Miscellaneous statistical

Operations Management

Data Analysis for Managers
with Microsoft® Excel

Duxbury Titles of Related Interest

To order copies contact your local bookstore or call 1-800-354-9706. For more information go to: www.duxbury.com

Data Analysis for Managers
with Microsoft® Excel, 2nd Edition

S. Christian Albright
Kelley School of Business, Indiana University

Wayne L. Winston
Kelley School of Business, Indiana University

Christopher Zappe
Bucknell University

With Case Studies by

Mark Broadie
Graduate School of Business, Columbia University

Peter Kolesar
Graduate School of Business, Columbia University

THOMSON

BROOKS/COLE

Australia • Canada • Mexico • Singapore • Spain • United Kingdom • United States

THOMSON

BROOKS/COLE

Publisher: Curt Hinrichs

Assistant Editor: Ann Day

Development Editor: Carol Pritchard-Martinez

Editorial Assistant: Katherine Brayton

Technology Project Manager: Burke Taft

Marketing Manager: Joseph Rogove

Marketing Assistant: Jessica Perry

Advertising Project Manager: Tami Strang

Print/Media Buyer: Jessica Reed

Permissions Editor: Bob Kauser

Production Service: Susan L. Reiland

Text Designer: Carolyn Deacy

Cover Designer: Joan Greenfield

Photo Researcher: Sue C. Howard

Copy Editor: Christine M. Levesque

Illustrator: Lori Heckelman

Cover Illustration: Todd Damen

Cover Printer: Phoenix Color

Compositor: ATLIS Graphics

Printer: Quebecor World, Taunton

For more information about our products, contact us at:
Thomson Learning Academic Resource Center
1-800-423-0563
For permission to use material from this text, contact us by:
Phone: 1-800-730-2214   **Fax:** 1-800-730-2215
Web: http://www.thomsonrights.com

Library of Congress Control Number 2002116788

Student Edition with InfoTrac College Edition:
  ISBN 0-534-38366-1
Instructor's Edition: ISBN 0-534-39721-2

Brooks/Cole—Thomson Learning
10 Davis Drive
Belmont, CA 94002
USA

Asia
Thomson Learning
5 Shenton Way #01-01
UIC Building
Singapore 068808

Australia/New Zealand
Thomson Learning
102 Dodds Street
Southbank, Victoria 3006
Australia

Canada
Nelson
1120 Birchmount Road
Toronto, Ontario M1K 5G4
Canada

Europe/Middle East/Africa
Thomson Learning
High Holborn House
50/51 Bedford Row
London WC1R 4LR
United Kingdom

Latin America
Thomson Learning
Seneca, 53
Colonia Polanco
11560 Mexico D.F.
Mexico

Spain/Portugal
Paraninfo Thomson Learning
Calle/Magallanes, 25
28015 Madrid, Spain

About the Authors

S. Christian Albright

Chris Albright got his B.S. degree in Mathematics from Stanford in 1968 and his Ph.D. in Operations Research from Stanford in 1972. Since then he has been teaching in the Operations & Decision Technologies Department in the Kelley School of Business at Indiana University. He has taught courses in management science, computer simulation, statistics, and computer programming to all levels of business students: undergraduates, MBAs, and doctoral students. In addition, he has taught simulation modeling at General Motors and Whirlpool, and he has taught database analysis for the Army. He has published over 20 articles in leading operations research journals in the area of applied probability, and he has authored the books *Statistics for Business and Economics, Practical Management Science, Data Analysis and Decision Making,* and *VBA for Modelers.* He is also currently working with the Palisade Corporation on a commercial version of his statistical StatPro add-in for Excel. His current interests are in spreadsheet modeling, the development of VBA applications in Excel, and programming in the .NET environment.

On the personal side, Chris has been married for 31 years to his wonderful wife, Mary, who recently retired after teaching 7th grade English for 30 years and is now working as a career counselor at the Kelley School. They have one son, Sam, who is currently working in New York City as a saxophone player and tours with the Monkees. Chris has many interests outside the academic area. They include activities with his family (especially traveling with Mary), going to cultural events at Indiana University, playing golf and tennis, running and power walking, and reading. And although he earns his livelihood from statistics and management science, his *real* passion is for playing the piano and listening to classical music. ●

Wayne Winston

Wayne L. Winston is Professor of Operations & Decision Technologies in the Kelley School of Business at Indiana University, where he has taught since 1975. Wayne received his B.S. degree in mathematics from MIT and his Ph.D. degree in operations research from Yale. He has written the successful textbooks *Operations Research: Applications and Algorithms, Mathematical Programming: Applications and Algorithms, Simulation Modeling with @RISK, Practical Management Science,* and *Financial Models Using Simulation and Optimization.* Wayne has published over 20 articles in leading journals and has won many teaching awards, including the school-wide MBA award four times. His current interest is in showing how spreadsheet models can be used to solve business problems in all disciplines, particularly in finance and marketing.

Wayne enjoys swimming and basketball, and his passion for trivia won him an appearance several years ago on the television game show *Jeopardy,* where he won two games. He is married to the lovely and talented Vivian. They have two children, Gregory and Jennifer. ●

Christopher J. Zappe

Chris earned his B.A. in mathematics from DePauw University in 1983 and his M.B.A. and Ph.D. in decision sciences from Indiana University in 1987 and 1988, respectively. Between 1988 and 1993, he performed research and taught various courses in the decision sciences area at the University of Florida in the College of Business Administration. Since 1993, Chris has been serving as an associate professor in the Department of Management at Bucknell University. He currently teaches undergraduate courses in business statistics, decision analysis, and computer simulation. Moreover, Chris teaches a graduate seminar in applied game theory. He has published articles in various journals including *Managerial and Decision Economics, OMEGA, Naval Research Logistics,* and *Interfaces.* His current scholarly interests focus on mathematical programming models of performance appraisal processes and innovative pedagogies in operations research/management science.

Chris has been married to his wonderful wife, Jeannie, for nearly four years now. Recently, Chris and Jeannie were blessed with the birth of their first child, Matthew. Beyond spending many long days with his students and colleagues at Bucknell, Chris enjoys playing with his new son, traveling to exciting faraway locations with his wife, reading great books of American history, watching major league baseball and college basketball games, and serving on the board of the local community center in Lewisburg, PA. ●

Dedication

To my main supporters

Mary, Sam, Tami, Ruth, and, of course, Charlie. And to Sam Senior, who is up there watching it all—S.C.A.

To my wonderful family

Vivian, Jennifer, Gregory—W.L.W.

To my wonderful family

Jeannie and Matthew—C.J.Z.

Brief Contents

Contents

Chapter 4: Getting the Right Data 129

Part ② **Probability, Uncertainty, and Decision Making**

Chapter 5: Probability and Probability Distributions 189

Chapter 6: Normal, Binomial, Poisson, and Exponential Distributions 239

Part ④ Regression, Forecasting, and Time Series

Chapter 11: Regression Analysis: Estimating Relationships 547

Chapter 12: Regression Analysis: Statistical Inference 619

Chapter 13: Time Series Analysis and Forecasting 689

Part ⑤

Design of Experiments, Quality Control, and Data Mining

Chapter 14: Analysis of Variance and Experimental Design 761

Chapter 15: Data Mining Techniques: Discriminant Analysis, Logistic Regression, and OLAP 813

Chapter 16: Statistical Process Control 865

Preface

With today's technology, companies are able to *collect* tremendous amounts of data with relative ease. Indeed, many companies now have more data than they know what to do with. However, the data are usually meaningless until they are analyzed for trends, patterns, relationships, and other useful *information*. This book illustrates in a practical way a variety of statistical methods, from simple to complex, to help you analyze data sets and uncover important information. In many business contexts, data analysis is only the first step in the solution of a problem. Acting on the solution and the information it provides to make good decisions is a critical next step. Therefore, there is a heavy emphasis throughout this book on analytical methods that are useful in decision making. Again, the methods vary considerably, but the objective is always the same—to equip you with data analysis and decision-making tools you can *apply* in your business careers.

We recognize that the vast majority of students in this type of course are *not* majoring in a quantitative area. They are typically *business* majors in finance, marketing, operations management, or some other business discipline who will need to analyze data and make quantitative-based decisions in their jobs. We offer a hands-on, example-based approach and introduce fundamental concepts as they are practically applied. Our vehicle is spreadsheet software, something with which most students are already familiar and will undoubtedly use in their careers. Our MBA students at Indiana University are so turned on by the required course that is based upon this book that *well over 50%* of them (mostly finance and marketing majors) take our follow-up *elective* course in spreadsheet modeling (our management science course). We believe that students see value in statistics and quantitative analysis when the course is taught in a practical and example-based approach.

Rationale for Writing This Book

Data Analysis for Managers is different from the many fine textbooks written for statistics and data analysis. Our rationale for writing this book is based on three fundamental objectives.

1. **Practical in approach**   We want the emphasis to be placed upon realistic business examples and the process by which a manager might analyze a problem from data acquisition to implementation of a solution—rather than exclusively on computational methods.

2. **Spreadsheet-based** We want the book to provide students with the skills to analyze business problems with tools they have access to and will use in their careers. To this end, we have adopted Excel and commercial spreadsheet add-ins.

3. **Relevant content** We have found from our teaching in industry that certain analytical tools are widely used in business yet seldom, if ever, covered in existing textbooks. For example, most data analyzed in business are obtained from company data warehouses, databases, or intranets rather than surveys. Chapter 4 discusses some of these widely used methods of data collection.

Practical in Approach

Taking a cue from our *Practical Management Science* book, we wanted this book to be very example-based and practical. We strongly believe that students learn best by working through examples, and they appreciate the material most when the examples are realistic and interesting business examples. Therefore, our approach in this book differs in two important ways.

First, there is just enough conceptual development to give students an understanding and appreciation for the issues raised in the examples. We often introduce important concepts (such as multicollinearity in regression) in the context of examples, rather than discussing them in the abstract. Our experience is that students gain greater intuition and understanding of the concepts and applications through this approach. This second edition features several new pedagogical features to help students grasp important concepts.

Second, we place little emphasis on hand calculations. The second edition contains just enough hand calculation and additional discussion to better understand how concepts are applied. We believe it is more important for students to understand why they are conducting an analysis and what it means. Therefore, we illustrate how powerful software can be used to create graphical and numerical outputs in a matter of seconds, freeing the rest of the time for in-depth interpretation of the output, sensitivity analysis, and alternative modeling approaches. In our own courses, we move directly into a discussion of examples, where we focus almost exclusively on interpretation and modeling issues and let the computer software perform the number crunching.

Spreadsheet-Based

As we have demonstrated in our other texts, we are strongly committed to teaching spreadsheet-based, example-driven courses, regardless of whether the basic area is statistics or management science. We have found tremendous enthusiasm for this approach, both from students and from faculty around the world who have used the book. The students appear to learn (and remember) more, and they *appreciate* the material more; instructors typically enjoy teaching more, and they usually receive immediate reinforcement through better teaching evaluations.

Relevant Content

There are many extremely useful concepts and techniques in business statistics. Indeed, properly utilized and applied, the concepts contained in this book will help you run your business smarter, more efficiently, and profitably. Our experience suggests that some of these methods are more widely used than others, and still others are widely used in business yet seldom, if ever, taught in a business program. We have sought to bridge this gap by including topics that are widely used and placing emphasis on central ideas of greatest relevance to students' future coursework and careers.

What We Hope to Accomplish in This Book

Condensing the ideas in the above paragraphs, we hope to:

- Reverse negative student attitudes about statistics and quantitative methods by making them real, accessible, and interesting.

- Teach problem-solving and reasoning skills. Give students lots of hands-on experience with real problems and challenge them to develop their intuition, logic, and problem-solving skills.

- Increase the appeal of statistics to a broad range of business students' interests by including real examples from a wide variety of business functions. Expose students to real problems in many business disciplines and show them how these problems can be attacked with analytical methods.

- Emphasize practical and marketable skills. Develop spreadsheet skills, including experience with powerful spreadsheet add-ins, that will add immediate value in other courses and in their future careers.

New in This Edition

The feedback we have received from users of the first edition has been very gratifying and informative. Before any writing of this second edition occurred, we combined this user feedback with the feedback from dozens of surveys and reviews to develop a plan that would make this second edition more effective as a teaching and learning resource. The following is a list of changes and enhancements made in this second edition of *Data Analysis for Managers*.

- **Simplified** We aimed to streamline the second edition by removing more difficult ideas and adding additional pedagogical tools.

- **More emphasis on concepts** An important change is the inclusion of many new pedagogical features for explaining concepts. These features are in the form of margin notes, boxed-in definitions (and formulas) in the text, enhanced explanations in the text itself, and stated objectives for the examples. The primary pedagogical tool is still the examples, but there is now more "up-front" material that explains the concepts.

- **New Conceptual Exercises** There are new conceptual problems at the end of each chapter. These either test the concepts or ask more open-ended questions—as opposed to the many data analysis problems already in the book.

- **New Review Summaries** There is a list of key terms and formulas at the end of each chapter. Together with the pedagogical "concept" features noted above, these should make it easier for students to study and review, and they should make it easier for instructors to teach.

- **Title change**   Perhaps the most obvious change is that the title of the book is *Data Analysis for Managers,* not *Managerial Statistics.* We believe the term "statistics" has an overly technical connotation. We want to stress the idea that this book is aimed at business people who need accessible tools to analyze their data and make decisions.

- **New modular organization**   The book is now divided into five parts: Part I: Getting, Describing, and Summarizing Data (Chapters 2–4); Part II: Probability, Uncertainty, and Decision Making (Chapters 5–7); Part III: Statistical Inference (Chapters 8–10); Part IV: Regression, Time Series, and Forecasting (Chapters 11–13); and Part V: Design of Experiments, Quality Control, and Data Mining (Chapters 14–16). This provides a clearer view of the organization of the book and allows instructors to more easily customize related groups of topics for their course.

- **New section on OLAP**   Chapter 15 now includes a section on online analytical processing (OLAP). We describe this very powerful data mining tool and show how it can be implemented in Excel.

- Other chapter by chapter changes:
  - Chapter 5, Probability and Probability Distributions, has been revised significantly. The more advanced examples have been simplified or eliminated.
  - Chapter 6, Normal, Binomial, Poisson, and Exponential Distributions, now contains a brief discussion of the exponential distribution.
  - In Chapter 7, Decision Making under Uncertainty, the section on influence diagrams with PrecisionTree has been deleted. We thought the explanation of this tool was too difficult to follow, and we didn't want to include the many extra pages that would have been necessary to explain it adequately.
  - Chapter 8, Sampling and Sampling Distributions, has been reorganized significantly. This is a chapter where conceptual development was lacking in the first edition. The concepts are now explained in a more complete and logical manner.
  - Chapter 13, Time Series Analysis and Forecasting, has been entirely reorganized. For example, there is now an introductory section on the various components of a time series (level, trend, seasonality, and noise), and the methods for handling seasonality are all contained in a single section.
  - Chapter 14, Analysis of Variance and Experimental Design, now *follows* the regression chapters. This allows us to include a section on implementing ANOVA with regression.
  - Chapter 15 has been renamed Data Mining Techniques: Discriminant Analysis, Logistic Regression, and OLAP.
  - The chapter on statistical process control has been moved to the end of the book. It is now Chapter 16.
  - The former Chapter 17 on statistical report writing has been made into an appendix.

- **New and updated data**   The data for many of the problems have been updated to be as timely as possible. This is specifically true for time series data. Rather than ending in 1996 or so (when we were writing the first edition), many of the data sets now end in 2000 or 2001.

Software

This book is based entirely on Microsoft Excel, the spreadsheet package that has become the standard analytical tool in business. Excel is an extremely powerful package, and one of our goals is to convert *casual* users into *power* users who can take full advantage of its features. If we accomplish no more than this, we will be imparting a valuable skill for the business world. However, Excel has many specific analytical limitations. Therefore, this book includes several Excel add-ins that greatly enhance Excel's capabilities. As a group, these add-ins comprise what is arguably the most impressive assortment of spreadsheet-based software in any book on the market.

StatPro Statistical add-in   When we began teaching this course in Excel, it quickly became obvious that Excel's inherent statistical capabilities were limited. Therefore, we provide an Excel add-in, StatPro™, that accompanies this book. We think you will find it to be quite powerful and extremely easy to use. StatPro™ does not attempt to do what Excel already does well (pivot tables, for example), but it performs most statistical analyses, even those as complex as stepwise regression, in a matter of seconds. (To see a summary of its capabilities, see the file StatProHelp.htm on the CD-ROM.) If you have been using Excel's built-in Analysis ToolPak, we think you will be very pleasantly surprised with the functionality of StatPro™. StatPro also contains a number of functions (all ending in an underscore, as in Discrete_) for simulating random numbers from a variety of probability distributions, including the common distributions (normal, binomial, uniform, and so on) and some not so common (multivariate normal and multinomial, for example).

DecisionTools add-in   The CD-ROM accompanying this book contains the powerful DecisionTools™ Suite by Palisade Corporation—the first time this software has been included in a textbook. This suite includes five separate Excel add-ins:

- @Risk, the popular add-in for simulation;
- PrecisionTree, a graphical-based add-in for creating and analyzing decision trees;
- TopRank, a powerful add-in for performing what-if analyses;
- BestFit, an add-in for fitting probability distributions to observed data;
- RiskView, a graphical add-in for drawing probability distributions.

The DecisionTools™ Suite included in this book is a special version for students only. It is only slightly scaled down from the professional version that sells for hundreds of dollars and is used by many leading companies. Specifically, the DecisionTools software will function for 2 years when properly installed and authorized and has modest limitations to the size of model that can be analyzed. We make extensive use of PrecisionTree in the chapter on decision making under uncertainty. We place less emphasis on @Risk, TopRank, BestFit, and RiskView, but we illustrate them in several examples.

Possible Sequences of Topics

Although we intend to use this book for our own required one-semester course, we admit that there is more material than can be covered adequately in one semester. We have tried to make the book as modular as possible. However, due to the natural progression of statistical topics, the basic topics in the early chapters must be covered before the more advanced topics (regression and time series analysis) in the later chapters. With this in mind, here are several possible ways to cover the topics.

- A one-semester required course, with no statistics prerequisite (or where MBA students have forgotten whatever statistics they might have learned years ago). Chapters 2, 3, 5, 6, 8–12, and possibly 4 should be covered. Depending on the time remaining, any of the topics in Chapters 7 (decision making under uncertainty), 13 (time series analysis), 14 (ANOVA), 15 (discriminant analysis, logistic regression, and OLAP), or 16 (statistical process control) can be covered in practically any order.

- A one-semester required course, with a statistics prerequisite. Assuming that students know the basic elements of statistics (up through hypothesis testing, say), then the material in Chapters 2, 3, 5, 6, and 8–10 can be reviewed *quickly,* primarily to illustrate how Excel and add-ins can be used to advantage. Then the instructor can choose between any of the topics in Chapters 7, 11–12, 13, 14, 15, or 16 (in practically any order) to fill up the remainder of the course.

- A two-semester required sequence. Given the luxury of spreading the topics over two semesters, the entire book can be covered. The statistics topics in Chapters 2, 3, 5, 6, and 8–10 should be covered in chronological order, but the remaining chapters can be covered in practically any order.

Custom Publishing

Should you wish to use only a subset of the text, or add chapters from the authors' other texts, or your own materials, you may do so through Thomson Custom Publishing. Contact your local Duxbury/Thomson Learning representative for more details.

Ancillaries for Adopting Faculty

The CD-ROM that accompanies this book contains:

- StatPro and DecisionTools Suite, add-ins described above
- Excel files for examples in the chapters (usually two versions of each—a template, or data-only version, and a finished version)
- Data files required for the problems and cases
- A file TUTORIAL.DOC that contains a brief tutorial in the basic elements of Excel
- A file README.HTM that contains instructions for installing and using the add-ins
- A file BORDERMACROS.DOC that explains how to develop a macro to create borders for color-coding your worksheets

In addition, adopting instructors may obtain *Instructors Suite* CD-ROM that includes:

- Solution files (in Excel and Word formats) for all of the problems and cases in the book
- PowerPoint presentation files for all of the examples in the book
- Test items in Word format
- Completed Excel files for all of the examples in the book

Finally, adopting instructors will have access to the following:

- The AWZ (authors' initials) Book Companion website that includes software updates, errata, additional problems and solutions, and additional resources for both students and faculty (accessible through www.duxbury.com by selecting "Online Book Companions")
- A Student Solutions manual with Hints for students (ISBN 0-534-39911-8).

Acknowledgments

The authors would like to thank several people who helped make this book a reality. First, the authors are indebted to Peter Kolesar and Mark Broadie of the Columbia Business School for contributing many excellent case studies that appear throughout the book. In addition, the manuscript went through several stages of surveys and review, where many valuable ideas emerged. We wish to thank the following reviewers and survey participants for their comments and suggestions:

James D. Behel, Harding University; Richard Bernhard, North Carolina State University; Cheryl Dale, William Carey College; Dinesh Dave, Appalachian State University; Abe Feinberg, California State University, Northridge; Soumen Ghosh, Georgia Institute of Technology; Irwin Greenberg, George Mason University; Ching-Chung Kuo, Pennsylvania State University at Harrisburg; Vedran Lelas, University of Connecticut; Shreevardhan Lele, University of Maryland; John Leschke, University of Virginia; Karl Majeske, University of Michigan; Leslie Marx, University of Rochester; Mike Middleton, University of San Francisco; Tyra Anne Mitchell, Georgia Institute of Technology; Michael H. Morris, University of Notre Dame; Richard Morris, Winthrop University; Herbert Moskowitz, Purdue University; Kelly Nichols, Gonzaga University; Tom Obremski, University of Denver; Paul Paschke, Oregon State University; David W. Pentico, Duquesne University; Tim Riggle, Baldwin Wallace College; Werner Schenk, University of Rochester; Catherine Shenoy, University of Kansas; Bala Shetty, Texas A&M University; William E. Stein, Texas A & M University; Donald N. Stengel, California State University, Fresno; Ralph E. Steuer, University of Georgia; Robert Stoll, Baldwin Wallace College; Joe Sullivan, Mississippi State University; Mustafa Yilmaz, Northeastern University; and Joe Zhu, Worcester Polytechnic Institute.

There are more people who helped to produce this book than we can list here. However, there are a few special people whom we were happy (and lucky) to have on our team. Carol Pritchard-Martinez made many important contributions to this second edition. Managing the development of this second edition and the many new pedagogical elements that it contains was seamless under Carol's direction. We think the text is much improved thanks to her contributions. There may be a few remaining errors, but they are certainly not the fault of our project editor, Susan Reiland. Susan is a perfectionist, and her influence clearly shows throughout the book. Not only did she capture the typos, but she helped us improve our writing styles immeasurably. If you want to learn how to write well, write a book with Susan! The driving force behind this project from day one has been our editor, Curt Hinrichs. There were nights and weekends when we were in no mood to thank Curt, but even when he pushed us to our limits, we knew that he was one-hundred percent behind us. No author can ask for more than an editor who knows the market and really cares about a project. We got that consistently from Curt. Any success this book has in the market is due largely to his efforts.

We are also grateful to many of the professionals at Duxbury who worked behind the scenes to make this book a success: Joseph Rogove, Marketing Manager; Hal Humphrey, Project Manager; Ann Day, Assistant Editor; and Katherine Brayton, Editorial Assistant.

S. Christian Albright
Wayne L. Winston
Christopher Zappe

1

Introduction to Managerial Statistics

Statistics from Washington, D.C.

There is almost certainly no organization that does more to collect, analyze, and disseminate statistical data than the federal government in Washington, D.C. However, most of you are probably unaware that there is a post called Chief Statistician of the U.S. Office of Management and Budget, and we would be willing to bet that none of you knows who has held that position since 1993. Her name is Katherine K. Wallman, and in a recent article in *Chance* (Citro, 1999), a magazine of the American Statistical Association (ASA), she discussed some of the goals and recent initiatives of her organization. We summarize some of this discussion here.

Perhaps her primary point is the importance of federally gathered statistics to the everyday lives of millions of Americans. Federal statistics govern cost-of-living adjustments for many wage earners and Social Security payments for retirees, they define who qualifies for food stamps and school lunches, and they determine where services such as schools and hospitals will be provided. However, despite the fact that statistics have a bearing on people's daily lives, Ms. Wallman laments the fact that so many of these same people are unequipped to understand even the most basic statistical information. As she quotes from the ASA's *Guidelines for the Teaching of Statistics K–12 Mathematics Curriculum*, "Raw data, graphs, charts, rates, percentages, probabilities, averages, forecasts, and trend lines are an inescapable part of our everyday lives. They affect decisions on health, citizenship, parenthood, employment, financial concerns, sports, and many other matters . . . Informed citizens should understand the latest news polls, the consumer price indices, and unemployment rates."

She defines "statistical literacy" as the "ability to understand and critically evaluate statistical results that permeate our daily lives—coupled with the ability to appreciate the contributions that statistical thinking can make in public and private, professional and personal decisions," and she states that one of her principal goals is to find ways for the U.S. federal statistical system to contribute more directly to improving the nation's statistical literacy and the people's ability to use federal statistics wisely. Interestingly, she notes that policy makers and the public have traditionally raised issues not so much with the scientific merit of the statistical methods that statisticians use, but rather with matters such as confidentiality of data and the burden placed on the public by data gatherers. But recently, with the political importance of such statistical instruments as the Consumer Price Index and the national decennial census, even the statistical *methods* have come under political debate. At the very least, the public should understand enough about statistics to judge whether politicians are attacking, say, the proposed sampling scheme for the U.S. census for statistical reasons or purely for political gain.

Ms. Wallman then discusses some of the initiatives she has championed in her position as Chief Statistician. Probably the most important, especially to readers of this book, has been the development of an interagency World Wide Web site (www.fedstats.gov) that permits easy access to the wide array of federal statistics available to the public. She calls this site "one-stop shopping." It has a Subjects A–Z listing, where users can find up-to-date data on such topics as AIDS/HIV, births, children, consumer credit, crime, education, environment, foreign trade, immigration, life expectancy, nuclear energy, poverty, productivity, tobacco use, transportation, unemployment, wages, and many others. As Wallman states, "The site's primary objective is to help users find the information they need without having to know and understand in advance how the decentralized federal statistical system is organized or which agency or agencies may produce the data they are seeking." For anyone who has searched through the maze of links on some statistical Web sites, this ability to one-stop shop will be most welcome.

In 1997, the first year it was released, the FedStats Web site hosted over 800,000 sessions and gained enthusiastic public support. (It is undoubtedly well above that level by now.) It has been featured in the *Wall Street Journal, The Washington Post,* Lycos, USA TODAY Online, and Yahoo. After exploring FedStats, one member of the public commented, "I was offended when the census taker came around and asked a lot of seemingly useless questions. I will have a different attitude the next time." As for the future, Wallman promises that the Interagency Council on Statistical Policy, the developers of FedStats, will continue to improve the site as suggestions and comments are received. It already provides a gateway to the Web sites of over 70 federal agencies, and it is currently developing enhanced search capabilities by indexing many of these Web sites. When we discuss sources of external data, including Web sites, in Chapter 4, we expect that the FedStats site will be one of the first you explore. ●

1.1 Introduction

We are living in the age of technology. This has two important implications for everyone entering the business world. First, technology has made it possible to collect huge amounts of data. Retailers collect point-of-sale data on products and customers every time a transaction occurs; credit agencies have all sorts of data on people who have or would like to obtain credit; investment companies have a limitless supply of data on the historical patterns of stocks, bonds, and other securities; and government agencies have data on economic trends, the environment, social welfare, consumer product safety, and virtually everything else we can imagine. It has become relatively *easy* to collect the data. As a result, data are plentiful. However, as many organizations are now beginning to discover, it is quite a challenge to analyze and make sense of all the data they have collected.

A second important implication of technology is that it has given many more people the power and responsibility to analyze data and make decisions on the basis of quantitative analysis. Those entering the business world can no longer pass all of the quantitative analysis to the "quant jocks," the technical specialists who have traditionally done the number crunching. The vast majority of employees now have a desktop or laptop computer at their disposal, they have access to relevant data, and they have been trained in easy-to-use software, particularly spreadsheet and database software. For these employees, statistical methods are no longer forgotten topics they once learned in college. Statistical analysis is now an integral part of their daily jobs.

Huge quantities of data already exist, and they are only going to expand in the future. Many companies already complain of swimming in a sea of data. However, enlightened companies are seeing this expansion as a source of competitive advantage. By using quantitative methods to uncover the *information* in the data and then acting on this information—again guided by statistical analysis—they are able to gain advantages that their less enlightened competitors are not able to gain. Several pertinent examples of this follow.

- Direct marketers analyze enormous customer databases to see which customers are likely to respond to various products and types of promotions. This allows them to target different classes of customers in different ways to maximize profits—and gives their customers what the customers want.

- Hotels and airlines also analyze enormous customer databases to see what their customers want and are willing to pay for. By doing this, they have been able to devise very clever pricing strategies, where not everyone pays the same price for the same accommodations. For example, a business traveler typically makes a plane reservation closer to the time of travel than a vacationer. The airlines know this. Therefore, they reserve seats for these business travelers and charge them a higher price (for the same seats). The airlines profit, and the customers are happy.

- Financial planning services have a virtually unlimited supply of data about security prices, and they have customers with widely differing preferences for various types of investments. Trying to find a match of investments to customers is a very challenging problem. However, customers can easily take their business elsewhere if good decisions are not made on their behalf. Therefore, financial planners are under extreme competitive pressure to analyze masses of data so that they can make informed decisions for their customers.

- We all know about the pressures U.S. manufacturing companies have faced from foreign competition in the past couple of decades. The automobile companies, for example, have had to change the way they produce and market automobiles to stay in business. They have had to improve quality and cut costs by orders of magnitude. Although the struggle continues, much of the success they have had can be attributed to continual data analysis and wise decision making. Starting on the shop floor and moving up through the organization, they now measure almost everything they do, they analyze these measurements, and then they act on the information from these measurements.

We talk about companies analyzing data and making decisions. However, *companies* don't really do this; *people* do it. And who will these people be in the future? They will be *you*! We know from experience that students in all areas of business, at both the undergraduate and graduate level, will soon be *required* to describe large complex data sets, run regression analyses, and make quantitative forecasts. You are the people who will soon be analyzing data and making important decisions to help gain your companies a competitive advantage. And if you are *not* willing or able to do so, there will be plenty of other technically trained people who will be more than happy to replace you.

Our goal in this book is to teach you how to use a variety of statistical methods to analyze data and make decisions. We plan to do so in a very hands-on way. We will discuss a number of statistical methods and illustrate their use in a large variety of realistic business problems. As you will see, this book is very "example driven," with examples from finance, marketing, operations, accounting, and other areas of business. To analyze these examples, we will take advantage of the Microsoft Excel spreadsheet package, together with a number of powerful Excel add-ins. In each example we will provide step-by-step details of the method and its implementation in Excel.

This is *not* a "theory" book. It is also not a book where you can lean comfortably back in your chair, prop your legs up on a table, and read about how *other* people perform statistical analysis. It is a "get your hands dirty" book, where you will learn best by actively following the examples throughout the book at your own PC. In short, you will learn by doing. By the time you have finished, you will have acquired some very useful skills for today's business world.

1.2 An Overview of the Book

This book is packed with statistical methods and examples, possibly more than can be covered in any single course. Therefore, we purposely intend to keep this introductory chapter brief so that you can get on with the analysis. Nevertheless, it is useful to introduce the methods you will be learning and the tools you will be using. In this section we will provide an overview of the methods covered in this book and the software that will be used to implement them. Then in the next section we will briefly discuss the pros and cons of using Excel-based add-ins versus the many stand-alone statistical software packages. Finally, in Section 1.4 we will preview some of the examples we will cover in much more detail in later chapters. Our primary purpose at this point is to stimulate your interest in what follows.

1.2.1 The Methods

Three important themes run through this book: **data analysis, dealing with uncertainty, and decision making.** Each of these themes has subthemes. Data analysis includes data **description,** data **inference,** and the search for **relationships** in data. Dealing with uncertainty includes **measuring** uncertainty and **modeling** uncertainty explicitly into the analysis. Decision making includes **decision analysis** for problems with uncertainty and structured **sensitivity analysis.** There are obvious overlaps between these themes and subthemes. When we make inferences from data and search for relationships in data, we must deal with uncertainty. When we use decision trees to help make decisions, uncertainty is the key driver. When we analyze data statistically, we typically do so with the ultimate goal of making good decisions.

Figure 1.1 shows where you will find these themes and subthemes in the remaining chapters of this book. In the next few paragraphs, we will discuss the book's contents in more detail.

We begin in Chapters 2 and 3 by illustrating a number of ways to summarize the information in data sets. These include graphical and tabular summaries, as well as numerical summary measures such as means, medians, and standard deviations. The material in these two chapters is elementary from a mathematical point of view, but it is extremely important. As we stated at the beginning of this chapter, organizations are now able to collect huge amounts of raw data. The question then becomes, What does it all mean? Although there are very sophisticated methods for analyzing data sets, many of which we will cover in later chapters, the "simple" methods in Chapters 2 and 3 are crucial for obtaining an initial understanding of the data. Fortunately, Excel and available add-ins now make what was once a very tedious task quite easy. For example, Excel's pivot table tool for "slicing and dicing" data is an analyst's dream come true. You'll be amazed at the complex analysis it will enable you to perform—with almost no effort!

After the analysis in Chapters 2 and 3, we step back for a moment in Chapter 4 to see how we get the data we need in the first place. We know from experience that many students and business people are able to perform appropriate statistical analysis once they have the data in a suitable form. Often the most difficult part, however, is getting the right data, in the right form, into a software package for analysis. Therefore, in Chapter 4 we

FIGURE 1.1
Themes and Sub-themes

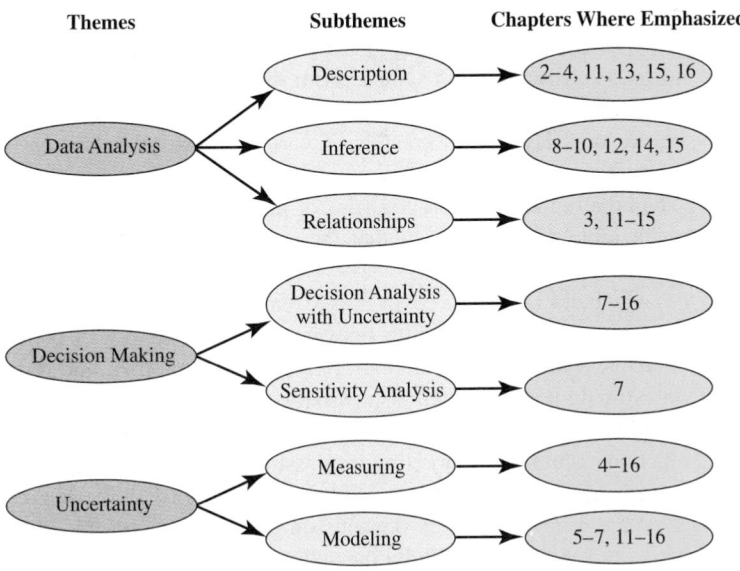

present a number of extremely useful methods for doing this within Excel. Specifically, we discuss methods for using Excel's built-in filtering tools to perform queries on Excel data sets, using Microsoft Query (part of Microsoft Office) to perform queries on *external* databases (such as Access) and bring the resulting data into Excel, writing queries to import data directly into Excel from Web sites, and "cleaning" data sets (getting rid of "bad" data values). This chapter provides tools that most analysts need but are usually not even aware of.

Uncertainty is a key aspect of most business problems. To deal with uncertainty, we need a basic understanding of probability. We provide this understanding in Chapters 5 and 6. Chapter 5 covers basic rules of probability and then discusses the extremely important concept of probability distributions in some generality. Chapter 6 follows up this discussion by focusing on two of the most important probability distributions, the normal and binomial distributions. It also briefly discusses the Poisson and exponential distributions, which have many applications in probability models but are not used as extensively in this book.

We have found that one of the best ways to make probabilistic concepts "come alive" and easier to understand is by using computer simulation. Therefore, simulation is a common theme that runs through this book, beginning in Chapter 5. We will use simulation to illustrate difficult probabilistic and statistical concepts.

In Chapter 7 we apply our knowledge of probability to decision making under uncertainty. These types of problems—faced by all companies on a continual basis—are characterized by the need to make a decision *now,* even though important information (such as demand for a product or returns from investments) will not be known until later. The material in Chapter 7 provides a rational basis for making such decisions. The methods we illustrate do not guarantee perfect outcomes—the future could unluckily turn out differently than we had expected—but they do enable us to proceed rationally and make the best of the given circumstances. Additionally, the software we will use to implement these methods allows us, with very little extra work, to see how sensitive the optimal decisions are to inputs. This is crucial, because the inputs to many business problems are at best educated guesses. Finally, we will examine the role of risk aversion in these types of decision problems.

In Chapters 8, 9, and 10 we discuss sampling and statistical inference. Here the basic problem is to estimate one or more characteristics of a population. If it is too expensive or time-consuming to learn about the *entire* population—and it usually is—we instead select a random sample from the population and then use the information in the sample to *infer* the characteristics of the population. We see this continually on news shows that describe the results of various polls. We also see it in many business contexts. For example, auditors typically sample only a fraction of a company's records. Then they infer the characteristics of the entire population of records from the results of the sample to conclude whether the company has been following acceptable accounting standards.

Chapters 11 and 12 cover the extremely important topic of regression analysis, which is used to study relationships between variables. Its power is its generality. Every part of a business has variables that are related to one another, and regression can often be used to estimate possible relationships between these variables. In managerial accounting, regression is used to estimate how overhead costs depend on direct labor hours and production volume. In marketing, regression is used to estimate how sales volume depends on advertising and other marketing variables. In finance, regression is used to estimate how the return of a stock depends on the "market" return. In real estate studies, regression is used to estimate how the selling price of a house depends on the assessed valuation of the house and characteristics such as the number of bedrooms and square footage. Regression analysis finds perhaps as many uses in the business world as any method in this book.

We discuss time series analysis and forecasting in Chapter 13. This topic is particularly important for providing inputs into business decision problems. For example, manufacturing companies must forecast demand for their product to make sensible decisions about order quantities from their suppliers. As another example, fast-food restaurants must forecast customer arrivals, sometimes down to the level of 15-minute intervals, so that they can staff their restaurants appropriately.

There are many approaches to forecasting, ranging from simple to complex. Some involve regression-based methods, in which one or more time series variables are used to forecast the variable of interest, whereas other methods are based on extrapolation. In an extrapolation method the historical patterns of a time series variable, such as product demand or customer arrivals, are studied carefully and are then "extrapolated" into the future to obtain forecasts. A number of extrapolation methods are available. We will study both regression and extrapolation methods for forecasting in Chapter 13.

In Chapter 14 we discuss a group of statistical methods that generally come under the headings of analysis of variance (ANOVA) or (more generally) design of experiments. Typically, ANOVA is used to see whether several experimental treatments, or combination of experimental treatments, produce different mean results in some variable of interest. This general topic is quite broad—indeed, entire books have been written about design of experiments—so we present only an introduction to its key concepts in this chapter.

In Chapter 15 we discuss two related statistical methods, discriminant analysis and logistic regression. The basic objective of both of these methods is to use available variables to explain or predict which of several groups observations are in. For example, these methods are used by market researchers to predict which consumers will purchase their product and which will not, based on demographic and other data on the consumers. Financial analysts have used these methods to predict which existing companies will remain solvent and which will go bankrupt, based on a number of financial variables. Although discriminant analysis and logistic regression are generally classified as more advanced "multivariate" methods, we have included an introduction to them here because of their increasing importance in the business world.

In Chapter 15 we also briefly discuss one of the increasingly important data mining methods, online analytical processing (OLAP). Data mining methods, which are used to search for useful information in today's large corporate databases, are typically complex and require specialized software. However, OLAP is basically just a generalization of pivot tables from Chapters 2 and 3, and fortunately a version of it can be implemented in Excel—without any specialized add-ins.

Chapter 16 presents an introduction to statistical process control, also known as quality control. Quality has become a buzzword in many companies, in both manufacturing and service industries, for the simple reason that it is almost impossible to survive in today's business world without paying close attention to quality. As we discuss in Chapter 16, a "quality program" requires a number of initiatives, some quantitative and some not. For example, most quality advocates argue that lasting quality will not occur unless management takes an active role in helping workers to do their best work, including extensive job training. These are management issues that fall outside of the scope of this book. However, an important part of any quality program involves gathering data on company processes and using statistical techniques to analyze these data.

We will focus on some of these statistical techniques in Chapter 16. Specifically, we will discuss control charts for checking whether a process is "in control," and we will discuss methods for checking whether a process is capable of producing outputs that meet specifications.

Finally, Appendix A presents a few ideas about statistical report writing. Even the best statistical analysis can be wasted unless it is reported accurately and succinctly, using appropriate tables and charts. A prerequisite of good statistical report writing is the ability to write well in general. Although we cannot make any claims that this appendix will transform poor writers into good writers, we will illustrate a few key elements of good statistical report writing by drawing on examples from previous chapters.

1.2.2 The Software

The topics we have just discussed are very important. Together, they can be used to solve a wide variety of business problems. However, they are not of much practical use unless we have the software to do the number crunching. Very few business problems are small enough to be solved with pencil and paper. They require powerful software.

The software included in this book, together with Microsoft Excel, provides you with a powerful software combination that you will not use for one course and then discard. It is software that is being used—and will continue to be used—by leading companies all over the world to solve large, complex problems. We firmly believe that the experience you obtain with this software, through working the examples and problems in this book, will give you a key competitive advantage in the marketplace.

It all begins with Excel. All of the quantitative methods that we discuss are implemented in Excel. It is obviously impossible to forecast the state of computer software into the long-term or even medium-term future, but as we are writing this book, Excel is *the* most heavily used spreadsheet package on the market, and there is every reason to believe that this state will persist for many years. Most companies use Excel, most employees and most students have been trained in Excel, and Excel is a *very* powerful, flexible, and easy-to-use package.

Although the tutorial is presented in a Word file, it contains "embedded" Excel spreadsheets that allow you to practice spreadsheet techniques within Word.

Built-in Excel Features Virtually everyone in the business world knows the basic features of Excel, but relatively few know many of its more powerful features. In short, relatively few people are the "power users" we expect you will become by working through this book. To get you started, the file TUTORIAL.DOC on the CD-ROM explains some of the "intermediate" features of Excel—features that we expect you to be able to use. These include the SUMPRODUCT, VLOOKUP, IF, NPV, and COUNTIF functions. They also include range names, the Data Table command, the Paste Special command, the Goal Seek command, and a few others. Finally, although we assume you can perform routine spreadsheet tasks such as copying and pasting, we include a few tips to help you perform these tasks more efficiently.

In the body of the book we describe several of Excel's advanced features in more detail. In Chapters 2 and 3 we introduce pivot tables, the Excel tool that enables you to summarize data sets in an almost endless variety of ways. (Excel has a lot of useful tools, but we personally believe that pivot tables are the most ingenious and powerful of all.) In Chapter 4 we discuss Excel's methods for querying databases, including Excel databases, databases in external database packages (such as Access), and even data from the Web. Beginning in Chapter 5, we introduce Excel's RAND function for generating random numbers. This function is used in all spreadsheet simulations (at least those that don't take advantage of an add-in). Finally, throughout the book, we will mention some of Excel's more exotic features. Our purpose is not to "wow" you with all the neat things Excel can do; rather, it is to explore how to get a job done in the most efficient way.

StatPro Add-in Because the topic of this book is statistical analysis, we were in a quandary as we developed the book. There are a number of excellent statistical software packages on the market, including Minitab, SPSS, SAS, StatGraphics, and others. Although there are now user-friendly Windows versions of these packages, they are *not* spreadsheet-based. We have found through our own experience that students resist the use of non-spreadsheet packages, regardless of their inherent quality, so we wanted to use Excel as our "statistics package." Unfortunately, Excel's built-in statistical tools are rather limited, and the Analysis ToolPak (developed by a third party) that ships with Excel also has significant limitations.

Therefore, we developed our own add-in called StatPro that accompanies this book.[1] Our goal when developing StatPro was to create an add-in that would be powerful, easy to use, and capable of generating output quickly in an easily interpretable form. We do *not* believe you should have to spend hours each time you want to produce some statistical output. This might be a good learning experience the first time through, but after that it acts as a strong incentive *not* to perform the analysis at all! We believe you should be able to generate output quickly and easily. This gives you the time to *interpret* the output—and possibly redo the analysis if it didn't turn out right the first time.

A good illustration involves the construction of histograms, scatterplots, and time series plots, discussed in Chapter 2. All of these extremely useful graphs can be created in a straightforward way with Excel's built-in tools. But by the time you perform all the necessary steps and "dress up" the charts exactly as you want them, you are not going to be very anxious to repeat the whole process again. StatPro does it all quickly and easily. (You still might want to "dress up" the resulting charts, but that's up to you.) Therefore, if we advise you in a later chapter, say, to look at several scatterplots as a prelude to a regression analysis, you can do so in a matter of seconds.

Decision Tools Suite Besides StatPro and built-in Excel add-ins, we have also contracted with Palisade Corporation to include a slightly scaled-down version of its powerful Decision Tools suite in this book. All items in this suite are Excel add-ins—so the learning curve isn't very steep. There are five separate add-ins in this suite: @Risk, PrecisionTree, TopRank, BestFit, and RiskView. The first two are the most important for our purposes, but all are useful for certain tasks.

@Risk The simulation add-in @Risk enables us to run as many replications of a spreadsheet simulation as we like. As the simulation runs, @Risk automatically keeps track of

---

[1]StatPro is also being developed as a separate product by Palisade Corporation.

the outputs we select, and it then displays the results in a number of tabular and graphical forms. It also enables us to perform a sensitivity analysis, so that we can see which inputs have the most effect on the outputs.

PrecisionTree The PrecisionTree add-in is used in Chapter 7 to analyze decision problems with uncertainty. The primary method for performing this type of analysis is to draw a decision tree. Decision trees are inherently graphical, and they were always difficult to implement in spreadsheets, which are based on rows and columns. However, Precision-Tree does this in a very clever and intuitive way. Equally important, once the basic decision tree has been built, it is easy to use PrecisionTree to perform a sensitivity analysis on the model inputs.

TopRank Although the other Palisade add-ins are probably not as useful as @Risk and PrecisionTree, they are all worth investigating. TopRank is the most general of them. It starts with any spreadsheet model, where a set of inputs is used, along with a number of spreadsheet formulas, to produce an output. TopRank then performs a sensitivity analysis to see which inputs have the largest effect on the output. For example, it might tell us which affects after-tax profit the most: the tax rate, the risk-free rate for investing, the inflation rate, or the price charged by a competitor. Unlike @Risk, TopRank is used when uncertainty is not *explicitly* built into a spreadsheet model. However, it considers uncertainty implicitly by performing sensitivity analysis on the important model inputs.

Palisade Corporation originally marketed BestFit and RiskView as separate products. Although they still exist as separate products, their functionality is now included in @Risk.

BestFit BestFit is used to determine the most appropriate probability distribution for a spreadsheet model when we have data on some uncertain quantity. For example, a simulation might model each week's demand for a product as a random variable. What probability distribution should we use for weekly demand: the well-known normal distribution or possibly some skewed distribution? If we have historical data on weekly demands for the product, we can feed them into BestFit and let it recommend the distribution that best fits the data. This is a very useful tool in real applications. Instead of guessing a distribution that we think might be relevant, we can let BestFit point us to a distribution that fits historical data well. We discuss BestFit in Chapter 6.

RiskView Finally, RiskView is a drawing tool that complements @Risk. A number of probability distributions are available with @Risk and can be used in simulations. Each has an associated @Risk function, such as RiskNormal, RiskBinomial, and so on. Before selecting any of these distributions, however, it is useful (especially for beginners) to see what these distributions look like. RiskView performs this task easily. For any selected probability distribution (and any selected parameters of this distribution), it creates a graph of the distribution.

Together with Excel and the add-ins included in this book, you have a wealth of software at your disposal. The examples and step-by-step instructions throughout this book will help you to become a power user of this software. Admittedly, this takes plenty of practice and a willingness to experiment, but it is certainly within your grasp. When you are finished, we won't be surprised if you rate improved software skills as the most valuable thing you've learned from this book.

1.3 Excel versus Stand-alone Statistical Software

Some readers (especially instructors) may be concerned about our exclusive use of Excel to perform statistical analysis. Why are we not using Minitab, SPSS, SAS, StatGraphics, or any of the many other fine stand-alone statistical packages now on the market? We have taught statistics for a number of years, and initially we were very disappointed with Excel's statistical capabilities. At the same time, we were very impressed with some of the emerging Windows-based statistical software packages. These packages are relatively easy to learn, they perform a wide array of statistical methods, and their graphical capabilities are tremendous.

However, we were eventually persuaded by the overwhelming majority of our students to use Excel. The students make at least three very good arguments for Excel. First, they want to learn a software package only if there is some assurance that their eventual employers will own it. The vast majority of companies own and use Microsoft Office, but the same cannot be said for specialized statistical packages. Second, specialized statistical packages are almost *too* good, in the sense that they contain loads of advanced features most students will never use (or understand). This wealth of functionality can be intimidating to students who are, after all, business majors, not statistics majors. Third and most important, students already know how to use Excel, and they are anxious to learn more about its possibilities because they know this is valuable knowledge in the workplace. Besides, they usually have their data in Excel worksheets, and if any pre-analysis data manipulation is necessary, they are most comfortable doing this in Excel. If they were then forced to export their data to another package—not always a simple chore!—they would regard this as a definite non-value-added step in the process.

We know from our experience of trying to force specialize statistical software on students that it just does not work—no one is happy in the end. On the other hand, we also knew that Excel's statistical capabilities are sadly lacking. The Analysis ToolPak that ships with Excel is not sufficiently good to build a course around. Therefore, we created StatPro to fill a clear gap. We have gotten very positive reviews about StatPro from our students, both for its ease of use and its power to perform all of the standard statistical analyses. We are the first to admit that StatPro does not have nearly as much functionality as SPSS, SAS, and the others, but it meets our intentions of providing students with Excel capabilities they can use to obtain immediate results—numerical outputs and charts—in an environment in which they are comfortable and with almost no learning required. In addition, StatPro is continually evolving (stay tuned to the www.indiana.edu/~mgtsci Web site), and we will attempt to add features (or modify the interface) in response to our users' wishes.

This argument over whether to use Excel or a specialized statistics package will undoubtedly continue, but we always come back to our intended audience. We believe the vast majority of our students will use statistical procedures from time to time in their work, maybe even on a daily basis, but they will *not* be professional statisticians. For these users, we believe Excel, along with an add-in such as StatPro, offers the best environment for performing data analysis, all things considered. Our recommendation for professional statisticians is completely different. They might consider Excel for relatively simple tasks, but they will definitely benefit from the additional functionality offered by specialized statistical packages.

1.4 A Sampling of Examples

Perhaps the best way to illustrate what you will be learning in this book is to preview a few examples from the remaining chapters. Our intention here is not to teach you any methods; that will come later. We only want to indicate the types of problems you will learn how to solve. Each example below is numbered as in the chapter where it appears.

Example 3.9

The Spring Mills Company produces and distributes a wide variety of manufactured goods. Because of its variety, it has a large number of customers. It classifies these customers as small, medium, and large, depending on the volume of business each does with Spring Mills. Recently, Spring Mills has noticed a problem with its accounts receivable. It is not getting paid back by its customers in as timely a manner as it would like. This obviously costs Spring Mills money. If a customer delays a payment of $300 for 20 days, say, then the company loses potential interest on this amount. The company has gathered data on 280 customer accounts. For each of these accounts, the data set lists three variables: Size, the size of the customer (coded 1 for small, 2 for medium, 3 for large); Days, the number of days since the customer was billed; and Amount, the amount the customer owes. What information can we obtain from these data?

Solution

This example from Chapter 3 is a typical example of trying to make sense out of a large data set. Spring Mills has 280 observations on each of three variables. By realistic standards, this is not a huge data set, but it still presents a challenge. We examine the data from a number of angles and present several tables and charts. For example, the scatterplots in Figures 1.2–1.4 clearly indicate that there is a *positive* relationship between the amount owed and the number of days since billing for the medium-size and large

FIGURE 1.2
Scatterplot of Amount versus Days for Small Customers

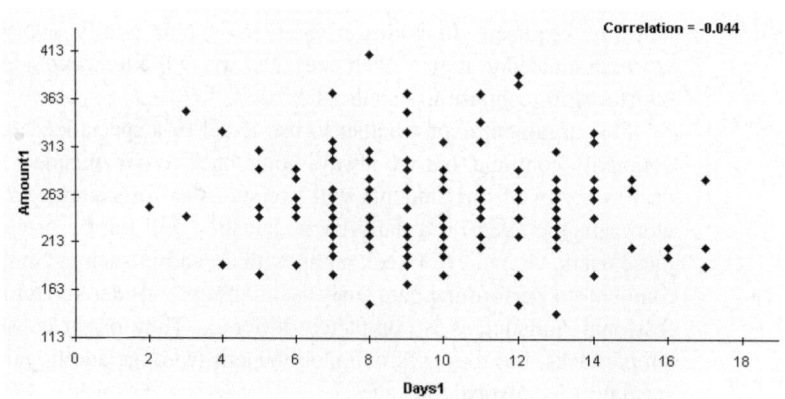

FIGURE 1.3
Scatterplot of Amount versus Days for Medium Customers

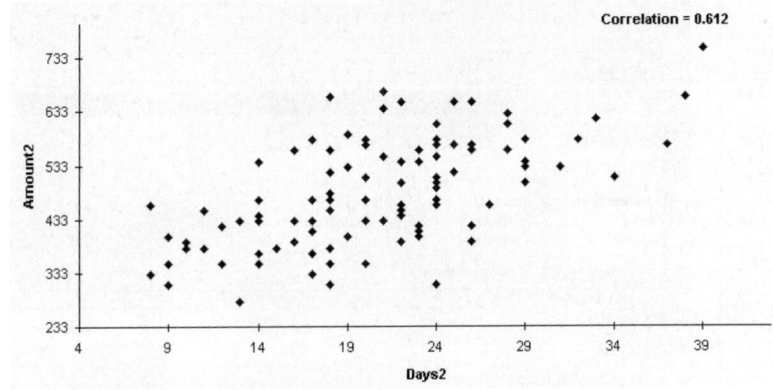

FIGURE 1.4
Scatterplot of Amount versus Days for Large Customers

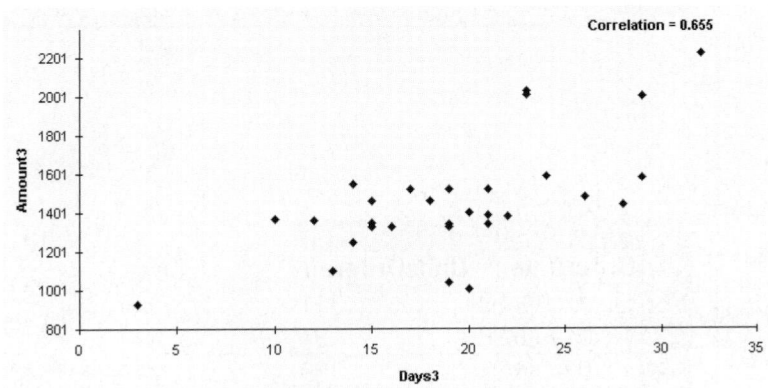

customers, but that no such relationship exists for the small customers. As we will see, graphs such as these are very easy to construct in Excel, regardless of the size of the data set. ●

Example 4.3

The Fine Shirt Company creates and sells shirts to its customers. These customers are retailers who sell the shirts to consumers. The company has created an Access database file that has information on sales to its customers over the past five years (1995 through 1999). There are three related tables in this database: Customers, Orders, and Products. The Customers table has the following information on the company's 7 customers: CustomerNum (an index for the customer, from 1 to 7), Name, Street, City, State, Zip, and Phone. The Products table has the following information on the company's 10 products (types of shirts): ProductNum (an index for the product, from 1 to 10), Description, Gender (whether the product is made for females, males, or both), and UnitPrice (the price to the retailer). Finally, the bulk of the data is in the Orders table. This table has a record

FIGURE 1.5
Microsoft Query
Dialog Box

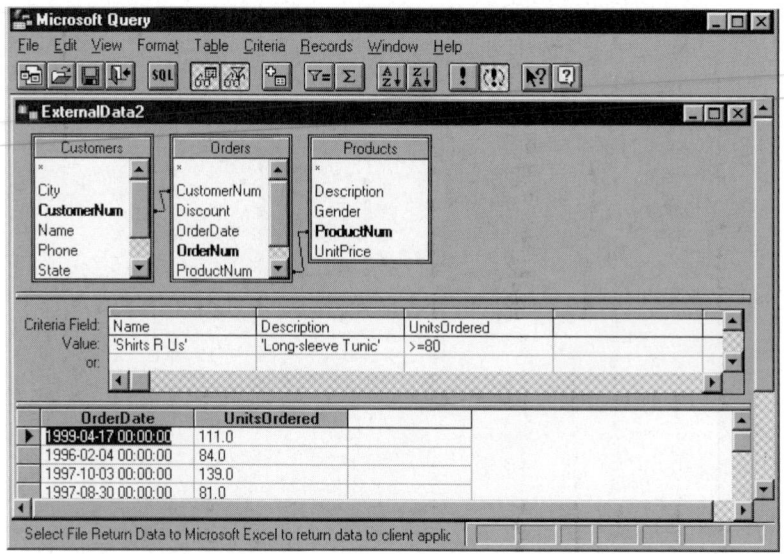

FIGURE 1.6
Data from Access
Returned to Excel

	H	I
1	**OrderDate**	**UnitsOrdered**
2	17-Apr-99	111
3	04-Feb-96	84
4	03-Oct-97	139
5	30-Aug-97	81
6	20-Feb-98	142
7	11-Nov-96	87
8	29-Feb-96	125
9	22-Sep-98	117
10	01-Jun-98	83
11	18-Jun-95	95
12	26-Dec-97	114
13	26-Jul-98	146
14	13-Jan-95	98
15	02-May-98	93
16	26-Apr-98	100
17	04-Jun-97	92
18	22-Jan-95	142

for each product ordered by each customer on each date during the 5-year period. There are 2245 records in this table. Note that if a customer ordered more than one product on a particular date, there is a separate record for each product ordered. The fields in the Orders table are: OrderNum (an index for the order, from 1 to 2245), CustomerNum (to link to the Customers table), ProductNum (to link to the Products table), OrderDate, UnitsOrdered (number of shirts of this type ordered), and Discount (percentage discount, if any, for this order).

The company wants to perform a statistical analysis of the data on orders within Excel. How can it use Microsoft Query to import that data from Access into Excel?

Solution

This example from Chapter 4 explains how to use Microsoft Query, software that is included in Microsoft Office, to import selected data from an Access database into Excel. For example, if we want to find all of the records in the Orders table that correspond to orders for at least 80 units made by the customer Shirts R Us (customer number 3) for the product Long-sleeve Tunic (product number 6), we need to fill in a Query dialog box as illustrated in Figure 1.5. This locates all of the order records that satisfy the specified conditions and returns the data to an Excel worksheet, as shown in Figure 1.6. ●

Example 7.1

SciTools Incorporated, a company that specializes in scientific instruments, has been invited to make a bid on a government contract. The contract calls for a specific number of these instruments to be delivered during the coming year. The bids must be sealed (so that no company knows what the others are bidding), and the low bid wins the contract. SciTools estimates that it will cost $5000 to prepare a bid and $95,000 to supply the instruments if it wins the contract. On the basis of past contracts of this type, SciTools believes that the possible low bids from the competition, if there is any competition, and the associated probabilities are those shown in Table 1.1. In addition, SciTools believes there is a 30% chance that there will be no competing bids.

Table 1.1

Data for Bidding Example

Low Bid	Probability
Less than $115,000	0.2
Between $115,000 and $120,000	0.4
Between $120,000 and $125,000	0.3
Greater than $125,000	0.1

FIGURE 1.7
Decision Tree for
SciTools

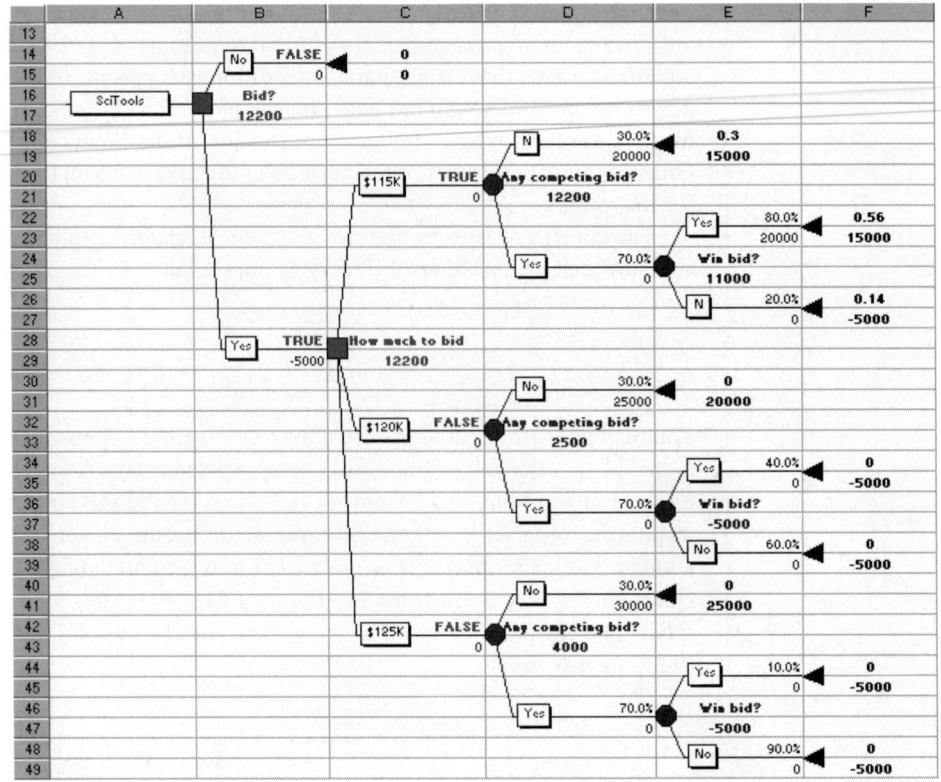

Solution

This is a typical example of decision making under uncertainty, the topic of Chapter 7. SciTools has to make decisions *now* (whether to bid and if so, how much to bid), without knowing what the competition is going to do. The company can't *assure* itself of a perfect outcome, but it can make a rational decision in light of the uncertainty it faces. We will see how decision trees, produced easily with the PrecisionTree add-in to Excel, not only lay out all of the elements of the problem in a logical manner, but also indicate the best solution. The completed tree for this problem appears in Figure 1.7. It indicates that SciTools should indeed prepare a bid, for the amount $115,000. ●

Example 9.4

An auditor wants to determine the proportion of invoices that contain price errors—that is, prices that do not agree with those on an authorized price list. He checks 93 randomly sampled invoices and finds that 2 of them include price errors. What can he conclude, in terms of a 95% one-sided confidence interval, about the proportion of *all* invoices with price errors?

Solution

This is an important application of statistical inference in the auditing profession. Auditors try to determine what is true about a population (in this case, all of a company's invoices) by examining a relatively small sample from the population. Here the auditor wants an upper limit so that he is 95% confident that the overall proportion of invoices with errors is no greater than this upper limit. We show the spreadsheet solution in Figure 1.8. This shows that the auditor can be 95% confident that the overall proportion of invoices with errors is no greater than 6.6%.

FIGURE 1.8
Analysis of Auditing Example

	A	B	C	D	E	F
1	Auditing example for an exact one-sided confidence interval					
2				**Range names**		
3	Confidence level	95%		ConfLev: B3		
4	Number of errors	2		NErrors: B4		
5	Sample size	93		SampProp: B7		
6				SampSize: B5		
7	Sample proportion	0.0215		UpLimit: B10		
8						
9	Exact upper confidence limit for p			Goal seek condition		
10	Upper	0.066		0.050	=	0.05
11						
12	Large-sample upper confidence limit for p					
13	Upper	0.046				

Example 11.2

The Bendrix Company manufactures various types of parts for automobiles. The manager of the factory wants to get a better understanding of overhead costs. These overhead costs include supervision, indirect labor, supplies, payroll taxes, overtime premiums, depreciation, and a number of miscellaneous items such as charges for building depreciation, insurance, utilities, and janitorial and maintenance expenses. Some of these overhead costs are "fixed" in the sense that they don't vary appreciably with the volume of work being done, whereas others are "variable" and do vary directly with the volume of work. The fixed overhead costs tend to come from the supervision, depreciation, and miscellaneous categories, whereas the variable overhead costs tend to come from the indirect labor, supplies, payroll taxes, and overtime premiums categories. However, it is not easy to draw a clear line between the fixed and variable overhead components.

The Bendrix manager has tracked total overhead costs over the past 36 months. To help "explain" these, he has also collected data on two variables that are related to the amount of work done at the factory. These variables are:

- MachHrs: number of machine hours used during the month
- ProdRuns: number of separate production runs during the month

The first of these is a direct measure of the amount of work being done. To understand the second, we note that Bendrix manufactures parts in fairly large batches. Each

FIGURE 1.9
Scatterplot of
Overhead versus
Machine Hours

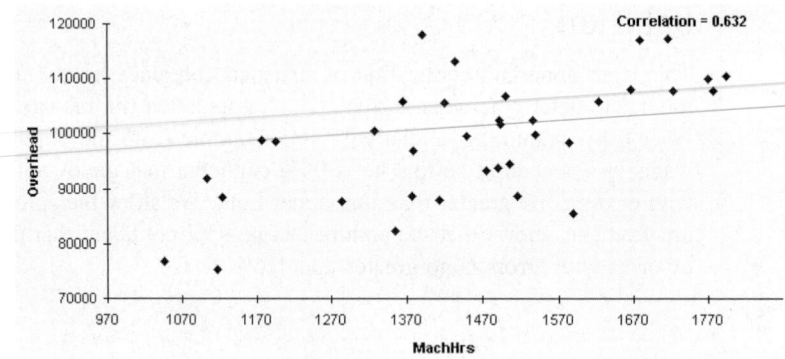

FIGURE 1.10
Scatterplot of
Overhead versus
Production Runs

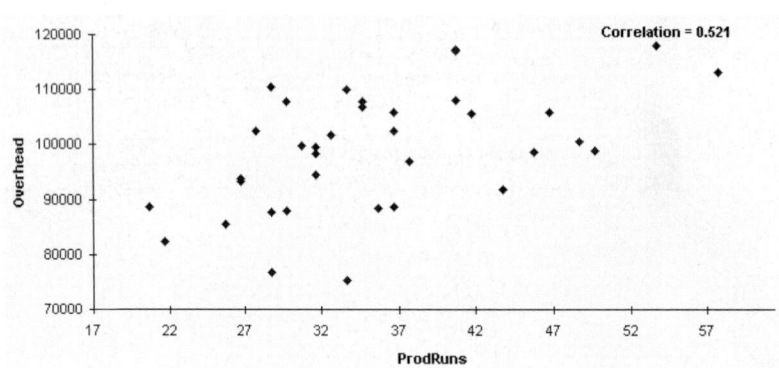

FIGURE 1.11
Regression Output

	A	B	C	D	E	F	G
1	*Results of multiple regression for Overhead*						
2							
3	*Summary measures*						
4		Multiple R	0.9308				
5		R-Square	0.8664				
6		Adj R-Square	0.8583				
7		StErr of Est	4108.9932				
8							
9	*ANOVA Table*						
10		Source	df	SS	MS	F	p-value
11		Explained	2	3614020652.0000	1807010326.0000	107.0261	0.0000
12		Unexplained	33	557166208.0000	16883824.4848		
13							
14	*Regression coefficients*						
15			Coefficient	Std Err	t-value	p-value	
16		Constant	3996.6782	6603.6509	0.6052	0.5492	
17		MachHrs	43.5364	3.5895	12.1289	0.0000	
18		ProdRuns	883.6179	82.2514	10.7429	0.0000	

batch corresponds to a production run. Once a production run is completed, the factory must "set up" for the next production run. During this setup there is typically some downtime while the machinery is reconfigured for the part type scheduled for production in the next batch. Therefore, the manager believes both of these variables might be responsible (in different ways) for variations in overhead costs. Do scatterplots support this belief?

Solution

This is a typical regression example, here in a cost accounting setting. The manager is trying to see what type of relationship, if any, there is between overhead costs and the two explanatory variables, number of machine hours and number of production runs. The scatterplots requested appear in Figures 1.9 and 1.10. They do indeed indicate a positive and linear relationship between overhead and the two explanatory variables.

However, regression goes well beyond scatterplots. It estimates an equation relating the variables. This equation can be determined from regression output such as that shown in Figure 1.11. This output implies the following equation for predicted overhead as a function of machine hours and production runs:

$$\text{Predicted Overhead} = 3997 + 43.54\text{MachHrs} + 883.62\text{ProdRuns}$$

The positive coefficients of MachHrs and ProdRuns indicate the effects these variables have on overhead. We will not take the example any further at this point but will simply indicate that it is easy to generate the output in Figure 1.11 with StatPro. The difficult part is learning how to interpret it. We will spend plenty of time in Chapters 11 and 12 on interpretation issues. ●

Example 13.3

The file INTEL.XLS contains quarterly sales data (in millions of dollars) for the chip manufacturing firm Intel from the beginning of 1986 through the first quarter of 2001. Are Intel's sales growing exponentially through the entire period?

Solution

This example illustrates a regression-based trend curve, one of several possible forecasting techniques for a time series variable. A time series plot of Intel's quarterly sales through 1996 appears in Figure 1.12. It indicates that sales have been increasing steadily at an increasing rate. This is basically what an exponential trend curve implies. To estimate this curve we use regression analysis to obtain the following equation for predicted quarterly sales as a function of time:

$$\text{Predicted Sales} = 295.377e^{0.0657\text{Time}}$$

This equation implies that Intel's sales are increasing by approximately 6.6% per quarter during this period, which translates to an annual percentage increase of about 29%! As we'll see in Chapter 13, this is the typical approach used in forecasting. We look at a time series plot to discover trends or other patterns in historical data, and then use one of a

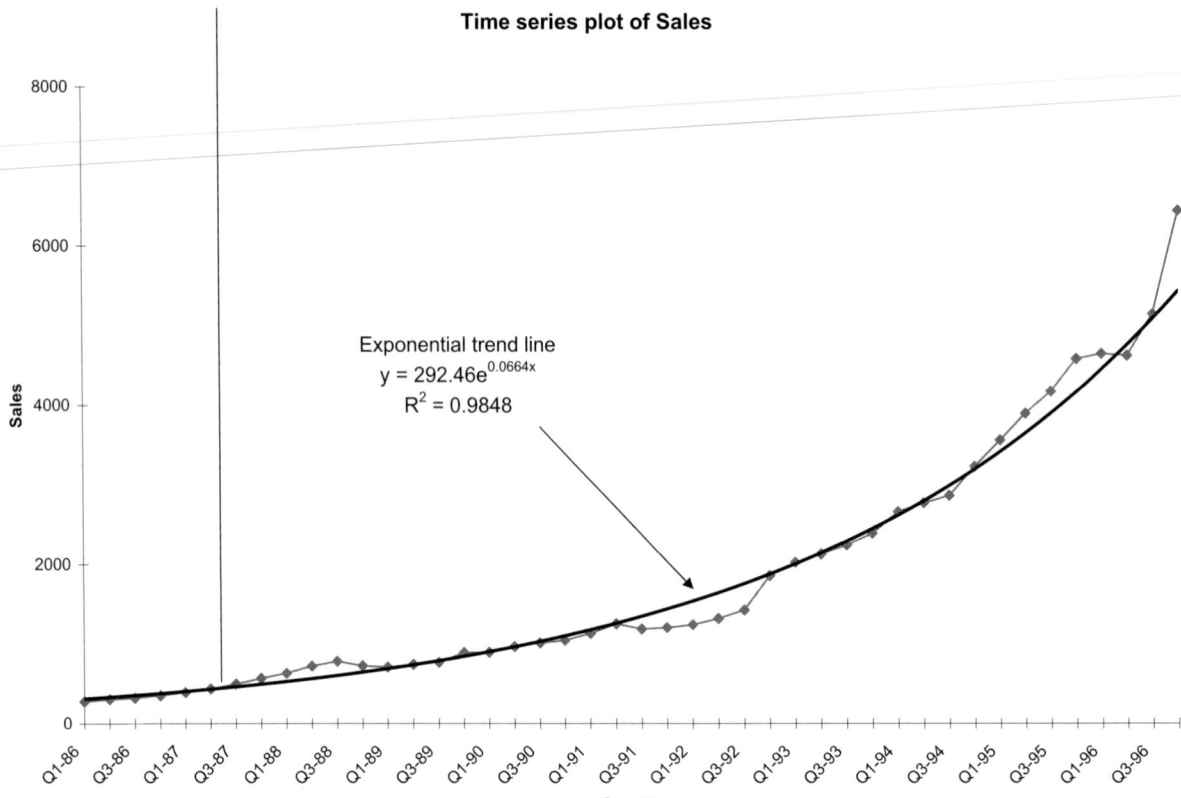

Time series plot of Sales

Exponential trend line
$y = 292.46e^{0.0664x}$
$R^2 = 0.9848$

FIGURE 1.12 Time Series Plot of Quarterly Sales at Intel

variety of techniques to fit the observed patterns and extrapolate them into the future. As we will see, this exponential trend did not extrapolate well into the period past 1996. Intel's sales still tended to increase, but not nearly at an annual rate of 29%. ●

Example 14.3

If you are a golfer, or even if you have ever seen golf ball commercials on television, you know that a number of golf ball manufacturers claim to have the "longest ball," that is, the ball that goes the farthest on drives. This example illustrates how these claims might be tested. We assume that there are five major brands, labeled A through E. A consumer testing service runs an experiment in which 60 balls of each brand are driven under three temperature conditions. The first 20 are driven in cool weather (about 40 degrees), the next 20 are driven in mild weather (about 65 degrees), and the last 20 are driven in warm weather (about 90 degrees). The goal is to see whether some brands differ significantly, on average, from other brands and what effect temperature has on mean differences between brands. For example, it is possible that brand A is the longest ball in warm weather but some other brand is longest in cool temperatures.

Solution

This is an example of an analysis of variance (ANOVA). We run an experiment as described, where two "factors," brand of golf ball and temperature, are varied in a specific manner and a "response variable," distance of drive, is recorded for each golf ball driven under each combination of factor levels. Although the details of the statistical output are somewhat complex, we can get the essence of the results in Figures 1.13–1.15.

FIGURE 1.13
Table of Means

	L	M	N	O	P
1	**Table of sample mean yardages**				
2					
3	Average of Yards	Temp ▼			
4	Brand ▼	Cool	Mild	Warm	Grand Total
5	A	218.8	236.5	258.4	237.9
6	B	224.1	245.1	258.3	242.5
7	C	228.0	242.7	263.0	244.6
8	D	215.0	237.6	256.1	236.2
9	E	224.8	255.7	270.9	250.5
10	Grand Total	222.1	243.5	261.4	242.3

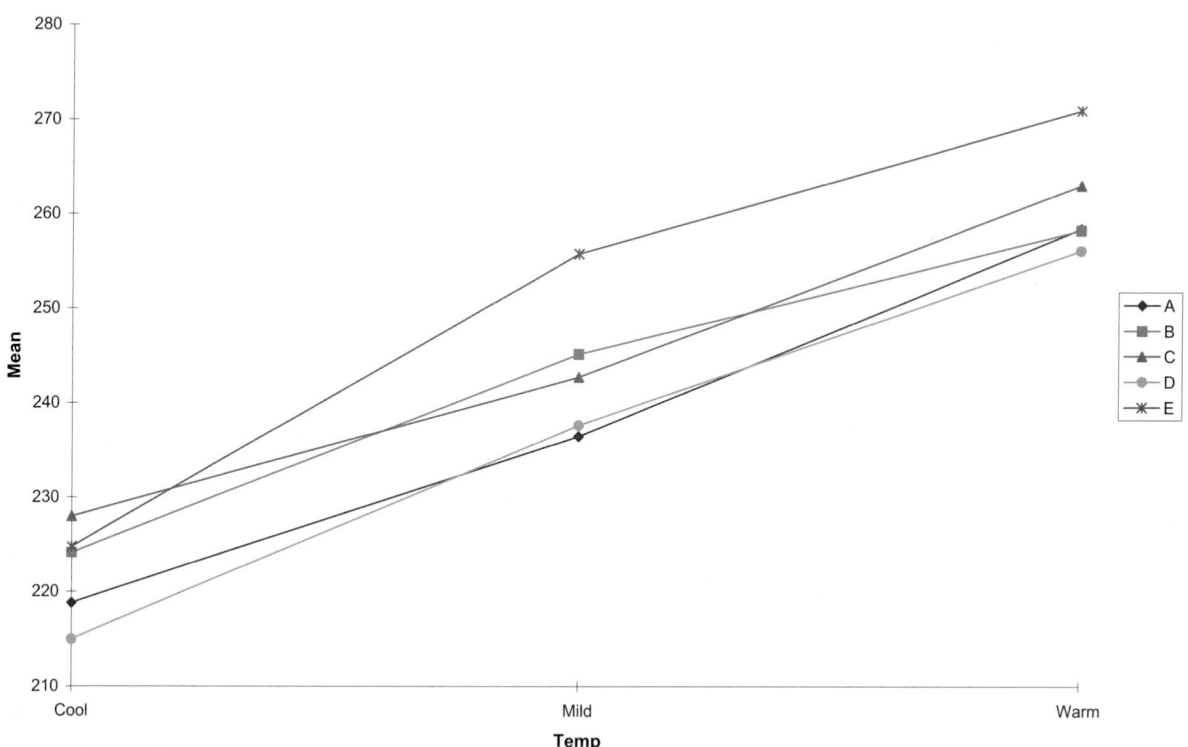

Chart of Cell Means to Check for Interactions

FIGURE 1.14 One View of Interactions

Chart of Cell Means to Check for Interactions

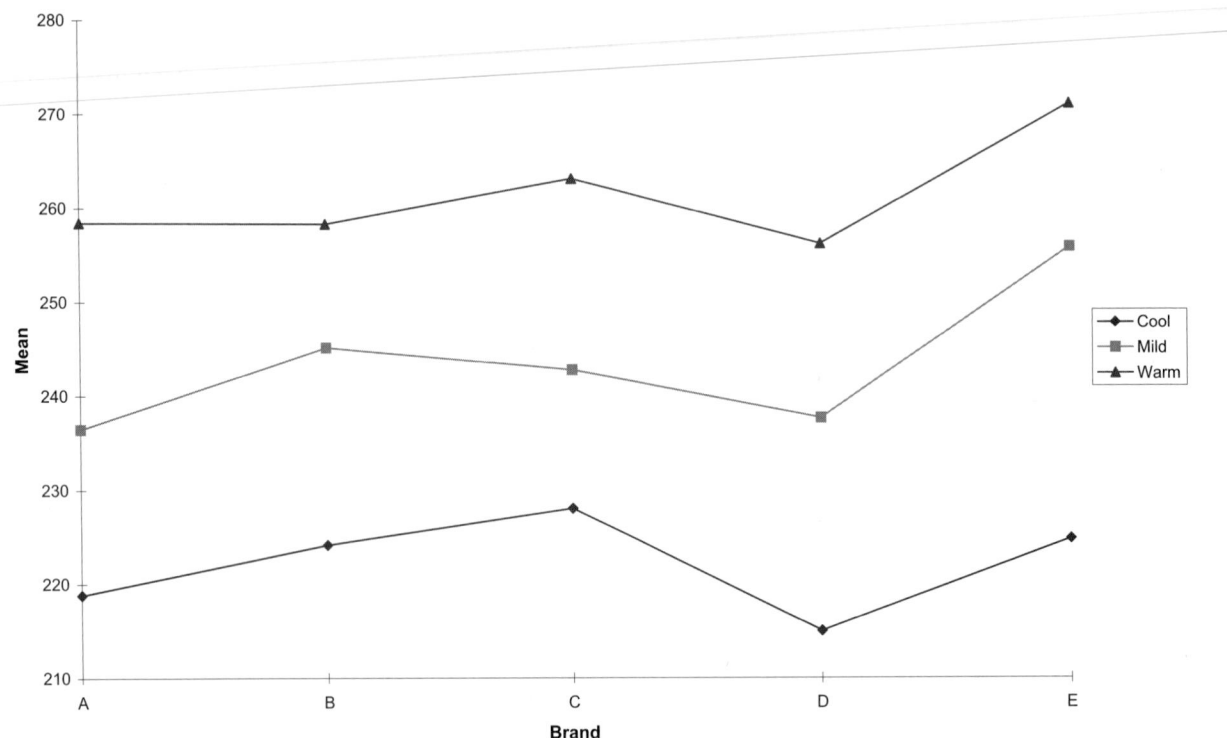

FIGURE 1.15 Another View of Interactions

The table in Figure 1.13 contains the sample mean distances for each combination of brand and temperature. For example, the 20 balls of brand A driven in cool temperature averaged 218.8 yards. In addition, this table includes row and column averages. For example, the 60 balls of brand A averaged 237.9 yards, whereas the 100 balls driven in cool temperature averaged 222.1 yards. The column averages indicate that temperature definitely makes a difference: balls go farther when it is warmer. The more important averages, however, are the row (brand) averages. They are closer together, and although brand E has the largest average, it is not clear whether brand E is *really* a longer ball or the differences across brands are due to chance. The ANOVA table (not shown here but explained in Chapter 14) resolves this issue.

Even if brand E is deemed the longest brand on average, it might not be the longest in all temperatures. The interaction charts in Figures 1.14 and 1.15 allow us to examine the data in more detail. They plot the sample means from Figure 1.13 in two equivalent ways. Figure 1.14 shows a separate line of means for each brand, where temperature appears on the horizontal axis. We see that brand E does indeed dominate in mild and warm temperatures, but brand C is slightly longer in cool temperatures. Figure 1.15 shows the same information in a slightly different way. There is a separate line of means for each temperature, and the brand is depicted on the horizontal axis. As their captions indicate, these charts allow us to see whether brand and temperature *interact* in some way. We will have more to say about these types of interactions in Chapter 14. ●

Example 15.3

The file WSJ1.XLS contains observations on 84 people, each of whom either subscribes or does not subscribe to *The Wall Street Journal* (WSJ). These are the two groups. The variables that we believe might be useful for discriminating between these two groups are Income (the person's annual income) and InvestAmt (the total amount the person has invested in stocks and bonds). A partial listing of the data appears in Figure 1.16. How well do these two variables discriminate between subscribers and nonsubscribers? How well can we classify these 84 people on the basis of these two variables?

Solution

This example illustrates the technique of *discriminant analysis*, which attempts to categorize members of some population into two or more groups, based on their values of various explanatory variables. For two groups and two explanatory variables, as in this example, the method is equivalent to drawing a scatterplot of Income versus InvestAmt, using different shapes of points for the subscribers and nonsubscribers, and seeing whether any line can "separate" the subscriber points from the nonsubscriber points. This scatterplot and corresponding line are shown in Figure 1.17. We then classify all points on one

FIGURE 1.16
Data for WSJ Example

	A	B	C	D	E	F
1	Discriminant analysis: two groups and two explanatory variables					
2						
3	Person	Income	InvestAmt	WSJSubscriber		
4	1	66400	26900	No		
5	2	68000	7100	No		
6	3	54900	21500	No		
7	4	50600	19300	No		
8	5	54100	16700	No		
56	53	60900	25800	No		
57	54	88900	28600	No		
58	55	68200	12300	No		
59	56	88400	34500	No		
60	57	66600	32200	No		
61	58	77800	48500	Yes		
62	59	86600	66600	Yes		
63	60	72900	39400	Yes		
64	61	90900	63800	Yes		
65	62	64300	50100	Yes		
66	63	53900	36400	Yes		
84	81	74100	36700	Yes		
85	82	78500	46000	Yes		
86	83	75200	51100	Yes		
87	84	100700	58800	Yes		

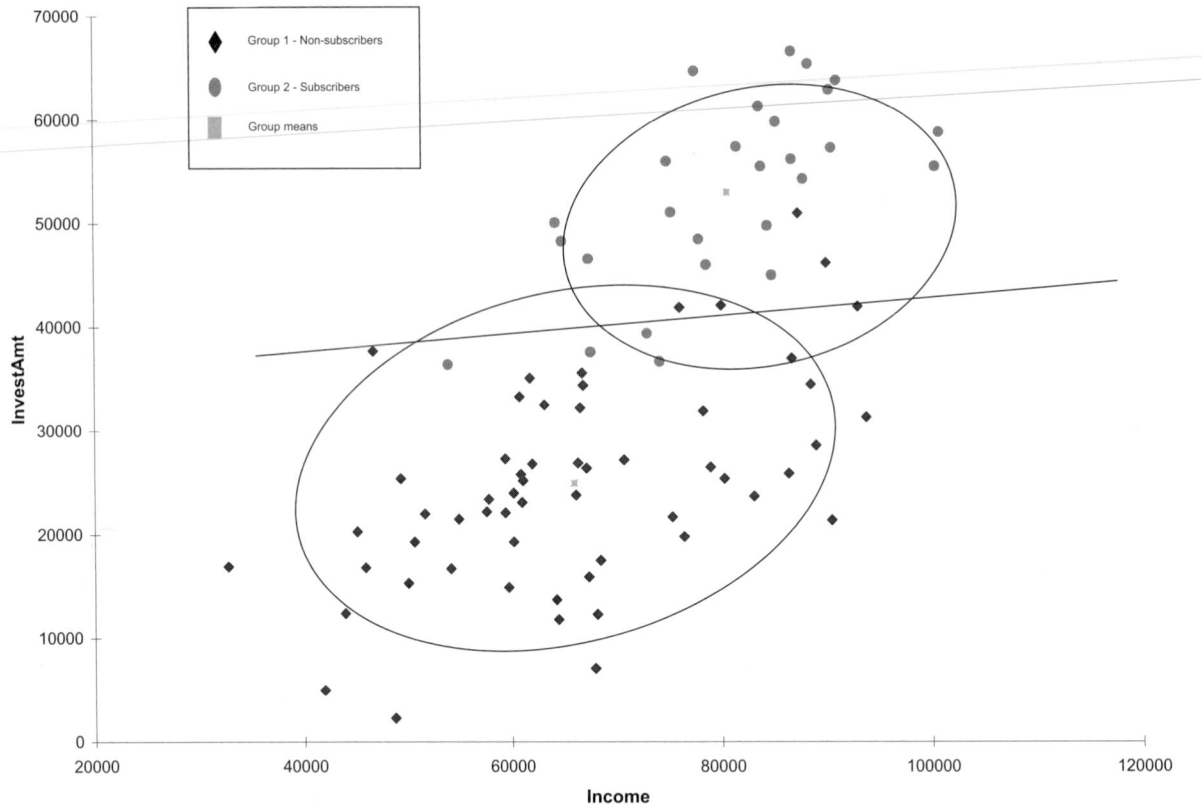

FIGURE 1.17 The Line of Maximal Separation

side of the line as subscribers and those on the other side as nonsubscribers. Our hope is that most of the classifications are correct.

There is an equivalent algebraic approach to discriminant analysis, which calculates a "score" for each member of the population and then classifies each member into one of the two groups based on whether the member's score is less than or greater than some cutoff score. In this particular example, the method correctly classifies 52 out of 57 of the nonsubscribers and 25 out of 27 of the subscribers. We will discuss this method in greater detail in Chapter 15, where we will also discuss another somewhat similar classification method, logistic regression. ●

Example 16.3

In a manufacturing process for gaskets, there are two parallel production machines that produce identical types of gaskets. A crucial dimension of the gaskets is their thickness, measured in millimeters. Every 15 minutes, four gaskets were sampled, two from each machine. What can we learn from the $\overline{X}$ and R charts?

Solution

This example illustrates the control charts that are used in many manufacturing and service companies to learn how their processes are behaving. Although it is probably not obvious at this point, the control charts in Figures 1.18 and 1.19 indicate some rather suspicious behavior. Upon closer examination, we discover that the two machines are producing quite different outputs—the average thickness of one is well above the average for the other. Armed with this knowledge, a machine operator can take immediate action to bring the machines back into line with one another. Fortunately, the StatPro add-in makes it easy to produce these types of control charts in Excel, as we discuss in Chapter 16.

FIGURE 1.18
$\bar{X}$ Chart for Gasket Data from Both Machines

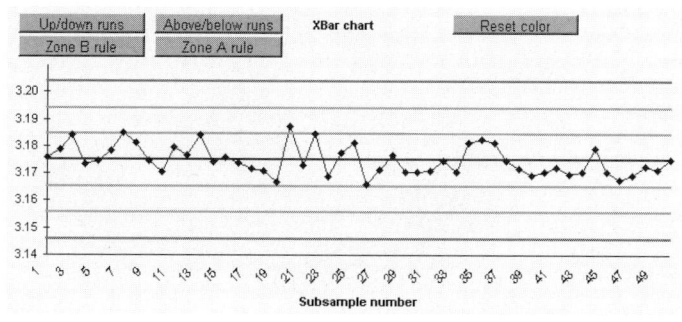

FIGURE 1.19
R Chart for Gasket Data from Both Machines

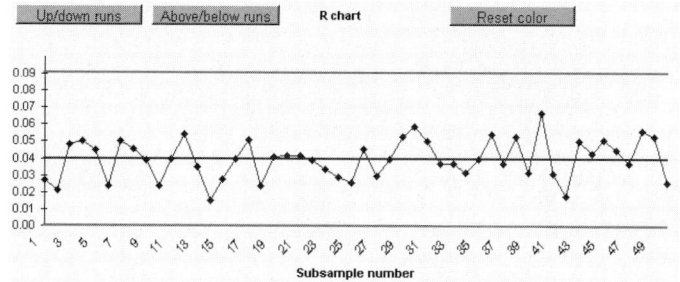

1.5 Conclusion

In this chapter we have attempted to convince you that the skills in this book are important for *you* to know as you enter the business world. The statistical methods we discuss are no longer the sole province of the "quant jocks." By having a PC on your desk that is loaded with powerful software, you incur a responsibility to use this software to solve business problems. We have described the types of problems you will learn to solve in this book, along with the software you will use to solve these problems. Now it's time for you to get started!

1.1 Entertainment on a Cruise Ship

Cruise ship traveling has become big business. Many cruise lines are now competing for customers of all age groups and socioeconomic status levels. They offer all types of cruises, from relatively inexpensive 3–4-day cruises in the Caribbean to 12–15-day cruises in the Mediterranean to several-month around-the-world cruises. Cruises have several features that attract customers, many of whom book six months or more in advance: (1) they offer a relaxing, everything-done-for-you way to travel, (2) the food they serve is plentiful, usually excellent, and included in the price of the cruise, (3) they stop at a number of ports that are interesting and offer travelers a way to see the world, and (4) they provide a wide variety of entertainment, particularly in the evening.

This last feature, the entertainment, presents a difficult problem for a ship's staff. A typical cruise might have well over a thousand customers, including elderly singles and couples, middle-aged people with or without children, and young people, often honeymooners. These different types of passengers have varied tastes in terms of their after-dinner preferences in entertainment. Some want traditional dance music, some want comedians, some want rock music, some want movies, some want to go back to their cabins and read, and so on. Obviously, cruise entertainment directors want to provide the variety of entertainment their customers desire—within a reasonable budget—because satisfied customers tend to be repeat customers. The question is how to provide the right mix of entertainment.

On a cruise one of the authors and his wife recently took, the entertainment was of high quality and there was plenty of variety. A seven-piece show band played dance music nightly in the largest lounge, two other small musical combos played nightly at two smaller lounges, a pianist played nightly at a piano bar in an intimate lounge, a group of professional singers and dancers played Broadway-type shows about twice weekly, and various professional singers and comedians played occasional single-night performances.[2] Although this entertainment was free to all of the passengers, much of it had embarrassingly low attendance. The nightly show band and musical combos, who were contracted to play nightly until midnight, often had less than a half dozen people in the audience—sometimes literally none. The professional singers, dancers, and comedians attracted larger audiences, but there were still plenty of empty seats. In spite of this, the cruise staff posted a weekly schedule, and they stuck to it regardless of attendance. In a short-term financial sense, it didn't make much difference. The performers got paid the same whether anyone was in the audience or not, the passengers had already paid (indirectly) for the entertainment as part of the cost of the cruise, and the only possible opportunity cost to the cruise line (in the short run) was the loss of liquor sales from the lack of passengers in the entertainment lounges. The morale of the entertainers was not great—entertainers love packed houses—but they usually argued, philosophically, that their hours were relatively short and they were still getting paid to see the world.

If you were in charge of entertainment on this ship, how would you describe the problem with entertainment—is it a problem with deadbeat passengers, low-quality entertainment, or a mismatch between the entertainment offered and the entertainment desired? How might you try to solve the problem? What constraints might you have to work within? Would you keep a strict schedule such as the one followed by this cruise director, or would you play it more "by ear"? Would you gather data to help solve the problem? What data would you gather? How might you analyze these data? How much would financial considerations dictate your decisions? Would they be long-term or short-term considerations?

---

[2] There was also a moderately large onboard casino, but it tended to attract the same people every night, and it was always closed when the ship was in port.

Describing Data: Graphs and Tables

Graphical Analysis of the Challenger Disaster

On the morning of January 28, 1986, the U.S. space shuttle *Challenger* exploded a few minutes after takeoff, killing all seven crew members. The physical cause of this tragic accident was found to be a failure in one of the O-rings located in a joint on the right-hand-side solid rocket booster. The Presidential Commission investigating the disaster concluded that the decision-making process leading up to the launch of *Challenger* was seriously flawed. Prior to the *Challenger* disaster, engineers from NASA and Morton Thiokol (the manufacturer of the shuttle's solid rocket motors) had observed evidence of in-flight damage to these O-rings. In spite of this evidence and the cold weather that adversely affected the performance of the O-ring seals, the *Challenger* was launched—and the outcome was disastrous.

This tragedy provides a dramatic example of how well-chosen graphs can make—or could have made—a huge difference. Data were available from previous shuttle flights on the ambient temperature of the solid rocket motor joints at launch and the number of joints observed to have suffered some form of damage. (These are included for your interest in the file CHALLENGER.XLS.) One set of data lists this information only for those 7 previous flights where at least one joint suffered damage. Another set lists this information for all 23 previous flights. A scatterplot—one of the graph types we will examine in this chapter—of the first set shows how 2 previous flights, one with a relatively cool temperature and one with a relatively warm temperature, each had two joints damaged. The other 5 flights, all with intermediate temperatures, each had a single joint damaged. This scatterplot appears in Figure 2.1. (Two of the points are identical, which explains why there appear to be only 6 points.) It contains virtually no evidence that damage is related to temperature. Perhaps the decision makers referred to this plot on launch day to confirm their go-ahead decision.

© 2002 Harris Welles

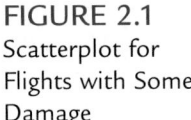

FIGURE 2.1
Scatterplot for Flights with Some Damage

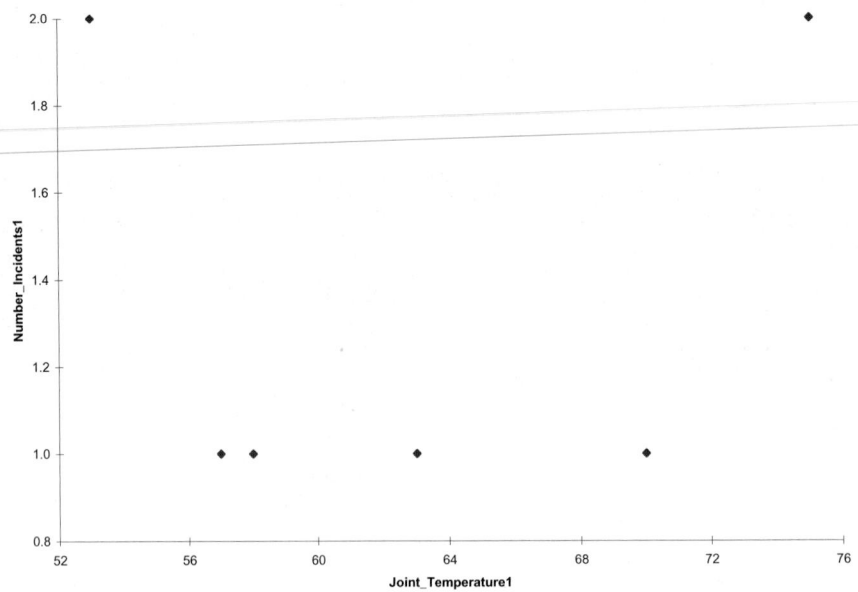

However, a scatterplot of the data from all 23 previous flights, shown in Figure 2.2, tells a somewhat different story. (Again, several points are identical, so there appear to be fewer than 23 points.) Note that the extra 16 points on this plot are all on the horizontal axis—that is, they all correspond to flights with no damage—and all correspond to relatively warm temperatures. In addition, it now becomes apparent that *most* of the flights with some damage occurred at relatively cool temperatures. The only exception is the point marked as a possible "outlier"—that is, a point outside the general pattern. If we ignore this potential outlier, a fairly clear pattern emerges: More damage tends to occur at low temperatures. Of course, this plot does not provide conclusive evidence that a shuttle launched at near-freezing tempera-

FIGURE 2.2
Scatterplot for All Previous Flights

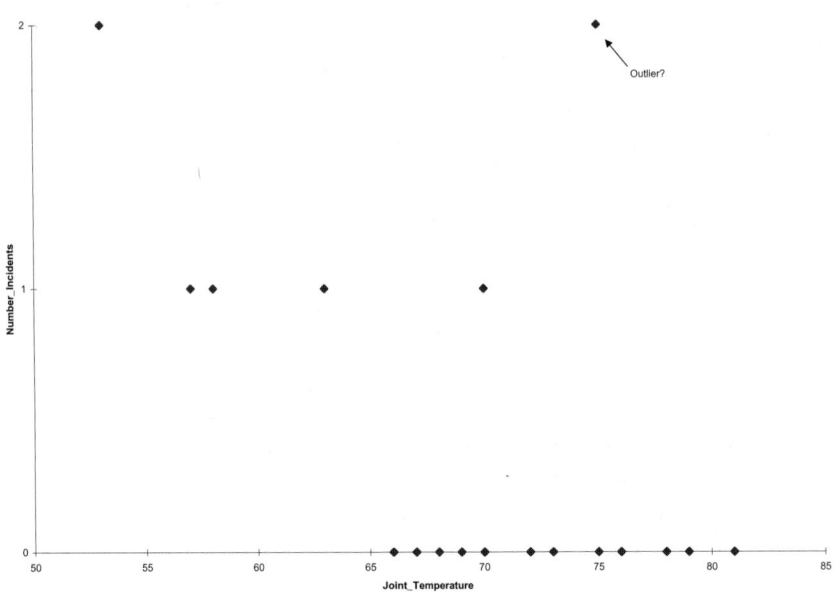

tures, which the fateful launch experienced, was doomed to disaster. However, it provides a clear warning. If you had observed this plot and the freezing temperature on that January morning, would you have decided to go ahead with the launch? ●

2.1 Introduction

The goal of this chapter and the next is very simple: to make sense out of data by constructing appropriate summary measures, tables, and graphs. Our purpose here is to take a set of data that at first glance has little meaning and to present the data in a form that makes sense to people. There are numerous ways to do this, limited only by our imagination, but there are several tools used most often: (1) a variety of graphs, including bar charts, pie charts, histograms, scatterplots, and time series plots, (2) tables of summary measures (such as totals, averages, etc.) grouped by categories, and (3) numerical summary measures such as counts, percentages, averages, and measures of variability. These terms might not all be familiar at this point, but you've undoubtedly seen examples of them in newspapers, magazine articles, and books.

The material in these two chapters is _simple, complex,_ and _important._ It is simple because there are no difficult mathematical concepts. With the possible exception of variance, covariance, and correlation, all of the numerical measures, tables, and graphs are natural and easy to understand. It used to be a tedious chore to produce them, but with the advances in statistical software, including add-ins for spreadsheet packages such as Excel, they can now be produced easily and quickly.

If it's so easy, why do we also claim that the material in this chapter is complex? The data sets available to companies in today's computerized world tend to be extremely large and filled with "unstructured" data. As we will see, even in data sets that are quite small in comparison to those real companies face, it is a challenge to summarize the data in such a way that the important _information_ stands out clearly. It is easy to produce summary measures, tables, and graphs, but the issue is one of producing the most _appropriate_ measures, tables, and graphs.

The typical employees of today—not just the managers and technical specialists—have a wealth of easy-to-use tools at their disposal, and it is frequently up to them to summarize data in a way that is both meaningful and useful to their constituents: people within their company, their company's suppliers, and their company's customers. It takes some training and practice to do this effectively.

Because today's companies are inundated with data, and because virtually every employee in the company must summarize data to some extent, the material in this chapter and Chapters 3 and 4 is arguably the most important material in this book. There has sometimes been a tendency to race through the "descriptive statistics" chapters to get to the more "interesting" material in later chapters as quickly as possible. We want to resist this tendency. The material covered in these three chapters deserves close examination, and this takes some time.

Most of the material in this chapter and the next two could be covered in any order. This chapter involves graphs and tables. The next chapter covers numerical summary measures and another type of graph, the _boxplot,_ which utilizes several of the summary measures. Chapter 3 concludes with several examples that put the descriptive tools from both chapters to good use. Finally, in Chapter 4 we discuss ways of getting data from various sources, such as Access databases, into a format in Excel where they can be analyzed by the methods discussed in Chapters 2 and 3 (and later chapters).

It is customary to refer to the raw numbers as _data_ and the output of a statistical analysis as _information._ We start with the data, and we hope to end with information that an organization can use for competitive advantage.

2.2 Basic Concepts

We begin with a short discussion of several important concepts: populations and samples, variables and observations, and types of data.

2.2.1 Populations and Samples

First, we distinguish between a *population* and a *sample.* A **population** includes all of the objects of interest, whether they be people, households, machines, or whatever. The following are three typical populations:

- All potential voters in a presidential election
- All subscribers to cable television
- All invoices submitted for Medicare reimbursement by a nursing home

In these situations and many others it is virtually impossible to obtain information about all members of the population. For example, it is far too costly to ask all potential voters which presidential candidates they prefer. Therefore, we often try to gain insights into the characteristics of a population by examining a **sample,** or subset, of the population. In later chapters we will examine populations and samples in some depth, but for now, it is enough to know that we typically want samples to be *representative* of the population so that observed characteristics of the sample can be generalized to the population as a whole.

Populations and Samples	A **population** includes all of the objects of interest in a study, whether they be people, households, machines, or whatever. A **sample** is a subset of the population, often randomly chosen and preferably representative of the population as a whole.

An example where a sample was *not* representative is found in the case of the *Literary Digest* fiasco of 1936. In the 1936 presidential election, subscribers to the *Literary Digest,* a highbrow literary magazine, were asked to mail in a ballot with their preference for president. Overwhelmingly, these ballots favored the Republican candidate, Alf Landon, over the Democratic candidate, Franklin D. Roosevelt. Despite this, FDR was a landslide winner. The discrepancy arose because the readers of the *Literary Digest* were not at all representative of most voters in 1936. Most voters in 1936 could barely make ends meet, let alone subscribe to a literary magazine. Thus the typical lower-middle-income voter had almost no chance of being chosen in this sample.

Today, Gallup, Harris, and other pollsters make a conscious effort to ensure that their samples—which usually include about 1500 people—are representative of the population. (It is truly remarkable, for example, that a sample of 1500 voters can almost surely predict a candidate's actual percentage of votes correct to within 3%. We will explain why this is possible in Chapters 8 and 9.) The important point is that a representative sample of reasonable size can give us a lot of important information about the population of interest.

We will use the terms *population* and *sample* a few times in this chapter, which is why we have defined them here. However, the distinction is not too important until later chapters. Our intent in this chapter is to focus entirely on the data in a given data set, not to generalize beyond it. Therefore, the given data set could be a population or a sample from a population. At this point, the distinction is largely irrelevant.

2.2.2 Variables and Observations

To standardize data analysis, especially on a computer, it is customary to present the data in rows and columns. Each column represents a **variable,** and each row corresponds to an **observation,** that is, a member of the population or sample. The numbers of variables and observations vary widely from one data set to another, but they can all be put in this row–column format.

The terms *variables* and *observations* are fairly standard. However, you might hear alternative terms. First, many people refer to **cases** instead of observations; that is, each row is a case. Second, if the data are stored in database packages such as Microsoft Access, the terms **fields** and **records** are typically used. Fields are the same as variables; each column corresponds to a field. Records are the same as observations (or cases); each row corresponds to a record.

Variables and Observations	A **variable** (or **field**) is an attribute, or measurement, on members of a population, such as height, gender, or salary. An **observation** (or **case** or **record**) is a list of all variable values for a single member of a population. A variable is usually listed in a column; an observation is usually listed in a row.

Example (2.1) Data from an Opinion Survey

The data set shown in Figure 2.3 represents 30 responses from a questionnaire concerning the president's environmental policies. (See the file CODING.XLS.) Identify the variables and observations.

FIGURE 2.3
Data from
Environmental
Survey

	A	B	C	D	E	F
1	Data from a questionnaire on environmental policy					
2						
3	Age	Gender	State	Children	Salary	Opinion
4	35	Male	Minnesota	1	$65,400	5
5	61	Female	Texas	2	$62,000	1
6	35	Male	Ohio	0	$63,200	3
7	37	Male	Florida	2	$52,000	5
8	32	Female	California	3	$81,400	1
9	33	Female	New York	3	$46,300	5
10	65	Female	Minnesota	2	$49,600	1
11	45	Male	New York	1	$45,900	5
12	40	Male	Texas	3	$47,700	4
13	32	Female	Texas	1	$59,900	4
14	57	Male	New York	1	$48,100	4
15	38	Female	Virginia	0	$58,100	3
16	37	Female	Illinois	2	$56,000	1
17	42	Female	Virginia	2	$53,400	1
18	38	Female	New York	2	$39,000	2
19	48	Male	Michigan	1	$61,500	2
20	40	Male	Ohio	0	$37,700	1
21	57	Female	Michigan	2	$36,700	4
22	44	Male	Florida	2	$45,200	3
23	40	Male	Michigan	0	$59,000	4
24	21	Female	Minnesota	2	$54,300	2
25	49	Male	New York	1	$62,100	4
26	34	Male	New York	0	$78,000	3
27	49	Male	Arizona	0	$43,200	5
28	40	Male	Arizona	1	$44,500	3
29	38	Male	Ohio	1	$43,300	1
30	27	Male	Illinois	3	$45,400	2
31	63	Male	Michigan	2	$53,900	1
32	52	Male	California	1	$44,100	3
33	48	Female	New York	2	$31,000	4

Objective To illustrate variables and observations in a typical data set.

Solution

This data set provides observations on 30 people who responded to the questionnaire. Each observation lists the person's age, gender, state of residence, number of children, annual salary, and opinion of the president's environmental policies. These six pieces of information represent the variables. It is customary to include a row (row 3) that gives variable names. These variable names should be meaningful—and no longer than necessary. ●

2.2.3 Types of Data

Three variables that appear to be numerical but are usually treated as categorical are phone numbers, zip codes, and Social Security numbers. Do you see why? Can you think of others?

There are several ways to categorize data, as we will explain in the context of Example 2.1. We distinguish, for example, between *numerical* and *categorical* data. The basic distinction here is whether we intend to do any arithmetic on the data. It makes sense to do arithmetic on numerical data, but not on categorical data. Clearly, the Gender and State variables are categorical, and the Children and Salary variables are numerical. The Age and Opinion variables are more difficult to categorize. Age is expressed numerically, and we *might* want to perform some arithmetic on age (such as calculating the average age of the respondents). However, age might be treated as a categorical variable, as we will see shortly.

Numerical and Categorical Variables	A variable is **numerical** if meaningful arithmetic can be performed on it. Otherwise, the variable is **categorical.**

The Opinion variable is expressed numerically, on a 1-to-5 **Likert** scale. These numbers are only "codes" for the categories "strongly disagree," "disagree," "neutral," "agree," and "strongly agree." We never intend to perform arithmetic on these numbers; in fact, it is not really appropriate to do so.[1] Here, then, we treat the Opinion variable as categorical. Note, too, that there is a definite ordering of its categories, whereas there is no natural ordering of the categories for the Gender or State variables. When there is a natural ordering of categories, we classify the variable as **ordinal.** If there is no natural ordering, as with the Gender and State variables, we classify the variables as **nominal.** However, both ordinal and nominal variables are categorical.

Ordinal and Nominal Variables	A categorical variable is **ordinal** if there is a natural ordering of its possible values. If there is no natural ordering, it is **nominal.**

Excel Tip *How do you remember, for example, that 1 stands for "strongly disagree" in the Opinion variable? You can enter a comment—a reminder to yourself and others—in any cell. This feature appears under the Insert menu. Alternatively, you can right-click on a cell and select the Insert Comment menu item. A small red tag appears in any cell with*

---

[1]Some people do take averages, for example, of numbers such as these, but there are conceptual reasons for *not* doing so; the resulting averages can be misleading.

a note. Moving the cursor over that cell causes the comment to appear. You will see numerous comments in the files that accompany this book.

Categorical variables can be *coded* numerically or left uncoded. In Figure 2.3, Gender has not been coded, whereas Opinion has been coded. This is largely a matter of taste—so long as you realize that coding a truly categorical variable does not make it numerical and open to arithmetic operations. An alternative is shown in Figure 2.4. Now Gender has been coded (1 for males, 2 for females), and Opinion has not been coded. In addition, we have categorized the Age variable as "young" (34 or younger), "middle-aged" (from 35 to 59), and "elderly" (60 or older). The purpose of the study dictates whether age should be treated numerically or categorically; there is no right or wrong way.

Numerical variables can be classified as either *discrete* or *continuous*. The basic distinction is whether the data arise from counts or continuous measurements. The variable Children is clearly a count (that is, *discrete*), whereas the variable Salary is best treated as continuous. This distinction between discrete and continuous variables is sometimes important, as it dictates the type of analysis that is most natural.

Discrete and Continuous Variables	A numerical variable is **discrete** if its possible values can be counted. A **continuous** variable is the result of an essentially continuous measurement.

FIGURE 2.4
Environmental Data Using a Different Coding

	A	B	C	D	E	F
1	Data from a questionnaire on environmental policy					
2						
3	Age	Gender	State	Children	Salary	Opinion
4	Middle-aged	1	Minnesota	1	$65,400	Strongly agree
5	Middle-aged	2	Texas	2	$62,000	Strongly disagree
6	Elderly	1	Ohio	0	$63,200	Neutral
7	Middle-aged	1	Florida	2	$52,000	Strongly agree
8	Young	2	California	3	$81,400	Strongly disagree
9	Young	2	New York	3	$46,300	Strongly agree
10	Elderly	2	Minnesota	2	$49,600	Strongly disagree
11	Middle-aged	1	New York	1	$45,900	Strongly agree
12	Middle-aged	1	Texas	3	$47,700	Agree
13	Young	2	Texas	1	$59,900	Agree
14	Middle-aged	1	New York	1	$48,100	Agree
15	Middle-aged	2	Virginia	0	$58,100	Neutral
16	Middle-aged	2	Illinois	2	$56,000	Strongly disagree
17	Middle-aged	2	Virginia	2	$53,400	Strongly disagree
18	Middle-aged	2	New York	2	$39,000	Disagree
19	Middle-aged	1	Michigan	1	$61,500	Disagree
20	Middle-aged	1	Ohio	0	$37,700	Strongly disagree
21	Middle-aged	2	Michigan	2	$36,700	Agree
22	Middle-aged	1	Florida	2	$45,200	Neutral
23	Middle-aged	1	Michigan	0	$59,000	Agree
24	Young	2	Minnesota	2	$54,300	Disagree
25	Middle-aged	1	New York	1	$62,100	Agree
26	Young	1	New York	0	$78,000	Neutral
27	Middle-aged	1	Arizona	0	$43,200	Strongly agree
28	Middle-aged	1	Arizona	1	$44,500	Neutral
29	Middle-aged	1	Ohio	1	$43,300	Strongly disagree
30	Young	1	Illinois	3	$45,400	Disagree
31	Elderly	1	Michigan	2	$53,900	Strongly disagree
32	Middle-aged	1	California	1	$44,100	Neutral
33	Middle-aged	2	New York	2	$31,000	Agree

Finally, data can be categorized as *cross-sectional* or *time series*. The opinion data in Example 2.1 are cross-sectional. A pollster evidently sampled a cross section of people at one particular point in time. In contrast, time series data occur when we track one or more variables through time. A typical example of a time series variable is the series of daily closing values of the Dow Jones Index. Very different types of analysis are appropriate for cross-sectional and time series data, as will become apparent in this and later chapters.

Cross-sectional and Time Series Data	**Cross-sectional** data are data on a population at a distinct point in time. **Time series** data are data collected across time.

(2.3) Frequency Tables and Histograms

A good place to start building a "toolkit" of descriptive methods is with *frequency tables* and their graphical analog, *histograms*. A frequency table indicates how many observations fall in various categories. A histogram shows this same information graphically. We'll construct a frequency table and a histogram in the following example.

Frequency Tables and Histograms	A **frequency table** lists the number of observations of some variable that fall in various categories. A **histogram** is a bar chart of these frequencies.

Example (2.2) Data on Famous Movie Stars

The file ACTORS.XLS contains information on 66 movie stars. (See Figure 2.5.) This data set contains the name of each actor and the following four variables:

- Gender
- DomesticGross: average domestic gross of star's last few movies (in $ millions)
- ForeignGross: average foreign gross of star's last few movies (in $ millions)
- Salary: current amount the star asks for a movie (in $ millions)[2]

   We are interested in summarizing the 66 salaries in a frequency table and a histogram.

Objective To use StatPro's histogram procedure to plot the distribution of actors' salaries.

Solution

To obtain a frequency table for data that are essentially continuous, such as the Salary variable, we must first choose appropriate categories. There is no set rule here. We want to have enough categories so that we can see a meaningful distribution, but we don't want so many categories that there are only a few observations per category. A good rule of thumb is to divide the range of values into 8 to 15 equally spaced categories, plus a pos-

---

[2]We realize that this quantity is not a salary in the usual sense of the term, but we'll use the word *salary* throughout this example.

FIGURE 2.5
Data on Famous
Actors and Ac-
tresses

	A	B	C	D	E
1	**Famous actors and actresses**				
2					
3	Note: All monetary values are in $ millions.				
4					
5	Name	Gender	DomesticGross	ForeignGross	Salary
6	Angela Bassett	F	32	17	2.5
7	Jessica Lange	F	21	27	2.5
8	Winona Ryder	F	36	30	4
9	Michelle Pfeiffer	F	66	31	10
10	Whoopi Goldberg	F	32	33	10
11	Emma Thompson	F	26	44	3
12	Julia Roberts	F	57	47	12
13	Sharon Stone	F	32	47	6
14	Meryl Streep	F	34	47	4.5
15	Susan Sarandon	F	38	49	3
16	Nicole Kidman	F	55	51	4
17	Holly Hunter	F	51	53	2.5
18	Meg Ryan	F	43	55	8.5
19	Andie Macdowell	F	26	75	2
59	Clint Eastwood	M	55	94	12.5
60	Mel Gibson	M	91	95	19
61	Bruce Willis	M	55	99	16.5
62	Bill Pullman	M	38	103	6
63	Liam Neeson	M	29	108	3
64	Samuel Jackson	M	40	122	4.5
65	Jim Carrey	M	122	123	15
66	Morgan Freeman	M	77	123	6
67	Arnold Scharz	M	108	124	20
68	Brad Pitt	M	57	124	10
69	Michael Douglas	M	68	137	18
70	Robin Williams	M	92	180	15
71	Tom Hanks	M	166	182	17.5

Although this is not a universal convention, when we indicate a category such as 2–4, we mean that this includes the right endpoint, 4, and excludes the left endpoint, 2.

sible open-ended category at either end of the range. For this data set we choose the categories 0–2, 2–4, 4–6, 6–8, 8–10, 10–12, 12–14, 14–16, 16–18, 18–20, and over 20. (All of these are in millions of dollars.)

To create the histogram in Excel, we use the StatPro add-in that accompanies this book.

Steps for creating a histogram and the associated frequency table

❶ Place the cursor anywhere within the data set. This is a common first step in the StatPro add-in. If you can imagine a big rectangular range that contains the data set, including the variable names at the top, the cursor should be somewhere—anywhere—within this range. If you forget to place it there, the add-in will prompt you to do so before continuing.

Excel Tip *There is a little-known key combination for selecting the rectangular region "surrounding" any cell in a data set: Ctrl-Shift-*. Try it. With the Actors file open, put your cursor in any cell inside the data set, such as D12. Press Ctrl-Shift-* (all three keys at once) to highlight the data range A5:E71. This is what StatPro does for you when it makes its guess in Step 1.*

❷ Select the StatPro/Charts/Histogram(s) menu item.

FIGURE 2.6
Histogram Dialog
Box in StatPro
Add-In

Defining histogram categories

Notes on inputs below

Left-most category extends up to 'Minimum value'.
'Number of categories' includes open-ended categories on left and right.
'Category length' applies to all other (non-open-ended) categories.

OK

Cancel

Information of variable being plotted

Here is some useful information about Salary

Minimum: 2.000   5th percentile: 2.500

Maximum: 20.000   95th percentile: 19.950

Category settings for variable Salary

Minimum value: 2

Number of categories: 11

Category length: 2

Example of settings

Entering 500, 6, and 100 would
produce categories: <=500, 500
to 600, 600 to 700, 700 to 800,
800 to 900, and >900.

☐ Test normal fit

③ A list of *numerical* variables in the data set appears. You can select one or more of these and obtain a frequency table and histogram for each variable you select. For now, select the Salary variable.

④ Now comes the important part, where you specify the categories. (See Figure 2.6.) You need to enter (1) the upper limit of the first (leftmost) category, (2) the total number of categories, and (3) the typical length of a category. For this example, enter 2, 11, and 2.

⑤ The histogram is placed on a separate "chart" sheet with the name Hist-Salary. If a sheet with this name already exists, it is replaced. Another sheet with the name Hist-SalaryData is also added. It contains the frequencies for the histogram.

FIGURE 2.7
Histogram of
Actor and Actress
Salaries

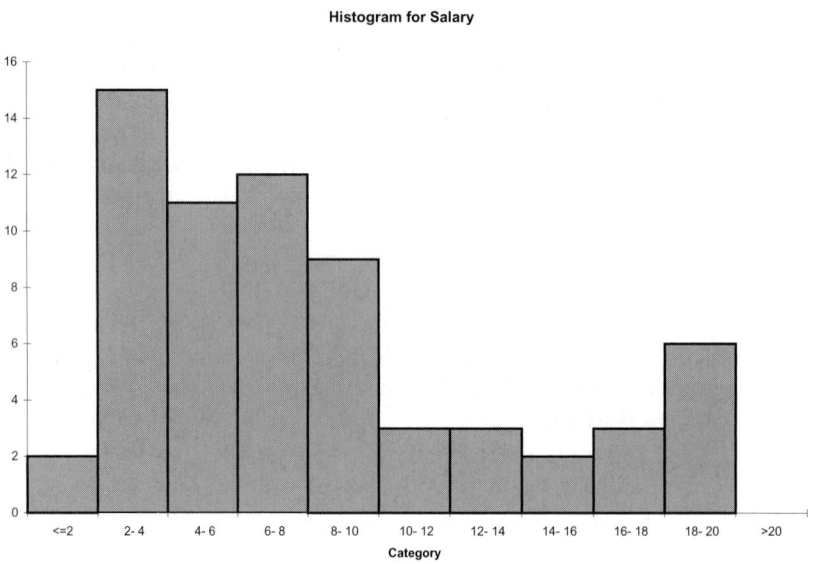

Histogram for Salary

FIGURE 2.8
Frequency Table
for Actor and
Actress Salaries

	A	B	C
1	**Frequency table for Salary**		
2			
3	Upper limit	Category	Frequency
4	2	<=2	2
5	4	2- 4	15
6	6	4- 6	11
7	8	6- 8	12
8	10	8- 10	9
9	12	10- 12	3
10	14	12- 14	3
11	16	14- 16	2
12	18	16- 18	3
13	20	18- 20	6
14		>20	0

The resulting histogram and frequency table for the Salary variable appear in Figures 2.7 and 2.8. It is clear that most salaries are in the $2–$10 million range, but a few are considerably larger. Specifically, 17 of the 66 actors, or about 25%, make more than $10 million, and 6 of the 66, or about 9%, are in the highest salary bracket, $18–$20 million.

If you aren't satisfied with this histogram (you want fewer categories, for example), just repeat the procedure. You will overwrite the previous output. ●

Excel Tip *StatPro automatically puts all charts on separate "chart sheets." If you would rather have a chart on the same sheet as the data, it is easy to move it. Go to the chart sheet, use the Chart/Location menu item, and select the "As object in" option, with the desired data sheet selected in the box. The chart sheet will no longer exist, and the chart will automatically appear in the data sheet. This process can also be reversed by selecting the "As new sheet" option under the Chart/Location menu item.*

To see how the choice of categories can make a difference, try selecting the following settings, instead of our 2, 11, and 2, for the salaries histogram: (1) 6, 3, 6; (2) 0.5, 40, 0.5; and (3) 10, 10, 2. None of these is "wrong," but none of them produces a very informative histogram.

Creating a histogram can be a tedious task, but an add-in such as StatPro makes it relatively easy. However, you must be prepared to fill in the dialog box in Figure 2.6—that is, you must be ready to specify the categories. This might take some trial and error. The following guidelines should be helpful.

Planning a histogram

- Usually, choose about 8–15 categories. The more observations you have, the more categories you can afford to have.

- Try to select categories that "fill" the range of data. For example, it wouldn't make sense to have 10 categories of length $1000, starting with the category "less than $30,000," if most of the observations are in the $20,000 to $50,000 range. In this case all of the data would fall in the first few categories, virtually no data would fall in the higher categories, and the histogram would not tell a very interesting story.

- It is customary to plan the categories so that the "breakpoints" between categories are nice round numbers.

- The leftmost and rightmost categories are often open-ended categories such as "less than or equal to $20,000" and "greater than $100,000." (In this case, the first entry in the dialog box in Figure 2.6 would be 20000.) The "category length" requested in the dialog box is for the middle categories, not these open-ended categories. If a typical middle category is from \$30,000 to \$40,000, then 10000 should be entered as the category length.

- Above all, remember that there is not a single "right" answer. If your initial entries in the dialog box don't produce a very interesting histogram, try it again with new entries. Note that the dialog box in Figure 2.6 indicates the range of the data. This should help you choose reasonable categories.

An Alternative Method for Creating Histograms in Excel It is also possible to create histograms in Excel *without* using StatPro or any other statistical add-in. This requires a combination of the FREQUENCY function—a good function to know in its own right—and the Chart Wizard.

Steps to create a histogram

1. Define the categories as in column A of Figure 2.8. That is, specify the *upper limit* of each category: 2 through 20. Excel refers to this range of upper limits (A4:A13 in the figure) as the "bin range," so we suggest giving it the range name Bins. The "greater than 20" is then treated as an implied extra category.

2. Enter *labels* to describe the categories in column B, again as in Figure 2.8. (This step is not necessary, but it will make the histogram more readable.)

3. Select the range C4:C14 (one cell longer than the Bins range), type the formula

$$=FREQUENCY(Salary,Bins)$$

and press Ctrl-Shift-Enter (press all three keys at once). Here, Salary is the range name for the Salary variable data (E6:E71 in Figure 2.5).

> **Excel Tip** *The FREQUENCY function is an **array** function in Excel, which means that it fills a whole range in one step. If you look at the formula bar, you'll notice curly brackets around this formula. You should **not** type these curly brackets. They appear automatically when you press Ctrl-Shift-Enter, and they indicate that this is an array formula. Note that the value in the "extra" cell, C14, is the number of observations greater than the last bin value, 20. For this example it just happens that there are no such observations.*

4. Form the histogram from columns B and C by using the Chart Wizard with the column chart type. You should obtain a chart essentially like the one shown in Figure 2.7, which can then be modified to your taste with Excel's many chart options.

In summary, keep in mind that when you create a histogram, you must first define the categories. To do this sensibly, you should first take a look at the data—how many observations are there, and how small and large are the data values? Of course, if you create a histogram and it doesn't appear to provide much information, you can always redefine the categories and try again.

The term **distribution** refers to the way the data are distributed in the various categories. It is common to refer to a positively skewed distribution, say, rather than a positively skewed histogram. However, either term can be used.

2.3.1 Shapes of Histograms

Four different shapes of histograms are commonly observed: symmetric, positively skewed, negatively skewed, and bimodal. A histogram is **symmetric** if it has a single peak and looks approximately the same to the left and right of the peak. For reasons that will become apparent in a later chapter, symmetric histograms are very common. One is illustrated in the following example.

Example 2.3

Distribution of Diameters of Elevator Rails at Otis Elevator

Otis Elevator has measured the diameter (in inches) of 400 elevator rails. (See the file OTIS1.XLS.) The diameters range from a low of approximately 0.449 inch to a high of approximately 0.548 inch. Check that these diameters follow a symmetric distribution.

Objective To illustrate a symmetric distribution of part diameters.

Solution

To create a histogram, we must first decide on categories. Given the range of the diameters, we've chosen to define the categories "less than 0.455," "0.455 to 0.465," and so on, up to "greater than 0.545." Of course, other choices are certainly possible. Now we will use StatPro's Histogram procedure to create the histogram in Figure 2.9. [The settings required to obtain these categories are: (1) upper limit of first category: 0.455; (2) number of categories: 11; and (3) length of typical category: 0.01.]

Clearly, the most likely diameters are between 0.495 and 0.505. Also, we see that the distribution of diameters is fairly symmetric. For example, 70 rails have diameters between 0.485 and 0.495, whereas nearly the same number (67) have diameters between 0.505 and 0.515. The diameters appear to follow the bell-shaped "normal" distribution. We will discuss the normal distribution in detail in Chapter 6.

FIGURE 2.9
Symmetric Distribution of Diameters

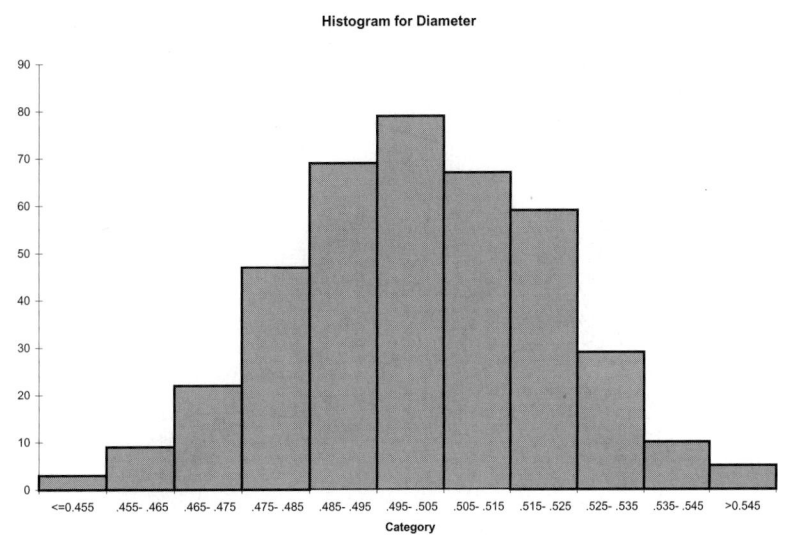

A histogram is **skewed to the right** (or **positively** skewed) if it has a single peak and the values of the distribution extend much farther to the right of the peak than to the left of the peak. One common example of positively skewed data is the following.

Example 2.4

Distribution of Times Between Arrivals to a Bank

The file BANK.XLS lists the time between customer arrivals—called interarrival times—for all customers arriving at a bank on a given day. Do these interarrival times appear to be positively skewed?

Objective To illustrate a positively skewed distribution of times between customer arrivals to a bank.

Solution

We subdivide the data into intervals of 2.5 seconds; the categories are "0 to 2.5," "2.5 to 5.0," and so on, up to "greater than 27.5." The resulting histogram appears in Figure 2.10. The interarrival times are clearly positively skewed: There is a "long tail" to the right of the peak, and none to the left. We also see that values over 15 minutes are quite unlikely. Evidently, there is usually very little time between consecutive customer arrivals. Now and then, however, there is a fairly large gap between arrivals. These occasional large gaps could be important for the bank. They could help enable tellers to clear out long waiting lines.

FIGURE 2.10
Positively Skewed Distribution of Interarrival Times

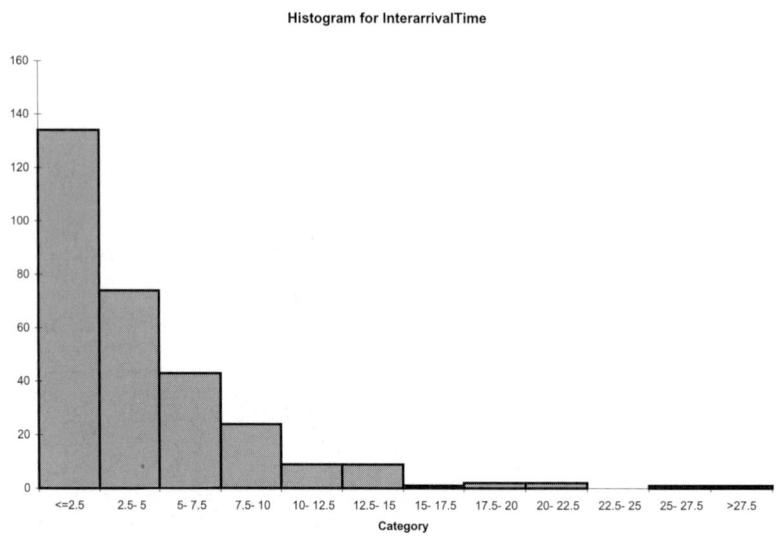

Histogram for InterarrivalTime

A histogram is **skewed to the left** (or **negatively** skewed) if its longer tail is on the left. Negatively skewed distributions are probably less common than positively skewed distributions, but the following example illustrates one situation where negative skewness often occurs.

Example **2.5**

Distribution of Accounting Midterm Scores

The file MIDTERM.XLS lists the midterm scores for a large class of accounting students. Does the histogram indicate a negatively skewed distribution?

Objective To illustrate a negatively skewed distribution of scores on a midterm exam.

Solution

For this example we divide the range into 5% intervals, creating categories "less than or equal to 45," "45 to 50," and so on, up to "greater than 95" to create the histogram in Figure 2.11. This histogram shows that the most likely score is between 85 and 90. Clearly, the right tail can extend only to 100, while the left tail tapers off gradually. This leads to the obvious negative skewness. You've probably been in such classes, where most students do reasonably well but a few pull down the class average.

FIGURE 2.11
Negatively Skewed
Distribution of
Midterm Scores

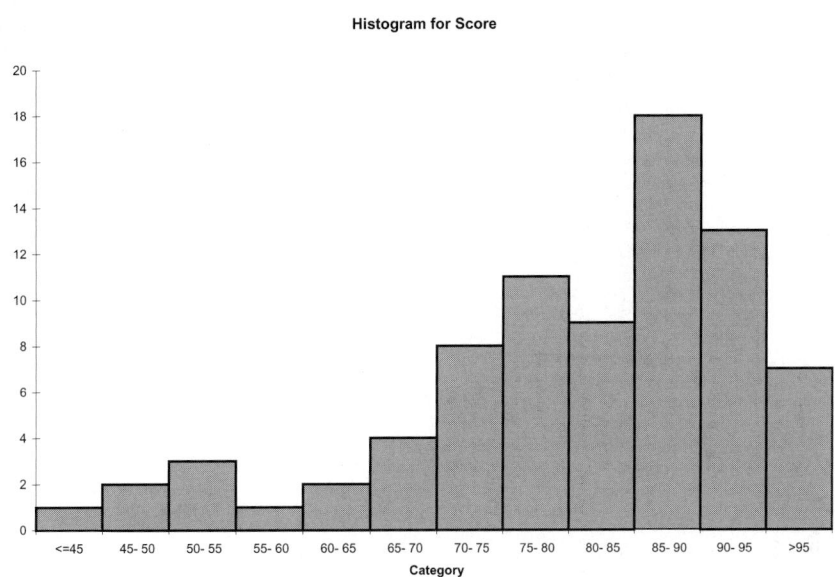

Some histograms have two or more peaks. This is often an indication that the data come from two or more distinct populations. Example 2.6 illustrates a situation where there are exactly *two* peaks, called a **bimodal distribution.**

Example **2.6**

Distribution of Diameters of Elevator Rails Produced on Two Machines

The file OTIS2.XLS lists the diameters of all elevator rails produced on a single day at Otis Elevator. Otis uses two machines to produce elevator rails. What do we learn from a histogram of these data?

Objective To illustrate a bimodal distribution of part diameters manufactured by two separate machines.

Solution

The diameters from the individual machines are listed in columns A and B of the OTIS2.XLS file. We merge these into a single variable in column C. Figure 2.12 shows the histogram on the merged data in column C. This is an obvious bimodal distribution, and it provides clear evidence of two distinct populations. Evidently, rails from one machine average about 0.5 inch in diameter, while rails from the other average about 0.6 inch in diameter. (A closer look at the data confirms this.) In such a case it is better to construct a single histogram for each machine's production, as in Figures 2.13 and 2.14. (These are based on the data in columns A and B, respectively.) These show that each machine's distribution is reasonably symmetric, although the scales on their horizontal axes are quite different.

FIGURE 2.12
Bimodal Distribution of Diameters from Both Machines

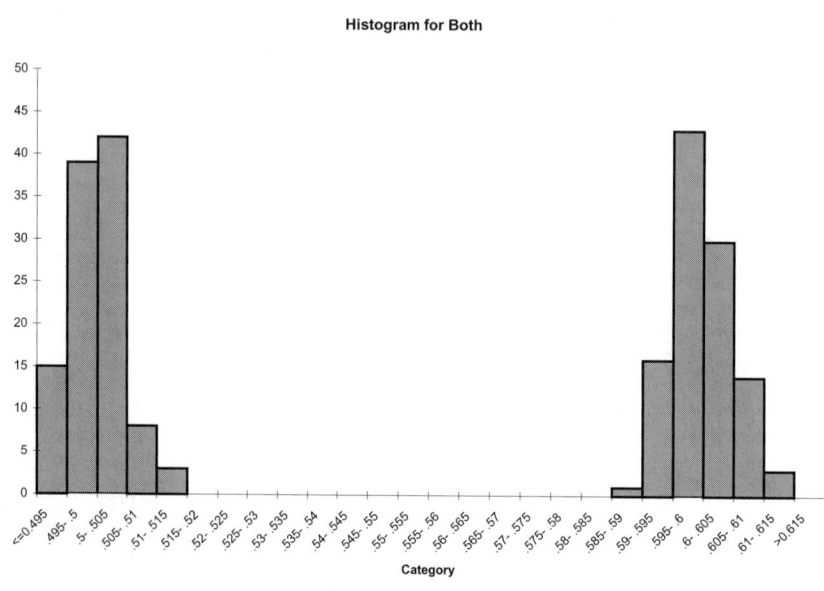

FIGURE 2.13
Distribution of Diameters from Machine 1

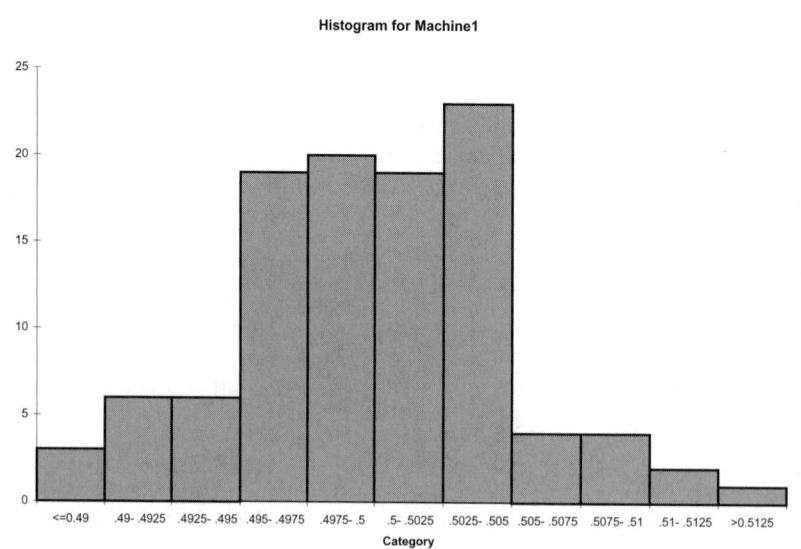

FIGURE 2.14
Distribution of Di-
ameters from Ma-
chine 2

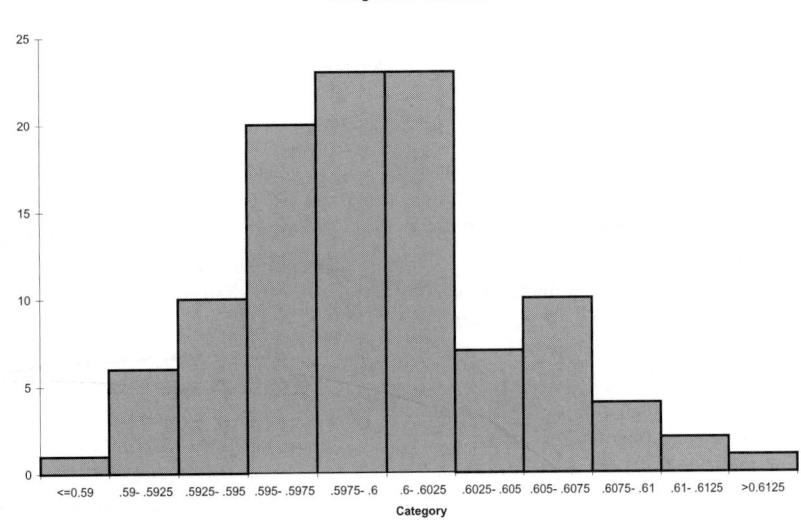

Histogram for Machine2

The bimodal histogram in Figure 2.12 provides a clear sign to management that there *are* two separate processes—and that possibly at least one of the two machines should be adjusted. ●

PROBLEMS

Level A

1. A human resources manager at Beta Technologies, Inc., has collected current annual salary figures and related data for 52 of the company's full-time employees. The data are in the file P02_01.XLS. In particular, these data include each selected employee's gender, age, number of years of relevant work experience prior to employment at Beta, the number of years of employment at Beta, the number of years of post-secondary education, and annual salary.
   a. Indicate the type of data for each of the six variables included in this set.
   b. Construct a frequency table and histogram for the ages of the employees included in this sample. How would you characterize the age distribution in this case?
   c. Construct a frequency table and histogram for the salaries of the employees included in this sample. How would you characterize the salary distribution in this case?

2. A production manager is interested in determining the proportion of defective items in a typical shipment of one of the computer components that her company manufactures. The proportion of defective components is recorded for each of 500 randomly selected shipments collected during a 1-month period. The data are

in the file P02_02.XLS. Construct a frequency table and histogram that will enable this production manager to begin to understand the variation of the proportion of defective components in the company's shipments.

3. *Business Week's Guide to the Best Business Schools* provides enrollment data on graduate business programs that they rate as the best in the United States. Specifically, this guide reports the percentages of women, minority, and international students enrolled in each of the top programs, as well as the total number of full-time students enrolled in each of these distinguished programs. The data are in the file P02_03.XLS.
   a. Generate frequency tables and histograms for the distributions of each of the four variables included in this data set.
   b. Compare the distributions of the proportions of women, minority, and international students enrolled in *Business Week*'s top graduate business schools. What general conclusions can be drawn by comparing the histograms of these three distributions?

4. The manager of a local fast-food restaurant is interested in improving the service provided to customers who use the restaurant's drive-up window. As a first step in this process, the manager asks his assistant to

record the time (in minutes) it takes to serve 200 different customers at the final window in the facility's drive-up system. The given 200 customer service times are all observed during the busiest hour of the day for this fast-food operation. The data are in the file P02_04.XLS.

a. Construct a frequency table and histogram for the distribution of observed customer service times.

b. Are shorter or longer service times more likely in this case?

5. A finance professor has just given a midterm examination in her corporate finance course. In particular, she is interested in learning how her class of 100 students performed on this exam. The data are in the file P02_05.XLS. Generate a histogram of this distribution of exam scores (where the maximum possible score is 100). Based on the histogram and associated frequency table, how would you characterize the group's performance on this test?

6. Five hundred households in a middle-class neighborhood were recently surveyed as a part of an economic development study conducted by the local government. Specifically, for each of the 500 randomly selected households, the survey requested information on the following variables: family size, approximate location of the household within the neighborhood, an indication of whether those surveyed owned or rented their home, gross annual income of the first household wage earner, gross annual income of the second household wage earner (if applicable), monthly home mortgage or rent payment, average monthly expenditure on utilities, and the total indebtedness (excluding the value of a home mortgage) of the household. The data are in the file P02_06.XLS.

a. Indicate the type of data for each of the eight variables included in this survey.

b. For each of the categorical variables in this survey, indicate whether the identified variable is *nominal* or *ordinal*. Explain your reasoning in each case.

c. Construct a frequency table and histogram for each of the numerical variables in this data set. Indicate whether each of these distributions is approximately symmetric or skewed. Which, if any, of these distributions are skewed to the right? Which, if any, of these distributions are skewed to the left?

7. A real estate agent has gathered data on 150 houses that were recently sold in a suburban community. Included in this data set are observations for each of the following variables: the appraised value of each house (in thousands of dollars), the selling price of each house (in thousands of dollars), the size of each house (in hundreds of square feet), and the number of bedrooms in each house. The data are in the file P02_07.XLS.

a. Indicate whether each of these four variables is *continuous* or *discrete*.

b. Generate frequency tables and histograms for both the appraised values and selling prices of the 150 houses included in the given sample. In what ways are these two distributions similar? In what ways are they different?

8. In a recent ranking of top graduate business schools in the United States published by *U.S. News & World Report,* data were provided on a number of attributes of recognized graduate programs. Specifically, the following variables were considered by *U.S. News & World Report* in establishing its overall ranking: each program's reputation rank by academics, each program's reputation rank by recruiters, each program's student selectivity rank, each program's placement success rate, the average GMAT score for students enrolled in each program, the average undergraduate grade-point average for students enrolled in each program, each program's recent acceptance rate, the typical starting base salary for recent graduates from each program, the proportion of recent graduates from each program who were employed within 3 months of completing their graduate studies, and the out-of-state tuition paid by affected full-time students in each program. The data are in the file P02_08.XLS.

a. Indicate the type of data for each of the ten variables considered in the formulation of the overall ranking.

b. For each of the categorical variables in this set, indicate whether the identified variable is *nominal* or *ordinal.* Explain your reasoning in each case.

c. Construct a frequency table and histogram for each of the numerical variables in this data set. Indicate whether each of these distributions is approximately symmetric or skewed. Which, if any, of these distributions are skewed to the right? Which, if any, of these distributions are skewed to the left?

9. The operations manager of a toll booth, located at a major exit of a state turnpike, is trying to estimate the average number of vehicles that arrive at the toll booth during a 1-minute period during the peak of rush-hour traffic. In an effort to estimate this average throughput value, he records the number of vehicles that arrive at the toll booth over a 1-minute interval commencing at the same time for each of 365 normal weekdays. The data are in the file P02_09.XLS.

a. Generate a histogram of the number of vehicles that arrive at this toll booth over the period of this study.

b. Characterize this observed arrival distribution. Specifically, is it equally likely for smaller and larger numbers of vehicles to arrive during the chosen 1-minute period?

10. The SAT test score includes both verbal and mathematical components. The average scores on both the verbal and mathematical portions of the SAT have been computed for students taking this standardized test in each of the 50 states and the District of Columbia. Also, the proportion of high school graduates taking the test in each of the 50 states and the District of Columbia is recorded. The data are in the file P02_10.XLS.
   a. Construct a histogram for each of these three distributions of numerical values. Are these distributions essentially symmetric or are they skewed?
   b. Compare the distributions of the average verbal scores and average mathematical scores. In what ways are these two distributions similar and in what ways are they different?

11. In ranking metropolitan areas in the United States, David Savageau and Geoffrey Loftus, the authors of *Places Rated Almanac,* consider the average time (in minutes) it takes a citizen of each metropolitan area to travel to work and back home each day. The data are in the file P02_11.XLS. Generate a histogram for this distribution of daily commute times. Are shorter or longer average daily commute times generally more likely for citizens residing in these metropolitan areas?

12. The U.S. Department of Transportation regularly publishes the *Air Travel Consumer Report,* which provides a variety of performance measures of major U.S. commercial airlines. One dimension of performance reported is each airline's percentage of domestic flights arriving within 15 minutes of the scheduled arrival time at major reporting airports throughout the country. The data are in the file P02_12.XLS.
   a. Construct a frequency table and histogram for each airline's distribution of percentage of on-time arrivals at the reporting airports. Indicate whether each distribution is skewed or not.
   b. Visually compare the histograms you constructed in part **a.** What general conclusions emerge from your visual comparisons regarding the on-time performance of these major U.S. air carriers?

13. According to a survey conducted by New York compensation consultants William M. Mercer Inc. and published in *The Wall Street Journal* (April 9, 1998), chief executive officers from 350 of the nation's biggest businesses gained an 11.7% increase in salaries and bonuses in 1997. The data are in the file P02_13.XLS. This dramatic increase came on the heels of an 8.9% jump in corporate profit. Construct frequency tables and histograms to gain a clearer understanding of both the distributions of annual base salaries and of bonuses earned by the surveyed CEOs in fiscal 1997.

2.4 Analyzing Relationships with Scatterplots

We are often interested in the relationship between two variables. A useful way to picture this relationship is to plot a point for each observation, where the coordinates of the point represent the values of the two variables. The resulting graph is called a *scatterplot.* By examining the scatter of points, we can usually see whether there is any relationship between the two variables, and if so, what type of relationship it is. We illustrate this method in the following example.

Scatterplot	A **scatterplot** contains a point for each observation, based on the values of two selected variables. The resulting plot indicates the relationship, if any, between these two variables.

Example 2.7 Relationship Between Salaries and Domestic Gross for Movie Stars

The data in the ACTORS.XLS file may lead us to suppose that stars whose movies gross large amounts have the largest salaries. Is this actually true?

Objective To use StatPro's scatterplot procedure to illustrate the relationship between movie gross and actor salary.

Solution

To analyze this, plot each star's salary on the vertical axis and the corresponding domestic gross on the horizontal axis. The resulting scatterplot appears in Figure 2.15. This graph can be obtained with StatPro add-in. To do so, use the StatPro/Charts/Scatterplot(s) menu item, select Salary and DomesticGross as the variables to plot, and (in the next dialog box) indicate that you want Salary on the vertical axis.

FIGURE 2.15
Scatterplot of Salary versus DomesticGross

StatPro's scatterplot procedure automatically calculates the *correlation* between the two variables and shows it in a text box at the top right of the chart. If you don't want the correlation to show, simply click on the text box and press the Delete key. Correlation will be explained in the next chapter.

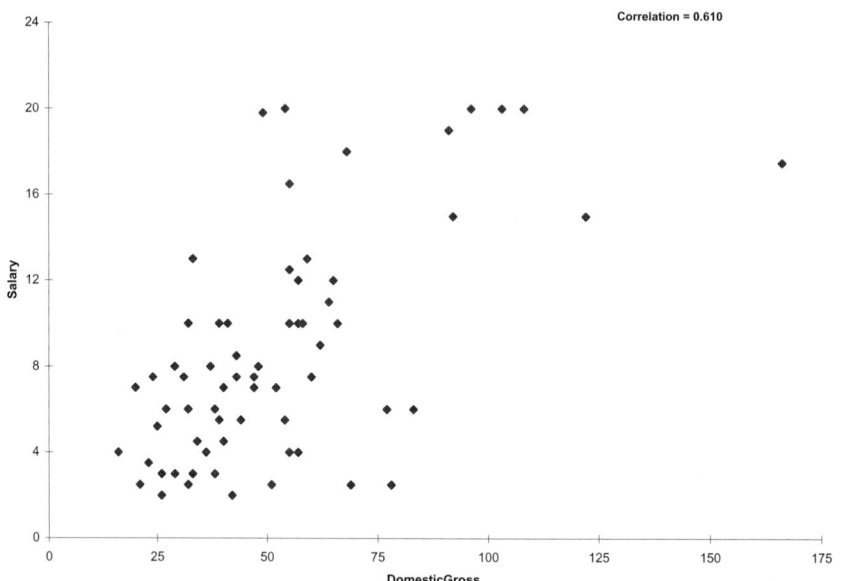

Alternatively, you can create the graph with Excel's Chart Wizard, using an "X-Y" type of chart. However, this method has a drawback. When you select the ranges for the two variables, say, the variables in columns B and E, Excel *automatically* puts the variable in the leftmost column (here column B) on the horizontal axis. You can't get around this Excel convention except by physically moving the data in column B to the right of the data in column E (or by doing something clever we haven't yet discovered!). There is no such limitation in StatPro.

The message from Figure 2.15 is fairly clear. First, the points tend to move up and to the right. This means that stars in films with large domestic grosses tend to make the largest salaries. The correlation of 0.61 shown in the chart supports this conclusion. As we will see in the next chapter, this implies a reasonably strong positive *linear* relationship between the two variables. ●

For the sake of contrast, we now consider a relationship that is quite different.

Example 2.8 Relationship Between Sales Productivity and Years of Experience for Salespeople

Suppose we are interested in the relationship between sales productivity and the number of years a salesperson has worked the territory. We collect the data in Figure 2.16. (See the file SALES.XLS.) Describe the relationship between sales and experience.

FIGURE 2.16
Sales versus Experi-
ence Data

	A	B	C	D	E
1	**Sales productivity versus years of experience**				
2					
3	Note: All monetary values are in $ thousands.				
4					
5	YrsExper	Sales			
6	24	54			
7	8	57			
8	2	45			
9	12	61			
10	8	57			
11	4	50			
12	6	54			
13	6	54			
14	11	60			
15	10	60			
16	11	60			
17	16	62			
18	14	62			
19	10	60			
20	18	61			
21	22	57			
22	20	60			

Objective To use a scatterplot to identify a nonlinear relationship between years of experience and sales.

Solution

Using StatPro's Scatterplot procedure, we construct the scatterplot shown in Figure 2.17. Note that as experience increases to around 14 years, sales increases, but at a decreasing rate. This indicates that there is a *nonlinear* relationship between sales and experience. Then beyond 14 years, additional experience appears to result in a sales decrease. Can

FIGURE 2.17
Scatterplot Illus-
trating a Nonlinear
Relationship

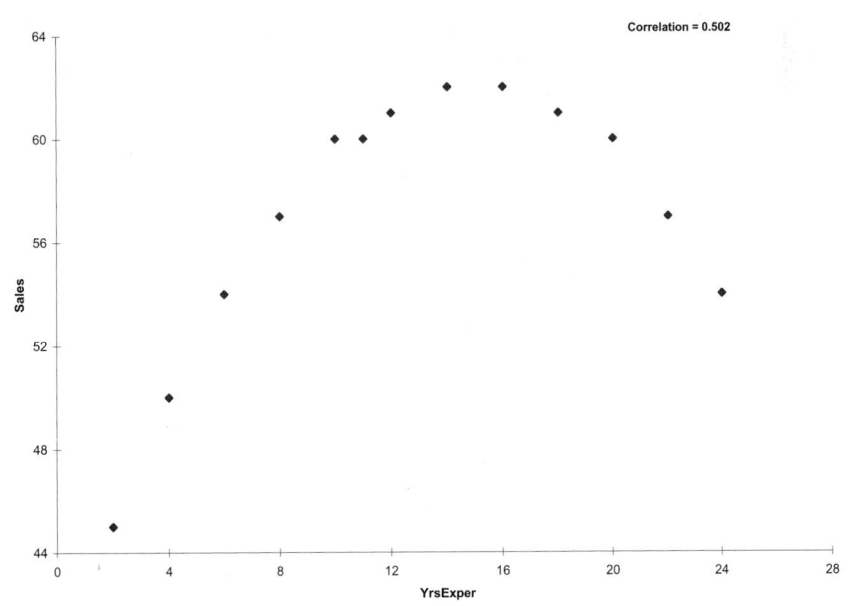

you think of a reason why this might be the case? In Chapter 11 we will learn how to estimate nonlinear relationships. ●

PROBLEMS

Level A

14. Explore the relationship between the selling prices and the appraised values of the 150 homes in the file P02_07.XLS by generating a scatterplot.
    a. Is there evidence of a *linear* relationship between the selling price and appraised value in this case? If so, characterize the relationship (i.e., indicate whether the relationship is a positive or negative one).
    b. For which of the two remaining variables, the size of the home and the number of bedrooms in the home, is the relationship with the home's selling price *stronger?* Justify your choice.

15. A human resources manager at Beta Technologies, Inc., is trying to determine the variable that best explains the variation of employee salaries using the previously gathered sample of 52 full-time employees in the file P02_01.XLS. Generate scatterplots to help this manager identify whether the employee's (a) gender, (b) age, (c) number of years of relevant work experience prior to employment at Beta, (d) the number of years of employment at Beta, or (e) the number of years of post-secondary education has the *strongest* linear relationship with annual salary.

16. Consider the enrollment data for *Business Week*'s top U.S. graduate business programs in the file P02_03.XLS. Specifically, generate scatterplots to assess whether there is a systematic relationship between the total number of full-time students and each of the following: (a) the proportion of female students, (b) the proportion of minority students, and (c) the proportion of international students enrolled at these distinguished business schools.

17. What is the relationship between the number of short-term general hospitals and the number of general or family physicians in metropolitan areas? Explore this question by producing a scatterplot for these two variables using the data in the file P02_17.XLS. Interpret your results.

18. Motorco produces electric motors for use in home appliances. One of the company's production managers is interested in examining the relationship between the dollars spent per month in inspecting finished motor products and the number of motors produced during that month that were returned by dissatisfied customers. He has collected the data in the file P02_18.XLS to explore this relationship for the past 36 months. Produce a scatterplot for these two variables and interpret it for this production manager.

19. The *ACCRA Cost of Living Index* provides a useful and reasonably accurate measure of cost of living differences among many urban areas. Items on which the index is based have been carefully chosen to reflect the different categories of consumer expenditures. The data are in the file P02_19.XLS. Generate scatterplots to explore the relationship between the composite index and each of the various expenditure components.
    a. Which expenditure component has the *strongest* relationship with the composite index?
    b. Which expenditure component has the *weakest* relationship with the composite index?

20. Consider the proportions of U.S. domestic airline flights arriving within 15 minutes of the scheduled arrival times at the Philadelphia and Pittsburgh airports. The data are in the file P02_20.XLS.
    a. Do you expect these two sets of performance measures to be *positively* or *negatively* associated with each other? Explain your reasoning.
    b. Compare your expectation to the actual relationship revealed by a scatterplot of the given data.

21. Examine the relationship between the average scores on the verbal and mathematical components of the SAT test across the 50 states and the District of Columbia by generating a scatterplot. The data are in the file P02_10.XLS. Also, explore the relationship between each of these variables and the proportion of high school graduates taking the SAT. Interpret each of these scatterplots.

22. Is there a strong relationship between a chief executive officer's annual compensation and her or his organization's recent profitability? Explore this question by generating relevant scatterplots for the survey data in the file P02_13.XLS. In particular, generate and interpret scatterplots for the change in the company's net income from 1996 to 1997 (see *Comp_NetInc96* column) and the CEO's 1997 base salary, as well as for the change in the company's net income from 1996 to 1997 and the CEO's 1997 bonus. Summarize your findings.

23. In response to a recent ranking of top graduate business schools in the United States published by *U.S. News & World Report,* the director of one of the recognized programs would like to know which variables are most strongly associated with a school's overall

score. Ideally, she hopes to use an enhanced understanding of the ranking scheme to improve her program's score in the forthcoming years and thus please both external and internal constituents. The data are in the file P02_08.XLS.

a. Use scatterplots to provide her with an indication of the measures that are most strongly related to the overall score in the *U.S. News & World Report* ranking.

b. Generally, how can she and her administrative colleagues proceed to improve their program's ranking in forthcoming publications?

24. Consider the relationship between the size of the population and the average household income level for residents of U.S. towns. What do you expect the relationship between these two variables to be? Using the data in the file P02_24.XLS, produce and interpret the scatterplot for these two variables.

25. Based on the annual data in the file P02_25.XLS from the U.S. Department of Agriculture, explore the relationship between the number of farms and the average size of a farm in the United States during the given time period. Specifically, generate a scatterplot and interpret it.

2.5 Time Series Plots

When we are interested in forecasting future values of a time series, it is helpful to create a time series plot. This is essentially a scatterplot, with the time series variable on the vertical axis and time itself on the horizontal axis. To make patterns in the data more apparent, the points are usually connected with lines.

When we look at a time series plot, we usually look for two things:

- Is there an observable trend? That is, do the values of the series tend to increase (an upward trend) or decrease (a downward trend) over time?

- Is there a seasonal pattern? For example, do the peaks or valleys for quarterly data tend to occur every fourth observation? Or do soft drink sales peak in the summer months?

The following example illustrates the construction and interpretation of a time series plot.

Example 2.9 Time Series Pattern of Quarterly Sales Revenue for Toys "R" Us

The file TOYS.XLS lists quarterly sales revenues (in $ millions) for Toys "R" Us during the years 1992–1995. The data are shown in Figure 2.18. Display these sales data in a time series plot and comment on whether trend and/or seasonality is present.

Objective To use StatPro's time series plot procedure to identify seasonality and an upward trend in sales.

Solution

To obtain the time series plot shown in Figure 2.19, we use the StatPro/Charts/Time Series Plot(s) menu item. This procedure allows us to plot one or more time series variables (on the same chart). In this example there is only one variable, Revenue, to plot. We also have the option of selecting a "date" variable for labeling the horizontal axis. Here we select Quarter (in column A) as the date variable. (If there were no date variable, the horizontal axis would be labeled with consecutive integers, starting with 1.)

FIGURE 2.18
Revenue Data for
Toys "R" Us

	A	B	C	D
1	**Toys "R" Us revenues**			
2				
3	Note: All monetary values are in $ millions.			
4				
5	Quarter	Revenue		
6	Q1-92	1026		
7	Q2-92	1056		
8	Q3-92	1182		
9	Q4-92	2861		
10	Q1-93	1172		
11	Q2-93	1249		
12	Q3-93	1346		
13	Q4-93	3402		
14	Q1-94	1286		
15	Q2-94	1317		
16	Q3-94	1449		
17	Q4-94	3893		
18	Q1-95	1462		
19	Q2-95	1452		
20	Q3-95	1631		
21	Q4-95	4200		

You can probably spot the upward trend and the seasonal pattern in this data set by looking directly at the numbers in column B. Such patterns are not usually so easy to detect, however, which is exactly why time series plots are so useful.

FIGURE 2.19
Time Series Plot of
Revenue

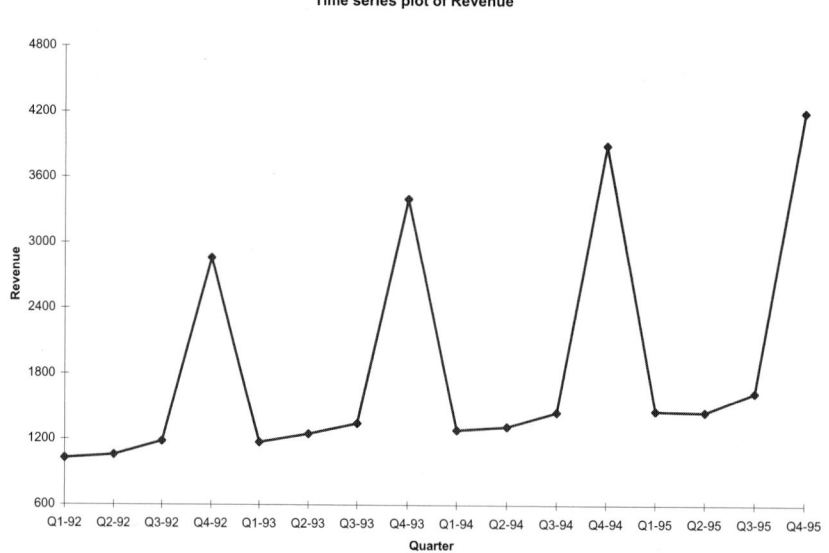

The time series plot exhibits an obvious seasonal pattern. Fourth quarter sales each year are much larger than the sales for the first three quarters. Of course, this is due to holiday sales. Focusing on the first three quarters of each successive year, we see a small upward trend in sales. This upward trend is also visible in the pattern of fourth quarters. The lesson from this graph is that if we want to forecast future quarterly sales for Toys "R" Us, we need to estimate the upward trend and seasonality of quarterly sales. ●

Sometimes it is useful to plot two time series variables on the same chart to compare their time series behavior. This is simple to do in StatPro by selecting *two* time series variables to plot. However, if the data for these two variables are of completely different magnitudes, the graph will be dominated by the variable with the largest data; the graph of the other variable will barely be visible. Therefore, StatPro provides the option of using the *same* vertical scale for both variables or using *different* vertical scales. The following example illustrates the second option. (Note that this latter option is not available if you select to plot *more* than two series on the same chart.)

Example 2.10 Time Series Pattern of Sales of Two Products with Different Magnitudes

Consider a company that sells two products. Product 1 is a much better seller than product 2. (See the file TWOVARS.XLS.) The monthly revenues from product 1 are typically above $100,000, whereas the revenues from product 2 are typically around $5000. How can the time series behavior of these revenues be shown on a single chart in a meaningful way?

Objective To illustrate how sales of two products with very different scales can be plotted in a single time series chart with StatPro.

Solution

The desired graph is shown in Figure 2.20. Note that it has two vertical scales. The scale on the left is appropriate for product 1, and the scale on the right is appropriate for product 2. We produced this chart in StatPro by filling in the key dialog box in the Time Series Plot procedure as shown in Figure 2.21 on page 54. (There is no "date" variable in this file, which explains why the second box is not checked.)

FIGURE 2.20
Plotting Time Series Variables on Two Different Scales

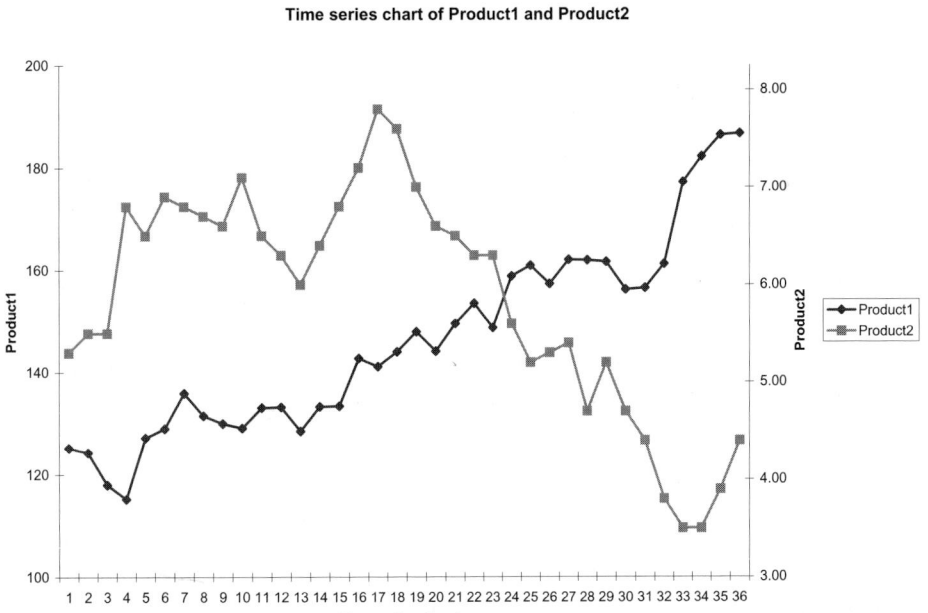

FIGURE 2.21
StatPro Dialog Box for a Time Series Plot with Two Separate Scales

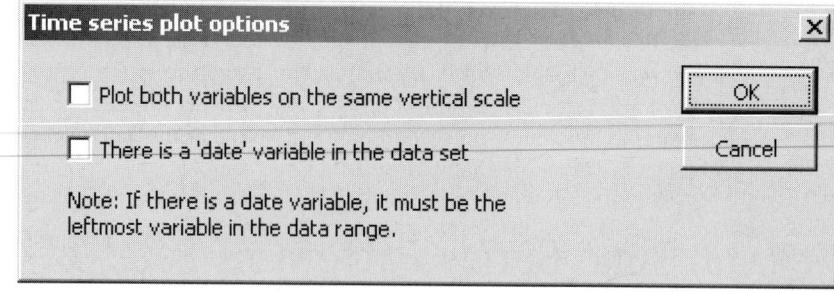

There are no obvious patterns (trend or seasonality) for the revenues of these two products. However, when they are both plotted on the same chart, we can see whether they move with one another. For example, in the last few months, revenues for product 1 increased, whereas revenues for product 2 decreased. Management might want to explore whether there is a reason for this. ●

PROBLEMS

Level A

26. Consider the Consumer Price Index (CPI), which provides the annual percentage change in consumer prices. Annual data are in the file P02_26.XLS.
  a. Construct a time series plot for these data.
  b. What, if any, trend do you see in the CPI for the given time period?

27. Compare the trends of the percentage changes of annual new orders for all manufacturing, durable goods, and nondurable goods industries in the United States. Annual data are in the file P02_27.XLS.
  a. Are the trends in these three times series similar?
  b. Can you explain the variation in each of these series over the given time period?

28. The Consumer Confidence Index (CCI) attempts to measure people's feelings about general business conditions, employment opportunities, and their own income prospects.
  a. Generate a time series plot for the annual average values of the CCI for the data in the file P02_28.XLS.
  b. Is it possible to say that U.S. consumers are more or less confident as we move forward in the present decade?
  c. How would you explain recent variations in the overall trend of the CCI?

29. Consider the proportion of Americans under the age of 18 living below the poverty level for each of the years in the file P02_29.XLS.
  a. Generate and interpret a time series plot for these data.

  b. How successful have Americans been recently in their efforts to win "the war against poverty" for the nation's children?

30. Examine the trends in the annual average values of the discount rate, the federal funds rate, and the prime rate for the data in the file P02_30.XLS. Can you discern any cyclical or other patterns in the times series plots of these three key interest rates?

31. Consider the U.S. balance of trade for both goods and services (measured in millions of U.S. dollars) for the data in the file P02_31.XLS.
  a. Produce a times series plot for each of the three given time series.
  b. Characterize recent trends in the U.S. balance of trade figures using your time series plots.

32. What is the recent trend in the number of mergers and acquisitions in the United States? Confirm your knowledge of this feature of U.S. business activity by generating a time series plot for the data in the file P02_32.XLS.

33. The Federal Deposit Insurance Corporation provides annual data on the number of insured commercial banks and the number of commercial failures in the United States. The file P02_33.XLS contains these data from 1980.
  a. Explore the relationship between these two time series by first producing and interpreting a scatterplot. Is the revealed relationship consistent with your expectations?
  b. Next, generate time series plots for each of these variables and comment on recent trends in their behavior.

34. Is cigar consumption in the United States on the rise? Explore this question by producing time series plots for each of the variables in the file P02_34.XLS. Comment on any observed trends in annual cigar consumption of the general U.S. population and of the U.S. male population over the given period.

35. Examine the trend of average annual interest rates on 30-year fixed mortgages in the United States for the data in the file P02_35.XLS. What conclusion(s) can be drawn from an analysis of the time series plot generated with the given data?

36. What has happened to the total *number* and average *size* of farms in the United States during the second half of the 20th century? Respond to this question by producing a time series plot of the data from the U.S. Department of Agriculture in the file P02_25.XLS. Is the observed result consistent with your knowledge of the structural changes within the U.S. farming economy?

37. Consider the file P02_37.XLS, which contains total monthly U.S. retail sales data for several recent years.
    a. Generate a plot of this time series and comment on any observable trends, including a possible seasonal pattern, in the data.
    b. Based on your time series plot, make a qualitative projection about the total retail sales levels for the next 12 months. Specifically, in which months of the subsequent year do you expect retail sales levels to be *highest?* In which months of the subsequent year do you expect retail sales levels to be *lowest?*

38. Are there certain times of the year at which Americans typically purchase greater quantities of liquor? Investigate this question by generating a time series plot for the monthly retail sales data from U.S. liquor stores listed in the file P02_38.XLS. Interpret the seasonal pattern revealed by your graph.

39. Examine the provided monthly time series data for total U.S. retail sales of building materials (which includes retail sales of building materials, hardware and garden supply stores, and mobile home dealers). The data are in the file P02_39.XLS.
    a. Is there an observable trend in these data? That is, do the values of the series tend to increase or decrease over time?
    b. Is there a seasonal pattern in these data? If so, how do you explain this seasonal pattern?

40. In which months of the calendar year do U.S. gasoline service stations typically have their *lowest* retail sales levels? In which months of the calendar year do U.S. gasoline service stations typically have their *highest* retail sales levels? Produce a time series plot for the monthly data in the file P02_40.XLS to respond to these two questions.

2.6 Exploring Data with Pivot Tables

We now look at one of Excel's most powerful—and easy-to-use—tools, **pivot tables.** This tool provides an incredible amount of useful information about a data set. **Pivot tables** allow us to "slice and dice" data in a variety of ways. That is, they break the data down into subpopulations so that we can, for example, see average salaries for male actors and female actresses separately. Statisticians often refer to the resulting tables as **contingency tables** or **crosstabs.** However, Excel provides more variety and flexibility with its pivot table tool than most other statistical software packages provide with their "crosstab" options. In particular, contingency tables list only counts, whereas pivot tables can list counts, averages, sums, and other summary measures.

It is easiest to understand pivot tables by means of examples, so we will illustrate several possibilities with the data in the ACTORS.XLS file. However, these only begin to show the power of pivot tables. The examples in Section 3.9 will show their real power.

Example 2.11 Comparison of Male and Female Movie Stars' Salaries

Female actresses claim they are being underpaid relative to male actors. Do the data support this claim?

Objective To use a pivot table to break down data on actors by gender.

Solution

We first need to segregate data on males and females. Although there are other ways to count the number of males and females, the following steps show how it can be done with a pivot table.

How to count data using a pivot table

❶   Position the cursor anywhere in the data range. (Pivot tables are similar to the StatPro add-in in expecting you to perform this step first.)

❷   Select the Data/PivotTable and PivotChart Report menu item. This takes you to a three-step PivotTable Wizard that leads you through the process.[3]

❸   The first step asks for two pieces of information: (1) the source of the data, and (2) whether you want an accompanying chart. For (1), make sure the Microsoft Excel list or database option is selected. (We will explore the External data source in Chapter 4.) For (2), there is absolutely no extra work in creating a chart, so you might as well select the PivotChart option. Then click on Next.

❹   In the second step, specify the range of the data set. Assuming that you placed the cursor somewhere in the data set, the Wizard correctly guesses the range of the data, so click on Next.

❺   The final step (which is a combination of the final *two* steps in Excel 97) asks you where you want to place the pivot table. (See Figure 2.22.) We suggest placing it on a new worksheet. Then there are two ways to proceed in this step. First, you can click on Finish to obtain a "blank" pivot table, as illustrated in Figure 2.23. Alternatively, you can click on the Layout button in Figure 2.22 to obtain the dialog box in Figure 2.24. We favor this approach, although it is a matter of taste. You can either construct the pivot table

> Asking for a pivot table (on a new worksheet) and a pivot chart will create two new sheets: one for the pivot table and one for the chart. Clicking on Finish in the last step of the wizard produces the chart sheet by default. From there you can navigate to the pivot table sheet.

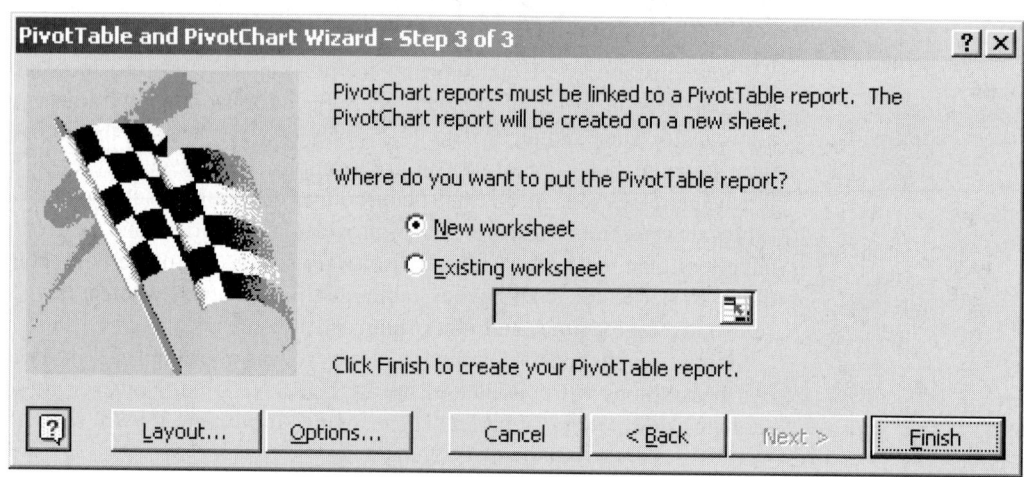

FIGURE 2.22 Step 3 dialog box for PivotTable Wizard

[3]If you have used pivot tables in Excel 97 or earlier versions, you will see two differences in Excel 2000 and more recent versions. First, the wizard is now a three-step procedure, not four. However, the basic functionality is still the same. Second, pivot charts now accompany pivot tables automatically if you request them.

FIGURE 2.23
Blank Pivot Table

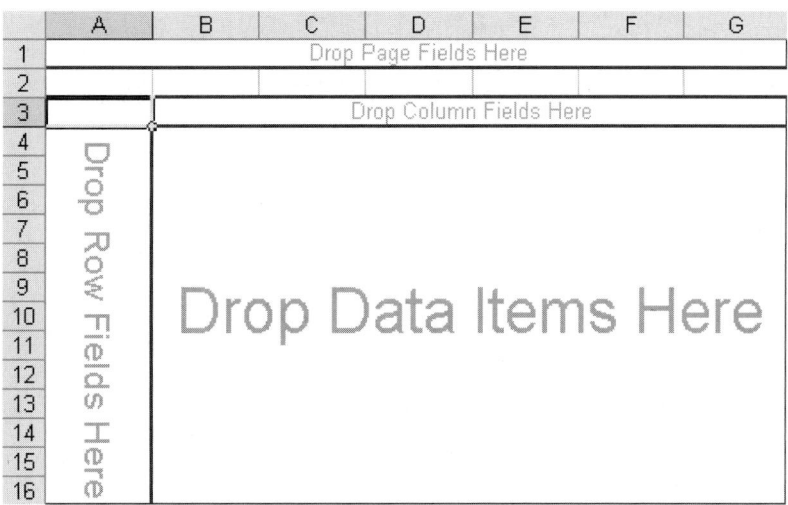

FIGURE 2.24
PivotTable layout
dialog box

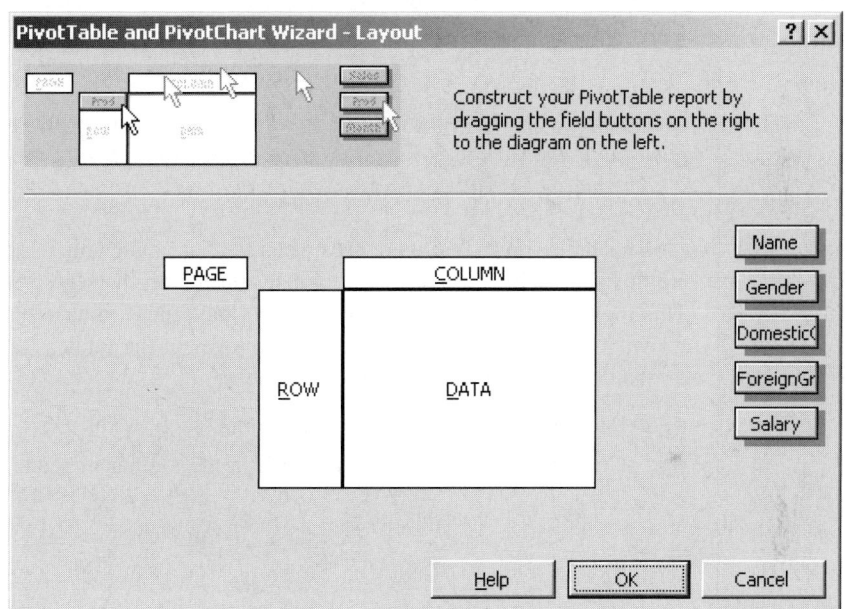

now, through this layout dialog box, or you can construct it later, by adding variables to the blank pivot table in Figure 2.23.

6 Assuming you clicked on the Layout button to produce the dialog box in Figure 2.24, you now drag variable buttons to the four areas: row, column, data, or page. Essentially, the row, column, and page areas allow you to break the data down by the categories of the variables in these areas. The "data" area specifies the data you want to calculate. For this example, drag Gender to the row area and Gender to the data area. The screen should appear as shown in Figure 2.25 (page 58). Click on OK to return to Figure 2.22, and then click on Finish to create the pivot table in Figure 2.26.

Excel Tip *In this example we want* **counts,** *the number of men and the number of women. To get these we place Gender in the row area (it could instead be in the column area),*

FIGURE 2.25
Layout dialog box
with selected fields

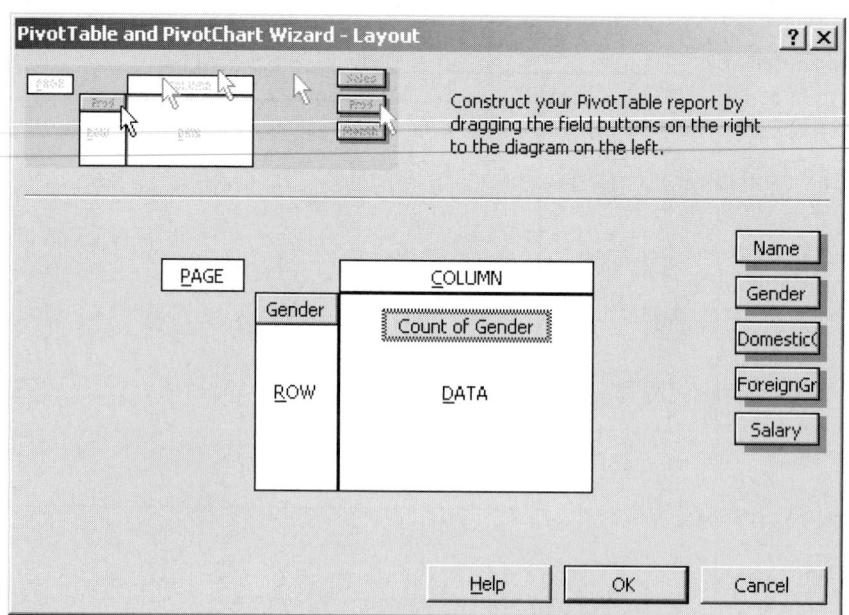

and we place **any nonnumerical** variable in the data area. When we want counts from a pivot table, the variable in the data area is irrelevant as long as it provides counts. When a nonnumerical variable is placed in the data area, we get counts by default.

This table shows that there are 66 stars; 48 are male and 18 are female. When we create this pivot table, a pivot table toolbar appears on the screen as shown in Figure 2.26. This toolbar allows us to modify the pivot table in various ways. For example, suppose we want to express these counts as percentages of the total.

To do so, go through the following steps:

❶ Put the cursor on any cell with a count, such as cell B5.

❷ Click on the PivotTable drop-down in the toolbar to see a menu.

FIGURE 2.26
Pivot Table of Gender Counts

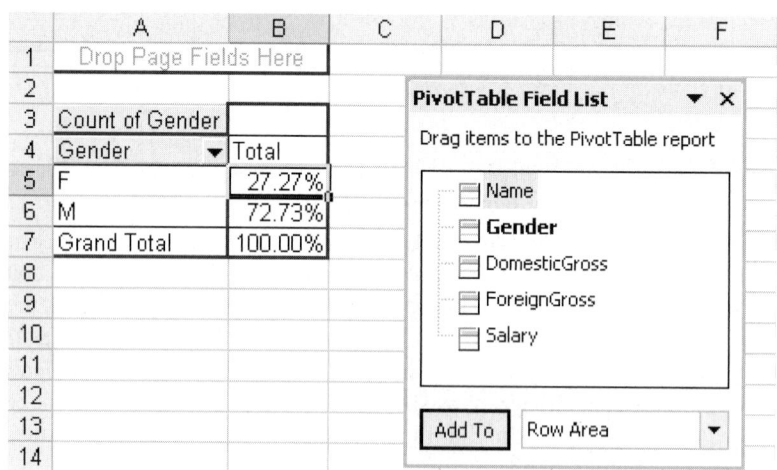

FIGURE 2.27
Field Settings Dialog Box

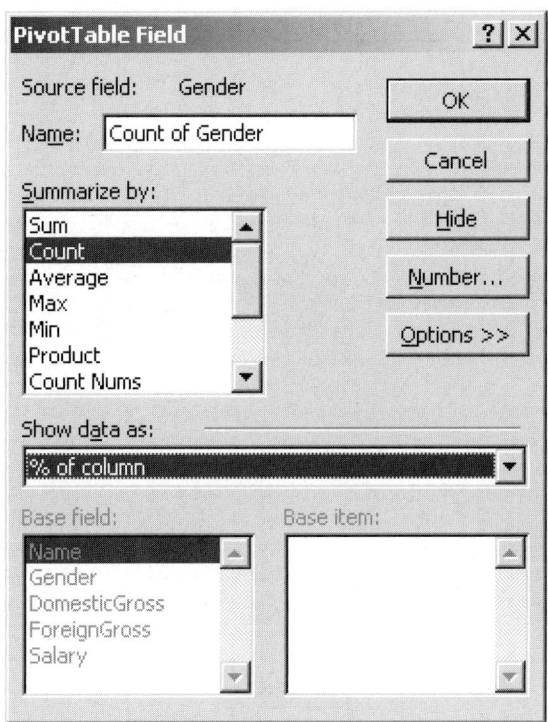

❸   Select Field settings in this menu to bring up a dialog box.

❹   Click on the Options button of this dialog box and select % of column in the Show data as drop-down. The resulting dialog box should appear as in Figure 2.27. Click on OK to see the counts as percentages.

The new pivot table is shown in Figure 2.28. It still contains the same information, but expressed differently.

To see the results graphically, look at the accompanying chart sheet, which should look like the one in Figure 2.29. As with the numbers, this chart shows that almost 75% of the actors are male. Is this an important finding? Without knowing exactly how these 66 actors and actresses were selected from the population of *all* actors and actresses, we can't make very definitive conclusions. If we knew, for example, that these 66 were today's highest paid actors and actresses, then it might be surprising that 75% of them were males.

The remarkable property of pivot charts is that they are linked entirely to the associated pivot tables, and vice versa. If you change something in the pivot table, the pivot chart changes automatically; if you change something in the pivot chart, the pivot table

FIGURE 2.28
Pivot Table with Counts Expressed as Percentages

	A	B
1	Drop Page Fields Here	
2		
3	Count of Gender	
4	Gender	Total
5	F	27.27%
6	M	72.73%
7	Grand Total	100.00%

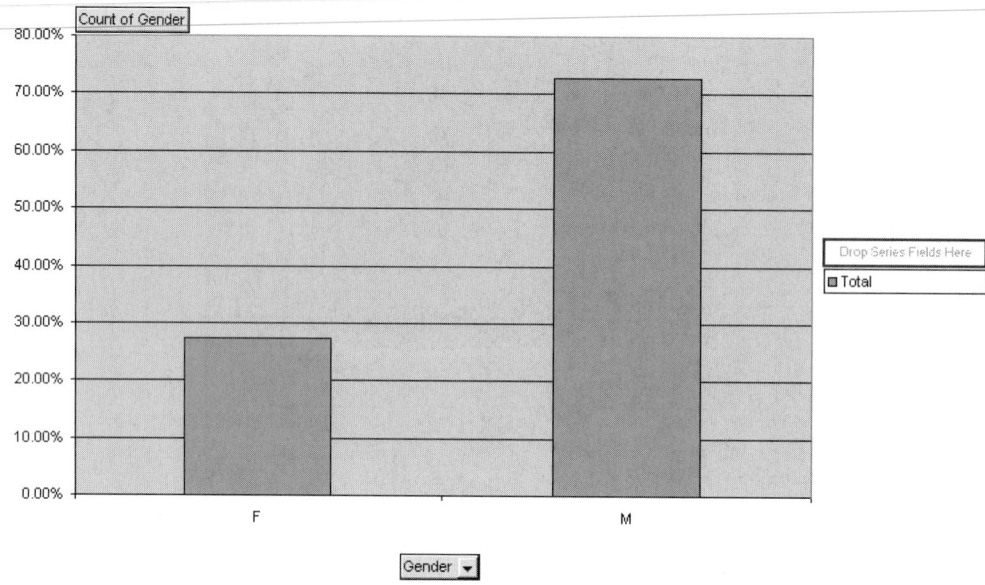

FIGURE 2.29 PivotChart for Gender Breakdown

changes automatically. Finally, you can manipulate pivot charts just like any other Excel charts. The easiest way is to click on the Chart Wizard on the PivotTable toolbar and then select the settings you want.

We still don't know whether women are underpaid, so we will create other pivot tables to examine the distribution of salaries, classified by gender. The following steps accomplish this.

① Place the cursor anywhere in the data range (on the Data sheet).

② Select the Data/PivotTable and PivotChart Report menu item, and click on Next in the first two steps to accept the defaults.[4]

③ In the Layout dialog box, drag the Salary variable to the Row area, drag Gender to the column area, drag Gender to the data area, and click on OK. (Again, we want to create counts in the pivot table, so *any* nonnumerical variable could be placed in the data area.)

④ Click on Finish to accept the defaults on the final screen.

The completed pivot table appears in Figure 2.30. It shows the number of men and women making each possible salary. (It also shows row and column totals.) This is a bit too much detail, although it is already apparent that men typically earn more than women. It would be better to show these counts for broader categories.

---

[4]When you build additional pivot tables in a workbook, the wizard asks you (after step 2) if you want to base the new pivot table on the data from an existing pivot table. This is a memory issue, and it is best to respond "yes." It then asks you which of the previous pivot tables to base this one on. Assuming they all came from the same data set, it doesn't matter which previous pivot table you choose.

FIGURE 2.30
Pivot Table Showing Distribution of Salary by Gender

	A	B	C	D
1	Drop Page Fields Here			
2				
3	Count of Gender	Gender ▼		
4	Salary ▼	F	M	Grand Total
5	2	1	1	2
6	2.5	4	1	5
7	3	2	2	4
8	3.5		1	1
9	4	2	3	5
24	15		2	2
25	16.5		1	1
26	17.5		1	1
27	18		1	1
28	19		1	1
29	19.8		1	1
30	20		4	4
31	Grand Total	18	48	66

We can easily "group" the Salary categories as follows.

❶ Right-click on any cell in the Salary column of the pivot table, such as cell A5.

❷ Select the Group and Outline/Group menu item. (As you can guess, the Ungroup menu item lets you get back to where you started.)

❸ In the resulting dialog box, specify that the groups should start at 2, end at 20, and use increments of 3. Then click on OK.

The modified pivot table should look essentially like the one in Figure 2.31. Actually, to make yours look exactly like ours, you'll have to express counts as percentages of column (as we did earlier). Also, we made three changes to the chart:

❶ With the chart sheet selected, use the Chart/Location menu item to locate the chart on the same sheet as the pivot table.

❷ With the chart selected, click on the Chart Wizard toolbar button, and select the chart subtype you prefer. (We selected the first subtype of a column chart.)

❸ With the chart selected, click on the PivotChart drop-down in the PivotTable toolbar, and select the Hide PivotChart Field Buttons menu item. This makes the chart more suitable for printing.

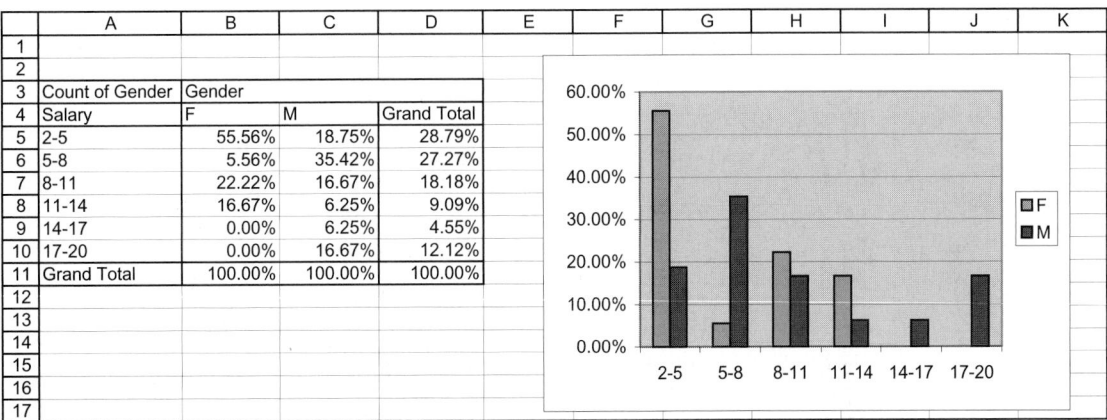

FIGURE 2.31 Using the Pivot Table Group Option

Figure 2.31 makes it clear that over half the women in this data set are in the lowest salary category, whereas only 19% of the men are in this category. Also, no women are in the highest two salary categories, whereas 23% of the men are in these categories. Evidently, male actors make considerably more money from movies than female actresses.

Another way to compare salaries of men and women is to look at the *average* salary by gender. We can also do this with a pivot table using the following steps.

❶   Proceed as before to get to the Layout dialog box.

❷   Drag Gender to the row area and Salary to the data area. Note that the data area now shows "Sum of Salary." When a *numerical* variable is dragged to the data area, the default is to show its sum in the pivot table. Therefore, if we finished now, we'd see the sum of all male salaries and the sum of all female salaries. However, we want averages, not sums. Therefore, go to step 3.

❸   Double-click on the Sum of Salary button in the Data area to bring up a PivotTable Field dialog box, and select Average in the Summarize By list. As long as you are there, click on the Number button, and select Currency with two decimals. This will format all of the averages nicely in the pivot table.

The result appears in Figure 2.32. Clearly, the male actors make considerably more on average than the female actresses. The corresponding 3-D column chart of the average salaries shows this discrepancy graphically.

The analysis so far appears to indicate that the movie industry discriminates against women. However, it is possible that women are paid less because movies with female leads gross less money than movies with male leads. To analyze this further, we look at the average salary of men and women for each domestic gross level. (We could also take into account the influence of foreign grosses, but we won't do so here.)

The pivot table for doing this appears in Figure 2.33. By this time you should be able to create the pivot table on your own. But if you need help, here are the basic steps: Drag DomesticGross to the row area, Gender to the column area, and Salary to the data area; summarize salaries by average (not sum); and in the resulting pivot table group the domestic gross values in increments of 20.

First, note the blanks in four of the cells. The reason is that no movies with female leads had domestic grosses in these highest four categories, so it is impossible to calculate averages for them. Now it's fair to ask whether men average more than women, after controlling for the domestic gross. Clearly, they do so in the two lowest domestic gross categories, but only barely in the third. Beyond the third category, it is hard to tell because no females were leads in the real blockbusters. In any case, we can now say with

FIGURE 2.32
Pivot Table of Average Salary by Gender

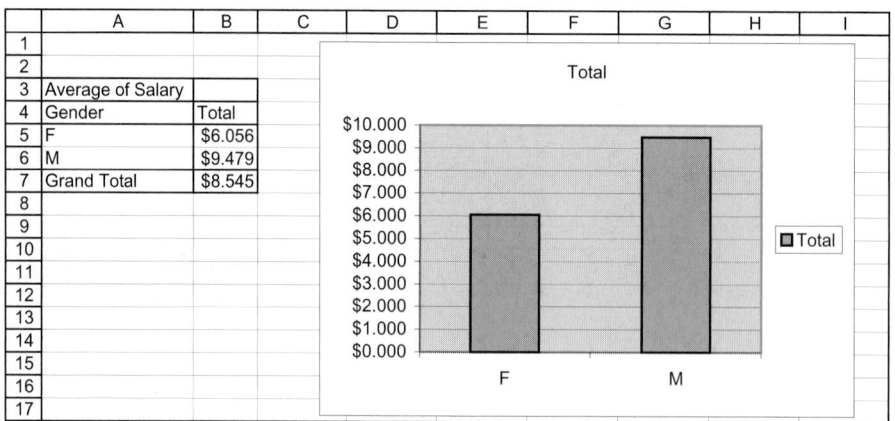

	A	B	C	D	E	F	G	H	I
1									
2									
3	Average of Salary								
4	Gender	Total							
5	F	$6.056							
6	M	$9.479							
7	Grand Total	$8.545							

FIGURE 2.33
Pivot Table of Average Salary by Domestic Gross and Gender

	A	B	C	D
1				
2				
3	Average of Salary	Gender		
4	DomesticGross	F	M	Grand Total
5	16-35	$4.357	$6.155	$5.456
6	36-55	$4.400	$8.990	$8.072
7	56-75	$9.417	$9.500	$9.462
8	76-95		$9.700	$9.700
9	96-115		$20.000	$20.000
10	116-135		$15.000	$15.000
11	156-175		$17.500	$17.500
12	Grand Total	$6.056	$9.479	$8.545

more assurance that the industry *does* appear to discriminate against women in terms of salary. ●

The next example provides further illustration of the possibilities of pivot tables.

Example 2.12 Effects of Operator and Machine on Elevator Rail Diameters

The file OTIS3.XLS lists the diameters (in inches) of elevator rails produced by Otis Elevator's two machines and two operators. Each diameter corresponds to a particular machine/operator combination, as shown in Figure 2.34. What effects, if any, do the operator and machine have on elevator rail diameters?

FIGURE 2.34
Diameters at Otis by Machine and Operator

	A	B	C	D	E
1	Diameters of elevator rails at Otis Elevator				
2					
3	Note: All diameters are expressed in fractions of inches.				
4					
5	Machine	Operator	Diameter		
6	2	1	0.5184		
7	2	2	0.5357		
8	1	2	0.5290		
9	1	2	0.5298		
10	2	1	0.5423		
11	1	1	0.5188		
12	2	2	0.5387		
13	1	1	0.5207		
14	2	1	0.5095		
15	2	2	0.5438		
16	2	1	0.5187		
17	1	1	0.5179		
133	1	1	0.5419		
134	1	1	0.5160		
135	2	2	0.5413		

Objective To use pivot tables to break rail diameters down by machine and operator.

Solution

The relevant pivot table and an associated column chart appear in Figure 2.35. Here we dragged Machine to the row area, Operator to the column area, and Diameter (expressed as an average) to the data area. The numbers and graph indicate several interesting results:

- On average, the diameters for machine 2 are about 0.01 inch larger than those for machine 1 (see column D). This pattern is approximately the same whether operator 1 or operator 2 is operating the machines (see columns B and C).

- On average, the diameters for operator 2 are about 0.01 inch larger than those for operator 1 (see row 7). This pattern is approximately the same whether operators are operating machine 1 or machine 2 (see rows 5 and 6).

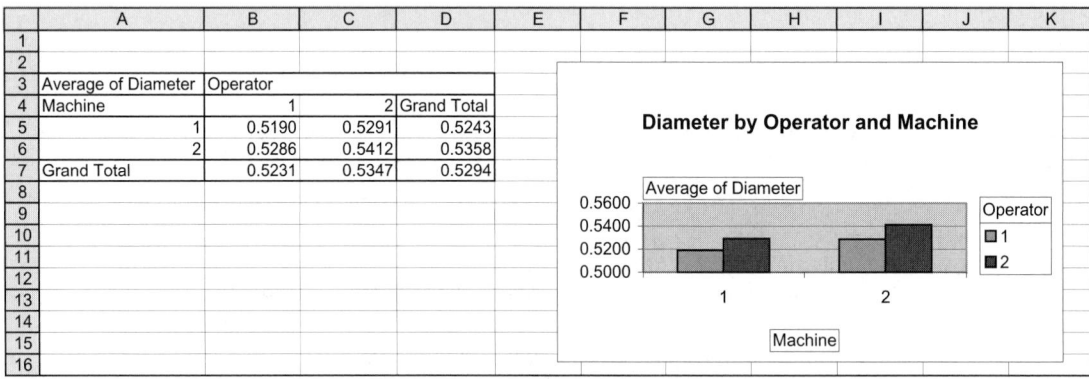

FIGURE 2.35 Diameters by Operator and Machine

What might this mean to a manager? Suppose, for example, that the desired diameter is 0.53 inch. Then the following decisions might be made: (1) machine 1's setting could be increased by about 0.006 inch, whereas machine 2's setting could be decreased by about 0.006 inch, and (2) operator 1 could be coached to increase his diameters by about 0.007 inch, whereas operator 2 could be coached to decrease her diameters by about 0.005 inch. ●

Final Remarks on Pivot Tables We have just begun to see the power and flexibility of pivot tables. Other available options include the following.

- The data used to create a pivot table can come from an external database (such as Microsoft Access) as well as a database within an Excel spreadsheet. We will explore this possibility in Chapter 4.

- We can use a "page" field for added flexibility. For example, if we drag Gender to the page field, we can then see three separate pivot tables with a click of the mouse: one for all people, one for males only, and one for females only. We will explore this option in the next chapter.

- We can drag multiple fields to any of the areas (row, column, page, or data) of a pivot table. Also, once the pivot table is visible, we can drag any row, column, or page area to another area—without going back to the Wizard—and the pivot table automatically readjusts itself. We can even drag an existing field's button to a blank space in the spreadsheet to remove this field from the pivot table.[5]

- By double-clicking on any pivot table entry, we see all of the data that were used to compute that entry. For example, by double-clicking on cell B5 in Figure 2.35, we see the diameters of all rods that were produced on machine 1 by operator 1.

- The pivot table retains a link to the original data, so that if the original data change, the pivot table recalculates automatically. To get it to do so, click on the exclamation point button of the Pivot Table toolbar.

- As stated previously, we can skip the Layout dialog box when we first create the pivot table. We then get a blank pivot table, as shown in Figure 2.23, to which we can drag fields from the list in the PivotTable toolbar to the row, column, page, and data areas.

PROBLEMS

Level A

41. A human resources manager at Beta Technologies, Inc., has collected current annual salary figures and related data for 52 of the company's full-time employees. In particular, these data include each selected employee's gender, age, number of years of relevant work experience prior to employment at Beta, number of years of employment at Beta, number of years of post-secondary education, and annual salary. The data are in the file P02_01.XLS. Use Excel to construct pivot tables or other descriptive graphs to answer the following questions:
   a. What proportion of these full-time Beta employees are female?
   b. Is there evidence of salary discrimination against women at Beta Technologies? What are the limitations of the conclusion that you have drawn in answering this question?
   c. Is additional post-secondary education positively associated with higher average salaries at Beta?
   d. Is there evidence of salary discrimination against older employees at Beta Technologies? What are the limitations of the conclusion that you have drawn in answering this question?

42. Consider *Business Week's Guide to the Best Business Schools* enrollment data for top-rated graduate business programs in the United States. Specifically, this guide reports the percentages of women, minority, and international students enrolled in each of the top programs, as well as the total number of full-time students enrolled in each. The data are in the file P02_03.XLS. Use Excel to construct pivot tables or other descriptive graphs to answer the following questions:
   a. Do graduate business programs with higher proportions of female student enrollments tend, on average, to have higher proportions of minority student enrollments?
   b. Do graduate business programs with higher proportions of female student enrollments tend, on average, to have higher proportions of international student enrollments?
   c. What, if any, are the limitations of the conclusions you have reached in responding to each of the previous questions?

43. Who is most likely to access the Internet today? Consider the survey data collected from 1000 randomly selected Internet users, given in the file P02_43.XLS. Construct pivot tables to answer each of the following questions.
   a. What proportion of these Internet users are men under the age of 30?
   b. What proportion of these Internet users are single with no formal education beyond high school?
   c. What proportion of these Internet users are currently employed? What is the average salary of the employed Internet users in this sample?

44. Is there a relationship between a state's number of classroom teachers and the average salary of classroom teachers in the state? Generate one or more pivot tables using Excel to answer this question for the data in the file P02_44.XLS. In addition to exploring the relationship between a state's total number of classroom teachers and their average

---

[5]These and many other pivot table features are explained in more detail in Chapter 25 of Walkenbach (1999). However, we have found that you learn best about pivot tables by experimenting with them: dragging fields around, changing field settings, grouping, and so on. Don't be afraid to experiment with this remarkably flexible tool.

annual salary, consider the same relationships for both elementary and secondary teachers.

45. Using the first given data set in file P02_45.XLS, which was collected in 1994, construct a single pivot table that breaks down the 1000 randomly selected U.S. workers by sex *and* race. Develop a similar breakdown for those workers randomly selected in 1982. How have the proportions of U.S. workers in these various cross sections changed between 1982 and 1994?

46. Given data in the file P02_13.XLS from a recent survey of chief executive officers from the nation's biggest businesses, generate pivot tables using Excel to determine whether the levels of the 1997 annual salaries and bonuses earned by CEOs are related somewhat to the *types* of companies in which they serve.

47. Consider the data in the file P02_12.XLS on various performance measures for the largest U.S. airlines. In particular, employ a pivot table and/or descriptive graphs to determine whether the level of consumer complaints about an airline is associated with the number of reports of mishandled baggage filed by the airline's passengers. Summarize your findings.

Level B

48. What influences a metropolitan area's vulnerability to recession? Construct pivot tables or other descriptive graphs to explore this issue using the given job statistics for selected towns in the United States in the file P02_48.XLS. Be sure to consider the relative mix of new blue-collar jobs versus new white-collar jobs in attempting to explain an area's unemployment threat level.

49. As a part of an economic development study conducted by the local government, 500 households in a middle-class neighborhood were recently surveyed. In particular, for each of the randomly selected households, the survey requested information on the following variables: family size, approximate location of the household within the neighborhood, an indication of whether those surveyed owned or rented their home, gross annual income of the first household wage earner, gross annual income of the second household wage earner (if applicable), monthly home mortgage or rent payment, average monthly expenditure on utilities, and the total indebtedness (excluding the value of a home mortgage) of the household. The data are in the file P02_06.XLS. Use Excel to construct pivot tables or other descriptive graphs to answer the following questions:

    a. What proportion of households in each of the four locations within this neighborhood own their homes?

    b. What relationship, if any, exists between the primary household income level and family size?

    c. What relationship, if any, exists between the primary household income level and the household's location within the neighborhood?

    d. What relationship, if any, exists between the primary household income level and whether the household owns or rents their home?

50. Consider the relationship between the population size of selected metropolitan areas in the United States and the location's average annual rates for various forms of violent crime, including murder, rape, robbery, and aggravated assault. The data are in the file P02_50.XLS. Use pivot tables or other descriptive methods to explore the relationship between a metro area's size and the area's level of violent criminal acts. Summarize your findings.

51. Using cost-of-living data from the *ACCRA Cost of Living Index* in the file P02_19.XLS, examine the relationship between the geographical *location* of an urban area within the United States (e.g., northeast, southeast, midwest, northwest, or southwest) and its *composite* cost-of-living index. In other words, is the overall cost of living higher or lower for urban areas in particular geographical regions of the country? You will need to assign the given urban areas to one of any number of such geographical regions before you can produce pivot tables or other graphical tools in responding to this question.

52. Consider the relationship between the size of the population and the average time (in minutes) it takes citizens of selected American communities to travel to work and back home each day. Using the data in the file P02_11.XLS, produce a pivot table with Excel to determine whether the population size is useful in explaining the variation of average commute times.

53. In a recent ranking of top graduate business schools in the United States published by *U.S. News & World Report,* data were provided on numerous attributes of recognized graduate programs. Specifically, we are interested in understanding the variation of the average starting base salaries for recent graduates from these programs. The data are in the file P02_08.XLS. Use Excel to construct pivot tables or other descriptive graphs to answer the following questions:

    a. Are higher average GMAT scores for enrolled students associated with higher average starting base salaries for recent graduates from these programs?

    b. Are higher average undergraduate grade-point averages for enrolled students associated with higher average starting base salaries for recent graduates from these programs?

    c. Are lower program acceptance rates associated with higher starting base salaries for recent graduates from these programs?

    d. What are the limitations of the conclusions you have reached in answering these three questions?

2.7 Conclusion

The graphs and tables we have discussed in this chapter are extremely useful for describing data sets. The graphs show at a glance how a single variable is distributed, how two variables are related, or how a variable varies over time. The tables are also useful, not only in their own right, but for providing the data needed to create graphs. We have paid special attention to the pivot table feature available in Excel. Pivot tables allow us to see relationships in a data set that would be very difficult to see in any other way. In fact, they might be the best kept secret in Excel, but their popularity will surely grow as business managers learn to take advantage of their power and flexibility.

Summary of Key Terms

Term	Explanation	Excel	Pages
Population	Includes all objects of interest in a study—people, households, machines, etc.		32
Sample	Representative subset of population, usually chosen randomly.		32
Variable (or Field)	Attribute or measurement on members of a population, such as height, gender, or salary.		33
Case (or Observation or Record)	List of all variable values for a single member of a population.		33
Data type	Several categorizations are possible: numerical versus categorical (with subcategories nominal, ordinal); discrete versus continuous; cross-sectional versus time series.		34, 35, 36
Frequency table	Contains counts of observations in specified categories.		36
Histogram	Bar chart of frequencies from a frequency table.	StatPro/Charts/ Histogram(s)	36
Distribution	General term for the way data are distributed, as indicated by a frequency table or histogram.		40
Skewed distribution	Indicated by nonsymmetric shape of histogram. Skewed to the right (or positively skewed) means a long tail to the right. Skewed to the left (or negatively skewed) is the opposite.		41, 42
Bimodal distribution	Indicated by histogram with two peaks, often due to data from two distinct populations.		43
Scatterplot	Chart showing relationship between two variables, with a point for each observation.	StatPro/Charts/ Scatterplot(s)	47
Time series plot	Chart showing behavior over time of a time series variable.	StatPro/Charts/ Time Series Plot(s)	51
Pivot table (and Pivot table chart)	Table (and corresponding chart) in Excel that summarizes data broken down by one or more categorical variables.	Data/PivotTable and PivotChart Report	55
Contingency table (or Crosstabs)	Traditional statistical terms for pivot tables that list counts.		55

Conceptual Exercises

C.1. An airline analyst seeks to estimate the proportion of all American adults who are currently afraid to fly in the wake of the terrorist attacks of September 11, 2001. To estimate this percentage, the analyst decides to survey 1500 Americans from across the nation. Identify the sample and the population in this case.

C.2. The number of children living in each of a large number of randomly selected households is an example of which data type? Be as specific as possible.

C.3. Does it make sense to construct a histogram for the state of residence of randomly selected individuals in a sample? Explain why or why not.

C.4. Characterize the likely shape of a histogram of the distribution of scores on a fairly easy midterm exam in a graduate statistics course.

C.5. Suppose that a researcher is interested in determining whether there is a relationship between the number of room air conditioning units sold each week in a given year and the time of year (in terms of weeks). What type of descriptive graph would be most useful in performing this analysis? Explain your choice.

Level A

54. An economic development researcher wants to understand the relationship between the size of the monthly home mortgage or rent payment for households in a particular middle-class neighborhood and each of the following household variables: family size, approximate location of the household within the neighborhood, an indication of whether those surveyed owned or rented their home, gross annual income of the first household wage earner, gross annual income of the second household wage earner (if applicable), average monthly expenditure on utilities, and the total indebtedness (excluding the value of a home mortgage) of the household. The data are in the file P02_06.XLS.
   a. Generate a scatterplot for each pairing of variables with the size of the household's monthly home mortgage or rent payment.
   b. Which of the variables have a *positive* linear relationship with the size of the household's monthly home mortgage or rent payment?
   c. Which of the variables have a *negative* linear relationship with the size of the household's monthly home mortgage or rent payment?
   d. Which of the variables have essentially *no* linear relationship with the size of the household's monthly home mortgage or rent payment?

55. The *Places Rated Almanac* ranked many metropolitan areas in the United States with consideration of the following aspects of life in each area: cost of living, transportation, jobs, education, climate, crime, arts, health, and recreation. The data are in the file P02_55.XLS.
   a. Generate scatterplots to discern the relationship between the metropolitan area's overall score and each of these numerical factors.
   b. Are the relationships revealed by the scatterplots consistent with your expectations? If not, can you explain any discrepancies between your findings and expectations?

56. The U.S. Bureau of Labor Statistics provides data on the year-to-year percentage changes in the wages and salaries of workers in private industries, including both "white-collar" and "blue-collar" occupations. Consider the data in the file P02_56.XLS.
   a. Is there evidence of a strong relationship between the yearly changes in the wages and salaries of "white-collar" and "blue-collar" workers in the United States over the given time period? Describe the nature of any observed systematic relationship between these two variables.
   b. Construct graphs for each of the three given time series and comment on any observed trends in these data.

57. The file P02_57.XLS contains three years (1999–2001) of sales data for the Sky's the Limit Women's Apparel store. Each observation represents sales during a 4-week period. Thus the first observation is sales during the first 4 weeks of 1999, and so on.
   a. Does there appear to be any trend in sales?
   b. Do sales appear to be seasonal? If so, discuss the nature of the seasonality.

58. For approximately 170 companies, the file P02_58.XLS contains two pieces of information: a measure of the strength of the corporate culture (from 1 = High to 5 = Low), and the percentage net income growth from 1977 to 1988. Do these data indicate that a strong corporate culture is associated with financial success or weakness?

59. The file P02_59.XLS contains quarterly sales revenues for Wal-Mart for several years. Construct a time series plot and discuss the trend and seasonal characteristics of Wal-Mart's sales.

60. Consider the data in the file P02_60.XLS. In particular, columns B–D contain the following information about a sample of Bloomington residents: education level (completed high school only or completed college), income level (low or high), and whether the last purchased car was financed.

a. Using the data in columns B–D, determine how education and income influence the likelihood that a family finances a car.

b. Column A of this file contains the exact salaries of these Bloomington residents. Using categories of length $10,000, construct a histogram of these salaries. (You can experiment with the appropriate leftmost and rightmost categories.) Does the histogram appear to be bell shaped?

61. The file P02_61.XLS contains the following information about a sample of Bloomington families: family size (large or small), number of cars owned by family (1, 2, 3, or 4), and whether family owns a foreign car.

a. Use these data to determine how family size and number of cars influence the likelihood that a family owns a foreign car.

b. Construct a histogram with four bars, using the obvious categories, for the number of cars owned by a family. Interpret the height of the second bar.

62. The file P02_62.XLS contains monthly returns on Barnes and Noble stock for several years. Do monthly stock returns appear to be skewed or symmetric?

63. The file P02_63.XLS contains annual returns for firms grouped by size. For example, in 1926, firms that ranked in the top 10% by size of sales returned an average of 14.9%, whereas firms that ranked in the bottom 10% by size returned an average of −6.1%. What do these data tell you about the relationship between firm size and average stock return? What are possible investment implications of this information?

64. The file P02_64.XLS contains monthly returns on Mattel stock for several years. Plot a histogram of these data and summarize what you learn from it.

65. It has been hypothesized that a reduction in the average length of the workweek in a country will reduce the unemployment rate. The theory is that of job-sharing—if everybody works 5% less, the unemployed workers can pick up the reduced hours. Table 2.1

gives the percentage decrease in annual working hours per employee from 1975 to 1994 as well as the increase in unemployment rate for nine countries. Do these data support the hypothesis that job-sharing reduces unemployment? (Source: *The Economist,* November 25, 1995)

66. Do countries with high rates of home ownership have higher or lower unemployment rates? The file P02_66.XLS lists the 1996 home ownership percentage and unemployment rate for various countries. Discuss the relationship between home ownership and unemployment. Do you have any explanation for this relationship?

67. The SoftBus Company sells PC equipment and customized software to small companies to help them manage their day-to-day business activities. To understand its customers better, SoftBus recently sent questionnaires to a large number of prospective customers. Key personnel—those who would be using the software—were asked to fill out the questionnaire. SoftBus received 82 usable responses, as shown in the file P02_67.XLS. The variables include the following information on the key person who filled out the questionnaire: gender, years of experience with this company, education level, whether the person owns a home PC, and the person's self-reported level of computer knowledge. Construct pivot tables (as many as you believe are necessary) to help SoftBus understand these customers, then write a short report that summarizes your findings.

68. The Comfy Company sells medium-priced patio furniture through a mail-order catalog. The company has operated primarily in the East but is now expanding to the Southwest. To get off to a good start, it plans to send potential customers a catalog with a discount coupon. However, Comfy is not sure how large a discount is necessary to entice customers to buy, so it experiments by sending catalogs to selected residents in

Table **2.1**

Data on Workweeks and Unemployment

Country	% Decrease in Annual Working Hours per Employee	% Increase in Unemployment
United States	3.0%	−2.0%
Italy	5.3	5.0
Japan	7.0	0.0
Canada	6.6	5.0
Britain	9.4	5.0
Spain	10.8	15.0
Holland	12.5	0.5
Germany	12.5	3.0
France	12.8	9.0

six cities. Tucson and San Diego receive coupons for 5% off any furniture within the next 2 months, Phoenix and Santa Fe receive coupons for 10% off, and Riverside and Albuquerque receive coupons for 15% off. The data are in the file P02_68.XLS. Develop one or more pivot tables to help the company determine the effect of the size of the coupon and whether this effect depends on the particular city. Write a short report that summarizes your findings.

69. You are a local Coca-Cola bottler. You want to determine whether sales of Coke and Diet Coke are more sensitive than the competition to changes in price. The file P02_69.XLS contains weekly data on the price per can of Coke, Diet Coke, Pepsi, and Diet Pepsi, and the number of cans (in hundreds) sold of each product. Which of these products exhibits more price sensitivity?

70. Use the data in the file P02_70.XLS to determine how the type of school (public or Catholic) that students attend affects their chance of graduating from high school.

71. It is well known that stock prices are a leading indicator of a recession. This means that several months before a recession begins, stock prices usually drop (foreshadowing a drop in the economy), and several months before a recession ends, stock prices usually increase (foreshadowing the end of the recession). Use the data in the file P02_71.XLS to argue that the Dow Jones Index was a leading indicator for both the April 1960–February 1961 recession and the December 1969–November 1970 recession.

72. The file P02_72.XLS contains information on monthly stock prices and trading volume for Wal-Mart. The data include date, low price for the month, high price for the month, closing price for the month, and number of shares traded during the month.
   a. Create a time series plot of High, Low, and Close, all on the same graph. Are there any obvious time series patterns?
   b. Create a time series plot of Volume. Are there any obvious time series patterns?
   c. Create a scatterplot of Volume versus Close. Does there appear to be any relationship between these two variables?

73. The file P02_73.XLS contains expense account data on your company's seven sales representatives for the past 4 months. Each row in the database includes a single expense record, which contains the rep's name, the month, the category (trip, entertaining client, or miscellaneous supplies), the amount claimed, and the amount reimbursed. (Only "legitimate" expenses are reimbursed.)
   a. Create a pivot table to tabulate the number of expense records of each category by each representative for the entire 4-month period. (For example, it will list the number of trips taken by Smith.) Use

the data in the resulting pivot table to construct an appropriate bar chart. (You can decide on the exact form of the chart.)
   b. Create a pivot table to show the total amount spent each month by each rep. Use the resulting pivot table to create time series plots, one for each rep.
   c. Create a histogram of reimbursed amounts, for the entertaining clients and miscellaneous supplies categories only.

Level B

74. The annual base salaries for 200 students graduating from a reputable MBA program this year are of interest to those in the admissions office who are responsible for marketing the program to prospective students. The data are in the file P02_74.XLS.
   a. Generate a frequency table and histogram for the given distribution of starting salaries. What does the histogram suggest about this distribution of starting salaries?
   b. Is it possible to separate these salaries into two or more subgroups? If so, generate a frequency distribution and histogram for each subset of starting salaries. Also, characterize the shape of each subgroup's distribution.
   c. As an admissions officer of this MBA program, how would you proceed to use these findings to market the program to prospective students?

75. The percentage of private-industry jobs that are managerial has steadily declined in recent years as companies have found middle management a ripe area for cutting costs. How have women and various minority groups fared in gaining management positions during this period of corporate downsizing of the management ranks? Relevant data are listed in the file P02_75.XLS. Generate scatterplots and/or time series plots using these data to make general comparisons across the various groups included in the set.

76. Chandler Enterprises produces Pentium chips. Five types of defects (labeled 1–5) have been known to occur. Chips are manufactured by two operators (A and B). Four machines (1–4) are used to manufacture chips. The file P02_76.XLS contains data for a sample of defective chips including the type of defect, operator, machine, and day of the week. Use the data in this file to chart a course of action that would lead, as quickly as possible, to improved product quality. You should use the Pivot Table Wizard to "stratify" the defects with respect to type of defect, day of the week, machine used, and operator working. You might even want to break the data down by machine and operator (or in some other way). Assume that each operator and machine made an equal number of products.

77. You own a local McDonald's and have done some market research in an attempt to better understand your customers. For a random sample of Bloomington

residents, the file P02_77.XLS contains the income, gender, and number of days per week the resident goes to McDonald's. Use this information to determine how gender and income influence the frequency with which Bloomington residents attend McDonald's.

78. Students at Faber College apply to study either English or science. You have been assigned to determine whether Faber College discriminates against women in admitting students to the school of their choice. The file P02_78.XLS contains the following data on Faber's students: gender, major applied for (English or science), and admission decision (yes or no). Assuming that women and men are equally qualified for each major, do the data indicate that the college discriminates against women? Make sure you use all available information.

79. You have been assigned to evaluate the quality of care given to heart attack patients at Emergency Room (ER) and Chicago Hope (CH). For the last month the file P02_79.XLS contains the following patient data: hospital where patient was admitted (ER or CH), risk category (high or low, where high-risk people are less likely to survive than low-risk people), and patient's outcome (lived or died). Use the data to determine which hospital is doing a better job of caring for heart attack patients. Use all of the data.

80. The file P02_80.XLS contains the monthly level of the Dow Jones Index for several decades. Do these data indicate any unusual seasonal patterns in stock returns? [*Hint:* You can extract the month (January, February, etc.) with the formula =TEXT(A4,"mmm") copied down any column.]

81. You sell station wagons and want to know how family size and salary influence the likelihood that a family will purchase a station wagon. You have surveyed some local families and found out whether they own a station wagon, the size of the family (Large means at least five people, Small means no more than four people), and the family's salary (High means at least

$80,000, Low means less than $80,000). The data are in the file P02_81.XLS. Analyze these data to determine how salary and family size influence the likelihood that a family will purchase a station wagon.

82. The file P02_82.XLS contains data on the diameter of an elevator rail, the operator who built the elevator rail, and the machine used to build the elevator rail. What can you learn from these data?

83. The file P02_83.XLS contains daily returns and the daily level of the Standard and Poor's 500 stock index. Describe what you learn from these data.

84. Viscerex, a small chemical company, wants to determine how viscosity of liquid supplied to the company influences the level of impurities. The file P02_84.XLS contains the following information: firm supplying the liquid to Viscerex, viscosity level of the liquid, and level of impurities in the liquid. Describe how viscosity affects the level of impurities.

85. Two major awards are given to daytime soap operas and their actors and actresses: the Daytime Emmys and the Soap Opera Digest Awards. The Daytime Emmys are voted on by members of the TV industry. The Soap Opera Digest Awards are voted on by soap opera viewers who are readers of *Soap Opera Digest.* The file P02_85.XLS contains the number of awards of each type won by each daytime soap. Use these data to determine whether voters for Daytime Emmys and Soap Opera Digest Awards appear to be looking for the same qualities when they vote for awards.

86. In recent years economists and others have debated whether investment in information technology is good or bad for employment growth. Table 2.2 contains information technology (IT) investment as a percentage of total investment for eight countries during the 1980s. It also contains the average annual percentage change in employment during the 1980s. Explain how these data shed light on the question of whether IT investment creates or costs jobs. (Source: *The Economist,* September 28, 1996)

Table **2.2**

Data on IT Investment

Country	IT Investment as % of Total Investment (1980s)	Annual Average % Change in Employment (1980–1989)
Netherlands	2.0%	1.3%
Italy	3.6	1.9
Germany	4.0	1.7
France	5.6	1.5
Canada	7.8	2.4
Japan	7.8	2.4
Britain	7.8	3.0
United States	11.9	3.4

Table **2.3**		
Data on Income Inequality	**Change in Wage Inequality Ratio**	**Change in Unemployment Rate**
Country		
Germany	−6.0%	6.0%
France	−3.5	5.6
Italy	1.0	5.2
Japan	0.0	0.6
Australia	5.0	2.4
Sweden	4.0	5.9
Canada	5.5	2.0
New Zealand	9.5	4.0
Britain	15.6	2.5
United States	15.8	−1.8

87. Do countries with more income inequality have lower unemployment rates? Table 2.3 contains the following information for ten countries during the 1980–1995 time period: change from 1980 to 1995 in ratio of the average wage of the top 10% of all wage earners to the median wage, and change from 1980 to 1995 in unemployment rate. (Source: *The Economist,* August 17, 1996)

   a. Explain why the ratio of the average wage of the top 10% of all wage earners to the median measures income inequality.

   b. Do these data help to confirm or contradict the hypothesis that increased wage inequality leads to lower unemployment levels?

   c. What other data would you need to be more confident that increased income inequality leads to lower unemployment?

88. One magazine reported that a man's weight at birth has a significant impact on the chance that the man will suffer a heart attack during his life. Analyze the data in the file P02_88.XLS to determine how birth weight influences the chances that a man will have a heart attack.

89. When Staples and Office Depot proposed merging in 1997, the Federal Trade Commission (FTC) rejected the merger. In analyzing the impact of the merger, the FTC looked at the prices of the following quantities:

- A Pentium 166 computer with 16 meg of RAM, 10-speed CD-ROM, and 2g hard drive (labeled Computer)
- A desk, filing cabinet, 10 reams of computer paper, and a laser printer (labeled Office)
- 10 notebooks, 5 boxes of pencils, a book bag, 10 boxes of crayons (labeled School)

The file P02_89.XLS contains data on these variables for several cities. For example, the first city had both a Staples and an Office Depot, and a computer cost $1979.31. The second city had only a Staples, and the school supplies cost $179.86. Based on these data, can you explain why the FTC rejected the merger? (*Note:* The data in this file are fictitious but are consistent with the conclusions of the article.) (Source: Based on *The Economist,* May 3, 1998)

90. An important question in finance is whether the stock market is efficient. The market is efficient if knowledge of past changes in a stock's price tells us nothing about future changes in the stock's price. Here you'll check whether daily price changes in IBM stock are consistent with efficient markets, using the data in the file P02_90.XLS. Define an "up" day for IBM as a day when the return is greater than 0. A "down" day is when the return is less than or equal to 0. Does it appear that knowledge of whether IBM went up or down yesterday can help us predict whether it will go up or down today?

91. You work for a small travel agency and are about to do a mass mailing of a travel brochure. Your funds are limited, so you want to mail to the people who spend the most money on travel. The file P02_91.XLS contains data for a random sample of 925 residents. These data include their gender, age, and amount spent on travel last year. Use these data to determine how gender and age influence a person's travel expenditures. Also make recommendations on the type of person to whom you should mail your brochure.

92. A question of great interest is how the distribution of family income has changed in the United States during the last 20 years. The file P02_92.XLS contains data for a sample of 499 family incomes (in real 1995 dollars). For each family, the 1975 and 1995 incomes are listed. (Although these data are fictitious, they are consistent with what has actually happened to U.S. family income during these years.) Based on these data, discuss as completely as possible how the distribution of family income in the United States changed from 1975 to 1995.

CASE

2.1 Customer Arrivals at Bank98

Bank98 operates a main location and three branch locations in a medium-size city. All four locations perform similar services, and customers typically do business at the location nearest them. The bank has recently had more congestion—long waiting lines—than it (or its customers) would like. As part of a study to learn the causes of these long lines and to suggest possible solutions, all locations have kept track of customer arrivals during 1-hour intervals for the past 10 weeks. All branches are open Monday through Friday from 9 A.M. until 5 P.M. and on Saturday from 9 A.M. until noon. For each location, the file BANK98.XLS contains the number of customer arrivals during each hour of a 10-week period. The manager of Bank98 has hired you to make some sense out of these data. Specifically, your task is to present charts and/or tables that indicate how customer traffic into the bank locations varies by day of week and hour of day. There is also interest in whether any daily or hourly patterns you observe are stable across weeks. Although you don't have full information about the way the bank currently runs its operations—you know only its customer arrival pattern and the fact that it is currently experiencing long lines—you are encouraged to append any suggestions for improving operations, based on your analysis of the data.

CASE

2.2 Automobile Production and Purchases

Are people in the United States buying more cars than in the past? Are they buying more foreign cars relative to domestic cars? Are auto sales seasonal, with more sales occurring during some months than others? Does automobile production mirror sales very closely? These are some questions you have been asked to answer, using the data in the file AUTOS.XLS. This file contains monthly data on U.S. sales of domestic and foreign cars since 1967. It also shows monthly production of domestic cars since 1993. The data are shown in two forms: not seasonally adjusted—the raw data—and seasonally adjusted. (Although you will learn more about seasonal adjustment of time series data in Chapter 13, the basic idea is that this is a method for smoothing out seasonal ups and downs so that underlying trends can be seen more clearly.) You have been asked to prepare a report that explains any important patterns you observe in these data. Of course, your report should contain relevant charts and/or tables. Make sure your report indicates whether you are using seasonally adjusted data or raw data (or both) and why.

2.3 Saving, Spending, and Social Climbing

The recent best-selling book *The Millionaire Next Door* by Thomas J. Stanley and William D. Danko (Longstreet Press, 1996) presents some very interesting data on the characteristics of millionaires. We tend to believe that people with expensive houses, expensive cars, expensive clothes, country club memberships, and other outward indications of wealth are the millionaires. The authors define wealth, however, in terms of savings and investments, not consumer items. In this sense, they argue that people with a lot of expensive *things* and even large incomes often have surprisingly little wealth. These people tend to spend much of what they make on consumer items, often trying to keep up with, or impress, their peers. In contrast, the real millionaires, in terms of savings and investments, frequently come from "unglamorous" professions (particularly teaching!), own unpretentious homes and cars, dress in in-expensive clothes, and otherwise lead rather ordinary lives.

Consider the (hypothetical) data in the file SOCIAL_CLIMBERS.XLS. For several hundred couples, it lists their education level, their annual combined salary, the market value of their home and cars, the amount of savings they have accumulated (in savings accounts, stocks, retirement accounts, and so on), and a self-reported "social climber index" on a scale of 1 to 10 (with 1 being very unconcerned about social status and material items and 10 being very concerned about these). Prepare a report based on these data, supported by relevant charts and/or tables, that might be used in a book such as *The Millionaire Next Door*. Although your report might be used in such a book, your conclusions can either support or contradict those of Stanley and Danko.

Describing Data: Summary Measures

Predictors of Successful Movies

The movie industry is a high-profile industry with highly variable revenue stream. In 1998, U.S. moviegoers spent close to $7 billion at the box office alone. With this much money at stake, it is not surprising that movie studios are interested in knowing what variables are useful for predicting a movie's financial success. The article by Simonoff and Sparrow (2000) examines this issue for 311 movies released in 1998 and late 1997. (They obtained their data from a public Web site at www.imdb.com.) Although it is preferable to examine movie *profits,* the costs of making movies are virtually impossible to obtain. Therefore, the authors focused instead on revenues—specifically, the total U.S. domestic gross revenue for each film.

Simonoff and Sparrow obtained prerelease information on a number of variables that were thought to be possible predictors of gross revenue. (By "prerelease," we mean that this information is known about a film *before* the film is actually released.) These variables include: (1) the genre of the film, categorized as action, children's, comedy, documentary, drama, horror, science fiction, or thriller; (2) the Motion Picture Association of America (MPAA) rating of the film, categorized as G (general audiences), PG (parental guidance suggested), PG-13 (possibly unsuitable for children under 13), R (children not admitted unless accompanied by an adult), NC-17 (no one under 17 admitted), or U (unrated); (3) the country of origin of the movie, categorized as United States, English-speaking but non–United States, or non–English-speaking; (4) number of actors and actresses in the movie who were listed in *Entertainment Weekly*'s lists of the 25 Best Actors and 25 Best Actresses, as of 1998; (5) number of actors and actresses in the movie who were among the top 20 actors and top 20 actresses in average box office gross per movie in their careers; (6) whether the movie was a sequel; (7) whether the movie was released before a holiday weekend; (8) whether the movie was released during the

© 2002 Harris Welles

Christmas season; and (9) whether the movie was released during the summer season.

To get a sense of whether these variables are related to gross revenue, we could calculate a lot of summary measures and create numerous tables. However, we agree with Simonoff and Sparrow that the information is best presented in a series of *side-by-side boxplots,* a type of graph we will introduce in this chapter. (See Figure 3.1.) These boxplots are slightly different from the versions we describe in this book, but they accomplish exactly the same purpose. (There are two differences: First, their boxplots are vertical; ours are horizontal. Second, their boxplots capture an extra piece of information—the *widths* of their boxes are proportional to the square roots of the sample sizes, so that wide boxes correspond to categories with more movies. In contrast, the *heights* of our boxes carry no information about sample size.) Basically, each box and the lines and points extending above and below it indicate the distribution of gross revenues for any category. The box itself, from bottom to top, captures the middle 50% of the revenues in the category, the line in the middle of the box represents the median revenue, and the lines and dots indicate possible skewness and outliers.

These particular boxplots indicate some interesting and possibly surprising information about the movie business. First, almost all of the boxplots indicate a high degree of variability and positive skewness, where there are a few movies that gross extremely large amounts compared to the "typical" movies in the category. Second, genre certainly makes a difference. There are more comedies and dramas (wider boxes), but they typically gross considerably less than action, children's, and science fiction films. Third, the same is true of R-rated movies compared to movies rated G, PG, or PG-13—there are more of them, but they typically gross much less. Fourth, U.S. movies do considerably better than foreign movies. Fifth, it helps to have stars, although there are quite a few "sleepers" that succeed without having big-name stars. Sixth, sequels do better, presumably reflecting the success of the earlier films. Finally, the release date makes a big difference. Movies released before holidays, during the Christmas season, or during the summer season tend to have larger gross revenues. Indeed, as Simonoff and Sparrow discuss, movie studios compete fiercely for the best release dates.

Are these prerelease variables sufficient to predict gross revenues accurately? As we might expect from the amount of variability in most of the boxplots in Figure 3.1, the answer is "no." Many intangible factors evidently determine the ultimate success of a movie, so that some, such as *There's Something About Mary,* do much better than expected, and others, such as *Godzilla,* do worse than expected. We will revisit this movie data set in the chapter opener to Chapter 12. There, we will see how Simonoff and Sparrow use "multiple regression" to predict gross revenue—with only limited success. ●

Introduction

In the previous chapter we summarized data mainly with tables and graphs. It is often useful to summarize data even further with a few well-chosen numbers. In this chapter we will learn the most frequently used numerical summary measures. These include measures for describing a single variable, such as the mean, median, and standard deviation, plus a couple of measures, correlation and covariance, for describing the potential relationship between two variables. Using these numerical summary measures, we will then discuss an additional graph called a boxplot. Boxplots are useful for describing a single variable or comparing two or more related variables.

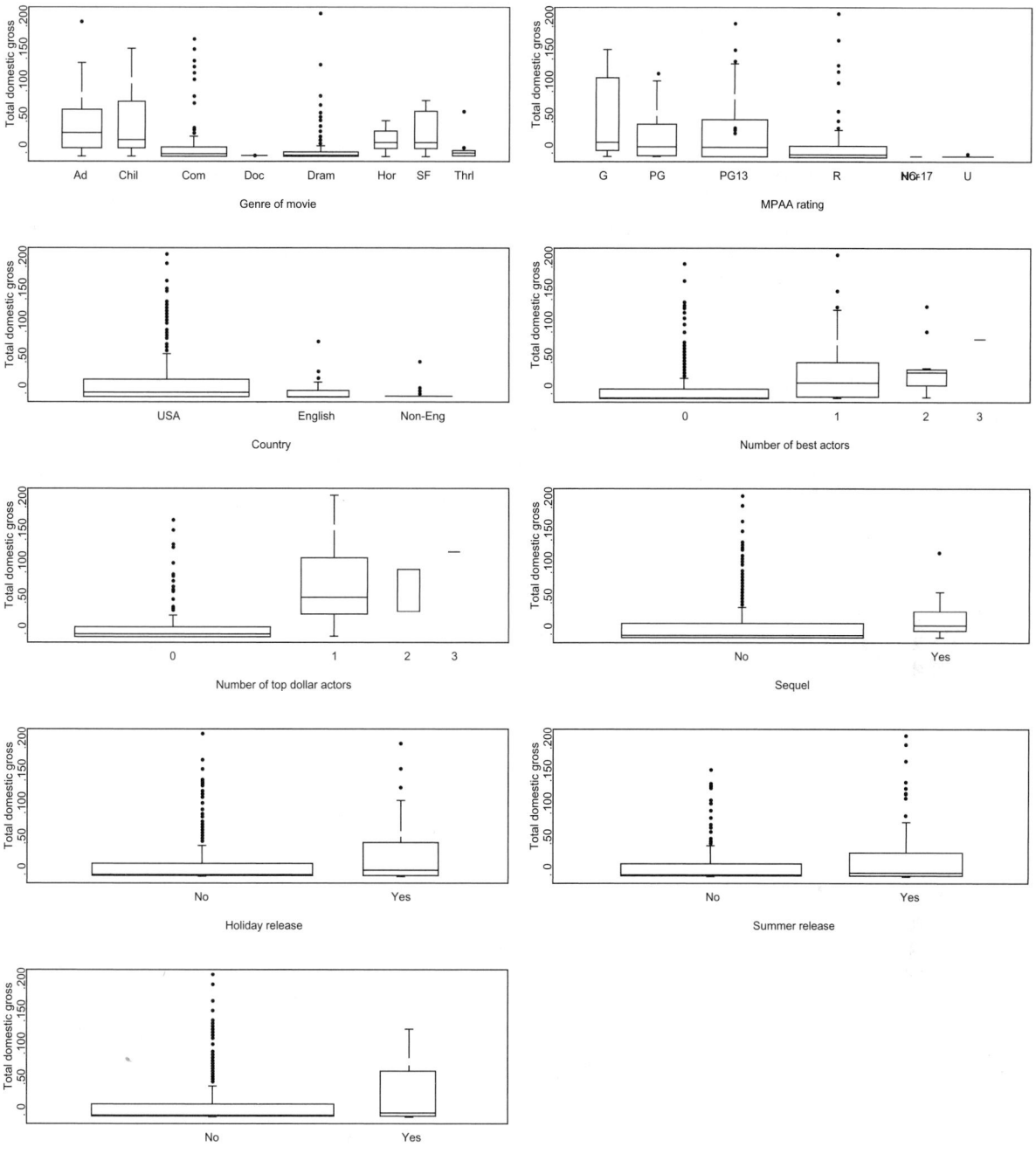

FIGURE 3.1 Boxplots of Domestic Gross Revenues for 1998 Movies

Source: From "Predicting Movie Grosses, Winners and Losers, Blockbusters and Sleepers" by J. S. Simonoff & I. R. Sparrow, *Chance*, pp. 15–22, Vol. 13, No. 3, Summer 2000. Used by permission.

We conclude this chapter by examining three relatively complex examples. Here we are able to put all of the descriptive tools we have learned to good use. These examples are typical of the large-scale data sets business managers face on a continual basis. Only by looking at the data from a number of points of view can we discover the information and patterns hidden in the data.

3.2 Measures of Central Location

Most of the numerical summary measures we consider here are for a single variable. Each describes some aspect of the distribution of the variable. That is, each number describes one feature of the distribution that we see graphically in a histogram. We begin with measures of central location. The three most commonly used measures are the *mean, median,* and *mode,* each of which gives a slightly different interpretation to the term "central location."

3.2.1 The Mean

The summation sign denoted by the Greek symbol sigma (Σ) is simply a shortcut for writing a sum, in this case $X_1 + X_2 + \cdots + X_n$. It is used extensively in statistical formulas.

The **mean** is the average of all values of a variable. If the data represent a sample from some larger population, we call this measure the **sample mean** and denote it by $\overline{X}$ (pronounced "x-bar"). If the data represent the entire population, we call it the **population mean** and denote it by μ. This distinction is not important in this chapter, but it will become relevant in later chapters when we discuss statistical inference. In either case the formula for the mean is given by equation (3.1).

Formula for the Mean		
	$\text{Mean} = \dfrac{\sum_{i=1}^{n} X_i}{n}$	**(3.1)**

Here n is the number of observations and X_i is the value of observation i. Equation (3.1) simply says to add all the observations and divide by n, the number of observations.

To obtain the mean in Excel, we use the AVERAGE function on the appropriate range, as illustrated in the following example.

Example 3.1 Summarizing Starting Salaries for Business Undergraduates

The file SALARY.XLS lists starting salaries for 190 graduates from an undergraduate school of business. The data are in the range named Salary on a sheet called Data. Find the average of all salaries.

Objective To use Excel's AVERAGE function to summarize salaries.

Solution

Figure 3.2 includes a number of summary measures produced by Excel's built-in functions. In particular, we calculate the mean salary by entering the formula

=AVERAGE(Salary)

in cell B6. It is nearly $30,000. (The other summary measures in Figure 3.2 are obtained using other Excel functions, as will be discussed subsequently.)

FIGURE 3.2

Selected Summary
Measures of
Salary Data

	A	B	C
1	**Summary measures using Excel functions**		
2			
3	Count	190	
4	Minimum	$17,100	
5	Maximum	$38,200	
6	Average	$29,762	
7	Median	$29,850	
8	Lower quartile	$27,325	
9	Upper quartile	$32,300	
10	5-percentile	$23,690	
11	95-percentile	$35,810	
12	Range	$21,100	
13	Standard deviation	$3,707	
14	Variance	13743424	

When working with a set of data, the terms *mean* and *average* are synonymous.

The mean in Example 3.1 is a "representative" measure of center because the distribution of salaries is nearly symmetric. (You can check this statement by constructing a histogram of salaries.) However, the mean is often misleading because of skewness. For example, if a few of the undergraduates got abnormally high salaries (over $100,000, say), these large values would tend to inflate the mean and make it unrepresentative of the majority of the salaries. In such cases, the median may be a more appropriate measure of center.

Think of the difference between mean and median for some common variables: (1) all final exam scores for a large college course; (2) all household incomes in your state; (3) total yearly medical expenses for all households in your state. For these variables, which measure would probably be more "representative," the mean or median? Why?

3.2.2 The Median

The **median** is the "middle" observation when the data are listed from smallest to largest. If there is an odd number of observations, the median is the middle observation. For example, if there are nine observations, the median is the fifth smallest (or fifth largest) observation. If there is an even number of observations, we take the median to be the average of the two middle observations. For example, if there are ten observations, the median is the average of the fifth and sixth smallest values.

We calculate the median salary in Example 3.1 by entering the formula

$$=\text{MEDIAN(Salary)}$$

in cell B7. (See Figure 3.2.) Its value is again approximately $30,000, almost the same as the mean. This is typical of symmetric distributions, but it is not true for skewed distributions. For example, if a few graduates received abnormally large salaries, the mean would be affected by them, but the median would not be affected at all. It would still represent the "middle" of the distribution.

Median

The **median** is the middle observation (for an odd number of observations) or the average of the middle two observations (for an even number of observations) after the observations have been sorted from low to high.

3.2.3 The Mode

The **mode** is the most frequently occurring value. If the values are essentially continuous, as with the salaries in Example 3.1, then the mode is essentially irrelevant. There is typically no *single* value that occurs more than once, or there are at best a few ties for the most frequently occurring value. In either case the mode is not likely to provide much information. However, the following example illustrates where the mode is useful.

Mode	The **mode** is the most frequently occurring observation.

Example 3.2 Shoe Size Purchased at a Shoe Store

The file SHOES.XLS lists shoe sizes purchased at a shoe store. What is the store's best-seller?

Objective To use Excel's MODE function to find the best-selling shoe size.

Solution

Shoe sizes come in discrete increments, rather than a continuum, so it makes sense to find the mode, the size that is requested most often. This can be done with Excel's MODE function, which shows that size 11 is the most frequently purchased size. This is also apparent from the histogram in Figure 3.3, where the category for size 11 corresponds to the highest bar. (By default, StatPro lists the categories on the horizontal axis as, for example, "10.5–11," which in this case really means size 11. You can easily change the axis labels by changing the labels in column B of the "Data" sheet that StatPro creates for the histogram.)

FIGURE 3.3
Distribution of Shoe Sizes for Example 3.2

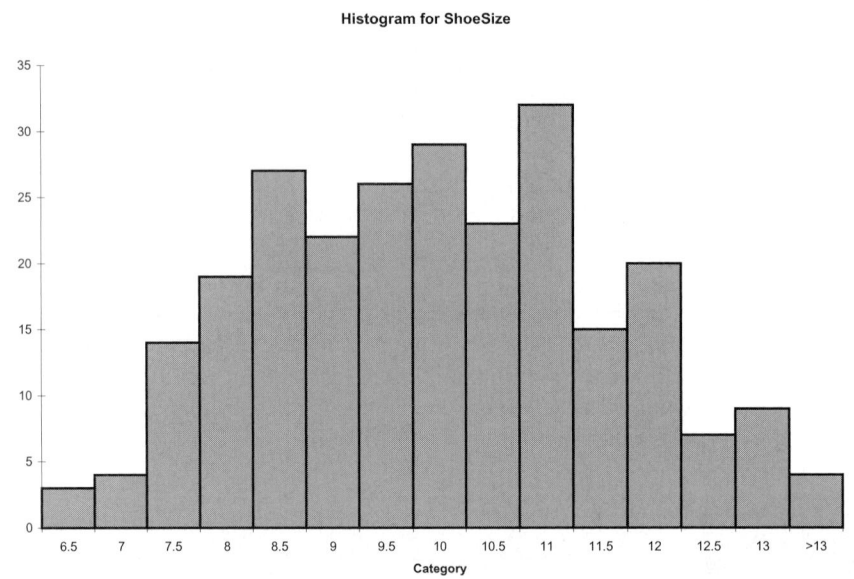

3.3 Quartiles and Percentiles

The median splits the data in half. It is sometimes called the 50th **percentile** because (approximately) half of the data are below the median. It is also called the second **quartile,** because if we divide the data into four parts, then the median separates the lower two parts from the upper two. We can also find other percentiles and quartiles. Some of these appear in Figure 3.2, which summarizes the salary data from Example 3.1.

For example, the 5th and 95th percentiles appear in cells B10 and B11. They are calculated with the formulas

$$=\text{PERCENTILE(Salary,.05)}$$

and

$$=\text{PERCENTILE(Salary,.95)}$$

They say that 5% of all salaries are below \$23,690 (so that 95% are above \$23,690), and 95% of all salaries are below \$35,810 (so that 5% are above \$35,810). These two percentiles (and sometimes others) are frequently quoted.

Similarly, Figure 3.2 lists the lower and upper quartiles in cells B8 and B9. These are the 25th and 75th percentiles, so they can be calculated with the PERCENTILE function. They can also be calculated with the formulas

> The second argument of the PERCENTILE function must be a *decimal* between 0 and 1. In contrast, the second argument of the QUARTILE function must be an *integer:* 1, 2, or 3.

$$=\text{QUARTILE(Salary,1)}$$

and

$$=\text{QUARTILE(Salary,3)}$$

That is, they are the first and third quartiles. (The median is the second quartile.) In short, 25% of the salaries are below \$27,325, 25% are above \$32,300, and the other 50% are in between.

The difference between the first and third quartiles is called the **interquartile range** (IQR). It measures the spread between the largest and smallest of the middle half of the data. In the salary example the IQR is \$4,975 (= \$32,300 − \$27,325). We will come back to the IQR when we discuss boxplots later in this chapter.

Excel Tip *Excel uses a rather obscure interpolation formula (which is not spelled out in its online help) for determining percentiles and quartiles. This can be particularly disconcerting—and wrong—for small data sets. For example, Excel reports that the 25th percentile (and the first quartile) of the ten numbers 1, 2, 3, 4, 5, 6, 7, 8, 9, 10 is 3.25. This does not appear to be consistent with our definition of the 25th percentile, which states that 25% of the numbers are below it and 75% are above it. As another example, it always reports the 25th percentile of 13 numbers as the fourth smallest of these numbers, which seems strange. Therefore, we recommend not putting too much reliance on Excel's PERCENTILE and QUARTILE functions for small data sets.*

3.4 Minimum, Maximum, and Range

Three other descriptive measures of a variable are its *minimum, maximum,* and *range.* The **minimum** is the smallest value, the **maximum** is the largest value, and the **range** is the difference between the maximum and minimum. These are listed in Figure 3.2 for the salary data. The minimum and maximum are calculated in cells B4 and B5 with the formulas

$$=\text{MIN(Salary)}$$

and

$$=\text{MAX(Salary)}$$

The range is then calculated in cell B12 with the formula

$$=\text{B5-B4}$$

We see that no salary is below \$17,100, no salary is above \$38,200, and all salaries are contained within an interval of length \$21,100.

The minimum and maximum provide bounds on the data set, and the range provides a crude measure of the variability of the data. These measures are often worth reporting, but they can obviously be affected by one or two extreme values. The range, in particular, is usually not as good a measure of variability as the measures discussed next.

3.5 Measures of Variability: Variance and Standard Deviation

To really understand a data set, we need to know more than measures of central location; we also need measures of variability. To see this, consider the following example.

Example 3.3 Variability of Elevator Rail Diameters at Otis Elevator

Suppose Otis Elevator is going to stop manufacturing elevator rails. Instead, it is going to buy them from an outside supplier. Otis would like each rail to have a diameter of 1 inch. The company has obtained samples of ten elevator rails from each supplier. These are listed in columns A and B of Figure 3.4. (See the file OTIS4.XLS.) Which supplier should Otis prefer?

Objective To calculate the variability for two suppliers and choose the one with the least variability.

FIGURE 3.4
Two Samples with
Different Amounts
of Variability

	A	B	C	D	E	F
1	**Diameters from two suppliers**					
2						
3	**Data**			**Summary measures**		
4	Supplier1	Supplier2			Supplier1	Supplier2
5	1.00	0.96		Mean	1.00	1.00
6	0.98	1.05		Median	1.00	1.00
7	1.02	1.00		Mode	1	1
8	1.01	0.97		Variance	0.000133	0.001200
9	1.00	1.00		Standard deviation	0.0115	0.0346
10	0.99	1.03				
11	0.99	0.98				
12	1.00	1.02				
13	1.01	0.95				
14	1.00	1.04				

Solution

Observe that the mean, median, and mode are all exactly 1 inch for each supplier. Based on these measures, the two suppliers are equally good and both are right on the mark. It is clear from a glance at the data, however, that supplier 1 is somewhat better than supplier 2. The reason is that supplier 2's rails exhibit more variability about the mean than supplier 1's rails. If we want rails to have a diameter of 1 inch, then variability around the mean is bad! ●

The most commonly used measures of variability are the *variance* and *standard deviation*. The **variance** is essentially the average of the squared deviations from the mean. We say "essentially" because there are two versions of variance: the **population variance,** usually denoted by σ^2, and the **sample variance,** usually denoted by s^2. The formulas for them are given by equations (3.2) and (3.3). These formulas differ only in their denominators. Also, their numerical values are practically the same when n, the number of observations, is large.

Formula for Population Variance	$$\sigma^2 = \frac{\sum_{i=1}^{n}(X_i - \overline{X})^2}{n}$$	**(3.2)**

Formula for Sample Variance	$$s^2 = \frac{\sum_{i=1}^{n}(X_i - \overline{X})^2}{n-1}$$	**(3.3)**

We say that each term in these sums is a squared deviation from the mean, where *deviation* means difference. Therefore, you can remember variance as the *average of the squared deviations from the mean*. This is not quite true for the sample variance (because of dividing by $n - 1$, not n), but it is nearly true. If we use the supplier 2 data in Figure 3.4 for illustration, the deviations from the mean are -0.04, 0.05, 0.00, -0.03, 0.00, 0.03, -0.02, 0.02, -0.05, and 0.04; the squares of these deviations are 0.0016, 0.0025, 0.0000, 0.0009, 0.0000, 0.0009, 0.0004, 0.0004, 0.0025, and 0.0016; and the sum of these squares is 0.0108. Therefore, the population variance is $0.0108/10 = 0.0011$, and the sample variance (the one quoted in the figure) is $0.0108/9 = 0.0012$.

It is more common to quote the sample variance and the sample standard deviation than their population counterparts. However, for large n, the difference is practically irrelevant.

As their names imply, σ^2 is relevant if the data set includes the entire population, whereas s^2 is relevant for a sample from a population. Excel has a built-in function for each. To obtain σ^2 we use the VARP function; to obtain s^2 we use the VAR function. In this chapter we will illustrate only the sample variance s^2.

The important part about either variance formula is that the variance tends to increase when there is more variability around the mean. Indeed, large deviations from the mean contribute heavily to the variance because they are *squared*. One consequence of this is that the variance is expressed in squared units (squared dollars, for example) rather than original units. Therefore, a more intuitive measure is the **standard deviation,** defined as the square root of the variance. The standard deviation is measured in original units, such as dollars, and, as we will discuss shortly, it is much easier to interpret.

Relationship Between Standard Deviation and Variance	$$\text{Standard deviation} = \sqrt{\text{Variance}}$$	**(3.4)**

Of course, depending on which variance measure we use, we obtain the corresponding standard deviation, either σ (population) or s (sample). Excel has built-in functions for each of these; we use STDEVP for σ and STDEV for s. In this chapter we will illustrate only s.

The variances and standard deviations of the diameters from the two suppliers in Example 3.3 appear in Figure 3.4. To obtain them, enter the formulas

$$=VAR(A5:A14)$$

and

$$=STDEV(A5:A14)$$

in cells E8 and E9, and enter similar formulas for supplier 2 in cells F8 and F9. Because of the relationship between variance and standard deviation, we could also have used the formula

$$=SQRT(E8)$$

in cell E9, but we instead took advantage of the STDEV function.

As we mentioned previously, it is difficult to interpret these variances numerically because they are expressed in squared inches, not inches. All we can say is that the variance from supplier 2 is considerably larger than the variance from supplier 1. The standard deviations, on the other hand, are expressed in inches. The standard deviation for supplier 1 is approximately 0.012 inch, and supplier 2's standard deviation is approximately three times this large. This is a considerable disparity, as we explain next.

Interpretation of the Standard Deviation: Empirical Rules

These rules are based on a theoretical distribution called the *normal distribution,* which will be discussed in detail in Chapter 6. Remarkably, many data sets are at least approximately normally distributed, so that these empirical rules often provide good approximations.

Many data sets follow certain "empirical rules." In particular, suppose that a histogram of the data is approximately symmetric and "bell shaped." That is, heights of the bars rise to some peak and then decline. Such behavior is quite common, and it allows us to interpret the standard deviation intuitively. Specifically, we can state that

- Approximately 68% of the observations are within 1 standard deviation of the mean, that is, within the interval $\overline{X} \pm s$;

- Approximately 95% of the observations are within 2 standard deviations of the mean, that is, within the interval $\overline{X} \pm 2s$; and

- Approximately 99.7%—almost all—of the observations are within 3 standard deviations of the mean, that is, within the interval $\overline{X} \pm 3s$.

We illustrate these empirical rules with the following example.

Example 3.4 Distribution of Monthly Dow Returns

The file DOW.XLS contains monthly closing prices for the Dow Jones index from January 1947 through January 1993. The monthly returns from the index are also shown, starting with the February 1947 value. Each return is the monthly percentage change (expressed as a decimal) in the index. How well do the empirical rules work for these data?

Objective To illustrate the empirical rules for Dow Jones monthly returns.

Solution

Figures 3.5 and 3.6 show time series plots of the index itself and the monthly returns. Clearly, the index has been increasing fairly steadily over the period, whereas the returns

FIGURE 3.5
Time Series Plot
of Dow Closing
Index

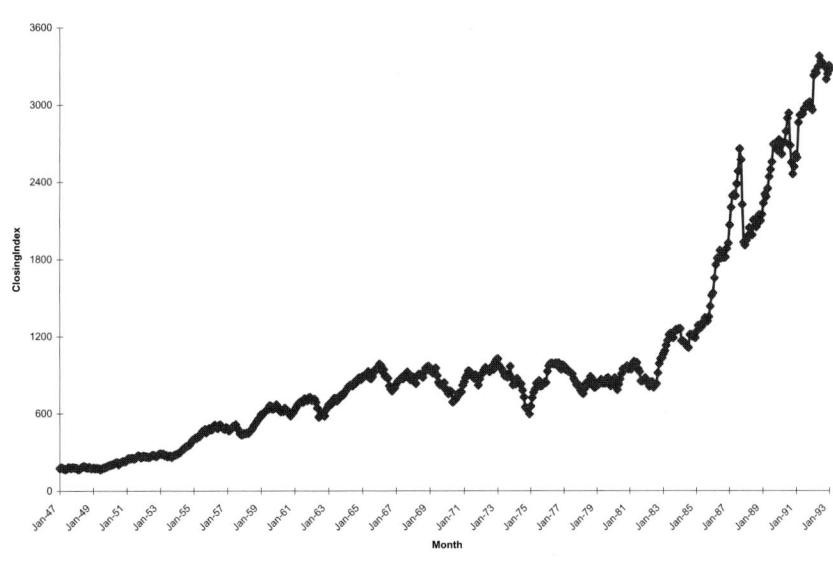

FIGURE 3.6
Time Series Plot
of Dow Returns

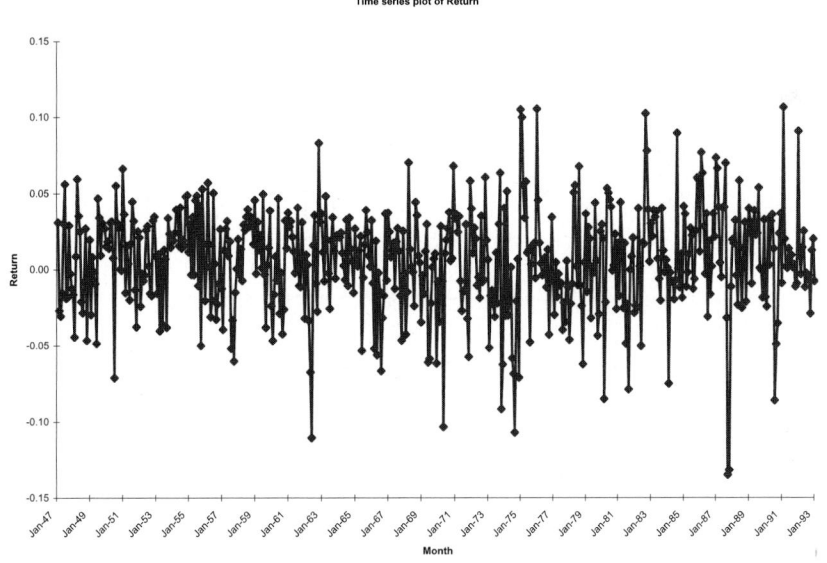

exhibit no obvious trend. Whenever a series indicates a clear trend, most of the measures we have been discussing are less relevant. For example, the mean closing index for this period has at most historical interest. We are probably more interested in predicting the *future* of the Dow, and the historical mean (or standard deviation or variance) has little relevance for predicting the future.

In contrast, the measures we have been discussing are relevant for the series of returns, which fluctuates around a stable mean. In Figure 3.7 we first calculate the mean and standard deviation of returns with the AVERAGE and STDEV functions in cells B4 and B5. These indicate an average return of about 0.59% and a standard deviation of about 3.37%. Therefore, the empirical rules (if they apply) imply, for example, that about 2/3 of all returns are within the interval 0.59% ± 3.37%, that is, from −2.78% to 3.95%.

We can use a frequency table to check whether the empirical rules apply to these returns. We first enter the upper limits of suitable categories in the range A8:A15. Although

FIGURE 3.7
Empirical Rules for Dow Jones Data

	A	B	C	D
1	**Checking empirical rules for returns**			
2				
3	Summary measures of returns			
4	Mean	0.0059		
5	Stdev	0.0337		
6				
7	Category	Upper limit	Frequency	
8	More than 3 stdevs below mean	-0.0951	5	
9	Between 2 and 3 stdevs below mean	-0.0614	13	
10	Between 1 and 2 stdevs below mean	-0.0278	57	
11	Between mean and 1 stdev below mean	0.0059	194	
12	Between mean and 1 stdev above mean	0.0395	217	
13	Between 1 and 2 stdevs above mean	0.0732	55	
14	Between 2 and 3 stdevs above mean	0.1069	11	
15	More than 3 stdevs above mean		0	
16				
17	Percentages within k stdevs of mean			
18	k	1	2	3
19	Actual	74.5%	94.7%	99.1%
20	From empirical rules	68.0%	95.0%	99.7%

any categories could be chosen, it is convenient to choose values of the form $\overline{X} \pm ks$ as breakpoints for the categories, where the open-ended categories on either end are "more than 3 standard deviations from the mean." Here, k is a typical multiple of the standard deviation s. We let k have the values 1, 2, and 3. Then the upper limits of the categories can be calculated in column B by entering the formulas

$$=B4-3*B5$$

and

$$=B8+B5$$

in cells B8 and B9, and then copying this latter formula to the range B10:B14. In words, each breakpoint is 1 standard deviation higher than the previous one.

Next, we use the FREQUENCY function to fill in column C. Specifically, we highlight the range C8:C15, type the formula

$$=FREQUENCY(Return,Bins)$$

and press Ctrl-Shift-Enter. (Here, Return is the range name for the Returns variable, and Bins is the range name of the range B8:B14.)

As an alternative to using the FREQUENCY function, the same counts can be found by creating a histogram of the returns with StatPro, using −0.0951 (3 standard deviations below the mean) as the "minimum" value, eight categories, and typical category length 0.0337 (the standard deviation).

Finally, we use the frequencies in column C to calculate the actual percentages of returns within k standard deviations of the mean for $k = 1$, $k = 2$, and $k = 3$, and we compare these with percentages from the empirical rules. (See rows 19 and 20.) The agreement between these percentages is not perfect—there are a few more observations within 1 standard deviation of the mean than the empirical rules predict—but in general the empirical rules work quite well. ●

The FREQUENCY function is an "array function" in Excel. We enter it with the Ctrl-Shift-Enter key combination.

3.6 Obtaining Summary Measures with Add-Ins

In the past few sections we have used Excel's built-in functions (AVERAGE, STDEV, and so on) to calculate a number of summary measures. A quicker way is to use the Stat-Pro add-in or Excel's Analysis ToolPak add-in. We illustrate the StatPro add-in in the following example. You can compare it with the Analysis ToolPak method if you like. To do so, use the Tools/Data Analysis menu item and select the Descriptive Statistics option.

Example 3.5 Summarizing Movie Star Salaries

Refer again to the SALARY.XLS data set used in Example 3.1, and find a set of useful summary measures for the salaries.

Objective To use StatPro's Summary Stats procedure to calculate summary measures of salaries.

Solution

This is easy with StatPro's Summary Stats procedure. As usual, place the cursor anywhere within the data range, then select the StatPro/Summary Stats/One-Variable Summary Stats menu item, select all variables you want to summarize, and select the summary measures you want from the dialog box shown in Figure 3.8. Note that four common measures (mean, median, standard deviation, and count) are checked by default, but you can override these defaults.

A typical output appears in Figure 3.9. It includes many of the summary measures we have discussed, plus a few more. The **mean absolute deviation** is similar to the variance, except that it is an average of the *absolute* (not squared) deviations from the mean. The **kurtosis** and **skewness** indicate the relative peakedness of the distribution and its skewness. These are relatively technical measures that we will not discuss here.

FIGURE 3.8
Summary Stats
Dialog Box

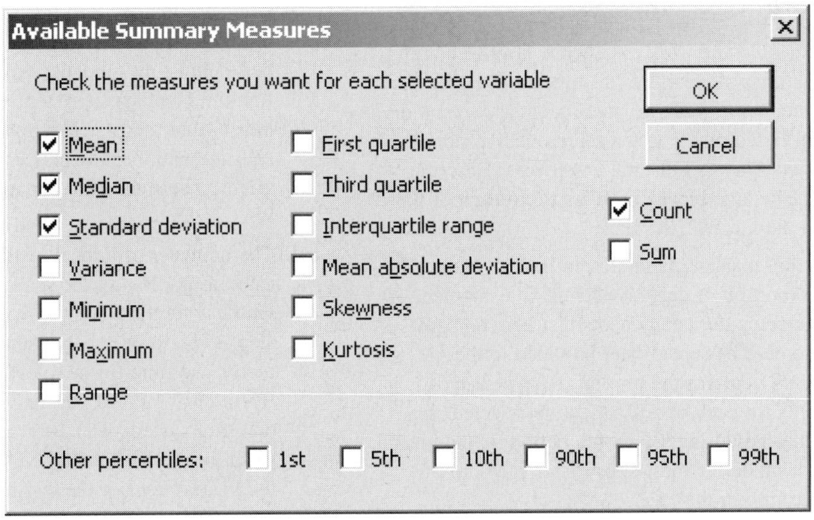

FIGURE 3.9

Selected Summary
Measures Using
StatPro Add-In

	A	B	C
1		**Summary measures for selected variables**	
2			Salary
3		Count	190.000
4		Mean	29762.105
5		Median	29850.000
6		Standard deviation	3707.212
7		Minimum	17100.000
8		Maximum	38200.000
9		Range	21100.000
10		Variance	13743424.116
11		First quartile	27325.000
12		Third quartile	32300.000
13		Interquartile range	4975.000
14		Mean absolute deviation	2967.767
15		Skewness	-0.166
16		Kurtosis	-0.071
17		5th percentile	23690.000
18		95th percentile	35810.000

By clicking on any of the cells in column B of Figure 3.9, you'll see that StatPro provides *formulas* for the outputs. (Excel's Analysis ToolPak does not do so.) The effect is that if any of the original data change, the summary measures we just produced change automatically. Finally, note that all outputs are formatted as "Number" with three decimal places by default. You might want to reformat them in a more appropriate manner. ●

PROBLEMS

Level A

1. A human resources manager at Beta Technologies, Inc., is interested in compiling some statistics on the *typical* full-time Beta employee. In particular, she is interested in finding the typical age, number of years of relevant full-time work experience prior to coming to Beta, number of years of full-time work experience at Beta, number of years of post-secondary education, and salary based on the given representative sample of 52 of the company's full-time employees. Describe the typical Beta employee with regard to each of these factors in the file P02_01.XLS.

2. A production manager is interested in determining the typical proportion of defective items in a shipment of one of the computer components that her company manufactures. The spreadsheet provided in the file P02_02.XLS contains the proportion of defective components for each of 500 randomly selected shipments collected during a 1-month period. Is the mean, median, or mode the most appropriate measure of central location in this case?

3. *Business Week's Guide to the Best Business Schools* provides enrollment data on graduate business programs that it rates as the best in the United States. Specifically, this guide reports the percentages of women, minority, and international students enrolled in each of the top programs, as well as the total number of full-time students enrolled in each program. Use the most appropriate measure(s) of central location to find the typical percentage of women, minority, and international students enrolled in these elite programs. These data are contained in the file P02_03.XLS.

4. The manager of a local fast-food restaurant is interested in improving the service provided to customers who use the restaurant's drive-up window. As a first step in this process, the manager asks his assistant to record the time (in minutes) it takes to serve 200 different customers at the final window in the facility's drive-up system. The 200 customer service times given in the file P02_04.XLS are all observed during the busiest hour of the day for this fast-food operation.

a. Compute the mean, median, and mode of this sample of customer service times.

b. Which of these measures do you believe is the most appropriate one in describing this distribution? Explain the reasoning behind your choice.

5. A finance professor has just given a midterm examination in her corporate finance course. In particular, she is interested in learning how her class of 100 students performed on this exam. The 100 exam scores are given in the file P02_05.XLS.

a. What are the mean and median scores (out of 100 possible points) on this exam?

b. Explain why the mean and median values are different in this case.

6. Compute the mean, median, and mode of the given set of average annual household income levels of citizens from selected U.S. metropolitan areas in the file P03_06.XLS. What can you infer about the shape of this particular income distribution from the computed measures of central location?

7. The operations manager of a toll booth, located at a major exit of a state turnpike, is trying to estimate the typical number of vehicles that arrive at the toll booth during a 1-minute period during the peak of rush-hour traffic. In an effort to estimate this typical throughput value, he records the number of vehicles that arrive at the toll booth over a 1-minute interval commencing at the same time for each of 365 normal weekdays. These data are contained in the file P02_09.XLS.

a. Find the most appropriate measure of the given distribution's central location.

b. Is this distribution of arrivals *skewed* somewhat? Explain.

8. The proportions of high school graduates annually taking the SAT test in each of the 50 states and the District of Columbia are provided in the file P02_10.XLS.

a. Compute the mean, median, and mode of this set of proportions.

b. Which of these measures do you believe is the most appropriate one in describing this distribution? Explain the reasoning behind your choice.

9. Consider the average time (in minutes) it takes a citizen of each metropolitan area to travel to work and back home each day. Refer to the data given in the file P02_11.XLS.

a. Find the most representative average commute time across this distribution.

b. Does it appear that this distribution of average commute times is *approximately* symmetric? Explain why or why not.

10. Five hundred households in a middle-class neighborhood were recently surveyed as a part of an economic development study conducted by the local government. Specifically, for each of the 500 randomly selected households, the survey requested information on several variables, including the household's level of indebtedness (excluding the value of any home mortgage). These data are provided in the file P02_06.XLS.

a. Find the maximum and minimum debt levels for the households in this sample.

b. Find the indebtedness levels at each of the 25th, 50th, and 75th percentiles.

c. Compute and interpret the interquartile range in this case.

11. A real estate agent has gathered data on 150 houses that were recently sold in a suburban community. Included in this data set are observations for each of the following variables: the appraised value of each house (in thousands of dollars), the selling price of each house (in thousands of dollars), the size of each house (in hundreds of square feet), and the number of bedrooms in each house. Refer to the file P02_07.XLS in answering the following questions.

a. Find the house(s) at the 80th percentile of all sample houses with respect to *appraised value.*

b. Find the house(s) at the 80th percentile of all sample houses with respect to *selling price.*

c. Find the maximum and minimum sizes (measured in *square footage*) of all sample houses.

d. What is the typical number of bedrooms in a recently sold house in this suburban community?

12. The U.S. Department of Transportation regularly publishes the *Air Travel Consumer Report,* which provides a variety of performance measures of major U.S. commercial airlines. One dimension of performance reported is each airline's percentage of domestic flights arriving within 15 minutes of the scheduled arrival time at major reporting airports throughout the country. Use these data, given in the file P02_12.XLS, to answer the following questions:

a. Which major U.S. airline has the *highest* third quartile on-time arrival percentage?

b. Which major U.S. airline has the *lowest* first quartile on-time arrival percentage?

c. Which major U.S. airline has the *largest* range of on-time arrival percentages?

d. Which major U.S. airline has the *smallest* range of on-time arrival percentages?

13. Having computed measures of central location for various numerical attributes of full-time employees, a human resources manager at Beta Technologies, Inc., is now interested in compiling some statistics on the variability of sample data values about their respective means. Using the data provided in the file P02_01.XLS, assist this manager by computing sample standard deviations for each of the following numerical variables: age, number of years of relevant full-time work experience prior to coming to Beta,

number of years of full-time work experience at Beta, number of years of post-secondary education, and salary. Do the empirical rules apply for any of these variables? In each case explain why the empirical rules apply or do not apply.

14. A production manager is interested in determining the variability of the proportion of defective items in a shipment of one of the computer components that her company manufactures. The spreadsheet provided in the file P02_02.XLS contains the proportion of defective components for each of 500 randomly selected shipments collected during a 1-month period. Compute the sample standard deviation of these data and use this value in interpreting the empirical rules in this case.

15. In an effort to provide more consistent customer service, the manager of a local fast-food restaurant would like to know the dispersion of customer service times about their average value for the facility's drive-up window. The file P02_04.XLS contains 200 customer service times, all of which were observed during the busiest hour of the day for this fast-food operation.
    a. Find and interpret the variance and standard deviation of these sample values.
    b. Are the empirical rules applicable in this case? If so, apply these rules and interpret your results. If not, explain why the empirical rules are not applicable here.

16. Compute the standard deviation of the given set of average annual household income levels of citizens from selected U.S. metropolitan areas in the file P03_06.XLS. Is it appropriate to apply the empirical rules in this case? Explain.

17. The file P02_26.XLS contains annual percentage changes in consumer prices. Are the empirical rules applicable in this case? If so, apply these rules and interpret your results.

18. The diameters of 100 rods produced by Rodco are listed in the file P03_18.XLS. Based on these data, you can be approximately 99.7% sure that the diameter of a typical rod will be between what two numbers?

19. The file P03_19.XLS contains the thickness (in centimeters) of some mica pieces. A piece meets specifications if it is between 7 and 15 centimeters in thickness.
    a. What fraction of mica pieces meet specifications?
    b. Do the empirical rules appear to be valid for this data set?

Level B

20. A finance professor has just given a midterm examination in her corporate finance course. In particular, she is now interested in assigning letter grades to the scores earned by her 100 students who took this exam. The top 10% of all ordered scores should receive a grade of A. The next 10% of all ordered scores should be assigned a grade of B. The third 10% of all ordered scores should receive a grade of C. The next 10% of all ordered scores should be assigned a grade of D. All subsequent scores should be considered to be failing (i.e., equivalent to F grades). The 100 exam scores are given in the file P02_05.XLS. Assist this instructor in assigning letter grades to each of the given finance exam scores.

21. The operations manager of a toll booth, located at a major exit of a state turnpike, is trying to estimate the variability of the number of vehicles that arrive at the toll booth during a 1-minute period during the peak of rush-hour traffic. In an effort to estimate this measure of dispersion, he records the number of vehicles that arrive at the toll booth over a 1-minute interval commencing at the same time for each of 365 normal weekdays. These data are contained in the file P02_09.XLS.
    a. Is the sample variance (or sample standard deviation) a reliable measure of dispersion in this case? Explain why or why not.
    b. If the sample variance (or sample standard deviation) is not a reliable measure of dispersion in this case, how can this operations manager most appropriately measure the variability of the values in the given sample?

22. The file P02_10.XLS contains the proportions of high school graduates annually taking the SAT test in each of the 50 states and the District of Columbia. Approximately 95% of these proportions fall between what two fractions?

23. The file P03_23.XLS contains the salaries of all Indiana University business school professors.
    a. If you increased every professor's salary by $1000, what would happen to the mean and median salary?
    b. If you increased every professor's salary by $1000, what would happen to the sample standard deviation of the salaries?
    c. If you increased everybody's salary by 5%, what would happen to the sample standard deviation of the salaries?

24. The file P03_24.XLS contains a sample of family incomes (in thousands of 1980 dollars) for a set of families sampled in 1980 and 1990. Assume that these families are representative of the whole United States. The Republicans claim that the country was better off in 1990 than 1980, because average income increased. Do you agree?

25. According to the Educational Testing Service, the scores of people taking the SAT were as listed in Table 3.1.

Table **3.1**

Data on SAT Scores

Range	Number Having Family Income ≥$18,000	Number Having Family Income ≤$6000
200–250	325	1638
251–300	4,212	7980
301–350	13,896	9622
351–400	26,175	8973
401–450	37,213	8054
451–500	41,412	6663
501–550	37,400	4983
551–600	28,151	3119
601–650	17,992	1626
651–700	9,284	686
701–750	3,252	239
751–800	415	17

a. Estimate the average and standard deviation of SAT scores for students whose families made at least $18,000 and for those whose families made no more than $6000. (*Hint:* Assume all scores in a group are concentrated at the group's midpoint.)

b. Do these results have any implications for college admissions?

26. The file P03_26.XLS lists the fraction of U.S. men and women of various heights. Use these data to estimate the mean and standard deviation of the height of American men and women. (*Hint:* Assume all heights in a group are concentrated at the group's midpoint.)

27. The file P03_27.XLS lists the fraction of U.S. men and women of various weights. Use these data to estimate the mean and standard deviation of the weights of U.S. men and women. (*Hint:* Assume all weights in a group are concentrated at the group's midpoint.)

3.7 Measures of Association: Covariance and Correlation

All of the summary measures to this point involve a single variable. It is also useful to summarize the relationship between two variables. Specifically, we would like to summarize the type of behavior often observed in a scatterplot. Two such measures are *covariance* and *correlation*. We will discuss them briefly here and in more depth in later chapters. Each measures the strength (and direction) of a *linear* relationship between two numerical variables. Intuitively, the relationship is "strong" if the points in a scatterplot cluster tightly around some straight line. If this straight line rises from left to right, then the relationship is "positive" and the measures are positive numbers. If it falls from left to right, then the relationship is "negative" and the measures are negative numbers.

If we want to measure the covariance or correlation between two variables X and Y— indeed, even if we just want to form a scatterplot of X versus Y— then X and Y must be "paired" variables. That is, they must have the same number of observations, and the X and Y values for any observation should be naturally paired. For example, each observation could be the height and weight for a particular person, the time in a store and the amount purchased for a particular customer, and so on.

With this in mind, let X_i and Y_i be the paired values for observation i, and let n be the number of observations. Then the covariance between X and Y, denoted by Cov(X, Y), is given by equation (3.5):

Formula for Covariance	$$\text{Cov}(X, Y) = \frac{\sum_{i=1}^{n}(X_i - \overline{X})(Y_i - \overline{Y})}{n - 1}$$	(3.5)

Scatterplots that rise from lower left to upper right will tend to have positive covariance and correlation. Those that fall from upper left to lower right will tend to have negative covariance and correlation.

You probably will never have to use this formula directly—Excel has a built-in COVAR function that does it for you—but the formula does indicate what covariance is all about. It is essentially an average of products of deviations from means. If X and Y vary in the *same* direction, then when X is above (or below) its mean, Y will also tend to be above (or below) its mean. In either case, the product of deviations will be positive—a positive times a positive or a negative times a negative—so the covariance will be positive. The opposite is true when X and Y vary in *opposite* directions. Then the covariance will be negative.

The limitation of covariance as a descriptive measure is that it is affected by the *units* in which X and Y are measured. For example, we can inflate the covariance by a factor of 1000 simply by measuring X in dollars rather than in thousands of dollars. The correlation, denoted by Corr(X, Y), remedies this problem. It is a *unitless* quantity defined by equation (3.6), where Stdev(X) and Stdev(Y) denote the standard deviations of X and Y. Again, you'll probably never have to use this formula for calculations—Excel does it for you with the built-in CORREL function—but it does show that to produce a unitless quantity, we need to divide the covariance by the product of the standard deviations.

Formula for Correlation	$$\text{Corr}(X, Y) = \frac{\text{Cov}(X, Y)}{\text{Stdev}(X) \times \text{Stdev}(Y)}$$	(3.6)

Statisticians use the symbol r for the correlation based on sample data. If they want to denote the correlation based on the entire population, they use the letter ρ (rho).

The correlation is unaffected by the units of measurement of the two variables, and it is *always* between -1 and $+1$. The closer it is to either of these two extremes, the closer the points in a scatterplot are to some straight line, either in the negative or positive direction. On the other hand, if the correlation is close to 0, then the scatterplot is typically a "cloud" of points with no apparent relationship. However, it is also possible that the points are close to a *curve* and have a correlation close to 0. This is because correlation is relevant only for measuring linear relationships.

When there are more than two variables in a data set, it is often useful to create a table of covariances and/or correlations. Each value in the table then corresponds to a particular pair of variables. The StatPro add-in allows you to do this easily, as illustrated in the following example.

Example (3.6) # Household Spending for Various Categories of Items

A survey questions members of 100 households about their spending habits. The data in the file EXPENSES.XLS represent the salary, expenses for cultural activities, expenses for sports-related activities, and expenses for dining out for each household over the past year. Do these variables appear to be related linearly?

Objective To use StatPro's correlation procedure to measure the relationship between expenses in various categories.

Solution

Scatterplots of each variable versus each other variable answer the question quite nicely, but six scatterplots are required, one for each pair. To get a quick indication of possible linear relationships, we can use StatPro to obtain a table of correlations and/or covariances. To do so, place the cursor anywhere in the data range, use the StatPro/Summary Stats/Correlations, Covariances menu item, select all four variables of interest, and choose to obtain correlations *and* covariances. Otherwise, accept all of StatPro's default settings. The tables of correlations and covariances appear in Figure 3.10.

The only relationships that stand out are the positive relationships between salary and cultural expenses and between salary and dining expenses, and the negative relationship between cultural and sports-related expenses. In contrast, there is very little linear relationship between salary and sports expenses or between dining expenses and either culture or sports expenses. To confirm these graphically, we show scatterplots of Salary versus Culture and Culture versus Sports in Figures 3.11 and 3.12. These indicate more intuitively what a correlation of approximately ±0.5 really means.

FIGURE 3.10
Table of Correlations and Covariances

	A	B	C	D	E	F
1		**Table of correlations**				
2			Salary	Culture	Sports	Dining
3		Salary	1.000			
4		Culture	0.506	1.000		
5		Sports	-0.081	-0.520	1.000	
6		Dining	0.558	0.170	0.266	1.000
7						
8		**Table of covariances (variances on the diagonal)**				
9			Salary	Culture	Sports	Dining
10		Salary	91130278.788			
11		Culture	1094786.800	52315.394		
12		Sports	-219026.400	-33607.520	81427.232	
13		Dining	2564694.800	18748.640	36461.280	236187.667

FIGURE 3.11
Scatterplot Indicating a Positive Relationship

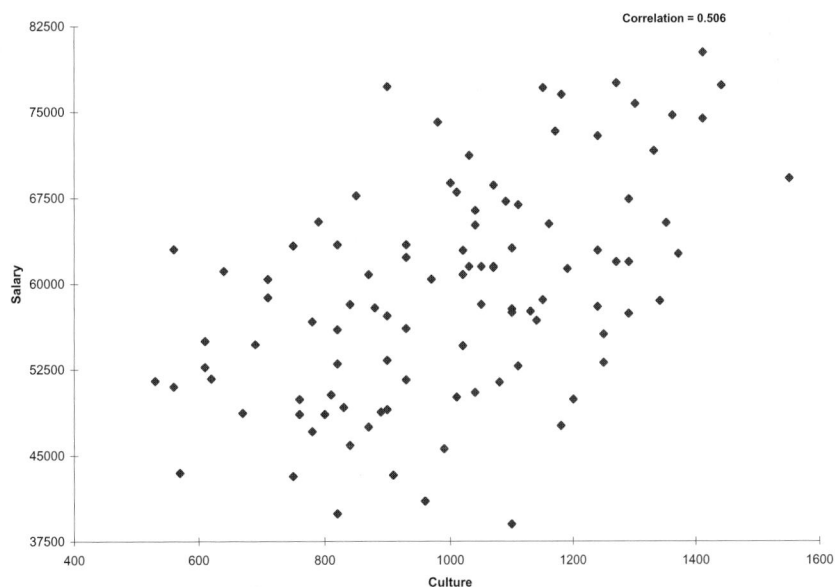

FIGURE 3.12
Scatterplot Indicating a Negative Relationship

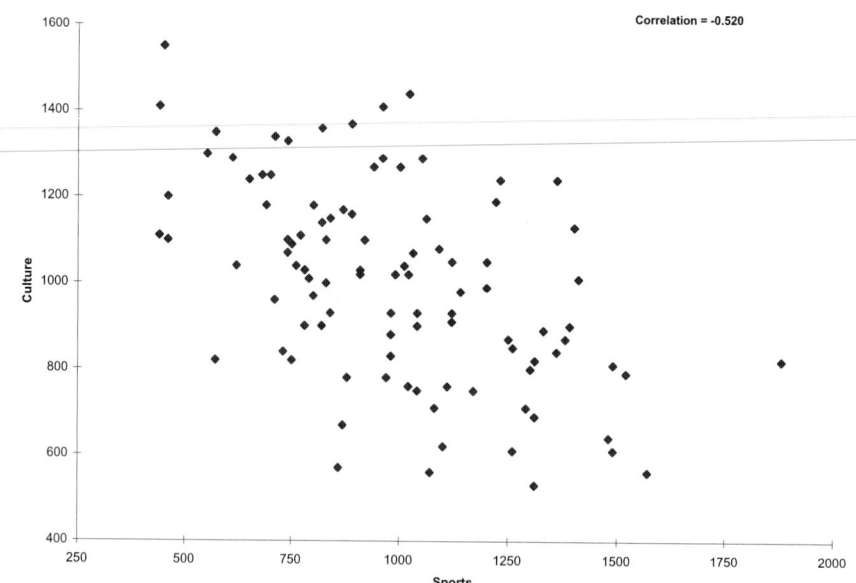

Excel Tip *The correlations in StatPro's correlation (or covariance) tables are linked by Excel formulas to the data. However, the correlation label in one of StatPro's scatterplots (in the upper right corner) is* not *linked to the data. It will* not *change if the data change.*

In general, we point out the following properties that are evident from Figure 3.10:

- The correlation between a variable and itself is always 1.

- The correlation between X and Y is the same as the correlation between Y and X. Therefore, it is sufficient to list the correlations below (or above) the diagonal in the table. (The same is true for covariances.) StatPro provides these options.

- The covariance between a variable and itself is the *variance* of that variable. We indicate this in the heading of the covariance table.

- It is difficult to interpret the magnitudes of the covariances. These depend on the fact that the data are measured in dollars rather than, say, thousands of dollars. It is much easier to interpret the magnitudes of the correlations because they are scaled to be between -1 and $+1$. ●

P R O B L E M S

Level A

28. Explore the relationship between the selling prices and the appraised values of the 150 homes in the file P02_07.XLS by computing a correlation.
   a. Is there evidence of a *linear* relationship between the selling price and appraised value in this case? If so, characterize the relationship (i.e., indicate whether the relationship is a positive or negative one).
   b. For which of the two remaining variables, the size of the home and the number of bedrooms in the

home, is the relationship with the home's appraised value *stronger*? Justify your choice.

29. A human resources manager at Beta Technologies, Inc., is trying to determine the variable that best explains the variation of employee salaries using the sample of 52 full-time employees in the file P02_01.XLS. Generate a table of correlations to help this manager identify whether the employee's (a) gender, (b) age, (c) number of years of relevant work experience prior to employment at Beta, (d) the number of years of employment at Beta, or (e) the number of years of post-secondary education has the *strongest* linear relationship with annual salary.

30. Consider the enrollment data for *Business Week*'s top U.S. graduate business programs in the file P02_03.XLS. Specifically, compute correlations to assess whether there is a linear relationship between the total number of full-time students and each of the following: (a) the proportion of female students, (b) the proportion of minority students, and (c) the proportion of international students enrolled at these distinguished business schools.

31. What is the relationship between the number of short-term general hospitals and the number of medical specialists in metropolitan areas? Explore this question by producing a correlation measure for these two variables using the data in the file P02_17.XLS. Interpret your result.

32. An economic development researcher wants to understand the relationship between the average monthly expenditure on utilities for households in a particular middle-class neighborhood and each of the following household variables: family size, approximate location of the household within the neighborhood, an indication of whether those surveyed owned or rented their home, gross annual income of the first household wage earner, gross annual income of the second household wage earner (if applicable), size of the monthly home mortgage or rent payment, and the total indebtedness (excluding the value of a home mortgage) of the household. The data are in the file P02_06.XLS.
   a. Generate a correlation for each pairing of variables with the household's average monthly expenditure on utilities.
   b. Which of the variables have a *positive* linear relationship with the household's average monthly expenditure on utilities?
   c. Which of the variables have a *negative* linear relationship with the household's average monthly expenditure on utilities?
   d. Which of the variables have essentially *no* linear relationship with the household's average monthly expenditure on utilities?

33. Motorco produces electric motors for use in home appliances. One of the company's production managers is interested in examining the relationship between the dollars spent per month in inspecting finished motor products and the number of motors produced during that month that were returned by dissatisfied customers. He has collected the data in the file P02_18.XLS to explore this relationship for the past 36 months. Produce a correlation measure for these two variables and interpret it for this production manager.

34. A large number of metropolitan areas in the United States have been ranked with consideration of the following aspects of life in each area: cost of living, transportation, jobs, education, climate, crime, arts, health, and recreation. The data are in the file P02_55.XLS.
   a. Generate a table of correlations to discern the relationship between the metropolitan area's overall score and each of these numerical factors.
   b. Which variables are most strongly associated with the overall score? Are you surprised by any of the results here?

35. The *ACCRA Cost of Living Index* provides a useful and reasonably accurate measure of cost-of-living differences among many urban areas. Items on which the index is based have been carefully chosen to reflect the different categories of consumer expenditures. The data are in the file P02_19.XLS. Compute correlation measures to explore the relationship between the composite index and each of the various expenditure components.
   a. Which expenditure component has the *strongest* linear relationship with the composite index?
   b. Which expenditure component has the *weakest* linear relationship with the composite index?

36. Based on the annual data in the file P02_25.XLS from the U.S. Department of Agriculture, determine whether a linear relationship exists between the number of farms and the average size of a farm in the United States during these years. Specifically, generate a correlation measure and interpret it.

 3.8 Describing Data Sets with Boxplots

We now introduce the *boxplot,* a very useful graphical method for summarizing data. We saved boxplots for this chapter because they are based on a number of summary measures we discussed in Section 3.2. **Boxplots** can be used in two ways: either to describe a single variable in a data set or to compare two (or more) variables. We illustrate these uses in the following examples.

Example (3.7) Distribution of Monthly Dow Returns

Recall that the DOW.XLS file lists the monthly returns on the Dow from February 1947 through January 1993. Use a boxplot to summarize the distribution of these returns.

Objective To use StatPro's boxplot procedure to describe the distribution of monthly Dow returns.

Solution

Excel has no boxplot option, but we have included this option in the StatPro add-in. To create a boxplot for Dow returns, place the cursor anywhere within the data set, use the StatPro/Charts/Boxplot(s) menu item, select the Single boxplot option, and select the Return variable. Eventually, two sheets will be added. One has the boxplot chart, whereas the other contains summary measures used to form the boxplot, as explained below.[1] The resulting boxplot appears in Figure 3.13 and the summary measure sheet appears in Figure 3.14.

Understanding a Boxplot

- The right and left of the box are at the third and first quartiles. Therefore, the length of the box equals the interquartile range (IQR), and the box itself represents the middle 50% of the observations. The height of the box has no significance. (Our boxplot conventions are not the only ones, as we saw in the chapter opener about movie revenues. However, the variations are primarily cosmetic.)

- The vertical line inside the box indicates the location of the median. The point inside the box indicates the location of the mean.

- Horizontal lines are drawn from each side of the box. They extend to the most extreme observations that are no farther than 1.5 IQRs from the box. They are useful for indicating variability and skewness.[2]

FIGURE 3.13
Boxplot of Dow
Returns

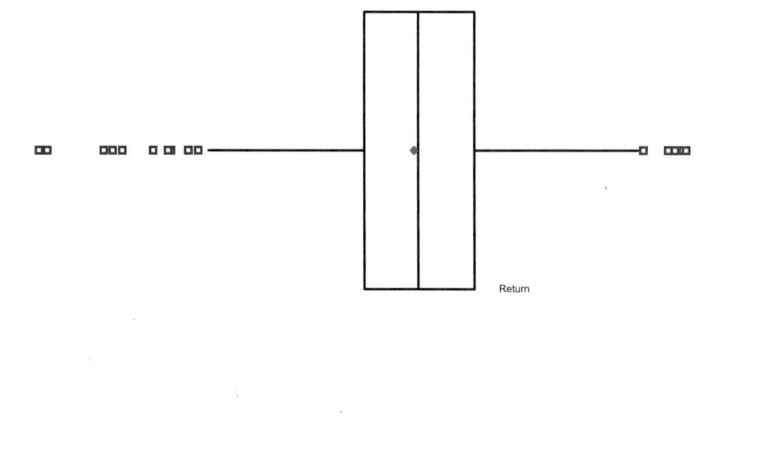

[1]The second sheet, which includes the summary data for the boxplot, is actually hidden by StatPro.
[2]Some statistical software packages create vertical boxplots rather than the horizontal type StatPro creates. However, the information is identical.

FIGURE 3.14
Summary Measures for Boxplot

	A	B
1	**Summary measures for boxplots**	
2		Return
3	Mean	0.00587967
4	Median	0.00741727
5	Q1	-0.01256551
6	Q3	0.02836001
7	IQR	0.04092552
8		
9	Outer lower fence	-0.13534207
10	Outer upper fence	0.15113656
11		
12	Inner lower fence	-0.07395379
13	Inner upper fence	0.08974828
14		
15	Lower adjacent value	-0.07112479
16	Upper adjacent value	0.08942125
17		
18	# of extreme outliers	0
19	# of mild outliers	16
20		
21	# of low outliers	10
22	# of high outliers	6

If you want to see this hidden sheet, use Excel's Format/ Sheet/Unhide menu item.

- Observations farther than 1.5 IQRs from the box are shown as individual points. If they are between 1.5 IQRs and 3 IQRs from the box, they are called **mild outliers** and are hollow. Otherwise, they are called **extreme outliers** and are solid.[3]

The boxplot implies that the Dow returns are approximately symmetric on each side of the median, although the mean is slightly below the median. In addition, there are a few mild outliers but no extreme outliers. ●

Boxplots are probably most useful for comparing two populations graphically, as we illustrate in the following example.

Example 3.8 Graphical Comparison of Male and Female Movie Stars' Salaries

Recall that the salaries of famous actors and actresses are listed in the file ACTORS.XLS. Use side-by-side boxplots to compare the salaries of male and female actors and actresses.

Objective To use StatPro's boxplot procedure to compare actors' salaries across gender.

---

[3]These conventions, along with the rather quaint terminology of fences and adjacent values listed in the summary table, are due to the statistician John Tukey.

Solution

The data setup for this type of "comparison" problem can be in one of two forms: stacked or unstacked. The data are stacked if there is a "code" variable such as Gender that designates which gender each observation is in, and there is a single "measurement" variable Salary that lists the salaries for both genders. The data are unstacked if there is a *separate* Salary column for each gender (one for males and one for females). As Figure 3.15 indicates, the data in the ACTORS.XLS file are in stacked form. Therefore, to obtain side-by-side boxplots of male and female salaries, use the StatPro/Charts/Boxplot(s) menu item, and, after the opening dialog box, check the "stacked" option. Then choose Gender as the code variable and Salary as the measurement variable.

FIGURE 3.15
Actor Data in
Stacked Form

	A	B	C	D	E
1	**Famous actors and actresses**				
2					
3	Note: All monetary values are in $ millions.				
4					
5	Name	Gender	DomesticGross	ForeignGross	Salary
6	Angela Bassett	F	32	17	2.5
7	Jessica Lange	F	21	27	2.5
8	Winona Ryder	F	36	30	4
9	Michelle Pfeiffer	F	66	31	10
10	Whoopi Goldberg	F	32	33	10
11	Emma Thompson	F	26	44	3
12	Julia Roberts	F	57	47	12
13	Sharon Stone	F	32	47	6
14	Meryl Streep	F	34	47	4.5
15	Susan Sarandon	F	38	49	3
16	Nicole Kidman	F	55	51	4
17	Holly Hunter	F	51	53	2.5
18	Meg Ryan	F	43	55	8.5
19	Andie Macdowell	F	26	75	2
20	Jodie Foster	F	62	85	9
21	Rene Russo	F	69	85	2.5
22	Sandra Bullock	F	64	104	11
23	Demi Moore	F	65	125	12
24	Danny Glover	M	42	4	2
25	Billy Crystal	M	52	14	7

Excel Tip *Many statistical software packages require data to be in stacked form. StatPro allows data to be in either stacked or unstacked form. In addition, StatPro can convert from stacked to unstacked form, or vice versa.*

The resulting side-by-side boxplots appear in Figure 3.16. It is clear that the female salary box is considerably to the left of the male salary box, although both have about the same IQR. Each boxplot has three indications that the salary distributions are skewed to the right: (1) the means are larger than the medians, (2) the medians are closer to the left sides of the boxes than to the right sides, and (3) the horizontal lines extend farther to the right than to the left of the boxes. However, there are no outliers—not even the big stars like Harrison Ford or Sylvester Stallone!

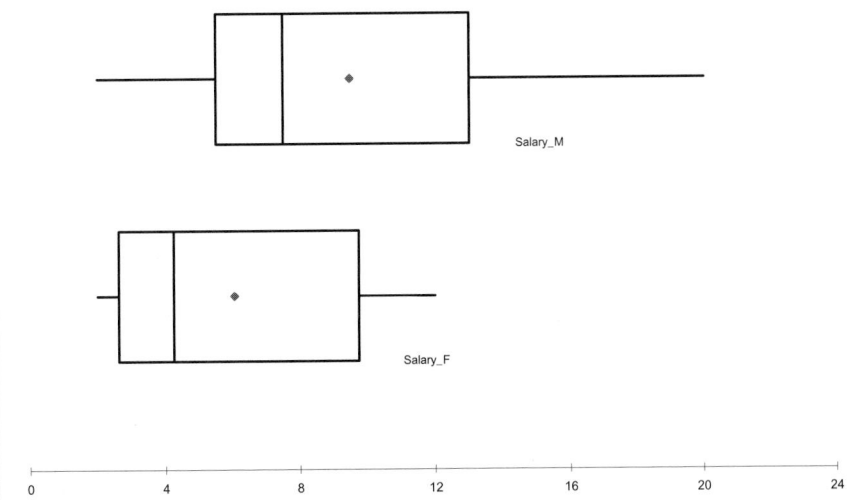

FIGURE 3.16
Side-by-side Box-
plots of Female
and Male Salaries

PROBLEMS

Level A

37. Consider the data in the file P03_37.XLS on various performance measures for the largest U.S. airlines. In particular, generate boxplots to summarize each of the given measures of airline performance. Are these four performance measures linearly associated with one another? Explain your findings.

38. In a recent ranking of top graduate business schools in the United States published by *U.S. News & World Report*, the average starting base salaries for recent graduates from recognized graduate programs were provided. These data are given in the file P02_08.XLS. Construct a boxplot to characterize this distribution of average starting MBA salaries. In particular, is this distribution essentially symmetric or skewed?

39. Consider the average annual rates for various forms of violent crime (including murder, rape, robbery, and aggravated assault) and the average annual rates for various forms of property-related crime (including burglary, larceny-theft, and motor vehicle theft) in selected U.S. metropolitan areas. The data are provided in the file P02_50.XLS. Use side-by-side boxplots to compare the rates of these two general classes of crimes. Summarize your findings.

40. The annual average values of the Consumer Confidence Index are given in the file P02_28.XLS. Construct a boxplot to find the middle 50% of this distribution of values. Characterize the nature and amount of the variability about the interquartile range in this case.

41. Consider the given set of average annual household income levels of citizens from selected U.S. metropolitan areas in the file P03_06.XLS. What can you infer about the shape of this particular income distribution from a boxplot of the given data? If applicable, note the presence of any outliers in this data set.

42. Using cost-of-living data from the *ACCRA Cost of Living Index* in the file P02_19.XLS, generate a boxplot to summarize the *composite* cost-of-living index values.

Level B

43. In 1970 a lottery was held to determine who would be drafted (and sent to Vietnam). For each date of the year, a ball was put into an urn. For instance, January 1 was number 305 and February 14 was number 4. Thus a person born on February 14 would be drafted before a person born on January 1. The file P03_43.XLS contains the "draft number" for each date for the 1970 and 1971 lotteries. Do you notice anything unusual about the results of either lottery? What do you think might have caused this result? (*Hint:* Use a boxplot for each month's numbers.)

3.9 Applying the Tools

Now that you are equipped with a collection of tools for describing data, it's time to apply these tools to some serious data analysis. We will examine three data sets in this section. Each of these is rather small by comparison with the data sets real companies often face, but they are large enough to make the analysis far from trivial. In each example we will illustrate some of the output that might be obtained by the company involved, but you should realize that we are never really finished. With data sets as rich as these, there are always more numbers that could be calculated, more tables that could be formed, and more charts that could be created. We encourage you to take each analysis a few steps beyond what we present there.

Each example has a decision problem lurking behind it. If these data belonged to real companies, the companies would not only want to describe the data, but they would want to use the information from their data analysis as a basis for decision making. We are not yet in a position to perform this decision making, but you should appreciate that the data analysis we perform here is really just the first step in an overall business analysis.

Example 3.9 Accounts Receivable at Spring Mills

The Spring Mills Company produces and distributes a wide variety of manufactured goods. Due to this variety, it has a large number of customers. The company classifies these customers as small, medium, and large, depending on the volume of business each does with Spring Mills. Recently, Spring Mills has noticed a problem with its accounts receivable. It is not getting paid back by its customers in as timely a manner as it would like. This obviously costs Spring Mills money. If a customer delays a payment of $300 for 20 days, say, then the company loses potential interest on this amount. The company has gathered data on 280 customer accounts. For each of these accounts, the data set lists three variables: Size, the size of the customer (coded 1 for small, 2 for medium, 3 for large); Days, the number of days since the customer was billed; and Amount, the amount the customer owes. (See the file RECEIVE.XLS.) What information can we obtain from these data?

Objective To use charts, summary measures, and pivot tables to understand data on accounts receivable at Spring Mills.

Solution

It is always a good idea to get a rough sense of the data first. We do this by calculating several summary measures for Days and Amount, a histogram of Amount, and a scatterplot of Amount versus Days in Figures 3.17, 3.18, and 3.19. Figure 3.17 indicates posi-

Appropriate charts let us see the big picture. StatPro makes it easy to create charts quickly and painlessly.

FIGURE 3.17
Summary Measures for the Combined Data

	A	B	C	D
1	**Summary measures for selected variables**			
2			Days	Amount
3		Count	280.000	280.000
4		Sum	4102.000	130000.000
5		Mean	14.650	464.286
6		Median	13.000	320.000
7		Standard deviation	7.221	378.055
8		Minimum	2.000	140.000
9		Maximum	39.000	2220.000

FIGURE 3.18
Histogram of All
Amounts Owed

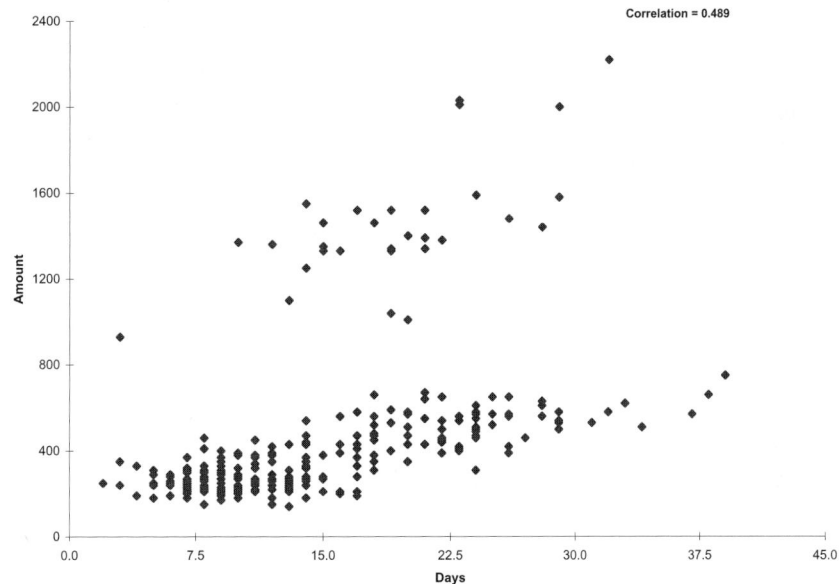

Histogram for Amount

FIGURE 3.19
Scatterplot of
Amount versus
Days for All
Customers

Correlation = 0.489

tive skewness in the Amount variable—the mean is considerably larger than the median, probably because of some large amounts due. Also, the standard deviation of Amount is quite large. This positive skewness is confirmed by the histogram. The scatterplot suggests some suspicious behavior, with two distinct groups of points. (You can check that the upper group of points in the scatterplot correspond to the large customers, whose amounts owed are uniformly greater than those in the other two groups.)

The next logical step is to see whether the different customer sizes have any effect on either Days, Amount, or the relationship between Days and Amount. To do this, it is useful to "unstack" the Days and Amount variables—that is, to create a new Days and Amount variable for *each* group of customer sizes. For example, the Days and Amount variables for customers of size 1 are named Days1 and Amount1. (We used StatPro's

	A	B	C	D	E	F	G	H
1	Summary measures for selected variables							
2			Days1	Amount1	Days2	Amount2	Days3	Amount3
3		Count	150.000	150.000	100.000	100.000	30.000	30.000
4		Mean	9.800	254.533	20.550	481.900	19.233	1454.333
5		Median	10.000	250.000	20.000	470.000	19.000	1395.000
6		Standard deviation	3.128	49.285	6.622	99.155	6.191	293.888
7		Minimum	2.000	140.000	8.000	280.000	3.000	930.000
8		Maximum	17.000	410.000	39.000	750.000	32.000	2220.000

FIGURE 3.20 Summary Measures Broken Down by Size

Unstack procedure, which is quite straightforward, to accomplish this, but copying and pasting also works.) Summary measures and a variety of charts based on these unstacked variables appear in Figures 3.20–3.28 (pages 102–105).

There is obviously a lot going on here, and most of it is clear from the figures. We point out the following: (1) there are far fewer large customers than small or medium customers; (2) the large customers tend to owe considerably more than small or medium customers; (3) the small customers do not tend to be as long overdue as the medium or large customers; and (4) there is no relationship between Days and Amount for the small customers, but there is a definite positive relationship between these variables for the medium and large customers.

We've now done the "obvious" analysis. There is still much more we can do, however. For example, suppose Spring Mills wants a breakdown of customers who owe at least $500. We first create a new variable called "Large?" next to the original variables that equals 1 for all amounts greater than $500 and equals 0 otherwise.[4] (See Figure 3.29 on page 106 for some of the data.) We do this by entering the formula

$$=IF(C6>=\$B\$3,1,0)$$

FIGURE 3.21
Histogram of
Amount for Small
Customers

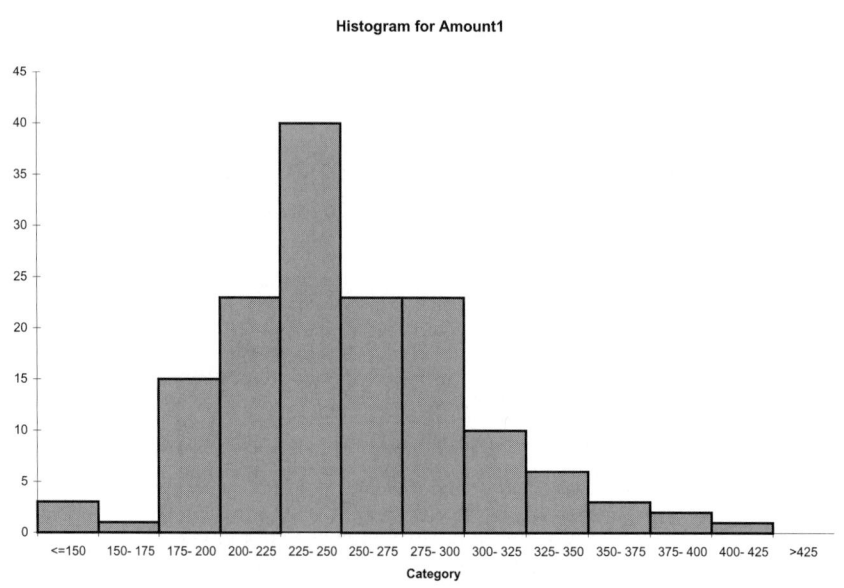

[4]We could just as well enter the labels "yes" and "no" in the Large? column. However, it is common to use 0–1 values for such a variable.

FIGURE 3.22
Histogram of
Amount for
Medium
Customers

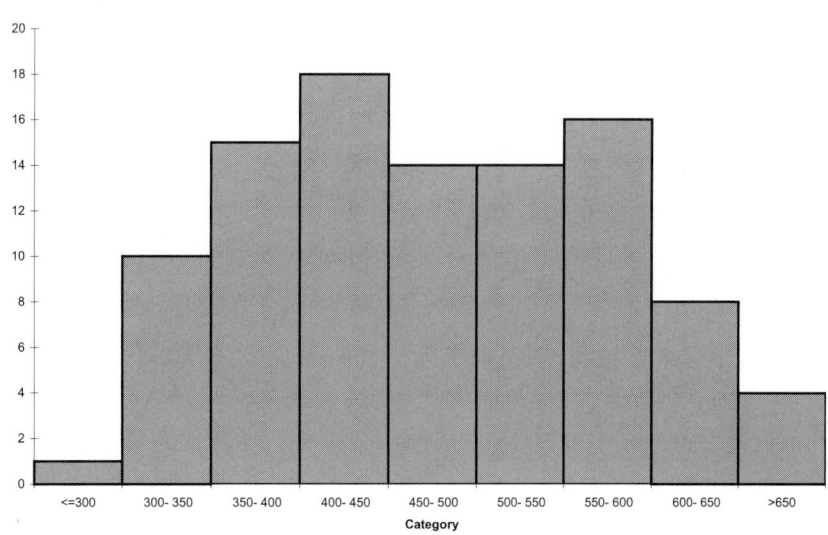

FIGURE 3.23
Histogram of
Amount for Large
Customers

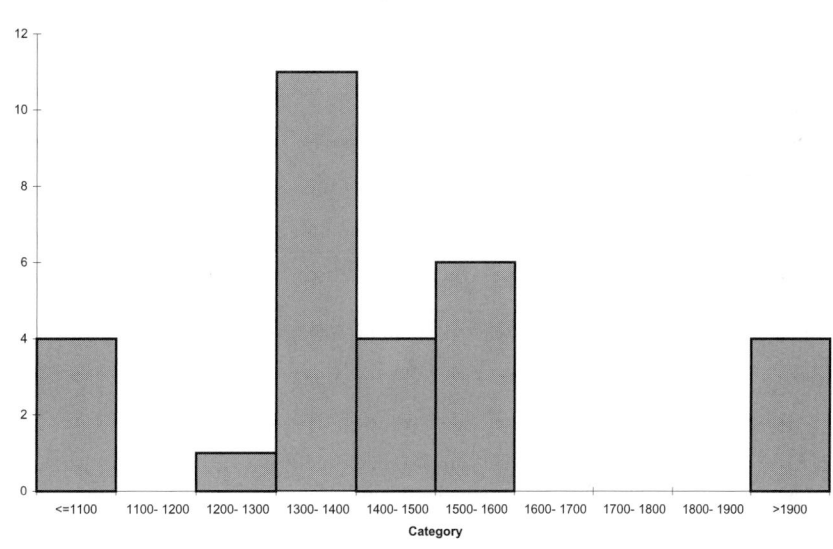

FIGURE 3.24
Boxplots of Days
Owed by Different
Size Customers

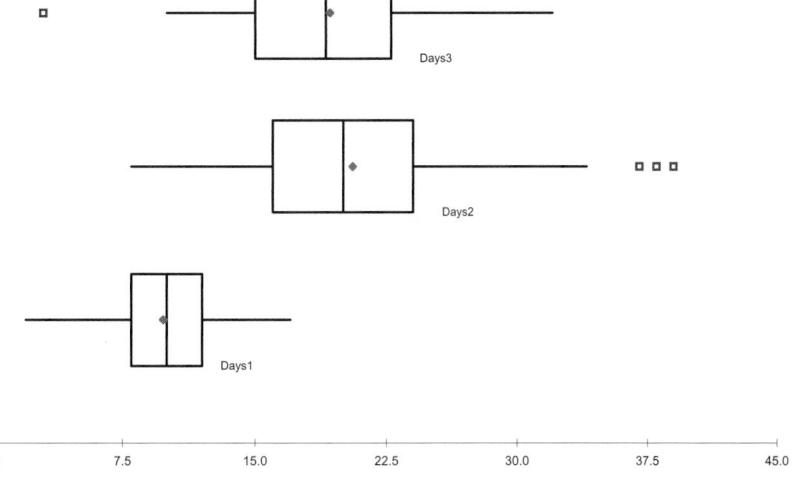

FIGURE 3.25
Boxplots of
Amounts Owed
by Different Size
Customers

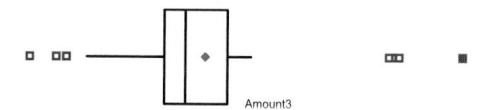

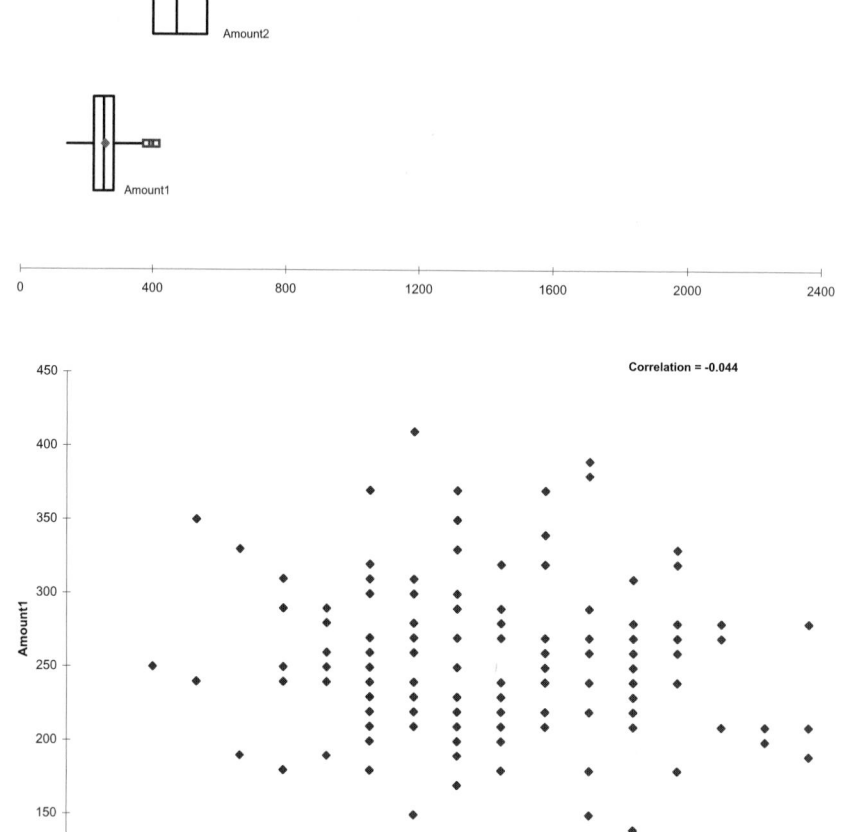

FIGURE 3.26
Scatterplot of
Amount versus
Days for Small
Customers

in cell D6 and copying down. We then use a pivot table to create a *count* of the number of 1's in this new variable for each value of the Size variable.

Figure 3.30 on page 106 shows the results. Actually, we created this pivot table twice, once (on top) showing counts as percentages of each column, and once showing them as percentages of each row. The top table shows, for example, that about 73% of all customers with amounts less than $500 are small customers. The bottom table shows, for example, that 45% of all medium-size customers owe at least $500. When you hear the expression "slicing and dicing the data," this is what it means. These two pivot tables are based on the *same* counts, but they portray them in slightly different ways. Neither is better than the other; each provides useful information.

Finally, we investigate the amount of interest Spring Mills is losing by the delays in its customers' payments. We assume that the company can make 12% annual interest on

FIGURE 3.27
Scatterplot of
Amount versus
Days for Medium
Customers

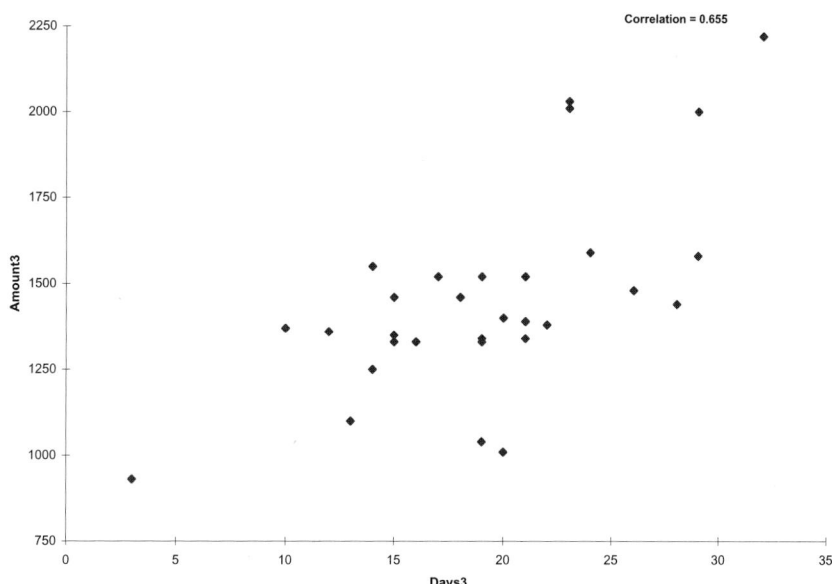

FIGURE 3.28
Scatterplot of
Amount versus
Days for Large
Customers

excess cash. Then we create a Lost variable for each customer size that indicates the amount of interest Spring Mills loses on each customer group. (See Figure 3.31 on page 106.) The typical formula for lost interest in cell C10 is

$$=B10*A10*\$C\$7/365$$

This is the amount owed multiplied by the number of days owed multiplied by the interest rate, divided by the number of days in a year. Then we calculate sums of these amounts in row 5. Although Spring Mills is losing more per customer from the large customers, it is losing more in *total* from the medium-size customers—because there are

FIGURE 3.29
Checking for
Amounts at Least
$500

	A	B	C	D
1	**Distribution of receivables**			
2				
3	Value to check for	$500		
4				
5		Size	Days	Amount	Large?
6		1	7	$180	0
7		1	8	$210	0
8		1	10	$210	0
9		1	8	$150	0
10		1	9	$300	0
11		1	5	$240	0
12		1	4	$330	0
13		1	10	$290	0
14		1	5	$240	0
15		1	13	$270	0
16		1	12	$220	0
17		1	11	$260	0

FIGURE 3.30
Pivot Tables
for Counts of
Customers Who
Owe More Than
$500

	A	B	C	D
1				
2	Count of Large?	Large? ▼		
3	Size ▼	0	1	Grand Total
4	1	73.17%	0.00%	53.57%
5	2	26.83%	60.00%	35.71%
6	3	0.00%	40.00%	10.71%
7	Grand Total	100.00%	100.00%	100.00%
8				
9	Count of Large?	Large? ▼		
10	Size ▼	0	1	Grand Total
11	1	100.00%	0.00%	100.00%
12	2	55.00%	45.00%	100.00%
13	3	0.00%	100.00%	100.00%
14	Grand Total	73.21%	26.79%	100.00%

	A	B	C	D	E	F	G	H	I
1	**Interest Lost**								
2									
3	**Summary measures for selected variables**								
4		Lost1	Lost2	Lost3					
5	Sum	$122.68	$338.65	$287.25					
6									
7	Annual interest rate	12%							
8									
9	Days1	Amount1	Lost1	Days2	Amount2	Lost2	Days3	Amount3	Lost3
10	7	$180.00	$0.41	17	$470.00	$2.63	19	$1,330.00	$8.31
11	8	$210.00	$0.55	22	$540.00	$3.91	20	$1,400.00	$9.21
12	10	$210.00	$0.69	28	$560.00	$5.16	14	$1,550.00	$7.13
13	8	$150.00	$0.39	24	$470.00	$3.71	15	$1,460.00	$7.20
14	9	$300.00	$0.89	26	$650.00	$5.56	23	$2,030.00	$15.35
15	5	$240.00	$0.39	29	$530.00	$5.05	19	$1,520.00	$9.49
16	4	$330.00	$0.43	21	$550.00	$3.80	15	$1,330.00	$6.56
17	10	$290.00	$0.95	33	$620.00	$6.73	17	$1,520.00	$8.50
18	5	$240.00	$0.39	16	$430.00	$2.26	21	$1,390.00	$9.60
19	13	$270.00	$1.15	27	$460.00	$4.08	24	$1,590.00	$12.55

FIGURE 3.31 Summary Measures of Lost Interest

FIGURE 3.32
Pie Chart of
Lost Interest by
Customer Size

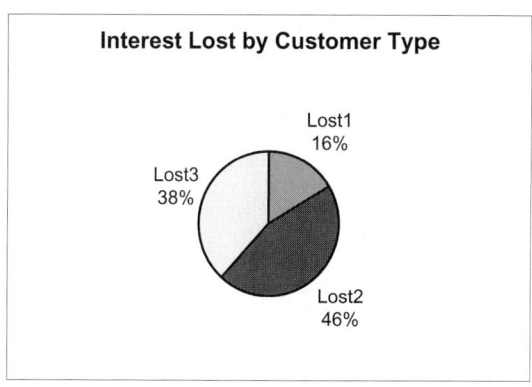

more of them. This is shown graphically in Figure 3.32 by a pie chart of the sums in row 5. This pie chart shows, for example, that 46% of the lost interest is due to the medium-size customers.

If Spring Mills really wants to decrease its receivables, it might want to target the medium-size customer group, from which it is losing the most interest. Or it could target the large customers because they owe the most on average. The most appropriate action depends on the cost and effectiveness of targeting any particular customer group. However, the analysis presented here gives the company a much better picture of what's currently going on. ●

Example 3.10 Customer Arrival and Waiting Patterns at R&P Supermarket

The R&P Supermarket is open 24 hours a day, 7 days a week. Lately, it has been receiving a lot of complaints from its customers about excessive waiting in line for checking out. R&P has decided to investigate this situation by gathering data on arrivals, departures, and line lengths at the checkout stations. It has collected data in half-hour increments for an entire week—336 observations—starting at 8 A.M. on Monday morning and ending at 8 A.M. the following Monday.

Specifically, it has collected data on the following variables: InitialWaiting, the number waiting or being checked out at the beginning of a half-hour period; Arrivals, the number of arrivals to the checkout stations during a period; Departures, the number finishing the checkout process during a period; and Checkers, the number of checkout stations open during a period. (See the file CHECKOUT.XLS.)

The data set also includes time variables: Day, day of week; StartTime, clock time at the beginning of each half-hour period; and TimeInterval, a descriptive term for the time of day, such as Lunch rush for 11:30 A.M. to 1:30 P.M. (The comment in cell C3 of the Data sheet spells these out.) Finally, the data set includes the *calculated* variable End-Waiting, the number waiting or being checked out at the end of a half-hour period. For any time period, it equals InitialWaiting plus Arrivals minus Departures; it also equals InitialWaiting for the *next* period. A partial listing of the data appears in Figure 3.33.

The manager of R&P wants to analyze these data to discover any trends, particularly in the pattern of arrivals throughout a day or across the entire week. Also, the store currently uses a "seat-of-the-pants" approach to opening and closing checkout stations each half hour. The manager would like to see how well the current approach is working. Of course, she would love to know the "best" strategy for opening and closing checkout stations—but this is beyond her (and our) capabilities at this point.

	A	B	C	D	E	F	G	H	I
1	Supermarket checkout efficiency								
2									
3	Day	StartTime	TimeInterval	InitialWaiting	Arrivals	Departures	EndWaiting	Checkers	TotalCustomers
4	Mon	8:00 AM	Morning rush	2	21	22	1	3	23
5	Mon	8:30 AM	Morning rush	1	25	18	8	3	26
6	Mon	9:00 AM	Morning	8	27	28	7	3	35
7	Mon	9:30 AM	Morning	7	21	23	5	3	28
8	Mon	10:00 AM	Morning	5	20	23	2	5	25
9	Mon	10:30 AM	Morning	2	36	31	7	5	38
10	Mon	11:00 AM	Morning	7	30	36	1	5	37
11	Mon	11:30 AM	Lunch rush	1	34	29	6	5	35
12	Mon	12:00 PM	Lunch rush	6	56	48	14	7	62
13	Mon	12:30 PM	Lunch rush	14	58	64	8	7	72
14	Mon	1:00 PM	Lunch rush	8	53	52	9	7	61
15	Mon	1:30 PM	Afternoon	9	30	36	3	5	39
16	Mon	2:00 PM	Afternoon	3	34	31	6	5	37
17	Mon	2:30 PM	Afternoon	6	36	37	5	5	42
18	Mon	3:00 PM	Afternoon	5	30	28	7	5	35
19	Mon	3:30 PM	Afternoon	7	29	34	2	5	36
20	Mon	4:00 PM	Afternoon	2	35	33	4	5	37
21	Mon	4:30 PM	Afternoon rush	4	32	25	11	5	36

FIGURE 3.33 A Partial Listing of the Supermarket Checkout Data

Objective To use charts, summary measures, and pivot tables to understand time patterns of arrivals and congestion at R&P Supermarket.

Solution

Obviously, time plays a crucial role in this example, so a good place to start is to create one or more time series plots. The graph in Figure 3.34 shows the time series behavior of InitialWaiting (the lower line) and Arrivals during the entire week. (This looks much better on a PC monitor, where the two lines are in different colors.) There is almost *too*

FIGURE 3.34
Time Series Plot
of InitialWaiting
and Arrivals
Variables

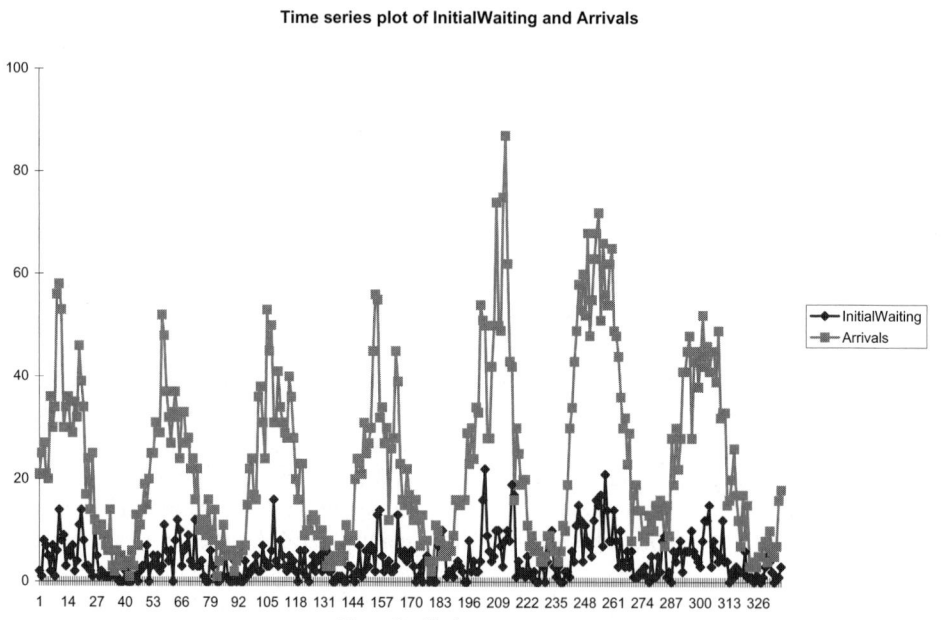

FIGURE 3.35
Time Series Plot
of Arrivals and
Departures
Variables

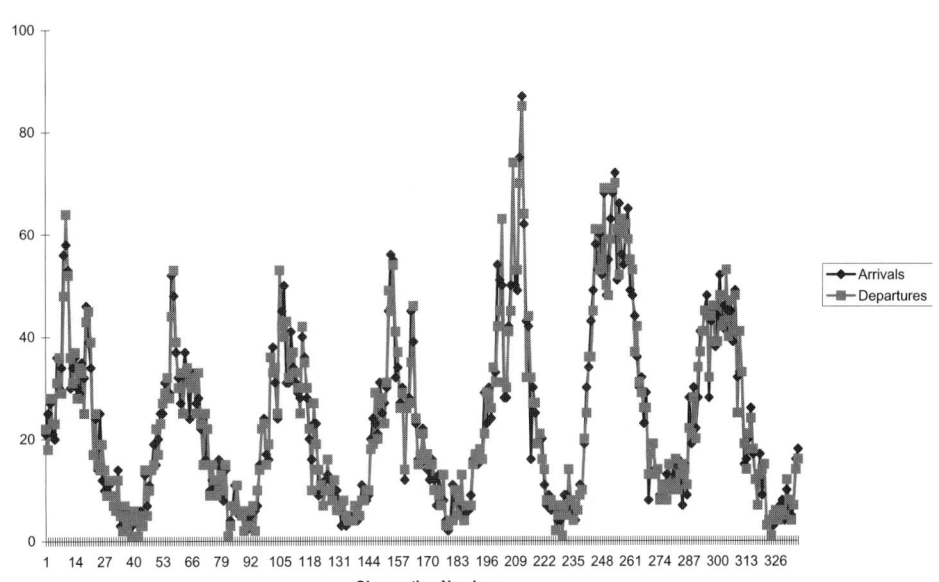

Time series plot of Arrivals and Departures

much clutter in this graph to see exactly what's happening, but it is clear that (1) Fridays and Saturdays are the busiest days; (2) the time pattern of arrivals is somewhat different—more spread out—during the weekends than during the weekdays; (3) there are fairly regular peak arrival periods during the weekdays; and (4) the number waiting is sometimes as large as 10 or 20, and the largest of these tend to be around the peak arrival times. You can decide whether it might be better (less clutter) to separate this plot into two time series plots, one for InitialWaiting and one for Arrivals. The advantage of the current plot is that we can match up time periods for the two variables.

A similar time series plot appears in Figure 3.35. This shows Arrivals and Departures, although it is difficult to separate the two time series—they are practically on top of one another. Perhaps this is not so bad. It means that for the most part, the store is checking out customers approximately as quickly as they are arriving.

A somewhat more efficient way to obtain this time series behavior is with pivot tables. Figure 3.36 shows one possibility. To create this pivot table, we drag the Initial-Waiting variable to the Data area, express it as an average, drag the StartTime variable to the Row area, and drag the Day variable to the Page area. (We also condensed the information from half-hour periods to hour-long periods by using the grouping option on the Pivot Table toolbar.) Finally, we create a time series plot from the data in the pivot table. Note how the variable in the Page area works. For example, we obtain the graph in Figure 3.36 if we choose Monday in the Page area (the dropdown list in cell B1). However, if we choose another day in cell B1, the data in the pivot table and the graph change automatically. So we can make comparisons across the days of the week with a couple of clicks of the mouse!

Similarly, the pivot table and corresponding column chart in Figure 3.37 indicate the average number of arrivals per half-hour period for each interval in the day. To obtain this output, we drag the Arrival variable to the Data area, express it as an average, drag the TimeInterval variable to the Row area, and drag the Day variable to the Page area. You can check that the pattern shown here for Friday is a bit different than for the other days— it has a significant bulge during the afternoon rush period.

You can either create this chart automatically with the pivot chart option, or you can create it with StatPro from the data in the pivot table.

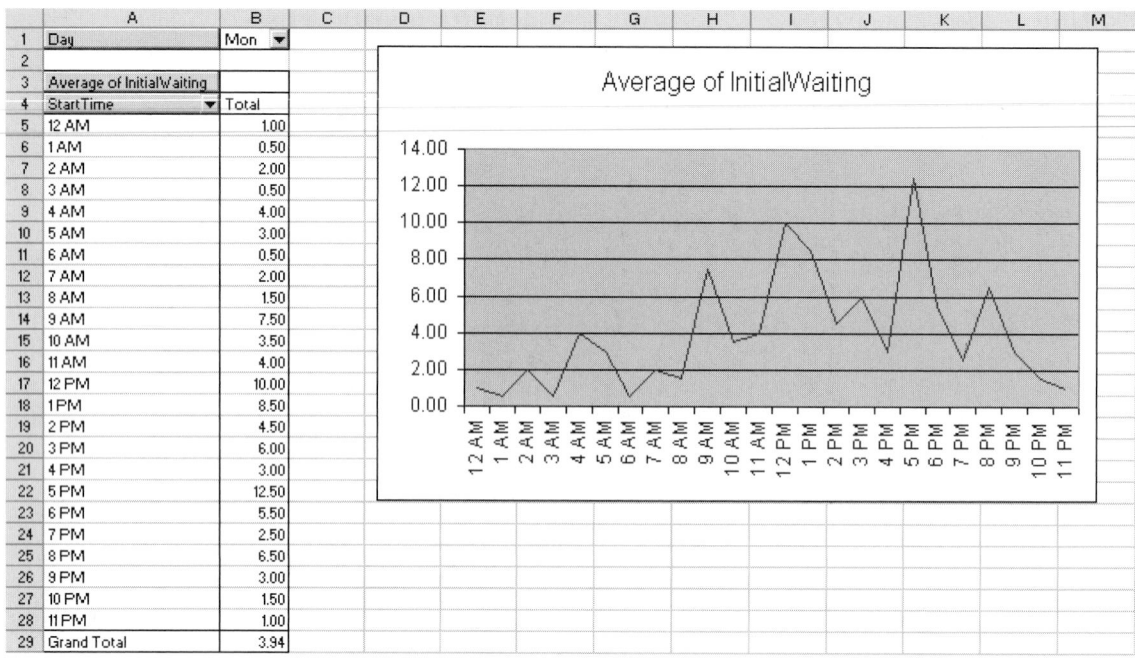

FIGURE 3.36 Average InitialWaiting by Hour of Day

FIGURE 3.37
Average Arrivals
by TimeInterval of
Day

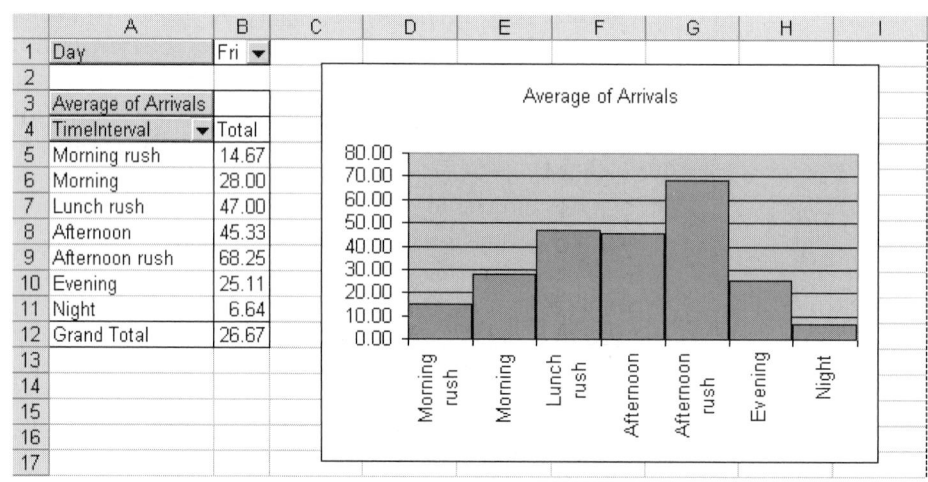

Excel Tip *If you try to create this pivot table on your own, you'll no doubt wonder how we got the time intervals in column A in the correct chronological order. The trick is to create a* **custom sort list.** *To do so, use the Tools/Options menu item, and select the Custom Lists tab. Then type a list of items in the List Entries box in the order you want them, and click on the Add button. To sort in this list order in the pivot table, place the cursor on any item in column A, select the Data/Sort menu item, click on Options, and select the new list from the dropdown list. This custom list will then be available in this or any other workbook you develop.*

FIGURE 3.38
Scatterplot of
Checkers versus
TotalCustomers

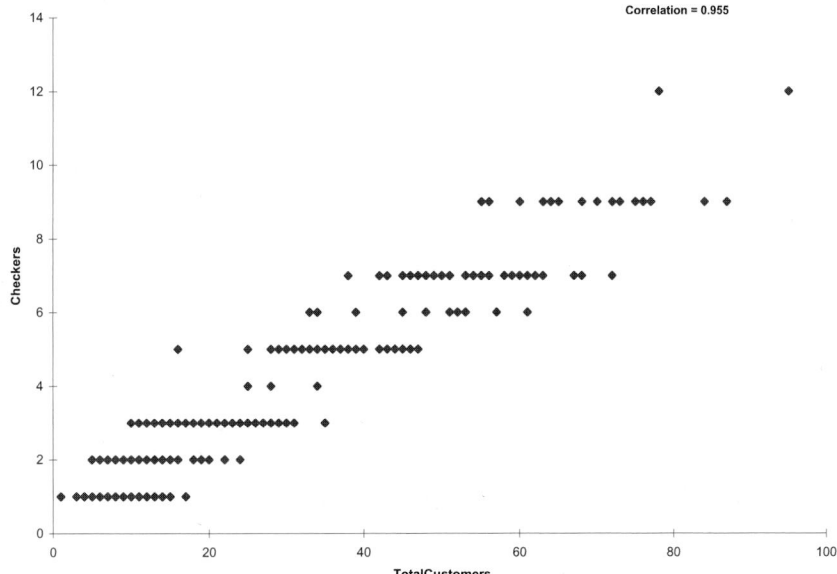

FIGURE 3.39
Scatterplot of
EndWaiting versus
Checkers

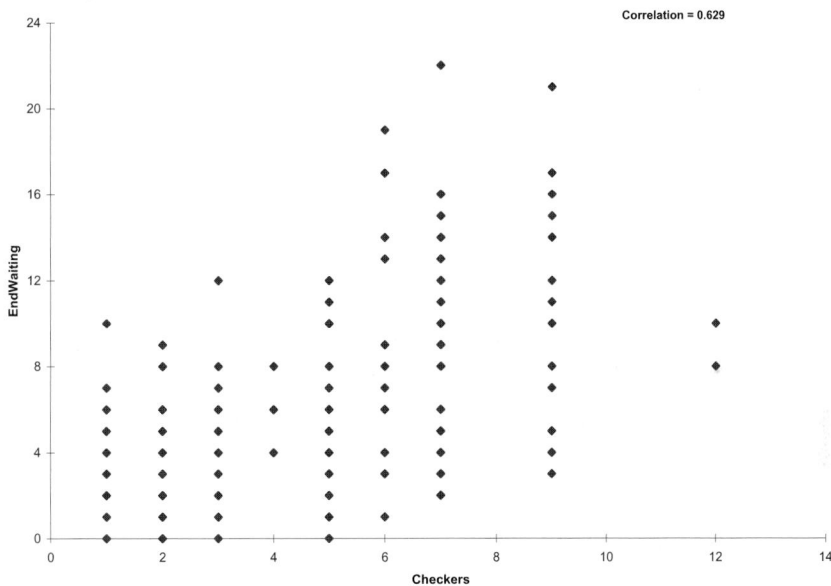

The manager of R&P is ultimately interested in whether the "right" number of check-out stations are available throughout the day. Figures 3.38 and 3.39 provide some evidence. The first of these is a scatterplot of Checkers versus TotalCustomers. (We calculated the TotalCustomers variable as the sum of the InitialWaiting and the Arrivals variables to measure the total amount of work presented to the checkout stations in any half-hour period.) There is an obvious positive relationship between these two variables. Evidently, management is reacting as it should—it is opening more checkout stations when there is more traffic. The second scatterplot shows EndWaiting versus Checkers. There is again a definite upward trend. Periods when more checkout stations are open tend to be associated with periods where more customers still remain in the checkout process.

In real situations like this one, company analysts often use a technique called discrete-event simulation to study the effect on waiting times and costs of using various policies for opening and closing checkout counters.

Presumably, management is reacting with more open checkout stations in busy periods, but it is not reacting strongly enough.

If you think you can solve the manager's problem just by fiddling with the numbers in the Checkers column, think again. There are two problems. First, there is a trade-off between the "cost" of having customers wait in line and the cost of paying extra checkout people. This is a difficult trade-off for any supermarket manager. Second, the number of departures is clearly related to the number of checkout stations open. (The relationship is a complex one that requires mathematical queueing theory to quantify.) Therefore, it doesn't make sense to change the numbers in the Checkers column without changing the numbers in the Departures (and hence the InitialWaiting and EndWaiting) columns in an appropriate way. This is *not* an easy problem! ●

Example 3.11 Demographic and Catalog Mailing Effects on Sales at HyTex

The HyTex Company is a direct marketer of stereophonic equipment, personal computers, and other electronic products. HyTex advertises entirely by mailing catalogs to its customers, and all of its orders are taken over the telephone. The company spends a great deal of money on its catalog mailings, and it wants to be sure that this is paying off in sales. Therefore, HyTex has collected data on 1000 customers at the end of the current year. (See the file CATALOGS.XLS.) For each customer the company has data on the following variables:

- Age: coded as 1 for 30 or younger, 2 for 31 to 55, 3 for 56 or older
- Gender: coded as 1 for males, 0 for females
- OwnHome: coded as 1 if customer owns a home, 0 otherwise
- Married: coded as 1 if customer is currently married, 0 otherwise
- Close: coded as 1 if customer lives reasonably close to a shopping area that sells similar merchandise, 0 otherwise
- Salary: combined annual salary of customer and spouse (if any)
- Children: number of children living with customer
- History: coded as "NA" if customer had no dealings with the company before this year, 1 if customer was a low-spending customer last year, 2 if medium-spending, 3 if high-spending
- Catalogs: Number of catalogs sent to the customer this year
- AmountSpent: Total amount of purchases made by the customer this year

HyTex wants to analyze these data carefully to understand its customers better. Also, the company wants to see whether it is sending the catalogs to the right customers. Currently, each customer receives either 6, 12, 18, or 24 catalogs through the mail each year. However, who receives how many has not really been thought out carefully. Is the current distribution of catalogs effective? Is there room for improvement?

Objective To use charts, summary measures, and pivot tables to understand the demographics of HyTex's customers, and to understand how these demographics, as well as the number of catalogs mailed, affect amounts spent.

Solution

This is the most difficult example we've faced so far, but it pales in comparison to the difficulty *real* direct marketing companies face. They have all sorts of data on millions of customers. How can they make sense of all these data? Using our relatively small data set, we will get the ball rolling. We'll let you discover additional patterns in the data that might exist. Furthermore, we'll only see an indication of whether the current distribution of catalog mailings is effective. It is well beyond our abilities at this point to find a more effective catalog distribution policy.

HyTex is primarily interested in the AmountSpent variable so it makes sense to create scatterplots of AmountSpent versus selected "explanatory" variables. We do this in Figures 3.40–3.42. Figure 3.40 shows AmountSpent versus Salary. It is clear that

FIGURE 3.40
Scatterplot of AmountSpent versus Salary

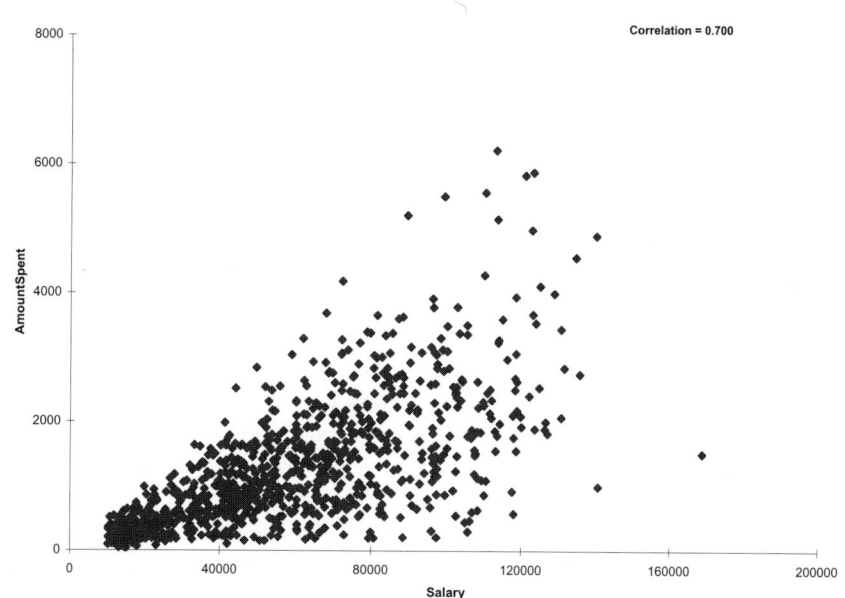

FIGURE 3.41
Scatterplot of AmountSpent versus Catalogs

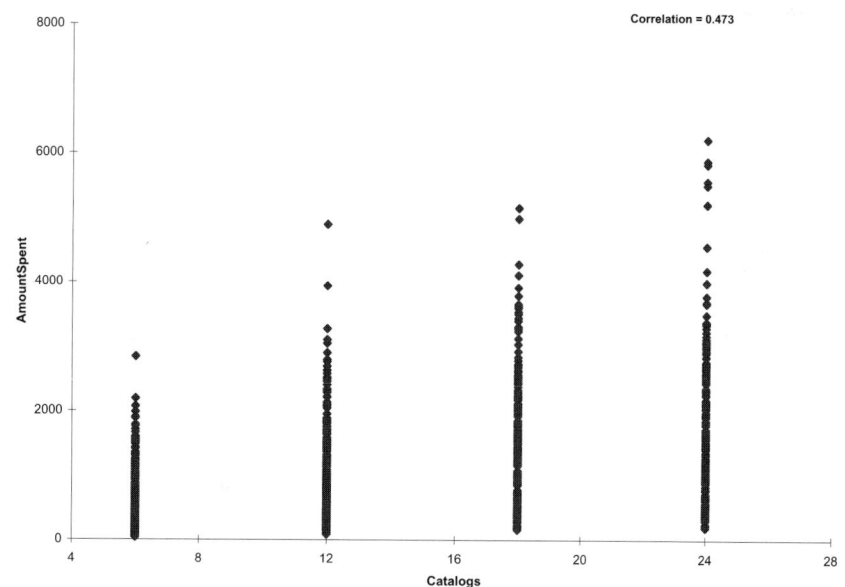

FIGURE 3.42
Scatterplot of
AmountSpent
versus Children

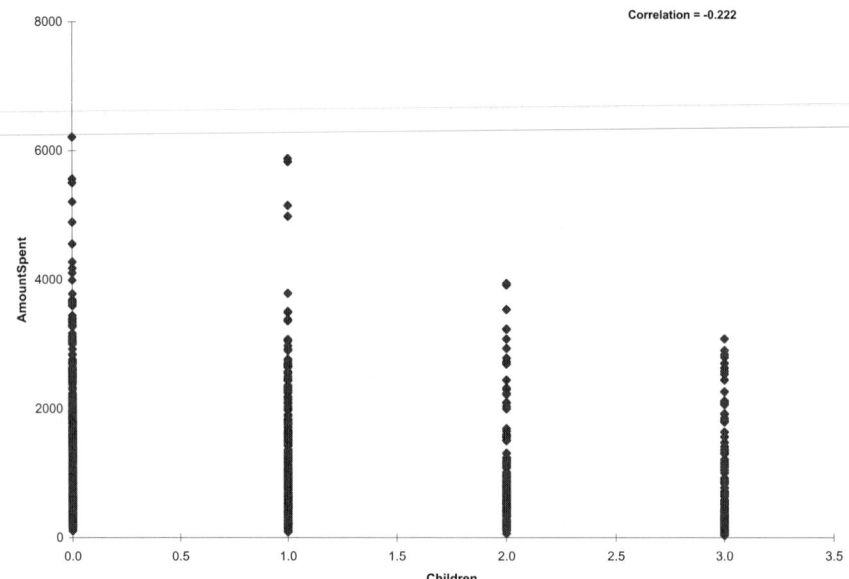

Correlation = -0.222

customers with higher salaries tend to spend more, although the variability in amounts spent increases significantly as salary increases. Figure 3.41 shows that there is some tendency toward higher spending among customers who receive more catalogs. But do the catalogs *cause* more spending, or are more catalogs sent to customers who would tend to spend more anyway? There is no way to answer this question with the data the company has collected. Figure 3.42 shows the interesting tendency of customers with more children to spend less. Perhaps customers with more children are already spending so much on $100-plus athletic shoes that they have little left to spend on electronic equipment!

Pivot tables and accompanying charts are very useful in this type of situation. We show several. First, Figures 3.43 and 3.44 can be used to better understand the demographics of the customers. Each row of Figure 3.43 shows the percentages of an age group who own homes. By changing the page variables Gender and Married in cells B5 and B6, we can see how these percentages change for married women, unmarried men, and so on.

	A	B	C	D	E	F	G	H	I	J	K
1											
2						Married (All) ▼ Gender 0 ▼					
3											
4											
5	Married	(All) ▼									
6	Gender	0 ▼					Count				
7						100.00%					
8	Count	OwnHome ▼				80.00%				OwnHome ▼	
9	Age ▼	0	1	Grand Total		60.00%				■ 0	
10	1	83.04%	16.96%	100.00%		40.00%				■ 1	
11	2	41.26%	58.74%	100.00%		20.00%					
12	3	30.23%	69.77%	100.00%		0.00%					
13	Grand Total	52.57%	47.43%	100.00%			1	2	3		
14											
15											
16							Age ▼				
17											
18											

FIGURE 3.43 Percent Home Owners versus Age, Married, and Gender

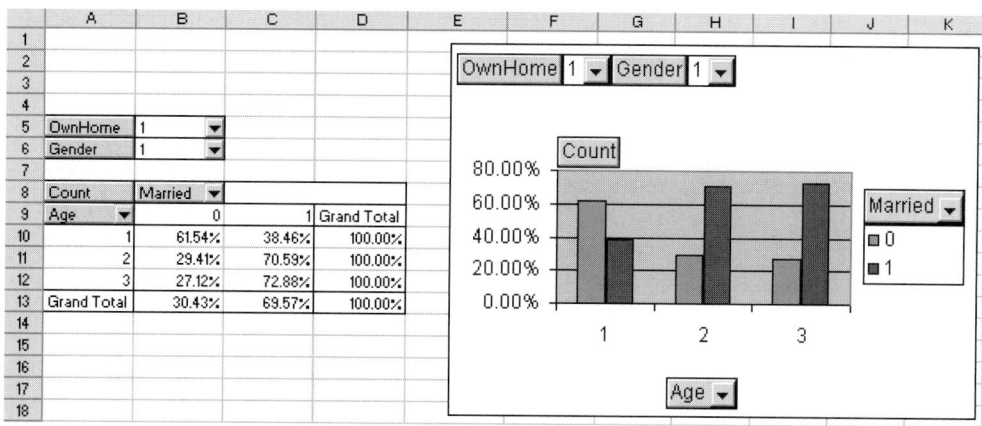

FIGURE 3.44 Percent Married versus Age, OwnHome, and Gender

You can check that these percentages remain relatively stable for the various groups. Specifically, a small percentage of the younger people own their own home, regardless of marital status or gender.

Figure 3.44 is similar. It shows the percentages of each age group who are married, for any combination of the Gender and OwnHome variables. Here the percentages change considerably for different settings of the page variables. For example, you can check that the married/unmarried split is quite different for women who don't own a home than for the male home owners shown in the figure.

Excel Tip *We do* not *have to create a new pivot table and pivot chart to get those shown in Figure 3.44. Starting with the pivot table shown in Figure 3.43, we can simply drag the gray OwnHome button to just above the gray Married button, and then drag the Married button to where the OwnHome button was. Everything updates automatically, including the chart! (This assumes that we created the chart as a pivot chart in the first place.) The pivot table/pivot chart combinations for the rest of this example were created from the original pivot table/pivot chart combination in Figure 3.43 in the same way— that is, by dragging field buttons to the appropriate locations in the pivot table. What a time-saver!*

Figure 3.45 provides more demographic information. Now we show the average Salary broken down by Age and Gender, with page variables for OwnHome and Married.

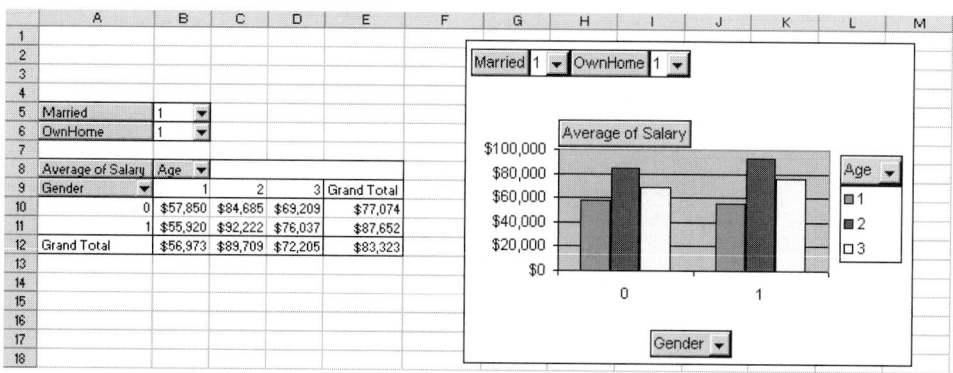

FIGURE 3.45 Average Salary versus Age, Gender, Married, and OwnHome

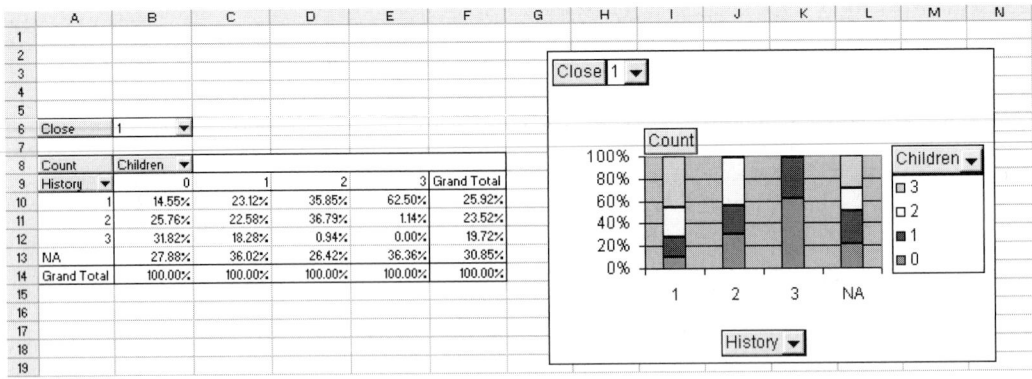

FIGURE 3.46 Percentages in History Categories versus Children and Close

Count	Children				
History	0	1	2	3	Grand Total
1	14.55%	23.12%	35.85%	62.50%	25.92%
2	25.76%	22.58%	36.79%	1.14%	23.52%
3	31.82%	18.28%	0.94%	0.00%	19.72%
NA	27.88%	36.02%	26.42%	36.36%	30.85%
Grand Total	100.00%	100.00%	100.00%	100.00%	100.00%

You can check that the *shape* of the resulting chart is practically the same for any combination of the page variables. However, the heights of the bars change appreciably. For example, the average salaries are considerably larger for the married home owners shown in the figure than for unmarried customers who are not home owners.

The pivot table and pivot chart in Figure 3.46 break the data down in another way. Each column in the pivot table shows the percentages in the various History categories for a particular number of children. Each of these columns corresponds to one of the bars in the "stacked" bar chart. Also, we have used Close as a page variable. Two interesting points emerge. First, customers with more children tend to be more heavily represented in the low-spending History category (and less heavily represented in the high-spending category). Also, as you can check by changing the setting of the Close variable from 1 to 0, the percentage of high-spenders among customers who live far from electronics stores is much higher than for those who live close to such stores.

Although we are not told exactly how HyTex determined its catalog mailing distribution, Figure 3.47 provides an indication. Each row of the pivot table shows the percentages of a particular History category that were sent 6, 12, 18, or 24 catalogs. The company's distribution policy is still somewhat unclear—and there is probably hope for improvement—but it *did* evidently send more catalogs to high-spending customers and fewer to low-spending customers.

Finally, Figure 3.48 shows the average AmountSpent versus History and Catalogs, with a variety of demographic variables in the page area. There are so many possible combinations that it is difficult to discover all the existing patterns. However, one thing stands

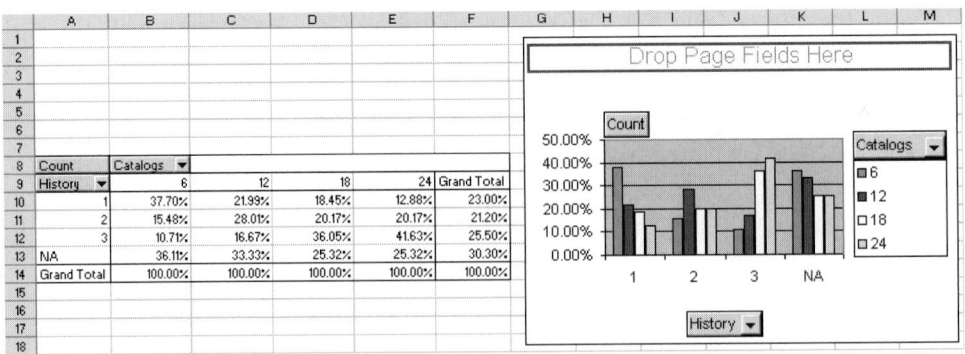

Count	Catalogs				
History	6	12	18	24	Grand Total
1	37.70%	21.99%	18.45%	12.88%	23.00%
2	15.48%	28.01%	20.17%	20.17%	21.20%
3	10.71%	16.67%	36.05%	41.63%	25.50%
NA	36.11%	33.33%	25.32%	25.32%	30.30%
Grand Total	100.00%	100.00%	100.00%	100.00%	100.00%

FIGURE 3.47 Catalog Distribution versus History

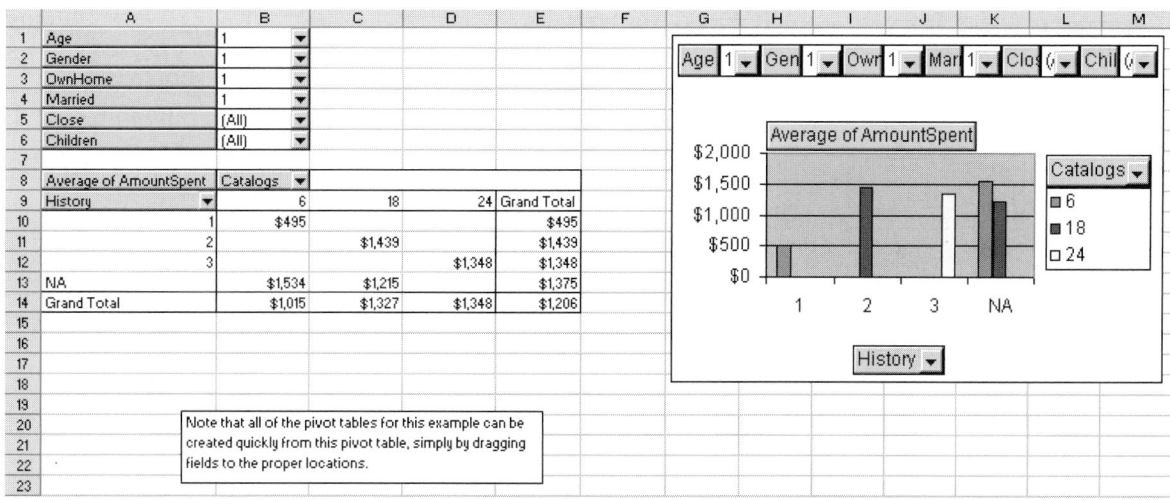

FIGURE 3.48 Average AmountSpent versus History, Catalogs, and Demographic Variables

out loud and clear from the graph: the more catalogs customers receive, the more they tend to spend. In addition, if they were large spenders last year, they tend to be large spenders this year.

In a pivot table with this many combinations, there will almost certainly be some combinations with no observations. For example, it turns out that there are no young married males who were low-spenders last year and received 18 catalogs. In this case you'll see a blank in the corresponding pivot table cell. In addition, there are no young married male home owners who received 12 catalogs. If you try this combination, the whole "12" column of the pivot table will disappear—and the corresponding bars will disappear from the chart. ●

3.10 Conclusion

This chapter and the previous chapter have illustrated the tremendous variety of descriptive measures we can obtain with Excel's built-in tools and add-ins such as StatPro. The *concepts* in these chapters are relatively simple. The key, therefore, is to have simple-to-use tools available to produce tables, graphs, and numerical summary measures in a matter of minutes. This is now possible, not only with statistical software packages but with spreadsheet packages, particularly Excel. These tools allow us to concentrate on presenting the data in the most appropriate way, so that interesting information hidden in the data is brought to the surface.

Summary of Key Terms

Term	Explanation	Excel	Pages	Equation Number
One-variable summary statistics	General term for measures that summarize distribution of a variable.	StatPro/Summary Stats/ One-Variable Summary Stats	78–88	
Mean $\bar{X}$ (for sample) μ (for population)	Average of observations.	=AVERAGE(*range*)	78	3.1

Summary of Key Terms (continued)

Term	Explanation	Excel	Pages	Equation Number
Median	Middle observation after observations are sorted from low to high.	=MEDIAN(*range*)	79	
Mode	Most frequently occurring observation.	=MODE(*range*)	80	
Percentile	Value such that a specified percentage of observations are below it.	=PERCENTILE(*range,pct*), where *pct* is decimal between 0 and 1	81	
Quartile	First quartile has 25% of observations below it. Third quartile has 25% of observations above it. Second quartile is the median.	=QUARTILE(*range,n*), where *n* is 1, 2, or 3	81	
Interquartile range	Difference between third and first quartiles.	=QUARTILE(*range*,3) − QUARTILE(*range*,1)	81	
Minimum	Smallest observation.	=Min(*range*)	81	
Maximum	Largest observation.	=MAX(*range*)	81	
Range	Difference between largest and smallest observations.	=MAX(*range*) − MIN(*range*)	81	
Variance s^2 (for sample) σ^2 (for population)	Measure of variability. Basically, average of squared deviations from mean.	=VAR(*range*) (for sample) =VARP(*range*) (for population)	83	3.2, 3.3
Standard deviation s (for sample) σ (for population)	Measure of variability in same units as observations. Square root of variance.	=STDEV(*range*) (for sample) =STDEVP(*range*) (for population)	83	3.4
Empirical rules	They specify approximate percentages of observations within 1, 2, or 3 standard deviations of mean for bell-shaped distributions.		84	
Measures of association	Measures of strength of linear relationship between two variables.	StatPro/Summary Stats/ Correlations, Covariances	91–94	
Covariance	Measure of association, affected by units of measurement.	=COVAR(*range1,range2*)	92	3.5
Correlation r (for sample) ρ (for population)	Measure of association, unaffected by units of measurement, always between −1 and +1.	=CORREL(*range1,range2*)	92	3.6
Boxplot	Chart that indicates the distribution of one or more variables. Box captures middle 50% of data, lines and points indicate possible skewness and outliers.	StatPro/Charts/Boxplot(s)	96	

Conceptual Exercises

C.1. Suppose that the histogram of a given income distribution is positively skewed. What does this fact imply about the relationship between the mean and median of the given distribution?

C.2. "The midpoint of the line segment joining the first quartile and third quartile of any distribution is the median." Indicate whether this statement is true or false. Explain your choice.

C.3. Explain why the standard deviation would likely *not* be a reliable measure of variability for a distribution of data that includes at least one extreme outlier.

C.4. "Two numerical variables are unrelated if the correlation measure between them is close to 0." Indicate whether this statement is true or false. Explain your choice.

C.5. Explain how a boxplot could be used to determine whether the associated distribution of values is essentially symmetric.

Level A

44. The annual base salaries for 200 students graduating from a reputable MBA program this year (see the file P02_74.XLS) are of interest to those in the admissions office who are responsible for marketing the program to prospective students. What salary level is *most* indicative of those earned by students graduating from this MBA program this year?

45. In its annual ranking of top graduate business schools in the United States, *U.S. News & World Report* provides data on a number of attributes of recognized graduate programs (refer to the file P02_08.XLS). One variable of interest is the annual out-of-state tuition paid by affected full-time students in each program.
 a. Find the annual out-of-state tuition levels at each of the 25th, 50th, and 75th percentiles.
 b. Identify the schools with the largest and smallest annual out-of-state tuitions. Does there appear to be a relationship between the program's overall ranking and its out-of-state tuition level?

46. Consider the Consumer Price Index, which provides the annual percentage change in consumer prices, for the period in the file P02_26.XLS. Find and interpret the interquartile range of these annual percentage changes.

47. The Consumer Confidence Index (CCI) attempts to measure people's feelings about general business conditions, employment opportunities. and their own income prospects. The annual average values of the CCI are given in the file P02_28.XLS.

 a. Fifteen percent of all years in this sample have annual CCI values that exceed what value?
 b. Forty percent of all years in this sample have annual CCI values that are less than or equal to what value?
 c. In which year of the given sample period were U.S. consumers *most* confident, as measured by the CCI?
 d. In which year of the given sample period were U.S. consumers *least* confident, as measured by the CCI?

48. Consider the proportion of Americans under the age of 18 living below the poverty level for the data in the file P02_29.XLS.
 a. In which years of the sample has the poverty rate for American children exceeded the rate that defines the third quartile of these data?
 b. In which years of the sample has the poverty rate for American children fallen below the rate that defines the first quartile of these data?
 c. What is the typical poverty rate for American children during this period?

49. The annual averages of the discount rate, federal funds rate, and the prime rate are given in the file P02_30.XLS. For each of these three key interest rates, determine the following:
 a. Thirty percent of all years in the sample period have annual average rates that exceed what value?
 b. Twenty-five percent of all years in the sample period have annual average rates that are less than or equal to what value?
 c. What is the most typical annual average rate over the given sample period?

50. Given data in the file P02_13.XLS from a recent survey of chief executive officers from 352 of the nation's biggest businesses, respond to the following questions.
 a. Find the annual salary below which 75% of all given CEO salaries fall.
 b. Find the annual bonus above which 55% of all given CEO bonuses fall.
 c. Determine the range of the middle 50% of all given total annual compensation figures (i.e., of the amounts found in column *Sum97*).

51. The file P02_44.XLS contains the number of classroom teachers and the average salary of classroom teachers for each of the 50 states and the District of Columbia. Which of the states paid their teachers average salaries that exceeded approximately 90% of all average salaries? Which of the states paid their teachers average salaries that exceeded only about 10% of all average salaries?

52. The annual base salaries for 200 students graduating from a reputable MBA program this year are given in the file P02_74.XLS.
   a. Is it appropriate to apply the empirical rules in this case? Explain.
   b. If it is appropriate to apply the empirical rules here, between what two numbers can we be about 68% sure that the salary of any one of these 200 students will fall?

53. Refer to the data given in the file P02_11.XLS. Consider the average time (in minutes) it takes a citizen of each metropolitan area to travel to work and back home each day.
   a. Find a reliable measure of the dispersion of these average commute times around the overall sample mean.
   b. Between what two numbers can we be approximately 99.7% sure that any one of these average travel times will fall?

54. Is there a strong relationship between a chief executive officer's annual compensation and her or his organization's recent profitability? Explore this question by generating correlations for the survey data in the file P02_13.XLS. In particular, compute and interpret correlation measures for the change in the company's net income from 1996 to 1997 (see *Comp_NetInc96* column) and the CEO's 1997 base salary, as well as for the change in the company's net income from 1996 to 1997 and the CEO's 1997 bonus. Summarize your findings.

55. Construct boxplots to compare recent job growth rates with forecasted job growth rates for selected towns in the United States. These growth rates are all provided in the file P03_55.XLS. Do you detect the presence of any outliers in either of these two distributions? Also, compute a correlation measure for these two sets of job growth rates and interpret it.

56. The percentage of private-industry jobs that are managerial has steadily declined in recent years as companies have found middle management a ripe area for cutting costs. How have women and various minority groups fared in gaining management positions during this period of corporate downsizing of the management ranks? Relevant data are given in the file P02_75.XLS. Generate boxplots using these data to make general comparisons across the various groups included in the set.

57. The U.S. Bureau of Labor Statistics provides data on the year-to-year percentage changes in the wages and salaries of workers in private industries, including both "white-collar" and "blue-collar" occupations. Consider the data in the file P02_56.XLS. Develop side-by-side boxplots to summarize these distributions of annual percentage changes. In particular, state and interpret the interquartile range for each of three given distributions.

58. Explore the given distribution of the numbers of beds in short-term general hospitals in selected U.S. metropolitan areas by producing a boxplot (refer to the data in the file P02_17.XLS). In particular, use your boxplot to answer the following questions:
   a. Is it more likely for these metropolitan areas to have larger or smaller numbers of hospital beds?
   b. Is the mean or median the more accurate measure of central location in this case? Explain.
   c. Characterize the variation of these values around the center of the data.
   d. Do you detect the presence of any *extreme* outliers in this case?

59. The file P03_59.XLS contains the proportion of annual revenue spent on research and development (R&D) activities for each of 200 randomly selected high-technology firms. Characterize this distribution by computing numerical summary measures and constructing a boxplot diagram. In particular, comment on the typical proportion of revenue dollars spent on R&D by these firms and the variation about the typical proportion value.

60. Consider various characteristics of the U.S. civilian labor force provided in the file P03_60.XLS. In particular, examine the given unemployment rates across the United States.
   a. Characterize the distribution of total unemployment rates. What is the most typical value? How are the other total unemployment rates distributed about the typical rate?
   b. Compare the distribution of total unemployment rate to those of the male unemployment rate and the female unemployment rate. How are these distributions similar? How are they different?
   c. Which is more strongly associated with the total unemployment rate in the United States: the male unemployment rate or the female unemployment rate?

61. The file P03_61.XLS contains the sale price of gasoline in each of the 50 states.
   a. Compare the distributions of gasoline sale price data (one for each year). Specifically, do you find the mean and standard deviation of these distributions to be changing over time? If so, how do you explain the trends?
   b. In which regions of the country have gasoline prices changed the most?
   c. In which regions of the country have gasoline prices remained relatively stable?

62. Examine life expectations (in years) at birth for various countries across the world. These data are in the file P03_62.XLS.
   a. Generate an estimate of the *typical* human's life span at birth using the 1997 data. What are the

limitations of the method you have employed in estimating this world population parameter?

b. Characterize the *variability* of the life spans at birth using the 1997 data. Is this distribution fairly symmetric or skewed? How do you know?

c. How strongly are the 1997 life expectations associated with projections for births in 2000 and 2010? Explain why the degree of linear association between the 1997 data and each set of projections diminishes somewhat over time.

63. This problem focuses on the per capita circulation of daily newspapers in the United States. The file P03_63.XLS contains these data.

a. Compare the yearly distributions of daily newspaper per capita circulation over the period.

b. Note any clear trends, both nationally and regionally, in the average value of and variability of per capita newspaper circulation during the given years.

64. Have the proportions of Americans receiving public aid changed in recent years? Explore this question through a careful examination of the data provided in the file P03_64.XLS. In particular, generate numerical summary measures to respond to each of the following.

a. Report any observed changes in the overall mean or median rates during the given time period.

b. Can you find evidence of regional changes in the proportions of Americans receiving public aid? If so, summarize your specific findings.

65. The file P03_65.XLS contains the measured weight (in ounces) of a particular brand of ready-to-eat breakfast cereal placed in each of 500 randomly selected boxes by one of five different filling machine operators. Quality assurance personnel at this company are interested in determining how well these five operators are performing their assigned task of *consistently* placing 15 ounces of cereal in each box.

a. Employ descriptive graphs and summary measures to ascertain whether some or all of these operators are consistently missing the target weight of 15 ounces per box.

b. If you were charged with selecting the "Outstanding Employee of the Month" from this set of filling machine operators, which operator would you select based on the given data? Defend your choice.

66. Electro produces voltage-regulating equipment in New York and ships the equipment to Chicago. The voltage held is measured in New York before each unit is shipped to Chicago. The voltage held by each unit is also measured when the unit arrives in Chicago. The file P03_66.XLS contains a sample of voltage measurements at each city. A voltage regulator is considered acceptable if it can hold a voltage of between 25 and 75 volts.

a. Using boxplots and descriptive statistics, what can you learn about the voltage held by units before shipment and after shipment?

b. What percentage of units are acceptable before and after shipping?

c. Do you have any suggestions about how to improve the quality of Electro's regulators?

d. Ten percent of all New York regulators have a voltage exceeding what value?

e. Five percent of all New York regulators have a voltage less than or equal to what value?

67. The file P03_67.XLS contains the individual scores of students in two different accounting sections who took the same exam. Comment on the differences between exam scores in the two sections.

68. The file P03_68.XLS contains the monthly interest rates on 3-month government T-bills. For example, in January 1985, 3-month T-bills yielded 7.76% annual interest. To succeed in investments, it is important to understand the characteristics of the monthly changes in T-bill rates.

a. Construct a histogram of the monthly changes in interest rates. Try to choose categories so that you get an "interesting" histogram.

b. Do the empirical rules hold for changes in monthly interest rates?

c. Based on the given data, there is a 5% chance that during a given month T-bill rates will increase by less than what value? (A negative number is allowed here.)

d. Based on the given data, there is a 10% chance that during a given month T-bill rates will increase by at least what number?

e. Based on the given data, estimate the chances that T-bill rates during a given month will increase by more than 0.5%.

Level B

69. Data on the numbers of insured commercial banks in the United States are given in the file P03_69.XLS.

a. Compare these distributions of the numbers of U.S. commercial banks (one for each year). Do you find the mean and standard deviation of these numbers to be changing over time? If so, how do you explain the trends?

b. What trends do you notice in the numbers of commercial banks *by region?* For example, how do the numbers of commercial banks appear to be changing in the northeastern United States over the given period? Summarize your findings for each region of the country.

70. Educational attainment in the United States is the focus of this problem. Employ descriptive methods with the data provided in the file P03_70.XLS to characterize the educational achievements of Americans in the given year. Do your findings surprise you in some way?

71. Data on U.S. homeownership rates are given in the file P03_71.XLS.

    a. Employ numerical summary measures to characterize the changes in homeownership rates across the country during this period.

    b. Do the trends appear to be uniform across the United States or are they unique to certain regions of the country? Explain.

72. Data on community hospital average daily cost are provided in the file P03_72.XLS. Do the yearly distributions of average cost figures tend to become more or less variable over the given time period? Justify your answer with descriptive graphs and/or relevant summary measures.

73. The median sales price of existing one-family homes in selected metropolitan areas is the variable of interest in this exercise. Using the data contained in the file P03_73.XLS, characterize the distribution of median sales prices of existing single-family homes in 1996. How is this distribution different from that for the median sales prices of such homes in 1992? Carefully summarize the essential differences.

74. Are U.S. traffic fatalities related to the speed limit and/or road type? Consider the data found in the file P03_74.XLS.

    a. Do the average number and/or variability in the number of traffic fatalities occurring on interstate highways tend to increase as the speed limit is raised above 55 miles per hour? Explain your answer.

    b. Do the average number and/or variability in the number of traffic fatalities occurring on noninterstate roads tend to increase as the speed limit rises above 35 miles per hour? Explain your answer.

    c. Do the average number and/or variability in the number of traffic fatalities occurring on a road with a posted speed limit of 55 miles per hour tend to change with the road type (i.e., interstate versus noninterstate highway)? Explain your answer.

    d. Based on these data, which combination of speed limit and road type appears to be most lethal for U.S. drivers?

    e. Based on these data, which combination of speed limit and road type appears to be safest for U.S. drivers?

75. Have greater or lesser proportions of Americans joined labor unions during the past decade? Respond to this question by applying descriptive summary measures and graphical tools to the data provided in the file P03_75.XLS. Interpret your output. What conclusions can you draw from an analysis of these data?

76. Consider the percentage of the U.S. population without health insurance coverage. The file P03_76.XLS contains such percentages by state for both 1998 and 1999.

    a. Describe the distribution of state percentages of Americans without health insurance coverage in 1999. Be sure to employ both measures of central location and dispersion in developing your characterization of this sample.

    b. Compare the 1999 distribution with the corresponding set of percentages taken in 1998. How are these two sets of figures similar? In what ways are they different?

    c. Compute a correlation measure for the two given sets of percentages. What does the correlation coefficient tell you in this case?

    d. Based on your answers in parts **b** and **c** above, what would you expect to find upon analyzing similar data for 2000?

77. Given data in the file P02_13.XLS from a recent survey of 352 chief executive officers of the nation's biggest businesses, apply your knowledge of numerical summary measures to determine whether the typical levels and variances of the 1997 annual salaries and bonuses earned by CEOs depend in part on the *types* of companies in which they serve.

78. Consider survey data collected from 1000 randomly selected Internet users, given in the file P02_43.XLS.

    a. Use these data to formulate a profile of the typical *female* Internet user. Consider such attributes as age, education level, marital status, annual income, and family size in formulating your profile.

    b. Use these data to formulate a profile of the typical *married* Internet user. Consider such attributes as gender, age, education level, annual income, and family size in formulating your profile.

    c. Use these data to formulate a profile of the typical *high-income* (say, with an annual income in excess of $80,000) Internet user. Consider such attributes as gender, age, education level, marital status, and family size in formulating your profile.

79. As a part of an economic development study conducted by the local government, 500 households in a middle-class neighborhood were recently surveyed. In particular, for each of the randomly selected households, the survey requested information on the following variables: family size, approximate location of the household within the neighborhood, an indication of whether those surveyed owned or rented their home, gross annual income of the first household wage earner, gross annual income of the second household wage earner (if applicable), monthly home mortgage or rent payment, average monthly expenditure on utilities, and the total indebtedness (excluding the value of a home mortgage) of the household. These data are provided in the file P02_06.XLS. Compute and interpret relevant numerical summary measures to complete the following.

    a. Use these data to formulate a profile of the typical household residing within each of the four neighborhood locations. Consider such attributes as family size, home ownership status, gross annual income(s) of household wage earner(s), monthly home mortgage or rent payment, average monthly expenditure on utilities, and the total indebtedness (excluding the value of a home mortgage) of the household in formulating your profile.

    b. Do differences arise in the mean or median income levels of those wage earners from households located in different quadrants of this neighborhood? If so, summarize these differences.

    c. Do differences arise in the mean or median monthly home mortgage or rent payment paid by households located in different quadrants of this neighborhood? If so, summarize these differences.

    d. Do differences arise in the mean or median debt levels of households located in different quadrants of this neighborhood? If so, summarize these differences.

80. A human resources manager at Beta Technologies, Inc., is interested in developing a profile of the highest paid full-time Beta employees based on the given representative sample of 52 of the company's full-time workers in the file P02_01.XLS. In particular, she is interested in determining the typical age, number of years of relevant full-time work experience prior to coming to Beta, number of years of full-time work experience at Beta, and number of years of postsecondary education for those employees in the *highest quartile* with respect to annual salary. Employ appropriate descriptive methods to help the human resources manager develop this desired profile.

81. Using cost-of-living data from the *ACCRA Cost of Living Index* (see the file P02_19.XLS), examine the relationship between the geographical *location* of an urban area within the United States (e.g., northeast, southeast, midwest, northwest, or southwest) and its *composite* cost-of-living index. In other words, is the overall cost of living higher or lower and more or less variable for urban areas in particular geographical regions of the country? You must assign the given urban areas systematically to one of several geographical regions before you can apply appropriate summary measures in responding to this question. Summarize your findings in detail.

82. The file P03_82.XLS contains monthly interest rates on bonds that pay money a year after the day they are bought. It is often suggested that interest rates are more volatile (tend to change more) when interest rates are high. Do these data support this statement?

83. The file P03_83.XLS contains data on 1000 of Marvak's best customers. Marvak is a direct-marketing firm that sells electronic items. It has collected these data to learn more about its customers. The variables are self-explanatory, although a few cell notes have been added in row 3. Your boss at Marvak would like you, the Excel guru, to do the following.

    a. She wants a breakdown of gender by age group. That is, she wants a pivot table that lists, for each age group, the percentages of females and males. She wants to be able to access this information easily (with a couple of clicks) for any subcategories of Home and Married.

    b. She guesses that customers with larger salaries tend to spend more at Marvak. To check this, she wants you to append a new variable called SalaryCat to the data set that contains the four category labels in column J. A customer with salary below $30,000 is categorized as "LowSal"; between $30,000 and $70,000 as "MedSal"; between $70,000 and $120,000 as "HighSal"; and over $120,000 as "HugeSal." (By the way, no incomes are exactly equal to $30,000, $70,000, or $120,000.) You can use the lookup table in columns I and J to form this new *SalaryCat* column. Then create a pivot table that shows the average amount spent for each of the four salary categories, and comment briefly (on that sheet) whether your boss's conjecture appears to be correct.

84. The file P03_84.TXT contains the largest 100 public companies in the world, as listed in *The Wall Street Journal* on September 24, 1992, ranked by market value. The rankings are shown for 1992, as well as for 1991. The following variables are included:

- Company: name of company
- Location: 1 for United States, 2 for Japan/Australia, 3 for Europe
- Bank: 1 if bank (or savings institution), 0 otherwise
- Rank92: rank according to market value in 1992
- Rank91: rank according to market value in 1991
- MarketVal: market value in millions of U.S. dollars (12/31/91 exchange rates used)
- Sales91: sales in 1991 in millions of U.S. dollars
- PctChSales: percent change in sales from 1990, based on home currency
- Profit91: profit in 1991 in millions of U.S. dollars
- PctChProfit: percent change in profit from 1990, based on home currency

    a. This file is in ASCII (text) form, with blanks between the items and names (nonnumerical data) in double quotes. Open this file in Excel. (There is no "import" command; you simply open the file.) Excel will recognize that this is a text file, and a "wizard" will lead you through the steps to open it properly into Excel format. The key is that it is "delimited" with blanks. Once the file is opened, look at it to make sure everything is lined up correctly. Then use the Save As command to save the file as an .XLS file. (Once you do this, the .TXT version is no longer needed.)

b. Note that percentage changes from 1990 to 1991 are given for sales and profits. Use formulas to create variables Sales90 and Profit90, the sales and profits for 1990. [*Hint:* For example, the formula for Sales90 is given by 100*Sales91/(100+PctChSales).]

c. Create a scattergram of Profit91 (vertical axis) versus Profit90. Most of the points lie close to a line. For this problem, consider an outlier to be any point that is obviously not very close to this line. Which companies are the worst outliers in this sense? In business terms, what makes these companies outliers?

d. The companies designated "banks" are clearly different from the other companies in that their sales figures are much larger. For this question, consider only the subset of *nonbanks*. Define an outlier with respect to any variable as an observation that is at least 1.5 IQRs above the third quartile or below the first quartile. (This is the boxplot definition of outliers.) How many outliers are there with respect to Sales91?

e. Notice that there is a variable that codes the location of the company: 1 for United States, 2 for Japan/Australia, 3 for Europe. With regard to Profit91, are there any obvious differences among these? (Use summary statistics and/or charts.)

f. The data in this file are somewhat dated. See if you can find similar, but up-to-date, data (possibly on the Web). Then do the same analysis on the new data.

85. The file P03_85.XLS contains 1993 compensation data on the top 200 CEOs. The data include the CEO's name, company, total compensation, and the company's 5-year total return to shareholders (expressed as an annual rate). The data are sorted in descending order according to the 5-year return. (Source: *Fortune,* July 25, 1994)

a. How large must a total compensation be to qualify as an outlier on the high side according to the boxplot definition of outliers? In column A of the spreadsheet, highlight the names of all CEOs whose total compensations are outliers. (You can highlight them by making them boldface, italiciz-

ing them, or painting them a different color, for example.)

b. Form a scatterplot of total compensation versus 5-year return. (Put total compensation on the horizontal axis.) Do the CEOs in this database appear to be worth their pay?

c. The data in this file are somewhat dated. See if you can find similar, but up-to-date, data (possibly on the Web). Then do the same analysis on the new data.

86. The file P03_86.XLS contains data on close to 1800 professional NFL football players. (In case you are not a pro football fan, there are two conferences in the NFL: the NFC and the AFC. Also, each player plays either on offense or defense.) The data for each player include his base salary, any bonus, and the total of the base salary and the bonus. For each part, do the requested analysis for the total salary.

a. Create histograms of the salaries for (i) all of the players, (ii) all of the NFC players, and (iii) all of the AFC players. Also, create tables of summary statistics for these three groups.

b. Proceed as in part **a,** but now make the distinction between offensive and defensive players instead of which conference they are in.

c. Repeat parts **a** and **b,** but now eliminate the quarterbacks (QB), who are often the highest paid players. Does it make a difference?

d. Use one or more pivot tables to break the data down in other interesting ways, such as by position. Your goal is to understand the salary structure in the NFL. Write up your results in a brief report.

87. The file P03_87.XLS contains questionnaire data from a random sample of 200 TV viewers. (The variable name headings actually begin in row 25.) The questionnaire was taken by the local station XYZ, an affiliate of one of the three main networks. Like the local affiliates of the other two networks, XYZ's local dinnertime news program follows the national news program. The purpose of the questionnaire was to discover characteristics of the viewing public, presumably with the intention of doing something to increase XYZ's ratings. Your assignment is very open-ended—purposely so. Summarize any aspects of the data that you think are relevant—find means, proportions, scatterplots, pivot tables, whatever—to help XYZ management understand these viewers. (All 200 people watch both national and local news.)

The Dow Jones Industrial Average (DJIA) is a composite index of 30 of the largest "blue-chip" companies in the United States. It is probably the most quoted index from Wall Street, partly because it is old enough that many generations of investors have become accustomed to quoting it, and partly because the U.S. stock market is the world's largest. Besides longevity, two other factors play a role in the Dow's widespread popularity: It is understandable to most people, and it reliably indicates the market's basic trend. As this edition was going to press, the Dow was languishing at around 10,000, after having been above 11,000. There is no predicting where it might be when you read this.

Unlike most other market indexes that are weighted indexes (usually by market capitalization, that is, price times shares outstanding), the DJIA is an unweighted index. It was originally an average, namely, the sum of the stock prices divided by the number of stocks. In fact, the very first average price of industrial stocks, on May 26, 1896, was 40.94. However, because of stock splits, the DJIA is now calculated somewhat differently to preserve historical continuity. To calculate the DJIA, the prices of the 30 stocks in the index are summed, and this sum is divided by the "divisor," which is currently slightly greater than 0.33. The effect is to multiply the sum by approximately 3. Therefore, if the combined price of the 30 stocks rises by $10, say, the DJIA rises by about 30 points.

The Dow originally consisted of 12 stocks in 1896 and increased to 20 in 1916. The 30-stock average made its debut in 1928, and the number has remained constant ever since. However, the 30 stocks comprising the Dow do not remain the same. Since the first edition of this book, SBC Communications, Home Depot, Honeywell International, Intel, Microsoft, and Citigroup have replaced Union Carbide, Sear Roebuck, Aluminum Company of America, Allied Signal, Chevron, and Traveler's Group in the select 30. The editors of *The Wall Street Journal* select the components of the DJIA. They take a broad view of what "industrial" means. In essence, it is almost any company that isn't in the transportation business and isn't a utility. In choosing a new company for the DJIA,

they look among substantial industrial companies with a history of successful growth and wide interest among investors. The components of the DJIA are not changed often. It isn't a "hot stock" index, and the *Journal* editors believe that stability of composition enhances the trust that many people have in the averages. The most frequent reason for changing a stock is that something is happening to one of the components (for example, a company is being acquired).

Some people make predictions about where the stock market is headed based in part on their interpretation of DJIA movements, as well as movements of the transportation and utilities averages. But indexes don't predict anything. They are doing their job if they accurately reflect where the market has been. However, there is a great deal of common ground between the economy and the market. Stock investors try to anticipate future profits, and corporate profits are a prime fuel for the U.S. economy. So, not surprisingly, the market frequently rises ahead of economic expansion and falls prior to economic slowdown or contraction. The trouble is that this relationship isn't perfectly correlated; there are other factors that move markets and still others that affect the economy. Moreover, many people make the mistake of calibrating their economic expectations to the DJIA's movements. The result is, as Nobel-laureate economist Paul A. Samuelson put it, "The market has predicted nine of the last five recessions."

As indicated previously, the DJIA is not the only market index. There are two other Dow Jones indexes, one for transportation (DJTA), consisting of 20 stocks, and one for utilities (DJUA), consisting of 15 stocks. An elaborate analytical system dubbed Dow Theory holds that the DJTA must "confirm" the movement of the industrial average for a market trend to have staying power. If the industrials reach a new high, the transportations would need to reach a new high to "confirm" the broad trend. The trend reverses when both averages experience sharp downturns at around the same time. If they diverge—for example, if the industrial average keeps climbing while the transportations decline—watch out! The underlying fundamentals of the Dow Theory hold that the industrials make and the transportations take. If the transportations aren't taking what the industrials are making, it portends economic weakness and market problems. Similarly, according to analysts who study the averages, a rise in utility stock prices indicates that investors anticipate falling interest rates, because utilities are big borrowers and their profits are enhanced by lower interest costs. But the utility average tends to decline when investors expect rising interest rates. Because of this interest-rate sensitivity, the utility average is regarded by some as a leading indicator for the stock market as a whole.

This information and other interesting facts about the Dow Jones averages are available at the www.dowjones.com Web site. Other Web sites have data on the averages themselves. We used one such site (chart.yahoo.com/t) to download two years' worth of daily data for the DJIA, DJTA, and DJUA, as well as for the stocks comprising these averages. The data are in the files DJI.XLS, DJT.XLS, and DJU.XLS. Each file contains a sheet for the Dow Jones average and a sheet for each stock in the average. (The sheet names for the latter indicate the ticker symbols for the stocks, such as HWP for Hewlett-Packard Co.)

Use the tools you've learned in the past two chapters to analyze these data sets (or more recent Dow Jones data sets if you can download them). Here are some suggested directions for analysis.

1. The return for any period is the percentage change in the price over that period. That is, the return is

$$\frac{p_{\text{end}} - p_{\text{beg}}}{p_{\text{beg}}} \times 100$$

where p_{end} is the ending price and p_{beg} is the beginning price. (The return also includes dividends, but you can ignore these here.) Are the daily returns for the 30 stocks in the DJIA highly correlated with each other? Are the daily returns correlated with the DJIA itself? What about weekly returns? What about monthly returns? Answer the same questions for the DJTA; for the DJUA.

2. How would you evaluate portfolios of any of these stocks over the 2-year period? How much better (or worse) are some portfolios?

3. As stated previously, some analysts believe that the DJUA is a leading indicator of the market. Do the data bear this out, assuming we identify the DJIA as "the market"? One way to answer this is with "cross-correlations," such as the correlation between the DJIA today and the DJUA a week ago. Alternatively, we could compare a time series graph of the DJIA with a "shifted" version of the DJUA.

4. Similarly, is there any relationship between the DJIA and the DJTA?

Following up on Case Study 3.1, there are many market indexes other than the Dow Jones averages. These include broad U.S. indexes such as the Nasdaq Composite Index (mainly technology stocks), the NYSE Composite Index (an index of many stocks on the New York Stock Exchange), the S&P 500 index (an index of 500 of the largest U.S. companies), and others. There are also U.S. indexes for particular industries, such as the AMEX Biotechnology Index, and foreign indexes, such as the AMEX Japan Index. Daily data (from the same source as in the previous case) for several of these indexes are listed in the file OTHERINDEXES.XLS. Formulate and answer any interesting questions relating to these data and the Dow Jones data. In particular, do all of these indexes tend to move together, or do they tend to move in opposite directions?

3.3 Correct Interpretation of Means

A mean, as defined in this chapter, is a pretty simple concept—it is the average of a set of numbers. But even this simple concept can cause confusion if we aren't careful. The data in Figure 3.49 are typical of data presented by marketing researchers for a type of product, in this case, beer. Each value is an average of the number of six-packs of beer purchased per customer during a month. For the individual brands, the value is the average only for the customers who purchased at least one six-pack of that brand. For example, the value for Miller is the average number of six-packs purchased of *all* of these brands for customers who purchased at least one six-pack of Miller. In contrast, the "Any" average is the average number of six-packs purchased of these brands for all customers in the population.

Is there a paradox in these averages? On first glance, it might appear unusual, or even impossible, that the "Any" average is less than each brand average. Make up your own (small) data set, where you list a number of customers, along with the numbers of six-packs of each brand of beer each customer purchased, and calculate the averages for your data that correspond to those in Figure 3.49. Do you get the same result (that the "Any" average is lower than all of the others)? Are you *guaranteed* to get this result? Does it depend on the amount of brand loyalty in your population, where brand loyalty is greater when customers tend to stick to the same brand, rather than buying multiple brands? Write up your results in a concise report.

	A	B	C	D	E	F	G	H
1	Miller	Budweiser	Coors	Michelob	Heineken	Old Milwaukee	Rolling Rock	Any
2	6.77	6.62	6.64	7.11	7.29	7.30	7.17	4.71

FIGURE 3.49 Average beer purchases

4

Getting the Right Data

Finding Information with Data Mining

The types of data analysis we discuss in this and other chapters of this book are crucial to the success of most companies in today's data-driven business world. However, the sheer volume of available data often defies traditional methods of data analysis. Therefore, a whole new set of methods—and accompanying software—have recently been developed under the name of *data mining*. **Data mining** attempts to discover the patterns, trends, and relationships among data, especially nonobvious and unexpected patterns. For example, the analysis might discover that people who purchase skim milk also tend to purchase whole wheat bread, or that cars built on Mondays before 10 A.M. on production line #5 using parts from suppliers ABC and XYZ have significantly more defects than average. This new knowledge can then be used for more effective management of a business.

© 2002 Harris Welles

A good introductory account of data mining appears in the article by Pass (1997). As he states, the place to start is with a *data warehouse*. Typically, a **data warehouse** is a huge database that is designed specifically to study patterns in data and is *not* the same as the databases companies use for their day-to-day operational activities. A data warehouse should (1) combine data from multiple sources to discover as many interrelationships as possible, (2) contain accurate and consistent data, (3) be structured to enable quick and accurate responses to a variety of queries, and (4) allow follow-up responses to specific, newly relevant questions. In short, a data warehouse represents a relatively new type of database, one that is specifically structured to enable data mining.

Once a data warehouse is in place, analysts can begin to mine the data with a collection of methodologies, techniques, and accompanying software. Some of the primary methodologies are *cluster analysis, linkage analysis, time series analysis,* and *categorization analysis.* Cluster analysis is used to identify associations among data points. For example, data mining software might search through credit card purchases to discover that meals charged on business-issued Gold Cards are typically purchased on weekdays and have an average value of more than $200. Linkage analysis is used to link two or more events together. It attempts to find items that are typically purchased together as part of a "market basket," such as beer and pretzels, yogurt and skim milk, or less obvious pairs. Time series analysis is used to relate events in time. Financial analysts, for example, might try to relate interest rate fluctuations or stock performance to a series of preceding events. Categorization analysis, which contains elements of the preceding three methodologies and is probably the most broadly applicable to different types of business problems, attempts to explain the influence that numerous factors have on one specific outcome. For example, given all information on a loan applicant, categorization analysis might attempt to predict whether the applicant will pay back a loan promptly.

In his article, Pass describes one successful application of data mining at Allders International, a company that operates duty-free outlets throughout Europe. Like many companies, Allders was deluged by paper-based reports and spreadsheets of data. In fact, meaningful information was usually obtained too late to be useful for day-to-day decision making. The introduction of data mining made an immediate impact, both on the bottom line and on employee morale. As one manager stated, "In one store we've been able to move the margin up by four points, by being able to identify why it wasn't performing as well as other outlets. We took out the lower margin lines, even though they might sell well, substituting them or adjusting their positioning." Data mining has enabled Allders to fine-tune its product line by identifying and eliminating the low-performing SKUs (stock keeping units). However, it has also identified apparently unprofitable items that still have an important role in pulling shoppers into the stores. The data warehouse is continually being made available to new users, and existing users expect to find new ways to exploit its power for competitive advantage.

We will not discuss the specific data mining tools mentioned above in this book. However, the methods we will discuss in this chapter are frequently necessary steps in data mining. Before we can mine the data for useful information and insights, we have to be able to get the data into a form suitable for analysis. This is exactly what we will learn how to do here. ●

 # Introduction

We have introduced several numerical and graphical methods for analyzing data statistically in the past two chapters, and we will examine many more statistical methods in later chapters. However, any statistical analysis, whether in Excel or any other software package, presumes that we have the appropriate data. This is a big presumption. Indeed, the majority of the time spent in many real-world statistical projects is devoted to getting the right data in the first place. Unfortunately, this aspect of data analysis is given very little, if any, attention in most statistics textbooks. We believe it is of sufficient importance that we have devoted this entire chapter to methods for getting the data we need. The rest of this book will then present methods for *analyzing* the data.

Our basic assumption throughout most of this chapter is that the appropriate data exist somewhere. In particular, we will not cover methods for collecting data from scratch, using opinion polls, for example. This is a topic in itself and is better left to a specialized textbook in sampling methods. We will assume that data already exist, either in an Excel file, in a database file (such as a Microsoft Access file), or on the Web. In the first case, where the data already reside in an Excel file, we might need to **filter** the data, that is, extract a subset from the entire data set that satisfies specified conditions. For example, we might have customer data on all customers who have ordered from our company in the past year. However, we might want to analyze the subset of these customers who live in the East and have ordered at least three times with a total order amount of at least $500. Therefore, we will examine Excel's built-in capabilities for filtering the data to find only those customers who meet certain conditions. These tools are surprisingly easy to use. Once you know they exist, we expect that you will use them routinely.

Most of the large databases that companies collect are not stored in Excel. Instead, they reside in database packages, such as Microsoft Access, SQL Server, Oracle, and others. These packages are constructed to do certain tasks such as data updating and report writing very well. However, they are not nearly as good as Excel at statistical data analysis—number crunching. Therefore, we will show how to import data from a typical database package into Excel. The key here is to form a *query*, using the Microsoft Query package that ships with Office, that specifies exactly which data we want to import. This package not only presents a friendly user interface for creating the query, but it also finds the appropriate data from the database file and automatically imports it into Excel. Again, the entire process is surprisingly easy, even if you know practically nothing about database packages and database design.

Next, we will briefly examine the possibility of importing data directly from the Web into Excel. Given that the amount of data on the Web is already enormous and is constantly growing, the ability to get it into Excel is valuable. As with importing data from a database file, we import data from the Web by creating a query and then running it in Excel. Unfortunately, the Web is still fairly new, and sophisticated, easy-to-use tools for interfacing between the Web and Excel are just being developed. Nevertheless, we will illustrate that the current possibilities are already powerful and relatively straightforward. If you think that querying from a Web site is something only expert programmers can do, we hope to change your mind.

Often data sets are available on the Web, but the Web queries discussed in the previous paragraph cannot be used to get the data into Excel. In optional Section 4.8, we will illustrate one such situation, where the data from a large government survey are available to download from the Web, but only into *another* statistical package (SAS or SPSS). We do not cover either of these packages in this book, but if you really must obtain these data, you have no choice but to learn some fundamentals about these other statistical packages.

Finally, we cannot always assume that the data we obtain, from the Web or elsewhere, are "clean." There can be (and often are) many instances of "wrong" values—which can occur for all sorts of reasons—and unless we fix these at the beginning, the resulting statistical analysis can be seriously flawed. Therefore, we will conclude this chapter by discussing a few techniques for data cleansing.

4.2 Sources of Data

There are numerous sources of data, more now than ever before. These include sources of *existing* data, as well as methods for creating *new* data. In this section we will discuss these data sources in some generality. In the rest of this chapter, we will examine specific methods for getting the data we need.

We begin by discussing sources of existing data, including (1) data stored in printed form (books, magazines, newspapers, and reports), (2) data stored in spreadsheet files, (3) data stored in database files, such as Access files, and (4) data available from Web sites. Of course, some of these overlap. For example, it is less common today to have a printed version of data that is not stored electronically in some form.

Some of these data sources are easy to obtain, and some require considerable work. Indeed, much of this chapter attempts to unravel the mysteries behind obtaining existing data from various sources. However, we cannot cover all situations, and the burden will often be on you, the analyst, to learn how to obtain data from existing sources. For example, most university libraries have access to online databases. These data are available to all students, but you might need to read some rather obscure manuals (or get help from a reference librarian) to obtain the data in analyzable form.

Some data sources are freely available to everyone (over the Web, say), whereas some contain proprietary company data. Proprietary data are frequently stored by companies in **data warehouses,** huge databases that selected employees can obtain, say, over the company's intranet. These data are often unavailable to nonemployees—at any price. Other data sets are available, often over the Internet, for a fee. We have frequently found Web sites that contain exactly the data we need, only to be asked on the next screen for a credit card! As you will probably discover, some of these data sets are quite expensive.

Even if you find the data you need and are allowed to access it, getting the data into a form suitable for analysis is often a real challenge. Data are stored in a variety of formats, including plain text files (possibly delimited by tabs, commas, spaces, or some other symbol), Excel spreadsheets, relational databases (Access, SQL Server, and others), HTML tables, binary format readable only by specific software packages (SPSS and SAS, for example), and others. It would be nice if all datasets were available in your favorite format (as Excel spreadsheets, say), but the world is not nearly so accommodating. As a data analyst, you will often be forced to learn new skills, including those discussed in this chapter, so that you can obtain the particular data you need.

In addition to the problem of getting data into the appropriate format, there is often a problem of *cleansing* the data. The simple fact is that you cannot always trust the integrity of the data you obtain from external sources. For example, there are often missing values in survey data, where respondents have refused to answer certain questions. To make things worse, these missing values are often not left blank but are instead coded as 9999 or some such value. Suppose you blindly accept these 9999 codes as "real" values and calculate, say, averages and standard deviations. You can only imagine how a few 9999's can affect the results! Data cleansing is tedious, especially for large data sets, but it is an absolute necessity when dealing with externally obtained data. We will discuss some data cleansing techniques in Section 4.9.

More data sets are available today than ever before, partly because of the Web and partly because of the relative ease with which companies can collect customer data (with point-of-sale scanners, for example). However, many companies (and academic researchers) continue to generate *new* data through *surveys* and *controlled experiments.* The techniques for doing this properly—both for design and implementation—take us well beyond the scope of this book. Indeed, entire books have been written about designing and implementing these data collection methods. We will limit our discussion to a brief overview.

4.2.1 Data from Surveys

In today's world, you can hardly exist without being intimately aware of surveys. We hear them discussed on the nightly news almost every day, we read about them in the newspapers, and most of us are asked to take part in them increasingly often through (uninvited) e-mail messages. Simply put, there are many organizations out there that want to know what we think, what products we buy, and what we do with our money. How do they design and implement these surveys?

Survey design is an art in itself. Part of it is common sense—don't ask poorly worded questions, and don't ask questions where the answers could be of no possible use. (We've seen plenty of both.) But phrasing questions in just the right way, and asking the right questions to elicit exactly the information we need is not easy. If you plan to conduct your own survey, we suggest that you read a book on survey design (or get help from a seasoned veteran) and perhaps run a pilot test on a small sample before you launch your survey on a large audience. Just remember that (1) people are reluctant to respond to one additional survey and probably do not consider your study as important as you do, and (2) you usually get only one chance—if your results come back as garbage, you probably won't have a chance to conduct the survey again.

When professional pollsters such as Gallup conduct surveys, whom do they survey? This is an extremely important issue. If the people selected are not "representative," the results of the survey can be biased in one direction or another. Typically, a *random* sample of some type is required. It is usually not sufficient, for example, to survey the first 50 people entering a supermarket on a given day. The rules for choosing random samples can be quite complex. For example, we discuss a large survey on substance abuse in Section 4.8 that was conducted by an agency within the U.S. Department of Health and Human Services. In an abstract to the study, the agency spelled out its sampling technique, about a third of which follows. (The full explanation can be found by following links from www.icpsr.umich.edu/SAMHDA.)

> Multistage area probability sample design involving five selection stages: (1) primary sampling unit (PSU) areas (e.g., counties), (2) subareas within primary areas (e.g., blocks or block groups), (3) listing units within subareas, (4) age domains within sampled listing units, and (5) eligible individuals within sampled age domains. The 1998 NHSDA used the same 115 PSUs selected for the 1995 through 1997 NHSDAs, 6 supplemental PSUs from Arizona and California, and an additional 16 noncertainty PSUs from 13 purposely selected states. The 115 PSUs were selected to represent the nation's total eligible population, including areas of high Hispanic concentration.

This quote illustrates the complexity of sample selection. We will discuss several basic random sampling schemes in Chapter 8, but we will only scratch the surface. To learn more about this topic—to see how the "pros" do it—you need to consult a book on survey sampling such as Levy and Lemeshow (1999).

4.2.2 Data from Controlled Experiments

Controlled experiments represent another popular method of obtaining new data. In a **controlled experiment,** a researcher purposely holds several variables (called *factors*) constant at prescribed levels and then sees how one or more selected variables vary as the experiment is run. For example, a tire manufacturer might run an experiment where selected tread designs are used at selected air pressures and selected outside temperatures. The experiment might be run by driving several cars with each tread design at each combination of pressure and temperature for 10,000 miles and recording the amount of tread deterioration. The objective is to see whether some tread designs perform better than others, and whether the answer to this depends on air pressure and external temperature. We say that the company *controls* for air pressure and temperature by explicitly incorporating them into the experiment.

Controlled experiments have long been used in the natural sciences. For example, we are all aware of medical experiments, involving animals or even human subjects, that attempt to measure the effectiveness of various drugs. More recently, many businesses, particularly those in manufacturing, have become aware of the usefulness of conducting controlled experiments such as the tire experiment mentioned previously. These experiments have frequently led to higher quality in manufactured products and lower manufacturing costs. Companies have learned that they can gain a lot of information about their products or processes through a well-designed experiment. As with survey design, the topic of controlled experiments is too large and complex to be covered in this book. However, we will introduce the topic when we study analysis of variance (ANOVA) later in the book.

4.3 Using Excel's AutoFilter

In the next few sections we will examine several methods for getting data into Excel in a form suitable for analysis. Most of these methods use built-in Excel functionality that many of you are probably unaware of. However, once we illustrate them, you will be surprised how simple they are to use. The first, Excel's autofilter tool that we will discuss in this section, is a little-known gem. It enables us to perform simple queries on an existing Excel database with almost no effort. First, however, we define the term *query* that we will be using throughout this chapter. This is a database term. Given a set of related data—that is, a database—a **query** is a command that asks for a subset of this database that satisfies specified conditions. A typical query on a customer database, for example, might be: Find all of the unmarried female customers over the age of 35 who have purchased at least one major appliance over the past 5 years.

Query	A **query** is an instruction to a database to return a subset of the data that satisfies specified conditions.

In short, a query on a large database produces a smaller database that satisfies certain conditions. Once we obtain this smaller database, we can then use any of the methods from the previous two chapters (or from later chapters) to analyze the data statistically. Before proceeding, we mention that it is customary in database terminology to speak of *fields* and *records* rather than variables and rows. Each **field** of a database corresponds to a variable (or column), and each **record** refers to a row. We will adopt this database terminology throughout most of this chapter. In these terms, a query specifies which fields and records the query should return.

The following example illustrates some simple queries we can perform with Excel's autofilter tool.

Example 4.1 Filtering HyTex's Customer Data

The file CATALOGS.XLS contains the same database that we discussed in Example 3.11 of the previous chapter. It contains data on 1000 customers of HyTex, a direct marketing company, for the current year. For convenience, we repeat the variable—or field—definitions here:

- Age: coded as 1 for 30 or younger, 2 for 31 to 55, 3 for 56 or older
- Gender: coded as 1 for males, 0 for females

- OwnHome: coded as 1 if customer owns a home, 0 otherwise
- Married: coded as 1 if customer is currently married, 0 otherwise
- Close: coded as 1 if customer lives reasonably close to a shopping area that sells similar merchandise, 0 otherwise
- Salary: combined annual salary of customer and spouse (if any)
- Children: number of children living with customer
- History: coded as "NA" if customer had no dealings with the company before this year, 1 if customer was a low-spending customer last year, 2 if medium-spending, 3 if high-spending
- Catalogs: Number of catalogs sent to the customer this year
- AmountSpent: Total amount of purchases made by the customer this year

In the previous chapter we obtained information from this database through a variety of charts and pivot tables. Here, we will see how HyTex can perform simple queries on the data by using Excel's autofilter tool.

Objective To illustrate how Excel's autofilter tool can be used to execute relatively simple queries on the Catalogs database.

Solution

To use the autofilter tool, make sure the cursor is *anywhere* within the database, and select the Data/Filter/AutoFilter menu item. A dropdown arrow immediately appears next to each field name in the database. Clicking on any of these dropdown arrows produces a list similar to the one shown in Figure 4.1. The first three items on the list are always (All), (Top 10...), and (Custom...). The other items are then the distinct values in that field. For example, the list for Children shows that customers in this database all have from 0 to 3 children. In contrast, the list for an essentially continuous variable such as Salary is quite long. It includes each individual salary in the database.

By clicking on any value in any field's dropdown list, we automatically see only the records where that field equals the selected value. All other records are temporarily hidden from view. (They are *not* deleted!) For example, if we click on 2 in the list

	A	B	C	D	E	F	G	H	I	J
1	Direct Marketing Data									
2										
3	A	Genc	OwnHor	Marri	Clo	Sala	Childr	Histc	Catalo	AmountSpe
4	1	0	0	0	1	$16,	(All)	1	12	$218
5	2	0	1	1	0	$108,	(Top 10...)	3	18	$2,632
6	2	1	1	1	1	$97,	(Custom...)	NA	12	$3,048
7	3	1	1	1	1	$26,	0	1	12	$435
8	1	1	0	0	1	$11,	1	NA	6	$106
9	2	0	0	0	1	$42,600	2	2	12	$759
10	2	0	0	0	1	$34,700	0	NA	18	$1,615
11	3	0	1	1	0	$80,000	0	3	6	$1,985
12	2	1	1	0	1	$60,300	0	NA	24	$2,091
13	3	1	1	1	0	$62,300	0	3	24	$2,644
14	2	1	0	1	1	$94,200	1	3	18	$1,211
15	2	1	1	1	0	$73,800	0	3	24	$3,120
16	2	1	1	0	1	$45,900	2	1	12	$416
17	2	1	0	0	0	$52,600	1	NA	18	$1,773
18	2	0	1	1	1	$82,200	1	NA	12	$1,517

FIGURE 4.1 Database After Invoking AutoFilter

	A	Gend	OwnHor	Marri	Clo	Sala	Childr	Histo	Catalo	AmountSpe
1	**Direct Marketing Data**									
3	A ▾	Gend ▾	OwnHor ▾	Marri ▾	Clo ▾	Sala ▾	Childr ▾	Histc ▾	Catalo ▾	AmountSpe ▾
16	2	1	1	0	1	$45,900	2	1	12	$416
19	2	0	1	1	1	$76,700	2	2	6	$534
36	1	0	0	0	0	$13,700	2	NA	24	$414
46	2	1	1	1	1	$107,300	2	2	18	$1,566
55	2	1	1	0	1	$56,700	2	1	18	$536
58	2	1	1	1	0	$96,800	2	3	18	$2,299
62	2	0	1	1	1	$63,300	2	2	24	$679
64	1	0	0	0	1	$11,000	2	NA	18	$393
88	2	0	1	0	1	$43,900	2	1	6	$358
94	1	0	1	0	1	$25,900	2	NA	12	$449
118	2	0	1	0	0	$64,100	2	3	24	$1,595
123	2	1	1	1	1	$83,600	2	2	6	$538
129	2	0	1	1	1	$93,400	2	NA	24	$3,079

FIGURE 4.2 Customers with Two Children

	A	Gend	OwnHor	Marri	Clo	Sala	Childr	Histo	Catalo	AmountSpe
1	**Direct Marketing Data**									
3	A ▾	Gend ▾	OwnHor ▾	Marri ▾	Clo ▾	Sala ▾	Childr ▾	Histc ▾	Catalo ▾	AmountSpe ▾
16	2	1	1	0	1	$45,900	2	1	12	$416
46	2	1	1	1	1	$107,300	2	2	18	$1,566
55	2	1	1	0	1	$56,700	2	1	18	$536
58	2	1	1	1	0	$96,800	2	3	18	$2,299
123	2	1	1	1	1	$83,600	2	2	6	$538
130	2	1	1	1	1	$83,100	2	2	6	$586
134	2	1	0	0	1	$59,100	2	1	18	$580
143	2	1	0	0	1	$55,100	2	NA	6	$826
157	2	1	1	1	1	$70,400	2	NA	12	$1,127
162	2	1	1	1	1	$63,300	2	NA	6	$633
177	2	1	0	1	1	$106,700	2	2	6	$607
188	2	1	1	1	1	$109,900	2	2	12	$870
197	2	1	0	1	1	$88,000	2	2	6	$524
224	2	1	0	1	1	$65,800	2	2	18	$578

FIGURE 4.3 Male Customers with Two Children

in Figure 4.1, only those customers with exactly 2 children are visible. See Figure 4.2. You can tell a query has been performed by looking at the row numbers—a lot are hidden. (Also, in the Excel screen, these row numbers are colored blue, as is the dropdown arrow for Children.) We can now specify a value for another variable, such as 1 (male) for Gender. Then we see only those records for males with exactly 2 children, as in Figure 4.3.[1] To see only the males, we can click on the (All) item of the Children field. To see the entire database, we would then click on the (All) item of the Gender field. Alternatively, to return the database to its original form at any point, we can select the Data/Filter/Show All menu item.

We can perform more complex queries by using the (Custom...) item on any dropdown list. This allows us to enter up to two conditions for any field. These two conditions can be of the "and" or "or" variety, where an "and" type means to return only records satisfying *both* conditions, and an "or" type means to return records satisfying *either* of the conditions. Also, we can choose from a number of types of conditions, such

[1]Because of space limitations, the full results of the queries are not shown in Figures 4.2 and 4.3.

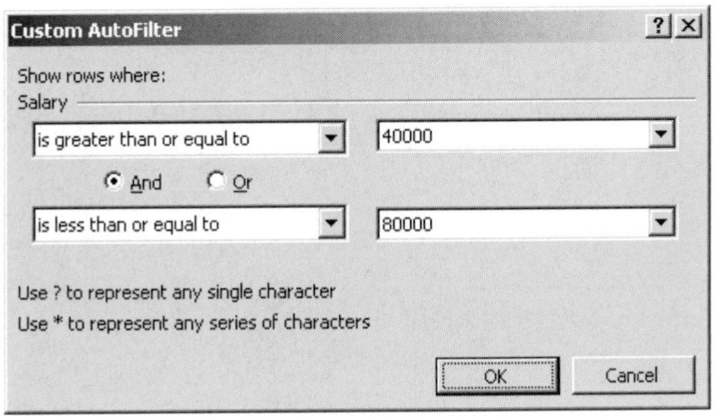

FIGURE 4.4
"And" Conditions
for Salary

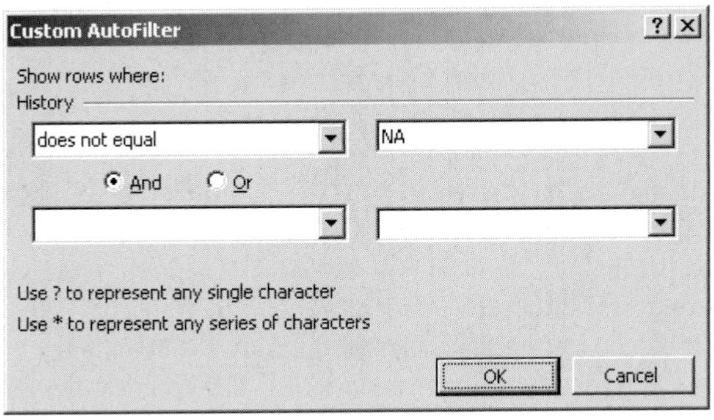

FIGURE 4.5
"Not Equal"
Condition for
History

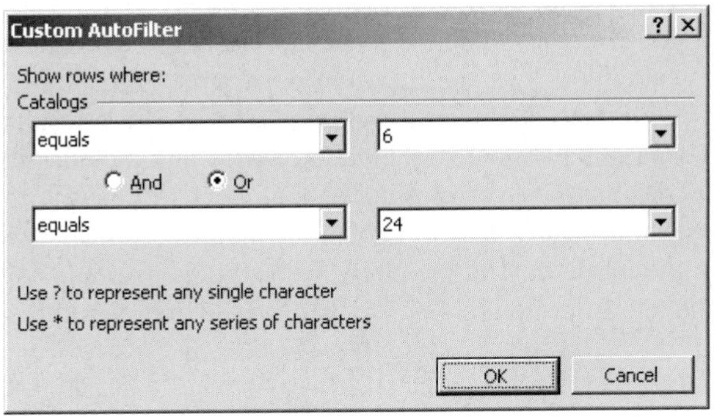

FIGURE 4.6
"Or" Conditions
for Catalogs

as "greater than some value," "contains some value," "does not equal some value," and others. We can also use the "wildcard" characters ? and *, which are especially useful with text data. The character ? stands for any *single* character, so that the condition "equals Bro?n" could return Brown, Broan, and so on. The character * stands for any *series* of characters, so that the condition "equals Sm*" could return Smith, Small, Smithsonian, and so on.

As an example, we performed a query where Salary must be between $40,000 and $80,000, History must not be "NA", and Catalogs can be either 6 or 24. To perform this query, we click on the (Custom...) item in each of these fields' lists and fill out the resulting dialog boxes as in Figures 4.4, 4.5, and 4.6. The first few records obtained appear in Figure 4.7.

Of course, we can mix and match these types of queries. For example, we could click on 1 in the Children list and use the above custom query for Salary to return all families

	A	B	C	D	E	F	G	H	I	J
1	Direct Marketing Data									
2										
3	A ▼	Genc ▼	OwnHor ▼	Marri ▼	Clo ▼	Sala ▼	Childr ▼	Histc ▼	Catalo ▼	AmountSpe ▼
11	3	0	1	1	0	$80,000	0	3	6	$1,985
13	3	1	1	1	0	$62,300	0	3	24	$2,644
15	2	1	1	1	0	$73,800	0	3	24	$3,120
19	2	0	1	1	1	$76,700	2	2	6	$534
20	2	0	1	1	1	$79,400	3	1	6	$200
28	2	1	0	0	1	$50,700	3	1	6	$157
30	2	0	1	0	1	$51,300	0	2	24	$1,424
48	2	1	1	1	0	$74,400	0	3	6	$1,893
62	2	0	1	1	1	$63,300	2	2	24	$679
71	2	1	1	0	1	$63,200	0	3	24	$1,932
72	1	1	0	1	0	$41,600	0	3	24	$1,690
77	2	0	0	1	1	$73,000	1	2	6	$490
87	3	1	1	1	1	$45,900	0	2	24	$1,645

FIGURE 4.7 Results from Custom Query

with 1 child and a salary between $40,000 and $80,000. We can also use the (Top 10. . .) item in any list to good advantage. For example, if we click on this item in the Salary list, we see the dialog box in Figure 4.8. This is actually more general than it appears. In the left dropdown, either Top or Bottom can be chosen; in the middle dropdown, any integer can be chosen; and in the right dropdown, either Items or Percent can be chosen. As it shows here, we will obtain records for the 10 customers with the highest salaries. But if we select the options in Figure 4.9, we will obtain the 5% of customers with the smallest salaries. Because there are 1000 records total, this query will return 50 of them.

Once we obtain the results of a query, we might want to summarize these results in some way. For example, we might want to use StatPro to create a table of summary measures or a scatterplot. This is a bit tricky. If we put the cursor inside the database that is the result of the query and then run StatPro, it will return results for the *entire* database. That is, it senses the temporarily hidden records and includes them in the analysis. The trick is to copy and paste the results of the query to a new location, below the original database (separated by at least one blank row) or on a new sheet, say. Then StatPro can be used in the usual way on the copy. We show an example of this in Figure 4.10. The query is the one from Figure 4.8. We then copied and pasted these ten records (and field

FIGURE 4.8
Query for the Top
10 Salaries

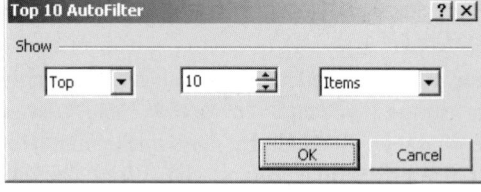

FIGURE 4.9
Query for the Bottom 5% of
Salaries

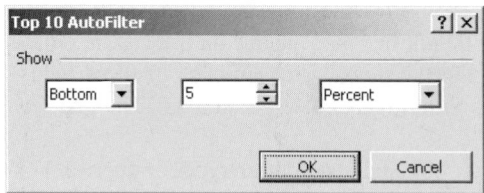

	A	B	C	D	E	F	G	H
	Age	Gender	OwnHome	Married	Close	Salary	Children	History
1005								
1006	3	0	1	1	0	$128,700	0	3
1007	2	0	1	1	1	$130,700	3	NA
1008	2	1	1	1	0	$130,600	0	3
1009	2	0	0	1	1	$135,700	1	3
1010	2	1	1	1	1	$140,700	3	2
1011	3	1	1	1	1	$134,500	0	3
1012	2	1	1	1	1	$127,000	0	NA
1013	2	0	1	1	1	$131,500	3	NA
1014	3	1	1	1	1	$168,800	0	3
1015	2	1	1	1	0	$140,000	0	3
1016								

Summary measures for selected variables

	Salary	AmountSpent
Count	10.000	10.000
Mean	136820.000	2887.880
Median	133000.000	2792.987
Standard deviation	12116.729	1317.779

FIGURE 4.10 StatPro Analysis of Query Results

names) to an area below the original database (separated by a blank row) and used StatPro's Summary procedure to analyze them. (Note that when we copy and paste, the dropdown arrows next to the field names are *not* copied.) ●

The AutoFilter menu item is a "toggle." This means that when it is checked, the dropdown arrows next to the field names appear, and queries can be run. If this menu item is selected again, AutoFilter is disabled, so that the arrows disappear, and the entire database is restored. In addition, even if AutoFilter is enabled (the arrows are visible), the entire database can be restored by selecting the Data/Filter/Show All menu item.

PROBLEMS

Level A

1. The file P04_01.XLS contains a data set that represents 30 responses from a questionnaire concerning the president's environmental policies. Each observation lists the person's age, gender, state of residence, number of children, annual salary, and opinion of the president's environmental policies.
   a. Use Excel's autofilter tool to identify all respondents who are female, middle-age, and who have two children. What is the average salary of these respondents?
   b. Use Excel's autofilter tool to identify all respondents who are elderly and who strongly disagree with the president's environmental policies. What is the average salary of these respondents?
   c. Use Excel's autofilter tool to identify all respondents who strongly agree with the president's environmental policies. What proportion of these individuals are young?

2. A human resources manager at Beta Technologies, Inc., has collected current annual salary figures and related data for 52 of the company's full-time employees. The data are in the file P04_02.XLS.
   a. Use Excel's autofilter tool to identify all employees who are male and who have exactly 4 years of postsecondary education. What is the average salary of these employees?
   b. Find the average salary of all *female* employees who have exactly 4 years of postsecondary education. How does this mean salary compare to the one obtained in part **a**?
   c. Use Excel's autofilter tool to identify all employees who have more than 4 years of postsecondary education. What proportion of these employees are male?

3. Five hundred households in a middle-class neighborhood were recently surveyed as part of an economic development study conducted by the local government.

The data are in the file P04_03.XLS. Use Excel's autofilter tool to answer the following questions:

a. What are the average monthly home mortgage payment, average monthly utility bill, and average total debt (excluding the home mortgage) of all homeowners residing in the southeast sector of the city?

b. What are the average monthly home mortgage payment, average monthly utility bill, and average total debt (excluding the home mortgage) of all homeowners residing in the northwest sector of the city? How do these results compare to those found in part **a**?

c. What is the average annual income of the first household wage earners who rent their home (i.e., house or apartment)? How does this statistic compare to the average annual income of the first household wage earners who own their home?

d. What proportion of households surveyed contain a single person who owns his or her home?

4. The file P04_04.XLS contains information on 66 movie stars. In particular, the data set contains the name of each actor and the following four variables: gender, domestic gross (average domestic gross of the star's last few movies, in millions of dollars), foreign gross (average foreign gross of the star's last few movies, in millions of dollars), and salary (current amount the star asks for a movie, in millions of dollars). Use Excel's autofilter tool to answer the following questions:

a. Identify all stars whose average domestic gross exceeds 75 million dollars and whose average foreign gross exceeds 75 million dollars. Find the average salary of these stars. What proportion of these stars are men?

b. Identify all stars whose average domestic gross is between 50 and 75 million dollars (inclusive) and whose average foreign gross is between 50 and 75 million dollars (inclusive). Find the average salary of these stars. What proportion of these stars are women?

c. Identify all stars whose average domestic gross is less than 50 million dollars and whose average foreign gross is less than 50 million dollars. Find the average salary of these stars. What proportion of these stars are men?

4.4 Complex Queries with the Advanced Filter

The autofilter tool is useful for quick and simple queries, but it limits us to "and" queries across fields. That is, it limits us to queries of the form: Find all records where Field 1 satisfies certain conditions *and* Field 2 satisfies certain conditions *and* Field 3 satisfies certain conditions, and so on. It cannot handle the following query: Find all customers who are either male with salary above $40,000 or female with at least 2 children. Here there are two "and" queries (male *and* salary above $40,000, and female *and* at least 2 children), but they are combined with an "or." To perform a more complex query of this type, we need Excel's advanced filter tool, found under the Data/Filter/Advanced Filter menu item. However, we pay a price for being able to perform more complex queries. The price is that we must first construct a "criteria range." This criteria range essentially spells out the query. It consists of a top row of field names and one or more rows of conditions. Each row of conditions becomes an "or" part of the query. For example, the first row (right below the field names) might indicate that we want males with salary above $40,000, and the second row might indicate that we want females with at least 2 children. Then the query will return all records that match the conditions in *either* (or both) of these rows. There is no limit to the number of rows—sets of conditions—we can put in the criteria range, although it is typically a small number such as 1 or 2.

When we use the advanced filter tool, we must specify the original data range, the criteria range, and (optionally) a range where the results of the query will be placed. Unlike autofilter, the query is not automatically done "in-place," where the original database is replaced by the results of the query by hiding some of the rows. This is still an option, but we can also request that the results of the query be placed in any range we select. We demonstrate the procedure in the following continuation of Example 4.1.

Example (4.1) Filtering HyTex's Customer Data (continued)

The Hytex Company would now like to perform more advanced queries on the data in the CATALOGS.XLS file by using Excel's advanced filter tool. How might it proceed?

Objective To illustrate how Excel's advanced filter tool can be used to execute more complex "or" queries on the Catalogs database.

Solution

We begin by copying the row of field names to any unused area of the Data sheet. This might be right above the database (by first inserting some blank rows) or just to the right of the database. We chose the latter. This row becomes the top row of the criteria range. Then we (manually) enter conditions in the cells just below these field names. The key is that the conditions in a given row are "and" conditions, whereas conditions across rows are treated in an "or" manner, as described above. An example appears in Figure 4.11. The first row specifies that we want all customers who are married *and* have salary at least $80,000 *and* have at least 2 children. The second row specifies that we want all customers who have salary $100,000 *and* received at least 12 catalogs. Using the range L3:U5 as the criteria range, the query will return records that match the conditions in either (or both) of rows 4 and 5.

Here are several example customers and an indication of whether they will be included in the query results:

❶ Married, salary $85,000, 3 children, received 6 catalogs: included (satisfies conditions in row 4 but not row 5)

❷ Married, salary $105,000, 2 children, received 18 catalogs: included (satisfies conditions in row 4 and row 5)

❸ Unmarried, salary $120,000, 1 child, received 6 catalogs: not included (does not satisfy the conditions in either row)

❹ Married, salary $120,000, 1 child, received 18 catalogs: included (satisfies the conditions in row 5 but not row 4)

Once the criteria range is created, we run the query by using the Data/Filter/ Advanced Filter menu item. This brings up a dialog box, which we fill in as shown in Figure 4.12 (page 142). If we select the top option (Filter the list, in-place), the query acts just like an autofilter query, so that records that do not match the conditions are temporarily hidden from view. We favor the second option, which places the query results in a separate output range and keeps the original database intact. Two other points are worth mentioning. First, we need specify only the top left cell of the output range, here, cell W3. Indeed, it would be difficult to specify the entire output range because we do not know how many records will match the query conditions! Second, if we check the box at the bottom, then any customers who meet the query conditions and are identical on all fields are listed only once in the output range. This is sometimes appropriate when

FIGURE 4.11
Criteria Range with Two Sets of Conditions

	L	M	N	O	P	Q	R	S	T	U
1	Criteria range (starts in row 3)									
2										
3	Age	Gender	OwnHome	Married	Close	Salary	Children	History	Catalogs	AmountSpent
4				1		>=80000	>=2			
5						>=100000			>=12	

FIGURE 4.12
Dialog Box for
Advanced Filter

If range names are
given to the data
and criteria ranges,
they can be used
instead of cell
addresses in the
dialog box.

FIGURE 4.13
Selected Query
Results in the
Output Range

	W	X	Y	Z	AA	AB	AC	AD	AE	AF
1	Query results									
2										
3	Age	Gender	OwnHome	Married	Close	Salary	Children	History	Catalogs	AmountSpent
4	2	0	1	1	0	$108,100	3	3	18	$2,632
5	2	1	1	1	1	$95,800	3	1	12	$678
6	2	1	1	1	1	$107,300	2	2	18	$1,566
7	2	1	1	1	0	$90,700	3	3	24	$2,265
8	2	0	1	1	1	$81,700	3	1	24	$879
9	2	1	1	1	0	$96,800	2	3	18	$2,299
10	2	1	0	1	1	$117,700	1	3	12	$2,104
11	3	0	1	1	1	$118,000	3	1	12	$581
12	3	0	1	1	0	$110,000	0	3	24	$5,564
13	3	1	1	1	1	$124,900	0	3	18	$4,109
14	3	0	1	1	0	$128,700	0	3	24	$3,995
15	2	0	0	1	1	$118,300	1	NA	18	$2,681
16	2	1	1	1	0	$120,000	3	3	18	$1,926
17	2	1	1	1	1	$83,600	2	2	6	$538
18	2	0	1	1	1	$130,700	3	NA	6	$2,070
19	2	0	1	1	1	$93,400	2	NA	24	$3,079
20	2	1	1	1	1	$83,100	2	2	6	$586
21	2	0	1	1	1	$102,500	3	1	12	$548
22	2	0	1	1	1	$85,500	2	2	18	$895

we want to avoid duplicate records in the query results. Some of the results of this particular query appear in Figure 4.13.

As another example, suppose we want the customers who are either (1) male with salary between $40,000 and $50,000, or (2) female with salary over $70,000. The problem here is that condition (1) includes an "and" condition (greater than $40,000 *and* less than $50,000) in the same field, Salary. How should this condition be entered in the criteria range? It is tempting to enter the label ">40000,<50000" in a cell under Salary, where we include both conditions, separated by a comma. However, this doesn't work! (We tried to see whether it would work, or what *would* work, in Excel's online help, but unfortunately there is very little we could find about specifying conditions for queries.) One solution—maybe you can find another—is to enter *two* Salary fields in the criteria range, as shown in Figure 4.14. There is no rule that every field name must be included in the criteria range. Only those names involved in the query are required. In addition, the same field name can be included more than once, evidently to deal with the situation we have posed. The criteria range, as set up in Figure 4.14, will return exactly the records we seek.

It is even possible to base the criteria on a formula. This is called a **computed query.** For example, suppose we want to locate all customers with salary at least $1000 greater

FIGURE 4.14
An "And"
Condition in
the Salary Field

	L	M	N
3	Gender	Salary	Salary
4	1	>40000	<50000
5	0	>60000	

FIGURE 4.15
A Computed
Query

	L
3	HighSalary
4	FALSE
5	
6	
7	MedianSalary+1000
8	$54,700

than the median salary for all customers. Then in the criteria range, we can enter the formula

$$=F4>(MEDIAN(\$F\$4:\$F\$1003)+1000) \tag{4.1}$$

under any field name such as HighSalary. (We use F4 in this formula because it is the *first* cell with data in the Salary column.) Better yet, we can calculate the median salary plus $1000 in some unused cell, cell L8, say, and replace formula (4.1) with the formula

$$=F4>\$L\$8 \tag{4.2}$$

(Note that the expression to the right of the first equals sign in either formula (4.1) or formula (4.2) is a *condition*. Therefore, the result of either formula will be TRUE or FALSE.) The setup for this is shown in Figure 4.15. Cell L4 contains formula (4.2), cell L8 contains the formula

$$=MEDIAN(F4:F1003)+1000$$

and the criteria range is L3:L4. The resulting query returns 482 records (slightly less than 50% of all records)—exactly those with a salary greater than $54,700.

Guidelines for computed queries

- The column heading above a computed criterion must *not* be the same as a field name in the database. This is why we used the name HighSalary, not Salary, in the criteria range.

- References to cells outside the database range should be *absolute*. This is why we put dollar signs around L8 in formula (4.2).

- References to cells inside the database range should be *relative*. This is why we made the leftmost F4 in formulas (4.1) and (4.2) relative. However, there is an exception to this rule, as shown in formula (4.1), where we made the range F4:F1003 absolute.

Once you understand the underlying logic, these last two rules—and the exception—make sense. As we see in Figure 4.15, the first salary, the one in cell F4, does *not* meet the criterion. This is why we see FALSE in cell L4. However, when we run the query, Excel recognizes that the cell reference F4 in formula (4.2) is relative. Therefore, it substitutes *each* salary (first the one in cell F4, then the one in cell F5, and so on) into the formula in cell L4 to check whether it meets the condition. This is why we want the left side of the inequality to be relative. However, because the median salary plus 1000 should remain fixed, we want the right side to be absolute. Finally, the reason we prefer formula (4.2) to formula (4.1) is that it is much faster. When we use formula (4.1), the median plus 1000 must be calculated 1000 times, once for each record in the database. When we use formula (4.2), the median plus 1000 must be calculated only once. ●

4.4.1 Tips for Forming Criteria

As this example has illustrated, Excel's advanced filter tool is very useful and relatively easy to use, provided that we know how to enter the conditions in the criteria range correctly. Unfortunately, this is not discussed in much detail in any online help we have been able to find. Here are some tips that might come in handy.

- For text fields like last names or cities, entering a single letter such as M will return any text that starts with that letter. Similarly, entering any sequence of letters such as Mon will return any text that starts with this sequence—Monday, Montana, and so on. In addition, it is *not* case sensitive. We could enter Mon or mon with exactly the same results.

- A *formula* of the form ="=Smith" can be entered under a text field. This returns all records that match Smith exactly. Why might we do this? The reason is that if we enter only the name Smith (as a label, not a formula), it will return any name that starts with Smith, such as Smithsonian. So for an exact match, it is best to use a formula.

- To specify a "not equal" condition, use the characters < > (less than followed by greater than), as in <>10.

- Wildcards are permitted, exactly as with Autofilter. The character ? stands for any single character, and * stands for any series of characters.

- Be careful of putting the criteria range just to the right of the database range (as we did). If you then run the advanced filter with the default option of showing the results in-place (hiding the rows that don't match the criteria), the rows of your criteria range might be hidden as well! Many experts insist on putting the criteria range directly *above* the database range (with at least one blank row between them).

- Remember that only the fields involved in the conditions need to be entered in the criteria range. Also, as we saw in the example, the same field name can be entered more than once.

- Because it is so important, we state once more how Excel decides which records to return. For each row in the criteria range, Excel finds all records that match *all* of the conditions in that row. Then, if a record is a match for *any* of the rows in the criteria range, it is returned in the query results.

4.4.2 Database Functions

We have already worked with Excel's summary functions, including COUNT, COUNTA, SUM, AVERAGE, and STDEV.[2] There are similar functions for summarizing results from a database query. They all begin with the letter D, as in DCOUNT, DCOUNTA, DSUM, DAVERAGE, and DSTDEV. Now that we have discussed criteria ranges, these database functions are easy to describe. They all take three arguments, as in

$$=\text{Dfunction}(\textit{database range,field name,criteria range})$$

Here, Dfunction is any of the database functions, such as DAVERAGE; *database range* is the range of the database, including the field labels at the top; *field name* is the name of a field we want to summarize, enclosed in double quotes; and *criteria range* is the criteria range, exactly as we discussed earlier in this section. We illustrate these functions in the following continuation of Example 4.1.

---

[2]Remember that COUNT returns the number of *numerical* values in a range, whereas COUNTA returns the number of *all nonblank* cells in a range.

Example (4.1) **Filtering HyTex's Customer Data (continued)**

For HyTex's database of 1000 customers, we would like to calculate summary measures regarding the amount spent for all customers who are male, have a salary above $50,000, had a previous history with HyTex, and received at least 18 catalogs, or are female, have a salary above $60,000, had a previous history with HyTex, and received at least 12 catalogs.

Objective To illustrate Excel's database summary functions on the Catalogs database.

Solution

The solution appears in Figure 4.16. The criteria range is formed in the usual way. It includes two criteria rows because of the "or" condition in the statement of the problem. For convenience, we gave range names Data and Criteria to the entire database and the criteria range, respectively. Then we entered the database functions for count, sum, average, and standard deviation in cells M9–M12 (and spelled them out as labels to the right for your convenience). There are two things to note. First, we must enter the field name AmountSpent inside double quotes. Second, the query itself does not need to be performed explicitly. That is, we do not need to use the Data/Filter/Advanced Filter menu item as a first step. The database functions perform the query implicitly and report only the summary results. Alternatively, we could perform the query explicitly with the advanced filter tool and then use the *usual* Excel functions COUNT, SUM, AVERAGE, and STDEV on the results of the query. The summary results would be identical.

FIGURE 4.16
Excel's Database Functions

	L	M	N	O	P	Q	R
1	Criteria range (range named Criteria)						
2							
3	Gender	Salary	History	Catalogs			
4	1	>50000	<>NA	>=18			
5	0	>60000	<>NA	>=12			
6							
7	Database functions (all summarizing AmountSpent)						
8							
9	Count		247	=DCOUNT(Database,"AmountSpent",Criteria)			
10	Sum		504042.5	=DSUM(Database,"AmountSpent",Criteria)			
11	Average		2040.658	=DAVERAGE(Database,"AmountSpent",Criteria)			
12	Stdev		1067.994	=DSTDEV(Database,"AmountSpent",Criteria)			

PROBLEMS

Level A

5. Recall that the file P04_01.XLS contains a data set that represents 30 responses from a questionnaire concerning the president's environmental policies. Each observation lists the person's age, gender, state of residence, number of children, annual salary, and opinion of the president's environmental policies.

a. Find all respondents who are either (1) middle-age men with at least one child and an annual salary of at least $50,000, or (2) middle-age women with two or fewer children and an annual salary of at least $30,000.

b. Find the mean and median salaries of the respondents who meet the conditions specified above.

c. What proportion of the respondents who satisfy the conditions specified above agree or strongly agree with the president's environmental policies?

6. Recall that the file P04_04.XLS contains information on 66 movie stars. In particular, the data set contains the name of each actor and the following four variables: gender, domestic gross (average domestic gross of the star's last few movies, in millions of dollars), foreign gross (average foreign gross of the star's last few movies, in millions of dollars), and salary (current amount the star asks for a movie, in millions of dollars).

   a. Find all movie stars who are either (1) females with domestic gross between 40 and 80 million dollars (inclusive) and foreign gross between 40 and 80 million dollars (inclusive), or (2) males with domestic gross between 50 and 90 million dollars (inclusive) and foreign gross between 50 and 90 million dollars (inclusive).

   b. Find the mean and median salaries of the movie stars who meet the conditions specified in part **a.**

   c. What proportion of the stars identified in part **a** earn salaries in excess of 10 million dollars per movie?

7. A human resources manager at Beta Technologies, Inc., has collected current annual salary figures and related data for 52 of the company's full-time employees. The data are in the file P04_02.XLS.

   a. Identify all full-time employees who are either (1) females between the ages of 30 and 50 (inclusive) who have at least 5 years of prior work experience, at least 10 years of prior work experience at Beta, and at least 4 years of post-secondary education; or (2) males between the ages of 40 and 60 (inclusive) who have at least 6 years of prior work experience, at least 12 years of prior work experience at Beta, and at least 4 years of post-secondary education.

   b. For those employees who meet the conditions specified in part **a,** compare the mean salary of the females with that of the males. Also, compare the median salary of the female employees with that of the male employees.

   c. What proportion of the full-time employees identified in part **a** earn less than $50,000 per year?

8. Five hundred households in a middle-class neighborhood were recently surveyed as part of an economic development study conducted by the local government. The data are in the file P04_03.XLS. Identify all of the households in the given data set that satisfy each of the following conditions:

   a. The household owns their home and their monthly home mortgage payment is in the top quartile of the monthly payments for all households.

   b. The household's typical monthly expenditure on utilities is within 2 standard deviations of the mean monthly expenditure on utilities for all households.

   c. The household's total indebtedness (excluding home mortgage) is less than 10% of the household's primary annual income level.

 # Importing External Data from Access

To this point, we have worked only with databases that already exist in Excel. Often, however, the data we need to analyze reside in an external source. In this section we will discuss the situation where the data were created in a database package. Specifically, we will consider data in Microsoft Access format. (This is the database package that is bundled with Microsoft Office.) Database packages such as Access, FoxPro, Oracle, and many others are extremely complex and powerful packages, and for database creation, querying, manipulation, and reporting, they have many advantages over spreadsheets. However, they are not nearly as powerful as spreadsheets for statistical analysis. Therefore, it is often necessary to import data from a database package—either all of it or just a subset of it, based on a query—into Excel, where we can then perform the statistical analysis. Fortunately, Microsoft has included a software package called Microsoft Query in its Office suite that makes the importing relatively easy. We will describe the process in this section.

4.5.1 A Brief Introduction to Relational Databases

First, we present some general concepts about database structure. The Excel "databases" we have discussed so far in this book are often called **flat files** or, more simply, *lists.* They are also called *single-table* databases, where *table* is the usual database terminology for a rectangular range of data, with columns corresponding to fields and rows corresponding to records.[3] For example, the data in the file CATALOGS.XLS that we used in Example 4.1 reside in a single table. This table consists of 10 fields and 1000 records arranged in a rectangular range. Flat files are fine for relatively simple database applications, but they are not powerful enough for more complex applications. For the latter we need *relational databases.* A **relational database** is a related set of tables, where each table is a rectangular arrangement of fields and records, and the tables are linked explicitly.

As a simple example, suppose you would like to keep track of information on all of the books you own. Specifically, you would like to keep track of data on each book (title, author, copyright date, whether you have read it, when you bought it, and so on), as well as data on each author (name, birthdate, awards won, number of books written, and so on). Now suppose you store *all* of these data in a flat file. Then if you own 10 books by Danielle Steele, say, you must fill in the identical personal information on Ms. Steele for *each* of the 10 records associated with her books. This is not only a waste of time, but it increases the chance of introducing errors as you enter the same information over and over.

A better solution is to create a Books table and an Authors table. In the Books table, each record would contain the data, including author name, for a particular book. It might also include an AuthorIndex field, where a unique number is associated with each author. Danielle Steele might have index 001, John Grisham might have index 002, and so on. The Authors table would have a *single* record for each author, and it would include the same AuthorIndex field. In this way, personal data on Danielle Steele would be entered only once. Similarly, for maintenance purposes, if any of her personal data changed, it would need to be updated in only one place: in her record of the Authors table.

The linked fields are called *keys.* Specifically, the AuthorIndex field in the Authors table is called a *primary key,* and the AuthorIndex field in the Books table is called a *foreign key.* A primary key must contain *unique* values, whereas a foreign key can contain duplicate values.

The key to relating these two tables is the AuthorIndex field. In a database package such as Access, we explicitly draw a link between the AuthorIndex fields in the two tables.[4] This link allows a user to find data from the two tables easily. For example, suppose you see in the Authors table that John Updike's index is 035. Then you can search through the Books table for all records with AuthorIndex 035. These correspond to the books you own by John Updike. Going the other way, if you see in the Books table that you own *The World According to Garp* by John Irving, who happens to have AuthorIndex 021, you can look up the (unique) record in the Authors table with AuthorIndex 021 to find personal information about John Irving.

The theory and implementation of relational databases is both lengthy and complex. Indeed, many books have been written about the topic. However, this brief introduction is probably enough for our purposes. As we will see in examples, an Access database file (recognizable by the .MDB extension) typically contains several related tables. They are related in the same basic way as the Books and Authors were related in the previous paragraph—through links of certain fields. These links will be apparent when we use Microsoft Query to import data from Access into Excel. Just keep in mind that we will not actually create Access databases. This would take us too far afield, given the goals of this book. In fact, we do not even require you to own Access. We will simply assume that (1) an Access database exists, (2) we know the type of data it contains, and (3) we want to query it for information that we can import into Excel for eventual statistical analysis.

---

[3]We have found the term *list* used in most of the how-to Excel books on the market. Database experts tend to reserve the term *database* for the *relational,* or *multitable,* databases we will discuss in this section.
[4]They do not actually have to have the same field name, such as AuthorIndex, but the indexes must match. For example, if 001 is Danielle Steele's index in one table, it must be her index in the other table.

4.5.2 Using Microsoft Query to Import Data from a Database Package

The Microsoft Query package allows us to import all or part of the data from many database packages into Excel—with very little work. You probably do not know you own this package. For example, if you click on the Windows Start button and then choose Programs, you will not find Microsoft Query on the list. However, it comes with Office, and you can use it. The only question is whether you installed it when you installed Office. To check, open a blank spreadsheet in Excel and select the Data/Get External Data menu item. (Microsoft has changed this menu item to Data/New Database Query in Excel 2002. We will use the Excel 2000 name.) If the Run Database Query is grayed out, then Microsoft Query was not installed. You will have to go through the Add/Remove part of the Office Setup program (with your Office CD-ROM) to install it. (You'll find it under the Data Access group.)

Once Microsoft Query is installed, importing data from Access (or any other supported database package) is essentially a three-step process:

1. Define the source, so that Excel knows what type of database the data are in.
2. Use Microsoft Query to define a query.
3. Return the data to Excel.

We illustrate these three steps in the following example.

Example 4.2 Fine Shirt Company's Relational Data

The Fine Shirt Company creates and sells shirts to its customers. These customers are retailers who sell the shirts to consumers. The company has created an Access database file SHIRTORDERS.MDB that has information on sales to its customers during the period (1995–1999). There are three related tables in this database: Customers, Orders, and Products. The Customers table has the following information on the company's 7 customers:

- CustomerID (an index for the customer, from 1 to 7)
- Name
- Street
- City
- State
- Zip
- Phone

The Products table has the following information on the company's 10 products (types of shirts):

- ProductID (an index for the product, from 1 to 10)
- Description
- Gender (whether the product is made for females, males, or both)
- UnitPrice (the price to the retailer)

Finally, the bulk of the data are in the Orders table. This table has a record for each product ordered by each customer on each date during the 5-year period. There are 2245

FIGURE 4.17
Relational
diagram

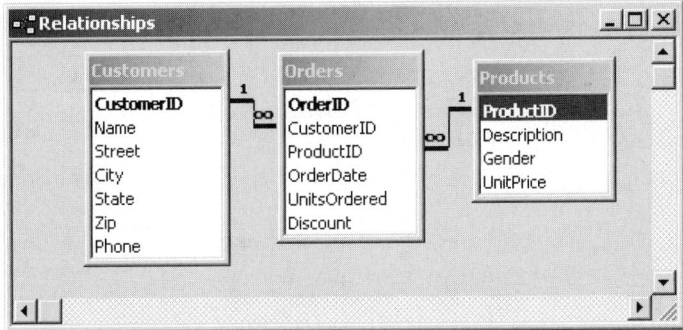

records in this table. If a customer ordered more than one product on a particular date, there is a separate record for each product ordered. The fields in the Orders table are:

- OrderID (an index for the order, from 1 to 2245)
- CustomerID (to link to the Customers table)
- ProductID (to link to the Products table)
- OrderDate
- UnitsOrdered (number of shirts of this type ordered)
- Discount (percentage discount, if any, for this order)

The Access file has a link between the CustomerID fields in the Customers and Orders tables, and a link between the ProductID fields in the Products and Orders tables. This way, the detailed information on customers and products must be entered only once. If we need any of this information for a particular order, we can find it through the links. For example, if a particular order shows that CustomerID and ProductID are 2 and 7, we can look up information about customer 2 and product 7 in the Customers and Products tables.

Access allows us to diagram the relationships between tables, as shown in Figure 4.17. This diagram clearly shows the links involving the CustomerID and ProductID fields. The 1 and ∞ signs on the links imply "many-to-one" relationships. Specifically, a given customer is included only once in the Customers table, but this same customer can be responsible for many orders in the Orders table. Similarly, a given product is included only once in the Products table, but it can be included in many orders in the Orders table.

The company wants to perform a statistical analysis of the data on orders within Excel. How can it use Microsoft Query to import the data from Access into Excel?

Objective To illustrate how Microsoft Query can be used to return the results of queries on the ShirtOrders database back into Excel.

Solution

Before going into the details, it is important to realize that the entire procedure is done within Excel and Microsoft Query, *not* Access. You need not even own Access to make the procedure work. All you need is the Access database file, in this case SHIRTORDERS.MDB.[5]

---

[5]We also note that this procedure can be done in the same way with databases from other database packages, such as FoxPro. We just decided to illustrate the procedure with Access.

FIGURE 4.18
Choose Data
Source Dialog Box

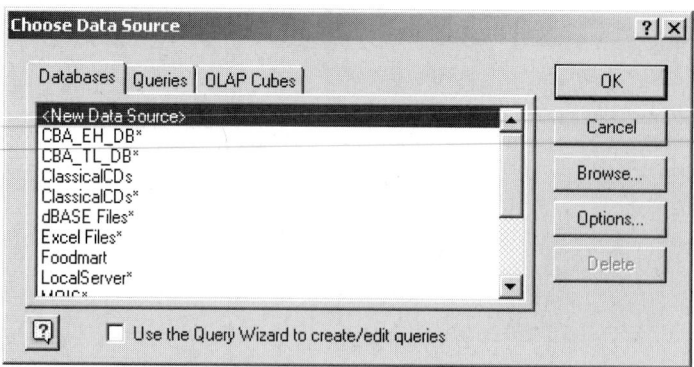

FIGURE 4.19
Create New Data
Source Dialog Box

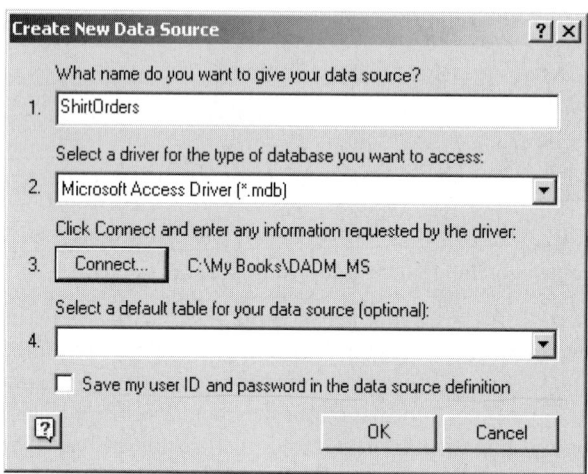

The first step of the procedure is to tell Excel what type of data you have. In its terminology, you must define a "data source." To do so, open a blank spreadsheet in Excel and select the Data/Get External Data/New Database Query menu item. This takes you to the Choose Data Source dialog box in Figure 4.18. Note that the list you see might not be the same as the one shown here. Each time you tell Excel about a new data source, it is added to the list shown. In any case, we want to add a new data source, so make sure the top item is highlighted. Also, make sure the bottom box is *not* checked. (We prefer *not* to use the Query Wizard, although you can experiment with it if you like.) Then click on OK.

This takes you to the Create New Data Source dialog. It should eventually be filled in as shown in Figure 4.19. Actually, there are three steps to filling it in. First, enter a descriptive title in line 1. (This does *not* need to be the same name as the Access file name.) Next, use the dropdown list in line 2 to select the appropriate driver, in this case the Microsoft Access Driver. (This is where you could specify another database package, such as FoxPro.) Finally, click on the Connect button in line 3 to bring up the ODBC Microsoft Access Setup dialog box in Figure 4.20, where you indicate which database *file* you want to use. To choose it, click on its Select button and browse for the SHIRTORDERS.MDB file. (Your file will almost certainly be in a different location than ours.) Once you have located this file, click on OK a couple of times to see the completed Create New Data Source dialog box, and click on OK once more to get back to the Choose Data Source dialog box, with your data source, ShirtOrders, now on the list. (See Figure 4.21.)

FIGURE 4.20
Dialog Box for Se-
lecting the Appro-
priate Database
File

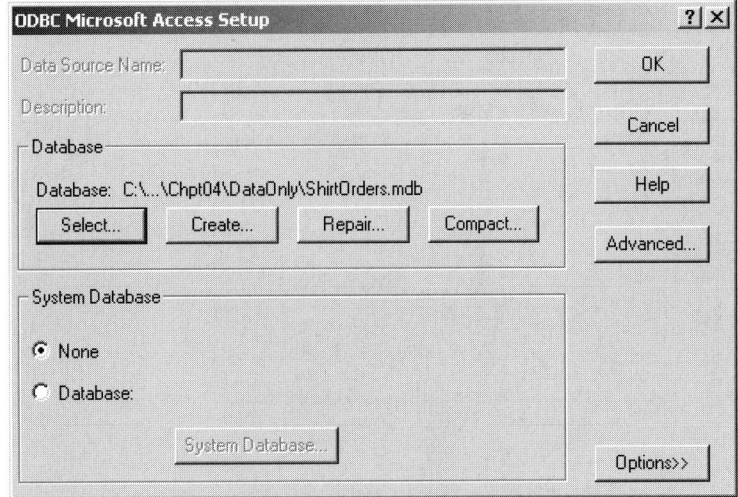

FIGURE 4.21
Choose Data
Source Dialog Box
with the New
Entry

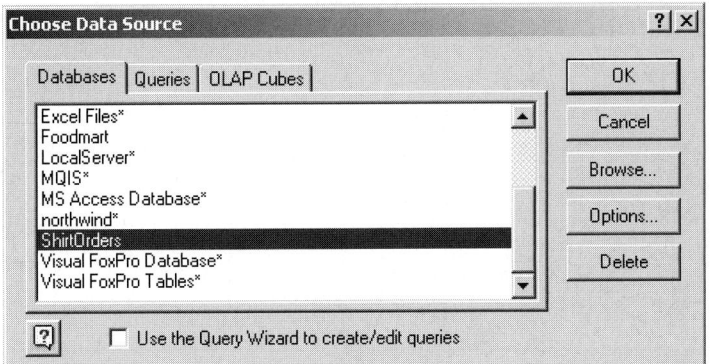

This completes step 1 of the overall procedure. You have defined a data source that you can now query. It is important to realize that once you have created this ShirtOrders source, you will not have to create it again. Specifically, if you want to run another query on this database at a later time, you can select the Tools/Get External Data/New Database Query menu item in Excel, select the ShirtOrders source from the list, and proceed directly to the query itself, bypassing the steps described above.

At this point, you should be looking at the Choose Data Source dialog box in Figure 4.21—with ShirtOrders on the list. Make sure the ShirtOrders item is selected and the bottom checkbox is *unchecked,* and click on OK.[6] This brings up the Add Tables dialog box shown in Figure 4.22 (page 152), in front of the Microsoft Query screen in Figure 4.23. This begins the second step of the overall procedure, where the query is defined. Essentially, we need to specify which tables are relevant for the query, which fields we want to return to Excel, and which records meet the criteria we spell out.

To get started, let's try a relatively easy single-table query. We will find all of the records from the Orders table where the order date is during the years 1997 or 1998, the product number is 3 or 5, and the number of units ordered is at least 100, and we will

---

[6]If the bottom checkbox is checked, the Query Wizard will be launched when you click on OK. You can try this if you like, but we find it confusing and less useful than the method we describe here.

FIGURE 4.22
Add Tables
Dialog Box

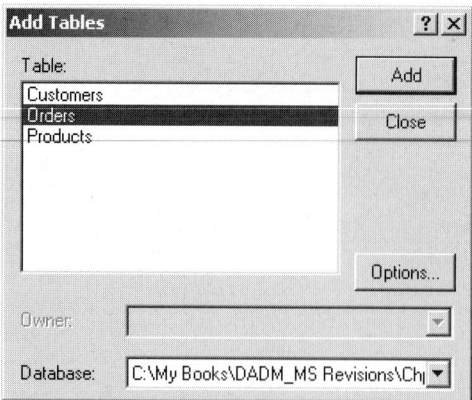

FIGURE 4.23
Microsoft Query
Screen

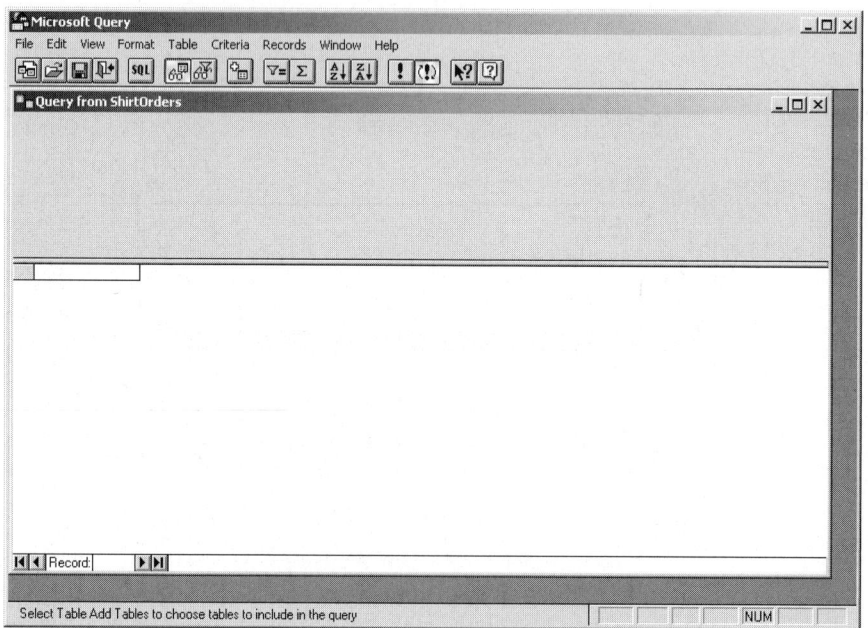

return to Excel all fields in the Orders table for these records. First, if the Add Tables dialog box is still showing, select the Orders table, click on Add, and then click on Close. (If the Add Tables dialog box is not showing, select the Table/Add Tables menu item to make it show.) The table appears in the top pane of the screen. (See Figure 4.24). You can double-click on any of the fields in this table to add fields that will be returned by the query. If you double-click on the top item (the asterisk), all fields will be returned. For this query, double-click on the asterisk, and you should see a sampling of the data that will be returned in the bottom pane of the screen. Finally, click on the Show/Hide Criteria button on the toolbar (the button with the glasses and the funnel). This opens a middle pane on the screen, where you can enter criteria. The screen should now appear as in Figure 4.24.

Now you enter the criteria for the query. Essentially, you fill in the middle pane of the Query screen like you filled in the criteria range in Excel for the advanced filter tool. Any conditions in a given row are "and" conditions, whereas those across rows are treated as "or" conditions. You can either type the conditions directly into the small "spreadsheet"

FIGURE 4.24
Query Screen
Before Entering
Criteria

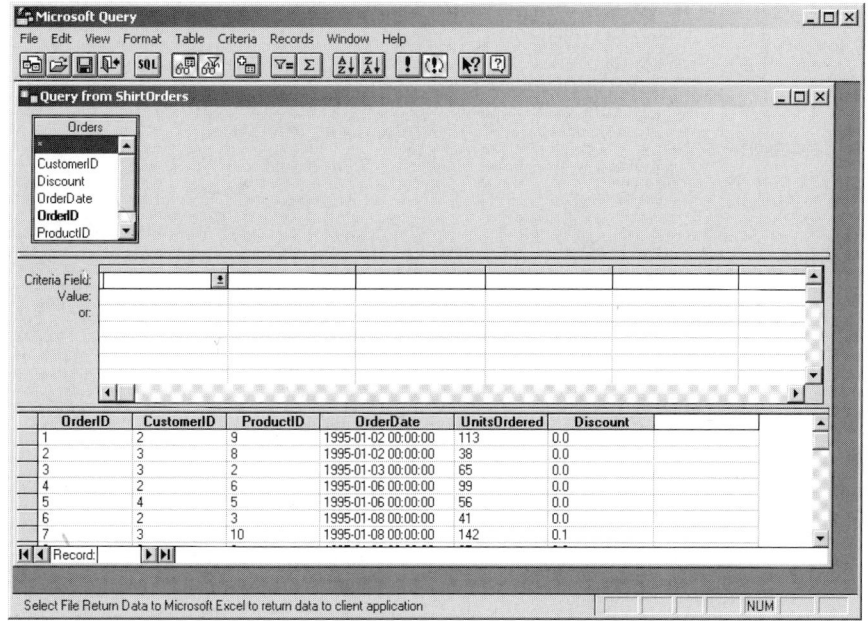

FIGURE 4.25
Add Criteria
Dialog Box

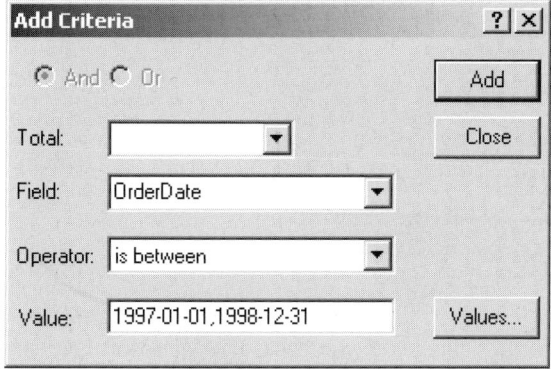

in the middle pane—if you know the correct syntax—or select the Criteria/Add Criteria menu item. This latter option brings up the dialog box shown in Figure 4.25. After a bit of experimenting, you'll see how to enter conditions in this dialog box. Then when you click on the Add button, the condition appears in the middle pane of the screen. By examining the syntax of the conditions that are entered, you can quickly learn how to type in your own conditions directly. The final conditions for our query appear in Figure 4.26 on page 154. (Note how dates are enclosed in # signs, and how the key words Between and In are used.)

If you scroll down the records in the bottom pane of the screen, you will see that this query returns 69 of the 2245 records in the Orders table. The final step in our three-step process is to get these data back into Excel. This is easy. Simply select the File/Return Data to Microsoft Excel menu item. This takes you back to Excel and brings up the dialog box in Figure 4.27, where you can specify where you want the results. When you click on OK, the results appear in a few seconds, and you can now analyze them statistically using any tools we have discussed. However, there is more—these data are still linked to the query. With the cursor anywhere in the data, select the Data/Get External Data/Edit

FIGURE 4.26
Criteria for Single-
Table Query

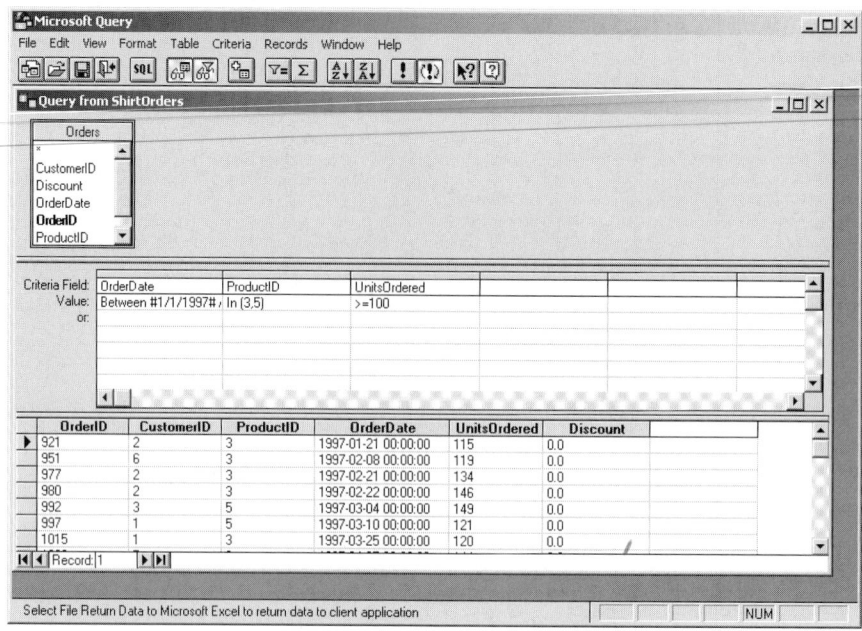

FIGURE 4.27
Location of Query
Results Dialog
Box

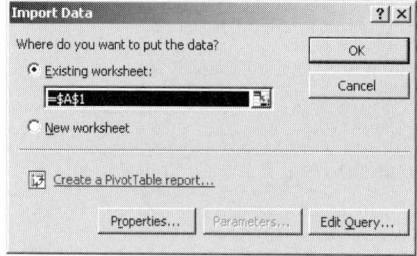

Query menu item.[7] This takes you back into Microsoft Query, with your criteria showing. If you change the criteria and select the File/Return Data to Microsoft Excel menu
item, the updated data will replace the original data in Excel.

One more possibility is to save the query itself. To do so, use the File/Save menu item
in the Microsoft Query screen with some suggestive file name such as OrdersQuery1. The
extension .DQY is added by default. This allows you to run this query at any time from
within Excel by using the Data/Get External Data/Run Saved Query menu item.

Let's now try a more ambitious query. We will find all of the records in the Orders
table that correspond to orders for at least 80 units made by the customer Shirts R Us
(customer number 3) for the product Long-sleeve Tunic (product number 6), and we will
return the dates and units ordered for these orders. The main difference is that we now
have to base the query on all three tables in the database. The reason is that the Orders
table does not have "Shirts R Us"—it contains only customer *numbers*. Similarly, it
doesn't know about "Long-sleeve Tunic." The trick is to use the links between the tables.

Starting in Excel (with the cursor *not* inside the data previously returned), select the
Data/Get External Data/New Database Query menu item. This time, however, simply
click on the ShirtOrders data source that is already there—you do not need to create it

---

[7]You can also make choices with the handy External Data toolbar that appears with the returned data.

FIGURE 4.28
Query Based on
All Three Tables

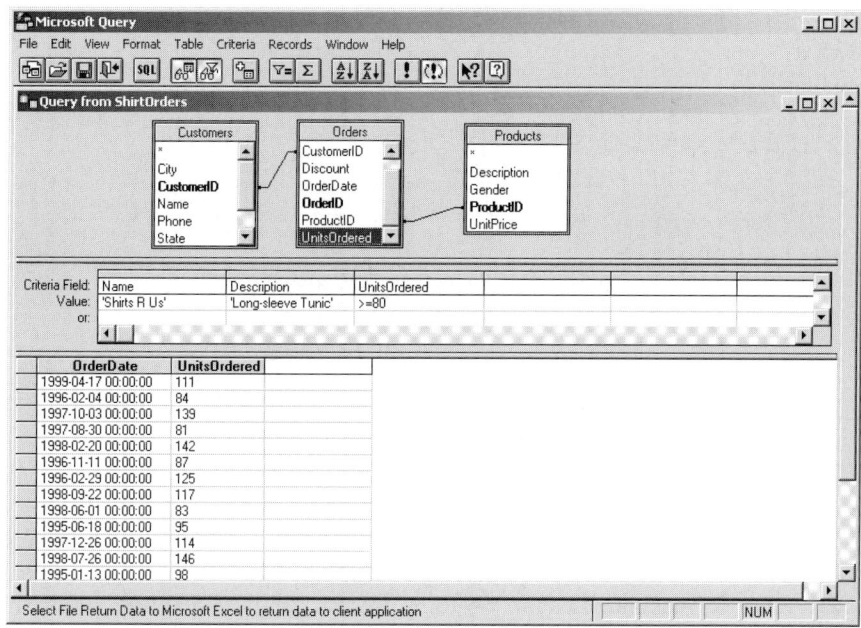

FIGURE 4.29
Data Returned to
Excel

	A	B
1	**OrderDate**	**UnitsOrdered**
2	17-Apr-99	111
3	4-Feb-96	84
4	3-Oct-97	139
5	30-Aug-97	81
6	20-Feb-98	142
7	11-Nov-96	87
8	29-Feb-96	125
9	22-Sep-98	117
10	1-Jun-98	83
11	18-Jun-95	95
12	26-Dec-97	114
13	26-Jul-98	146
14	13-Jan-95	98
15	2-May-98	93
16	26-Apr-98	100
17	4-Jun-97	92
18	22-Jan-95	142

Make sure that
Microsoft Query is
closed before trying
to create a new
query.

again. This takes you directly into the Microsoft Query screen. Inside this screen, first add all three tables to the top pane of the Query screen by using the Tables/Add Tables menu item. Next, double-click on the OrderDate and UnitsOrdered fields in the Orders table (because we want data in these two fields to be returned to Excel). Finally, fill out the criteria as shown in Figure 4.28. Note that the field names for the three criteria are from different tables. The Name field is from the Customers table, the Description field is from the Products table, and the UnitsOrdered field is from the Orders table. A good exercise is to think through the logic that Microsoft Query uses. From the Customers table, Microsoft Query finds that Shirts R Us corresponds to customer number 3. From the Products table, it finds that Long-sleeve Tunic corresponds to product number 6. Therefore, it searches the Orders table for all records where CustomerID is 3, ProductID is 6, and UnitsOrdered is at least 80. This returns 17 records, as shown in Figure 4.29.

FIGURE 4.30
Query with a
Calculated Field

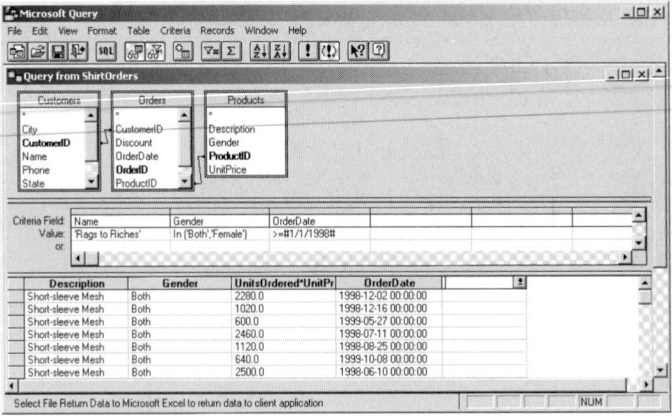

One last possibility we will illustrate is returning *calculated fields.* Suppose we want to return the *revenues* for all orders during 1998 or 1999 from Rags to Riches for shirts sold to females, where revenue is calculated as units ordered times unit price times 1 minus the discount. We form the query in the usual way, but in the bottom pane, we type the *expression* UnitsOrdered*UnitPrice*(1-Discount) as one of the field names. (Note that unlike Excel, there is no equals sign to the left of the expression.) The resulting Query screen, assuming we want to return the fields Description, Gender, and OrderDate in addition to revenue, should appear as in Figure 4.30. When we return the data to Excel, the field name for revenue will be something like Expr1001. This can then be changed to Revenue.

We reiterate that once the results of the query data are returned to Excel, we can then begin the statistical analysis of the data—creating summary measures, scatterplots, pivot tables, and so on. ●

4.5.3 SQL Statements

Queries represent a large part of the power behind relational databases. Regardless of the particular database package, whether it be Access, FoxPro, or any of the others, the *types* of queries we create are all basically the same. We typically base the query on one or more tables and ask it to return selected fields with records that satisfy certain conditions. To standardize queries across packages, SQL (structured query language and pronounced "S-Q-L" or "sequel") was developed. Sitting behind each query you develop in a user-friendly interface such as the Microsoft Query screen is a SQL statement. Although these statements are beyond the scope of this book, you might like to take a look at them, just to see how the experts create queries. This is easy to do. Once you have created a query, click on the SQL button in the Query toolbar.

As an example, if you form the query shown in Figure 4.26 and click on the SQL button, you see the SQL statement in Figure 4.31. At first, this is probably intimidating. However, if you break it down into its parts, it isn't that bad. SQL has a number of keywords that are capitalized. This statement includes the keywords SELECT, FROM, WHERE, and AND. The SELECT part of the statement specifies which fields to return (where, in the case of multiple tables, the table name and a period precede the field name). The FROM part specifies which tables to base the query on. Finally, the WHERE part spells out the criteria, separated by ANDs. If you want to learn more about SQL, the best way is to create a query through the interface and then look at the corresponding

FIGURE 4.31
SQL Statement

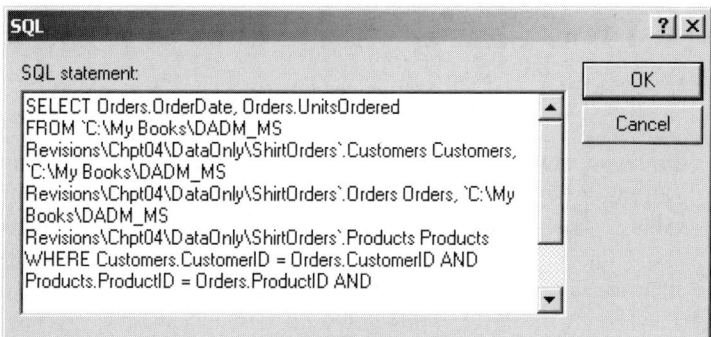

SQL statement.[8] Once you get used to SQL statements, you can edit a query by editing its SQL equivalent. If you get really proficient, you can even create a query from scratch by typing the appropriate SQL statement directly.

PROBLEMS

Level A

9. The Fine Shirt Company creates and sells shirts to its customers. These customers are retailers who sell shirts to customers. The company has created an Access database file P04_09.MDB that has information on sales to its customers over the past 5 years (i.e., 1995–1999). There are three related tables in this database: Customers, Orders, and Products. These tables are described in detail in Example 4.2. Find all of the records from the Orders table where the order was placed in 1998 or 1999, the product number is 1 or 10, the customer number is not 7, and the number of units ordered is at least 75. Return to Excel all fields in the Orders table for each of these records.

10. Continuing with the Fine Shirt Company database found in the file P04_09.MDB, find all of the records from the Orders table that correspond to orders for between 50 and 100 items made by the customer Rags to Riches for the product Short-sleeve Polo. Return to Excel the dates, units ordered, and discounts for each of these orders.

11. Continuing with the Fine Shirt Company database found in the file P04_09.MDB, find all of the records from the Orders table that correspond to orders for more than 75 items made by the customer Threads for products designed to be worn by women. Return to Excel the dates, units ordered, and product description for each of these orders.

Level B

12. Returning to the Fine Shirt Company, use the three tables contained in file P04_09.MDB to perform the following:
   a. Find all of the records from the Orders table that correspond to orders placed in 1998 by the customer The Shirt on Your Back for shirts designed to be worn by *both* men and women. Return to Excel the fields OrderDate, Description, Gender, UnitsOrdered, UnitPrice, Discount, and a *calculated field* revenue. Note that revenue equals

   $$UnitsOrdered * UnitPrice * (1-Discount)$$

   b. Analyze the distribution of revenues associated with order records identified in part **a**. Be sure to consider measures of central location, variability, and skewness in characterizing this distribution.
   c. Repeat parts **a** and **b** with the same criteria except that the analysis should now focus on the orders placed in 1999. Summarize the differences between the revenue distributions for 1998 and 1999.

13. Write the SQL statement to perform the query given in Problem 4.10.

14. Write the SQL statement to perform the query given in Problem 4.11.

---

[8]If you are really ambitious about learning database queries and SQL, the book by Kauffman et al. (2001) is a very good reference.

Creating Pivot Tables from External Data

In the previous section we learned how to import data from external databases by using Microsoft Query. We now briefly discuss how external data can be used to create pivot tables.[9] The procedure is nearly the same as for creating pivot tables from an existing Excel database—the procedure we discussed in the previous two chapters. However, the data we now base the pivot table on are the result of a query on an external database. Fortunately, to develop this query, we do not have to learn anything new. We do it exactly as in the previous section. The following continuation of Example 4.2 illustrates the procedure.

Example 4.2 | Fine Shirt Company's Relational Data (continued)

The Fine Shirt Company would like to break down revenue from its various customers and products by using pivot tables. How should it proceed?

Objective To illustrate how a pivot table can be created directly from data in the ShirtOrders database, using Microsoft Query.

Solution

We use the pivot table wizard in the usual way, except that steps 1 and 2 are slightly different. Starting with a blank spreadsheet in Excel, use the Data/PivotTable and

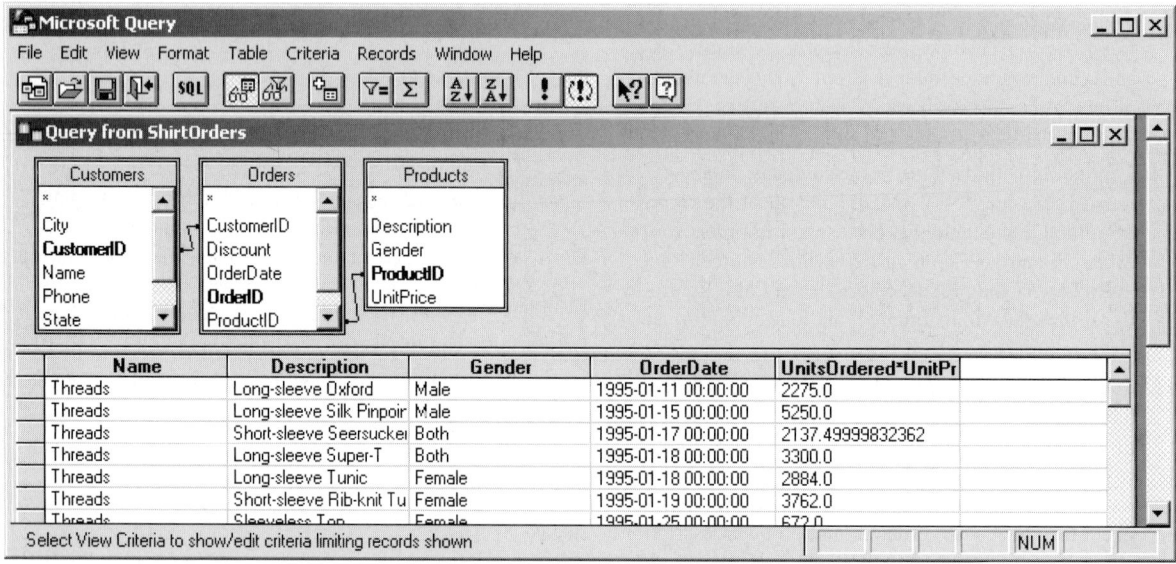

FIGURE 4.32 Specification of the Query

---

[9]This section assumes you have read the previous section and know how to create pivot tables, as discussed in Chapters 2 and 3.

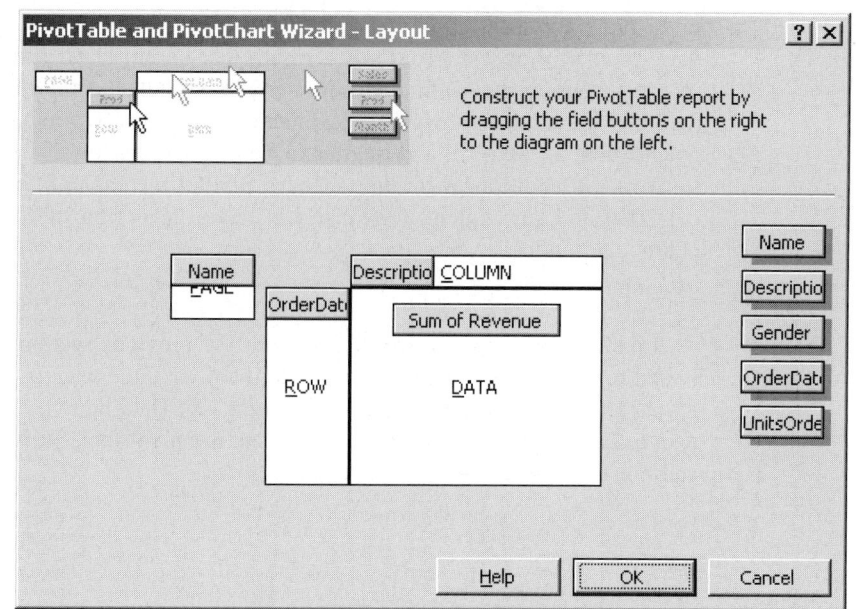

FIGURE 4.33 Layout Dialog Box

	A	B	C	D	E	F	G	H	I
1	Name	(All)							
2									
3	Sum of Revenue		Description						
4	Years	OrderDate	Long-sleeve Broadcloth	Long-sleeve Oxford	Long-sleeve Silk Pinpoint	Long-sleeve Super-T	Long-sleeve Tunic	Short-sleeve Mesh	Short-sleeve Polo
5	1995	Qtr1	$25,396	$27,034	$40,161	$23,530	$39,487	$19,277	$28,287
6		Qtr2	$33,880	$41,350	$29,148	$29,461	$27,930	$11,505	$48,025
7		Qtr3	$23,598	$13,417	$41,501	$17,629	$20,649	$20,238	$33,991
8		Qtr4	$24,550	$34,635	$58,964	$21,746	$17,478	$28,353	$45,331
9	1996	Qtr1	$14,915	$21,072	$40,455	$31,373	$33,597	$18,688	$24,922
10		Qtr2	$21,122	$14,295	$29,797	$10,927	$26,810	$20,517	$25,077
11		Qtr3	$26,136	$24,861	$45,570	$25,582	$17,305	$20,134	$32,161
12		Qtr4	$31,369	$22,721	$28,826	$27,405	$38,828	$14,245	$37,308
13	1997	Qtr1	$17,485	$20,825	$30,940	$31,287	$29,445	$33,897	$45,160
14		Qtr2	$24,461	$19,824	$45,288	$10,098	$26,502	$34,831	$35,910
15		Qtr3	$33,758	$28,865	$43,526	$25,970	$33,400	$23,521	$37,938
16		Qtr4	$26,419	$18,799	$36,050	$14,214	$19,901	$20,700	$26,859
17	1998	Qtr1	$14,215	$24,275	$35,821	$22,101	$27,462	$24,111	$33,115
18		Qtr2	$33,257	$21,422	$25,056	$34,108	$19,270	$13,529	$24,186
19		Qtr3	$33,965	$40,511	$28,360	$15,268	$29,863	$17,779	$25,410
20		Qtr4	$19,678	$22,902	$48,365	$18,356	$18,231	$15,715	$30,526
21	1999	Qtr1	$26,198	$21,837	$21,791	$21,905	$14,335	$19,660	$38,242
22		Qtr2	$35,510	$30,240	$24,001	$32,432	$31,437	$17,293	$31,996
23		Qtr3	$10,827	$16,650	$43,568	$23,113	$30,974	$14,273	$21,237
24		Qtr4	$25,369	$27,687	$34,207	$19,147	$17,122	$20,917	$18,882
25	Grand Total		$502,108	$493,225	$731,397	$455,653	$520,024	$409,183	$644,568

FIGURE 4.34 Pivot Table Results After Grouping on OrderDate

PivotChart Report menu item and select the External Data Source option in step 1. For step 2, click on the Get Data button. This takes you through the same query procedure as in the previous section. Specifically, select the ShirtOrders data source and then define the query. We defined it as shown in Figure 4.32 with *no* criteria—just a set of fields to return, one of which is calculated revenue—but you can impose criteria if you like. When you use the File/Return Data to Microsoft Excel menu item, you go back to step 2 of the pivot table wizard. At this point, click on Next and you should be in familiar territory.

From here, you can create any pivot tables you desire. We filled out the Layout dialog box as shown in Figure 4.33 to obtain our pivot table results, some of which are shown in Figure 4.34. The only trick here involves the OrderDate field. The original pivot table contains a row for each date—over 1000 rows. We decided to group the data by quarter of year. To do this, right-click on any date in the original pivot table, select the Group and Outline/Group menu item, and select *both* Quarter and Year. The resulting pivot table shows total revenue broken down by product, customer (using the "page" area at the top), and quarter of year. This is a lot of useful data with very little work! In addition, with the enhanced pivot table capabilities now available in Excel 2000 and subsequent versions, you have the option of obtaining corresponding charts automatically. ●

Like the query results we discussed in the previous section, pivot table results are linked to the query we create in step 2 of the pivot table wizard. This means that we can go back to step 2 of the wizard, click on the Get Data button, edit the query, return the data to Excel, and click on Finish to update the pivot table. It is an amazingly intuitive and powerful tool!

PROBLEMS

Level A

15. The Fine Shirt Company would like to know how many units of each of its products were sold to each customer during each of the past 5 years (i.e., 1995–1999). Using the database given in the file P04_09.MDB, construct one or more pivot tables that provide Fine Shirt with the desired information.

16. The Fine Shirt Company would also like to know how many units of its products designed for each gender subset (i.e., men, women, and both genders) were sold to each customer during each quarter of the past 5 years (i.e., from the first quarter of 1995 through the fourth quarter of 1999). Using the database given in the file P04_09.MDB, construct one or more pivot ta-

bles that provide Fine Shirt with the desired information.

Level B

17. The Fine Shirt Company would like to know what proportion of each customer's total dollar purchases in 1999 came from buying Short-sleeve Seersucker shirts. Furthermore, it would like to compare this proportion to that of the most popular product, as measured by 1999 total dollar purchases, for each customer. Using the database given in the file P04_09.MDB, construct one or more pivot tables that provide Fine Shirt with the desired information. Summarize your findings.

 # Web Queries

The chances are good that all of you have found interesting data on the Web that are amenable to statistical analysis in Excel. The question is how to import the data from the Web into Excel. Fortunately, this is now possible with Excel's Web queries, a feature that was added to Excel in Office 97. It is still relatively primitive and will undoubtedly change as Office and the Web develop, but it provides powerful capabilities most users are completely unaware of. (We have already seen considerable changes in Excel's Web query feature, both from Excel 97 to Excel 2000, and from Excel 2000 to Excel XP. It is now better than ever—almost user-friendly!) We will discuss Web queries briefly in this section, just to provide a glimpse of the possibilities. Hopefully, this will inspire you to try some things on your own.

To understand how it is possible to query a Web site from Excel, you should first understand at least a little of how Web pages are constructed. They are created with HTML (hypertext markup language), a text language that includes "tags" for displaying the various items you see on a typical Web page. One tag that is particularly useful for our purposes is the TABLE tag. When this tag is used as part of an HTML document, followed by data, it puts these data in a readable tabular form. Of course, the table might be surrounded by a lot of text and graphics, but the chances are that when we query a Web page from Excel, we are most interested in the table data and would like to ignore the surrounding stuff. Web queries allow us to do exactly this. They search for TABLE tags, find the corresponding data, and bring them into Excel in the usual row and column format. (If there are no TABLE tags, the Web queries bring in the entire HTML page, which could look pretty strange in a spreadsheet!)

As you know from Web experience, some Web sites are just there to read, whereas others ask you for information by using HTML "forms." There are two corresponding types of Web queries: *static* and *dynamic*. A **static query** simply asks for the data on a Web page. A **dynamic query,** on the other hand, prompts the user for parameters that define the query. As an example, a Web query we will see below imports stock price data into Excel, but it first prompts the user for one or more stock symbols, so that it knows which stock prices to import. To add to the confusion, when someone creates a Web page with an HTML form, she specifies that certain input parameters from the user should be sent to a server and then processed appropriately. There are two ways to specify this—the *GET* method and the *POST* method. Although the details are fairly technical, the essential difference is that the **GET** method is used for sending small amounts of information, whereas the **POST** method is used for larger amounts. It helps to be aware of the method used when you create a Web query, even if you do not entirely understand the technical details behind the methods.

We will begin with a simple static Web query. We (the authors) have a Web server called bl-bus-dotnet1.ads.iu.edu that we control. (This means that unlike other ever-changing Web sites, this one will continue to behave as we describe here—probably!) There is an HTML page SCORES.HTM on this site, created just for this example, that contains a heading and a table of course scores for students in a fictitious course. To get the data in this table into Excel, use the following steps:

1. Make sure you have an active connection to the Web, and open a new workbook in Excel.
2. Select the Data/Get External Data/New Web Query menu item.
3. Fill in this dialog box as shown in Figure 4.35 on page 162 (for Excel 2000) or Figure 4.36 (for Excel XP). The most important part is the URL (the address of the page) at the top, which is

http://bl-bus-dotnet1.ads.iu.edu:81/scores.htm

You have to know this or browse the Web for it. We find it easiest to browse to the intended Web site, copy its URL, and paste it into the dialog box. In Excel 2000 you then indicate whether you want to import the whole page, the tables only, or specific

FIGURE 4.35
Web Query Dialog
Box for Excel 2000

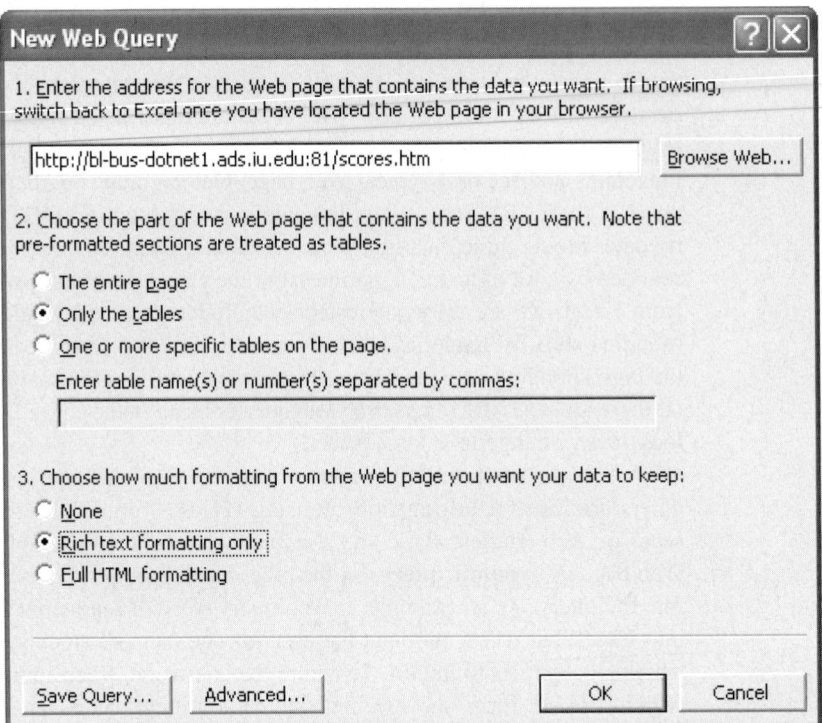

FIGURE 4.36
Web Query Dialog
Box for Excel XP

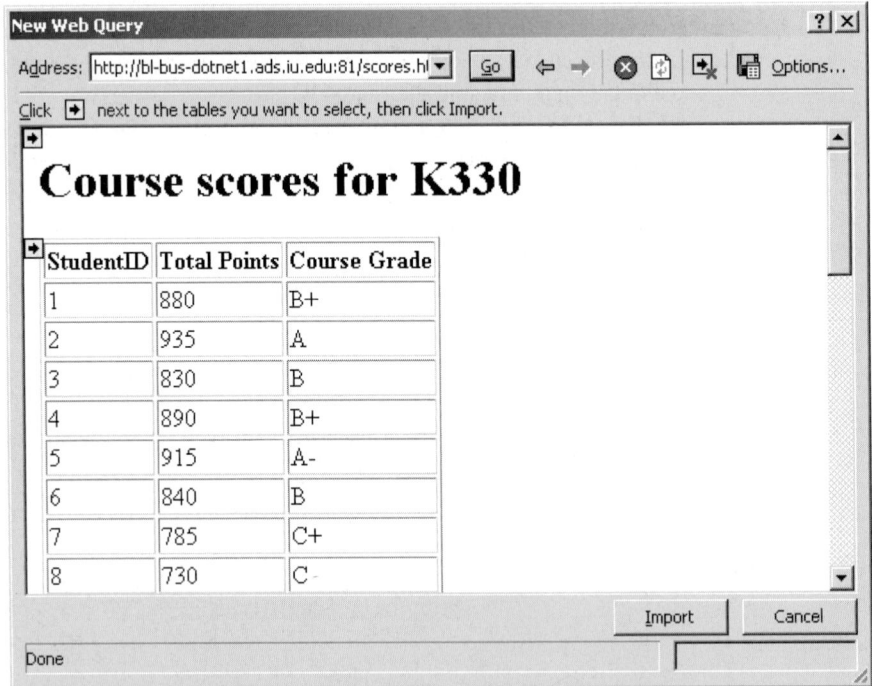

tables only, and you can also choose formatting options. (We chose all tables with rich text formatting.) Excel XP makes the process much easier. Once you enter the URL and click on Go, it shows you the Web page with yellow arrows next to all of the tables. (Some of these will probably not look like "tables," but they all have the HTML <Table> tag.) You can then click on any of these yellow arrows to import any desired tables into Excel.

4. After you click on OK, you will be asked where to place the results. We specified cell A1 of the blank worksheet.

The results then appear as shown in Figure 4.37. This Web page has only one table, and its contents have been imported into Excel and formatted nicely. In addition, a link to the Web page remains. This means that if the data on the Web page change, as they often do, we can update the Excel table to obtain the latest data. To do so, put the cursor anywhere inside the Excel table and select the Data/Refresh Data menu item. (This item is enabled only if the cursor is inside a table imported from an external data source.)

FIGURE 4.37
Results of Web Query

	A	B	C
1	StudentID	Total Points	Course Grade
2	1	880	B+
3	2	935	A
4	3	830	B
5	4	890	B+
6	5	915	A-
7	6	840	B
8	7	785	C+
9	8	730	C
10	9	810	B-
11	10	905	A-
12	11	865	B
13	12	720	C-
14	13	895	B+
15	14	835	B
16	15	965	A

You can also save the definition of the query in an .IQY file. (You might want to save it so that you could give it to a friend or use it on a different PC.) To save it, make sure the cursor is inside the Excel table, select the Data/Get External Data/Edit Query menu item to bring back the New Web Query dialog box, and click on the Save Query button. By default, Microsoft stores such queries in the C:\Windows\Application Data\ Microsoft\Queries folder, although this might depend on your operating system. In any case, you can override this default. Then you can run this query later on by selecting the Data/Get External Data/Run Saved Query (in Excel 2000) or Data/Get External Data/ Import Data (in Excel XP) menu item and selecting your saved query file.

These saved query file are simply text files—very short text files, in fact. You can open one of them in Notepad to see how it is constructed. The one we saved (as SCORES.IQY) for the previous query has the following lines:

```
WEB

1

http://b1-bus-dotnet1.ads.iu.edu:81/scores.htm

Selection=AllTables

Formatting=RTF

PreFormattedTextToColumns=True

ConsecutiveDelimitersAsOne=True

SingleBlockTextImport=False

DisableDateRecognition=False
```

The only required line in this file is the third one, which lists the URL of the Web page. The first two lines are optional, and the last six, which indicate the settings from the New Web Query dialog box (Figure 4.35 or 4.36), including the advanced options, are also optional. We point this out because it is possible to create your Web query directly in Notepad as an .IQY file and then run it in Excel with the Data/Get External Data/Run Saved Query menu item.

In fact, Microsoft has included several .IQY files with Office to indicate some of the possibilities. On our PC, these files are located in the C:\Program Files\Microsoft Office\Office\Queries folder, although this may differ depending on your operating system. One of these is MICROSOFT INVESTOR CURRENCY RATES.IQY. Its contents (as seen in Notepad) follow:

```
WEB

1

http://investor.msn.com/external/excel/quotes.asp?symbol=/ADY,/BPY,/CDY
,/ZEY,/DMY,/SFY

Selection=EntirePage

Formatting=All

PreFormattedTextToColumns=True

ConsecutiveDelimitersAsOne=True

SingleBlockTextImport=False
```

The third and fourth lines, the ones containing the URL, are really one long line that has been broken into two lines here to fit on the page. It is the key to the query. But what does it mean, and how would you know how to write it? Unfortunately, this is where Web queries get a bit complex, as we now explain.

This particular query is a static query, in that there are no prompts for information from the user. However, it can be changed easily to ask for different values. Specifically, the values after "symbol=" are called *parameter* values. For this query, each parameter value is a symbol for a currency. You can use these particular symbols (for American dollar, British pound, and so on), or you can obtain data for other currencies by including other symbols in this line.

As another example, the sample file MICROSOFT INVESTOR STOCK QUOTES.IQY is a dynamic query. It includes the URL

```
http://investor.msn.com/external/excel/quotes.asp?SYMBOL=["QUOTE0",
"Enter stock, fund or other MS Investor symbols separated by commas."]
```

(Again, this should be entered as one long line in Notepad.) In this case, Quote0 is the name of a parameter, and the sentence following it is a prompt to the user. If you run this saved query in Excel, you will see the dialog box in Figure 4.38. You can enter any stock symbols you like, and the Web query will return data for these stocks.[10] Figure 4.39 shows the data returned for our choices. The data in rows 3–7 are probably what we really want, in which case we delete the other information below (which, in spite of its appearance, could be HTML tables).

FIGURE 4.38
Parameter Dialog Box

	A	B	C	D	E	F	G	H	I	J	K	L	M	N	O	P
1	Stock Quotes Provided by MSN MoneyCentral Investor															
2	Click here to visit MSN MoneyCentral Investor															
3				Last	Previous Close	High	Low	Volume	Change	% Change	52 Wk High	52 Wk Low	Market Cap	EPS	P/E Ratio	# Shares Out
4	Microsoft Corporation	Chart	News	56.29	56.45	57.3	55.95	20,496,400	-0.16	-0.28%	76.15	47.5	326,592,039,000	1.09	55.3	5,415,222,000
5	International Business Machines Corporation	Chart	News	98.41	100.84	101.25	97.37	7,686,700	-2.43	-2.41%	126.39	87.49	168,964,268,508	4.35	23.9	1,716,942,000
6	Sears, Roebuck and Co.	Chart	News	50.06	49.76	50.55	49.4	1,136,300	0.3	0.60%	54.29	29.9	16,037,972,940	2.24	22.9	320,375,000
7	Coca-Cola Company	Chart	News	52.18	52.02	52.28	51.61	2,268,300	0.16	0.31%	52.68	42.37	129,652,429,458	1.6	32.7	2,484,715,000
8																
9	Symbol Lookup		MSN MoneyCentral Investor Home				Microsoft Office Update									
10	Find stocks, mutual funds, options, indices, and currencies.		Discover Investor's tools, columns, and more!				Get the latest from Microsoft Office									
11																
12	Terms of Use. © 2000 Microsoft Corporation and/or its suppliers. All rights reserved.															
13	Stock and fund data provided by Media General Financial Services.															
14	Quotes supplied by Standard & Poor's ComStock, Inc. and are delayed at least 20 minutes. NYSE, AMEX, and NASDAQ index data are provided real time.															
15	Fund data provided by Morningstar, Inc. © 2000. All rights reserved.															
16	Source Standard & Poor's Fund Services © Micropal Ltd 2000 All rights reserved www.funds-sp.com															
17	Source © Standard & Poor's International Ratings Ltd 2000 www.ifisinc.com [+81-3-3593-8681] All rights reserved. Errors and omissions excepted															

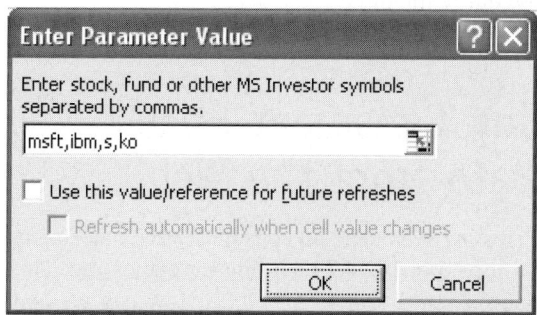

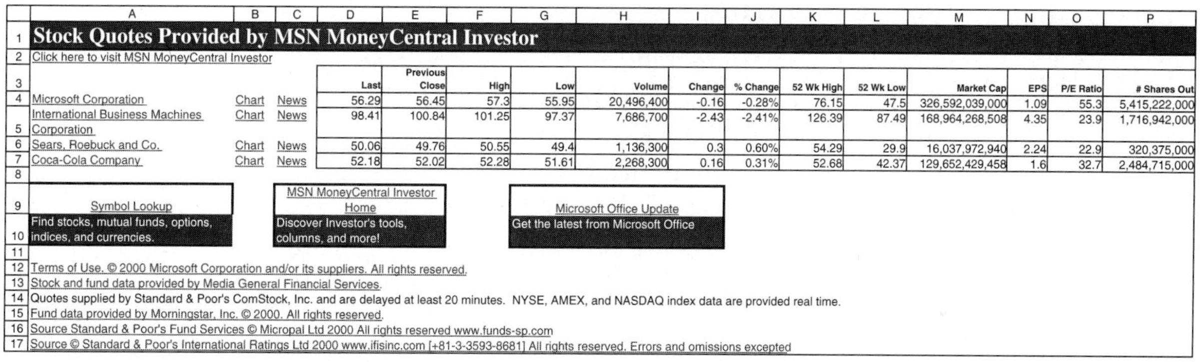

FIGURE 4.39 Results of Web Query

You can experiment with the sample queries Microsoft has supplied. However, what if you want to create your own Web query? What are the rules? The short answer is, It depends. We will not try to go through all of the details, but we will get you started. (Even Excel's online help on Web queries is pretty sketchy on the details.)

If you are lucky, you will find a page, such as our SCORES.HTM page referred to previously, that is static. There is no form to fill out, and the data are just there waiting to be downloaded to Excel. If you find such a page, you can proceed exactly as we did for the scores page, by entering its URL into the New Web Query dialog box in Figure 4.35 or 4.36.

[10]If you check the enabled box in Figure 4.38, you won't have to retype the stock symbols when you rerun or refresh this query. It will continue to use your initial choices. Also, as usual, we warn you that the screens you see might not match ours exactly, because Web pages are in a continual state of flux.

More frequently, however, you will browse to a site that has a form you must fill out to indicate exactly what data you want. This form might have a text box where you type in a stock symbol, option buttons where you choose which years you are interested in, or other means of capturing user choices. Then when you submit this form, you get the data. In this type of situation, how do you know what URL to include in the New Web Query dialog box (or in an .IQY file)?

There are many possibilities, so you will need to experiment. Of course, you should also check out Excel's online help for Web queries, but as we stated previously, many of the details we would like to see there are missing. The following example illustrates how one possible Web query might be created.

Example 4.3) Importing Consumer Price Index Data from the Web

We found an interesting Web page on consumer price indexes for various commodities at the URL address http://146.142.4.24/cgi-bin/surveymost?ap. This page presents us with a number of choices, as shown in Figure 4.40. How can we construct a Web query that gives us a choice of which data to obtain and then downloads the requested data?

FIGURE 4.40
Consumer Price
Index Web Page

Average Price Data
(Select from list below)

☐ 500 kwh Electricity – APU000072621
☐ Utility Gas, 40 Therms – APU000072601
☐ Utility GAs, 100 Therms – APU000072611
☐ Fuel Oil, Per Gallon – APU000072511
☐ Gasoline, All Types – APU00007471A
☐ Gasoline, Unleaded Regular – APU000074714
☐ Bread, White – APU0000702111
☐ Ground Beef, All Types, Per Pound – APU0000703111
☐ Whole Chicken, Per Pound – apu0000706111
☐ Eggs, Large, Per Dozen – APU0000708111
☐ Milk, All Types, Per Gallon – APU0000709111
☐ Red Delicious Apples – APU0000711111
☐ Navel Oranges – APU0000711311
☐ Bananas – APU0000711211
☐ Tomatoes – APU0000712311
☐ Orange Juice, Frozen – APU0000713111
☐ Coffee, Ground Roast, All Sizes – APU0000717311
☐ Iceberg Lettuce – APU0000712211

| Retrieve data | Reset form |

Objective To illustrate how a Web query can be created to obtain dynamic data from the consumer price index site.

Solution

The key is to look at the HTML source code for the Web page. (This can be done with the View/Source menu item in Internet Explorer or the View/Page Source menu item in Netscape.) Somewhere in this page there is a <FORM> tag with the following line:

<FORM ACTION=/cgi-bin/surveymost METHOD=POST>

As stated previously, there are two methods for sending a user's choices from a form to a Web server for processing: the POST method and the GET method. This form uses the POST method. For our Web query, this means that the information from the form—the parameter values—should be placed on a separate line in the .IQY file, right below the URL line. (With the GET method, they are placed on the *same* line as the URL, following a question mark.)

Regarding user inputs, we can see from the Web page itself that the user needs to specify the commodity (through a coded APU number). To get the proper syntax for the parameters line of the query, we search the HTML source code for INPUT tags. The following is a typical INPUT tag:

<INPUT TYPE=checkbox NAME=series_id VALUE=APU000072621> 500 kwh
    Electricity

This indicates that the name of the commodity input is "series_id" and a typical value for this input is one of the APU numbers shown on the Web page. Another line is

<INPUT SIZE=0 TYPE=hidden NAME=survey VALUE=ap>

There are several other <INPUT> tags in the source code, but they are not necessary for our purposes.

Using this information, we type the following query into Notepad and save it in the file CONSUMER PRICE INDEXES.IQY.

```
WEB

1

http://146.142.4.24/cgi-bin/surveymost?ap

series_id=["series_id","Select APU#"]&survey=ap

Selection=AllTables

Formatting=RTF

PreFormattedTextToColumns=True

ConsecutiveDelimitersAsOne=True

SingleBlockTextImport=False

DisableDateRecognition=False
```

The "series_id=" line is the tricky part. It is a sequence of "parameter name = parameter value" items (such as survey=ap) separated by ampersand (&) symbols. We know these parameter names from the <INPUT> lines in the source code. If we want to prompt the user for a parameter value, then we include a parameter name and a prompt to the right of the equals sign, enclosed in square brackets, as in =["series_id", "select APU#"].

FIGURE 4.41
Prompt for APU
Number

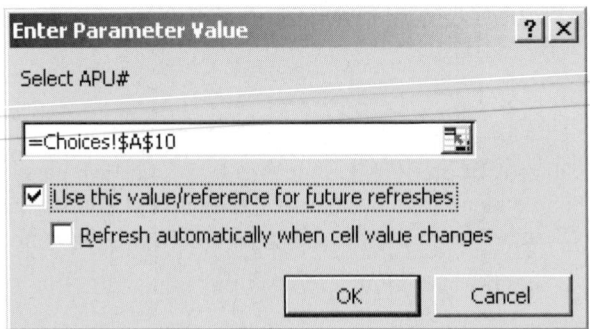

Enter Parameter Value	? X
Select APU#	

=Choices!$A$10

☑ Use this value/reference for future refreshes
☐ Refresh automatically when cell value changes

[OK] [Cancel]

FIGURE 4.42
Possible Choices
(Entered
Manually)

	A	B	C	D	E
1	Possible APU #'s				
2	APU000072621	500 kwh Electricity			
3	APU000072601	Utility Gas, 40 Therms			
4	APU000072611	Utility GAs, 100 Therms			
5	APU000072511	Fuel Oil, Per Gallon			
6	APU00007471A	Gasoline, All Types			
7	APU000074714	Gasoline, Unleaded Regular			
8	APU0000702111	Bread, White			
9	APU0000703111	Ground Beef, All Types, Per Pound			
10	APU0000706111	Whole Chicken, Per Pound			
11	APU0000708111	Eggs, Large, Per Dozen			
12	APU0000709111	Milk, All Types, Per Gallon			
13	APU0000711111	Red Delicious Apples			
14	APU0000711311	Navel Oranges			
15	APU0000711211	Bananas			
16	APU0000712311	Tomatoes			
17	APU0000713111	Orange Juice, Frozen			
18	APU0000717311	Coffee, Ground Roast, All Sizes			
19	APU0000712211	Iceberg Lettuce			

If we run this query from Excel, using the Data/Get External Data/Run Saved Query menu item, we are presented with a dialog box prompting for the APU number, as shown in Figure 4.41. Unfortunately, users will typically not know what to enter in this dialog box. Who could remember which APU number corresponds to white bread, for example? Therefore, a nice touch is to manually enter the APU numbers and corresponding commodities in an Excel range.[11] We did this, as shown in Figure 4.42. Now a user can respond to a dialog box by clicking on the appropriate APU-number cell, as we indicate in Figure 4.41.

Some of the results from one particular query (for whole chicken) appear in Figure 4.43. When we run this query, we keep our fingers crossed. We are usually not entirely

[11]If we are lucky, we might be able to cut and paste this information from the Web page into Excel instead of typing it.

Series Id: apu0000706111
Area: U.S. city average
Item: Chicken, fresh, whole, per lb. (453.6 gm)

Year	Jan	Feb	Mar	Apr	May	Jun	Jul	Aug	Sep	Oct	Nov	Dec
1992	0.878	0.849	0.859	0.861	0.854	0.861	0.878	0.882	0.881	0.865	0.885	0.879
1993	0.875	0.871	0.879	0.874	0.888	0.891	0.89	0.896	0.897	0.893	0.918	0.91
1994	0.899	0.888	0.891	0.899	0.902	0.918	0.92	0.905	0.901	0.889	0.904	0.895
1995	0.897	0.893	0.917	0.911	0.891	0.908	0.919	0.908	0.939	0.93	0.948	0.939
1996	0.941	0.94	0.933	0.947	0.949	0.968	0.972	0.99	1.006	1.021	1.003	1.002
1997	1.016	1.008	1.009	1.002	1.007	0.993	1.005	0.988	0.992	0.984	1.018	1.001
1998	1.022	1.007	1.032	1.029	1.032	1.016	1.033	1.059	1.074	1.076	1.084	1.06
1999	1.072	1.064	1.057	1.057	1.026	1.041	1.045	1.043	1.08	1.055	1.078	1.053
2000	1.059	1.046	1.064	1.069	1.052	1.069	1.089	1.086	1.087	1.09	1.065	1.078
2001	1.091	1.09	1.103	1.101	1.095	1.106	1.102	1.11	1.107	1.115	1.133	1.109
2002	1.091	1.111	1.11									

FIGURE 4.43 Web Query Results

sure what kind of data we will get because we are at the mercy of the Web site creator. The data in Figure 4.43 look just about right, except possibly for some formatting. ●

We are not sure how to rate the usefulness of Web queries at this stage. On the one hand, they are somewhat difficult to master, and the data we obtain in Excel might or might not be in a form useful for statistical analysis. On the other hand, the Web itself is only about 10 years old, so the fact that we can get live data into Excel with a query file that contains only a few lines is pretty amazing. We suspect that the situation will only improve in the future, especially as Microsoft creates better tools for interfacing between Excel and the Web.

Many Web sites have data that can be downloaded as a text file (probably with a .TXT extension). This is quite different from what we have been describing so far. If you find one of these sites, you will have the option of downloading the file to a directory of your choice on your hard drive. Then you can open the file in Excel. Just make sure that the option in the "Files of type" box in the Open File dialog box is either All files or Text files. You will then be led through a Text Import Wizard that helps you get the data in the proper format in Excel. It is a fairly straightforward process, but you should always examine the resulting Excel data carefully to ensure that the columns are lined up correctly.

PROBLEMS

Level A

18. Import data of interest to you from the Web site at URL address http://146.142.4.24/cgi-bin/surveymost?eb

19. Import data of interest to you from the Web site at URL address http://venus.census.gov/cdrom/lookup

20. Import data of interest to you from the Web site at URL address http://wonder.cdc.gov/

21. Import data of interest to you from the Web site at URL address http://www.ers.usda.gov/db/fatus/

22. Import data of interest to you from the Web site at URL address http://nces.ed.gov/nationsreportcard/naepdata/

23. Import data of interest to you from the Web site at URL address http://www.bts.gov/ntda/oai/search.html

4.8 Other Data Sources on the Web[12]

In the previous section we saw how it is sometimes possible to use Excel's Web query tool to import data on the Web into Excel. This works when we are lucky enough to find data *displayed* on a Web page in table form. However, there are many other types of data sources available on the Web—sometimes free and sometimes for a charge—and the number of these sources increases daily. It can often be quite a challenge to get these data into a form where they can be analyzed by the methods discussed in this book. We cannot hope in this section to discuss all of the possible data formats and available methods for extracting data from the Web. Instead, we will illustrate one possibility in the following example. As you read this example, you should imagine that your job depends on getting (and then analyzing) these data. Therefore, quitting because the process is too complex or because you don't know the required software is not an option!

Example 4.4 Acquiring Data on Substance Abuse

An interesting article by Kovar (2000) discusses whether adolescents smoke as much as we tend to hear in the news media. To make her arguments, she analyzes data from a large national survey, the National Household Survey of Drug Abuse, funded by the Office of Applied Studies of the Substance Abuse and Mental Health Services Administration. As she indicates, the data are freely available from the Substance Abuse and Mental Health Data Archive (SAMHDA) Web site www.icpsr.umich.edu/SAMHDA. Suppose you would like to analyze these survey data on your own. How should you proceed?

Objective To illustrate how to get the survey data from the SAMHDA Web site into a software package in a form suitable for statistical analysis.

Solution

The instructions we give in this example work correctly *now*. However, because the Web is in constant flux, we can only hope that nothing substantial will change by the time you try them. First, we visit the SAMHDA Web site at www.icpsr.umich.edu/SAMHDA, and click on the Download Data button. This takes us to a page, where we click on the National Household Survey on Drug Abuse (NHSDA) link. This takes us to a page that briefly describes the purpose and history of the survey. From there, we click on the Download Data & Documentation link that takes us to the page that lists yearly surveys (1979–1999). There are two links for each survey: Description and Downloads. The first of these takes us to an abstract of the survey. The second takes us to a page where we can download data. We now discuss each of these possibilities.

In a large survey such as this, it is extremely important to know the details of the survey: who did it, when it was done, and how it was done. Many of these details are listed in the abstract. For example, for the 1998 survey (the one we accessed), the investigator was the U.S. Department of Health and Human Services, Substance Abuse and Mental Health Services Administration, Office of Applied Studies. The intended population was "the civilian, noninstitutionalized population of the United States aged 12 and older, including residents of noninstitutional group quarters such as college dormitories, group

---

[12]This section can be omitted without any loss of continuity.

FIGURE 4.44
Download Page
from Web Site

Compressed (Using Gzip)	Uncompressed Size in KB	Filename (Filetype)	Dataset #: Name	Search Codebook
Compressed	Uncompressed 918 KB	cb2934.pdf.gz (Codebook, PDF format file)	DS1: National Household Survey on Drug Abuse, 1998	(Warning: PDF searches are **slow**) search

Data and Non-Documentation Files: Freely available

Compressed (Using Gzip)	Uncompressed Size in KB	Filename (Filetype)	# of Records	Record Length	Dataset #: Name
Compressed	Uncompressed 64224 KB	da2934.gz (Data file)	25500	2579	1: National Household Survey on Drug Abuse, 1998
Compressed	Uncompressed 1027 KB	sa2934.gz (SAS data def. statements file)	13648	77	1: National Household Survey on Drug Abuse, 1998
Compressed	Uncompressed 965 KB	sp2934.gz (SPSS data def. statements file)	13174	75	1: National Household Survey on Drug Abuse, 1998

homes, shelters, rooming houses, and civilians dwelling on military installations." The data source was "personal interviews and self-enumerated answer sheets (drug use)." Beyond this information, the abstract includes detailed paragraphs entitled Summary (the basic objectives of the survey), Collect.Note (technical details dealing with the data), and Sampling (the precise way the samples were selected). Finally, the abstract includes the cryptic line "1 data file + machine-readable documentation (PDF) + SAS data definition statements + SPSS data definition statements" under an Extent.Collect heading. (You'll understand this line shortly.) If you plan to do any serious analysis on these data, you should read the information in this abstract carefully.

Having read the abstract, we now follow the <u>Downloads</u> link to the download page. This page appears as in Figure 4.44. Unfortunately, it now becomes a bit complex. What do we need to download, and what will we get? As usual on the Web, some experimentation is required.

First, there is a link to a "codebook." This is a huge document, viewable in Adobe's .pdf format, that describes the survey in minute detail. (This file and the other available files can be downloaded in "zipped" format or in uncompressed format. We recommend the former. After downloading, they can then be unzipped.) This codebook is the "machine-readable documentation" listed in the Extent.Collect heading referred to previously. At some point, you might need to look at this document, but it can probably be skipped for now.

In the bottom section of Figure 4.44, we see three lines. The first line contains the data file. We first downloaded the zipped file and then unzipped it to a file that is called da2934. (You can change the name if you like. We changed ours to AbuseData.txt.) This file is huge (over 64 MB), and at first glance it appears to be virtually useless. It is a plain text file with nothing but long lines of numbers—not even separated by delimiters such as tabs, spaces, or commas. How can anyone analyze a data set in this form!

This is where the bottom two entries in Figure 4.44 come into play. The survey agency has written *commands* for importing the data in the text file into two very common and powerful statistical packages: SAS and SPSS. The middle entry in the figure is for SAS; the bottom entry is for SPSS. Given that we are not covering these packages in this book, you might imagine that we have hit a dead end at this stage. However, remember that your job might depend on analyzing such data sometime in the near future. We cannot afford to stop yet, so we will describe how to get the data into SPSS. (You or your instructor might want to try SAS instead.) If you have access to SPSS and plenty of hard drive space, you can follow along. Otherwise, you can read the following instructions to get the gist of the procedure.

SPSS is a Windows package with the usual menus and toolbars we are used to seeing in Windows packages. (The current version is 10.0.) However, SPSS still retains a command-driven language for performing various tasks, such as importing text data into the package. This is exactly what the download in the bottom line of Figure 4.44 contains: SPSS command lines for importing the data in our AbuseData.txt text file into SPSS. If you download this command file, unzip it (the unzipped version is called sp2934 by default, but we changed ours to AbuseSPSSCommands.txt), and load it in a text editor such as WordPad, you will see a boxed-in explanation at the top, followed by a few command lines and interspersed with many data lines. The next few lines were taken from this file. We will briefly explain each of them.

```
*  SPSS FILE HANDLE, DATA LIST COMMANDS.
FILE HANDLE DATA / NAME="data-filename" LRECL=2579.
DATA LIST FILE=DATA /
    RESPID 1-6    ENCPSU 7-9    ENCSEG 10-13
```

Any line preceded by an asterisk is a "comment," which can be ignored. The next line indicates where the data are coming from. You should substitute the path and name of the data set for "data-filename." (We used c:\statbook\chpt4\AbuseData.txt.) The next two lines (and many lines below these) describe the data setup. We mentioned previously that the data file contains long lines of digits. The command lines indicate how to "chop up" these digits. The first variable is called RESPID and contains the first 6 digits in each line. The second is called ENCPSU and contains the next 3 digits. The third is called ENCSEG and contains the next 4 digits. This continues on and on. It turns out that there are 1405 variables, and the number of digits on each line is 2579.

```
*  SPSS VARIABLE LABELS COMMAND.
VARIABLE LABELS
    RESPID "RESPONDENT IDENTIFICATION NUMBER"
    ENCPSU "PRIMARY SAMPLING UNIT (ENCRYPTED)"
    ENCSEG "SEGMENT IDENTIFICATION NUMBER (ENCRYPTED"
```

These lines give variable labels (or nicknames) to the cryptic variable names. When SPSS generates statistical output, it uses these nicknames instead of the variable names to label the output. The line for the variable ENCSEG suggests that there is a limit on the length of variable labels in SPSS.

```
*  SPSS MISSING VALUES COMMAND.
*  MISSING VALUES
    ACRDALC (81 THRU HI)    POUNDS2 (981 THRU HI)
    ACRDANL (81 THRU HI)    PPRES (81 THRU HI)
```

Missing values are an extremely important issue, especially in survey data. Many respondents leave questions blank or respond in some unintended way. These command lines indicate which responses should be considered "missing." For example, any response to the variable ACRDALC that is coded 81 or higher should be considered "missing." However, because there is an asterisk next to the MISSING VALUES command, this command will currently be ignored. If you want the command to be active, you can just delete the asterisk—we did.

```
* SPSS VALUE LABELS COMMAND.
VALUE LABELS
  SKPMARTL
  1 "Respondent is 12-14 years old - SKIP"
  2 "Respondent is 15 years old or older"
  3 "12-14 years old LOGICALLY ASSIGNED - SKI"
  4 "15 years or older LOGICALLY ASSIGNED" /
  MARITAL
  1 "Married"
  2 "Widowed"
  3 "Divorced or separated"
  4 "Never been married"
  85 "BAD DATA Logically assigned"
  89 "LEGITIMATE SKIP Logically assigned"
  96 "MULTIPLE RESPONSE"
  97 "REFUSED"
  98 "BLANK (NO ANSWER)"
  99 "LEGITIMATE SKIP" /
```

The value labels explain the coding used. For example, the SKPMARTL variable had four possible responses, coded 1–4. The value labels tell us what these codes really mean. Note the large codes for the MARITAL variable. Codes of 85 or greater are evidently "unusual" responses and are candidates for being treated as "missing."

```
* Create SPSS system file
* SAVE outfile="spss-filename.sav".
```

These final two lines allow us to save the imported data in a special binary format with the SPSS extension .sav. (This is similar to saving Excel files in a binary format with the extension .xls.) Once the file is saved in this format, it is much easier to open in later SPSS sessions. You should replace spss-filename.sav with the path and filename you prefer. (We used c:\statbook\chpt4\AbuseData.sav.) Also, this command is currently "commented out." Again, you should remove the asterisk in front of SAVE to make it active.

We're almost there. We have the data in a huge text file, and we have another text file of SPSS commands that will be used to import the data into SPSS. We now explain how to run these commands with the following four-step procedure.

Importing the data into SPSS

❶ Open SPSS. (The survey documentation indicates that version 9.0 or higher should be used.)

❷ Select the File/New/Syntax menu item. This opens a blank "syntax" window where you can enter SPSS commands.

❸ Copy the contents of the DataSPSSCommands.txt file (or whatever you named it), and paste them into the syntax window. If you have not already done so, make the changes we indicated previously to some of these command lines. (Change the file names appropriately, and delete the asterisk next to the MISSING VALUES and SAVE commands if you like.) See Figure 4.45 on page 174.

❹ Select the Run/All menu item.

FIGURE 4.45
SPSS Syntax
Window with
Pasted
Commands

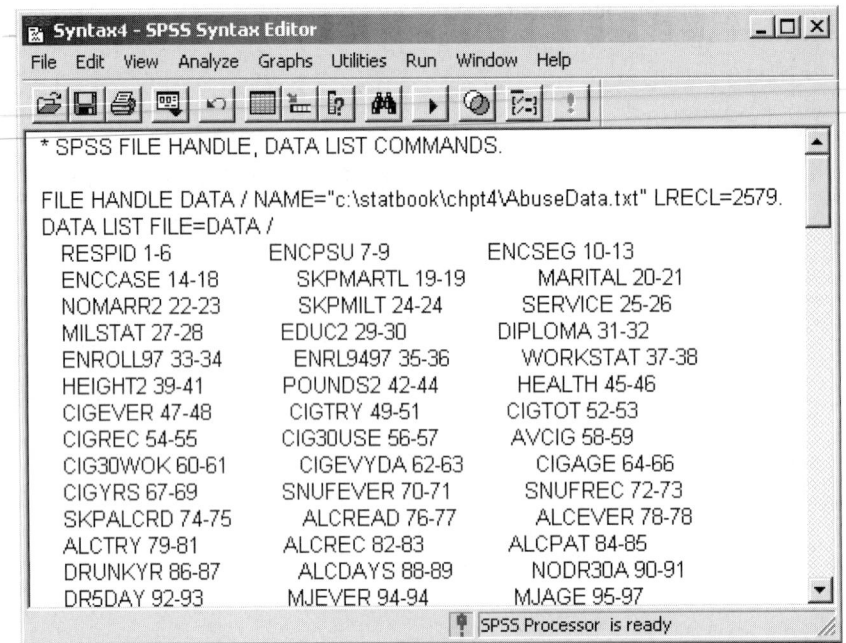

That's all there is to it. SPSS will now run the commands and import the data into a spreadsheet-like interface. It even saves the file in .sav format automatically for you. (Be prepared to wait a minute or more, depending on the speed of your computer.) We will not pursue this example any further because we don't want to get into a lot of SPSS interface details. However, the following points should be helpful if you want to experiment on your own.

Helpful Hints for Using SPSS

- SPSS has two "sheets" with tabs just like Excel sheets. The Data View sheet, shown in Figure 4.46, allows you to look at the data in the usual row–column format. The Variable View sheet, shown in Figure 4.47, provides detailed information on all of the variables—a "data dictionary."

- The Analyze menu contains all of the statistical procedures. For example, the Analyze/Descriptive Statistics/Descriptives menu item is functionally similar to the StatPro/Summary Stats/One-Variable Summary Stats menu item in StatPro. However, SPSS contains *many* more statistical procedures than StatPro.

- The results from all SPSS procedures are placed in an output window. They can then be stored in an output file, with extension .spo, if desired.

We conclude this example by putting everything in perspective. You might be annoyed at this point to have to learn about a new package (SPSS or SAS), but you have little choice if you really need to access these survey data. You might argue that we could avoid SPSS by importing the data text file directly into Excel. However, there are two problems: First, the data wouldn't fit. This data set has 1405 variables and over 25,000 cases, well beyond Excel's capacity (at least on a single sheet). Second, even if size were not an issue, we would need to use Excel's wizard for importing data from a text file. The wizard would ask us how to break these long lines into individual pieces, and it would also ask us for variable names. This would not be impossible, but it would be *extremely* tedious. We are better off taking advantage of what the survey agency has given us— SPSS (or SAS) command lines for quick and easy importing into a heavy-duty package.

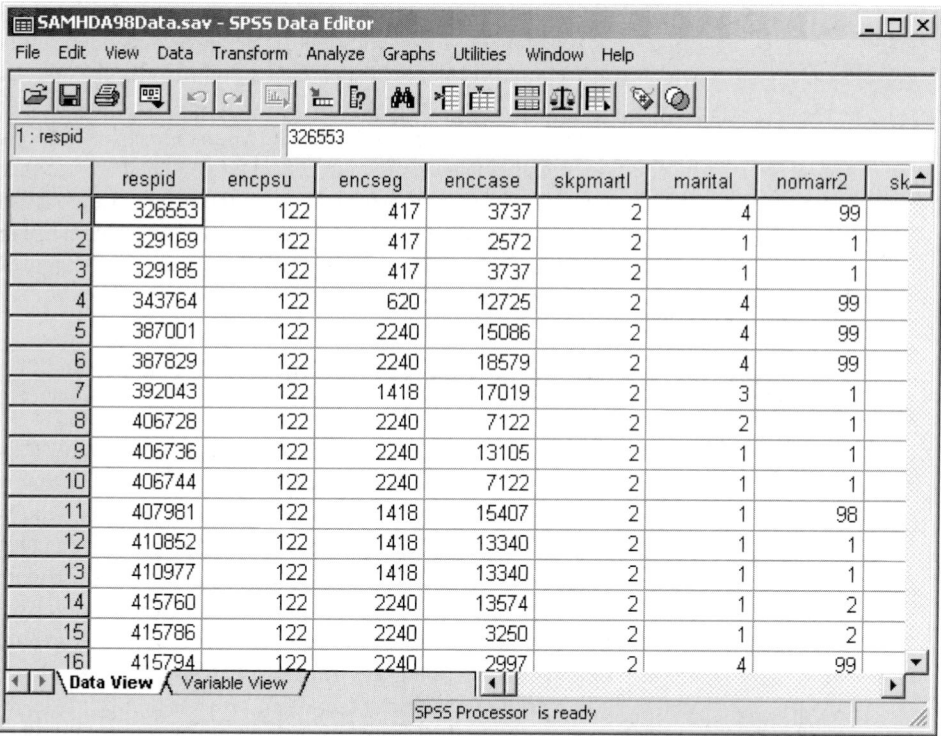

FIGURE 4.46 SPSS Data View Sheet

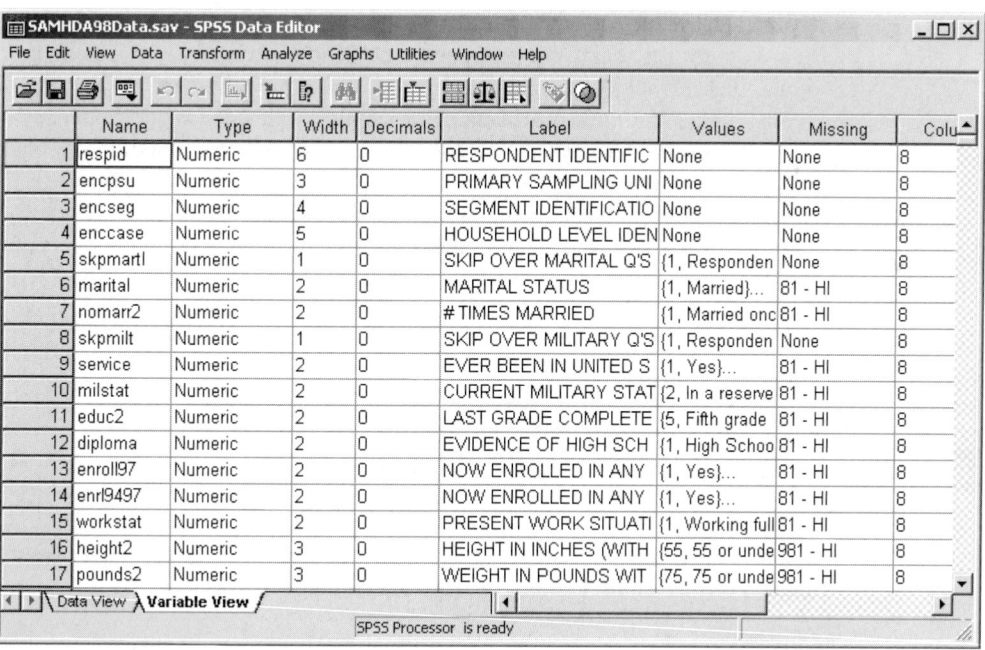

FIGURE 4.47 SPSS Variable View Sheet

The issues in this example are somewhat specific, but they are also quite general. They are specific in the sense that the details we have discussed apply only to data sets for which the Web author has furnished SPSS command lines. We have no idea how many Web sites are set up this way, but we wouldn't be surprised if there are quite a few. The issues are general, however, in the sense that, as a user of Web data, you will frequently be confronted with a "new" situation. The data you need are out there, and you have access to them, but you must learn how to import them into some statistical package, Excel or otherwise, in a form suitable for analysis. The Web has suddenly made available a wealth of data for us to analyze, but obtaining the data often poses real challenges.

4.9 Cleansing the Data

When you study statistics in a course, the data sets you analyze have usually been carefully prepared by the textbook author or your instructor. For that reason, they are usually in good shape—that is, they usually contain exactly the data you need, there are no missing data, and there are no "bad" entries (that might have been caused by keypunch errors, for example). Unfortunately, you cannot count on real-world data sets to be so perfect. This is especially the case when you obtain data from external sources such as the Web. There can be all sorts of problems with the data, and it is your responsibility to correct these problems before doing any serious analysis. This initial step, called *cleansing* the data, can be very tedious, but it can often prevent totally misleading results later on.

In this section we will examine one data set that has a number of errors, all of which could very possibly occur in real data sets. We will discuss methods for *finding* the problems and for *correcting* them. However, you should be aware of two things. First, the "errors" we consider here are only a few of those that could occur. Cleansing data requires real detective work to uncover all possible errors that might be present. Second, once an error is found, it is not always clear how to correct it. A case in point is missing data. For example, some respondents to a questionnaire, when asked for their annual income, might leave this box blank. How should we treat these questionnaires when we perform the eventual analysis? Should we delete them entirely, should we replace their blank incomes with the *average* income of all who responded to this question, or should we use a more complex rule to estimate the missing incomes? All three of these options have been suggested by statisticians, and all three have their pros and cons. Perhaps the safest method is to delete any questionnaires with missing data, so that we don't have to "guess" at the missing values, but then we might be throwing away a lot of potentially useful data. Our point is that some subjectivity and common sense must often be used when cleansing data sets.

Example 4.5 Customer Data with Errors

The file CLEANSING.XLS has data on 1500 customers of a particular company. A portion of these data appears in Figure 4.48, where many of the rows have been hidden. How much of this data set is usable? How much needs to be cleansed?

Objective To find and fix data errors in this company's data set.

Solution

We purposely constructed this data set to have a number of "problems," all of which you might encounter in real data sets. We begin with the Social Security Number (SSN).

	A	B	C	D	E	F	G	H	I
1	A customer database with some "bad" data								
2									
3	Customer	SSN	Birthdate	Age	Region	CredCardUser	Income	Purchases	AmtSpent
4	1	539-84-9599	10/26/44	57	East	0	62900	4	2080
5	2	444-05-4079	01/01/32	67	West	1	23300	0	0
6	3	418-18-5649	08/17/73	25	East	1	48700	8	3990
7	4	065-63-3311	08/02/47	51	West	1	137600	2	920
8	5	059-58-9566	10/03/48	50	East	0	101400	2	1000
9	6	443-13-8685	03/24/60	39	East	0	139700	1	550
10	7	638-89-7231	12/02/43	55	South	1	50900	3	1400
11	8	202-94-6453	11/08/74	24	South	1	50500	0	0
12	9	266-29-0308	09/28/67	31	North	0	151400	2	910
13	10	943-85-8301	07/05/65	33	West	0	88300	2	1080
14	11	047-07-5332	11/13/64	34	North	0	120300	3	1390
1496	1493	542-84-9996	11/03/75	23	South	1	29700	3	1640
1497	1494	880-57-0607	03/14/48	51	West	0	102700	5	2480
1498	1495	632-29-6841	02/06/45	54	West	1	89700	2	1000
1499	1496	347-70-0762	09/28/65	33	West	0	71800	2	970
1500	1497	638-19-2849	07/31/30	68	South	0	121100	5	2540
1501	1498	670-57-4549	07/21/54	44	North	1	64000	4	2160
1502	1499	166-84-2698	10/30/66	32	South	0	91000	6	2910
1503	1500	366-03-5021	09/23/34	64	South	0	121400	1	530

FIGURE 4.48 Data Set That Needs Cleansing

Presumably, all 1500 customers are distinct people, so all 1500 SSNs should be different. How can we tell if they are? One simple way is as follows. First, sort on the SSN column. (An easy way to do this is to select any SSN and click on the "AZ" button on the top toolbar. If this button isn't on your toolbar, you can use the Data/Sort menu item instead.) Once the SSNs are sorted, enter the formula

$$=IF(B5=B4,1,0)$$

in cell J5 and copy this formula down column J. This formula checks whether two adjacent SSNs are equal. Then enter the formula

$$=SUM(J5:J1503)$$

You might think that a visual scan of column B (after sorting) would find the duplicates. However, with 1500 entries, it's easy to miss something. That's why we recommend entering the formulas in column J and using Excel's Find tool.

in cell J4 to see if there are any duplicate SSNs. (See Figure 4.49 on page 178.) As we see, there are two pairs of duplicate SSNs. To find them, highlight the range from cell J5 down and use the Edit/Find menu item, with the resulting dialog box filled in as shown in Figure 4.50. In particular, make sure the bottom box has Values selected. Then click on the Find Next button two times to find the offenders. Customers 369 and 618 each have SSN 283-42-4994, and customers 159 and 464 each have SSN 680-00-1375. At this point, the company should check the SSNs of these four customers (hopefully available from another source) and enter them correctly here. (You can now delete column J and sort on column A to bring the data set back to its original form.)

The Birthdate and Age columns present two interesting problems. When the birthdates were entered, they were entered in exactly the form shown (10/26/44, for example). Then the age was calculated by a somewhat complex formula, just as you would calculate your own age.[13] Are there any problems? First, sort on Birthdate. You'll see that the first 18 customers all have birthdate 05/17/27—quite a coincidence! (See Figure 4.51.) It turns

---

[13]In case you are interested in some of Excel's date functions, we left the formula for age in cell D4. (We replaced this formula by its values in the rest of column D; otherwise, Excel takes quite a while to recalculate it 1500 times!) This formula uses Excel's TODAY, YEAR, MONTH, and DAY functions. Check online help to learn more about these functions.

	A	B	C	D	E	F	G	H	I	J
1	A customer database with some "bad" data									
2										
3	Customer	SSN	Birthdate	Age	Region	CredCardUser	Income	Purchases	AmtSpent	DupSSN
4	681	001-05-3748	03/24/36	63	North	0	159700	1	530	2
5	685	001-43-2336	08/21/63	35	North	0	149300	4	1750	0
6	62	001-80-6937	12/27/54	44	West	1	44000	4	2020	0
7	787	002-23-4874	01/31/76	23	North	0	153000	3	1330	0
8	328	004-10-8303	10/19/76	22	West	1	49800	4	1940	0
9	870	004-39-9621	10/13/57	41	South	0	138900	2	1010	0
10	156	004-59-9799	06/12/38	60	North	0	79700	2	980	0
11	1481	005-06-4020	06/16/52	46	South	1	42700	6	2890	0
12	127	005-83-0032	09/13/50	48	West	1	122400	3	1460	0
13	1441	006-14-1331	03/07/64	35	West	0	77400	2	930	0
14	1225	006-22-5891	08/16/51	47	East	0	125400	4	2070	0
15	1319	006-48-5947	08/19/31	67	North	0	157000	4	1920	0
16	41	006-59-5729	11/07/42	56	West	1	59100	7	460	0
17	293	007-25-4693	09/01/56	42	South	0	152200	1	540	0
18	699	007-29-6368	08/10/62	36	South	1	103100	4	1990	0

FIGURE 4.49 Checking for Duplicate SSNs

FIGURE 4.50
Dialog Box
for Locating
Duplicates

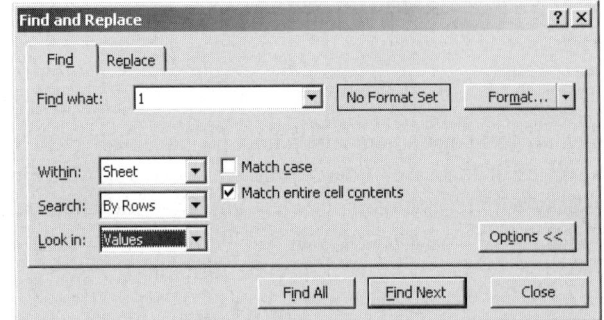

	A	B	C	D	E	F	G	H	I
1	A customer database with some "bad" data								
2									
3	Customer	SSN	Birthdate	Age	Region	CredCardUser	Income	Purchases	AmtSpent
4	494	085-32-5438	05/17/27	71	East	0	103700	1	480
5	645	086-39-4715	05/17/27	71	North	0	155300	5	2480
6	1068	110-67-7322	05/17/27	71	North	0	138500	3	1400
7	730	142-06-2339	05/17/27	71	West	1	38200	1	510
8	782	183-25-0406	05/17/27	71	West	0	51600	0	0
9	1179	183-40-5102	05/17/27	71	East	0	44800	4	1940
10	64	205-84-3572	05/17/27	71	East	0	50500	1	490
11	661	212-01-7062	05/17/27	71	West	0	147900	5	2450
12	429	279-23-7773	05/17/27	71	South	0	120300	4	2100
13	813	338-58-7652	05/17/27	71	East	1	47500	2	1020
14	486	364-94-9180	05/17/27	71	West	0	116500	2	1040
15	466	365-18-7407	05/17/27	71	East	0	155400	4	1900
16	1131	602-63-2343	05/17/27	71	North	1	67800	3	1520
17	463	619-94-0553	05/17/27	71	East	0	62300	2	930
18	607	626-04-1182	05/17/27	71	South	1	75900	3	1540
19	1329	678-19-0332	05/17/27	71	West	0	83900	5	2710
20	1045	715-28-2884	05/17/27	71	South	0	82400	4	1850
21	754	891-12-9133	05/17/27	71	North	0	77300	4	1980
22	174	240-78-9827	01/09/30	69	East	0	29900	2	960
23	43	053-13-0416	01/21/30	69	West	1	146300	1	480

FIGURE 4.51 Suspicious Duplicate Birthdates

	A	B	C	D	E	F	G	H	I
1	A customer database with some "bad" data								
2									
3	Customer	SSN	Birthdate	Age	Region	CredCardUser	Income	Purchases	AmtSpent
4	148	237-88-3817	08/11/29	-31	South	0	63800	8	3960
5	324	133-99-5496	05/13/28	-30	North	0	142500	2	1000
6	426	968-16-0774	09/29/28	-30	North	0	68400	2	1100
7	1195	806-70-0226	10/14/28	-30	West	0	40600	4	1960
8	1310	380-84-2860	10/17/28	-30	West	0	91800	2	980
9	440	618-84-1169	10/19/28	-30	West	1	113600	1	470
10	589	776-44-8345	04/16/27	-29	West	1	59300	2	1030
11	824	376-25-7809	11/02/27	-29	North	1	9999	2	1070
12	922	329-51-3208	03/21/28	-29	East	1	35400	6	3000
13	229	964-27-4755	01/29/27	-28	East	0	26700	1	450
14	1089	808-29-7482	02/28/27	-28	South	0	90000	5	2580
15	1037	594-47-1955	08/10/25	-27	East	1	128300	3	1510

FIGURE 4.52 Negative Ages: A Y2K Problem

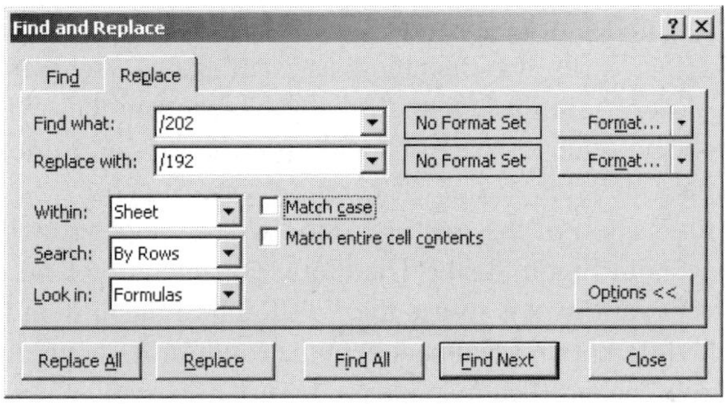

FIGURE 4.53
Dialog Box for Correcting the Y2K Problem

The "code" used by analysts to denote missing data is not at all standard. Some use 9999, others leave the entry blank, and others use some other code.

out that Excel's dates are stored internally as integers, which you can see by formatting dates as numbers. So highlight these 18 birthdates and use the Format/Cells menu item with the Number option (and zero decimals) to see what number they correspond to. It turns out to be 9999, the "code" many analysts use for missing values. Therefore, it is likely that these 18 customers were not born on 05/17/27 after all. Their birthdates were probably *missing* and simply entered as 9999, which were then formatted as dates. If birthdate is important for further analysis, these 18 customers should probably be deleted from the data set.

It gets even more interesting if you sort on the Age variable. You'll see that the first 12 customers after sorting have *negative* ages. (See Figure 4.52.) You have just run into a Y2K (year 2000) problem! These 12 customers were all born before 1930. Excel guesses that any two-digit year from 00 to 29 corresponds to the 21st century, whereas those from 30 to 99 correspond to the 20th century.[14] Obviously, this guess was a bad one for these 12 customers, and we should change their birthdates to the 20th century. An easy way to do so is to highlight these 12 birthdates, choose the Edit/Replace menu item, fill out the resulting dialog box as shown in Figure 4.53, and click on the Replace All button. This replaces any year that starts 202, as in 2028, with a year that starts 192. (Always be careful with the Replace All option. For example, if we had entered /20 and /19 in the "Find what:" and "Replace with:" boxes, we would not only have replaced the years, but the 20th *day* of any month would also have been replaced by the 19th day!) If

[14]To make matters even worse, a *different* rule was used in earlier versions of MS Office. So if you are running Excel 95, you might see a different result. In addition, there is no guarantee that Microsoft will continue to use this same rule in future editions of Office.

FIGURE 4.54

Pivot Table
with Too Many
Categories

	A	B
3	Sum of AmtSpent	
4	Region	Total
5	East	48550
6	North	36530
7	South	67289
8	West	35140
9	East	555909
10	North	514135
11	South	563585
12	West	520294
13	Grand Total	2341432

you copy the formula for Age that was originally in cell D4 to all of column D, the ages should recalculate automatically as *positive* numbers.

The Region variable presents a problem that can be very hard to find—because you usually are not looking for it. There are four regions: North, South, East, and West. If you sort on Region and starting scrolling down, you'll find a few Easts, a few Norths, a few Souths, and a few Wests, and then the Easts start again. Why aren't the Easts all together? If you look closely, you'll see that a few of the labels in these cells—those at the top after sorting—begin with a space. Whoever typed them inadvertently entered a space before the name. Does this matter? It certainly can. Suppose you create a pivot table, for example, with Region in the row area. You will get eight row categories, not four. (An example appears in Figure 4.54.) Therefore, you should get rid of the extra spaces. The most straightforward way is to use the Edit/Replace menu item in the obvious way.

A slightly different problem occurs in the CredCardUser column, where 1 corresponds to credit card users and 0 corresponds to nonusers. A typical use of these numbers might be to find the proportion of credit card users, which we can find by entering the formula

$$=\text{AVERAGE}(F4{:}F1503)$$

in cell F2, say. This *should* give the proportion of 1's, but instead it gives an error (#DIV/0!). What's wrong? A clue is that the numbers in column F are left-justified, whereas numbers in Excel are usually right-justified. Here is what might have happened. Data on users and nonusers might initially have been entered as the labels Yes and No. Then to convert them to 1 and 0, someone might have entered the formula

$$=\text{IF}(F4=\text{``Yes''},\text{``1''},\text{``0''})$$

The double quotes around 1 and 0 cause them to be interpreted as *text,* not *numbers,* and no arithmetic can be done on them. (In addition, text is typically left-justified, the telltale sign we observed.) Fortunately, Excel has a function called VALUE that converts text entries that look like numbers to numbers. So we should form a new column that uses this VALUE function on the entries in column F to convert them to numbers. (Specifically, we could create these VALUE formulas in a new column, then do a Copy and Paste-Special/Values to replace the formulas by their values, and finally cut and paste these values over the original text in column F.)

Next we turn to the Income column. If you sort on it, you'll see that most incomes go from $20,000 to $160,000. However, there are a few at the top that are much smaller, and there are a few 9999's. (See Figure 4.55.) By this time, you can guess that the 9999's correspond to missing values, so these customers should probably be deleted if Income is crucial to the analysis. The small numbers at the top take some educated guesswork. Because they range from 22 to 151, we might guess (and hopefully we could confirm) that the person who entered these data thought of them as "thousands" and sim-

The newest release of Office, Version 2002 (or XP), puts a comment in such cells, warning that numbers have been formatted as text.

The moral is to omit double quotes around numbers in IF statements. Use double quotes only around text.

	A	B	C	D	E	F	G	H	I
1	A customer database with some "bad" data								
2									
3	Customer	SSN	Birthdate	Age	Region	CredCardUser	Income	Purchases	AmtSpent
4	439	390-77-9781	06/03/70	28	West	0	22	8	4160
5	593	744-30-0499	05/04/60	38	East	0	25	5	2460
6	1343	435-02-2521	08/24/42	56	West	1	43	5	2600
7	925	820-65-4438	11/12/32	66	North	0	55	6	2980
8	1144	211-02-9333	08/13/34	64	North	0	71	9999	9999
9	460	756-41-9393	05/14/71	27	East	0	81	3	1500
10	407	241-86-3823	07/03/59	39	East	1	88	4	2000
11	833	908-76-1846	09/17/60	38	West	0	104	4	1970
12	233	924-59-1581	05/12/31	67	South	0	138	6	2950
13	51	669-39-4544	10/05/33	65	West	0	149	2	1010
14	816	884-27-5089	03/05/62	37	North	1	151	2	900
15	824	376-25-7809	11/02/27	-29	North	1	9999	2	1070
16	518	378-83-7998	11/02/74	24	West	1	9999	2	940
17	570	758-72-4033	11/07/70	28	South	1	9999	3	1520
18	605	600-05-9780	07/10/58	40	North	1	9999	0	0
19	796	918-32-8454	03/22/54	45	North	0	9999	3	1570
20	1235	470-28-5741	07/01/52	46	North	0	9999	1	480

FIGURE 4.55 Suspicious Incomes

ply omitted the trailing 000's. If this is indeed correct, we can fix them by multiplying each by 1000.

Finally, we examine the Purchases (number of separate purchases by a customer) and AmtSpent (total spent on all purchases) columns. First, sort on Purchases. You'll see the familiar 9999's at the bottom. In fact, each 9999 for Purchases has a corresponding 9999 for AmtSpent. This makes sense. If the number of purchases is unknown, the total amount spent is probably also unknown. We can effectively delete these 9999 rows by inserting a blank row right above them. Excel then automatically senses the boundary of the data. Essentially, a blank row or column imposes a separation from the "active" data. (See Figure 4.56.)

	A	B	C	D	E	F	G	H	I
1485	1427	182-48-9138	05/18/40	58	East	0	105000	9	4450
1486									
1487	1144	211-02-9333	08/13/34	64	North	0	71	9999	9999
1488	287	133-53-5943	09/22/35	63	North	1	20000	9999	9999
1489	1298	552-06-0509	10/12/37	61	North	0	23700	9999	9999
1490	375	867-63-6238	09/17/71	27	West	0	29900	9999	9999
1491	250	586-87-0627	06/24/52	46	East	1	53300	9999	9999
1492	14	614-59-6703	08/01/72	26	South	1	54400	9999	9999
1493	1106	102-74-2447	03/14/30	69	West	0	59300	9999	9999
1494	1121	637-23-3846	06/14/54	44	South	0	64000	9999	9999
1495	153	048-55-8930	09/05/34	64	West	1	64400	9999	9999
1496	980	967-97-4228	07/04/63	35	South	1	76800	9999	9999
1497	1061	377-29-0406	10/08/51	47	West	1	93000	9999	9999
1498	858	819-34-4450	05/26/59	39	South	1	101300	9999	9999
1499	432	572-79-9529	01/21/67	32	West	1	104500	9999	9999
1500	1438	452-69-6883	01/16/74	25	South	0	116400	9999	9999
1501	1125	394-20-9464	10/20/75	23	North	1	129400	9999	9999
1502	469	797-55-3419	09/16/61	37	North	1	132800	9999	9999
1503	443	087-21-2053	07/02/52	46	West	0	141200	9999	9999
1504	317	865-85-3875	12/19/31	67	South	0	149900	9999	9999

FIGURE 4.56 Separating Rows with Missing Data from the Rest

FIGURE 4.57
Scatterplot with
Suspicious Out-
liers

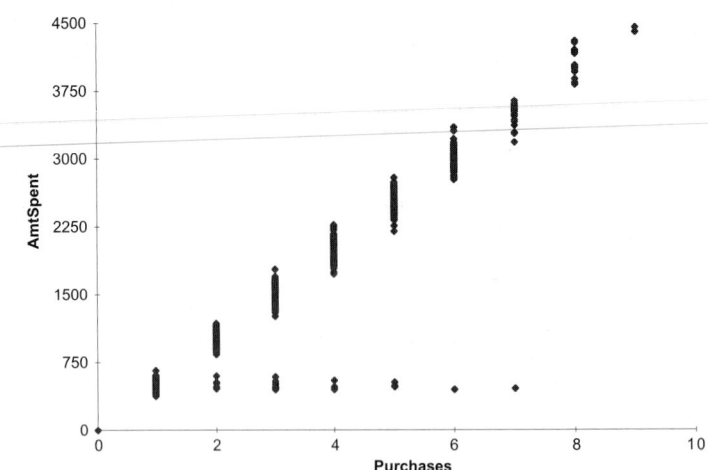

	A	B	C	D	E	F	G	H	I
641	708	682-62-7862	10/25/52	46	West	1	43100	2	1160
642	227	580-03-8512	10/12/53	45	East	0	48400	2	1170
643	1455	169-31-5478	06/19/45	53	North	1	144600	2	1170
644	777	820-27-6346	07/04/36	62	West	0	155000	2	1180
645	259	731-52-6832	02/05/51	48	East	1	41700	3	450
646	121	345-16-5545	07/08/59	39	West	1	112700	3	450
647	109	280-07-3023	08/04/43	55	West	0	24300	3	460
648	1469	719-98-9028	03/15/69	30	North	1	91300	3	470
649	1331	745-63-6259	07/22/58	40	South	0	63700	3	480
650	1313	041-74-0192	12/04/59	39	East	0	25900	3	510
651	501	156-39-5201	08/15/38	60	East	0	111000	3	540
652	936	261-74-3204	10/01/37	61	West	0	65000	3	590
653	921	601-98-9218	05/06/38	60	South	1	131000	3	1260
654	294	728-06-3395	07/12/66	32	West	0	159800	3	1300
655	568	375-92-1009	01/13/59	40	North	1	73600	3	1310
656	597	062-42-6963	03/11/58	41	East	1	67800	3	1320

FIGURE 4.58 Suspicious Values of AmtSpent

Now we examine the remaining data for these two variables. Presumably, there is a relationship between these variables, where the amount spent increases with the number of purchases. We can check this with a scatterplot of the (nonmissing) data, as shown in Figure 4.57. There is a clear upward trend for most of the points, but there are some suspicious outliers at the bottom of the plot. Again, we take an educated guess. Perhaps the *average spent per purchase,* rather than the total amount spent, was entered for a few of the customers. This would explain the abnormally small values. (It would also explain why these outliers are all at about the same height in the plot.) If we can locate these outliers on the Data sheet, we should multiply each by the corresponding number of purchases. How do we find them on the Data sheet? First, sort on AmtSpent, then sort on Purchases. This will arrange the amounts spent in increasing order for each value of Purchases. Then, using the scatterplot as a guide, scroll through each value of Purchases (starting with 2) and locate the abnormally low values of AmtSpent (which are all together). For example, Figure 4.58 indicates the suspicious values for 3 purchases. This procedure is a bit tedious, but it beats working with invalid data. ●

Level A

24. The file P04_01.XLS contains a data set that represents 30 responses from a questionnaire concerning the president's environmental policies. Each observation lists the person's age, gender, state of residence, number of children, annual salary, and opinion of the president's environmental policies. How much of this data set is usable? Cleanse all usable data.

25. The file P04_05.XLS contains information on 66 movie stars. In particular, the data set contains the name of each actor and the following four variables: gender, domestic gross (average domestic gross of the star's last few movies, in millions of dollars), foreign gross (average foreign gross of the star's last few movies, in millions of dollars), and salary (current amount the star asks for a movie, in millions of dollars). Are all of these data usable? Explain why or why not.

26. A human resources manager at Beta Technologies, Inc., has collected current annual salary figures and related data for 52 of the company's full-time employees. The data are in the file P04_02.XLS. Specifically,

these data include each selected employee's gender, age, number of years of relevant work experience prior to employment at Beta, the number of years of employment at Beta, the number of years of post-secondary education, and annual salary. How much of this data set is usable? Cleanse all usable data.

27. Five hundred households in a middle-class neighborhood were recently surveyed as part of an economic development study conducted by the local government. Specifically, for each of the 500 randomly selected households, the survey requested information on the following variables: family size, approximate location of the household within the neighborhood, an indication of whether those surveyed owned or rented their home, gross annual income of the first household wage earner, gross annual income of the second household wage earner (if applicable), monthly home mortgage or rent payment, average monthly expenditure on utilities, and the total indebtedness (excluding the value of a home mortgage) of the household. The data are in the file P04_03.XLS. Cleanse all usable data in this set.

 # Conclusion

This chapter has covered some very powerful tools for getting the right data into Excel. As with many other features of Excel, the tools we have discussed are fairly easy to use—once you know they exist. We believe that once you know that something can be done and have a general idea of how to do it, you can figure out the rest of the details. Indeed, as the software changes, you will be forced to learn the details on your own through experimenting and consulting online help. Therefore, as you look back on this chapter, focus on what *can* be done, not the nitty-gritty details. It *is* possible to create queries in Excel ranges so that we can find subsets of an Excel database that satisfy certain conditions. It *is* possible to create queries in Microsoft Query so that we can import data from many database packages into Excel. It *is* even possible to create text file queries so that we can import data from Web pages into Excel. Once you realize these possibilities, you will be able to accomplish tasks that the majority of Excel users have never even tried.

Summary of Key Terms

Term	Explanation	Excel	Page
Data warehouse	A type of database used by companies to store large quantities of historical data for later statistical analysis.		132
Survey	A questionnaire used to gather information from a sample of a population.		132
Controlled experiment	An experiment where certain variables are deliberately set at specified levels to learn about their effects on one or more other variables.		133
Query	An instruction to a database to return a subset of the data that satisfies specified conditions.		134
Autofilter	A simple way to query an Excel database.	Data/Filter/ AutoFilter	134
Advanced filter	A more general way to query an Excel database, where a combination of "and" and "or" conditions can be used.	Data/Filter/ Advanced Filter	140
Flat file	A single-table database, called a list in Excel.		146
Relational database	A database where the data are stored in related tables, which are related by primary and foreign key fields.		147
Microsoft Query	Software that is packaged with Microsoft Office, used to get data from external databases and return the data to Excel.	Data/Get External Data/ New Database Query, or Data/PivotTable and PivotChart Report	148
SQL	Structured Query Language, a general language used to specify database queries.		156
Web query	A method for importing tables from selected Web pages into Excel.	Data/Get External Data/ New Web Query	160
SPSS, SAS	Two heavy-duty statistical software packages favored by many statisticians.		171
Cleansing data	The process of removing errors—keypunch errors, Y2K errors, or any other types of errors—from a data set.		176

PROBLEMS

Conceptual Exercises

C.1. What is the difference between a *survey* and a *census?*

C.2. An organizational behavior professor wonders whether the use of role-playing exercises will help her students learn how groups make decisions. She decides to conduct an experiment in which some students engage in role-playing exercises while others engage in a more traditional discussion about group decision making during recitation sections. This pro- fessor teaches two sections of her OB course: one at 8:00–9:30 A.M. and another at 1:00–2:30 P.M. on Tuesdays and Thursdays. To keep things simple, she decides that all students enrolled in the 8:00 A.M. section will be subjected to the role-playing peda- gogy and those students enrolled in the 1:00 P.M. section will learn about how groups make decisions through one or more class discussions. Does the pro- fessor's plan for conducting this experiment appear to be sound? Explain why or why not.

C.3. Assume that a national insurance company has randomly selected 1500 of its customers to assess their attitudes toward the service they receive from the company's agents. Provide an example of how an analyst from this company might perform a query on the customer database in conducting this investigation.

C.4. Identify all flat files in the relational database described in Example 4.2.

C.5. Suppose that you collect a random sample of 250 salaries for the salespersons employed by a large PC manufacturer. Furthermore, assume that you find that two of these salaries are considerably higher than the others in the sample. In cleansing this data set, should you delete the unusual observations? Explain why or why not.

Level A

28. Consider the given survey data collected from 1000 randomly selected Internet users. The data are in the file P04_28.XLS. Use Excel's AutoFilter tool to answer the following questions:
   a. What proportion of those surveyed are females who are married, employed, and have achieved more than a high school education?
   b. What proportion of those surveyed are males who are single, unemployed, and have achieved a high school education or less?
   c. Find the average annual income of the females who are married, employed, and have achieved more than a high school education.
   d. Find the average annual income of the males who are married, employed, and have achieved more than a high school education. How does this result compare to the average found in part **c?**

29. Consider the given survey data collected from 1000 randomly selected Internet users. The data are in the file P04_28.XLS.
   a. Find all Internet users in the sample who are either (1) married men between the ages of 21 and 40 (inclusive) who are employed and have more than a high school education and two or fewer children, or (2) married women between the ages of 21 and 40 (inclusive) who are employed and have more than a high school education and at least one child.
   b. Characterize the distribution of annual incomes for the individuals who meet the conditions specified in part **a.** In particular, report the mean, median, and standard deviation for the resulting income distribution. Is this distribution skewed?
   c. For those Internet users who satisfy the conditions specified in part **a,** compare the mean salary of the men with that of the women. Also, compare the standard deviation of the salaries earned by the men with that of the salaries earned by the women.

30. ShirtCo is a direct competitor of the Fine Shirt Company described previously in Example 4.2. Like its rival, ShirtCo makes and sells shirts to its customers.

The main difference is that ShirtCo focuses its efforts on the creation and production of specialty T-shirts. The company has created an Access database file P04_30.MDB that contains information on sales to its customers from 1995 through 1998. There are two related tables in this database: Sales and Customer. Each of the 2245 records in the Sales table contains the order number, customer number (1–7), order data, channel of sale (wholesale or retail), type of T-shirt product (Art, Dinosaurs, Environment, Humorous, Kids, Political, or Sports), units sold, list price, total invoice amount, and amount paid by customer. Each of the 7 records in the Customer table contains the customer number, the customer's name, street address, city, state, zip code, country, phone number, and the date of first contact. Find all of the records from the Sales table where the order was placed in 1997 or 1998, the sale channel was retail, the product type was not Kids, and the number of units ordered was at least 400. Return to Excel all fields in the Sales table for each of these records.

31. Continuing with the ShirtCo database found in the file P04_30.MDB, find all of the records from the Sales table that correspond to orders for over 500 items made by the customer Shirts R Us for the Environment, Humorous, and Political products. Return to Excel the dates, sale channel, product type, units ordered, and amounts paid for each of these orders.

32. ShirtCo would like to know the total amount spent by each of its customers on each of its products during each of the years 1995–1998. Using the database given in the file P04_30.MDB, construct one or more pivot tables that provide ShirtCo with the desired information.

33. ShirtCo would also like to know the proportions sold through each channel (i.e., wholesale versus retail) for each of its products during each quarter of the years 1995–1998. Using the database given in the file P04_30.MDB, construct one or more pivot tables that provide ShirtCo with the desired information.

34. Who is most likely to access the Internet today? Consider the given survey data collected from 1000 randomly selected Internet users. The data are in the file P04_28.XLS. Are all of these data usable? Explain why or why not.

Level B

35. The file P04_35.XLS contains 1997 compensation data for chief executive officers from 350 of the nation's biggest businesses.
   a. Find all executives whose annual salary in fiscal 1997 was at least $1,000,000 and whose company type was either Energy or Cyclical. Find the average bonus earned by these chief executive officers in fiscal 1997.
   b. Find all executives whose annual salary in fiscal 1997 was less than $500,000 and whose company

type was either Technology or Noncyclical. Find the average bonus earned by these chief executive officers in fiscal 1997.

c. Find all executives whose annual salary in fiscal 1997 was between $500,000 and $1,000,000 (inclusive) and whose company type was either Basic Materials or Noncyclical. Find the average bonus earned by these chief executive officers in fiscal 1997.

36. Recall that the HyTex Company is a direct marketer of stereophonic equipment, personal computers, and other electronic products. The file P04_36.XLS contains recent data on 1000 HyTex customers.

a. Identify all customers in the sample who are 55 years of age or younger, female, single, and who have had at least some dealings with HyTex before this year. Find the average number of catalogs sent to these customers and the average amount spent by these customers this year. How strongly correlated are the numbers of catalogs sent and the amounts spent on HyTex purchases for these customers?

b. Do any of the customers who satisfy the conditions stated in part **a** have salaries that fall in the bottom 10% of all 1000 combined salaries in the sample? If so, how many?

c. Identify all customers in the sample who are more than 30 years of age or younger, male, homeowners, married, and who have had little if any dealings with HyTex before this year. Find the average combined household salary and the average amount spent by these customers this year. How strongly correlated are the combined household salaries and the amounts spent on HyTex purchases for these customers?

d. Do any of the customers who satisfy the conditions stated in part **a** have salaries that fall in the top 10% of all 1000 combined salaries in the sample? If so, how many?

37. Recall that the HyTex Company is a direct marketer of stereophonic equipment, personal computers, and other electronic products. The file P04_36.XLS contains recent data on 1000 HyTex customers.

a. Identify all customers in the given sample who are either (1) homeowners between the ages of 31 and 55 who live reasonably close to a shopping area that sells similar merchandise, and who have a combined salary between $40,000 and $90,000 (inclusive) and a history of being a medium- or high-spender at HyTex; or (2) homeowners greater than the age of 55 who live reasonably close to a shopping area that sells similar merchandise, and who have a combined salary between $40,000 and $90,000 (inclusive) and a history of being a medium- or high-spender at HyTex.

b. Characterize the subset of customers who satisfy the conditions specified in part **a**. In particular, what proportion of these customers are women? What proportion of these customers are married? On average, how many children do these cus-

tomers have? Finally, how many catalogs do these customers typically receive, and how much do they typically spend each year at HyTex?

c. In what ways are the customers who satisfy condition (1) in part **a** different from those who satisfy condition (2) in part **a**? Be as specific as possible.

38. Refer to Problem 37 with the data provided in the file P04_36.XLS. Find all of the customers in the given sample who satisfy each of the following conditions:

a. AmountSpent is at least $1000 greater than the median of AmountSpent for all customers.

b. AmountSpent is more than two standard deviations above the mean of AmountSpent for all customers.

c. Salary is no less than the 90th percentile of salaries for all customers.

39. ShirtCo is trying to determine who was its biggest customer, as measured by total units sold, in 1998. Once ShirtCo determines which customer was responsible for the maximum level of total unit sales, the company would then like to know the breakdown of this customer's 1998 total *expenditures* by product and channel. Using the database given in the file P04_30.MDB, construct pivot tables that provide ShirtCo with the desired information. Summarize your findings.

40. According to a survey conducted by New York compensation consultants William M. Mercer Inc. and published recently in *The Wall Street Journal* (April 9, 1998), chief executive officers from 350 of the nation's biggest businesses gained an 11.7% increase in salaries and bonuses in 1997. The data are in the file P04_35.XLS. Cleanse all usable data in this set.

41. The HyTex Company is a direct marketer of stereophonic equipment, personal computers, and other electronic products. HyTex advertises entirely by mailing catalogs to its customers, and all of its orders are taken over the telephone. The company spends a great deal of money on its catalog mailings, and wants to be sure that this is paying off in sales. Therefore, it has collected data on 1000 customers at the end of the current year. For each customer it has data on the following variables: Age (coded as 1 for 30 or younger, 2 for 31 to 55, 3 for 56 or older), Gender (coded as 1 for males, and 2 for females), OwnHome (coded as 1 if customer owns a home, and 2 otherwise), Married (coded as 1 if customer is currently married, and 2 otherwise), Close (coded as 1 if customer lives reasonably close to a shopping area that sells similar merchandise, and 2 otherwise), Salary (combined annual salary of customer and spouse, if applicable), Children (number of children living with the customer), History (coded as "NA" if customer had no dealings with the company before this year, 1 if customer was a low-spending customer last year, 2 if medium-spending, and 3 if high-spending), Catalogs (number of catalogs sent to the customer this year), and AmountSpent (total amount of purchases made by the customer this year). These data are provided in the file P04_36.XLS. Cleanse all usable data in this file.

4.1 EduToys, Inc.

EduToys, Inc., sells a wide variety of educational toy products to its customers through its Web site. Jeannie Dobson, director of information services at EduToys, recently developed a relational database to store critical information that the management team needs to more effectively serve EduToys' customers. The database, which is provided in the file P4_EduToys.MDB, consists of five related tables: Company, Customer, Inventory, Orders, and Toys.

The Company table consists of the following information on each of the 159 companies that manufacture and supply products to EduToys: identification number, name, and telephone number. The Customer table maintains the following data on each of the 307 customers who purchased at least one item from EduToys' electronic store during the first 10 months of operation (i.e., January–October 1998): identification number, last name, first name, age, gender, street address, city, state, zip code, and telephone number. The Inventory table consists of the following information on each of the 201 products that EduToys purchases from its various suppliers: identification number, name, quantity in current inventory, quantity on order, and expected delivery date of order. The Orders table records the following information for each of the customer transactions that took place during the first 10 months of 1998: transaction identification number, date, customer identification number, customer credit card number, product identification number, and quantity purchased. Finally, the Toys table maintains the following data on each of the products sold by EduToys: product identification number, company (i.e., supplier) identification number, product name, type of product, appropriate age group for product, unit price, and detailed product description.

As part of your internship with EduToys, you have been asked by your supervisor to prepare a memorandum that responds to the following questions. Your supervisor encourages you to make extensive use of the database in completing this assignment. Also, she wants you to retain copies of all Excel spreadsheets that you prepare to generate the needed information.

1. How do EduToys' past customers break down by age and gender?

2. Which of EduToys' past customers have spent amounts that fall in the top 20% of all transactions (as measured in dollars)? Report the first name, last name, street address, city, state, and zip code for each of these customers.

3. Which products have generated sales revenues (in dollars) that fall in the top 25% of all such revenue contributions? Report the current inventory level, quantity on order, and supplier of each of these best-selling products.

4. How do the given 1998 sales (in dollars) break down by product type and product age group?

5. What proportion of all given transactions were conducted through the use of each type of credit card (including American Express, Discover, MasterCard, and Visa)?

6. What changes or additions would you recommend making to the present database? Provide the reasoning behind each of your recommendations.

Probability and Probability Distributions

© Peter Samuels/CORBIS

Game at McDonald's

Several years ago McDonald's ran a campaign in which it gave game cards to its customers. These game cards made it possible for customers to win hamburgers, french fries, soft drinks, and other fast-food items, as well as cash prizes. Each card had 10 covered spots that could be uncovered by rubbing them with a coin. Beneath three of these spots were "zaps." Beneath the other seven spots were names of prizes, two of which were identical. (Some cards had variations of this pattern, but we'll use this type of card for purposes of illustration.) For example, one card might have two pictures of a hamburger, one picture of a Coke, one of french fries, one of a milk shake, one of $5, one of $1000, and three zaps. For this card the customer could win a hamburger. To win on any card, the customer had to uncover the two matching spots (which showed the potential prize for that card) before uncovering a zap; any card with a zap uncovered was automatically void. Assuming that the two matches and the three zaps were arranged randomly on the cards, what is the probability of a customer winning?

We'll label the two matching spots M_1 and M_2, and the three zaps Z_1, Z_2, and Z_3. Then the probability of winning is the probability of uncovering M_1 *and* M_2 before uncovering Z_1, Z_2, *or* Z_3. In this case the relevant set of outcomes is the set of all orderings of M_1, M_2, Z_1, Z_2, and Z_3, shown in the order they are uncovered. As far as the outcome of the game is concerned, the other five spots on the card are irrelevant. Then an outcome such as M_2, M_1, Z_3, Z_1, Z_2 is a winner, whereas M_2, Z_2, Z_1, M_1, Z_3 is a loser. Actually, the first of these would be declared a winner as soon as M_1 were uncovered, and the second would be declared a loser as soon as Z_2 were uncovered. However, we show the whole sequence of M's and Z's so that we can count outcomes correctly. We then find the probability of winning using the argument of equally likely probabilities. Specifically, we divide the number of outcomes that are winners by the total number of outcomes. It can be shown that the number of outcomes that are winners is 12, whereas the total number of outcomes is 120. Therefore, the probability of a winner is 12/120 = 0.1.

This calculation, which showed that on the average, 1 out of 10 cards could be winners, was obviously important for McDonald's. Actually, this provides only an upper bound on the fraction of cards where a prize was awarded. The fact is that many customers threw their cards away without playing the game, and even some of the winners neglected to claim their prizes. So, for example, McDonald's knew that if they made 50,000 cards where a milk shake was the winning prize, somewhat less than 5000 milk shakes would be given away. Knowing approximately what their expected "losses" would be from winning cards, McDonald's was able to design the game (how many cards of each type to print) so that the expected extra revenue (from customers attracted to the game) would cover the expected losses. ●

5.1 Introduction

A large part of the subject of statistics deals with uncertainty. Demands for products are uncertain, times between arrivals to a supermarket are uncertain, stock price returns are uncertain, changes in interest rates are uncertain, and so on. In these examples and many others, the uncertain quantity—demand, time between arrivals, stock price return, change in interest rate—is a numerical quantity. In the language of statistics and probability, such a numerical quantity is called a *random variable*. More formally, a **random variable** associates a numerical value with each possible outcome of a random phenomenon.

Associated with each random variable is a **probability distribution** that lists all of the possible values of the random variable and their corresponding probabilities. A probability distribution provides very useful information. It not only tells us the possible values of the random variable, but also how likely they are. For example, it is useful to know that the possible demands for a product are, say, 100, 200, 300, and 400, but it is even more useful to know that the probabilities of these four values are 0.1, 0.2, 0.4, and 0.3. Now we know, for example, that there is a 70% chance that demand will be at least 300.

It is often useful to summarize the information from a probability distribution with several well-chosen numerical summary measures. These include the mean, variance, and standard deviation, and, for distributions of more than one random variable, the covariance and correlation. As their names imply, these summary measures are much like the summary measures in Chapter 3. However, they are not identical. The summary measures in this chapter are based on probability distributions, not an observed data set. We will use numerical examples to explain the difference between the two—and how they are related.

The purpose of this chapter is to explain the basic concepts and tools necessary to work with probability distributions and their summary measures. We will begin by briefly discussing the basic rules of probability, which we will need in this chapter and in several later chapters. We will also introduce *computer simulation,* an extremely useful tool for illustrating important concepts in probability and statistics.

We conclude this chapter with a discussion of *weighted sums of random variables.* These are particularly useful in investment analysis, where we want to analyze portfolios of stocks or other securities. In this case the weights are the relative amounts invested in the securities, and the weighted sum represents the portfolio return. Our goal is to investigate the probability distribution of a weighted sum, particularly its summary measures.

5.2 Probability Essentials

We begin with a brief discussion of probability. The concept of probability is one that we all encounter in everyday life. When a weather forecaster states that the chance of rain is 70%, she is making a probability statement. When we hear that the odds of the Los Angeles Lakers winning the NBA Championship are 2 to 1, this is also a probability statement. The *concept* of probability is quite intuitive. However, the *rules* of probability are not always as intuitive or easy to master. We will examine the most important of these rules in this section.

Probability	A **probability** is a number between 0 and 1 that measures the likelihood that some event will occur. An event with probability 0 cannot occur, whereas an event with probability 1 is certain to occur. An event with probability greater than 0 and less than 1 involves uncertainty, but the closer its probability is to 1, the more likely it is to occur.

As the examples in the preceding paragraph illustrate, we often express probabilities as percentages or odds. However, these can easily be converted to probabilities on a 0–1 scale. If the chance of rain is 70%, then the probability of rain is 0.7. Similarly, if the odds of the Lakers winning are 2 to 1, then the probability of the Lakers winning is 2/3 (or 0.6667).

5.2.1 Rule of Complements

The simplest probability rule involves the *complement* of an event. If A is any event, then the **complement of A,** denoted by $\overline{A}$ (or in some books by A^c), is the event that A does *not* occur. For example, if A is the event that the Dow Jones Index will finish the year at or above the 11,000 mark, then the complement of A is that the Dow will finish the year below 11,000.

If the probability of A is $P(A)$, then the probability of its complement, $P(\overline{A})$, is given by equation (5.1). Equivalently, the probability of an event and the probability of its complement sum to 1. For example, if we believe that the probability of the Dow finishing at or above 11,000 is 0.55, then the probability that it will finish the year below 11,000 is $1 - 0.55 = 0.45$.

Rule of Complements	$$P(\overline{A}) = 1 - P(A) \qquad \text{(5.1)}$$

5.2.2 Addition Rule

We say that events are **mutually exclusive** if at most one of them can occur. That is, if one of them occurs, then none of the others can occur. For example, consider the following three events involving a company's annual revenue in 2002: (1) revenue is less than \$1 million, (2) revenue is at least \$1 million but less than \$2 million, and (3) revenue is at least \$2 million. Clearly, only one of these events can occur. Therefore, they are mutually exclusive. They are also **exhaustive,** which means that they exhaust all possibilities—one of these three events *must* occur.

Let A_1 through A_n be any n events. Then the *addition rule* of probability involves the probability that at least one of these events will occur. In general, this probability is quite complex, but it simplifies considerably when the events are mutually exclusive. In this case the probability that at least one of the events will occur is the sum of their individual probabilities, as shown in equation (5.2). Of course, when the events are mutually exclusive, "at least one" is equivalent to "exactly one." In addition, if the events A_1 through A_n are exhaustive, then the probability above is 1. In this case we are certain that one of the events will occur.

Addition Rule for Mutually Exclusive Events	$P(\text{at least one of } A_1 \text{ through } A_n) = P(A_1) + P(A_2) + \cdots + P(A_n)$ **(5.2)**

In a typical application, the events A_1 through A_n are chosen to partition the set of all possible outcomes into a number of mutually exclusive events. For example, in terms of a company's annual revenue, define A_1 as "revenue is less than \$1 million," A_2 as "revenue is at least \$1 million but less than \$2 million," and A_3 as "revenue is at least \$2 million." As we discussed previously, these three events are mutually exclusive and exhaustive. Therefore, their probabilities must sum to 1. Suppose these probabilities are $P(A_1) = 0.5$, $P(A_2) = 0.3$, and $P(A_3) = 0.2$. (Note that these probabilities *do* sum to 1.) Then the additive rule enables us to calculate other probabilities. For example, the event that revenue is at least \$1 million is the event that either A_2 or A_3 occurs. From the addition rule, its probability is

$$P(\text{revenue is at least \$1 million}) = P(A_2) + P(A_3) = 0.5$$

Similarly,

$$P(\text{revenue is less than \$2 million}) = P(A_1) + P(A_2) = 0.8$$

and

$$P(\text{revenue is less than \$1 million } or \text{ at least \$2 million}) = P(A_1) + P(A_3) = 0.7$$

5.2.3 Conditional Probability and the Multiplication Rule

Probabilities are always assessed relative to the information currently available. As new information becomes available, probabilities often change. For example, if you read that Kobe Bryant pulled a hamstring muscle, your assessment of the probability that the Lakers will win the NBA Championship would obviously change. A formal way to revise probabilities on the basis of new information is to use *conditional probabilities*.

Let A and B be any events with probabilities $P(A)$ and $P(B)$. Typically, the probability $P(A)$ is assessed without knowledge of whether B occurs. However, if we are *told* that B has occurred, then the probability of A might change. The new probability of A is called the **conditional probability** of A given B. It is denoted by $P(A|B)$. Note that there is still uncertainty involving the event to the left of the vertical bar in this notation; we do not know whether it will occur. However, there is no uncertainty involving the event to the right of the vertical bar; we *know* that it has occurred.

The **conditional probability formula** enables us to calculate $P(A|B)$ as shown in equation (5.3). The numerator in this formula is the probability that *both* A and B occur. This probability must be known to find $P(A|B)$. However, in some applications $P(A|B)$ and $P(B)$ are known. Then we can multiply both sides of the conditional probability formula by $P(B)$ to obtain the **multiplication rule** for $P(A$ and $B)$ in equation (5.4).

Conditional Probability	$$P(A\mid B) = \dfrac{P(A \text{ and } B)}{P(B)}$$	**(5.3)**

Multiplication Rule	$$P(A \text{ and } B) = P(A\mid B)P(B)$$	**(5.4)**

The conditional probability formula and the multiplication rule are both valid; in fact, they are equivalent. The one we use depends on which probabilities we know and which we want to calculate, as illustrated in the following example.

Example 5.1 Assessing Uncertainty at the Bendrix Company

The Bendrix Company supplies contractors with materials for the construction of houses. The company currently has a contract with one of its customers to fill an order by the end of July. However, there is some uncertainty about whether this deadline can be met, due to uncertainty about whether Bendrix will receive the materials it needs from one of its suppliers by the middle of July. Right now it is July 1. How can the uncertainty in this situation be assessed?

Objective To apply several of the essential probability rules in determining the likelihood that Bendrix will meet its end-of-July deadline, given the information the company has at the beginning of July.

Solution

Let A be the event that Bendrix meets its end-of-July deadline, and let B be the event that Bendrix receives the materials from its supplier by the middle of July. The probabilities Bendrix is best able to assess on July 1 are probably $P(B)$ and $P(A\mid B)$. At the beginning of July, Bendrix might estimate that the chances of getting the materials on time from its supplier are 2 out of 3, that is, $P(B) = 2/3$. Also, thinking ahead, Bendrix estimates that *if* it receives the required materials on time, the chances of meeting the end-of-July deadline are 3 out of 4. This is a conditional probability statement, namely, that $P(A\mid B) = 3/4$. Then we can use the multiplication rule to obtain

$$P(A \text{ and } B) = P(A\mid B)P(B) = (3/4)(2/3) = 0.5$$

That is, there is a 50–50 chance that Bendrix will get its materials on time *and* meet its end-of-July deadline.

The ∩ symbols in this figure mean "intersection." The event $A \cap B$ is the event where both A and B occur.

This uncertain situation is depicted graphically in the form of a *probability tree* in Figure 5.1 on page 194. Note that Bendrix initially faces (at the leftmost branch of the tree diagram) the uncertainty of whether event B or its complement will occur. Regardless of whether event B takes place, Bendrix must next confront the uncertainty regarding event A. This uncertainty is reflected in the set of two parallel pairs of branches that model the possibility that either event A or its complement could occur next. Hence, there are four mutually exclusive outcomes regarding the two uncertain events in this situation, as shown on the right-hand side of Figure 5.1. Initially, we are interested in the first possible outcome, the joint occurrence of events A and B, found at the top of the probability tree diagram. Another way to compute the probability of *both* events B and A occurring

FIGURE 5.1
Probability Tree
for Example 5.1

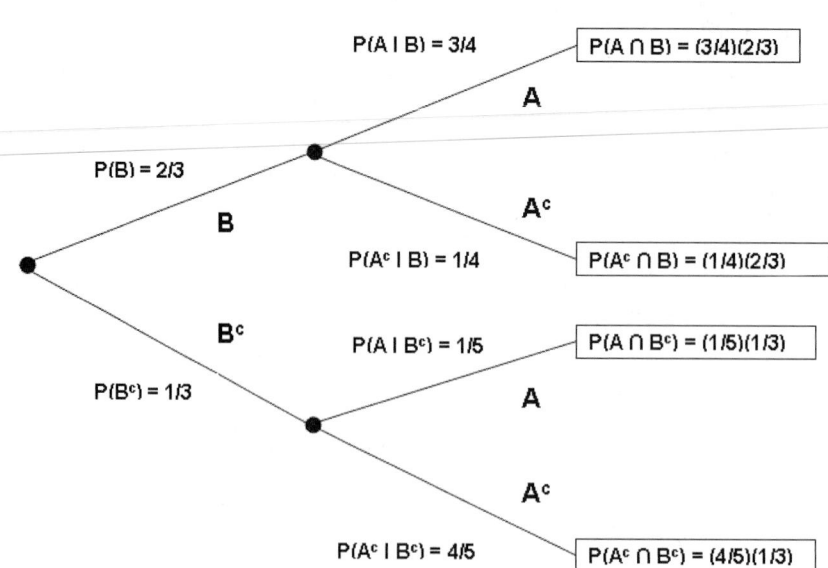

$P(A \mid B) = 3/4$

$P(A \cap B) = (3/4)(2/3)$

A

$P(B) = 2/3$

B

A^c

$P(A^c \mid B) = 1/4$

$P(A^c \cap B) = (1/4)(2/3)$

B^c

$P(A \mid B^c) = 1/5$

$P(A \cap B^c) = (1/5)(1/3)$

A

$P(B^c) = 1/3$

A^c

$P(A^c \mid B^c) = 4/5$

$P(A^c \cap B^c) = (4/5)(1/3)$

is to multiply the probabilities associated with the branches along the path from the root of the tree (on the left-hand side) to the desired terminal point or outcome of the tree (on the right-hand side). In this case, we multiply the probability of B, corresponding to the first branch along the path of interest, by the conditional probability of A given B, associated with the second branch along the path of interest.

There are several other probabilities of interest in this example. First, let $\overline{B}$ be the complement of B; it is the event that the materials from the supplier do *not* arrive on time. We know that $P(\overline{B}) = 1 - P(B) = 1/3$ from the rule of complements. However, we do not yet know the conditional probability $P(A|\overline{B})$, the probability that Bendrix will meet its end-of-July deadline, given that it does not receive the materials from the supplier on time. In particular, $P(A|\overline{B})$ is *not* equal to $1 - P(A|B)$. (Can you see why?) Suppose Bendrix estimates that the chances of meeting the end-of-July deadline are 1 out of 5 if the materials do not arrive on time, that is, $P(A|\overline{B}) = 1/5$. Then a second use of the multiplication rule gives

$$P(A \text{ and } \overline{B}) = P(A|\overline{B})P(\overline{B}) = (1/5)(1/3) = 0.0667$$

In words, there is only 1 chance out of 15 that the materials will not arrive on time *and* Bendrix will meet its end-of-July deadline.

Again, we can use the probability tree for Bendrix in Figure 5.1 to compute the probability of the joint occurrence of the complement of event B and event A. This outcome is the third (from the top of the diagram) terminal point of the tree. To find the desired probability, we multiply the probabilities corresponding to the two branches included in this path from the left-hand side of the tree to the right-hand side. Of course, we confirm that the probability of interest is the product of the two relevant probabilities, namely, 1/5 and 1/3. Simply stated, probability trees can be quite useful in modeling and assessing such uncertain outcomes in real-life situations.

The bottom line for Bendrix is whether it will meet its end-of-July deadline. After mid-July, this probability is either $P(A|B) = 3/4$ or $P(A|\overline{B}) = 1/5$ because by this time, Bendrix will *know* whether the materials arrived on time. But on July 1, the relevant probability is $P(A)$—there is still uncertainty about whether B or $\overline{B}$ will occur. Fortunately, we can calculate $P(A)$ from the probabilities we already know. The logic is that A consists of the two mutually exclusive events (A and B) and (A and $\overline{B}$). That is, if A is to occur, it must occur with B or with $\bar{B}$. Therefore, using the *addition* rule for mutually exclusive events, we obtain

$$P(A) = P(A \text{ and } B) + P(A \text{ and } \bar{B}) = 1/2 + 1/15 = 17/30 = 0.5667$$

The chances are 17 out of 30 that Bendrix will meet its end-of-July deadline, given the information it has at the beginning of July. ●

5.2.4 Probabilistic Independence

A concept that is closely tied to conditional probability is *probabilistic independence*. We just saw how the probability of an event A can depend on whether another event B has occurred. Typically, the probabilities $P(A)$, $P(A|B)$, and $P(A|\bar{B})$ are all different, as in Example 5.1. However, there are situations where all of these probabilities are equal. In this case we say that the events A and B are independent. This does *not* mean they are mutually exclusive. Rather, **probabilistic independence** means that knowledge of one event is of no value when assessing the probability of the other.

The main advantage to knowing that two events are independent is that the multiplication rule simplifies to equation (5.5). This follows by substituting $P(A)$ for $P(A|B)$ in the multiplication rule, which we are allowed to do because of independence. In words, the probability that both events occur is the product of their individual probabilities.

Multiplication Rule for Independent Events		
	$P(A \text{ and } B) = P(A)P(B)$	**(5.5)**

How can we tell whether events *are* probabilistically independent? Unfortunately, this issue usually cannot be settled with mathematical arguments; typically, we need empirical data to decide whether independence is reasonable. As an example, let A be the event that a family's first child is male, and let B be the event that its second child is male. Are A and B independent? We would argue that they aren't independent if we believe, say, that a boy is more likely to be followed by another boy than by a girl. We would argue that they are independent if we believe the chances of the second child being a boy are the same, regardless of the gender of the first child. (Note that neither argument has anything to do with boys and girls being equally likely.)

In any case, the only way to settle the argument is to observe many families with at least two children and see what occurs. If we observe, say, that 55% of all families with first child male also have the second child male, and only 45% of all families with first child female have the second child male, then we can make a good case for *nonindependence* of A and B.

5.2.5 Equally Likely Events

Much of what you know about probability is probably based on situations where outcomes are equally likely. These include flipping coins, throwing dice, drawing balls from urns, and other random mechanisms that are often discussed in books on probability. For example, suppose an urn contains 20 red marbles and 10 blue marbles. We plan to randomly select 5 marbles from the urn, and we are interested, say, in the probability of selecting at least 3 red marbles. To find this probability, we argue that because of randomness, every possible group of 5 marbles is equally likely to be chosen. Then we *count* the number of groups of 5 marbles that contain at least 3 red marbles, we count the total number of groups of 5 marbles that could be selected, and we set the desired probability equal to the ratio of these two counts.

Let us put this method of calculating probabilities into proper perspective. It is true that many probabilities, particularly in games of chance, *can* be calculated by using an equally likely argument. It is also true that probabilities calculated in this way satisfy all of the rules of probability, including the rules we have already discussed. However, many probabilities, especially those in business situations, *cannot* be calculated by equally likely arguments, simply because the possible outcomes are not equally likely. For example, just because we are able to identify five possible scenarios for a company's future, there is probably no reason whatsoever to conclude that each scenario has probability 1/5.

The bottom line is that we will have almost no need in this book to discuss counting rules for equally likely outcomes. If you dreaded learning about probability in terms of balls and urns, rest assured that you will *not* do so here!

5.2.6 Subjective Versus Objective Probabilities

In this section we ask a very basic question: Where do the probabilities in a probability distribution come from? A complete answer to this question could lead to a chapter by itself, so we will only briefly discuss the issues involved. There are essentially two distinct ways to assess probabilities, *objectively* and *subjectively*. **Objective probabilities** are those that can be estimated from long-run proportions, whereas **subjective probabilities** cannot be estimated from long-run proportions. Some examples will make this distinction clearer.

Consider throwing two dice and observing the sum of the two sides that face up. What is the probability that the sum of these two sides is 7? We might argue as follows. Because there are $6 \times 6 = 36$ ways the two dice can fall, and because exactly 6 of these result in a sum of 7, the probability of a 7 is $6/36 = 1/6$. This is the equally likely argument we discussed previously that reduces probability to counting. Let's look at the fundamental principle involved.

What if the dice are weighted in some way? Then the equally likely argument is no longer valid. We can, however, toss the dice many times and record the proportion of tosses that result in a sum of 7. This proportion is called a *relative frequency*.

Relative Frequency	The **relative frequency** of an event is the proportion of times the event occurs out of the number of times the random experiment is run. A relative frequency can be recorded as a proportion or a percentage.

A famous result called the **law of large numbers** states that this relative frequency, in the long run, will get closer and closer to the "true" probability of a 7. This is exactly what we mean by an objective probability. It is a probability that can be estimated as the long-run proportion of times an event occurs in a sequence of many identical experiments.

When it comes to flipping coins, throwing dice, and spinning roulette wheels, objective probabilities are certainly relevant. We don't need a person's *opinion* of the probability that a roulette wheel, say, will end up pointing to a red number; we can simply spin it many times and keep track of the proportion of times it points to a red number. However, there are many situations, particularly in business, that cannot be repeated many times—or even more than once—under identical conditions. In these situations objective probabilities make no sense (and equally likely arguments usually make no sense either), so we must resort to subjective probabilities. A subjective probability is one person's assessment of the likelihood that a certain event will occur. We assume that the person making the assessment uses all of the information available to make the most rational assessment possible.

This definition of subjective probability implies that one person's assessment of a certain probability might differ from another person's assessment of the *same* probability. For example, consider the probability that the Dallas Cowboys will win the next Super Bowl. If we ask a casual football observer to assess this probability, we'll get one answer, but if we ask a person with a lot of inside information about injuries, team cohesiveness, and so on, we might get a very different answer. Because these probabilities are *subjective,* people with different information typically assess probabilities in different ways.

Subjective probabilities are usually relevant for unique, one-time situations. However, most situations are not completely unique; we usually have some history to guide us. That is, historical relative frequencies can be assimilated into subjective probabilities. For example, suppose a company is about to market a new product. This product might be quite different in some ways from any products the company has marketed before, but it might also share some features with the company's previous products. If the company wants to assess the probability that the new product will be a success, it will certainly analyze the unique features of this product and the current state of the market to obtain a subjective assessment. However, the company will also look at its past successes and failures with reasonably similar products. If the proportion of successes with past products was 40%, then this value might be a starting point in the search for a subjective estimate of *this* product's probability of success.

All of the "given" probabilities in this chapter and later chapters can be placed somewhere on the objective-to-subjective continuum, usually closer to the subjective end. An important implication of this placement is that these probabilities are not cast in stone; they are only educated guesses. Therefore, it is always a good idea to run a **sensitivity analysis** (especially on a spreadsheet, where this is easy to do) to see how any "bottom-line" answers depend on the "given" probabilities. Sensitivity analysis is especially important in Chapter 7, when we study decision making under uncertainty.

PROBLEMS

Level A

1. In a particular suburb, 30% of the households have installed electronic security systems.
   a. If a household is chosen at random from this suburb, what is the probability that this household has not installed an electronic security system?
   b. If two households are chosen at random from this suburb, what is the probability that *neither* has installed an electronic security system?

2. Several major automobile producers are competing to have the largest market share for sport utility vehicles in the coming quarter. A professional automobile market analyst assesses that the odds of General Motors *not* being the market leader are 6 to 1. The odds against Chrysler and Ford having the largest market share in the coming quarter are similarly assessed to be 12 to 5 and 8 to 3, respectively.

   a. Find the probability that General Motors will have the largest market share for sport utility vehicles in the coming quarter.
   b. Find the probability that Chrysler will have the largest market share for sport utility vehicles in the coming quarter.
   c. Find the probability that Ford will have the largest market share for sport utility vehicles in the coming quarter.
   d. Find the probability that some other automobile manufacturer will have the largest market share for sport utility vehicles in the coming quarter.

3. The publisher of a popular financial periodical has decided to undertake a campaign in an effort to attract new subscribers. Market research analysts in this company believe that there is a 1 in 4 chance that the increase in the number of new subscriptions resulting

from this campaign will be less than 3000, and there is a 1 in 3 chance that the increase in the number of new subscriptions resulting from this campaign will be between 3000 and 5000. What is the probability that the increase in the number of new subscriptions resulting from this campaign will be less than 3000 *or* more than 5000?

4. Suppose that 18% of the employees of a given corporation engage in physical exercise activities during the lunch hour. Moreover, assume that 57% of all employees are male, and 12% of all employees are males who engage in physical exercise activities during the lunch hour.
   a. If we choose an employee at random from this corporation, what is the probability that this person is a female who engages in physical exercise activities during the lunch hour?
   b. If we choose an employee at random from this corporation, what is the probability that this person is a female who does not engage in physical exercise activities during the lunch hour?

5. In a study designed to gauge married women's participation in the workplace today, the data provided in the file P05_05.XLS are obtained from a sample of 750 randomly selected married women. Consider a woman selected at random from this sample in answering each of the following questions.
   a. What is the probability that this randomly selected woman has a job outside the home?
   b. What is the probability that this randomly selected woman has at least one child?
   c. What is the probability that this randomly selected woman has a full-time job and no more than one child?
   d. What is the probability that this randomly selected woman has a part-time job or at least one child but not both?

6. Suppose that we draw a single card from a standard deck of 52 playing cards.
   a. What is the probability that a diamond *or* club is drawn?
   b. What is the probability that the drawn card is not a 4?
   c. Given that a black card has been drawn, what is the probability that a spade has been drawn?
   d. Let E_1 be the event that a black card is drawn. Let E_2 be the event that a spade is drawn. Are E_1 and E_2 independent events? Why or why not?
   e. Let E_3 be the event that a heart is drawn. Let E_4 be the event that a 3 is drawn. Are E_3 and E_4 independent events? Why or why not?

Level B

7. In a large accounting firm, the proportion of accountants with MBA degrees and at least 5 years of pro-

fessional experience is 75% as large as the proportion of accountants with no MBA degree and less than 5 years of professional experience. Furthermore, 35% of the accountants in this firm have MBA degrees, and 45% have less than 5 years of professional experience. If one of the firm's accountants is selected at random, what is the probability that this accountant has an MBA degree or at least 5 years of professional experience but not both?

8. A local beer producer sells two types of beer, a regular brand and a light brand with 30% fewer calories. The company's marketing department wants to verify that its traditional approach of appealing to local white-collar workers with light beer commercials and appealing to local blue-collar workers with regular beer commercials is indeed a good strategy. A randomly selected group of 400 local workers are questioned about their beer-drinking preferences, and the data shown in the file P05_08.XLS are obtained.
   a. If a blue-collar worker is chosen at random from this group, what is the probability that she or he prefers light beer (to regular beer or no beer at all)?
   b. If a white-collar worker is chosen at random from this group, what is the probability that she or he prefers light beer (to regular beer or no beer at all)?
   c. If we restrict our attention to workers who like to drink beer, what is the probability that a randomly selected blue-collar worker prefers to drink light beer?
   d. If we restrict our attention to workers who like to drink beer, what is the probability that a randomly selected white-collar worker prefers to drink light beer?
   e. Does the company's marketing strategy appear to be appropriate? Explain why or why not.

9. Suppose that two dice are tossed. For each die, it is equally likely that 1, 2, 3, 4, 5, or 6 dots will show.
   a. What is the probability that the sum of the dots on the uppermost faces of the two dice will be 5 or 7?
   b. What is the probability that the sum of the dots on the uppermost faces of the two dice will be some number other than 4 or 8?
   c. Let E_1 be the event that the first die shows a 3. Let E_2 be the event that the sum of the dots on the uppermost faces of the two dice is 6. Are E_1 and E_2 independent events?
   d. Again, let E_1 be the event that the first die shows a 3. Let E_3 be the event that the sum of the dots on the uppermost faces of the two dice is 7. Are E_1 and E_3 independent events?
   e. Given that the sum of the dots on the uppermost faces of the two dice is 7, what is the probability that the first die showed 4 dots?
   f. Given that the first die shows a 3, what is the probability that the sum of the dots on the uppermost faces of the two dice is an even number?

5.3 Distribution of a Single Random Variable

We now discuss the topic of most interest in this chapter, probability distributions. In this section we examine the probability distribution of a single random variable. In later sections we will discuss probability distributions of two or more related random variables.

There are really two types of random variables: *discrete* and *continuous.* A **discrete random variable** has only a finite number of possible values, whereas a **continuous random variable** has a continuum of possible values.[1] As an example, consider demand for televisions. Is this discrete or continuous? Strictly speaking, it is discrete because the number of televisions demanded must be an integer. However, because the number of possible demand values is probably quite large—all integers between 1000 and 5000, say—it might be easier to treat demand as a continuous random variable. On the other hand, for reasons of simplicity we might go the other direction and treat demand as discrete with only a few possible values, such as 1000, 2000, 3000, 4000, and 5000. This is obviously an approximation to reality, but it might suffice for all practical purposes.

Mathematically, there is an important difference between discrete and continuous probability distributions. Specifically, a proper treatment of continuous distributions, analogous to the treatment we will provide in this chapter, requires calculus—which we do not presume for this book. Therefore, we will discuss only discrete distributions in this chapter. In later chapters we will often *use* continuous distributions, particularly the bell-shaped normal distribution, but we will simply state their properties, without trying to derive them mathematically.

The essential properties of a discrete random variable and its associated probability distribution are quite simple. We will discuss them in general and then analyze a numerical example. Let X be a random variable. (Usually, capital letters toward the end of the alphabet, such as X, Y, and Z, are used to denote random variables.)

To specify the probability distribution of X, we need to specify its possible values and their probabilities. We assume that there are k possible values, denoted $v_1, v_2, \ldots, v_k$. The probability of a typical value v_i is denoted in one of two ways, either $P(X = v_i)$ or $p(v_i)$. The first reminds us that this is a probability involving the random variable X, whereas the second is a simpler "shorthand" notation. Probability distributions must satisfy two criteria: (1) be nonnegative, and (2) sum to 1. In symbols, we must have

$$\sum_{i=1}^{k} p(v_i) = 1, \quad p(v_i) \geq 0$$

This is basically all there is to it: a list of possible values and a list of associated probabilities that sum to 1. Although this list of probabilities completely determines a probability distribution, it is sometimes useful to calculate *cumulative* probabilities. A **cumulative probability** is the probability that the random variable is *less than or equal to* some particular value. For example, assume that 10, 20, 30, and 40 are the possible values of a random variable X, with corresponding probabilities 0.15, 0.25, 0.35, and 0.25. Then a typical cumulative probability is $P(X \leq 30)$. From the addition rule it can be calculated as

$$P(X \leq 30) = P(X = 10) + P(X = 20) + P(X = 30) = 0.75$$

It is often convenient to summarize a probability distribution with two or three well-chosen numbers. The first of these is the *mean,* usually denoted μ. It is also called the

---

[1]Actually, a more rigorous discussion allows a discrete random variable to have an infinite number of possible values, such as all the positive integers. The only time this occurs in this book is when we discuss the Poisson distribution in Chapter 6.

expected value of X and denoted $E(X)$ (for *expected* X). The **mean** is a weighted sum of the possible values, weighted by their probabilities, as shown in equation (5.6). In much the same way that an average of a set of numbers indicates "central location," the mean indicates the center of the probability distribution. We will see this more clearly when we analyze a numerical example.

Mean of a Probability Distribution, μ	$$\mu = E(X) = \sum_{i=1}^{k} v_i p(v_i)$$	(5.6)

To measure the variability in a distribution, we calculate its *variance* or *standard deviation*. The **variance,** denoted by σ^2 or $\text{Var}(X)$, is a weighted sum of the squared deviations of the possible values from the mean, where the weights are again the probabilities. This is shown in equation (5.7).

As in Chapter 3, the variance is expressed in the *square* of the units of X, such as dollars squared. Therefore, a more natural measure of variability is the **standard deviation,** denoted by σ or $\text{Stdev}(X)$. It is the square root of the variance, as shown in equation (5.8).

Variance of a Probability Distribution, σ^2	$$\sigma^2 = \text{Var}(X) = \sum_{i=1}^{k} (v_i - E(X))^2\, p(v_i)$$	(5.7)

Standard Deviation of a Probability Distribution, σ	$$\sigma = \text{Stdev}(X) = \sqrt{\text{Var}(X)}$$	(5.8)

We now consider a typical example.

Example 5.2 Market Return Scenarios for the National Economy

An investor is concerned with the market return for the coming year, where the market return is defined as the percentage gain (or loss, if negative) over the year. The investor believes there are five possible scenarios for the national economy in the coming year: rapid expansion, moderate expansion, no growth, moderate contraction, and serious contraction. Furthermore, she has used all of the information available to her to estimate that the market returns for these scenarios are, respectively, 0.23, 0.18, 0.15, 0.09, and 0.03. That is, the possible returns vary from a high of 23% to a low of 3%. Also, she has assessed that the probabilities of these outcomes are 0.12, 0.40, 0.25, 0.15, and 0.08. Use this information to describe the probability distribution of the market return.

Objective To compute the mean, variance, and standard deviation of the probability distribution of the market return for the coming year.

FIGURE 5.2
Probability Distribution of Market Returns

	A	B	C	D
1	**Mean, variance, and standard deviation of a random variable**			
2				
3	Economic outcome	Probability	Market	Sq dev from mean
4	Rapid Expansion	0.12	0.23	0.005929
5	Moderate Expansion	0.40	0.18	0.000729
6	No Growth	0.25	0.15	0.000009
7	Moderate Contraction	0.15	0.09	0.003969
8	Serious Contraction	0.08	0.03	0.015129
9				
10	Mean return	0.153	**Range names**	
11	Variance of return	0.002811	Mean: B10	
12	Stdev of return	0.053	Probs: B4:B8	
13			Returns: C4:C8	
14			SqDevs: D4:D8	
15			Var: B11	
16				

Solution

To make the connection between the general notation and this particular example, we let X denote the market return for the coming year. Then each possible economic scenario leads to a possible value of X. For example, the first possible value is $v_1 = 0.23$, and its probability is $p(v_1) = 0.12$. These values and probabilities appear in columns B and C of Figure 5.2.[2] (See the file MARKETRETURN.XLS.) Note that the five probabilities sum to 1, as they should. This probability distribution implies, for example, that the probability of a market return at least as large as 0.18 is $0.12 + 0.40 = 0.52$, because it could occur as a result of rapid or moderate expansion of the economy. Similarly, the probability that the market return is 0.09 or less is $0.15 + 0.08 = 0.23$, because this could occur as a result of moderate or serious contraction of the economy.

The summary measures of this probability distribution appear in the range B10:B12. They can be calculated with the following steps.

Procedure for Calculating Summary Measures

❶ Mean return. Calculate the mean return in cell B10 with the formula

$$=\text{SUMPRODUCT(Returns,Probs)}$$

This formula illustrates the general rule in equation (5.6): The mean is the sum of products of possible values and probabilities.

❷ Squared deviations. To get ready to compute the variance, calculate the squared deviations from the mean by entering the formula

$$=(\text{C4-Mean})\char`\^2$$

in cell D4 and copying it down through cell D8.

❸ Variance. Calculate the variance of the market return in cell B11 with the formula

$$=\text{SUMPRODUCT(SqDevs,Probs)}$$

This illustrates the general formula for variance in equation (5.7): The variance is always a sum of products of squared deviations from the mean and probabilities.

As always, range names are not required, but they make the Excel formulas easier to read. You can use them or omit them, as you wish.

[2]From here on, we will often shade the given inputs in the spreadsheet figures gray. This way you can see which values are given and which are calculated. We will also box in the range names used.

4 **Standard deviation.** Calculate the standard deviation of the market return in cell B12 with the formula

$$=SQRT(Var)$$

We see that the mean return is 15.3% and the standard deviation is 5.3%. What do these measures really mean? First, the mean, or *expected,* return does not imply that the most likely return is 15.3%, nor is this the value that the investor "expects" to occur. In fact, the value 15.3% is not even a possible market return (at least not according to the model). We can understand these measures better in terms of long-run averages. That is, if we could imagine the coming year being repeated many times, each time using the probability distribution in columns B and C to generate a market return, then the average of these market returns would be close to 15.3%, and their standard deviation—calculated as in Chapter 3—would be close to 5.3%. ●

PROBLEMS

Level A

10. A fair coin (i.e., heads and tails are equally likely) is tossed three times. Let X be the number of heads observed in three tosses of this fair coin.
  a. Find the probability distribution of X.
  b. Compute the probability that two or fewer heads are observed in three tosses.
  c. Compute the probability that at least one head is observed in three tosses.
  d. Find the expected value of X.
  e. Find the standard deviation of X.

11. Consider a random variable with the following probability distribution: $P(X = 0) = 0.1$, $P(X = 1) = 0.2$, $P(X = 2) = 0.3$, $P(X = 3) = 0.3$, and $P(X = 4) = 0.1$.
  a. Find $P(X \leq 2)$.
  b. Find $P(1 < X \leq 3)$.
  c. Find $P(X > 0)$.
  d. Find $P(X > 3 | X > 2)$.
  e. Find the expected value of X.
  f. Find the standard deviation of X.

12. A study has shown that the probability distribution of X, the number of customers in line (including the one being served, if any) at a checkout counter in a department store, is given by $P(X = 0) = 0.25$, $P(X = 1) = 0.25$, $P(X = 2) = 0.20$, $P(X = 3) = 0.20$, and $P(X \geq 4) = 0.10$. Consider a newly arriving customer to the checkout line.
  a. What is the probability that this customer will not have to wait behind anyone?
  b. What is the probability that this customer will have to wait behind at least one customer?
  c. On average, behind how many other customers will the newly arriving customer have to wait?

13. A construction company has to complete a project no later than 3 months from now or there will be significant cost overruns. The manager of the construction company believes that there are four possible values for the random variable X, the number of months from now it will take to complete this project: 2, 2.5, 3, and 3.5. The manager currently thinks that the probabilities of these four possibilities are in the ratio 1 to 2 to 4 to 2. That is, $X = 2.5$ is twice as likely as $X = 2$, $X = 3$ is twice as likely as $X = 2.5$, and $X = 3.5$ is half as likely as $X = 3$.
  a. Find the probability distribution of X.
  b. What is the probability that this project will be completed in less than 3 months from now?
  c. What is the probability that this project will *not* be completed on time?
  d. What is the expected completion time (in months) of this project from now?
  e. How much variability (in months) exists around the expected value you found in part **d?**

14. A corporate executive officer is attempting to arrange a meeting of his three vice presidents for tomorrow morning. He believes that each of these three busy individuals, independently of the others, has about a 60% chance of being able to attend the meeting.
  a. Find the probability distribution of X, the number of vice presidents who can attend the meeting.
  b. What is the probability that none of the three vice presidents can attend the meeting?
  c. If the meeting will be held tomorrow morning only if everyone can attend, what is the probability that the meeting will take place at that time?
  d. How many of the vice presidents should the CEO expect to be available for tomorrow morning's meeting?

Level B

15. Several students enrolled in a finance course subscribe to *Money* magazine. If two students are selected at random from this class, the probability that neither of the chosen students subscribes to *Money* is 0.81. Furthermore, the probability of selecting one student who subscribes and one student who does not subscribe to this magazine is 0.18. Finally, the probability of selecting two students who subscribe to *Money* is 0.01. Let *X* be the number of students who subscribe to *Money* magazine from the two selected at random. Find the mean and standard deviation of *X*.

16. The "house edge" in any game of chance is defined as

$$\frac{E(\text{player's loss on a bet})}{\text{Size of player's loss on a bet}}$$

For example, if a player wins $10 with probability 0.48 and loses $10 with probability 0.52 on any bet, then the house edge is

$$\frac{-[10(0.48) - 10(0.52)]}{10} = 0.04$$

Give an interpretation to the house edge that relates to how much money the house is likely to win on average. Which do you think has a larger house edge: roulette or sports gambling? Why?

An Introduction to Simulation

In the previous section we asked you to "imagine" many repetitions of an event, with each repetition resulting in a different random outcome. Fortunately, we can do more than *imagine;* we can make it happen with computer *simulation.* Simulation is an extremely useful tool that can be used to incorporate uncertainty explicitly into spreadsheet models. As we will see, a simulation model is the same as a regular spreadsheet model except that some cells include random quantities. Each time the spreadsheet recalculates, new values of the random quantities occur, and these typically lead to different "bottom-line" results. By forcing the spreadsheet to recalculate many times, a business manager is able to discover the results that are most likely to occur, those that are least likely to occur, and best-case and worst-case results. We will use simulation in several places in this book to help explain difficult concepts in probability and statistics. We begin in this section by using simulation to explain the connection between summary measures of probability distributions and the corresponding summary measures from Chapter 3.

We continue to use the market return distribution in Figure 5.2 from Example 5.2. Because this is our first discussion of computer simulation in Excel, we proceed in some detail. Our goal is to simulate many returns (we arbitrarily choose 400) from this distribution and analyze the resulting returns. We want each simulated return to have probability 0.12 of being 0.23, probability 0.40 of being 0.18, and so on. Then, using the methods for summarizing data from Chapter 3, we will calculate the average and standard deviation of the 400 simulated returns.

The method for simulating many market returns is straightforward once we know how to simulate a *single* market return. The key to this is Excel's RAND function, which generates a random number between 0 and 1. The RAND function has no arguments, so every time we call it, we enter RAND(). (Although there is nothing inside the parentheses next to RAND, the parentheses cannot be omitted.) That is, to generate a random number between 0 and 1 in any cell, we enter the formula

$$=\text{RAND}()$$

in that cell. The RAND function can also be used as part of another function. For example, we can simulate the result of a single flip of a fair coin by entering the formula

$$=\text{IF(RAND()}<=0.5,\text{"Heads","Tails")}$$

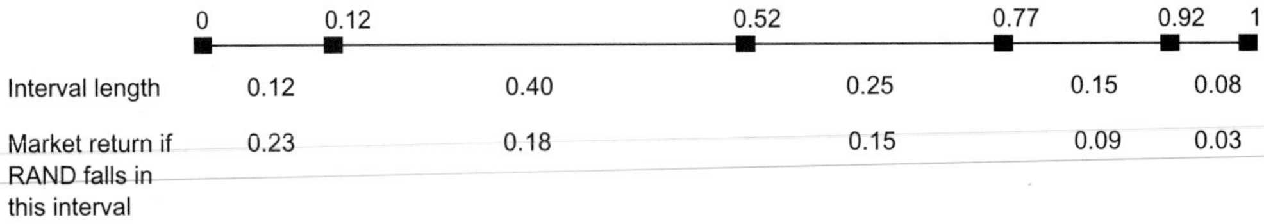

	0	0.12		0.52	0.77	0.92	1
Interval length	0.12	0.40		0.25	0.15	0.08	
Market return if RAND falls in this interval	0.23	0.18		0.15	0.09	0.03	

FIGURE 5.3 Associating RAND Values with Market Returns

Random numbers generated with Excel's RAND function are said to be **uniformly distributed** between 0 and 1 because all decimal values between 0 and 1 are equally likely. These uniformly distributed random numbers can then be used to generate numbers from any discrete distribution such as the market return distribution in Figure 5.2. To see how this is done, note first that there are five possible values in this distribution. Therefore, we divide the interval from 0 to 1 into five parts with lengths equal to the probabilities in the probability distribution. Then we see which of these parts the random number from RAND falls into and generate the associated market return. If the random number is between 0 and 0.12 (of length 0.12), we generate 0.23 as the market return; if the random number is between 0.12 and 0.52 (of length 0.40), we generate 0.18 as the market return; and so on. See Figure 5.3.

This procedure is accomplished most easily in Excel through the use of a *lookup table*. Lookup tables are useful when we want to compare a particular value to a set of values, and depending on where the particular value falls, assign a given "answer" or value from an associated list of values. In this case we want to compare a generated random number to values (between 0 and 1) falling in each of the five intervals shown in Figure 5.3, and then report the corresponding market return. This process is made relatively simple in Excel by applying the VLOOKUP function, as explained in the following steps.[3] (Refer to Figure 5.4 and the MARKETRETURN.XLS file.)

FIGURE 5.4
Simulation of Market Returns

	A	B	C	D	E
1	Simulating market returns				
2					
3	Summary statistics from simulation below				
4	Average return	0.154			
5	Stdev of returns	0.054			
6					
7	Exact values from previous sheet (for comparison)				
8	Average return	0.153			
9	Stdev of returns	0.053			
10					
11	Simulation			Lookup table	
12	Random #	Market return		CumProb	Return
13	0.662320	0.15		0	0.23
14	0.851650	0.09		0.12	0.18
15	0.267725	0.18		0.52	0.15
16	0.537825	0.15		0.77	0.09
17	0.190522	0.18		0.92	0.03
18	0.330463	0.18			
19	0.686100	0.15	Range names		
20	0.401385	0.18	Ltable: D13:E17		
21	0.213011	0.18	SimReturns: B13:B412		
22	0.187867	0.18			
410	0.023228	0.23			
411	0.330127	0.18			
412	0.386493	0.18			

[3]This could also be accomplished with nested IF functions, but the resulting formula would be much more complex.

Procedure for Generating Random Market Returns in Excel

1. Lookup table. Copy the possible returns to the range E13:E17. Then enter the *cumulative* probabilities next to them in the range D13:D17. To do this, enter the value 0 in cell D13. Then enter the formula

$$=D13+Market!B4$$

in cell D14 and copy it down through cell D17. (Note that the Market!B4 in this formula refers to cell B4 in the Market sheet, that is, cell B4 in Figure 5.2.) Each value in column D is the current probability plus the previous value. The table in this range, D13:E17, becomes the lookup range. For convenience, we have named this range LTable.

2. Random numbers. Enter random numbers in the range A13:A412. An easy way to do this is to highlight the range, then type the formula

$$=RAND()$$

and finally press Ctrl-Enter. Note that these random numbers are "live." That is, each time you do any calculation in Excel or press the recalculation key (the F9 key), these random numbers change.

3. Market returns. Generate the random market returns by referring the random numbers in column A to the lookup table. Specifically, enter the formula

$$=VLOOKUP(A13,LTable,2)$$

in cell B13 and copy it down through cell B412. This formula compares the random number in cell A13 to the cumulative probabilities in the first column of the lookup table and sees where it "fits," as illustrated in Figure 5.3. Then it returns the corresponding market return in the second column of the lookup table. (It uses the *second* column because we set the third argument of the VLOOKUP function to 2.)

Excel Tip *In general, the VLOOKUP function takes three arguments: (1) the value to be compared, (2) a table of lookup values, with the values to be compared against always in the leftmost column, and (3) the column number of the lookup table where we find the "answer."*

4. Summary statistics. Summarize the 400 market returns by entering the formulas

$$=AVERAGE(SimReturns)$$

and

$$=STDEV(SimReturns)$$

in cells B4 and B5. For comparison, copy the average and standard deviation from the Market sheet in Figure 5.2 to cells B8 and B9.

Now let's step back and see what we've accomplished. The following points are relevant.

- Simulations such as this are very common, and we will continue to use them to illustrate concepts in probability and statistics.

- The numbers you obtain will be different from the ones in Figure 5.4 because of the nature of simulation. The results depend on the particular random numbers that happen to be generated.

- The way we entered cumulative probabilities and then used a lookup table is generally the best way to generate random numbers from a discrete probability distribution. However, there is an easier way if a simulation add-in is available.

- Each generated market return in the SimReturns range is one of the five possible market returns. If you count the number of times each return appears and then divide by 400, the number of simulated values, you'll see that the resulting fractions are *approximately* equal to the original probabilities. For example, the fraction of times the highest return 0.23 appears is about 0.12. This is the essence of what it means to simulate from a given probability distribution.

- The average and standard deviation in cells B4 and B5, calculated from the formulas in Chapter 3, are very close to the mean and standard deviation of the probability distribution in cells B8 and B9. Note, however, that these measures are calculated in entirely different ways. For example, the average in cell B4 is a simple average of 400 numbers, whereas the mean in cell B8 is a weighted sum of the possible market returns, weighted by their probabilities.

This last point allows us to interpret the summary measures of a probability distribution. Specifically, the mean and standard deviation of a probability distribution are approximately what we would obtain if we calculated the average and standard deviation, using the formulas from Chapter 3, of many simulated values from this distribution. In other words, the mean is the long-run average of the simulated values. Similarly, the standard deviation measures their variability.

You might ask whether this long-run average interpretation of the mean is relevant if the situation is going to occur only once. For example, the market return in the example is for "the coming year," and the coming year is going to occur only once. So what is the use of a long-run average? In this type of situation, the long-run average interpretation is probably *not* very relevant, but fortunately, there is another use of the expected value that we will exploit in Chapter 7—namely, when a decision maker must choose among several actions that have uncertain outcomes, the preferred decision is often the one with the largest expected (monetary) value. This makes the expected value of a probability distribution extremely important in decision-making contexts.

PROBLEMS

Level A

17. A personnel manager of a large manufacturing plant is investigating the number of reported on-the-job accidents at the facility over the past several years. Let X be the number of such accidents reported during a 1-month period. Based on past records, the manager has established the probability distribution for X as shown in the file P05_17.XLS.
    a. Generate 400 values of this random variable X with the given probability distribution using computer simulation.
    b. Compare the distribution of simulated values to the given probability distribution. Is the simulated distribution indicative of the given probability distribution? Explain why or why not.

18. A quality inspector picks a sample of 15 items at random from a manufacturing process known to produce 10% defective items. Let X be the number of defective items found in the random sample of 15 items. Assume that the condition of each item is independent of that of each of the other items in the sample. The probability distribution of X is provided in the file P05_18.XLS.
    a. Generate 500 values of this random variable with the given probability distribution using computer simulation.
    b. Compute the mean and standard deviation of the distribution of simulated values. How do these summary measures compare to the mean and standard deviation of the given probability distribution?

19. The file P05_19.XLS gives the probability distribution for the number of job applications processed at a small employment agency during a typical week.

a. Generate 400 values of this random variable with the given probability distribution using computer simulation.

b. Compute the mean and standard deviation of the distribution of simulated values. How do these summary measures compare to the mean and standard deviation of the given probability distribution?

c. Use your simulated distribution to find the probability that the weekly number of job applications processed will be within two standard deviations of the mean.

20. Consider a random variable with the following probability distribution: $P(X = 0) = 0.1$, $P(X = 1) = 0.2$, $P(X = 2) = 0.3$, $P(X = 3) = 0.3$, and $P(X = 4) = 0.1$.

a. Generate 400 values of this random variable with the given probability distribution using computer simulation.

b. Compare the distribution of simulated values to the given probability distribution. Is the simulated distribution indicative of the given probability distribution? Explain why or why not.

c. Compute the mean and standard deviation of the distribution of simulated values. How do these summary measures compare to the mean and standard deviation of the given probability distribution?

21. The probability distribution of X, the number of customers in line (including the one being served, if any) at a checkout counter in a department store, is given by $P(X = 0) = 0.25$, $P(X = 1) = 0.25$, $P(X = 2) = 0.20$, $P(X = 3) = 0.20$, and $P(X = 4) = 0.10$.

a. Generate 500 values of this random variable with the given probability distribution using computer simulation.

b. Compare the distribution of simulated values to the given probability distribution. Is the simulated distribution indicative of the given probability distribution? Explain why or why not.

c. Compute the mean and standard deviation of the distribution of simulated values. How do these summary measures compare to the mean and standard deviation of the given probability distribution?

Level B

22. Betting on a football point spread works as follows. Suppose Michigan is favored by 17.5 points over Indiana. If you bet a "unit" on Indiana and Indiana loses by 17 or less, you win $10. If Indiana loses by 18 or more points, you lose $11. Find the mean and standard deviation of your winnings on a single bet. Assume that there is a 0.5 probability that you will win your bet and a 0.5 probability that you will lose your bet. Also simulate 1600 "bets" to estimate the average loss per bet. (*Note:* Do not be too disappointed if you are off by up to 50 cents. It takes many, say 10,000, simulated bets to get a really good estimate of the mean loss per bet because there is a lot of variability on each bet.)

5.5 Distribution of Two Random Variables: Scenario Approach

We now turn to the distribution of two related random variables. In this section we discuss the situation where the two random variables are related in the sense that they both depend on which of several possible scenarios occurs. In the next section we will discuss a second way of relating two random variables probabilistically. These two methods differ slightly in the way they assign probabilities to different outcomes. However, for both methods there are two summary measures, *covariance* and *correlation,* that measure the relationship between the two random variables. As with the mean, variance, and standard deviation, covariance and correlation are similar to the measures with the same names from Chapter 3, but they are conceptually different. In Chapter 3, correlation and covariance were calculated from data; here they are calculated from a probability distribution.

If the random variables are X and Y, then we denote the covariance and correlation between X and Y by Cov(X, Y) and Corr(X, Y). These are defined by equations (5.9) and (5.10). Here, $p(x_i, y_i)$ in equation (5.9) is the probability that X and Y equal the values x_i and y_i; it is called a **joint probability.**

Formula for Covariance	$$\text{Cov}(X, Y) = \sum_{i=1}^{k} (x_i - E(X))(y_i - E(Y))p(x_i, y_i)$$	**(5.9)**

Formula for Correlation	$$\text{Corr}(X, Y) = \frac{\text{Cov}(X, Y)}{\text{Stdev}(X) \times \text{Stdev}(Y)}$$	**(5.10)**

Although covariance and correlation based on a joint probability distribution are calculated differently than for known data, their interpretation is essentially the same as that discussed in Chapter 3. Each indicates the strength of a linear relationship between X and Y. That is, if X and Y tend to vary in the *same* direction, then both measures are positive. If they vary in *opposite* directions, both measures are negative. As before, the magnitude of the covariance is more difficult to interpret because it depends on the units of measurement of X and Y. However, the correlation is always between -1 and $+1$.

The following example illustrates the scenario approach, as well as covariance and correlation. Simulation is used to explain the relationship between the covariance and correlation as defined here and the similar measures from Chapter 3.

Example (5.3) Analyzing a Portfolio of Investments in GM Stock and Gold

An investor plans to invest in General Motors (GM) stock and in gold. He assumes that the returns on these investments over the next year depend on the general state of the economy during the year. To keep things simple, he identifies four possible states of the economy: depression, recession, normal, and boom. Also, given the most up-to-date information he can obtain, he assumes that these four states have probabilities 0.05, 0.30, 0.50, and 0.15. For each state of the economy, he estimates the resulting return on GM stock and the return on gold. These appear in the shaded section of Figure 5.5. (See the file GMGOLD.XLS.) For example, if there is a depression, the investor estimates that GM stock will decrease by 20% and the price of gold will increase by 5%. The investor wants to analyze the joint distribution of returns on these two investments. He also wants to analyze the distribution of a portfolio of investments in GM stock and gold.

Objective To obtain the relevant joint distribution and use it to calculate the covariance and correlation between returns on the two given investments.

Solution

To obtain the joint distribution, we use the distribution of GM return, defined by columns B and C of the shaded region in Figure 5.5, and the distribution of gold return, defined by columns B and D. The scenario approach applies because a given state of the economy determines *both* GM and gold returns, so that only four pairs of returns are possible. For example, -0.20 is a possible GM return and 0.09 is a possible gold return, but they cannot occur simultaneously. The only possible *pairs* of returns are -0.20 and 0.05, 0.10 and 0.20, 0.30 and -0.12, and 0.50 and 0.09. These possible pairs have the joint probabilities shown in column B.

To calculate means, variances, and standard deviations, we treat GM and gold returns separately. For example, the formula for the mean GM return in cell B10 is

=SUMPRODUCT(GMReturns,Probs)

	A	B	C	D	E	F	G	H
1	Calculating covariance and correlation between two random variable							
2								
3	Economic outcome	Probability	GM Return	Gold Return				
4	Depression	0.05	-0.20	0.05		Selected range names		
5	Recession	0.30	0.10	0.20		Covar: B23		
6	Normal	0.50	0.30	-0.12		GMDevs: B14:B17		
7	Boom	0.15	0.50	0.09		GMMean: B10		
8						GMReturns: C4:C7		
9		GM	Gold			GMSqDevs: D14:D17		
10	Means	0.245	0.016			GMStdev: B21		
11						GoldDevs: C14:C17		
12		Deviations from means		Sq devs from means		GoldMean: C10		
13		GM	Gold	GM	Gold	GoldReturns: D4:D7		
14	Depression	-0.45	0.03	0.1980	0.0012	GoldSqDevs: E14:E17		
15	Recession	-0.15	0.18	0.0210	0.0339	GoldStdev: C21		
16	Normal	0.06	-0.14	0.0030	0.0185	Probs: B4:B7		
17	Boom	0.26	0.07	0.0650	0.0055			
18								
19		GM	Gold					
20	Variances	0.0275	0.0203					
21	Stdevs	0.166	0.142					
22								
23	Covariance	-0.0097						
24	Correlation	-0.410						

FIGURE 5.5 Distribution of GM and Gold Returns

We don't necessarily advise using this many range names, but we have used them here to help clarify the meaning of the Excel formulas.

The only new calculations in Figure 5.5 involve the covariance and correlation between GM and gold returns. To obtain these, we use the following steps.

Procedure for Calculating the Covariance and Correlation

❶ **Deviations between means.** The formula for covariance [equation (5.9)] is a weighted sum of deviations from means (not squared deviations), so we first need to calculate these deviations. To do this, enter the formula

$$=C4-GMMean$$

in cell B14 and copy it down through cell B17. Calculate the deviations for gold similarly in column C.

❷ **Covariance.** Calculate the covariance between GM and gold returns in cell B23 with the formula

$$=SUMPRODUCT(GMDevs,GoldDevs,Probs)$$

Note the use of the SUMPRODUCT function in this formula. It usually takes two range arguments, but it is allowed to take more than two, all of which must have exactly the same dimension. This function multiplies corresponding elements from each of the three ranges and sums these products—exactly as prescribed by equation (5.9).

❸ **Correlation.** Calculate the correlation between GM and gold returns in cell B24 with the formula

$$=Covar/(GMStdev*GoldStdev)$$

as prescribed by equation (5.10).

The negative covariance indicates that GM and gold returns tend to vary in opposite directions, although it is difficult to judge the strength of the relationship between them by the magnitude of the covariance. The correlation of −0.410, on the other hand, is also

FIGURE 5.6
Simulation of GM
and Gold Returns

	A	B	C	D	E	F	G
1	**Simulating GM and Gold returns**						
2							
3	**Summary measures from simulation below**						
4		GM	Gold				
5	Means	0.248	0.024		Selected range names		
6	Stdevs	0.171	0.142		LTable: E21:G24		
7					SimGM: B21:B420		
8	Covariance	-0.0092			SimGold: C21:C420		
9	Correlation	-0.382					
10							
11	**Exact results from previous sheet (for comparison)**						
12		GM	Gold				
13	Means	0.245	0.016				
14	Stdevs	0.166	0.142				
15							
16	Covariance	-0.0097					
17	Correlation	-0.410					
18							
19	**Simulation results**				Lookup table for generating returns		
20	Random #	GM return	Gold return		CumProb	GM return	Gold return
21	0.4082230	0.30	-0.12		0.00	-0.20	0.05
22	0.3696801	0.30	-0.12		0.05	0.10	0.20
23	0.8393543	0.30	-0.12		0.35	0.30	-0.12
24	0.5103072	0.30	-0.12		0.85	0.50	0.09
25	0.9379595	0.50	0.09				
26	0.9114591	0.50	0.09				
418	0.1803136	0.10	0.20				
419	0.0709424	0.10	0.20				
420	0.9369361	0.50	0.09				

This simulation is
not necessary for
the calculation of
the covariance and
correlation, but it
provides some in-
sight into their
meaning.

negative and indicates a moderately strong relationship. We can't rely too much on this correlation, however, because the relationship between GM and gold returns is *not* linear. From the values in the range C4:D7, we see that GM does better and better as the economy improves, whereas gold does better, then worse, then better.

A simulation of GM and gold returns helps to explain the covariance and correlation measures. This simulation is shown in Figure 5.6. There are two keys to this simulation. First, we simulate the states of the economy, not—at least not directly—the GM and gold returns. For example, any random number between 0.05 and 0.35 implies a recession. The returns for GM and gold from a recession are then known to be 0.10 and 0.20. We implement this idea by entering a RAND function in cell A21 and then entering the formulas

$$=VLOOKUP(A21,LTable,2)$$

and

$$=VLOOKUP(A21,LTable,3)$$

in cells B21 and C21. These formulas are then copied down through row 420. This way, the *same* random number—hence the same scenario—is used to generate both returns in a given row, and the effect is that only four *pairs* of returns are possible.

Second, once we have the simulated returns in the range B21:C420, we can calculate the covariance and correlation of these numbers in cells B8 and B9 with the formulas[4]

$$=COVAR(SimGM,SimGold)$$

and

$$=CORREL(SimGM,SimGold)$$

---

[4]These formulas implement the covariance and correlation definitions from Chapter 3, not equations (5.9) and (5.10) of this chapter because these formulas are based on given data.

	A	B	C	D	E	F	G	H	I	J
1	Analyzing a portfolio of GM and Gold									
2										
3	Total to invest	$10,000								
4										
5	Investments	GM	Gold							
6	Fraction of total	0.60	0.40							
7	Dollar value	$6,000	$4,000							
8										
9	Distribution of portfolio returns				Sq devs from means					
10	Economic outcome	Per dollar	Total		Per dollar	Total				
11	Depression	-0.1	($1,000)		0.064212	6421156				
12	Recession	0.14	$1,400		0.00018	17956				
13	Normal	0.132	$1,320		0.000458	45796				
14	Boom	0.336	$3,360		0.033343	3334276				
15										
16	Summary measures of portfolio returns									
17		Per dollar	Total							
18	Mean return	0.153	$1,534.00							
19	Variance of return	0.008495	849484							
20	Stdev of return	0.092167	$921.67							
21										
22	Data table for mean and stdev of portfolio return as a function of GM investment									
23	GM investment	Mean	StDev							
24		$1,534.00	$921.67							
25	0.0	$160.00	$1,424.22							
26	0.1	$389.00	$1,223.28							
27	0.2	$618.00	$1,048.16							
28	0.3	$847.00	$913.81							
29	0.4	$1,076.00	$840.04							
30	0.5	$1,305.00	$842.90							
31	0.6	$1,534.00	$921.67							
32	0.7	$1,763.00	$1,059.57							
33	0.8	$1,992.00	$1,236.97							
34	0.9	$2,221.00	$1,439.34							
35	1.0	$2,450.00	$1,657.56							

FIGURE 5.7 Distribution of Portfolio Return

Here, COVAR and CORREL are the built-in Excel functions discussed in Chapter 3 for calculating the covariance and correlation between pairs of numbers. A comparison of cells B8 and B9 with B16 and B17 shows that there is a reasonably good agreement between the covariance and correlation of the probability distribution [from equations (5.9) and (5.10)] and the measures based on the simulated values. This agreement is not perfect, but it typically improves as we simulate more pairs (because of the law of large numbers).

The final question in this example involves a portfolio consisting of GM stock and gold. The analysis appears in Figure 5.7. We assume that the investor has $10,000 to invest. He puts some fraction of this in GM stock (see cell B6) and the rest in gold. Of course, these fractions determine the total dollar values invested in row 7. The key to the analysis is the following. Because there are only four possible scenarios, there are only four possible portfolio returns. For example, if there is a recession, the GM and gold returns are 0.10 and 0.20, so the portfolio return (per dollar) is a weighted average of these returns, weighted by the fractions invested:

Portfolio return in recession = 0.6(0.10) + 0.4(0.20) = 0.14

In this way, we can calculate the entire portfolio return distribution—either per dollar or total dollars—and then calculate its summary measures in the usual way. The details, which are similar to other spreadsheet calculations in this chapter, can be found in the GMGOLD.XLS file. In particular, the possible returns are listed in the ranges B11:B14 and C11:C14 of Figure 5.7, and the associated probabilities are the same as those used previously in this example. These lead to the summary measures in the range B18:C20.

In particular, the investor's expected return per dollar invested is 0.153 and the standard deviation is 0.092. Based on a $10,000 investment, these translate to an expected total dollar return of $1534 and a standard deviation of $921.67.

Let's see how the expected portfolio return and the standard deviation of portfolio return change as the amount the investor puts into GM stock changes. To do this, we make sure that the value in cell B6 is a constant and that *formulas* are entered in cells C6, B7, and C7. In this way, these last three cells update automatically when the value in cell B6 changes—and the total investment amount remains $10,000. Then we form a data table in the range A24:C35 that calculates the mean and standard deviation of the total dollar portfolio return for each of several GM investment proportions in column A. (To do this, enter the formulas =C18 and =C20 in cells B24 and C24, highlight the range A24:C35, select the Data/Table command, and enter cell B6 as the column input cell. No row input cell is necessary.)

The graph of the means and standard deviations from this data table is in Figure 5.7. It shows that the expected portfolio return steadily increases as more and more is put into GM (and less is put into gold). However, the standard deviation, often used as a measure of risk, first decreases, then increases. This means there is a trade-off between expected return and risk as measured by the standard deviation. The investor could obtain a higher expected return by putting more of his money into GM, but past a fraction of approximately 0.4, the risk also increases. ●

PROBLEMS

Level A

23. The quarterly sales levels (in millions of dollars) of two U.S. retail giants are dependent on the general state of the national economy in the forthcoming months. The file P05_23.XLS provides the probability distribution for the projected sales volume of each of these two retailers in the forthcoming quarter.
    a. Find the mean and standard deviation of the quarterly sales volume for each of these two retailers. Compare these two sets of summary measures.
    b. Compute covariance and correlation measures for the given quarterly sales volumes. Interpret your numerical findings.

24. The possible annual percentage returns of the stocks of Alpha, Inc., and Beta, Inc., share a common probability distribution, given in the file P05_24.XLS.
    a. What is the expected annual return of Alpha's stock? What is the expected annual return of Beta's stock?
    b. What is the standard deviation of the annual return of Alpha's stock? What is the standard deviation of the annual return of Beta's stock?
    c. On the basis of your answers to the questions in parts a and b, which of these two stocks would you prefer to buy? Defend your choice.
    d. Are the annual returns of these two stocks positively or negatively associated with each other? How might the answer to this question influence your decision to purchase shares of one or both of these companies?

25. The annual bonuses awarded to members of the management team and assembly line workers of an automobile manufacturer depend largely on the corporation's sales performance during the preceding year. The file P05_25.XLS contains the probability distribution associated with possible bonuses awarded (measured in hundreds of dollars) to white-collar and blue-collar employees at the end of the company's fiscal year.
    a. How much do a manager and an assembly line worker expect to receive in their bonus check at the end of a typical year?
    b. For which group of employees within this organization does there appear to be more variability in the distribution of possible annual bonuses?
    c. How strongly associated are the bonuses awarded to the white-collar and blue-collar employees of this company at the end of the year? What are some possible implications of this result for the relations between members of the management team and the assembly line workers in the future?

26. Consumer demand for small, economical automobiles depends somewhat on recent trends in the average price of unleaded gasoline. For example, consider the information given in the file P05_26.XLS on the distributions of average annual sales of the Honda Civic

and the Saturn SL in relation to the trend of the average price of unleaded fuel over the past 2 years.

a. Find the annual mean sales levels of the Honda Civic and the Saturn SL.

b. For which of these two models are sales levels more sensitive to recent changes in the average price of unleaded gasoline?

c. Given the available information, how strongly associated are the typical annual sales volumes of these two popular compact cars? Provide a qualitative explanation of your quantitative measure here.

Level B

27. Upon completing their respective homework assignments, marketing majors and accounting majors at a large state university enjoy hanging out at the local tavern in the evenings. The file P05_27.XLS contains the distribution of number of hours spent by these students at the tavern in a typical week, along with typical cumulative grade-point averages (on a 4-point scale) for marketing and accounting students with common social habits.

a. Compare the means and standard deviations of the grade-point averages of the two groups of students. Does one of the two groups consistently perform better academically than the other? Explain.

b. Does academic performance, as measured by cumulative GPA, seem to be associated with the amount of time students typically spend at the local tavern? If so, characterize the observed relationship.

c. Compute the covariance and correlation between the typical grade-point averages earned by the two subgroups of students. What do these measures of association indicate in this case?

 5.6 # Distribution of Two Random Variables: Joint Probability Approach

The previous section illustrated the scenario approach for specifying the joint distribution of two random variables. We first identify several possible scenarios, next specify the value of each random variable that will occur under each scenario, and then assess the probability of each scenario. For people who think in terms of scenarios—and this includes many business managers—this is a very appealing approach.

In this section we illustrate an alternative method for specifying the probability distribution of two random variables X and Y. We first identify the possible values of X and the possible values of Y. Let x and y be any two such values. Then we *directly* assess the joint probability of the pair (x, y) and denote it by $P(X = x$ and $Y = y)$ or more simply by $p(x, y)$. This is the probability of the joint event that $X = x$ and $Y = y$ both occur. As always, the joint probabilities must be nonnegative and sum to 1.

A joint probability distribution, specified by all probabilities of the form $p(x, y)$, provides a tremendous amount of information. It indicates not only how X and Y are related, but also how each of X and Y is distributed in its own right. In probability terms, the joint distribution of X and Y determines the *marginal distributions* of both X and Y, where each **marginal distribution** is the probability distribution of a *single* random variable. The joint distribution also determines the *conditional distributions* of X given Y, and of Y given X. The **conditional distribution** of X given Y, for example, is the distribution of X, given that Y is known to equal a certain value.

These concepts are best explained by means of an example, as we do next.

Example (5.4) # Understanding the Relationship Between Demands for Substitute Products

A company sells two products, product 1 and product 2, that tend to be substitutes for one another. That is, if a customer buys product 1, she tends not to buy product 2, and vice

FIGURE 5.8
Joint Probability
Distribution of
Demands

	A	B	C	D	E	F
1	**Probability distribution of demands for substitute products**					
2						
3				Demand for product 1		
4			100	200	300	400
5		50	0.015	0.040	0.050	0.035
6	Demand	100	0.030	0.080	0.075	0.025
7	for	150	0.050	0.100	0.100	0.020
8	product 2	200	0.045	0.100	0.050	0.010
9		250	0.060	0.080	0.025	0.010

versa. The company assesses the joint probability distribution of demand for the two products during the coming month. This joint distribution appears in the shaded region of Figure 5.8. (See the Demand sheet of the file SUBSTITUTES.XLS.) The left and top margins of this table show the possible values of demand for the two products. Specifically, the company assumes that demand for product 1 can be from 100 to 400 (in increments of 100) and demand for product 2 can be from 50 to 250 (in increments of 50). Furthermore, each possible value of demand 1 can occur with each possible value of demand 2, with the joint probability given in the table. For example, the joint probability that demand 1 is 200 *and* demand 2 is 100 is 0.08. Given this joint probability distribution, describe more fully the probabilistic structure of demands for the two products.

Objective To use the given joint probability distribution of demands to find the conditional distribution of demand for each product (given the demand for the other product) as well as to calculate the covariance and correlation between demands for these substitutes.

Solution

Let D_1 and D_2 denote the demands for products 1 and 2. We first find the marginal distributions of D_1 and D_2. These are the row and column sums of the joint probabilities in Figure 5.9. An example of the reasoning is as follows. Consider the probability $P(D_1 = 200)$. If demand for product 1 is to be 200, it must be accompanied by *some* value of D_2; that is, exactly one of the joint events ($D_1 = 200$ and $D_2 = 50$) through ($D_1 = 200$ and $D_2 = 250$) must occur. Using the addition rule for probability, we find the total probability of these joint events by summing the corresponding joint probabilities. The result is $P(D_1 = 200) = 0.40$, the column sum corresponding to $D_1 = 200$. Similarly, marginal probabilities for D_2 such as $P(D_2 = 150) = 0.27$ are the row sums in column G. Note that the marginal probabilities, either those in row 10 or those in column G, sum to 1, as they should. These marginal probabilities indicate how the demand for either product behaves in its own right, aside from any considerations of the *other* product.

The marginal distributions indicate that "in-between" values of D_1 or of D_2 are most likely, whereas extreme values in either direction are less likely. However, these marginal distributions tell us nothing about the *relationship* between D_1 and D_2. After all, products 1 and 2 are supposedly substitute products. The joint probabilities spell out this relationship, but they are rather difficult to interpret. A better way is to calculate the conditional distributions of D_1 given D_2, or of D_2 given D_1. We do this in rows 12–29 of Figure 5.9.

We first focus on the conditional distribution of D_1 given D_2, shown in rows 12–19. In each row of this table (rows 15–19), we fix the value of D_2 at the value in column B and calculate the conditional probabilities of D_1 given this fixed value of D_2. The condi-

	A	B	C	D	E	F	G	H	I	J	K
1	Probability distribution of demands for substitute products										
2											
3				Demand for product 1							
4			100	200	300	400			Selected range names		
5		50	0.015	0.040	0.050	0.035	0.14		CovDem: B47		
6	Demand	100	0.030	0.080	0.075	0.025	0.21		Demands1: C4:F4		
7	for	150	0.050	0.100	0.100	0.020	0.27		Demands2: B5:B9		
8	product 2	200	0.045	0.100	0.050	0.010	0.21		JtProbs: C5:F9		
9		250	0.060	0.080	0.025	0.010	0.18		MeanDem1: B32		
10			0.20	0.40	0.30	0.10			MeanDem2: C32		
11									Probs1: C10:F10		
12	Conditional distribution of demand for product 1, given demand for product 2								Probs2: G5:G9		
13				Demand for product 1					ProdDevsDem: C37:F41		
14			100	200	300	400			SqDevsDem1: C36:F36		
15		50	0.11	0.29	0.36	0.25	1		SqDevsDem2: B37:B41		
16	Demand	100	0.14	0.38	0.36	0.12	1		StdevDem1: B45		
17	for	150	0.19	0.37	0.37	0.07	1		StdevDem2: C45		
18	product 2	200	0.22	0.49	0.24	0.05	1				
19		250	0.34	0.46	0.14	0.06	1				
20											
21	Conditional distribution of demand for product 2, given demand for product 1										
22				Demand for product 1							
23			100	200	300	400					
24		50	0.08	0.10	0.17	0.35					
25	Demand	100	0.15	0.20	0.25	0.25					
26	for	150	0.25	0.25	0.33	0.20					
27	product 2	200	0.23	0.25	0.17	0.10					
28		250	0.30	0.20	0.08	0.10					
29			1	1	1	1					
30											
31		Product 1	Product 2								
32	Means	230.00	153.25								
33											
34	Squared deviations from means (along left and top)										
35	and products of deviations from mean (in body)										
36			16900.0	900.0	4900.0	28900.0					
37		10660.6	13422.5	3097.5	-7227.5	-17552.5					
38		2835.6	6922.5	1597.5	-3727.5	-9052.5					
39		10.6	422.5	97.5	-227.5	-552.5					
40		2185.6	-6077.5	-1402.5	3272.5	7947.5					
41		9360.6	-12577.5	-2902.5	6772.5	16447.5					
42											
43		Product 1	Product 2								
44	Variances	8100.00	4176.94								
45	Stdevs	90.00	64.63								
46											
47	Covariance	-1647.50									
48	Correlation	-0.283									

FIGURE 5.9 Marginal and Conditional Distributions and Summary Measures

tional probability is the joint probability divided by the marginal probability of D_2. For example, the conditional probability that D_1 equals 200, given that D_2 equals 150, is

$$P(D_1 = 200|D_2 = 150) = \frac{P(D_1 = 200 \text{ and } D_2 = 150)}{P(D_2 = 150)} = \frac{0.10}{0.27} = 0.37$$

These conditional probabilities can be calculated all at once by entering the formula

$$=C5/\$G5$$

in cell C15 and copying it to the range C15:F19. (Make sure you see why only column G, not row 5, is held absolute in this formula.) We can also check that each row of this table is a probability distribution in its own right by summing across rows. The row sums shown in column G are all equal to 1, as they should be.

Similarly, the conditional distribution of D_2 given D_1 is in rows 21–29. Here, each column represents the conditional probability distribution of D_2 given the fixed value of D_1 in row 23. These probabilities can be calculated by entering the formula

$$=\text{C5/C\$10}$$

in cell C24 and copying it to the range C24:F28. Now the column sums shown in row 29 are 1, indicating that each column of the table represents a probability distribution.

Various summary measures can now be calculated. We show some of them in Figure 5.9. The following steps present the details.

Procedure for Calculating Summary Measures

❶   Expected values. The expected demands in cells B32 and C32 follow from the marginal distributions. To calculate these, enter the formulas

$$=\text{SUMPRODUCT(Demands1,Probs1)}$$

and

$$=\text{SUMPRODUCT(Demands2,Probs2)}$$

in these two cells. Note that each of these is based on equation (5.6) for an expected value, that is, a sum of products of possible values and their (marginal) probabilities.

❷   Variances and standard deviations. These measures of variability are also calculated from the marginal distributions by appealing to equation (5.7). We first find squared deviations from the means, then calculate the weighted sum of these squared deviations, weighted by the corresponding marginal probabilities. For example, to find the variance of D_1, enter the formula

$$=\text{(C4-MeanDem1)\textasciicircum 2}$$

in cell C36 and copy it across to cell F36. Then enter the formula

$$=\text{SUMPRODUCT(SqDevsDem1,Probs1)}$$

in cell B44, and take its square root in cell B45.

❸   Covariance and correlation. The formulas for covariance and correlation are the same as before [see equations (5.9) and (5.10)]. However, we proceed somewhat differently than in Example 5.3. Now we form a complete table of products of deviations from means in the range C37:F41. To do so, enter the formula

$$=\text{(C\$4-MeanDem1)*(\$B5-MeanDem2)}$$

in cell C37 and copy it to the range C37:F41. Then calculate the covariance in cell B47 with the formula

$$=\text{SUMPRODUCT(ProdDevsDem,JtProbs)}$$

Finally, calculate the correlation in cell B48 with the formula

$$=\text{CovDem/(StdevDem1*StdevDem2)}$$

Now let's step back and see what we have. If we are interested in the behavior of a single demand only, say, D_1, then the relevant quantities are the marginal probabilities in row 10 and the mean and standard deviation of D_1 in cells B32 and B45. However, we are often more interested in the joint behavior of D_1 and D_2. The best way to see this behavior is in the conditional probability tables. For example, compare the probability distributions in rows 15–19. As the value of D_2 increases, the probabilities for D_1 tend to shift to the left. That is, as demand for product 2 increases, demand for product 1 tends

FIGURE 5.10
Conditional
Distributions of
Demand 1 Given
Demand 2

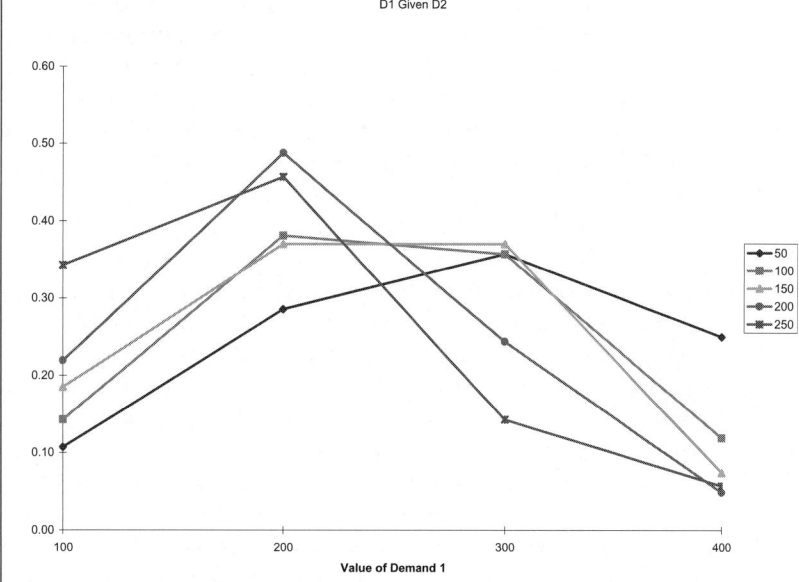

to decrease. This is only a *tendency*. When D_2 equals its largest value, there is still some chance that D_1 will be large, but this probability is fairly small.

This behavior can be seen more clearly from the graph in Figure 5.10. Each line in this graph corresponds to one of the rows 15–19. The legend represents the different values of D_2. We see that when D_2 is large, D_1 tends to be small, although again, this is only a tendency, not a perfect relationship. When we say that the two products are substitutes for one another, this is the type of behavior we imply.

By symmetry, the conditional distribution of D_2 given D_1 shows the same type of behavior. This is illustrated in Figure 5.11, where each line represents one of the columns C–F in the range C24:F28 and the legend represents the different values of D_1.

FIGURE 5.11
Conditional
Distributions of
Demand 2 Given
Demand 1

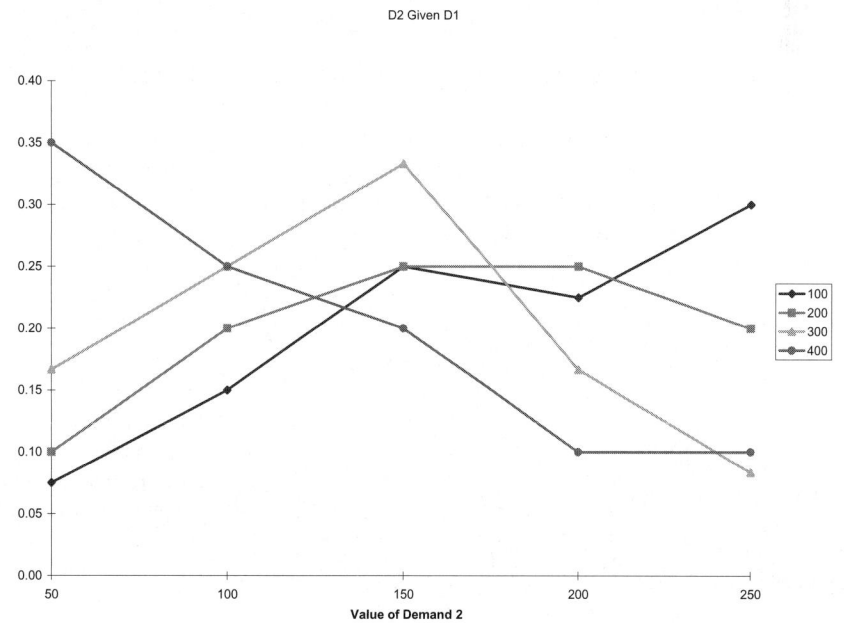

The information in these graphs is confirmed—to some extent, at least—by the co-variance and correlation between D_1 and D_2. In particular, their negative values indicate that demands for the two products tend to move in opposite directions. Also, the rather small magnitude of the correlation, -0.283, indicates that the relationship between these demands is far from perfect. When D_1 is large, there is still a reasonably good chance that D_2 will be large, and when D_1 is small, there is still a reasonably good chance that D_2 will be small. ●

How to Assess Joint Probability Distributions In the scenario approach from Section 5.5, only one probability for each scenario has to be assessed. In the joint probability approach, a whole table of joint probabilities must be assessed. This can be quite difficult, especially when there are many possible values for each of the random variables. In Example 5.4 it requires $4 \times 5 = 20$ joint probabilities that not only sum to 1 but imply the "substitute product" behavior we want.

One approach is to proceed *backward* from the way we proceeded in the example. Instead of specifying the joint probabilities and then deriving the marginal and conditional distributions, we can specify either set of marginal probabilities and either set of conditional probabilities, and then use these to *calculate* the joint probabilities. The reasoning is based on the multiplication rule for probability in the form given by equation (5.11).

Joint Probability Formula	$P(X = x \text{ and } Y = y) = P(X = x \mid Y = y)P(Y = y)$	**(5.11)**

In words, the joint probability on the left is the conditional probability that $X = x$ given $Y = y$, multiplied by the marginal probability that $Y = y$. Of course, the roles of X and Y can be reversed, yielding the alternative formula in equation (5.12).

Alternative Joint Probability Formula	$P(X = x \text{ and } Y = y) = P(Y = y \mid X = x)P(X = x)$	**(5.12)**

We choose the formula that makes the probabilities on the right-hand side easiest to assess.

The advantage of this procedure over assessing the joint probabilities directly is that it is probably *easier* and *more intuitive* for a business manager. The manager has more control over the relationship between the two random variables, as determined by the conditional probabilities she assesses.

PROBLEMS

Level A

28. Let X and Y represent the number of Dell and Compaq desktop computers, respectively, sold per month at a local computer store. The file P05_28.XLS contains the probabilities of various combinations of monthly sales volumes of these competitors.

a. Find the marginal distributions of X and Y. Interpret your findings.
b. Calculate the expected monthly desktop computer sales volumes for Dell and Compaq at this computer store.

c. Calculate the standard deviations of the monthly desktop computer sales volumes for Dell and Compaq at this computer store.

d. Construct and interpret the conditional distribution of X given Y.

e. Construct and interpret the conditional distribution of Y given X.

f. Find and interpret the correlation between X and Y.

29. The joint probability distribution of the weekly demand for two brands of diet soda is provided in the file P05_29.XLS. In particular, let D_1 and D_2 represent the weekly demand (in hundreds of 2-liter bottles) for brand 1 and brand 2, respectively, in a small town in central Pennsylvania.

a. Find the mean and standard deviation of this community's weekly demand for each brand of diet soda.

b. What is the probability that the weekly demand for each brand will be at least 1 standard deviation above its mean?

c. What is the probability that at least one of the two weekly demands will be at least 1 standard deviation above its mean?

d. What is the correlation between the weekly demands for these two brands of diet soda? What does this measure of association tell you about the relationship between these two products?

30. A local pharmacy has two checkout stations available to its customers: a regular checkout station and an express checkout station. Customers with six or fewer items are assumed to join the express line. Let X and Y be the numbers of customers in the regular checkout line and the express checkout line, respectively, at the busiest time of a typical day. Note that these numbers include the customer(s) being served, if any. The joint distribution for X and Y is given in the file P05_30.XLS.

a. Find the marginal distributions of X and Y. What does each of these distributions tell you?

b. Calculate the conditional distribution of X given Y. What is the practical benefit of knowing this conditional distribution?

c. What is the probability that no one is waiting or being served in the regular checkout line?

d. What is the probability that no one is waiting or being served in the express checkout line?

e. What is the probability that no more than two customers are waiting in both lines combined?

f. On average, how many customers would we expect to see in each of these two lines during the busiest time of day at the pharmacy?

31. Suppose that the manufacturer of a particular product assesses the joint distribution of price P per unit and demand D for its product in the upcoming quarter as provided in the file P05_31.XLS.

a. Find the expected price and demand level in the upcoming quarter.

b. What is the probability that the price of this product will be above its mean in the upcoming quarter?

c. What is the probability that the demand for this product will be below its mean in the upcoming quarter?

d. What is the probability that demand for this product will exceed 2500 units during the upcoming quarter, given that its price will be less than $40?

e. What is the probability that demand for this product will be less than 3500 units during the upcoming quarter, given that its price will be greater than $30?

f. Compute the correlation between price and demand. Is the result consistent with your expectations? Explain.

Level B

32. The recent weekly trends of two particular stock prices, P_1 and P_2, can best be described by the joint probability distribution shown in the file P05_32.XLS.

a. What is the probability that the price of stock 1 will not increase in the upcoming week?

b. What is the probability that the price of stock 2 will change in the upcoming week?

c. What is the probability that the price of stock 1 will not decrease, given that the price of stock 2 will remain constant in the upcoming week?

d. What is the probability that the price of stock 2 will change, given that the price of stock 1 will change in the upcoming week?

e. Why is it impossible to find the correlation between the typical weekly movements of these two stock prices from the information given? Nevertheless, does it appear that they are positively or negatively related? Why? What are the implications of this result for choosing an investment portfolio that may or may not include these two particular stocks?

33. Two service elevators are used in parallel by employees of a three-story hotel building. At any point in time when both elevators are stationary, let X_1 and X_2 be the floor numbers at which elevators 1 and 2, respectively, are currently located. The joint probability distribution of X_1 and X_2 is given in the file P05_33.XLS.

a. What is the probability that these two elevators are not stationed on the same floor?

b. What is the probability that elevator 2 is located on the third floor?

c. What is the probability that elevator 1 is not located on the first floor?

d. What is the probability that elevator 2 is located on the first floor, given that elevator 1 is not stationed on the first floor?

e. What is the probability that a hotel employee approaching the first-floor elevators will find at least one available for service?

f. Repeat part **e** for a hotel employee approaching each of the second- and third-floor elevators.

g. How might this hotel's operations manager respond to your findings for each of the previous questions?

Independent Random Variables

A very important special case of joint distributions is when the random variables are **independent.** Intuitively, this means that any information about the values of any of the random variables is worthless in terms of predicting any of the others. In particular, if there are only two random variables X and Y, then information about X is worthless in terms of predicting Y, and vice versa. Usually, random variables in real applications are *not* independent; they are usually related in some way, in which case we say they are **dependent.** However, we often make an assumption of independence in mathematical models to simplify the analysis.

The most intuitive way to express independence of X and Y is to say that their conditional distributions are equal to their marginals. For example, the conditional probability that X equals any value x, given that Y equals some value y, equals the marginal probability that X equals x—and this statement is true for *all* values of x and y. In words, knowledge of the value of Y has no effect on probabilities involving X. Similarly, knowledge of the value of X has no effect on probabilities involving Y.

An equivalent way of stating the independence property is that for all values x and y, the events $X = x$ and $Y = y$ are probabilistically independent in the sense of Section 5.2.4. This leads to the important property that *joint probabilities equal the product of the marginals,* as shown in equation (5.13). This follows from equation (5.11) and also because conditionals equal marginals under independence. Equation (5.13) might not be as intuitive, but it is very useful, as illustrated in the following example.

Joint Probability Formula for Independent Random Variables	$P(X = x \text{ and } Y = y) = P(X = x)P(Y = y)$	**(5.13)**

Example 5.5 Analyzing the Sales of Two Popular Personal Digital Assistants

A local office supply and equipment store, Office Station, sells several different brands of personal digital assistants (PDAs). One of the store's managers has studied the daily sales of its two most popular personal digital assistants, the Palm M505 and the Palm Vx, over the past quarter. In particular, she has used historical data to assess the joint probability distribution of the sales of these two products on a typical day. The assessed distribution is shown in Figure 5.12 (see the file PDA.XLS). The manager would like to use this distribution to determine whether there is support for the claim that the sales of the Palm Vx are often made at the expense of Palm M505 sales, and vice versa.

FIGURE 5.12
Joint Probability
Distribution of
Sales

	A	B	C	D	E	F
1	Assessed probability distribution of sales of two popular PDAs					
2						
3				Daily sales of Palm Vx		
4			0	1	2	3
5		0	0.01	0.03	0.06	0.09
6	Daily sales of	1	0.02	0.06	0.12	0.09
7	Palm M505	2	0.03	0.12	0.06	0.09
8		3	0.04	0.09	0.06	0.03

Objective To use the given assessed joint probability distribution to find the conditional distribution of daily sales of each PDA (given the sales of the other PDA), and to determine whether the daily sales of these two products are *independent* random variables.

Solution

As in the solution of Example 5.4, we begin by applying the addition rule for probability to find the marginal probability distribution of sales for each of the two personal digital assistants, as shown in Figure 5.13. By summing the corresponding joint probabilities recorded in the various rows and columns of C5:F8, we find the marginal probability that Office Station sells exactly 0, 1, 2, or 3 units of a particular PDA per day. For instance, the sum of the joint probabilities in the range C5:F5 indicates that P(daily sales of Palm M505 = 0) = 0.19, and the sum of the joint probabilities in the range C5:C8 indicates that the P(daily sales of Palm Vx = 0) = 0.10. The marginal probabilities for Palm M505 sales calculated in G5:G8, as well as the marginal probabilities for Palm Vx sales calculated in C9:F9, sum exactly to 1, as they should.

FIGURE 5.13
Marginal and
Conditional Distributions of Sales

	A	B	C	D	E	F	G
1	Assessed probability distribution of sales of two popular PDAs						
2							
3				Daily sales of Palm Vx			
4			0	1	2	3	
5		0	0.01	0.03	0.06	0.09	0.19
6	Daily sales of	1	0.02	0.06	0.12	0.09	0.29
7	Palm M505	2	0.03	0.12	0.06	0.09	0.30
8		3	0.04	0.09	0.06	0.03	0.22
9			0.10	0.30	0.30	0.30	
10							
11	Conditional distribution of sales of Palm Vx, given sales of Palm M505						
12				Daily sales of Palm Vx			
13			0	1	2	3	
14		0	0.05	0.16	0.32	0.47	1
15	Daily sales of	1	0.07	0.21	0.41	0.31	1
16	Palm M505	2	0.10	0.40	0.20	0.30	1
17		3	0.18	0.41	0.27	0.14	1
18							
19	Conditional distribution of sales of Palm M505, given sales of Palm Vx						
20				Daily sales of Palm Vx			
21			0	1	2	3	
22		0	0.10	0.10	0.20	0.30	
23	Daily sales of	1	0.20	0.20	0.40	0.30	
24	Palm M505	2	0.30	0.40	0.20	0.30	
25		3	0.40	0.30	0.20	0.10	
26			1	1	1	1	

Before constructing the conditional distribution of sales for each product, we check whether these two random variables are independent. Let M and V denote the daily sales for the Palm M505 and Palm Vx, respectively. We know from equation (5.13) that $P(M = m$ and $V = v) = P(M = m)P(V = v)$ for all values of m and v if M and V are independent. However, we have already discovered in computing the marginal probabilities that $P(M = 0)P(V = 0) = (0.19)(0.10) = 0.019$, whereas $P(M = 0$ and $V = 0) = 0.01$. In other words, the joint probability that Office Station sells no units of each product does not equal the product of the marginal probabilities that this office supply store sells no units of these products. This inequality rules out the possibility that M and V are independent random variables. If you are not yet convinced of this conclusion, compare the products of other marginal probabilities with corresponding joint probabilities in Figure 5.13. You can verify that equation (5.13) fails to hold for virtually all of the different combinations of sales levels in this case.

We now construct and interpret the conditional distribution of daily sales of each PDA. We first focus on the conditional distribution of V given M, shown in rows 11–17 of Figure 5.13. In each row of this table (located in rows 14–17), we fix the value of M at the value in column B and calculate the conditional probabilities of V given this fixed value of M. These conditional probabilities can be calculated all at once by entering the formula

$$= C5/\$G5$$

in cell C14 and copying it to the range C14:F17. We can verify that each row of this table is a probability distribution in its own right by summing across rows. The rows sums shown in the range G14:G17 are all equal to the required value of 1.

Similarly, the conditional distribution of M given V is in rows 19–26 of Figure 5.13. In each column of this table (located in columns C–F), we fix the value of V at the value in row 21 and calculate the conditional probabilities of M given this fixed value of V. All of these conditional probabilities can be calculated at once by entering the formula

$$= C5/C\$9$$

in cell C22 and copying it to the range C22:F25. Note that the column sums shown in row 26 are 1, indicating that each column of the table represents a probability distribution.

What can the Office Station manager infer by examining these conditional probability distributions? Observe in the first table (the one located in rows 14–17) that the likelihood of achieving the *highest* daily sales level of the Palm Vx *decreases* as the given daily sales level of the Palm M505 increases. This same table reveals that the probability of experiencing the *lowest* daily sales level of the Palm Vx *increases* as the given daily sales level of the Palm M505 increases. Furthermore, by closely examining the second table (the one located in rows 22–25), we see that the likelihood of achieving the *highest* daily sales level of the Palm M505 *decreases* as the given daily sales level of the Palm Vx increases. This same table reveals that the probability of experiencing the *lowest* daily sales level of the Palm M505 *increases* as the given daily sales level of the Palm Vx increases.

Thus, it appears that there is considerable support for the claim that the sales of the Palm Vx are often made at the expense of Palm M505 sales, and vice versa. This result makes sense in light of our previous finding that the daily sales of these two products are *not* independent of one another. In other words, by knowing the sales level of one of these PDAs, the manager has a better understanding of the likelihood of achieving particular results regarding sales of the other product. ●

Level A

34. The file P05_34.XLS shows the conditional distribution of the daily number of accidents at a given intersection during the winter months, X_2, given the amount of snowfall (in inches) for the day, X_1. The marginal distribution of X_1 is provided in the bottom row of the table.

   a. Are X_1 and X_2 independent random variables? Explain why or why not.

   b. What is the probability of observing no accidents at this intersection on a winter day with no snowfall?

   c. What is the probability of observing no accidents at this intersection on a randomly selected winter day?

   d. What is the probability of observing at least two accidents at this intersection on a randomly selected winter day on which the snowfall is at least 3 inches?

   e. What is the probability of observing less than 4 inches of snowfall on a randomly selected day in this area?

35. A sporting goods store sells two competing brands of exercise bicycles. Let X_1 and X_2 be the numbers of the two brands sold on a typical day at this store. Based on available historical data, the conditional probability distribution of X_1 given X_2 is assessed and provided in the file P05_35.XLS. The marginal distribution of X_2 is given in the bottom row of the table.

   a. Are X_1 and X_2 independent random variables? Explain why or why not.

   b. What is the probability of observing the sale of one brand 1 bicycle and one brand 2 bicycle on the same day at this sporting goods store?

   c. What is the probability of observing the sale of at least one brand 1 bicycle on a given day at this sporting goods store?

   d. What is the probability of observing no more than two brand 2 bicycles on a given day at this sporting goods store?

   e. Given that no brand 2 bicycles are sold on a given day, what is the likelihood of observing the sale of at least one brand 1 bicycle at this sporting goods store?

36. The file P05_28.XLS contains the probabilities of various combinations of monthly sales volumes of Dell (X) and Compaq (Y) desktop computers at a local computer store. Are the monthly sales of these two competitors independent of each other? Explain your answer.

37. Let D_1 and D_2 represent the weekly demand (in hundreds of 2-liter bottles) for brand 1 diet soda and brand 2 diet soda, respectively, in a small central Pennsylvania town. The joint probability distribution of the weekly demand for these two brands of diet soda is provided in the file P05_29.XLS. Are D_1 and D_2 independent random variables in this case? Explain why or why not.

38. The file P05_31.XLS contains the joint probability distribution of price P per unit and demand D for a particular product in the upcoming quarter.

   a. Are P and D independent random variables? Explain your answer.

   b. If P and D are *not* independent random variables, which joint probabilities result in the same *marginal* probabilities for P and D as given in the file but make P and D independent of each other?

Level B

39. You know that in 1 year you are going to buy a house. (In fact, you've already selected the neighborhood, but right now you're finishing your graduate degree, and you're engaged to be married this summer, so you're delaying the purchase for a year.) The annual interest rate for fixed-rate 30-year mortgages is currently 7.00%, and the price of the type of house you're considering is $120,000. However, things may change. Using your knowledge of the economy (and a crystal ball), you estimate that the interest rate may increase or decrease by as much as 1 percentage point. Also, the price of the house may increase by as much as $10,000—it certainly won't decrease! You assess the probability distribution of the interest rate change as shown in the file P05_39.XLS. The probability distribution of the increase in the price of the house is also shown in this file. Finally, you assume that the two random events (change in interest rate, change in house price) are probabilistically independent. This means that the probability of any joint event, such as an interest increase of 0.50% and a price increase of $5000, is the product of the individual probabilities.

   a. Use a spreadsheet and the PMT function to find the expected monthly house payment (using a 30-year fixed-rate mortgage) if there is no down payment. Find the variance and standard deviation of this monthly payment.

   b. Repeat part **a,** but assume that the down payment is 10% of the price of the house (so that you finance only 90%).

5.8 Weighted Sums of Random Variables

In this section we will analyze summary measures of weighted sums of random variables. An extremely important application of this topic is in financial investments. The example in this section illustrates such an application. However, there are many other applications of weighted sums of random variables, both in business and elsewhere. It is a topic well worth learning.

Before proceeding to the example, we lay out the main concepts and results. Let X_1, $X_2, \ldots, X_n$ be any n random variables (which could be independent or dependent), and let $a_1, a_2, \ldots, a_n$ be any n constants. We form a new random variable Y that is the weighted sum of the X's:

$$Y = a_1 X_1 + a_2 X_2 + \cdots + a_n X_n$$

It is common (but not required) to use capital letters to denote random variables and lowercase letters to denote constants.

In general, it is too difficult to obtain the complete probability distribution of Y, so we will be content to obtain its summary measures, namely, the mean $E(Y)$ and the variance $\text{Var}(Y)$. Of course, we can then calculate $\text{Stdev}(Y)$ as the square root of $\text{Var}(Y)$.

The mean is the easy part. We substitute the mean of each X into the formula for Y to obtain $E(Y)$ as follows in equation (5.14).

Expected Value of a Weighted Sum of Random Variables	$E(Y) = a_1 E(X_1) + a_2 E(X_2) + \cdots + a_n E(X_n)$	(5.14)

Using summation notation, we can write this more compactly as

$$E(Y) = \sum_{i=1}^{n} a_i E(X_i)$$

The variance is not as straightforward. Its value depends on whether the X's are independent or dependent. If they are independent, then $\text{Var}(Y)$ is a weighted sum of the variances of the X's, using the *squares* of the a's as weights, as shown in equation (5.15).

Variance of a Weighted Sum of Independent Random Variables	$\text{Var}(Y) = a_1^2 \text{Var}(X_1) + a_2^2 \, \text{Var}(X_2) + \cdots + a_n^2 \text{Var}(X_n)$	(5.15)

Using summation notation, this becomes

$$\text{Var}(Y) = \sum_{i=1}^{n} a_i^2 \text{Var}(X_i)$$

If the X's are not independent, the variance of Y is more complex and requires covariance terms. In particular, for every pair X_i and X_j, there is an extra term in equation (5.15): $2a_i a_j \text{Cov}(X_i, X_j)$. The general result is best written in summation notation, as shown in equation (5.16).

Variance of a Weighted Sum of Dependent Random Variables	$$\mathrm{Var}(Y) = \sum_{i=1}^{n} a_i^2 \mathrm{Var}(X_i) + \sum_{i<j} 2a_i a_j \mathrm{Cov}(X_i, X_j)$$	**(5.16)**

The first summation is the variance when the X's are independent. The second summation indicates that we should add the covariance term for all pairs of X's that have nonzero covariances. Actually, this equation is *always* valid, regardless of independence, because the covariance terms are all zero when the X's are independent.

Special Cases of Expected Value and Variance

- **Sum of independent random variables.** Here we assume the X's are independent and the weights are all 1, that is,

$$Y = X_1 + X_2 + \cdots + X_n$$

Then the mean of the sum is the sum of the means, and the variance of the sum is the sum of the variances:

$$E(Y) = E(X_1) + E(X_2) + \cdots + E(X_n)$$
$$\mathrm{Var}(Y) = \mathrm{Var}(X_1) + \mathrm{Var}(X_2) + \cdots + \mathrm{Var}(X_n)$$

- **Difference between two independent random variables.** Here we assume X_1 and X_2 are independent and the weights are $a_1 = 1$ and $a_2 = -1$, so that we can write Y as

$$Y = X_1 - X_2$$

Then the mean of the difference is the difference between means, but the variance of the difference is the *sum* of the variances (because $a_2^2 = (-1)^2 = 1$):

$$E(Y) = E(X_1) - E(X_2)$$
$$\mathrm{Var}(Y) = \mathrm{Var}(X_1) + \mathrm{Var}(X_2)$$

- **Sum of two dependent random variables.** In this case we make no independence assumption and set the weights equal to 1, so that $Y = X_1 + X_2$. Then the mean of the sum is again the sum of the means, but the variance of the sum includes a covariance term:

$$E(Y) = E(X_1) + E(X_2)$$
$$\mathrm{Var}(Y) = \mathrm{Var}(X_1) + \mathrm{Var}(X_2) + 2\mathrm{Cov}(X_1, X_2)$$

- **Difference between two dependent random variables.** This is the same as the second case, except that the X's are no longer independent. Again, the mean of the difference is the difference between means, but the variance of the difference now includes a covariance term, and because of the negative weight $a_2 = -1$, the sign of this covariance term is negative:

$$E(Y) = E(X_1) - E(X_2)$$
$$\mathrm{Var}(Y) = \mathrm{Var}(X_1) + \mathrm{Var}(X_2) - 2\mathrm{Cov}(X_1, X_2)$$

- **Linear Function of a Random Variable.** Suppose that Y can be written as

$$Y = a + bX$$

for some constants a and b. In this special case the random variable Y is said to be a *linear function* of another random variable X. Then the mean, variance, and standard deviation of Y can be calculated from the similar quantities for X with the following formulas:

$$E(Y) = a + bE(X)$$
$$\text{Var}(Y) = b^2 \text{Var}(X)$$
$$\text{Stdev}(Y) = b\, \text{Stdev}(X)$$

In particular, if Y is a constant multiple of X (that is, if $a = 0$), then the mean and standard deviation of Y are the same multiple of the mean and standard deviation of X.

We now put these concepts to use in an investment example.

Example 5.6 Describing Investment Portfolio Returns

An investor has $100,000 to invest, and she would like to invest it in a portfolio of eight stocks. She has gathered historical data on the returns of these stocks and has used the historical data to estimate means, standard deviations, and correlations for the stock returns. These summary measures appear in rows 12, 13, and 17–24 of Figure 5.14. (See the file INVEST.XLS.)

For example, the mean and standard deviation of stock 1 are 0.101 and 0.124. These imply that the historical annual returns of stock 1 averaged 10.1% and the standard deviation of the annual returns was 12.4%. Also, the correlation between the annual returns on stocks 1 and 2, for example, is 0.32 (see either cell C17 or B18, which necessarily contain the same value). This value, 0.32, indicates a moderate positive correlation between the historical annual returns of these stocks.

In fact, all of the correlations are positive, which probably indicates that each stock tends to vary in the same direction as some underlying economic indicator. Also, the diagonal entries in the correlation matrix are all 1 because any stock return is perfectly correlated with itself.

Although these summary measures have been obtained from historical data, the investor believes they are relevant for *future* returns. Now she would like to analyze a portfolio of these stocks, using the investment amounts shown in row 9. What is the mean annual return from this portfolio? What are its variance and standard deviation?

	A	B	C	D	E	F	G	H	I	J
1	Calculating mean, variance, and stdev for a weighted sum of random variables									
2										
3	Random variables are one-year returns from various stocks									
4	Weights are amounts invested in stocks									
5	Weighted sum is return from portfolio									
6										
7	Given quantities									
8		Stock1	Stock2	Stock3	Stock4	Stock5	Stock6	Stock7	Stock8	Total
9	Weights	$10,500	$16,300	$9,600	$9,300	$9,500	$15,400	$14,300	$15,100	$100,000
10										
11		Stock1	Stock2	Stock3	Stock4	Stock5	Stock6	Stock7	Stock8	
12	Means	0.101	0.073	0.118	0.099	0.118	0.091	0.096	0.123	
13	Stdevs	0.124	0.119	0.134	0.141	0.158	0.159	0.113	0.174	
14										
15	Correlations between stock returns									
16		Stock1	Stock2	Stock3	Stock4	Stock5	Stock6	Stock7	Stock8	
17	Stock1	1.000	0.320	0.370	0.610	0.800	0.610	0.550	0.560	
18	Stock2	0.320	1.000	0.410	0.780	0.430	0.800	0.950	0.480	
19	Stock3	0.370	0.410	1.000	0.330	0.860	0.380	0.340	0.700	
20	Stock4	0.610	0.780	0.330	1.000	0.680	0.500	0.500	0.670	
21	Stock5	0.800	0.430	0.860	0.680	1.000	0.580	0.420	0.540	
22	Stock6	0.610	0.800	0.380	0.500	0.580	1.000	0.920	0.340	
23	Stock7	0.550	0.950	0.340	0.500	0.420	0.920	1.000	0.650	
24	Stock8	0.560	0.480	0.700	0.670	0.540	0.340	0.650	1.000	

FIGURE 5.14 Input Data for Investment Example

Objective To determine the mean annual return of this investor's portfolio, and to quantify the risk associated with the total dollar return from the given weighted sum of annual stock returns.

Solution

This is a typical weighted sum model. The random variables, the X's, are the annual returns from the stocks; the weights, the a's, are the dollar amounts invested in the stocks; and the summary measures of the X's are given in rows 12, 13, and 17–24 of Figure 5.14. Be careful about units, however. Each X_i represents the return on a *single* dollar invested in stock i, whereas Y, the weighted sum of the X's, represents the *total* dollar return. So a typical value of an X might be 0.105, whereas a typical value of Y might be $10,500.

We can immediately apply equation (5.14) to obtain the mean return from the portfolio. This appears in cell B49 of Figure 5.15 (page 228), using the formula

$$=\text{SUMPRODUCT(Weights,Means)}$$

We are not quite ready to calculate the variance of the portfolio return. The reason is that the input data include standard deviations and correlations for the X's, not the variances and covariances required in equation (5.16) for Var(Y).[5] But the variances and covariances are related to standard deviations and correlations by

$$\text{Var}(X_i) = (\text{Stdev}(X_i))^2 \qquad (5.17)$$

and

$$\text{Cov}(X_i, X_j) = \text{Stdev}(X_i) \times \text{Stdev}(X_j) \times \text{Corr}(X_i, X_j) \qquad (5.18)$$

To calculate these in Excel, it is useful first to create a *column* of standard deviations in column L by using Excel's TRANSPOSE function. To do this, highlight the range L12:L19, type the formula

$$=\text{TRANSPOSE(Stdevs)}$$

and press Ctrl-Shift-Enter—all three keys at once.

Next, we form a table of variances and covariances of the X's in the range B28:I35, using equations (5.17) and (5.18). Actually, we can do this all at once by entering the formula

$$=\$L12*B\$13*B17$$

in cell B28 and copying it to the range B28:I35. Each diagonal element of this range is a variance, and the other elements off the diagonal are covariances. (As usual, make sure you understand the effect of the mixed references $L12 and B$13.)

Finally, we use equation (5.16) for Var(Y) to calculate the portfolio variance in cell B50. To do so, we form a table of the terms needed in equation (5.16) and then sum these terms, as in the following steps.

❶ Row of weights. Enter the weights in row 38 by highlighting the range B38:I38, typing the formula

$$=\text{Weights}$$

and pressing Ctrl-Enter.

❷ Column of weights. Enter these same weights as a *column* in the range A39:A46 by highlighting this range, typing the formula

$$=\text{TRANSPOSE(Weights)}$$

and pressing Ctrl-Shift-Enter.

This is a useful procedure any time you want to transform a row into a column or vice versa. But don't forget to highlight the range where the transpose will go before you type the formula.

---

[5]This was intentional. It is often easier for an investor to assess standard deviations and correlations because they are more intuitive measures.

	A	B	C	D	E	F	G	H	I	J	K	L
1	Calculating mean, variance, and stdev for a weighted sum of random variables											
2												
3	Random variables are one-year returns from various stocks											
4	Weights are amounts invested in stocks											
5	Weighted sum is return from portfolio											
6												
7	Given quantities											
8		Stock1	Stock2	Stock3	Stock4	Stock5	Stock6	Stock7	Stock8	Total		
9	Weights	$10,500	$16,300	$9,600	$9,300	$9,500	$15,400	$14,300	$15,100	$100,000		
10												
11		Stock1	Stock2	Stock3	Stock4	Stock5	Stock6	Stock7	Stock8		Standard deviations	
12	Means	0.101	0.073	0.118	0.099	0.118	0.091	0.096	0.123		Stock1	0.124
13	Stdevs	0.124	0.119	0.134	0.141	0.158	0.159	0.113	0.174		Stock2	0.119
14											Stock3	0.134
15	Correlations between stock returns										Stock4	0.141
16		Stock1	Stock2	Stock3	Stock4	Stock5	Stock6	Stock7	Stock8		Stock5	0.158
17	Stock1	1.000	0.320	0.370	0.610	0.800	0.610	0.550	0.560		Stock6	0.159
18	Stock2	0.320	1.000	0.410	0.780	0.430	0.800	0.950	0.480		Stock7	0.113
19	Stock3	0.370	0.410	1.000	0.330	0.860	0.380	0.340	0.700		Stock8	0.174
20	Stock4	0.610	0.780	0.330	1.000	0.680	0.500	0.500	0.670			
21	Stock5	0.800	0.430	0.860	0.680	1.000	0.580	0.420	0.540		Range names	
22	Stock6	0.610	0.800	0.380	0.500	0.580	1.000	0.920	0.340		Corrs: B17:I24	
23	Stock7	0.550	0.950	0.340	0.500	0.420	0.920	1.000	0.650		Covar: B28:I35	
24	Stock8	0.560	0.480	0.700	0.670	0.540	0.340	0.650	1.000		Means: B12:I12	
25											PortVarTerms: B39:I46	
26	Covariances between stock returns (variances of stock returns are on the diagonal)										Stdevs: B13:I13	
27		Stock1	Stock2	Stock3	Stock4	Stock5	Stock6	Stock7	Stock8		Var: B50	
28	Stock1	0.0154	0.0047	0.0061	0.0107	0.0157	0.0120	0.0077	0.0121		Weights: B9:I9	
29	Stock2	0.0047	0.0142	0.0065	0.0131	0.0081	0.0151	0.0128	0.0099			
30	Stock3	0.0061	0.0065	0.0180	0.0062	0.0182	0.0081	0.0051	0.0163			
31	Stock4	0.0107	0.0131	0.0062	0.0199	0.0151	0.0112	0.0080	0.0164			
32	Stock5	0.0157	0.0081	0.0182	0.0151	0.0250	0.0146	0.0075	0.0148			
33	Stock6	0.0120	0.0151	0.0081	0.0112	0.0146	0.0253	0.0165	0.0094			
34	Stock7	0.0077	0.0128	0.0051	0.0080	0.0075	0.0165	0.0128	0.0128			
35	Stock8	0.0121	0.0099	0.0163	0.0164	0.0148	0.0094	0.0128	0.0303			
36												
37	Weights (top row, left column) and terms for calculating portfolio variance											
38		10500	16300	9600	9300	9500	15400	14300	15100			
39	10500	1695204	808156.6	619710.3	1041461	1563442	1944727	1157146	1915690			
40	16300	808156.61	3762436	1023044	1983952	1251941	3799640	2977643	2446257			
41	9600	619710.34	1023044	1654825	556662.6	1660562	1196954	706755.9	2365921			
42	9300	1041460.7	1983952	556662.6	1719508	1338418	1605425	1059465	2308357			
43	9500	1563441.6	1251941	1660562	1338418	2253001	2131702	1018696	2129613			
44	15400	1944727.1	3799640	1196954	1605425	2131702	5995642	3640157	2187374			
45	14300	1157146	2977643	706755.9	1059465	1018696	3640157	2611133	2759650			
46	15100	1915689.9	2446257	2365921	2308357	2129613	2187374	2759650	6903231			
47												
48	Summary measures of portfolio											
49	Mean	$10,056.40										
50	Variance	124992021										
51	Stdev	$11,179.98										

FIGURE 5.15 Calculations for Investment Example

❸ Table of terms. Now use these weights and the covariances to fill in the table of terms required for the portfolio variance. To do so, enter the formula

$$=\$A39*B28*B\$38$$

in cell B39 and copy it to the range B39:I46. The terms on the diagonal of this table are squares of weights multiplied by variances. These are the terms in the first summation in equation (5.16). Each term off the diagonal is the product of two weights and a covariance. These are the terms needed in the second summation in equation (5.16).

❹ Portfolio variance and standard deviation. Calculate the portfolio variance in cell B50 with the formula

$$=SUM(PortVarTerms)$$

Then calculate the standard deviation of the portfolio return in cell B51 as the square root of the variance.

The results in Figure 5.15 indicate that the investor has an expected return of slightly more than $10,000 (or 10%) from this portfolio. However, the standard deviation of approximately $11,200 is sizable. This standard deviation is a measure of the portfolio's risk. Investors always want a large mean return, but they also want low risk. Moreover, they realize that the only way to obtain a higher mean return is usually to accept more risk. You can experiment with the spreadsheet for this example to see how the mean and standard deviation of portfolio return vary with the investment amounts. Just enter new weights in row 9 (keeping the sum equal to $100,000) and see how the values in B49–B51 change. ●

P R O B L E M S

Level A

40. A typical consumer buys a random number (X) of polo shirts when he shops at a men's clothing store. The distribution of X is given by the following probability distribution: $P(X = 0) = 0.30$, $P(X = 1) = 0.30$, $P(X = 2) = 0.20$, $P(X = 3) = 0.10$, and $P(X = 4) = 0.10$.
  a. Find the mean and standard deviation of X.
  b. Assuming that each shirt costs $35, let Y be the total amount of money (in dollars) spent by a customer when he visits this clothing store. Find the mean and standard deviation of Y.
  c. Compute the probability that a customer's expenditure will be more than 1 standard deviation above the mean expenditure level.

41. Based on past experience, the number of customers who arrive at a local gasoline station during the noon hour to purchase fuel is best described by the probability distribution given in the file P05_41.XLS.
  a. Find the mean, variance, and standard deviation of this random variable.
  b. Find the probability that the number of arrivals during the noon hour will be within 1 standard deviation of the mean number of arrivals.
  c. Suppose that the typical customer spends $15 on fuel upon stopping at this gasoline station during the noon hour. Compute the mean and standard deviation of the total gasoline revenue earned by this gas station during the noon hour.
  d. What is the probability that the total gasoline revenue will be less than the mean value found in part **c**?
  e. What is the probability that the total gasoline revenue will be more than 2 standard deviations above the mean value found in part **c**?

42. Let X be the number of defective items found by a quality inspector in a random batch of 15 items from a particular manufacturing process. The probability distribution of X is provided in the file P05_18.XLS. This firm earns $500 profit from the sale of each *acceptable* item in a given batch. In the event that an

item is found to be *defective*, it must be reworked at a cost of $100 before it can be sold, thus reducing its per-unit profitability to $400.
  a. Find the mean and standard deviation of the profit earned from the sale of all items in a given batch.
  b. What is the probability that the profit earned from the sale of all items in a given batch is within 2 standard deviations of the mean profit level? Is this result consistent with the empirical rules we learned in Chapter 3? Explain.

43. The probability distribution for the number of job applications processed at a small employment agency during a typical week is given in the file P05_19.XLS.
  a. Assuming that it takes the agency's administrative assistant 2 hours to process a submitted job application, on average how many hours in a typical week will the administrative assistant spend processing incoming job applications?
  b. Find an interval with the property that the administrative assistant can be approximately 95% sure that the total amount of time he spends each week processing incoming job applications will be in this interval.

44. Consider a financial services salesperson whose annual salary consists of both a fixed portion of $25,000 and a variable portion that is a commission based on her sales performance. In particular, she estimates that her monthly sales commission can be represented by a random variable with mean $5000 and standard deviation $700.
  a. What annual salary can this salesperson expect to earn?
  b. Assuming that her sales commissions in different months are independent random variables, what is the standard deviation of her annual salary?
  c. Between what two annual salary levels can this salesperson be approximately 95% sure that her true total earnings will fall?

45. A film processing shop charges its customers 18 cents per print, but customers may refuse to accept one or more of the prints for various reasons. Assume that this shop does not charge its customers for refused prints. The number of prints refused per 24-print roll is a random variable with mean 1.5 and standard deviation 0.5.

 a. What are the mean and standard deviation of the amount that customers pay for the development of a typical 24-print roll?

 b. Assume that this shop processes 250 24-print rolls of film in a given week. If the numbers of refused prints on these rolls are independent random variables, what are the mean and standard deviation of the weekly film processing revenue of this shop?

 c. Find an interval such that the manager of this film shop can be approximately 95% sure that the weekly processing revenue will be contained within the interval.

46. Suppose the monthly demand for Thompson televisions has a mean of 40,000 and a standard deviation of 20,000. Determine the mean and standard deviation of the annual demand for Thompson TVs. Assume that demand in any month is probabilistically independent of demand in any other month. (Is this assumption realistic?)

47. Suppose there are five stocks available for investment and each has an annual mean return of 10% and a standard deviation of 4%. Assume the returns on the stocks are independent random variables.

 a. If you invest 20% of your money in each stock, determine the mean, standard deviation, and variance of the annual dollar return on your investments.

 b. If you invest $100 in a single stock, determine the mean, standard deviation, and variance of the annual return on your investment.

 c. How do the answers to parts **a** and **b** relate to the phrase, "Don't put all your eggs in one basket"?

48. An investor puts $10,000 into each of four stocks, labeled A, B, C, and D. The file P05_48.XLS contains the means and standard deviations of the annual returns of these four stocks. Assuming that the returns of these four stocks are independent of each other, find the mean and standard deviation of the total amount that this investor earns in 1 year from these four investments.

Level B

49. Consider again the investment problem described in Problem 48. Now, assume that the returns of the four stocks are no longer independent of one another. Specifically, the correlations between all pairs of stock returns are given in the file P05_49.XLS.

 a. Find the mean and standard deviation of the total amount that this investor earns in 1 year from these four investments. Compare these results to those you found in the previous problem. Explain the differences in your answers.

 b. Suppose that this investor now decides to place $15,000 each in stocks B and D, and $5000 each in stocks A and C. How do the mean and standard deviation of the total amount that this investor earns in 1 year change from the allocation used in part **a?** Provide an intuitive explanation for the changes you observe here.

50. A supermarket chain operates five stores of varying sizes in Harrisburg, Pennsylvania. Profits (represented as a percentage of sales volume) earned by these five stores are 2.75%, 3%, 3.5%, 4.25%, and 5%, respectively. The means and standard deviations of the daily sales volumes at these five stores are given in the file P05_50.XLS. Assuming that the daily sales volumes are independent of each other, find the mean and standard deviation of the total *profit* that this supermarket chain earns in 1 day from the operation of its five stores in Harrisburg.

51. A manufacturing company constructs a 1-cm assembly by snapping together four parts that average 0.25 cm in length. The company would like the standard deviation of the length of the assembly to be 0.01 cm. Its engineer, Peter Purdue, believes that the assembly will meet the desired level of variability if each part has a standard deviation of $0.01/4 = 0.0025$ cm. Instead, show Peter that you can do the job by making each part have a standard deviation of $0.01/\sqrt{4} = 0.005$ cm. This could save the company a lot of money, because not as much precision is needed for each part.

52. The weekly demand function for one of a given firm's products can be represented by $Q = 200 - 5p$, for $p = 1, 2, \ldots, 40$, where Q is the number of units purchased (in hundreds) at a sales price of p (in dollars). Assume that the probability distribution of the sales price is given by $P(p = k) = .025$, for $k = 1, 2, \ldots, 40$. (In words, each price is equally likely.)

 a. Find the mean and standard deviation of p. Interpret these measures in this case.

 b. Find the mean and standard deviation of Q. Interpret these measures in this case.

 c. Assuming that it costs this firm $10 to manufacture and sell each unit of the given product, define π to be the firm's weekly contribution to profit from the sale of this product (measured in dollars). Express π as a function of the quantity purchased, Q.

 d. Find the expected weekly contribution to the firm's profit from the sale of this product. Also, compute the standard deviation of the weekly contribution to the firm's profit from the sale of this product.

53. A retailer purchases a batch of 1000 fluorescent lightbulbs from a wholesaler at a cost of $2 per bulb. The wholesaler agrees to replace each defective bulb with one that is guaranteed to function properly for a charge of $0.20 per bulb. The retailer sells the bulbs at a price of $2.50 per bulb and gives his customers free replacements if they bring defective bulbs back to the store. Let X be the number of defective bulbs in a typical batch, and assume that the mean and standard deviation of X are 50 and 10, respectively.

a. Find the mean and standard deviation of the profit (in dollars) the retailer makes from selling a batch of lightbulbs.

b. Find an interval with the property that the retailer can be approximately 95% sure that his profit will be in this interval.

 5.9 # Conclusion

This chapter has introduced some very important concepts, including the basic rules of probability, random variables, probability distributions, and summary measures of probability distributions. We have also shown how computer simulation can be used to help explain some of these concepts. Many of the concepts presented in this chapter will be used in later chapters, so it is important to learn them now. In particular, we will rely heavily on probability distributions in Chapter 7 when we discuss decision making under uncertainty. There we will learn how the expected value of a probability distribution is the primary criterion for making decisions. We will also continue to use computer simulation in later chapters to help explain complex statistical concepts.

Summary of Key Terms

Term	Explanation	Excel	Pages	Equation Number
Random variable	Associates a numerical value with each possible outcome of a random phenomenon.		190	
Probability	A number between 0 and 1 that measures the likelihood that some event will occur.		191	
Rule of complements	The probability of any event A and the probability of its complement—the event that A does *not* occur—sum to 1.	Must be done manually	191	5.1
Mutually exclusive events	Events where only one of them can occur.		191	
Exhaustive events	Events where at least one of them must occur.		191	
Addition rule for mutually exclusive events	The probability that at least one of a set of mutually exclusive events will occur is the sum of their probabilities.	Must be done manually	192	5.2
Conditional probability formula	Updates the probability of an event, given the knowledge that another related event has occurred.	Must be done manually	193	5.3
Multiplication rule for two events	Formula for the probability that two events *both* occur.	Must be done manually	193	5.4

Summary of Key Terms (continued)

Term	Explanation	Excel	Pages	Equation Number
Probabilistic independence	Events where knowledge that one of them has occurred is of no value when assessing the probability that the other will occur; allows for simplification of the multiplication rule.	Must be done manually	195	5.5
Relative frequency	The proportion of times the event occurs out of the number of times the random experiment is run.		196	
Mean (or expected value) of a probability distribution	A measure of central tendency—the weighted sum of the possible values of a random variable, weighted by their probabilities.	Must be done manually	200	5.6
Variance of a probability distribution	A measure of variability: the weighted sum of the squared deviations of the possible values of a random variable from the mean, weighted by the probabilities.	Must be done manually	200	5.7
Standard deviation of a probability distribution	A measure of variability: the square root of the variance.	Must be done manually	200	5.8
Simulation	An extremely useful tool that can be used to incorporate uncertainty explicitly into spreadsheet models.		203	
Uniformly distributed random numbers	Random numbers such that all decimal values between 0 and 1 are equally likely.	=RAND()	204	
Covariance for a joint probability distribution	A measure of the relationship between two jointly distributed random variables.	Must be done manually	208	5.9
Correlation for a joint probability distribution	A measure of the relationship between two jointly distributed random variables, scaled to be between -1 and $+1$.	Must be done manually	208	5.10
Multiplication rule for random variables	Formula for a joint probability as the product of a marginal probability and a conditional probability.	Must be done manually	218	5.11, 5.12
Independent random variables	Random variables where information about one of them is of no value in terms of predicting the other.		220	
Multiplication rule for independent random variables	The joint probability is the product of the marginal probabilities.	Must be done manually	220	5.13
Expected value of a weighted sum of random variables	Useful for finding the expected value of Y, where $Y = a_1X_1 + a_2X_2 + \cdots + a_nX_n$.	Must be done manually	224	5.14

Summary of Key Terms (continued)

Term	Explanation	Excel	Pages	Equation Number
Variance of a weighted sum of independent random variables	Useful for finding the variance of Y, where $Y = a_1X_1 + a_2X_2 + \cdots + a_nX_n$ and the X's are independent of one another.	Must be done manually	224	5.15
Variance of a weighted sum of dependent random variables	Useful for finding the variance of Y, where $Y = a_1X_1 + a_2X_2 + \cdots + a_nX_n$ and the X's are *not* independent of one another.	Must be done manually	225	5.16
Covariance in terms of standard deviations and correlation	Used to calculate covariance when only information on correlations and standard deviations is given.	Must be done manually	227	5.18

PROBLEMS

Conceptual Exercises

C.1. Suppose that you want to find the probability that event A or event B will occur in the case where these two events are *not* mutually exclusive. Explain how you would proceed to compute the probability that at least one of these two events will occur.

C.2. "If two events are mutually exclusive, then they must *not* be independent events." Is this statement true or false? Explain your choice.

C.3. Is the number of passengers who show up for a particular commercial airline flight a discrete or a continuous random variable? Is the time between flight arrivals at a major airport a discrete or a continuous random variable? Explain your answers.

C.4. Suppose that officials in the federal government are trying to determine the likelihood of a major small-pox epidemic in the United States within the next 12 months. Is this an example of an objective probability or a subjective probability? Explain your choice.

C.5. What is another term for the covariance between a random variable and itself?

Level A

54. A business manager who needs to make many phone calls has estimated that when she calls a client, the probability that she will reach the client right away is 60%. If she does not reach the client on the first call, the probability that she will reach the client with a subsequent call in the next hour is 20%.
   a. Find the probability that the manager will reach her client in two or fewer calls.

   b. Find the probability that the manager will reach her client on the second call but not on the first call.
   c. Find the probability that the manager will be unsuccessful on two consecutive calls.

55. Suppose that a marketing research firm sends questionnaires to two different companies. Based on historical evidence, the marketing research firm believes that each company, independently of the other, will return the questionnaire with probability 0.40.
   a. What is the probability that *both* questionnaires will be returned?
   b. What is the probability that *neither* of the questionnaires will be returned?
   c. Now, suppose that this marketing research firm sends questionnaires to *ten* different companies. Assuming that each company, independently of the others, returns its completed questionnaire with probability 0.40, how do your answers to parts **a** and **b** change?

56. Based on past sales experience, an appliance store stocks five window air conditioner units for the coming week. No orders for additional air conditioners will be made until next week. The weekly consumer demand for this type of appliance has the probability distribution given in the file P05_56.XLS.
   a. Let X be the number of window air conditioner units left at the end of the week (if any), and let Y be the number of special stockout orders required (if any), assuming that a special stockout order is required each time there is a demand and no unit is available in stock. Find the probability distributions of X and Y.

b. Find the expected value of X and the expected value of Y.

c. Assume that this appliance store makes a $40 profit on each air conditioner sold from the weekly available stock, but the store loses $10 for each unit sold on a special stockout order basis. Let Z be the profit that the store earns in the upcoming week from the sale of window air conditioners. Find the probability distribution of Z.

d. Find the expected value of Z.

57. Simulate 400 weekly consumer demands for window air conditioner units with the probability distribution given in the file P05_56.XLS. How does your simulated distribution compare to the given probability distribution? Explain any differences between these two distributions.

58. The probability distribution of the weekly demand (in hundreds of reams) of copier paper used in the duplicating center of a corporation is provided in the file P05_58.XLS.

a. Find the mean and standard deviation of this distribution.

b. Find the probability that weekly copier paper demand will be at least 1 standard deviation above the mean.

c. Find the probability that weekly copier paper demand will be within 1 standard deviation of the mean.

59. Consider the probability distribution of the weekly demand (in hundreds of reams) of copier paper used in a corporation's duplicating center, as shown in the file P05_58.XLS.

a. Generate 500 values of this random variable with the given probability distribution using computer simulation.

b. Compute the mean and standard deviation of the simulated values.

c. Use your simulated distribution to find the probability that weekly copier paper demand will be within 1 standard deviation of the mean.

60. The probability distribution of the weekly demand (in hundreds of reams) of copier paper used in the duplicating center of a corporation is provided in the file P05_58.XLS. Assuming that it costs the duplicating center $5 to purchase a ream of paper, find the mean and standard deviation of the weekly copier paper cost for this corporation.

61. The instructor of an introductory organizational behavior course believes that there might be a relationship between the number of writing assignments (X) she makes in the course and the final grades (Y) earned by students enrolled in this class. She has taught this course with varying numbers of writing assignments for many semesters now. She has compiled relevant historical data in the file P05_61.XLS for you to review.

a. Convert the given frequency table to a table of conditional probabilities of final grades (Y) earned by students enrolled in this class, given the number of writing assignments (X) made in the course. Comment on your constructed table of conditional probabilities. Generally speaking, what does this table tell you?

b. Given that this instructor makes only one writing assignment in the course, what is the expected final grade earned by the typical student?

c. How much variability exists around the conditional mean grade you found in part **b**? Furthermore, what proportion of all relevant students earn final grades within 2 standard deviations of this conditional mean?

d. Given that this instructor makes more than one writing assignment in the course, what is the expected final grade earned by the typical student?

e. How much variability exists around the conditional mean grade you found in part **d**? Moreover, what proportion of all relevant students earn final grades within 2 standard deviations of this conditional mean?

f. Compute the covariance and correlation of X and Y in this case. What does each of these measures tell you? In particular, is this organizational behavior instructor correct in believing that there is a systematic relationship between the number of writing assignments made and final grades earned in her classes?

62. The file P05_62.XLS contains the joint probability distribution of recent weekly trends of two particular stock prices, P_1 and P_2.

a. Are P_1 and P_2 independent random variables? Explain why or why not.

b. If P_1 and P_2 are *not* independent random variables, which joint probabilities result in the same *marginal* probabilities for P_1 and P_2 as given in this file but make P_1 and P_2 independent of each other?

63. Consider two service elevators used in parallel by employees of a three-story hotel building. At any point in time when both elevators are stationary, let X_1 and X_2 be the floor numbers at which elevators 1 and 2, respectively, are currently located. The file P05_33.XLS contains the joint probability distribution of X_1 and X_2.

a. Are X_1 and X_2 independent random variables? Explain your answer.

b. If X_1 and X_2 are *not* independent random variables, which joint probabilities result in the same *marginal* probabilities for X_1 and X_2 as given in this file but make X_1 and X_2 independent of each other?

64. A roulette wheel contains the numbers 0, 00, and 1, 2, . . . , 36. If you bet $1 on a single number coming up, you earn $35 if the number comes up and lose $1 if the number does not come up. Find the mean and standard deviation of your winnings on a single bet.

65. Assume that there are four equally likely states of the economy: boom, low growth, recession, and depression. Also, assume that the percentage annual return you obtain when you invest a dollar in gold or the stock market is shown in the file P05_65.XLS.
    a. Find the covariance and correlation between the annual return on the market and the annual return on gold. Interpret your answers.
    b. Suppose you invest 40% of your money in the market and 60% of your money in gold. Determine the mean and standard deviation of the annual return on your portfolio.
    c. Obtain your part **b** answer by determining the actual return on your portfolio in each state of the economy and determining the mean and variance directly without using any formulas involving covariances or correlations.
    d. Suppose you invested 70% of your money in the market and 30% in gold. Without doing any calculations, determine whether the mean and standard deviation of your portfolio would increase or decrease from your answer in part **b**. Give an intuitive explanation to support your answers.

66. You are considering buying a share of Ford stock with the possible returns for the next year as shown in the file P05_66.XLS. Determine the mean, variance, and standard deviation of the annual return on Ford stock.

67. Suppose there are three states of the economy: boom, moderate growth, and recession. The annual return on GM and Ford stock in each state of the economy is shown in the file P05_67.XLS.
    a. Calculate the mean, standard deviation, and variance of the annual return on each stock assuming the probability of each state is 1/3.
    b. Calculate the mean, standard deviation, and variance of the annual return on each stock assuming the probabilities of the three states are 1/4, 1/4, and 1/2.
    c. Calculate the covariance and correlation between the annual return on GM and Ford stocks assuming the probability of each state is 1/3.
    d. Calculate the covariance and correlation between the annual return on GM and Ford stocks assuming the probabilities of the three states are 1/4, 1/4, and 1/2.
    e. You have invested 25% of your money in GM and 75% in Ford. Assuming that each state is equally likely, determine the mean and variance of your portfolio's return.
    f. Now check your answer to part **e** by directly computing for each state the return on your portfolio and use the formulas for mean and variance of a random variable. For example, in the Boom state, your portfolio earns 0.25(0.25) + 0.75(0.32).

68. You have placed 30% of your money in investment A and 70% of your money in investment B. The annual returns on investments A and B depend on the state of the economy as shown in the file P05_68.XLS. Determine the mean and standard deviation of the annual return on your investments.

69. There are three possible states of the economy during the next year (states 1, 2, and 3). The probability of each state of the economy and the percentage annual return on IBM and Disney stocks are as shown in the file P05_69.XLS.
    a. Find the correlation between the annual return on IBM and Disney. Interpret this correlation.
    b. If you put 80% of your money in IBM and 20% in Disney, find the mean and standard deviation of your annual return.

70. The return on a portfolio during a period is defined by

$$\frac{PV_{end} - PV_{beg}}{PV_{beg}}$$

where PV_{beg} is the portfolio value at the beginning of a period and PV_{end} is the portfolio value at the end of the period. Suppose there are two stocks in which we can invest, stock 1 and stock 2. During each year there is a 50% chance that each dollar invested in stock 1 will turn into $2 and a 50% chance that each dollar invested in stock 1 will turn into $0.50. During each year there is a 50% chance that each dollar invested in stock 2 will turn into $2 and a 50% chance that each dollar invested in stock 2 will turn into $0.50.
    a. If you invest all your money in stock 1, determine the expected value, variance, and standard deviation of your 1-year return.
    b. Assume the returns on stocks 1 and 2 are independent random variables. If you put half your money into each stock, determine the expected value, variance, and standard deviation of your 1-year return.
    c. Can you give an intuitive explanation of why the variance and standard deviation in part **b** are smaller than the variance and standard deviation in part **a**?
    d. Use simulation to check your answers to part **b**. Use at least 1000 trials.

71. Each year the employees at Zipco receive a $0, $2000, or $4500 salary increase. They also receive a merit rating of 0, 1, 2, or 3, with 3 indicating outstanding performance and 0 indicating poor performance. The joint probability distribution of salary increase and merit rating is listed in the file P05_71.XLS. For example, 20% of all employees receive a $2000 increase and have a merit rating of 1. Find the correlation between salary increase and merit rating. Then interpret this correlation.

72. Suppose X and Y are independent random variables. The possible values of X are -1, 0, and 1; the possible values of Y are 10, 20, and 30. You are given that $P(X = -1$ and $Y = 10) = 0.05$, $P(X = 0$ and $Y = 30) = 0.20$, $P(Y = 10) = 0.20$, and $P(X = 0) = 0.50$. Determine the joint probability distribution of X and Y.

73. You are involved in a risky business venture where three outcomes are possible: (1) you will lose not only your initial investment ($5000) but an additional $3000; (2) you will just make back your initial investment (for a net gain of $0); or (3) you will make back your initial investment plus an extra $10,000. The probability of (1) is half as large as the probability of (2), and the probability of (3) is one-third as large as the probability of (2).

    a. Find the individual probabilities of (1), (2), and (3). (They should sum to 1.)

    b. Find the expected value of your net gain (or loss) from this venture. Find its variance and standard deviation.

Level B

74. Imagine that you are trying to predict what will happen to the price of gasoline (regular unleaded) and the price of natural gas for home heating during the next month. Assume you believe that the price of either will stay the same, go up by 5%, or go down by 5%. Assess the joint probabilities of these possibilities, that is, assess nine probabilities that sum to 1 and are "realistic." Do you believe it is easier to assess the marginal probabilities of one and the conditional probabilities of the other, or to assess the joint probabilities directly? (*Note:* There is no "correct" answer, but there are unreasonable answers, those that do not reflect reality.)

75. Consider an individual selected at random from a sample of 750 married women (see the data in the file P05_05.XLS) in answering each of the following questions.

    a. What is the probability that this woman does not work outside the home, given that she has at least one child?

    b. What is the probability that this woman has no children, given that she works part-time?

    c. What is the probability that this woman has at least two children, given that she does not work full-time?

76. Suppose that 8% of all managers in a given company are African-American, 13% are women, and 17% have earned an MBA degree from a top-10 graduate business school. Let A, B, and C be, respectively, the events that a randomly selected individual from this population is African-American, is a woman, and has earned an MBA from a top-10 graduate business school.

    a. Would you expect A, B, and C to be independent events? Explain why or why not.

    b. Assuming that A, B, and C *are* independent events, compute the probability that a randomly selected manager from this company is a white male and has earned an MBA degree from a top-10 graduate business school.

    c. If A, B, and C are *not* independent events, could you calculate the probability requested in part **b** from the information given? What further information would you need?

77. Consider again the supermarket chain described in Problem 50. Now, assume that the daily sales of the five stores are no longer independent of one another. In particular, the file P05_77.XLS contains the correlations between all pairs of daily sales volumes.

    a. Find the mean and standard deviation of the total profit that this supermarket chain earns in 1 day from the operation of its five stores in Harrisburg. Compare these results to those you found in Problem 50. Explain the differences in your answers.

    b. Find an interval such that the regional sales manager of this supermarket chain can be approximately 95% sure that the total daily profit earned by its stores in Harrisburg will be contained within the interval.

78. A manufacturing plant produces two distinct products, X and Y. The cost of producing one unit of X is $18 and that of Y is $22. Assume that this plant incurs a weekly setup cost of $24,000 regardless of the number of units of X or Y produced. The means and standard deviations of the weekly production levels of X and Y are given in the P05_78.XLS.

    a. Assuming that the weekly production levels of X and Y are independent, compute the mean and standard deviation of this plant's total weekly production cost. Between what two total cost figures can we be about 68% sure that this plant's actual total weekly production cost will fall?

    b. How do your answers in part **a** change when you discover that the correlation between the weekly production levels of X and Y is actually 0.29? Explain the differences in the two sets of results.

79. The typical standard deviation of the annual return on a stock is 20% and the typical mean return is about 12%. The typical correlation between the annual returns of two stocks is about 0.25. Mutual funds often put an equal percentage of their money in a given number of stocks. By choosing a large number of stocks, they hope to diversify away the risk involved with choosing particular stocks. How many stocks does an investor need to own to diversify away the risk associated with individual stocks? To answer this question, use the above information about "typical" stocks to determine the mean and standard deviation for the following portfolios:

- Portfolio 1: Half your money in each of 2 stocks
- Portfolio 2: 20% of your money in each of 5 stocks
- Portfolio 3: 10% of your money in each of 10 stocks
- Portfolio 4: 5% of your money in each of 20 stocks
- Portfolio 5: 1% of your money in each of 100 stocks

What do your answers tell you about the number of stocks a mutual fund needs to invest in to diversify? (*Hint:* You will need to consider a square range. For portfolio 2, for example, it will be a square range with 25 cells, 5 on the diagonal and 20 off the diagonal. Each diagonal term makes the same contribution to the variance and each off-diagonal term makes the same contribution to the variance.)

80. You are ordering milk for Mr. D's and are determined to please! Milk is delivered once a week (at midnight Sunday). The mean and standard deviation of the number of gallons of milk demanded each day are given in the file P05_80.XLS. Determine the mean and standard deviation of the weekly demand for milk. What assumption must you make to determine the weekly standard deviation? Presently you are ordering 1000 gallons per week. Is this a sensible order quantity? Assume all milk spoils after 1 week.

81. At the end of 1995, Wall Street's best estimates of the means, standard deviations, and correlations for the 1996 returns on stocks, bonds, and T-bills were as shown in the file P05_81.XLS. Stocks have the highest average return and the most risk, whereas T-bills have the lowest average return and the least risk. Find the mean and standard deviation for the return on your 1996 investments for the three asset allocations listed in this file. For example, portfolio 1 allocates 53% of all assets to stocks, 6% to bonds, and 41% to T-bills. (This is what most Wall Street firms did.) Based on your results, can you explain why nobody in 1996 should have allocated all their assets to bonds? (*Note:* T-bills are 90-day government issues; the bonds are 10-year government bonds.)

82. The annual returns on stocks 1 and 2 for three possible states of the economy are given in the file P05_82.XLS.
    a. Find and interpret the correlation between stocks 1 and 2.
    b. Consider another stock (stock 3) that always yields an annual return of 10%. Suppose you invest 60% of your money in stock 1, 10% in stock 2, and 30% in stock 3. Determine the standard deviation of the annual return on your portfolio. (*Hint:* You do not need Excel to compute the variance of stock 3 and the covariance of stock 3 with the other two stocks!)

83. The application at the beginning of this chapter describes the campaign McDonald's used several years ago, where customers could win various prizes.
    a. Verify the figures that are given in the description. That is, argue why there are 10 winning outcomes and 120 total outcomes.
    b. Suppose McDonald's had designed the cards so that each card had 2 zaps and 3 pictures of the winning prize (and again 5 pictures of other irrelevant prizes). The rules are the same as before: To win, the customer must uncover all 3 pictures of the winning prize before uncovering a zap. Would there be more or fewer winners with this design? Argue by calculating the probability that a card is a winner.
    c. Going back to the original game (as in part **a**), suppose McDonald's printed 1 million cards, each of which was eventually given to a customer. Assume that the (potential) winning prizes on these were: 500,000 Cokes worth $0.40 each, 250,000 french fries worth $0.50 each, 150,000 milk shakes worth $0.75 each, 75,000 hamburgers worth $1.50 each, 20,000 cards with $1 cash as the winning prize, 4000 cards with $10 cash as the winning prize, 800 cards with $100 cash as the winning prize, and 200 cards with $1000 cash as the winning prize. Find the expected amount (the dollar equivalent) that McDonald's gave away in winning prizes, assuming everyone played the game and claimed the prize if they won. Find the standard deviation of this amount.

5.1 Simpson's Paradox

The results we obtain when we work with conditional probabilities can be quite unintuitive, even paradoxical. This case is similar to one described in an article by Blyth (1972) and is usually referred to as Simpson's paradox. [Two other examples of Simpson's paradox are described in the articles by Westbrooke (1998) and Appleton et al. (1996).] Essentially, Simpson's paradox says that even if one treatment has a better effect than another on *each* of two separate subpopulations, it can have a *worse* effect on the population as a whole.

Suppose that the population is the set of managers in a large company. We categorize the managers as those with an MBA degree (the B's) and those without an MBA degree (the $\bar{B}$'s). These categories are the two "treatment groups." We also categorize the managers as those who were hired directly out of school by this company (the C's) and those who worked with another company first (the $\bar{C}$'s). These two categories form the two "subpopulations." Finally, we use as a measure of effectiveness those managers who have been promoted within the past year (the A's).

Assume the following conditional probabilities are given:

$$P(A|B \text{ and } C) = 0.10, \quad P(A|\bar{B} \text{ and } C) = 0.05 \quad (5.19)$$

$$P(A|B \text{ and } \bar{C}) = 0.35, \quad P(A|\bar{B} \text{ and } \bar{C}) = 0.20 \quad (5.20)$$

$$P(C|B) = 0.90, \quad P(C|\bar{B}) = 0.30 \quad (5.21)$$

Each of these can be interpreted as a proportion. For example, the probability $P(A|B \text{ and } C)$ implies that

10% of all managers who have an MBA degree and were hired by the company directly out of school were promoted last year. Similar explanations hold for the other probabilities.

Joan Seymour, the head of personnel at this company, is trying to understand these figures. From the probabilities in equation (5.19), she sees that among the subpopulation of workers hired directly out of school, those with an MBA degree are twice as likely to be promoted as those without an MBA degree. Similarly, from the probabilities in equation (5.20), she sees that among the subpopulation of workers hired after working with another company, those with an MBA degree are *almost* twice as likely to be promoted as those without an MBA degree. The information provided by the probabilities in equation (5.21) is somewhat different. From these, she sees that employees with MBA degrees are three times as likely as those without MBA degrees to have been hired directly out of school.

Joan can hardly believe it when a whiz-kid analyst uses these probabilities to show—correctly—that

$$P(A|B) = 0.125, \quad P(A|\bar{B}) = 0.155 \quad (5.22)$$

In words, those employees *without* MBA degrees are more likely to be promoted than those with MBA degrees. This appears to go directly against the evidence in equations (5.19) and (5.20), both of which imply that MBAs have an advantage in being promoted. Can you derive the probabilities in equation (5.22)? Can you shed any light on this "paradox"?

6

Normal, Binomial, Poisson, and Exponential Distributions

© Ed Eckstein/CORBIS

Challenging Claims of *The Bell Curve*

One of the most controversial books of the past decade is *The Bell Curve* (Herrnstein and Murray, 1994). The authors are the late Richard Herrnstein, a psychologist, and Charles Murray, an economist, both of whom had extensive training in statistics. The book is a scholarly treatment of differences in intelligence, measured by IQ, and its effect on socioeconomic status (SES). The authors argue, by appealing to many past studies and presenting many statistics and graphs, that there are significant differences in IQ among different groups of people, and that these differences are at least partially responsible for differences in SES. Specifically, their basic claims are that (1) there is a quantity, intelligence, that can be measured by an IQ test, (2) the distribution of IQ scores is essentially a symmetric bell-shaped curve, (3) IQ scores are highly correlated with various indicators of success, (4) IQ is determined predominantly by genetic factors and less so by environmental factors, and (5) African-Americans score significantly lower—about 15 points lower—on IQ than whites.

Although the discussion of this latter point takes up a relatively small part of the book, it has generated by far the most controversy. Many criticisms of the authors' racial thesis have been based on emotional arguments. However, it can also be criticized on entirely statistical grounds, as Barnett (1995) has done.[1] Barnett never states that the analysis by Herrnstein and Murray is *wrong*. He merely states that (1) the assumptions behind some of the analysis are at best questionable, and (2) some of the crucial details are not made as explicit as they should have been. As he states, "The issue is not that *The Bell Curve* is demonstrably wrong,

[1] Arnold Barnett is a professor in operations research at MIT's Sloan School of Management and specializes in data analyses about issues of health and safety.

but that it falls so far short of being demonstrably right. The book does not meet the burden of proof we might reasonably expect of it."

For example, Barnett takes issue with the claim that the genetic component of IQ is, in the words of Herrnstein and Murray, "unlikely to be smaller than 40 percent or higher than 80 percent." Barnett asks what it would mean if genetics made up, say, 60 percent of IQ. His only clue from the book is in an endnote, which implies this definition: If a large population of genetically identical newborns grew up in randomly chosen environments, and their IQs were measured once they reached adulthood, then the variance of these IQs would be 60 percent less than the variance for the entire population. The key word is *variance*. As Barnett notes, however, this statement implies that the corresponding drop in *standard deviation* is only 37 percent. That is, even if all members of the population were exactly the same genetically, differing environments would create a standard deviation of IQs 63 percent as large as the standard deviation that exists today. If this is true, it is hard to argue, as Herrnstein and Murray have done, that environment plays a minor role in determining IQ.

Because the effects of different racial environments are so difficult to disentangle from genetic effects, Herrnstein and Murray try at one point to bypass environmental influences on IQ by matching blacks and whites from similar environments. They report that blacks in the top decile of SES have an average IQ of 104, but that whites within that decile have an IQ 1 standard deviation higher. Even assuming that they have their facts straight, Barnett criticizes the vagueness of their claim. What standard deviation are they referring to: the standard deviation of the entire population or the standard deviation of only the people in the upper decile of SES? The latter is certainly much smaller than the former. Should we assume that the "top-decile blacks" are in the top decile of the black population or of the overall population? If the latter, then the matched comparison between blacks and whites is flawed because the wealthiest 10 percent of whites have far more wealth than the wealthiest 10 percent of blacks. Moreover, even if the reference is to the pooled national population, the matching is imperfect. It is possible that the blacks in this pool could average around the ninth percentile, whereas the whites could average around the fourth percentile, with a significant difference in income between the two groups.

The problem is that Herrnstein and Murray never state these details explicitly. Therefore, we have no way of knowing—without collecting and analyzing all of the data ourselves—whether their results are essentially correct. As Barnett concludes his article, "I believe that *The Bell Curve*'s statements about race would have been better left unsaid even if they were definitely true. And they are surely better left unsaid when, as we have seen, their meaning and accuracy [are] in doubt." ●

6.1 Introduction

In the previous chapter we discussed probability distributions in general. In this chapter we investigate several specific distributions that commonly occur in a variety of management science applications. The first of these is a continuous distribution called the *normal* distribution, which is characterized by a symmetric, bell-shaped curve and is the cornerstone of statistical theory. The second distribution is a discrete distribution called the *binomial* distribution. It is relevant when we sample from a population with only two types of members or when we perform a series of independent, identical "experiments" with only two possible outcomes. The other two distributions we will discuss are the *Poisson* and *exponential* distributions. These are often used when we are counting events of some type through time, such as arrivals to a bank. In this case, the Poisson distribution, which is discrete, describes the *number* of arrivals in any period of time, whereas the exponential distribution, which is continuous, describes the *times* between arrivals.

The main goals in this chapter are to present the properties of these distributions, give some examples of when they apply, and see how to perform calculations involving them. Regarding this last objective, analysts have traditionally used special tables to look up probabilities or values for the distributions in this chapter. However, we will see how these tasks can be simplified with the statistical functions available in Excel. Given the availability of these Excel functions, the traditional tables are no longer necessary.

We cannot overemphasize the importance of these distributions. Almost all of the statistical results we will learn in later chapters are based on either the normal distribution or the binomial distribution. The Poisson and exponential distributions play a less important role in this book, but they are nevertheless extremely important in many management science applications. Therefore, it is essential that you become thoroughly familiar with these distributions before proceeding.

6.2 The Normal Distribution

The single most important distribution in statistics is the normal distribution. It is a *continuous* distribution and is associated with the familiar symmetric, bell-shaped curve. The normal distribution is defined by its mean and standard deviation. By changing the mean, we can shift the normal curve to the right or left. By changing the standard deviation, we can make the curve more or less spread out. Therefore, there are really many normal distributions, not just a single normal distribution.

6.2.1 Continuous Distributions and Density Functions

We first take a moment to discuss continuous probability distributions in general. In the previous chapter we dealt with discrete distributions, characterized by a list of possible values and their probabilities. The same idea holds for continuous distributions such as the normal distribution, but the mathematics becomes more complex. Now instead of a list of possible values, there is a *continuum* of possible values, such as all values between 0 and 100 or all values greater than 0. Instead of assigning probabilities to each individual value in the continuum, we "spread" the total probability of 1 over this continuum. The key to this spreading is called a *probability density function,* which acts like a histogram. The higher the value of the density function, the more likely this region of the continuum is.

Probability Density Function	A **probability density function,** usually denoted by $f(x)$, specifies the probability distribution of a continuous random variable X. The higher $f(x)$ is, the more likely x is. Also, the total area between the graph of $f(x)$ and the horizontal axis, which represents the total probability, is equal to 1. Finally, $f(x)$ is nonnegative for all possible values of X.

As an example, consider the density function—*not* a normal density function—shown in Figure 6.1 on page 242. It indicates that all values in the continuum from 25 to 100 are possible, but that the values near 70 are most likely. (This density function might

FIGURE 6.1
A Skewed Density
Function

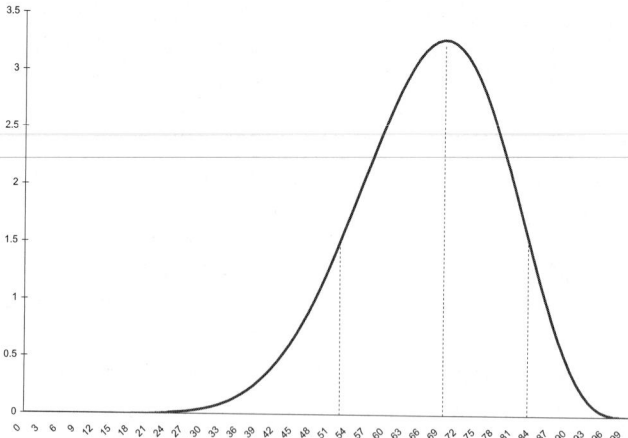

correspond to scores on an exam.) To be a bit more specific, because the height of the density at 70 is approximately twice the height of the curve at 84 or 53, a value near 70 is approximately twice as likely as a value near 84 or a value near 53. In this sense, the height of the density function indicates *relative* likelihoods.

To find probabilities from a density function, we need to calculate areas under the curve. For example, the area of the designated region in Figure 6.2 represents the probability of a score between 65 and 75. Also, the area under the *entire* curve is 1, because the total probability of all possible values is always 1. Unfortunately, this is about as much as we can say without calculus. Integral calculus is necessary to find areas under curves. Fortunately, statistical tables have been constructed to find such areas for a number of well-known density functions, including the normal. Even better, special Excel functions have been developed to find these areas—without the need for bulky tables. We will take advantage of these Excel functions as we study the normal distribution (and other distributions).

What about the mean and standard deviation (or variance) of a continuous distribution? As before, the mean is a measure of central tendency of the distribution, and the standard deviation (or variance) measures the variability of the distribution. Again, however, calculus is generally required to calculate these quantities. We will simply list their values (which *were* obtained through calculus) for the normal distribution and any other continuous distributions where we need them. By the way, the mean for the density in Figure 6.1 is slightly *less* than 70—it is always to the left of the peak for a left-skewed

FIGURE 6.2
Probability as the
Area Under the
Density

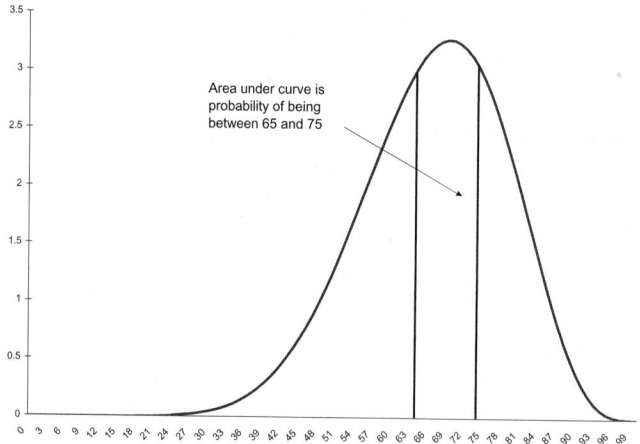

Area under curve is
probability of being
between 65 and 75

distribution and to the right of the peak for a right-skewed distribution—and the standard deviation is approximately 15.

6.2.2 The Normal Density

The normal distribution is a continuous distribution with possible values ranging over the *entire* number line—from "minus infinity" to "plus infinity." However, only a relatively small range has much chance of occurring. The normal density function is actually quite complex, in spite of its "nice" bell-shaped appearance. For the sake of completeness, we list the formula for the normal density function in equation (6.1). Here, μ and σ are the mean and standard deviation of the distribution.

Normal Probability Density Function

$$f(x) = \frac{1}{\sqrt{2\pi}\sigma}\, e^{-(x-\mu)^2/(2\sigma^2)} \quad \text{for } -\infty < x < +\infty \qquad \textbf{(6.1)}$$

The curves in Figure 6.3 illustrate several normal density functions for different values of μ and σ. The mean μ can be any number: negative, positive, or zero. As we see, the effect of increasing or decreasing the mean μ is to shift the curve to the right or the left. On the other hand, the standard deviation σ must be a *positive* number. It controls the spread of the normal curve. When σ is small, the curve is more peaked; when σ is large, the curve is more spread out. For shorthand, we use the notation $N(\mu, \sigma)$ to refer to the normal distribution with mean μ and standard deviation σ. For example, $N(-2, 1)$ refers to the normal distribution with mean -2 and standard deviation 1.

FIGURE 6.3
Several Normal Density Functions

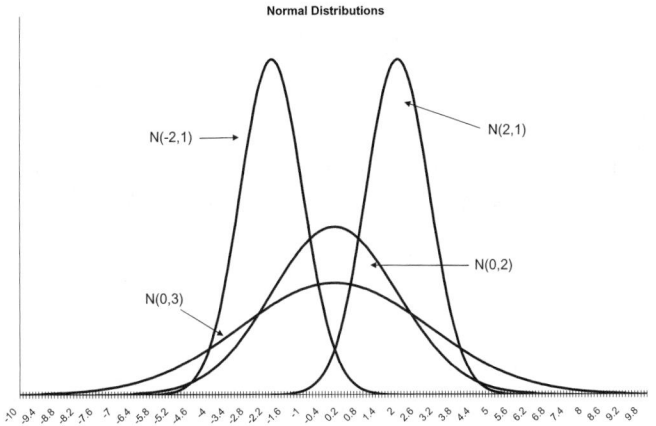

6.2.3 Standardizing: *Z*-Values

There are infinitely many normal distributions, one for each pair μ and σ. We single out one of these for special attention, the *standard normal* distribution. The **standard normal** distribution has mean 0 and standard deviation 1, so we can denote it by $N(0,1)$. It is also referred to as the **Z distribution.** Suppose the random variable X is normally distributed with mean μ and standard deviation σ. We define the random variable Z by equation (6.2). This operation is called *standardizing*. That is, to **standardize** a variable, we subtract its mean and then divide the difference by the standard deviation. When X is normally distributed, the standardized variable is $N(0, 1)$.

Standardizing a Normal Random Variable	$$Z = \frac{X - \mu}{\sigma}$$	(6.2)

One reason for standardizing is to measure variables with different means and/or standard deviations on a single scale. For example, suppose several sections of a college course are taught by different instructors. Because of differences in teaching methods and grading procedures, the distributions of scores in these sections might differ, possibly by a wide margin. However, if each instructor calculates his or her mean and standard deviation and then calculates a Z-value for each student, the distributions of the Z-values should be approximately the same in each section.

It is also easy to interpret a Z-value. It is the number of standard deviations to the right or the left of the mean. If Z is positive, the original value (in this case, the original score) is to the *right* of the mean; if Z is negative, the original score is to the *left* of the mean. For example, if the Z-value for some student is 2, then this student's score is 2 standard deviations above the mean. If the Z-value for another student is -0.5, then this student's score is 0.5 standard deviation below the mean. We illustrate Z-values in the following example.

Example 6.1 Standardizing Returns from Mutual Funds

The annual returns for 30 mutual funds appear in Figure 6.4. (See the file FUNDS.XLS.) Find and interpret the Z-values of these returns.

Objective To use Excel to standardize annual returns of various mutual funds.

Solution

The 30 annual returns appear in column B of Figure 6.4. Their mean and standard deviation are calculated in cells B4 and B5 with the AVERAGE and STDEV functions. The corresponding Z-values are calculated in column C by entering the formula

=(B8-MeanReturn)/StdevReturn

in cell C8 and copying it down column C.

There is an equivalent way to calculate these Z-values in Excel. We do this in column D, using Excel's STANDARDIZE function directly. To use this function, enter the formula

=STANDARDIZE(B8,MeanReturn,StdevReturn)

in cell D8 and copy it down column D.

The Z-values in Figure 6.4 range from a low of -1.80 to a high of 2.19. Specifically, the return for stock 1 is about 1.80 standard deviations below the mean, whereas the return for fund 17 is about 2.19 standard deviations above the mean. As we will see shortly, these values are typical: Z-values are usually in the range from -2 to $+2$ and values beyond -3 or $+3$ are very uncommon (recall the *empirical* rules that we first discussed in Chapter 3). Also, the Z-values automatically have mean 0 and standard deviation 1, as we see in cells C5 and C6 by using the AVERAGE and STDEV functions on the Z-values in column C (or D).

FIGURE 6.4
Mutual Fund
Returns and
Z-Values

	A	B	C	D	E	F	G
1	**Annual returns from mutual funds**						
2							
3	Summary statistics from returns below						
4	Mean return	0.091	0		Calculated two different ways - the second with the Standardize function		
5	Stdev of returns	0.047	1				
6							
7	Fund	Annual return	Z value	Z value			
8	1	0.007	-1.8047	-1.8047			
9	2	0.080	-0.2363	-0.2363			
10	3	0.082	-0.1934	-0.1934			
11	4	0.123	0.6875	0.6875			
12	5	0.022	-1.4824	-1.4824	Range names:		
13	6	0.054	-0.7949	-0.7949	MeanReturn: B4		
14	7	0.109	0.3867	0.3867	StdevReturn: B5		
15	8	0.097	0.1289	0.1289	Returns: B8:B37		
16	9	0.047	-0.9453	-0.9453	ZValues: C8:C37		
17	10	0.021	-1.5039	-1.5039			
18	11	0.111	0.4297	0.4297			
19	12	0.180	1.9121	1.9121			
20	13	0.157	1.4180	1.4180			
21	14	0.134	0.9238	0.9238			
22	15	0.140	1.0528	1.0528			
23	16	0.107	0.3438	0.3438			
24	17	0.193	2.1914	2.1914			
25	18	0.156	1.3965	1.3965			
26	19	0.095	0.0859	0.0859			
27	20	0.039	-1.1172	-1.1172			
28	21	0.034	-1.2246	-1.2246			
29	22	0.064	-0.5801	-0.5801			
30	23	0.071	-0.4297	-0.4297			
31	24	0.079	-0.2578	-0.2578			
32	25	0.088	-0.0645	-0.0645			
33	26	0.077	-0.3008	-0.3008			
34	27	0.125	0.7305	0.7305			
35	28	0.094	0.0645	0.0645			
36	29	0.078	-0.2793	-0.2793			
37	30	0.066	-0.5371	-0.5371			

6.2.4 Normal Tables and Z-Values

A common use for Z-values and the standard normal distribution is in calculating probabilities and percentiles by the "traditional" method.[2] This method is based on a table of the standard normal distribution found in most statistics textbooks. Such a table is given in Figure 6.5 (page 246). The body of the table contains probabilities. The left and top margins contain possible values. Specifically, suppose we want to find the probability that a standard normal random variable is less than 1.35. We locate 1.3 along the left and .05—for the second decimal in 1.35—along the top, and then read into the table to find the probability 0.9115. In words, the probability is about 0.91 that a standard normal random variable is less than 1.35.

Alternatively, if we are given a probability, we can use the table to find the value with this much probability to the left of it under the standard normal curve. We call this a *percentile* calculation. For example, if the probability is 0.75, we can find the 75th percentile by locating the probability in the table closest to 0.75 and then reading to the left and up. With interpolation, the required value is approximately 0.675. In words, the probability of being to the left of 0.675 under the standard normal curve is approximately 0.75.

---

[2]If you intend to rely on Excel functions for normal calculations, you can omit this subsection.

z	0.00	0.01	0.02	0.03	0.04	0.05	0.06	0.07	0.08	0.09
0.0	0.5000	0.5040	0.5080	0.5120	0.5160	0.5199	0.5239	0.5279	0.5319	0.5359
0.1	0.5398	0.5438	0.5478	0.5517	0.5557	0.5596	0.5636	0.5675	0.5714	0.5753
0.2	0.5793	0.5832	0.5871	0.5910	0.5948	0.5987	0.6026	0.6064	0.6103	0.6141
0.3	0.6179	0.6217	0.6255	0.6293	0.6331	0.6368	0.6406	0.6443	0.6480	0.6517
0.4	0.6554	0.6591	0.6628	0.6664	0.6700	0.6736	0.6772	0.6808	0.6844	0.6879
0.5	0.6915	0.6950	0.6985	0.7019	0.7054	0.7088	0.7123	0.7157	0.7190	0.7224
0.6	0.7257	0.7291	0.7324	0.7357	0.7389	0.7422	0.7454	0.7486	0.7517	0.7549
0.7	0.7580	0.7611	0.7642	0.7673	0.7704	0.7734	0.7764	0.7794	0.7823	0.7852
0.8	0.7881	0.7910	0.7939	0.7967	0.7995	0.8023	0.8051	0.8078	0.8106	0.8133
0.9	0.8159	0.8186	0.8212	0.8238	0.8264	0.8289	0.8315	0.8340	0.8365	0.8389
1.0	0.8413	0.8438	0.8461	0.8485	0.8508	0.8531	0.8554	0.8577	0.8599	0.8621
1.1	0.8643	0.8665	0.8686	0.8708	0.8729	0.8749	0.8770	0.8790	0.8810	0.8830
1.2	0.8849	0.8869	0.8888	0.8907	0.8925	0.8944	0.8962	0.8980	0.8997	0.9015
1.3	0.9032	0.9049	0.9066	0.9082	0.9099	0.9115	0.9131	0.9147	0.9162	0.9177
1.4	0.9192	0.9207	0.9222	0.9236	0.9251	0.9265	0.9279	0.9292	0.9306	0.9319
1.5	0.9332	0.9345	0.9357	0.9370	0.9382	0.9394	0.9406	0.9418	0.9429	0.9441
1.6	0.9452	0.9463	0.9474	0.9484	0.9495	0.9505	0.9515	0.9525	0.9535	0.9545
1.7	0.9554	0.9564	0.9573	0.9582	0.9591	0.9599	0.9608	0.9616	0.9625	0.9633
1.8	0.9641	0.9649	0.9656	0.9664	0.9671	0.9678	0.9686	0.9693	0.9699	0.9706
1.9	0.9713	0.9719	0.9726	0.9732	0.9738	0.9744	0.9750	0.9756	0.9761	0.9767
2.0	0.9772	0.9778	0.9783	0.9788	0.9793	0.9798	0.9803	0.9808	0.9812	0.9817
2.1	0.9821	0.9826	0.9830	0.9834	0.9838	0.9842	0.9846	0.9850	0.9854	0.9857
2.2	0.9861	0.9864	0.9868	0.9871	0.9875	0.9878	0.9881	0.9884	0.9887	0.9890
2.3	0.9893	0.9896	0.9898	0.9901	0.9904	0.9906	0.9909	0.9911	0.9913	0.9916
2.4	0.9918	0.9920	0.9922	0.9925	0.9927	0.9929	0.9931	0.9932	0.9934	0.9936
2.5	0.9938	0.9940	0.9941	0.9943	0.9945	0.9946	0.9948	0.9949	0.9951	0.9952
2.6	0.9953	0.9955	0.9956	0.9957	0.9959	0.9960	0.9961	0.9962	0.9963	0.9964
2.7	0.9965	0.9966	0.9967	0.9968	0.9969	0.9970	0.9971	0.9972	0.9973	0.9974
2.8	0.9974	0.9975	0.9976	0.9977	0.9977	0.9978	0.9979	0.9979	0.9980	0.9981
2.9	0.9981	0.9982	0.9982	0.9983	0.9984	0.9984	0.9985	0.9985	0.9986	0.9986
3.0	0.9987	0.9987	0.9987	0.9988	0.9988	0.9989	0.9989	0.9989	0.9990	0.9990
3.1	0.9990	0.9991	0.9991	0.9991	0.9992	0.9992	0.9992	0.9992	0.9993	0.9993
3.2	0.9993	0.9993	0.9994	0.9994	0.9994	0.9994	0.9994	0.9995	0.9995	0.9995
3.3	0.9995	0.9995	0.9995	0.9996	0.9996	0.9996	0.9996	0.9996	0.9996	0.9997
3.4	0.9997	0.9997	0.9997	0.9997	0.9997	0.9997	0.9997	0.9997	0.9997	0.9998

FIGURE 6.5 Normal Probabilities

We can perform the same kind of calculations for *any* normal distribution if we first standardize. As an example, suppose that X is normally distributed with mean 100 and standard deviation 10. We will find the probability that X is less than 115 and the 85th percentile of this normal distribution. To find the probability that X is less than 115, we first standardize the value 115. The corresponding Z-value is

$$Z = (115 - 100)/10 = 1.5$$

Now we look up 1.5 in the table (1.5 row, .00 column) to obtain the probability 0.9332. For the percentile question we first find the 85th percentile of the standard normal distribution. Interpolating, we obtain a value of approximately 1.037. Then we set this value equal to a standardized value:

$$Z = 1.037 = (X - 100)/10$$

Finally, we solve for X to obtain 110.37. In words, there is a probability 0.85 of being to the left of 110.37 in the $N(100, 10)$ distribution.

There are some obvious drawbacks to using the standard normal table for probability calculations. The first is that there are holes in the table—we often have to interpolate. A second drawback is that the standard normal table takes different forms in different text-books. These differences are rather minor, but they can easily cause errors. Finally, the table requires us to perform calculations. For example, we might have to standardize. More importantly, we often have to use the symmetry of the normal distribution to find probabilities that are not in the table. As an example, to find the probability that Z is less than -1.5, we must go through some mental gymnastics. First, by symmetry this is the same as the probability that Z is greater than 1.5. Then since only left-tail probabilities are tabulated, we must find the probability that Z is less than 1.5 and subtract this probability from 1. The chain of reasoning is

$$P(Z < -1.5) = P(Z > 1.5) = 1 - P(Z < 1.5) = 1 - .9332 = .0668$$

This is not too difficult, given a bit of practice, but it is easy to make a mistake. Spreadsheet functions make the whole procedure much easier and less prone to errors.

6.2.5 Normal Calculations in Excel

Two types of calculations are typically made with normal distributions: finding probabilities and finding percentiles. Excel makes each of these fairly simple. The functions used for normal probability calculations are NORMDIST and NORMSDIST. The main difference between these is that the one with the "S" (for standardized) applies only to $N(0, 1)$ calculations, whereas NORMDIST applies to *any* normal distribution. On the other hand, percentile calculations, where we supply a probability and require a value, are often called *inverse* calculations. Therefore, the Excel functions for these are named NORMINV and NORMSINV. Again, the "S" in the second of these indicates that it applies only to the standard normal distribution.

The NORMDIST and NORMSDIST functions give left-tail probabilities, such as the probability that a normally distributed variable is *less than* 35. The syntax for these functions is

$$=\text{NORMDIST}(x, \mu, \sigma, 1)$$

and

$$=\text{NORMSDIST}(x)$$

Here, x is a number we supply, and μ and σ are the mean and standard deviation of the normal distribution. The last argument "1" in the NORMDIST function is used to obtain the *cumulative* normal probability, the only kind we'll ever need. (This 1 is a bit of a nuisance to remember, but it's necessary.) Note that NORMSDIST takes only one argument (because μ and σ are known to be 0 and 1), so it is easier to use—when it applies.

The NORMINV and NORMSINV functions return values for user-supplied probabilities. For example, if we supply the probability 0.95, these functions return the 95th percentile. Their syntax is

$$=\text{NORMINV}(p, \mu, \sigma)$$

and

$$=\text{NORMSINV}(p)$$

where p is a probability we supply. These are analogous to the NORMDIST and NORMSDIST functions (except there is no fourth argument "1" necessary in the NORMINV function).

We illustrate these Excel functions in the following example.[3]

---

[3]Actually, we already illustrated the NORMSDIST function; it was used to create the body of Figure 6.5.

Example (6.2) Becoming Familiar with Normal
Calculations in Excel

Use Excel to calculate the following probabilities and percentiles for the standard normal distribution: (a) $P(Z < -2)$, (b) $P(Z > 1)$, (c) $P(-0.4 < Z < 1.6)$, (d) the 5th percentile, (e) the 75th percentile, and (f) the 99th percentile. Then for the $N(75, 8)$ distribution, find the following probabilities and percentiles: (a) $P(X < 70)$, (b) $P(X > 73)$, (c) $P(75 < X < 85)$, (d) the 5th percentile, (e) the 60th percentile, and (f) the 97th percentile.

Objective To calculate probabilities and percentiles for standard normal and nonstandard normal random variables in Excel.

Solution

The solution appears in Figure 6.6. (See the file NORMFNS.XLS.) The $N(0, 1)$ calculations are in rows 7–14; the $N(75, 8)$ calculations are in rows 23–30. For your convenience, the formulas used in column B are spelled out in column D. Note that the standard normal calculations use the normal functions with the "S" in the middle; the rest use the normal functions without the "S"—and require more arguments.

Note the following for normal *probability* calculations:

- For "less than" probabilities, use NORMDIST or NORMSDIST directly. (See rows 7 and 23.)

	A	B	C	D	E	F	G	H	I
1	**Using Excel's normal functions**								
2									
3	**Examples with standard normal**								
4									
5	**Probability calculations**								
6	Range	Probability		Formula					
7	Less than -2	0.0228		NORMSDIST(-2)					
8	Greater than 1	0.1587		1-NORMSDIST(1)					
9	Between -.4 and 1.6	0.6006		NORMSDIST(1.6)-NORMSDIST(-0.4)					
10									
11	**Percentiles**								
12	5th	-1.645		NORMSINV(0.05)					
13	75th	0.674		NORMSINV(0.75)					
14	99th	2.326		NORMSINV(0.99)					
15									
16	**Examples with nonstandard normal**								
17				Range names					
18	Mean	75		Mean: B18					
19	Stdev	8		Stdev: B19					
20									
21	**Probability calculations**								
22	Range	Probability		Formula					
23	Less than 70	0.2660		NORMDIST(70,Mean,Stdev,1)					
24	Greater than 73	0.5987		1-NORMDIST(73,Mean,Stdev,1)					
25	Between 75 and 85	0.3944		NORMDIST(85,Mean,Stdev,1)-NORMDIST(75,Mean,Stdev,1)					
26									
27	**Percentiles**								
28	5th	61.841		NORMINV(0.05,Mean,Stdev)					
29	60th	77.027		NORMINV(0.6,Mean,Stdev)					
30	97th	90.046		NORMINV(0.97,Mean,Stdev)					

FIGURE 6.6 Normal Calculations with Excel Functions

- For "greater than" probabilities, subtract the NORMDIST or NORMSDIST function from 1. (See rows 8 and 24.)

- For "between" probabilities, subtract the two NORMDIST or NORMSDIST functions. For example, in row 9 the probability of being between -0.4 and 1.6 is the probability of being less than 1.6 minus the probability of being less than -0.4.

The percentile calculations are even more straightforward. In most percentile problems we want to find the value with a certain probability to the *left* of it. In this case we use the NORMINV or NORMSINV function with the specified probability as the first argument. See rows 12–14 and 28–30. ●

There are a couple of variations of percentile calculations. First, suppose we want the value with probability 0.05 to the *right* of it. This is the same as the value with probability 0.95 to the left of it, so we use NORMINV or NORMSINV with probability argument 0.95. For example, the value with probability 0.4 to the right of it in the $N(75, 8)$ distribution is 77.027 (see cell B29 in Figure 6.6).

As a second variation, suppose we want to find an interval of the form $-x$ to x, for some positive number x, with (1) probability 0.025 to the left of $-x$, (2) probability 0.025 to the right of x, and (3) probability 0.95 between $-x$ and x. This is a very common problem in statistical inference. In general, we want a probability (such as 0.95) to be in the middle of the interval so that half of the remaining probability (0.025) is in each of the tails. (See Figure 6.7.) Then the required x can be found with NORMINV or NORMSINV, using probability argument 0.975, because there must be a total probability of 0.975 to the left of x.

For example, if the relevant distribution is the standard normal, then the required value of x is 1.96, found with the function NORMSINV(0.975). Similarly, if we want probability 0.90 in the middle and probability 0.05 in each tail, the required x is 1.645, found with the function NORMSINV(0.95). Remember these two numbers, 1.96 and 1.645. They occur frequently in statistical applications.

6.2.6 Empirical Rules Revisited

Chapter 3 introduced the empirical rules that apply to many data sets. Namely, about 68% of the data fall within 1 standard deviation of the mean, about 95% fall within 2 standard

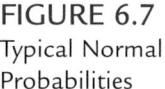

FIGURE 6.7
Typical Normal
Probabilities

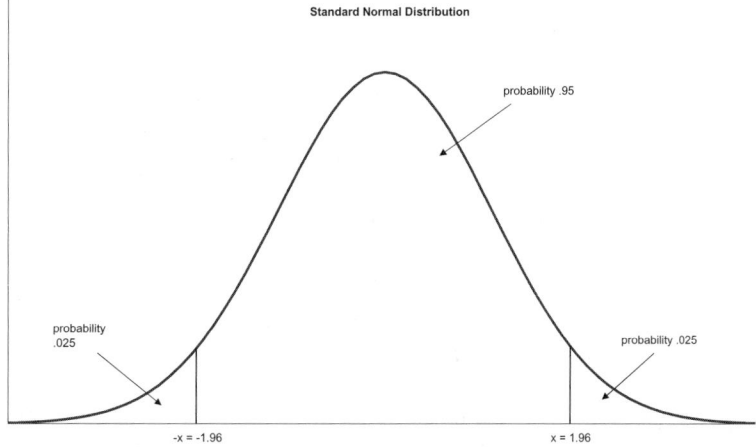

deviations of the mean, and almost all fall within 3 standard deviations of the mean. For these rules to hold with real data, the distribution of the data must be at least approximately symmetric and bell shaped. Let's look at these rules more closely.

Let X be normally distributed with mean μ and standard deviation σ. To perform a probability calculation on X, we can first standardize X and then perform the calculation on the standardized variable Z. Specifically, we will find the probability that X is within k standard deviations of the mean for $k = 1$, $k = 2$, and $k = 3$. In general, this probability is $P(\mu - k\sigma < X < \mu + k\sigma)$. But by standardizing the values $\mu - k\sigma$ and $\mu + k\sigma$, we obtain the equivalent probability $P(-k < Z < k)$, where Z has a $N(0, 1)$ distribution. This latter probability can be calculated in Excel with the formula

The normal distribution is the basis for the empirical rules introduced in Chapter 3.

$$=\text{NORMSDIST}(k)\text{-NORMSDIST}(-k)$$

By substituting the values 1, 2, and 3 for k, we find the following probabilities:

$$P(-1 < Z < 1) = .6827$$
$$P(-2 < Z < 2) = .9545$$
$$P(-3 < Z < 3) = .9973$$

As we see, there is virtually no chance of being beyond 3 standard deviations from the mean, the chances are about 19 out of 20 of being within 2 standard deviations of the mean, and the chances are about 2 out of 3 of being within 1 standard deviation of the mean. These probabilities are the basis for the empirical rules in Chapter 3. These rules more closely approximate reality as the histograms of observed data become more bell shaped.

6.3 Applications of the Normal Distribution

In this section we apply the normal distribution to a variety of business problems.

Example 6.3 Personnel Testing at ZTel

The personnel department of ZTel, a large communications company, is reconsidering its hiring policy. Each applicant for a job must take a standard exam, and the hire or no-hire decision depends at least in part on the result of the exam. The scores of all applicants have been examined closely. They are approximately normally distributed with mean 525 and standard deviation 55.

The current hiring policy occurs in two phases. The first phase separates all applicants into three categories: automatic accepts, automatic rejects, and "maybes." The automatic accepts are those whose test scores are 600 or above. The automatic rejects are those whose test scores are 425 or below. All other applicants (the "maybes") are passed on to a second phase where their previous job experience, special talents, and other factors are used as hiring criteria. The personnel manager at ZTel wants to calculate the percentage of applicants who are automatic accepts or rejects, given the current standards. She also wants to know how to change the standards to automatically reject 10% of all applicants and automatically accept 15% of all applicants.

Objective To determine test scores that can be used to accept or reject job applicants at ZTel.

FIGURE 6.8
Calculations for Personnel Example

	A	B	C	D	E	F
1	**Personnel Accept/Reject Example**					
2						
3	Mean of test scores	525		**Range names**		
4	Stdev of test scores	55		Mean: B3		
5				Stdev: B4		
6	**Current Policy**					
7	Automatic accept point	600				
8	Automatic reject point	425				
9						
10	Percent accepted	8.63%		1-NORMDIST(B7,Mean,Stdev,1)		
11	Percent rejected	3.45%		NORMDIST(B8,Mean,Stdev,1)		
12						
13	**New Policy**					
14	Percent accepted	15%				
15	Percent rejected	10%				
16						
17	Automatic accept point	582		NORMINV(1-B14,Mean,Stdev)		
18	Automatic reject point	455		NORMINV(B15,Mean,Stdev)		

Solution

Let X be the test score of a typical applicant. Then the distribution of X is $N(525, 55)$. If we find a probability such as $P(X \leq 425)$, we can interpret this as the probability that a typical applicant is an automatic reject, or we can interpret it as the percentage of *all* applicants who are automatic rejects. Given this observation, the solution to ZTel's problem appears in Figure 6.8. (See the file PERSONNEL.XLS.) The probability that a typical applicant is automatically accepted is 0.0863, found in cell B10 with the formula

$$=1\text{-NORMDIST(B7,Mean,Stdev,1)}$$

Similarly, the probability that a typical applicant is automatically rejected is .0345, found in cell B11 with the formula

$$=\text{NORMDIST(B8,Mean,Stdev,1)}$$

Therefore, ZTel automatically accepts about 8.6% and rejects about 3.5% of all applicants under the current policy.

To find new cutoff values that reject 10% and accept 15% of the applicants, we need the 10th and 85th percentiles of the $N(525, 55)$ distribution. These are 455 and 582 (rounded to the nearest integer), respectively, found in cells B17 and B18 with the formulas

$$=\text{NORMINV(1-B14,Mean,Stdev)}$$

and

$$=\text{NORMINV(B15,Mean,Stdev)}$$

To accomplish its objective, ZTel needs to raise the automatic rejection point from 425 to 455 and lower the automatic acceptance point from 600 to 582. ●

Example 6.4 Quality Control at PaperStock Company

The PaperStock Company runs a manufacturing facility that produces a paper product. The fiber content of this product is supposed to be 20 pounds per 1000 square feet. (This is typical for the type of paper used in grocery bags, for example.) Because of random variations in the inputs to the process, however, the fiber content of a typical 1000-square-foot roll varies according to a $N(\mu, \sigma)$ distribution. The mean fiber content (μ) can be

	A	B	C	D	E	F	G	H	I
1	**Paper Machine Setting Problem**								
2									
3	Mean	20							
4	Stdev in good case	0.1		Range names					
5	Stdev in bad case	0.15		Mean: B3					
6				Stdev1: B4					
7	Reject region			Stdev2: B5					
8	Lower limit	19.8							
9	Upper limit	20.3							
10									
11	Probability of reject								
12	in good case	0.024		NORMDIST(B8,Mean,Stdev1,1)+(1-NORMDIST(B9,Mean,Stdev1,1))					
13	in bad case	0.114		NORMDIST(B8,Mean,Stdev2,1)+(1-NORMDIST(B9,Mean,Stdev2,1))					
14									
15	Data table of rejection probability as a function of the mean and good standard deviation								
16					Standard deviation				
17		0.024	0.1	0.11	0.12	0.13	0.14	0.15	
18		19.7	0.841	0.818	0.798	0.779	0.762	0.748	
19		19.8	0.500	0.500	0.500	0.500	0.500	0.500	
20		19.9	0.159	0.182	0.203	0.222	0.240	0.256	
21	Mean	20	0.024	0.038	0.054	0.072	0.093	0.114	
22		20.1	0.024	0.038	0.054	0.072	0.093	0.114	
23		20.2	0.159	0.182	0.203	0.222	0.240	0.256	
24		20.3	0.500	0.500	0.500	0.500	0.500	0.500	
25		20.4	0.841	0.818	0.798	0.779	0.762	0.748	

FIGURE 6.9 Calculations for Paper Quality Example

controlled—that is, it can be set to any desired level by adjusting an instrument on the machine. The variability in fiber content, as measured by the standard deviation σ, is 0.1 pound when the process is "good," but it sometimes increases to 0.15 pound when the machine goes "bad." A given roll of this product must be rejected if its actual fiber content is less than 19.8 pounds or greater than 20.3 pounds. Calculate the probability that a given roll is rejected, for a setting of $\mu = 20$, when the machine is "good" and when it is "bad."

Objective To determine the machine settings that result in paper of acceptable quality at PaperStock Company.

Solution

Let X be the fiber content of a typical roll. The distribution of X will be either $N(20, 0.1)$ or $N(20, 0.15)$, depending on the status of the machine. In either case, the probability that the roll must be rejected can be calculated as shown in Figure 6.9. (See the file PAPER.XLS.) The formula for rejection in the "good" case appears in cell B12:

$$=\text{NORMDIST(B8,Mean,Stdev1,1)}+(1-\text{NORMDIST(B9,Mean,Stdev1,1))}$$

It is the sum of two probabilities: the probability of being to the left of the lower limit and the probability of being to the right of the upper limit. These probabilities of rejection are represented graphically in Figure 6.10. A similar formula for the "bad" case appears in cell B13, using Stdev2 in place of Stdev1.

We see that the probability of a rejected roll in the "good" case is 0.024; in the "bad" case it is 0.114. That is, when the standard deviation increases by 50% from 0.1 to 0.15, the percentage of rolls rejected more than quadruples, from 2.4% to 11.4%.

It is certainly possible that the true process mean and "good" standard deviation will not always be equal to the values we've assumed in cells B3 and B4. Therefore, it is

FIGURE 6.10
Rejection Regions
for Paper Quality
Example

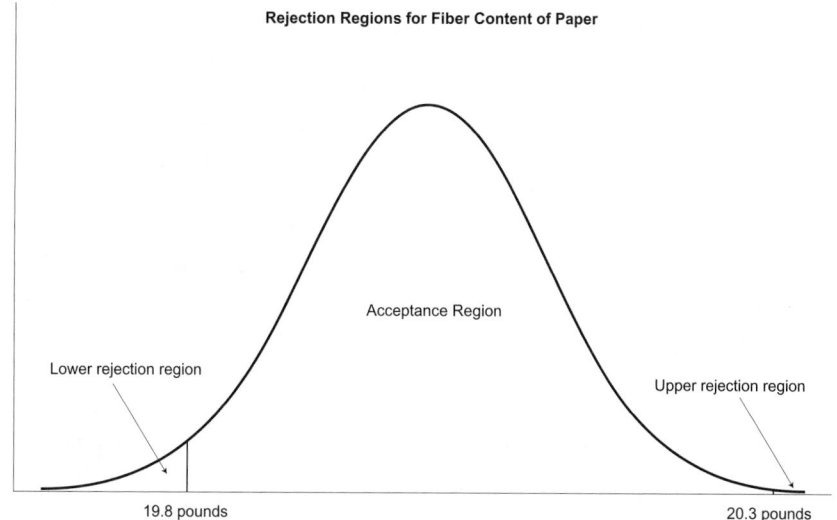

Rejection Regions for Fiber Content of Paper

Acceptance Region

Lower rejection region

Upper rejection region

19.8 pounds

20.3 pounds

To form this data table, enter the formula =B12 in cell B17, highlight the range B17:H25, and use the Data/Table menu item with row input cell B4 and column input cell B3.

useful to see how sensitive the rejection probability is to these two parameters. We do this with a two-way data table, as shown in Figure 6.9. The tabulated values show that the probability of rejection varies greatly even for small changes in the key inputs. In particular, a combination of a badly centered mean and a large standard deviation can make the probability of rejection very large. ●

Example 6.5 Analyzing an Investor's After-Tax Profit

Howard Davis invests $10,000 in a certain stock on January 1, 2003. By examining past movements of this stock and consulting with his broker, Howard estimates that the annual return from this stock, X, is normally distributed with mean 10% and standard deviation 4%. Here X (when expressed as a decimal) is the profit Howard receives per dollar invested. It means that on December 31, 2003, his $10,000 will have grown to $10,000(1 + X)$ dollars. Because Howard is in the 33% tax bracket, he will then have to pay the Internal Revenue Service 33% of his profit. Calculate the probability that Howard will have to pay the IRS at least $400. Also, calculate the dollar amount such that Howard's after-tax profit is 90% certain to be less than this amount; that is, calculate the 90th percentile of his after-tax profit.

Objective To determine the after-tax profit Howard Davis can be 90% certain of earning.

Solution

Howard's before-tax profit is $10,000X$ dollars, so the amount he pays the IRS is $0.33(10,000X)$, or $3300X$ dollars. We want the probability that this is at least $400. Since $3300X > 400$ is the same as $X > 4/33$, the probability of this outcome can be found as in Figure 6.11 on page 254. (See the file STOCKTAX.XLS.) It is calculated with the formula

$$=1-NORMDIST(400/(InvestAmt*TaxRate),Mean,Stdev,1)$$

in cell B8. As we see, Howard has about a 30% chance of paying at least $400 in taxes.

To answer the second question, note that the after-tax profit is 67% of the before-tax profit, or $6700X$ dollars, and we want its 90th percentile. If this percentile is x, then we know that $P(6700X < x) = 0.90$, which is the same as $P(X < x/6700) = 0.90$. In words,

	A	B	C	D	E	F	G	H
1	**Tax on Stock Return Problem**							
2				**Range names**				
3	Amount invested	$10,000		InvestAmt: B3				
4	Mean return	10%		Mean: B4				
5	StDev of return	4%		Stdev: B5				
6	Tax rate	33%		TaxRate: B6				
7								
8	Probability he pays at least $400 in taxes	0.298		1-NORMDIST(400/(InvestAmt*TaxRate),Mean,Stdev,1)				
9								
10	90th percentile of stock return	15.13%		NORMINV(0.9,Mean,Stdev)				
11	90th percentile of after-tax return	$1,013		(1-TaxRate)*InvestAmt*B10				

FIGURE 6.11 Calculations for Taxable Returns Example

we want the 90th percentile of the X distribution to be $x/6700$. From cell B10 of Figure 6.11, we see that the 90th percentile is 15.13%, so the required value of x is $1,013. Note that the *mean* after-tax profit is $670 (67% of the mean before-tax profit of 0.10 multiplied by $10,000). Of course, Howard might get lucky and make more than this, but he is 90% certain that his after-tax profit will be no greater than $1,013. ●

It is sometimes tempting to model every continuous random variable with a normal distribution. This can be dangerous for at least two reasons. First, not all random variables have a *symmetric* distribution. Some are skewed to the left or the right, and for these the normal distribution can be a poor approximation. The second problem is that many random variables in real applications must be *nonnegative*, and the normal distribution allows the possibility of negative values. The following example shows how assuming normality can get us into trouble if we aren't careful.

Example 6.6 Predicting Future Demand for Microwave Ovens at Highland Company

The Highland Company is a retailer that sells microwave ovens. The company wants to model its demand for microwaves over the next 12 years. Using historical data as a guide, it assumes that demand in year 1 is normally distributed with mean 5000 and standard deviation 1500. It assumes that demand in every subsequent year is normally distributed with mean equal to the *actual* demand from the previous year and standard deviation 1500. For example, if demand in year 1 turns out to be 4500, then the *mean* demand in year 2 is 4500. This assumption appears plausible because it leads to correlated demands. For example, if demand is high one year, it will tend to be high the next year. Investigate the ramifications of this model, and suggest models that might be more realistic.

Objective To construct and analyze a spreadsheet model for microwave oven demand over the next 12 years using Excel's NORMINV function, and to show how "normal" models can lead to nonsensical outcomes unless we are careful.

Solution

The best way to analyze this model is with simulation, much as we did in Chapter 5. To do this, we must be able to simulate normally distributed random numbers in Excel. We can do this with the NORMINV function. Specifically, to generate a normally distributed number with mean μ and standard deviation σ, we use the formula

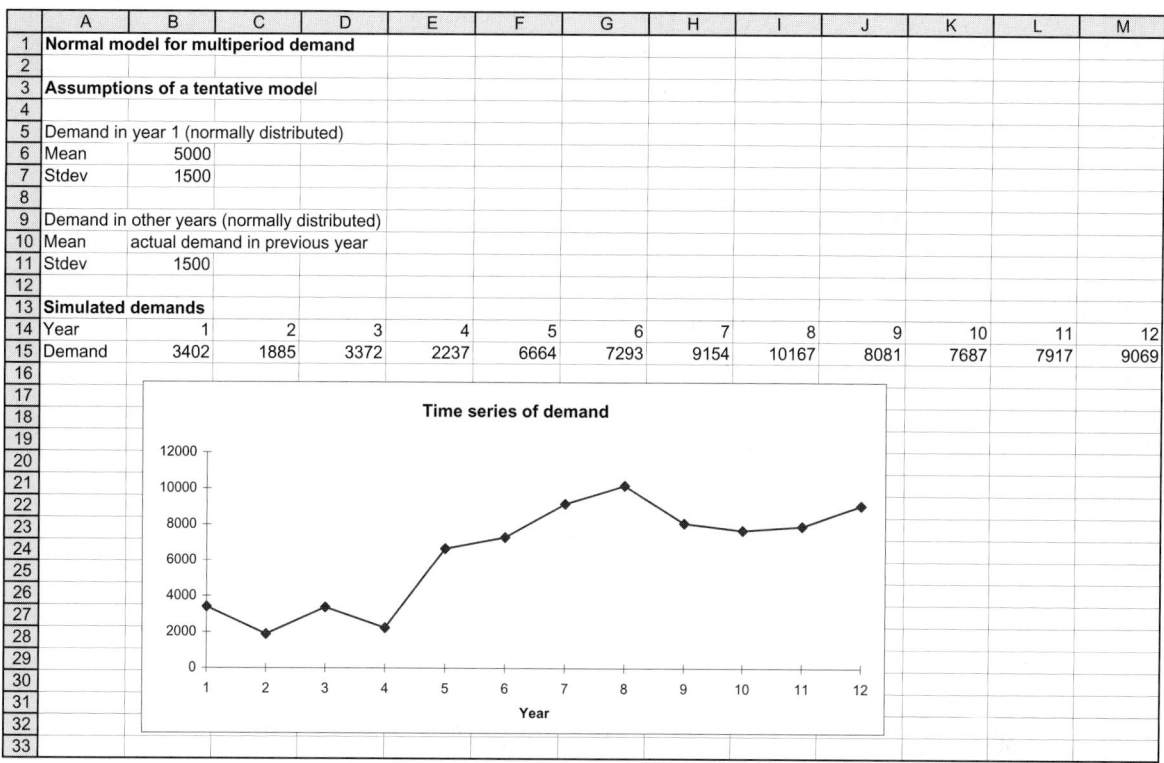

FIGURE 6.12 One Set of Demands for Model 1 in the Microwave Example

$$=\text{NORMINV}(\text{RAND}(),\mu,\sigma)$$

Because this formula uses the RAND function, it generates a *different* random number each time it is used—and each time the spreadsheet recalculates.[4]

The spreadsheet in Figure 6.12 shows a simulation of yearly demands over a 12-year period. To simulate the demands in row 15, we enter the formula

$$=\text{NORMINV}(\text{RAND}(),\text{B6},\text{B7})$$

in cell B15. Then we enter the formula

$$=\text{NORMINV}(\text{RAND}(),\text{B15},\$\text{B}\$11)$$

in cell C15 and copy it across row 15. As the accompanying time series graph of these demands indicates, the model seems to be performing well.

However, the simulated demands in Figure 6.12 are only one set of possible demands. Remember that each time the spreadsheet recalculates, all of the random numbers change.[5] Figure 6.13 (page 256) shows a different set of random numbers generated by the *same* formulas. Clearly, the model is not working well in this case—some demands are negative, which makes no sense. The problem is that if the actual demand is low in one year,

---

[4]To see why this formula makes sense, note that the RAND function in the first argument generates a uniformly distributed random value between 0 and 1. Therefore, the effect of the function is to generate a random *percentile* from the normal distribution.

[5]The usual way to get Excel to recalculate is to press the F9 key. However, this makes all of the data tables in the workbook recalculate, which can take forever. Because there is a data table in another sheet of the OVEN-DEMAND.XLS file, we suggest a different way to recalculate. Simply position the cursor on any blank cell and press the Delete key.

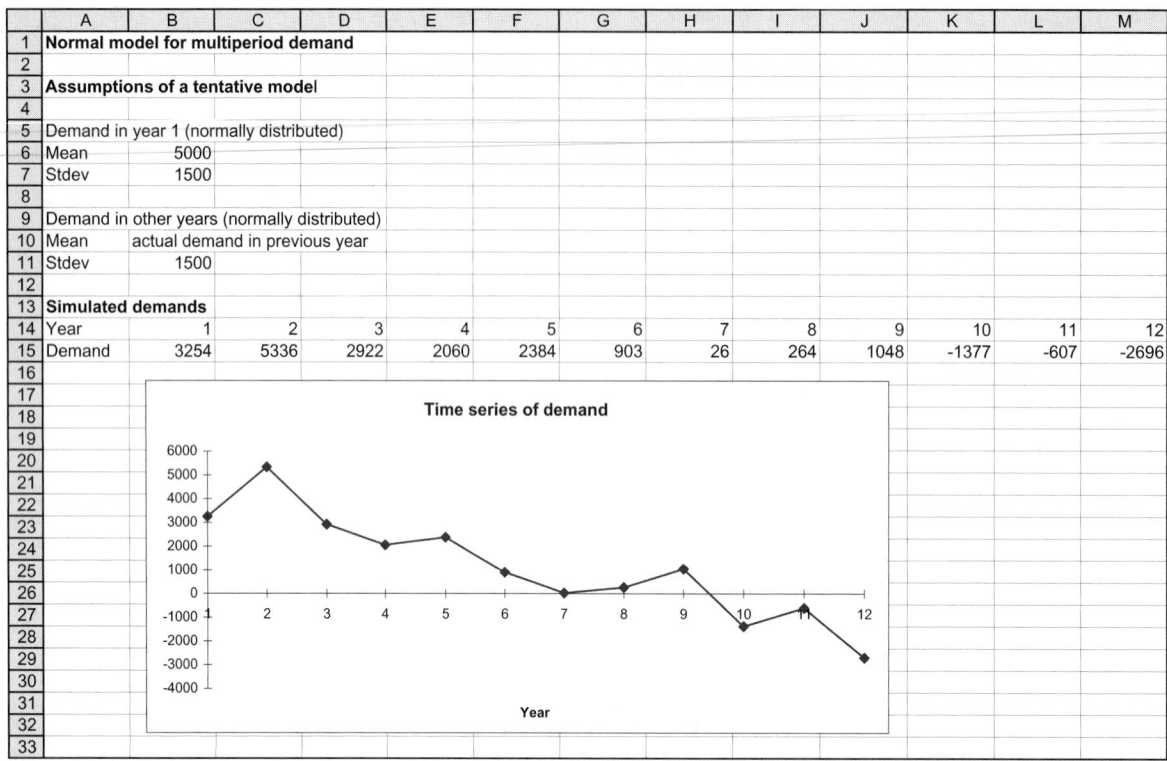

The spreadsheet shown contains:

Normal model for multiperiod demand

Assumptions of a tentative model

Demand in year 1 (normally distributed)
Mean	5000
Stdev	1500

Demand in other years (normally distributed)
Mean	actual demand in previous year
Stdev	1500

Simulated demands

Year	1	2	3	4	5	6	7	8	9	10	11	12
Demand	3254	5336	2922	2060	2384	903	26	264	1048	-1377	-607	-2696

Time series of demand

FIGURE 6.13 Another Set of Demands for Model 1 in the Microwave Example

there is a fair chance that the next normally distributed demand will be negative. You can check (by recalculating many times) that the demand sequence is *usually* all positive, but every now and then you'll get a nonsense sequence as in Figure 6.13. We need a new model!

One way to modify the model is to let the standard deviation and mean move together. That is, if the mean is low, then the standard deviation will also be low. This minimizes the chance that the *next* random demand will become negative. Besides, this type of model is probably more realistic. If demand in one year is low, there is likely to be less variability in next year's demand. Figure 6.14 illustrates one way to model this changing standard deviation.

We let the standard deviation of demand in any year (after year 1) be the original standard deviation, 1500, multiplied by the ratio of the expected demand for this year to the expected demand in year 1. For example, if demand in some year is 500, then the expected demand next year is 500, and the standard deviation of next year's demand is reduced to 1500(500/5000) = 150. The only change to the spreadsheet model is in row 15, where we enter

$$=NORMINV(RAND(),B15,\$B\$7*B15/\$B\$6)$$

in cell C15 and copy it across row 15. Now the chance of a negative demand is practically negligible because this would require a value more than 3 standard deviations below the mean.

The model in Figure 6.14 is still not foolproof. By recalculating many times, we can still generate a negative demand now and then. To be even safer, we can "truncate" the demand distribution at some nonnegative value such as 250, as shown in Figure 6.15.

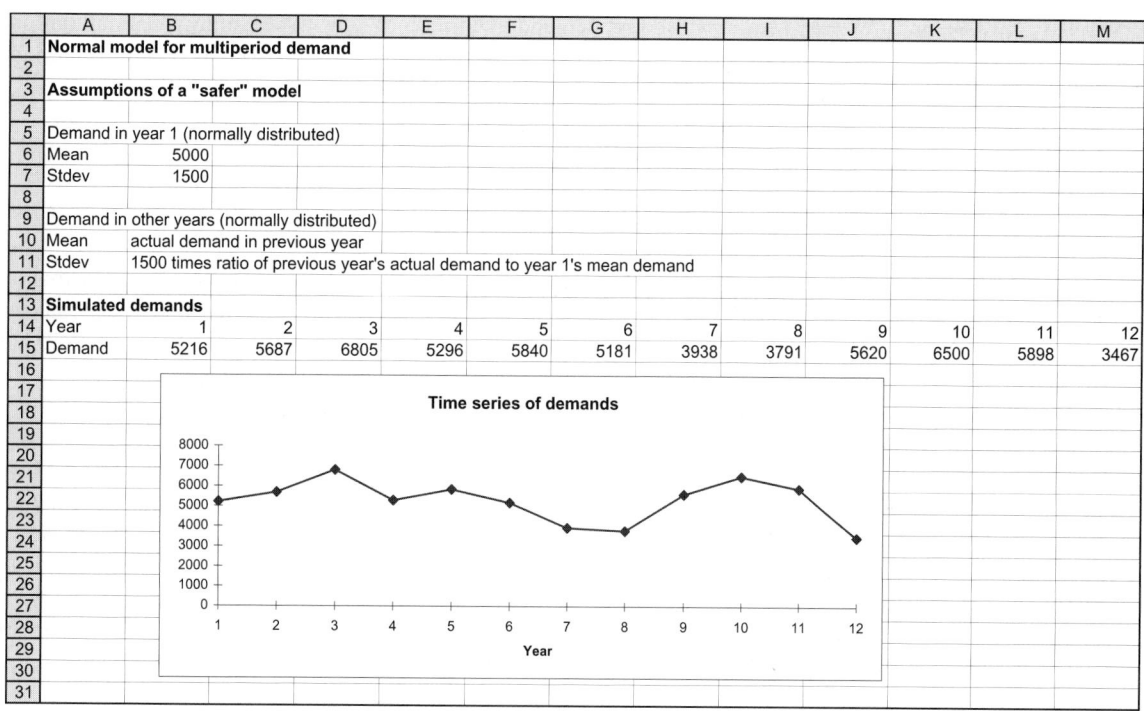

	A	B	C	D	E	F	G	H	I	J	K	L	M
1	**Normal model for multiperiod demand**												
2													
3	**Assumptions of a "safer" model**												
4													
5	Demand in year 1 (normally distributed)												
6	Mean	5000											
7	Stdev	1500											
8													
9	Demand in other years (normally distributed)												
10	Mean	actual demand in previous year											
11	Stdev	1500 times ratio of previous year's actual demand to year 1's mean demand											
12													
13	**Simulated demands**												
14	Year	1	2	3	4	5	6	7	8	9	10	11	12
15	Demand	5216	5687	6805	5296	5840	5181	3938	3791	5620	6500	5898	3467

FIGURE 6.14 Generated Demands for Model 2 in Microwave Example

	A	B	C	D	E	F	G	H	I	J	K	L	M
1	**Normal model for multiperiod demand**												
2													
3	**Assumptions of an even "safer" model**												
4													
5	Minimum demand in any year			250									
6													
7	Demand in year 1 (truncated normal)												
8	Mean	5000											
9	Stdev	1500											
10													
11	Demand in other years (truncated normal)												
12	Mean	actual demand in previous year											
13	Stdev	1500 times ratio of previous year's actual demand to year 1's mean demand											
14													
15	**Simulated demands**												
16	Year	1	2	3	4	5	6	7	8	9	10	11	12
17	Demand	6025	3289	2967	3272	1972	2150	1502	1436	1972	2818	3028	2460

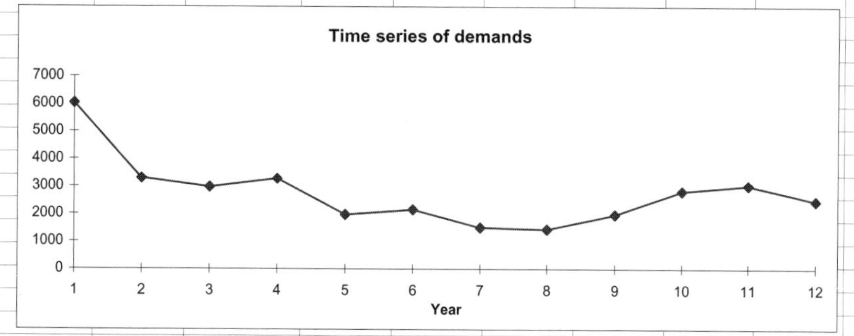

FIGURE 6.15 Generated Demands for a Truncated Model in Microwave Example

Now we generate a random demand as in the previous model, but if this randomly generated value is below 250, we set the demand equal to 250. This is done by entering the formulas

$$=\text{MAX(NORMINV(RAND(),B8,B9),D5)}$$

and

$$=\text{MAX(RAND(),B17,\$B\$9*B17/\$B\$8),\$D\$5)}$$

in cells B17 and C17 and copying this latter formula across row 17. Whether this is the way the demand process works for Highland's microwaves is an open question, but at least we have prevented demands from ever becoming negative—or even falling below 250. Moreover, this type of truncation is a common way of modeling when we want to use a normal distribution but for physical reasons cannot allow the random quantities to become negative.

Before leaving this example, we challenge your intuition. In the final model in Figure 6.15, the demand in any year (say, year 6) is, aside from the truncation, normally distributed with a mean and standard deviation that depend on the previous year's demand. Does this mean that if we recalculate many times and keep track of the year 6 demand each time, the resulting histogram of these year 6 demands will be normally distributed? Perhaps surprisingly, the answer is a clear "no." We show the evidence in Figures 6.16 and 6.17. In Figure 6.16 we use a data table to obtain 400 replications of demand in year 6 (in column B). Then we use StatPro's histogram procedure to create a histogram of these simulated demands in Figure 6.17. It is clearly skewed to the right and *nonnormal*.

What causes this distribution to be nonnormal? It is *not* the truncation. Truncation has a relatively minor effect because most of the demands don't need to be truncated anyway. The real reason is that the distribution of year 6 demand is only normal *conditional* on the demand in year 5. That is, if we fix the demand in year 5 at any level and then repli-

FIGURE 6.16
Replication of
Demand in Year 6

	A	B	C	D	E
36	Replication	Demand			
37		1699		Average	4930
38	1	2868		Stdev	4153
39	2	3256			
40	3	1307			
41	4	7615			
42	5	2695			
43	6	8890			
44	7	4772			
45	8	2002			
46	9	5245			
47	10	8341			
48	11	1103			
49	12	3983			
50	13	8656			
51	14	5622			
52	15	13672			
53	16	9564			
54	17	5432			
55	18	4544			
56	19	4838			
433	396	7432			
434	397	2498			
435	398	6582			
436	399	6874			
437	400	2379			

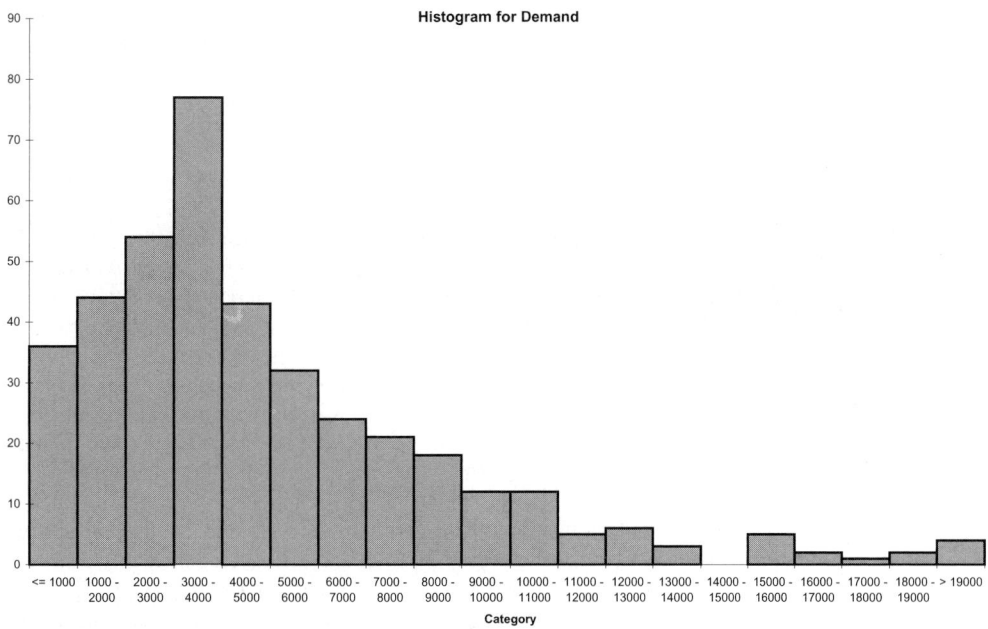

FIGURE 6.17 Histogram of Year 6 Demands

cate year 6 demand many times, the resulting histogram *is* normally shaped. But we don't fix the year 5 demand. It varies from replication to replication, and this variation causes the skewness in Figure 6.17. Admittedly, the reason for this skewness is not obvious from an intuitive standpoint, but simulation makes it easy to demonstrate. ●

PROBLEMS

Level A

1. The grades on the midterm examination given in a large managerial statistics class are normally distributed with mean 75 and standard deviation 9. The instructor of this class wants to assign an A grade to the top 10% of the scores, a B grade to the next 10% of the scores, a C grade to the next 10% of the scores, a D grade to the next 10% of the scores, and an F grade to all scores below the 60th percentile of this distribution. For each possible letter grade, find the lowest acceptable score within the established range. For example, the lowest acceptable score for an A is the score at the 90th percentile of this normal distribution.

2. Suppose it is known that the distribution of purchase amounts by customers entering a popular retail store is approximately normal with mean $25 and standard deviation $8.

a. What is the probability that a randomly selected customer spends less than $35 at this store?

b. What is the probability that a randomly selected customer spends between $15 and $35 at this store?

c. What is the probability that a randomly selected customer spends more than $10 at this store?

d. Find the dollar amount such that 75% of all customers spend no more than this amount.

e. Find the dollar amount such that 80% of all customers spend at least this amount.

f. Find two dollars amounts, equidistant from the mean of $25, such that 90% of all customer purchases are between these values.

3. A machine used to regulate the amount of a certain chemical dispensed in the production of a particular type of cough syrup can be set so that it discharges an average of μ milliliters (ml) of the chemical in each bottle of cough syrup. The amount of chemical placed into each bottle of cough syrup is known to have a normal distribution with a standard deviation of 0.250 ml. If this machine discharges more than 2 ml of the chemical when preparing a given bottle of this cough syrup, the bottle is considered to be un-acceptable by industry standards. Determine the set-ting for μ so that no more than 1% of the bottles of cough syrup prepared through the use of this machine will be rejected.

4. The weekly demand for Ford car sales follows a nor-mal distribution with mean 50,000 cars and standard deviation 14,000 cars.
   a. There is a 1% chance that Ford will sell more than what number of cars during the next year?
   b. What is the probability that Ford will sell between 2.4 and 2.7 million cars during the next year?

5. Warren Dinner has invested in nine different invest-ments. The returns on the different investments are probabilistically independent, and each return follows a normal distribution with mean $500 and standard deviation $100.
   a. There is a 1% chance that the total return on the nine investments is less than what value? (Use the fact that the sum of independent normal random variables is normally distributed, with mean equal to the sum of the individual means, and variance equal to the sum of the individual variances.)
   b. What is the probability that Warren's total return is between $4000 and $5200?

6. Scores on an exam appear to follow a normal distribu-tion with $\mu = 60$ and $\sigma = 20$. The instructor wishes to give a grade of D to students scoring between the 10th and 30th percentiles on the exam. For what range of scores should a D be given?

7. Suppose the weight of a typical American male follows a normal distribution with $\mu = 180$ lb and $\sigma = 30$ lb. Also, suppose 91.92% of all American males weigh more than I weigh.
   a. What fraction of American males weigh more than 225 pounds?
   b. How much do I weigh?

8. Assume that the length of a typical televised baseball game, including all the commercial timeouts, is nor-mally distributed with mean 2.45 hours and standard deviation 0.37 hour. Consider a televised baseball game that begins at 2:00 in the afternoon. The next regularly scheduled broadcast is at 5:00.
   a. What is the probability that the game will cut into the next show, that is, go past 5:00?
   b. If the game is over before 4:30, another half-hour show can be inserted into the 4:30–5:00 slot. What is the probability of this occurring?

9. The amount of a soft drink that goes into a typical 12-ounce can varies from can to can. It is normally dis-tributed with an adjustable mean μ and a fixed stan-dard deviation of 0.05 ounce. (The adjustment is made to the filling machine.)
   a. If regulations require that cans have at least 11.9 ounces, what is the smallest mean μ that

can be used so that at least 99.5% of all cans meet the regulation?
   b. If the mean setting from part a is used, what is the probability that a typical can has at least 12 ounces?

Level B

10. The manufacturer of a particular bicycle model has the following costs associated with the management of this product's inventory. In particular, the company currently maintains an inventory of 1000 units of this bicycle model at the beginning of each year. If X units are demanded each year and X is less than 1000, the excess supply, $1000 - X$ units, must be stored until next year at a cost of $50 per unit. If X is greater than 1000 units, the excess demand, $X - 1000$ units, must be produced separately at an extra cost of $80 per unit. Assume that the annual demand (X) for this bi-cycle model is normally distributed with mean 1000 and standard deviation 75.
    a. Find the expected annual cost associated with managing potential shortages or surpluses of this product.
    b. Find two annual total cost levels, equidistant from the expected value found in part a, such that 95% of all costs associated with managing potential shortages or surpluses of this product are between these values.
    c. Comment on this manufacturer's annual produc-tion policy for this bicycle model in light of your findings in part b.

11. Matthew's Bakery prepares peanut butter cookies for sale every morning. It costs the bakery $0.25 to bake each peanut butter cookie, and each cookie is sold for $0.50. At the end of the day, leftover cookies are dis-counted and sold the following day at $0.10 per cookie. The daily demand (in dozens) for peanut but-ter cookies at this bakery is known to be normally distributed with mean 50 and standard deviation 15. The manager of Matthew's Bakery is trying to deter-mine how many dozen peanut butter cookies to make each morning to maximize the product's contribution to bakery profits. Use computer simulation to find a very good, if not optimal, production plan in this case.

12. Suppose that a particular production process fills detergent in boxes of a given size. Specifically, this process fills the boxes with an amount of detergent (in ounces) that is adequately described by a normal distribution with mean 50 and standard deviation 0.5.
    a. Simulate this production process for the filling of 500 boxes of detergent. Compute the mean and standard deviation of your simulated sample weights. How do your sample statistics compare to the theoretical population parameters in this case? How well do the empirical rules

apply in describing the variation in the weights of the detergent in your simulated detergent boxes?

b. A box of detergent is rejected by quality control personnel if it is found to contain less than 49 ounces or more than 51 ounces of detergent. Given these quality standards, what proportion of all boxes are rejected? What step(s) could the supervisor of this production process take to reduce this proportion to 1%?

13. It is widely known that many drivers on interstate highways in the United States do not observe the posted speed limit. Assume that the actual rates of speed driven by U.S. motorists are normally distributed with mean μ mph and standard deviation 5 mph. Given this information, answer each of the following independent questions:

a. If 40% of all U.S. drivers are observed traveling at 65 mph or more, what is the mean μ?

b. If 25% of all U.S. drivers are observed traveling at 50 mph or less, what is the mean μ?

c. Suppose now that the mean μ and standard deviation σ of this distribution are both unknown. Furthermore, it is observed that 40% of all U.S. drivers travel at less than 55 mph and 10% of all U.S. drivers travel at more than 70 mph. What must μ and σ be?

14. The lifetime of a certain manufacturer's washing machine is normally distributed with mean 4 years. Only 15% of all these washing machines last at least 5 years. What is the standard deviation of the lifetime of a washing machine made by this manufacturer?

15. You have been told that the distribution of regular unleaded gasoline prices over all gas stations in Indiana is normally distributed with mean $1.25 and standard deviation $0.075, and you have been asked to find two dollar values such that 95% of all gas stations charge somewhere between these two values. Why is each of the following an acceptable answer: between $1.076 and $1.381, or between $1.103 and $1.397? Can you find any other acceptable answers? Which of the many possible answers would you prefer if you are asked to obtain the *shortest* such interval?

16. When we create boxplots, we place the sides of the "box" at the first and third quartiles, and the difference between these (the length of the box) is called the interquartile range (IQR). A mild outlier is then defined as an observation that is between 1.5 and 3 IQRs from the box, and an extreme outlier is defined as an observation that is more than 3 IQRs from the box.

a. If the data are normally distributed, what percentage of values will be mild outliers? What percent-

age will be extreme outliers? Why don't the answers depend on the mean and/or standard deviation of the distribution?

b. Check out your answers in part **a** with simulation. Simulate a large number of normal random numbers (you can choose any mean and standard deviation), and count the number of mild and extreme outliers with appropriate IF functions. Do these match, at least approximately, your answers to part **a**?

17. A fast-food restaurant sells hamburgers and chicken sandwiches. On a typical weekday the demand for hamburgers is normally distributed with mean 313 and standard deviation 57; the demand for chicken sandwiches is normally distributed with mean 93 and standard deviation 22.

a. How many hamburgers must the restaurant stock to be 98% sure of not running out on a given day?

b. Answer part **a** for chicken sandwiches.

c. If the restaurant stocks 400 hamburgers and 150 chicken sandwiches for a given day, what is the probability that it will run out of hamburgers or chicken sandwiches (or both) that day? Assume that the demand for hamburgers and the demand for chicken sandwiches are probabilistically independent.

d. Why is the independence assumption in part **c** probably not realistic? Using a more realistic assumption, do you think the probability requested in part **c** would increase or decrease?

18. Suppose that the demands for a company's product in weeks 1, 2, and 3 are each normally distributed. The means are 50, 45, and 65. The standard deviations are 10, 5, and 15. Assume that these three demands are probabilistically independent. This means that if you observe one of them, it doesn't help you to predict the others. Then it turns out that total demand for the 3 weeks is also normally distributed. Its mean is the sum of the individual means, and its variance is the sum of the individual variances. (Its standard deviation, however, is not the sum of the individual standard deviations; square roots don't work that way.)

a. Suppose that the company currently has 180 units in stock, and it will not be receiving any further shipments from its supplier for at least 3 weeks. What is the probability that it will stock out during this 3-week period?

b. How many units should the company currently have in stock so that it can be 98% certain of not stocking out during this 3-week period? Again assume that it won't receive any further shipments during this period.

6.4 The Binomial Distribution

The normal distribution is undoubtedly the most important probability distribution in statistics. Not far behind in order of importance is the *binomial* distribution. The binomial distribution is a discrete distribution that can occur in two situations: (1) whenever we sample from a population with only two types of members (males and females, for example), and (2) whenever we perform a sequence of identical experiments, each of which has only two possible outcomes.

Imagine any experiment that can be repeated many times under identical conditions. It is common to refer to each repetition of the experiment as a *trial*. We assume that the outcomes of successive trials are probabilistically independent of one another and that each trial has only two possible outcomes. We label these two possibilities generically as success and failure. In any particular application the outcomes might be Democrat/Republican, defective/nondefective, went bankrupt/remained solvent, and so on. The probability of a success on each trial is p, and the probability of a failure is $1 - p$. The number of trials is n.

Binomial Distribution	Consider a situation in which there are n independent, identical trials, where the probability of a success on each trial is p and the probability of a failure is $1 - p$. Define X to be the random number of successes in the n trials. Then X has a **binomial** distribution with parameters n and p.

For example, the binomial distribution with parameters 100 and 0.3 is the distribution of the number of successes in 100 trials when the probability of success is 0.3 on each trial. A simple example that you can keep in mind throughout this section is the number of heads you would see if you flipped a coin n times. Assuming the coin is well balanced, the relevant distribution is binomial with parameters n and $p = 0.5$. This coin-flipping example is often used to illustrate the binomial distribution because of its simplicity, but we will see that the binomial distribution also applies to many important business situations.

To understand how the binomial distribution works, consider the coin-flipping example with $n = 3$. If X represents the number of heads in three flips of the coin, then the possible values of X are 0, 1, 2, and 3. We can find the probabilities of these values by considering the eight possible outcomes of the three flips: (T,T,T), (T,T,H), (T,H,T), (H,T,T), (T,H,H), (H,T,H), (H,H,T), and (H,H,H). Because of symmetry (the well-balanced property of the coin), each of these eight possible outcomes must have the same probability, so each must have probability 1/8. Next, note that one of the outcomes has $X = 0$, three outcomes have $X = 1$, three outcomes have $X = 2$, and one outcome has $X = 3$. Therefore, the probability distribution of X is

$$P(X = 0) = 1/8, \quad P(X = 1) = 3/8, \quad P(X = 2) = 3/8, \quad P(X = 3) = 1/8$$

This is a special case of the binomial distribution, with $n = 3$ and $p = 0.5$. In general, where n can be any positive integer and p can be any probability between 0 and 1, there is a rather complex formula for calculating $P(X = k)$ for any integer k from 0 to n. Instead of presenting this formula, we will discuss how to calculate binomial probabilities in Excel. We do this with the BINOMDIST function. The general form of this function is

$$=\text{BINOMDIST}(k,n,p,cum)$$

The middle two arguments are as stated previously: the number of trials n and the probability of success p on each trial. The first parameter k is an integer number of successes that we specify. The last parameter, *cum*, is either 0 or 1. It is 1 if we want the probability of *less than or equal to k* successes, and it is 0 if we want the probability of *exactly k* successes. We illustrate typical binomial calculations in the following example.

Example 6.7 Battery Life Experiment

Suppose 100 identical batteries are inserted in identical flashlights. Each flashlight takes a single battery. After 8 hours of continuous use, we assume that a given battery is still operating with probability 0.6 and has failed with probability 0.4. Let *X* be the number of successes in these 100 trials, where a success means that the battery is still functioning. Find the probabilities of the following events: (a) exactly 58 successes, (b) no more than 65 successes, (c) less than 70 successes, (d) at least 59 successes, (e) greater than 65 successes, (f) between 55 and 65 successes (inclusive), (g) exactly 40 failures, (h) at least 35 failures, and (i) less than 42 failures. Then find the 95th percentile of the distribution of *X*.

Objective To use Excel's BINOMDIST and CRITBINOM functions for calculating binomial probabilities and percentiles in the context of batteries in flashlights.

Solution

Figure 6.18 shows the solution to all of these problems. (See the file BINOMFNS.XLS.) The probabilities requested in parts (a)–(f) all involve the number of successes *X*. The key to these is the wording of phrases such as "no more than," "greater than," and so on. In particular, we have to be careful to distinguish between probabilities such as $P(X < k)$ and $P(X \leq k)$. The latter includes the possibility of having $X = k$ and the former does not.

	A	B	C	D	E	F	G	H	I
1	**Binomial Probability Calculations**								
2				Range names					
3	Number of trials	100		NTrials: B3					
4	Probability of success on each trial	0.6		PSucc: B4					
5									
6	**Event**	**Probability**		Formula					
7	Exactly 58 successes	0.0742		BINOMDIST(58,NTrials,PSucc,0)					
8	No more than 65 successes	0.8697		BINOMDIST(65,NTrials,PSucc,1)					
9	Less than 70 successes	0.9752		BINOMDIST(69,NTrials,PSucc,1)					
10	At least 59 successes	0.6225		1-BINOMDIST(58,NTrials,PSucc,1)					
11	Greater than 65 successes	0.1303		1-BINOMDIST(65,NTrials,PSucc,1)					
12	Between 55 and 65 successes (inclusive)	0.7386		BINOMDIST(65,NTrials,PSucc,1)-BINOMDIST(54,NTrials,PSucc,1)					
13									
14	Exactly 40 failures	0.0812		BINOMDIST(40,NTrials,1-PSucc,0)					
15	At least 35 failures	0.8697		1-BINOMDIST(34,NTrials,1-PSucc,1)					
16	Less than 42 failures	0.6225		BINOMDIST(41,NTrials,1-PSucc,1)					
17									
18	**Finding the 95th percentile (trial and error)**								
19	Trial values	CumProb							
20	65	0.8697		BINOMDIST(A20,NTrials,PSucc,1)					
21	66	0.9087		(Copy down)					
22	67	0.9385							
23	68	0.9602							
24	69	0.9752							
25	70	0.9852							
26				Formula in cell A27:					
27	68	0.95		CRITBINOM(NTrials,PSucc,B27)					

FIGURE 6.18 Typical Binomial Calculations

With this in mind, we can translate the probabilities requested in (a)–(f) to the following:

a. $P(X = 58)$

b. $P(X \leq 65)$

c. $P(X < 70) = P(X \leq 69)$

d. $P(X \geq 59) = 1 - P(X < 59) = 1 - P(X \leq 58)$

e. $P(X > 65) = 1 - P(X \leq 65)$

f. $P(55 \leq X \leq 65) = P(X \leq 65) - P(X < 55) = P(X \leq 65) - P(X \leq 54)$

Note how we have converted each of these so that it includes only terms of the form $P(X = k)$ or $P(X \leq k)$ (for suitable values of k). These are the types of probabilities that can be handled directly by the BINOMDIST function. The answers appear in the range B7:B12, and the corresponding formulas are shown in column D.

The probabilities requested in (g)–(i) involve *failures* rather than successes. But because each trial results in either a success or a failure, the number of failures is also binomially distributed, with parameters n and $1 - p = 0.4$. So in rows 14–16, we calculate the requested probabilities in exactly the same way, except that we substitute 1-PSucc for PSucc in the third argument of the BINOMDIST function.

Finally, to calculate the 95th percentile of the distribution of X, we proceed by trial and error. For each value k from 65 to 70, we have calculated the probability $P(X \leq k)$ in column B with the BINOMDIST function. Note that there is no value k such that $P(X \leq k) = 0.95$ exactly. We see that $P(X \leq 67)$ is slightly less than 0.95, and $P(X \leq 68)$ is slightly greater than 0.95. Therefore, the meaning of the "95th percentile" is a bit ambiguous. If we want the largest value k such that $P(X \leq k) \leq 0.95$, then this k is 67. If instead we want the smallest value k such that $P(X \leq k) \geq 0.95$, then this value is 68. The latter interpretation is the one usually accepted for binomial percentiles.

In fact, Excel has another built-in function, CRITBINOM, for finding this value of k. We illustrate it in row 27 of Figure 6.18. Now we enter the requested probability, 0.95, in cell B27 and the formula

$$=\text{CRITBINOM(NTrials,PSucc,B27)}$$

in cell A27. It returns 68, the smallest value k such that $P(X \leq k) \geq 0.95$ for this binomial distribution. ●

6.4.1 Mean and Standard Deviation of the Binomial Distribution

It can be shown that the mean and standard deviation of a binomial distribution with parameters n and p are given by the following equations.

Mean and Standard Deviation of a Binomial Distribution	
$$E(X) = np$$	(6.3)
$$\text{Stdev}(X) = \sqrt{np(1 - p)}$$	(6.4)

The formula for the mean is quite intuitive. For example, if you observe 100 trials, each with probability of success 0.6, what is your best guess for the number of successes? It is clearly $100(0.6) = 60$. The standard deviation is less obvious but still very useful. It indicates how far the actual number of successes might deviate from the mean. In this case the standard deviation is $\sqrt{100(0.6)(0.4)} = 4.90$.

Fortunately, the empirical rules discussed in Chapter 3 also apply, at least approximately, to the binomial distribution. That is, there is about a 95% chance that the actual number of successes will be within 2 standard deviations of the mean, and there is almost no chance that the number of successes will be more than 3 standard deviations from the mean. So for this example, it is very likely that the number of successes will be in the range of approximately 50 to 70, and it is very unlikely that there will be fewer than 45 or more than 75 successes.

This reasoning is extremely useful. It gives us a rough estimate of the number of successes we are likely to observe. Suppose we randomly sample 1000 parts from an assembly line and, based on historical performance, we know that the percentage of parts with some type of defect is about 5%. Translated into a binomial model, we assume that each of the 1000 parts, independently of the others, has some type of defect with probability 0.05. Would we be surprised to see, say, 75 parts with a defect? The mean is $1000(0.05) = 50$ and the standard deviation is $\sqrt{1000(0.05)(0.95)} = 6.89$. Therefore, the number of parts with defects is 95% certain to be within $50 \pm 2(6.89)$, or approximately from 35 to 65. Because 75 is slightly beyond 3 standard deviations from the mean, it is highly unlikely that we would observe 75 (or more) parts with defects.

6.4.2 The Binomial Distribution in the Context of Sampling

We now consider how the binomial distribution applies to sampling from a population with two types of members. Let's say these two types are men and women, although in applications they might be Democrats versus Republicans, users of our product versus nonusers, and so on. We will assume that the population has N members, of whom N_M are men and N_W are women (where $N_M + N_W = N$). If we sample n of these randomly, we are typically interested in the composition of the sample. The question is whether the number of men in the sample is binomially distributed with parameters n and $p = N_M/N$, the fraction of men in the population. The answer depends on how the sampling is performed.

If sampling is done **without replacement,** then each member of the population can be sampled only once. That is, once a person is sampled, his or her name is struck from the list and cannot be sampled again. If sampling is done **with replacement,** then it is possible, although maybe not likely, to select a given member of the population any number of times. Most real-world sampling is performed *without* replacement. There is no point in obtaining information from the same person more than once. However, *the binomial model applies only to sampling with replacement.* Because the composition of the remaining population keeps changing as the sampling progresses, the binomial model can provide only an approximation of the population if sampling is done without replacement. If there is no replacement, the value of p, the proportion of men in this case, does *not* stay constant, a requirement of the binomial model. The appropriate distribution for sampling without replacement is the *hypergeometric* distribution, a distribution we will not discuss in detail here.

If n is small relative to N, however, the binomial model is a very good approximation to the hypergeometric model and can be used even if sampling is performed without replacement. A rule of thumb is that if n is no greater than 10% of N, that is, no more than 10% of the population is sampled, then the binomial model can be used safely. Of course, as you probably know, many national polls sample considerably less than 10% of the population. In fact, they often sample only a few thousand people from the hundreds of millions in the entire population. The bottom line is that in most real-world sampling contexts, the binomial model is perfectly adequate.

6.4.3 The Normal Approximation to the Binomial

If we graph the binomial probabilities, we see an interesting phenomenon—namely, the graph begins to look symmetric and bell shaped when n is fairly large and p is not too close to 0 or 1. An example is illustrated in Figure 6.19 with the parameters $n = 30$ and $p = 0.4$. Generally, if $np > 5$ and $n(1 - p) > 5$, the binomial distribution can be approximated well by a normal distribution with mean np and standard deviation $\sqrt{np(1 - p)}$.

One practical consequence of the normal approximation to the binomial is that the empirical rules can be applied. That is, when the binomial distribution is approximately symmetric and bell shaped, we know the chances are about 2 out of 3 that the number of successes will be within 1 standard deviation of the mean. Similarly, there is about a 95% chance that the number of successes will be within 2 standard deviations of the mean, and the number of successes will almost surely be within 3 standard deviations of the mean. Here, the mean is np and the standard deviation is $\sqrt{np(1 - p)}$.

FIGURE 6.19
Bell-shaped
Binomial
Distribution

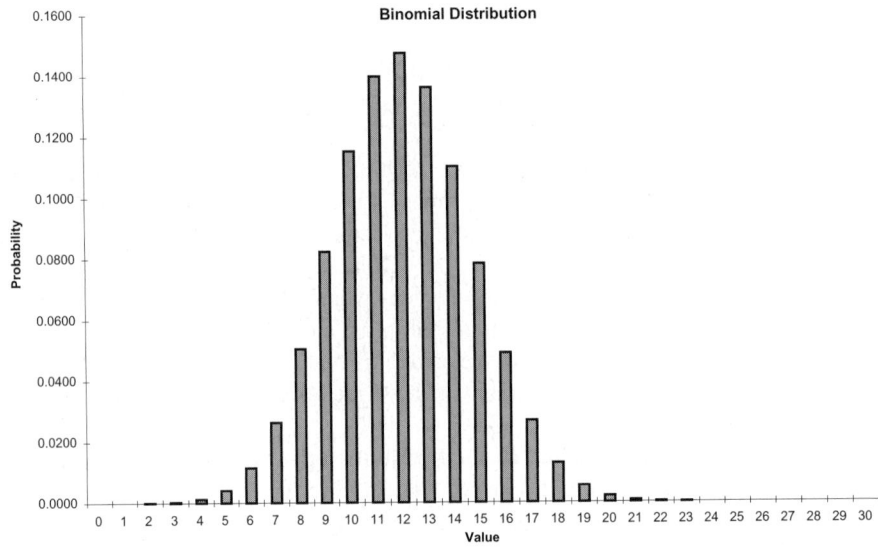

6.5 Applications of the Binomial Distribution

The binomial distribution finds many applications in the business world and elsewhere. We discuss a few such applications in this section.

Example 6.8 Is This Mutual Fund Really a Winner?

An investment broker at the Michaels & Dodson Company claims that he has found a real winner. He has tracked a mutual fund that has beaten a standard market index in 37 of the past 52 weeks. Could this be due to chance, or has he *really* found a winner?

	A	B	C	D	E	F	G	H	I
1	**Michaels & Dodson mutual fund problem**								
2									
3	Weeks beating market index	37						**Range names**	
4	Total number of weeks	52						NBeatMkt: B3	
5								NFunds: B8	
6	Probability of doing at least this well by chance	0.00159		1-BINOMDIST(NBeatMkt-1,NWeeks,0.5,1)				NWeeks: B4 PAtLeast37: B6	
7								PStar: B9	
8	Number of mutual funds	400							
9	Probability of at least one doing at least this well	0.471		1-BINOMDIST(0,NFunds,PAtLeast37,1)					
10									
11	Two-way data table of the probability in B9 as a function of values in B3 and B8								
12			Number of weeks beating the market index						
13		0.471	36	37	38	39	40		
14	Number of mutual funds	200	0.542	0.273	0.113	0.040	0.013		
15		300	0.690	0.380	0.164	0.060	0.019		
16		400	0.790	0.471	0.213	0.079	0.025		
17		500	0.858	0.549	0.258	0.097	0.031		
18		600	0.904	0.616	0.301	0.116	0.038		

FIGURE 6.20 Binomial Calculations for Investment Example

Objective To determine the probability of a mutual fund outperforming a standard market index at least 37 out of 52 weeks.

Solution

The broker is no doubt tracking a lot of mutual funds, and he is probably reporting on the best of these. Therefore, we will check whether the best of *many* mutual funds could do at least this well purely by chance. To do this, we first specify what we mean by "purely by chance." This means that each week, a given fund has a 50–50 chance of beating the market index, independently of performance in other weeks. In other words, the number of weeks where a given fund outperforms the market index is binomially distributed with $n = 52$ and $p = 0.5$. With this in mind, cell B6 of Figure 6.20 shows the probability that a given fund does at least as well—beats the market index at least 37 out of 52 weeks— as the reported fund. (See the MUTUALFUND.XLS file.) Because $P(X \geq 37) = 1 - P(X \leq 36)$, the relevant formula is

$$=1\text{-BINOMDIST(NBeatMkt-1,NWeeks,0.5,1)}$$

Obviously, this probability, 0.00159, is quite small. A single fund isn't likely to beat the market this often purely by chance.

However, the probability that the *best* of many mutual funds does at least this well is much larger. To calculate this probability, let's assume that 400 funds are being tracked, and let Y be the number of these that beat the market at least 37 of 52 weeks. Then Y is also binomially distributed, with parameters $n = 400$ and $p = 0.00159$, the probability calculated previously. We want to know whether *any* of the 400 funds beats the market at least 37 of 52 weeks, so we calculate $P(Y \geq 1) = 1 - P(Y = 0)$. We do this in cell B9 with the formula

$$=1\text{-BINOMDIST(0,NFunds,PAtLeast37,1)}$$

(Can you see why the fourth argument could be 0 *or* 1?) The resulting probability is nearly 0.5—that is, there is nearly a 50–50 chance that at least one of 400 funds will do as well as the reported fund. This certainly casts doubt on the broker's claim that he found a real winner. Perhaps his star fund just got lucky and will perform no better than average in succeeding weeks.

To see how the probability in cell B9 depends on the level of success of the reported fund (the value in cell B3) and the number of mutual funds being tracked (in cell B8), we create a two-way data table in the range B13:G18. (The formula in cell B13 is =PStar, the row input cell is B3, and the column input cell is B8.) As we saw, beating the market 37 times out of 52 is no big deal with 400 funds, but beating it 40 times out of 52, even with 600 funds, is something worth reporting. The probability of this happening purely by chance is only 0.038, or less than 1 out of 25. ●

The next example requires a normal calculation to find a probability p, which is then used in a binomial calculation.

Example 6.9 Analyzing Daily Sales at Diggly Wiggly Supermarket

Customers at the Diggly Wiggly Supermarket spend varying amounts. Historical data show that the amount spent per customer is normally distributed with mean $85 and standard deviation $30. If 500 customers shop in a given day, calculate the mean and standard deviation of the number who spend at least $100. Then calculate the probability that at least 30% of all customers spend at least $100.

Objective To use the normal *and* binomial distributions to calculate the typical number of customers who spend at least $100 per day and the probability that at least 30% of all 500 daily customers spend at least $100.

Solution

Both questions involve the number of customers who spend at least $100. Because the amounts spent are normally distributed, the probability that a typical customer spends at least $100 is found with the NORMDIST function. This probability, 0.309, appears in cell B7 of Figure 6.21. (See the file SUPERMARKET.XLS.) We calculate it with the formula

$$=1\text{-NORMDIST}(100, \text{NormMean}, \text{NormStdev}, 1)$$

	A	B	C	D	E	F	G
1	**Supermarket problem**			Range names			
2				NCusts: B10			
3	Amount spent per customer (normally distributed)			NormMean: B4			
4	Mean	$85		NormStdev: B5			
5	StDev	$30		PAtLeast100: B8			
6							
7	Probability that a customer spends at least $100	0.309		1-NORMDIST(100,NormMean,NormStdev,1)			
8							
9							
10	Number of customers	500					
11							
12	Mean and stdev of number who spend at least $100						
13	Mean	154.27		NCusts*PAtLeast100			
14	StDev	10.33		SQRT(NCusts*PAtLeast100*(1-PAtLeast100))			
15							
16	Probability at least 30% spend at least $100	0.676		1-BINOMDIST(0.3*NCusts-1,NCusts,PAtLeast100,1)			

FIGURE 6.21 Calculations for Supermarket Example

This probability is then used as the parameter p in a binomial model. The mean and standard deviation of the number who spend at least \$100 are calculated in cells B13 and B14 as np and $\sqrt{np(1-p)}$, using $n = 500$, the number of shoppers, and $p = 0.309$. The expected number who spend at least \$100 is slightly greater than 154, and the standard deviation of this number is slightly greater than 10.

To answer the second question, note that 30% of 500 customers is 150 customers. Then the probability that at least 30% of the customers spend at least \$100 is the probability that a binomially distributed random variable, with $n = 500$ and $p = 0.309$, is at least 150. We calculate this binomial probability, which turns out to be about 2/3, in cell B16 with the formula

$$=1\text{-BINOMDIST}(0.3*\text{NCusts-}1,\text{NCusts},\text{PAtLeast}100,1)$$

Note that the first argument calculates to 149. This is because the probability of *at least* 150 customers is 1.0 minus the probability of less than or equal to 149 customers. ●

Example 6.10 Overbooking by Airlines

This example presents a simplified version of calculations used by airlines when they overbook flights. They realize that a certain percentage of ticketed passengers will cancel at the last minute. Therefore, to avoid empty seats, they sell more tickets than there are seats, hoping that just about the right number of passengers show up. We will assume that the no-show rate is 5%. In binomial terms, we are assuming that each ticketed passenger, independently of the others, shows up with probability 0.95 and cancels with probability 0.05.

For a flight with 200 seats, the airline wants to find how sensitive various probabilities are to the number of tickets it issues. In particular, it wants to calculate (a) the probability that more than 205 passengers show up, (b) the probability that more than 200 passengers show up, (c) the probability that at least 195 seats will be filled, and (d) the probability that at least 190 seats will be filled. The first two of these are "bad" events from the airline's perspective; they mean that some customers will be bumped from the flight. The last two events are "good" in the sense that the airline wants most of the seats to be occupied.

Objective To assess the benefits and drawbacks of issuing various numbers of tickets on an airline flight with 200 seats.

Solution

To solve the airline's problem, we use the BINOMDIST function and a data table. The solution appears in Figure 6.22 on page 270. (See the file OVERBOOK.XLS.) We first enter a possible number of tickets issued in cell B6 and, for this number, calculate the required probabilities in row 10. For example, the formulas in cells B10 and D10 are

$$=1\text{-BINOMDIST}(205,\text{NTickets},1\text{-PNoShow},1)$$

and

$$=1\text{-BINOMDIST}(194,\text{NTickets},1\text{-PNoShow},1)$$

Note that the condition "more than" requires a slightly different calculation than "at least." The probability of more than 205 is 1.0 minus the probability of less than or equal to 205, whereas the probability of at least 195 is 1.0 minus the probability of less than or equal to 194. Also, note that we are treating a "success" as a passenger who shows up.

FIGURE 6.22
Binomial Calculations for Overbooking Example

	A	B	C	D	E
1	**Airline Overbooking Problem**				
2					
3	Number of seats	200			
4	Probability of no-show	0.1	**Range names**		
5			NTickets: B6		
6	Number of tickets issued	215	PNoShow: B4		
7					
8	**Required probabilities**				
9		More than 205 show up	More than 200 show up	At least 195 seats filled	At least 190 seats filled
10		0.001	0.050	0.421	0.820
11					
12	**Data table showing sensitivity of probabilities to number of tickets issued**				
13	Number of tickets issued	More than 205 show up	More than 200 show up	At least 195 seats filled	At least 190 seats filled
14		0.001	0.050	0.421	0.820
15	206	0.000	0.000	0.012	0.171
16	209	0.000	0.001	0.064	0.384
17	212	0.000	0.009	0.201	0.628
18	215	0.001	0.050	0.421	0.820
19	218	0.013	0.166	0.659	0.931
20	221	0.064	0.370	0.839	0.978
21	224	0.194	0.607	0.939	0.995
22	227	0.406	0.802	0.981	0.999
23	230	0.639	0.920	0.995	1.000
24	233	0.822	0.974	0.999	1.000

Therefore, the third argument of each BINOMDIST function is 1.0 minus the no-show probability.

To see how sensitive these probabilities are to the number of tickets issued, we create a one-way data table at the bottom of the spreadsheet. It is *one-way* because there is only one *input*, the number tickets issued, even though four output probabilities are tabulated. (To create the data table, list several possible numbers of tickets issued along the side in column A and transfer the probabilities from row 10 to row 14. That is, enter the formula =B10 in cell B14 and copy it across row 14. Then form a data table using the range A14:E24, no row input cell, and column input cell B6.)

The results are as expected. As the airline issues more tickets, there is a larger chance of having to bump passengers from the flight, but there is also a larger chance of filling most seats. In reality, the airline has to make a trade-off between these two, taking its various costs and revenues into account. ●

The following is another simplified example of a real problem that occurs every time we watch election returns on TV. This problem is of particular interest in light of the highly unusual events that took place during election night television coverage of the U.S. presidential election on November 7, 2000. The basic question is how soon the networks can declare one of the candidates the winner, based on early voting returns. Our example is somewhat unrealistic because it ignores the possibility that early tabulations might be biased one way or the other. For example, the earliest reporting precincts might be known to be more heavily in favor of the Democrat than the population in general. Nevertheless, the example explains why the networks are able to make conclusions based on such seemingly small amounts of data.

Example 6.11 Projecting Election Winners from Early Returns

We assume that there are N voters in the population, of whom N_R will vote for the Republican and N_D will vote for the Democrat. The eventual winner will be the Republican if $N_R > N_D$ and will be the Democrat otherwise, but we won't know which until all of the votes are tabulated. (To simplify the example, we assume there are only two candidates and that the election will *not* end in a tie.) Let's suppose that a small percentage of the votes have been counted and the Republican is currently ahead 540 to 460. On what basis can the networks declare the Republican the winner?

Objective To use a binomial model to determine whether early returns reflect the eventual winner of an election between two candidates.

Solution

Let $n = 1000$ be the total number of votes that have been tabulated. If X is the number of Republican votes so far, $X = 540$. Now we pose the following question. If the Democrat were going to be the eventual winner, that is, $N_D > N_R$, and we randomly sampled 1000 voters from the population, how likely is it that at least 540 of these voters would be in favor of the Republican? If this is very *unlikely*, then the only reasonable conclusion is that the Democrat will *not* be the eventual winner. This is the reasoning the networks use to declare the Republican the winner.

We use a binomial model to see how unlikely is the event "at least 540 out of 1000," assuming that the Democrat will be the eventual winner. We need a value for p, the probability that a typical vote is for the Republican. This probability should be the proportion of voters in the entire population who favor the Republican. All we know is that this probability is less than 0.5, because we have assumed that the Democrat will win. In Figure 6.23, we show how the probability of at least 540 out of 1000 varies with values of p less than, but close to, 0.5. (See the file VOTING.XLS.)

FIGURE 6.23
Binomial Calculations for Voting Example

	A	B	C	D	E
1	**Election Voting Example**				
2					
3	Population proportion for Republican	0.49	**Range names**		
4			PRepub: B3		
5	Votes tabulated so far	1000	NVotes: B5		
6	Votes for Republican so far	540	NVoteRepub: B6		
7					
8	Binomial probability of at least this many votes for Republican				
9		0.0009			
10					
11	Data table showing sensitivity of this probability to population proportion for Republican				
12	Population proportion for Republican	Probability			
13		0.0009			
14	0.490	0.0009			
15	0.492	0.0013			
16	0.494	0.0020			
17	0.496	0.0030			
18	0.498	0.0043			
19	0.499	0.0052			

We enter a trial value of 0.49 for p in cell B3 and then calculate the required probability in cell B9 with the formula

$$=1-\text{BINOMDIST(NVotesRepub-1,NVotes,PRepub,1)}$$

Then we use this to create the data table at the bottom of the spreadsheet. This data table tabulates the probability of the given lead (at least 540 out of 1000) for various values of p less than 0.5. As shown in the last few rows, even if the eventual outcome were going to be a virtual tie—with the Democrat slightly ahead—there would still be very little chance of the Republican being at least 80 votes ahead so far. But because the Republican *is* currently ahead by 80 votes, the networks feel safe in declaring the Republican the winner. ●

The final example in this section challenges the two assumptions of the binomial model. So far, we have assumed that the outcomes of successive trials (1) have the same probability p of success and (2) are probabilistically independent. There are many situations where either or both of these assumptions are questionable. For example, consider successive items from a production line, where each item either meets specifications (a success) or doesn't (a failure). If the process deteriorates over time, at least until it receives maintenance, then the probability p of success could slowly decrease. Even if p remains constant, defective items could come in bunches (because of momentary inattentiveness on the part of a worker, say), which would invalidate the independence assumption.

If an analyst believes that the binomial assumptions are invalid, then an alternative model must be specified that reflects reality more closely. This is not easy—all kinds of *nonbinomial* assumptions can be imagined. Furthermore, even when we make such assumptions, there are probably no simple formulas to use, such as the BINOMDIST formulas we have been using. Simulation might be the only alternative, as we illustrate in the following example.

Example 6.12 Streak Shooting in Basketball

Do basketball players shoot in streaks? This question has been debated by thousands of basketball fans, and it has even been studied statistically by several academic researchers. Most fans believe the answer is "yes," arguing that players clearly alternate between hot streaks where they can't miss and cold streaks where they can't hit the broad side of a barn. This situation does not fit a binomial model where, say, a "450 shooter" has a 0.450 probability of making each shot and a 0.550 probability of missing, independently of other shots. If the binomial model does not apply, what model might be appropriate, and how could it be used to calculate a probability such as the probability of making at least 13 shots out of 25 attempts?[6]

Objective To formulate a nonbinomial model of basketball shooting, and to use it to find the probability of a 0.450 shooter making at least 13 out of 25 shots.

Solution

This problem is quite open ended. There are numerous alternatives to the binomial model that could capture the "streakiness" most fans believe in, and the one we suggest here is by no means definitive. We challenge you to develop others.

The model we propose assumes that this shooter makes 45% of his shots in the long run. The probability that he makes his first shot in a game is 0.45. In general, consider

[6]There are obviously a lot of extenuating circumstances surrounding any shot: the type of shot (layup versus jump shot), the type of defense, the score, the time left in the game, and so on. For this example we focus on a pure jump shooter who is more or less unaffected by the various circumstances in the game.

	A	B	C	D	E	F	G	H	I
1	**Basketball shooting example**								
2									
3	Long-run average	0.45							
4	Increment d1 after a make	0.015							
5	Increment d2 after a miss	0.015							
6									
7	Number of shots	25							
8									
9	Binomial probability of at least 13 out of 25	0.306							
10									
11	Summary statistics from simulation below			Compare these		Fraction of reps with at least 13 from table below			
12	Number of makes	10				0.328			
13	At least 13 makes?	0							
14									
15	Simulation of makes and misses using nonbinomial model					Data table to replicate 25 shots many times			
16		Shot	Streak	P(make)	Make?		Rep	At least 13?		
17		1	NA	0.45	0			0		
18		2	-1	0.435	0		1	0		
19		3	-2	0.42	1		2	0		
20		4	1	0.465	1		3	0		
21		5	2	0.48	1		4	1		
22		6	3	0.495	0		5	1		
23		7	-1	0.435	0		6	1		
24		8	-2	0.42	0		7	0		
25		9	-3	0.405	0		8	1		
26		10	-4	0.39	1		9	1		
27		11	1	0.465	0		10	0		
28		12	-1	0.435	1		11	1		
29		13	1	0.465	1		12	0		
30		14	2	0.48	1		13	1		
31		15	3	0.495	0		14	0		
32		16	-1	0.435	0		15	0		
33		17	-2	0.42	0		16	0		
34		18	-3	0.405	1		17	1		
35		19	1	0.465	1		18	1		
36		20	2	0.48	1		19	0		
37		21	3	0.495	0		20	1		
38		22	-1	0.435	0		21	0		
39		23	-2	0.42	0		22	0		
40		24	-3	0.405	0		23	0		
41		25	-4	0.39	0		24	0		
42							25	0		
43							26	0		
265							248	0		
266							249	0		
267							250	1		

Range names
AtLeast13: B13
Inc_d1: B4
Inc_d2: B5
NMakes: B12
NShots: B7
PMake: B3

FIGURE 6.24 Simulation of Basketball Shooting Model

his nth shot. If he has made his last k shots, then the probability of making shot n is $0.45 + kd_1$. On the other hand, if he has missed his last k shots, the probability of making shot n is $0.45 - kd_2$. Here, d_1 and d_2 are small values (0.01 and 0.02, for example) that indicate how much the shooter's probability of success increases or decreases depending on his current streak. The model implies that the shooter gets better the more shots he makes and worse the more he misses.

To implement this model, we use simulation as shown in Figure 6.24. (See the file BASKETBALL.XLS.) Actually, we first do a "baseline" binomial calculation in cell B9, using the parameters $n = 25$ and $p = 0.450$. The formula in cell B9 is

$$=1-\text{BINOMDIST}(12,\text{NShots},\text{PMake},1)$$

If the player makes each shot with probability 0.45, independently of the other shots, then the probability that he will make over half of his 25 shots is 0.306—about a 30% chance.

The simulation in the range A17:D41 shows the results of 25 random shots according to the *nonbinomial* model we have assumed. Column B indicates the length of the current streak, where a negative value indicates a streak of misses and a positive value indicates a streak of makes. Column C indicates the probability of a make on the current shot, and column D contains 1's for makes and 0's for misses. Here are step-by-step instructions for developing this range.

❶ **First shot.** Enter the formulas

$$=PMake$$

and

$$=IF(RAND()<C17,1,0)$$

in cells C17 and D17 to determine the outcome of the first shot.

❷ **Second shot.** Enter the formulas

$$=IF(D17=0,-1,1)$$
$$=IF(B18<0,PMake+B18*Inc\_d2,PMake+B18*Inc\_d1)$$

and

$$=IF(RAND()<C18,1,0)$$

in cells B18, C18, and D18. The first of these indicates that by the second shot, the shooter will have a streak of one make or one miss. The second formula is the important one. It indicates how the probability of a make changes depending on the current streak. The third formula simulates a make or a miss, using the probability in cell C18.

❸ **Length of streak on third (and succeeding) shots.** Enter the formula

$$=IF(AND(B18<0,D18=0),B18-1, IF(AND(B18<0,D18=1),1,$$
$$IF(AND(B18>0,D18=0),-1,B18+1)))$$

in cell B19 and copy it down column B. This nested IF formula checks for all four combinations of the previous streak (negative or positive, indicated in cell B18) and the most recent shot (make or miss, indicated in cell D18) to see whether the current streak continues by one or a new streak starts.

❹ **Results of remaining shots.** The logic for the formulas in columns C and D is the same for the remaining shots as for shot 2, so copy the formulas in cells C18 and D18 down their respective columns.

❺ **Summary of 25 shots.** Enter the formulas

$$=SUM(D17:D41)$$

and

$$=IF(NMakes>=13,1,0)$$

in cells B12 and B13 to summarize the results of the 25 simulated shots. In particular, the value in cell B13 is 1 only if at least 13 of the shots are successes.

What about the *probability* of making at least 13 shots with this nonbinomial model? So far, we have simulated one set of 25 shots and have reported whether at least 13 of the shots are successes. We need to replicate this simulation many times and report the fraction of the replications where at least 13 of the shots are successes. We do this with a data table in columns F and G.

To create this table, enter the replication numbers 1–250 (you could use any number of replications) in column F, using the Edit/Fill/Series menu item. Then transfer the value in B13 to cell G17 by entering =AtLeast13 in this cell. Essentially, we are recalculating this value 250 times, each with different random numbers. To do this, highlight the range F17:G267, use the Data/Table menu item, leave the row input cell blank, and enter *any blank cell* (such as H17) as the column input cell. This causes Excel to recalculate the basic simulation 250 times, each time with different random numbers. Finally, enter the formula

$$=\text{AVERAGE(G18:G267)}$$

in cell F12 to calculate the fraction of the replications with at least 13 makes out of 25 shots.

After finishing all of this, you'll note that the spreadsheet is "live" in the sense that if you press the F9 recalculation key, all of the simulated quantities change—new random numbers. In particular, the estimate in cell F12 of the probability of at least 13 makes out of 25 shots changes. It is sometimes less than the binomial probability in cell B9 and sometimes greater. In general, the two probabilities are roughly the same. The bottom line? Even if the world doesn't behave exactly as the binomial model indicates, probabilities of various events can often be approximated fairly well by binomial probabilities—which saves us the trouble of developing and working with more complex models! ●

PROBLEMS

Level A

19. In a typical month, an insurance agent presents life insurance plans to 40 potential customers. Historically, one in four such customers chooses to buy life insurance from this agent. Based on the relevant binomial distribution, answer the following questions:

  a. What is the probability that exactly 5 customers will buy life insurance from this agent in the coming month?

  b. What is the probability that no more than 10 customers will buy life insurance from this agent in the coming month?

  c. What is the probability that at least 20 customers will buy life insurance from this agent in the coming month?

  d. Determine the mean and standard deviation of the number of customers who will buy life insurance from this agent in the coming month.

  e. What is the probability that the number of customers who buy life insurance from this agent in the coming month will lie within 2 standard deviations of the mean?

  f. What is the probability that the number of customers who buy life insurance from this agent in the coming month will lie within 3 standard deviations of the mean?

20. Continuing the previous exercise, use the normal approximation to the binomial to answer each of the questions posed in parts **a**–**f** above. How well does the normal approximation perform in this case? Explain.

21. Many vehicles used in space travel are constructed with redundant systems to protect flight crews and their valuable equipment. In other words, backup systems are included within many vehicle components so that if one or more systems fail, backup systems will assure the safe operation of the given component and thus the entire vehicle. For example, consider one particular component of the U.S. space shuttle that has n duplicated systems (i.e., one original system and $n - 1$ backup systems). Each of these systems functions, independently of the others, with probability 0.98. This shuttle component functions successfully provided that *at least* one of the n systems functions properly.

  a. Find the probability that this shuttle component functions successfully if $n = 2$.

  b. Find the probability that this shuttle component functions successfully if $n = 4$.

  c. What is the minimum number n of duplicated systems that must be incorporated into this shuttle component to ensure at least a 0.9999 probability of successful operation?

22. Suppose that a popular hotel for vacationers in Orlando, Florida, has a total of 300 identical rooms. Like many major airline companies, this hotel has adopted an overbooking policy in an effort to maximize the usage of its available lodging capacity. Assume that each potential hotel customer holding a room reservation, independently of other customers, cancels the reservation or simply does not show up at the hotel on a given night with probability 0.15.

   a. Find the largest number of room reservations that this hotel can book and still be at least 95% sure that everyone who shows up at the hotel will have a room on a given night.

   b. Given that the hotel books the number of reservations found in answering part **a**, find the probability that at least 90% of the available rooms will be occupied on a given night.

   c. Given that the hotel books the number of reservations found in answering part **a**, find the probability that at most 80% of the available rooms will be occupied on a given night.

   d. How does your answer to part **a** change as the required assurance rate increases from 95% to 97%? How does your answer to part **a** change as the required assurance rate increases from 95% to 99%?

   e. How does your answer to part **a** change as the cancellation rate varies between 5% and 25% (in increments of 5%)? Assume now that the required assurance rate is held fixed at 95%.

23. A production process manufactures items with weights that are normally distributed with mean 15 pounds and standard deviation 0.1 pound. An item is considered to be defective if its weight is less than 14.8 pounds or greater than 15.2 pounds. Suppose that these items are currently produced in batches of 1000 units.

   a. Find the probability that at most 5% of the items in a given batch will be defective.

   b. Find the probability that at least 90% of the items in a given batch will be acceptable.

   c. How many items would have to be produced in a batch to guarantee that a batch consists of no more than 1% defective items?

24. Past experience indicates that 30% of all individuals entering a certain store decide to make a purchase. Using (i) the binomial distribution and (ii) the normal approximation to the binomial, find that probability that 10 or more of the 30 individuals entering the store in a given hour will decide to make a purchase. Compare the results obtained using the two different approaches. Under what conditions will the normal approximation to this binomial probability become even more accurate?

25. Suppose that the number of ounces of soda put into a Pepsi can is normally distributed with $\mu = 12.05$ ounces and $\sigma = 0.03$ ounce.

   a. Legally, a can must contain at least 12 ounces of soda. What fraction of cans will contain at least 12 ounces of soda?

   b. What fraction of cans will contain less than 11.9 ounces of soda?

   c. What fraction of cans will contain between 12 and 12.08 ounces of soda?

   d. One percent of all cans will weigh more than what value?

   e. Ten percent of all cans will weigh less than what value?

   f. Pepsi controls the mean weight in a can by setting a timer. For what mean should the timer be set so that only 1 in 1000 cans will be underweight?

   g. Every day Pepsi produces 10,000 cans. The government inspects 10 randomly chosen cans each day. If at least two are underweight, Pepsi is fined $10,000. Given that $\mu = 12.05$ ounces and $\sigma = 0.03$ ounce, what is the probability that Pepsi will be fined on a given day?

26. Suppose that 52% of all registered voters prefer George Bush to Al Gore. (You may substitute the names of the current presidential candidates!)

   a. In a random sample of 100 voters, what is the probability that the sample will indicate that Bush will win the election (that is, there will be more votes in the sample for Bush)?

   b. In a random sample of 100 voters, what is the probability that the sample will indicate that Gore will win the election?

   c. In a random sample of 100 voters, what is the probability that the sample will indicate a dead heat (50–50)?

   d. In a random sample of 100 voters, what is the probability that between 40 and 60 (inclusive) voters will prefer Bush?

27. Assume that, on average, 95% of all ticket-holders show up for a flight. If a plane seats 200 people, how many tickets should be sold to make the chance of an overbooked flight as close as possible to 5%?

28. Suppose that 60% of all people prefer Coke to Pepsi. We randomly choose 500 people and ask them if they prefer Coke to Pepsi. What is the probability that our survey will (erroneously) indicate that Pepsi is preferred by more people than Coke?

29. A firm's office contains 150 PCs. The probability that a given PC will not work on a given day is 0.05.

   a. On a given day what is the probability that exactly one computer will not be working?

   b. On a given day what is the probability that at least two computers will not be working?

   c. What assumptions do your answers in parts **a** and **b** require?

30. Suppose that 4% of all tax returns are audited. In a group of n tax returns, consider the probability that at

most two returns are audited. How large must n be before this probability will be less than 0.01?

31. The height of a typical American female is normally distributed with $\mu = 64$ inches and $\sigma = 4$ inches. We observe the height of 10 American females.
  a. What is the probability that exactly half the women will be under 58 inches tall?
  b. Let X be the number of the 10 women who are under 58 inches tall. Determine the mean and standard deviation of X.

32. Consider a large population of shoppers, each of whom spends a certain amount during their current shopping trip; the distribution of these amounts is normally distributed with mean $55 and standard deviation $15. We randomly choose 25 of these shoppers. What is the probability that at least 15 of them spend between $45 and $75?

Level B

33. Many firms utilize sampling plans to control the quality of manufactured items ready for shipment. To illustrate the use of a sampling plan, suppose that a particular company produces and ships electronic computer chips in lots, each consisting of 1000 chips. This company's sampling plan specifies that quality control personnel will randomly sample 50 chips from each lot and accept the lot for shipping if the number of defective chips is less than 5. Of course, the given lot will be rejected if the number of defective chips is found to be 5 or more.
  a. Find the probability of accepting a lot as a function of the actual fraction of defective chips. In particular, let the actual fraction of defective chips in a given lot equal any of 0.1, 0.2, . . . , 0.8, 0.9. Then compute the lot acceptance probability for each of these lot defective fractions.
  b. Construct a graph showing the probability of lot acceptance for each of the 9 lot defective fractions. Interpret your graph.

34. Continuing the previous exercise, repeat parts **a** and **b** under a revised sampling plan that calls for accepting a given lot if the number of defective chips found in the random sample of 50 chips is *not greater than* 5. Summarize any notable differences between the two graphs you have constructed in completing part **b** of this and the previous exercise.

35. Dell Computer receives computer chips from Chipco. Each batch sent by Chipco is inspected as follows: 35 chips are tested and the batch passes inspection if at most one defective chip is found in the set of 35 tested chips. Past history indicates an average of 1% of all chips produced by Chipco are defective. Dell has received 10 batches this week. What is the probability that at least 9 of the batches will pass inspection?

36. A standardized test consists entirely of multiple-choice questions, each with 5 possible choices. You want to ensure that a student who randomly guesses on each question will obtain an expected score of zero. How would you accomplish this?

37. In the current tax year, suppose that 5% of the millions of individual tax returns are fraudulent. That is, they contain errors that were purposely made to cheat the government. Although these errors are often well concealed, let's suppose that a thorough IRS audit will uncover them.
  a. If a random 250 tax returns are audited, what is the probability that the IRS will uncover at least 15 fraudulent returns?
  b. Answer the same question as in part **a,** but this time assume there is only a 90% chance that a given fraudulent return will be spotted as such if it is audited.

38. Suppose you work for a survey research company. In a typical survey, you mail questionnaires to 150 companies. Of course, some of these companies might decide not to respond. We'll assume that the nonresponse rate is 45%; that is, each company's probability of not responding, independently of the others, is 0.45. If your company requires at least 90 responses for a "valid" survey, what is the chance that it will get this many? Use a data table to see how your answer varies as a function of the nonresponse rate (for a reasonable range of response rates surrounding 45%).

39. Continuing the previous problem, suppose your company does this survey in two "waves." It mails the 150 questionnaires and waits a certain period for the responses. As above, we assume that the nonresponse rate is 45%. However, after this initial period, your company follows up (by telephone, say) on the nonrespondents, asking them to please respond. Suppose that the nonresponse rate on this second "wave" is 70%; that is, each original nonrespondent now responds with probability 0.3, independently of the others. Your company now wants to find the probability of obtaining at least 110 responses total. It turns out that this is a very difficult probability to calculate directly. So instead, approximate it with simulation.

40. A person claims that she is a fortune teller. Specifically, she claims that she can predict the direction of the change (up or down) in the Dow Jones Industrial Average for the next 10 days (such as U, U, D, U, D, U, U, D, D, D). (You can assume that she makes all 10 predictions right now, although that won't affect your answer to the question.) Obviously, you're skeptical, thinking that she's just guessing, so you'll be surprised if her predictions are accurate. Which would surprise you more: (1) she predicts at least 8 out of 10 correctly, or (2) she predicts at least 6 out of 10 correctly on each of 4 separate occasions? Answer by assuming that (1) she really is guessing and (2) each day the Dow is equally likely to go up or down.

6.6 The Poisson and Exponential Distributions

The final two distributions in this chapter are called the *Poisson* and *exponential* distributions. In most statistical applications, including those in the rest of this book, these distributions play a much less important role than the normal and binomial distributions. For this reason we will not analyze them in as much detail. However, in many applied management science models, the Poisson and exponential distributions are as important as any other distributions, discrete or continuous. For example, much of the study of probabilistic inventory models, queueing models, and reliability models relies heavily on these two distributions.

6.6.1 The Poisson Distribution

The Poisson distribution is a discrete distribution. It usually applies when we are interested in the *count* of events occurring within a specified period of time or space. Its possible values are all of the nonnegative integers: 0, 1, 2, and so on—there is no upper limit. Even though there is an infinite number of possible values, this causes no real problems because the probabilities of all sufficiently large values are essentially 0.

The Poisson distribution is characterized by a single parameter, usually labeled λ (Greek lambda), which must be positive. By adjusting the value of λ, we are able to produce different Poisson distributions, all of which have the same basic shape as in Figure 6.25. That is, they first increase, then decrease. It turns out that λ is easy to interpret. It is both the mean and the variance of the Poisson distribution. Therefore, the standard deviation is $\sqrt{\lambda}$.

Typical Examples of the Poisson Distribution

1. A bank manager is studying the arrival pattern to the bank. Then the events are customer arrivals, the number of arrivals in an hour is Poisson distributed, and λ represents the expected number of arrivals per hour.

2. An engineer is interested in the lifetime of a type of battery. A device that uses this type of battery is operated continuously. When the first battery fails, it is replaced by

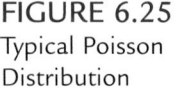

FIGURE 6.25
Typical Poisson Distribution

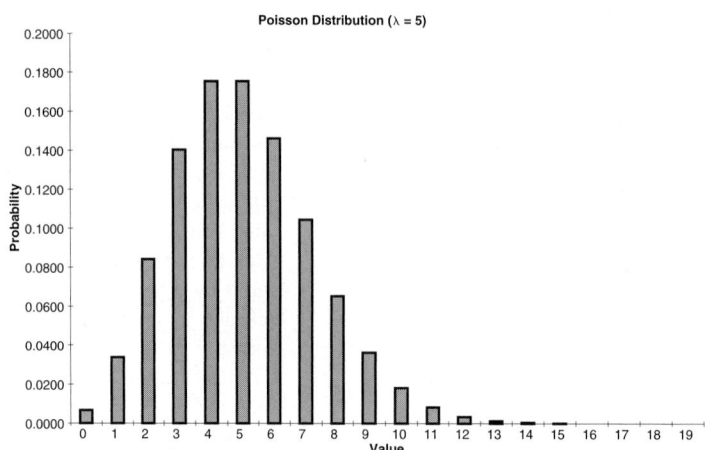

a second; when the second fails, it is replaced by a third, and so on. The events are battery failures, the number of failures that occur in a month is Poisson distributed, and λ represents the expected number of failures per month.

3. A retailer is interested in the number of units of a product demanded in a particular unit of time such as a week. Then the events are customer demands, the number of units demanded in a week is Poisson distributed, and λ is the expected number of units demanded per week.

4. In a quality control setting, the Poisson distribution is often relevant for describing the number of defects in some unit of space. For example, when paint is applied to the body of a new car, any minor blemish is considered a defect. Then the number of defects on the hood, say, might be Poisson distributed. In this case, λ is the expected number of defects per hood.

These examples are representative of the many situations where the Poisson distribution has been applied. For the obvious reason, the parameter λ is often called a rate—arrivals per hour, failures per month, and so on. If we change the unit of time, we simply modify the rate accordingly. For example, if the number of arrivals to a bank in a single hour is Poisson distributed with rate $\lambda = 30$, then the number of arrivals in a half-hour period is Poisson distributed with rate $\lambda = 15$.

We can use Excel to calculate Poisson probabilities much as we did with binomial probabilities. The relevant function is the POISSON function. It takes the form

$$=\text{POISSON}(k, \lambda, cum)$$

The third argument *cum* works exactly as in the binomial case. If it is 0, the function returns $P(X = k)$; if it is 1, the function returns $P(X \le k)$. As examples, if $\lambda = 5$, POISSON(7,5,0) returns the probability of exactly 7, POISSON(7,5,1) returns the probability of less than or equal to 7, and 1-POISSON(3,5,1) returns the probability of greater than 3.

The following example shows how a manager or consultant might use the Poisson distribution.

Example 6.13 Managing Inventory of Televisions at Kriegland

Kriegland is a department store that sells various brands of color television sets. One of the manager's biggest problems is to decide on an appropriate inventory policy for stocking television sets. On the one hand, he wants to have enough in stock so that customers receive their requests right away, but on the other hand, he does not want to tie up too much money in inventory that sits on the storeroom floor.

Most of the difficulty results from the unpredictability of customer demand. If this demand were constant and known, the manager could decide on an appropriate inventory policy fairly easily. But the demand varies widely from month to month in a random manner. All the manager knows is that the historical average demand per month is approximately 17. Therefore, he decides to call in a consultant. The consultant immediately suggests using a probability model. Specifically, she attempts to find the probability distribution of demand in a typical month. How might she proceed?

Objective To model the probability distribution of monthly demand for color television sets with a particular Poisson distribution.

Solution

Let X be the demand in a typical month. The consultant knows that there are many possible values of X. For example, if historical records show that monthly demands have always been between 0 and 40, the consultant knows that almost all of the probability should be assigned to the values 0 through 40. However, she does not relish the thought of finding 41 probabilities, $P(X = 0)$ through $P(X = 40)$, that sum to 1 and reflect historical frequencies. Instead, she discovers from the manager that the histogram of demands from previous months is shaped much like the graph in Figure 6.25. That is, it rises to some peak, then falls.

Knowing that a Poisson distribution has this same basic shape, the consultant decides to model the monthly demand with a Poisson distribution. To choose a particular Poisson

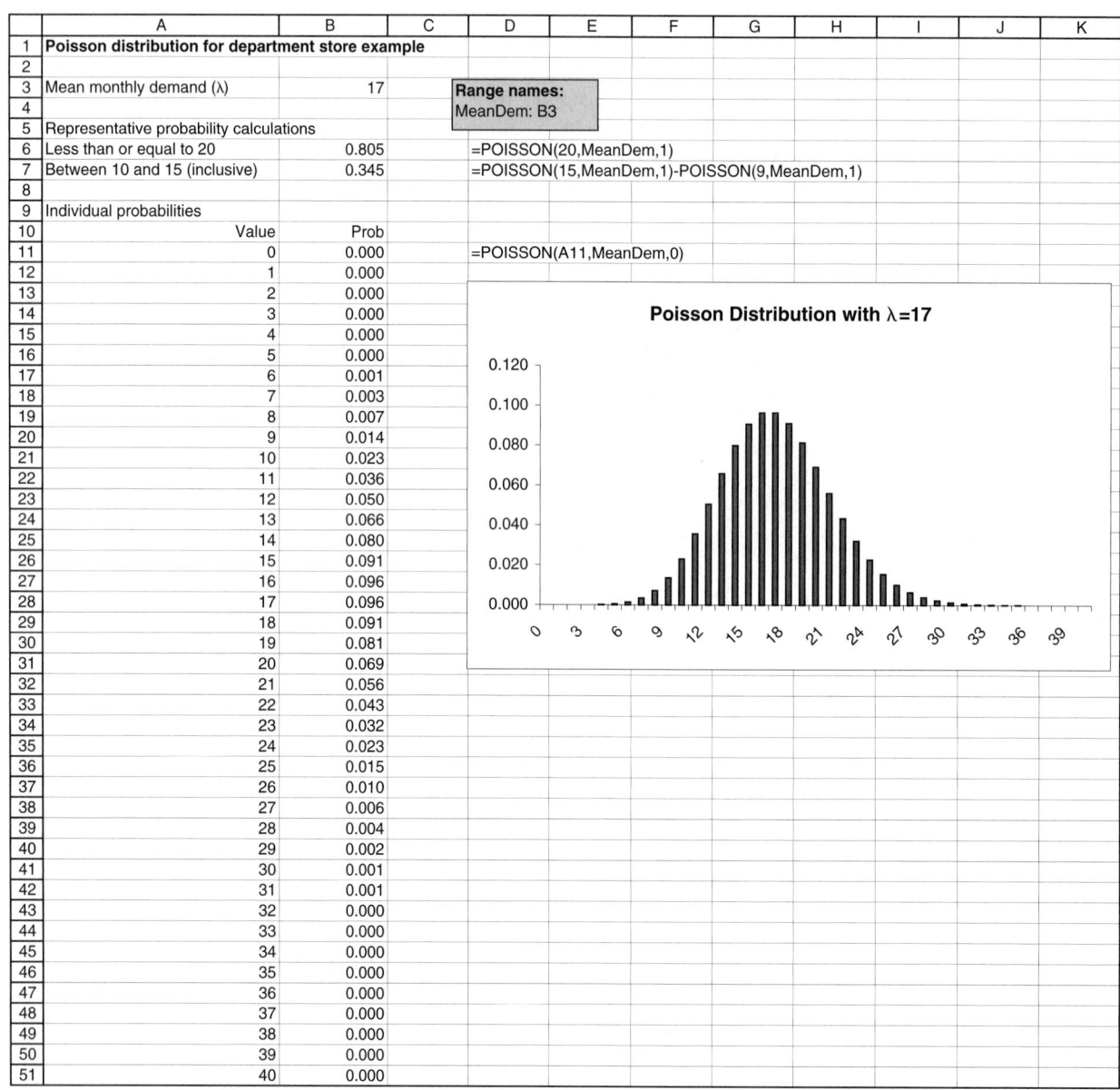

	A	B	C	D	E	F	G	H	I	J	K
1	Poisson distribution for department store example										
2											
3	Mean monthly demand (λ)	17		Range names:							
4				MeanDem: B3							
5	Representative probability calculations										
6	Less than or equal to 20	0.805		=POISSON(20,MeanDem,1)							
7	Between 10 and 15 (inclusive)	0.345		=POISSON(15,MeanDem,1)-POISSON(9,MeanDem,1)							
8											
9	Individual probabilities										
10	Value	Prob									
11	0	0.000		=POISSON(A11,MeanDem,0)							
12	1	0.000									
13	2	0.000									
14	3	0.000									
15	4	0.000									
16	5	0.000									
17	6	0.001									
18	7	0.003									
19	8	0.007									
20	9	0.014									
21	10	0.023									
22	11	0.036									
23	12	0.050									
24	13	0.066									
25	14	0.080									
26	15	0.091									
27	16	0.096									
28	17	0.096									
29	18	0.091									
30	19	0.081									
31	20	0.069									
32	21	0.056									
33	22	0.043									
34	23	0.032									
35	24	0.023									
36	25	0.015									
37	26	0.010									
38	27	0.006									
39	28	0.004									
40	29	0.002									
41	30	0.001									
42	31	0.001									
43	32	0.000									
44	33	0.000									
45	34	0.000									
46	35	0.000									
47	36	0.000									
48	37	0.000									
49	38	0.000									
50	39	0.000									
51	40	0.000									

FIGURE 6.26 Poisson Calculations for Television Example

distribution, all she has to do is choose a value of λ, the mean demand per month. Because the historical average is approximately 17, she chooses λ = 17. Now she can test the Poisson model by calculating probabilities of various events and asking the manager whether these probabilities are a reasonable approximation to reality.

For example, the Poisson probability that monthly demand is less than or equal to 20, $P(X \leq 20)$, is 0.805 (using the Excel function POISSON(20,17,1)), and the probability that demand is between 10 and 15 inclusive, $P(10 \leq X \leq 15)$, is 0.345 (using POISSON(15,17,1)-POISSON(9,17,1)). Figure 6.26 illustrates various probability calculations and shows the graph of the individual Poisson probabilities. (See the file DPTSTORE.XLS.)

If the manager believes that these probabilities and other similar probabilities are reasonable, then the *statistical* part of the consultant's job is finished. Otherwise, she must try a different Poisson distribution—a different value of λ—or perhaps a different type of distribution altogether. ●

6.6.2 The Exponential Distribution

Suppose that a bank manager is studying the pattern of customer arrival at her branch location. As indicated previously in this section, the number of arrivals in an hour at a facility such as a bank is often well described by a Poisson distribution with parameter λ, where λ represents the expected number of arrivals per hour. An alternative way to view the uncertainty in the arrival process is to consider the *times* between customer arrivals. The most common probability distribution used to model these times, often called *interarrival times,* is the *exponential* distribution.

In general, the *continuous* random variable X has an **exponential** distribution with parameter λ (with λ > 0) if the probability density function for X has the form $f(x) = \lambda e^{-\lambda x}$ for x > 0. This exact form is not as important as the shape of the graph it implies, as shown in Figure 6.27. Because this density function decreases continually from left to right, its most likely value is x = 0. Alternatively, if we collect many observations from an exponential distribution and draw a histogram of the observed values, then we expect it to resemble the smooth curve shown in Figure 6.27, with the tallest bars to the left. The mean and standard deviation of this distribution are easy to remember. They are both equal to the *reciprocal* of the parameter λ. For example, an exponential distribution with parameter λ = 0.1 has mean and standard deviation both equal to 10.

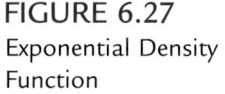
FIGURE 6.27
Exponential Density Function

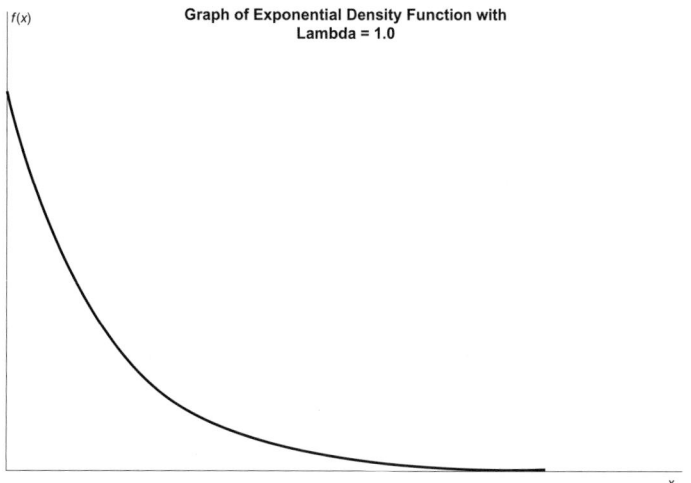

As with the normal distribution, we usually want probabilities to the left or right of a given value. For any exponential distribution, we can calculate the probability to the left of a given value $x > 0$ with Excel's EXPONDIST function. In particular, this function takes the form

$$=\text{EXPONDIST}(x, \lambda, 1)$$

For example, if $x = 0.5$ and $\lambda = 5$ (so that the mean equals $1/5 = 0.2$), then the probability of being less than 0.5 can be found with the formula $=\text{EXPONDIST}(0.5, 5, 1)$. This returns the probability 0.918. Of course, the probability of being greater than 0.5 is then $1 - 0.918 = 0.082$.

Returning to the bank manager's analysis of customer arrival data, when the times between arrivals are exponentially distributed, we sometimes hear that "arrivals occur according to a Poisson process." This is because there is a close relationship between the exponential distribution, which measures *times* between events such as arrivals, and the Poisson distribution, which counts the *number* of events in a certain length of time. The details of this relationship are beyond the level of this book, so we will not explore the topic further. It is sufficient for our purposes to say, for example, that if customers arrive at a facility according to a Poisson process with rate 6 per hour, then we know the corresponding times between arrivals are exponentially distributed with mean $1/\lambda = 1/6$ hour.

PROBLEMS

Level A

41. The annual number of industrial accidents occurring in a particular manufacturing plant is known to follow a Poisson distribution with mean 12.
    a. What is the probability of observing exactly 12 accidents at this plant during the upcoming year?
    b. What is the probability of observing no more than 12 accidents at this plant during the upcoming year?
    c. What is the probability of observing at least 15 accidents at this plant during the upcoming year?
    d. What is the probability of observing between 10 and 15 accidents (inclusive) at this plant during the upcoming year?
    e. Find the smallest integer k such that we can be at least 99% sure that the annual number of accidents occurring at this plant will be less than k.

42. Suppose that the number of customers arriving each hour at the only checkout counter in a local pharmacy is approximately Poisson distributed with an expected arrival rate of 20 customers per hour.
    a. Find the probability that exactly 10 customers arrive at this checkout counter in a given hour.
    b. Find the probability that at least 5 customers arrive at this checkout counter in a given hour.
    c. Find the probability that no more than 25 customers arrive at this checkout counter in a given hour.
    d. Find the probability that between 10 and 30 customers (inclusive) arrive at this checkout counter in a given hour.

    e. Find the largest integer k such that we can be at least 95% sure that the number of customers arriving at this checkout counter in a given hour will be greater than k.
    f. Recalling the relationship between the Poisson and exponential distributions, find the probability that the time between two successive customer arrivals is more then 4 minutes. Find the probability that it is less than 2 minutes.

43. Suppose the number of points scored by the Indiana University basketball team in 1 minute follows a Poisson distribution with $\lambda = 1.5$. In a 10-minute span of time, what is the probability that Indiana University scores exactly 20 points? (Use the fact that if the rate per minute is λ, then the rate in t minutes is λt.)

44. Suppose that the times between arrivals at a bank during the peak period of the day are exponentially distributed with a mean of 45 seconds. If you just observed an arrival, what is the probability that you will need to wait for more than a minute before observing the next arrival? What is the probability you will need to wait at least 2 minutes?

Level B

45. Consider a Poisson random variable X with parameter $\lambda = 2$.
    a. Find the probability that X is within 1 standard deviation of its mean.

b. Find the probability that X is within 2 standard deviations of its mean.

c. Find the probability that X is within 3 standard deviations of its mean.

d. Do the empirical rules we learned previously seem to be applicable in working with the Poisson distribution where $\lambda = 2$? Explain why or why not.

e. Repeat parts **a–d** for the case of a Poisson random variable where $\lambda = 20$.

46. Based on historical data, the probability that a major league pitcher pitches a no-hitter in a game is about 1/1300.

a. Use the binomial distribution to determine the probability that in 650 games 0, 1, 2, or 3 no-hitters will be pitched. (Find the separate probabilities of these four events.)

b. Repeat part **a** using the Poisson approximation to the binomial. This approximation says that if n is large and p is small, a binomial distribution with parameters n and p is approximately Poisson with $\lambda = np$.

Fitting a Probability Distribution to Data: BestFit

The normal, binomial, Poisson, and exponential distributions are three of the most commonly used distributions in real applications. However, many other discrete and continuous distributions are also used. These include the uniform, triangular, Erlang, lognormal, gamma, Weibull, and others. How do we know which to choose for any particular application? Often we can answer this by seeing which of several potential distributions fits a given set of data most closely. Essentially, we compare a histogram of the data with the theoretical probability distributions available and see which gives the best fit.

BestFit, one of the add-ins in the Decision Tools suite, makes this fairly easy. Before seeing how BestFit works, we note that it is actually a stand-alone program that can be run *independently* of Excel. However, versions 4.0 and 4.5 of @Risk, an Excel add-in, incorporate all of BestFit's functionality. [This is simply a new marketing strategy by Palisade. Formerly, customers had to purchase BestFit separately. Now they can essentially obtain BestFit (and RiskView) by purchasing @Risk.] Therefore, you can fit a distribution to data using the stand-alone BestFit program, or you can do it through @Risk. To keep things simple for now, we will illustrate only BestFit's stand-alone capabilities in the following example.

Example 6.14 Assessing a Distribution of Supermarket Checkout Times

A supermarket has collected checkout times on over 100 customers. (See the file CHECKOUTTIMES.XLS.) As shown in Figure 6.28 on page 284, the times vary from 40 seconds to 279 seconds, with the mean and median right around 2 minutes.

The supermarket manager would like to check whether these data are normally distributed or whether some other distribution fits them better. How can he tell?

Objective To determine which probability distribution fits the given data best.

Solution

The easiest way to open BestFit (regardless of whether Excel is open) is to click on the Start button and navigate to the BestFit item in the Palisade Decision Tools suite. Once

FIGURE 6.28
Supermarket
Checkout Times

	A	B	C	D	E	F	G
1	Checkout times (seconds) at a supermarket						
2							
3	Customer	Time			**Summary measures for selected variables**		
4	1	131				Time	
5	2	101			Count	113.000	
6	3	178			Mean	159.239	
7	4	246			Median	155.000	
8	5	207			Standard deviation	52.609	
9	6	155			Minimum	40.000	
10	7	95			Maximum	279.000	
11	8	105					
12	9	168					
13	10	92					
113	110	138					
114	111	279					
115	112	90					
116	113	155					

BestFit is open, we see a blank column called Samples that looks like a spreadsheet. We can enter data values in this column manually or paste them from an Excel spreadsheet. For this example we do the latter, copying the checkout times from Figure 6.28 to this BestFit window. (See Figure 6.29, which shows the first few data values.) We can then go in several directions. Specifically, we can test the fit of a *given* distribution or we can find the distribution that fits the data best.

Since the supermarket manager wants to know whether the data could come from a normal distribution, we check this possibility first. To do so, select the Fitting/Specify Distribution To Fit menu item, click on the Fit to Predefined Distributions button, fill out the left side of the resulting dialog box as shown in Figure 6.30, click on the Add button, and click on OK. (To get the best possible normal fit, we use the sample mean and the sample standard deviation of the data in this dialog box.) To see how well this normal distribution fits the data, select the Fitting/Run Fit menu item. This produces the graph and data in Figure 6.31, with a normal curve superimposed on the histogram of the data. A visual examination of this graph is often sufficient to tell whether the fit is any "good." (We would judge this fit to be "fair," but not great.)

In addition, BestFit provides plenty of numerical measures of the goodness of fit in its GOF tab, shown in the figure. Probably the most important of these are the test values in

FIGURE 6.29
Data in BestFit
Window

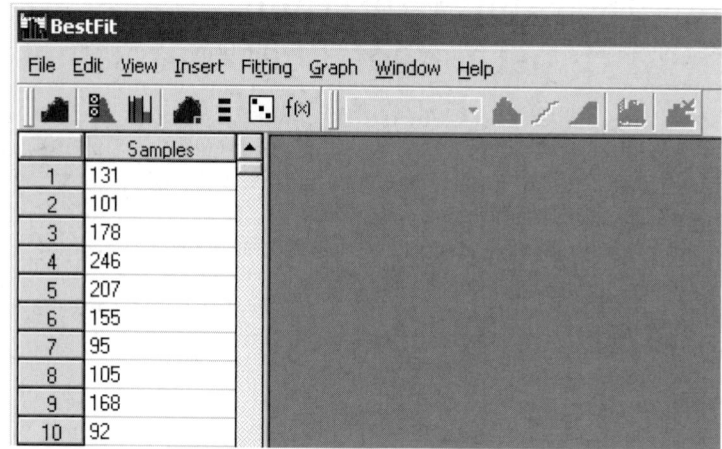

FIGURE 6.30
Selecting a Predefined Distribution in BestFit

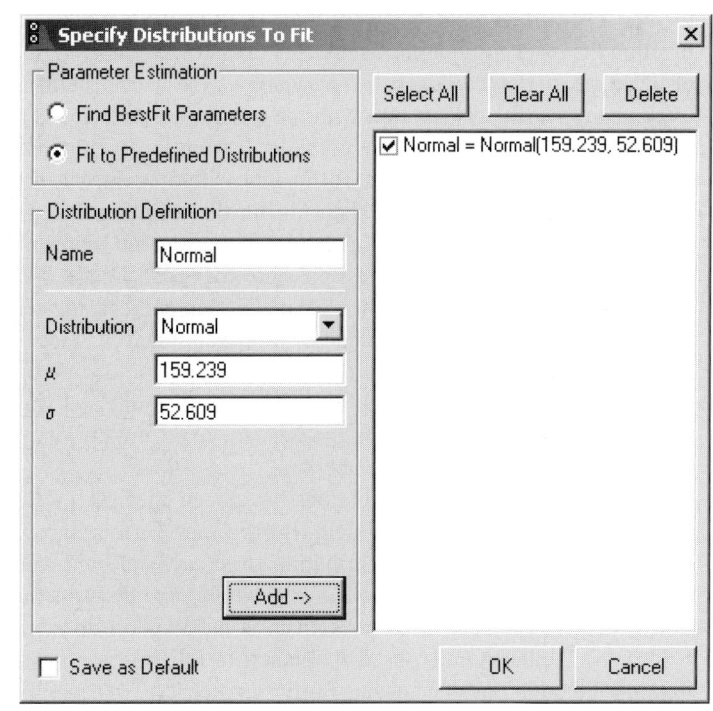

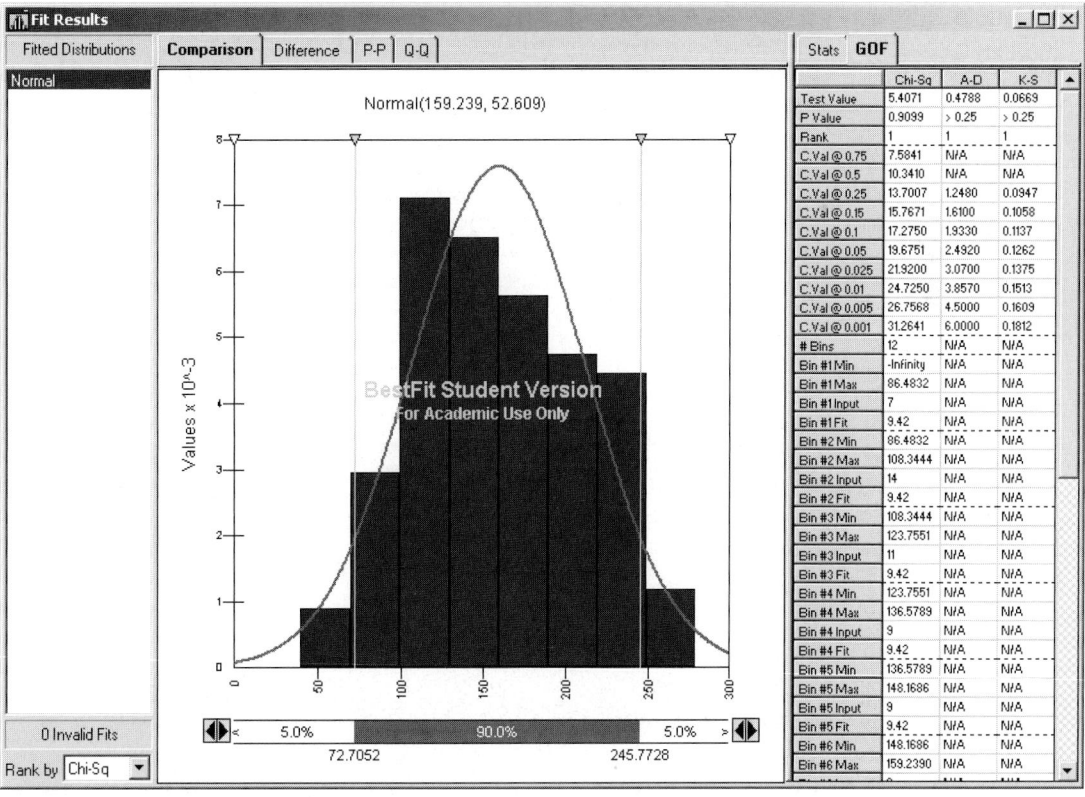

FIGURE 6.31 Normal Fit to the Data

the first row of this table. We won't pursue all of the technical details, but we will simply mention that each column measures "goodness of fit" in a slightly different way. For each of these measures, the larger the test value is, the *worse* the fit is. They can then be used to compare fits; the distribution with the lowest test statistics is the winner.

We next see which of many possible distributions fit the data best. To do this, select the Fitting/Specify Distribution To Fit menu item again, but this time click on the Find BestFit Parameters button and then fill out the resulting dialog box as shown in Figure 6.32. On the left we have made a couple of "reasonable" choices about the checkout data. We have specified that the lowest possible checkout time is 0 and that there is no upper limit on the checkout times. When we make such choices, the set of possible distributions that are checked on the right changes. For example, the selected list here contains only distributions with a lower limit of 0 and no upper limit. (Note that the normal distribution does *not* satisfy these conditions.) We can then uncheck any distributions we do not want included in the search for the best fit. (You might want to uncheck distributions you've never heard of, for instance!)

Once these candidate distributions have been specified, we close this dialog box and select the Fitting/Run Fit menu item. BestFit then performs an extensive numerical algorithm to find the best-fitting distribution from each selected distribution family (the best gamma of all gamma distributions, for example) and displays them in ranked order, from best to worst. The best fit for these data is the Weibull distribution, as shown in Figure 6.33. (The Weibull family, used frequently in reliability models, includes skewed distributions, although this one appears fairly symmetric.) We can also click on any of the "runner up" distributions to see how well they fit. For example, the gamma fit is shown in Figure 6.34. The gamma distribution appears to fit the data just as well as the Weibull.

FIGURE 6.32
Candidate Distributions for BestFit

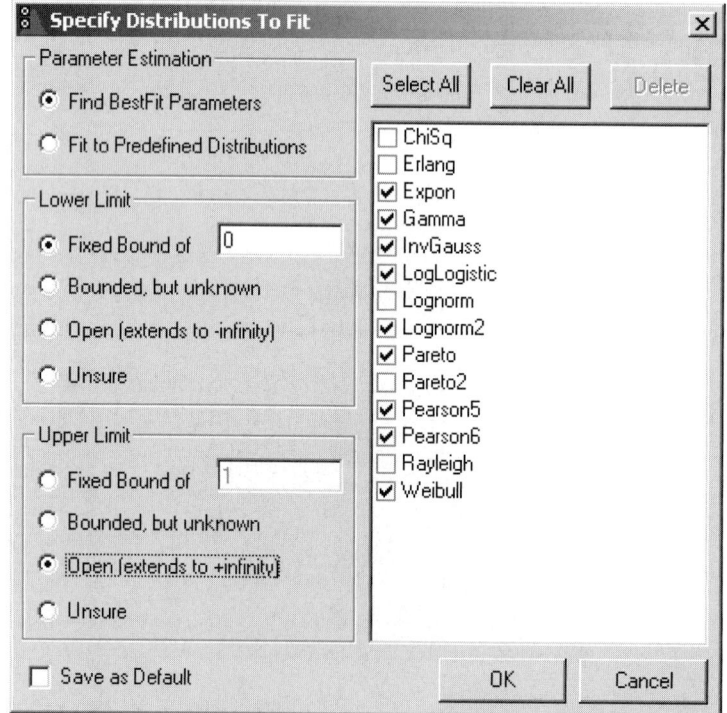

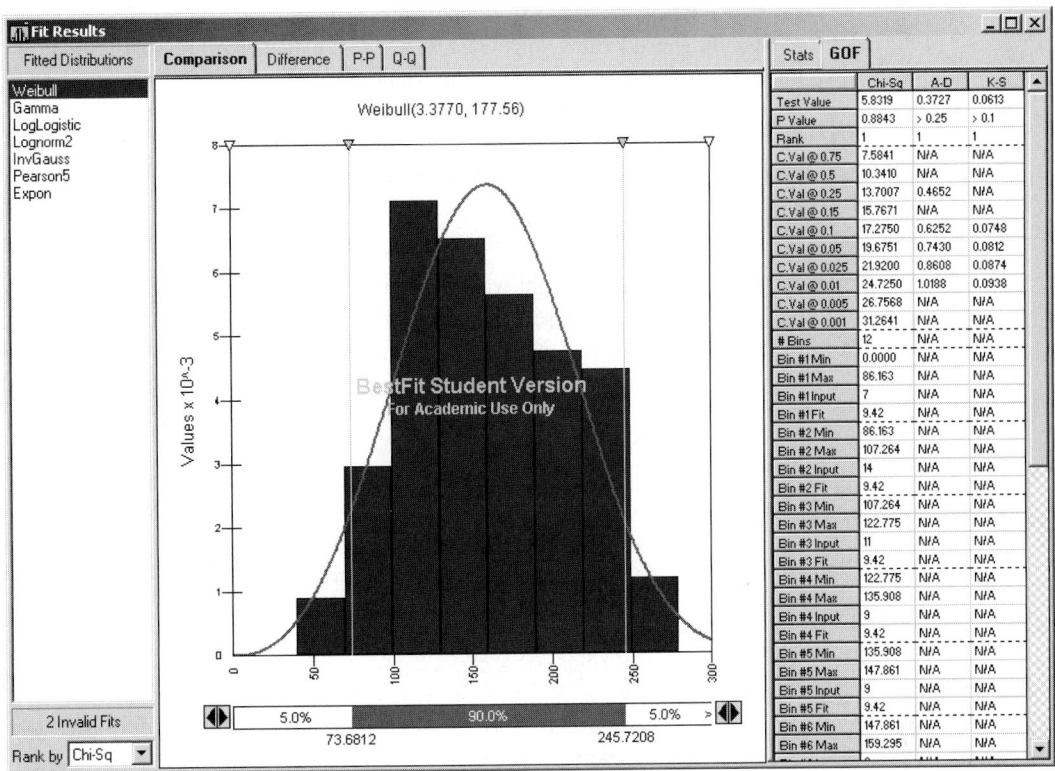

FIGURE 6.33 Weibull Fit to the Data

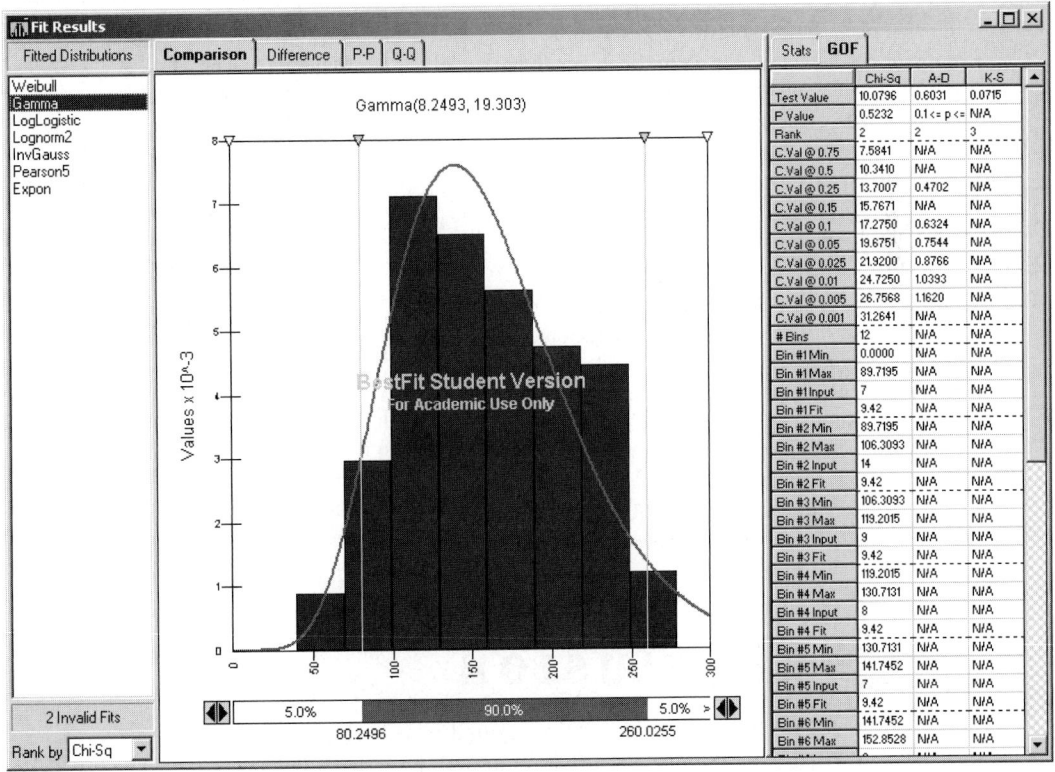

FIGURE 6.34 Gamma Fit to the Data

It is not always easy to "eyeball" these graphs and judge which fit is best. This is the reason for the goodness-of-fit measures. Comparing Figures 6.33 and 6.34, we see that the gamma fit is considerably worse than the Weibull—its test values are all much larger. Surprisingly, the test values for the normal fit in Figure 6.31 are very comparable to those for the Weibull. The only downside to the normal distribution, in this example, is that checkout times cannot possibly be negative, as the normal allows. But the probability of a negative value for this particular normal distribution might be so low that the manager would decide to use it anyway. ●

At this point, you might wonder why we bother fitting a distribution to a set of data in the first place. The usual reason is given in the following scenario. Suppose a manager needs to make a decision, but there is at least one source of uncertainty. If the manager wants to develop a decision model or perhaps a simulation model to help solve his problem, probability distributions of all uncertain outcomes are typically required. The manager could always choose one of the "well-known" distributions, such as the normal, for all uncertain outcomes, but these might not reflect reality. Instead, the manager could gather historical data, such as those in the preceding example, find the distribution that fits these data best, and then use this distribution in the decision or simulation model. Of course, as this example has illustrated, it helps to know a few distributions other than the normal—the Weibull and the gamma, for instance. Although we will not pursue these in this book, the more distributions you have in your "tool kit," the more effectively you can model.

PROBLEMS

Level A

47. A production manager is interested in determining the proportion of defective items in a typical shipment of one of the computer components that her company manufactures. The proportion of defective components is recorded for each of 500 randomly selected shipments collected during a 1-month period. The data are in the file P02_02.XLS. Use the BestFit add-in to determine which probability distribution best fits these data.

48. The manager of a local fast-food restaurant is interested in improving the service provided to customers who use the restaurant's drive-up window. As a first step in this process, the manager asks his assistant to record the time (in minutes) it takes to serve 200 different customers at the final window in the facility's drive-up system. The given 200 customer service times are all observed during the busiest hour of the day for this fast-food operation. The data are in the file P02_04.XLS. Use the BestFit add-in to determine which probability distribution best fits these data.

49. The operations manager of a toll booth, located at a major exit of a state turnpike, is trying to estimate the average number of vehicles that arrive at the toll booth during a 1-minute period during the peak of rush-hour traffic. In an effort to estimate this average throughput value, he records the number of vehicles that arrive at the toll booth over a 1-minute interval commencing at the same time for each of 365 normal weekdays. The data are in the file P02_09.XLS. Use the BestFit add-in to determine which probability distribution best fits these data.

50. A finance professor has just given a midterm examination in her corporate finance course and is interested in learning how her class of 100 students performed on this exam. The data are in the file P02_05.XLS. Use the BestFit add-in to determine which probability distribution best fits these data.

6.8 Conclusion

We have covered a lot of ground in this chapter, and much of the material, especially that on the normal distribution, will be used in later chapters. The normal distribution is the cornerstone for much of statistical theory. As we will see when we study statistical inference and regression, an assumption of normality is behind most of the procedures we use. Therefore, it is important to understand the properties of the normal distribution and how to work with it in Excel. The binomial, Poisson, and exponential distributions, while not used as frequently as the normal distribution in this book, are also extremely important. The examples we have discussed indicate how these distributions can be used in a variety of business situations.

Although we have attempted to stress *concepts* in this chapter, we have also described the details necessary to work with these distributions in Excel. Fortunately, these details are not too difficult to master once you understand Excel's built-in functions such as NORMDIST, NORMINV, and BINOMDIST. Figures 6.6 and 6.18 provide typical examples of these functions. We suggest that you keep a copy of these figures handy.

Summary of Key Terms

Term	Explanation	Excel	Page	Equation Number
Probability density function	Specifies the probability distribution of a continuous random variable.		241	
Normal distribution	A continuous distribution with possible values ranging over the *entire* number line; its density function is a symmetric bell-shaped curve.		243	6.1
Standardizing a normal random variable	Transforms any normal distribution with mean μ and standard deviation σ to the *standard* normal distribution with mean 0 and standard deviation 1.	STANDARDIZE	243	6.2
Normal calculations in Excel	Useful for finding probabilities and percentiles for nonstandard and standard normal distributions.	NORMDIST, NORMSDIST, NORMINV, NORMSINV	247	
Empirical rules for normal distribution	About 68% of the data fall within 1 standard deviation of the mean, about 95% of the data fall within 2 standard deviations of the mean, and almost all fall within 3 standard deviations of the mean.		249	
Binomial distribution	The distribution of the number of successes in n independent, identical trials, where each trial has probability p of success.	BINOMDIST CRITBINOM	262	
Mean and standard deviation of a binomial distribution	The mean and standard deviation of a binomial distribution with parameters n and p are np and $[np(1-p)]^{1/2}$, respectively.		264	6.3, 6.4
Normal approximation to the binomial distribution	If $np > 5$ and $n(1-p) > 5$, the binomial distribution can be approximated well by a normal distribution with mean np and standard deviation $[np(1-p)]^{1/2}$.		266	

Summary of Key Terms (continued)

Term	Explanation	Excel	Page	Equation Number
Poisson distribution	A discrete probability distribution that often describes the number of events occurring within a specified period of time or space; mean and variance both equal the parameter λ.	POISSON	278	
Exponential distribution	A continuous probability distribution useful for measuring *times* between events such as customer arrivals to a service facility; mean and standard deviation both equal $1/\lambda$.	EXPONDIST	281	
Relationship between Poisson and exponential distributions	Exponential distribution measures *times* between events; Poisson distribution counts the *number* of events in a certain period of time.		282	
BestFit	A software package for finding how well a specified distribution fits a set of data, or for finding the distribution that best fits a set of data.	Excel not required	283	

PROBLEMS

Conceptual Exercises

C.1 Explain why the probability that a continuous random variable takes on any particular value must equal zero.

C.2 What is the relationship between the mean and standard deviation of any normal distribution? Explain.

C.3 A New York Yankees fan would like to determine the probability that his beloved team will sweep the National League Championship team in the first four games of this year's World Series. Explain, in words, how the fan should proceed to assess this likelihood. Be sure to state all assumptions that the fan needs to make to proceed with the approach you prescribe.

C.4 A production manager would like to determine the probability that a particular production process yields more than 3 defective items per hour. Explain, in words, how the manager should proceed to assess this likelihood. Be sure to state all assumptions that the manager needs to make to proceed with the approach you prescribe.

C.5 State the major similarities and differences between the *normal* distribution and the *exponential* distribution.

Level A

51. Suppose the annual return on XYZ stock follows a normal distribution with mean 0.12 and standard deviation 0.30.

  a. What is the probability that XYZ's value will decrease during a year?

  b. What is the probability that the return on XYZ during a year will be at least 20%?

  c. What is the probability that the return on XYZ during a year will be between −6% and 9%?

  d. There is a 5% chance that the return on XYZ during a year will be greater than or equal to what value?

  e. There is a 1% chance that the return on XYZ during a year will be less than what value?

  f. There is a 95% chance that the return on XYZ during a year will be between what two values (equidistant from the mean)?

52. Assume the annual mean return on Walt Disney stock is around 15% and the annual standard deviation is around 25%. Assume the annual and daily returns on Disney stock are normally distributed.

  a. What is the probability that Disney will lose money during a year?

b. There is a 5% chance that Disney will earn a return of at least what value during a year?

c. There is a 10% chance that Disney will earn a return of less than or equal to what value during a year?

d. What is the probability that Disney will earn at least 35% during a year?

e. Assume there are 252 trading days in a year. What is the probability that Disney will lose money on a given day? [*Hint:* Let Y be the annual return on Disney and X_i be the return on Disney on day i. Then (approximately) $Y = X_1 + X_2 + \cdots + X_{252}$. Also, use the fact that the sum of independent normal random variables is normally distributed, with mean equal to the sum of the individual means, and variance equal to the sum of the individual variances.]

53. Suppose Dell Computer receives its disk drives from Diskco. On average, 4% of all floppy disk drives received by Dell are defective.

a. Dell has adopted the following policy. It samples 50 disk drives in each shipment and accepts the shipment if all disk drives in the sample are non-defective. What fraction of batches will Dell accept?

b. Suppose instead that the batch is accepted if at most 1 disk drive in the sample is defective. What fraction of batches will Dell accept?

c. What is the probability that a sample of size 50 will contain at least 10 defectives?

54. A family is considering a move from a midwestern city to a city in California. The distribution of housing costs where the family currently lives is normal with mean $105,000 and standard deviation $18,200. The distribution of housing costs in the California city is normal with mean $135,000 and standard deviation $20,400. The family's current house is valued at $110,000.

a. What percentage of houses in the family's current city cost less than theirs?

b. If the family buys a $110,000 house in the new city, what percentage of houses there will cost less than theirs?

c. What price house will the family need to buy to be in the same percentile (of housing costs) in the new city as they are in the current city?

55. The number of traffic fatalities in a typical month in a given state has a normal distribution with mean 125 and standard deviation 31.

a. If a person in the highway department claims that there will be at least m fatalities in the next month with probability 0.95, what value of m makes this claim true?

b. If the claim is that there will be no more than n fatalities in the next month with probability 0.98, what value of n makes this claim true?

56. It can be shown that a sum of independent normally distributed random variables is also normally distributed. Do *all* functions of normal random variables lead to normal random variables? Consider the following.

SuperDrugs is a chain of drugstores with three similar-size stores in a given city. The sales in a given week for any of these stores is normally distributed with mean $15,000 and standard deviation $3000. At the end of each week, the sales figure for the store with the largest sales among the three stores is recorded. Is this maximum value normally distributed? Find out with simulation. Simulate a weekly sales figure at each of the three stores and calculate the maximum. Then replicate this maximum 500 times with a data table and create a histogram of the 500 maximum values. Does it appear to be normally shaped? Whatever this distribution looks like, use your simulated values to estimate its mean and standard deviation.

Level B

57. When we sum 30 or more independent random variables, the sum of the random variables will usually be approximately normally distributed, even if each individual random variable is not normally distributed. Use this fact to estimate the probability that a casino will be behind after 90,000 roulette bets, given that it wins $1 or loses $35 on each bet with probabilities 37/38 and 1/38.

58. The daily demand for six-packs of Coke at Mr. D's follows a normal distribution with mean 120 and standard deviation 30. Every Monday the Coke delivery driver delivers Coke to Mr. D's. If Mr. D's wants to have only a 1% chance of running out of Coke by the end of the week, how many should Mr. D's order for the week? Assume orders are placed Sunday at midnight. (Assume also that demands on different days are probabilistically independent. Use the fact that the sum of independent normal random variables is normally distributed, with mean equal to the sum of the individual means, and variance equal to the sum of the individual variances.)

59. Many companies use sampling to determine whether a batch should be accepted. An (n, c) sampling plan consists of inspecting n randomly chosen items from a batch and accepting the batch if c or fewer sampled items are defective. Suppose a company uses a (100, 5) sampling plan to determine whether a batch of 10,000 computer chips is acceptable.

a. The "producer's risk" of a sampling plan is the probability that an acceptable batch will be rejected by a sampling plan. Suppose the customer considers a batch with 3% defectives acceptable. What is the producer's risk for this sampling plan?

b. The "consumer's risk" of a sampling plan is the probability that an unacceptable batch will be accepted by a sampling plan. Our customer says that a batch with 9% defectives is unacceptable. What is the consumer's risk for this sampling plan?

60. Suppose that if a presidential election were held today, 52% of all voters would vote for Bush over Gore. (You may substitute the names of the current presidential candidates!) This problem shows that even if there are 100 million voters, a sample of several thousand is enough to determine the outcome, even in a fairly close election.

a. If we were to randomly sample 1500 voters, what is the chance that the sample would indicate (correctly) that Bush is preferred to Gore?

a. If we were to randomly sample 6000 voters, what is the chance that the sample would indicate (correctly) that Bush is preferred to Gore?

61. The Coke factory fills bottles of soda by setting a timer on a filling machine. It has generally been observed that the distribution of the number of ounces the machine puts into a bottle is normal with standard deviation 0.05 oz. The company wants 99.9% of all its bottles to have at least 16 oz of soda. To what amount should the mean amount put in each bottle be set? (The company does not want to put in any more than is necessary!)

62. The time it takes me to swim 100 yards in a race is normally distributed with mean 62 seconds and standard deviation 2 seconds. In my next five races, what is the chance I will swim under a minute exactly twice?

63. We assemble a large part by joining two smaller parts together. In the past, the smaller parts we have produced have a mean length of 1 inch and a standard deviation of 0.01 inch. Assume that the lengths of the smaller parts are normally distributed.

a. What fraction of the larger parts are longer than 2.05 inches? (Use the fact that the sum of independent normal random variables is normally distributed, with mean equal to the sum of the individual means, and variance equal to the sum of the individual variances.)

b. What fraction of the larger parts are between 1.96 inches and 2.02 inches long?

64. (Suggested by Sam Kaufmann, Indiana University MBA who runs Harrah's Lake Tahoe Casino) A high roller has come to the casino to play 300 games of craps. For each game of craps played there is a 0.493 probability that the high roller will win $1 and a 0.507 probability that the high roller will lose $1. After 300 games of craps, what is the probability that the casino will be behind more than $10?

65. (Suggested by Sam Kaufmann, Indiana University MBA who runs Harrah's Lake Tahoe Casino) A high roller comes to the casino intending to play 500 hands of blackjack for $1 a hand. On each hand, the high roller will win $1 with probability 0.48 and lose $1 with probability 0.52. After the 500 hands, what is the probability that the casino has lost more than $40?

66. Bottleco produces 100,000 12-oz bottles of soda per year. By adjusting a timer, Bottleco can adjust the mean number of ounces placed in a bottle. No matter what the mean, the standard deviation of the number of ounces in a bottle is 0.05. Soda costs $0.05 per ounce. Any bottle weighing less than 12 ounces will incur a $10 fine for being underweight. Determine a setting for the mean number of ounces per bottle of soda that will minimize the expected cost per year of producing soda. Your answer should be accurate within 0.001 ounce. Does the number of bottles produced per year influence your answer?

67. The weekly demand for televisions at Lowland Appliance is normally distributed with mean 400 and standard deviation 100. Each time an order for TVs is placed, it arrives exactly 4 weeks later. That is, TV orders have a 4-week lead time. Lowland doesn't want to run out of TVs during any more than 1% of all lead times. How low should Lowland let its TV inventory drop before it places an order for more TVs? (*Hint:* How many standard deviations above the mean lead-time demand must the reorder point be for there to be a 1% chance of a stockout during the lead time? Also, use the fact that the sum of independent normal random variables is normally distributed, with mean equal to the sum of the individual means, and variance equal to the sum of the individual variances.)

68. An elevator rail is assumed to meet specifications if its diameter is between 0.98 and 1.01 inches. Each year we make 100,000 elevator rails. For a cost of $10/\sigma^2$ per year we can rent a machine that produces elevator rails whose diameters have a standard deviation of σ. Any machine will produce rails having a mean diameter of 1 inch. Any rail that does not meet specifications must be reworked (at a cost of $12) to meet specifications. Assume that the diameter of an elevator rail follows a normal distribution.

a. What standard deviation (within 0.001 inch) will minimize our annual cost of producing elevator rails? You need not try standard deviations in excess of 0.02 inch.

b. For your answer in part **a,** one elevator rail in 1000 will be at least how many inches in diameter?

69. A 20-question true–false examination is given. Each correct answer is worth 5 points. Consider an unprepared student who randomly guesses on each question.

a. If no points are deducted for incorrect answers, what is the probability that the student will score at least 60 points?

b. If 5 points are deducted for each incorrect answer, what is the probability that the student will score at least 60 points?

c. If 5 points are deducted for each incorrect answer, what is the probability that the student will receive a negative score?

70. The percentage of examinees who took the GMAT (Graduate Management Admission) exam from June 1992 to March 1995 and scored below each total score is given in the file P06_70.XLS. For example, 96% of all examinees scored 690 or below. The mean GMAT score for this time period was 497 and the standard deviation was 105. Does it appear that GMAT scores can accurately be approximated by a normal distribution? (Source: 1995 GMAT Examinee Interpretation Guide)

71. What caused the crash of TWA Flight 800? Physics Professors Hailey and Helfand of Columbia University believe there is a reasonable possibility that a meteor hit Flight 800. They reason as follows. On a given day, 3000 meteors of a size large enough to destroy an airplane hit the earth's atmosphere. Around 50,000 flights per day, averaging 2 hours in length, have been flown from 1950 to 1996. (Flight 800 flew in 1996.) This means that at any given point in time, planes in flight cover approximately two-billionths of the world's atmosphere. Determine the probability that at least one plane in the last 47 years has been downed by a meteor. (*Hint:* Use the Poisson approximation to the binomial. This approximation says that if n is large and p is small, a binomial distribution with parameters n and p is approximately Poisson with $\lambda = np$.)

72. In the decade 1982–1991, ten employees working at the Amoco Company chemical research center were stricken with brain tumors. The average employment at the center was 2000 employees. Nationwide, the average incidence of brain tumors in a single year is 20 per 100,000 people. If the incidence of brain tumors at the Amoco chemical research center were the same as the nationwide incidence, what is the probability that at least 10 brain tumors would be observed among Amoco workers during the decade 1982–1991? What do you conclude from your analysis? (Source: AP wire service report, March 12, 1994)

73. Claims arrive at random times to an insurance company. The daily amount of claims is normally distributed with mean $1570 and standard deviation $450. Total claims on different days each have this distribution, and they are probabilistically independent of one another.

a. Find the probability that the amount of total claims over a period of 100 days is at least $150,000. (Use the fact that the sum of independent normally distributed random variables is normally distributed, with mean equal to the sum of the individual

means and variance equal to the sum of the individual variances.)

b. If the company receives premiums totaling $165,000, find the probability that the company will net at least $10,000 for the 100-day period.

74. A popular model for stock prices is the following. If p_0 is the current stock price, then the price, p_k, k periods from now (where a period could be a day, week, or any other convenient unit of time, and k is any positive integer) is given by

$$p_k = p_0 \exp((\mu - 0.5\sigma^2)k + sZ\sqrt{k})$$

where exp is the exponential function (EXP in Excel), μ is the mean percentage growth rate per period of the stock, σ is the standard deviation of the growth rate per period, and Z is a normally distributed random variable with mean 0 and standard deviation 1. Both μ and σ are typically estimated from actual stock price data, and they are typically expressed in decimal form, such as $\mu = 0.01$ for a 1% mean growth rate. Suppose a period is defined as a month, the current price of the stock (as of the end of December 2000) is $75, $\mu = 0.006$, and $\sigma = 0.028$. Use simulation to obtain 500 possible stock price changes from the end of December 2000 to the end of December 2003. (Note that you can simulate a given change in one line and then copy it down.) Draw a histogram of these changes to see whether the stock price change is at least approximately normally distributed. Also, use the simulated data to estimate the mean price change and the standard deviation of the change.

75. Continuing the previous problem (with the same parameters), use simulation to generate the ending stock prices for each month in 2001. (Use $k = 1$ to get January's price from December's, use $k = 1$ again to get February's price from January's, and so on.) Then use a data table to replicate the ending December 2001 stock price 500 times. Draw a histogram of these 500 values. Do they appear to resemble a normal distribution?

76. Your company is running an audit on the Sleaze Company. Since Sleaze has a bad habit of overcharging its customers, the focus of your audit is on checking whether the billing amounts on its invoices are correct. We'll assume that each invoice is for too high an amount with probability 0.06 and for too low an amount with probability 0.01 (so that the probability of a correct billing is 0.93). Also, we assume that the outcome for any invoice is probabilistically independent of the outcomes for other invoices.

a. If you randomly sample 200 of Sleaze's invoices, what is the probability that you will find at least 15 invoices that overcharge the customer? What is the probability you won't find any that undercharge the customer?

b. Find an integer k such that the probability is at least 0.99 that you will find at least k invoices that overcharge the customer. (*Hint:* Use trial and error with the BINOMDIST function to find *k*.)

77. Continuing the previous problem, suppose that when Sleaze overcharges a customer, the distribution of the amount overcharged (expressed as a percentage of the correct billing amount) is normally distributed with mean 15% and standard deviation 4%.
   a. What percentage of overbilled customers are charged at least 10% more than they should pay?
   b. What percentage of *all* customers are charged at least 10% more than they should pay?
   c. If your auditing company samples 200 randomly chosen invoices, what is the probability that it will find at least 5 where the customer was over-charged by at least 10%?

78. Let *X* be normally distributed with a given mean and standard deviation. Sometimes you want to find two values *a* and *b* such that $P(a < X < b)$ is equal to some specific probability such as 0.90 or 0.95. There are many answers to this problem, depending on how much probability you put in each of the two tails. For this question, assume the mean and standard deviation are $\mu = 100$ and $\sigma = 10$, and that we want to find *a* and *b* such that $P(a < X < b) = 0.90$.
   a. Find *a* and *b* so that there is probability 0.05 in each tail.
   b. Find *a* and *b* so that there is probability 0.025 in the left tail and 0.075 in the right tail.
   c. The "usual" answer to the general problem is the answer from part **a,** that is, where you put equal probability in the two tails. It turns out that this is the answer that minimizes the length of the interval from *a* to *b*. That is, if you solve the problem: min (*b* − *a*), subject to $P(a < X < b) = 0.90$, you'll get the same answer as in part **a.** Verify this using Excel's Solver.

79. Your manufacturing process makes parts such that each part meets specifications with probability 0.98. You need a batch of 250 parts that meet specifications. How many parts must you produce to be at least 99% certain of producing at least 250 parts that meet specifications?

80. The Excel functions discussed in this chapter are useful for solving a lot of probability problems, but there are other problems that, even though they are similar to normal or binomial problems, cannot be solved with these functions. In cases like this, computer simulation can often be used. Here are a couple of such problems for you to simulate. For each example, use 500 replications of the experiment.
   a. You observe a sequence of parts from a manufacturing line. These parts use a component that is supplied by one of two suppliers. The probability that a given part uses a component supplied by

supplier 1 is 0.6; it is supplied by supplier 2 with probability 0.4. Each part made with a component from supplier 1 works properly with probability 0.95, and each part made with a component from supplier 2 works properly with probability 0.98. Assuming that 30 of these parts are made, we want the probability that at least 29 of them work properly.
   b. Here we look at a more generic example such as coin flipping. That is, there is a sequence of trials where each trial is a success with probability *p* and a failure with probability 1 − *p*. A "run" is a sequence of consecutive successes or failures. For most of us, intuition says that there should not be "long" runs. Test this by finding the probability that there is at least 1 run of length at least 6 in a sequence of 15 trials. (The run could be of 0's or 1's.) You can use any value of *p* you like—or try different values of *p*.

81. As any credit-granting agency knows, there are always some customers who default on credit charges. Typically, customers are grouped into relatively homogeneous categories, so that customers within any category have approximately the same chance of defaulting on their credit charges. Here we'll look at one particular group of customers. We'll assume each of these customers has (1) probability 0.07 of defaulting on his or her current credit charges, and (2) total credit charges that are normally distributed with mean $350 and standard deviation $100. We'll also assume that if a customer defaults, 20% of his or her charges can be recovered. The other 80% are written off as bad debt.
   a. What is the probability that a typical customer in this group will default and produce a write-off of more than $250 in bad debt?
   b. If there are 500 customers in this group, what are the mean and standard deviation of the number of customers who will meet the description in part **a?**
   c. Again assuming there are 500 customers in this group, what is the probability that at least 25 of them will meet the description in part **a?**
   d. Suppose now that nothing is recovered from a default—the whole amount is written off as bad debt. Show how to simulate the total amount of bad debt from 500 customers in just two cells, one with a binomial calculation, the other with a normal calculation.

82. You have a device that uses a single battery, and you operate this device continuously, never turning it off. Whenever a battery fails, you replace it with a brand new one immediately. Suppose the lifetime of a typical battery has an exponential distribution with mean 205 minutes. If you operate the device, starting with a

new battery, until you have observed 25 battery failures, what is the probability that at least 15 of these 25 batteries lived at least 3.5 hours?

83. In the previous problem, we ran the "experiment" until there are a certain number of failures and then answered a question about the times between failures. In this problem, we take a different point of view. We run the experiment for a certain amount of time and then ask a question about the number of failures during this time. Specifically, suppose you operate the device from the previous problem continuously for 3 days, making battery changes when necessary. Find the probability that you will observe at least 25 failures. (*Hint:* Do a Poisson calculation using an appropriate λ for the number of failures in a 3-day period.)

CASE

6.1

EuroWatch Company

The EuroWatch Company assembles expensive wristwatches and then sells them to retailers throughout Europe. The watches are assembled at a plant with two assembly lines. These lines are intended to be identical, but line 1 uses somewhat older equipment than line 2 and is typically less reliable. Historical data have shown that each watch coming off line 1, independently of the others, is free of defects with probability 0.98. The similar probability for line 2 is 0.99. Each line produces 500 watches per hour. The production manager has asked you to answer the following questions.

1. She wants to know how many defect-free watches each line is likely to produce in a given hour. Specifically, find the smallest integer k (for each line separately) such that you can be 99% sure that the line will not produce more than k defective watches in a given hour.

2. EuroWatch currently has an order for 500 watches from an important customer. The company plans to fill this order by packing slightly more than 500 watches, all from line 2, and sending this package off to the customer. Obviously, EuroWatch wants to send as few watches as possible, but it wants to be 99% sure that when the customer opens the package, there are at least 500 defect-free watches. How many watches should be packed?

3. EuroWatch has another order for 1000 watches. Now it plans to fill this order by packing slightly more than one hour's production from each line. This package will contain the *same* number of watches from each line. As in the previous question, EuroWatch wants to send as few watches as possible, but it again wants to be 99% sure that when the customer opens the package, there are at least 1000 defect-free watches. The question of how many watches to pack is unfortunately quite difficult because the total number of defect-free watches is *not* binomially distributed. (Why not?) Therefore, the manager asks you to solve the problem with simulation (and some trial and error). (*Hint:* It turns out that it's much faster to simulate

small numbers than large numbers, so simulate the number of watches with defects, not the number without defects.)

4. Finally, EuroWatch has a third order for 100 watches. The customer has agreed to pay $50,000 for the order—that is, $500 per watch. If Euro-Watch sends more than 100 watches to the customer, its revenue doesn't increase; it can never exceed $50,000. Its unit cost of producing a watch is $450, regardless of which line it is assembled on. The order will be filled entirely from a single line, and EuroWatch plans to send slightly more than 100 watches to the customer.

If the customer opens the shipment and finds that there are fewer than 100 defect-free watches (which we'll assume the customer has the ability to do), then he'll pay only for the defect-free watches—EuroWatch's revenue will decrease by $500 per watch short of the 100 required—and on top of this, EuroWatch will be required to make up the difference at an expedited cost of $1000 per watch. The customer won't pay a dime for these expedited watches. (If expediting is required, EuroWatch will make sure that the expedited watches are defect-free. It doesn't want to lose this customer entirely!)

You have been asked to develop a spreadsheet model to find EuroWatch's expected profit for any number of watches it sends to the customer. You should develop it so that it responds correctly, regardless of which assembly line is used to fill the order and what the shipment quantity is. (*Hints:* Use the BINOMDIST function, with last argument 0, to fill up a column of probabilities for each possible number of defective watches. Next to each of these, calculate EuroWatch's profit. Then use a SUMPRODUCT to obtain the expected profit. Finally, you can assume that EuroWatch will never send more than 110 watches. It turns out that this large a shipment is not even close to optimal.)

6.2 Cashing in on the Lottery

Many states supplement their tax revenues with state-sponsored lotteries. Most of them do so with a game called lotto. Although there are various versions of this game, they are all basically as follows. People purchase tickets that contain r distinct numbers from 1 to m, where r is generally 5 or 6 and m is generally around 50. For example, in Virginia, the state discussed in this case, $r = 6$ and $m = 44$. Each ticket costs $1, about 39 cents of which is allocated to the total jackpot.[7] There is eventually a drawing of $r = 6$ distinct numbers from the $m = 44$ possible numbers. Any ticket that matches these 6 numbers wins the jackpot.

There are two interesting aspects of this game. First, the current jackpot includes not only the revenue from this round of ticket purchases but any jackpots carried over from previous drawings because of no winning tickets. Therefore, the jackpot can build from one drawing to the next, and in celebrated cases it has become huge. Second, if there is more than one winning ticket—a distinct possibility—the winners share the jackpot equally. (This is called the "parimutuel" effect.) So, for example, if the current jackpot is $9 million and there are three winning tickets, then each winner receives $3 million.

It can be shown that for Virginia's choice of r and m, there are approximately 7 million possible tickets (7,059,052 to be exact). Therefore, any ticket has about one chance out of 7 million of being a winner. That is, the probability of winning with a single ticket is $p = 1/7,059,052$—not very good odds! If n people purchase tickets, then the number of winners is binomially distributed with parameters n and p. Because n is typically very large and p is small, the number of winners has approximately a Poisson distribution with rate $\lambda = np$. (This makes ensuing calculations somewhat easier.) For example, if 1 million tickets are purchased, then the number of winning tickets is approximately Poisson distributed with $\lambda = 1/7$.

In 1992, an Australian syndicate purchased a huge number of tickets in the Virginia lottery in an attempt to assure itself of purchasing a winner. It worked! Although the syndicate wasn't able to purchase all 7 million possible tickets (it was about 1.5 million shy of this), it did purchase a winning ticket, and there were no other winners. Therefore, the syndicate won a 20-year income stream worth approximately $27 million, with a net present value of approximately $14 million. This easily covered the cost of the tickets it purchased. Two questions come to mind: (1) Is this "hogging" of tickets unfair to the rest of the public? (2) Is it a wise strategy on the part of the syndicate (or did it just get lucky)?

To answer the first question, consider how the lottery changes for the general public with the addition of the syndicate. To be specific, suppose the syndicate can invest $7 million and obtain *all* of the possible tickets, making itself a sure winner. Also, suppose n people from the general public purchase tickets, each of which has 1 chance out of 7 million of being a winner. Finally, let R be the jackpot carried over from any previous lotteries. Then the total jackpot on this round will be $[R + 0.39(7,000,000 + n)]$, because 39 cents from every ticket goes toward the jackpot. The number of winning tickets for the public will be Poisson distributed with $\lambda = n/7,000,000$. However, any member of the public who wins will *necessarily* have to share the jackpot with the syndicate, which is a sure winner. Use this information to calculate the expected amount the public will win. Then do the same calculation when the syndicate does *not* play. (In this case the jackpot will be smaller, but the public won't have to share any winnings with the syndicate.) For values of n and R that you can select, is the public better off with or without the syndicate? Would you, as a general member of the public, support a move to outlaw syndicates from "hogging" the tickets?

The second question is whether the syndicate is wise to buy so many tickets. Again assume that the syndicate can spend $7 million and purchase each possible ticket. (Would this be possible in reality?) Also, assume that n members of the general public purchase tickets, and that the carryover from the

[7]Of the remaining 61 cents, the state takes about 50 cents. The other 11 cents is used to pay off lesser prize winners whose tickets match some, but not all, of the winning 6 numbers. To keep this case relatively simple, however, we will ignore these lesser prizes and concentrate only on the jackpot.

previous jackpot is R. The syndicate is thus assured of having a winning ticket, but is it assured of covering its costs? Calculate the expected net benefit (in terms of net present value) to the syndicate, using any reasonable values of n and R, to see whether the syndicate can expect to come out ahead.

Actually, the analysis suggested in the previous paragraph is not complete. There are at least two complications to consider. The first is the effect of taxes. Fortunately for the Australian syndicate, it did not have to pay federal or state taxes on its winnings, but a U.S. syndicate wouldn't be so lucky. Second, the jackpot from a $20 million jackpot, say, is actually paid in 20 annual $1 million payments. The Lottery Commission pays the winner $1 million immediately and then purchases 19 "strips" (bonds with the interest not included) maturing at 1-year intervals with face value of $1 million each. Unfortunately, the lottery prize does not offer the liquidity of the Treasury issues that back up the payments. This lack of liquidity could make the lottery less attractive to the syndicate.

7

Decision Making Under Uncertainty

© Getty Images

Decision Analysis at Du Pont

Formal decision analysis in the face of uncertainty frequently occurs at the most
strategic levels of a company's planning
process and typically involves teams of high-
level managers from all areas of the company.
This is certainly the case with Du Pont, as reported by two internal decision analysis experts,
Krumm and Rolle (1992), in their article "Management and Application of Decision and Risk Analysis
in Du Pont." Du Pont's formal use of decision analysis
began in the 1960s, but because of a lack of computing power
and distrust of the method by senior-level management, it never really got
a foothold. However, by the mid-1980s things had changed considerably.
The company was involved in a faster-moving, more uncertain environment, more people throughout the company were empowered to make
decisions, and these decisions had to be made more quickly. In addition,
the computing power had arrived to make large-scale quantitative analysis feasible. Since that time, Du Pont has embraced formal decision-
making analysis in all of its businesses, and the trend is almost certain
to continue.

The article describes a typical example of decision analysis within the
company. One of Du Pont's businesses, Business Z (so-called for reasons
of confidentiality), was stagnating. It was not set up to respond quickly to
changing customer demands, and its financial position was declining due
to lower prices and market share. A decision board and a project team
were empowered to turn things around. The project team developed a detailed timetable to accomplish three basic steps: frame the problem, assess uncertainties and perform the analysis, and implement the recommended decision. The first step involved setting up a "strategy table" to
list the possible strategies and the factors that would affect or be affected
by them. The three basic strategies were (1) a base-case strategy (continue operating as is), (2) a product differentiation strategy (develop new

products), and (3) a cost leadership strategy (shut down the plant and streamline the product line).

In the second step the team asked a variety of experts throughout the company for their assessments of the likelihood of key uncertain events. In the analysis step they then used all of the information gained to determine the strategy with the largest expected net present value. Two important aspects of this analysis step were the extensive use of sensitivity analysis (many what-if questions) and the emergence of new "hybrid" strategies that dominated the strategies that had been considered to that point. In particular, the team finally decided on a product differentiation strategy that also decreased costs by shutting down some facilities in each plant.

By the time of the third step, implementation, the decision board needed little convincing. Since all of the key people had been given the opportunity to provide input to the process, everyone was convinced that the right strategy had been selected. All that was left was to put the plan in motion and monitor its results. The results were impressive. Business Z made a complete turnaround, and its net present value increased by close to $200 million. Besides this tangible benefit, there were definite intangible benefits from the overall process. As Du Pont's vice president for finance said, "The D&RA [decision and risk analysis] process improved communication within the business team as well as between the team and corporate management, resulting in rapid approval and execution. As a decision maker, I highly value such a clear and logical approach to making choices under uncertainty and will continue to use D&RA whenever possible." ●

7.1 **Introduction**

In this chapter we will provide a formal framework for analyzing decision problems that involve uncertainty. We will discuss the most frequently used criteria for choosing among alternative decisions, how probabilities are used in the decision-making process, how decisions made at an early stage can influence decisions made at a later stage, how a decision maker can quantify the value of information, and how attitudes toward risk can affect the analysis. Throughout, we will employ graphical tools—decision trees—to guide the analysis. These enable the decision maker to view all important aspects of the problem at once: the decision alternatives, the uncertain outcomes and their probabilities, the economic consequences, and the chronological order of events. We will show how to implement these graphical tools in Excel by taking advantage of a very powerful and flexible add-in from Palisade called PrecisionTree.

Many examples of decision making under uncertainty exist in the business world. Here are several examples.

Examples of Decision Making Under Uncertainty

1. Companies routinely place bids for contracts to complete a certain project within a fixed time frame. Often these are sealed bids, where each of several companies presents in a sealed envelope a bid for completing the project; then the envelopes are opened, and the low bidder is awarded the bid amount to complete the project. Any particular company in the bidding competition must deal with the possible uncertainty of its *actual* cost of completing the project (should it win the bid), as well as the uncertainty involved in what the other companies will bid. The trade-off is between bidding low to win the bid and bidding high to make a profit.

2. Whenever a company contemplates introducing a new product into the market, there are a number of uncertainties that affect the decision, probably the most important being the customers' reaction to this product. If the product generates high customer demand, then the company will make a large profit. But if demand is low (and, after all, the vast majority of new products do poorly), then the company might not even recoup its development costs. Because the level of customer demand is critical, the company might try to gauge this level by test marketing the product in one region of the country. If this test market is a success, the company can then be more optimistic that a full-scale national marketing of the product will also be successful. But if the test market is a failure, the company can cut its losses by abandoning the product.

3. Should athletes be required to undergo drug testing? And if drug testing is required, should an athlete who tests positive be banned from the sport? Several sources of uncertainty, as well as several "costs," are relevant here. The uncertainties involve the proportion of the total athlete population who use drugs and the reliability of the tests. The costs include the obvious cost of the tests, but also the less obvious "costs" of invading an athlete's privacy and of declaring an athlete a drug user when in fact he or she is not (because of a faulty test).

4. A recent *Interfaces* article (Borison, 1995) describes an application of formal decision analysis by Oglethorpe Power Corporation (OPC), a Georgia-based electricity supplier. The basic decision OPC faced was whether to build a new transmission line to supply large amounts of electricity to parts of Florida and, if they decided to build it, how to finance this project. OPC had to deal with several sources of uncertainty: the cost of building new facilities, the demand for power in Florida, and various market conditions, such as the spot price of electricity.

5. Another *Interfaces* article (Ulvila, 1987) describes the decision analysis performed by the U.S. Postal Service regarding the purchase of automation equipment. One of the investment decisions was which type of OCR (optical character recognition) equipment the Postal Service should purchase (or convert) for reading single- and/or multiple-line addresses on packages. An important factor in this decision was the level of use by businesses of the "zip+4" (nine-digit zip codes). Zip+4 usage had been recommended for some time but was used only sporadically. The Postal Service was uncertain about the future level of business zip+4 usage. If businesses used the nine-digit codes heavily in the future, then a certain type of (expensive) OCR equipment would be most economical. If business use of zip+4 did not increase, then purchasing this equipment would be a waste of money. The decision was an extremely important one, given the expense of the proposed equipment and the fact that the Postal Service would have to live with whatever equipment it purchased for a number of years.

6. Utility companies must make many decisions that have significant environmental and economic consequences. [A good discussion of such consequences appears in an *Interfaces* article by Balson et al. (1992).] For these companies it is not necessarily enough to conform to federal or state environmental regulations. Recent court decisions have found companies liable—for huge settlements—when accidents occurred, even though the companies followed all existing regulations. Therefore, when utility companies decide, say, whether to replace equipment or mitigate the effects of environmental pollution, they must take into account the possible environmental consequences (such as injuries to people) as well as economic consequences (such as lawsuits). An aspect of these situations that makes decision analysis particularly difficult is that the potential "disasters" are often extremely improbable; hence, their likelihoods are very difficult to assess accurately.

Elements of a Decision Analysis

Although decision making under uncertainty occurs in a wide variety of contexts, all problems have three elements in common: (1) the set of decisions (or strategies) available to the decision maker, (2) the set of possible outcomes and the probabilities of these outcomes, and (3) a value model that prescribes results, usually monetary values, for the various combinations of decisions and outcomes. Once these elements are known, the decision maker can find an "optimal" decision, depending on the optimality criterion chosen. The following example illustrates these concepts.

Example 7.1 Competitive Bidding by SciTools Incorporated

SciTools Incorporated, a company that specializes in scientific instruments, has been invited to make a bid on a government contract. The contract calls for a specific number of these instruments to be delivered during the coming year. The bids must be sealed (so that no company knows what the others are bidding), and the low bid wins the contract. SciTools estimates that it will cost $5000 to prepare a bid and $95,000 to supply the instruments if it wins the contract. On the basis of past contracts of this type, SciTools believes that the possible low bids from the competition, if there is any competition, and the associated probabilities are those shown in Table 7.1. In addition, SciTools believes there is a 30% chance that there will be *no* competing bids.

Objective To determine the probability of SciTools winning the government contract and its associated payoff for each possible bidding decision.

Table 7.1

Probabilities of Low Bids from Competition	Low Bid	Probability
	Less than $115,000	0.2
	Between $115,000 and $120,000	0.4
	Between $120,000 and $125,000	0.3
	Greater than $125,000	0.1

Solution

Let's discuss the three elements of SciTools' problem. First, SciTools has two basic strategies: submit a bid or do not submit a bid. If SciTools submits a bid, then it must decide how much to bid. Based on SciTools' cost to prepare the bid and its cost to supply the instruments, there is obviously no point in bidding less than $100,000—SciTools wouldn't make a profit even if it won the bid. Although any bid amount over $100,000 might be considered, the data in Table 7.1 might persuade SciTools to limit its choices to $115,000, $120,000, and $125,000.[1]

---

[1]The problem with a bid such as $117,000 is that the data in Table 7.1 make it impossible to calculate the probability of SciTools winning the contract if it bids this amount. Other than this, however, there is nothing that rules out such an "in-between" bid.

The next element of the problem involves the uncertain outcomes and their probabilities. We have assumed that SciTools knows exactly how much it will cost to prepare a bid and how much it will cost to supply the instruments if it wins the bid. (In reality, these are probably estimates of the actual costs.) Therefore, the only source of uncertainty is the behavior of the competitors—will they bid, and if so, how much? From SciTools' standpoint, this information is difficult to obtain. The behavior of the competitors depends on (1) how many competitors are likely to bid and (2) how the competitors assess *their* costs of supplying the instruments. However, we will assume that SciTools has been involved in similar bidding contests in the past and can, therefore, predict competitor behavior from past experience. The result of such prediction is the assessed probability distribution in Table 7.1 and the 30% estimate of the probability of no competing bids.

The last element of the problem is the value model that transforms decisions and outcomes into monetary values for SciTools. The value model is straightforward in this example, but it can become quite complex in other applications, especially when the time value of money is involved and some quantities (such as the costs of environmental pollution) are difficult to quantify. If SciTools decides right now not to bid, then its monetary value is $0—no gain, no loss. If it makes a bid and is underbid by a competitor, then it loses $5000, the cost of preparing the bid. If it bids B dollars and wins the contract, then it makes a profit of (B − $100,000), that is, B dollars for winning the bid, less $5000 for preparing the bid, less $95,000 for supplying the instruments. For example, if it bids $115,000 and the lowest competing bid, if any, is greater than $115,000, then SciTools makes a profit of $15,000.

It is often convenient to list the monetary outcomes in a *payoff table,* as shown in Table 7.2.

> **Payoff Table**
>
> For each possible decision and each possible outcome, the **payoff table** lists the monetary value earned by an organization, where a positive value represents a profit and a negative value represents a loss. The probabilities of the various outcomes are listed at the bottom of the table.

For example, the probability that the competitors' low bid is less than $115,000 is 0.7 (the probability of at least one competing bid) multiplied by 0.2 (the probability that the lowest competing bid is less than $115,000, given that there is at least one competing bid).

Table 7.2

Payoff Table for SciTools Bidding Example

		Competitors' Low Bid ($1000s)				
		No Bid	<115	>115, <120	>120, <125	>125
	No Bid	0	0	0	0	0
SciTools'	115	15	−5	15	15	15
Bid	120	20	−5	−5	20	20
($1000s)	125	25	−5	−5	−5	25
	Probability	0.3	0.7(0.2)	0.7(0.4)	0.7(0.3)	0.7(0.1)

It is sometimes possible to simplify payoff tables to better understand the essence of the problem. In the present example, if SciTools bids, then the only necessary information about the competitors' bid is whether it is lower or higher than SciTools' bid. That is, SciTools cares only whether it wins the contract. Therefore, an alternative way of presenting the payoff table is shown in Table 7.3 on page 304.

Table 7.3

Alternative Payoff Table for SciTools Bidding Example		Monetary Value		Probability That
		SciTools Wins	SciTools Loses	SciTools Wins
	No Bid	NA	0	0.00
SciTools'	115	15	−5	0.86
Bid	120	20	−5	0.58
($1000s)	125	25	−5	0.37

The third and fourth columns of this table indicate the payoffs to SciTools, depending on whether it wins or loses the bid. The rightmost column shows the probability that SciTools wins the bid for each possible decision. For example, if SciTools bids $120,000, then it wins the bid if there are no competing bids (probability 0.3) or if there are competing bids but the lowest of these is greater than $120,000 (probability 0.7(0.3 + 0.1)). In this case the total probability that SciTools wins the bid is 0.3 + 0.28 = 0.58.

7.2.1 Risk Profiles

From Table 7.3 we can obtain *risk profiles* for each of SciTools' decisions.

Risk Profile	A **risk profile** for any decision lists all possible monetary values and their corresponding probabilities.

For example, if SciTools bids $120,000, there are two monetary values possible, a profit of $20,000 or a loss of $5000, and their probabilities are 0.58 and 0.42, respectively. On the other hand, if SciTools decides not to bid, there is a sure monetary value of $0—no profit, no loss.

A risk profile can also be illustrated graphically as a bar chart. There is a bar above each possible monetary value with height proportional to the probability of that value. For example, the risk profile for a $120,000 bid decision is a bar chart with two bars, one above −$5000 with height 0.42 and one above $20,000 with height 0.58. The risk profile for the "no bid" decision is even simpler: It has a single bar above $0 with height 1. We don't show these bar charts for this example because they are so simple, but in more complex examples they can provide insight.

7.2.2 Expected Monetary Value (EMV)

From the information we have discussed so far, it is not at all obvious which decision SciTools should make. The "no bid" decision is certainly safe, but it is certain to make zero profit. If SciTools decides to bid, the probability that it will lose $5000 is smallest with the $115,000 bid, but this bid has the smallest potential profit. Of course, if SciTools knew what the competitors were going to do, its decision would be easy. However, this uncertainty is the defining aspect of the problems in this chapter. The decision must be made *before* the uncertainty is resolved.

The most common way to make the choice is to calculate the *expected monetary value* (EMV) of each alternative and then choose the alternative with the largest EMV.

| Expected Monetary Value | The **expected monetary value (EMV)** is a weighted average of the possible monetary values, weighted by their probabilities. Formally, if v_i is the monetary value corresponding to outcome i and p_i is its probability, then expected monetary value is defined as |

$$EMV = \Sigma \, v_i p_i \qquad (7.1)$$

Actually, this is nothing new. The EMV is the mean of the probability distribution of possible monetary outcomes, as defined in Chapter 5.

The EMVs for SciTools' problem are listed in Table 7.4. They indicate that if SciTools uses the EMV criterion for making its decision, it should bid $115,000, as this yields the largest EMV.

Table **7.4**

EMVs for SciTools Bidding Example

Alternative	EMV Calculation	EMV
No bid	0(1)	$0
Bid $115,000	15,000(0.86) + (−5000)(0.14)	$12,200
Bid $120,000	20,000(0.58) + (−5000)(0.42)	$9,500
Bid $125,000	25,000(0.37) + (−5000)(0.63)	$6,100

Don't confuse an EMV with a sure amount that we "expect" to make.

It is very important to understand what an EMV implies and what it does not imply. If SciTools bids $115,000, then its EMV is $12,200. This does *not* mean SciTools will earn a profit of $12,200. It will earn $15,000 or it will lose $5000. So what does the EMV of $12,200 really mean? It means that if SciTools could enter many "gambles" like this, where on each gamble it would win $15,000 with probability 0.86 or lose $5000 with probability 0.14, then *on average* it would win $12,200 per gamble. In other words, the EMV can be interpreted as a long-term average.

It might seem peculiar, then, to base a one-time decision on EMV, which represents a long-term average. There are two ways to explain this apparent inconsistency. First, most companies make frequent decisions under uncertainty. Although each decision might have its own unique characteristics, it seems reasonable that if the company plans to make many such decisions, it should be willing to "play the averages," as it does when it uses EMV as the decision criterion. Second, even if this is the only such decision the company is *ever* going to make, decision theorists have proven that under certain conditions, maximizing EMV is a rational basis for making this decision. These "certain conditions" relate to the decision maker's attitude toward risk. As we will discuss later in this chapter, if the decision maker is risk averse and the possible monetary payoffs or losses are large relative to her wealth, then EMV is *not* the appropriate decision criterion to use. However, the EMV criterion has proved useful in the vast majority of decision-making applications, so we will use it throughout most of this chapter.

7.2.3 Decision Trees

By now, we have gone through most of the steps of solving SciTools' problem. We have listed the decision alternatives, the uncertain outcomes and their probabilities, and the profits and losses from all combinations of decisions and outcomes. We have then calculated the EMV for each alternative and have chosen the alternative with the largest EMV.

FIGURE 7.1
Decision Tree for
SciTools Bidding
Example

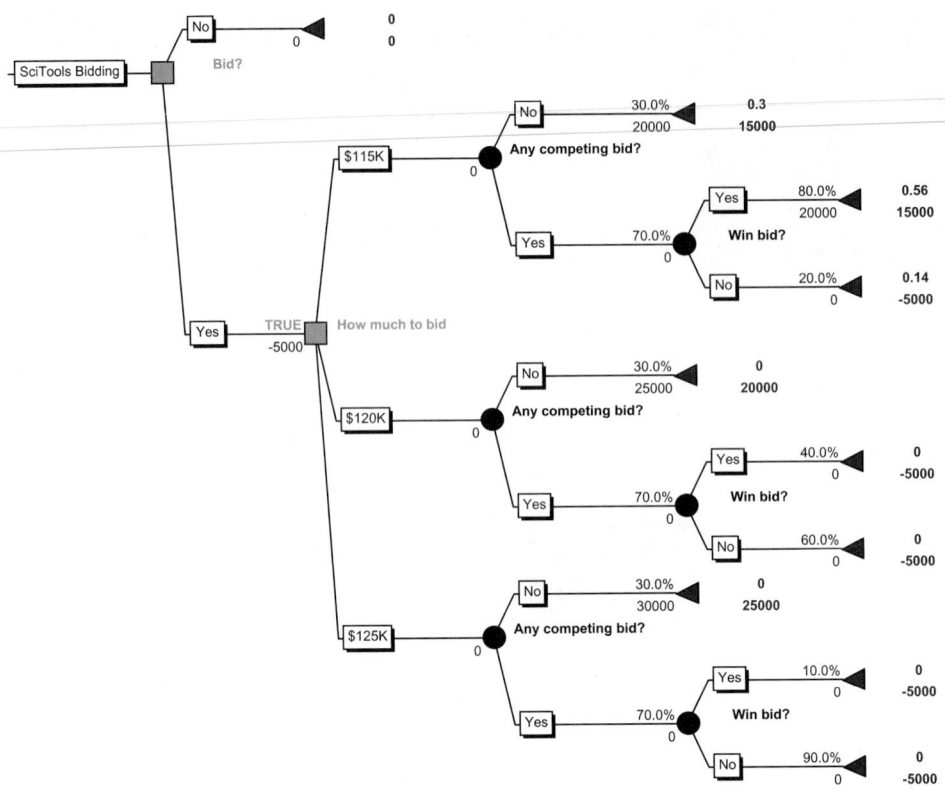

All of this can be done efficiently using a graphical tool called a *decision tree*. The decision tree that corresponds to SciTools' problem appears in Figure 7.1. (This figure is actually part of an Excel spreadsheet and was created with the PrecisionTree add-in. We'll explain how it was created shortly.)

To understand Figure 7.1, we need to know the following conventions that have been established for decision trees.

Attributes of Decision Trees

1. Decision trees are composed of **nodes** (circles, squares, and triangles) and **branches** (lines).

2. The nodes represent points in time. A **decision node** (a square) is a time when the decision maker makes a decision. A **probability node** (a circle) is a time when the result of an uncertain event becomes known. An **end node** (a triangle) indicates that the problem is completed—all decisions have been made, all uncertainty has been resolved, and all payoffs/costs have been incurred.

3. Time proceeds *from left to right*. This means that any branches leading into a node (from the left) have already occurred. Any branches leading out of a node (to the right) have not yet occurred.

4. Branches leading out of a decision node represent the possible decisions; the decision maker can choose the preferred branch. Branches leading out of probability nodes represent the possible outcomes of uncertain events; the decision maker has no control over which of these will occur.

5. Probabilities are listed on probability branches. These probabilities are *conditional* on the events that have already been observed (those to the left). Also, the probabilities on branches leading out of any particular probability node must sum to 1.

6. Individual monetary values are shown on the branches where they occur, and cumulative monetary values are shown to the right of the end nodes. (Actually, Precision-Tree shows two values to the right of each end node. The top one is the probability of getting to that end node, and the bottom one is the associated monetary value.)

The decision tree in Figure 7.1 illustrates these conventions for a *single-stage* decision problem, the simplest type of decision problem. In a single-stage problem all decisions are made *first,* and then all uncertainty is resolved. Later in this chapter we will see *multistage* decision problems, where decisions and outcomes alternate. That is, a decision maker makes a decision, then some uncertainty is resolved, then the decision maker makes a second decision, then some further uncertainty is resolved, and so on. Because these multistage decisions problems are inherently more complex, we will focus initially on single-stage problems.

Single-Stage and Multistage Decision Problems	In a **single-stage** decision problem all decisions are made *first,* and then all uncertainty is resolved. In a **multistage** decision problem decisions and outcomes *alternate.*

Once a decision tree has been drawn and labeled with probabilities and monetary values, it can be solved easily. The solution for the decision tree in Figure 7.1 is shown in Figure 7.2 on page 308. Among other things, it shows that the decision to bid $115,000 is optimal (follow the decision branches with "True" above them), with a corresponding EMV of $12,200 (the value under "Bid?" at the left of the tree). This is consistent with what we saw previously for this example.

The solution procedure used to develop Figure 7.2 is called *folding back* on the tree. Starting at the right of the tree and working back to the left, the procedure consists of two types of calculations. (These calculation are performed automatically by PrecisionTree.)

1. At each probability node, we calculate the EMV (sum of monetary values times probabilities) and write it below the name of the node. For example, consider the node (top right) after SciTools' decision to bid $115,000 and after it learns that there will be a competing bid. From that point, SciTools will either win $15,000 with probability 0.8 or lose $5000 with probability 0.2. The corresponding EMV is

$$0.8(15,000) + 0.2(-5000) = 11,000$$

and this value is entered below the node name "Win bid?".

Now, back up a step and consider the preceding probability node (the one to the left of the "Win bid?" node). At this point, SciTools has bid $115,000 and is about to discover whether there will be a competing bid. If there is none, with probability 0.3, then SciTools will win $15,000. But if there is a competing bid, with probability 0.7, the EMV from that point on is the $11,000 we just calculated. Essentially, this $11,000 summarizes the consequences of being at the "Win bid?" node, and SciTools acts the same as if it were going to receive $11,000 *for certain.* Therefore, the EMV for the "Any competing bid?" node is

$$0.3(15,000) + 0.7(11,000) = 12,200$$

This EMV is written below the node name.

2. Decision nodes are much easier. At each decision node we find the maximum of the EMVs and write it below the node name. PrecisionTree indicates the winner by placing "True" on the decision branch with the maximum EMV and "False" on all other branches emanating from this node. For example, consider the node where SciTools

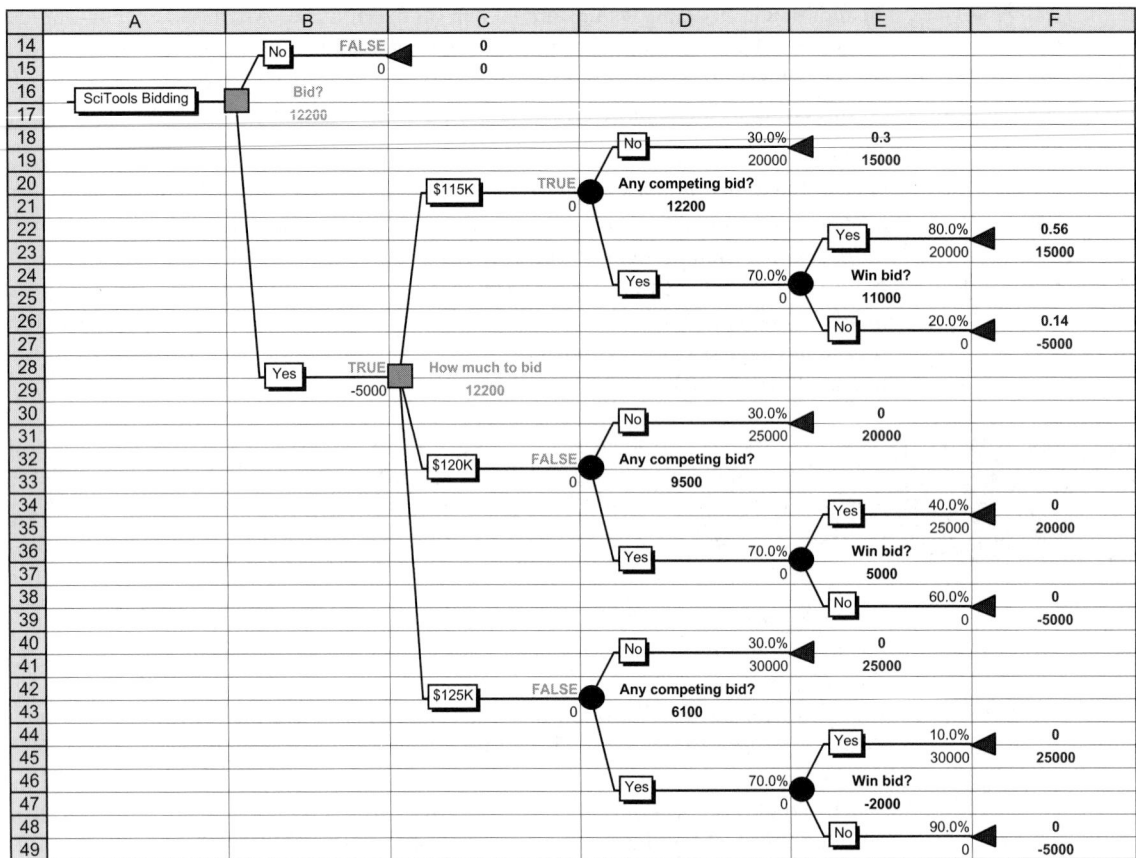

FIGURE 7.2 Result of Folding Back to Obtain Optimal Decision

is deciding how much to bid (after having already decided to place a bid). The EMVs under the three succeeding *probability* nodes are $12,200, $9500, and $6100. Since the maximum of these is $12,200, the EMV for the "How much to bid" node is $12,200 and is written below the node name.

After the folding-back process is completed—that is, after we have calculated EMVs for all nodes—we can trace the "True" labels from left to right to see the optimal strategy. In this case SciTools should place a bid, and it should be for $115,000. The EMV written below the leftmost decision node, $12,200, indicates SciTools' EMV for this strategy. If SciTools is truly willing to use the EMV criterion, that is, if it is willing to play the averages, then the company should be indifferent between receiving $12,200 *for certain* and bidding $115,000—with the associated risk of winning $15,000 or losing $5000.

PROBLEMS

Level A

1. The SweetTooth Candy Company knows it will need 10 tons of sugar 6 months from now to implement its production plans. Jean Dobson, SweetTooth's purchasing manager, has essentially two options for acquiring the needed sugar. She can either buy the sugar at the going market price when she needs it, 6 months from now, or she can buy a forward contract now. The contract guarantees delivery of the sugar in 6 months but the cost of purchasing it will be based on today's mar-

ket price. Assume that possible sugar forward contracts available for purchase are for 5 tons or 10 tons only. No forward contracts can be purchased or sold in the intervening months. Thus, SweetTooth's possible decisions are: (1) purchase a forward contract for 10 tons of sugar now, (2) purchase a forward contract for 5 tons of sugar now and purchase 5 tons of sugar in 6 months, or (3) purchase all 10 tons of needed sugar in 6 months. The price of sugar bought now for delivery in 6 months is $0.0851 per pound. The transaction costs for 5-ton and 10-ton forward contracts are $65 and $110, respectively. Finally, Ms. Dobson has assessed the probability distribution for the possible prices of sugar 6 months from now (in dollars per pound). The file P07_01.XLS contains these possible prices and their corresponding probabilities.

    a. Given that SweetTooth wants to acquire the needed sugar in the least costly way, formulate a payoff table that specifies the cost (in dollars) associated with each possible decision and possible sugar price in the future.

    b. Generate a risk profile for each of SweetTooth's possible decisions in this problem.

    c. Construct a decision tree to identify the course of action that minimizes SweetTooth's *expected* cost of meeting its sugar demand.

2. Carlisle Tire and Rubber, Inc. is considering expanding production to meet potential increases in the demand for one of its tire products. Carlisle's alternatives are to construct a new plant, expand the existing plant, or do nothing in the short run. The market for this particular tire product may expand, remain stable, or contract. Carlisle's marketing department estimates the probabilities of these market outcomes as 0.25, 0.35, and 0.40, respectively. The file P07_02.XLS contains Carlisle's estimated payoff (in dollars) table.

    a. Generate a risk profile for each of Carlisle's possible decisions in this problem.

    b. Construct a decision tree to identify the course of action that maximizes Carlisle's expected profit.

3. A local energy provider offers a landowner $180,000 for the exploration rights to natural gas on a certain site and the option for future development. This option, if exercised, is worth an additional $1,800,000 to the landowner, but this will occur only if natural gas is discovered during the exploration phase. The landowner, believing that the energy company's interest in the site is a good indication that gas is present, is tempted to develop the field herself. To do so, she must contract with local experts in natural gas exploration and development. The initial cost for such a contract is $300,000, which is lost forever if no gas is found on the site. If gas is discovered, however, the landowner expects to earn a net profit of $6,000,000. Finally, the landowner estimates the probability of finding gas on this site to be 60%.

    a. Formulate a payoff table that specifies the landowner's payoff (in dollars) associated with each possible decision and each outcome with respect to finding natural gas on the site.

    b. Generate a risk profile for each of the landowner's possible decisions in this problem.

    c. Construct a decision tree to identify the course of action that maximizes the landowner's expected gain (in dollars) from this opportunity.

4. Techware Incorporated is considering the introduction of two new software products to the market. In particular, the company has four options regarding these two proposed products: introduce neither product, introduce product 1 only, introduce product 2 only, or introduce both products. Research and development costs for products 1 and 2 are $180,000 and $150,000, respectively. Note that the first option entails no costs because research and development efforts have not yet begun. The success of these software products depends on the trend of the national economy in the coming year and on the consumers' reaction to these products. The company's revenues earned by introducing product 1 only, product 2 only, or both products in various states of the national economy are given in the file P07_04.XLS. The probabilities of observing a strong, fair, and weak trend in the national economy in the coming year are 0.30, 0.50, and 0.20, respectively.

    a. Formulate a payoff table that specifies Techware's profit (in dollars) for each possible decision and each outcome with respect to the trend in the national economy.

    b. Generate a risk profile for each of Techware's possible decisions in this problem.

    c. Construct a decision tree to identify the course of action that maximizes Techware's expected profit (in dollars) from these marketing opportunities.

5. Consider an investor with $10,000 available to invest. He has the following options regarding the allocation of his available funds: (1) he can invest in a risk-free savings account with a guaranteed 3% annual rate of return; (2) he can invest in a fairly safe stock, where the possible annual rates of return are 6%, 8%, or 10%; or (3) he can invest in a more risky stock where the possible annual rates of return are 1%, 9%, or 17%. Note that the investor can place all of his available funds in any one of these options, or he can split his $10,000 into two $5000 investments in any two of these options. The joint probability distribution of the possible return rates for the two stocks is given in the file P07_05.XLS.

    a. Formulate a payoff table that specifies this investor's earnings (in dollars) in one year for each possible decision and each outcome with respect to the two stock returns.

    b. Generate a risk profile for each of this investor's possible decisions in this problem.

c. Construct a decision tree to identify the course of action that maximizes this investor's expected earnings (in dollars) in one year from these investment opportunities.

6. A buyer for a large department store chain must place orders with an athletic shoe manufacturer 6 months prior to the time the shoes will be sold in the department stores. In particular, the buyer must decide on November 1 how many pairs of the manufacturer's newest model of tennis shoes to order for sale during the upcoming summer season. Assume that each pair of this new brand of tennis shoes costs the department store chain $45. Furthermore, assume that each pair of these shoes can then be sold to the chain's customers for $70. Any pairs of these shoes remaining unsold at the end of the summer season will be sold in a closeout sale next fall for $35 each. The probability distribution of consumer demand for these tennis shoes (in hundreds of pairs) during the upcoming summer season has been assessed by market research specialists and is provided in the file P07_06.XLS. Finally, assume that the department store chain must purchase these tennis shoes from the manufacturer in lots of 100 pairs.

    a. Formulate a payoff table that specifies the contribution to profit (in dollars) from the sale of the tennis shoes by this department store chain for each possible purchase decision (in hundreds of pairs) and each outcome with respect to consumer demand.

    b. Generate a risk profile for each of the buyer's possible decisions in this problem.

    c. Construct a decision tree to identify the buyer's course of action that maximizes the expected profit (in dollars) earned by the department store chain from the purchase and subsequent sale of tennis shoes in the coming year.

Level B

7. In designing a new space vehicle, NASA must decide whether to provide 0, 1, or 2 backup systems for a crucial component of the vehicle. The first backup system, if included, comes into use only if the original system fails. The second backup system, if included, comes into use only if the original system and the first backup system both fail. NASA engineers claim that each system, independently of the others, has a 1% chance of failing if called into use. Each backup system costs $70,000 to produce and install within the vehicle. Once the vehicle is in flight, the mission will be scrubbed only if the original system and all backups fail. The cost of a scrubbed mission, in addition to production costs, is assessed to be $8,000,000.

    a. Generate a risk profile for each of NASA's possible decisions in this problem.

    b. Construct a decision tree to identify the course of action that minimizes NASA's expected total cost in this case.

8. Mr. Maloy has just bought a new $30,000 sport utility vehicle. As a reasonably safe driver, he believes that there is only about a 5% chance of being in an accident in the coming year. If he is involved in an accident, the damage to his new vehicle depends on the severity of the accident. The probability distribution for the range of possible accidents and the corresponding damage amounts (in dollars) are given in the file P07_08.XLS. Mr. Maloy is trying to decide whether he is willing to pay $170 each year for collision insurance with a $300 deductible. Note that with this type of insurance, he pays the *first* $300 in damages if he causes an accident and the insurance company pays the remainder.

    a. Formulate a payoff table that specifies the cost (in dollars) associated with each possible decision and type of accident.

    b. Generate a risk profile for each of Mr. Maloy's possible decisions in this problem.

    c. Construct a decision tree to identify the course of action that minimizes Mr. Maloy's annual expected total cost, including the possible insurance premium, deductible payment, and damage payment.

9. The purchasing agent for a microcomputer manufacturer is currently negotiating a purchase agreement for a particular electronic component with a given supplier. This component is produced in lots of 1000, and the cost of purchasing a lot is $30,000. Unfortunately, past experience indicates that this supplier has occasionally shipped defective components to its customers. Specifically, the proportion of defective components supplied by this supplier is well described by the probability distribution given in the file P07_09.XLS. Although the microcomputer manufacturer can repair a defective component at a cost of $20 each, the purchasing agent is intrigued to learn that this supplier will now assume the cost of replacing defective components in excess of the first 100 faulty items found in a given lot. This guarantee may be purchased by the microcomputer manufacturer prior to the receipt of a given lot at a cost of $1000 per lot. The purchasing agent is interested in determining whether it is worthwhile for her company to purchase the supplier's guarantee policy.

    a. Formulate a payoff table that specifies the microcomputer manufacturer's total cost (in dollars) of purchasing and repairing (if necessary) a complete lot of components for each possible decision and each outcome with respect to the proportion of defective items.

    b. Generate a risk profile for each of the purchasing agent's possible decisions in this problem.

    c. Construct a decision tree to identify the purchasing agent's course of action that minimizes the expected total cost (in dollars) of achieving a complete lot of satisfactory components.

7.3 The PrecisionTree Add-In

Decision trees present a challenge for Excel. We must somehow take advantage of Excel's calculating capabilities (to calculate EMVs, for example) and its graphical capabilities (to depict the decision tree). Fortunately, there is now a powerful add-in, PrecisionTree developed by Palisade Corporation, that makes the process relatively straightforward. This add-in not only enables us to build and label a decision tree, but it performs the folding-back procedure automatically and then allows us to perform sensitivity analysis on key input parameters.

The first thing you must do to use PrecisionTree is to "add it in." The easiest way to do this is from the Windows Start button: navigate to Programs, then to the Palisade Decision Tools group, and select PrecisionTree. This can be done even if Excel is not running. In this case, Excel will be launched and PrecisionTree will be opened.[2] You'll know that PrecisionTree is ready for use when you see its toolbar (shown in Figure 7.3) and a PrecisionTree menu to the left of the Help menu. (You'll probably also see a "DTools" toolbar that allows you to switch between any of the add-ins in the Palisade suite.)

PrecisionTree is quite easy to use—at least its most basic items are—but it can be confusing at first. We'll lead you through the steps for the SciTools example. (The file SCITOOLS.XLS lists the inputs. You should work through the following steps on your own.)

Using PrecisionTree

1. **Inputs.** Check that the inputs are as shown in columns A and B of Figure 7.4. In fact, we strongly recommend *always* having an "inputs" section in your Excel decision models. This section should include the given numerical inputs from the problem statement. Then you can refer to these input cells in your decision tree.

FIGURE 7.3
PrecisionTree Toolbar

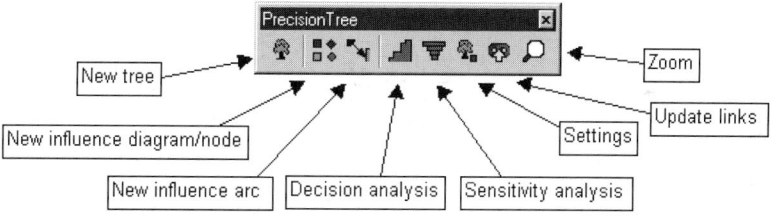

FIGURE 7.4
Inputs for SciTools Bidding Example

	A	B	C
1	**SciTools Bidding Example**		
2			
3	**Inputs**		
4	Cost to prepare a bid	$5,000	
5	Cost to supply instruments	$95,000	
6			**Range names**
7	Probability of no competing bid	0.3	BidCost: B4
8	Comp bid distribution (if they bid)		PrNoBid: B7
9	<$115K	0.2	ProdCost: B5
10	$115K to $120K	0.4	
11	$120K to $125K	0.3	
12	>$125K	0.1	

[2] The version of PrecisionTree on the CD-ROM limits the number of nodes total in any tree. See the Readme.htm file on the CD-ROM for details.

FIGURE 7.5
Beginnings of a
New Tree

	A	B
11	$120K to $125K	0.3
12	>$125K	0.1
13		
14	SciTools Bidding	1
15		0
16		

FIGURE 7.6
Dialog Box for
Adding a New De-
cision Node and
Branches

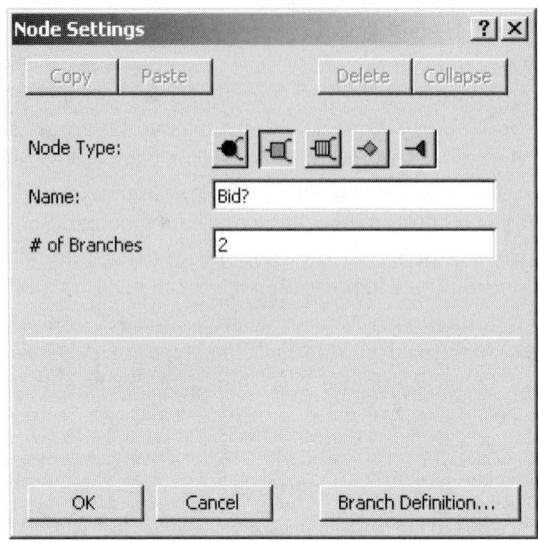

2. **New tree.** Click on the new tree button (the far left button) on the PrecisionTree tool-bar, and then click on any cell (say, cell A14) below the input section to start a new tree. Click on the name box of this new tree (it probably says "tree #1") to open a dialog box. Type in a descriptive name for the tree, such as SciTools Bidding, and click on OK. You should now see the beginnings of a tree, as shown in Figure 7.5.

3. **Decision nodes and branches.** From here on, keep the finished tree in Figure 7.2 in mind. This is the finished product toward which we're building. To obtain decision nodes and branches, click on the (only) triangle end node to open the dialog box in Figure 7.6. Click on the green square to indicate that this is a decision node, and fill in the dialog box as shown. We're calling this decision "Bid?" and specifying that there are two possible decisions. The tree expands as shown in Figure 7.7. The boxes that say "branch" show the default labels for these branches. Click on either of these boxes to open another dialog box where you can provide a more descriptive name for the branch. We label the two branches "No" and "Yes." Also, you can enter the immediate payoff/cost for either branch right below it. Since there is a $5000 cost of bidding, enter the formula

$$= -\text{BidCost}$$

FIGURE 7.7
Tree with Initial De-
cision Node and
Branches

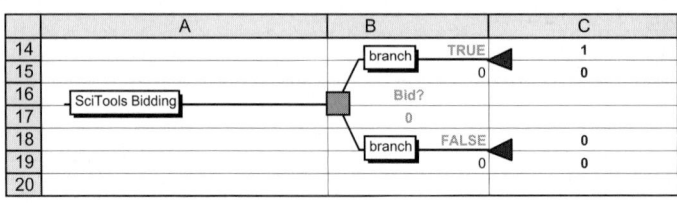

FIGURE 7.8

Decision Tree with
Decision Branches
Labeled

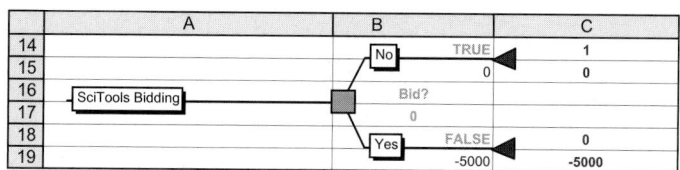

right below the "Yes" branch in cell B19. (It is negative to reflect a *cost.*) The tree should now appear as in Figure 7.8.

4. **More decision branches.** The top branch is completed; if SciTools does not bid, there is nothing left to do. If SciTools decides to bid, click on the bottom end node, and proceed as in the previous step to add and label the decision node and three decision branches for the amount to bid. (Refer to Figure 7.2.) The tree to this point should appear as in Figure 7.9. Note that there are no monetary values below these decision branches because no *immediate* payoffs or costs are associated with the bid amount decision.

5. **Probability nodes and branches.** We now need a probability node and branches from the rightmost end nodes to capture whether the competition bids. Click on the top one of these end nodes to bring up the same dialog box as in Figure 7.6. Now, however, click on the red circle box to indicate that this is a probability node. Label it "Any competing bid?", specify two branches, and click on OK. Then label the two branches "No" and "Yes." Next, repeat this procedure to form another probability node (with two branches) following the "Yes" branch, call it "Win bid?", and label its branches as shown in Figure 7.10 on page 314.

6. **Copying probability nodes and branches.** You could now repeat the same procedure from the previous step to build probability nodes and branches following the other bid amount decisions, but because they're structurally equivalent, you can save a lot of work by using PrecisionTree's copy and paste feature. Click on the leftmost probability node to open a dialog box (Figure 7.6) and click on Copy. Then click on either end node to bring up the same dialog box and click on Paste. Do this again with the other end node. Decision trees can get very "bushy," but this copy and paste feature can make them much less tedious to construct.

7. **Labeling probability branches.** You should now have the decision tree shown in Figure 7.11. It is structurally the same as the completed tree in Figure 7.2, but the probabilities and monetary values on the probability branches are not correct. Note that each probability branch has a value above and below the branch. The value above is the probability (the default values make the branches equally likely), and the value below is the monetary value (the default values are 0). We can enter any values or

FIGURE 7.9

Tree with All Deci-
sion Nodes and
Branches

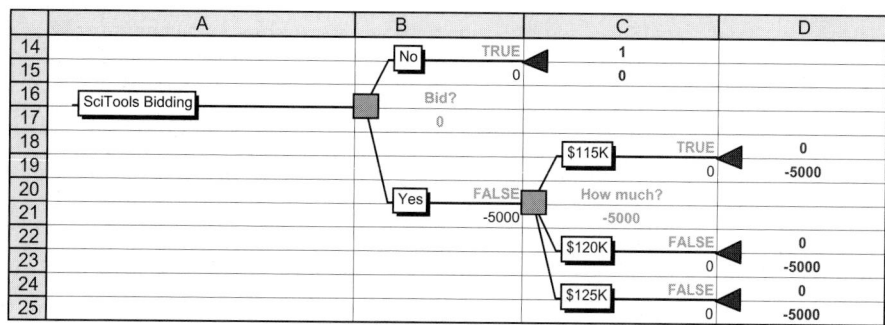

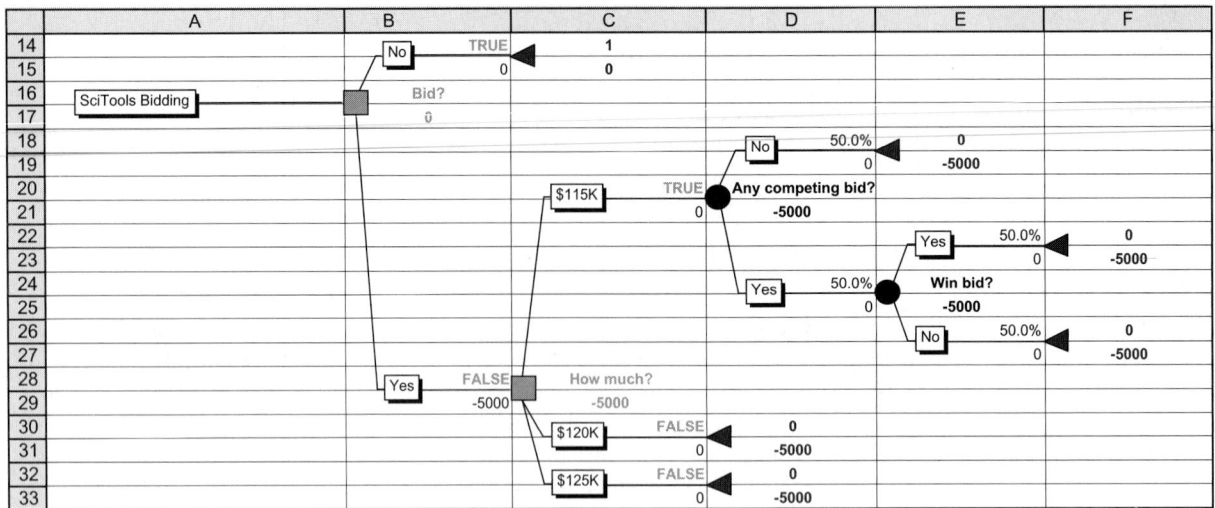

FIGURE 7.10 Decision Tree with One Set of Probability Nodes and Branches

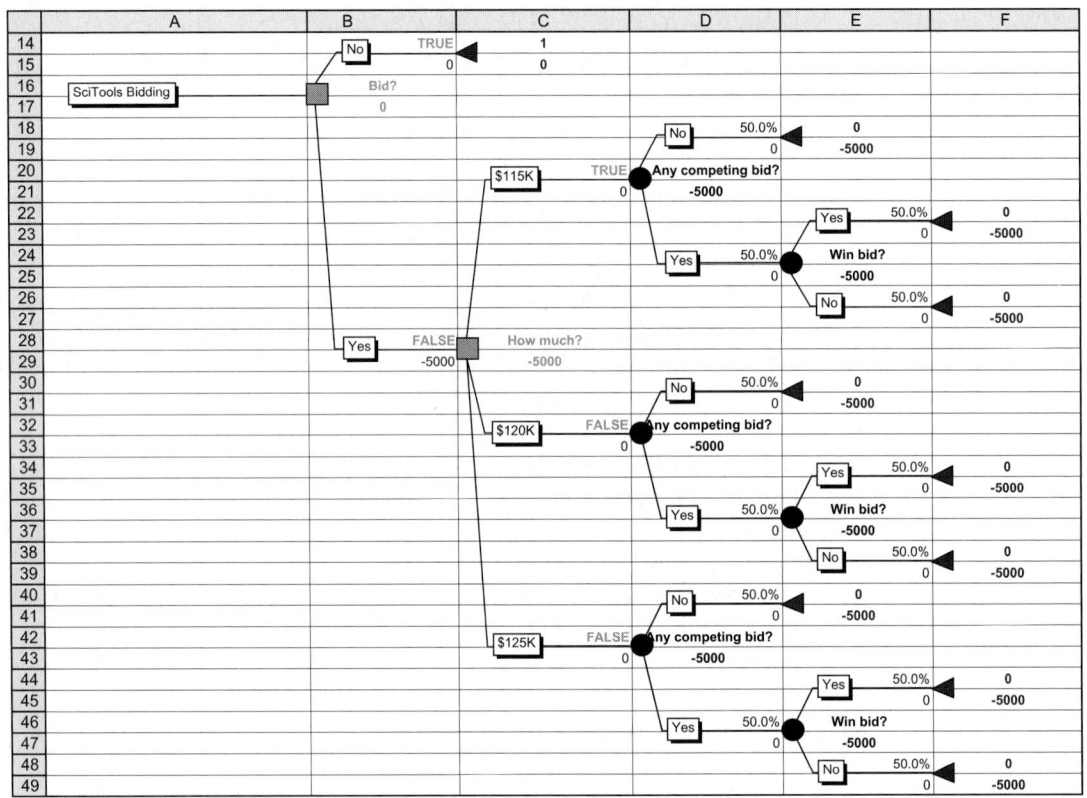

FIGURE 7.11 Structure of Completed Tree

formulas in these cells, exactly as we do in typical Excel worksheets. As usual, it is a good practice to refer to input cells in these formulas whenever possible. We'll get you started with the probability branches following the decision to bid $115,000. First, enter the probability of no competing bid in cell D18 with the formula

$$=PrNoBid$$

and enter its complement in cell D24 with the formula

$$=1-D18$$

Next, enter the probability that SciTools wins the bid in cell E22 with the formula

$$=SUM(B10:B12)$$

and enter its complement in cell E26 with the formula

$$=1-E22$$

(Remember that SciTools wins the bid only if the competitor bids higher, and in this part of the tree, SciTools is bidding $115,000.) For the monetary values, enter the formula

$$=115000-ProdCost$$

in the two cells, D19 and E23, where SciTools wins the contract. Because we already subtracted the cost of the bid (cell B29), we shouldn't do so again. This would be double-counting, and it should always be avoided in decision problems.

8. **Enter the other formulas on probability branches.** Using the previous step and Figure 7.2 as a guide, enter formulas for the probabilities and monetary values on the other probability branches, that is, those following the decision to bid $120,000 or $125,000. Note that PrecisionTree doesn't "help" ensure that the probabilities emanating from any probability node sum to 1. You have to make sure that they do. If they do not, you see errors in the decision tree.

We're finished! The completed tree in Figure 7.2 shows the best strategy and its associated EMV, as we discussed previously. Note that we never have to perform the folding-back procedure manually. PrecisionTree does it for us. In fact, the tree is completed as soon as we finish entering the relevant inputs. In addition, if we change any of the inputs, the tree reacts automatically. For example, try changing the bid cost in cell B4 from $5000 to some large value such as $20,000. You'll see that the tree calculations update automatically, and the best decision is then *not* to bid, with an associated EMV of $0.

7.3.1 Risk Profile of Optimal Strategy

Once the decision tree is completed, we can use PrecisionTree's tools for decision analysis. We can obtain, for example, a risk profile and other information about the *optimal* decision. To do so, click on the fourth button from the left on the PrecisionTree toolbar (it looks like a staircase) and fill in the resulting dialog box as shown in Figure 7.12 on page 316. (You can experiment with other options. The Policy Suggestion option, which allows us to see only that part of the tree that corresponds to the best decision, might not be available in the academic version of PrecisionTree.)

The Risk Profile option presents a graphical risk profile of the optimal decision. The Statistics Report shows this information numerically. In either form, we see the probability distribution of payoffs when using the optimal decision. As the statistics report in

FIGURE 7.12
Dialog Box for Information About Optimal Decision

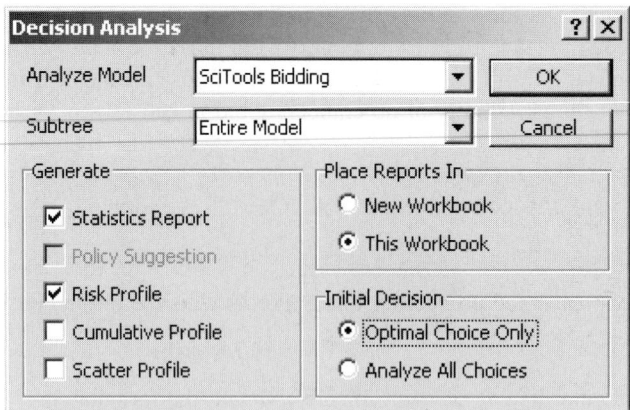

FIGURE 7.13
Statistics Report for Optimal Decision

	A	B	C	D
1	PrecisionTree Statistics Report			
2	For SciTools Bidding of SciTools.xls			
3	Created on 8/25/2001 at 11:29:49 AM			
4				
5				
6				
7	STATISTICS			
8	Mean	12200		
9	Minimum	-5000		
10	Maximum	15000		
11	Mode	15000		
12	Std Dev	6939.741		
13	Skewness	-2.075006		
14	Kurtosis	5.305648		
15				
16	PROFILE:			
17	#	X	P	
18	1	-5000	0.14	
19	2	15000	0.86	

Figure 7.13 and the risk profile in Figure 7.14 show, there are only two possible monetary outcomes if SciTools bids $115,000. It either earns $15,000 or loses $5000, and the former is much more likely. (The associated probabilities are 0.86 and 0.14.) This graphical information is even more useful when there are a larger number of possible monetary outcomes. We can see what they are and how likely they are.

7.3.2 Sensitivity Analysis

We have already stressed the importance of a follow-up sensitivity analysis for any decision problem, and PrecisionTree makes this relatively easy to perform. First, we can enter any values into the input cells and watch how the tree changes. But we can get more systematic information by clicking on PrecisionTree's sensitivity button, the fifth from the left on the toolbar (it looks like a tornado). This brings up the dialog box in Figure 7.15. It requires an EMV cell (and an optional descriptive name) to analyze at the top and one or more input cells in the middle. The specifications for these input cells are actually entered at the bottom of the dialog box.

FIGURE 7.14
Risk Profile of Opti-
mal Decision

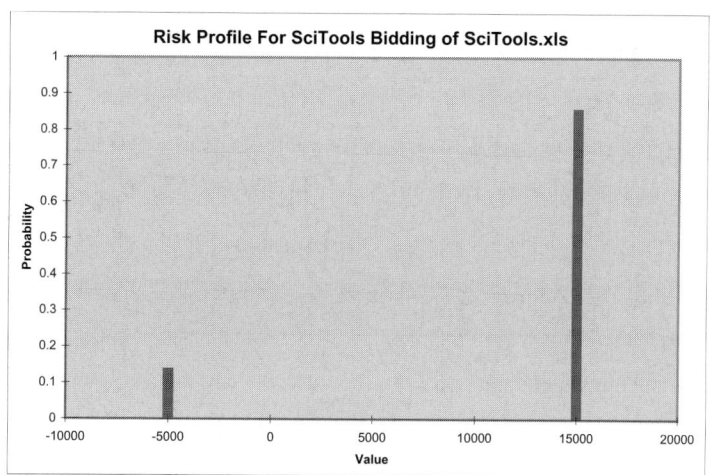

FIGURE 7.15
Sensitivity Analysis
Dialog Box

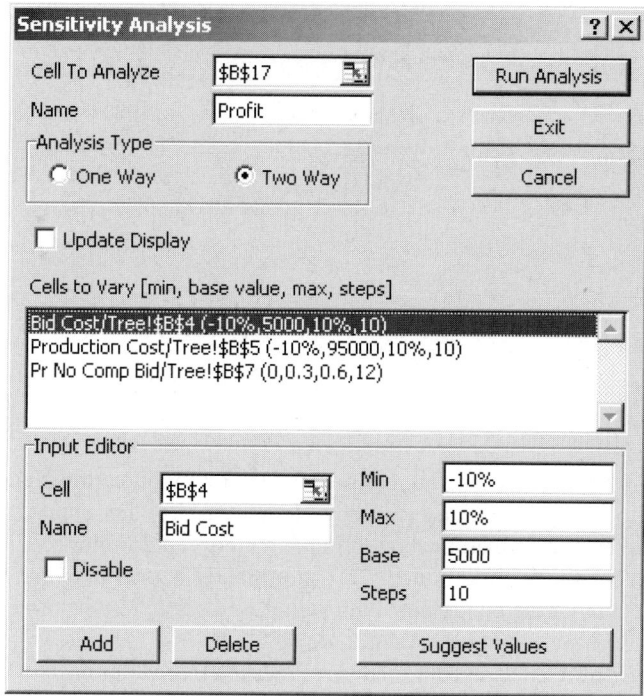

The cell to analyze (at the top) is usually the EMV cell at the far left of the decision tree—this is the cell shown in the figure—but it can be any EMV cell. For example, if we *assume* SciTools will prepare a bid and we want to see how sensitive the EMV from that point on is to inputs, we could select cell C29 (refer to Figure 7.2) to analyze. Next, for any input cell such as the production cost cell (B5), we enter a minimum value, a maximum value, a base value (probably the original value in the model), and a step size. For example, to specify these for the production cost, we clicked on the Suggest Values button. This default setting varies the production cost by as much as 10% from the original value in either direction in a series of 10 steps. We can also enter our own desired values. We did so for the probability of no competing bids, varying its value from 0 to 0.6 in a sequence of 12 steps.

FIGURE 7.16
EMV versus Production Cost for the
Optimal Decision

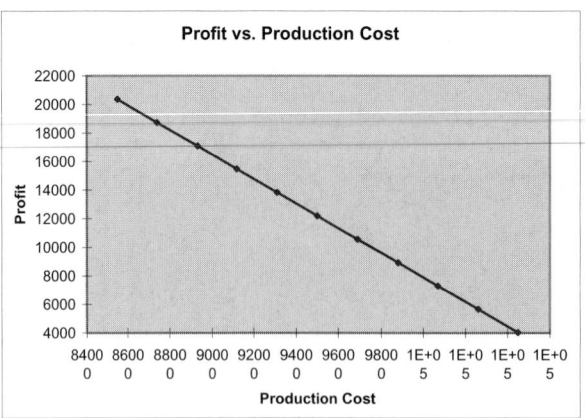

FIGURE 7.17
Tornado Chart for
SciTools Example

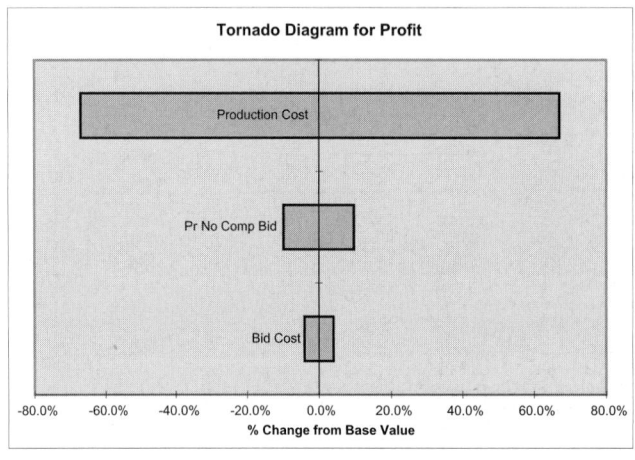

We have found the dialog box in Figure 7.15 a bit confusing to fill in, because we are used to filling in from top to bottom. In contrast, for any input cell you want to vary, you should enter information about it in the bottom Input Editor section of the dialog box and then click on the Add button. This information is then *automatically* placed in the middle section of the dialog box. Although it is possible to type directly into this middle section, we have found it best not to do so.

When we click on Run Analysis, PrecisionTree varies each of the specified inputs (one at a time if we select the One Way option) and presents the results in several ways in a *new* Excel file with Sensitivity, Tornado, and Spider Graph sheets. The Sensitivity sheet includes several charts, a typical one of which appears in Figure 7.16. This shows how the EMV varies with the production cost for the optimal decision. Sometimes the optimal decision changes over the specified range of the input parameter. This often shows up as a "kink" in the sensitivity chart. For example, if the graph showed a downward-sloping line and suddenly flattened out at 0, we would suspect that a change in the optimal decision occurred at that point. (No such behavior can be seen in Figure 7.16.)

The Tornado sheet shows how sensitive the EMV of the *optimal* decision is to each of the selected inputs over the ranges selected. (See Figure 7.17.) The length of each bar shows the percentage change in the EMV in either direction, so the longer the bar, the more sensitive this EMV is to the particular input. The bars are always arranged from longest on top to shortest on the bottom—hence the name *tornado* chart. Here we see that production cost has the largest effect on EMV, and bid cost has the smallest effect.

You can copy the sheets in this new file back to the original file if you like. To do so, use Excel's Edit/Move or Copy Sheet menu item.

FIGURE 7.18
Spider Chart for
SciTools Example

FIGURE 7.18
Spider Chart for
SciTools Example

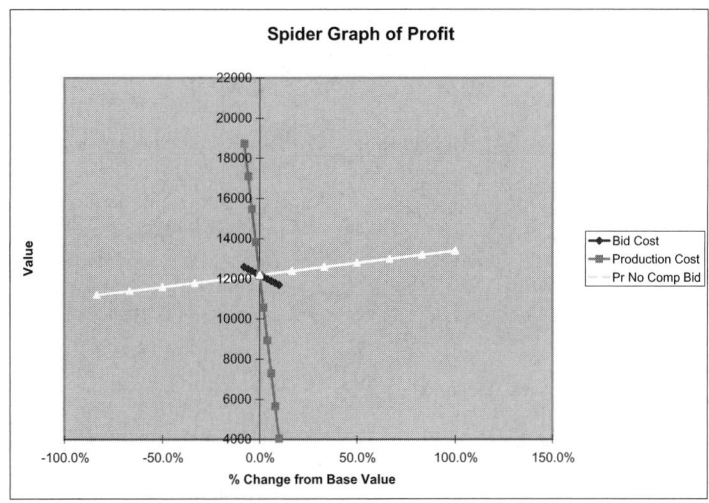

FIGURE 7.19
Two-Way Sensitivity
Analysis

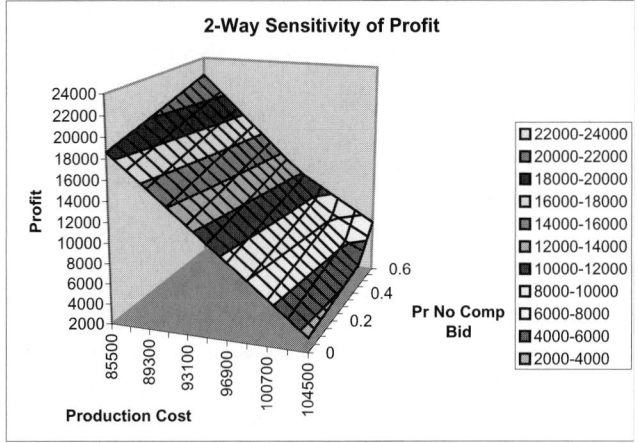

Finally, the Spider Chart sheet contains the chart in Figure 7.18. It shows how much the optimal EMV varies in magnitude for various percentage changes in the input variables. The steeper the slope of the line, the more the EMV is affected by a particular input. We again see that the production cost has a relatively large effect, whereas the other two inputs have relatively small effects.

Each time we click on the sensitivity button, we can run a different sensitivity analysis. An interesting option is to run a two-way analysis (by clicking on the Two Way button in Figure 7.15). Then we see how the selected EMV varies as each *pair* of inputs vary simultaneously. We analyzed the EMV in cell B17 with this option, using the same inputs as before. A typical result appears in Figure 7.19. This chart shows how the EMV from the optimal decision varies as the production cost and the probability of no competing bid *both* vary over their selected ranges. (The legend—and color coding on the Excel chart—indicates the values of EMV.)

We reiterate that a sensitivity analysis is always an important aspect in real decision analyses. If we had to construct decision trees by hand—with paper and pencil—a sensitivity analysis would be virtually out of the question. We would have to recompute everything each time through. Therefore, one of the most valuable features of the PrecisionTree add-in is that it enables us to perform sensitivity analyses in a matter of seconds.

Level A

10. Consider again SweetTooth Candy Company's decision problem described in Problem 1. Use the PrecisionTree add-in to identify the strategy that minimizes SweetTooth's expected cost of meeting its sugar demand. Also, perform sensitivity analysis on the optimal decision and summarize your findings. In response to which model inputs is the expected cost value most sensitive?

11. Consider again Carlisle Tire and Rubber's decision problem described in Problem 2. Use the PrecisionTree add-in to identify the strategy that maximizes this tire manufacturer's expected profit. Also, perform sensitivity analysis on the optimal decision and summarize your findings. In response to which model inputs is the expected profit value most sensitive?

12. Consider again the landowner's decision problem described in Problem 3. Use the PrecisionTree add-in to identify the strategy that maximizes the landowner's expected net earnings from this opportunity. Also, perform sensitivity analysis on the optimal decision and summarize your findings. In response to which model inputs is the expected net earnings value most sensitive?

13. Consider again Techware's decision problem described in Problem 4. Use the PrecisionTree add-in to identify the strategy that maximizes Techware's expected profit from the given marketing opportunities. Also, perform sensitivity analysis on the optimal decision and summarize your findings. In response to which model inputs is the expected profit value most sensitive?

14. Consider again the investor's decision problem described in Problem 5. Use the PrecisionTree add-in to identify the strategy that maximizes the investor's expected earnings in one year from the given investment opportunities. Also, perform sensitivity analysis on the optimal decision and summarize your findings. In response to which model inputs is the expected earnings value most sensitive?

15. Consider again the department store buyer's decision problem described in Problem 6. Use the PrecisionTree add-in to identify the strategy that maximizes the department store chain's expected profit earned by purchasing and subsequently selling pairs of the new tennis shoes. Also, perform sensitivity analysis on the optimal decision and summarize your findings. In response to which model inputs is the expected profit value most sensitive?

Level B

16. Consider again NASA's decision problem described in Problem 7. Use the PrecisionTree add-in to identify the strategy that minimizes NASA's expected total cost. Also, perform sensitivity analysis on the optimal decision and summarize your findings. In response to which model inputs is the expected total cost value most sensitive?

17. Consider again Mr. Maloy's decision problem described in Problem 8. Use the PrecisionTree add-in to identify the strategy that minimizes Mr. Maloy's annual expected total cost. Also, perform sensitivity analysis on the optimal decision and summarize your findings. In response to which model inputs is the expected total cost value most sensitive?

18. Consider again the purchasing agent's decision problem described in Problem 9. Use the PrecisionTree add-in to identify the strategy that minimizes the expected total cost of achieving a complete lot of satisfactory microcomputer components. Also, perform sensitivity analysis on the optimal decision and summarize your findings. In response to which model inputs is the expected total cost value most sensitive?

 # More Single-Stage Examples

All applications of decision making under uncertainty follow the procedures discussed so far. We first identify the possible decision alternatives, assess relevant probabilities, and calculate monetary values. Then we use a decision tree (or influence diagram) to identify the alternative with the largest EMV and follow this up with a thorough sensitivity analysis. We can also examine the risk profiles for the various alternatives. This is particularly useful if criteria other than straight EMV maximization are considered, as we will discuss in Section 7.7. In this section we will illustrate the process with several single-stage examples, where the decision maker makes one decision and then learns which of several uncertain outcomes occurs. In this next section we will examine multistage examples, where two or more sequential decisions must be made.

The following example illustrates a decision problem most of us face on an annual basis, although most of us probably don't go to the trouble of analyzing it formally.

Example 7.2 Choosing an Employee's Health Insurance Payment Plan

Each year employees at State University are asked to decide on one of three health care plans. The terms of these are as follows:[3]

Plan 1: The monthly cost is $24. There is a $500 deductible. The participant pays all expenses until payments for the year equal $500. After that, 90% of remaining expenses are paid by the insurer.

Plan 2: This is the same as plan 1, except that the monthly cost is $1 and the deductible amount is $1000.

Plan 3: The monthly cost is $20. There is no deductible. The employee pays 30% of all medical expenses. The rest is paid by the insurer.

Which of these three plans should an employee choose?

Objective To use a decision tree model to identify the optimal health care premium plan that minimizes an employee's expected cost.

Solution

Clearly, the solution will vary from one employee to another, depending on the assessed probability distribution of medical expenses. To illustrate, however, we will consider an employee who assesses the distribution of yearly medical expenses shown in Table 7.5. These expenses include hospital visits, surgery, office visits, and prescriptions, all of which are covered under the terms of the plans. As in the previous example, this distribution is only an approximation of the real distribution, which would contain a continuum of expenses. However, it is adequate for making a decision among the three plans.

Table 7.5

Distribution of Medical Expenses for Insurance Example	Total Medical Expense	Probability
	$200	0.30
	$600	0.50
	$1000	0.15
	$5000	0.03
	$15,000	0.02

The next step is to determine the employee's cost for each plan and each outcome. For example, suppose that the employee chooses plan 1 and incurs $600 in expenses.

---

[3]We assume that these terms apply only to the employee; that is, these are not family plans.

Then the total cost is the cost of the insurance plus the full amount of the first $500 in expenses plus 10% of the last $100 in expenses, that is,

$$24(12) + 500 + 0.1(100) = \$798$$

However, if this employee's medical expenses are only $200, then the cost is

$$24(12) + 200 = \$488$$

The costs for the other plans and other outcomes can be calculated in a similar manner. We list all of the costs in Table 7.6.

Table **7.6**

Employee Cost Table for Insurance Example	Medical Expense	Plan 1	Plan 2	Plan 3
	$200	$488	$212	$300
	$600	$798	$612	$420
	$1000	$838	$1012	$540
	$5000	$1238	$1412	$1740
	$15,000	$2238	$2412	$4740

The choice is certainly not clear from this table. The plan with the lowest premium, plan 2, looks good if the year's medical expenses are low. This is also true for the no-deductible plan, plan 3, although its cost is quite large in case of a disaster. For moderate medical expenses, plan 1 is obviously inferior, but it is the best for guarding against a disaster. These trade-offs could be illustrated by risk profiles, which you might want to examine. Instead, we turn directly to the decision tree.

Using PrecisionTree

1 **Inputs.** Enter the inputs for the three plans and the probabilities from Table 7.5 in the top left portion of the spreadsheet (down to row 15). (See Figure 7.20 and the file MEDICAL.XLS.)

A cost table is more natural than a pay-off table in this example because all monetary values are costs.

2 **Cost table.** For later use in the decision tree, calculate the costs to the employee (not counting insurance premiums) in the range B19:D23. To do this, enter the formula

$$=IF(\$A19<=B\$6,\$A19,B\$6+B\$7*(\$A19-B\$6))$$

in cell B19 and copy this to the range B19:D23. This IF function says that if the medical expense is less than the deductible, the employee pays it all. Otherwise, the employee pays the deductible amount plus a percentage of the remainder.

3 **Decision tree.** Use PrecisionTree to create the decision tree shown in Figure 7.21 (page 324). Here are some tips. First, create the decision node and decision branches, and enter formulas for their values as 12 times the relevant monthly premiums. Then create a single probability node and its branches, label the branches, and enter formulas for the probabilities with *absolute* references. For example, enter the formula

$$=\$C\$11$$

for the probability of the top branch. Next, copy the probability node to the end nodes below it. (Do you see the effect of the absolute references?) Finally, link the values for all of the probability branches to the cells in the cost table. (We know of no quick way

FIGURE 7.20

Inputs and Cost
Table for Medical
Example

	A	B	C	D
1	**Medical insurance problem**			
2				
3	Inputs for plans			
4		Plan1	Plan2	Plan3
5	Monthly cost	$24	$1	$20
6	Deductible	$500	$1,000	$0
7	Copay Pct	10%	10%	30%
8				
9	Distribution of medical expenses			
10		Expense	Prob	
11		$200	0.3	
12		$600	0.5	
13		$1,000	0.15	
14		$5,000	0.03	
15		$15,000	0.02	
16				
17	Out of pocket cost table (plan along top, expense along side), not including premiums			
18		Plan1	Plan2	Plan3
19	$200	$200	$200	$60
20	$600	$510	$600	$180
21	$1,000	$550	$1,000	$300
22	$5,000	$950	$1,400	$1,500
23	$15,000	$1,950	$2,400	$4,500

to do this. We entered 15 separate formulas, one for each branch. However, it is much easier to create a cost table and link branch formulas to it than to create the branch formulas directly from input values.)

❹ Minimize costs. If we quit here, we would mistakenly choose the *worst* of the three plans. This is because PrecisionTree *maximizes* EMV by default, and in this problem we want to *minimize* the EMV of the costs. However, this is simple to change. Click on the name box at the far left in the decision tree. This brings up a dialog box (not shown here) where we can select the Minimize option.

Remember to check
the Minimize option
if you really want to
minimize. It's easy
to forget!

As we see from Figure 7.21, the optimal plan is plan 3. Its EMV—an expected *cost*—is $528. The EMVs for plans 1 and 2 are $753 and $612, respectively. Evidently, this employee's chances of large medical expenses (where plan 3 is at its worst) are not large enough to outweigh plan 3's no-deductible benefit. However, we might want to experiment with various inputs, either the properties of the plans or the employee's medical expense distribution, to see whether plan 3 continues to be the preferred plan. For example, if the probabilities in Table 7.5 change to 0.30, 0.40, 0.15, 0.10, and 0.05, so that large expenses are more probable, the EMVs for the three plans become $827, $722, and $750. Now plan 2 is preferred, although the difference in EMV between plans 2 and 3 is quite small.

We can use this insurance example to illustrate one *nonmonetary* aspect of decision problems that is difficult to incorporate into a decision tree. At the university where we teach, there is another insurance plan in addition to the types in the example. Its premiums are low, and there are *no* copayments—the insurer pays all medical expenses. This plan is clearly the cheapest of all plans offered, but it is not chosen by many employees. Why? The plan is through an HMO, where all employees must go to a specified set of physicians; otherwise, the plan does not pay their expenses. Evidently, many employees believe that the "cost" of having to go to physicians they would not choose otherwise outweighs the dollar savings from the plan.

FIGURE 7.21
Decision Tree for
Medical Insurance
Example

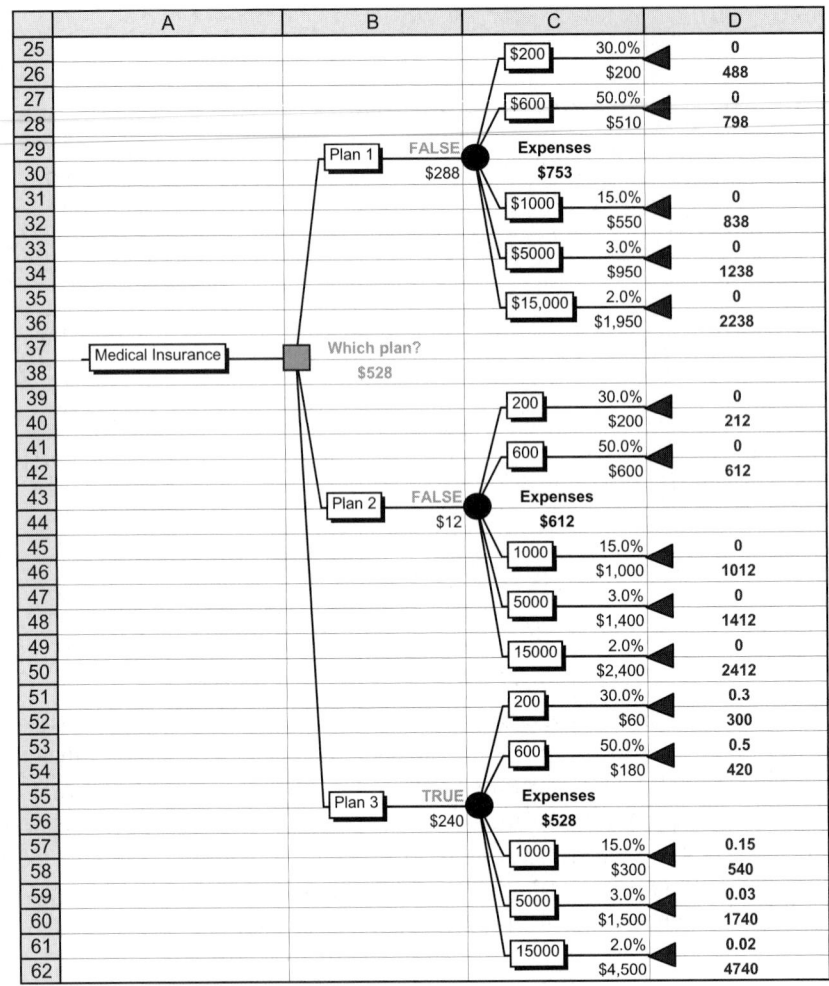

The following example illustrates one method for using a *continuous* probability distribution in a decision tree model.

Example 7.3 Selecting a Supplier of Lightbulbs for FreshWay

FreshWay, a chain of supermarkets, requires 24,000 fluorescent lightbulbs for its stores. There are two suppliers of these lightbulbs. Supplier A offers them at $4.00 per bulb and will replace the first 900 defective bulbs with guaranteed good ones for $3.00 each. It will replace all defectives after the first 900 for nothing. Supplier B is similar. It will charge $4.15 per bulb, replace the first 1200 defectives for $1.00 each, and replace all defectives after the first 1200 for nothing. FreshWay plans to sell these lightbulbs for $4.40 apiece and charge its customers nothing for replacement of defectives. The only uncertainty is the number of defective bulbs from either supplier. Based on historical data from each supplier, FreshWay believes that the percentage of defectives is normally distributed with mean 4% and standard deviation 1% from supplier A, and mean 4.2% and standard deviation 1.2% from supplier B. Which supplier should be chosen to maximize FreshWay's EMV?

FIGURE 7.22
Midpoints of Five
Equal Probability
Regions

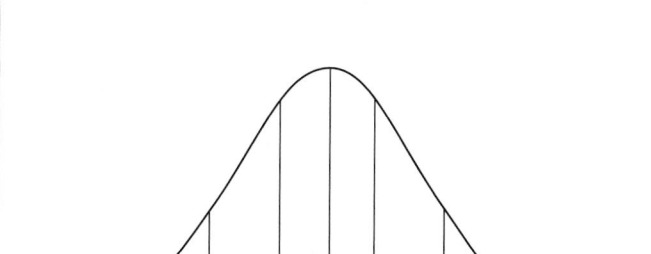

Critical Percentiles for FreshWay Illustrative Calculations

Objective To use a decision tree model to identify the lightbulb supplier that maximizes FreshWay's expected monetary value.

Solution

Let p be the percentage of lightbulbs that are defective. Then the profit to FreshWay from buying from supplier A is

$$\text{Profit} = \begin{cases} 24{,}000(4.40 - 4.00) - (24{,}000p)(3.00) & \text{if } p \le 900/24{,}000 \\ 24{,}000(4.40 - 4.00) - (900)(3.00) & \text{if } p > 900/24{,}000 \end{cases}$$

A similar expression holds for supplier B. The only random quantity in this expression is p, which is normally distributed. The question is how we can model the *continuous* distribution of p in a *discrete* decision tree—that is, a tree with a discrete number of probability branches. The method usually used is to approximate the continuous normal distribution by a discrete distribution with a relatively small, say 5, number of equally likely values.

The idea is to divide the normal distribution into an equal number of equal probability regions and take the midpoint (in a probability sense) of each region as a value for the decision tree. For example, if we use five points, then each region has probability 0.2. The probability halfway between 0 and 0.2 is 0.1, so the first point on the tree is the 10th percentile of the normal distribution. Similarly, the next point is the 30th percentile, the next is the 50th, the next is the 70th, and the last is the 90th. A graphical display of these five midpoints is provided in Figure 7.22.

Figure 7.23 (page 326) illustrates the calculations. (See the file LIGHTBULB.XLS.) Through row 13 we enter the given inputs for the problem. Then in rows 17–26 we enter the information we'll use in the decision tree regarding the percentage defective for each supplier. This information is based on the five-point approximation to the normal distribution. For example, the 10th percentile of the normal distribution for supplier A is found in cell C17 with the formula

$$=\text{NORMINV(B17,\$B\$12,\$C\$12)}$$

and this is copied down to cell C21. Then the cost to FreshWay from defectives, assuming the value in C17 is the percentage of defectives, is calculated in cell D17 with the formula

$$=\text{\$C\$7*IF(C17<=\$D\$7/Quantity,Quantity*C17,\$D\$7)}$$

and it is copied down to cell D21. Similar formulas are used for supplier B.

FIGURE 7.23
Inputs and Calculations for Lightbulb Example

	A	B	C	D
1	**FreshWay lightbulb purchasing example**			
2			Range names	
3	Quantity	24000	Quantity: B3	
4	Selling price	$4.40	SellingPrice: B4	
5				
6		UnitCost	ReplaceCost	Charge for first:
7	Supplier A	$4.00	$3.00	900
8	Supplier B	$4.15	$1.00	1200
9				
10	Distribution of percent defective: normal			
11		Mean	Stdev	
12	Supplier A	4.0%	1.0%	
13	Supplier B	4.2%	1.2%	
14				
15	Percentages to use on decision tree			
16		Midpoint probability	Percentile	FreshWay's cost
17	Supplier A	0.1	2.72%	$1,957.28
18		0.3	3.48%	$2,502.43
19		0.5	4.00%	$2,700.00
20		0.7	4.52%	$2,700.00
21		0.9	5.28%	$2,700.00
22	Supplier B	0.1	2.66%	$638.91
23		0.3	3.57%	$856.97
24		0.5	4.20%	$1,008.00
25		0.7	4.83%	$1,159.03
26		0.9	5.74%	$1,200.00

FIGURE 7.24
Decision Tree for Lightbulb Example

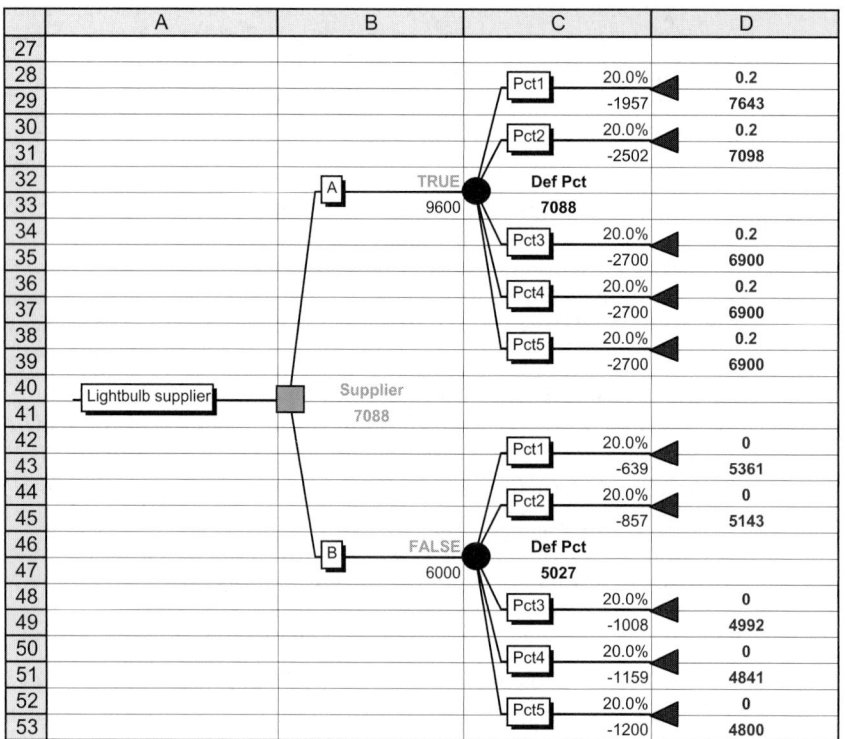

FIGURE 7.25
Tornado Chart to Analyze the EMV for Supplier B

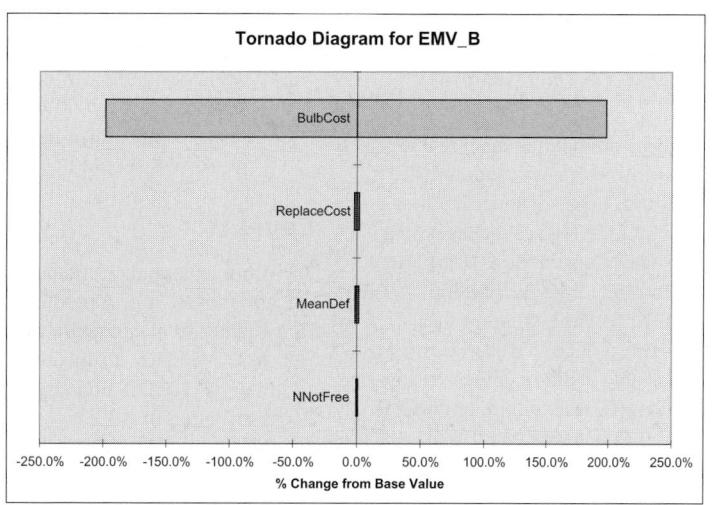

It is then straightforward to construct the decision tree shown in Figure 7.24. We have entered the revenue from selling the bulbs and the cost of purchasing them in cells B33 and B47. For example, the formula in cell B33 is

$$=Quantity*(SellingPrice-B7)$$

Then we have linked the monetary values below the probability branches to the relevant cells in the D17:D26 range.

The EMVs for suppliers A and B are $7088 and $5027, so supplier A is the clear choice. Evidently, the higher price charged by supplier B and its slightly higher mean percentage of defects outweigh its better deal on replacing defectives. Of course, if supplier B really wants to get FreshWay's business, it could attempt to sweeten its deal in a number of ways. Sensitivity analysis is useful to see how the EMV for supplier B (in cell C47) is affected by the various input parameters. We tried this, varying the inputs in cells B8, C8, D8, and B13 by PrecisionTree's default values (10% in either direction) and keeping track of the change in the EMV for supplier B. The tornado chart in Figure 7.25 makes it very clear that the most important input is the unit purchase cost. The effects of the other three inputs are practically negligible in comparison. If supplier B wants Fresh-Way's business, it will have to lower its unit purchase cost. ●

Modeling Issue The discrete approximation used in Example 7.3 can be used in *any* decision tree with continuous probability distributions, regardless of whether they are normal. We first need to decide how many values to have in the discrete approximation. The usual choices are 5 or 3. (Surprisingly, a three-point approximation does an adequate job in many situations.) Then we need to use the "inverse" function—in the previous example it was the NORMINV function—to find the values to use in the decision tree. The appropriate inverse function is available in Excel for a number of widely used continuous distributions.

PROBLEMS

Level A

19. Each day the manager of a local bookstore must decide how many copies of the community newspaper to order for sale in her shop. She pays the newspaper's publisher $0.40 for each copy and sells the newspapers to local residents for $0.50 each. Newspapers that are unsold at the end of day are considered worthless. The probability distribution of the number of copies of the newspaper purchased daily at her shop is provided in the file P07_19.XLS. Employ a decision tree to find the book store manager's profit-maximizing daily order quantity.

20. Two construction companies are bidding against one another for the right to construct a new community center building in Lewisburg, Pennsylvania. The first construction company, Fine Line Homes, believes that its competitor, Buffalo Valley Construction, will place a bid for this project according to the distribution shown in the file P07_20.XLS. Furthermore, Fine Line Homes estimates that it will cost $160,000 for its own company to construct this building. Given its fine reputation and long-standing service within the local community, Fine Line Homes believes that it will likely be awarded the project in the event that it and Buffalo Valley Construction submit exactly the same bids. Employ a decision tree to identify Fine Line Homes' profit-maximizing bid for the new community center building.

21. Suppose that you have sued your employer for damages suffered when you recently slipped and fell on an icy surface that should have been treated by your company's physical plant department. Specifically, your injury resulting from this accident was sufficiently serious that you, in consultation with your attorney, decided to sue your company for $500,000. Your company's insurance provider has offered to settle this suit with you out of court. If you decide to reject the settlement and go to court, your attorney is confident that you will win the case but is uncertain about the amount the court will award you in damages. He has provided his assessment of the probability distribution of the court's award to you in the file P07_21.XLS. Let S be the insurance provider's proposed out-of-court settlement (in dollars). For which values of S will you decide to accept the settlement? For which values of S will you choose to take your chances in court? Of course, you are seeking to maximize the expected payoff from this litigation.

22. Suppose that one of your colleagues has $2000 available to invest. Assume that all of this money must be placed in one of three investments: a particular money market fund, a certain stock, or gold. Each dollar your colleague invests in the money market fund earns a virtually guaranteed 12% annual return. Each dollar he invests in the stock earns an annual return characterized by the probability distribution provided in the file P07_22.XLS. Finally, each dollar he invests in gold earns an annual return characterized by the probability distribution given in this same file.
   a. If your colleague must place all of his available funds in a single investment, which investment should he choose to maximize his expected earnings over the next year?

   b. Suppose now that your colleague can place all of his available funds in one of these three investments as before, or he can invest $1000 in one alternative and $1000 in another. Assuming that he seeks to maximize his expected total earnings in 1 year, how should he allocate his $2000?

Level B

23. A home appliance company is interested in marketing an innovative new product. The company must decide whether to manufacture this product essentially on its own or employ a subcontractor to manufacture it. The file P07_23.XLS contains the estimated probability distribution of the cost of manufacturing one unit of this new product (in dollars) under the alternative that the home appliance company produces the item on its own. The same file contains the estimated probability distribution of the cost of purchasing one unit of this new product (in dollars) under the alternative that the home appliance company commissions a subcontractor to produce the item.
   a. Assuming that the home appliance company seeks to minimize the expected unit cost of manufacturing or buying the new product, should the company make the new product or buy it from a subcontractor?
   b. Perform sensitivity analysis on the optimal expected cost. Under what conditions, if any, would the home appliance company select an alternative different from the one you identified in part a?

24. A grapefruit farmer in central Florida is trying to decide whether to take protective action to limit damage to his crop in the event that the overnight temperature falls to a level well below freezing. He is concerned that if the temperature falls sufficiently low and he fails to make an effort to protect his grapefruit trees, he runs the risk of losing his entire crop, which is worth approximately $75,000. Based on the latest forecast issued by the National Weather Service, the farmer estimates that there is a 60% chance that he will lose his entire crop if it is left unprotected. Alternatively, the farmer can insulate his fruit by spraying water on all of the trees in his orchards. This action, which would likely cost the farmer C dollars, would prevent total devastation but might not completely protect the grapefruit trees from incurring some damage as a result of the unusually cold overnight temperatures. The file P07_24.XLS contains the assessed distribution of possible damages (in dollars) to the insulated fruit in light of the cold weather forecast. Of course, this farmer seeks to minimize the expected total cost of coping with the threatening weather.
   a. Find the maximum value of C below which the farmer will choose to insulate his crop in hopes of

limiting damage as result of the unusually cold weather.

b. Set C equal to the value identified in part **a.** Perform sensitivity analysis to determine under what conditions, if any, the farmer might be better off not spraying his grapefruit trees and taking his chances in spite of the threat to his crop.

25. Consider again the department store buyer's decision problem described in Problem 6. Assume now that consumer demand for the new tennis shoe model (in hundreds of pairs) during the upcoming summer season is *normally* distributed with mean 6 and standard deviation 1.5.

    a. Formulate a payoff table that specifies the contribution to profit (in dollars) from the sale of the tennis shoes by this department store chain for each possible purchase decision (in hundreds of pairs) and each outcome with respect to consumer demand. Use an appropriate discrete approximation of the given normal demand distribution.

    b. Construct a decision tree to identify the buyer's course of action that maximizes the expected profit (in dollars) earned by the department store chain from the purchase and subsequent sale of tennis shoes in the coming year.

26. Consider again the purchasing agent's decision problem described in Problem 9. Assume now that the proportion of defective components supplied by this supplier is well described by the *triangular* distribution with parameters 0, 0, and 1. Its density rises linearly from 0 to 1.

    a. Formulate a payoff table that specifies the microcomputer manufacturer's total cost (in dollars) of

purchasing and repairing (if necessary) a complete lot of components for each possible decision and each outcome with respect to the proportion of defective items. Use an appropriate discrete approximation of the given triangular distribution for the proportion of defective items.

    b. Construct a decision tree to identify the purchasing agent's course of action that minimizes the expected total cost (in dollars) of achieving a complete lot of satisfactory components.

27. A retired partner from Goldman Sachs has $1 million available to invest in particular stocks or bonds. Each investment's annual rate of return depends on the state of the economy in the coming year. The file P07_27.XLS contains the distribution of returns for these stocks and bonds as a function of the economy's state in the coming year. This investor wants to allocate her $1 million to maximize her expected total return 1 year from now.

    a. If $X = Y = 15\%$, find the optimal investment strategy for this investor.

    b. For which values of X (where $10\% < X < 20\%$) and Y (where $12.5\% < Y < 17.5\%$), if any, will this investor prefer to place all of her available funds in the given stocks to maximize her expected total return 1 year from now?

    c. For which values of X (where $10\% < X < 20\%$) and Y (where $12.5\% < Y < 17.5\%$), if any, will this investor prefer to place all of her available funds in the given bonds to maximize her expected total return 1 year from now?

(7.5) Multistage Decision Problems

So far, all of the examples have required a single decision. We now examine a problem where the decision maker must make at least two decisions that are separated in time, such as when a company must decide whether to buy information that will help it make a second decision. The following example illustrates the typical situation.

Example (7.4) Selecting the Best Marketing Strategy at the Acme Company

The Acme Company is trying to decide whether to market a new product. As in many new-product situations, there is considerable uncertainty about whether the new product will eventually "catch on." Acme believes that it might be prudent to introduce the product in a regional test market before introducing it nationally. Therefore, the company's first decision is whether to conduct the test market. Acme estimates that the fixed cost of the test market is $3 million. If it decides to conduct the test market, it must then wait for the results. Based on the results of the test market, it can then decide whether to market the product nationally, in which case it will incur a fixed cost of $90 million. On the other hand, if the original decision is *not* to run a test market, then the final decision—whether to market the product nationally—can be made without further delay. Acme's unit margin, the difference between its selling price and its unit variable cost, is $18 (in the test market and in the national market).

Acme classifies the results in either the test market or the national market as great, fair, or awful. Each of these is accompanied by a forecast of total units sold. These sales volumes (in 1000s of units) are 200, 100, and 30 for the test market and 6000, 3000, and 900 for the national market. Based on previous test markets for similar products, Acme estimates that probabilities of the three test market outcomes are 0.3, 0.6, and 0.1. Then, based on historical data from previous products that were test marketed and eventually marketed nationally, it assesses the probabilities of the national market outcomes given each possible test market outcome. If the test market is great, the probabilities for the national market outcomes are 0.8, 0.15, and 0.05. If the test market is fair, these probabilities are 0.3, 0.5, and 0.2. If the test market is awful, they are 0.05, 0.25, and 0.7. (Note how the probabilities of the national market outcomes tend to mirror the test market outcomes.)

The company wants to use a decision tree approach to find the best strategy.

Objective To construct a multi-stage decision tree model to identify the marketing contingency plan that maximizes Acme's expected profit.

Solution

We begin by discussing the three basic elements of this decision problem: the possible strategies, the possible outcomes and their probabilities, and the value model. The possible strategies are clear. Acme must first decide whether to conduct a test market. Then it must decide whether to introduce the product nationally. However, it is important to realize that if Acme decides to conduct a test market, it can base the national market decision on the results of the test market. In this case its final strategy will be a **contingency plan,** where it conducts the test market, then introduces the product nationally if it receives sufficiently positive test market results or abandons the product if it receives sufficiently negative test market results. The optimal strategies from many multistage decision problems involve similar contingency plans.

Regarding the uncertain outcomes and their probabilities, we note that the given probabilities—probabilities of test market outcomes and *conditional* probabilities of national market outcomes given test market outcomes—are exactly the ones we need in the decision tree. This is because the test market outcome is known *before* the national market outcome will occur. However, suppose Acme decides not to run a test market and then decides to market nationally. Then what are the probabilities of the national market outcomes?

It is important to realize that we cannot simply assess three *new* probabilities for this situation. These probabilities are *implied* by the given probabilities. This follows from the rules of conditional probability we first encountered in Section 5.2 of Chapter 5. If we let T_1, T_2, and T_3 be the test market outcomes, and N be any of the national market outcomes, then by the addition rule for probability and the conditional probability formula, we obtain equations (7.2). (These equivalent equations are sometimes called the **law of total probability.**) For example, if N_1 represents a great national market, then from equations (7.2),

$$P(N_1) = (0.8)(0.3) + (0.3)(0.6) + (0.05)(0.1) = 0.425$$

Similarly, we find that $P(N_2) = 0.37$ and $P(N_3) = 0.205$. These are the probabilities we need to use for the probability branches when no test market is used.

| Law of Total Probability | $P(N) = P(N \text{ and } T_1) + P(N \text{ and } T_2) + P(N \text{ and } T_3)$
 $= P(N|T_1)P(T_1) + P(N|T_2)P(T_2) + P(N|T_3)P(T_3)$ | (7.2) |

Finally, the monetary values in the tree are straightforward. There are fixed costs of test marketing or marketing nationally, and these are incurred as soon as these "go ahead" decisions are made. From that point, we observe the sales volumes and multiply them by the unit margin to obtain the profits.

The inputs for the decision tree appear in Figure 7.26. (See the file ACME.XLS.) In this part of the spreadsheet the only calculated values, which follow from equations (7.2), are in row 28. Specifically, the formula in cell B28 is

$$=\text{SUMPRODUCT(B22:B24,\$B\$16:\$B\$18)}$$

which is copied across row 28. The tree is then straightforward to build and label, as shown in Figure 7.27 on page 332. Note how the fixed costs of test marketing and marketing nationally appear on the decision branches where they occur, so that only the selling profits need to be placed on the probability branches. Also, the probabilities on the various probability branches are exactly those listed in Figure 7.26.

The interpretation of this tree is fairly straightforward if we realize that each value just below each node name is an EMV. For example, the 807 in cell B43 is the EMV for the entire decision problem, which means that Acme's best EMV is $807,000. As another example, the 5910 in cell D47 means that if Acme ever gets to that point—the test market has been conducted and it has been great—the EMV for ACME is $5,910,000. Each

FIGURE 7.26

Inputs for Acme Marketing Example

	A	B	C	D
1	**Acme marketing example**			
2				
3	Fixed costs ($1000s)		**Range names:**	
4	Test mkt	$3,000	NatlMktCost: B5	
5	National mkt	$90,000	TestMktCost: B4	
6			UnitMargin: B18	
7	Unit margin	$18		
8				
9	Possible quantities sold (1000s)			
10		Test mkt	Natl mkt	
11	Great	200	6000	
12	Fair	100	3000	
13	Awful	30	900	
14				
15	Probabilities of test outcomes			
16	Great	0.30		
17	Fair	0.60		
18	Awful	0.10		
19				
20	Probabilities of natl mkt outcomes, given test mkt outcomes			
21		Natl great	Natl fair	Natl awful
22	Test great	0.80	0.15	0.05
23	Test fair	0.30	0.50	0.20
24	Test awful	0.05	0.25	0.70
25				
26	Probabilities of natl mkt outcomes without test mkt (calculated from above inputs)			
27		Natl great	Natl fair	Natl awful
28		0.425	0.370	0.205

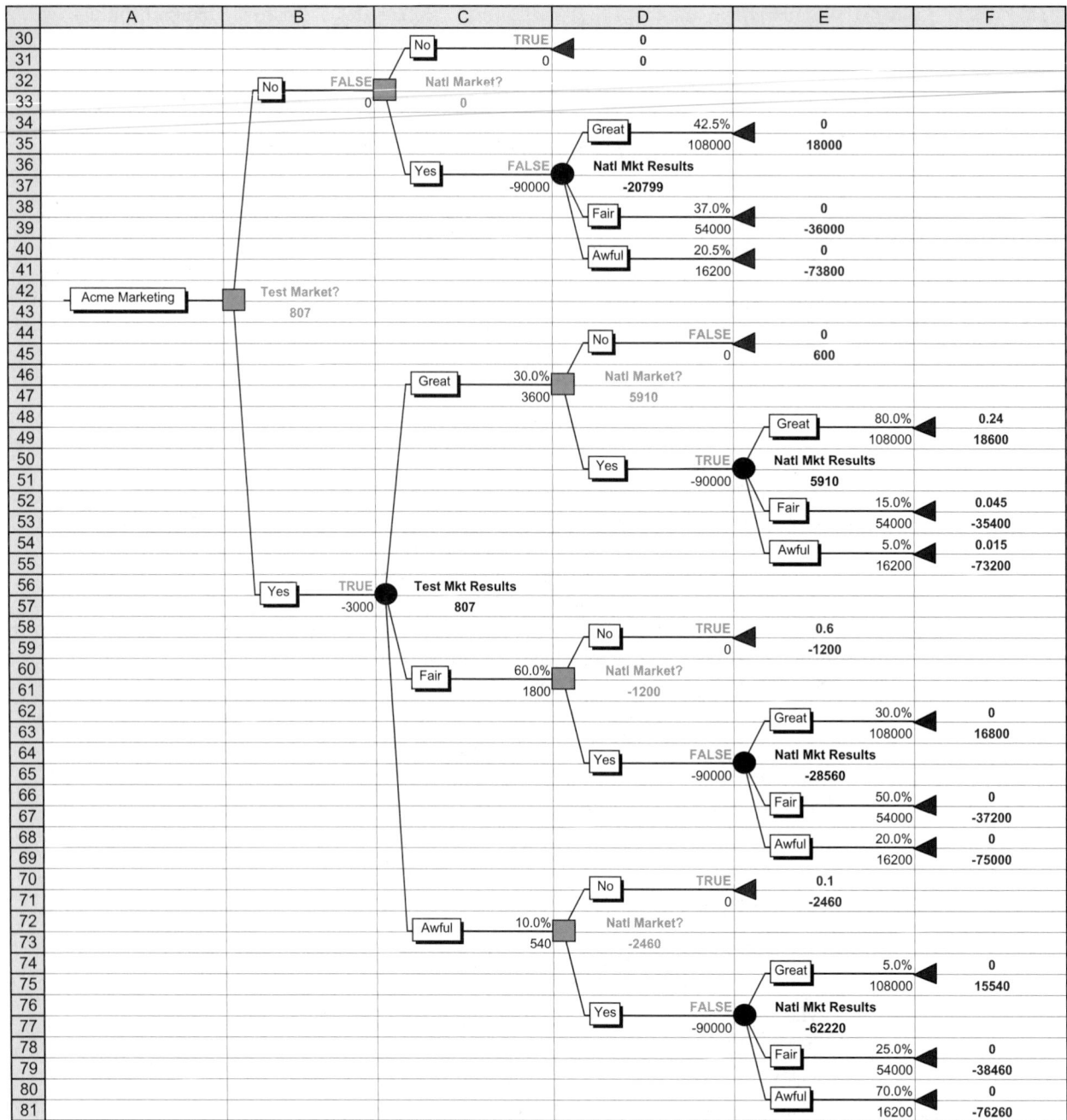

FIGURE 7.27 Decision Tree for Acme Marketing Example

of these EMVs has been calculated by the folding-back procedure we discussed previously, starting from the right and working back toward the left. PrecisionTree takes EMVs at probability nodes and maximums at decision nodes.

We can also see Acme's optimal strategy by following the "TRUE" branches from left to right. Acme should first run a test market. If the test market results are great, then the product should be marketed nationally. However, if the test market results are only fair or awful, the product should be abandoned. In these cases the prospects from a national

FIGURE 7.28
Risk Profile of
Optimal Strategy

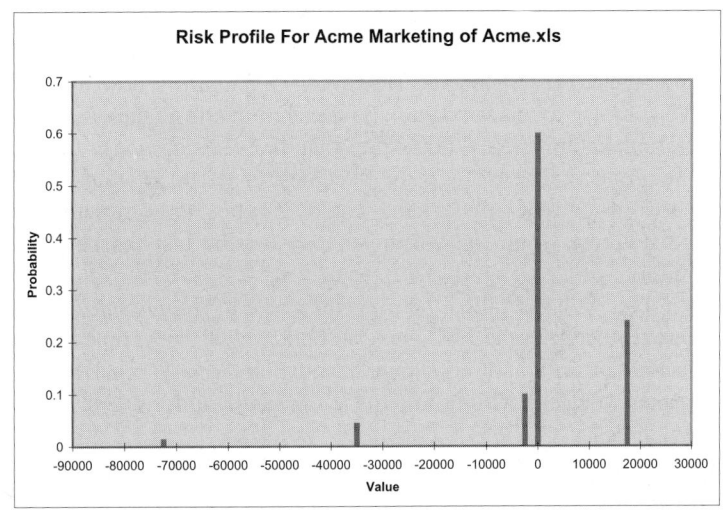

FIGURE 7.29
Distribution of
Profit/Loss from
the Optimal
Strategy

	A	B	C
16	**PROFILE:**		
17	#	X	P
18	1	-73200	0.015
19	2	-35400	0.045
20	3	-2460	0.1
21	4	-1200	0.6
22	5	18600	0.24

market look bleak, so Acme should cut its losses. (And there *are* losses. In these latter two cases, Acme has spent $3,000,000 on the test market and has recouped only $1,800,000 or $540,000 on test market sales.)

The risk profile from the optimal strategy appears in Figure 7.28, which is based on the data in Figure 7.29. (These were obtained by clicking on PrecisionTree's Decision Analysis button—which resembles a staircase—and selecting the Statistics and Risk Profile options.) We see that there is a small chance of two possible large losses (approximately $73 million and $35 million), there is a 70% chance of a moderate loss of about $1 or $2 million, and there is a 24% chance of an $18.6 million profit. Of course, the net effect is an EMV of $807,000.

You might argue that the large potential losses and the slightly higher than 70% chance of *some* loss should persuade Acme to abandon the product right away—without a test market. However, this is what "playing the averages" with EMV is all about. Since the EMV of this optimal strategy is greater than 0, the EMV of abandoning the product right away, Acme should go ahead with this optimal strategy if the company is indeed an EMV maximizer. In Section 7.7 we will see how this reasoning can change if Acme is a risk-averse decision maker—as it might be with multimillion dollar losses looming in the future! ●

7.5.1 Expected Value of Sample Information

The role of the test market in the Acme marketing example is to provide information in the form of more accurate probabilities of national market results. Information usually costs something, as it does in Acme's problem. Currently, the fixed cost of the test market is $3 million, which is evidently not too much to pay because Acme's best strategy is to conduct the test market. However, we might ask how much this test market is worth. This is easy to answer. From the decision tree in Figure 7.27, we see that the EMV from test marketing is $807,000 better than the decision *not* to test market (and then abandon the product). Therefore, if the fixed cost of test marketing were any more than $807,000 above its current value, Acme would be better not to run a test market. Equivalently, the most Acme would be willing to pay for the test market (as a fixed cost) is $3.807 million.

This value is called the *expected value of sample information,* or *EVSI*. It is the difference between the EMV we can obtain with the sample information and the EMV we can obtain without it, as shown in equation (7.3). Alternatively, it is the most the decision maker would be willing to pay for the sample information.

Expected Value of Sample Information (EVSI)	EVSI = EMV with *free* information − EMV without information (7.3)

In Acme's problem, the EMV with free information is $3.807 million (just don't charge for the test market fixed cost), and the EMV without any test market information is $0 (because Acme abandons the product when there is no test market available). Therefore,

$$\text{EVSI} = \$3.807 - \$0 = \$3.807 \text{ million}$$

7.5.2 Expected Value of Perfect Information

The term *sample* implies that the information does not remove all uncertainty about the future. That is, even after the test market results are in, there is still uncertainty about the national market results. Therefore, we might go one step further and ask how much *perfect* information is worth. We can imagine perfect information as an envelope that contains the true final outcome (of the national market). That is, either "the national market will be great," "the national market will be fair," or "the national market will be awful" is written inside the envelope. Admittedly, no such envelope exists, but if it did, how much would Acme be willing to pay for it?

We can answer this question with the simple decision tree in Figure 7.30. Now the probability node on the left corresponds to opening the envelope. Its probabilities are the same as before (when there is no test market available). Note the reasoning here. Acme doesn't know what the contents of the envelope will be, so we need a probability node. However, once the envelope is opened, the true national market outcome will be revealed. At that point Acme's decision is fairly obvious. If Acme learns that a national market will be great, it knows the product will be profitable and will market it. Otherwise, if Acme learns that the national market will be fair or poor, it knows that there will be a loss from marketing nationally, so it will abandon the product. Folding back in the usual way produces an EMV of $7.65 million.

Now compare this $7.65 million with the EMV in the top part of Figure 7.27 that results from no test market, namely, $0. The difference, $7.65 million, is called the *expected value of perfect information,* or *EVPI*. It represents the maximum amount the company would pay for perfect information about the final outcome (of the national market). The EVPI is the difference between the EMV with perfect information and the EMV with no additional information, as shown in equation (7.4).

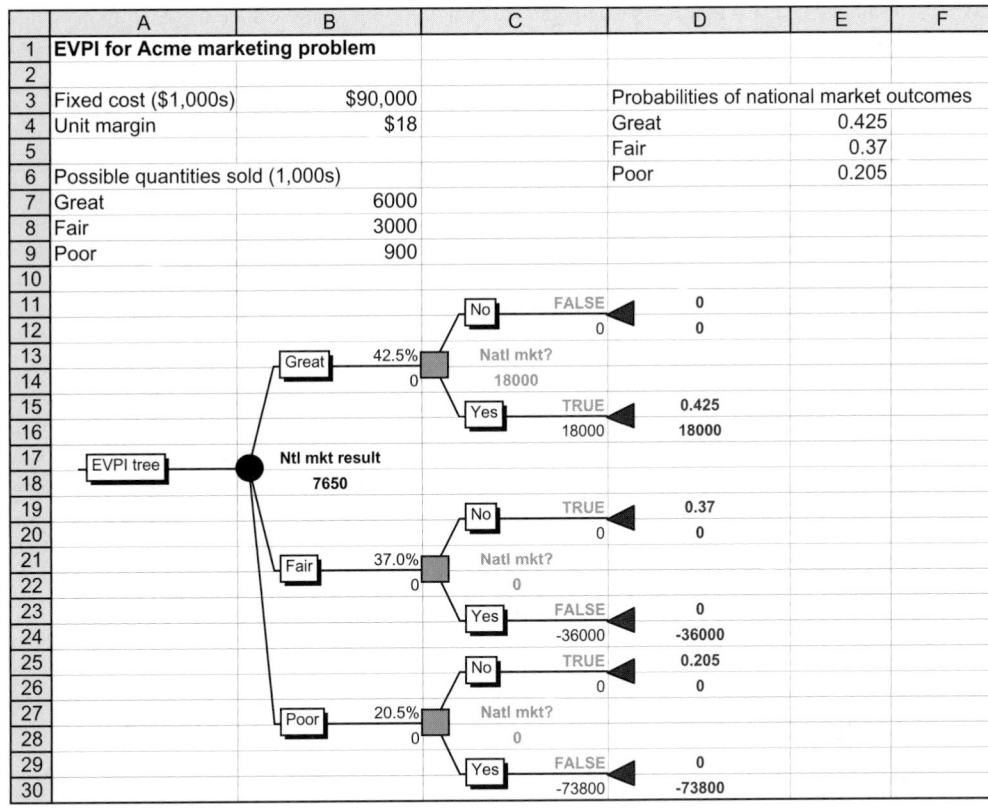

The table in the figure:

	A	B	C	D	E	F
1	EVPI for Acme marketing problem					
2						
3	Fixed cost ($1,000s)	$90,000		Probabilities of national market outcomes		
4	Unit margin	$18		Great	0.425	
5				Fair	0.37	
6	Possible quantities sold (1,000s)			Poor	0.205	
7	Great	6000				
8	Fair	3000				
9	Poor	900				

FIGURE 7.30 EVPI for Acme Marketing Example

Expected Value of Perfect Information (EVPI)

$$\text{EVPI} = \text{EMV with } \textit{free} \text{ perfect information} - \text{EMV with no information} \quad (7.4)$$

In Acme's case this expression becomes

$$\text{EVPI} = \$7.65 - \$0 = \$7.65 \text{ million}$$

The EVPI may appear to be an irrelevant concept since perfect information is almost never available—at *any* price. However, it is often useful because it represents an *upper bound* on the EVSI for any potential sample information. That is, no sample information can ever be worth more than the EVPI. For example, if Acme is contemplating an expensive test market with an anticipated fixed cost of more than $8 million, then there is really no point in pursuing it any further. The information gained from this test market, no matter how reliable it is, cannot possibly justify its cost because its cost is greater than the EVPI.

PROBLEMS

Level A

28. The senior executives of an oil company are trying to decide whether to drill for oil in a particular field in the Gulf of Mexico. It costs the company $300,000 to drill in the selected field. Company executives believe that if oil is found in this field its estimated value will be $1,800,000. At present, this oil company believes that there is a 50% chance that the selected field actually contains oil. Before drilling, the company can hire a geologist at a cost of $30,000 to prepare a report that contains a recommendation regarding drilling in the selected field. There is a 55% chance that the geologist will issue a favorable recommendation and a 45% chance that the geologist will issue an unfavorable recommendation. Given a favorable recommendation from the geologist, there is a 75% chance that the field actually contains oil. Given an unfavorable recommendation from the geologist, there is a 15% chance that the field actually contains oil.

a. Assuming that this oil company wishes to maximize its expected net earnings, determine its optimal strategy through the use of a decision tree.

b. Compute and interpret the expected value of sample information (EVSI) in this decision problem.

c. Compute and interpret the expected value of perfect information (EVPI) in this decision problem.

29. A local certified public accountant must decide which of two copying machines to purchase for her expanding business. The cost of purchasing the first machine is $4500, and the cost of maintaining the first machine each year is uncertain. The CPA's office manager believes that the annual maintenance cost for the first machine will be $0, $150, or $300 with probabilities 0.35, 0.35, and 0.30, respectively. The cost of purchasing the second machine is $3000, and the cost of maintaining the second machine through a guaranteed maintenance agreement is $225 per year.

Before the purchase decision is made, the CPA can hire an experienced copying machine repairman to evaluate the quality of the first machine. Such an evaluation would cost the CPA $60. If the repairman believes that the first machine is satisfactory, there is a 65% chance that its annual maintenance cost will be $0 and a 35% chance that its annual maintenance cost will be $150. If, however, the repairman believes that the first machine is unsatisfactory, there is a 60% chance that its annual maintenance cost will be $150 and a 40% chance that its annual maintenance cost will be $300. The CPA's office manager believes that the repairman will issue a satisfactory report on the first machine with probability 0.50.

a. Provided that the CPA wishes to minimize the expected total cost of purchasing and maintaining one of these two machines for a 1-year period, which machine should she purchase? When, if ever, would it be worthwhile for the CPA to obtain the repairman's review of the first machine?

b. Compute and interpret the expected value of sample information (EVSI) in this decision problem.

c. Compute and interpret the expected value of perfect information (EVPI) in this decision problem.

30. Upjohn is developing a new product to promote hair growth in cases of male pattern baldness. If Upjohn markets the new product and it is successful, the com-

pany will earn $500,000 in additional profit. If the marketing of this new product proves to be unsuccessful, the company will lose $350,000 in development and marketing costs. In the past, similar products have been successful 60% of the time. At a cost of $50,000, the effectiveness of the new restoration product can be thoroughly tested. If the results of such testing are favorable, there is an 80% chance that the marketing efforts of this new product will be successful. If the results of such testing are not favorable, there is a mere 30% chance that the marketing efforts of this new product will be successful. Upjohn currently believes that the probability of receiving favorable test results is 0.60.

a. Identify the strategy that maximizes Upjohn's expected net earnings in this situation.

b. Compute and interpret the expected value of sample information (EVSI) in this decision problem.

c. Compute and interpret the expected value of perfect information (EVPI) in this decision problem.

31. Hank is considering placing a bet on the upcoming showdown between the Penn State and Michigan football teams in State College. The winner of this contest will represent the Big Ten Conference in the Rose Bowl on New Year's Day. Without any additional information, Hank believes that each team has an equal chance of winning this big game. If he wins the bet, he will win $500; if he loses the bet, he will lose $550. Before placing his bet, he may decide to pay his friend Al, who happens to be a football sportswriter for the *Philadelphia Enquirer*, $50 for Al's expert prediction on the game. Assume that Al predicts that Penn State will win similar games 55% of the time, and that Michigan will win similar games 45% of the time. Furthermore, Hank knows that when Al predicts that Penn State will win, there is a 70% chance that Penn State will indeed win the football game. Finally, when Al predicts that Michigan will win, there is a 20% chance that Penn State will proceed to win the upcoming game.

a. To maximize his expected profit from this betting opportunity, how should Hank proceed in this situation?

b. Compute and interpret the expected value of sample information (EVSI) in this decision problem.

c. Compute and interpret the expected value of perfect information (EVPI) in this decision problem.

32. A product manager at Procter & Gamble seeks to determine whether her company should market a new brand of toothpaste. If this new product succeeds in the marketplace, P&G estimates that it could earn $1,800,000 in future profits from the sale of the new toothpaste. If this new product fails, however, the company expects that it could lose approximately $750,000. If P&G chooses not to market this new brand, the product manager believes that there would

be little, if any, impact on the profits earned through sales of P&G's other products. The manager has estimated that the new toothpaste brand will succeed with probability 0.55. Before making her decision regarding this toothpaste product, the manager can spend $75,000 on a market research study. Such a study of consumer preferences will yield either a positive recommendation with probability 0.50 or a negative recommendation with probability 0.50. Given a positive recommendation to market the new product, the new brand will eventually succeed in the marketplace with probability 0.75. Given a negative recommendation regarding the marketing of the new product, the new brand will eventually succeed in the marketplace with probability 0.25.

a. To maximize expected profit in this case, what course of action should the P&G product manager take?

b. Compute and interpret the expected value of sample information (EVSI) in this decision problem.

c. Compute and interpret the expected value of perfect information (EVPI) in this decision problem.

Level B

33. A publishing company is trying to decide whether to publish a new business law textbook. Based on a careful reading of the latest draft of the manuscript, the publisher's senior editor in the business textbook division assesses the distribution of possible payoffs earned by publishing this new book. The file P07_33.XLS contains this probability distribution. Before making a final decision regarding the publication of the book, the editor can learn more about the text's potential for success by thoroughly surveying business law instructors teaching at universities across the country. Historical frequencies based on similar surveys administered in the past are provided in the same file.

a. Find the strategy that maximizes the publisher's expected payoff (in dollars).

b. What is the most (in dollars) that the publisher should be willing to pay to conduct a new survey of business law instructors?

c. If the actual cost of conducting the given survey is less than the amount identified in part **a,** what should the publisher do?

d. Assuming that a survey could be constructed that provides "perfect information" to the publisher, how much should the company be willing to pay to acquire and implement such a survey?

34. Lands' End Direct Merchants is trying to decide whether to ship some customer orders now via UPS or wait until after the threat of another UPS strike is over. If Lands' End decides to ship the requested merchandise now and the UPS strike takes place, the company will incur $60,000 in delay and shipping costs. If Lands' End decides to ship the customer orders via UPS and no strike occurs, the company will incur $4000 in shipping costs. If Lands' End decides to postpone shipping its customer orders via UPS, the company will incur $10,000 in delay costs regardless of whether or not UPS goes on strike. Let p represent the probability that UPS will go on strike and affect Lands' End's shipments.

a. For which values of p, if any, does Lands' End minimize its expected total cost by choosing to postpone shipping its customer orders via UPS?

b. Suppose now that at a cost of $1000, Lands' End can purchase information regarding the likelihood of a UPS strike in the near future. Based on similar strike threats in the past, the probability that this information indicates the occurrence of a UPS strike is 27.5%. If the purchased information indicates the occurrence of a UPS strike, the chance of a strike actually occurring is 0.105/0.275. If the purchased information does not indicate the occurrence of a UPS strike, the chance of a strike not occurring is 0.680/0.725. Provided that $p = 0.15$, what strategy should Lands' End pursue to minimize its expected total cost?

c. Continuing part **b,** compute and interpret the expected value of sample information (EVSI) when $p = 0.15$.

d. Continuing part **b,** compute and interpret the expected value of perfect information (EVPI) when $p = 0.15$.

7.6 Bayes' Rule

In multistage decision problems we typically have alternating sets of decision nodes and probability nodes. The decision maker makes a decision, some uncertain outcomes are observed, the decision maker makes another decision, more uncertain outcomes are observed, and so on. In the resulting decision tree, all probability branches at the *right* of the tree are conditional on outcomes that have occurred earlier, to their left. Therefore, the probabilities on these branches are of the form $P(A|B)$, where B is an event that occurs *before* event A in time. However, it is sometimes more natural to assess conditional probabilities in the opposite order, that is, $P(B|A)$. Whenever this is the case, we require *Bayes' rule* to obtain the probabilities we need on the tree. Essentially, Bayes' rule is a mechanism for updating probabilities as new information becomes available. We illustrate the mechanics of Bayes' rule in the following example. [See Feinstein (1990) for a real application of this example.]

Example 7.5 Testing Athletes for Illegal Drug Usage

If an athlete is tested for a certain type of drug usage (steroids, say), then the test will come out either positive or negative. However, these tests are never perfect. Some athletes who are drug free test positive, and some who are drug users test negative. The former are called false positives; the latter are called false negatives. We will assume that 5% of all athletes use drugs, 3% of all tests on drug-free athletes yield false positives, and 7% of all tests on drug users yield false negatives. The question then is, What we can conclude from a positive or negative test result?

Objective To illustrate Bayes' rule for updating the probability of drug usage after observing a positive or negative drug test result.

Solution

Let D and ND denote that a randomly chosen athlete is or is not a drug user, and let $T+$ and $T-$ indicate a positive or negative test result. We are given the following probabilities. First, since 5% of all athletes are drug users, we know that $P(D) = 0.05$ and $P(ND) = 0.95$. These are called **prior probabilities** because they represent the chance that an athlete is or is not a drug user *prior* to the results of a drug test. Second, from the information on drug test accuracy, we know the conditional probabilities $P(T+ |ND) = 0.03$ and $P(T- |D) = 0.07$. But a drug-free athlete either tests positive or negative, and the same is true for a drug user. Therefore, we also have the probabilities $P(T- |ND) = 0.97$ and $P(T+ |D) = 0.93$. These four conditional probabilities of test results given drug user status are often called the **likelihoods** of the test results.

Given these priors and likelihoods, we want *posterior probabilities* such as $P(D|T+)$, the probability that an athlete who tested positive is a drug user, or $P(ND|T-)$, the probability that an athlete who tested negative is drug free. They are called **posterior probabilities** because they are assessed *after* the drug test results. This is where Bayes' rule enters. We will develop Bayes' rule generally and then apply it to the present example.

Let A be any "information" event, such as the result of a drug test, and let B_1, B_2, ..., B_n be any mutually exclusive and exhaustive set of events. That is, exactly one of the B_i's must occur. To apply Bayes' rule, we assume that the prior probabilities $P(B_1)$, $P(B_2)$, ..., $P(B_n)$ are given, as are the likelihoods $P(A|B_i)$ for each i. Then we want the posterior probabilities $P(B_i|A)$ for each i. Bayes' rule, shown in equation (7.5), indicates how to calculate these posterior probabilities.

Bayes' Rule	

$$P(B_i|A) = \frac{P(A|B_i)P(B_i)}{P(A|B_1)P(B_1) + \cdots + P(A|B_n)P(B_n)} \tag{7.5}$$

This formula says that a typical posterior probability is a ratio. The numerator is a likelihood times a prior, and the denominator is the sum of likelihoods times priors.

Before illustrating Bayes' rule numerically, we make two other observations about the terms in Bayes' rule. First, we can use the multiplication rule of probability to write any product of a likelihood and a prior as

$$P(A|B_i)P(B_i) = P(A \text{ and } B_i)$$

The probability on the right, that *both* A and B_i occur, is called a **joint probability.** Second, we can use the definition of conditional probability directly to write

$$P(B_i|A) = \frac{P(A \text{ and } B_i)}{P(A)}$$

Therefore, the probability in the denominator of Bayes' rule is really just the probability of A:

$$P(A) = P(A|B_1)P(B_1) + \cdots + P(A|B_n)P(B_n)$$

As we will see shortly, this natural by-product of Bayes' rule will come in very handy in decision trees.

It is fairly easy to implement Bayes' rule in a spreadsheet, as illustrated in Figure 7.31 for the drug example. Here A corresponds to either test result, and B_1 and B_2 correspond to D and ND. (See the file DRUGBAYES.XLS.[4]) In words, we want to see how the chances of D and ND change after seeing the results of the drug test.

The given priors and likelihoods are listed in the ranges B5:C5 and B9:C10. We then calculate the products of likelihoods and priors in the range B15:C16. The formula in cell B15 is

$$=B\$5*B9$$

FIGURE 7.31

Bayes' Rule for Drug Testing Example

	A	B	C	D
1	**Illustration of Bayes' rule using drug example**			
2				
3	Prior probabilities of drug user status			
4		User	Non-user	
5		0.05	0.95	1
6				
7	Likelihoods of test results, given drug user status			
8		User	Non-user	
9	Test positive	0.93	0.03	
10	Test negative	0.07	0.97	
11		1	1	
12				
13	Joint probabilities of drug user status and test results			
14		User	Non-user	Unconditional
15	Test positive	0.0465	0.0285	0.075
16	Test negative	0.0035	0.9215	0.925
17				1
18				
19	Posterior probabilities of drug user status			
20		User	Non-user	
21	Test positive	0.620	0.380	1
22	Test negative	0.004	0.996	1

[4]The Bayes2 sheet in the finished version of this file illustrates how Bayes' rule can be used when there are more than two possible test results and/or drug user categories.

and this is copied to the rest of the B15:C16 range. Their row sums are calculated in the range D15:D16. These represent the unconditional probabilities of the two possible results. They are also (as we saw above) the denominators of Bayes' rule. Finally, we calculate the posterior probabilities in the range B21:C22. The formula in cell B21 is

$$=B15/\$D15$$

and this is copied to the rest of the B21:C22 range. The various 1's in the margins of Figure 7.31 are row sums or column sums that must equal 1. We show them only as checks of our logic.

Note that a negative test result leaves little doubt that the athlete is drug free. The posterior probability that the athlete is drug free, given a negative test result, is 0.996. However, there is still a lot of doubt about an athlete who tests positive. The posterior probability that the athlete uses drugs, given a positive test result, is only 0.620. This asymmetry occurs because of the prior probabilities. We are fairly certain that a randomly selected athlete is drug free because only 5% of all athletes use drugs. It takes a lot of evidence to convince us otherwise. This initial bias, plus the fact that the test produces a few false positives, means that athletes with positive test results still have a decent chance (probability 0.380) of being drug free. Is this a valid argument for not requiring drug testing of athletes? We explore this question in the following continuation of the drug-testing example. It all depends on the "costs." (It might also depend on whether there is a second type of test that could help confirm the findings of the first test. However, we won't consider such a test.) ●

Example ⑦.⑥ Deciding Whether to Implement Mandatory Drug Testing

The administrators at State University are trying to decide whether to institute mandatory drug testing for the athletes. They have the same information about priors and likelihoods as in the previous example, but now they want to use a decision tree approach to see whether the benefits outweigh the costs.[5]

Objective To use a decision tree model to help administrators at State University decide whether to implement mandatory drug testing for all athletes.

Solution

We have already discussed the uncertain outcomes and their probabilities. Now we need to discuss the decision alternatives and the monetary values—the other two elements of a decision analysis. We will assume that there are only two alternatives: perform drug testing on all athletes or don't perform any drug testing. In the former case we assume that if an athlete tests positive, this athlete is barred from sports.

The "monetary" values are more difficult to assess. They include

- the benefit B from correctly identifying a drug user and barring him or her from sports
- the cost C_1 of the test itself for a single athlete (materials and labor)
- the cost C_2 of falsely accusing a nonuser (and barring him or her from sports)
- the cost C_3 of not identifying a drug user (either by not testing at all or by obtaining a false negative)
- the cost C_4 of violating a nonuser's privacy by performing the test

[5]Again, see Feinstein (1990) for an enlightening discussion of this drug-testing problem at a real university.

It is clear that only C_1 is a direct monetary cost that is easy to measure. However, the other "costs" and the benefit B are real, and they must be compared on some scale to enable administrators to make a rational decision. We will do so by comparing everything to the cost C_1, to which we will assign value 1. (This does not mean that the cost of testing an athlete is necessarily \$1; it just means that we will express all other costs as multiples of C_1.) Clearly, there is a lot of subjectivity involved in making these comparisons, so sensitivity analysis on the final decision tree is a must.

Before developing this decision tree, it is useful to form a benefit–cost table for both alternatives and all possible outcomes. Because we will eventually maximize expected net *benefit,* all benefits in this table have a positive sign and all costs have a negative sign. These net benefits appear in Table 7.7. The first two columns are relevant if no tests are performed; the last four are relevant when testing is performed. For example, if a positive test is obtained for a nonuser, there are three costs: the cost of the test (C_1), the cost of falsely accusing the athlete (C_2), and the cost of violating the nonuser's privacy (C_4). The other entries are obtained similarly.

Table **7.7**

Net Benefit for Drug-Testing Example	Don't Test		Perform Test			
	D	*ND*	*D* and *T+*	*ND* and *T+*	*D* and *T−*	*ND* and *T−*
	$-C_3$	0	$B - C_1$	$-(C_1 + C_2 + C_4)$	$-(C_1 + C_3)$	$-(C_1 + C_4)$

The solution with PrecisionTree shown in Figure 7.32 (page 342) is now fairly straightforward. (See the file DRUG.XLS.) We first enter all of the benefits and costs in an input section. These, together with the Bayes' rule calculations from before, appear at the top of the spreadsheet. Then we use PrecisionTree in the usual way to build the tree and enter the links to the values and probabilities.

Before we interpret this solution, we discuss the timing (from left to right). If drug testing is performed, the result of the drug test is observed first (a probability node). Each test result leads to an action (bar from sports or don't), and then the eventual benefit or cost depends on whether the athlete uses drugs (again a probability node). You might argue that the university never knows for certain whether the athlete uses drugs, but we must include this information in the tree to get the benefits and costs correct. If no drug testing is performed, then there is no intermediate test result node or branches.

Now let's interpret our results. First, we discuss the benefits and costs shown in Figure 7.32. These were chosen fairly arbitrarily, but with some hope of reflecting reality. They say that the largest cost is falsely accusing (and barring) a nonuser. This is 50 times as large as the cost of the test. The benefit of identifying a drug user is only half this large, and the cost of not identifying a user is 40% as large as barring a nonuser. The violation of privacy of a nonuser is twice as large as the cost of the test. Based on these values, the decision tree implies that drug testing should *not* be performed. The EMVs for testing and for not testing are both negative, indicating that the costs outweigh the benefits for each, but the EMV for not testing is slightly *less* negative.[6]

What would it take to change this decision? We'll start with the assumption, probably accepted by most people in our society, that the cost of falsely accusing a nonuser (C_2) ought to be the largest of the benefits or costs in the range B4:B10. In fact, because of possible legal costs, we might argue that C_2 should be *more* than 50 times the cost of the test. But if we increase C_2, the scales are tipped even farther in the direction of not

[6]The university in the Feinstein (1990) study came to the same conclusion.

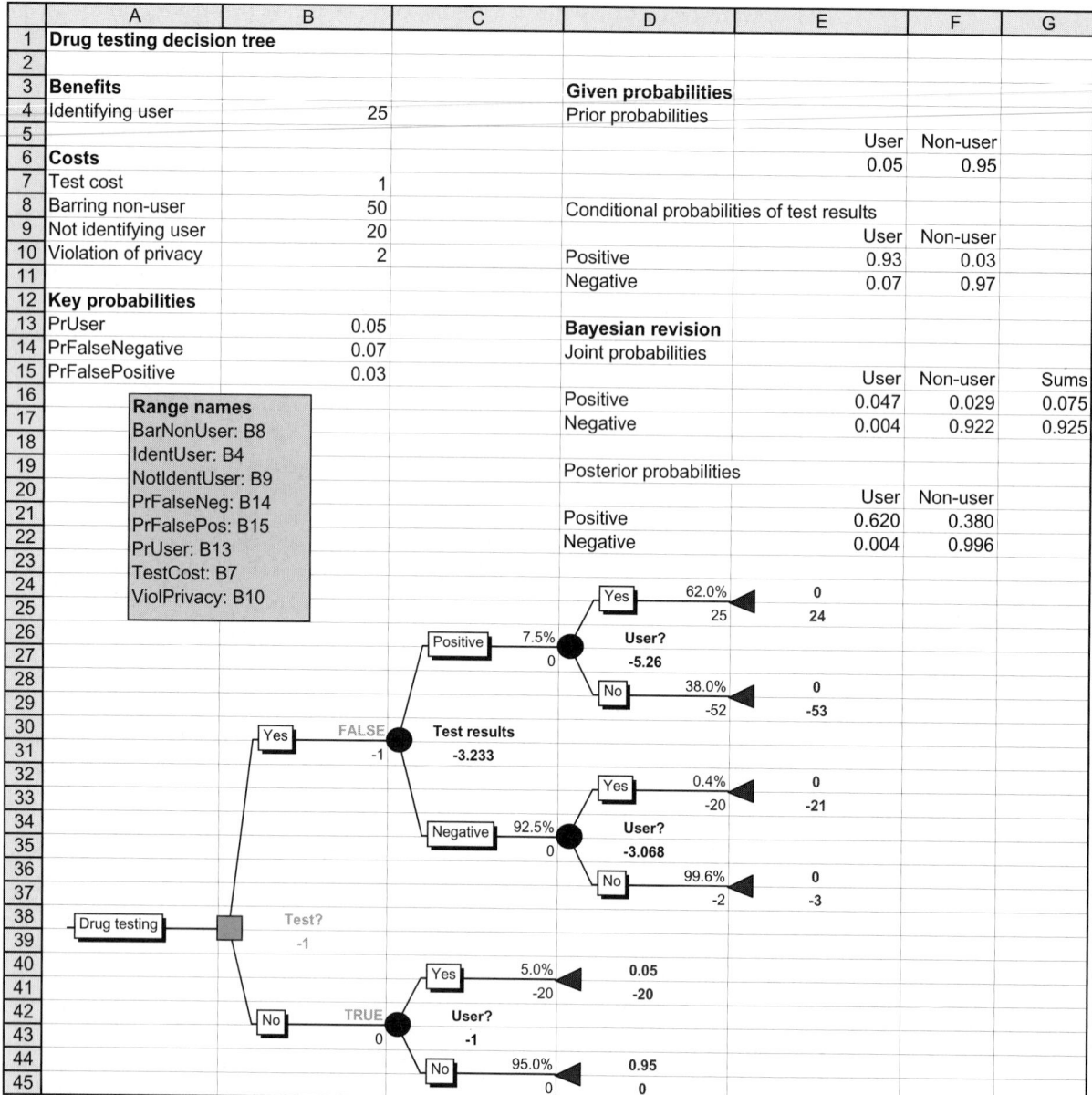

FIGURE 7.32 Decision Tree for Drug-Testing Example

testing. On the other hand, if the benefit B from identifying a user and/or the cost C_3 for not identifying a user increase, then testing might be the preferred alternative. We tried this, keeping C_2 constant at 50. When B and C_3 both had value 45, no testing was still optimal, but when they both increased to 50—the same magnitude as C_2—then testing won out by a small margin. However, it would be difficult to argue that B and C_3 should be of the same magnitude as C_2.

Other than the benefits and costs, the only other thing we might vary is the accuracy of the test, measured by the error probabilities in cells B14 and B15. Presumably, if the test makes fewer false positives and false negatives, testing might be a more attractive alternative. We tried this, keeping the benefits and costs the same as those shown in Figure

7.32 but changing the error probabilities. Even when each error probability was decreased to 0.01, however, the no-testing alternative was still optimal—by a fairly wide margin.

In summary, based on a number of reasonable assumptions and parameter settings, this example has shown that it is difficult to make a case for mandatory drug testing. ●

PROBLEMS

Level A

35. Consider a population of 2000 individuals, 800 of whom are women. Assume that 300 of the women in this population earn at least $60,000 per year, and 200 of the men earn at least $60,000 per year.
  a. What is the probability that a randomly selected individual from this population earns less than $60,000 per year?
  b. If a randomly selected individual is observed to earn less than $60,000 per year, what is the probability that this person is a man?
  c. If a randomly selected individual is observed to earn at least $60,000 per year, what is the probability that this person is a woman?

36. Yearly automobile inspections are required for residents of the state of Pennsylvania. Suppose that 18% of all inspected cars in Pennsylvania have problems that need to be corrected. Unfortunately, Pennsylvania state inspections fail to detect these problems 12% of the time. Consider a car that is inspected and is found to be free of problems. What is the probability that there is indeed something wrong that the inspection has failed to uncover?

37. Consider again the landowner's decision problem described in Problem 3. Suppose now that, at a cost of $90,000, the landowner can request that a soundings test be performed on the site where natural gas is believed to be present. The company that conducts the soundings concedes that 30% of the time the test will indicate that no gas is present when it actually is. When natural gas is not present in a particular site, the soundings test is accurate 90% of the time.
  a. Given that the landowner pays for the soundings test and the test indicates that gas is present, what is the landowner's revised estimate of the probability of finding gas on this site?
  b. Given that the landowner pays for the soundings test and the test indicates that gas is not present, what is the landowner's revised estimate of the probability of not finding gas on this site?
  c. Should the landowner request the given soundings test at a cost of $90,000? Explain why or why not. If not, when (if ever) would the landowner choose to obtain the soundings test?

38. The chief executive officer of a firm in a highly competitive industry believes that one of her key employ-

ees is providing confidential information to the competition. She is 90% certain that this informer is the vice president of finance, whose contacts have been extremely valuable in obtaining financing for the company. If she decides to fire this vice president and he is the informer, she estimates that the company will gain $500,000. If she decides to fire this vice president but he is not the informer, the company will lose his expertise and still have an informer within the staff; the CEO estimates that this outcome would cost her company about $2.5 million. If she decides not to fire this vice president, she estimates that the firm will lose $1.5 million regardless of whether he actually is the informer (because in either case the informer is still with the company).

Before deciding whether to fire the vice president for finance, the CEO could order lie detector tests. To avoid possible lawsuits, the lie detector tests would have to be administered to all company employees, at a total cost of $150,000. Another problem she must consider is that the available lie detector tests are not perfectly reliable. In particular, if a person is lying, the test will reveal that the person is lying 95% of the time. Moreover, if a person is not lying, the test will indicate that the person is not lying 85% of the time.
  a. To minimize the expected total cost of managing this difficult situation, what strategy should the CEO adopt?
  b. Should the CEO order the lie detector tests for all of her employees? Explain why or why not.
  c. Determine the maximum amount of money that the CEO should be willing to pay to administer lie detector tests.

39. A customer has approached a bank for a $10,000 1-year loan at a 12% interest rate. If the bank does not approve this loan application, the $10,000 will be invested in bonds that earn a 6% annual return. Without additional information, the bank believes that there is a 4% chance that this customer will default on the loan, assuming that the loan is approved. If the customer defaults on the loan, the bank will lose $10,000. At a cost of $100, the bank can thoroughly investigate the customer's credit record and supply a favorable or unfavorable recommendation. Past experience indicates that in cases where the customer did not default on the approved loan, the probability of receiving a favorable recommendation on the basis of the credit investigation was 77/96. Furthermore, in cases where the customer defaulted on the approved loan, the probability of receiving a favorable recommendation on the basis of the credit investigation was 1/4.

  a. What course of action should the bank take to maximize its expected profit?
  b. Compute and interpret the expected value of sample information (EVSI) in this decision problem.
  c. Compute and interpret the expected value of perfect information (EVPI) in this decision problem.

40. A company is considering whether to market a new product. Assume, for simplicity, that if this product is marketed, there are only two possible outcomes: success or failure. The company assesses that the probabilities of these two outcomes are p and $1 - p$, respectively. If the product is marketed and it proves to be a failure, the company will lose \$450,000. If the product is marketed and it proves to be a success, the company will gain \$750,000. Choosing not to market the product results in no gain or loss for the company.

The company is also considering whether to survey prospective buyers of this new product. The results of the consumer survey can be classified as favorable, neutral, or unfavorable. In similar cases where proposed products proved to be market successes, the likelihoods that the survey results were favorable, neutral, and unfavorable were 0.6, 0.3, and 0.1, respectively. In similar cases where proposed products proved to be market failures, the likelihoods that the survey results were favorable, neutral, and unfavorable were 0.1, 0.2, and 0.7, respectively. The total cost of administering this survey is C dollars.

  a. Let $p = 0.4$. For which values of C, if any, would this company choose to conduct the consumer survey?
  b. Let $p = 0.4$. What is the largest amount that this company would be willing to pay for perfect information about the potential success or failure of the new product?
  c. Let $p = 0.5$ and $C = \$15,000$. Find the strategy that maximizes the company's expected earnings in this situation. Does the optimal strategy involve conducting the consumer survey? Explain why or why not.

41. The U.S. government is attempting to determine whether immigrants should be tested for a contagious disease. Let's assume that the decision will be made on a financial basis. Furthermore, assume that each immigrant who is allowed to enter the United States and has the disease costs the country \$100,000. Also, each immigrant who is allowed to enter the United States and does not have the disease will contribute \$10,000 to the national economy. Finally, assume that x percent of all potential immigrants have the disease. The U.S. government can choose to admit all immi-

grants, admit no immigrants, or test immigrants for the disease before determining whether they should be admitted. It costs T dollars to test a person for the disease; the test result is either positive or negative. A person who does not have the disease *always* tests negative. However, 20% of all people who *do* have the disease test negative. The government's goal is to maximize the expected net financial benefits per potential immigrant.

  a. Let $x = 10$ (i.e., 10%). What is the largest value of T at which the U.S. government will choose to test potential immigrants for the disease?
  b. How does your answer to the question in part **a** change when x increases to 15?
  c. Let $x = 10$ and $T = \$100$. Find the government's optimal strategy in this case.
  d. Let $x = 10$ and $T = \$100$. Compute and interpret the expected value of perfect information (EVPI) in this decision problem.
  e. Comment on the ethical aspects of the U.S. government's conduct in this case.

Level B

42. A city in Ohio is considering replacing its fleet of gasoline-powered automobiles with electric cars. The manufacturer of the electric cars claims that this municipality will experience significant cost savings over the life of the fleet if it chooses to pursue the conversion. If the manufacturer is correct, the city will save about \$1.5 million dollars. If the new technology employed within the electric cars is faulty, as some critics suggest, the conversion to electric cars will cost the city \$675,000. A third possibility is that less serious problems will arise and the city will break even with the conversion. A consultant hired by the city estimates that the probabilities of these three outcomes are 0.30, 0.30, and 0.40, respectively.

The city has an opportunity to implement a pilot program that would indicate the potential cost or savings resulting from a switch to electric cars. The pilot program involves renting a small number of electric cars for 3 months and running them under typical conditions. This program would cost the city \$75,000. The city's consultant believes that the results of the pilot program would be significant but not conclusive; she submits the data in the file P07_42.XLS, a compilation of probabilities based on the experience of other cities, to support her contention. For example, the first row of her table indicates that, given that a conversion to electric cars actually results in a savings of \$1.5 million, the conditional probabilities that the pilot program will indicate that the city saves money, loses money, and breaks even are 0.6, 0.1, and 0.3, respectively.

  a. What actions should this city take to maximize the expected savings?

b. Should the city implement the pilot program at a cost of $75,000?

c. Compute and interpret the expected value of sample information (EVSI) in this decision problem.

43. A manufacturer must decide whether to extend credit to a retailer who would like to open an account with the firm. Past experience with new accounts indicates that 45% are high-risk customers, 35% are moderate-risk customers, and 20% are low-risk customers. If credit is extended, the manufacturer can expect to lose $60,000 with a high-risk customer, make $50,000 with a moderate-risk customer, and make $100,000 with a low-risk customer. If the manufacturer decides not to extend credit to a customer, the manufacturer neither makes nor loses any money.

Prior to making a credit extension decision, the manufacturer can obtain a credit rating report on the retailer at a cost of $2000. The credit agency concedes that its rating procedure is not completely reliable. In particular, the credit rating procedure will rate a low-risk customer as a moderate-risk customer with probability 0.10 and as a high-risk customer with probability 0.05. Furthermore, the given rating procedure will rate a moderate-risk customer as a low-risk customer with probability 0.06 and as a high-risk customer with probability 0.07. Finally, the rating procedure will rate a high-risk customer as a low-risk customer with probability 0.01 and as a moderate-risk customer with probability 0.05.

a. Find the strategy that maximizes the manufacturer's expected net earnings.

b. Should the manufacturer routinely obtain credit rating reports on those retailers who seek credit approval? Why or why not?

c. Compute and interpret the expected value of sample information (EVSI) in this decision problem.

44. A television network earns an average of $1.6 million each season from a hit program and loses an average of $400,000 each season on a program that turns out to be a flop. Of all programs picked up by this network in recent years, 25% turn out to be hits and 75% turn out to be flops. At a cost of *C* dollars, a market research firm will analyze a pilot episode of a prospective program and issue a report predicting whether the given program will end up being a hit. If the program is actually going to be a hit, there is a 90% chance that the market researchers will predict the program to be a hit. If the program is actually going to be a flop, there is a 20% chance that the market researchers will predict the program to be a hit.

a. Assuming that $C = \$160,000$, identify the strategy that maximizes this television network's expected profit in responding to a newly proposed television program.

b. What is the maximum value of *C* that this television network should be willing to incur in choosing to hire the market research firm?

c. Compute and interpret the expected value of perfect information (EVPI) in this decision problem.

 7.7 # Incorporating Attitudes Toward Risk

Rational decision makers are sometimes willing to violate the EMV maximization criterion when large amounts of money are at stake. These decision makers are willing to sacrifice some EMV to reduce risk. Are you ever willing to do so personally? Consider the following scenarios.

1. You have a chance to enter a lottery where you will win $100,000 with probability 0.1 or win nothing with probability 0.9. Alternatively, you can receive $5000 for certain. How many of you—truthfully—would take the certain $5000, even though the EMV of the lottery is $10,000? Change the $100,000 to $1,000,000 and the $5000 to $50,000 and ask yourself whether you'd prefer the sure $50,000!

2. You can either buy collision insurance on your expensive new car or not buy it, where the insurance costs a certain premium and carries some deductible provision. If you decide to pay the premium, then you are essentially paying a certain amount to avoid a gamble—the possibility of wrecking your car and not having it insured. You can be sure that the premium is greater than the expected cost of damage; otherwise, the insurance company would not stay in business. Therefore, from an EMV standpoint you should not purchase the insurance. But how many of you drive without this type of insurance?

These examples, the second of which is certainly realistic, illustrate situations where rational people do not behave as EMV maximizers. Then how do they act? This question has been studied extensively by many researchers, both mathematically and behaviorally. Although the answer is still not agreed upon universally, most researchers believe that if certain basic behavioral assumptions hold, people are **expected utility** maximizers—that is, they choose the alternative with the largest expected utility. Although we will not go deeply into the subject of expected utility maximization, the discussion in this section will acquaint you with the main ideas.

7.7.1 Utility Functions

We begin by discussing an individual's *utility function.* This is a mathematical function that transforms monetary values—payoffs and costs—into *utility values.*

Utility Function	An individual's **utility function** specifies the individual's preferences for various monetary payoffs and costs and, in doing so, it automatically encodes the individual's attitudes toward risk.

Most individuals are **risk averse,** which means intuitively that they are willing to sacrifice some EMV to avoid risky gambles. In terms of the utility function, this means that every extra dollar of payoff is worth slightly less to the individual than the previous dollar, and every extra dollar of cost is considered slightly more costly (in terms of utility) than the previous dollar. The resulting utility functions are shaped as shown in Figure 7.33. Mathematically, these functions are said to be **increasing** and **concave.** The increasing part means that they go uphill—everyone prefers more money to less money. The concave part means that they increase at a decreasing rate. This is the risk-averse preference.

There are two problems involved in implementing utility maximization in a real decision analysis. The first is obtaining an individual's (or company's) utility function; we will discuss this next. The second is using the resulting utility function to find the best decision. This second step is straightforward. We simply substitute utility values for monetary values in the decision tree and then fold back as usual. That is, we calculate expected *utilities* at probability branches and take maximums (of expected utilities) at de-

FIGURE 7.33
Risk-Averse Utility Function

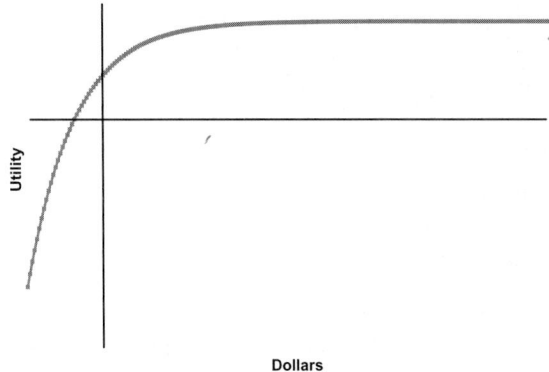

cision branches. We will look at a numerical example later in this section. So the real work involves finding an individual's (or company's) utility function in the first place.

7.7.2 Assessing a Utility Function

We will outline a method that can be used to estimate a person's utility function. There are two things we must understand about this method. First, it asks the person to make a series of trade-offs. Because each of us has different attitudes toward risk, we will not all make the trade-offs in the same way. Therefore, each of us will obtain our own utility function. Second, even a particular person's utility function is not unique. If $U(x)$ represents a person's utility function, then it turns out that $aU(x) + b$ gives the same preferences, for any constants a and b with $a > 0$. They are equivalent in the sense that they lead to exactly the same decisions.

We take advantage of this nonuniqueness by specifying two points on the utility function. Specifically, we begin by asking the person for two monetary values that represent the worst possible loss and the best possible gain imaginable. Let's say these values are $-A$ and B. Then we *arbitrarily* assign utility values 0 and 1 to these two monetary values, that is, $U(-A) = 0$ and $U(B) = 1$. Don't worry about the absolute magnitudes, 0 and 1, we've assigned—we could assign any other values, such as 14 and 320. The important thing is to use these as "anchors" and then obtain other utility values in terms of them.

The procedure is as follows. Given any two known utility values, say, $U(x)$ and $U(y)$, where x and y are monetary values, we present the person with a choice between the following two options:

- Option 1: Obtain a certain payoff of z.
- Option 2: Obtain a payoff of either x or y, depending on the flip of a fair coin.

Then we ask the person to select the monetary value z in option 1 so that he or she is *indifferent* to the two options. If the person is indifferent, then the expected utilities from the two options must be equal. We will call the resulting value of z the **indifference value.** This leads to equation (7.6) for $U(z)$.

Formula for Generating the Utility of an Indifference Value	$$U(z) = 0.5U(x) + 0.5U(y) \qquad (7.6)$$

In words, we have generated a new utility value from two known utility values. This process continues until we have enough utility values to approximate a utility curve. (Note that if any of x, y, and z are negative, then "payoff" really means "cost.") We'll illustrate this procedure with an example.

Example 7.7 Assessing an Entrepreneur's Utility Function

John Jacobs owns his own business. Because he is about to make an important decision where large losses or large gains are at stake, he wants to use the expected utility criterion to make his decision. He knows that he must first assess his own utility function, so he hires a decision analysis expert, Susan Schilling, to help him out. How might the session between John and Susan proceed?

Objective To illustrate the process of assessing an individual's utility function by repeatedly generating a new utility value from two known utility values.

Solution

Susan first asks John for the largest loss and largest gain he can imagine. He answers with the values $200,000 and $300,000, so she assigns utility values $U(-200000) = 0$ and $U(300000) = 1$ as anchors for the utility function. Now she asks John to choose between two options:

- Option 1: Obtain a payoff of z (really a loss if z is negative).
- Option 2: Obtain a loss of $200,000 or a payoff of $300,000, depending on the flip of a fair coin.

Susan reminds John that the EMV of option 2 is $50,000 (halfway between $-$200,000 and $300,000). He realizes this, but because he is quite risk averse, he would far rather have $50,000 for certain than take the gamble in option 2. Therefore, the indifference value of z must be less than $50,000. Susan then poses several values of z to John. Would he rather have $10,000 for sure or take option 2? He says he'd rather take the $10,000. Would he rather *pay* $5000 for sure or take the gamble in option 2? (This is like an insurance premium.) He says he'd rather take option 2. By this time, we know the indifference value of z must be less than $10,000 and greater than $-$5000. With a few more questions of this type, John finally decides on $z = 5000 as his indifference value. He is indifferent between obtaining $5000 for sure and taking the gamble in option 2. We can substitute these values into equation (7.6):

$$U(5000) = 0.5U(-200000) + 0.5U(300000) = 0.5(0) + 0.5(1) = 0.5$$

Note that John is giving up $45,000 in EMV because of his risk aversion. The EMV of the gamble in option 2 is $50,000, and he is willing to accept a *sure* $5000 in its place.

The process would then continue. For example, since she now knows $U(5000)$ and $U(300000)$, Susan could ask John to choose between these options:

- Option 1: Obtain a payoff of z.
- Option 2: Obtain a payoff of $5000 or a payoff of $300,000, depending on the flip of a fair coin.

If John decides that his indifference value is now $z = $130,000$, then with equation (7.6) we know that

$$U(130000) = 0.5U(5000) + 0.5U(300000) = 0.5(0.5) + 0.5(1) = 0.75$$

Note that John is now giving up $22,500 in EMV because the EMV of the gamble in option 2 is $152,500. By continuing in this manner, Susan can help John assess enough utility values to approximate a continuous utility curve. ●

As this example illustrates, utility assessment is tedious. Even in the best of circumstances, when a trained consultant attempts to assess the utility function of a single person, the process requires the person to make a series of choices between hypothetical alternatives involving uncertain outcomes. Unless the person has some training in probability, these choices can be difficult to understand, let alone make, and it is unlikely that the person will answer *consistently* as the questioning proceeds. The process is even more difficult when a company's utility function is being assessed. Because company executives involved typically have different attitudes toward risk, it is difficult for these people to reach a consensus on a common utility function.

7.7.3 Exponential Utility

For these reasons there are classes of "ready-made" utility functions that have been developed. One important class is called *exponential utility* and has been used in many financial investment analyses. An **exponential utility function** has only one adjustable numerical parameter, and there are straightforward ways to discover the most appropriate value of this parameter for a particular individual or company. So the advantage of using an exponential utility function is that it is relatively easy to assess. The drawback is that exponential utility functions do not capture all types of attitudes toward risk. Nevertheless, their ease of use has made them popular.

An exponential utility function has the form in equation (7.7). Here x is a monetary value (a payoff if positive, a cost if negative), $U(x)$ is the utility of this value, and $R > 0$ is an adjustable parameter called the *risk tolerance*. Basically, the **risk tolerance** measures how much risk the decision maker will tolerate. The larger the value of R, the less risk averse the decision maker is. That is, a person with a large value of R is more willing to take risks than a person with a small value of R.

Exponential Utility Function	$$U(x) = 1 - e^{-x/R} \qquad (7.7)$$

To assess a person's (or company's) exponential utility function, we need only to assess the value of R. There are a couple of tips for doing this. First, it has been shown that the risk tolerance is approximately equal to that dollar amount R such that the decision maker is indifferent between the following two options:

- Option 1: Obtain no payoff at all.
- Option 2: Obtain a payoff of R dollars or a loss of $R/2$ dollars, depending on the flip of a fair coin.

For example, if you are indifferent between a bet where you win $1000 or lose $500, with probability 0.5 each, and not betting at all, then your R is approximately $1000. From this criterion it certainly makes intuitive sense that a wealthier person (or company) ought to have a larger value of R. This has been found in practice.

A second tip for finding R is based on empirical evidence found by Ronald Howard, a prominent decision analyst. Through his consulting experience with several large companies, he discovered tentative relationships between risk tolerance and several financial variables—net sales, net income, and equity. [See Howard (1992).] Specifically, he found that R was approximately 6.4% of net sales, 124% of net income, and 15.7% of equity for the companies he studied. For example, according to this prescription, a company with net sales of $30 million should have a risk tolerance of approximately $1.92 million. Howard admits that these percentages are only guidelines. However, they do indicate that larger and more profitable companies tend to have larger values of R, which means that they are more willing to take risks involving given dollar amounts.

We illustrate the use of the expected utility criterion, and exponential utility in particular, with the following example.

Example 7.8 Evaluating Risky Ventures at Venture Limited

Venture Limited is a company with net sales of $30 million. The company currently must decide whether to enter one of two risky ventures or do nothing. The possible outcomes of the less risky venture are a $0.5 million loss, a $0.1 million gain, and a $1 million gain. The probabilities of these outcomes are 0.25, 0.50, and 0.25. The possible outcomes of the more risky venture are a $1 million loss, a $1 million gain, and a $3 million gain. The probabilities of these outcomes are 0.35, 0.60, and 0.05. If Venture Limited can enter at most one of the two risky ventures, what should it do?

Objective To use a decision tree model to assist Venture Limited's management in identifying the course of action that maximizes the company's *expected utility* for a given risk tolerance.

Solution

We will assume that Venture Limited has an exponential utility function. Also, based on Howard's guidelines, we will assume that the company's risk tolerance is 6.4% of its net sales, or $1.92 million. (We'll do a sensitivity analysis on this parameter later on.) We can substitute into the exponential utility formula (7.7) to find the utility of any monetary outcome. For example, the gain from doing nothing is $0, and its utility is

$$U(0) = 1 - e^{-0/1.92} = 1 - 1 = 0$$

As another example, the utility of a $1 million loss is

$$U(-1) = 1 - e^{-(-1)/1.92} = 1 - 1.683 = -0.683$$

These are the values we use (instead of monetary values) in the decision tree.

Fortunately, PrecisionTree takes care of all the details. After we build a decision tree and label it (with monetary values) in the usual way, we click on the name of the tree (the box on the far left of the tree) to open the dialog box in Figure 7.34. We then fill in the utility function information as shown in the upper right section of the dialog box. This says to use an exponential utility function with risk tolerance 1.92. The dialog box also indicates that we want expected utilities (as opposed to EMVs) to appear in the decision tree.

The completed tree for this example appears in Figure 7.35. (See the file VENTURE.XLS.) We build it in exactly the same way as usual and link probabilities and monetary values to its branches in the usual way. For example, there is a link in cell C22

Don't forget the extra minus sign when calculating the exponential utility of a negative number (a cost).

FIGURE 7.34
Dialog Box for Specifying the Exponential Utility Criterion

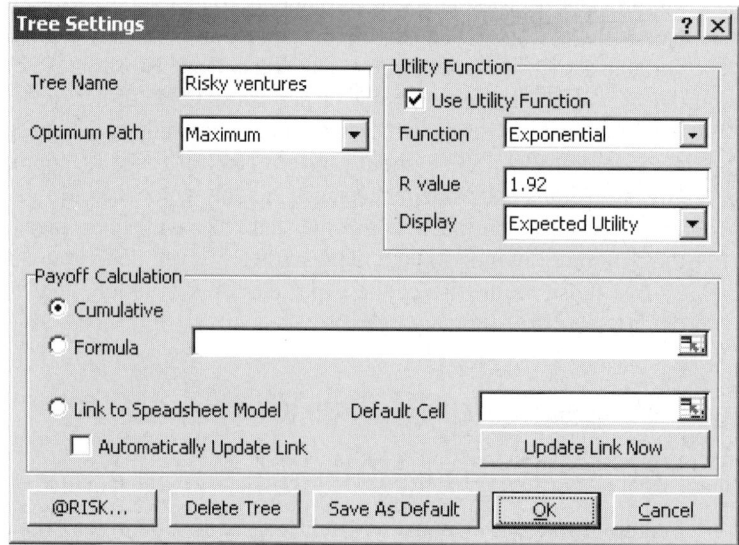

FIGURE 7.35

Decision Tree for Risky Venture Example

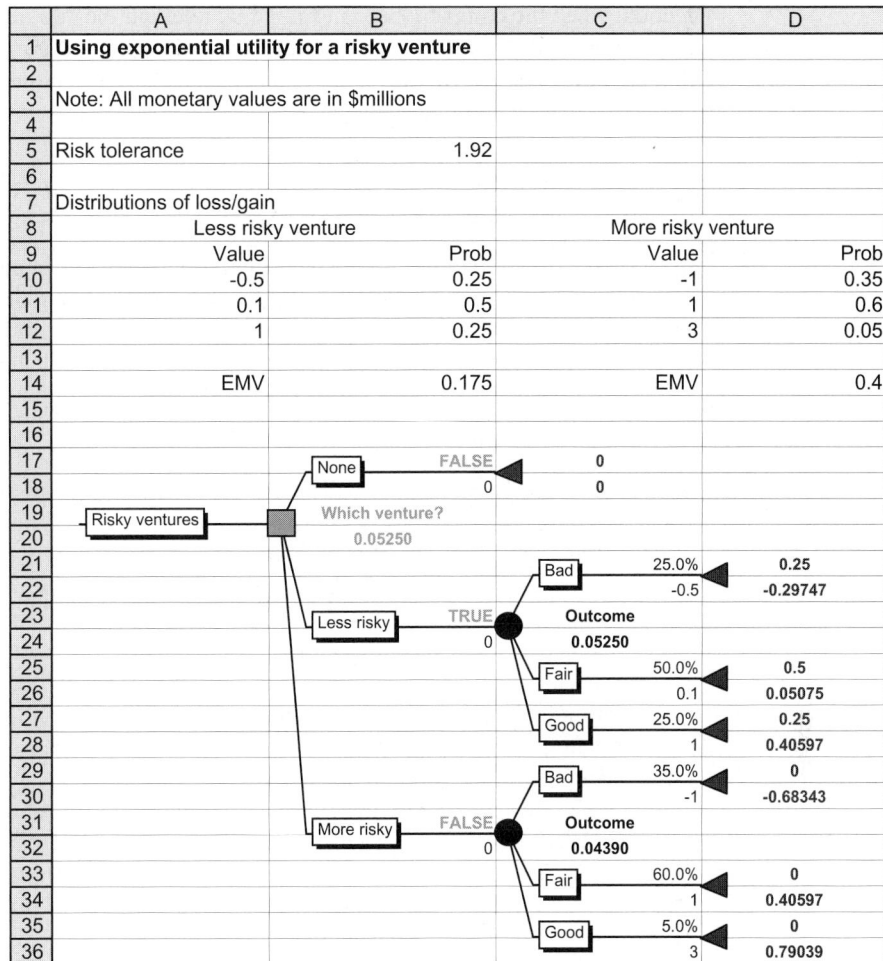

	A	B	C	D
1	**Using exponential utility for a risky venture**			
2				
3	Note: All monetary values are in $millions			
4				
5	Risk tolerance	1.92		
6				
7	Distributions of loss/gain			
8	Less risky venture		More risky venture	
9	Value	Prob	Value	Prob
10	-0.5	0.25	-1	0.35
11	0.1	0.5	1	0.6
12	1	0.25	3	0.05
13				
14	EMV	0.175	EMV	0.4

Decision tree (rows 17–36):

- None — FALSE — 0 / 0 / 0
- Risky ventures → Which venture? 0.05250
  - Less risky — TRUE — 0
    - Bad 25.0% — 0.25 / -0.5 / -0.29747
    - Outcome 0.05250
    - Fair 50.0% — 0.5 / 0.1 / 0.05075
    - Good 25.0% — 0.25 / 1 / 0.40597
  - More risky — FALSE — 0
    - Bad 35.0% — 0 / -1 / -0.68343
    - Outcome 0.04390
    - Fair 60.0% — 0 / 1 / 0.40597
    - Good 5.0% — 0 / 3 / 0.79039

to the monetary value in cell A10. However, the expected values shown in the tree (those shown in color on your screen) are expected *utilities,* and the optimal decision is the one with the largest expected utility. In this case the expected utilities for doing nothing, investing in the less risky venture, and investing in the more risky venture are 0, 0.0525, and 0.0439, respectively. Therefore, the optimal decision is to invest in the less risky venture.

Note that the EMVs of the three decisions are $0, $0.175 million, and $0.4 million. The latter two of these are calculated in row 14 as the usual "sumproduct" of monetary values and probabilities. So from an EMV point of view, the more risky venture is definitely best. However, Venture Limited is sufficiently risk averse, and the monetary values are sufficiently large, that the company is willing to sacrifice EMV to reduce its risk.

How sensitive is the optimal decision to the key parameter, the risk tolerance? We can answer this by changing the risk tolerance (through the dialog box in Figure 7.34) and watching how the decision tree changes.[7] You can check that when the company becomes *more* risk tolerant, the more risky venture eventually becomes optimal. In fact, this occurs when the risk tolerance increases to approximately $2.075 million. In the other di-

---

[7]We show the risk tolerance in cell B5, but the values in the decision tree are not linked to that cell. We need to go through the dialog box to change the risk tolerance.

rection, when the company becomes *less* risk tolerant, the "do nothing" decision eventually becomes optimal. This occurs when the risk tolerance decreases to approximately $0.715 million. So the "optimal" decision depends heavily on the attitudes toward risk of Venture Limited's top management. ●

7.7.4 Certainty Equivalents

Now suppose that Venture Limited has only two options. It can either enter the less risky venture or receive a *certain* dollar amount x and avoid the gamble altogether. We want to find the dollar amount x such that the company is indifferent between these two options. If it enters the risky venture, its expected utility is 0.0525, calculated previously. If it receives x dollars for certain, its (expected) utility is

$$U(x) = 1 - e^{-x/1.92}$$

To find the value x where it is indifferent between the two options, we set $1 - e^{-x/1.92}$ equal to 0.0525, or $e^{-x/1.92} = 0.9475$, and solve for x. Taking natural logarithms of both sides and multiplying by -1.92, we obtain

$$x = -1.92 \ln(0.9475) \approx \$0.104 \text{ million}$$

This value is called the **certainty equivalent** of the risky venture. The company is indifferent between entering the less risky venture and receiving $0.104 million to avoid it. Although the EMV of the less risky venture is $0.175 million, the company acts as if it is equivalent to a sure $0.104 million. In this sense, the company is willing to give up the difference in EMV, $71,000, to avoid a gamble.

By a similar calculation, the certainty equivalent of the more risky venture is approximately $0.086 million. That is, the company acts as if this more risky venture is equivalent to a sure $0.086 million, when in fact its EMV is a hefty $0.4 million! So in this case it is willing to give up the difference in EMV, $314,000, to avoid this particular gamble. Again, the reason is that the company dislikes risk. We can see these certainty equivalents in PrecisionTree by adjusting the Display box in Figure 7.34 to show Certainty Equivalent. The resulting tree is shown in Figure 7.36. The certainty equivalents we just discussed appear in cells C24 and C32.

FIGURE 7.36
Decision Tree with Certainty Equivalents

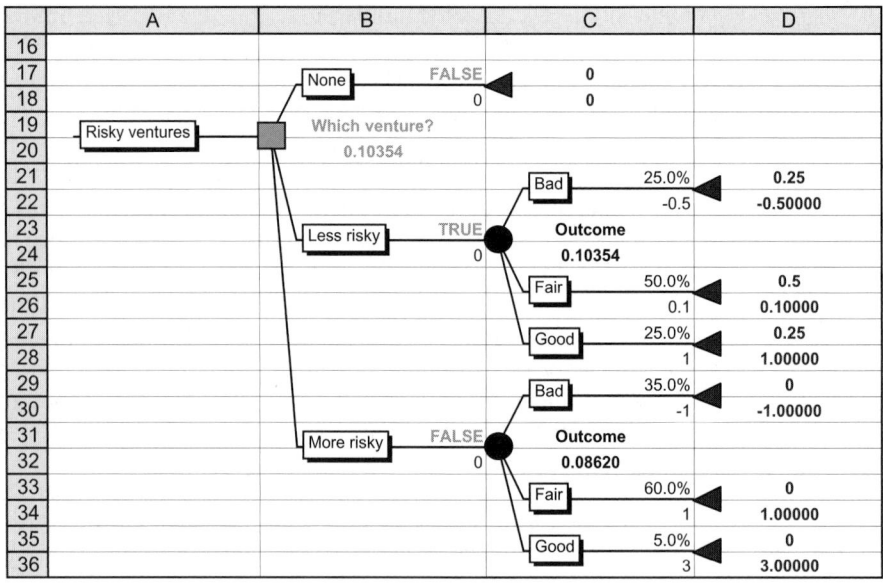

7.7.5 Is Expected Utility Maximization Used?

The previous discussion indicates that utility maximization is a fairly involved task. The bottom line, then, is whether the difficulty is worth the trouble. Theoretically, expected utility maximization might be interesting to researchers, but is it really used? The answer appears to be: not very often. For example, one recent article on the practice of decision making [see Kirkwood (1992)] quotes Ronald Howard—the same person we quoted previously—as having found risk aversion to be of practical concern in only 5% to 10% of business decision analyses. This same article quotes the president of a Fortune 500 company as saying, "Most of the decisions we analyze are for a few million dollars. It is adequate to use expected value (EMV) for these."

With these comments in mind, it is clear that knowledge of expected utility maximization is an important requirement for anyone intending to specialize in the field. In some of the greatest success stories, expected utility maximization was indeed implemented. For nonspecialists, however, a passing knowledge of the concepts is sufficient.

PROBLEMS

Level A

45. Suppose that a decision maker's utility as a function of his wealth, x, is given by $U(x) = \ln x$ (the natural logarithm of x).
   a. Is this decision maker risk averse? Explain why or why not.
   b. The decision maker now has \$10,000 and two possible decisions. For decision 1, he loses \$500 for certain. For decision 2, he loses \$0 with probability 0.9 and loses \$5000 with probability 0.1. Which decision maximizes the expected utility of his net wealth?

46. An investor has \$10,000 in assets and can choose between two different investments. If she invests in the first investment opportunity there is an 80% chance that she will increase her assets by \$590,000 and a 20% chance that she will increase her assets by \$190,000. If she invests in the second investment opportunity there is a 50% chance that she will increase her assets by \$1.19 million and a 50% chance that she will increase her assets by \$1000. This investor has an exponential utility function for final assets with a risk tolerance parameter equal to \$600,000. Which investment opportunity will she prefer?

47. Consider again FreshWay's decision problem described in Example 7.3. Suppose now that FreshWay's utility function of profit π, earned from the acquisition and sale of the 24,000 fluorescent lightbulbs, is $U(\pi) = \ln(\pi)$. Find the course of action that maximizes FreshWay's expected utility. How does this optimal decision compare to the optimal decision with an EMV criterion? Explain any difference between the two decisions.

48. Consider again the landowner's decision problem described in Problem 3. Suppose now that the landowner's utility function of financial gain is exponential with risk tolerance \$1 million. Find the course of action that maximizes the landowner's expected utility. How does this optimal decision compare to the optimal decision with an EMV criterion? Explain any difference between the two decisions.

49. Consider again Techware's decision problem described in Problem 4. Suppose now that Techware's utility function of profit r (measured in dollars), earned from the given marketing opportunities, is $U(r) = 1 - e^{-r/350,000}$.
   a. Find the course of action that maximizes Techware's expected utility. How does this optimal decision compare to the optimal decision with an EMV criterion? Explain any difference between the two optimal decisions.
   b. Repeat part **a** when Techware's utility function is $U(r) = 1 - e^{-r/50,000}$.

50. Consider again the bank's customer loan decision problem in Problem 39. Suppose now that the bank's utility function of profit π (in dollars) is $U(\pi) = 1 - e^{-\pi/10,000}$. Find the strategy that maximizes the bank's expected utility in this case. How does this optimal strategy compare to the optimal decision with an EMV criterion? Explain any difference between two optimal strategies.

Level B

51. Suppose that a decision maker has a utility function for monetary gains x given by $U(x) = (x + 10000)^{0.5}$.
   a. Show that this decision maker is indifferent between gaining nothing (i.e., \$0) and entering a risky situation where she gains \$80,000 with probability 1/3 and loses \$10,000 with probability 2/3.

b. If there is a 10% chance that one of the decision maker's family heirlooms, valued at $5000, will be stolen during the next year, what is the most that she would be willing to pay each year for an insurance policy that completely covers the potential loss of her cherished item?

52. A decision maker is going to invest $2000 for a period of 6 months. Two potential investments are available to him: U.S. Treasury bills and gold. If this decision maker invests the $2000 in T-bills, he is sure to end the 6-month period with $2592. If this decision maker invests in gold, there is a 75% chance that he will end the 6-month period with $800 and a 25% chance that he will end up with $20,000. The decision maker's utility function of ending up with x dollars is $U(x) = x^{0.5}$.

a. Should this decision maker invest in gold or T-bills?

b. Suppose the decision maker invests a proportion y of his $2000 in T-bills and the remaining fraction $(1 - y)$ of his available funds in gold. In this case his gain or loss from either investment is reduced proportionally. For example, if he invests half of his money in gold, he will either lose $600 with probability 0.75 or gain $9000 with probability 0.25. Given the same utility function $U(x) = x^{0.5}$, find the investor's optimal choice of y.

 # Conclusion

In this chapter we have discussed methods that can be used in decision-making problems in which future uncertainty is a key element. Perhaps the most important skill we can gain from this chapter is the ability to approach decision problems that include uncertainty in a systematic manner. This systematic approach requires the decision maker to list all possible decisions or strategies, list all possible uncertain outcomes, assess the probabilities of these outcomes (possibly with the aid of Bayes' rule), calculate all necessary monetary values, and finally do the calculations necessary to obtain the best decision. If large dollar amounts are at stake, it might also be necessary to perform a utility analysis, where the decision maker's feelings toward risk are taken into account. Once the basic analysis has been completed, using "best guesses" for the various parameters of the problem, a sensitivity analysis should be conducted to see whether the best decision continues to be best within a range of problem parameters.

Summary of Key Terms

Term	Explanation	Excel	Page	Equation Number
Payoff table	A list of payoffs (or costs, if negative) for each decision/outcome combination.		303	
Risk profile	A graphical representation of the probability distribution of all possible outcomes from a particular decision.	PrecisionTree	304	
Expected monetary value (EMV)	The weighted average of all possible monetary outcomes, weighted by their probabilities.	=SUMPRODUCT(values, probs)	305	7.1
Decision tree	Graphical representation of all elements of a decision analysis.	PrecisionTree	305–308	
Decision node	Has branches corresponding to possible decisions at some point in time.	PrecisionTree	306	
Probability node	Has branches corresponding to possible outcomes, of which exactly one will occur.	PrecisionTree	306	

Summary of Key Terms (continued)

Term	Definition	Tool	Page	Box
Single-stage decision problems	One where a single decision is made and then all uncertainty is resolved.		307, 320–327	
Multistage decision problems	One where decisions and observation of uncertain outcomes alternate.		307, 329–335	
Folding back	Process of taking maximums at decision nodes and EMVs at probability nodes to find optimal decision and its EMV.	PrecisionTree	307	
Sensitivity analysis	Process of seeing how optimal decision and EMV vary when one or more inputs vary.	PrecisionTree	316	
Contingency plan	Strategy in a multistage decision problem that specifies which decision to make for each possible outcome.		330	
Law of total probability	Formula for finding a probability by conditioning on all possible outcomes.		331	7.2
Expected value of sample information (EVSI)	Difference between EMV with sample information and EMV without any information, the most a decision maker should pay for the sample information.		333	7.3
Expected value of perfect information (EVPI)	Difference between EMV with perfect information and EMV without any information, the most a decision maker should pay for the perfect information.		334	7.4
Bayes' rule	Formula for updating probabilities as new information becomes available.		337	7.5
Prior, posterior probabilities	Probabilities before and after new information becomes available.		338	
Likelihoods	Probabilities of possible sample information.		338	
Utility function	Function that encodes a person's (or company's) feelings toward risk.		346	
Risk aversion	Tendency to prefer a sure thing to a gamble with the same EMV.		346	
Assessing a utility function	Process of asking questions about trade-offs to learn a person's attitude toward risk.		347	7.6
Exponential utility function	Particular type of utility function that requires only one parameter to be specified.		349	7.7
Risk tolerance	Parameter of exponential utility function that indicates how much risk the person will tolerate.		349	
Certainty equivalent	Dollar value such that a person is indifferent between receiving this value for sure and entering a gamble.		352	

PROBLEMS

Conceptual Exercises

C.1. What is the maximum number of monetary outcomes that could appear in a payoff table associated with a single-stage decision problem that includes x possible decisions and y possible uncertain outcomes? Explain.

C.2. Is it possible for the bar chart of a risk profile to consist of *one* bar only? Explain your answer.

C.3. A marketing manager constructs and solves a decision tree model to determine whether he should recommend the introduction of a highly innovative product into the marketplace. The optimal strategy yields an expected monetary value of $350 million. He prepares a memo to his boss (the vice-president for marketing) that indicates the company is guaranteed to earn $350 million by pursuing the given optimal strategy. Assuming that the marketing manager's decision tree model is valid, how should the vice-president for marketing respond to the manager's memo?

C.4. Under what conditions, if any, would the expected value of sample information equal the expected value of perfect information for a given multistage decision problem?

C.5. In applying Bayes' rule, what is the difference between a *likelihood* probability and a *posterior* probability?

Level A

53. Ford is going to produce a new vehicle, the Pioneer, and wants to determine the amount of annual capacity it should build. Ford's goal is to maximize the profit from this vehicle over the next 10 years. Each vehicle will sell for $13,000 and incur a variable production cost of $10,000. Building one unit of annual capacity will cost $3000. Each unit of capacity will also cost $1000 per year to maintain, even if the capacity is unused. Demand for the Pioneer is unknown but marketing estimates the distribution of annual demand to be as shown in the file P07_53.XLS. Assume that unit sales during a year is the minimum of capacity and annual demand.
  a. Explain why a capacity of 1,300,000 is not a good choice.
  b. Which capacity level should Ford choose?

54. You are CEO of the venture capital firm D&D. Billy comes to you with an investment proposition. You estimate that your distribution of cash flows from this investment is as shown in the file P07_54.XLS.
  a. If you are trying to maximize the expected value of the firm's cash flows, should you take the project?

b. Suppose you assess your firm to be risk averse, with an exponential utility function. You also use the rule of thumb that the firm's risk tolerance is about 6.4% of its annual revenues, which are $30 million. Determine whether D&D should enter the venture.

55. Pizza King (PK) and Noble Greek (NG) are competitive pizza chains. Pizza King believes there is a 25% chance that NG will charge $6 per pizza, a 50% chance that NG will charge $8 per pizza, and a 25% chance that NG will charge $10 per pizza. If PK charges price p_1 and NG charges price p_2, PK will sell $100 + 25(p_2 - p_1)$ pizzas. It costs PK $4 to make a pizza. PK is considering charging $5, $6, $7, $8, or $9 per pizza. To maximize its expected profit, what price should PK charge for a pizza?

56. Sodaco is considering producing a new product: Chocovan soda. Sodaco estimates that the annual demand for Chocovan, D (in thousands of cases), has the following probability distribution: $P(D = 30) = 0.30$, $P(D = 50) = 0.40$, $P(D = 80) = 0.30$. Each case of Chocovan sells for $5 and incurs a variable cost of $3. It costs $800,000 to build a plant to produce Chocovan. Assume that if $1 is received every year (forever), this is equivalent to receiving $10 at the present time. If Sodaco decides to build the plant and produce Chocovan, find the expected net present value of its profit.

57. Many decision problems have the following simple structure. A decision maker has two possible decisions, 1 and 2. If decision 1 is made, a *sure* cost of c is incurred. If decision 2 is made, there are two possible outcomes, with costs c_1 and c_2 and probabilities p and $1 - p$. We assume that $c_1 < c < c_2$. The idea is that decision 1, the riskless decision, has a "moderate" cost, whereas decision 2, the risky decision, has a "low" cost c_1 or a "high" cost c_2.
  a. Find the decision maker's cost table—that is, the cost for each possible decision and each possible outcome.
  b. Calculate the expected cost from the risky decision.
  c. List as many scenarios as you can think of that have this structure. (Here'a an example to get you started. Think of insurance, where you pay a sure premium to avoid a large possible loss.)

58. During the summer, Olympic swimmer Adam Johnson swims every day. On sunny summer days he goes to an outdoor pool, where he may swim for no charge. On rainy days he must go to a domed pool. At the beginning of the summer, he has the option of purchasing a $15 season pass to the domed pool, which allows him use for the entire summer. If he doesn't buy the season pass, he must pay $1 each time he goes there. Past meteorological records indicate that there is a 60% chance that the summer will be sunny (an average of 6 rainy days during the summer) and a 40% chance the summer will be rainy (an average of 30 rainy days during the summer).

Before the summer begins, Adam has the option of purchasing a long-range weather forecast for $1. The forecast predicts a sunny summer 80% of the time and a rainy summer 20% of the time. If the forecast predicts a sunny summer, there is a 70% chance that the summer will actually be sunny. If the forecast predicts a rainy summer, there is an 80% chance that the summer will actually be rainy. Assuming that Adam's goal is to minimize his total expected cost for the summer, what should he do? Also find the EVSI and the EVPI.

59. Erica is going to fly to London on August 5 and return home on August 20. It is now July 1. On July 1, she can buy a one-way ticket (for $350) or a round-trip ticket (for $660). She can also wait until August 1 to buy a ticket. On August 1, a one-way ticket will cost $370, and a round-trip ticket will cost $730. It is possible that between July 1 and August 1, her sister (who works for the airline) will be able to obtain a free one-way ticket for Erica. The probability that her sister will obtain the free ticket is 0.30. If Erica has bought a round-trip ticket on July 1 and her sister has obtained a free ticket, she can return "half" of her round-trip to the airline. In this case, her total cost will be $330 plus a $50 penalty. Use a decision tree approach to determine how to minimize Erica's expected cost of obtaining round-trip transportation to London.

60. A nuclear power company is deciding whether to build a nuclear power plant at Diablo Canyon or at Roy Rogers City. The cost of building the power plant is $10 million at Diablo and $20 million at Roy Rogers City. If the company builds at Diablo, however, and an earthquake occurs at Diablo during the next 5 years, construction will be terminated and the company will lose $10 million (and will still have to build a power plant at Roy Rogers City). Without further information, the company believes there is a 20% chance that an earthquake will occur at Diablo during the next 5 years. For $1 million, a geologist can be hired to analyze the fault structure at Diablo Canyon. She will either predict that an earthquake will occur or that an earthquake will not occur. The geologist's past record indicates that she will predict an earthquake on 95% of the occasions for which an earthquake will occur and no earthquake on 90% of the occasions for which an earthquake will not occur. Should the power company hire the geologist? Also find the EVSI and the EVPI.

61. Joan's utility function for her asset position x (for x between 0 and $160,000) is given by $U(x) = \sqrt{x}/400$.

a. Is Joan risk averse? Explain.

b. Currently, Joan's assets consist of $10,000 in cash and a $90,000 home. During a given year, there is a 0.001 probability that Joan's home will be destroyed by fire or other causes. How much should Joan be willing to pay for insurance that covers her home completely from this type of destruction?

62. My current annual income is $40,000. I believe that I owe $8000 in taxes. For $500, I can hire a CPA to review my tax return. There is a 20% chance she will save me $4000 in taxes and an 80% chance she won't save me anything. If x is my disposable income for the current year, my utility function is given by $U(x) = \sqrt{x}/200$.

a. Am I risk averse or risk seeking?

b. Should I hire the accountant?

Level B

63. City officials in Ft. Lauderdale, Florida, are trying to decide whether to evacuate coastal residents in anticipation of a major hurricane that may make landfall near their city within the next 48 hours. Based on previous studies, it is estimated that it will cost approximately 1 million dollars to evacuate the residents living along the coast of this major metropolitan area. However, if city officials choose not to evacuate their residents and the storm strikes Ft. Lauderdale, there would likely be some deaths as a result of the hurricane's storm surge along the coast. While city officials are reluctant to place an economic value on the loss of human life resulting from such a storm, they realize that it may ultimately be necessary to do so to make a sound judgment in this situation. Prior to making the evacuation decision, city officials consult hurricane experts at the National Hurricane Center in Coral Gables regarding the accuracy of past predictions. They learn that in similar past cases, hurricanes that were *predicted* to make landfall near a particular coastal location actually did so 60% of the time. Moreover, they learn that in past similar cases hurricanes that were predicted *not* to make landfall near a particular coastal location actually did so 20% of the time. Finally, in response to similar threats in the past, weather forecasters have issued predictions of a major hurricane making landfall near a particular coastal location 40% of the time.

a. Let L be the economic valuation of the loss of human life resulting from a coastal strike by the hurricane in this case. Employ a decision tree to help these city officials make a decision that minimizes the expected cost of responding to the threat of the impending storm as a function of L. To proceed, you might begin by choosing an initial value of L and then perform sensitivity analysis on the optimal decision by varying this model parameter. Summarize your findings.

b. For which values of *L* will these city officials *always* choose to evacuate the coastal residents, regardless of the Hurricane Center's prediction?

64. A homeowner wants to decide whether he should install an electronic heat pump in his home. Given that the cost of installing a new heat pump is fairly large, the homeowner would like to do so only if he can count on being able to recover the initial expense over *five* consecutive years of cold winter weather. Upon reviewing historical data on the operation of heat pumps in various kinds of winter weather, he computes the expected annual costs of heating his home during the winter months with and without a heat pump in operation. These cost figures are shown in the file P07_64.XLS. The probabilities of experiencing a mild, normal, colder than normal, and severe winter are $0.2(1 - x)$, $0.5(1 - x)$, $0.3(1 - x)$, and x, respectively.

a. Given that $x = 0.1$, what is the most that the homeowner is willing to pay for the heat pump?

b. If the heat pump costs $500, how large must x be before the homeowner decides it is economically worthwhile to install the heat pump?

c. Given that $x = 0.1$, compute and interpret the expected value of perfect information (EVPI) when the heat pump costs $500.

d. Repeat part **c** when $x = 0.15$.

65. Consider a company that manufactures computer memory chips in lots of ten chips. From past experience, the company knows that 80% of all lots contain 10% defective chips, and 20% of all lots contain 50% defective chips. If an *acceptable* (that is, 10% defective) batch of chips is sent on to the next stage of production, processing costs of $10,000 are incurred. If an *unacceptable* (that is, 50% defective) batch is sent on to the next stage of production, processing costs of $40,000 are incurred. This company also has the option of reworking a batch of chips at a cost of $10,000. A reworked batch is guaranteed to be acceptable. Alternatively, at a cost of $1000, the company can test one memory chip from each batch in an attempt to determine whether the given batch is unacceptable. If a randomly selected chip is found to be defective, the batch from which the chip came is acceptable with probability 8/18. If a randomly selected chip is found *not* to be defective, the batch from which the chip came is acceptable with probability 72/82.

a. Determine how this company can minimize the expected total cost per batch of computer memory chips.

b. Compute and interpret the expected value of sample information (EVSI) in this decision problem.

c. Compute and interpret the expected value of perfect information (EVPI) in this decision problem.

d. Suppose now that this manufacturer's utility function tion of cost *c* per batch is $U(c) = -(c)^{0.5}$. Find the strategy that maximizes the manufacturer's expected utility. How does this optimal strategy compare to the optimal decision with an EMV criterion? Explain any difference between the two optimal strategies.

66. Patty is trying to determine whether to take management science or statistics. If she takes management science, she believes there is a 10% chance she will receive an A, a 40% chance she will receive a B, and a 50% chance she will receive a C. If Patty takes statistics, she has a 70% chance of receiving a B, a 25% chance of a C, and a 5% chance of a D. Patty is indifferent between the following two options:

• Option 1: Receiving a B for certain
• Option 2: A 70% chance at an A and a 30% chance at a D

Patty is also indifferent between the following two options:

• Option 3: Receiving a C for certain
• Option 4: A 25% chance at an A and a 75% chance at a D

To maximize the expected utility associated with her final grade, which course should Patty take?

67. Many men over 50 take the PSA blood test. The purpose of the PSA test is to catch prostate cancer early. Dr. Rene Labrie of Quebec conducted a study to determine whether the PSA test can actually prevent cancer deaths. In 1989 Dr. Labrie randomly divided all male registered voters between 45 and 80 in Quebec City into two groups. Two-thirds of the men were asked to be tested for prostate cancer and one-third were not asked. Eventually, 8137 men were screened for prostate cancer (PSA plus digital rectal exam) in 1989; 38,056 men were not screened. By 1997 only 5 of the screened men had died of prostate cancer whereas 137 of the men who were not screened had died of prostate cancer. (Source: *New York Times*, May 19, 1998)

a. Discuss why this study seems to indicate that screening for prostate cancer saves lives.

b. Despite the results of this study, many doctors are not convinced that early screening for prostate cancer saves lives. Can you see why they doubt the conclusions of the study?

68. You have just been chosen to appear on *Hoosier Millionaire*! The rules are as follows: There are four hidden cards. One says "STOP" and the other three have dollar amounts of $150,000, $200,000, and $1,000,000. You get to choose a card. If the card says "STOP," you win no money. At any time you may quit and keep the largest amount of money that has appeared on any card you have chosen, or you may continue. If you continue and choose the STOP card, however, you win no money. As an example, you might first choose the $150,000 card, then the $200,000 card, and then choose to quit and receive $200,000.

    a. If your goal is to maximize your expected payoff, what strategy should you follow?

    b. Suppose your utility function for an increase in cash satisfies $U(0) = 0$, $U(\$40,000) = 0.25$, $U(\$120,000) = 0.50$, $U(\$400,000) = 0.75$, and $U(\$1,000,000) = 1$. Are you risk averse? Explain.

    c. After drawing a curve through the points in part **b**, determine a strategy that maximizes your expected utility. (Alternatively, you might want to assess and use your *actual* utility function.)

69. You are trying to determine how much money to put in your Tax Saver Benefit (TSB) plan. At the beginning of the calendar year, a TSB allows you to put money into an account. The money in the account can be used to pay for medical expenses incurred during the year. Once the TSB is exhausted, you must pay the medical expenses out of pocket. The benefit of the TSB is that money placed in the TSB is not subject to federal taxes. The catch is that any money left in the TSB at the end of the year is lost to you. Suppose the federal tax rate is 40% and your current annual salary is $50,000. You believe that it is equally likely that your medical expenses during the current year will be $3000, $4000, $5000, $6000, or $7000.

    a. If you are risk neutral and want to maximize your expected disposable income, how much should you put in your TSB?

    b. Suppose you assess a utility function for disposable income given by $U(x) = 0.000443x^{0.713595}$. (Who said they all have to have nice round numbers?) Are you risk averse? How much should you put in the TSB?

70. Peter is thinking of purchasing an advertising company from Amanda. At present, only Amanda (not Peter) knows the current value of the company. Peter knows, however, that there is an equal chance that the company is worth 10, 20, 30, 40, 50, 60, 70, 80, 90, or 100 million dollars. Amanda will accept an offer from Peter only if Peter bids at least the value of the company. For example, if Amanda knows the company is worth $20 million, she will accept any bid of $20 million or higher. As soon as Peter purchases the company, his reputation as a skilled businessman immediately increases the actual value of the company by 80%.

    a. Suppose Peter is risk neutral and is considering bidding 10, 20, 30, 40, 50, 60, 70, 80, 90, or 100 million dollars. What should he bid?

    b. Suppose Peter's utility function for financial gains or losses (in millions of dollars) is given by $U(x) = ((x + 82)/144)^{1.7}$. Determine whether Peter is risk averse or risk seeking and determine Peter's optimal decision.

71. Sarah Chang is the owner of a small electronics company. In 6 months a proposal is due for an electronic timing system for the 2008 Olympic Games. For several years, Chang's company has been developing a new microprocessor, a critical component in a timing system that would be superior to any product currently on the market. However, progress in research and development has been slow, and Chang is unsure about whether her staff can produce the microprocessor in time. If they succeed in developing the microprocessor (probability p_1), there is an excellent chance (probability p_2) that Chang's company will win the $1 million Olympic contract. If they do not, there is a small chance (probability p_3) that she will still be able to win the same contract with an alternative, inferior timing system that has already been developed.

    If she continues the project, Chang must invest $200,000 in research and development. In addition, making a proposal (which she will decide whether to do after seeing whether the R&D is successful or not) requires developing a prototype timing system at an additional cost of $50,000. Finally, if Chang wins the contract, the finished product will cost an additional $150,000 to produce.

    a. Develop a decision tree that can be used to solve Chang's problem. You can assume in this part that she is using EMV (of her net profit) as a decision criterion. Build the tree so that she can enter any values for p_1, p_2, and p_3 (in input cells) and automatically see her optimal EMV and optimal strategy from the tree.

    b. If $p_2 = 0.8$ and $p_3 = 0.1$, what value of p_1 makes Chang indifferent between abandoning the project and going ahead with it?

    c. How much would Chang be willing to pay the Olympic organization (now) to guarantee her the contract in the case where her company is successful in developing the contract? (This guarantee is in force only if she is successful in developing the product.) Assume $p_1 = 0.4$, $p_2 = 0.8$, and $p_3 = 0.1$.

    d. Suppose now that this a "big" project for Chang. Therefore, she decides to use expected utility as her criterion, with an exponential utility function. Using some trial and error, see which risk tolerance changes her initial decision from "go ahead" to "abandon" when $p_1 = 0.4$, $p_2 = 0.8$, and $p_3 = 0.1$.

72. Suppose an investor has the opportunity to buy the following contract, a stock call option, on March 1. The contract allows him to buy 100 shares of ABC stock at the end of March, April, or May at a guaranteed price of $50 per share. He can "exercise" this option at most once. For example, if he purchases the stock at the end of March, he cannot purchase more in April or May at the guaranteed price. The current price of the stock is $50. Each month, we assume the stock price either goes up by a dollar (with probability 0.6) or down by a dollar (with probability 0.4). If the investor buys the contract, he is hoping that the stock price will go up. The reasoning is that if he buys the contract, the price goes up to $51, and he buys the stock (that is, he exercises his option) for $50, he can turn around and sell the stock for $51 and make a profit of $1 per share. On the other hand, if the stock price goes down, he doesn't have to exercise his option; he can just throw the contract away.

a. Use a decision tree to find the investor's optimal strategy (that is, when he should exercise the option), *assuming* he purchases the contract.

b. How much should he be willing to pay for such a contract?

73. The Ventron Engineering Company has just been awarded a $2 million development contract by the U.S. Army Aviation Systems Command to develop a blade spar for its Heavy Lift Helicopter program. The blade spar is a metal tube that runs the length of and provides strength to the helicopter blade. Due to the unusual length and size of the Heavy Lift Helicopter blade, Ventron is unable to produce a single-piece blade spar of the required dimensions, using existing extrusion equipment and material.

The engineering department has prepared two alternatives for developing the blade spar: (1) sectioning or (2) an improved extrusion process. Ventron must decide which process to use. (Backing out of the contract at any point is not an option.) The risk report has been prepared by the engineering department. An explanation of the information in this report follows.

The sectioning option involves joining several shorter lengths of extruded metal into a blade spar of sufficient length. This work will require extensive testing and rework over a 12-month period at a total cost of $1.8 million. Although this process will definitely produce an adequate blade spar, it merely represents an extension of existing technology.

To improve the extrusion process, on the other hand, it will be necessary to perform two steps: (1) improve the material used, at a cost of $300,000, and (2) modify the extrusion press, at a cost of $960,000. The first step will require 6 months of work, and if this first step is successful, the second step will require another 6 months of work. If both steps are successful, the blade spar will be available at that time—that is, 1 year from now. The engineers estimate that the probabilities of succeeding in steps 1 and 2 are 0.9 and 0.75, respectively. However, if either step is unsuccessful (which will be known only in 6 months for step 1 and in 1 year for step 2), Ventron will have no alternative but to switch to the sectioning process—and incur the sectioning cost on top of any costs already incurred.

Development of the blade spar must be completed within 18 months to avoid holding up the rest of the contract. If necessary, the sectioning work can be done on an accelerated basis in a 6-month period, but the cost of sectioning will then increase from $1.8 million to $2.4 million.

Frankly, the Director of Engineering, Dr. Smith, wants to try developing the improved extrusion process. This is not only cheaper (if successful) for the current project, but its expected side benefits for future projects could be sizable. Although these side benefits are difficult to gauge, Dr. Smith's best guess is an additional $2 million. (Of course, these side benefits are obtained only if both steps of the modified extrusion process are completed successfully.)

a. Develop a decision tree to maximize Ventron's EMV. This includes the revenue from this project, the side benefits (if applicable) from an improved extrusion process, and relevant costs. You don't need to worry about the time value of money—that is, no discounting or NPVs are required. Summarize your findings in words in the spreadsheet.

b. What value of side benefits would make Ventron indifferent between the two alternatives?

c. How much would Ventron be willing to pay, right now, for perfect information about both steps of the improved extrusion process? (This information would tell Ventron, right now, the ultimate success/failure outcomes of both steps.)

74. Ligature, Inc. is a company that does contract work for publishing companies. It specializes in writing textbooks for secondary schools. Because states such as Texas and California typically adopt only about four to eight textbooks for any given subject and grade level (from which individual schools can choose), the potential for large profits is great.

Ligature is currently negotiating a contract with Brockway and Coates (B&C), a large publishing company, to write a Social Studies series for grades 9–12. Actually, the development of the books is already well under way, and the only details not yet worked out concern the fee B&C will pay Ligature. Ligature has always operated on a fixed-fee basis. Under this arrangement, B&C would pay Ligature its costs, in this case $4.15 million, plus 25%. Ligature would receive this payment in 6 months, at the beginning of year 1. Although this is still an option, the companies have also been discussing a royalty arrangement as an alternative.

Under the royalty plan, B&C would still pay Ligature its $4.15 million costs at the beginning of year 1, but Ligature would then receive yearly royalty payments at the ends of years 1–5. These payments would depend on (1) total sales over the 5 years, (2) the timing of sales, and (3) the negotiated royalty rate—that is, Ligature's percentage of each sales dollar. As for timing, both parties agree that 10% of total sales will be in year 1, 20% will be in each of years 2 and 3, 30% will be in year 4, and 20% will be in year 5. They also estimate that the probability distribution of total sales is discrete, with possible values $25 million, $30 million, $50 million, and $70 million, and corresponding probabilities 0.10, 0.45, 0.30, and 0.15.

To guard its interests, B&C has imposed the following restriction to any royalty agreement. It places a cap on the amount Ligature can earn through the royalty scheme. Specifically, the royalties, discounted back to the beginning of year 1 at a 10% discount rate, cannot exceed 33% of Ligature's $4.15 million costs. Obviously, this limits B&C's downside exposure, regardless of the negotiated royalty rate or how well the books sell.

Ligature is interested in maximizing the NPV of its profit from this project (discounted back to the beginning of year 1), using a 10% discount rate. The following steps lead you through the required calculations to "solve" the problem. No decision tree is required for this problem.

a. The file P07_74.XLS supplies the inputs in an input section (blue border), and it has a calculation section (red border). First, calculate the upper part of the calculation section. To do so, enter any trial value for total sales in cell G8 and do the necessary calculations to eventually find (in cell G17) the NPV to Ligature from the royalty agreement. At this point, you can use any royalty rate in the RoyRate cell (C28).

b. Using the calculations from part **a,** complete the data table in the middle part of the calculation section. It should show the NPV to Ligature for any potential value of total sales. Then use these NPVs to calculate the expected NPV to Ligature in the ExpNPV cell (G27).

c. Suppose the current "offer on the table" is a 3% royalty rate. In the bottom part of the calculation section, use IF comparisons to see which arrangement, fixed fee or royalty, each party would favor.

d. Continuing part **c** (with the 3% offer on the table), what do you think the two parties will eventually agree upon? That is, will they stick with the 3% royalty rate, move to a different royalty rate, or settle on the fixed-fee arrangement? Answer below cell B36.

75. The American chess master Jonathan Meller is playing the Soviet expert Yuri Gasparov in a two-game exhibition match. Each win earns a player one point, and each draw earns half a point. The player who has the most points after two games wins the match. If the players are tied after two games, they play until one wins a game; then the first player to win a game wins the match. During each game, Meller has two possible strategies: to play a daring strategy or to play a con-

servative strategy. His probabilities of winning, losing, and drawing when he follows each strategy are shown in the file P07_75.XLS. To maximize his probability of winning the match, what should the American do?

76. [Based on Balson et al. (1992)] An electric utility company is trying to decide whether to replace its PCB transformer in a generating station with a new and safer transformer. To evaluate this decision, the utility needs information about the likelihood of an incident, such as a fire, the cost of such an incident, and the cost of replacing the unit. Suppose that the total cost of replacement as a present value is $75,000. If the transformer is replaced, there is virtually no chance of a fire. However, if the current transformer is retained, the probability of a fire is assessed to be 0.0025. If a fire occurs, then the cleanup cost could be high ($80 million) or low ($20 million). The probability of a high cleanup cost, given that a fire occurs, is assessed at 0.2.

a. If the company uses EMV as its decision criterion, should it replace the transformer?

b. Perform a sensitivity analysis on the key parameters of the problem that are difficult to assess, namely, the probability of a fire, the probability of a high cleanup cost, and the high and low cleanup costs. Does the optimal decision from part **a** remain optimal for a "wide" range of these parameters?

c. Do you believe EMV is the correct criterion to use in this type of problen involving environmental accidents?

77. [Based on Mellichamp et al. (1993)] Construction equipment managers typically have many large pools of engines, transmissions, and other equipment units to maintain. One approach to this maintenance is to use *oil analysis,* where the oil from any of these is subjected periodically to an inspection. These inspections can sometimes signal an impending failure (for example, too much iron in the oil), and preventive maintenance can then be performed (at a relatively low cost), eliminating the risk of failure (failure would result in a relatively high cost). However, oil analysis costs money, and it is not perfect. That is, it can indicate that a unit is defective when in fact it is not about to fail, and it can indicate that a unit is nondefective when in fact it is about to fail. As a possible substitute for oil analysis, the company could simply change the oil periodically, thereby reducing the probability of a failure.

Suppose the company has four alternatives: (1) do nothing, (2) use oil analysis only, (3) replace oil only, or (4) replace oil and do oil analysis. For option (1) the probability of a failure is p_1, and the cost of a failure is C_1. For option (2), the probability of a failure remains at p_1. If the unit is about to fail, the oil analysis will indicate this with probability $1 - \alpha$; if the unit is not about to fail, the oil analysis will indicate this with probability $1 - \beta$. (Therefore, α and β are the error probabilities of the oil analysis.) The oil analysis itself costs C_2, and if it indicates that a failure is about to occur, the oil will be changed, at cost C_3, and preventive maintenance will be performed. The cost of maintenance to restore a unit that is about to fail is C_4, whereas the cost of maintenance for a unit that is not about to fail is C_5. The only difference between options (3) and (4) is that the probability of a failure decreases to p_2 after changing the oil. The values of these parameters for a particular class of units (engines in light trucks, say) are given in the file P07_77.XLS.

a. For these parameters, develop a decision tree to find the company's best decision and the corresponding expected cost.

b. If the company has 500 units, what should it do? What is the expected cost for the entire fleet?

c. Suppose that the company has different types of units. For example, the cost of an oil change might be higher for some, or the cost of a failure might be higher or lower. Run a sensitivity analysis on any of the parameters you believe might be "key" parameters and see whether the optimal decision changes in ways you would anticipate.

78. [Based on Hess (1993)] A company that is heavily involved in R&D projects believes it might have the potential to develop a very lucrative commercial product that would (if successful) reduce pulp mill water pollution. At the current stage, however, everything is quite uncertain, and the company is trying to decide whether to go ahead with its R&D or abandon the product. The following are the primary risks:

- Would market tests confirm that there is a significant market for the product?
- Could the company develop a new process for making this product—that is, is it technically feasible?
- Even if there is a significant market and the process is technically feasible, would the company's board sanction the new plant capital necessary to produce the product on a commercial scale?
- Assuming the answers to the above questions are all yes and the plant is built, would the venture turn out to be successful?

We assume that each of these questions has a yes or no answer. The probabilities of yes answers are shown in the file P07_78.XLS. The plus-or-minus value indicates the company's uncertainty about the true probabilities.

The primary economic factors are the following:
- the research expenses to identify a new production process for the product
- the marketing development cost to determine whether there is a significant market
- the process development costs, including presanction engineering
- the commercial development costs, both before and after the board's sanction
- the venture value (net present value) if successful

The estimates of these values are also shown in the file P07_78.XLS. Again, the plus-or-minus values indicate the company's considerable uncertainty about the values. All dollar values are in millions of dollars.

The timing of events is as follows:
- Decide whether to abandon product now. (This is really the only nontrivial decision the company will make.) If not, then:
- Spend on research and marketing development. If marketing development indicates an insignificant market for the product *or* research indicates that the process is technically infeasible, cut expenses and quit. Otherwise:
- Spend on process and commercial development. If company board then declines to sanction money for plant, cut expenses and quit. Otherwise:
- Spend on further commercial development. By this time, the company has made all of its decisions. If the venture turns out to be a commercial success, then it gains the venture value for a success (less expenses so far). Otherwise, the company has lost the money spent so far, but that is all.

Analyze the company's problem. Obviously, with the high degree of uncertainty, sensitivity analysis is the key. Note that there are many uncertainties about the input parameters in the file P07_78.XLS. In fact, there are far too many to allow you to try every combination. Therefore, just try a few combinations that you believe might be the most important. Write a report of your findings.

CASE

7.1 Jogger Shoe Company

The Jogger Shoe Company is trying to decide whether to make a change in its most popular brand of running shoes. The new style would cost the same to produce, and it would be priced the same, but it would incorporate a new kind of lacing system that (according to its marketing research people) would make it more popular. There is a fixed cost of $300,000 of changing over to the new style. The unit contribution to before-tax profit for either style is $8. The tax rate is 35%. Also, because the fixed cost can be depreciated and will therefore affect the after-tax cash flow, we need a depreciation method. We assume it is straight-line depreciation.

The current demand for these shoes is 190,000 pairs annually. The company assumes this demand will continue for the next 3 years if the current style is retained. However, there is uncertainty about demand for the new style, if it is introduced. The company models this uncertainty by assuming a normal distribution in year 1, with mean 220,000 and standard deviation 20,000. The company also assumes that this demand, whatever it is, will remain constant for the next 3 years. However, if demand in year 1 for the new style is sufficiently low, the company can always switch back to the current style and realize an annual demand of 190,000. The company wants a strategy that will maximize the expected net present value (NPV) of total cash flow for the next 3 years, where a 15% interest rate is used for the purpose of calculating NPV.

The Westhouser Paper Company in the state of Washington currently has an option to purchase a piece of land with good timber forest on it. It is now May 1, and the current price of the land is $2.2 million. Westhouser does not actually need the timber from this land until the beginning of July, but its top executives fear that another company might buy the land between now and the beginning of July. They assess that there is 1 chance out of 20 that a competitor will buy the land during May. If this does not occur, they assess that there is 1 chance out of 10 that the competitor will buy the land during June. If Westhouser does not take advantage of its current option, it can attempt to buy the land at the beginning of June or the beginning of July, provided that it is still available.

Westhouser's incentive for delaying the purchase is that its financial experts believe there is a good chance that the price of the land will fall significantly in one or both of the next two months. They assess the possible price decreases and their probabilities in Tables 7.8 and 7.9. Table 7.8 shows the probabilities of the possible price decreases during May. Table 7.9 shows the *conditional* probabilities of the possible price decreases in June, *given* the price decrease in May. For example, if the price decrease in May is $60,000, then the possible price decreases in June are $0, $30,000, and $60,000 with respective probabilities 0.6, 0.2, and 0.2.

If Westhouser purchases the land, it believes that it can gross $3 million. (This does not count the cost of purchasing the land.) But if it does not purchase the land, it believes that it can make $650,000 from alternative investments. What should the company do?

Table 7.8

Distribution of Price Decrease in May

Price Decrease	Probability
$0	0.5
$60,000	0.3
$120,000	0.2

Table 7.9

Distribution of Price Decrease in June

Price Decrease in May					
$0		$60,000		$120,000	
June Decrease	Probability	June Decrease	Probability	June Decrease	Probability
$0	0.3	$0	0.6	$0	0.7
$60,000	0.6	$30,000	0.2	$20,000	0.2
$120,000	0.1	$60,000	0.2	$40,000	0.1

8

Sampling and Sampling Distributions

© 2002 Harris Welles

Choosing Samples of Customers to Receive Mailings

In the first half of this chapter, we discuss methods for selecting random samples. The purpose of these samples is to discover characteristics of a population, such as the proportion who favor the President's economic policy. By selecting a *random* sample of perhaps 1000 people out of a population of millions, we can make fairly accurate inferences about the population as a whole, at savings of much time and money.

A different type of sample has recently become the focus of many direct response marketers, companies that mail advertisements for their products directly to prospective customers. The experience of one such company, the Franklin Mint (FM) of Philadelphia, is described in Zahavi (1995). The FM markets expensive collectibles, ranging from famous Precision Car models to the Sword of Francis Drake, to a relatively small, but avid, collector population. The FM relies entirely on its mailings to prospective customers for sales. However, it is important for the company to mail ads for any particular products to the right customers; otherwise, mailing costs can seriously erode profits. This is especially the case when, on average, the response rate to products in this type of market is less than one-half percent—that is, no more than 1 person out of every 200 who receive a mailing actually purchases a product.

Until recently, companies such as the FM used relatively subjective rules to choose the sample of customers to receive mailings. However, these companies now have an abundance of data about their customers, and they are beginning to use sophisticated statistical methods to locate the customers who are most likely to purchase any particular type of product. In essence, they build a probability model that relates the probability of purchasing to (1) the customer's purchase history; (2) demographic variables (many of which can be acquired from outside vendors and appended to the customer's record); and (3) the product attributes, such as theme, material, artist, sponsor, and product code. The most challenging part of

365

the model is to identify the best predictor variables from the hundreds available, but techniques (and software) are now available to perform this task efficiently.

Direct marketers such as the FM have found that this is a situation where even a small amount of explanatory power from a statistical model can make a big difference in the bottom line. No model can correctly identify *exactly* who will respond positively to an ad and who will not, but if the model can identify customer samples where the response rate to mailings is even a *little* higher than it was, the ratio of mailing costs to eventual sales can decrease significantly. For example, the FM installed its system (called AMOS) in 1992 and realized an increase in profit of approximately 7.5% in 3 years; undoubtedly, it has increased even more since then. As Zahavi states, "Looking beyond the FM, the implications of using AMOS-like systems to support the decision-making process in the database marketing industry are likely to be quite substantial, which, given the size of the industry, could run well in excess of several hundred million dollars a year!" ●

8.1 Introduction

In a typical statistical inference problem we want to discover one or more characteristics of a given population. For example, we might want to know the proportion of toothpaste customers who have tried, or intend to try, a particular brand. Or we might want to know the average amount owed on credit card accounts for a population of customers at a shopping mall. Generally, the population is large and/or spread out, and it is difficult, maybe even impossible, to contact each member. Therefore, we identify a sample of the population and then obtain information from the members of the sample.

There are two main objectives of this chapter. The first is to discuss the sampling schemes that are generally used in real sampling applications. We will focus on several types of *random* samples and see why these are preferable to nonrandom samples. The second objective is to see how the information from a sample of the population—for example, 1% of the population—can be used to infer the properties of the entire population. The key here is the concept of *sampling distributions*. We will focus on the sampling distribution of the sample mean, and we will see how a famous mathematical result called the central limit theorem is the key to the analysis.

8.2 Sampling Terminology

We begin by introducing some of the terminology that is used in sampling. In any sampling problem there is a relevant *population*. A **population** is the set of all members about which a study intends to make inferences, where an **inference** is a statement about a numerical characteristic of the population, such as an average income or the proportion of incomes below $50,000. It is important to realize that a population is defined in relationship to any particular study. Any analyst planning a survey should first decide which population the conclusions of the study will concern, so that a sample can be chosen from *this* population.

Population	The relevant **population** contains all members about which a study intends to make inferences.

For example, if a marketing researcher plans to use a questionnaire to infer consumers' reactions to a new product, she must first decide which population of consumers is of interest—all consumers, consumers over 21 years old, consumers who do most of their shopping in shopping malls, or others. Once the relevant consumer population has been designated, a sample from this population can then be surveyed. However, inferences made from the study pertain only to this *particular* population.

Before we can choose a sample from a given population, we typically need a list of all members of the population. This list is called a **frame,** and the potential sample members are called **sampling units.** Depending on the context, sampling units could be individual people, households, companies, cities, or others.

Frame	A **frame** is a list of all members, called **sampling units,** in the population.

In this chapter we will assume that the population is finite and consists of N sampling units. We also assume that a frame of these N sampling units is available. Unfortunately, there are situations where a complete frame is practically impossible to obtain. For example, if we want to survey the attitudes of all unemployed teenagers in Chicago, it is practically impossible to obtain a complete frame of them. In this situation all we can hope to obtain is a partial frame, from which the sample can be selected. If the partial frame omits any significant segments of the population—which a complete frame would include—then the resulting sample could be biased. For instance, if we use the Yellow Pages of a Los Angeles telephone book to choose a sample of restaurants, we automatically omit all restaurants that do not advertise in the Yellow Pages. Depending on the purposes of the study, this might be a serious omission.

There are two basic types of samples, *probability samples* and *judgmental samples.* A **probability sample** is a sample in which the sampling units are chosen from the population by means of a random mechanism such as a random number table. In contrast, no formal random mechanism is used to select a **judgmental sample.** In this case the sampling units are chosen according to the sampler's judgment.

Probability Sample	The members of a **probability sample** are chosen according to a random mechanism, whereas the members of a **judgmental sample** are chosen according to the sampler's judgment.

We will not discuss judgmental samples. The reason is very simple—there is no way to measure the accuracy of judgmental samples because the rules of probability do not apply to them. In other words, if we estimate some population characteristic from the observations in a judgmental sample, there is no way to tell how accurate this estimate is. In addition, it is very difficult to choose a representative sample from a population *without* using some random mechanism. Because our judgment is usually not as good as we think, judgmental samples are likely to contain our own built-in biases. Therefore, we will focus exclusively on probability samples from here on.

8.3 Methods for Selecting Random Samples

In this section we discuss the types of random samples that are used in real sampling applications. Different types of sampling schemes have different properties. There is typically a trade-off between cost and accuracy. Some sampling schemes are cheaper and easier to administer, whereas others cost more but provide more accurate information. We will discuss some of these issues, but anyone who intends to make a living in survey sampling needs to learn much more about the topic than we can cover here.

8.3.1 Simple Random Sampling

The simplest type of sampling scheme is appropriately called *simple random sampling*. Consider a population of size N and suppose we want to sample n units from this population. Then a **simple random sample** of size n has the property that every possible sample of size n has the same probability of being chosen. Simple random samples are the easiest to understand, and their statistical properties are fairly straightforward. Therefore, we will focus primarily on simple random samples in the rest of this book. However, as we will discuss shortly, simple random samples are typically *not* used in real applications.

Simple Random Sample	A **simple random sample** is one where each member of the population has the same chance of being chosen.

We illustrate a simple random sample for a small population. Suppose the population size is $N = 5$, and we label the five members of the population as a, b, c, d, and e. Also, suppose we want to sample $n = 2$ of these members. Then the possible samples are (a, b), (a, c), (a, d), (a, e), (b, c), (b, d), (b, e), (c, d), (c, e), and (d, e). That is, there are 10 possible samples—the number of ways two members can be chosen from five members. Then a *simple* random sample of size $n = 2$ has the property that each of these 10 possible samples has an equal probability, 1/10, of being chosen.

One other property of simple random samples can be seen from this example. If we focus on any member of the population, say, member b, we note that b is a member of 4 of the 10 samples. Therefore, the probability that b is chosen in a simple random sample is 4/10, or 2/5. In general, any member has probability n/N of being chosen in a simple random sample. If you are one of 100,000 members of a population, then the probability that you will be selected in a simple random sample of size 100 is 100/100,000, or 1 out of 1000.

There are several ways simple random samples can be chosen, all of which involve random numbers. One approach that works well for our small example with $N = 5$ and $n = 2$ is to generate a single random number with the RAND function in Excel. We divide the interval from 0 to 1 into 10 equal subintervals of length 1/10 each and see which of these subintervals the random number falls into. We then choose the corresponding sample. For example, suppose the random number is 0.465. This is in the fifth subinterval, that is, the interval from 0.4 to 0.5, so we choose the fifth sample, (b, c).

Clearly, this method is consistent with simple random sampling—each of the samples has the same chance of being chosen—but it is prohibitive when n and N are large. In this case there are too many possible samples to list. Fortunately, there is another method that can be used. The idea is simple. We sort the N members of the population randomly, using Excel's RAND function to generate random numbers for the sort. Then we include the first n members from the sorted sequence in the random sample. (If you haven't covered the simulation sections of previous chapters, the RAND function in Excel generates numbers that are distributed randomly and uniformly between 0 and 1.) We illustrate this procedure in the following example.

Example (8.1) **Selecting a Sample of Families to Analyze Annual Incomes**

Consider the frame of 40 families with annual incomes shown in column B of Figure 8.1. (See the file RANDSAMP.XLS.) We want to choose a simple random sample of size 10 from this frame. How can this be done? And how do summary statistics of the chosen families compare to the corresponding summary statistics of the population?

Objective To illustrate how Excel's random number function, RAND, can be used to generate simple random samples.

Solution

The idea is very simple. We first generate a column of random numbers in column C. Then we sort the rows according to the random numbers and choose the first 10 families in the sorted rows. The following procedure produces the results in Figure 8.2 on page 370. (See the Manual sheet in the finished version of the file RANDSAMP.XLS.)

❶ **Random numbers.** Enter the formula

$$=RAND()$$

in cell C10 and copy it down column C.

❷ **Replace with values.** To enable sorting we must first "freeze" the random numbers—that is, replace their formulas with values. To do this, select the range C10:C49,

FIGURE 8.1
Population Income Data

	A	B	C	D
1	Illustration of simple random sampling			
2				
3	Summary statistics			
4		Mean	Median	Stdev
5	Population	$39,985	$38,500	$7,377
6	Sample			
7				
8	Population			
9	Family	Income		
10	1	$43,300		
11	2	$44,300		
12	3	$34,600		
13	4	$38,000		
14	5	$44,700		
15	6	$45,600		
16	7	$42,700		
17	8	$36,900		
18	9	$38,400		
19	10	$33,700		
20	11	$44,100		
21	12	$51,500		
22	13	$35,900		
23	14	$35,600		
24	15	$43,000		
47	38	$46,900		
48	39	$37,300		
49	40	$41,000		

FIGURE 8.2
Selecting a Simple
Random Sample

	A	B	C	D	E	F	G
1	Illustration of simple random sampling						
2							
3	Summary statistics						
4		Mean	Median	Stdev			
5	Population	$39,985	$38,500	$7,377			
6	Sample	$38,750	$38,200	$5,922			
7							
8	Population				Random sample		
9	Family	Income	Random #		Family	Income	Random #
10	1	$43,300	0.6835		32	$31,700	0.0513
11	2	$44,300	0.5237		34	$39,300	0.0742
12	3	$34,600	0.8313		4	$38,000	0.0814
13	4	$38,000	0.0814		14	$35,600	0.0922
14	5	$44,700	0.1955		22	$33,600	0.1162
15	6	$45,600	0.2007		40	$41,000	0.1219
16	7	$42,700	0.8300		12	$51,500	0.1231
17	8	$36,900	0.8036		10	$33,700	0.1677
18	9	$38,400	0.1788		9	$38,400	0.1788
19	10	$33,700	0.1677		5	$44,700	0.1955
20	11	$44,100	0.2638		6	$45,600	0.2007
21	12	$51,500	0.1231		16	$38,600	0.2316
22	13	$35,900	0.7965		21	$56,400	0.2571
23	14	$35,600	0.0922		36	$36,300	0.2585
24	15	$43,000	0.6672		11	$44,100	0.2638
47	38	$46,900	0.7248		31	$33,000	0.9277
48	39	$37,300	0.3655		20	$31,900	0.9489
49	40	$41,000	0.1219		37	$28,400	0.9491

use Edit/Copy menu item, and then use the Edit/Paste Special menu item with the Values option.

③ Copy to a new range. Copy the range A10:C49 to the range E10:G49.

④ Sort. Select the range E10:G49 and use the Data/Sort menu item. Sort according to the Random # column (column G) in ascending order. Then the 10 families with the 10 smallest random numbers are the ones in the sample. (These are enclosed in a border.)

⑤ Means. Use the AVERAGE, MEDIAN, and STDEV functions in row 6 to calculate summary statistics of the first 10 incomes in column F. These similar summary statistics for the population have already been calculated in row 5. (Cell D5 uses the STDEVP function, because this is the population standard deviation.)

To obtain more random samples of size 10 (for comparison), we would need to go through this process repeatedly. To save you the trouble of doing so, we wrote a macro to automate the process. (See the Automated sheet in the RANDSAMP.XLS file.) This sheet looks essentially the same as the sheet in Figure 8.2, except that there is a button to run the macro and only the required data remain on the spreadsheet. Try clicking on this button. Each time you do so, you'll get a different random sample—and different summary measures in row 6. By doing this many times and keeping track of the sample summary data, you can see how the summary measures vary from sample to sample. We will have much more to say about this type of variation later in this chapter. ●

The procedure described in Example 8.1 can be used in Excel to select a simple random sample of any size from any population. All we need is a frame, a list of the population values. Then it's just a matter of inserting random numbers, freezing them, and sorting on the random numbers.

Perhaps surprisingly, simple random samples are almost never used in real applications. There are several reasons for this.

Despite this, most of the statistical analysis in this book assumes simple random samples. The analysis is considerably more complex for other types of random samples and is discussed in more advanced books on sampling.

- Because each sampling unit has the same chance of being sampled, simple random sampling can result in samples that are spread over a large geographical region. This can make sampling extremely expensive, especially if personal interviews are used.

- Simple random sampling requires that all sampling units be identified prior to sampling. Sometimes this is infeasible.

- Simple random sampling can result in underrepresentation or overrepresentation of certain segments of the population. For example, if the primary—but not sole—interest is in the graduate student subpopulation of university students, a simple random sample of *all* university students might not provide enough information about the graduate students.

8.3.2 Using StatPro to Generate Simple Random Samples

The method described in Example 8.1 is simple but somewhat tedious, especially if we want to generate more than one random sample. (Even the macro described at the end of the example works only for that particular file.) Therefore, we developed a more general method as part of StatPro. This procedure generates any number of simple random samples of any specified sample size from a given data set. It can be found under the StatPro/Statistical Inference/Generate Random Samples menu item.

Actually, this procedure returns only the *indexes* of the members selected in the samples. For example, from a data set with 500 members, one random sample of size 5 might have the indexes 413, 22, 310, 156, and 209. This indicates that these particular members are included in the sample. To get the *data* for these members, we can use a lookup command, as illustrated in the following example.

Example 8.2 Sampling from Accounts Receivable at Spring Mills Company

The file RECEIVE.XLS contains 280 accounts receivable for the Spring Mills Company (the same data we discussed in Example 3.9). There are three variables:

- Size: customer size (small, medium, large), depending on its volume of business with Spring Mills
- Days: number of days since the customer was billed
- Amount: amount of the bill

Generate 50 random samples of size 15 each from the small customers only, calculate the average amount owed in each random sample, and construct a histogram of these 50 averages.

Objective To illustrate StatPro's method of choosing simple random samples, and how the associated data can be found with Excel's lookup functions.

Solution

The original file (from Chapter 3) contains only the Size, Days, and Amount variables, and the data are sorted by Size. We append an extra variable Account in column A that indexes the accounts: 1 to 280. (Use the Edit/Fill menu item to do this quickly.) To

	A	B	C	D	E	AV	AW	AX	AY	AZ	BA	BB
1	*Indices of members in samples*										Averages shown as a column	
2		Sample1	Sample2	Sample3	Sample4	Sample47	Sample48	Sample49	Sample50			
3		86	124	139	44	16	101	122	77		Averages	
4		124	84	81	91	76	104	84	111		250.67	
5		85	12	32	82	49	41	50	16		244.00	
6		42	28	46	142	120	103	71	20		238.67	
7		88	88	71	45	27	89	65	150		255.33	
8		30	50	148	47	119	102	12	66		242.67	
9		127	59	50	24	103	30	5	70		248.00	
10		128	136	132	17	88	21	35	99		261.33	
11		15	30	24	26	67	63	86	85		246.00	
12		2	43	29	7	74	12	108	94		251.33	
13		90	115	124	143	3	34	7	142		258.00	
14		108	94	88	68	32	14	133	1		252.67	
15		150	18	70	112	79	23	28	32		260.00	
16		61	142	143	16	48	95	139	68		246.67	
17		20	55	37	39	97	15	30	102		256.00	
18											259.33	
19	Amounts owed for accounts selected in samples										254.67	
20		Sample1	Sample2	Sample3	Sample4	Sample47	Sample48	Sample49	Sample50		251.33	
21		280	270	210	210	270	310	220	220		269.33	
22		270	280	300	320	270	320	280	240		258.67	
23		180	260	200	290	250	230	190	270		270.00	
24		280	240	230	240	240	240	240	250		246.67	
25		260	260	240	320	220	220	200	370		242.67	
26		140	190	240	270	250	280	260	280		260.00	
27		190	320	190	230	240	140	300	240		258.67	
28		300	170	260	260	260	240	200	320		267.33	
29		350	140	230	210	390	190	280	180		259.33	
30		210	310	260	330	370	260	260	240		260.00	
31		180	220	270	210	210	330	330	240		261.33	
32		260	240	260	260	200	320	290	180		239.33	
33		370	240	240	220	190	290	240	200		264.00	
34		240	240	210	270	240	260	210	260		258.00	
35		250	280	240	190	290	350	140	280		241.33	
36	Averages										263.33	
37		250.67	244.00	238.67	255.33	259.33	265.33	242.67	251.33		239.33	
38											232.00	

FIGURE 8.3 Randomly Generated Samples

select small accounts only, insert a blank row after account 150 (the last small account). Then, with the cursor anywhere in the small account data set, use the StatPro/Statistical Inference/Generate Random Samples menu item, enter 50 and 15 as the number of samples and the sample size, and put the results in a new sheet. The top part of Figure 8.3 should appear. (We have hidden many of the intermediate samples.)

To find the amounts owed for the sampled accounts, enter the formula

$$=VLOOKUP(B3,Data!Data,4)$$

in cell B21 and copy it to the range B21:AY35. (Note that Data is the range name StatPro automatically gives to the original data range in the Data sheet. We use this range here as the lookup table range.) Then calculate the averages in row 37 with the AVERAGE function, and transpose this *row* of averages to a *column* of averages in column BA by highlighting the range BA4:BA53, entering the formula

$$=TRANSPOSE(B37:AY37)$$

and pressing Ctrl-Shift-Enter. (A column is needed by StatPro to create the histogram.) Finally, use StatPro's histogram procedure with appropriate settings to create a histogram similar to that shown in Figure 8.4. (Yours might look different because the random numbers you use to generate the random samples will be different from ours.)

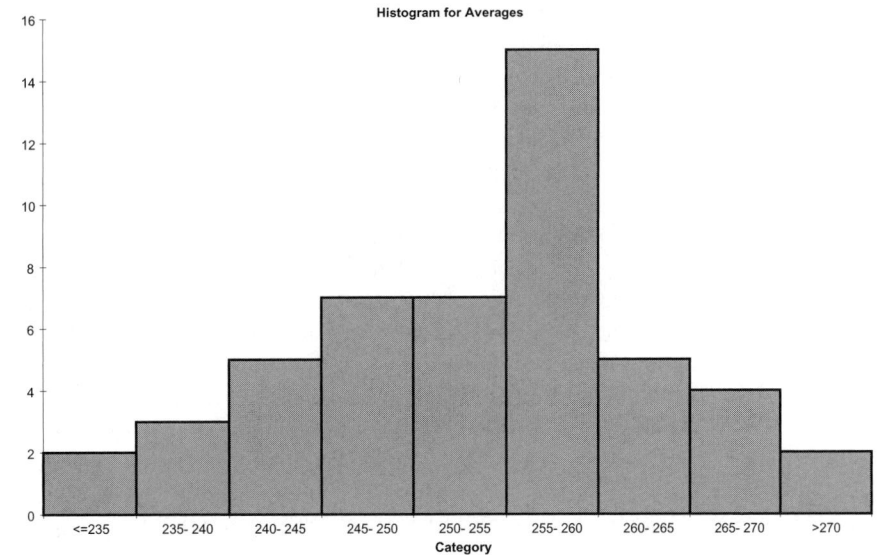

FIGURE 8.4
Histogram of 50
Sample Averages

The histogram in Figure 8.4 indicates the variability of sample means we might obtain by selecting many *different* random samples of size 15 from this particular population of small customer accounts. This histogram, which is approximately bell shaped, approximates the sampling distribution of the sample mean. We will come back to this important idea when we study sampling distributions in Section 8.4. ●

In the next several subsections we will describe sampling plans that are often used. These plans differ from simple random sampling both in the way the samples are chosen and in the way the data analysis is performed. However, we will barely touch on this latter issue. The details are quite complicated and are better left to a book devoted entirely to sampling. [See, for example, the excellent book by Levy and Lemeshow (1999).]

8.3.3 Systematic Sampling

Suppose you are asked to select a random sample of 250 names from the white pages of a telephone book. Let's also say that there are 55,000 names listed in the white pages. A *systematic sample* provides a convenient way to choose the sample, as follows. First, we calculate the **sampling interval** as the population size divided by the sample size: 55,000/250 = 220. Conceptually, we can think of dividing the book into 250 "blocks" with 220 names per block. Next, we use a random mechanism to choose a number between 1 and 220. Say this number is 131. Then we choose the 131st name and every 220th name thereafter. So we would choose the 131st name, the 351st name, the 571st name, and so on. The result is a systematic sample of size $n = 250$.

Systematic
Sampling

In **systematic sampling,** one of the first k members is selected randomly, and then every kth member after this one is selected. The value k is called the **sampling interval** and equals the ratio N/n, where N is the population size and n is the desired sample size.

Clearly, systematic sampling is different from simple random sampling because not every sample of size 250 has a chance of being chosen. In fact, there are only 220 different samples possible (depending on the first number chosen), and each of these is equally likely. Nevertheless, systematic sampling is generally similar to simple random sampling in its statistical properties. The key is the relationship between the ordering of the sampling units in the frame (the white pages of the telephone book in this case) and the purpose of the study.

If the purpose of the study is to analyze personal incomes, say, then there is probably no relationship between the alphabetical ordering of names in the telephone book and personal income. However, there are situations where the ordering of the sampling units is not random, which can make systematic sampling more or less appealing. For example, suppose that a company wants to sample randomly from its customers, and its customer list is in decreasing order of order volumes. That is, the largest customers are at the top of the list and the smallest are at the bottom. Then systematic sampling might be more representative than simple random sampling because it guarantees a wide range of customers in terms of order volumes.

Systematic samples are typically chosen because of their convenience.

However, some type of cyclical ordering in the list of sampling units can lead to very *unrepresentative* samples. As an extreme, suppose a company has a list of daily transactions and it decides to draw a systematic sample with the sampling interval equal to 7. Then if the first sampled day is Monday, all other days in the sample will be Mondays! This could clearly bias the sample. Nevertheless, except for obvious examples like this one, systematic sampling can be an attractive alternative to simple random sampling and is often used because of its convenience.

8.3.4 Stratified Sampling

Suppose we can identify various subpopulations within the total population. We call these subpopulations **strata.** Then instead of taking a simple random sample from the entire population, it might make more sense to select a simple random sample from each stratum separately. This sampling method is called **stratified sampling.** It is a particularly useful approach when there is considerable variation *between* the various strata but relatively little variation *within* a given stratum.

Stratified Sampling	In **stratified sampling,** the population is divided into relatively homogeneous subsets called **strata,** and then random samples are taken from each of the strata.

There are several advantages to stratified sampling. One obvious advantage is that we obtain separate estimates within each stratum—which we would not obtain if we took a simple random sample from the entire population. Even if we eventually plan to pool the samples from the individual strata, it cannot hurt to have the total sample broken down into separate samples initially.

Stratified samples are typically chosen because they provide more accurate estimates of population parameters for a given sampling cost.

A more important advantage of stratified sampling is that the accuracy of the resulting population estimates can be increased by using appropriately defined strata. The trick is to define the strata so that there is less variability within the individual strata than in the population as a whole. We want strata such that there is relative homogeneity within the strata, but relative heterogeneity among the strata, with respect to the variable(s) being analyzed. By choosing the strata in this way, we can generally obtain more accuracy for a given sampling cost than we could obtain from a simple random sample at the same cost. Alternatively, we can achieve the same level of accuracy at a lower sampling cost.

The key to using stratified sampling effectively is selecting the appropriate strata. Suppose a company that advertises its product on television wants to estimate the reaction of viewers to the advertising. Here the population consists of all viewers who have seen the advertising. But what are the appropriate strata? The answer depends on the company's objectives and its product. The company could stratify the population by gender, by income, by amount of television watched, by the amount of the product class consumed, and probably others. Without knowing more specific information about the company's objectives, it is impossible to say which of these stratification schemes is most appropriate.

Suppose that we have identified I nonoverlapping strata in a given population. Let N be the total population size, and let N_i be the population size of stratum i, so that

$$N = N_1 + N_2 + \cdots + N_I$$

To obtain a stratified random sample, we must choose a total sample size n, and we must choose a sample size n_i from each stratum i, such that

$$n = n_1 + n_2 + \cdots + n_I$$

We can then select a simple random sample of the specified size from *each* stratum exactly as in Example 8.1.

However, how do we choose the individual sample sizes n_1 through n_I, given that the total sample size n has been chosen? For example, if we decide to sample 500 customers in total, how many should come from each stratum? There are many ways that we could choose numbers n_1 through n_I that sum to n, but probably the most popular method is to use *proportional sample sizes*. The idea is very simple. If one stratum has, say, 15% of the total population, then we select 15% of the total sample from this stratum. For example, if the total sample size is $n = 500$, we select $0.15(500) = 75$ members from this stratum.

Proportional Sample Sizes	With **proportional sample sizes,** the proportion of each stratum selected is the same from stratum to stratum.

The advantage of proportional sample sizes is that they are very easy to determine. The disadvantage is that they ignore differences in variability among the strata. To illustrate, suppose that we are attempting to estimate the population mean amount paid annually per student for textbooks at a large university. We identify three strata: undergraduates, master's students, and doctoral students. Their population sizes are 20,000, 4000, and 1000, respectively. Therefore, the proportions of students in these strata are $20,000/25,000 = 0.80$, $4000/25,000 = 0.16$, and $1000/25,000 = 0.04$. If the total sample size is $n = 150$, then the sample should include 120 undergraduates, 24 master's students, and 6 doctoral students if proportional sample sizes are used.

However, let σ_i be the standard deviation of annual textbook payments in stratum i, and suppose that $\sigma_1 = \$50$, $\sigma_2 = \$120$, and $\sigma_3 = \$180$. Thus, there is considerably more variation in the amounts paid by doctoral students than by undergraduates, with the master's students in the middle. If we are interested in estimating the mean amount spent per student, then despite its small sample size, the doctoral sample is likely to have a large effect on the accuracy of our estimate of the mean. This is because of its relatively large standard deviation. In contrast, we might not need to sample as heavily from the undergraduate population because of its relatively small standard deviation. In general, strata with less variability can afford to be sampled less heavily than proportional sampling calls for, and the opposite is true for strata with larger variability. In fact, there are *optimal* sample size formulas that take the σ_i's into account, but we will not present them here.

In the following example we illustrate how stratified sampling can be accomplished with Excel by using random numbers.

Example 8.3 Stratified Sampling from the Smalltown Population of Sears Credit Card Holders

The file STRATIFIED.XLS contains a frame of all 1000 people in the city of Smalltown who have Sears credit cards. Sears is interested in estimating the average number of *other* credit cards these people own, as well as other information about their use of credit. The company decides to stratify these customers by age, select a stratified sample of size 100 with proportional sample sizes, and then contact these 100 people by phone. How might Sears proceed?

Objective To illustrate how stratified sampling, with proportional sample sizes, can be implemented in Excel.

Solution

First, Sears has to decide exactly how to stratify by age. Their reasoning is that different age groups probably have different attitudes and behavior regarding credit. After some preliminary investigation, they decide to use three age categories: 18–30, 31–62, and 63–80. (No one in the population is younger than 18 or older than 80.)

Figure 8.5 shows how the calculations might then proceed. We begin with the following inputs: (1) the total sample size in cell C3, (2) the definitions of the strata in rows 6–8, and (3) the customer data in the range A11:B1010. To see which age category each customer is in, we enter the formula

$$=IF(B11<=\$D\$6,1,IF(B11<=\$D\$7,2,3))$$

in cell C11 and then copy it down column C.

	A	B	C	D	E	F	G	H	I	J	K	L	M	N
1	Stratified sampling by Sears													
2														
3	Total sample size		100											
4									Range names:					
5	Strata based on age					Counts	SampSize		TotSampSize: C3					
6	Stratum 1	18	to	30		132	13							
7	Stratum 2	31	to	62		766	77							
8	Stratum 3	63	to	80		102	10							
9														
10	Cust	Age	Category		Cust_1	Age_1	Cust_2	Age_2	Cust_3	Age_3		Cust_1	Age_1	Rand #
11	1	49	2		11	23	1	49	4	66		835	24	0.002255
12	2	39	2		13	24	2	39	12	63		412	26	0.04736
13	3	55	2		15	30	3	55	38	75		67	27	0.054435
14	4	66	3		20	29	5	52	40	64		163	30	0.057802
15	5	52	2		26	30	6	37	42	71		874	23	0.063774
16	6	37	2		34	26	7	34	53	63		637	23	0.092121
17	7	34	2		43	25	8	34	64	71		513	28	0.092475
18	8	34	2		55	25	9	33	68	66		409	27	0.096296
19	9	33	2		56	28	10	36	95	64		56	28	0.098075
20	10	36	2		60	21	14	47	102	76		531	29	0.119406
21	11	23	1		62	28	16	59	117	63		250	29	0.133212
22	12	63	3		67	27	17	31	127	67		622	28	0.137265
23	13	24	1		79	30	18	34	128	71		215	30	0.147072
24	14	47	2		80	29	19	35	130	74		99	26	0.157176

FIGURE 8.5 Selecting a Stratified Sample

Next, it is useful to "unstack" the data into three groups, one for each age category, as shown in columns E–J. For example, columns E and F list the customer numbers and ages for all customers in the first age category. It is easy to unstack the data in columns A–C with StatPro. With the cursor anywhere in the A10:C1010 range, use the StatPro/Data Utilities/Unstack Variables menu item, select Category as the Code variable, select Cust and Age as the variables to unstack, and accept the default location for the unstacked variables.

Once the variables are unstacked, we can calculate the information in the range F6:G8 by entering the formulas

$$=COUNT(E11:E142)$$

and

$$=ROUND(TotSampSize*F6/1000,0)$$

in cells F6 and G6, with similar formulas for the other two categories. In words, if Sears wants to use proportional sample sizes, then it should sample 13, 77, and 10 customers from the three age categories. The formula for proportional sample sizes might lead to fractional values, so we use Excel's ROUND function to round to integers.

Finally, we can proceed as in Example 8.1 for each of the three categories separately. Figure 8.5 illustrates the selection of 13 customers from age category 1. We copy the data in columns E and F to columns L and M, append a column of random numbers with the RAND function in column N, freeze these random numbers (with the Copy and Paste Special/Values commands), sort on the random number column, and choose the first 13 customers. (The finished version of the STRATIFIED.XLS file shows similar calculations for the other two age categories to the right of column N.) Note that if we wanted a *different* sample of 13 from age category 1, all we would need to do is generate new random numbers in column N with the RAND function, freeze them, and sort again. ●

Many of the sampling schemes used in real applications use some form of stratification. The following example describes one of these applications.

Stratified Sampling at the OTA

When congressional committees and other federal agencies in Washington contemplate changes in tax laws, they frequently ask the Office of Tax Analysis (OTA) to provide information on how the tax changes will affect various groups of taxpayers. The OTA finds this information by simulating the changes on a randomly selected representative sample of taxpayers. That is, it discovers how the tax returns for these sampled taxpayers would be affected by the proposed changes. This is a time-consuming process, and the OTA's job is made even more difficult because (1) the information is sought quickly, often within a day, and (2) the OTA has to run literally thousands of these simulations per year. Therefore, it is essential that the sample of taxpayers be as small as possible, in addition to being representative.

One researcher, John Mulvey, reported a methodology for choosing this sample. [See Mulvey (1980).] Actually, he explains that the OTA already had a representative subset of approximately 155,000 taxpayers. For each of these there was information on 192 attributes such as adjusted gross income, taxes paid, salary and wages, total tax credits, pensions, and so on. This represents an enormous amount of data, so the OTA wanted to reduce the sample size to about 75,000 without losing any of the representativeness of the original large sample.

One idea was to stratify the sample of 155,000 taxpayers in some way and then choose representative members from each stratum to be in the smaller sample. The question,

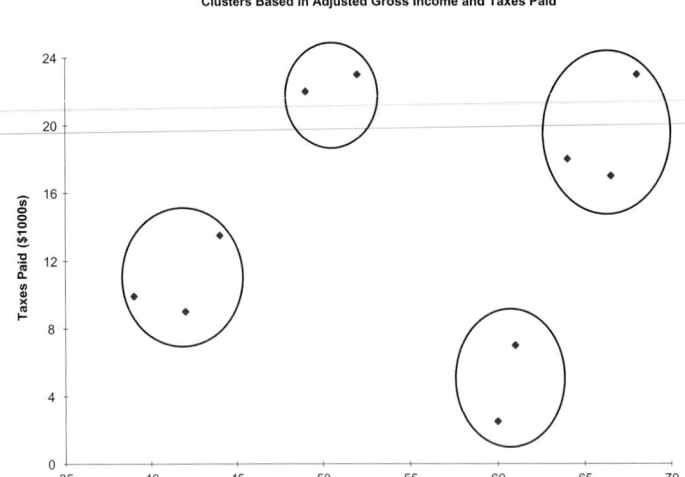

FIGURE 8.6
Clusters for Collecting Tax Information

though, was how to define the strata. For example, if the strata were based on only a single attribute, such as adjusted gross income, the smaller sample might resemble the larger sample very closely with respect to adjusted gross income, but the resemblance might not be very good for other important attributes.

Therefore, Mulvey decided to define strata in a more sophisticated manner. He used an intricate computer program to separate the 155,000 taxpayers into strata such that the taxpayers within each stratum were similar with respect to two attributes, adjusted gross income and taxes paid. His idea is shown graphically in Figure 8.6, where each point represents a single taxpayer. (For clarity, we show only ten taxpayers from the thousands available.) For the ten points shown, the four strata, or groupings, are shown by circles. Mulvey then chose representative taxpayers from each grouping so that the total sample size was approximately 75,000. As he reports, the resemblance between the original large sample and the newly created small sample was very good. In particular, the percentage differences between means and standard deviations for the two samples were less than 2% for most of the important attributes. This was good enough for the OTA, which began using the reduced sample in 1980. ●

8.3.5 Cluster Sampling

Suppose that a company is interested in various characteristics of households in a particular city. The sampling units are households. We could select a random sample of households by one of the sampling methods already discussed. However, it might be more convenient to proceed somewhat differently. We could first divide the city into city blocks and consider the city blocks as sampling units. We could then select a simple random sample of city blocks and then sample *all* of the households in the chosen blocks. In this case the city blocks are called **clusters** and the sampling scheme is called **cluster sampling.**

Cluster Sampling	In **cluster sampling,** the population is separated into clusters, such as cities or city blocks, and then a random sample of the clusters is selected.

The primary advantage of cluster sampling is sampling convenience (and possibly less cost). If an agency is sending interviewers out to interview heads of household, say, it is

Cluster analysis is typically more convenient and less costly than other random sampling methods.

much easier for them to concentrate on particular city blocks than to contact households throughout the city. The downside, however, is that the inferences drawn from a cluster sample can be less accurate, for a given sample size, than for other sampling plans.

Consider the following scenario. A nationwide company wants to survey its salespeople with regard to management practices. It decides to randomly select several sales districts (the clusters) and then interview all salespeople in the selected districts. It is likely that in any particular sales district the attitudes toward management are somewhat similar. This overlapping information means that the company is probably not getting the maximum amount of information per sampling dollar spent. Instead of sampling 20 salespeople from a given district who all have similar attitudes, it might be better to sample 20 salespeople from different districts who have a wider variety of attitudes. Nevertheless, the relative convenience of cluster sampling sometimes outweighs these statistical considerations.

Selecting a cluster sample is straightforward. The key is to define the sampling units as the *clusters*—the city blocks, for example. Then we can select a simple random sample of clusters exactly as in Example 8.1. Once the clusters are selected, we typically sample all of the population members in each selected cluster.

8.3.6 Multistage Sampling Schemes

The cluster sampling scheme just described, where a sample of clusters is chosen and then all of the sampling units within each chosen cluster are taken, is called a **single-stage** sampling scheme. Real applications are often more complex than this, resulting in **multistage** sampling schemes. For example, the Gallup organization uses multistage sampling in its nationwide surveys. A random sample of approximately 300 locations is chosen in the first stage of the sampling process. City blocks or other geographical areas are then randomly sampled from the first-stage locations in the second stage of the process. This is followed by a systematic sampling of households from each second-stage area. A total of about 1500 households comprise a typical Gallup poll.

The following presents another application of multistage sampling. It shows how complex actual sampling plans can be.

Multistage Sampling in Market Research

A marketing research company—the "supplier" in this discussion—has the job of supplying a large client with monthly random samples of its customers.[1] The client company, which is in the sales/service business, considers these random samples extremely important. The samples consist of selected customers from the entire nation. The sampled customers are asked about the quality of service from their respective salespeople, and this information is then used by the client to rate its salespeople. Because important decisions such as raises and promotions are based at least partially on the results of the samples, the client considers it crucial that the sampling be done according to well-established statistical guidelines.

Each month, the client gives the supplier a randomly selected sample of its customers. This list is chosen from several successive hierarchical (geographical) levels. First, the entire nation is divided into four areas: North, Central, South, and West. Then each of these areas is subdivided into several regions. For example, the Central area is subdivided into the Midwest, South Central, and Southwest regions. Then each region is subdivided into divisions, which are defined by states. For example, the Midwest region might include

---

[1] This discussion describes the consulting experience of one of the authors. For proprietary reasons, all company names are withheld in this discussion.

specified parts of Iowa, Minnesota, and Michigan. Finally, each division is subdivided into districts. A district is at the lowest end of the hierarchy.

The client randomly samples a certain number of its customers from each district and furnishes this list (one for each district) to the supplier. It also sets a target sample size for each district. The job of the supplier is to survey the specified number of customers from each list during a given month. For example, the list for one district might include 320 customers, from which the supplier is required to survey 10 customers.

Although the procedural details followed by the supplier are quite complex and will not be described here, the supplier basically selects a systematic sample from each district list that is about 5 times as large as the required number of contacts. So if the supplier needs 10 contacts from 320 customers, it chooses a systematic sample of about 50 customers from that district's list. It then "releases" the customers in this random sample to its phone-calling headquarters, and callers begin calling customers in the release sample until the required number of contacts is made.

Because of the importance of the sample information to the client, the client is understandably critical of all phases of the supplier's procedure. One particular cause of concern is the following. In each district the customers come from approximately 25 different calling zones, defined by their area code and the first three digits of their phone number. Some of these calling zones have up to 30 customers on the district list given to the supplier, whereas other zones have only one or two customers on the list. Now, the supplier is not instructed to stratify by calling zones, which would ensure proportional representation from the various calling zones among the release sample and/or the final contacts. However, the client *expects* that random sampling will guarantee such representativeness.

Unfortunately, it does not. For example, if one district list has 320 customers, 5% of whom come from a given calling zone, there is no guarantee that a random sample taken from these 320 customers will contain exactly 5% from that particular zone. In one month the sample might have more than 5%, and in another month it might have less than 5%. Neither the client nor the supplier understood exactly how much variation to expect. The client suspected that the procedure was producing too much variation (more than would be expected under a "valid" random sampling procedure), whereas the supplier defended its scheme as being valid.

This is where a consultant entered the picture. He was hired by the supplier to provide statistical evidence that the supplier's random sampling scheme was indeed valid and that the variation just described was within acceptable limits. The consultant agreed that the supplier was acting appropriately. He used computer simulation and appropriate probability models to illustrate the amount of variation that could be expected (between percentage available and percentage released/contacted in a given calling zone) under true random sampling and showed that the supplier's observed variation was of a similar magnitude. This demonstration satisfied both the supplier and the client that the sampling procedures being used were statistically valid. ●

PROBLEMS

Level A

1. Consider the frame of 52 full-time employees of Beta Technologies, Inc. Beta's human resources manager has collected current annual salary figures and related data for these employees. The data are in the file P02_01.XLS. In particular, these data include each selected employee's gender, age, number of years of relevant work experience prior to employment at Beta, the number of years of employment at Beta, the number of years of post-secondary education, and annual salary.

   a. Compute the mean, median, and standard deviation of the annual salaries for the 52 employees in the given frame.

b. Use Excel to choose a simple random sample of size 15 from this frame.

c. Compute the mean, median, and standard deviation of the annual salaries for the 15 employees included in your simple random sample. Compare these statistics with your computed descriptive measures for the frame obtained in part **a.** Is your simple random sample representative of the frame with respect to the annual salary variable?

2. A manufacturing company's quality control personnel have recorded the proportion of defective items for each of 500 monthly shipments of one of the computer components that the company produces. The data are in the file P02_02.XLS. The quality control department manager does not have sufficient time to review all of these data. Rather, she would like to examine the proportions of defective items for a simple random sample of 50 shipments.

a. Use Excel to generate such a random sample from the given frame.

b. What are the advantages and disadvantages of the sampling method requested by this quality control manager?

3. The manager of a local fast-food restaurant is interested in improving the service provided to customers who use the restaurant's drive-up window. As a first step in this process, the manager asks his assistant to record the time (in minutes) it takes to serve a large number of customers at the final window in the facility's drive-up system. The given frame of 200 customer service times are all observed during the busiest hour of the day for this fast-food operation. The data are in the file P02_04.XLS.

a. Compute the mean, median, and standard deviation of the customer service times in the given frame.

b. Use Excel to choose a simple random sample of size 20 from this frame.

c. Compute the mean, median, and standard deviation of the service times for the 20 customers included in your simple random sample. Compare these statistics with your computed descriptive measures for the frame obtained in part **a.**

4. A finance professor has just given a midterm examination in her corporate finance course. In particular, she is interested in learning how her large class of 100 students performed on this exam. The data are in the file P02_05.XLS.

a. Using these 100 students as the frame, generate a simple random sample of size 10 with Excel.

b. Compare the mean of the scores in the frame with that of the scores contained in the simple random sample.

5. Consider a frame consisting of 500 households in a middle-class neighborhood that was the recent focus of an economic development study conducted by the

local government. Specifically, for each of the 500 households, information was gathered on each of the following variables: family size, location of the household within the neighborhood, an indication of whether those surveyed owned or rented their home, gross annual income of the first household wage earner, gross annual income of the second household wage earner (if applicable), monthly home mortgage or rent payment, average monthly expenditure on utilities, and the total indebtedness (excluding the value of a home mortgage) of the household. The data are in the file P02_06.XLS.

a. Compute the mean, median, and standard deviation of the monthly home mortgage or rent payments of all households in the given frame.

b. Use Excel to choose a simple random sample of size 25 from this frame.

c. Compute the mean, median, and standard deviation of the monthly home mortgage or rent payments for the 25 households included in your simple random sample. Compare these statistics with your computed descriptive measures for the frame obtained in part **a.**

6. A real estate agent has received data on 150 houses that were recently sold in a suburban community. Included in this data set are observations for each of the following variables: the appraised value of each house (in thousands of dollars), the selling price of each house (in thousands of dollars), the size of each house (in hundreds of square feet), and the number of bedrooms in each house. The data are in the file P02_07.XLS. Suppose that this real estate agent wishes to examine a representative subset of these 150 houses. Use Excel to assist her by finding a simple random sample of size 10 from this frame.

7. Consider the given set of average annual household income levels of citizens of selected U.S. metropolitan areas in the file P03_06.XLS. Use Excel to obtain a simple random sample of size 15 from this frame.

8. The operations manager of a toll booth located at a major exit of a state turnpike is trying to estimate the average number of vehicles that arrive at the toll booth during a 1-minute period during the peak of rush-hour traffic. In an effort to estimate this average throughput value, he records the number of vehicles that arrive at the toll booth over a 1-minute interval commencing at the same time for each of 365 normal weekdays. The data are provided in the file P02_09.XLS. Choose a simple random sample of size 20 from the given frame of 365 values to help the operations manager estimate the average throughput value.

9. In ranking metropolitan areas in the United States, the *Places Rated Almanac* considers the average time (in minutes) it takes a citizen of each metropolitan area to travel to work and back home each day. The data are

in the file P02_11.XLS. Use Excel to obtain a simple random sample of 20 average commute times from the given set of such values.

10. Given data in the file P02_13.XLS from a recent survey of chief executive officers from the nation's biggest businesses (*The Wall Street Journal*, April 9, 1998), choose a simple random sample of 25 executives and find the mean, median, and standard deviation of the bonuses awarded to them in 1997. How do these sample statistics compare to the mean, median, and standard deviation of the bonuses given to all executives included in the frame?

11. A lightbulb manufacturer wants to know the number of defective bulbs contained in a typical box shipped by the company. Production personnel at this company have recorded the number of defective bulbs found in each of the 1000 boxes shipped during the past week. These data are provided in P08_11.XLS. Using this shipment of boxes as a frame, select a simple random sample of 50 boxes and compute the mean number of defective bulbs found in a box.

12. Consider the frame of 52 full-time employees of Beta Technologies, Inc. Beta's human resources manager has collected current annual salary figures and related data for these employees. The data are in the file P02_01.XLS.
    a. Compute the mean, median, and standard deviation of the annual salaries for the 52 employees in the given frame.
    b. Use Excel to choose a systematic sample of size 13 from this frame.
    c. Compute the mean, median, and standard deviation of the annual salaries for the 13 employees included in your systematic sample. Compare these statistics with your computed descriptive measures for the frame obtained in part **a**. Is your systematic sample representative of the frame with respect to the annual salary variable?

13. A manufacturing company's quality control personnel have recorded the proportion of defective items for each of 500 monthly shipments of one of the computer components that the company produces. The data are in the file P02_02.XLS. The quality control department manager does not have sufficient time to review all of these data. Rather, she would like to examine the proportions of defective items for a systematic sample of 50 shipments.
    a. Use Excel to generate such a systematic sample from the given frame.
    b. What are the advantages and disadvantages of the sampling method requested by this quality control manager?

14. The manager of a local fast-food restaurant is interested in improving the service provided to customers who use the restaurant's drive-up window. As a first step in this process, the manager asks his assistant to record the time (in minutes) it takes to serve a large number of customers at the final window in the facility's drive-up system. The given frame of 200 customer service times are all observed during the busiest hour of the day for this fast-food operation. The data are in the file P02_04.XLS.
    a. Compute the mean, median, and standard deviation of the customer service times in the given frame.
    b. Use Excel to choose a systematic sample of size 20 from this frame.
    c. Compute the mean, median, and standard deviation of the service times for the 20 customers included in your systematic sample. Compare these statistics with your computed descriptive measures for the frame obtained in part **a**.

15. A finance professor has just given a midterm examination in her corporate finance course. In particular, she is interested in learning how her large class of 100 students performed on this exam. The data are in the file P02_05.XLS.
    a. Using these 100 students as the frame, generate a systematic sample of size 10 with Excel.
    b. Compare the mean of the scores in the frame with that of the scores included in the systematic sample.

16. Consider a frame consisting of 500 households in a middle-class neighborhood that was the recent focus of an economic development study conducted by the local government. The data are in the file P02_06.XLS.
    a. Compute the mean, median, and standard deviation of the monthly home mortgage or rent payments of all households in the given frame.
    b. Use Excel to choose a systematic sample of size 25 from this frame.
    c. Compute the mean, median, and standard deviation of the monthly home mortgage or rent payments for the 25 households included in your systematic sample. Compare these statistics with your computed descriptive measures for the frame obtained in part **a**.

17. A real estate agent has received data on 150 houses that were recently sold in a suburban community. Included in this data set are observations for each of the following variables: the appraised value of each house (in thousands of dollars), the selling price of each house (in thousands of dollars), the size of each house (in hundreds of square feet), and the number of bedrooms in each house. The data are in the file P02_07.XLS. Suppose that this real estate agent wishes to examine a representative subset of these 150 houses. Use Excel to assist her by finding a systematic sample of size 10 from this frame.

18. Consider the given set of average annual household income levels of citizens of selected U.S. metropolitan areas in the file P03_06.XLS. Use Excel to obtain a systematic sample of size 25 from this frame.

19. The operations manager of a toll booth located at a major exit of a state turnpike is trying to estimate the average number of vehicles that arrive at the toll booth during a 1-minute period during the peak of rush-hour traffic. In an effort to estimate this average throughput value, he records the number of vehicles that arrive at the toll booth over a 1-minute interval commencing at the same time for each of 365 normal weekdays. The data are provided in the file P02_09.XLS. Choose a systematic sample of size 20 from the given frame of 365 values to help the operations manager estimate the average throughput value.

20. Consider the average time (in minutes) it takes citizens of each of the metropolitan areas across the United States to travel to work and back home each day. The data are in the file P02_11.XLS. Use Excel to obtain a systematic sample of 25 average commute times from the given set of such values.

21. Consider the frame of 52 full-time employees of Beta Technologies, Inc. Beta's human resources manager has collected current annual salary figures and related data for these employees. The data are in the file P02_01.XLS.
    a. Compute the mean, median, and standard deviation of the annual salaries for the 52 employees in the given frame.
    b. Assuming that the human resources manager wishes to stratify these employees by the number of years of post-secondary education, select such a stratified sample of size 15 with approximately proportional sample sizes.
    c. Compute the mean, median, and standard deviation of the annual salaries for the 15 employees included in your stratified sample. Compare these statistics with your computed descriptive measures for the frame obtained in part a. Is your stratified sample representative of the frame with respect to the annual salary variable?

22. Consider a frame consisting of 500 households in a middle-class neighborhood that was the recent focus of an economic development study conducted by the local government. Specifically, for each of the 500 households, information was gathered on each of the following variables: family size, location of the household within the neighborhood, an indication of whether those surveyed owned or rented their home, gross annual income of the first household wage earner, gross annual income of the second household wage earner (if applicable), monthly home mortgage or rent payment, average monthly expenditure on utilities, and the total indebtedness (excluding the value

of a home mortgage) of the household. The data are in the file P02_06.XLS.
    a. Compute the mean, median, and standard deviation of the gross annual income of the first wage earner of all households in the given frame.
    b. Given that researchers have decided to stratify the given households by location within the neighborhood, choose a stratified sample of size 25 with proportional sample sizes.
    c. Compute the mean, median, and standard deviation of the gross annual income of the first wage earner of the 25 households included in your stratified sample. Compare these statistics with your computed descriptive measures for the frame obtained in part a.
    d. Explain how economic researchers could apply cluster sampling in selecting a sample of size 25 from this frame. What are the advantages and disadvantages of employing cluster sampling in this case?

23. A real estate agent has received data on 150 houses that were recently sold in a suburban community. Included in this data set are observations for each of the following variables: the appraised value of each house (in thousands of dollars), the selling price of each house (in thousands of dollars), the size of each house (in hundreds of square feet), and the number of bedrooms in each house. The data are in the file P02_07.XLS.
    a. Suppose that this real estate agent wishes to examine a representative subset of these 150 houses that has been stratified by the number of bedrooms. Use Excel to assist her by finding such a stratified sample of size 15 with proportional sample sizes.
    b. Explain how the real estate agent could apply cluster sampling in selecting a sample of size 15 from this frame. What are the advantages and disadvantages of employing cluster sampling in this case?

24. Given data in the file P02_13.XLS from a recent survey of chief executive officers from the nation's biggest businesses (*The Wall Street Journal,* April 9, 1998), choose a sample of 25 executives stratified by company type with proportional sample sizes. Next, find the mean, median, and standard deviation of the salaries awarded to them in 1997. How do these sample statistics compare to the mean, median, and standard deviation of the salaries given to all executives included in the frame?

Level B

25. The employee benefits manager of a small private university would like to know the proportion of its full-time employees who prefer adopting each of three available health care plans in the forthcoming annual enrollment period. A reliable frame of the university's

employees and their tentative health care preferences are given in P08_25.XLS.

a. Compute the proportion of the employees in the given frame who favor *each* of the three plans (i.e., plans A, B, and C).

b. Use Excel to choose a sample of 45 employees stratified by employee classification with proportional sample sizes.

c. Compute the proportion of the 45 employees in the stratified sample who favor each health plan. Compare these sample proportions to the corresponding values obtained in part **a.** Explain any differences between the corresponding values.

d. What are the advantages and disadvantages of employing stratified sampling in this particular case?

e. Explain how the benefits manager could apply cluster sampling in selecting a sample of size 30 from this frame. What are the advantages and disadvantages of employing cluster sampling in this case?

26. The file P02_17.XLS reports the number of short-term general hospitals in each of a large number of U.S. metropolitan areas. Suppose that you are a sales manager for a major pharmaceutical producer and are interested in estimating the average number of such hospitals in *all* metropolitan areas across the entire country. Assuming that you do not have access to the data for each metropolitan location, you decide to select a sample that will be representative of all such areas.

a. Choose a simple random sample of 30 metropolitan areas from the given frame. Compute the mean number of short-term general hospitals for the metropolitan areas included in your sample.

b. Do you believe that simple random sampling is the best approach to obtaining a representative subset of the metropolitan areas in the given frame? Explain. If not, how might you proceed to select a better sample of size 30 using the data provided in the file? Compute the mean number of short-term general hospitals for the metropolitan areas included in your revised sample. How does this sample mean compare to that computed from your simple random sample in part **a?**

27. As human resources manager of a manufacturing plant, you are quite concerned about recent reports of sexual and racial harassment from the production workers within the organization. In an effort to gain a better understanding of the apparent problems, you decide that it would be wise to interview a cross section of your employees about this and other issues in the workplace.

a. Using the frame of employees provided in the file P08_27.XLS, select a subset of 30 production workers stratified by sex with proportional sample sizes.

b. Next, select another subset of 30 workers stratified by race with proportional sample sizes.

c. Finally, select one more subset of 30 workers stratified by *both* sex and race (e.g., Black women, White men, Asian women, Hispanic men, etc.) with proportional sample sizes.

d. Explain how the human resources manager could apply cluster sampling in selecting a sample of size 30 from this frame. What are the advantages and disadvantages of employing cluster sampling in this case?

28. Is the overall cost of living higher or lower for urban areas in particular geographical regions of the United States?

a. Begin by first selecting 40 urban areas from the given frame provided in the file P08_28.XLS. The urban areas you select should be stratified by geographical location within the United States (e.g., northeast, southeast, midwest, northwest, or southwest) and should reflect proportional sample sizes. Note that you will need to assign the given urban areas to one of any number of such geographical regions before you can generate a stratified sample.

b. Explain how you could apply cluster sampling in selecting a sample of size 40 from this frame. What are the advantages and disadvantages of employing cluster sampling in this case?

An Introduction to Estimation

The purpose of any random sample is to estimate properties of a population from the data observed in the sample. The following is a good example to keep in mind. Suppose a government agency wants to know the average household income, where this average is taken over the population of all households in Atlanta. Then this unknown average is the population parameter of interest, and the government is likely to estimate it by sampling several "representative" households in Atlanta and reporting the average of their incomes.

The mathematical procedures appropriate for performing this estimation depend on which properties of the population are of interest and which type of random sampling scheme is used. Because the details are considerably more complex when a more complex sampling scheme such as multistage sampling is used, we will focus on *simple* random samples, where the mathematics is relatively straightforward. Details for other sampling schemes such as stratified sampling can be found in Levy and Lemeshow (1999). However, even for more complex sampling schemes, the *concepts* are the same as those we will discuss here; only the details change.

Throughout most of this section, we will focus on the population mean of some variable such as household income. Our goal is to estimate this population mean by using the data in a randomly selected sample. We first discuss the types of errors that can occur in this estimation problem.

8.4.1 Sources of Estimation Error

There are two basic sources of errors that can occur when we sample randomly from a population: *sampling error* and all other sources, usually lumped together as *nonsampling error*. Sampling error results from "unlucky" samples. As such, the term *error* is somewhat misleading. Suppose, for example, that the mean grade-point average (GPA) in a large class of 400 students is 2.85. The instructor wants to know this mean but doesn't want to ask *each* student for his or her GPA. Therefore, the instructor asks a random sample of 20 students for their GPAs, and it turns out that the sample mean for these 20 students is 2.97. If the instructor then infers that the mean of *all* GPAs is 2.97, the resulting sampling error is the difference between the reported value and the true value, 0.12. Note that the instructor hasn't done anything "wrong." This sampling error is essentially due to bad luck.

Sampling Error	**Sampling error** is the inevitable result of basing an inference on a random sample rather than on the entire population.

We will see shortly how to measure the potential sampling error involved. The point here is that the resulting estimation error is not caused by anything we're doing wrong—we might just get unlucky.

Nonsampling error is quite different and can occur for a variety of reasons. We discuss a few of them.

- Perhaps the most serious type of nonsampling error is **nonresponse bias.** This occurs when a portion of the sample fails to respond to the survey. Anyone who has ever conducted a questionnaire, whether by mail, by phone, or any other method, knows that the percentage of nonrespondents can be quite large. The question is whether this introduces estimation error. If the nonrespondents *would* have responded similarly to the respondents, had they responded, we don't lose much by not hearing from them. However, because the nonrespondents don't respond, we typically have no way of knowing whether they differ in some important respect from the respondents. Therefore, unless we are able to persuade the nonrespondents to respond—through a follow-up phone call, for example—we must guess at the amount of nonresponse bias.

  By the way, it is interesting that we always learn (from statistics books!) to take *larger* samples to reduce sampling error. This makes intuitive sense. We can typically learn more about a population from a larger sample. However, if the potential for nonresponse bias is large, *smaller* samples might be preferable—they tend to decrease the amount of nonresponse bias.

- Another source of nonsampling error is **nontruthful responses.** This is particularly a problem when we ask sensitive questions in a questionnaire. For example, if the questions "Have you ever had an abortion?" or "Do you regularly use cocaine?" are asked, most people will answer "no," regardless of whether the true answer is "yes" or "no."

  There is a way of getting at such sensitive information, called the **randomized response** technique. Here the investigator presents each respondent with two questions, one of which is the sensitive question. The other is innocuous, such as, "Were you born in the summer?" The respondent is asked to decide randomly which of the two questions to answer—by flipping a coin, say—and then answer the chosen question truthfully. The investigator sees only the answer (yes or no), not the result of the coin flip. That is, the investigator doesn't know which question is being answered. However, by using probability theory, it is possible for the investigator to infer from many such responses the percentage of the population whose truthful answer to the sensitive question is "yes."

- Another type of nonsampling error is **measurement error.** This occurs when the responses to the questions do not reflect what the investigator had in mind. It might result from poorly worded questions, questions the respondents don't fully understand, questions that require the respondents to supply information they don't have, and so on. Undoubtedly, there have been times when you were filling out a questionnaire and said to yourself, "OK, I'll answer this as well as I can, but I know it's not what they want to know."

- One final type of nonsampling error is **voluntary response bias.** This occurs when the subset of people who respond to a survey differ in some important respect from all potential respondents. For example, suppose a population of students are surveyed to see how many hours they study per night. If the students who respond are predominantly those who get the best grades, the resulting sample mean number of hours will be biased on the high side.

From this discussion and your own experience with questionnaires, you should realize that the potential for nonsampling error is enormous. However, unlike sampling error, it cannot be measured with probability theory. It can only be controlled by using appropriate sampling procedures and designing good survey instruments. We will not pursue this topic any further here. If you are interested, however, you can learn about methods for controlling nonsampling error, such as proper questionnaire design, from books on marketing research.

8.4.2 Key Terms in Sampling

We now set the stage for the rest of this chapter, as well as for several later chapters. Suppose there is at least one population with some numerical parameter we would like to know. This parameter could be a population mean, a population proportion, the difference between two population means, the difference between two population proportions, or many others. Unless we measure each member of the population—that is, we take a complete census—we cannot learn the exact value of this population parameter. Therefore, we instead take a random sample of some type and try to *estimate* the population parameter from the data in the sample.

We typically begin by calculating a **point estimate** (or, simply, estimate) from the sample data, a "best guess" of the population parameter. The difference between the point estimate and the true value of the population parameter is called the **estimation error.** We then try to use probability theory to gauge the magnitude of the estimation error. The key to this is the **sampling distribution** of the point estimate, which is defined as the distribution of the point estimates we would see from *all* possible samples (of a given sample size) from the population. Often we report the accuracy of the point estimate with an accompanying *interval estimate* (which we will refer to as a *confidence interval* in succeeding chapters). An **interval estimate** is an interval around the point estimate, calculated from the sample data, where we strongly believe the true value of the population parameter lies.

Point Estimate	A **point estimate** is a single numeric value, a "best guess" of a population parameter, based on the data in a sample.

Estimation Error	The **estimation error** is the difference between the point estimate and the true value of the population parameter being estimated.

Sampling Distribution	The **sampling distribution** of any point estimate is the distribution of the point estimates we would see from *all* possible samples (of a given sample size) from the population.

Interval Estimate (or Confidence Interval)	An **interval estimate** (or **confidence interval**) is an interval around the point estimate, calculated from the sample data, where we strongly believe the true value of the population parameter lies.

Additionally, there are two other key terms you should know. First, consider the *mean* of the sampling distribution of a point estimate. It is simply the average value of the point estimates we would see from all possible samples. When this mean is equal to the true value of the population parameter, we say that the point estimate is **unbiased.** Otherwise, we say that it is **biased.** Naturally, we prefer unbiased estimates. They sometimes miss on the low side and sometimes on the high side, but on average, they tend to be right on.

Unbiased Estimate	An **unbiased estimate** is a point estimate such that the mean of its sampling distribution is equal to the true value of the population parameter being estimated.

Unbiased estimates are desirable because they average out to the correct value. However, this isn't enough. We do not want point estimates from different samples to vary wildly from sample to sample because we couldn't rely much on the estimate from any particular sample. Therefore, we measure the standard deviation of the sampling distribution of the estimate, which indicates how much point estimates from different samples vary. In the context of sampling, this standard deviation is called the **standard error** of the estimate. Ideally, we want estimates that have *small* standard errors.

Standard Error	The **standard error** of an estimate is the standard deviation of the sampling distribution of the estimate. It measures how much estimates from different samples vary.

The terms in this subsection are relevant for practically any population parameter we might want to estimate. In the following subsection we will discuss them in the context of estimating a population mean.

8.4.3 Sampling Distribution of the Sample Mean

In this section we discuss the estimation of the population mean from some population. For example, we might be interested in the mean household income for all families in a particular city, the mean diameter of all parts from a manufacturing process, the mean pollution count over all cities in the world with a population of at least 100,000, and so on. We label the unknown population mean by μ.

The point estimate of μ we will use, based on a sample from the population, is the sample mean $\overline{X}$, the average of the observations in the sample. Actually, you might not imagine that there are *other* possible point estimates for a population mean besides the sample mean. In fact, there are. We could use the sample median, the *trimmed mean* (where we average all but the few most extreme observations), and others. However, it turns out that the "natural" estimate, the sample mean, has very good theoretical properties, so it is usually the preferred point estimate.

How accurate is $\overline{X}$ in estimating μ? That is, how large does the estimation error $\overline{X} - \mu$ tend to be? As we discussed in the previous subsection, the sampling distribution of $\overline{X}$ provides the key. Before describing this sampling distribution in some generality, we provide some insight into it by revisiting the population of 40 incomes in Example 8.1. There we showed how to generate a single random sample of size 10. For the particular sample we generated (see Figure 8.2), the sample mean was $38,750. Since we know that the population mean of all 40 incomes is $39,985, the estimation error based on this particular sample is the difference $38,750 − $39,985, or $1235 on the low side.

However, this is only one of many possible samples. To see other possibilities, we used StatPro's procedure for generating random samples to generate 100 random samples of size 10 from the population of 40 incomes. We then calculated the sample mean for each random sample and created a histogram of these sample means, shown in Figure 8.7. Although this is not *exactly* the sampling distribution of the sample mean (because there are many more than 100 possible samples of size 10 from a population of size 40), it indicates how the possible sample means are distributed. They are most likely to be near the population mean ($39,985), they are very unlikely to be more than about $3000 from this population mean, and the distribution is approximately bell shaped.

The insights in the previous paragraph can be generalized. It turns out that the sampling distribution of the sample mean has the following properties, regardless of the underlying population. First, it is an unbiased estimate of the population mean, as indicated in equation (8.1). The sample means from some samples will be too low, and those from other samples will be too high, but on the average, they will be just right.

FIGURE 8.7

Approximate Sampling Distribution of Sample Mean

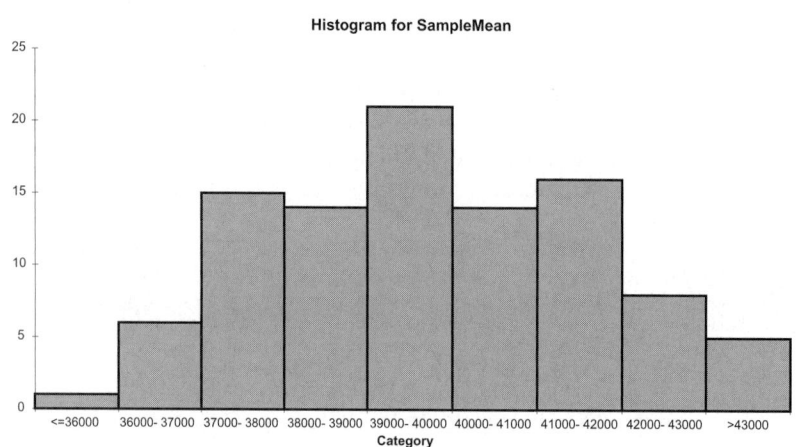

Unbiased Property of Sample Mean	$E(\bar{X}) = \mu$	**(8.1)**

The second property involves the variability of the $\bar{X}$ estimate. Recall that the standard deviation of an estimate, called the standard error, indicates how much the estimates vary from sample to sample. The standard error of $\bar{X}$ is given in equation (8.2). Here, SE($\bar{X}$) is our abbreviation for the standard error of $\bar{X}$, σ is the standard deviation of the population, and n is the sample size. We see that the standard error is large when the observations in the population are spread out (large σ), but that the standard error can be reduced by taking a larger sample.[2]

Standard Error of Sample Mean	$\text{SE}(\bar{X}) = \sigma/\sqrt{n}$	**(8.2)**

There is one problem with the standard error in equation (8.2). We need its value to gauge the amount of estimation error, but its value depends on another unknown population parameter, σ. Therefore, it is customary to approximate the standard error by substituting the *sample* standard deviation, s, for σ. This leads to equation (8.3).

Approximate Standard Error of Sample Mean	$\text{SE}(\bar{X}) = s/\sqrt{n}$	**(8.3)**

As we will discuss in the next subsection, the shape of the sampling distribution of $\bar{X}$ is approximately normal. Therefore, we can use the standard error exactly as we have used standard deviations in previous chapters to obtain interval estimates of the population mean. Specifically, if we go out 2 standard errors on either side of the sample mean, as shown in expression (8.4), we are 95% confident of capturing the population mean. Alternatively, we are 95% confident that the estimation error will be no greater than 2 standard errors in magnitude.

Interval Estimate of Population Mean	$\bar{X} \pm 2s/\sqrt{n}$	**(8.4)**

The following example illustrates a typical use of sample information.

Example 8.4 Estimating the Mean of Accounts Receivable for a Furniture Retailer

An internal auditor for a furniture retailer wants to estimate the average of all accounts receivable, where this average is taken over the population of all customer accounts.

---

[2]This formula for SE($\bar{X}$) assumes that the sample size n is small relative to the population size N. As a rule of thumb, we assume that n is no more than 5% of N. Later we will provide a "correction" to this formula when n is a larger percentage of N.

FIGURE 8.8
Sampling in Auditing Example

	A	B	C	D	E
1	Random sample of accounts receivable				
2					
3	Population size	10000			
4	Sample size	100			
5					
6	Sample of receivables			Summary measures from sample	
7	Account	Amount		Sample mean	$278.92
8	1	$85		Sample stdev	$419.21
9	2	$1,061		Std Error of mean	$41.92
10	3	$0			
11	4	$1,260		With fpc	$41.71
12	5	$924			
13	6	$129			
105	98	$657			
106	99	$86			
107	100	$0			

Because the company has approximately 10,000 accounts, an exhaustive enumeration of all accounts receivable is impossible. Therefore, the auditor randomly samples 100 of the accounts. The observed data appear in Figure 8.8. (See the file AUDIT.XLS.) What can the auditor conclude from this sample?

Objective To illustrate the meaning of standard error of the mean in a sample of accounts receivable.

Solution

The receivables for the 100 sampled accounts appear in column B. This is the only information available to the auditor, so he must base all conclusions on these sample data. We calculate the sample mean and sample standard deviation in cells E7 and E8 with the formulas

$$=AVERAGE(B8:B107)$$

and

$$=STDEV(B8:B107)$$

Then we use equation (8.3) to calculate the (approximate) standard error of the mean in cell B9 with the formula

$$=E8/SQRT(B4)$$

The auditor should interpret these values as follows. First, the sample mean $279 can be used to estimate the unknown population mean. It provides a best guess for the average of the receivables from all 10,000 accounts. In fact, because the sample mean is an unbiased estimate of the population mean, there is no reason to suspect that $279 either underestimates or overestimates the population mean. Second, the standard error $42 provides a measure of accuracy of the $279 estimate. Specifically, there is about a 95% chance that the estimate differs by no more than 2 standard errors ($84) from the true (but unknown) population mean. Therefore, the auditor can be 95% certain that the mean from all 10,000 accounts is within the interval $279 ± $84, that is, between $195 and $363. ●

It is important to distinguish between the sample standard deviation s and the standard error of the mean, approximated by $s/\sqrt{n}$. The sample standard deviation in the auditing example, $419, measures the variability in *individual* receivables in the sample (or in the population). By scrolling down column B, we see that there are some very low amounts (many zeros) and some fairly large amounts. This variability is indicated by the rather large sample standard deviation s. However, this value does not measure the accuracy of the sample mean as an estimate of the population mean. To judge *its* accuracy, we need to divide s by the square root of the sample size n. The resulting standard error, about $42, is much smaller than the sample standard deviation. It indicates that we can be about 95% certain that the sampling error is no greater than $84.

The Finite Population Correction We mentioned that equation (8.2) [or equation (8.3)] for the standard error of $\overline{X}$ is appropriate when the sample size n is small relative to the population size N. Generally, "small" means that n is no more than 5% of N. In most realistic samples this is certainly true. For example, political polls are typically based on samples of approximately 1000 people from the entire U.S. population.

There are situations, however, when we sample more than 5% of the population. In this case the formula for the standard error of the mean should be modified with a *finite population correction*, or *fpc*, factor. Then the modified standard error of the mean appears in equation (8.5), where the *fpc* is given by equation (8.6). Note that this factor is always less than 1 (when $n > 1$) and it decreases as n increases. Therefore, the standard error of the mean decreases—and the accuracy increases—as n increases.

Standard Error of Mean with Finite Population Correction Factor	$$\mathrm{SE}(\overline{X}) = \mathit{fpc} \times (s/\sqrt{n})$$	**(8.5)**

Finite Population Correction Factor	$$\mathit{fpc} = \sqrt{\dfrac{N-n}{N-1}}$$	**(8.6)**

To see how the *fpc* varies with n and N, consider the values in Table 8.1. Rather than listing n, we have listed the percentage of the population sampled, that is, $n/N \times 100\%$. It is clear that when 5% or less of the population is sampled, the *fpc* is very close to 1 and can safely be ignored. In this case we can use $s/\sqrt{n}$ as the standard error of the mean. Otherwise, we should use the modified formula in equation (8.5).

Table **8.1**

Finite Population Correction Factors	N	% Sampled	*fpc*
	100	5%	.980
	100	10%	.953
	10,000	1%	.995
	10,000	5%	.975
	10,000	10%	.949
	1,000,000	1%	.995
	1,000,000	5%	.975
	1,000,000	10%	.949

If less than 5% of the population is sampled, as is often the case, the *fpc* can safely be ignored.

In the auditing example, $n/N = 100/100,000 = 0.1\%$. This suggests that the *fpc* can safely be omitted. We illustrate this in cell B11 of Figure 8.8, which uses the formula from equation (8.5):

$$=SQRT((B3-B4)/(B3-1))*E9$$

Clearly, it makes no practical difference in this example whether we use the *fpc* or not. The standard error, rounded to the nearest dollar, is $42 in either case.

Virtually all standard error formulas used in sampling include an *fpc* factor. However, because it is rarely necessary—the sample size is usually very small relative to the population size—we will omit it from here on.

8.4.4 The Central Limit Theorem

Our discussion to this point has concentrated primarily on the mean and standard deviation of the sampling distribution of the sample mean. In this section we discuss this sampling distribution in more detail. Because of an important theoretical result called the central limit theorem, we know that this distribution is approximately *normal* with mean μ and standard deviation $\sigma/\sqrt{n}$. This theorem is the reason why the normal distribution appears in so many statistical results. We can state the theorem as follows.

Central Limit Theorem

For any population distribution with mean μ and standard deviation σ, the sampling distribution of the sample mean $\overline{X}$ is approximately normal with mean μ and standard deviation $\sigma/\sqrt{n}$, and the approximation improves as n increases.

The important part of this result is the normality of the sampling distribution. We know, without any conditions placed upon the sample size n, that the mean and standard deviation are μ and $\sigma/\sqrt{n}$. However, the central limit theorem also implies normality, provided that n is reasonably large.

How large must n be for the approximation to be valid? Most analysts suggest $n \geq 30$ as a rule of thumb. However, this depends on the population distribution. If the population distribution is very *nonnormal*—extremely skewed or bimodal, say—then the normal approximation might not be accurate unless n is considerably greater than 30. On the other hand, if the population distribution is already approximately symmetric, then the normal approximation is quite good for n considerably less than 30. In fact, in the special case where the population distribution itself is normal, the sampling distribution of $\overline{X}$ is *exactly* normal for *any* value of n.

The central limit theorem is not a simple concept to grasp. To help explain it, we employ simulation in the following example.

Example 8.5 Average Winnings from Spinning a Wheel of Fortune

Suppose you have the opportunity to play a game with a "wheel of fortune" (similar to a popular television game show). When you spin a large wheel, it is equally likely to stop in any position. Depending on where it stops, you win anywhere from $0 to $1000. Let's suppose your winnings are actually based on not one, but n spins of the wheel. For example, if $n = 2$, your winnings are based on the average of two spins. If the first spin results in $580 and the second spin results in $320, then you win the average, $450. How does the distribution of your winnings depend on n?

FIGURE 8.9
Uniform
Distribution

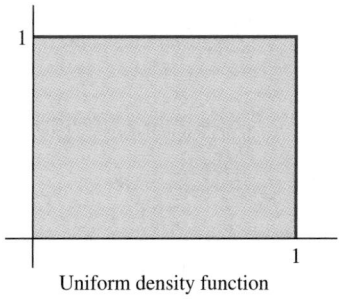

Uniform density function

Objective To illustrate the central limit theorem in the context of winnings in a game of chance.

Solution

First, we need to discuss what this experiment has to do with random sampling. Here, the population is the set of all outcomes we could obtain from a *single* spin of the wheel—that is, all dollar values from $0 to $1000. Each spin results in one randomly sampled dollar value from this population. Furthermore, because we have assumed that the wheel is equally likely to land in any position, all possible values in the continuum from $0 to $1000 have the same chance of occurring. The resulting population distribution is called the **uniform distribution** on the interval from $0 to $1000. (See Figure 8.9.) It can be shown (with calculus) that the mean and standard deviation of this uniform distribution are $\mu = \$500$ and $\sigma = \$289$.[3]

Before we go any further, take a moment to test your own intuition. If you play this game once and your winnings are based on the average of n spins, how likely is that you will win at least $600 if $n = 1$; if $n = 3$; if $n = 10$? (The answers are 0.4, 0.27, and 0.14, respectively, where the last two answers are approximate and are based on the central limit theorem or on our simulations. So you are much less likely to "win big" if your winnings are based on the average of many spins.)

Now we'll analyze the distribution of winnings based on the average of n spins. We do so by means of a sequence of simulations in Excel, for $n = 1$, $n = 2$, $n = 3$, $n = 6$, and $n = 10$. (See the files SPIN1.XLS, SPIN2.XLS, SPIN3.XLS, SPIN6.XLS, and SPIN10.XLS.) For each simulation we consider 1000 replications of an experiment. Each replication of the experiment simulates n spins of the wheel and calculates the average—that is, the winnings—from these n spins. Based on these 1000 replications, we can then calculate the average winnings, the standard deviation of winnings, and a histogram of winnings for each n. These will show clearly how the distribution of winnings depends on n.

The values in Figure 8.10 and the histogram in Figure 8.11 (page 394) show the results for $n = 1$. Here there is no averaging—we spin the wheel once and win the amount shown. To replicate this experiment 1000 times and collect statistics, we proceed as follows.

1 **Random outcomes.** To generate outcomes uniformly distributed between $0 and $1000, enter the formula

$$=\$B\$3+RAND()*(\$B\$4-\$B\$3)$$

in cell B11 and copy it down column B. The effect of this formula, given the values in cells B3 and B4, is to generate a random number between 0 and 1 and multiply it by $1000.

---

[3]In general, if a distribution is uniform on the interval from a to b, then its mean is the midpoint $(a + b)/2$ and its standard deviation is $(b - a)/\sqrt{12}$.

	A	B	C	D	E	F
1	**Wheel of fortune simulation**					
2						
3	Minimum winnings	$0		**Summary measures of winnings**		
4	Maximum winnings	$1,000		Mean	$505	
5				Stdev	$288	
6	Number of spins	1				
7						
8	**Simulation of spins**					
9						
10	Replication	Outcome				
11	1	$369				
12	2	$536				
13	3	$154				
14	4	$876				
15	5	$797				
16	6	$589				
17	7	$846				
18	8	$896				
19	9	$721				
20	10	$10				
21	11	$762				
1008	998	$973				
1009	999	$993				
1010	1000	$948				

FIGURE 8.10 Simulation of Winnings from a Single Spin

FIGURE 8.11
Histogram of
Simulated
Winnings from a
Single Spin

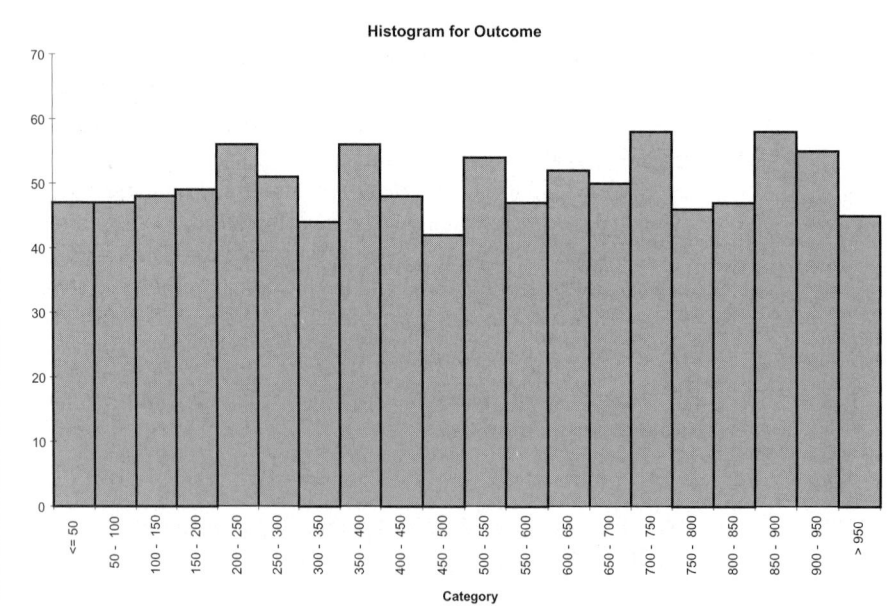

② **Summary measures.** Calculate the average and standard deviation of the 1000 winnings in column B with the AVERAGE and STDEV functions. These values appear in cells E4 and E5.

③ **Frequency table and histogram.** Use the StatPro histogram procedure to create a histogram of the values in column B.

Note the following from Figures 8.10 and 8.11.

- The sample mean of the winnings (cell E4) is very close to the population mean, $500.
- The standard deviation of the winnings (cell E5) is very close to the population standard deviation, $289.
- The histogram is nearly flat.

These properties should come as no surprise. When $n = 1$, the "sample mean" is a single observation—that is, no averaging takes place. Therefore, the sampling distribution of the sample mean is *equivalent* to the flat population distribution in Figure 8.9.

But what happens when $n > 1$? Figures 8.12 and 8.13 (page 396) show the results for $n = 2$. Now we form a second column of outcomes in column C, corresponding to the second spin in each experiment, and we average the values in columns B and C to obtain each of the winnings in column D. None of the rest of the spreadsheet changes. The average winnings is again very close to $500, but the standard deviation of winnings is much lower. In fact, it is close to $\sigma/\sqrt{2} = 289/\sqrt{2} = \$204$, exactly as theory predicts. In addition, the histogram of winnings is no longer flat. It is triangularly shaped—symmetric, but not yet bell shaped.

To develop similar simulations for $n = 3$, $n = 6$, $n = 10$, or any other n, we insert additional outcome columns and make sure that the AVERAGE formula in the Winnings

	A	B	C	D	E	F	G	H
1	Wheel of fortune simulation							
2								
3	Minimum winnings	$0				Summary measures of winnings		
4	Maximum winnings	$1,000				Mean	$490	
5						Stdev	$200	
6	Number of spins	2						
7								
8	Simulation of spins							
9								
10	Replication	Outcome1	Outcome2	Winnings				
11	1	$371	$601	$486				
12	2	$623	$430	$527				
13	3	$772	$262	$517				
14	4	$803	$733	$768				
15	5	$969	$392	$680				
16	6	$980	$634	$807				
17	7	$744	$849	$796				
18	8	$180	$270	$225				
19	9	$169	$90	$130				
20	10	$675	$601	$638				
21	11	$709	$260	$484				
1007	997	$363	$489	$426				
1008	998	$110	$455	$283				
1009	999	$335	$878	$607				
1010	1000	$770	$87	$429				

FIGURE 8.12 Simulation of Winnings from Two Spins

FIGURE 8.13
Histogram of Sim-
ulated Winnings
from Two Spins

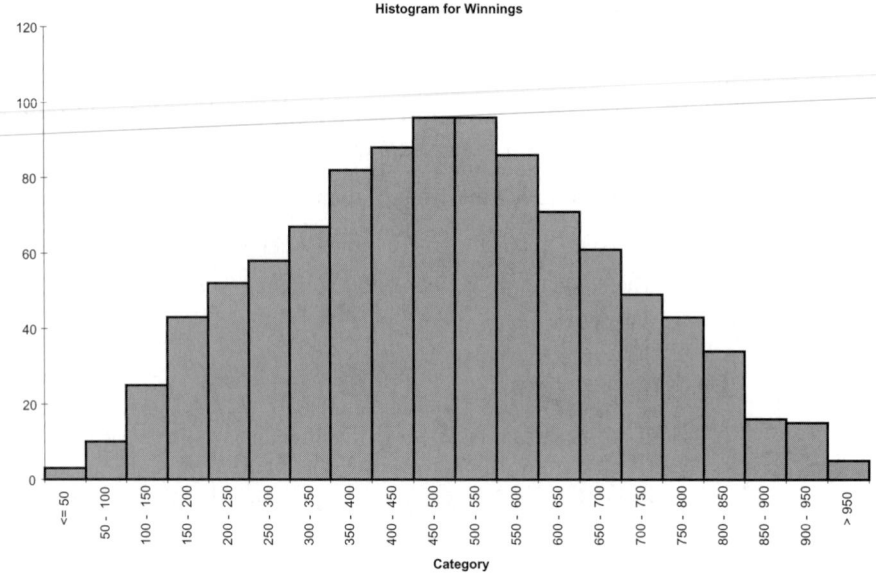

Histogram for Winnings

FIGURE 8.14
Histogram of Sim-
ulated Winnings
from Three Spins

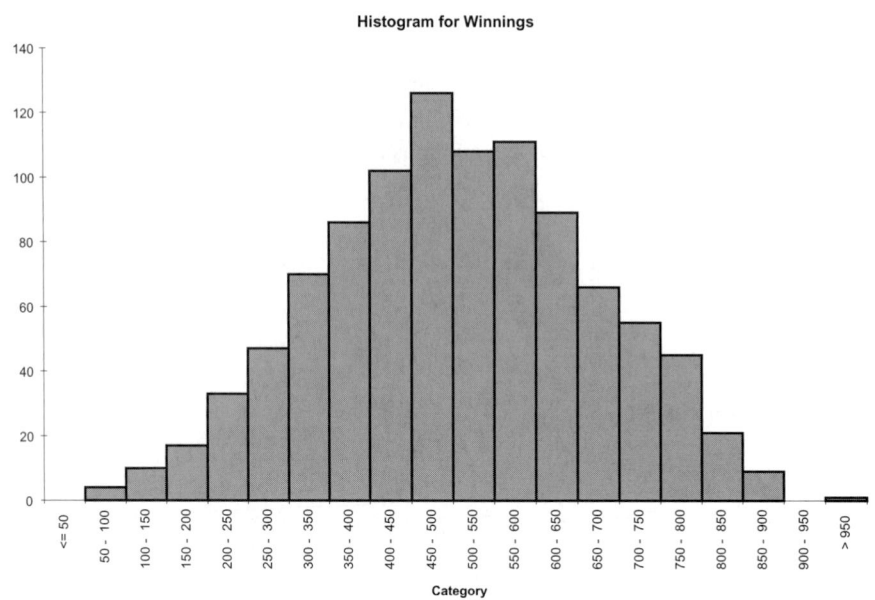

Histogram for Winnings

column averages all n outcomes to its left. The resulting histograms appear in Figures 8.14–8.16. They clearly show two effects of increasing n: (1) the histogram becomes more bell shaped, and (2) there is less variability. However, the mean stays right at $500. This behavior is exactly what the central limit theorem predicts. In fact, because the population distribution is symmetric in this example—it's flat—we see the effect of the central limit theorem for n much less than 30; it is already evident for n as low as 6.

Finally, although the numerical results are not shown here, we could find the answers to the question we posed previously: How does the probability of winning at least $600 depend on n? For any specific value of n, we would simply find the fraction of the 1000 replications where the average of n spins is greater than $600. (This information is given

FIGURE 8.15
Histogram of Simulated Winnings from Six Spins

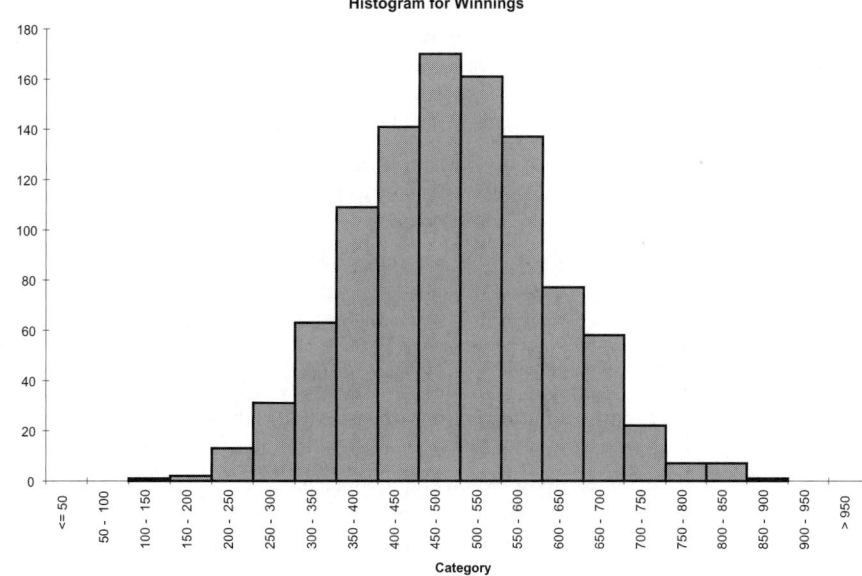

FIGURE 8.16
Histogram of Simulated Winnings from Ten Spins

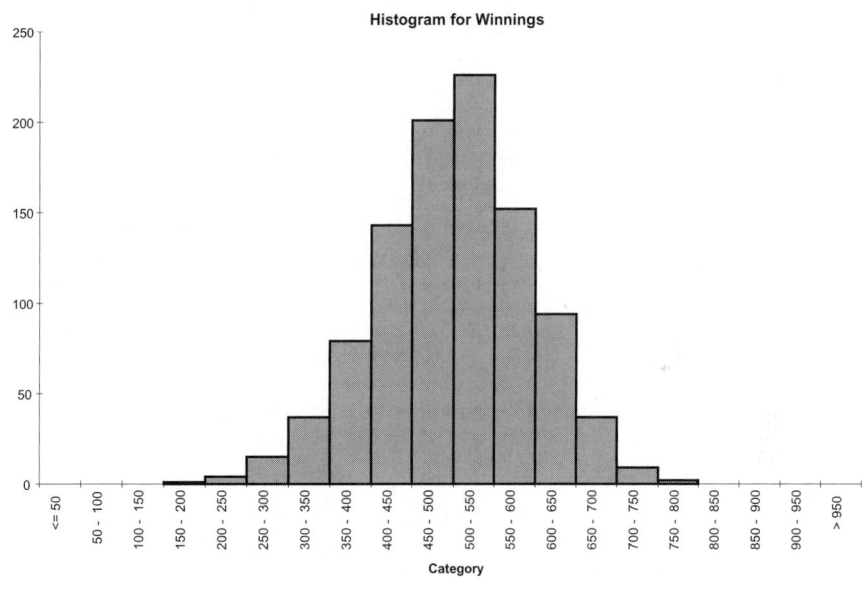

by StatPro as part of its histogram output.) You can check that the answers we gave previously are supported by the simulation results. ●

What are the main lessons from this example? For one, we see that the sampling distribution of the sample mean (winnings) is bell shaped when n is reasonably large. This is in spite of the fact that the population distribution is flat—far from bell shaped. Actually, the population distribution could have *any* shape, not just uniform, and the bell-shaped property would still hold (although n might have to be larger than in the example). This bell-shaped normality property allows us to perform probability calculations, as we will see in subsequent examples.

Equally important, this example demonstrates the *decreased variability* in the sample means as n increases. Why should an increased sample size lead to decreased variability? The reason is the averaging process. Think about obtaining a winnings of $750 based on the average of two spins. All we need is two lucky spins. In fact, one really lucky spin and an average spin will do. But think about obtaining a winnings of $750 based on the average of *ten* spins. Now we need a *lot* of really lucky spins—and virtually no unlucky ones. The point is that we are much less likely to obtain a really large (or really small) sample mean when n is large than when n is small. This is exactly what it means to say that the variability of the sample means decreases with increasing sample size.

This decreasing variability is predicted by the formula for the standard error of the mean, $\sigma/\sqrt{n}$. As n increases, the standard error obviously decreases. This is what drives the behavior in Figures 8.13–8.16. In fact, using $\sigma = \$289$, the standard errors for $n = 2$, $n = 3$, $n = 6$, and $n = 10$ are $204, $167, $118, and $91.

Finally, what does this decreasing variability have to do with estimating a population mean with a sample mean? Very simply, it means that the sample mean tends to be a more *accurate* estimate when the sample size is large. Because of the approximate normality from the central limit theorem, we know from Chapter 6 that there is about a 95% chance that the sample mean will be within 2 standard errors of the population mean. In other words, there is about a 95% chance that the sampling error will be no greater than two standard errors in magnitude. Therefore, because the standard error decreases as the sample size increases, the sampling error is likely to decrease as well.

To illustrate this, reconsider the auditor in Example 8.4. The standard error based on a sample of size $n = 100$ yielded a sample standard deviation of $419 and a standard error of about $42. Therefore, the sampling error has a 95% chance of being less than 2 standard errors, or $84, in magnitude. If the auditor believes that this sampling error is too large and therefore randomly samples 300 more accounts, then the new standard error will be $419/\sqrt{400} \approx \$21$. Now there is about a 95% chance that the sampling error will be no more than $42. Note that because of the square root, small standard errors come at a high price. To halve the standard error, we must quadruple the sample size!

8.4.5 Sample Size Determination

The problem of determining the appropriate sample size in any sampling context is not an easy one, but it must be faced in the planning stages, *before* any sampling is done. We focus here on the relationship between sampling error and sample size. As we discussed previously, the sampling error tends to decrease as the sample size increases, so the desire to minimize sampling error encourages us to select larger sample sizes. We should note, however, that several other factors encourage us to select *smaller* sample sizes. The ultimate sample size selection must achieve a trade-off between these opposing forces.

What are these other factors? First, there is the obvious cost of sampling. Larger sample sizes require larger costs. Sometimes, a company or agency might have a budget for a given sampling project. If the sample size required to achieve an "acceptable" sampling error is 500, say, but the budget allows for a sample size of only 300, budget considerations will probably prevail.

Another problem caused by large sample sizes is timely collection of the data. Suppose a retailer wants to collect sample data from its customers to decide whether to run an advertising blitz in the coming week. Obviously, the retailer needs to collect these data quickly if they are to be of any use, and a large sample could require too much time to collect.

Finally, a more subtle problem caused by large sample sizes is the increased chance of *nonsampling* error, such as nonresponse bias. As we discussed previously in this chapter, there are many potential sources of nonsampling error, and they are usually very difficult to quantify. However, it is likely that they tend to increase as the sample size increases. Arguably, the potential increase in *sampling* error from a smaller sample could be more than offset by a decrease in nonsampling error, especially if the cost saved by the smaller sample size is used to reduce the sources of nonsampling error—more follow-up of nonrespondents, for example.

Nevertheless, the determination of sample size is usually driven by sampling error considerations. If we want to estimate a population mean with a sample mean, then the key is the standard error of the mean, given by

$$SE(\overline{X}) = \sigma/\sqrt{n}$$

We know from the central limit theorem that if n is reasonably large, there is about a 95% chance that the magnitude of the sampling error will be no more than 2 standard errors. Because σ is fixed in the formula for $SE(\overline{X})$, we can choose n to make $2SE(\overline{X})$ acceptably small.

The usual procedure is to select an acceptable sampling error B, called the *maximum probable absolute error.* Then we set $2SE(\overline{X})$ equal to B and solve for n. After doing the required algebra, we obtain equation (8.7). The implication is that if we randomly sample this many members from the population, then there is a 95% chance that the resulting sampling error will be no greater than B in magnitude.[4]

Sample Size to Achieve a Given Maximum Probable Absolute Error	$$n = \frac{4\sigma^2}{B^2}$$ (8.7)

Maximum Probable Absolute Error	The **maximum probable absolute error** is the largest sampling error we are willing to accept. We typically choose the sample size n so that there is only a 5% chance of getting a sampling error larger than this maximum.

Equation (8.7) makes sense intuitively. Specifically, when the variability in the population is large, as measured by σ^2, we need a large sample size to achieve a given level of accuracy. In an extreme case, if all household incomes in the population are within $100 dollars of one another—very little variability—we don't need to sample many households to get a good estimate of the mean household income. But if the household incomes vary from a low of, say, $10,000, to a high of over a million dollars, we might need to sample quite a few households to achieve an accurate estimate of the mean household income.

Similarly, it makes sense that n should vary inversely with B. The value of B indicates the maximum sampling error we are willing to tolerate. The lower the value of B, the less error we will accept. Therefore, if we want really accurate estimates, we have to pay for them with large sample sizes.

From a practical point of view, the only problem in applying equation (8.7) is that we generally do not know σ, and we cannot simply approximate σ by the sample standard deviation s because we are still in the planning stages—we don't yet have a sample! The

---

[4]The sample size formula given here assumes the eventual sample size will be "small" relative to the population size. When this isn't true, there are more precise formulas that take the population size into account. See Levy and Lemeshow (1999), for example.

usual way to overcome this problem is to use a reasonable "guess" for σ, perhaps based on historical data or a pilot sample. We discuss this issue in the following example.

Example 8.6 Analyzing Videocassette Rentals

A marketing researcher has been hired by a videocassette rental company to estimate the average number of videocassettes rented annually by households in a particular metropolitan area. The researcher decides to determine the sample size that makes the maximum probable absolute error approximately equal to 10. Discuss how she should proceed.

Objective To find the sample size that achieves a given maximum probable absolute error for a marketing researcher.

Solution

The solution appears in Figure 8.17. (See the file SAMPSIZE.XLS.) The researcher has chosen the maximum probable absolute error criterion, with $B = 10$, as the value she is willing to tolerate. Therefore, she should use equation (8.7). To use this equation, she must estimate a value for σ. Based on her knowledge of the industry and available historical data, she uses a best guess of $σ = 50$. She then uses these values (see cells C7 and C8 of Figure 8.17) to find the required sample size in cell C10 with the formula

$$=4*C7\hat{}2/C8\hat{}2$$

Finally, she takes a sample of size 100, as prescribed in cell C10. We assume for the sake of illustration that she observes the sample values shown in column F. Based on this sample, we calculate summary measures in the usual way in the range C13:C16. Note that the standard error in cell C15 is the standard deviation in cell C14 divided by the square root of 100, and that the value in cell C16 is 2 times the standard error in cell C15. This actual absolute error is slightly greater than the maximum absolute error she specified (10, in cell C8) because she observed a larger standard deviation than she guessed (59.31 versus 50). In other words, the fact that there is evidently more variation in the population than she thought makes her sample mean based on 100 households slightly less accurate than she intended.

FIGURE 8.17
Sample Size
Determination

	A	B	C	D	E	F
1	Determining sample size for estimating a population mean					
2						
3	Assumption: Population size is very large					
4						
5					Sample results	
6	Inputs				Household	# of rentals
7		Pop stdev (guess)	50		1	44
8		Max absolute error	10		2	95
9					3	42
10		Resulting sample size	100.00		4	39
11					5	155
12	Outputs from sample				6	38
13		Sample mean	89.78		7	159
14		Sample stdev	59.31		8	86
15		Std Error	5.93		9	151
16		Absolute error	11.86		10	42
17					11	36
104					98	26
105					99	43
106					100	159

Sample size determination is extremely important in real sampling applications. It is one of the few "levers" the analyst has for controlling the amount of sampling error. We will have more to say about sample size determination in the next chapter when we study confidence interval estimation.

8.4.6 Summary of Key Ideas for Simple Random Sampling

To this point, we have covered some very important concepts. Because we will build on these concepts in later chapters, we summarize them here.

Key Concepts of Simple Random Sampling

- To estimate a population mean with a simple random sample, we use the sample mean as a "best guess." This estimate is called a point estimate. That is, $\overline{X}$ is a point estimate of μ.

- The accuracy of the point estimate is measured by its standard error. It is the standard deviation of the sampling distribution of the point estimate. The standard error of $\overline{X}$ is approximately $s/\sqrt{n}$, where s is the sample standard deviation.

- An interval estimate (with 95% confidence) for the population mean extends to 2 standard errors on either side of the sample mean.

- From the central limit theorem, the sampling distribution of $\overline{X}$ is approximately normal when n is reasonably large.

- There is approximately a 95% chance that any particular $\overline{X}$ will be within 2 standard errors of the population mean μ.

- The sampling error can be reduced by increasing the sample size n. Appropriate sample size formulas are available for achieving an acceptable level of the maximum probable absolute error.

PROBLEMS

Level A

29. A manufacturing company's quality control personnel have recorded the proportion of defective items for each of 500 monthly shipments of one of the computer components that the company produces. The data are in the file P02_02.XLS. The quality control department manager does not have sufficient time to review all of these data. Rather, she would like to examine the proportions of defective items for a sample of these shipments.
   a. Use Excel to generate a simple random sample of size 25 from the given frame.
   b. Compute a point estimate of the population mean from the sample selected in part **a.** What is the sampling error in this case? Assume that the population consists of the proportion of defective items for each of the given 500 monthly shipments.
   c. Determine a good approximation to the standard error of the mean in this case.

   d. Repeat parts **b** and **c** after generating a simple random sample of size 50 from the given frame.

30. The manager of a local fast-food restaurant is interested in improving the service provided to customers who use the restaurant's drive-up window. As a first step in this process, the manager asks his assistant to record the time (in minutes) it takes to serve a large number of customers at the final window in the facility's drive-up system. The given frame of 200 customer service times are all observed during the busiest hour of the day for this fast-food operation. The data are in the file P02_04.XLS.
   a. Use Excel to generate a simple random sample of size 10 from this frame.
   b. Compute a point estimate of the population mean from the sample selected in part **a.** What is the sampling error in this case? Assume that the population consists of the given 200 customer service times.

c. Determine a good approximation to the standard error of the mean in this case.

d. Repeat parts **b** and **c** after generating a simple random sample of size 20 from the given frame.

31. Consider the given set of average annual household income levels of citizens of selected U.S. metropolitan areas in the file P03_06.XLS.

   a. Use Excel to obtain a simple random sample of size 15 from this frame.

   b. Compute a point estimate of the population mean from the sample selected in part **a.** What is the sampling error in this case? Assume that the population consists of all average annual household income levels in the given frame.

   c. Determine a good approximation to the standard error of the mean in this case.

   d. Repeat parts **b** and **c** after generating a simple random sample of size 30 from the given frame.

32. The operations manager of a toll booth located at a major exit of a state turnpike is trying to estimate the average number of vehicles that arrive at the toll booth during a 1-minute period during the peak of rush-hour traffic. In an effort to estimate this average throughput value, he records the number of vehicles that arrive at the toll booth over a 1-minute interval commencing at the same time for each of 365 normal weekdays. The data are provided in the file P02_09.XLS.

   a. Choose a simple random sample of size 18 from the given frame to help the operations manager estimate the average throughput value.

   b. Compute a point estimate of the population mean from the sample selected in part **a.** What is the sampling error in this case? Assume that the population consists of the numbers of vehicle arrivals over the 365 weekdays in the given frame.

   c. Determine a good approximation to the standard error of the mean in this case.

   d. Repeat parts **b** and **c** after generating a simple random sample of size 36 from the given frame.

33. Continuing the previous problem with the same data file, answer the following questions.

   a. What sample size would be required for the operations manager to be approximately 95% sure that his estimate of the average throughput value is within 1 unit of the true mean? Assume that his best estimate of the population standard deviation σ is 1.7 arrivals per minute.

   b. How does the answer to part **a** change if the operations manager wants his estimate to be within 0.75 unit of the actual population mean? Explain the difference in your answers to parts **a** and **b.**

34. A lightbulb manufacturer wants to estimate the average number of defective bulbs contained in a box shipped by the company. Production personnel at this company have recorded the number of defective bulbs found in each of the 1000 boxes shipped during the past week. These data are provided in P08_11.XLS.

   a. What sample size would be required for the production personnel to be approximately 95% sure that their estimate of the average number of defective bulbs per box is within 0.25 unit of the true mean? Assume that their best estimate of the population standard deviation σ is 0.9 defective bulb per box.

   b. How does the answer to part **a** change if the production personnel want their estimate to be within 0.40 unit of the actual population mean? Explain the difference in your answers to parts **a** and **b.**

35. Senior management of a certain consulting services firm is concerned about a growing decline in the organization's productivity. In an effort to understand the depth and extent of this problem, management would like to estimate the average number of hours its employees spend on work-related activities in a typical week. The frame of virtually all of the firm's full-time employees, including the employees' self-reported amounts of time typically devoted to work activities each week, is provided in P08_35.XLS.

   a. What sample size would be required for management to be approximately 95% sure that its estimate of the average number of hours the employees spend on work-related activities in a typical week is within 6 hours of the true mean? Assume that management's best estimate of the population standard deviation σ is 10 hours per week.

   b. How does the answer to part **a** change if management want its estimate to be within 3 hours of the actual population mean? Explain the difference in your answers to parts **a** and **b.**

36. Elected officials in a small Florida town are preparing the annual budget for their community. Specifically, they would like to estimate how much their constituents living in this town are typically paying each year in real estate taxes. Given that there are over 3000 homeowners in this small community, officials have decided to sample a representative subset of taxpayers and thoroughly study their tax payments. The latest frame of homeowners is given in P08_36.XLS.

   a. What sample size would be required for elected officials to be approximately 95% sure that their estimate of the average annual real estate tax payment made by homeowners in their community is within $100 of the true mean? Assume that their best estimate of the population standard deviation σ is $535.

   b. Choose a simple random sample of the size found in part **a.**

   c. Compute the observed sampling error based on the sample you have drawn from the population given in the file. How does the actual sampling error compare to the maximum probable absolute error established in part **a?** Explain.

Level B

37. Continuing Problem 29, what proportion of the given 500 monthly shipments contain fractions of defective components within 1 standard error of the mean (based on the original simple random sample of size 25)? What proportion of the 500 monthly shipments contain fractions of defective components within 2 standard errors of the mean (again, based on the original simple random sample of size 25)?

38. Continuing Problem 30, what proportion of the given 200 customer service times are within 2 standard errors of the mean (based on the original simple random sample of size 10)? What proportion of the 200 customer service times are within 3 standard errors of the mean (again, based on the original simple random sample of size 10)?

39. Continuing Problem 31, what proportion of the given average annual household income levels are within 2 standard errors of the mean (based on the original simple random sample of size 15)? What proportion of the given average annual household income levels are within 2 standard errors of the mean (now, based on the second simple random sample of size 30)?

40. Continuing Problem 32, what proportion of the numbers of vehicle arrivals over the given 365 weekdays are within 2 standard errors of the mean (based on the original simple random sample of size 18)? What proportion of the given numbers of vehicle arrivals are within 2 standard errors of the mean (now, based on the second simple random sample of size 36)?

41. Wal-Mart buyers seek to purchase adequate supplies of various brands of toothpaste to meet the ongoing demands of its customers. In particular, Wal-Mart is interested in estimating the proportion of its customers who favor the country's leading brand of toothpaste, Crest. The file P08_41.XLS contains the toothpaste brand preferences of 2000 Wal-Mart customers, obtained recently through the administration of a customer survey.
  a. Use Excel to choose a simple random sample of size 100 from the given frame.
  b. Using the sample found in part **a,** compute a point estimate (called the sample proportion, $\hat{p}$) of the true proportion of Wal-Mart customers who prefer Crest toothpaste. What is the sampling error in this case? Assume that the population consists of the preferences of all customers in the given frame.
  c. Given that the standard error of the sampling distribution of the sample proportion $\hat{p}$ is approximately $\sqrt{\hat{p}(1-\hat{p})/n}$, compute a good approximation to the standard error of the sample proportion in this case.
  d. Repeat parts **b** and **c** after generating a simple random sample of size 50 from the given frame. How do you explain the differences in your computed results?

42. A finance professor has just given a midterm examination in her corporate finance course. The 100 scores are provided in P02_05.XLS.
  a. Generate an appropriate histogram for the given distribution of 100 examination scores. Characterize this distribution. Also, compute the mean and standard deviation of the given scores.
  b. Repeatedly choose simple random samples of size 2 from the original distribution given in the file. Record the sample mean for each of 500 sampling repetitions and generate an appropriate histogram of the resulting sampling distribution. Characterize this sampling distribution and compute its mean and standard deviation.
  c. Repeatedly choose simple random samples of size 5 from the original distribution given in the file. Record the sample mean for each of 500 sampling repetitions and generate an appropriate histogram of the resulting sampling distribution. Characterize this sampling distribution and compute its mean and standard deviation.
  d. Repeatedly choose simple random samples of size 10 from the original distribution given in the file. Record the sample mean for each of 500 sampling repetitions and generate an appropriate histogram of the resulting sampling distribution. Characterize this sampling distribution and compute its mean and standard deviation.
  e. Explain the changes in your constructed sampling distributions as the sample size was increased from $n = 2$ to $n = 10$. In particular, how does the sampling distribution you constructed in part **d** compare to the original distribution (where $n = 1$) you described in part **a?**

43. The annual base salaries for 200 students graduating from a reputable MBA program this year are of interest to those in the admissions office who are responsible for marketing the program to prospective students. These salaries are given in the file P02_74.XLS.
  a. Generate an appropriate histogram for the given distribution of 200 annual salaries. Characterize this distribution. Also, compute the mean and standard deviation of the given salaries.
  b. Repeatedly choose simple random samples of size 3 from the original distribution given in the file. Record the sample mean for each of 500 sampling repetitions and generate an appropriate histogram of the resulting sampling distribution. Characterize this sampling distribution and compute its mean and standard deviation.
  c. Repeatedly choose simple random samples of size 6 from the original distribution given in the file. Record the sample mean for each of 500 sampling repetitions and generate an appropriate histogram of the resulting sampling distribution. Characterize this sampling distribution and compute its mean and standard deviation.

d. Repeatedly choose simple random samples of size 12 from the original distribution given in the file. Record the sample mean for each of 500 sampling repetitions and generate an appropriate histogram of the resulting sampling distribution. Characterize this sampling distribution and compute its mean and standard deviation.

e. Explain the changes in your constructed sampling distributions as the sample size was increased from $n = 3$ to $n = 12$. In particular, how does the sampling distribution you constructed in part **d** compare to the original distribution (where $n = 1$) you described in part **a**?

44. A market research consultant hired by the Pepsi-Cola Co. is interested in determining the proportion of consumers who favor Pepsi-Cola over Coke Classic in a particular urban location. A frame of customers from the market under investigation is provided in P08_44.XLS.

a. What sample size would be required for the market research consultant to be approximately 95% sure that her estimate of the proportion of consumers who favor Pepsi-Cola in the given urban location is within 0.20 of the true proportion? Assume that her best estimate of the population proportion parameter p is 0.45. [*Hint:* The required sample size formula in this case is given by $n = 4p(1 - p)/B^2$, where p is the population proportion parameter and B is the familiar maximum probable absolute error.]

b. How does the answer to part **a** change if the market research consultant wants her estimate to be within 0.15 of the actual population proportion? Explain the difference in your answers to parts **a** and **b**.

8.5 Conclusion

This chapter has provided the fundamental concepts behind statistical inference. We have discussed ways to obtain random samples from a population, how to calculate a point estimate of a particular population parameter, the population mean, and how to measure the accuracy of this point estimate. The key idea is the sampling distribution of the estimate and specifically its standard deviation, called the standard error of the estimate. From the central limit theorem, we have seen that the sampling distribution of the sample mean is approximately normal, which implies that the sample mean will be within 2 standard errors of the population mean in approximately 95% of all random samples. In the next two chapters we will build on these important concepts.

Summary of Key Terms

Term	Symbol	Explanation	Excel	Page	Equation Number
Population		Contains all members about which a study intends to make inferences.		366	
Frame		A list of all members of the population.		367	
Sampling units		Potential members of a sample from a population.		367	
Probability sample		Any sample that is chosen by using a random mechanism.		367	
Judgmental sample		Any sample that is chosen according to a sampler's judgment rather a random mechanism.		367	
Simple random sample		A sample where each member of the population has the same chance of being chosen.	StatPro/ Statistical Inference/ Generate Random Samples	368	
Systematic sample		A sample where one of the first k members is selected randomly, and then every kth member after this one is selected.		373	
Stratified sample		A sample where the population is divided into relatively homogeneous subsets called strata, and then random samples are taken from each of the strata.		374	
Proportional sample sizes (in stratified sampling)		Occurs when the proportion of each stratum selected is the same from stratum to stratum.		375	
Cluster sampling		A sample where the population is separated into clusters, such as cities or city blocks, and then a random sample of the clusters is selected.		378	
Sampling error		The inevitable result of basing an inference on a sample rather than on the entire population.		385	
Nonsampling error		Any type of estimation error that is not sampling error, including nonresponse bias, nontruthful responses, measurement error, and voluntary response bias.		385	
Point estimate		A single numeric value, a "best guess" of a population parameter, based on the data in a sample.		387	
Estimation error		Difference between the estimate of a population parameter and the true value of the parameter.		387	

Summary of Key Terms (continued)

Term	Symbol	Explanation	Excel	Page	Equation Number
Sampling distribution		The distribution of the point estimates we would see from *all* possible samples (of a given sample size) from the population.		387	
Interval estimate		An interval around the point estimate, calculated from the sample data, where we strongly believe the true value of the population parameter lies.		387	
Unbiased estimate		An estimate where the mean of its sampling distribution equals the value of the parameter being estimated.		387	
Standard error of an estimate		The standard deviation of the sampling distribution of the estimate.		387	
Mean of sample mean	$E(\bar{X})$	Indicates property of unbiasedness of sample mean.		389	8.1
Standard error of sample mean	$SE(\bar{X})$	Indicates how sample means from different samples vary.		389	8.2, 8.3
Interval estimate of population mean		We are 95% confident that the population mean is within this interval.		389	8.4
Finite population correction factor	*fpc*	A correction for the standard error when the sample size is fairly large relative to the population size.		391	8.5, 8.6
Central limit theorem		States that the distribution of the sample mean is approximately normal for sufficiently large sample sizes.		392	
Maximum probable absolute error		The maximum error the analyst is willing to tolerate, achievable only by taking a large enough sample.		399	8.7

PROBLEMS

Conceptual Exercises

C.1. Suppose that you want to know the opinions of American secondary school teachers about establishing a national test for high school graduation. You obtain a list of the members of the National Education Association (the largest teachers' union) and mail a questionnaire to 3000 teachers chosen at random from this list. In all, 1529 teachers return the questionnaire. Identify the relevant *population* and *frame* in this case.

C.2. A sportswriter wants to know how strongly the residents of Indianapolis, Indiana, support the local minor league baseball team, the Indianapolis Indians. She stands outside the stadium before a game and interviews the first 30 people who enter the stadium. Suppose that the newspaper asks you to comment on the approach taken by this sportswriter in performing the survey of local opinion. How do you respond?

C.3. A large corporation has 4520 male and 1167 female employees. The organization's equal employment opportunity officer wants to poll the opinions of a random sample of employees. To give adequate attention to the opinions of female employees, exactly how should the EEO officer sample from the given population? Explain in detail.

C.4. Suppose that you want to estimate the mean monthly gross income of all households in your local community. You decide to estimate this population parameter by calling 150 randomly selected residents and asking each individual to report the household's monthly income. Assume that you use the local phone directory as the frame in selecting the households to be included in your sample. What are some possible sources of error that might arise in your effort to estimate this population mean?

C.5. What is the difference between a *standard deviation* and a *standard error?*

Level A

45. The annual base salaries for 200 students graduating from a reputable MBA program this year are of interest to those in the admissions office who are responsible for marketing the program to prospective students. The data are in the file P02_74.XLS. Use Excel to choose 10 simple random samples of size 15 from the given frame. For each simple random sample you obtain, compute the mean annual salary. Are these sample means equivalent? Explain why or why not.

46. A market research consultant hired by the Pepsi-Cola Co. is interested in determining who favors the Pepsi-Cola brand over Coke Classic in a particular urban location. A frame of customers from the market under investigation is provided in P08_44.XLS.
   a. Compute the proportion of the customers in the given frame who favor Pepsi.
   b. Use Excel to choose a simple random sample of size 30 from the given frame.
   c. Compute the proportion of the 30 customers in the random sample who favor Pepsi. Compare this sample proportion to the value obtained in part **a.** Explain any difference between the two values.
   d. What are the advantages and disadvantages of employing simple random sampling in this particular case?

47. The employee benefits manager of a small private university would like to know the proportion of its full-time employees who prefer adopting each of three available health care plans in the forthcoming annual enrollment period. A reliable frame of the university's employees and their tentative health care preferences are given in P08_25.XLS.
   a. Compute the proportion of the employees in the given frame who favor *each* of the three plans (i.e., plans A, B, and C).
   b. Use Excel to choose a simple random sample of size 45 from the given frame.
   c. Compute the proportion of the 45 employees in the random sample who favor each health plan. Compare these sample proportions to the corresponding values obtained in part **a.** Explain any differences between the corresponding values.
   d. What are the advantages and disadvantages of employing simple random sampling in this particular case?

48. Senior management of a certain consulting services firm is concerned about a growing decline in the organization's productivity. In an effort to understand the depth and extent of this problem, management would like to determine the average number of hours its employees spend on work-related activities in a typical week. The frame of virtually all of the firm's full-time employees, including the employees' self-reported amounts of time typically devoted to work activities each week, is provided in P08_35.XLS.
   a. Select a simple random sample of size 100 from the given frame.
   b. Compute the mean and standard deviation of the weekly number of hours worked by all employees in the frame. Also, compute the mean and standard deviation of the weekly number of hours worked by employees in the simple random sample. How do these two sets of descriptive measures compare?

49. Elected officials in a small Florida town are preparing the annual budget for their community. Specifically, they would like to know how much their constituents living in this town are typically paying each year in real estate taxes. Given that there are over 3000 homeowners in this small community, officials have decided to sample a representative subset of taxpayers and thoroughly study their tax payments. The latest frame of homeowners is given in P08_36.XLS. Note that this file contains the real estate tax payment made by each homeowner last year.
   a. Compute the average real estate tax payment made by the homeowners included in the frame. Is the overall mean a valid measure of central tendency in this case?
   b. Use Excel to choose a simple random sample of 150 homeowners from the given frame.
   c. Compute the average real estate tax payment for the 150 homeowners in the random sample. Compare this sample mean to the corresponding summary measure obtained in part **a.**

d. Is the sample mean computed in part **c** a good estimate of the average real estate tax payment made by homeowners living in this small town? Explain why or why not.

50. Auditors of a particular bank are interested in comparing the reported value of customer savings account balances with their own findings regarding the actual value of such assets. Rather than reviewing the records of each savings account at the bank, the auditors decide to examine a representative sample of savings account balances. The frame from which they will sample is given in the file P08_50.XLS.

a. Assist the bank's auditors by selecting a simple random sample of 100 savings accounts.

b. Explain how the auditors might use the simple random sample identified in part **a** to estimate the value of *all* savings accounts balances within this bank.

51. The manager of a local supermarket wants to know the average amount (in dollars) customers spend at his store on Fridays. He would like to study the buying behavior of each customer who makes a purchase at the store on a typical Friday. However, the manager's assistant, who is currently enrolled in a managerial statistics course at a local college, urges the manager to save his scarce time and money by studying a sample of customer purchases. The available frame of relevant customer purchases is provided in file P08_51.XLS.

a. Compute the average purchase amount made by the customers included in the given frame.

b. Use Excel to choose a simple random sample of 25 customers from the given frame.

c. Compute the average purchase amount made by the 25 customers in the random sample. Compare this sample mean to the corresponding summary measure obtained in part **a.**

d. Is the sample mean a good estimate of the overall population mean in this case? Explain why or why not.

52. The annual base salaries for 200 students graduating from a reputable MBA program this year are of interest to those in the admissions office who are responsible for marketing the program to prospective students. The data are in the file P02_74.XLS. Use Excel to choose 15 systematic samples of size 10 from the given frame. For each systematic sample you obtain, compute the mean annual salary. Are these sample means equivalent? Explain why or why not.

53. Given data in the file P02_13.XLS from a recent survey of chief executive officers from the nation's biggest businesses (*The Wall Street Journal*, April 9, 1998), choose a systematic sample of 25 executives and find the mean, median, and standard deviation of the bonuses awarded to them in 1997. How do these sample statistics compare to the mean, median, and

standard deviation of the bonuses given to all executives included in the frame?

54. A market research consultant hired by the Pepsi-Cola Co. is interested in determining who favors the Pepsi-Cola brand over Coke Classic in a particular urban location. A frame of customers from the market under investigation is provided in P08_44.XLS.

a. Compute the proportion of the customers in the given frame who favor Pepsi.

b. Use Excel to choose a systematic sample of size 30 from the given frame.

c. Compute the proportion of the 30 customers in the systematic sample who favor Pepsi. Compare this sample proportion to the value obtained in part **a.** Explain any difference between the two values.

d. What are the advantages and disadvantages of employing systematic sampling in this particular case?

55. A lightbulb manufacturer wants to know the number of defective bulbs contained in a typical box shipped by the company. Production personnel at this company have recorded the number of defective bulbs found in each of the 1000 boxes shipped during the past week. These data are provided in P08_11.XLS. Using this shipment of boxes as a frame, select a systematic sample of 50 boxes and compute the mean number of defective bulbs found in a box.

56. The employee benefits manager of a small private university would like to know the proportion of its full-time employees who prefer adopting each of three available health care plans in the forthcoming annual enrollment period. A reliable frame of the university's employees and their tentative health care preferences are given in P08_25.XLS.

a. Compute the proportion of the employees in the given frame who favor *each* of the three plans (i.e., plans A, B, and C).

b. Use Excel to choose a systematic sample of size 47 from the given frame.

c. Compute the proportion of the 47 employees in the systematic sample who favor each health plan. Compare these sample proportions to the corresponding values obtained in part **a.** Explain any differences between the corresponding values.

d. What are the advantages and disadvantages of employing systematic sampling in this particular case?

57. Senior management of a certain consulting services firm is concerned about a growing decline in the organization's productivity. In an effort to understand the depth and extent of this problem, management would like to determine the average number of hours their employees spend on work-related activities in a typical week. The frame of virtually all of the firm's full-time employees, including the employees' self-reported amounts of time typically devoted to work activities each week, is provided in P08_35.XLS.

a. Select a systematic sample of size 100 from the given frame.

b. Compute the mean and standard deviation of the weekly number of hours worked by all employees in the frame. Also, compute the mean and standard deviation of the weekly number of hours worked by employees in the systematic sample. How do these two sets of descriptive measures compare?

58. Elected officials in a small Florida town are preparing the annual budget for their community. Specifically, they would like to know how much their constituents living in this town are typically paying each year in real estate taxes. Given that there are over 3000 homeowners in this small community, officials have decided to sample a representative subset of taxpayers and thoroughly study their tax payments. The latest frame of homeowners is given in P08_36.XLS. Note that this file contains the real estate tax payment made by each homeowner last year.

a. Use Excel to choose a systematic sample of 150 homeowners from the given frame.

b. Compute the average real estate tax payment for the 150 homeowners in the systematic sample.

c. Is the sample mean computed in part **b** a good estimate of the average real estate tax payment made by homeowners living in this small town? Explain why or why not.

59. Auditors of a particular bank are interested in comparing the reported value of customer savings account balances with their own findings regarding the actual value of such assets. Rather than reviewing the records of each savings account at the bank, the auditors decide to examine a representative sample of savings account balances. The frame from which they will sample is given in the file P08_50.XLS.

a. Assist the bank's auditors by selecting a systematic sample of 151 savings accounts.

b. Explain how the auditors might use the systematic sample identified in part **a** to estimate the value of *all* savings accounts balances within this bank.

60. The manager of a local supermarket wants to know the average amount (in dollars) customers spend at his store on Fridays. He would like to study the buying behavior of each customer who makes a purchase at the store on a typical Friday. However, the manager's assistant, who is currently enrolled in a managerial statistics course at a local college, urges the manager to save his scarce time and money by studying a sample of customer purchases. The available frame of relevant customer purchases is provided in file P08_51.XLS.

a. Compute the average purchase amount made by the customers included in the given frame.

b. Use Excel to choose a systematic sample of 43 customers from the given frame.

c. Compute the average purchase amount made by the 43 customers in the systematic sample. Compare this sample mean to the corresponding summary measure obtained in part **a.**

d. Is the sample mean a good estimate of the overall population mean in this case? Explain why or why not.

61. Elected officials in a small Florida town are preparing the annual budget for their community. Specifically, they would like to know how much their constituents living in this town are typically paying each year in real estate taxes. Given that there are over 3000 homeowners in this small community, officials have decided to sample a representative subset of taxpayers and thoroughly study their tax payments. The latest frame of homeowners is given in P08_36.XLS. Note that this file contains the real estate tax payment made by each homeowner last year.

a. Compute the average real estate tax payment made by the homeowners included in the frame. Is the overall mean a valid measure of central tendency in this case?

b. Use Excel to choose a sample of 150 homeowners stratified by neighborhood with proportional sample sizes.

c. Compute the average real estate tax payment for the 150 homeowners in the stratified sample. Compare this sample mean to the corresponding summary measure obtained in part **a.**

d. Is the sample mean computed in part **c** a good estimate of the average real estate tax payment made by homeowners living in this small town? Explain why or why not.

e. Explain how the elected officials could apply cluster sampling in selecting a sample of size 150 from this frame. What are the advantages and disadvantages of employing cluster sampling in this case?

62. Auditors of a particular bank are interested in comparing the reported value of customer savings account balances with their own findings regarding the actual value of such assets. Rather than reviewing the records of each savings account at the bank, the auditors decide to examine a representative sample of savings account balances. The frame from which they will sample is given in the file P08_50.XLS.

a. What sample size would be required for the auditors to be approximately 95% sure that their estimate of the average savings account balance at this bank is within $100 of the true mean? Assume that their best estimate of the population standard deviation σ is $500.

b. Choose a simple random sample of the size found in part **a.**

c. Compute the observed sampling error based on the sample you have drawn from the population given in the file. How does the actual sampling error compare to the maximum probable absolute error established in part **a?** Explain.

63. The manager of a local supermarket wants to estimate the average amount customers spend at his store on Fridays. He would like to study the buying behavior of each customer who makes a purchase at the store on a typical Friday. However, the manager's assistant, who is currently enrolled in a managerial statistics course at a local college, urges the manager to save his scarce time and money by studying a sample of customer purchases. The available frame of relevant customer purchases is provided in file P08_51.XLS.
   a. What sample size would be required for the supermarket manager to be approximately 95% sure that his estimate of the average customer expenditure on Fridays is within $25 of the true mean? Assume that his best estimate of the population standard deviation σ is $72.
   b. Choose a simple random sample of the size found in part **a.**
   c. Compute the observed sampling error based on the sample you have drawn from the population given in the file. How does the actual sampling error compare to the maximum probable absolute error established in part **a?** Explain.

64. *The Hite Report* was Sheri Hite's survey of the attitudes of American women toward sexuality. She sent out over 100,000 surveys; each contained multiple-choice and open-ended questions. These surveys were given to women's groups and announced in church newsletters. Ads were also placed in women's magazines. A total of 3019 surveys were returned. Sheri Hite's findings challenged much conventional wisdom about sexuality. She found that most women were unhappy in their romantic relationships (some for reasons that are too graphic for this book!). How would you criticize Hite's methodology? A later poll, by the way, contradicted many of her findings.
   a. Give two criticisms of Hite's sampling methodology.
   b. Despite these criticisms, what value might you see in Hite's results?

65. A market research consultant hired by the Pepsi-Cola Co. is interested in determining who favors the Pepsi-Cola brand over Coke Classic in a particular urban location. A frame of customers from the market under investigation is provided in P08_44.XLS.
   a. Compute the proportion of the consumers in the given frame who favor Pepsi.
   b. Use Excel to choose a sample of size 30 stratified by gender with proportional sample sizes.
   c. Compute the proportion of the 30 consumers in the stratified sample who favor Pepsi. Compare this sample proportion to the value obtained in

part **a.** Explain any difference between the two values.
   d. What are the advantages and disadvantages of employing stratified sampling in this particular case?

Level B

66. Repeat Problem 65, but now stratify the consumers in the given frame by *age* rather than by gender. How does this modification affect your answers to the questions posed in parts **b** and **c?** Finally, stratify the consumers by both gender *and* age (e.g., all females over 60) with proportional sample sizes. How does this change affect your answers to the questions posed in parts **b** and **c?** Which approach to stratification appears to give the best results in estimating the actual proportion of the customers in the given frame who favor Pepsi?

67. Wal-Mart buyers seek to purchase adequate supplies of various brands of toothpaste to meet the ongoing demands of their customers. In particular, Wal-Mart is interested in knowing the proportion of its customers who favor such leading brands of toothpaste as Aquafresh, Colgate, Crest, and Mentadent. The file P08_41.XLS contains the toothpaste brand preferences of 2000 Wal-Mart customers, obtained recently through the administration of a customer survey.
   a. Determine the proportion of Wal-Mart customers who favor each major brand of toothpaste.
   b. Assuming that the given data constitute an appropriate frame, choose a simple random sample of 100 of these customers.
   c. Calculate the proportion of Wal-Mart customers in the random sample who favor each major brand of toothpaste. Compare these sample proportions to the corresponding values found in part **a.** How do you explain any disparities between corresponding proportions for customers included in the sample and those in the frame?

68. Suppose that you are an entrepreneur interested in establishing a new Internet-based sports information service. Furthermore, suppose that you have gathered basic demographic information on a large number of Internet users. Assume that these 1000 individuals were carefully selected through stratified sampling. These data are stored in the file P02_43.XLS.
   a. To assess potential interest in your proposed enterprise, you would like to conduct telephone interviews with a representative subset of the 1000 Internet users you surveyed previously. How would you proceed to stratify the given frame of 1000 individuals to choose 50 for telephone interviews? Explain your approach and implement it to select a useful sample of size 50.
   b. Explain how the entrepreneur could apply cluster sampling to obtain a sample of size 50 from this

frame. What are the advantages and disadvantages of employing cluster sampling in this case?

69. A market research consultant hired by the Pepsi-Cola Co. is interested in determining the proportion of consumers who favor Pepsi-Cola over Coke Classic in a particular urban location. A frame of customers from the market under investigation is provided in P08_44.XLS.

a. Use Excel to choose a simple random sample of size 30 from the given frame.

b. Using the sample found in part **a,** compute a point estimate (called the sample proportion, $\hat{p}$) of the true proportion of consumers who favor Pepsi-Cola in this market. What is the sampling error in this case? Assume that the population consists of the preferences of all consumers in the given frame.

c. Given that the standard error of the sampling distribution of the sample proportion $\hat{p}$ is approximately $\sqrt{\hat{p}(1 - \hat{p})/n}$, compute a good approximation to the standard error of the sample proportion in this case.

d. Repeat parts **b** and **c** after generating a simple random sample of size 15 from the given frame. How do you explain the differences in your computed results?

70. The employee benefits manager of a small private university would like to estimate the proportion of full-time employees who prefer adopting the first (i.e., plan A) of three available health care plans in the forthcoming annual enrollment period. A reliable frame of the university's employees and their tentative health care preferences are given in P08_25.XLS.

a. Use Excel to choose a simple random sample of size 45 from the given frame.

b. Using the sample found in part **a,** compute a point estimate (called the sample proportion, $\hat{p}$) of the true proportion of university employees who prefer plan A. What is the sampling error in this case? Assume that the population consists of the preferences of all employees in the given frame.

c. Given that the standard error of the sampling distribution of the sample proportion p is approximately $\sqrt{\hat{p}(1 - \hat{p})/n}$, compute a good approximation to the standard error of the sample proportion in this case.

d. Repeat parts **b** and **c** after generating a simple random sample of size 25 from the given frame. How do you explain the differences in your computed results?

71. Auditors of a particular bank are interested in comparing the reported value of customer savings account balances with their own findings regarding the actual value of such assets. Rather than reviewing the records of each savings account at the bank, the auditors decide to examine a representative sample of sav-ings account balances. The frame from which they will sample is given in the file P08_50.XLS.

a. Generate an appropriate histogram for the given distribution of savings account balances. Characterize this distribution. Also, compute the mean and standard deviation of the given scores.

b. Repeatedly choose simple random samples of size 2 from the original distribution given in the file. Record the sample mean for each of 500 sampling repetitions and generate an appropriate histogram of the resulting sampling distribution. Characterize this sampling distribution and compute its mean and standard deviation.

c. Repeatedly choose simple random samples of size 5 from the original distribution given in the file. Record the sample mean for each of 500 sampling repetitions and generate an appropriate histogram of the resulting sampling distribution. Characterize this sampling distribution and compute its mean and standard deviation.

d. Repeatedly choose simple random samples of size 10 from the original distribution given in the file. Record the sample mean for each of 500 sampling repetitions and generate an appropriate histogram of the resulting sampling distribution. Characterize this sampling distribution and compute its mean and standard deviation.

e. Explain the changes in your constructed sampling distributions as the sample size was increased from $n = 2$ to $n = 10$. In particular, how does the sampling distribution you constructed in part **d** compare to the original distribution (where $n = 1$) you described in part **a?**

72. A lightbulb manufacturer wants to estimate the number of defective bulbs contained in a typical box shipped by the company. Production personnel at this company have recorded the number of defective bulbs found in each of the 1000 boxes shipped during the past week. These data are provided in P08_11.XLS.

a. Generate an appropriate histogram for the given distribution of 1000 numbers of defective bulbs. Characterize this distribution. Also, compute the mean and standard deviation of the given numbers.

b. Repeatedly choose simple random samples of size 3 from the original distribution given in the file. Record the sample mean for each of 500 sampling repetitions and generate an appropriate histogram of the resulting sampling distribution. Characterize this sampling distribution and compute its mean and standard deviation.

c. Repeatedly choose simple random samples of size 6 from the original distribution given in the file. Record the sample mean for each of 500 sampling repetitions and generate an appropriate histogram of the resulting sampling distribution. Characterize this sampling distribution and compute its mean and standard deviation.

d. Repeatedly choose simple random samples of size 12 from the original distribution given in the file. Record the sample mean for each of 500 sampling repetitions and generate an appropriate histogram of the resulting sampling distribution. Characterize this sampling distribution and compute its mean and standard deviation.

e. Explain the changes in your constructed sampling distributions as the sample size was increased from $n = 3$ to $n = 12$. In particular, how does the sampling distribution you constructed in part **d** compare to the original distribution (where $n = 1$) you described in part **a?**

73. The employee benefits manager of a small private university would like to estimate the proportion of full-time employees who prefer adopting the first (i.e., plan A) of three available health care plans in the forthcoming annual enrollment period. A reliable frame of the university's employees and their tentative health care preferences are given in P08_25.XLS.

   a. What sample size would be required for the benefits manager to be approximately 95% sure that her estimate of the proportion of full-time university employees who prefer adopting plan A is within 0.15 of the true proportion? Assume that her best estimate of the population proportion parameter p is 1/3. [*Hint:* The required sample size formula in this case is given by $n = 4p(1 - p)/B^2$, where π is the population proportion parameter and B is the familiar maximum probable absolute error.]

   b. How does the answer to part **a** change if the benefits manager wants her estimate to be within 0.25 of the actual population proportion? Explain the difference in your answers to parts **a** and **b.**

74. Suppose the monthly unpaid balance on a Citicorp Mastercard is normally distributed with a mean of $1200 and standard deviation of $240. We want to show that the sample mean $\overline{X}$ is an unbiased estimate of the population mean μ, and the sample variance s^2 is an unbiased estimate of the population variance σ^2. Note that you can generate observations from CITICORP accounts by using the formula =NORMINV(RAND(),1200,240). Develop a simulation as follows:

   • Generate 50 samples of five credit card balances each. Then freeze the random numbers.

   • Calculate $\overline{X}$ and s^2 for each sample.
   • Show that the $\overline{X}$'s average to a value near the actual mean of $1200.
   • Show that the s^2's average to a value near the true value $\sigma^2 = 240^2$.

75. (Based on an actual case) Indiana audits nursing homes to see whether and how much the nursing home has overbilled Medicaid. Here is how they do it. Nurseco has 70 homes in Indiana. The state randomly samples one invoice per nursing home and determines Nurseco's liability as follows. Suppose nursing home 1 has billed Medicaid $100,000. If the one surveyed invoice at nursing home 1 indicates Nurseco has overbilled by 40%, then Nurseco would have to return 40% (or $40,000) that it has collected from the state. What is wrong with this approach? Assuming all nursing homes have overbilled at a similar rate, can you suggest a better plan to determine how much money should be returned to the state?

76. The central limit theorem states that when many independent random variables are summed, the result follows a normal distribution even if the individual random variables in the sum do not. To illustrate this idea, simulate 500 samples of size 15 from the uniform (0, 1) distribution (generated with the RAND function). Are the 500 sums (where each is a sum of 15 values) normally distributed? Show by constructing a histogram and also by checking the empirical rules.

77. Assume a very large normally distributed population of scores on a test with mean 70 and standard deviation 7.

   a. Find an interval that includes 95% of the population.

   b. Suppose you randomly sample a single member from this population. Find an interval so that you are 95% confident that this member's score will be in the interval.

   c. Now suppose you sample 30 members randomly from this population. Find an interval so that you are 95% confident that the average of these members' scores will be in the interval.

   d. Finally, suppose you sample 300 members randomly from this population. Find an interval so that you are 95% confident that the average of these members' scores will be in the interval.

   e. Explain intuitively why the answers to parts **a–d** are not all the same.

8.1 Sampling from Videocassette Renters

The file VIDEOS.XLS contains a large database on 10,000 customer transactions for a fictional chain of video stores in the United States. Each row corresponds to a different customer and lists (1) a customer ID number (1–10,000); (2) the state where the customer lives; (3) the city where the customer lives; (4) the customer's gender; (5) the customer's favorite type of movie (drama, comedy, science fiction, or action); (6) the customer's next favorite type of movie; (7) the number of times the customer has rented movies in the past year; and (8) the total dollar amount the customer has spent on movie rentals during the past year. The data are sorted by state, then city, then gender. We assume that this database represents the entire population of customers for this video chain. (Of course, national chains would have significantly larger customer populations, but this database is large enough to illustrate the ideas.)

Imagine that only the data in columns A–D are readily available for this population. The company is interested in summary statistics of the data in columns E–H, such as the percentage of customers whose favorite movie type is drama or the average amount spent annually per customer, but it will have to do some work to obtain the data in columns E–H for any particular customer. Therefore, the company wants to perform sampling. The question is: What form of sampling—simple random sampling, systematic sampling, stratified sampling, cluster sampling, or even some type of multistage sampling—is most appropriate?

Your job is to investigate the possibilities and to write a report on your findings. For any sampling method, any sample size, and any quantity of interest (such as average dollar amount spent annually), you should be concerned with sampling cost and accuracy. One way to judge the latter is to generate several random samples from a particular method and calculate the mean and standard deviation of your point estimates from these samples. For example, you might generate 10 systematic samples, calculate the average amount spent (an $\bar{X}$) for each sample, and then calculate the mean and standard deviation of these 10 $\bar{X}$'s. If your sampling method is accurate, the mean of the $\bar{X}$'s should be close to the population average, and the standard deviation should be small. By doing this for several sampling methods and possibly several sample sizes, you can experiment to see what is most cost efficient for the company. You can make any reasonable assumptions about the cost of sampling with any particular method.

Confidence Interval Estimation

Estimating a Company's Total Taxable Income

In Example 8.4 from the previous chapter, we illustrated how sampling can be used in auditing. We will see another illustration of sampling in auditing in Example 9.5 of this chapter. In both examples, the point of the sampling is to discover some property (such as a mean or a proportion) from a large population of a company's accounts by examining a small fraction of these accounts and projecting the results to the population. The article by Press (1995) offers an interesting variation on this problem. He poses the question of how a government revenue agency should assess a business taxpayer's income for tax purposes on the basis of a sample audit of the company's business transactions. A sample of the company's transactions will indicate a taxable income for each sampled transaction. The methods of this chapter will be applied to the sample information to obtain a confidence interval for the total taxable income owed by the company.

Suppose for the sake of illustration that this confidence interval extends from $1,000,000 to $2,200,000 and is centered at $1,600,000. In words, the government's best guess of the company's taxable income is $1,600,000, and it is fairly confident that the true taxable income is between $1,000,000 and $2,200,000. How much tax should it assess the company? Press argues that the agency would like to maximize its revenue while minimizing the risk that the company will be assessed more than it really owes. This last assumption, that the government does not want to *overassess* the company, is crucial. By making several reasonable assumptions, he is able to argue that the agency should base the tax on the *lower* limit of the confidence interval, in this case, $1,000,000.[1]

---

[1]In case this sounds overly generous on the government's part, the result is based on two important assumptions: (1) the confidence interval is a 90% confidence interval, and (2) the agency is 19 times more concerned about overassessing than about underassessing.

On the other hand, if the agency were indifferent between overcharging and undercharging, then it would base the tax on the midpoint, $1,600,000, of the confidence interval. Using this strategy, the agency would overcharge in about half the cases and undercharge in the other half. This would certainly be upsetting to companies—it would appear that the agency were flipping a coin to decide whether to overcharge or undercharge! (Today this strategy is most commonly used by state agencies, but it is being challenged, and the courts will have to make a decision.)

If the government agency does indeed decide to base the tax on the *lower* limit of the confidence interval, Press argues that it can still increase its tax revenue—by increasing the sample size of the audit. When the sample size increases, the confidence interval shrinks in width, and the lower limit, which governs the agency's tax revenue, almost surely increases. But there is some point at which larger samples are not warranted, for the simple reason that larger samples cost more money to obtain. Therefore, there is an optimal size that will balance the cost of sampling with the desire to obtain more tax revenue. ●

9.1 **Introduction**

This chapter expands on the ideas from the previous chapter. Given an observed data set, we want to make inferences to some larger population. Two typical examples follow:

- A mail-order company has accounts with thousands of customers. The company would like to infer the average time its customers take to pay their bills, so it randomly samples a relatively small number of its customers, sees how long these customers take to pay their bills, and draws inferences about the entire population of customers.

- A manufacturing company is considering two compensation schemes to implement for its workers. It believes that these two different compensation schemes might provide different incentives and hence result in different worker productivity. To see whether this is true, the company randomly assigns groups of workers to the two different compensation schemes for a period of 3 months and observes their productivity. Then it attempts to infer whether any differences observed in the experiment can be generalized to the overall worker population.

In each of these examples, there is an unknown population parameter a company would like to estimate. In the mail-order example the unknown parameter is the mean length of time customers take to pay their bills. Its true value could be discovered only by learning how long *every* customer in the entire population takes to pay its bills. This might not be possible, given the large number of customers. In the manufacturing example the unknown parameter is really a mean difference, the difference between the mean productivities with the two different compensation schemes. This mean difference could be discovered only by subjecting each worker to each compensation scheme and measuring their resulting productivities. This procedure would almost certainly be infeasible from a practical standpoint. Therefore, in these examples the companies involved are likely to select random samples and base their estimates of the unknown population means on sample data.

The inferences we draw in this chapter are always based on an underlying probability model, which means that some type of random mechanism must generate the given data. Two random mechanisms are generally used. The first involves sampling randomly from a larger population, as we discussed in the previous chapter. This is the mechanism responsible for generating the sample of customers in the mail-order example. Regardless of whether the sample is a simple random sample or a more complex random sample, such as a stratified sample, the fact that it is *random* allows us to use the rules of probability to make inferences about the population as a whole.

The second commonly used random mechanism is called a **randomized experiment.** The compensation scheme example just described is a typical randomized experiment. Here we select a set of subjects (employees), randomly assign them to two different **treatment groups** (compensation schemes), and then compare some quantitative measure (productivity) across the groups. The fact that the subjects are *randomly* assigned to the two treatment groups is useful for two reasons. First, it allows us to rule out a number of factors that might have led to differences across groups. For example, assuming that males and females are randomly spread across the two groups, we can rule out gender as the cause of observed group differences. Second, the random selection allows us to use the rules of probability to infer whether observed differences can be generalized to all employees.

Generally, statistical inferences are of two types, **confidence interval estimation** and **hypothesis testing.** The first of these is the subject of the current chapter; we will study hypothesis testing in the next chapter. They differ primarily in their point of view. For example, the mail-order company might sample 100 customers and find that they average 15.5 days before paying their bills. In confidence interval estimation, we use the data to obtain a point estimate and a confidence interval around this point estimate. In this example the point estimate is 15.5 days. It is a best guess for the mean bill-paying time in the entire customer population. Then, using the methods in this chapter, the company might find that a 95% confidence interval for the mean bill-paying time in the population is from 13.2 days to 17.8 days. The company is now 95% certain that the true mean bill-paying time in the population is within this interval.

Hypothesis testing takes a different point of view. Here we wish to check whether the observed data provide support for a particular hypothesis. In the compensation scheme example, suppose the manager believes that workers will have higher productivity if they are paid by salary than by an hourly wage. He runs the 3-month randomized experiment described previously and finds that the salaried workers produce on average eight more parts per day than the hourly workers. Now he must make one of two conclusions. Either salaried workers are in general no more productive than hourly workers and the ones in the experiment just got lucky, or salaried workers really *are* more productive. We will learn in the next chapter how to decide which of these conclusions is more reasonable.

There are only a few key ideas in this chapter, and the most important of these, sampling distributions, was introduced in Chapter 8. It is important to concentrate on these key ideas and not get bogged down in formulas or numerical calculations. Statistical software such as the StatPro add-in is generally available to take care of these calculations. The job of a business person is much more dependent on knowing which methods to use in which situations and how to interpret computer output than on memorizing and plugging into formulas.

Sampling Distributions

As we will soon learn, most confidence intervals are of the form in expression (9.1) on page 418. For example, when estimating a population mean, the point estimate is the sample mean, the standard error is the sample standard deviation divided by the square root of the sample size, and the multiple is approximately equal to 2. To learn why it works this way, we must first learn a bit about sampling distributions. Then in the next section we will put this knowledge to use.

Typical Form of Confidence Interval	Point Estimate $\pm$ Multiple $\times$ Standard Error	(9.1)

In the previous chapter we introduced the sampling distribution of the sample mean $\overline{X}$ and saw how it was related to the central limit theorem. In general, whenever we make inferences about one or more population parameters, such as a mean or the difference between two means, we always base this inference on the sampling distribution of a point estimate, such as the sample mean. Although the *concepts* of point estimates and sampling distributions are no different from those in the previous chapter, there are some new details we need to learn.

We again begin with the sample mean $\overline{X}$. From the central limit theorem we know that if the sample size n is reasonably large, then for *any* population distribution, the sampling distribution of $\overline{X}$ is approximately normally distributed with mean μ and standard deviation $\sigma/\sqrt{n}$, where μ and σ are the population mean and standard deviation. An equivalent statement is that the standardized quantity Z defined in equation (9.2) is approximately normal with mean 0 and standard deviation 1.

Standardized Z-Value	$$Z = \frac{\overline{X} - \mu}{\sigma/\sqrt{n}}$$	(9.2)

Typically, we use this fact to make inferences about an unknown population mean μ. There is one problem, however—we usually do not know the population standard deviation σ. This parameter, σ, is called a **nuisance parameter** because we need its value even though it is typically not the parameter of primary interest. The solution appears to be straightforward: Replace the nuisance parameter σ by its sample estimate s in the formula for Z and proceed from there. However, when we replace σ by the sample standard deviation s, we introduce a new source of variability, and the sampling distribution is no longer normal. It is instead called the *t* **distribution,** a close relative of the normal distribution that appears in a variety of statistical applications.

9.2.1 The *t* Distribution

We first set the stage for this new sampling distribution. We are interested in estimating a population mean μ with a sample of size n. We assume the population distribution is normal with unknown standard deviation σ. We intend to base inferences on the standardized value of $\overline{X}$, where σ is replaced by the sample standard deviation s. Then the standardized value in equation (9.3) has a *t* **distribution with** $n - 1$ **degrees of freedom.**

Standardized Value	$$t = \frac{\overline{X} - \mu}{s/\sqrt{n}}$$	(9.2)

The "degrees of freedom" is a numerical parameter of the *t* distribution that defines the precise shape of the distribution. Each time we encounter a *t* distribution, we will specify its degrees of freedom. In this particular sampling context, where we are basing inferences about μ on the sampling distribution of $\overline{X}$, the degrees of freedom turns out to be 1 less than the sample size n.

The *t* distribution looks very much like the standard normal distribution. It is bell shaped and is centered at 0. The only difference is that it is slightly more spread out, and

FIGURE 9.1
The *t* and Standard
Normal Distri-
butions

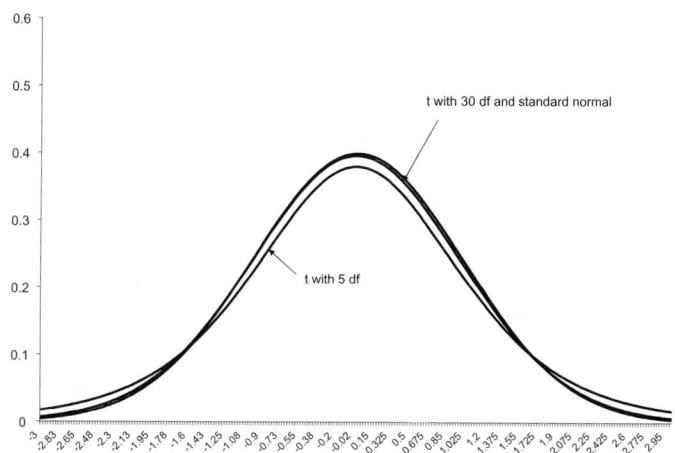

this increase in spread is greater for *small* degrees of freedom. In fact, when *n* is large, so that the degrees of freedom is large, the *t* distribution and the standard normal distribution are practically indistinguishable. This is illustrated in Figure 9.1. With 5 degrees of freedom, it is possible to see the increased spread in the *t* distribution. With 30 degrees of freedom, the *t* and standard normal curves are practically the same curve.

The *t*-value in equation (9.3) is very much like a typical *Z*-value such as in equation (9.2). That is, the *t*-value represents the number of standard errors by which the sample mean differs from the population mean. For example, if a *t*-value is 2.5, then the sample mean is 2.5 standard errors above the population mean. In contrast, if a *t*-value is −2.5, then the sample mean is 2.5 standard errors below the population mean. Also, *t*-values greater in magnitude than 3 are very unexpected because of the same property we saw with the normal distribution: It is very unlikely for a random value to be more than 3 standard deviations away from its mean. (In this case the random value is a sample mean, and the standard deviation is the standard error of the mean.)

Interpretation of a *t*-value	A **t-value** indicates the number of standard errors by which a sample mean differs from a population mean.

Because of this interpretation, *t*-values are perfect candidates for the "multiple" in expression (9.1), as we will soon see. First, however, we will briefly examine two Excel functions that help us to work with the *t* distribution in Excel.

In Chapter 6 we learned how to use Excel's NORMSDIST and NORMSINV functions to calculate probabilities or percentiles from the standard normal distribution. There are similar Excel functions for the *t* distribution: TDIST and TINV. Unfortunately, these functions are somewhat more difficult to master than their normal counterparts. To make the transition easier, it helps to know that these functions are generally used to find (1) the probability beyond a certain value, either in one or both tails, or (2) the value that has a certain (usually small) probability beyond it, in either one or both tails.

Example 9.1 Gasoline Prices in the United States

Suppose a government agency randomly samples 30 gas stations from the population of all gas stations in the United States. Its goal is to estimate the mean price for a gallon of premium unleaded gasoline. What is the probability that the sample mean price will be

One-tailed Probability

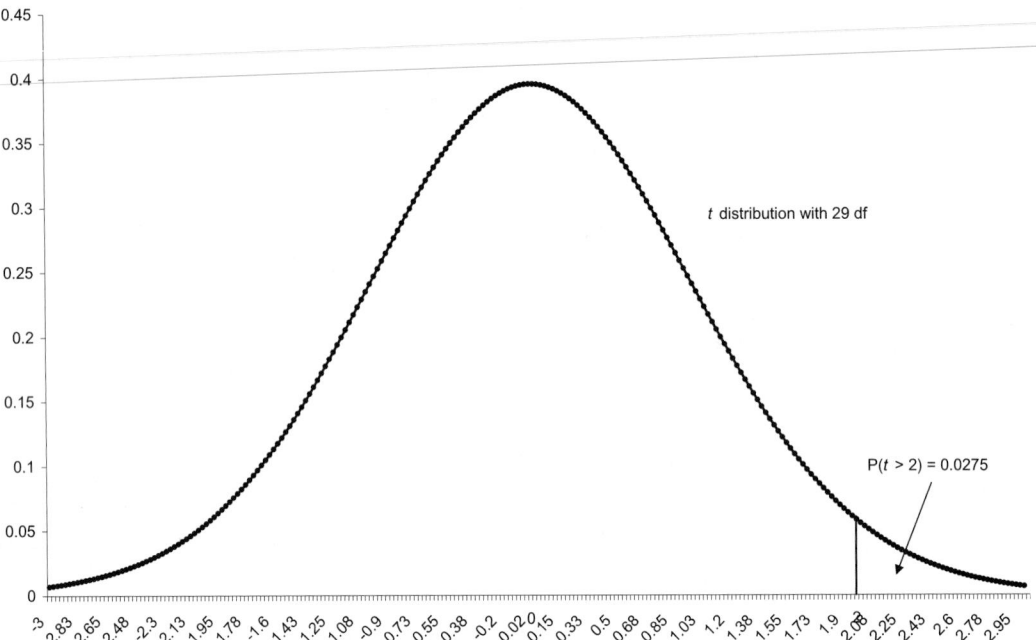

FIGURE 9.2 One-Tailed Probability for a *t* Distribution

Two-tailed Probability

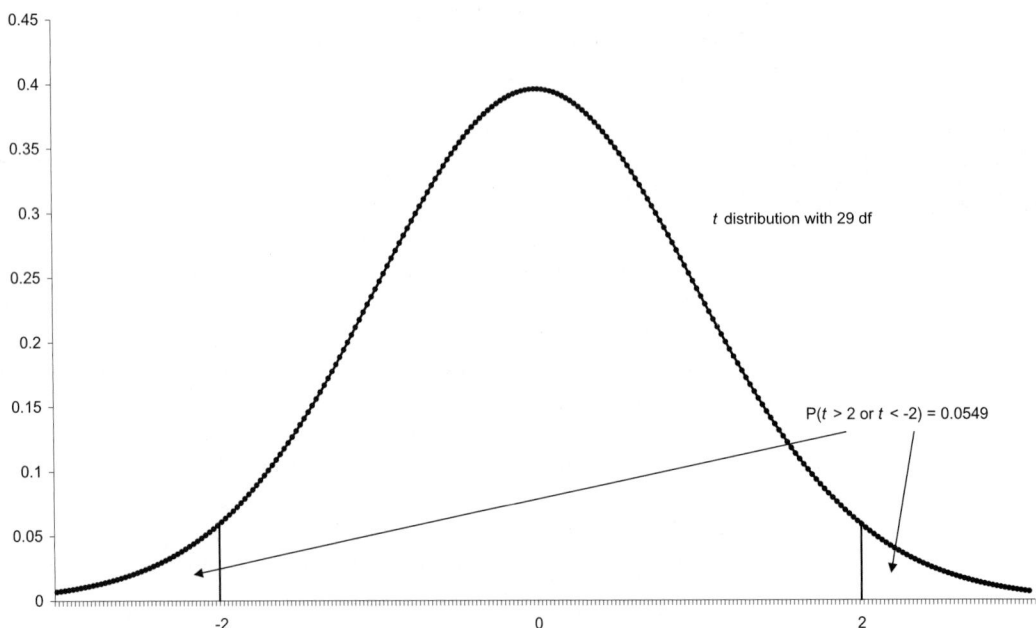

FIGURE 9.3 Two-Tailed Probability for a *t* Distribution

FIGURE 9.4

Excel Functions
for the *t* Distri-
bution

	A	B	C	D	E
1	Using the TDIST and TINV functions for the t distribution				
2					
3	Sample size	30			
4	Degrees of freedom	29			
5					
6	One-tailed probability				
7	t-value	2		Formula	
8	Probability in right tail	0.0275		=TDIST(B7,B4,1)	
9					
10	Two-tailed probability				
11	t-value	2		Formula	
12	Probability in both tails	0.0549		=TDIST(B11,B4,2)	
13					
14	TINV calculations				
15	Probability in right tail	0.05		Formula	
16	t-value	1.699		=TINV(2*B15,B4)	
17					
18	Probability in both tails	0.05		Formula	
19	t-value	2.045		=TINV(B18,B4)	

at least 2 standard errors above the population mean price? What is the probability that the sample mean price will differ by at least 2 standard errors from the population mean price?

Objective To use Excel's TDIST function to analyze differences between a sample mean and a population mean for gasoline prices.

Solution

First, we note that the answers to these questions do *not* depend on the values of the sample and population means or the standard error of the mean. They depend only on finding the probability that a "standardized" *t*-value is beyond some value, as illustrated in Figures 9.2 and 9.3. Figure 9.2 shows a one-tailed probability, where we are interested in whether a *t*-value exceeds some positive value, 2. Figure 9.3 shows a two-tailed probability, where we are interested in whether the *magnitude* of a *t*-value, positive or negative, exceeds 2. In each case the *t* distribution has $30 - 1 = 29$ degrees of freedom.

The calculations are in Figure 9.4. (See the file TDIST.XLS.) We answer the first question in rows 7 and 8. We want the probability that a *t*-value with 29 degrees of freedom exceeds 2. We find this with the formula spelled out in row 8. The first argument of TDIST is the value we want to exceed, the second is the degrees of freedom, and the third is the number of tails (1 or 2). We see that the probability of the sample mean exceeding the population mean by this much—2 standard errors—is only 0.0275. The answer to the second question is exactly twice this probability. We find it with the formula spelled out in row 12. The only difference is that the third argument is now 2. ●

In general, here are the technical details for using the TDIST function properly:

- Its first argument must be nonnegative.
- Unlike the NORMSDIST function, TDIST returns the probability to the *right* of the first argument (if the third argument is 1).
- Its third argument is either 1 or 2 and indicates the number of tails. By using 1 for this argument, we get the probability in the right-hand tail only. If we use 2 for the third argument, we obtain the probability of greater than the first argument or less than its negative.

Before proceeding, we mention the "inverse" problem, also illustrated in Figure 9.4. Instead of specifying a t-value and asking for the probability of exceeding this value, we can specify a probability and ask for the t-value that has this much probability beyond it. Again, this can be a one-tailed or a two-tailed problem. Rows 15 and 16 illustrate the one-tailed problem, whereas rows 18 and 19 illustrate the two-tailed problem. Each of these uses Excel's TINV function to find the appropriate t-value, as spelled out by the formulas in the figure. As we see, the t-value 1.699 has probability 0.05 to its right (when there are 29 degrees of freedom), whereas the t-values ± 2.045 have total probability 0.05 beyond them, or 0.025 in each tail.

The technical details for using the TINV function properly are as follows:

- The first argument is the total probability we want in both tails—half of this goes in the right-hand tail and half goes in the left-hand tail.

- Unlike the TDIST function, there is no third argument for the TINV function.

We agree that these differences between the t distribution and normal distribution functions are more complex than they ought to be, but this is the way Microsoft decided to program them. Fortunately, the StatPro add-in simplifies the process for most statistical inference applications.

9.2.2 Other Sampling Distributions

We have seen that the t distribution, a close relative of the normal distribution, is used when we want to make inferences about a population mean and the population standard deviation is unknown. Throughout this chapter (and later chapters) we will see other contexts where the t distribution appears. The theme is always the same—one or more means are of interest, and one or more standard deviations are unknown.

The t (and normal) distributions are not the only sampling distributions we will encounter. Two other close relatives of the normal distribution that appear in various contexts are the *chi-square* and *F distributions*. These are used primarily to make inferences about variances (or standard deviations), as opposed to means. We omit the details of these distributions for now, but you can look forward to seeing them in the near future.

PROBLEMS

Level A

1. Compute the following probabilities using Excel:
   a. $P(t_{10} \geq 1.75)$, where t_{10} has a t distribution with 10 degrees of freedom.
   b. $P(t_{100} \geq 1.75)$, where t_{100} has a t distribution with 100 degrees of freedom. How do you explain the difference between this result and the one obtained in part a?
   c. $P(Z \geq 1.75)$, where Z is a standard normal random variable. Compare this result to the results obtained in parts a and b. How do you explain the differences in these probabilities?
   d. $P(t_{20} \leq -0.80)$, where t_{20} has a t distribution with 20 degrees of freedom.
   e. $P(t_3 \leq -0.80)$, where t_3 has a t distribution with 3 degrees of freedom. How do you explain the dif-

ference between this result and the result obtained in part d?

2. Determine the following quantities using Excel:
   a. $P(-2.00 \leq t_{10} \leq 1.00)$, where t_{10} has a t distribution with 10 degrees of freedom.
   b. $P(-2.00 \leq t_{100} \leq 1.00)$, where t_{100} has a t distribution with 100 degrees of freedom. How do you explain the difference between this result and the one obtained in part a?
   c. $P(-2.00 \leq Z \leq 1.00)$, where Z is a standard normal random variable. Compare this result to the results obtained in parts a and b. How do you explain the differences in these probabilities?
   d. Find the 68th percentile of the t distribution with 20 degrees of freedom.

e. Find the 68th percentile of the t distribution with 3 degrees of freedom. How do you explain the difference between this result and the result obtained in part **d?**

3. Determine the following quantities using Excel:

a. Find the value of x such that $P(t_{10} > x) = 0.75$, where t_{10} has a t distribution with 10 degrees of freedom.

b. Find the value of y such that $P(t_{100} > y) = 0.75$, where t_{100} has a t distribution with 100 degrees of freedom. How do you explain the difference between this result and the result obtained in part **a?**

c. Find the value of z such that $P(Z > z) = 0.75$, where Z is a standard normal random variable. Compare this result to the results obtained in parts **a** and **b.** How do you explain the differences in the values of x, y, and z?

4. The NORMSDIST and NORMSINV functions in Excel give you probabilities and z-values, respectively, for the standard normal distribution. The analogous functions for the t distribution are TDIST and TINV. However, they do not work exactly the same as the normal functions. Try the following.

a. You have a t distribution with 15 degrees of freedom (df), and you want the probability to the right of the value 1.074. From t tables included in many statistics books, this probability is 0.15. Verify that you can get the answer in Excel with =TDIST(1.074,15,1), where the last "1" means one tail only.

b. This is the same as part **a,** but now you want the probability to the left of -1.074 or the right of 1.074, that is, the combined probability in both tails. Verify that you can get the answer in Excel with =TDIST(1.074,15,2), where the last "2" means two tails.

c. You have a t distribution with 15 df, and you want the t-value with probability 0.05 to the right of it. From t tables, this t-value is 1.753. Verify that you can get the answer in Excel with =TINV(.10,15). (The tricky part here is that the first argument, 0.10, is double the original probability you asked for. This is hard to remember!)

d. This is the same as part **c,** but now you want the t-value, call it t, such that probability 0.025 is to the right of t, and probability 0.025 is to the left of $-t$. The t tables show that t is 2.131. Verify that you can get the answer in Excel with =TINV(.05,15). (Here you can see the trick in part **c** better. You use the *combined* probability in both tails as the first argument in TINV.)

9.3 Confidence Interval for a Mean

We now come to the focal point of this chapter: using results about sampling distributions to construct confidence intervals. As explained in the introduction to this chapter, we assume that data have been generated by some random mechanism, either by observing a random sample from some population or by performing a randomized experiment. We want to use these data to infer the values of one or more population parameters such as the mean, the standard deviation, or a proportion. For each such parameter we will use the data to calculate a point estimate, which can be considered a "best guess" for the unknown parameter. We will also calculate a confidence interval around the point estimate to gauge its accuracy. This is directly analogous to the way we went out 2 standard errors on either side of the point estimate to form intervals in the previous chapter. However, we will expand on this procedure in this chapter.

We begin by deriving a confidence interval for a population mean μ, and we discuss its interpretation. Although the particular details pertain to a specific parameter, the mean, the same ideas carry over to estimation of other parameters as well, as we will see in later sections. As usual, we use $\overline{X}$, the sample mean, as the point estimate of μ.

To obtain a confidence interval for μ, we first specify a **confidence level,** usually 90%, 95%, or 99%. We then use the sampling distribution of the point estimate to determine the *multiple* of the standard error we need to go out on either side of the point estimate to achieve the given confidence level. If the confidence level is 95%, the value used most frequently in applications, then the multiple is approximately 2. More precisely, it is a t-value. That is, a typical confidence interval for μ is of the form in expression (9.4), where $SE(\overline{X}) = s/\sqrt{n}$.

Confidence Interval for Population Mean	$$\overline{X} \pm t\text{-multiple} \times SE(\overline{X}) \qquad \qquad \textbf{(9.4)}$$

To obtain the correct t-multiple, let α be 1 minus the confidence level (expressed as a decimal). For example, if the confidence level is 90%, then $\alpha = 0.10$. Then the appropriate t-multiple is the value that cuts off probability $\alpha/2$ in each tail of the t distribution with $n - 1$ degrees of freedom. For example, if $n = 30$ and the confidence level is 95%, we see from cell B19 of Figure 9.4 that the correct t-value is 2.045. The corresponding 95% confidence interval for μ is then

$$\overline{X} \pm 2.045(s/\sqrt{n})$$

If the confidence level is instead 90%, the appropriate t-value is 1.699 (to see this, change the probability in cell B18 to 0.10), and the resulting 90% confidence interval is

$$\overline{X} \pm 1.699(s/\sqrt{n})$$

If the confidence level is 99%, the appropriate t-value is 2.756 (to see this, change the probability in cell B18 to 0.01), and the resulting 99% confidence interval is

$$\overline{X} \pm 2.756(s/\sqrt{n})$$

Note that as the confidence level increases, the width of the confidence interval also increases. Since we naturally want confidence intervals to be as narrow as possible, this presents a trade-off. We can either have less confidence and a narrow interval, or we can have more confidence and a wide interval. However, we can also take a larger sample. As n increases, the standard error $s/\sqrt{n}$ decreases, and the length of the confidence interval tends to decrease for *any* confidence level. (Why won't it decrease for sure? The larger sample *might* result in a larger value of s that could offset the increase in n.)

The following example illustrates confidence interval estimation for a population mean. It uses the One-Sample procedure in the StatPro add-in to perform the calculations. However, by examining the resulting Excel formulas, you can check that all it is really doing is (1) calculating the sample mean, (2) calculating the standard error of the sample mean, $s/\sqrt{n}$, (3) finding the appropriate t-multiple with the TINV function, and (4) combining these to form the confidence interval via expression (9.4).

Example 9.2 Customer Response to a New Sandwich

A fast-food restaurant recently added a new sandwich to its menu. To estimate the popularity of this sandwich, a random sample of 40 customers who ordered the sandwich were surveyed. Each of these customers was asked to rate the sandwich on a scale of 1 to 10, 10 being the best. The results of this survey appear in column B of Figure 9.5. (See the file SANDWICH1.XLS.) The manager wants to estimate the mean satisfaction rating over the entire population of customers by using a 95% confidence interval. How should she proceed?

FIGURE 9.5
Analysis of New
Sandwich Data

	A	B	C	D	E	F	G
1	Customer satisfaction with new sandwich						
2							
3	Customer	Satisfaction		Results for one-sample analysis for Satisfaction			
4	1	7					
5	2	5		Summary measures			
6	3	5		Sample size		40	
7	4	6		Sample mean		6.250	
8	5	8		Sample standard deviation		1.597	
9	6	7					
10	7	6		Confidence interval for mean			
11	8	7		Confidence level		95.0%	
12	9	10		Sample mean		6.250	
13	10	7		Std error of mean		0.253	
14	11	9		Degrees of freedom		39	
15	12	5		Lower limit		5.739	
16	13	5		Upper limit		6.761	
17	14	8					
42	39	5					
43	40	4					

Objective To use StatPro's one-sample procedure to obtain a 95% confidence interval for the mean satisfaction rating of the new sandwich.

Solution

We use StatPro's One-Sample procedure on the Satisfaction variable. Although this procedure is capable of doing more than we're asking here, it is the appropriate procedure when we want to estimate the mean from a single population. To use it, place the cursor anywhere in the data set (cell B4, say) and select the StatPro/Statistical Inference/One-Sample Analysis menu item. In the succeeding dialog boxes, select Satisfaction as the variable that you want to analyze, check that you want a confidence interval for the mean, and then accept the defaults from there on. You should obtain the output shown in Figure 9.5.

The principal results are that (1) the best guess for the population mean rating is 6.250, the sample average in cell F12, and (2) a 95% confidence interval for the population mean rating extends from 5.739 to 6.761, as seen in cells F15 and F16. The manager can be 95% confident that the true mean rating over all customers who might try the sandwich is within this confidence interval.

To understand where these numbers come from, take a look at the formulas in column F. Note that the standard error in cell F13 corresponds to $s/\sqrt{n}$. It is calculated with the formula[2]

$$=\text{STDEV(Data!Satisfaction)/SQRT(COUNT(Data!Satisfaction))}$$

The degrees of freedom for the t distribution is 1 less than the sample size, as shown in cell F14. Finally, the t-multiple for the confidence interval, although not shown explicitly, is used in the formulas (via the TINV function) in cells F15 and F16. For example, the formula for the lower limit in cell F15 is

$$=\text{F12-TINV(1-F11,F14)*F13}$$

We stated previously that as the confidence level increases, the length of the confidence interval increases. You can convince yourself of this by entering different confidence

---

[2]The Data! in this formula is a reference to the sheet name that StatPro automatically attaches to ranges names.

levels, such as 90% or 99%, in cell F11 of the file SANDWICH1.XLS. The lower and upper limits of the confidence interval in cells F15 and F16 will change automatically, getting closer together for the 90% level and farther apart for the 99% level. Just remember that you, the analyst, can choose the confidence level you favor, although 95% is the level most commonly chosen.

Before leaving this example, we discuss the assumptions that lead to the confidence interval. First, we might question whether the sample is really a *random* sample—or whether it matters. Perhaps the manager used some random mechanism to select the customers to be surveyed. More likely, however, she simply surveyed 40 consecutive customers who tried the sandwich on a given day. This is called a **convenience sample** and is not really a random sample. However, unless there is some reason to believe that these 40 customers differ in some relevant aspect from the entire population of customers, it is probably safe to treat them as a random sample.

A second assumption is that the population distribution is *normal*. We made this assumption when we discussed the *t* distribution. Obviously, the population distribution *cannot* be exactly normal because it is concentrated on the ten possible satisfaction ratings, and the normal distribution describes a continuum. However, this is probably not a problem for two reasons. First, confidence intervals based on the *t* distribution are **robust** to violations of normality. This means that the resulting confidence intervals are valid for any populations that are *approximately* normal. Second, the normal population assumption is less crucial for larger sample sizes because of the central limit theorem. For *n* as large as 40, the results should be valid.

Finally, it is important to recognize what this confidence interval tells us and what it doesn't. In the entire population of customers who ordered this sandwich, there is a distribution of satisfaction ratings. Some fraction rate it as 1, some rate it as 2, and so on. All we are trying to determine here is the *average* of all these ratings, and based on our analysis, we can be 95% confident that this (still unknown) average is between 5.739 and 6.761. However, this confidence interval doesn't tell us other characteristics of the population of ratings that might be of interest, such as the proportion of customers who rate the sandwich 6 or higher. It tells us only about the *mean* rating. Later in this chapter, we will see how to find a confidence interval for a proportion, which allows us to analyze another important characteristic of a population distribution. ●

In the sandwich example we said that the manager can be 95% confident that the true mean rating is between 5.739 and 6.761. What does this statement really mean? Contrary to what you might expect, it does *not* mean that the true mean lies between 5.739 and 6.761 with probability 0.95. Either the true mean is inside this interval or it is not. The true meaning of a 95% confidence interval is based on the *procedure* used to obtain it. Specifically, if we use this procedure on a large number of random samples, all from the same population, then approximately 95% of the resulting confidence intervals will be "good" ones that include the true mean, and the other 5% will be "bad" ones that do not include the true mean. Of course, when we have only a single sample, as in the sandwich example, we have no way of knowing whether our confidence interval is one of the good ones or one of the bad ones, but we can be 95% confident that we obtained one of the good intervals.

Because this is such an important concept, we illustrate it in Figure 9.6 with simulation. (See the file CONFINT.XLS.) The data in column B are generated randomly from a normal distribution with the *known* values of μ and σ in cells B3 and B4. Next, we in-

	A	B	C	D	E	F	G	H	I	J	K
1	Interpretation of a "95% confidence interval"										
2											
3	Population mean	100									
4	Population stdev	20									
5											
6		Random sample			Results for one-sample analysis for Random sample						
7		113.90									
8		84.99			Summary measures						
9		93.71			Sample size	30					
10		114.72			Sample mean	99.503					
11		81.19			Sample standard deviation	21.818					
12		80.57									
13		98.53			Confidence interval for mean						
14		89.20			Confidence level	95.0%					
15		80.19			Sample mean	99.503					
16		172.12			Std error of mean	3.983					
17		109.59			Degrees of freedom	29					
18		95.19			Lower limit	91.356					
19		90.71			Upper limit	107.650					
20		72.20			Mean captured?	1					
21		130.92									
22		106.85			% of replications where mean is within confidence interval (from data table below)						
23		79.62			95.3%						
24		78.13									
25		72.31			Data table used to replicate confidence intervals						
26		86.96		Rep	Mean captured?						
27		101.51			1						
28		82.32		1	1						
29		124.36		2	1						
30		106.48		3	0						
31		99.64		4	1						
32		83.82		5	1						
33		139.34		6	1						
34		104.90		7	0						
35		101.11		8	1						
36		110.04		9	0						
37				10	1						
38				11	0						

FIGURE 9.6 Simulation Demonstration of Confidence Intervals

voke StatPro's One-Sample procedure to calculate a 95% confidence interval for the true value of μ, exactly as in the sandwich example. However, we now know whether the true value of μ is within the interval, so we record a 1 in cell F20 if it is and a 0 otherwise. This requires the formula

$$=IF(AND(B3>=F18,B3<=F19),1,0)$$

This simulation is performed only to illustrate the true meaning of a "95% confidence interval." In any real situation we will obtain only a single random sample and corresponding confidence interval.

Finally, we use a data table to replicate the simulated results 1000 times.[3] Specifically, we enter the formula

$$=F20$$

in cell E27 and build a data table (only a few rows of which are shown) in the range D27:E1027, leaving the row input cell box empty and using any blank cell as the column input cell. Then we calculate the fraction of values in the range E28:E1027 that are 1's in cell E23 with the AVERAGE function.

We see that 953 of the simulated confidence intervals (each based on a *different* random sample of size 30) contain the true mean 100. In theory, we would expect 950 of the 1000 intervals to cover the true mean, and this is almost exactly what we obtained. Of course, in a particular application you might unluckily obtain the third sample (in row 30). However, without knowing that the true mean is 100, you would have no way of knowing that you obtained a "bad" interval!

---

[3]It takes quite awhile to simulate 1000 samples of size 30 in this data table. Therefore, it's definitely a good idea to use the Tools/Options menu item to set the recalculation mode to "automatic except tables." That way, the data table will recalculate only if you explicitly tell it to (by pressing the F9 key).

PROBLEMS

Level A

5. A manufacturing company's quality control personnel have recorded the proportion of defective items for each of 500 monthly shipments of one of the computer components that the company produces. The data are in the file P02_02.XLS. The quality control department manager does not have sufficient time to review all of these data. Rather, she would like to examine the proportions of defective items for a sample of these shipments.

   a. Use Excel to generate a simple random sample of size 25 from the given frame.

   b. Using the sample generated in part **a,** construct a 95% confidence interval for the mean proportion of defective items over all monthly shipments. Assume that the population consists of the proportion of defective items for each of the given 500 monthly shipments.

   c. Interpret the 95% confidence interval constructed in part **b.**

   d. Does the 95% confidence interval contain the actual population mean in this case? If not, explain why not. What proportion of many similarly constructed confidence intervals should include the true population mean value?

6. Consider the given set of average annual household income levels of citizens of selected U.S. metropolitan areas in the file P03_06.XLS.

   a. Use Excel to obtain a simple random sample of size 15 from this frame.

   b. Using the sample generated in part **a,** construct a 99% confidence interval for the mean average annual household income level of citizens in the selected U.S. metropolitan areas. Assume that the population consists of all average annual household income levels in the given frame.

   c. Interpret the 99% confidence interval constructed in part **b.**

   d. Does the 99% confidence interval contain the actual population mean? If not, explain why not. What proportion of many similarly constructed confidence intervals should include the true population mean value?

7. The file P03_86.XLS contains data on all NFL players as of 2000. Because this file contains all players, you can calculate the *population* mean if we define "population" as all 2000 NFL salaries. However, proceed as in Chapter 8 to select a random sample of size 50 from this population. Based on this random sample, calculate a 95% confidence interval for the mean NFL salary in 2000. Does it contain the population mean? Repeat this procedure several times until you find a random sample where the population mean is *not* included in the confidence interval.

8. The file P09_08.XLS contains data on repetitive task times for each of two workers. John has been doing this task for months, whereas Fred has just started. Each time listed is the time (in seconds) to perform a routine task on an assembly line. The times shown are in chronological order.

   a. Find a 95% confidence interval for the mean time it takes John to perform the task. Do the same for Fred.

   b. Do you believe both of the confidence intervals in part **a** are valid and/or useful? Why or why not? Which of the two workers would you rather have (assuming time is the only issue)?

9. The manager of a local fast-food restaurant is interested in improving the service provided to customers who use the restaurant's drive-up window. As a first step in this process, the manager asks his assistant to record the time (in minutes) it takes to serve a large number of customers at the final window in the facility's drive-up system. The given frame of 200 customer service times are all observed during the busiest hour of the day for this fast-food operation. The data are in the file P02_04.XLS.

   a. Use Excel to generate a simple random sample of size 10 from this frame.

   b. Using the sample generated in part **a,** construct a 90% confidence interval for the mean service time of all customers arriving during the busiest hour of the day at this fast-food operation. Assume that the population consists of the given 200 customer service times.

   c. Interpret the 90% confidence interval constructed in part **b.**

Level B

10. Continuing the previous problem, use Excel to generate 100 simple random samples of size 10 from the frame given in the file P02_04.XLS. Then use each of these random samples to construct a 90% confidence interval for the mean service time of all customers arriving during the busiest hour of the day at this fast-food operation. How many of the 100 constructed confidence intervals actually contain the true value of the population mean in this case? Is this result consistent with your expectations? Explain.

9.4 Confidence Interval for a Total[4]

There are situations where a population mean is not the population parameter of most interest. A good example is the auditing example discussed in the previous chapter (Example 8.4). Rather than estimating the mean amount of receivables *per account,* the auditor might be more interested in the *total* amount of all receivables, summed over all accounts. In this section we will provide a point estimate and a confidence interval for a population total.

First, we introduce some notation. Let T be a population total we want to estimate, such as the total of all receivables, and let $\hat{T}$ be a point estimate of T based on a simple random sample of size n from a population of size N. We need a reasonable formula for $\hat{T}$; that is, we need to know how to calculate a point estimate of T. For the population total T, it is reasonable to sum all of the values in the sample, denoted T_S, and then "project" this total to the population with the formula (9.5), where the second equality follows because the sample total T_S divided by the sample size n is the sample mean $\overline{X}$.

Point Estimate for Population Total	$$\hat{T} = \frac{N}{n} T_S = N\overline{X}$$	(9.5)

Actually, equation (9.5) is quite intuitive. Suppose there are 1000 accounts in the population, we sample 50 of them, and we observe a sample total of $5000. Then, because we sampled only 1/20 of the population, a natural estimate of the population total is 20 × $5000 = $100,000.

Like the sample mean $\overline{X}$, the estimate $\hat{T}$ has a sampling distribution. The mean and standard deviation of this sampling distribution are given in equations (9.6) and (9.7), where σ is again the population standard deviation.

Mean and Standard Error of Point Estimate for Population Total	$E(\hat{T}) = T$	(9.6)
	$SE(\hat{T}) = N\sigma/\sqrt{n}$	(9.7)

Because σ is usually unknown, we use s instead of σ to obtain the approximate standard error of $\hat{T}$ given in formula (9.5). The second equality follows because $s/\sqrt{n}$ is the standard error of $\overline{X}$.

Approximate Standard Error of Point Estimate for Population Total	$SE(\hat{T}) = Ns/\sqrt{n} = N \times SE(\overline{X})$	(9.8)

Note from equation (9.6) that $\hat{T}$ is an unbiased estimate of the population total T. Therefore, it has no tendency to either overestimate or underestimate T.

[4]This section can be omitted without any loss of continuity.

From equations (9.5) and (9.8), we see that the point estimate of T is the point estimate of the mean multiplied by N, and that the standard error of this point estimate is the standard error of the sample mean multiplied by N. This has a very nice consequence. We can form a confidence interval for T with the following two-step procedure:

1. Find a confidence interval for the sample mean in the usual way.
2. Multiply each endpoint of the confidence interval by the population size N.

We illustrate this procedure in the following example.

Example 9.3 Estimating Total Tax Refunds

Suppose the Internal Revenue Service would like to estimate the total net amount of refund due to a particular set of 10,000 taxpayers. Each taxpayer will either receive a refund, in which case the net refund is positive, or will have to pay an amount due, in which case the net refund is negative. Therefore the *total* net amount of refund is a natural quantity of interest; it is the net amount the IRS will have to pay out (or receive, if negative). Find a 95% confidence interval for this total using the refunds from a random sample of 500 taxpayers in the file IRS.XLS.

Objective To use StatPro's one-sample procedure, with an appropriate modification, to find a 95% confidence interval for the total (net) amount the IRS must pay out to these 10,000 taxpayers.

Solution

The solution appears in Figure 9.7. Although there is no explicit StatPro procedure for dealing with population totals, we can take advantage of the close relationship between

FIGURE 9.7
Confidence Interval for a Population Total

	A	B	C	D	E	F	G
1	IRS tax refunds & payments						
2							
3	Population size	10000					
4							
5	Customer	Refund			Results for one-sample analysis for Refund		
6	1	$70					
7	2	$1,190			Summary measures		
8	3	$220			Sample size	500	
9	4	($280)			Sample mean	294.980	
10	5	$260			Sample standard deviation	581.312	
11	6	$370					
12	7	$450			Confidence interval for mean		
13	8	$210			Confidence level	95.0%	
14	9	$1,150			Sample mean	294.980	
15	10	$270			Std error of mean	25.997	
16	11	$470			Degrees of freedom	499	
17	12	($10)			Lower limit	243.903	
18	13	($160)			Upper limit	346.057	
19	14	$2,430					
20	15	$140			Confidence interval for population total		
21	16	($190)			Confidence level	95.0%	
22	17	($810)			Point estimate	$2,949,800	
23	18	($20)			Standard error	$259,970	
24	19	$300			Lower limit	$2,439,029	
25	20	($280)			Upper limit	$3,460,571	
26	21	($300)					
503	498	$190					
504	499	$1,840					
505	500	($20)					

the confidence interval for a mean and the confidence interval for a total. That is, we first use StatPro to find a 95% confidence interval for the population mean. This output appears in rows 5–18. The average refund per taxpayer in the sample is slightly less than $300 (cell F14), and the standard error of this sample mean (in cell F15) is about $26. The confidence interval for the mean (in cells F17 and F18) extends from $244 to $346. This part of the output analyzes the average refund for a single taxpayer.

Now all we need to do is project these results to the entire population. We do this in the range F22:F25 by multiplying each of the values in the previous paragraph by the population size, 10,000. The IRS can be 95% confident that it will need to pay out somewhere between 2.44 and 3.46 million dollars to these 10,000 taxpayers. ●

PROBLEMS

Level A

11. The operations manager of a toll booth located at a major exit of a state turnpike is trying to estimate the total number of vehicles that arrive at the toll booth during a 1-minute period during the peak of rush-hour traffic. In an effort to estimate this total throughput value, he records the number of vehicles that arrive at the toll booth over a 1-minute interval commencing at the same time for each of 50 normal weekdays. The data are provided in the file P09_11.XLS. Construct a 95% confidence interval for the total number of vehicles that arrive at the toll booth during a 1-minute period during the peak of rush-hour traffic for 1000 normal weekdays. What does this interval reveal about the actual throughput value of interest?

12. A lightbulb manufacturer wants to estimate the total number of defective bulbs contained in all of the boxes shipped by the company during the past week. Production personnel at this company have recorded the number of defective bulbs found in each of 50 randomly selected boxes shipped during the past week. These data are provided in the file P09_12.XLS. Construct a 99% confidence interval for the total number of defective bulbs contained in the 1000 boxes shipped by this company during the past week. Interpret this confidence interval for the production personnel at this company.

13. Auditors of a particular bank are interested in comparing the reported value of all 2265 customer savings account balances with their own findings regarding the actual value of such assets. Rather than reviewing the records of each savings account at the bank, the auditors decide to examine a representative sample of savings account balances. The frame from which they will sample is given in the file P08_50.XLS.
    a. Select a simple random sample consisting of 100 savings account balances from the given frame.
    b. Using the sample generated in part **a**, construct a 90% confidence interval for the total value of all savings account balances within this bank. Assume that the population consists of all savings account balances in the given frame.
    c. Interpret the 90% confidence interval constructed in part **b**.

Level B

14. Continuing the previous problem, use Excel to generate 50 simple random samples of size 100 from the frame given in P08_50.XLS. Then use each of these random samples to construct a 90% confidence interval for the total value of all 2265 savings account balances within this bank. How many of the 50 constructed confidence intervals actually contain the true total value in this case? Is this result consistent with your expectations? Explain.

9.5 Confidence Interval for a Proportion

How often have you heard on the evening news a survey finding such as, "52% of the public agree with the President's handling of the economy, with a sampling error of plus or minus 3%"? Surveys are often used to estimate proportions, such as the proportion of the public that agree with the President's handling of the economy. We will now see how to form a confidence interval for any population proportion p.

The basic procedure is very similar to what we described for a population mean. We find a point estimate, the standard error of this point estimate, and a multiple that depends on the confidence level. Then the confidence level has the same form as in expression (9.1):

$$\text{point estimate} \pm \text{multiple} \times \text{standard error}$$

In the news example the point estimate is 52% and the "multiple×standard error" is 3%. Therefore, the confidence interval extends from 49% to 55%. Although the news show doesn't state the confidence level explicitly, it is 95% by convention. In words, we can be 95% confident that the percentage of the public who agree with the President's handling of the economy is somewhere between 49% and 55%.

The theory that leads to this result is fairly straightforward. Let A be any property that members of a population either have or do not have. As examples, A might be the property that

- a person agrees with the President's handling of the economy;
- a person has purchased a company's product at least once within the past 3 months;
- the diameter of a part is with specification limits;
- a customer's account is at least 2 months overdue;
- a customer's rating of a new sandwich is at least 6 on a 10-point scale.

In each of these examples, we are interested in the proportion p of the population that have property A. We sample n members randomly and let $\hat{p}$ be the sample proportion of members with property A. For example, if 10 out of 50 sampled members have property A, then $\hat{p} = 10/50 = 0.2$. Then we use $\hat{p}$ as a point estimate of p.

It can be shown that for sufficiently large n, the sampling distribution of $\hat{p}$ is approximately normal with mean p and standard deviation $\sqrt{p(1 - p)/n}$. Because p is the unknown parameter, we substitute $\hat{p}$ for p in this standard deviation to obtain the following approximate standard error of $\hat{p}$:

Standard Error of Sample Proportion	$$\text{SE}(\hat{p}) = \sqrt{\dfrac{\hat{p}(1 - \hat{p})}{n}}$$	(9.9)

Finally, the multiple we use to obtain a confidence interval for p is a z-value. It is the standard normal value that cuts off an appropriate probability in each tail. For example, the z-multiple for a 95% confidence interval is 1.96 because this value cuts off probability 0.025 in each tail of the standard normal distribution. In general, the confidence interval has the form in expression (9.10):

Confidence Interval for a Proportion	$$\hat{p} \pm z\text{-multiple} \times \sqrt{\dfrac{\hat{p}(1 - \hat{p})}{n}}$$	(9.10)

This confidence interval is based on the assumption of a large sample size. A rule of thumb for checking the validity of this assumption is the following. Let p_L and p_U be the lower and upper limits of the confidence interval. Then the sample size is sufficiently large—and the confidence interval is valid—if $np_L > 5$, $n(1 - p_L) > 5$, $np_U > 5$, and $n(1 - p_U) > 5$.

We illustrate the procedure in the following example.

Example 9.4 Estimating the Response to a New Sandwich

The fast-food manager from Example 9.2 has already sampled 40 customers to estimate the population mean rating of its new sandwich. Recall that each rating is on a 1 to 10 scale, 10 being the best. The manager would now like to use the same sample to estimate the proportion of customers who rate the sandwich at least 6. Her thinking is that these are the customers who are likely to purchase the sandwich on subsequent visits.

Objective To illustrate the procedure for finding a confidence interval for the proportion of customers who rate the new sandwich at least 6 on a 10-point scale.

Solution

The solution appears in Figure 9.8. (See the file SANDWICH2.XLS.) StatPro does not yet have a procedure for calculating confidence intervals for proportions, but this example illustrates how a few Excel functions accomplish the job.) We first count the number of ratings that are at least 6 in cell E6. The easiest way to do this is with the formula

$$=COUNTIF(B4:B43,">=6")$$

(The quotes around the condition are required.) Then we calculate the sample proportion in cell E7 with the formula

$$=E6/E3$$

The rest is simply a matter of implementing equation (9.9) and expression (9.10). Specifically, the standard error formula in cell E11 is

$$=SQRT(E7*(1-E7)/E3)$$

and the formula for the lower limit of the confidence interval, in cell E13, is

$$=E7-E12*E11$$

Of course, the formula for the upper limit is the same except with a plus sign.

Then using the confidence interval limits, $p_L = 0.475$ and $p_U = 0.775$, we can check the assumption of large sample size. With $n = 40$, you can check that np_L, $n(1 - p_L)$, np_U,

FIGURE 9.8
Analysis of a Proportion for New Sandwich Data

	A	B	C	D	E	F
1	Customer satisfaction with new sandwich					
2						
3	Customer	Satisfaction		Sample size	40	
4	1	7				
5	2	5		Ratings at least 6		
6	3	5		Number	25	
7	4	6		Proportion	0.625	
8	5	8				
9	6	7		Confidence interval calculations		
10	7	6		Confidence level	95%	
11	8	7		Standard error	0.077	
12	9	10		z-multiple	1.960	
13	10	7		Lower limit	0.475	
14	11	9		Upper limit	0.775	
15	12	5				
16	13	5				
42	39	5				
43	40	4				

and $n(1 - p_U)$ are all well above 5, so that the validity of this confidence interval is established.

The output is fairly good news for the manager. Based on this sample of size 40, she can be 95% confident that the percentage of all customers who would rate the sandwich 6 or higher is somewhere between 47.5% and 77.5%. Of course, she realizes that this is a very wide interval, so there is still a lot of uncertainty about the *true* population proportion. To reduce the length of this interval, she would need to sample more customers—quite a few more customers. Typically, confidence intervals for proportions are fairly wide unless n is quite large. ●

We explore this final statement a bit more. Referring again to news shows, you have probably noticed that they almost always quote a sampling error of plus or minus 3%. In words, the "plus or minus" part of their 95% confidence interval is 3%, or 0.03. How large a sample size must they use to achieve this? We know that the "plus or minus" part of the confidence interval is 1.96 times the standard error of $\hat{p}$, so we must have

$$1.96 \times \sqrt{\hat{p}(1 - \hat{p})/n} = 0.03$$

Now, the quantity $\hat{p}(1 - \hat{p})$ is fairly constant for values of $\hat{p}$ between 0 and 1, provided that $\hat{p}$ isn't too close to 0 or 1. To get a reasonable estimate of the required n, we assume $\hat{p} = 0.5$. Then we have

$$1.96 \times \sqrt{(0.5)(0.5)/n} = 0.03$$

Solving for n, we obtain $n = [(1.96)(0.5)/0.03]^2 \approx 1067$.

This is a rather remarkable result. To obtain a 95% confidence interval of this length for a population proportion, where the population consists of millions of people, only about 1000 people need to be sampled. The remarkable fact is that this small a sample can provide such accurate information about such a large population.

One of many business applications of confidence intervals for proportions is in auditing. Auditors typically use **attribute sampling** to check whether certain procedures are being followed correctly. The term "attribute" implies that each item checked is done either correctly or incorrectly—there is no "in between." Examples of items not done correctly might include (1) an invoice copy that is not initialed by an accounting clerk, (2) an invoice quantity that does not agree with the quantity on the shipping document, (3) an invoice price that does not agree with the price on an authorized price list, and (4) an invoice with a clerical inaccuracy. Typically, an auditor focuses on one of these types of errors and then estimates the proportion of items with this type of error.

Because auditors are concerned primarily with how *large* the proportion of errors might be, they usually calculate 95% **one-sided** confidence intervals for proportions. Instead of using sample data to find lower and upper limits p_L and p_U of a confidence interval, they automatically use $p_L = 0$ and then determine an upper limit p_U such that the 95% confidence interval is from 0 to p_U. A simple modification of the confidence interval in formula (9.10) provides the result in equation (9.11), where the z-multiple is chosen so that the entire probability (0.05 for a 95% interval) is in the right-hand tail. For a 95% confidence level, the relevant z-multiple is 1.645.

Upper Limit of a One-Sided Confidence Interval for a Proportion	$p_U = \hat{p} + z\text{-multiple} \times \sqrt{\hat{p}(1 - \hat{p})/n}$ **(9.11)**

One further complication occurs, however. This formula for p_U relies on the large-sample approximation of the normal distribution to the binomial distribution. Auditors typically use an *exact* procedure to find p_U that is based directly on the binomial distribution. We illustrate how this is done in the following example.

Example 9.5 Auditing for Price Errors

An auditor wants to check the proportion of invoices that contain price errors—that is, prices that do not agree with those on an authorized price list. He checks 93 randomly sampled invoices and finds that two of them include price errors. What can he conclude, in terms of a 95% one-sided confidence interval, about the proportion of all invoices with price errors?

Objective To find the upper limit of a one-sided 95% confidence interval for the proportion of errors in the context of attribute sampling in auditing.

Solution

The results appear in Figure 9.9. (See the file AUDIT.XLS.) The sample proportion is $\hat{p} = 2/93 = 0.0215$ and the upper confidence limit based on the large-sample approximation is 0.046. This latter value is calculated in cell B14 with the formula

$$=B7+B13*SQRT(B7*(1-B7)/B5)$$

However, the fact that $np_U = 93(0.046) = 4.278$ is less than 5 indicates that the large-sample approximation might not be very accurate.

A more accurate procedure, based on the binomial distribution, appears in row 10. It turns out that if p_U is the appropriate upper confidence limit, then p_U satisfies the equation

$$P(X \leq k) = \alpha \tag{9.12}$$

Here, X is binomially distributed with parameters n and p_U, k is the observed number of errors, and α is 1 minus the confidence level. There is no way to find p_U directly (by means of a formula) from equation (9.12). However, we can use Excel's Goal Seek tool. First, we enter *any* trial value of p_U in cell B10 and the binomial formula

$$=BINOMDIST(B4,B5,B10,1)$$

in cell D10. (This formula calculates $P(X \leq k)$ from the trial value in cell B10.) Then we use the Tools/Goal Seek menu item, with cell B10 as the Set cell, 0.05 as the target value, and cell B10 as the Changing cell. (See Figure 9.10 on page 436.)

FIGURE 9.9
Analysis of Auditing Example

	A	B	C	D	E	F
1	Auditing example for an exact one-sided confidence interval					
2						
3	Confidence level	95%				
4	Number of errors	2				
5	Sample size	93				
6						
7	Sample proportion	0.0215				
8						
9	Exact upper confidence limit for p			Goal seek condition		
10	Upper limit	0.066		0.050	=	0.05
11						
12	Large-sample upper confidence limit for p					
13	z-multiple	1.645				
14	Upper limit	0.046				

FIGURE 9.10
Settings in Goal Seek Dialog Box

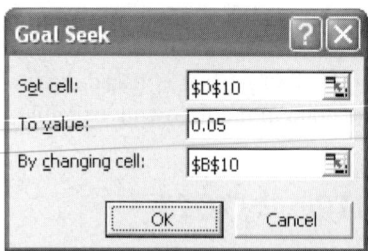

The resulting value of p_U is 0.066. This is considerably different (from the auditor's point of view) from the 0.046 value found from the large-sample approximation. It allows the auditor to state with 95% confidence that the percentage of invoices with price errors is no greater than 6.6%, based on the 2 errors out of 93 observed in the sample. ●

PROBLEMS

Level A

15. Wal-Mart buyers seek to purchase adequate supplies of various brands of toothpaste to meet the ongoing demands of its customers. In particular, Wal-Mart is interested in estimating the proportion of its customers who favor the country's leading brand of toothpaste, Crest. The file P09_15.XLS contains the toothpaste brand preferences of 200 Wal-Mart customers, obtained recently through the administration of a customer survey. Construct a 95% confidence interval for the proportion of all Wal-Mart customers who prefer Crest toothpaste. Interpret this confidence interval for the buyers at Wal-Mart.

16. The employee benefits manager of a small private university would like to estimate the proportion of full-time employees who prefer adopting the first (i.e., plan A) of three available health care plans in the coming annual enrollment period. A reliable frame of the university's employees and their tentative health care preferences are given in the file P08_25.XLS.
    a. Use Excel to choose a simple random sample of size 45 from the given frame.
    b. Using the sample found in part **a**, construct a 99% confidence interval for the proportion of university employees who prefer plan A. Assume that the population consists of the preferences of all employees in the given frame.
    c. Interpret the 99% confidence interval constructed in part **b**.

17. A market research consultant hired by the Pepsi-Cola Co. is interested in determining the proportion of consumers who favor Pepsi-Cola over Coke Classic in a particular urban location. A random sample of 250 consumers from the market under investigation is provided in P09_17.XLS. Construct a 90% confidence interval for the proportion of all consumers in this market who prefer Pepsi. Interpret this confidence interval for Pepsi-Cola's market researchers.

Level B

18. Continuing Problem 16, select simple random samples of 30 individuals from *each* of the given employee classifications (i.e., administrative staff, support staff, and faculty). Construct a 99% confidence interval for the proportion of employees who prefer adopting plan A for each of the three classifications. Do you see evidence of significant differences among these three interval estimates? Summarize your findings.

19. Continuing Problem 17, separate the given random sample of consumers (provided in the file P09_17.XLS) into two gender subgroups: *males* and *females*. Construct 90% confidence intervals for the proportion of male consumers who prefer Pepsi and the proportion of female consumers who prefer Pepsi. Do you see evidence of a significant difference between the preferences of males and females in this case? Repeat this same process with the *age* attribute of the given consumers. In other words, separate the given sample of consumers by age (i.e., *under 20, between 20 and 40, between 40 and 60,* and *over 60*). Construct a 90% confidence interval for the proportion of consumers in each age category favoring Pepsi. Summarize your findings.

9.6 Confidence Interval for a Standard Deviation[5]

In Section 9.3 we focused primarily on estimation of a population *mean*. Our concern with the population standard deviation σ was in its role as a nuisance parameter. That is, we needed an estimate of σ to estimate the standard error of the sample mean. However, there are cases where the variability in the population, measured by σ, is of interest in its own right. We briefly describe a procedure for obtaining a confidence interval for σ in this section.

The theory is somewhat more complex than for the case of the mean. As you might expect, we use the sample standard deviation s as a point estimate of σ. However, the sampling distribution of s is not symmetric—in particular, it is not the normal distribution or the t distribution. Rather, the appropriate sampling distribution is a right-skewed distribution called the **chi-square** distribution. Like the t distribution, the chi-square distribution has a degrees of freedom parameter, which (for this procedure) is again $n - 1$.

Tables of the chi-square distribution, for selected degrees of freedom, appear in many statistics books, but the necessary information can be obtained more easily with Excel's CHIDIST and CHIINV functions. The CHIDIST function takes the form

$$=\text{CHIDIST}(v,df)$$

This function returns the probability to the right of value v when the degrees of freedom parameter is df. Similarly, the CHIINV function takes the form

$$=\text{CHIINV}(p,df)$$

It returns the value with probability p to the right of it when the degrees of freedom parameter is df.

We will not present the rather complex confidence interval formulas for σ. However, we point out that because of the skewness of the sampling distribution of s, a confidence interval for σ is not centered at s. That is, the confidence interval is *not* the point estimate plus or minus a multiple of a standard error. Instead, s is always closer to the left endpoint of the confidence interval than to the right endpoint, as indicated in Figure 9.11.

The StatPro One-Sample procedure enables us to obtain a confidence interval for a population standard deviation as easily as for a mean. We illustrate this in the following example.

FIGURE 9.11
Confidence Interval for a Standard Deviation

Lower limit   Sample stdev s Upper limit

Example 9.6 Analyzing Variability in Diameters of Machine Parts

A machine produces parts that are supposed to have diameter 10 centimeters. However, due to inherent variability, some diameters are greater than 10 and some are less. The production supervisor is concerned about two things. First, he is concerned that the mean

[5]This section can be omitted without any loss of continuity.

	A	B	C	D	E	F	G	H	I	J
1	Measuring diameters of parts from a production process									
2										
3	Note: Each diameter is measured in cm									
4										
5	Part	Diameter			Results for one-sample analysis for Diameter					
6	1	10.031								
7	2	10.011			Summary measures					
8	3	10.003			Sample size	50				
9	4	10.025			Sample mean	9.996				
10	5	10.048			Sample standard deviation	0.034				
11	6	10.014								
12	7	10.030			Confidence interval for mean					
13	8	10.008			Confidence level	95.0%				
14	9	10.049			Sample mean	9.996				
15	10	9.995			Std error of mean	0.005				
16	11	9.965			Degrees of freedom	49				
17	12	10.003			Lower limit	9.986				
18	13	9.959			Upper limit	10.005				
19	14	10.013								
20	15	10.012			Confidence interval for standard deviation					
21	16	10.005			Confidence level	95.0%				
22	17	9.921			Sample standard deviation	0.034				
23	18	9.930			Degrees of freedom	49				
24	19	9.990			Lower limit	0.029				
25	20	9.948			Upper limit	0.043				
26	21	10.077								
27	22	9.959			Proportion of unusable parts					
28	23	10.000			Maximum deviation for unusability	0.065				
29	24	9.998			Assumed mean	10				
30	25	9.983			Assumed standard deviation	0.043				
31	26	9.995			Proportion unusable	0.131				
32	27	9.917								
33	28	9.934			Two-way data table for finding proportion unusuable as a function of mean and stdev					
34	29	10.044					Assumed standard deviation			
35	30	10.023				0.131	0.029	0.034	0.043	
36	31	9.997			Assumed mean	9.986	0.043	0.077	0.151	
37	32	10.020				9.996	0.026	0.058	0.132	
38	33	9.983				10.005	0.027	0.059	0.133	
39	34	9.998								
54	49	9.973								
55	50	9.970								

FIGURE 9.12 Analysis of Parts Data

diameter might not be 10 centimeters. Second, he is worried about the extent of variability in the diameters. Even if the mean is on target, excessive variability implies that many of the parts will fail to meet specifications. To analyze the process, he randomly samples 50 parts during the course of a day and measures the diameter of each part to the nearest millimeter. The results are shown in columns A and B of Figure 9.12. (See the file PARTS.XLS.) Should he be concerned about the results from this sample?

Objective To use StatPro's one-sample procedure to find a confidence interval for the standard deviation of part diameters, and to see how variability affects the proportion of unusable parts produced.

Solution

Because the manager is concerned about the mean *and* the standard deviation of diameters, we obtain 95% confidence intervals for both. This is easy to do with StatPro's One-Sample procedure. We go through the same dialog boxes as before, except that we now check the boxes for both confidence interval options—mean and standard deviation. The top part of the output in Figure 9.12 provides a 95% confidence interval for the mean. This confidence interval extends from 9.986 cm to 10.005 cm. Therefore, there is probably not too much cause for concern about the mean. The supervisor can be fairly confident that the mean diameter of all parts is close to 10 cm.

The bottom part of the output provides a 95% confidence interval for the standard deviation of diameters. This interval extends from 0.029 cm to 0.043 cm.[6] Is this good news or bad news? It depends. Let's say that a part is unusable if its diameter is more than 0.065 cm from the target. Let's also assume the true mean is right on target and that the standard deviation is at the *upper* end of the confidence interval, that is, $\sigma = 0.043$ cm. Finally, we assume that the population distribution of diameters is normal. Then the calculation in cell F31 shows that 13.1% of the parts will be unusable! The formula in cell F31 is

$$=\text{NORMDIST}(10\text{-}F28,F29,F30,1)+(1\text{-}\text{NORMDIST}(10+F28,F29,F30,1))$$

It adds the normal probabilities of being below or above the usable range.

To pursue this analysis one step further, we form a two-way data table in the range F35:I38. The assumed means we use in column F are the lower confidence limit, the sample mean, and the upper confidence limit. Similarly, the assumed standard deviations in row 35 are the lower confidence limit, the sample standard deviation, and the upper confidence limit. To form the table, enter the formula =F31 in cell F35, highlight the range F35:I38, use the Data/Table menu item, and enter F30 and F29 as the row and column input cells.

Each value in the body of the data table is the resulting proportion of unusable parts. Obviously, a mean close to the target and a small standard deviation are best, but even this best-case scenario results in 2.6% unusable parts (see cell G37). However, a mean off target and a large standard deviation can lead to as many as 15.1% unusable parts (see cell I36). In any case, the message for the supervisor should be clear—he must work to reduce the underlying variability in the process. This variability is hurting him much more than an off-target mean. ●

PROBLEMS

Level A

20. Consider a frame consisting of 500 households in a middle-class neighborhood that was the recent focus of an economic development study conducted by the local government. Specifically, for each of the 500 households, information was gathered on the total indebtedness (excluding the value of a home mortgage) of the household and each of several other variables. The data are in the file P02_06.XLS.
   a. Use Excel to choose a simple random sample of size 25 from this frame.
   b. Using the sample generated in part **a,** construct a 95% confidence interval for the standard deviation of the total indebtedness of all households in the given neighborhood.
   c. Interpret the 95% confidence interval constructed in part **b.**

   d. Does the 95% confidence interval contain the actual value of the population standard deviation in this case? If not, explain why not. What proportion of many similarly constructed confidence intervals should include the true population standard deviation?

21. Senior management of a certain consulting services firm is concerned about a growing decline in the organization's weekly number of billable hours. Ideally, the organization expects each professional employee to spend *at least* 40 hours per week on work. In an effort to understand this problem better, management would like to estimate the standard deviation of the number of hours their employees spend on work-related activities in a typical week. The frame of virtually all of the firm's full-time employees, including the employees' self-reported amounts of time typically

[6]You can check the spreadsheet formulas for the confidence interval limits to see how they use the CHIINV function.

devoted to work activities each week, is provided in the file P08_35.XLS.

a. Select a simple random sample of size 100 from the given frame.

b. Using the sample generated in part **a,** construct a 99% confidence interval for the standard deviation of the number of hours this organization's employees spend on work-related activities in a typical week.

c. Given the target range of 40–60 hours of work per week, should senior management be concerned about the number of hours their employees are currently devoting to work? Explain why or why not.

Level B

22. A manufacturing company's quality control personnel have recorded the proportion of defective items for each of 500 randomly selected shipments of one of the computer components that the company produces. The data are in the file P09_22.XLS. The quality control department manager would like to use this random sample to estimate the mean and standard deviation of the proportion of defective items in the company's shipments. She is concerned that some shipments of this computer component contain an unacceptably high proportion of defective items. In particular, her company cannot tolerate a defective rate higher than 5% for any of its shipments.

a. Use the given random sample to construct 95% confidence intervals for the mean and standard deviation of the proportion of defective items in the company's shipments. Interpret each of these interval estimates.

b. Assuming the proportion of defective computer components in a given shipment is *normally* distributed, what is the probability that a randomly selected shipment will be unacceptable? Based on information derived from the confidence intervals constructed in part **a,** compute this probability for various combinations of the mean and standard deviation. You might want to generate a two-way data table to compute this probability for various combinations of the mean and standard deviation of the defective rate in the company's shipments.

 9.7 Confidence Interval for the Difference Between Means

One of the most important applications of statistical inference is the comparison of two population means. There are many applications to business, as follows.

Applications of Statistical Inference to Business

- Men and women shop at a retail clothing store. The manager would like to know how much more (or less), on average, a woman spends on a typical purchase occasion than a man.

- Two airline companies fly similar routes. A consumer organization would like to check how much the average delay differs between the two airlines, where delay is defined as the actual arrival time at the destination minus the scheduled arrival time.

Statisticians call these general types of problems "comparison problems." They are among the most important types of problems attacked by statistical methods.

- A supermarket chain mails coupons for various products to its customers in one city. Its customers in another city receive no such coupons. The chain would like to check how much the average amount spent on these products differs between the two sets of customers over the next couple of months.

- A computer company has a customer service center that responds to customers' questions and complaints. The center employs two types of people: those who have had a recent course in dealing with customers (but little actual experience) and those with a lot of experience dealing with customers (but no formal course). The company would like to know how these two types of employees differ with respect to the average number of customer complaints of poor service in the last 6 months.

- A consulting company hires business students directly out of undergraduate school. The new hires all take a problem-solving test. They then go through an intensive 3-month training program, after which they take another similar problem-solving test.

The company wants to know how much the average test score improves after the training program.

- A car dealership often deals with husband–wife pairs shopping for cars. To check whether husbands react differently than their wives to the sales presentation, husbands and wives are asked (separately) to rate the quality of the sales presentation. The dealership wants to know how much husbands differ from their wives in terms of average ratings.

Each of these examples deals with a difference between means from two populations. However, the first four examples differ in one important respect from the last two. In the last two examples there is a natural *pairing* across the two samples. In the first of these, each employee takes a test before a course and then a test after the course, so that each employee is naturally paired with himself or herself. In the final example husbands and wives are naturally paired with one another. There is no such pairing in the first four examples. Instead, we assume that the samples in these examples are chosen *independently* of one another. For statistical reasons we need to distinguish these two cases, independent samples and paired samples, in the discussion that follows.

9.7.1 Independent Samples

The framework for this situation is the following. We are interested in some quantity, such as dollars spent or airplane delay, for each of two populations. The population means are μ_1 and μ_2, and the population standard deviations are σ_1 and σ_2. We take random samples of sizes n_1 and n_2 from the populations to estimate the difference between means, $\mu_1 - \mu_2$. A point estimate of this difference is the natural one, the difference between sample means, $\overline{X}_1 - \overline{X}_2$. Starting with this estimate, we want to form a confidence interval for the unknown population mean difference, $\mu_1 - \mu_2$.

It turns out that the appropriate sampling distribution of this estimate is again the t distribution, now with $n_1 + n_2 - 2$ degrees of freedom.[7] Therefore, a confidence interval for $\mu_1 - \mu_2$ is given by expression (9.13). The t-multiple is found as usual with the TINV function. It is the value that cuts off the appropriate probability (depending on the confidence level) in each tail of the t distribution with $n_1 + n_2 - 2$ degrees of freedom. For example, if the confidence level is 95% and $n_1 = n_2 = 30$, then the appropriate t-multiple is 2.002, found in Excel with the function TINV(0.05,58).

Confidence Interval for Difference Between Means	$$\overline{X}_1 - \overline{X}_2 \pm t\text{-multiple} \times \text{SE}(\overline{X}_1 - \overline{X}_2)$$	**(9.13)**

The standard error, $\text{SE}(\overline{X}_1 - \overline{X}_2)$, is more involved. We must first make the assumption that the population standard deviations are equal, that is, $\sigma_1 = \sigma_2$. (We will shortly present an alternative procedure to use when it is clear that the population standard deviations are *not* equal.) Then an estimate of this common standard deviation is provided by the "pooled" estimate from both samples, labeled s_p:

---

[7]This assumes that either the population distributions are normal or that the sample sizes are reasonably large, conditions that are at least approximately met in a wide variety of applications.

| Pooled Estimate of Common Standard Deviation | $$s_p = \sqrt{\dfrac{(n_1 - 1)s_1^2 + (n_2 - 1)s_2^2}{n_1 + n_2 - 2}}$$ |

Here, s_1 and s_2 are the sample standard deviations from the two samples. This pooled estimate is somewhere between s_1 and s_2, with the relative sample sizes determining its exact value. Then the standard error of $\bar{X}_1 - \bar{X}_2$ is given by equation (9.14):

| Standard Error of Difference Between Sample Means | $$\mathrm{SE}(\bar{X}_1 - \bar{X}_2) = s_p\sqrt{\dfrac{1}{n_1} + \dfrac{1}{n_2}}$$ | (9.14) |

Fortunately, the StatPro Two-Sample procedure takes care of all these calculations, as illustrated in the following example.

Example 9.7 Reliability of Treadmill Motors at the SureStep Company

The SureStep Company manufactures high-quality treadmills for use in exercise clubs. SureStep currently purchases its motors for these treadmills from supplier A. However, it is considering a change to supplier B, which offers a slightly lower cost. The only question is whether supplier B's motors are as reliable as supplier A's. To check this, SureStep installs motors from supplier A on 30 of its treadmills and motors from supplier B on another 30 of its treadmills. It then runs these treadmills under typical conditions and, for each treadmill, records the number of hours until the motor fails. The data from this experiment appear in Figure 9.13. (See the file MOTORS.XLS.) What can SureStep conclude?

FIGURE 9.13

Analysis of Treadmill Motors Data

	A	B	C	D	E	F	G
1	**Differences between treadmill motors**						
2							
3	Sample data (hours until motor fails)						
4							
5	Supplier A	Supplier B			*Two-sample analysis for Supplier A minus Supplier B*		
6	1358	658					
7	793	404			*Summary stats for two samples*		
8	587	735				Supplier A	Supplier B
9	608	457			Sample sizes	30	30
10	472	431			Sample means	748.800	655.667
11	562	658			Sample standard deviations	283.881	259.986
12	879	453					
13	575	488			*Confidence interval for difference between means*		
14	1293	522			Confidence level	95.0%	
15	1457	1247			Sample mean difference	93.133	
16	705	1095			Pooled standard deviation	272.196	NA
17	623	430			Std error of difference	70.281	70.281
18	725	726			Degrees of freedom	58	58
19	569	793			Lower limit	-47.549	-47.549
20	424	498			Upper limit	233.815	233.815
21	436	502					
22	1250	589			*Test of equality of variances*		
23	493	975			Ratio of sample variances	1.192	
24	485	808			p-value	0.319	
25	462	456					
35	666	507					

FIGURE 9.14
Boxplots for
Treadmill Motors
Data

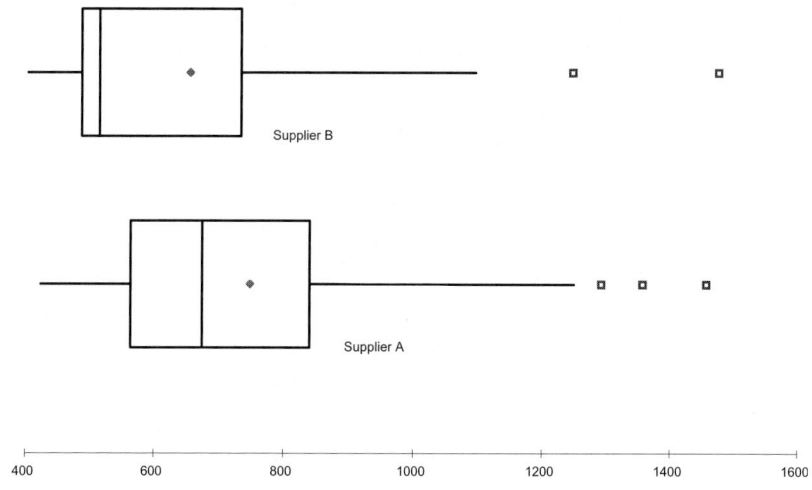

Objective To use StatPro's two-sample procedure to find a confidence interval for the difference between mean lifetimes of motors, and to see how this confidence interval can help SureStep choose the better supplier.

Solution

In any comparison problem it is a good idea to look initially at side-by-side boxplots of the two samples. These appear in Figure 9.14. These show that (1) the distributions of times until failure are skewed to the right for each supplier, (2) the mean for supplier A is somewhat greater than the mean for supplier B, and (3) there are several mild outliers. There seems to be little doubt that supplier A's motors will last longer on average than supplier B's—or is there? A confidence interval for the mean difference allows us to see whether the differences apparent in the boxplots can be generalized to *all* motors from the two suppliers.

We find this confidence interval by using the StatPro Two-Sample procedure. To do so, place the cursor anywhere in the data set, use the StatPro/Statistical Inference/ Two-Sample Analysis menu item, select the Unstacked option, select Supplier A and Supplier B as the variables to analyze, and indicate that you want a 95% confidence interval. We obtain the output in Figure 9.13. The top part of the output summarizes the two samples. It shows that the sample means differ by approximately 93 hours and that the sample standard deviations are of roughly the same magnitude.

Excel Tip *The data are "unstacked" because there are separate columns for supplier A's times and supplier B's times. StatPro's messages indicate the difference between stacked and unstacked data.*

The confidence interval calculations appear in the range F14:F20. Here we see that the difference between sample means is 93.133 hours, the pooled estimate of the common population standard deviation is 272.196 hours, the standard error of the sample mean difference is 70.281 hours, and a 95% confidence interval for the mean difference extends from −47.549 to 233.815 hours. Not only is this interval quite wide, but it extends from a negative value to a positive value. If SureStep had to make a guess, it would say that supplier A's motors last longer on average than supplier B's. But because of the negative part of the confidence interval, there is still a possibility that the opposite is true.

Should SureStep continue with supplier A? This depends on the trade-off between the cost of the motors and warranty costs (and any other relevant costs). Because the warranty probably depends on whether a motor lasts a certain amount of time, warranty costs probably depend on a *proportion* (the proportion that fail before 500 hours, say) rather than a mean. Therefore, we will postpone further discussion of this issue until we discuss differences between proportions in Section 9.8. ●

Equal-Variance Assumption

This two-sample analysis makes the strong assumption that the standard deviations (or variances) from the two populations are equal. How can we tell whether they are equal, and what do we do if they are not equal?

To check whether they are equal, we should first look at the two sample standard deviations. If they are of widely different magnitudes, this certainly casts doubt on the equal-variance assumption. The sample standard deviations in the treadmill example, 283.881 and 259.986, are of similar magnitudes and present no clear evidence of unequal population variances. A statistical test for equality of two population variances is automatically shown at the bottom of the StatPro Two-Sample output. Because we have not yet discussed hypothesis testing, however, we will postpone discussion of this test for now. Suffice it to say that it also presents no evidence of unequal variances for this example.

If we do have reason to believe that the population variances are unequal, then a slightly different procedure can be used to calculate a confidence interval for the difference between means. The appropriate standard error of $\overline{X}_1 - \overline{X}_2$ is now

$$\text{SE}(\overline{X}_1 - \overline{X}_2) = \sqrt{s_1^2/n_1 + s_2^2/n_2}$$

and the degrees of freedom used to find the t-multiple is given by a complex expression that we will not present here.

StatPro's Two-Sample procedure automatically calculates the confidence interval for this unequal-variance case. For the treadmill example you can see the results in the range G17:G20 of Figure 9.13. In this example they are exactly the same as the results (in column F) when we make the equal-variance assumption. This is a consequence of the equal sample sizes and roughly equal sample variances. In general, the two results will differ appreciably only when the sample sizes *and* the sample variances differ considerably across samples. In any case, the appropriate results to use are those on the right (column G) if we have reason to suspect unequal population variances and those on the left (column F) otherwise.

Excel Tip *StatPro always calculates the results in both columns. When they are nearly the same, as they often are, it makes no practical difference which you quote.*

We next revisit the R&P Supermarket data in Example 3.10 from Chapter 3. We again make a comparison between two means, this time the mean number of customers left in line during rush times versus normal times. There are two objectives in this example. First, it provides one more illustration of the two-sample procedure, now with unequal sample sizes. Perhaps more importantly, it illustrates that not all data sets come "ready-made" for performing a particular analysis. We have to do some data manipulation before we can invoke StatPro's Two-Sample procedure. Indeed, this is often the most time-consuming (and sometimes frustrating) part of statistical analysis in real applications—getting the data ready for the analysis.

Example 9.8 Analyzing Customer Waiting at R&P Supermarket

As in Example 3.10, the manager of the R&P Supermarket has collected a week's worth of data on customer arrivals, departures, and waiting. There are 48 observations per day, each taken at the end of a half-hour period. The data appear in the file SUPERMKT.XLS. The various times of day are listed in the TimeInterval variable. (See Figure 9.15.) They include Morning Rush, Morning, Lunch Rush, Afternoon, Afternoon Rush, Evening, and Night. (The comment in cell C3 explains exactly which time intervals these refer to.) There is also a variable, EndWaiting, that records the number of customers still being served or waiting in line at the end of each half-hour period.

The manager would like to check whether the average value of EndWaiting differs during rush periods from normal, non-night periods. She is concerned that there might be excessive waiting during rush periods, in which case she might need to add more checkout people during these times. She plans to exclude the night period from the analysis because she knows from experience that customers very seldom need to wait during the night.

Objective To use StatPro's Two-Sample procedure to find a confidence interval for the difference between mean waiting times during the supermarket's rush periods versus its normal periods.

Solution

Starting with the data set in its original form, we need to perform two main steps:

1 Rename the seven time intervals (Morning rush, Morning, and so on) so that there are only three: Rush, Normal, and Night.

2 Perform the statistical comparison between the EndWaiting variables for the Rush and Normal periods.

	A	B	C	D	E	F	G	H	I
1	Supermarket checkout efficiency								
2									
3	Day	StartTime	TimeInterval	InitialWaiting	Arrivals	Departures	EndWaiting	Checkers	TotalCustomers
4	Mon	8:00 AM	Morning rush	2	21	22	1	3	23
5	Mon	8:30 AM	Morning rush	1	25	18	8	3	26
6	Mon	9:00 AM	Morning	8	27	28	7	3	35
7	Mon	9:30 AM	Morning	7	21	23	5	3	28
8	Mon	10:00 AM	Morning	5	20	23	2	5	25
9	Mon	10:30 AM	Morning	2	36	31	7	5	38
10	Mon	11:00 AM	Morning	7	30	36	1	5	37
11	Mon	11:30 AM	Lunch rush	1	34	29	6	5	35
12	Mon	12:00 PM	Lunch rush	6	56	48	14	7	62
13	Mon	12:30 PM	Lunch rush	14	58	64	8	7	72
14	Mon	1:00 PM	Lunch rush	8	53	52	9	7	61
15	Mon	1:30 PM	Afternoon	9	30	36	3	5	39
16	Mon	2:00 PM	Afternoon	3	34	31	6	5	37
17	Mon	2:30 PM	Afternoon	6	36	37	5	5	42
18	Mon	3:00 PM	Afternoon	5	30	28	7	5	35
19	Mon	3:30 PM	Afternoon	7	29	34	2	5	36
20	Mon	4:00 PM	Afternoon	2	35	33	4	5	37
21	Mon	4:30 PM	Afternoon rush	4	32	25	11	5	36

FIGURE 9.15 Original Data for Supermarket Example

	A	B	C	D	E	F	G	H	I
1	Supermarket checkout efficiency								
2									
3	Day	StartTime	TimeInterval	InitialWaiting	Arrivals	Departures	EndWaiting	Checkers	TotalCustomers
4	Mon	8:00 AM	Rush	2	21	22	1	3	23
5	Mon	8:30 AM	Rush	1	25	18	8	3	26
6	Mon	9:00 AM	Normal	8	27	28	7	3	35
7	Mon	9:30 AM	Normal	7	21	23	5	3	28
8	Mon	10:00 AM	Normal	5	20	23	2	5	25
9	Mon	10:30 AM	Normal	2	36	31	7	5	38
10	Mon	11:00 AM	Normal	7	30	36	1	5	37
11	Mon	11:30 AM	Rush	1	34	29	6	5	35
12	Mon	12:00 PM	Rush	6	56	48	14	7	62
13	Mon	12:30 PM	Rush	14	58	64	8	7	72
14	Mon	1:00 PM	Rush	8	53	52	9	7	61
15	Mon	1:30 PM	Normal	9	30	36	3	5	39
16	Mon	2:00 PM	Normal	3	34	31	6	5	37
17	Mon	2:30 PM	Normal	6	36	37	5	5	42
18	Mon	3:00 PM	Normal	5	30	28	7	5	35
19	Mon	3:30 PM	Normal	7	29	34	2	5	36
20	Mon	4:00 PM	Normal	2	35	33	4	5	37

FIGURE 9.16 Supermarket Data with Time Categories Renamed

The finished version of the SUPERMKT.XLS file contains the results of step 1 in the NewData sheet and the results of step 2 in the Analysis sheet. If you want to follow along, hands-on, with the step-by-step procedure, you should use the "data only" version of the SUPERMKT.XLS file and perform the following steps.

1 **Copy sheet:** Create a copy of the OriginalData sheet by pressing the Ctrl key and dragging the OriginalData sheet tab to the right. Double-click on the new sheet tab and rename it NewData.

2 **Rename time intervals:** To rename the time intervals on the NewData sheet, use Excel's Find and Replace feature. Click on column C's tab to select the entire column, then select the Edit/Replace menu item. Type **Morning rush** in the "Find what:" box, type **Rush** in the "Replace with:" box, and click on the Replace All button. Repeat this for the other time intervals to be renamed. That is, replace Lunch rush and Afternoon rush by Rush, and replace Morning, Afternoon, and Evening by Normal. Figure 9.16 shows some of the results.

3 **Create boxplots:** Use StatPro's Boxplot procedure to create side-by-side boxplots of the EndWaiting variable. Select the Stacked option, and select TimeInterval as the code variable. (See Figure 9.17.)

4 **Perform Two-Sample Analysis:** Select the StatPro/Statistical Inference/Two-Sample Analysis menu item from the NewData sheet. Then select the Stacked option, select TimeInterval as the code variable, select Normal and Rush as the two categories to analyze, select EndWaiting as the measurement variable, and ask for a 95% confidence interval. (See Figure 9.18.)

Excel Tip *With the Stacked option, StatPro always asks you if it is OK to sort the data on the code variable. You should respond "Yes."*

The side-by-side boxplots in Figure 9.17 show that (1) the distribution of EndWaiting is definitely skewed to the right for each time interval, with a number of outliers, and (2) the mean value of EndWaiting is slightly larger for Rush than for Normal, with Night a distant third. Given the nature of the data, it should not really be surprising that the data are skewed to the right with a number of outliers. When the supermarket gets busy, wait-

FIGURE 9.17
Boxplots for
Supermarket
Example

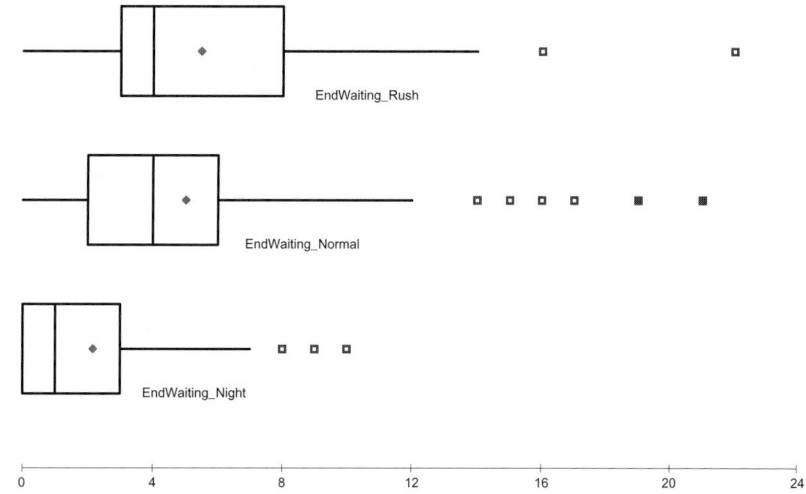

ing lines can really build. All it takes is one or two really long checkout times to produce an excessively large value of EndWaiting, and this is evidently what happened at R&P.

The output from the two-sample procedure appears in Figure 9.18. The sample means of EndWaiting are 5.480 and 5.014 for the Rush and Normal periods, the sample standard deviations are 4.284 and 4.293, and these are based on sample sizes of 98 and 140 half-hour periods. These summary statistics provide some evidence of a difference between population means but very little evidence of different population variances. This latter statement means that we can use the results in column M, not column N (although they are practically identical). We see that a point estimate for the mean difference (Normal minus Rush) is -0.465 and that a 95% confidence interval for this mean difference extends from -1.578 to 0.648.

What can the manager conclude from this analysis? Should she add extra checkout people during rush periods? This is difficult to answer because it obviously involves a trade-off between the cost of extra checkout people and the "cost" of making customers wait in line. Also, we have no way of knowing, at least not from the present analysis, how

FIGURE 9.18
Analysis of Supermarket Data

	K	L	M	N
3		**Two-sample analysis for EndWaiting_Normal minus EndWaiting_Rush**		
4				
5		**Summary stats for two samples**		
6			EndWaiting_Normal	EndWaiting_Rush
7		Sample sizes	140	98
8		Sample means	5.014	5.480
9		Sample standard deviations	4.293	4.284
10				
11		**Confidence interval for difference between means**		
12		Confidence level	95.0%	
13		Sample mean difference	-0.465	
14		Pooled standard deviation	4.290	NA
15		Std error of difference	0.565	0.565
16		Degrees of freedom	236	209
17		Lower limit	-1.578	-1.579
18		Upper limit	0.648	0.648
19				
20		**Test of equality of variances**		
21		Ratio of sample variances	1.004	
22		p-value	0.496	

much effect extra checkout people would have on waiting. However, the manager does know from this analysis that the mean difference between rush and normal periods is rather minor. Specifically, because the confidence interval extends from a negative value to a positive value, there is a good possibility that the *true* mean difference could be *positive*. That is, the mean for normal times could be *larger* than the mean for rush times. Therefore, the results of this analysis do not provide a strong incentive for the manager to change the current system. ●

9.7.2 Paired Samples

When the samples we want to compare are paired in some natural way, such as a pretest or posttest for each person or husband–wife pairs, there is a more appropriate form of analysis than the two-sample procedure we've been discussing. Consider the example where each new employee takes a test, then receives a 3-month training course, and finally takes another similar test. There is likely to be a fairly strong correlation between the pretest and posttest scores. Employees who score relatively low on the first test are likely to score relatively low on the second test, and employees who score relatively high on the first test are likely to score relatively high on the second test. The two-sample procedure does not take this correlation into account and therefore ignores important information. The paired procedure described in this section, on the other hand, uses this information to advantage.

The procedure itself is very straightforward. We do not directly analyze two separate variables (pretest scores and posttest scores, say); we analyze their *differences*. For each pair in the sample, we calculate the difference between the two scores for the pair. Then we perform a *one*-sample analysis, as in Section 9.3, on these differences. Actually, Stat-Pro's Paired-Sample procedure does the differencing *and* the ensuing one-sample analysis automatically, as described in the following example.

Example (9.9) Husband and Wife Reactions to Sales Presentations at Stevens Honda–Olds

The Stevens Honda–Olds automobile dealership often sells to husband–wife pairs. The manager would like to check whether the sales presentation is viewed any more or less favorably by the husbands than the wives. If it is, then some new training might be recommended for its salespeople. To check for differences, a random sample of husbands and wives are asked (separately) to rate the sales presentation on a scale of 1 to 10, 10 being the most favorable rating. The results appear in Figure 9.19. (See the AUTO.XLS file.) What can the manager conclude from these data?

Objective To use StatPro's Paired-Sample procedure to find a confidence interval for the mean difference between husbands' and wives' ratings of sales presentations.

Solution

We illustrate two ways to perform the analysis. Normally, we would only use the second of these, but the first sheds some light on the procedure. For the first method, make a copy of the Data sheet and call it OneSample. Then manually form a new variable in column D called Difference by entering the formula

$$=\text{B4-C4}$$

FIGURE 9.19
Data for Sales
Presentation
Example

	A	B	C
1	**Sales presentation ratings**		
2			
3	Pair	Husband	Wife
4	1	6	3
5	2	7	8
6	3	8	5
7	4	6	4
8	5	8	5
9	6	7	6
10	7	8	5
11	8	6	7
12	9	7	8
33	30	7	3
34	31	7	5
35	32	5	1
36	33	7	5
37	34	7	4
38	35	10	5

in cell D4 and copying it down column D. (See Figure 9.20.) This new variable is, for each couple, the husband's rating minus the wife's rating. Next, with the cursor anywhere in the resulting data set, select the StatPro/Statistical Inference/One-Sample Analysis menu item, select the variable Difference for analysis, and request a 95% confidence interval. This produces the output shown in Figure 9.20. We see that the sample mean Husband minus Wife difference is 1.629 and that a 95% confidence interval for this difference extends from 1.057 to 2.200. The standard error in cell H13, 0.281, refers to the standard error of the sample mean *difference*.

To perform this analysis more efficiently, again make a copy of the Data sheet and called it Paired. Then use the StatPro/Statistical Inference/Paired-Sample Analysis menu item, select the Husband and Wife variables for analysis, and accept the defaults in the other dialog boxes. We obtain the output in Figure 9.21 (page 450). The results are exactly the same as before. This is because StatPro's Paired-Sample procedure performs a one-sample analysis on the differences—and it saves you the work of creating the differences.

Figure 9.22 shows side-by-side boxplots of the husband and wife scores. These boxplots are not as useful here as in the two-sample procedure because we lose sight of which

	A	B	C	D	E	F	G	H	I
1	**Sales presentation ratings**								
2									
3	Pair	Husband	Wife	Difference			**Results for one-sample analysis for Difference**		
4	1	6	3	3					
5	2	7	8	-1			**Summary measures**		
6	3	8	5	3			Sample size	35	
7	4	6	4	2			Sample mean	1.629	
8	5	8	5	3			Sample standard deviation	1.664	
9	6	7	6	1					
10	7	8	5	3			**Confidence interval for mean**		
11	8	6	7	-1			Confidence level	95.0%	
12	9	7	8	-1			Sample mean	1.629	
13	10	7	5	2			Std error of mean	0.281	
14	11	6	3	3			Degrees of freedom	34	
15	12	5	4	1			Lower limit	1.057	
16	13	8	5	3			Upper limit	2.200	
17	14	7	8	-1					
37	34	7	4	3					
38	35	10	5	5					

FIGURE 9.20 One-Sample Analysis of Differences for Automobile Data

	A	B	C	D	E	F	G	H
1	Sales presentation ratings							
2								
3	Pair	Husband	Wife	Husband-Wife			**Paired-sample analysis for Husband minus Wife**	
4	1	6	3	3				
5	2	7	8	-1			**Summary measures for Husband-Wife**	
6	3	8	5	3			Sample size	35
7	4	6	4	2			Sample mean	1.629
8	5	8	5	3			Sample standard deviation	1.664
9	6	7	6	1				
10	7	8	5	3			**Confidence interval for mean**	
11	8	6	7	-1			Confidence level	95.0%
12	9	7	8	-1			Sample mean	1.629
13	10	7	5	2			Std error of mean	0.281
14	11	6	3	3			Degrees of freedom	34
15	12	5	4	1			Lower limit	1.057
16	13	8	5	3			Upper limit	2.200
17	14	7	8	-1				
37	34	7	4	3				
38	35	10	5	5				

FIGURE 9.21 Paired-Sample Analysis of Automobile Data

husbands are paired with which wives. A more useful boxplot is of the differences, shown in Figure 9.23. Here we see that the sample mean difference is positive, but even more importantly, we see that the vast majority of husband scores are greater than the corresponding wife scores. There is little doubt that most husbands tend to react more favorably to the sales presentations than their wives. Perhaps the salespeople need to be somewhat more sensitive to their female customers!

Before leaving this example, let's see what would have happened if we had used the two-sample procedure on the Husband and Wife variables. The results appear in Figure 9.24. Because there is a considerable difference between the sample standard deviations, we should probably use the confidence interval output in column H, not column G, although there is not much difference between the two. The important point is that the resulting confidence interval for the mean difference extends from 0.895 to 2.362, which is somewhat *wider* than the confidence interval from the paired-sample procedure. This is typical. When we use the two-sample procedure in a situation where the paired-sample procedure is more appropriate, we do not use the data as efficiently. The effect is that the

FIGURE 9.22
Side-by-side Boxplots for Automobile Data

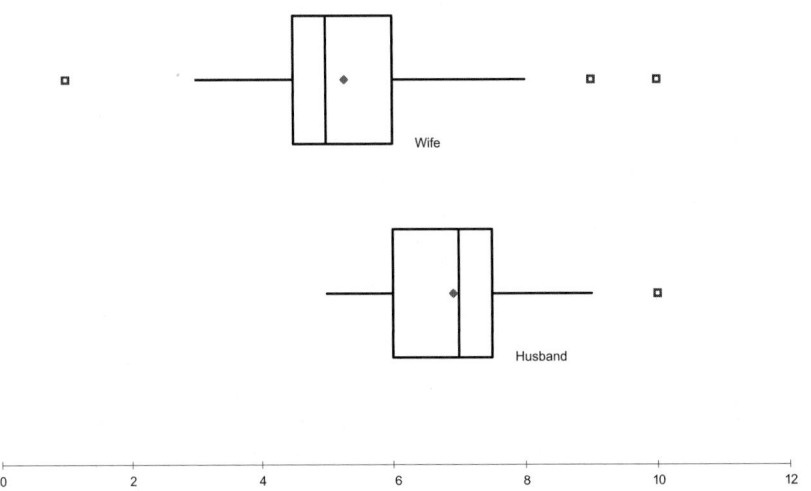

FIGURE 9.23
Single Boxplot of
Differences for
Automobile Data

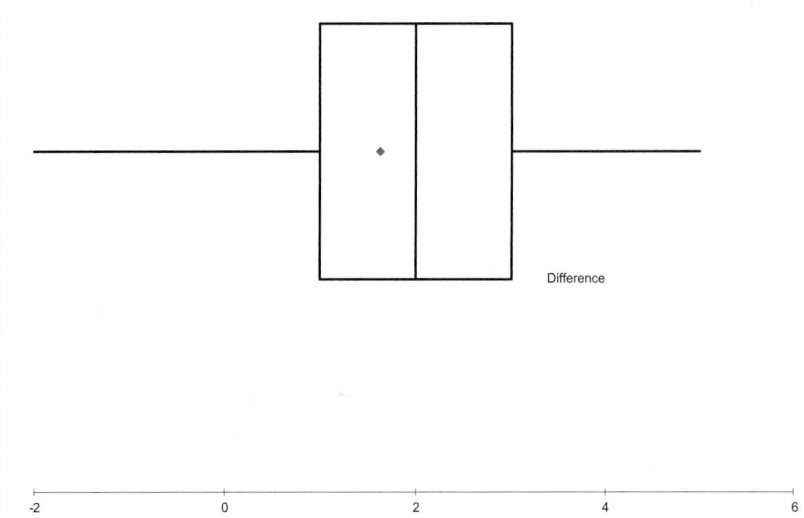

FIGURE 9.23
Single Boxplot of
Differences for
Automobile Data

	A	B	C	D	E	F	G	H
1	**Sales presentation ratings**							
2								
3	Pair	Husband	Wife			**Two-sample analysis for Husband minus Wife**		
4	1	6	3					
5	2	7	8			**Summary stats for two samples**		
6	3	8	5				Husband	Wife
7	4	6	4			Sample sizes	35	35
8	5	8	5			Sample means	6.914	5.286
9	6	7	6			Sample standard deviations	1.222	1.792
10	7	8	5					
11	8	6	7			**Confidence interval for difference between means**		
12	9	7	8			Confidence level	95.0%	
13	10	7	5			Sample mean difference	1.629	
14	11	6	3			Pooled standard deviation	1.533	NA
15	12	5	4			Std error of difference	0.367	0.367
16	13	8	5			Degrees of freedom	68	60
17	14	7	8			Lower limit	0.897	0.895
18	15	7	5			Upper limit	2.360	2.362
19	16	7	6					
20	17	6	5			**Test of equality of variances**		
21	18	5	4			Ratio of sample variances	2.151	
22	19	6	5			p-value	0.014	
23	20	9	10					
37	34	7	4					
38	35	10	5					

FIGURE 9.24 Two-Sample Analysis of Automobile Data

standard error of the difference tends to be larger, and the resulting confidence interval tends to be wider.

Why is the paired-sample procedure appropriate here? It is *not* just because husbands and wives naturally come in pairs. It is because they tend to react similarly to one another. You can check that the correlation between the husbands' scores and their wives' scores is 0.442. (Use Excel's CORREL function on the Husband and Wife variables.) This is far from perfect correlation, but it is large enough to warrant using the paired-sample procedure. ●

In general, the paired-sample procedure is appropriate when the samples are naturally paired in some way *and* there is a reasonably large positive correlation between the pairs. In this case the paired-sample procedure makes more efficient use of the data and generally results in narrower confidence intervals.

PROBLEMS

Level A

23. The director of a university's career development center is interested in comparing the starting annual salaries of male and female students who recently graduated from the university and commenced full-time employment. The director has formed pairs of male and female graduates with the same major and similar grade-point averages. Specifically, she has collected a random sample of 50 such pairs and has recorded the starting annual salary of each person. These data are provided in the file P09_23.XLS. Construct a 99% confidence interval for the mean difference between similar male and female graduates of this university. Interpret your result.

24. A real estate agent has collected a random sample of 75 houses that were recently sold in a suburban community. She is particularly interested in comparing the appraised value and recent selling price of the houses in this particular market. The values of these two variables for each of the 75 randomly chosen houses are provided in the file P09_24.XLS. Using the sample data, generate a 95% confidence interval for the mean difference between the appraised values and selling prices of the houses sold in this suburban community. Interpret the constructed interval estimate for the real estate agent.

Level B

25. A company employs two shifts of workers. Each shift produces a type of gasket where the thickness is the critical dimension. The average thickness and the standard deviation of thickness for shift 1, based on a random sample of 30 gaskets, are 10.53 mm and 0.14 mm. The similar figures for shift 2, based on a random sample of 25 gaskets, are 10.55 mm and 0.17 mm. Let $\mu_1 - \mu_2$ be the mean difference in thickness between shifts 1 and 2.
   a. Find a 95% confidence interval for $\mu_1 - \mu_2$.
   b. Based on your answer to part **a**, are you convinced that the gaskets from shift 2 are, on average, wider than those from shift 1? Why or why not?
   c. How would your answers to parts **a** and **b** change if the sample sizes were instead 300 and 250?

26. Consider a random sample of 100 households from a middle-class neighborhood that was the recent focus of an economic development study conducted by the local government. Specifically, for each of the 100 households, information was gathered on each of the following variables: family size, location of the household within the neighborhood, an indication of whether those surveyed owned or rented their home, gross annual income of the first household wage earner, gross annual income of the second household wage earner (if applicable), monthly home mortgage or rent payment, average monthly expenditure on utilities, and the total indebtedness (excluding the value of a home mortgage) of the household. The data are in the file P09_26.XLS.
   a. Separate the households in the sample by the *location* of their residence within the given community. For each of the four locations, use the sample information to generate a 90% confidence interval for the mean annual income of all relevant first household wage earners. Compare these four interval estimates. You might also consider generating boxplots of the primary wage earner variable for households in each of the four given locations.
   b. Generate a 90% confidence interval for the difference between the mean annual income levels of the first household wage earners in the first (i.e., SW) and second (i.e., NW) sectors of this community. Generate similar 90% confidence intervals for the differences between the mean annual income levels of primary wage earners from all other pairs of locations (i.e., first and third, first and fourth, second and third, second and fourth, and third and fourth). Summarize your findings.

27. Given data in the file P02_13.XLS from a recent survey of chief executive officers of the nation's biggest businesses (*The Wall Street Journal,* April 9, 1998), construct 95% confidence intervals for the differences in the mean levels of 1997 salaries awarded to executives from *Technology* companies and each of the other seven company types. For instance, construct a 95% confidence interval for the difference in the mean 1997 salaries of executives from *Technology* and *Basic Materials* companies. What conclusions can you draw from the seven 95% confidence intervals you have generated?

9.8 Confidence Interval for the Difference Between Proportions

The final confidence interval we examine is a confidence interval for the difference between two population proportions. As in the previous section, this "comparison" procedure finds many real applications. Several potential business applications are the following:

Applications of Confidence Interval Comparisons to Business

- When an appliance store is about to run a sale, it sometimes sends selected customers a mailing to notify them of the sale. On other occasions it includes a coupon for 5% off the sale price in these mailings. The store's manager would like to know whether the inclusion of coupons affects the proportion of customers who respond.

- A manufacturing company has two plants that produce identical products. The company wants to know how much the proportion of out-of-spec products differs across the two plants.

- A pharmaceutical company has developed a new over-the-counter sleeping pill. To judge its effectiveness, the company runs an experiment where one set of randomly chosen people takes the new pill and another set takes a placebo. (Neither set knows which type of pill they are taking.) The company judges the effectiveness of the new pill by comparing the proportions of people who fall asleep within a certain amount of time with the new pill and with the placebo.

- An advertising agency would like to check whether men are more likely than women to switch TV channels when a commercial comes on. The agency runs an experiment where the channel switching behavior of randomly chosen men and women can be monitored, and it collects data on the proportion of viewers who switch channels on at least half of the commercial times. The agency then compares these proportions across gender.

The basic form of analysis in each of these examples is the same as in the two-sample analysis for differences between means. However, instead of comparing two means, we now compare two proportions.

Formally, let p_1 and p_2 represent the two unknown population proportions, and let $\hat{p}_1$ and $\hat{p}_2$ be the two sample proportions, based on samples of sizes n_1 and n_2. Then the point estimate of the difference between proportions, $p_1 - p_2$, is the difference between sample proportions, $\hat{p}_1 - \hat{p}_2$. If we assume that the sample sizes are reasonably large, then the sampling distribution of $\hat{p}_1 - \hat{p}_2$ is approximately normal.[8]

Therefore, a confidence interval for $p_1 - p_2$ is given by expression (9.15). Here, the z-multiple is the usual value from the normal distribution that cuts off the appropriate probability in each tail (1.96 for a 95% confidence interval, for example). Also, the standard error of $\hat{p}_1 - \hat{p}_2$ is given by equation (9.16).

Confidence Interval for Difference Between Proportions		
	$\hat{p}_1 - \hat{p}_2 \pm z\text{-multiple} \times \text{SE}(\hat{p}_1 - \hat{p}_2)$	**(9.15)**

[8]This large-sample assumption is valid as long as $n_i \hat{p}_i > 5$ and $n_i(1 - \hat{p}_i) > 5$ for $i = 1$ and $i = 2$.

Standard Error of Difference Between Sample Proportions

$$\text{SE}(\hat{p}_1 - \hat{p}_2) = \sqrt{\frac{\hat{p}_1(1-\hat{p}_1)}{n_1} + \frac{\hat{p}_2(1-\hat{p}_2)}{n_2}} \quad \textbf{(9.16)}$$

The following example illustrates this procedure.

Example 9.10 Sales Response to Coupons for Discounts on Appliances

An appliance store is about to run a big sale. It selects 300 of its best customers and randomly divides them into two sets of 150 customers each. It then mails a notice of the sale to all 300 customers but includes a coupon for an extra 5% off the sale price to the second set of customers only. As the sale progresses, the store keeps track of which of these customers purchase appliances. The resulting data appear in Figure 9.25. (See the file COUPONS.XLS.) What can the store's manager conclude about the effectiveness of the coupons?

Objective To illustrate how to find a confidence interval for the difference between proportions of customers purchasing appliances with and without 5% discount coupons.

Solution

First, note that the data have been arranged in a "contingency" table, much like the pivot tables we discussed in Chapters 2 and 3.[9] Of the 150 customers who received coupons, 55 purchased an appliance. Of the 150 who did not receive coupons, only 35 purchased an appliance. These translate to the sample proportions 0.3667 and 0.2333, calculated in cells B8 and B9 with the formulas

=B4/D4

FIGURE 9.25
Analysis of Coupon Data

	A	B	C	D
1	**Effectiveness of coupons in promoting a sale**			
2				
3		Purchased	Didn't purchase	Total
4	Received coupon	55	95	150
5	Didn't receive coupon	35	115	150
6				
7	Sample proportions who purchased			
8	Received coupon	0.3667		
9	Didn't receive coupon	0.2333		
10				
11	Difference between sample proportions	0.1333		
12	Standard error of difference	0.0524		
13				
14	Confidence level	95%		
15	z-multiple	1.960		
16				
17	Confidence interval for difference between proportions			
18	Lower limit	0.0307		
19	Upper limit	0.2359		

[9]StatPro doesn't have a procedure for solving this problem when the data are in the form of a contingency table. However, the COUPONS.XLS file can be used as a "template" for all such problems.

and

$$=B5/D5$$

By subtraction, these lead directly to the sample difference between proportions, 0.1333, in cell B11. The standard error of this difference is calculated in cell B12 with the formula

$$=SQRT(B8*(1-B8)/D4+B9*(1-B9)/D5)$$

Also, the z-multiple for the confidence interval is calculated in cell B15 with the formula

$$=NORMSINV(B14+(1-B14)/2)$$

Finally, the limits of the confidence interval for the difference are calculated in cells B18 and B19 with the formulas

$$=B11-B15*B12$$

and

$$=B11+B15*B12$$

Because the confidence limits are both positive, we can conclude that the effect of coupons is almost surely to *increase* the proportion of buyers. Our best guess is that the increase is about 13% (from 23.3% to 36.7%). How can the store manager interpret this mean difference? He can use it to estimate the extra business he will receive by including coupons as opposed to not including them. The confidence interval implies that for every 100 customers, the coupons will probably induce an extra 3 to 23 customers to purchase an appliance who otherwise would not have made a purchase.

However, the difference between proportions does not directly indicate the difference in *profit* from including coupons. This is because the customers with coupons pay 5% less than the customers without them. Suppose, for example, the average purchase amount without a coupon is $400, $50 of which is profit for the store. For every 100 customers who receive a mailing with no coupon, the store can expect to make about

$$\$50(0.2333)(100) = \$1166.50$$

in profit. If these 100 customers receive coupons, the expected profit becomes

$$\$30(0.3667)(100) = \$1100.10$$

because these customers pay only $380 on average, $30 of which is profit to the store. Therefore, it appears that if the sample proportions in cells B8 and B9 are anywhere near the true proportions, the store will make *less* profit by including coupons than by not including them. ●

We now revisit Example 9.7, where the SureStep Company is trying to decide which of two suppliers to buy its treadmill motors from. We now compare the two suppliers with regard to warranty costs by analyzing the difference between relevant proportions.

Example 9.11 Analyzing Warranties on Treadmill Motors at SureStep Company

As before, the SureStep Company is trying to decide whether to switch from supplier A to supplier B for the motors in its treadmills. Let's suppose that each treadmill carries a 3-month warranty on the motor. If the motor fails within 3 months, SureStep will supply the customer with a new motor at no cost. This includes installation of the new motor at

FIGURE 9.26

Analysis of Tread-
mill Warranty
Data

	A	B	C	D	E	F	G
1	Differences between treadmill motors - warranty analysis						
2							
3	Sample data (hours until motor fails)			Failure within warranty period (500 hours)			
4	Supplier A	Supplier B			Supplier A	Supplier B	
5	1358	658		Sample size	30	30	
6	793	404		Number	6	11	
7	587	735		Proportion	0.200	0.367	
8	608	457					
9	472	431		Confidence interval for difference calculations			
10	562	658		Difference (B-A)	0.167		
11	879	453		StErr	0.114		
12	575	488					
13	1293	522		Confidence level	95%		
14	1457	1247		z-multiple	1.960		
15	705	1095					
16	623	430		Lower limit	-0.057		
17	725	726		Upper limit	0.391		
18	569	793					
19	424	498					
20	436	502					
21	1250	589					
22	493	975					
23	485	808					
24	462	456					
25	765	731					
26	854	491					
27	634	487					
28	1109	503					
29	800	465					
30	883	1475					
31	522	508					
32	791	846					
33	684	732					
34	666	507					

SureStep's expense. Based on the normal usage at most exercise clubs, SureStep translates the 3-month warranty period into approximately 500 hours of treadmill use. Therefore, using the data from Example 9.7 (in the WARRANTY.XLS file, the same data as in
the MOTORS.XLS file), it would like to compare the proportion of motors failing before
500 hours across the two suppliers.

Objective To illustrate how to find a confidence interval for the difference between proportions of motors failing within the warranty period for the two suppliers.

Solution

The analysis is almost exactly the same as in the previous example, so we will omit most
of the detailed formulas. However, in this case we must find the counts (numbers of failures within warranty) in cells E6 and F6 of Figure 9.26. The easiest way to do this is to
use Excel's COUNTIF function. For example, the formula in cell E6 is

$$=COUNTIF(A5:A34,"<500")$$

We see that the point estimate for the difference in proportions is 0.167 and that a
95% confidence interval for this difference extends from -0.057 to 0.391. Keep in mind
that this difference is the proportion for supplier B minus the proportion for supplier A.

This is fairly convincing, but not conclusive, evidence that a higher proportion of supplier B motors will fail under warranty. It says that if 100 motors from each supplier were
tested, as many as 39 more B motors than A motors might fail before 500 hours—but as
many as 5 or 6 more A motors than B motors might fail before 500 hours. In other words,

there is still some uncertainty about which supplier makes the more reliable motors, even though the weight of the evidence favors supplier A.

What does this mean in terms of costs? And should SureStep change suppliers? As we just saw, the confidence interval implies that more motors are likely to fail under warranty if SureStep changes to supplier B, but B's motors cost less. A cost analysis might go as follows. Suppose that each motor from supplier A costs SureStep $500, whereas supplier B offers them for $475 apiece. Let's follow 100 motors sent to exercise clubs for a period of 3 months. If these are from supplier A, they cost $500 apiece, and approximately 20% (see cell E7) will fail within the warranty period. Of course, each failure costs SureStep another $500. Therefore, the expected cost to SureStep is

$$\$500(100) + \$500(20\%)(100) = \$60,000$$

On the other hand, if these 100 motors come from supplier B, the unit cost is only $475, but approximately 36.7% of them will fail within the warranty period. Therefore, the expected cost is

$$= \$475(100) + \$475(36.7\%)(100) = \$64,933$$

Based on this analysis, the cheaper motors from supplier B are likely to cost more in the long run, so SureStep should probably not switch suppliers. (By the way, we omitted the cost of installing the motors from the analysis. This would have made supplier A look even better.) ●

PROBLEMS

Level A

28. A market research consultant hired by the Pepsi-Cola Co. is interested in estimating the difference between the proportions of female and male consumers who favor Pepsi-Cola over Coke Classic in a particular urban location. A random sample of 250 consumers from the market under investigation is provided in the file P09_17.XLS. After separating the 250 randomly selected consumers by *gender,* construct a 95% confidence interval for the difference between these two proportions. Of what value might this interval estimate be to marketing managers at the Pepsi-Cola Co.?

Level B

29. Continuing the previous problem, marketing managers at the Pepsi-Cola Co. have asked their market research consultant to explore further the difference between the proportions of women and men who prefer drinking Pepsi over Coke Classic. Specifically, Pepsi managers would like to know whether the difference between the proportions of female and male consumers who favor Pepsi varies by the *age* of the consumers. Use the random sample of 250 consumers provided in the file P09_17.XLS to assess whether estimates of this difference vary across the four given age categories: under 20, between 20 and 40, between 40 and 60, and over 60. Employ a 95% confidence

level in generating each of the *four* required interval estimates. Summarize your findings in detail. Finally, what recommendations would you make to the marketing managers in light of your statistical findings?

30. The employee benefits manager of a small private university would like to estimate differences in the proportions of various groups of full-time employees who prefer adopting the third (i.e., plan C) of three available health care plans in the forthcoming annual enrollment period. A reliable frame of the university's employees and their tentative health care preferences are given in the file P08_25.XLS.

   a. First, select a simple random sample of 25 employees from *each* of three employee classifications: administrative staff, support staff, and faculty.

   b. Use the three simple random samples obtained in part **a** to generate three 90% confidence intervals for the differences in the proportions of employees within respective classifications who favor plan C in the coming year. For instance, the first such confidence interval should estimate the difference between the proportion of administrative employees who favor plan C and the proportion of the support staff who prefer plan C.

   c. Interpret each of your constructed confidence intervals. How might the benefits manager use the information you have derived from the three random samples of university employees?

31. Consider a random sample of 100 households from a middle-class neighborhood that was the recent focus of an economic development study conducted by the local government. Specifically, for each of the 100 randomly selected households, information was gathered on each of the following variables: family size, location of the household within the neighborhood, an indication of whether those surveyed owned or rented their home, gross annual income of the first household wage earner, gross annual income of the second household wage earner (if applicable), monthly home mortgage or rent payment, average monthly expenditure on utilities, and the total indebtedness (excluding the value of a home mortgage) of the household. The data are provided in the file P09_26.XLS.

Researchers would like to use the available sample information to discern whether home ownership rates vary by household *location*. For example, is there a nonzero difference between the proportions of individuals who own their homes (as opposed to those who rent their homes) in households located in the first (i.e., SW) and second (i.e., NW) sectors of this community? Use the given sample to construct a 99% confidence interval that estimates this potential difference in home ownership rates as well as those of other combinations of household locations. Interpret and summarize your results. (*Hint:* To be complete, you should construct and interpret a total of *six* 99% confidence intervals.)

9.9 Controlling Confidence Interval Length

In this section we discuss the most widely used methods for achieving a confidence interval of a specified length. Confidence intervals are a function of three things: (1) the data in the sample, (2) the confidence level, and (3) the sample size(s). We briefly discuss the role of the first two in terms of their effect on confidence interval length and then discuss the effect of sample size in more depth.

The data in the sample directly affect the length of a confidence interval through their determination of the sample standard deviation(s). It might appear that because of *random* sampling, we have no control over the sample data, but this is not entirely true. In the case of surveys from a population, there are random sampling plans that can reduce the amount of variability in the sample and hence reduce confidence interval length. Indeed, this is the primary reason for using the stratified sampling procedure we discussed in the previous chapter.

Variance reduction is also possible in randomized experiments. There is a whole area of statistics called **experimental design** that suggests how to perform experiments to obtain the most information from a given amount of sample data. Although this is often aimed at scientific and medical research, it is also appropriate in business contexts. For example, the automobile dealership in Example 9.9 was wise to use *paired* husband–wife data rather than two independent samples of men and women. The pairing leads to a potential reduction in variability and hence a narrower confidence interval.

The confidence level has a clear effect on confidence interval length. As the confidence level increases, the length of the confidence interval increases as well. For example, a 99% confidence interval is always longer than a 95% confidence interval, assuming that they are both based on the same data. However, we rarely use the confidence level to control the length of the confidence interval. Instead, we usually choose the confidence level based on convention, and 95% is by far the most commonly used value. For example, it is the default level built into most software packages, including the StatPro add-in. We can override this default (by choosing 90% or 99%, for example), but we don't usually do so simply to control the confidence interval length.

The most obvious way to control confidence interval length is to choose the sample size(s) appropriately. In the rest of this section, we will see how this can be done.

FIGURE 9.27
Half-Length of a
Confidence Interval

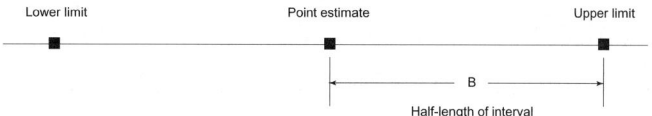

For each parameter we discuss, our goal is to make the length of a confidence interval sufficiently narrow. Because each confidence interval we have discussed (with the exception of the confidence interval for a standard deviation) is a point estimate plus or minus some quantity, we will focus on the "plus or minus" part, called the *half-length* of the interval. (See Figure 9.27.) The usual approach is to specify the half-length B we would like to obtain. Then we find the sample size(s) necessary to achieve this half-length.

9.9.1 Sample Size for Estimation of the Mean

We begin with a confidence interval for the mean. From Section 9.3, we know that its formula is

$$\overline{X} \pm t\text{-multiple} \times s/\sqrt{n}$$

We want to make the half-length of this interval equal to some prescribed value B. For example, if we want the confidence interval to be of the form $\overline{X} \pm 5$, we use $B = 5$. Actually, we won't be able to achieve this half-length B exactly, but we will be able to approximate it.

By setting

$$t\text{-multiple} \times s/\sqrt{n} = B$$

and solving for n, we obtain

$$n = \left(\frac{t\text{-multiple} \times s}{B}\right)^2$$

Unfortunately, sample size selection must be done *before* a sample is observed. Therefore, no value of s is yet available. Also, because the t-multiple depends on n (through the degrees of freedom parameter), it is not clear which t-multiple to use.

The usual way out of this dilemma is to replace s by some reasonable estimate σ_{est} of the population standard deviation σ, and to replace the t-multiple with the corresponding z-multiple from the standard normal distribution. The latter replacement is justified because z-values and t-values are practically equal unless n is very small. The resulting sample size formula is given in equation (9.17). This formula generally results in a noninteger value of n, in which case the practice is to round n *up* to the next larger integer.

Sample Size Formula for Estimating a Mean	$$n = \left(\frac{z\text{-multiple} \times \sigma_{est}}{B}\right)^2$$	**(9.17)**

The following example, an extension of Example 9.2, shows how to implement equation (9.17).

Example (9.12) Sample Size Selection for Estimating
Reaction to New Sandwich

The fast-food manager in Example 9.2 surveyed 40 customers, each of whom rated a
new sandwich on a scale 1 to 10. Based on the data, a 95% confidence interval for the
mean rating of all potential customers extended from 5.739 to 6.761, for a half-length
of $(6.761 - 5.739)/2 = 0.511$. How large a sample would be needed to reduce this half-
length to approximately 0.3?

Objective To find the sample size of customers required to achieve a sufficiently narrow
confidence for the mean rating of the new sandwich.

Solution

Formula (9.17) for n uses three inputs: the z-multiple, which is 1.96 for a 95% confidence
level; the prescribed confidence interval half-length B, which is 0.3; and an estimate σ_{est}
of the standard deviation. This final quantity must be guessed, but based on the given sample of size 40, we can use the observed sample standard deviation, 1.597. (This standard
deviation does not appear explicitly in Figure 9.5, but it can be found with the STDEV
function.) Therefore, formula (9.17) yields

$$n = \left(\frac{1.96(1.597)}{0.3}\right)^2 = 108.86$$

which we round up to $n = 109$. The claim, then, is that if the manager surveys 109 customers, a 95% confidence interval will have approximate half-length 0.3. Its *exact* half-
length will differ slightly from 0.3 because the standard deviation from the sample will
not equal 1.597 exactly.

The StatPro add-in has a Sample Size Selection procedure that performs this sample
size calculation. It can be used anywhere in a spreadsheet—the cursor doesn't have to be
placed inside a data set. There doesn't even have to be a data set. Just select the
StatPro/Statistical Inference/Sample Size Selection menu item, select the parameter to analyze (in this case the mean), and enter the requested values. In this case the requested
values are the confidence level (95), the half-length of the interval (0.3), and an estimate
of the standard deviation (1.597). This produces the message in Figure 9.28.

What if the manager were at the planning stage and didn't have a "preliminary" sample of size 40? What standard deviation estimate should she use for σ_{est} (since the value
1.597 is no longer available)? This is not an easy question to answer, but because of the
role of σ_{est} in equation (9.17), it is crucial for the determination of n. The manager basically has three choices: (1) she can base her estimate of the standard deviation on historical data, assuming relevant historical data are available; (2) she can take a small preliminary sample (of size 20, say) just to get an estimate of the standard deviation; or (3) she
can simply guess a value for the standard deviation. We do not recommend the third option, but there are cases where it is the only feasible option available.

FIGURE 9.28

Sample Size for a
Mean

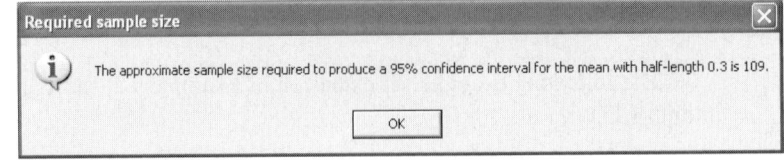

We have demonstrated the use of equation (9.17) for a sample mean. In the same way, it can also be used in the paired-sample procedure. In this case the resulting value of n refers to the number of *pairs* that should be included in the sample, and σ_{est} refers to an estimate of the standard deviation of the *differences* (husband scores minus wife scores, for example).

9.9.2 Sample Size for Estimation of Other Parameters

The sample-size analysis for the mean carries over with very few changes to other parameters. We discuss three other parameters in this section: a proportion, the difference between two means, and the difference between two proportions. In each case the required confidence interval can be obtained by setting the half-length equal to a prescribed value B and solving for n.

There are two points worth mentioning. First, the confidence interval for the difference between means uses a t-multiple. As we did for the mean, we replace this by a z-multiple, which is perfectly acceptable in most situations. Second, the confidence intervals for differences between means or proportions require *two* sample sizes, one for each sample. The formulas below assume that each sample uses the *same* sample size, denoted by n.

The sample size formula for a proportion p is given by equation (9.18). Here, p_{est} is an estimate of the population proportion p. A *conservative* value of n can be obtained by using $p_{est} = 0.5$. It is conservative in the sense that the sample size obtained by using $p_{est} = 0.5$ guarantees a confidence interval half-length no greater than B, regardless of the true value of p.

Sample Size Formula for Estimating a Proportion	$$n = \left(\frac{z\text{-multiple}}{B}\right)^2 p_{est}(1 - p_{est})$$	(9.18)

The sample size formula for the difference between means is given by equation (9.19). Here, σ_{est} is an estimate of the standard deviation of *each* population, where we again make the assumption (as in Section 9.7.1) that the two populations have a *common* standard deviation σ.

Sample Size Formula for Estimating the Difference Between Means	$$n = 2\left(\frac{z\text{-multiple} \times \sigma_{est}}{B}\right)^2$$	(9.19)

Finally, the sample size formula for the difference between proportions is given by equation (9.20). Here, p_{1est} and p_{2est} are estimates of the two unknown population proportions p_1 and p_2. As in the case of a single proportion, we obtain a conservative value of n by using the estimates $p_{1est} = p_{2est} = 0.5$.

$$n = \left(\frac{z\text{-multiple}}{B}\right)^2 [p_{1est}(1 - p_{1est}) + p_{2est}(1 - p_{2est})] \qquad (9.20)$$

Example 9.13 Sample Size Selection for Estimating the Proportion Who Have Tried a New Sandwich

Suppose that the fast-food manager from the previous example wants to estimate the proportion of customers who have tried its new sandwich. It wants a 90% confidence interval for this proportion to have half-length 0.05. For example, if the sample proportion turns out to be 0.42, then a 90% confidence interval should be (approximately) 0.42 ± 0.05. How many customers need to be surveyed?

Objective To find the sample size of customers required to achieve a sufficiently narrow confidence interval for the proportion of customers who have tried the new sandwich.

Solution

If the manager has "no idea" what the proportion is, then she can use $p_{est} = 0.5$ in equation (9.18) to obtain a conservative value of n. The appropriate z-multiple is now 1.645 because this value cuts off probability 0.05 in each tail of the standard normal distribution. Therefore, the required value of n is

$$n = \left(\frac{1.645}{0.05}\right)^2 (0.5)(1 - 0.5) \approx 271$$

On the other hand, if the manager is "pretty sure" that the proportion who have tried the new sandwich is around 0.3, she can use $p_{est} = 0.3$ instead. This time we use StatPro and enter the values 90 (confidence level), 0.05 (desired half-length), and 0.3 (estimate of the proportion). We receive the message in Figure 9.29.

Again remember that lower confidence levels result in narrower confidence intervals.

These calculations indicate that if we have more specific information about the unknown proportion, we can get by with a smaller sample size—in this case 228 rather than 271. Also, note that we selected a 90% confidence level rather than the usual 95% level. There is a trade-off here. Using 90% rather than 95% obviously gives us less confidence in the result, but it requires a smaller sample size. You can check that the required sample sizes for a 95% confidence level increase from 271 and 228 to 385 and 323.

FIGURE 9.29
Sample Size for a
Proportion

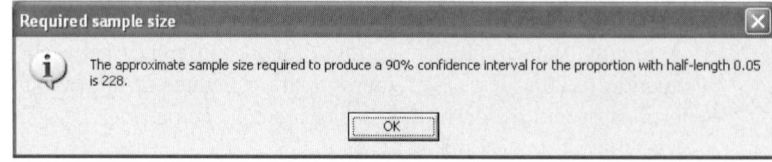

Required sample size

The approximate sample size required to produce a 90% confidence interval for the proportion with half-length 0.05 is 228.

OK

Example 9.14 Sample Size Selection for Analyzing Customer Complaints About Poor Service

A computer company has a customer service center that responds to customers' questions and complaints. The center employs two types of people: those who have had a recent course in dealing with customers (but little actual experience) and those with a lot of experience dealing with customers (but no formal course). The company wants to estimate the difference between these two types of employees in the average number of customer complaints regarding poor service in the last 6 months. The company plans to obtain information on a randomly selected sample of each type of employee, using equal sample sizes. How many employees should be in each sample to achieve a 95% confidence interval with approximate half-length 2?

Objective To see how many employees in each experimental group must be sampled to achieve a sufficiently narrow confidence interval for the difference between the mean numbers of complaints.

Solution

We use equation (9.19) with z-multiple 1.96 and $B = 2$. However, this formula also requires a value for σ_{est}, an estimate of the (assumed) common standard deviation for each group of employees, and there is no obvious estimate available. The manager might use the following argument. Based on a brief look at complaint data, he believes that some employees receive as few as 6 complaints over a 6-month period, whereas others receive as many as 36 (about 6 per month). Now he can estimate σ_{est} by arguing that all observations are likely to be within 3 standard deviations of the mean, so that the range of data—minimum to maximum—is about 6 standard deviations. Therefore, he sets

$$6\sigma_{est} = 36 - 6 = 30$$

and obtains $\sigma_{est} = 5$. Using this value in equation (9.19), the required sample size is

$$n = 2\left(\frac{1.96(5)}{2}\right)^2 \simeq 49$$

The StatPro Sample Size Selection procedure confirms this value. Here, we enter the values 95 (confidence level), 2 (desired half-length), and 5 (estimated standard deviation). We receive message in Figure 9.30.

Some analysts prefer the estimate

$$4\sigma_{est} = 36 - 6 = 30$$

that is, $\sigma_{est} = 7.5$, arguing that the quoted range (6–36) might not include "extreme" values and hence might extend to only 2 standard deviations on either side of the mean. By using this estimate of the standard deviation, you can check that the required sample size increases from 49 to 109. The important point here is that the estimate of the standard deviation can have a dramatic effect on the required sample size. (And don't forget that this size sample must be taken from *each* group of employees.)

FIGURE 9.30
Sample Size for a Difference Between Means

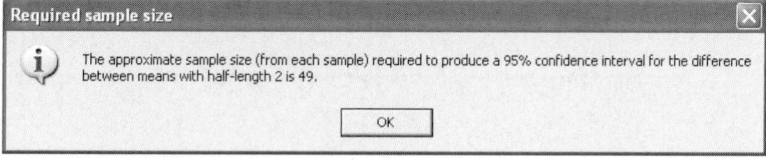

Required sample size

The approximate sample size (from each sample) required to produce a 95% confidence interval for the difference between means with half-length 2 is 49.

OK

The final example in this section illustrates what can happen when we ask for extremely accurate confidence intervals.

Example (9.15) Sample Size Selection for Analyzing Proportions of Out-of-Spec Products

A manufacturing company has two plants that produce identical products. The production supervisor wants to know how much the proportion of out-of-spec products differs across the two plants. He suspects that the proportion of out-of-spec products in each plant is in the range of 3% to 5%, and he wants a 99% confidence interval to have approximate half-length 0.005 (or 0.5%). How many items should he sample from each plant?

Objective To see how many products in each plant must be sampled to achieve a sufficiently narrow confidence interval for the difference between the proportions of out-of-spec products.

Solution

Here we use equation (9.20) with z-multiple 2.576 (the value that cuts off probability 0.005 in each tail of the standard normal distribution), $B = 0.005$, and $p_{1est} = p_{2est} = 0.05$. The reasoning for the latter is that the supervisor believes each proportion is around 3% to 5%, and we obtain the most conservative (largest) sample size by using the larger 5% figure. Then the required sample size is

$$n = \left(\frac{2.576}{0.005}\right)^2 [0.05(0.95) + 0.05(0.95)] \approx 25{,}213$$

This sample size (from *each* sample) is almost certainly prohibitive, so the supervisor decides he must lower his goals. One way is to decrease the confidence level, say, from 99% to 95%. Another way is to increase the desired half-length from 0.005 to, say, 0.025. We implemented both of these changes in the StatPro Sample Size Selection procedure by entering the values 95 (confidence level), 0.025 (desired half-length), and 0.05 and 0.05 (estimates of the proportions). The resulting message is in Figure 9.31. Even now the required sample size is 584. Obviously, narrow confidence intervals for differences between proportions can require very large sample sizes.

FIGURE 9.31

Sample Size for a Difference Between Proportions

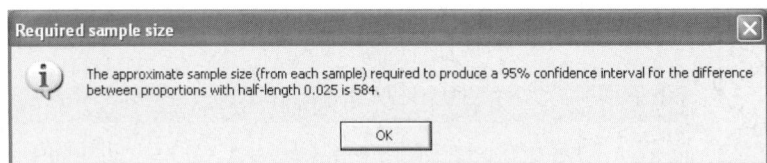

Required sample size

The approximate sample size (from each sample) required to produce a 95% confidence interval for the difference between proportions with half-length 0.025 is 584.

OK

Level A

32. Elected officials in a small Florida town are preparing the annual budget for their community. Specifically, they would like to estimate how much their constituents living in this town are typically paying each year in real estate taxes. Given that there are over 3000 homeowners in this small community, officials have decided to sample a representative subset of taxpayers and thoroughly study their tax payments. The latest frame of homeowners is given in the file P08_36.XLS.

    a. What sample size would be required to generate a 95% confidence interval for the community's mean annual real estate tax payment with a half-length of $100? Assume that the best estimate of the population standard deviation σ is $535.

    b. Choose a simple random sample of the size found in part **a** from the frame in the file. Construct a 95% confidence interval for the population mean. What is the half-length of this interval estimate? Is the half-length consistent with your expectations? Explain.

    c. Now suppose that elected officials want to construct a 95% confidence interval with a half-length of $75. What sample size would be required to achieve this objective? Again, assume that the best estimate of the population standard deviation σ is $535. Explain the difference between your result here and the result you obtained in part **a.**

33. You have been assigned to determine whether more people prefer Coke or Pepsi. Assume that roughly half the population prefers Coke and half prefers Pepsi. How large a sample would you need to take to ensure that you could estimate, with 95% confidence, the fraction of people preferring Coke within 2% of the actual value?

34. You are trying to estimate the average amount a family spends on food during a year. In the past the standard deviation of the amount a family has spent on food during a year has been approximately $1000. If you want to be 99% sure that you have estimated average family food expenditures within $50, how many families do you need to survey?

35. In past years, approximately 20% of all U.S. families purchased potato chips at least once a month. We are interested in determining the fraction of all U.S. families that currently purchase potato chips at least once a month. How many families must we survey if we want to be 99% sure that our estimate of the fraction of U.S. families currently purchasing potato chips at least once a month is accurate within 2%?

36. Continuing Problem 32, suppose that elected officials in this community would like to estimate the proportion of taxpayers whose annual real estate tax payments exceed $2000.

    a. What sample size would be required to generate a 99% confidence interval for this proportion with a half-length of 0.10? Assume for now that the relevant population proportion p is close to 0.50.

    b. Assume now that officials discover old tax records that suggest that approximately 30% of all property owners in this community pay more than $2000 annually in real estate taxes. What sample size would now be required to generate a 99% confidence interval for this proportion with a half-length of 0.10?

    c. Explain the difference in your answers to parts **a** and **b.**

    d. Choose a simple random sample of the size found in part **b** from the frame given in the file P08_36.XLS. Use this sample to evaluate the revised assumption that approximately 30% of all property owners in this community pay more than $2000 annually in real estate taxes.

Level B

37. Continuing the previous problem, suppose that elected officials in this town would like to estimate the difference between the proportions, labeled p_2 and p_6, of taxpayers living in the *second* neighborhood whose annual real estate tax payments exceed $2000 and those living in the *sixth* neighborhood whose annual real estate tax payments exceed $2000.

    a. What sample size (randomly selected from a frame of all taxpayers residing in each neighborhood) would be required to generate a 90% confidence interval for this difference between proportions with a half-length of 0.10? Assume for now that p_2 and p_6 are both close to 0.5.

    b. Choose a simple random sample of the size found in part **a** from the frame of all taxpayers residing in the second neighborhood. Also, choose a simple random sample of the size found in part **a** from the frame of all taxpayers residing in the sixth neighborhood. Use these samples to evaluate the assumption that p_2 and p_6 are both close to 0.5.

    c. Use the samples obtained in part **b** to generate revised estimates of p_2 and p_6. Repeat part **a** with these revised estimates of p_2 and p_6. Explain the difference between your original and revised responses to the question posed in part **a.**

9.10 Conclusion

When we want to estimate a population parameter from sample data, one of the most useful ways to do so is to report a point estimate and a corresponding confidence interval. This confidence interval gives us a quick sense of where the true parameter lies. It essentially quantifies the amount of uncertainty in the point estimate. Obviously, we prefer narrow confidence intervals. We have seen that the length of a confidence interval is determined by the variability in the data, the confidence level, usually set at 95%, and the sample size(s). We have also seen how sample size formulas can be used at the planning stage to achieve confidence intervals that are sufficiently narrow. Finally, we have seen how confidence intervals can be calculated from mathematical formulas or with statistical software such as the StatPro add-in. The advantage of software is that it enables us to concentrate on the important issues for business applications: which confidence intervals are appropriate, how to interpret them, and how to control their length.

Summary of Key Terms

Term	Explanation	Excel	Page	Equation Number
Confidence interval	An interval that, with a stated level of confidence, is likely to capture a population parameter.		418	9.1
t distribution	The sampling distribution of the standardized sample mean when the sample standard deviation is used in place of the population standard deviation.	=TDIST(*value, df,* 1 or 2) =TINV(*prob,df*)	418	9.3
Confidence level	Percentage (usually 90%, 95%, or 99%) that indicates how confident we are that the interval will capture the true population parameter.		423	
Confidence interval for a mean	Interval that is likely to capture a population mean.	StatPro/ Statistical Inference/One-Sample Analysis	424	9.4
Confidence interval for a total	Interval that is likely to capture the total of all observations in a population.	Can be derived from StatPro/ Statistical Inference/One-Sample Analysis	429	9.8
Confidence interval for a proportion	Interval that is likely to capture the proportion of all population members that satisfy a specified property.	Must be done manually	432, 434	9.10
Confidence interval for a standard deviation	Interval that is likely to capture a population standard deviation.	StatPro/ Statistical Inference/One-Sample Analysis	437	
Chi-square distribution	Skewed distribution useful for estimating standard deviations.	=CHIDIST(*value, df*) =CHIINV(*prob, df*)	437	
Confidence interval for difference between means with independent samples	Interval that is likely to capture the difference between two population means when the samples are independent of one another.	StatPro/ Statistical Inference/Two-Sample Analysis	441	9.13, 9.14
Confidence interval for difference between means with paired samples	Interval that is likely to capture the difference between two population means when the samples are paired in a natural way.	StatPro/ Statistical Inference/ Paired-Sample Analysis	448	
Confidence interval for difference between proportions	Interval that is likely to capture the difference between similarly defined proportions from two populations.	Must be done manually	453	9.15, 9.16
Sample size formulas	Formulas that specify the sample size(s) required to obtain sufficiently narrow confidence intervals.	StatPro/ Statistical Inference/ Sample Size Selection	459–462	9.17–9.20

PROBLEMS

Conceptual Exercises

C.1. Under what conditions, if any, is it *not* possible for us to assume that the sampling distribution of the sample mean is approximately normally distributed?

C.2. When, if ever, would it be appropriate to use the standard normal distribution as a substitute for the *t* distribution with $n - 1$ degrees of freedom in estimating a population mean?

C.3. "Assuming that all else remains constant, the width of a confidence interval for a population mean increases whenever the confidence level and sample size increase simultaneously." Is this statement true or false? Explain your choice.

C.4. Assuming that all else remains constant, what happens to the width of a 95% confidence interval for a population parameter when the sample size is reduced by half? You can assume that the resulting sample size is still quite large. Justify your answer.

C.5. "The probability is 0.99 that a 99% confidence interval contains the true value of the relevant population parameter." Is this statement true or false? Explain your choice.

Level A

38. A sample of 9 quality control managers with over 20 years experience have an average salary of $68,000 and a sample standard deviation of $19,000.
   a. You can be 95% confident that the mean salary for all quality managers with at least 20 years of experience is between what two numbers? What assumption are you making about the distribution of salaries?
   b. What size sample would be needed to ensure that we could estimate the true mean salary of all quality managers with more than 20 years of experience and have only 1 chance in 100 of being off by more than $500?

39. Political polls typically sample randomly from the U.S. population to investigate the percentage of voters who favor some candidate or issue. The number of people polled is usually on the order of 1000. Suppose that one such poll asks voters how they feel about the President's handling of environmental issues. The results show that 575 out of the 1280 people polled say they either "approve" or "strongly approve" of the President's handling. Find a 95% confidence interval for the proportion of the entire voter population who "approve" or "strongly approve" of the President's handling. If the same sample proportion were found in a sample twice as large—that is 1150 out of 2560—how would this affect the confi-

dence interval? How would the confidence interval change if the confidence level were 90% instead of 95%?

40. Referring to the previous problem, we often hear the results of such a poll in the news. In fact, the newscasters usually report something such as, "44.9% of the population approve or strongly approve of the President's handling of the environment. The margin of error in this result is plus or minus 3%." Where do you believe this 3% comes from? If the pollsters want the margin of error to be plus or minus 3%, how does this lead to a sample size of approximately 1000?

41. The widths of 100 elevator rails have been measured. The sample mean and standard deviation of the elevator rails are 2.05 inches and 0.01 inch.
   a. Construct a 95% confidence interval for the average width of an elevator rail. Do we need to assume that the width of elevator rails follows a normal distribution?
   b. How large a sample of elevator rails would we have to measure to ensure that we could estimate, with 95% confidence, the average diameter of an elevator rail within 0.01 inch?

42. We want to determine the percentage of Fortune 500 CEOs who think Indiana University deserves its current *Business Week* rating. We mail a questionnaire to all 500 CEOs and 100 respond. Exactly half of the respondents believe IU does deserve its ranking.
   a. Construct a 99% confidence interval for the fraction of Fortune 500 CEOs who believe IU deserves its ranking.
   b. Suppose again that we want to estimate the fraction of Fortune 500 CEOs who believe IU deserves its ranking. Our goal is to have only a 5% chance of having our estimate be in error by more than 0.02. What size sample do we need to take?

43. The SEC requires companies to file annual reports concerning their financial status. It is impossible to audit every account receivable. Suppose we audit a random sample of 49 accounts receivable invoices and find a sample average of $128 and a sample standard deviation of $53.
   a. Find a 99% confidence interval for the mean size of an accounts receivable invoice. Does your answer require that the sizes of the accounts receivable invoices follow a normal distribution?
   b. How large a sample do we need if we want to be 99% sure that we can estimate the mean invoice size within $5?

44. An opinion poll surveyed 900 people and reported that 52% believe the White House broke campaign financing laws.

a. Compute a 95% confidence interval for the population proportion of people who believe the White House broke campaign financing laws. Does the result of the poll convince you that a *majority* of citizens favor that viewpoint?

b. Suppose 10,000 (not 900) people are surveyed and 52% believe that the White House broke campaign financing laws. Would you now be convinced that a majority of citizens believe the White House broke campaign financing laws? Why might your answer be different than in part **a?**

c. How many people would you have to survey to be 99% confident that you can estimate the fraction of people who believe the White House has broken a campaign financing law to within 1%?

45. Sometimes you are given summary data, not the original data, and are asked for a confidence interval. In this case it is probably easier to calculate it using a hand calculator. Try it in the following examples.

a. A sample of 35 jazz CD recordings has been examined. The average playing time of these 35 recordings is 54.7 minutes, and the standard deviation is 6.8 minutes. Find a 95% confidence interval for the mean playing time of all jazz recordings in the population from which this was a sample.

b. You are told that a random sample of 130 people from Indiana has been given cholesterol tests, and 47 of these people had levels over the "safe" count of 180. Find a 95% confidence interval for the population proportion of people with cholesterol levels over 180.

Level B

46. We know that IQs are normally distributed with a mean of 100 and standard deviation of 15. Suppose we did not know this, and we took 100 random samples of four people's IQs and, for each sample, constructed a 95% confidence interval for the mean IQ. We expect that approximately 95 of these intervals would contain the true mean IQ (100) and approximately five of these intervals would not contain the true mean. Use simulation in Excel to see whether this is the case.

47. In Section 9.9, we gave a sample size formula for confidence interval estimation of a mean. If the confidence level is 95%, then (because the z-multiple is about 2), this formula is essentially

$$n = \frac{4\sigma^2}{B^2}$$

However, this formula is based on the assumption that the sample size n will be small relative to the population size N. If this is *not* the case, the appropriate formula is

$$n = \frac{N\sigma^2}{\sigma^2 + (N-1)B^2/4}$$

Now suppose we want to find a 95% confidence interval for a population mean. Based on preliminary (or historical) data, we believe that the population standard deviation is approximately 15. We want the confidence interval to have length 4. That is, we want the confidence interval to be of the form $\bar{X} \pm 2$. What sample size is required if $N = 400$; if $N = 800$; if $N = 10,000$; if $N = 100,000,000$? How would you summarize these findings in words?

48. The Ritter Manufacturing Company has kept track of machine hours and overhead costs at its main manufacturing plant for the past 52 weeks. The data appear in the file P09_48.XLS. Ritter has studied these data to understand the relationship between machine hours and overhead costs. Although the relationship is far from perfect, Ritter believes it can obtain a fairly accurate prediction of overhead costs from machine hours through the equation

Estimated Overhead = 746.5 + 3.32MachHrs

By substituting any observed value of MachHrs into this equation, Ritter obtains an estimated value of Overhead, which is always somewhat different from the true value of Overhead. The difference is called the prediction error.

a. Find a 95% confidence interval for the mean prediction error. Do the same for the *absolute* prediction error. (*Hint:* For example, the prediction error in week 1, actual overhead minus predicted overhead, is −94.5. The absolute prediction error is the absolute value, 94.5.)

b. A close examination of the data suggests that week 45 is a possible outlier. Illustrate this by creating a boxplot of the prediction errors. In what sense would you say week 45 is an outlier? See whether week 45 has much effect on the confidence intervals from part **a** by recalculating these confidence intervals, this time with week 45 deleted. Discuss your findings briefly.

Problems 49–58 are related to the data in the file P09_49.XLS. This file contains data on 400 customers' orders from ElecMart, a company that sells electronic appliances by mail order. The variables are:

- Date: date of order
- Day: day (Monday–Sunday) of order
- Time: time of day (morning, afternoon, evening) order was placed
- Region: region of customer (Northeast, Midwest, South, West) customer is from
- CardType: whether order is paid for by ElecMart's own credit card or another type of credit card
- Gender: gender of customer
- BuyCategory: level of customer's previous order volume from ElecMart (high, medium, low)
- ItemsOrdered: number of items ordered on this order

- TotalCost: total cost of this order
- HighItem: cost of most expensive item on this order

You can consider the data as a random sample from all of ElecMart's orders.

49. Find a 95% confidence interval for the mean total cost of all customer orders. Then do this separately for each of the four regions. Create side-by-side boxplots of total cost for the four regions. Does the positive skewness in these boxplots invalidate the confidence interval procedure used?

50. Find a 95% confidence interval for the proportion of all customers whose order is for more than $100. Then do this separately for each of three times of day.

51. Find a 95% confidence interval for the proportion of all customers whose orders contain at least 3 items *and* cost at least $100 total.

52. Find a 95% confidence interval for the difference between the mean amount of the highest cost item purchased for the High customer category and the similar mean for the Medium customer category. Do the same for the difference between the Medium and Low customer categories. Because of the way these customer categories are defined, you would probably expect these mean differences to be positive. Is this what the data indicate?

53. Of the subpopulation of customers who order in the evening, consider the proportion who are female. Similarly, of the subpopulation of customers who order in the morning, consider the proportion who are female. Find a 95% confidence interval for the difference between these two proportions.

54. Find a 95% confidence interval for the difference between the proportion of female customers who order during the evening and the proportion of male customers who order during the evening.

55. Find a 95% confidence interval for the difference between the mean total order cost for West customers and Northeast customers. Do the same for the other combinations: West versus Midwest, West versus South, Northeast versus South, Northeast versus Midwest, and South versus Midwest.

56. Find a 95% confidence interval for the difference between the mean cost per item for female orders and the similar mean for males.

57. Let $p_{E,F}$ be the proportion of female orders that are paid for with the ElecMart credit card, and let $p_{E,M}$ be the similar proportion for male orders.
  a. Find a 95% confidence interval for $p_{E,F}$; for $p_{E,M}$; for the difference $p_{E,F} - p_{E,M}$.
  b. Let $p_{E,F,Wd}$ be the proportion of female orders on weekdays that are paid for with the ElecMart credit card, and let $p_{E,F,We}$ be the similar proportion for weekends. Define $p_{E,M,Wd}$ and $p_{E,M,We}$

similarly for males. Find a 95% confidence interval for the difference $(p_{E,F,Wd} - p_{E,M,Wd}) - (p_{E,F,We} - p_{E,M,We})$. Interpret this difference in words. Might it be of any interest to ElecMart?

58. Suppose these 400 orders are a sample of the 4295 orders made during this time period, and suppose 2531 of these orders were placed by females. Find a 95% confidence interval for the total paid for all 4295 orders. Do the same for all 2531 orders placed by females. Do the same for all 1764 orders placed by males.

Problems 59–64 are related to the data in the file P09_59.XLS. This file contains data on 91 billings from Rebco, a company that sells plumbing supplies to retailers. The three variables in the file are:
- CustSize: small, medium, or large, depending on the volume of business the customer does with Rebco
- Days: number of days from when Rebco billed the customer until Rebco got paid
- Amount: amount of the bill

You can consider the data as a random sample from all of Rebco's billings.

59. Find a 95% confidence interval for the mean amount of all Rebco's bills. Do the same for each customer size separately.

60. Find a 95% confidence interval for the mean number of days it takes Rebco's customers (as a combined group) to pay their bills. Do the same for each customer size separately. Create a boxplot for the variable Days, based on all 91 billings. Also, create side-by-side boxplots for Days for the three separate customer sizes. Do any of these suggest problems with the validity of the confidence intervals?

61. Find a 95% confidence interval for the proportion of all large customers who pay bills of at least $1000 at least 15 days after they are billed.

62. Find a 95% confidence interval for the proportion of all bills paid within 15 days. Find a 95% confidence interval for the difference between the proportion of large customers who pay within 15 days and the similar proportion of medium-size customers. Find a 95% confidence interval for the difference between the proportion of medium-size customers who pay within 15 days and the similar proportion of small customers.

63. Suppose a bill is considered late if it is paid after 20 days. In this case its "lateness" is the number of days over 20. For example, a bill paid 23 days after billing has a lateness of 3, whereas a bill paid 18 days after billing has a lateness of 0. Find a 95% confidence interval for the mean amount of lateness for all customers. Find similar confidence intervals for each customer size separately.

64. Suppose Rebco can earn interest at the rate of 0.011% daily on excess cash. The company realizes that it could earn extra interest if its customers would pay their bills more promptly.

   a. Find a 95% confidence interval for the mean amount of interest it could gain if each of its customers would pay exactly 1 day more promptly. Find similar confidence intervals for each customer class separately.

   b. Suppose these 91 billings represent a random sample of the 965 billings Rebco generates during the year. Find a 95% confidence interval for the total amount of extra interest it could gain by getting each of these 965 billings to be paid 2 days more promptly.

65. The file P09_65.XLS contains data on the first 100 customers who entered a two-teller bank on Friday. All variables in this file are times, measured in minutes. These variables are:

   • ArriveTime: arrival time of customer (measured from the time the bank's doors opened)
   • ServiceTime: amount of time customer spent with a teller
   • WaitTime: amount of time customer spent waiting in line
   • BankTime: amount of time customer spent in the bank (waiting plus in service)

   a. Find a 95% confidence interval for the mean amount of time a customer spends in service with a teller.

   b. The bank is most interested in mean waiting times because customers get upset when they have to spend a lot of time waiting in line. Use the usual procedure to calculate a 95% confidence interval for the mean waiting time per customer.

   c. Your answer in part **b** is not valid! (It is much too narrow. It makes you believe you have a much more accurate estimate of the mean waiting time than you really have.) We made two implicit assumptions when we stated the confidence interval procedure for a mean: (1) The individual observations all come from the same distribution, and (2) the individual observations are probabilistically independent. Why are both of these, particularly (2), violated for the customer waiting times? [*Hint:* For (1), how do the first few customers differ from "typical" customers? For (2), if you are behind someone in line who has to wait a long time, what about your own waiting time?]

   d. Following up on (2) of part **c,** you might expect waiting times of successive customers to be autocorrelated, that is, correlated with each other. Large waiting times tend to be followed by large waiting times, and small by small. Check this with StatPro's Autocorrelation procedure, under the StatPro/Summary Stats/Autocorrelations menu

item. An autocorrelation of a certain lag, say, lag 2, is the correlation in waiting times between a customer and the customer two behind her. Do these successive waiting times appear to be autocorrelated? (A *valid* confidence interval for the mean waiting time takes autocorrelations into account—but it is considerably more difficult to calculate.)

Problems 66–68 are related to the data in the file P09_66.XLS. The SoftBus Company sells PC equipment and customized software to small companies to help them manage their day-to-day business activities. Although SoftBus spends time with all customers to understand their needs, the customers are eventually on their own to use the equipment and software intelligently. To understand its customers better, SoftBus recently sent questionnaires to a large number of prospective customers. Key personnel—those who would be using the software—were asked to fill out the questionnaire. SoftBus received 82 usable responses, as shown in the file. The variables are:

   • Gender: gender of key person
   • YrsExper: years of experience of key person with this company
   • Education: level of education of key person
   • OwnPC: whether key person owns his or her own home PC
   • PCKnowledge: key person's self-reported level of computer knowledge

   You can assume that these employees represent a random sample of all of SoftBus's prospective customers.

66. Construct a histogram of the PCKnowledge variable. (Since there are only five possible responses (1–5), this histogram should have only five bars.) Repeat this separately for those who own a PC and those who do not. Then find a 95% confidence interval for the mean value of PCKnowledge for all of SoftBus's prospective customers; of all its prospective customers who own PCs; of all its prospective customers who do not own PCs. The PCKnowledge variable obviously can't be exactly normally distributed because it has only five possible values. Do you think this invalidates the confidence intervals?

67. SoftBus believes it can afford to spend much less time with customers who own PCs and score at least 4 on PCKnowledge. We'll call these the "PC-savvy" customers. On the other hand, SoftBus believes it will have to spend a lot of time with customers who do not own a PC and score 2 or less on PCKnowledge. We'll call these the "PC-illiterate" customers.

   a. Find a 95% confidence interval for the proportion of all prospective customers who are PC-savvy. Find a similar interval for the proportion who are PC-illiterate.

b. Repeat part **a** twice, once for the subpopulation of customers who have at least 12 years of experience, and once for the subpopulation who have less than 12 years of experience.

c. Again repeat part **a** twice, once for the subpopulation of customers who have no more than a high school diploma, and once for the subpopulation who have more than a high school diploma.

d. Find a 95% confidence interval for the difference between the proportion of all customers with some college education who are PC-savvy and the similar proportion of all customers with no college education. Repeat this, substituting "PC-savvy" with "PC-illiterate."

e. Discuss any insights you gain from parts **a–d** that might be of interest to SoftBus.

68. Following up on the previous problem, SoftBus believes its profit from each prospective customer depends on the customer's level of PC knowledge. It divides the customers into three classes: PC-savvy, PC-illiterate, and all others (where the first two classes are as defined in the previous problem). As a rough guide, SoftBus figures it can gain profit P_1 from each PC-savvy customer, profit P_3 from each PC-illiterate company, and profit P_2 from each of the others.

a. What values of P_1, P_2, and P_3 seem "reasonable"? For example, would you expect $P_1 < P_2 < P_3$ or the opposite?

b. Using any reasonable values for P_1, P_2, and P_3, find a 95% confidence interval for the mean profit per customer that SoftBus can expect to obtain.

Problems 69–72 are related to the data in the file P09_69.XLS. The Comfy Company sells medium-priced patio furniture through a mail-order catalog. It has operated primarily in the East but is now expanding to the Southwest. To get off to a good start, it plans to send potential customers a catalog with a discount coupon. However, Comfy is not sure how large a discount is needed to entice customers to buy. It experiments by sending catalogs to selected residents in six cities. Tucson and San Diego receive coupons for 5% off any furniture within the next 2 months, Phoenix and Santa Fe receive coupons for 10% off, and Riverside and Albuquerque receive coupons for 15% off. The variables are:

- City: city where customer lives
- Discount: discount offered (5%, 10%, 15%)
- ItemsPurch: number of items purchased with the discount
- TotPaid: total paid (after subtracting the discount) for the items

69. Find a 95% confidence interval for the proportion of customers who will purchase at least one item if they receive a coupon for 5% off. Repeat for 10% off; for 15% off.

70. Find a 95% confidence interval for the proportion of customers who will purchase at least one item and pay at least $500 total if they receive a coupon for 5% off. Repeat for 10% off; for 15% off.

71. Comfy wonders whether the customers who receive larger discounts are buying more expensive items. Recalling that the value in the TotPaid column is *after* the discount, find a 95% confidence interval for the difference between the mean *original price per item* for customers who purchase something with the 5% coupon and the similar mean for customers who purchase something with the 10% coupon. Repeat with 5% and 10% replaced by 10% and 15%. What can you conclude?

72. Comfy wonders whether there are differences across cities that receive the *same* discount.

a. Find a 95% confidence interval for the difference between the mean amount spent in Tucson and the similar mean in San Diego. (These means should include the "0 purchases.") Repeat this for the difference between Phoenix and Santa Fe; between Riverside and Albuquerque. Does city appear to make a difference?

b. Repeat part **a,** but instead of analyzing differences between means, analyze differences between proportions of customers who purchase something. Does city appear to make a difference?

Problems 73–76 are related to the data in the files P09_73a.XLS and P09_73b.XLS. The Niyaki Company sells VCRs through a number of retail stores. On one popular model, there is a standard warranty that covers parts for the first 6 months and labor for the first year. Customers are always asked whether they wish to purchase an extended service plan for $25 that extends the original warranty 2 more years—that is, to 30 months on parts and 36 months on service. To get a better understanding of warranty costs, the company has gathered data on 70 VCRs purchased. The variables in the P09_73a.XLS file are:

- ExtendedPlan: whether customer purchased the extended service plan
- FailureTime: time (months) until the *first* failure of the unit
- PartsCost: cost of parts to repair the unit
- LaborCost: cost of labor to repair the unit

The latter two costs are tracked only for repairs covered by warranty. [Otherwise, the customer bears the cost(s).] The variables in the P09_73b.XLS file are similar, but they also include information of *subsequent* failures of the units (that occur during the warranty period).

73. Construct a histogram of the time until first failure for this type of VCR. Then find a 95% confidence interval for the mean time until failure for this type of

VCR. Does the shape of the histogram invalidate the confidence interval? Why or why not?

74. Find a 95% confidence interval for the proportion of customers who purchase the extended service plan. Find a 95% confidence interval for the proportion of all customers who would benefit by purchasing the extended service plan.

75. Find a 95% confidence interval for Niyaki's mean net warranty cost per unit sold (net of the $25 paid for the plan for those who purchase it). You can assume that this mean is for the *first* failure only; subsequent failures of the same units are ignored here.

76. This problem follows up on the previous two problems with the data in the P09_73b.XLS file. Here Niyaki did more investigation on the same 70 customers. It tracked subsequent failures and costs (if any) that occurred within the warranty period. (Note that only two customers had three failures within the warranty period, and parts weren't covered for either on the third failure. Also, no one had more than three failures within the warranty period.)

a. With these data, find the confidence intervals requested in the previous two problems.

b. Suppose that Niyaki sold this VCR model to 12,450 customers during the year. Find a 95% confidence interval for its total net cost due to warranties from all of these sales.

9.1 Harrigan University Admissions

Harrigan University is a liberal arts university in the Midwest that attempts to attract the highest quality students, especially from its region of the country. It has gathered data on 178 applicants who were accepted by Harrigan (a random sample from all acceptable applicants over the past several years). The data are in the file HARRIGAN.XLS. The variables are:

- Accepted: whether the applicant accepts Harrigan's offer to enroll

- MainRival: whether the applicant enrolls at Harrigan's main rival university

- HSClubs: number of high school clubs applicant served as an officer

- HSSports: number of varsity letters applicant earned

- HSGPA: applicant's high school GPA

- HSPctile: applicant's percentile (in terms of GPA) in his or her graduating class

- HSSize: number of students in applicant's graduating class

- SAT: applicant's combined SAT score

- CombinedScore: a combined score for the applicant used by Harrigan to rank applicants

The derivation of the combined score is a closely kept secret by Harrigan, but it is basically a weighted average of the various components of high school performance and SAT. Harrigan is concerned that it is not getting enough of the best students, and worse yet, it is concerned that many of these best students are going to Harrigan's main rival. Solve the following problems and then, based on your analysis, comment on whether Harrigan appears to have a legitimate concern.

1. Find a 95% confidence interval for the proportion of all acceptable applicants who accept Harrigan's invitation to enroll. Do the same for all acceptable applicants with a combined score less than 330; with a combined score between 330 and 375; with a combined score greater than 375. (Note that 330 and 375 are approximately the first and third quartiles of the Score variable.)

2. Find a 95% confidence interval for the proportion of all acceptable students with a combined score less than the median (356) who choose Harrigan's rival over Harrigan. Do the same for those with a combined score greater than the median.

3. Find 95% confidence intervals for the mean combined score, the mean high school GPA, and the mean SAT score of all acceptable students who accept Harrigan's invitation to enroll. Do the same for all acceptable students who choose to enroll elsewhere. Then find 95% confidence intervals for the differences between these means, where each difference is a mean for students enrolling at Harrigan minus the similar mean for students enrolling elsewhere.

4. Harrigan is interested (as are most schools) in getting students who are involved in extracurricular activities (clubs and sports). Does it appear to be doing so? Find a 95% confidence interval for the proportion of all students who decide to enroll at Harrigan who have been officers of at least two clubs. Find a similar confidence interval for those who have earned at least four varsity letters in sports.

5. The combined score Harrigan calculates for each student gives some advantage to students who rank highly in a *large* high school relative to those who rank highly in a small high school. Therefore, Harrigan wonders whether it is relatively more successful in attracting students from large high schools than from small high schools. Develop one or more confidence intervals for relevant parameters to shed some light on this issue.

D emand for systems analysts in the consulting industry is greater than ever. Graduates with a combination of business and computer knowledge—some even from liberal arts programs—are getting great offers from consulting companies. Once these people are hired, they frequently switch from one company to another as competing companies lure them away with even better offers. One consulting company, D&Y, has collected data on a sample of systems analysts they hired with an undergraduate degree several years ago. The data are in the file D&Y.XLS. The variables are:

- StartSal: employee's starting salary at D&Y
- OnRoadPct: percentage of time employee has spent on the road with clients
- StateU: whether the employee graduated from State University (D&Y's principal source of recruits)
- CISDegree: whether the employee majored in Computer Information Systems (CIS) or a similar computer-related area
- Stayed3Yrs: whether the employee stayed at least 3 years with D&Y
- Tenure: tenure of employee at D&Y (months) if he or she moved before 3 years

D&Y is trying to learn everything it can about retention of these valuable employees. You can help by solving the following problems and then, based on your analysis, presenting a report to D&Y.

1. Although starting salaries are in a fairly narrow band, D&Y wonders whether they have anything to do with retention.

a. Find a 95% confidence interval for the mean starting salary of all employees who stay at least 3 years with D&Y. Do the same for those who leave before 3 years. Then find a 95% confidence interval for the difference between these means.

b. Among all employees whose starting salary is above the median ($37,750), find a 95% confidence interval for the proportion who stay with D&Y for at least 3 years. Do the same for the employees with starting salaries above the median. Then find a 95% confidence interval for the difference between these proportions.

2. D&Y wonders whether the percentage of time on the road might influence who stays and who leaves. Repeat the previous problem, but now do the analysis in terms of percentage of time on the road rather than starting salary. (The median percentage of time on the road is 54%.)

3. Find a 95% confidence interval for the mean tenure (in months) of all employees who leave D&Y within 3 years of being hired. Why is it not possible with the given data to find a confidence interval for the mean tenure at D&Y among *all* systems analysts hired by D&Y?

4. State University's students, particularly those in its nationally acclaimed CIS area, have traditionally been among the best of D&Y's recruits. But are they relatively hard to retain? Find one or more relevant confidence intervals to help you make an argument one way or the other.

9.3 Delivery Times at SnowPea Restaurant

The SnowPea Restaurant is a Chinese carryout/delivery restaurant. Most of SnowPea's deliveries are within a 10-mile radius, but it occasionally delivers to customers more than 10 miles away. SnowPea employs a number of delivery people, four of whom are relatively new hires. The restaurant has recently been receiving customer complaints about excessively long delivery times. Therefore, SnowPea has collected data on a random sample of deliveries by its four new delivery people during the peak dinner time. The data are in the file SNOWPEA.XLS. The variables are:

- Deliverer: which person made the delivery
- PrepTime: time from when order was placed until delivery person started driving it to the customer
- TravelTime: time to drive from SnowPea to customer
- Distance: distance (miles) from SnowPea to customer

Solve the following problems and then, based on your analysis, write a report that makes reasonable recommendations to SnowPea management.

1. SnowPea is concerned that one or more of the new delivery people might be slower than others.
   a. Let μ_{Di} and μ_{Ti} be the mean delivery time and mean total time for delivery person i, where the total time is the sum of the delivery and prep times. Find 95% confidence intervals for each of these means for each delivery person. Although these might be interesting, give two reasons why they are not really fair measures for comparing the efficiency of the delivery people.
   b. Responding to the criticisms in part **a,** find a 95% confidence interval for the mean speed of delivery for each delivery person, where speed is measured as miles per hour during the trip from SnowPea to the customer. Then find 95% confidence intervals for the mean difference in speed between each pair of delivery people.

2. SnowPea would like to advertise that it can achieve a total delivery time of no more than M minutes for all customers within a 10-mile radius. On all orders that take more than M minutes,

SnowPea will give the customers a $10 certificate on their next purchase.
   a. Assuming for now that the delivery people in the sample are representative of all of SnowPea's delivery people, find a 95% confidence interval for the proportion of deliveries (within the 10-mile limit) that will be on time if $M = 25$ minutes; if $M = 30$ minutes; if $M = 35$ minutes.
   b. Suppose SnowPea makes 1000 deliveries within the 10-mile limit. For each of the values of M in part **a,** find a 95% confidence interval for the total dollar amount of certificates it will have to give out for being late.

3. The policy in the previous problem is simple to state and simple to administer. However, it is somewhat unfair to customers who live close to SnowPea—they will never get $10 certificates! A fairer, but more complex, policy is the following. SnowPea first analyzes the data and finds that total delivery times can be predicted fairly well with the equation

   Predicted Delivery Time $= 14.8 + 2.06$Distance

   (This is based on regression analysis, the topic of Chapters 11 and 12.) Also, most of these predictions are within 5 minutes of the actual delivery times. Therefore, whenever SnowPea receives an order over the phone, it looks up the customer's address in its computerized geographical database to find distance, calculates the predicted delivery time based on this equation, rounds this to the nearest minute, adds 5 minutes, and guarantees this delivery time or else a $10 certificate. It does this for *all* customers, even those beyond the 10-mile limit.
   a. Assuming again that the delivery people in the sample are representative of all of SnowPea's delivery people, find a 95% confidence interval for the proportion of all deliveries that will be within the guaranteed total delivery time.
   b. Suppose SnowPea makes 1000 deliveries. Find a 95% confidence interval for the total dollar amount of certificates it will have to give out for being late.

CASE

9.4 The Bodfish Lot Cruise[10]

Ralph Butts, Manager of Woodland Operations for Intergalactica Papelco's Southeastern Region, had to decide this morning whether to approve the Bodfish Lot logging contract that was sitting on his desk. Accompanying the contract was a cruise report that gave Mr. Butts the results of a sample survey of the timber on the Bodfish Lot. Was there enough timber to make logging operations worthwhile?

The Pluto Mill of Intergalactica Papelco is located on the River Styxx in Median, Michigan. The scale of operations at Pluto is enormous. Just one of its several $500 million, football-field-long, 4-story-high paper machines has the capability to produce a 20-mile-long, 16-foot-wide, 20-ton reel of paper every hour. Such a machine is run nonstop 24 hours a day for as many of the 365 days in the year that mill maintenance can keep it up and producing paper within specified quality levels. In total, the Pluto Mill produces about 400,000 tons of white paper a year. Because it takes about a ton of wood to produce a ton of paper, a huge quantity of cordwood logs suitable for chipping and pulping must be supplied continually to keep the mill operating. Intergalactica Papelco runs a large-scale logistics, planning, and procurement operation to provide the Pluto Mill with the requisite species, quantity, and quality of wood in a timely fashion.

The Pluto Mill sits on 500 acres of land in the midst of a region in which the huge Intergalactica Papelco owns over a quarter of a million acres of forest. Although this wholly-owned forest is the single largest supplier of wood to the mill, more than 60% of the wood used at Pluto is purchased from independent landowners and loggers under contract. Supplying contract wood dependably on such an enormous scale involves frequent purchasing decisions by the Intergalactica Woodlands Operations as to which independent woodlots have sufficient wood volume and quality to support economical logging operations. A prospective seller enters into a tentative agreement with Intergalactica on the basis of market price and a

visual scan of the woodlot. The final decision about whether to proceed with the logging is usually based on sampling estimates of the total wood volume on the lot.

A recent case in point was the Bodfish Lot in Henryville, Arkansas, whose owner approached Intergalactica with a proposal for logging during the 1991–1992 season. Aerial photographs indicated that the land was sufficiently promising to warrant a "cruise" to estimate the total volume of wood. (*Cruising* is a term used in the forestry industry to describe a systematic procedure for estimating the quantity, quality, variety, and value of the wood on a plot of land. Indeed, standard cruising methods have been developed and disseminated by the U.S. Department of Agriculture and Forestry Service.) Estimation based on limited sampling is essential. Even for the modest-size Bodfish Lot, with 586 acres of forested land, it would be practically impossible to measure every tree on the lot.

For the Bodfish Lot cruise it was decided to sample 89 distinct 1/7-acre plots for actual measurement. Although the plots were chosen "systematically," the sample was, Intergalactica hoped, still effectively "random." Indeed, *no* consistent attempt was made to select the plots from areas of heavy tree growth, large-diameter trees, heavy spruce concentration, and so on. In fact, the opposite was true: The regular spacing of the sampling grid more or less guaranteed a good cross section of the entire lot. This was what is called in forestry industry jargon a "standard line plot cruise." The total lot was 700 acres in area. The plots were spaced at 8-chain intervals apart on a rectangular grid drawn in advance at the Intergalactica Woodlands Field Office at One Rootmean Square in the town of Covariance, Illinois. The aerial photographs showed that, of the Bodfish Lot's 700 total acres, 586 acres were forested. The total volume estimate, to be done separately for each species, was to be based on the average for the 89 sampled plots on these 586 acres.

A circular area two-person cruise was then initiated. Typically, about ten plots could be cruised in one day. The foresters counted the entire number of cordwood trees over 6 inches in diameter within each

[10]This case was contributed by Peter Kolesar from Columbia University.

1/7-acre circle. Then, back in the office in Covariance, the number of trees on each plot was entered into a computer according to species, diameter, and possible end product. The file BODFISH.XLS contains this tabulation from the cruise notes of the counts for spruce, hard maple, and beech of the number of cordwood trees on the 89 sampled plots. (In the actual database, 13 different species of trees were recorded, and Intergalactica would have decided which trees were more suitable for lumber, plywood, or pulping applications.)

With these data, Intergalactica now had to decide whether to contract to log the lot. Ralph Butts, manager of Woodlands Operations, knew that even though Intergalactica would pay on the basis of the weight received at the mill, he needed at least 31,000 cordwood size trees on the lot to make operations economical. More detailed knowledge of the amount of timber by species would help the Pluto Mill make the crucial blending decisions that affect the cost and quality of the resulting wood pulp.

This was just one of several hundred similiar contracts to be made over the coming year. Butts was concerned with the rising cost of cruising in the Southeastern Region. Was the Bodfish Lot cruise excessive, he wondered? Could he get by in the future with considerably smaller samples? Suppose that only a half or a quarter of the plots on Bodfish had been cruised?

10

Hypothesis Testing

© 2002 Harris Welles

Official Sponsors of the Olympics

Hypothesis testing is one of the most frequently used tools in academic research, including research in the area of business. Many studies pose interesting questions, stated as hypotheses, and then test these with appropriate statistical analysis of experimental data. One such study is reported in McDaniel and Kinney (1996). They investigate the effectiveness of "ambush marketing" in prominent sports events such as the Olympic Games. Many companies pay significant amounts of money, perhaps $10 million, to become official sponsors of the Olympics. Ambushers are their competitors who pay no such fees but nevertheless advertise heavily during the Olympics, with the intention of linking their own brand image to the event in the minds of consumers. The question McDaniel and Kinney investigate is whether consumers are confused into thinking that the ambushers are the official sponsors.

At the time of the 1994 Winter Olympics in Lillehammer, Norway, the researchers ran a controlled experiment using 215 subjects ranging in age from 19 to 49 years old. Approximately half of the subjects—the "control group"—viewed a 20-minute tape of a women's skiing event in which several actual commercials for official sponsors in four product categories were interspersed. (The categories were fast food, automobile, credit card, and insurance; the official sponsors were McDonald's, Chrysler, VISA, and John Hancock.) The other half—the "treatment group"—watched the same tape but with commercials for competing ambushers. (The ambushers were Wendy's, Ford, American Express, and Northwestern Mutual, all of which advertised during the 1994 Olympics.) After watching the tape, each subject was asked to fill out a questionnaire. This questionnaire asked subjects to recall the official Olympics sponsors in each product category, to rate their attitudes toward the products, and to state their intentions to purchase the products.

McDaniel and Kinney tested several hypotheses. First, they tested the hypothesis that there would be no difference between the control and treatment groups in terms of which products they would recall as official Olympics sponsors. The experimental evidence allowed them to reject this hypothesis decisively. For example, the vast majority of the control group, who watched the McDonald's commercial, recalled McDonald's as being the official sponsor in the fast-food category. But a clear majority of the treatment group, who watched the Wendy's commercial, recalled Wendy's as being the official sponsor in this category. Evidently, Wendy's commercial was compelling.[1]

Because the ultimate objective of commercials is to increase purchases of a company's brand, the researchers also tested the hypothesis that viewers of official sponsor commercials would rate their intent to purchase that brand *higher* than viewers of ambusher commercials would rate their intent to purchase the ambusher brand. After all, isn't this why the official sponsors were paying large fees to be "official" sponsors? However, except for the credit card category, the data did *not* support this hypothesis. VISA viewers did indeed rate their intent to use VISA higher than American Express viewers rated their intent to use American Express. But in the other three product categories, the ambusher brand came out ahead of the official brand in terms of intent to purchase (although the differences were not statistically significant).

There are at least two important messages this research should convey to business. First, if a company is going to spend a lot of money to become an official sponsor of an event such as the Olympic Games, it must create a more vivid link in the mind of consumers between its product and the event. Otherwise, the company might be wasting its money. Second, ambush marketing is very possibly a wise strategy. By seeing enough of the ambushers' commercials during the event, consumers get confused into thinking that the ambusher is an "official" sponsor. In addition, previous research in the area suggests that consumers do not view ambushers negatively for using an ambushing strategy. ●

10.1 Introduction

When we want to make inferences to a population on the basis of sample data, we can perform the analysis in either of two ways. We can proceed as in the previous chapter, where we calculate a point estimate of a population parameter and then form a confidence interval around this point estimate. In this way we bring no preconceived ideas to the analysis but instead let the data "speak for themselves" in telling us where the true parameter is likely to be.

In contrast, an analyst often has a particular theory, or hypothesis, that he or she would like to test. This hypothesis might be that a new packaging design will produce more sales than the current design, that a new drug will have a higher cure rate for a given disease than any drug currently on the market, that people who smoke cigarettes are more susceptible to heart disease than nonsmokers, and so on. In this case the analyst typically collects sample data and checks whether the data provide enough evidence to support the hypothesis.

The hypothesis that the analyst is attempting to prove is called the **alternative hypothesis.** It is also frequently called the **research hypothesis.** The opposite of the alter-

---

[1]Whereas the McDonald's commercial featured the five-ringed Olympics logo and had an Olympics theme, the Wendy's commercial used a humorous approach built around the company's founder, Dave Thomas, and his dream of winning gold in Olympics bobsled competition.

Hypothesis testing is a form of decision making under uncertainty, where we decide which of two competing hypotheses to accept, based on sample data. However, in contrast to the methods discussed in Chapter 7, it is performed in a very specific way, as described in this chapter.

native hypothesis is called the **null hypothesis.** It usually represents the current thinking or status quo. That is, the null hypothesis is usually the accepted theory that the analyst is trying to *disprove*. In the previous examples the null hypotheses are:

- The new packaging design is no better than the current design.
- The new drug has a cure rate no higher than other drugs on the market.
- Smokers are no more susceptible to heart disease than nonsmokers.

The burden of proof is traditionally on the alternative hypothesis. It is up to the analyst to provide enough evidence in support of the alternative; otherwise, the null hypothesis will continue to be accepted. A slight amount of evidence in favor of the alternative is usually not enough. For example, if a slightly higher percentage of people are cured with a new drug in a sequence of clinical tests, this still might not be enough evidence to warrant introducing the new drug to the market. In general, we reject the null hypothesis—and accept the alternative—only if the results of a hypothesis test are "statistically significant," a concept we will explain in this chapter.

| Null and Alternative Hypotheses | The **null hypothesis** is usually the current thinking, or "status quo." The **alternative,** or **research, hypothesis** is usually the hypothesis a researcher wants to prove. The burden of proof is on the alternative hypothesis. |

As we will see in this chapter, confidence interval estimation and hypothesis testing use data in much the same way and they often report basically the same results, only from different points of view. There continues to be a debate (largely among academic researchers) over which of these two procedures is more useful. We believe that in a business context, confidence interval estimation is more useful and enlightening than hypothesis testing. However, hypothesis testing continues to be a key aspect of statistical analysis. Indeed, statistical software packages routinely include the elements of standard hypothesis tests in their outputs. We will see this, for example, when we study regression analysis in Chapters 11 and 12. Therefore, it is essential to understand the fundamentals of hypothesis testing so that we can interpret this output intelligently.

10.2 Concepts in Hypothesis Testing

Before we plunge into the details of specific hypothesis tests, it is useful to discuss the *concepts* behind hypothesis testing. There are a number of concepts and statistical terms involved, all of which lead eventually to the key concept of statistical significance. To make this discussion somewhat less abstract, we place it in the context of the following example.

Example 10.1 Experimenting with a New Pizza Style at the Pepperoni Pizza Restaurant

The manager of the Pepperoni Pizza Restaurant has recently begun experimenting with a new method of baking its pepperoni pizzas. He personally believes that the new method produces a better-tasting pizza, but he would like to base a decision on whether to switch from the old method to the new method on customer reactions. Therefore, he performs an experiment. For 100 randomly selected customers who order a pepperoni pizza for home delivery, he includes both an old-style and a free new-style pizza in the order. All he asks is that these customers rate the *difference* between pizzas on a -10 to $+10$ scale, where -10 means that they strongly favor the old style, $+10$ means they strongly favor the new style, and 0 means they are indifferent between the two styles. Once he gets the ratings from the customers, how should he proceed?

We begin by stating that Example 10.1 is used primarily to explain hypothesis-testing concepts. We do *not* want to imply that the manager would, or should, use a hypothesis-testing procedure to decide whether to switch from the old method to the new method. First, hypothesis testing does not take costs into account. If the new method of making pizzas uses more expensive cheese, for example, then hypothesis testing would ignore this important aspect of the decision problem. Second, even if the costs of the two pizza-making methods are equivalent, the manager might base his decision on a simple point estimate and possibly a confidence interval. For example, if the sample mean rating is 1.8 and a 95% confidence interval for the mean rating extends from 0.3 to 3.3, this in itself might be enough evidence to make the manager switch to the new method.

We will come back to these ideas—basically, that hypothesis testing is not necessarily the best procedure to use in a business decision-making context—throughout this chapter. However, with these caveats in mind, we discuss how the manager *might* proceed by using hypothesis testing. ●

10.2.1 Null and Alternative Hypotheses

As we stated in the introduction to this chapter, the hypothesis the manager is trying to prove is called the alternative, or research, hypothesis, whereas the "status quo" is called the null hypothesis. In this example the manager would personally like to prove that the new method provides better-tasting pizza, so this becomes the alternative hypothesis. The opposite, that the old-style pizzas are at least as good as the new-style pizzas, becomes the null hypothesis. We'll assume he judges which of these is true on the basis of the mean rating over the entire customer population, labeled μ. If it turns out that $\mu \leq 0$, then the null hypothesis is true. Otherwise, if $\mu > 0$, the alternative hypothesis is true. The hypotheses are summarized in the box.

Hypotheses for Pizza Example	Null hypothesis: $\mu \leq 0$ Alternative hypothesis: $\mu > 0$ where μ is the mean population rating.

Usually, the null hypothesis is labeled H_0 and the alternative hypothesis is labeled H_a. Therefore, in our example we can specify these as $H_0 : \mu \leq 0$ and $H_a : \mu > 0$. This is typical. The null and alternative hypotheses divide all possibilities into two nonoverlapping sets, exactly one of which must be true. In our case either the mean rating is less than or equal to 0, or it is positive. Exactly one of these possibilities *must* be true, and the manager intends to use sample data to learn which of them is true.

Traditionally, hypothesis testing has been phrased as a decision-making problem, where an analyst decides either to accept the null hypothesis or reject it, based on the sample evidence. In our example, accepting the null hypothesis means deciding that the new-style pizza is not really better than the old-style pizza and presumably discontinuing the new style. In contrast, rejecting the null hypothesis means deciding that the new-style pizza is indeed better than the old-style pizza and presumably switching to the new style.

10.2.2 One-Tailed Versus Two-Tailed Tests

The form of the alternative hypothesis can be either *one-tailed* or *two-tailed,* depending on what the analyst is trying to prove. The pizza manager's alternative hypothesis is **one-tailed** because he is hoping to prove that the customers' ratings are, on average, greater than 0. The only sample results that can lead to rejection of the null hypothesis are those in a particular direction, namely, those where the sample mean rating is *positive.* Of course, if the manager sets up his rating scale in the reverse order, so that *negative* ratings favor the new-style pizza, then the test is still one-tailed, but now only negative sample means lead to rejection of the null hypothesis.

In contrast, a **two-tailed** test is one where results in either of two directions can lead to rejection of the null hypothesis. A slight modification of the pizza example where a two-tailed alternative might be appropriate is the following. Suppose the manager currently uses two methods for producing pepperoni pizzas. He is thinking of discontinuing one of these methods if it appears that customers, on average, favor one method over the other. Therefore, he runs the same experiment as before, but now the hypotheses he tests are $H_0: \mu = 0$ versus $H_a: \mu \neq 0$, where μ is again the mean rating across the customer population. In this case *either* a large positive sample mean or a large negative sample mean will lead to rejection of the null hypothesis—and presumably to discontinuing one of the production methods. The difference between these two types of test is summarized in the box.

One-Tailed versus Two-Tailed Alternatives	A **one-tailed alternative** is one that is supported only by evidence in a single direction. A **two-tailed alternative** is one that is supported by evidence in either direction.

It is important to realize that the analyst, not the data, determines the type of alternative hypothesis. The hypothesis depends entirely on what the analyst wants to prove, and it should be formed *before* the data are collected.

Once the hypotheses are set up, it is easy to detect whether the test is one-tailed or two-tailed. One-tailed alternatives are phrased in terms of ">" or "<" whereas two-tailed alternatives are phrased in terms of "≠". The real question is whether to set up hypotheses for a particular problem as one-tailed or two-tailed. There is no *statistical* answer to this question. It depends entirely on what we are trying to prove. If the pizza manager is trying to prove that the new-style pizza is better than the old-style pizza—only results on "one side" will lead to a switch—a one-tailed alternative is appropriate. However, if he is trying to decide whether to discontinue either of two existing production methods—where results on "either side" will lead to a switch—then a two-tailed alternative is appropriate.

10.2.3 Types of Errors

Regardless of whether the manager decides to accept or reject the null hypothesis, it *might* be the wrong decision. He might incorrectly reject the null hypothesis when it is true ($\mu \leq 0$), and he might incorrectly accept the null hypothesis when it is false ($\mu > 0$). In the tradition of hypothesis testing, these two types of errors have acquired the names *type I* and *type II errors*. In general, we commit a **type I error** when we incorrectly *reject* a null hypothesis that is true. We commit a **type II error** when we incorrectly *accept* a null hypothesis that is false. These ideas appear graphically in Figure 10.1.

The pizza manager commits a type I error if he concludes, based on sample evidence, that the new-style pizza is better (and switches to it) when in fact the entire customer population would, on average, favor the old-style pizza. In contrast, he commits a type II error if he concludes, again based on sample evidence, that the new-style is no better (and

FIGURE 10.1
Types of Errors in
Hypothesis Testing

		Truth	
		H_0 is true	H_a is true
Decision	Reject H_0	Type I error	No error
	Do not reject H_0	No error	Type II error

discontinues it) when in fact the entire customer population would, on average, favor the new style.

Possible Errors in Pizza Example	Type I error: Switching to new style when it is no better than old style Type II error: Staying with old style when new style is better

Although we might be inclined to regard these two types of errors as equally serious or costly, type I errors have traditionally been regarded as the more serious of the two. Therefore, the hypothesis-testing procedure favors caution in terms of rejecting the null hypothesis. The thinking is that if we reject the null hypothesis and it is really true, then we commit a type I error—which is bad. Given this rather conservative way of thinking, we are inclined to accept the null hypothesis unless the sample evidence provides strong support for the alternative hypothesis. Unfortunately, we can't have it both ways. By accepting the null hypothesis, we risk committing a type II error.

Type I errors are usually considered more "costly," although this can lead to conservative decision making.

This is exactly the dilemma the pizza manager faces. If he wants to avoid a type I error (where he switches to the new style but really shouldn't), then he will require fairly convincing evidence from the survey that he *should* switch. If he observes *some* evidence to this effect, such as a sample mean rating of $+1.5$ and a 95% confidence interval that extends from -0.3 to $+3.3$, say, this evidence might not be strong enough to make him switch. However, if he decides not to switch, he risks committing a type II error.

10.2.4 Significance Level and Rejection Region

The analyst gets to choose the significance level α. It is traditionally chosen to be 0.05, but it is occasionally chosen to be 0.01 or 0.10.

The real question, then, is how strong the evidence in favor of the alternative hypothesis must be to reject the null hypothesis. Two approaches to this problem are commonly used. In the first, the analyst prescribes the probability of a type I error that he is willing to tolerate. This type I error probability is usually denoted by α and is most commonly set equal to 0.05, although $\alpha = 0.01$ and $\alpha = 0.10$ are also frequently used. The value of α is called the **significance level** of the test. Then, given the value of α, we use statistical theory to determine a *rejection region*. If the sample evidence falls into the **rejection region,** we reject the null hypothesis; otherwise, we accept it. The rejection region is chosen precisely so that the probability of a type I error is at most α. Sample evidence that falls into the rejection region is called **statistically significant at the α level**. For example, if $\alpha = 0.05$, we say that the evidence is statistically significant at the 5% level. These terms are summarized in the following boxes.

Rejection Region	The **rejection region** is the set of sample data that leads to the rejection of the null hypothesis.

Significance Level and Statistically Significant Results	The **significance level,** α, determines the size of the rejection region. Sample results in the rejection region are called **statistically significant** at the α level.

It is important to understand the effect of varying α. If α is small, such as 0.01, then the probability of a type I error is small. Therefore, we require a lot of sample evidence in favor of the alternative hypothesis to reject the null hypothesis. Equivalently, the rejection region in this case is small. In contrast, when α is larger, such as 0.10, the rejection region is larger, and it is easier to reject the null hypothesis.

10.2.5 Significance from *p*-values

A second approach, and one that is currently more popular, is to avoid the use of an α level and instead simply report "how significant" the sample evidence is. We do this by means of a *p-value*. The idea is quite simple—and very important. Suppose in the pizza example that the true mean rating (if it could be observed) is $\mu = 0$. In other words, the customer population, on average, judges the two styles of pizza to be equal. Now suppose that the sample mean rating is $+2.5$. The manager has two options at this point. (Remember that he doesn't know that $\mu = 0$; he only observes the sample.) He can conclude that (1) the null hypothesis is true—the new-style pizza is not preferred over the old style—and he just observed an unusual sample, or (2) the null hypothesis is *not* true—customers do prefer the new-style pizza—and the sample he observed is a typical one.

The *p*-value of the sample quantifies this. The **p-value** is the probability of seeing a random sample at least as extreme as the observed sample, given that the null hypothesis is true. Here, "extreme" is relative to the null hypothesis. For example, a sample mean rating of $+3.5$ from the pizza customers is more extreme evidence than a sample mean rating of $+2.5$. Each provides some evidence against the null hypothesis, but the former provides stronger, more extreme evidence.

> **p-value**
>
> The **p-value of a sample** is the probability of seeing a sample with at least as much evidence in favor of the alternative hypothesis as the sample actually observed. The smaller the *p*-value, the more evidence there is in favor of the alternative hypothesis.

Let's suppose that the pizza manager collects data from the 100 sampled customers and finds that the *p*-value for the sample is 0.03. This means that *if* the entire customer population, on average, judges the two types of pizza to be approximately equal, then only 3 random samples out of 100 would provide as much evidence in support of the new style as the observed sample. So should he conclude that the null hypothesis is true and he just happened to observe an unusual sample, or should he conclude that the null hypothesis is *not* true? There is no clear statistical answer to this question; it depends on how convinced the manager must be before switching. But we can say in general that smaller *p*-values indicate more evidence in support of the alternative hypothesis. If a *p*-value is sufficiently small, almost any decision maker will conclude that rejecting the null hypothesis (and accepting the alternative) is the most "reasonable" decision.

How small is a "small" *p*-value? This is largely a matter of semantics, but Figure 10.2 indicates the attitude of many analysts. If a *p*-value is less than 0.01, it provides "convincing" evidence that the alternative hypothesis is true. After all, fewer than 1 sample out of 100 would provide such support for the alternative hypothesis if it weren't true. If

FIGURE 10.2 Evidence in Favor of the Alternative Hypothesis

the p-value is between 0.01 and 0.05, there is "strong" evidence in favor of the alternative hypothesis. Unless the consequences of making a type I error are really serious, we are likely to reject the null hypothesis in this case.

The interval between 0.05 and 0.10 is a "gray area." If a scientific researcher were trying to prove a research hypothesis and observed a p-value between 0.05 and 0.10, she would probably be reluctant to publish her results as "proof" of the alternative hypothesis, but she would probably be encouraged to continue her research and collect more sample evidence. Finally, p-values larger than 0.10 are generally interpreted as weak or no evidence in support of the alternative.

There is a strong connection between the α-level approach and the p-value approach. Namely, we can reject the null hypothesis at a specified level of significance α only if the p-value from the sample is less than or equal to α. Equivalently, the sample evidence is statistically significant at a given α level only if its p-value is less than or equal to α. For example, if the p-value from a sample is 0.03, then we can reject the null hypothesis at the 10% and the 5% significance levels, but we cannot reject it at the 1% level. The p-value essentially states *how* significant a given sample is.

Relationship Between p-values and Statistically Significant Results	Sample evidence is statistically significant at the α level only if the p-value is less than α.

If you remember only one thing from this chapter, remember that a p-value measures how unlikely the observed sample results would be, given that the null hypothesis is true. Therefore, a low p-value provides evidence for rejecting the null hypothesis and accepting the alternative.

The advantage of the p-value approach is that the analyst doesn't have to choose a significance value α ahead of time. Since it is far from obvious what value of α we should choose in any particular situation, this is certainly an advantage. Another compelling advantage is that p-values for standard hypothesis tests are routinely included in most statistical software output. In addition, all p-values can be interpreted in basically the same way: a small p-value provides support for the alternative hypothesis.

10.2.6 Type II Errors and Power

A type II error occurs when the alternative hypothesis is true but there isn't enough evidence in the sample to reject the null hypothesis. This type of error is traditionally considered less important than a type I error, but it can lead to serious consequences in a real situation. For example, in medical trials on a proposed new cancer drug, a type II error would occur if the new drug is really superior to existing drugs but experimental evidence is not sufficiently conclusive to warrant marketing the new drug. For patients suffering from cancer, this is a serious error!

As we stated previously, the alternative hypothesis is typically the hypothesis a researcher wants to prove. If it is in fact true, the researcher wants to be able to reject the null hypothesis and hence avoid a type II error. The probability that she is able to do so is called the **power** of the test—that is, the power is 1 minus the probability of a type II error. There are several ways to achieve high power, the most popular of which is to increase sample size. By sampling more members of the population, we are better able to see whether the alternative is really true and hence avoid a type II error if it is true. As in the previous chapter, there are formulas that specify the sample size required to achieve a certain power for a given set of hypotheses. We will not pursue these in this book, but you should be aware that they exist. The definition of power is summarized in the box.

Power	The **power of a test** is 1 minus the probability of a type II error. It is the probability that we reject the null hypothesis when the alternative hypothesis is true.

10.2.7 Hypothesis Tests and Confidence Intervals

When we present the results of hypothesis tests, we often include confidence intervals in the output. This gives us two complementary ways to interpret the data. However, there is a more formal connection between the two, at least for two-tailed tests. Let α be the stated significance level of the test. We will state the connection for the most commonly used level, $\alpha = 0.05$, although it extends to any α value. The connection is that we can reject the null hypothesis at the 5% significance level if and only if a 95% confidence interval does *not* include the hypothesized value of the parameter.

Using a Confidence Interval to Perform a Two-Tailed Hypothesis Test	Reject the null hypothesis only if the hypothesized value does *not* lie inside a confidence interval for the parameter.

As an example, consider the test of $H_0: \mu = 0$ versus $H_a: \mu \neq 0$. Suppose a 95% confidence interval for μ extends from 1.35 to 3.42; that is, it does *not* include the hypothesized value 0. Then we can reject H_0 at the 5% significance level, and we know that the p-value from the sample must be less than 0.05. On the other hand, if a 95% confidence interval for μ extends, say, from -1.25 to 2.31 (negative to positive), then we can't reject the null hypothesis at the 5% significance level, and the p-value must be greater than 0.05.

There is also a correspondence between one-tailed hypothesis tests and "one-sided" confidence intervals, but we will not pursue this here.

10.2.8 Practical versus Statistical Significance

We have stated that statistically significant results are those that produce sufficiently small p-values. In other words, statistically significant results are those that provide strong evidence in support of the alternative hypothesis. We frequently hear about studies, particularly in the medical sciences, that produce statistically significant results. For example, we might hear that mice injected with one kind of drug develop "significantly more" cancer cells than mice injected with a second kind of drug.

The point of this section is that such results are not necessarily significant in the sense of being *important*. They might be significant only in the statistical sense. An example of what could happen is the following. An education researcher wants to see whether quantitative SAT scores differ, on average, across gender. He sets up the hypotheses $H_0: \mu_M = \mu_F$ versus $H_a: \mu_M \neq \mu_F$, where μ_M and μ_F are the mean quantitative SAT scores for males and females, respectively. He then randomly samples scores from 4000 males and 4000 females and finds the male and female sample averages to be 521 and 524. He also finds that the sample standard deviation for each group is about 50. Based on these numbers, the p-value for the sample data is approximately 0.007. (We'll see how to make this calculation later in this chapter.) Therefore, he claims that the results are "significant proof" that males do score differently (lower) than females.

If you read these results in a newspaper, your immediate reaction might be, "Who cares?" After all, the difference between 521 and 524 is certainly not very large from a practical point of view. So why does the education researcher get to make his claim? Here's what's going on. The chances are that the means μ_M and μ_F are not *exactly* equal. There is bound to be some difference between genders over the entire population. If the researcher takes large enough samples—and 4000 is plenty large—he is almost certain to obtain enough evidence to "prove" that the means are not equal. That is, he will almost surely obtain *statistically* significant results. However, the difference he finds, as in the numbers we quoted, might be of little *practical* significance. No one really cares whether females score 3 points higher or lower than males. If the difference were on the order of 30 to 40 points, then we might be interested.

As this example illustrates, there is always a possibility of statistical significance but not practical significance with large sample sizes. To be fair, we should also mention the opposite case, which typically occurs with small sample sizes. Here we fail to obtain statistical significance even though the truth about the population(s), if it were known, would be of practical significance. Let's assume that a medical researcher wants to test whether a new form of treatment produces a higher cure rate for a deadly disease than the best treatment currently on the market. Due to expenses, the researcher is able to run a controlled experiment on only a relatively small number of patients with the disease. Unfortunately, the results of the experiment are inconclusive. They show some evidence that the new treatment works better, but the *p*-value for the test is only 0.25.

In the scientific community these results would not be enough to warrant a switch to the new treatment. However, it is certainly possible that the new treatment, if it were used on a large number of patients, would provide a "significant" improvement in the cure rate—where "significant" now implies *practical* significance. In this type of situation, we could easily fail to discover practical significance because the sample sizes are not large enough to detect it statistically.

From here on, when we use the term "significant," we mean *statistically* significant. However, you should always keep the ideas in this section in mind. A statistically significant result is not necessarily of practical importance. Conversely, a result that fails to meet the criterion for statistical significance is not necessarily one that should be ignored.

> Extremely large samples can easily lead to significantly significant results that are not practically important. In contrast, small samples can fail to produce statistically significant results that might indeed be practically important.

(10.3) Hypothesis Tests for a Population Mean

Now that we have covered the general concepts behind hypothesis testing and the principal sampling distributions, the mechanics of hypothesis testing are fairly straightforward. We will discuss in some detail how the procedure works for a population mean. Then in later sections we will illustrate similar hypothesis tests for other parameters.

As with confidence intervals, the key to the analysis is the sampling distribution of the sample mean. We know that if we subtract the true mean μ from the sample mean and divide the difference by the standard error $s/\sqrt{n}$, the result has a t distribution with $n - 1$ degrees of freedom. In a hypothesis-testing context, the true mean to use is the null hypothesis value, specifically, the borderline value between the null and alternative hypotheses. This value is usually labeled μ_0, where the subscript reminds us that it is based on the *null* hypothesis.

To run the test, we calculate the "test statistic" in equation (10.1). This t-value indicates how many standard errors the sample mean is from the null value, μ_0. If the null hypothesis is true, or more specifically, if $\mu = \mu_0$, this test statistic has a t distribution with $n - 1$ degrees of freedom. The p-value for the test is the probability beyond the test statistic in both tails (for a two-tailed alternative) or in a single tail (for a one-tailed alternative) of the t distribution.

Test Statistic for Test of Mean	$$t\text{-value} = \frac{\overline{X} - \mu_0}{s/\sqrt{n}}$$	**(10.1)**

We illustrate the procedure by continuing the pizza manager's problem in Example 10.1.

Example 10.1 Experimenting with a New Pizza Style at the Pepperoni Pizza Restaurant (continued)

Recall that the manager of the Pepperoni Pizza Restaurant is running an experiment to test the hypotheses $H_0: \mu \leq 0$ versus $H_a: \mu > 0$, where μ is the mean rating in the entire customer population. Here, each customer rates the difference between an old-style pizza and a new-style pizza on a scale from -10 to $+10$, where negative ratings favor the old style and positive ratings favor the new style. The ratings for 40 randomly selected customers and several summary statistics appear in Figure 10.3. (See the file PIZZA.XLS.) Is there sufficient evidence from these sample data for the manager to reject H_0?

Objective To use a one-sample t test to see whether consumers prefer the new style pizza to the old style.

Solution

From the summary statistics, we see that the sample mean is $\overline{X} = 2.10$ and the sample standard deviation is $s = 4.717$. This positive sample mean provides some evidence in favor of the alternative hypothesis, but given the rather large value of s and the boxplot of ratings shown in Figure 10.4, which indicates a lot of negative ratings, does it provide *enough* evidence to reject H_0?

To run the test, we calculate the test statistic, using the borderline null hypothesis value $\mu_0 = 0$, and report how much probability is beyond it in the right tail of the appropriate t distribution. We use the *right* tail because the alternative is one-tailed of the "greater than" variety. The test statistic is

$$t\text{-value} = \frac{2.10 - 0}{4.717/\sqrt{40}} = 2.816$$

FIGURE 10.3

Data and Summary Measures for Pizza Example

	A	B	C	D	E	F
1	Testing the pizza manager's one-tailed hypothesis					
2						
3	Customer	Rating			Results for one-sample analysis for Rating	
4	1	-7				
5	2	7			Summary measures	
6	3	-2			Sample size	40
7	4	4			Sample mean	2.100
8	5	7			Sample standard deviation	4.717
9	6	6				

FIGURE 10.4

Boxplot for Pizza
Data

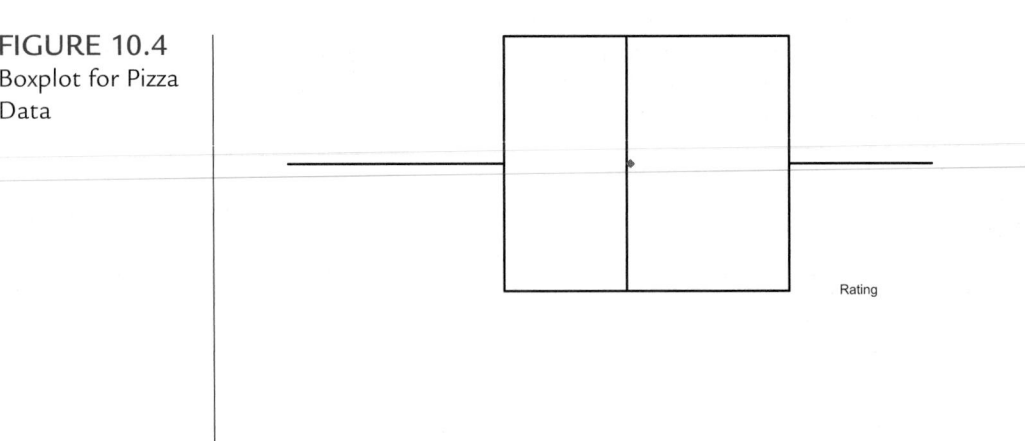

This *t*-value indicates that the sample mean is over 2.8 standard errors to the right of the null value, 0. Intuitively, this provides a lot of evidence in favor of the alternative—it is quite unlikely to see a sample mean 2.8 standard errors to right of a "true" mean. The probability beyond this value in the right tail of a *t* distribution with $n - 1 = 39$ degrees of freedom is approximately 0.004, which can be found in Excel with the function TDIST(2.816,39,1). (Recall that the first argument is the *t*-value, the second is the degrees of freedom, and the third is the number of tails.)

This probability, 0.004, is the *p*-value for the test. It indicates that these sample results would be *very* unlikely if the null hypothesis is true. The manager has two choices at this point. He can conclude that the null hypothesis is true and he obtained a very unlikely sample, or he can conclude that the alternative hypothesis is true—and presumably switch to the new-style pizza. This second conclusion certainly appears to be the more reasonable of the two.

Another way of interpreting the results of the test is in terms of traditional significance levels. We can reject H_0 at the 1% significance level because the *p*-value is less than 0.01. Of course, we can also reject H_0 at the 5% level or the 10% level because the *p*-value is also less than 0.05 and 0.10. But as we discussed previously, the *p*-value is a preferred way of reporting the results because it indicates exactly *how* significant these sample results are.

The StatPro One-Sample procedure can be used to perform this analysis easily, with the results shown in Figure 10.5. To use it, select the StatPro/Statistical Inference/One-

FIGURE 10.5

Hypothesis Test
for the Mean for
the Pizza Example

	D	E	F	G
3		*Results for one-sample analysis for Rating*		
4				
5		*Summary measures*		
6		Sample size	40	
7		Sample mean	2.100	
8		Sample standard deviation	4.717	
9				
10		*Test of mean<=0 versus one-tailed alternative*		
11		Hypothesized mean	0.000	
12		Sample mean	2.100	
13		Std error of mean	0.746	
14		Degrees of freedom	39	
15		t-test statistic	2.816	
16		p-value	0.004	

FIGURE 10.6
One-Sample Op-
tions Dialog Box

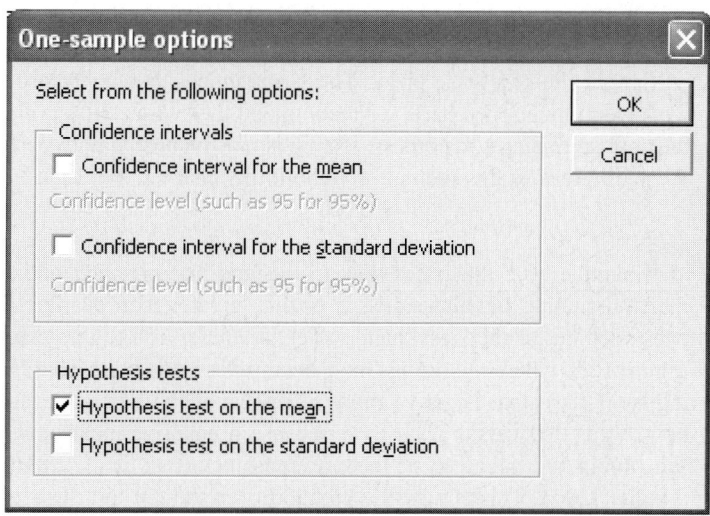

FIGURE 10.7
Hypothesis Test
Dialog Box

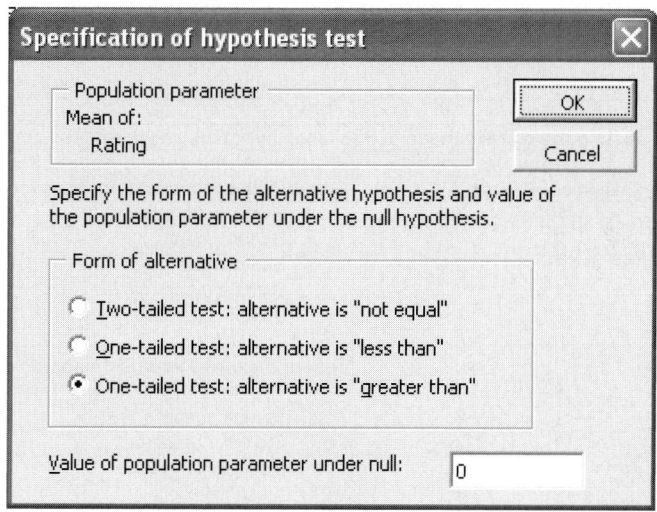

Sample Analysis menu item, and choose the Rating variable as the variable to analyze. Then fill out the next two dialog boxes as shown in Figures 10.6 and 10.7.

Most of the output in Figure 10.5 should be familiar: It mirrors the calculations we did previously, and you can check the formulas in the output cells to ensure that you understand the procedure. We note the following. First, the value in cell F11, 0, is the null hypothesis value μ_0 at the borderline between H_0 and H_a; it is the value specified in the dialog box in Figure 10.7. Second, take a look at the notes entered in cells D10 and F16. (These notes aren't visible in Figure 10.5, but they can be seen in the completed file.) The note in cell D10 reminds us that this test is based on normality of the underlying population distribution and/or a sufficiently large sample size. If these conditions are not satisfied (which is not a problem for this example), then other more appropriate tests are available. The note in cell F16 indicates that the results are significant at the 1% level. In general, StatPro compares the p-value to the three traditional significance levels, 1%, 5%, and 10%, and interprets significance in terms of these.

Before leaving this example, we ask one last question. Should the manager switch to the new-style pizza on the basis of these sample results? We would probably recommend "yes." There is no indication that the new-style pizza costs any more to make than the old-style pizza, and the sample evidence is fairly convincing that customers, on average, will prefer the new-style pizza. Therefore, unless there are reasons for not switching that we haven't mentioned here, we recommend the switch. However, if it costs more to make the new-style pizza, hypothesis testing is *not* the best way to perform a decision analysis. We will return to this theme throughout this chapter. ●

Example 10.1 illustrates how to run and interpret any one-tailed hypothesis for the mean, assuming the alternative is of the "greater than" variety. If the alternative is still one-tailed but of the "less than" variety, there is virtually no change. We illustrate this in Figure 10.8, where the ratings have been reversed in sign. That is, we multiplied each rating by -1, so that negative ratings now favor the new-style pizza. The hypotheses are now $H_0: \mu \geq 0$ versus $H_a: \mu < 0$ because a negative mean now supports the new style. We obtain the negative of the previous test statistic, -2.816, and exactly the same p-value, 0.004. This is now the probability in the *left* tail of the t distribution, but the interpretation of the results is exactly the same as before.

The analysis of two-tailed tests for the mean is also quite similar to the analysis in Example 10.1. We illustrate a typical two-tailed test in the following example.

FIGURE 10.8
Hypothesis Test with Reverse Coding

	A	B	C	D	E	F	G
1	Testing the pizza manager's one-tailed hypothesis with reverse coding						
2							
3	Customer	Rating			Results for one-sample analysis for Rating		
4	1	7			Hypothesized mean	0.000	
5	2	-7			Sample mean	-2.100	
6	3	2			Sample size	40	
7	4	-4			Sample mean	-2.100	
8	5	-7			Sample standard deviation	4.717	
9	6	-6					
10	7	0			Test of mean>=0 versus one-tailed alternative		
11	8	-2			Hypothesized mean	0.000	
12	9	-8			Sample mean	-2.100	
13	10	-2			Std error of mean	0.746	
14	11	-3			Degrees of freedom	39	
15	12	4			t-test statistic	-2.816	
16	13	-8			p-value	0.004	
17	14	5					

Example 10.2 Measuring Student Reaction to a New Textbook

A large required chemistry course at State University has been using the same textbook for a number of years. Over the years, the students have been asked to rate this textbook on 10-point scale, and the average rating has been stable at about 5.2. This year, the faculty decided to experiment with a new textbook. After the course, 50 randomly selected students were asked to rate this new textbook, also on a scale of 1 to 10. The results appear in column B of Figure 10.9. Can we conclude that the students like this new textbook any more or less than the previous textbook?

Objective To use a one-sample t test to see whether students like the new textbook any more or less than the old textbook.

FIGURE 10.9
Test of Two-Tailed Alternative

	A	B	C	D	E	F	G
1	Testing an instructor's two-tailed hypothesis about a new textbook						
2							
3	Mean rating of previous textbook (on a 1-10 scale)					5.2	
4							
5	Ratings of a new textbook (on a 1-10 scale)						
6							
7	Student	Rating			Results for one-sample analysis for Rating		
8	1	6					
9	2	3			Summary measures		
10	3	6			Sample size	50	
11	4	7			Sample mean	5.680	
12	5	6			Sample standard deviation	1.953	
13	6	10					
14	7	6			Confidence interval for mean		
15	8	8			Confidence level	95.0%	
16	9	7			Sample mean	5.680	
17	10	10			Std error of mean	0.276	
18	11	3			Degrees of freedom	49	
19	12	6			Lower limit	5.125	
20	13	4			Upper limit	6.235	
21	14	6					
22	15	8			Test of mean=5.2 versus two-tailed alternative		
23	16	10			Hypothesized mean	5.200	
24	17	5			Sample mean	5.680	
25	18	4			Std error of mean	0.276	
26	19	6			Degrees of freedom	49	
27	20	4			t-test statistic	1.738	
28	21	6			p-value	0.088	
29	22	6					
30	23	4					
56	49	6					
57	50	5					

Solution

The first question is whether the test should be one-tailed or two-tailed. Of course, the faculty have chosen the new textbook with the expectation that it will be preferred by the students, but it is very possible that students will like it *less* than the previous textbook. (Students are notoriously unpredictable in their acceptance of textbooks.) Therefore, we set this up as a two-tailed test—that is, the alternative hypothesis is that the mean rating of the new textbook is either less than *or* greater than the mean rating, 5.2, of the previous textbook. Formally, we write the hypotheses as H_0: $\mu = 5.2$ versus H_a: $\mu \neq 5.2$.

The test is run (and the StatPro One-Sample procedure can be used) almost exactly as with a one-tailed test. We calculate the *t*-distributed test statistic in the same way as before (see the output in Figure 10.9):

$$t\text{-value} = \frac{\overline{X} - 5.2}{s/\sqrt{n}} = \frac{5.680 - 5.2}{1.953/\sqrt{50}} = 1.738$$

The *p*-value is then the probability beyond -1.738 in the left tail *and* beyond $+1.738$ in the right tail of a *t* distribution with $n - 1 = 49$ degrees of freedom. The effect is to double the one-tailed *p*-value. From the output (cell F28) in Figure 10.9, we see that the two-tailed *p*-value is 0.088. (It results from the Excel function =TDIST(ABS(F27),F26,2). Note that the first argument of TDIST must always be nonnegative, and the absolute value function ABS ensures that it will be. The third argument is 2 because this is a two-tailed test.)

This moderately small *p*-value provides some evidence, but probably not convincing evidence, that the mean rating of the new textbook is different from the old mean rating of 5.2. If the *p*-value were lower (which might occur if more students were sampled), the evidence would be more conclusive. As in Example 10.1, we can now ask whether the faculty should continue to use the new textbook. Here again, it is probably not a decision that hypothesis testing, at least by itself, should determine. The students *appear* to favor the new textbook, if only by a small margin. If the faculty also favor it, we see no reason for not continuing to use it.

Because this is a two-tailed test, we can also perform the test by appealing to confidence intervals. A 95% confidence interval for the mean rating of the new textbook was also requested in the StatPro output in Figure 10.9. This interval extends from 5.125 to 6.235. Because this interval *does* include the old mean rating of 5.2, we cannot reject the null hypothesis at the 5% significance level. This is in agreement with the *p*-value of the test, which is greater than 0.05. However, we can check that a 90% confidence interval for the mean does *not* include 5.2. Therefore, we can reject the null hypothesis at the 10% level. This too is in agreement with the *p*-value, which is less than 0.10. ●

P R O B L E M S

Level A

1. Suppose a firm producing lightbulbs wants to know whether it can claim that its lightbulbs typically last more than 1000 burning hours. Hoping to find support for this claim, the firm collects a random sample of 100 lightbulbs and records the lifetime (in hours) of each. The sample data are contained in the file P10_01.XLS.
   a. Using a 5% significance level, can this lightbulb manufacturer claim that its bulbs typically last more than 1000 hours? Explain your answer.
   b. Using a 1% significance level, can this lightbulb manufacturer claim that its bulbs typically last more than 1000 hours? Explain your answer.

2. A manufacturer is interested in determining whether it can claim that the boxes of detergent it sells contain, on average, more than 500 grams of detergent. From past experience the manufacturer knows that the amount of detergent in the boxes is approximately normally distributed. The firm selects a random sample of 100 boxes and records the amount of detergent (in grams) in each box. These data are provided in the file P10_02.XLS. Formulate an appropriate hypothesis test and report a *p*-value. Do you find statistical support for the manufacturer's claim? Explain.

3. A producer of steel cables wants to know whether the steel cables it produces have an average breaking strength of 5000 pounds. An average breaking strength of less than 5000 pounds would not be adequate, and to produce steel cables with an average breaking strength in excess of 5000 pounds would unnecessarily increase production costs. The producer collects a random sample of 64 steel cable pieces. The breaking strength for each of these cable pieces is recorded in the file P10_03.XLS.
   a. Using a 5% significance level, what statistical conclusion can the producer reach regarding the aver-

age breaking strength of its steel cables? Explain your answer.
   b. Using a 1% significance level, what statistical conclusion can the producer reach regarding the average breaking strength of its steel cables? Explain your answer.

4. A U.S. Navy recruiting center knows from past experience that the heights of its recruits are normally distributed with mean 68 inches. The recruiting center wants to test the claim that the average height of this year's recruits is greater than 68 inches. To do this, recruiting personnel take a random sample of 64 recruits from this year and record their heights (in inches). The data are provided in the file P10_04.XLS.
   a. On the basis of the available sample information, do the recruiters find support for the given claim at the 1% significance level? Explain.
   b. Use the sample data to construct a 95% confidence interval for the average height of this year's recruits. Based on this confidence interval, what conclusion should recruiting personnel reach regarding the given claim?

5. Suppose that we wish to test $H_0: \mu = 10$ versus $H_a: \mu > 10$ at the $\alpha = 0.05$ significance level. Furthermore, suppose that we observe values of the sample mean and sample standard deviation when $n = 40$ that do *not* lead to the rejection of H_0. Is it true that we might reject H_0 if we observed the same values of the sample mean and sample standard deviation from a sample with $n > 40$? Why or why not?

Level B

6. A study is performed in a large southern town to determine whether the average weekly grocery bill per four-person family in the town is significantly different from the national average. A random sample of the

weekly grocery bills of four-person families in this town is given in the file P10_06.XLS.

a. Assume that the national average weekly grocery bill for a four-person family is $100. Is the sample evidence statistically significant? If so, at what significance levels can you reject the null hypothesis?

b. For which values of the sample mean (i.e., average weekly grocery bill) would you decide to reject the null hypothesis at the $\alpha = 0.01$ significance level? For which values of the sample mean would you decide to reject the null hypothesis at the $\alpha = 0.10$ level?

7. An aircraft manufacturer needs to buy aluminum sheets with an average thickness of 0.05 inch. The manufacturer knows that significantly thinner sheets would be unsafe and considerably thicker sheets would be too heavy. A random sample of 100 sheets from a potential supplier is collected. The thickness of each sheet in this sample is measured (in inches) and recorded in the file P10_07.XLS.

a. Based on the results of an appropriate hypothesis test, should the aircraft manufacturer buy aluminum sheets from this supplier? Explain why or why not.

b. For which values of the sample mean (i.e., average thickness) would the aircraft manufacturer decide to buy sheets from this supplier? Assume that $\alpha = 0.05$ in answering this question.

8. Suppose that we observe a random sample of size n from a normally distributed population. If we are able to reject $H_0:\mu = \mu_0$ in favor of a two-tailed alternative hypothesis at the 10% significance level, is it true that we can definitely reject H_0 in favor of the appropriate one-tailed alternative at the 5% significance level? Why or why not?

Hypothesis Tests for Other Parameters

Just as we developed confidence intervals for a variety of parameters, we can develop hypothesis tests for other parameters. They are based on the same sampling distributions we discussed in the previous chapter, and they are run and interpreted exactly as the tests for the mean in the previous section. In each case we use sample data to calculate a test statistic that has a well-known sampling distribution. Then we calculate a corresponding p-value to measure the support for the alternative hypothesis. Beyond this, only the details change, as we illustrate in this section.

10.4.1 Hypothesis Tests for a Population Proportion

To test a population proportion p, recall that the sample proportion $\hat{p}$ has a sampling distribution that is approximately normal when the sample size is reasonably large. Specifically, the standardized value

$$\frac{\hat{p} - p}{\sqrt{p(1 - p)/n}}$$

is approximately distributed as a standard normal random variable Z.

Let p_0 be the borderline value of p between the null and alternative hypotheses. Then we substitute p_0 for p to obtain the test statistic in equation (10.2). The p-value of the test is found by seeing how much probability is beyond this test statistic in the tail (or tails) of the standard normal distribution.[2] A rule of thumb for checking the large-sample assumption of this test is to check whether $np_0 > 5$ and $n(1 - p_0) > 5$.

---

[2]Do not confuse the unknown proportion p with the p-value of the test. They are logically different concepts and just happen to share the same letter p.

$$z\text{-value} = \frac{\hat{p} - p_0}{\sqrt{p_0(1 - p_0)/n}} \qquad (10.2)$$

We illustrate this test of proportion in the following example.

Example 10.3 Customer Complaints at Walpole Appliance Company

The Walpole Appliance Company has a customer service department that handles customer questions and complaints. This department's processes are set up to respond quickly and accurately to customers who phone in their concerns. However, there is a sizable minority of customers who prefer to write letters. Traditionally, the customer service department has not been very efficient in responding to these customers.

Letter writers first receive a mail-gram asking them to call customer service (which is exactly what letter writers wanted to avoid in the first place!), and when they do call, the customer service representative who answers the phone typically has no knowledge of the customer's problem. As a result, the department manager estimates that 15% of letter writers have not obtained a satisfactory response within 30 days of the time their letters were first received. The manager's goal is to reduce this value by at least half, that is, to 7.5% or less.

To do so, she changes the process for responding to letter writers. Under the new process, these customers now receive a prompt and courteous form letter that responds to their problem. (This is possible because the vast majority of concerns can be addressed by one of several form letters.) Each form letter states that if the customer still has problems, he or she can call the department. The manager also files the original letters so that if customers do call back, the representative who answers will be able to find their letters quickly and respond intelligently. With this new process in place, the manager has tracked 400 letter writers and has found that only 23 of them are classified as "unsatisfied" after a 30-day period. Does it appear that the manager has achieved her goal?

Objective To use a test for a proportion to see whether the new process of responding to complaint letters results in an acceptably low proportion of unsatisfied customers.

Solution

The manager's goal is to reduce the proportion of unsatisfied customers after 30 days from 0.15 to 0.075 or less. Because the burden of proof is on her to "prove" that she has accomplished this goal, we set up the hypotheses as $H_0: p \geq 0.075$ versus $H_a: p < 0.075$, where p is the proportion of all letter writers who are still unsatisfied after 30 days. The sample proportion she has observed is $\hat{p} = 23/400 = 0.0575$. This is obviously less than 0.075, but is it *enough* less to reject the null hypothesis?

The test statistic for the data, using the borderline value $p_0 = 0.075$, is

$$z\text{-value} = \frac{0.0575 - 0.075}{\sqrt{0.075(1 - 0.075)/400}} = -1.329$$

This value appears in cell B10 of Figure 10.10. You can already guess that the results will not be statistically significant. This z-value indicates that the observed proportion is only about 1.33 standard errors below the null proportion, which isn't all that large a difference. (See the file LETTERS.XLS. Although StatPro does not have a procedure for test-

FIGURE 10.10
Analysis of New Process for Letter Writers

	A	B
1	**Test of a proportion: responding to letter writers**	
2		
3	Target proportion with new procedure	0.075
4		
5	Number of unsatisfied customers after 30 days	23
6	Number of customers sampled	400
7	Sample proportion	0.0575
8	Standard error of sample proportion	0.01317
9		
10	z test statistic	-1.329
11	p value for a one-tailed test	0.092
12		
13	Confidence interval for true proportion	
14	Confidence level	95%
15	Standard error	0.012
16	z-multiple	1.960
17	Lower limit	0.035
18	Upper limit	0.080

ing proportions, this file serves as a "template" for all such tests.) Note that we first find the denominator (the standard error of $\hat{p}$) in cell B8 with the formula

$$=SQRT(B3*(1-B3)/B6)$$

The corresponding p-value, 0.092, is found with the formula

$$=NORMSDIST(B10)$$

in cell B11. It is the probability to the *left* of -1.329 in the standard normal distribution. Also, because $np_0 = 400(0.075) = 30 > 5$ and $n(1 - p_0) = 400(0.925) > 5$, this test is valid; that is, the sample size is large enough for the normal approximation to hold.

The p-value in cell B11 might not be as low as you expected—or as low as the manager would like. In spite of the fact that the sample proportion appears to be well below the target proportion of 0.075, the evidence in support of the alternative hypothesis is not overwhelming. In statistical terminology, the results are significant at the 10% level, but not at the 5% or 1% levels.

We also show a 95% confidence interval for the unknown proportion p in Figure 10.10. For example, the formula in cell B17 is

$$=B7-B16*SQRT(B7*(1-B7)/B6)$$

This confidence interval extends from 0.035 to 0.080. It includes the target value, 0.075, but just barely. In this sense it also provides some support for the argument that the manager has indeed achieved her goal.[3]

Analysts might disagree on whether a hypothesis test or a confidence interval is the more appropriate way to present these results. However, we see them as complementary and do not necessarily favor one over the other. The bottom line is that they both provide some, but not totally conclusive, evidence that the manager has achieved her goal. ●

---

[3]Note that the standard error in cell B8 for the hypothesis test uses the target proportion 0.075. In contrast, the standard error for the confidence interval uses the sample proportion 0.0575. The sampling distribution for a hypothesis test always uses the borderline value between H_0 and H_a. But because confidence intervals aren't connected to any hypotheses, their standard errors must rely on sample data. In most cases the two standard errors are practically the same.

10.4.2 Hypothesis Tests for Differences Between Population Means

This comparison problem—comparing two population means—is one of the most important problems analyzed with statistical methods. It can be analyzed with confidence intervals, hypothesis tests, or both.

We now discuss the comparison problem, where we test the difference between two population means. As in the previous chapter, the form of the analysis depends on whether the two samples are independent or paired.

If they are paired, then the test proceeds exactly as in Section 10.3, using the differences as the single variable of analysis. That is, if $\overline{D}$ is the sample mean difference between n pairs, D_0 is the hypothesized difference (the borderline value between H_0 and H_a), and s_D is the sample standard deviation of the differences, then the test is based on the test statistic in equation (10.3). If D_0 is the true mean difference, then this test statistic has a t distribution with $n - 1$ degrees of freedom. The validity of the test also requires that n be reasonably large and/or the population of *differences* be approximately normally distributed. The test statistic is defined in the box.

Test Statistic for Paired Samples Test of Difference Between Means

$$t\text{-value} = \frac{\overline{D} - D_0}{s_D/\sqrt{n}} \tag{10.3}$$

If the samples are independent and the population standard deviations are equal, then the two-sample theory discussed in Section 9.6.1 is relevant. It leads to the test statistic in equation (10.4). Here, $\overline{X}_1$ and $\overline{X}_2$ are the two sample means, D_0 is the hypothesized difference, n_1 and n_2 are the sample sizes, and s_p is the same pooled estimate of the common population standard deviation as in the previous chapter:

$$s_p = \sqrt{\frac{(n_1 - 1)s_1^2 + (n_2 - 1)s_2^2}{n_1 + n_2 - 2}}$$

If D_0 is the true mean difference, then this test statistic has a t distribution with $n_1 + n_2 - 2$ degrees of freedom. The validity of this test again requires that the sample sizes be reasonably large and/or the populations be approximately normally distributed.

Test Statistic for Independent Samples Test of Difference Between Means

$$t\text{-value} = \frac{(\overline{X}_1 - \overline{X}_2) - D_0}{s_p\sqrt{1/n_1 + 1/n_2}} \tag{10.4}$$

Fortunately, these formulas are implemented automatically by StatPro's procedures. We begin by illustrating an example of the paired-sample t test.

Example 10.4 Measuring the Effects of Traditional and New Styles of Soft-Drink Cans

Beer and soft-drink companies have recently become very concerned about the style of their cans. There are cans with fluted and embossed sides and cans with six-color graphics and holograms. Coca-Cola is even experimenting with a contoured can, shaped like the old-fashioned Coke bottle minus the neck. Evidently, these companies believe the style of the can makes a difference to consumers, which presumably translates into higher sales.

Assume that a soft-drink company is considering a style change to its current can, which has been the company's trademark for many years. To determine whether this new style is popular with consumers, the company runs a number of focus group sessions around the country. At each of these sessions, randomly selected consumers are allowed to examine the new and traditional styles, exchange ideas, and offer their opinions. Eventually, they fill out a form where, among other items, they are asked to respond to the following items, each on a scale of 1 to 7, 7 being the best:

- Rate the attractiveness of the traditional-style can (AO).
- Rate the attractiveness of the new-style can (AN).
- Rate the likelihood that you would buy the product with the traditional-style can (WBO).
- Rate the likelihood that you would buy the product with the new-style can (WBN).

The results over all focus groups are shown in Figures 10.11 and 10.12. (See the file CANS.XLS.) What can the company conclude from these data? Are hypothesis tests appropriate?

Objective To use paired-sample t tests for differences between means to see whether consumers rate the attractiveness, and their likelihood to purchase, higher for a new-style can than for the traditional-style can.

Solution

First, it is a good idea to examine summary statistics for the data. The averages from each survey item are shown at the bottom of Figure 10.11. They indicate some support for the new-style can. Also, we might expect the ratings for a given consumer to be correlated. This turns out to be the case, as shown by the relatively large positive correlations in Figure 10.12. These large positive correlations indicate that if we want to examine differences between survey items, a paired-sample procedure will make the most efficient use of the

FIGURE 10.11
Data on Soft-Drink Cans

	A	B	C	D	E
1	Focus group results on can styles				
2					
3	Consumer	AO	AN	WBO	WBN
4	1	5	7	4	1
5	2	7	7	6	6
6	3	6	7	7	6
7	4	1	3	1	1
8	5	3	4	1	1
9	6	7	7	7	7
10	7	5	7	4	6
11	8	6	7	6	7
12	9	5	7	6	6
13	10	5	4	4	6
14	11	1	3	1	1
15	12	2	1	1	3
16	13	6	6	6	6
17	14	4	5	3	3
18	15	2	5	1	1
19	16	6	7	7	7
20	17	4	5	2	1
181	178	5	4	4	3
182	179	3	4	1	3
183	180	3	5	6	7
184					
185	Averages	4.41	4.95	3.86	4.34

FIGURE 10.12

Correlations for
Soft-Drink Can
Data

	A	B	C	D	E	F
1	*Table of correlations*					
2			AO	AN	WBO	WBN
3		AO	1.000			
4		AN	0.740	1.000		
5		WBO	0.746	0.595	1.000	
6		WBN	0.594	0.401	0.774	1.000

data. Of course, a paired-sample procedure also makes sense because each consumer answers each item on the form. (If this is confusing, think about the following alternative setup. We have four *separate* groups of consumers. The first group responds to item 1 only, the second group responds to item 2 only, and so on. Then the responses to the various items are in no way paired, and an *independent-sample* procedure would be used instead.)

There are several differences of interest. The two most obvious are the difference between the attractiveness ratings of the two styles and the difference between the likelihoods of buying the two styles—that is, column B minus column C and column D minus column E. A third difference of interest is the difference between the attractiveness ratings of the new style and the likelihoods of buying the new can—that is, column C minus column E. This difference indicates whether perceptions of the new-style can are likely to translate into sales. Finally, a fourth difference that might be of interest is the difference between the third difference (column C minus column E) and the similar difference for the old style (column B minus column D). This checks whether the translation of perceptions into sales is any different for the two styles of cans.

All of these differences appear next to the original data in Figure 10.13. In terms of the original data, they are labeled as:

- Diff1: AO − AN
- Diff2: WBO − WBN
- Diff3: AN − WBN
- Diff4: AO − WBO
- Diff5: (AN − WBN) − (AD − WBO)

We generate these differences in columns F–J. (Actually, StatPro's Paired-Sample procedure generates the differences in columns F, G, H, and J automatically when it tests these differences. We manually inserted the difference in column I so that it could be used to generate the difference in column J.)

For each of the differences, Diff1, Diff2, Diff3, and Diff5, we test the mean difference over all potential consumers with a paired-sample analysis. (We actually run the one-sample procedure on the difference variables.) Exactly as in the previous chapter, we treat each difference variable as a *single* sample and run the same t test as in Section 10.3 on this sample. In each case the hypothesized difference, D_0, is 0. The only question is whether to run one-tailed or two-tailed tests. We propose that the tests for Diff1, Diff2, and Diff5 be two-tailed tests and that the test on Diff3 be a one-tailed test with the alternative of the "greater than" variety. The reasoning is that the company probably has little idea which way the differences Diff1, Diff2, and Diff5 will go (positive or negative), whereas it expects that Diff3 will be positive on average. That is, the company expects that consumers' ratings of the attractiveness of the new design will, on average, be larger than their likelihoods of purchasing the product. However, any of these hypotheses could be run as one-tailed or two-tailed tests. It depends on the prior beliefs of the company. In any case, to change a one-tailed p-value to a two-tailed p-value, all we need to do is multiply by 2. Similarly, we can change two-tailed p-values to one-tailed p-values by dividing by 2.

The results from the four tests appear in Figures 10.14 and 10.15. These outputs also include 99% confidence intervals for the corresponding mean differences. We obtained each

	A	B	C	D	E	F	G	H	I	J
1	Focus group results on can styles									
2										
3	Consumer	AO	AN	WBO	WBN	AO-AN	WBO-WBN	AN-WBN	AO-WBO	(AN-WBN)-(AO-WBO)
4	1	5	7	4	1	-2	3	6	1	5
5	2	7	7	6	6	0	0	1	1	0
6	3	6	7	7	6	-1	1	1	-1	2
7	4	1	3	1	1	-2	0	2	0	2
8	5	3	4	1	1	-1	0	3	2	1
9	6	7	7	7	7	0	0	0	0	0
10	7	5	7	4	6	-2	-2	1	1	0
11	8	6	7	6	7	-1	-1	0	0	0
12	9	5	7	6	6	-2	0	1	-1	2
13	10	5	4	4	6	1	-2	-2	1	-3
14	11	1	3	1	1	-2	0	2	0	2
15	12	2	1	1	3	1	-2	-2	1	-3
16	13	6	6	6	6	0	0	0	0	0
17	14	4	5	3	3	-1	0	2	1	1
18	15	2	5	1	1	-3	0	4	1	3
19	16	6	7	7	7	-1	0	0	-1	1
20	17	4	5	2	1	-1	1	4	2	2
181	178	5	4	4	3	1	1	1	1	0
182	179	3	4	1	3	-1	-2	1	2	-1
183	180	3	5	6	7	-2	-1	-2	-3	1

FIGURE 10.13 Original and Difference Variables for Soft-Drink Can Data

output for Diff1, Diff2, and Diff3 with the StatPro/Statistical Inference/Paired-Sample Analysis menu item, used on the appropriate pair of original variables (those in columns B–E in Figure 10.11). The output for Diff5 was based on the Diff3 and Diff4 variables.

Results of the analysis of soft-drink can style

- From the output for the Diff1 variable in Figure 10.14, there is overwhelming evidence that consumers, on average, rate the attractiveness of the new design higher than the attractiveness of the current design. The t-distributed test statistic is -5.351, calculated as

$$\frac{-0.539 - 0}{0.101} = -5.351$$

	L	M	N	O	P	Q	R	S
3		Paired-sample analysis for AO minus AN				Paired-sample analysis for WBO minus WBN		
4								
5		Summary measures for AO-AN				Summary measures for WBO-WBN		
6		Sample size	180			Sample size	180	
7		Sample mean	-0.539			Sample mean	-0.478	
8		Sample standard deviation	1.351			Sample standard deviation	1.347	
9								
10		Confidence interval for mean				Confidence interval for mean		
11		Confidence level	99.0%			Confidence level	99.0%	
12		Sample mean	-0.539			Sample mean	-0.478	
13		Std error of mean	0.101			Std error of mean	0.100	
14		Degrees of freedom	179			Degrees of freedom	179	
15		Lower limit	-0.801			Lower limit	-0.739	
16		Upper limit	-0.277			Upper limit	-0.216	
17								
18		Test of mean=0 versus two-tailed alternative				Test of mean=0 versus two-tailed alternative		
19		Hypothesized mean	0.000			Hypothesized mean	0.000	
20		Sample mean	-0.539			Sample mean	-0.478	
21		Std error of mean	0.101			Std error of mean	0.100	
22		Degrees of freedom	179			Degrees of freedom	179	
23		t-test statistic	-5.351			t-test statistic	-4.758	
24		p-value	0.000			p-value	0.000	

FIGURE 10.14 Analysis of Diff1 and Diff2 Variables

	T	U	V	W	X	Y	Z	AA	AB
3	**Paired-sample analysis for AN minus WBN**				**Paired-sample analysis for AN-WBN minus AO-WBO**				
4									
5	**Summary measures for AN-WBN**				**Summary measures for AN-WBN-AO-WBO**				
6		Sample size	180			Sample size	180		
7		Sample mean	0.611			Sample mean	0.061		
8		Sample standard deviation	2.213			Sample standard deviation	2.045		
9									
10	**Confidence interval for mean**				**Confidence interval for mean**				
11		Confidence level	99.0%			Confidence level	99.0%		
12		Sample mean	0.611			Sample mean	0.061		
13		Std error of mean	0.165			Std error of mean	0.152		
14		Degrees of freedom	179			Degrees of freedom	179		
15		Lower limit	0.182			Lower limit	-0.336		
16		Upper limit	1.041			Upper limit	0.458		
17									
18	**Test of mean<=0 versus one-tailed alternative**				**Test of mean=0 versus two-tailed alternative**				
19		Hypothesized mean	0.000			Hypothesized mean	0.000		
20		Sample mean	0.611			Sample mean	0.061		
21		Std error of mean	0.165			Std error of mean	0.152		
22		Degrees of freedom	179			Degrees of freedom	179		
23		t-test statistic	3.705			t-test statistic	0.401		
24		p-value	0.000			p-value	0.689		

FIGURE 10.15 Analysis of Diff3 and Diff5 Variables

and the corresponding p-value for a two-tailed test of the mean difference is (to three decimal places) 0.000. A 99% confidence interval for the mean difference extends from -0.801 to -0.277. Note that this 99% confidence interval does *not* include the hypothesized value 0. This is consistent with the fact that the two-tailed p-value is less than 0.01. (Recall the relationship between confidence intervals and two-tailed hypothesis tests from Section 10.2.7.)

- The results are basically the same for the difference between consumers' likelihoods of buying the product with the two styles. (See the output for the Diff2 variable in Figure 10.14.) Again, consumers are definitely more likely, on average, to buy the product with the new-style can. A 99% confidence interval for the mean difference extends from -0.739 to -0.216.

- The company's hypothesis that consumers' ratings of attractiveness of the new-style can are greater, on average, than their likelihoods of buying the product with this style can is confirmed. (See the output for the Diff3 variable in Figure 10.15.) The test statistic for this one-tailed test is 3.705 and the corresponding p-value is 0.000. A 99% confidence interval for the mean difference extends from 0.182 to 1.041.

- There is no evidence that the difference between attractiveness ratings and the likelihood of buying is any different for the new-style can than for the current-style can. (See the output for the Diff5 variable in Figure 10.15.) The test statistic for a two-tailed test of this difference is 0.401 and the corresponding p-value, 0.689, isn't close to any of the traditional significance levels. Furthermore, a 99% confidence interval for the mean difference extends from a negative value, -0.336, to a positive value, 0.458.

These results are further confirmed by histograms of the difference variables such as those shown in Figures 10.16 and 10.17.[4] (Boxplots could be used, but we prefer histograms when the variables include only a few possible integer values.) The histogram of the Diff1 variable in Figure 10.16 shows many more negative differences than positive

---

[4]To obtain the simplified horizontal axis labels (-3 instead of ≤ -3, for example), we simply change the labels in column B of the corresponding "Data" sheets produced by StatPro's Histogram procedure.

FIGURE 10.16
Histogram of the
Diff1 Variable

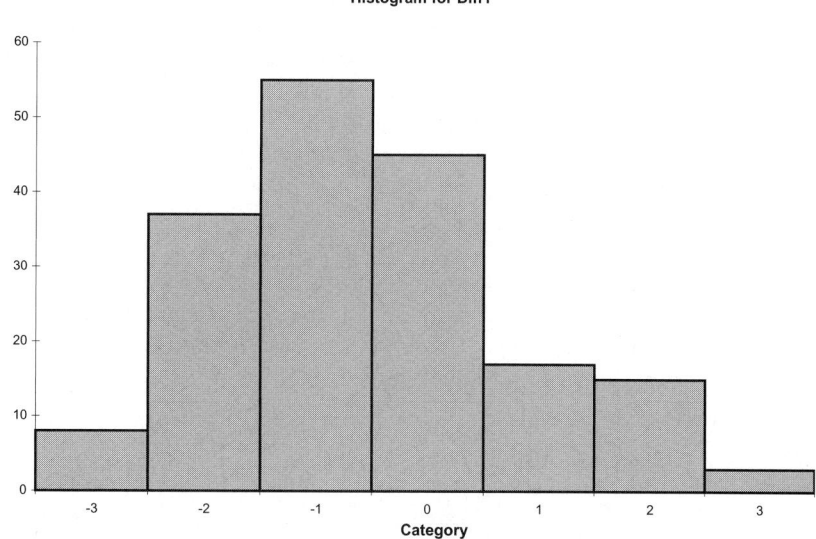

Histogram for Diff1

FIGURE 10.17
Histogram of the
Diff5 Variable

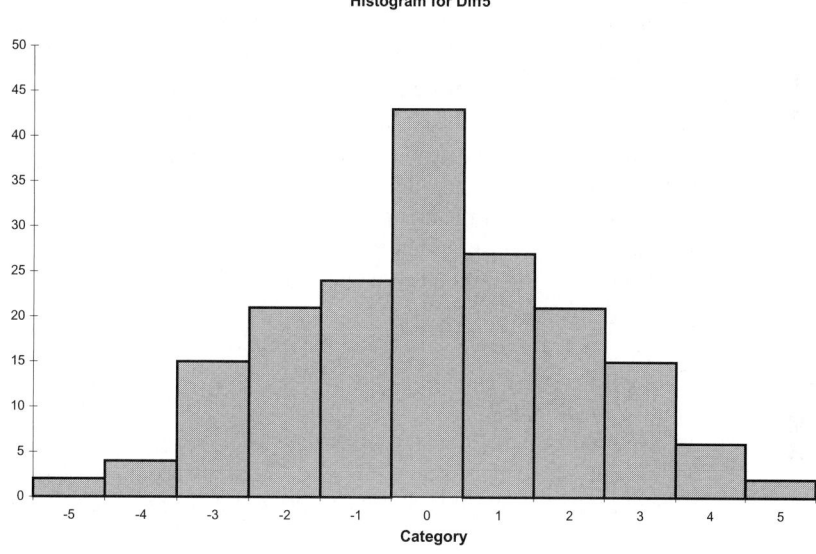

Histogram for Diff5

differences. This leads to the large negative test statistic and the all-negative confidence interval. In contrast, the histogram of the Diff5 variable in Figure 10.17 is almost perfectly symmetric around 0 and hence provides no evidence that the mean difference is nonzero.

This example illustrates once again how hypothesis tests and confidence intervals provide complementary information, although the confidence intervals are arguably the more useful of the two here. The hypothesis test for the first difference, for example, shows that the average rating for the new style is undoubtedly larger than for the current style. This is useful information, but it might be even more useful to know *how much* larger the average for the new style is. A confidence interval provides this information.

We conclude this example by recalling the distinction between practical significance and statistical significance. Due to the extremely low *p*-values, the results in Figure 10.14, for example, leave no doubt as to statistical significance. But this could be due to the large sample size. That is, if the true mean differences are even slightly different from 0, large samples will almost surely discover this and report small p-values. The soft-drink company, on the other hand, is more interested in knowing whether the observed differences are of any practical importance. This is not a statistical question. It is a question of what differences are important for the *business*. We suspect that the company would indeed be quite impressed with the observed differences in the sample—and might very well switch to the new-style can. ●

The following example illustrates the independent two-sample t test. We can tell that a paired-sample procedure is not appropriate because there is no attempt to match the observations in the two samples in any way. Indeed, this would be impossible because the sample sizes are not equal.

Example 10.5 Productivity Due to Exercise at Informatrix Software Company

Many companies are now installing exercise facilities at their plants. The goal is not only to provide a bonus (free use of exercise equipment) for their employees, but to make the employees more productive by getting them in better shape. One such company, the Informatrix Software Company, installed exercise equipment on site a year ago. To check whether it is having a beneficial effect on employee productivity, the company has gathered data on a sample of 80 randomly chosen employees, all between the ages of 30 and 40 and all with similar job titles and duties. The company observed which of these employees use the exercise facility regularly (at least three times per week on average). This group included 23 of the 80 employees in the sample. The other 57 employees were asked whether they exercise regularly elsewhere, and six of them replied that they do. The remaining 51, who admitted to being nonexercisers, were then compared to the combined group of 29 exercisers.

The comparison was based on the employees' productivity over the year, as rated by their supervisors. Each rating was on a scale of 1 to 25, 25 being the best. To increase the validity of the study, neither the employees nor the supervisors were told that a study was in progress. In particular, the supervisors did not know which employees were involved in the study or which were exercisers. The data from the study appear in Figure 10.18. (See the file EXERCISE.XLS.) Do these data support the company's (alternative) hypothesis that exercisers outperform nonexercisers on average? Can the company infer that any difference between the two groups is due to exercise?

Objective To use a two-sample t test for the difference between means to see whether regular exercise increases worker productivity.

Solution

Side-by-side boxplots are typically a good way to begin the analysis when comparing two populations.

To see whether there is any indication of a difference between the two groups, we create side-by-side boxplots of the Rating variable. These appear in Figure 10.19. Although there is a great deal of overlap between the two distributions, the distribution for the exercisers is somewhat to the right of that for the nonexercisers. Also, the variances of the two distributions appear to be roughly the same, although there is a bit more variation in the nonexerciser distribution.

A formal test on the mean difference uses the hypotheses $H_0: \mu_1 - \mu_2 \geq 0$ versus $H_a: \mu_1 - \mu_2 < 0$, where μ_1 and μ_2 are the mean ratings for the nonexerciser and exer-

FIGURE 10.18
Data for Study on
Effectiveness of
Exercise

	A	B	C	D
1	Study on the effect of regular exercise			
2				
3	Employee	Exercise	Rating	
4	30	0	6	
5	31	0	19	
6	32	0	19	
7	33	0	14	
8	34	0	20	
9	35	0	12	
10	36	0	15	
11	37	0	16	
12	38	0	16	
13	39	0	14	
73	19	1	15	
74	20	1	23	
75	21	1	20	
76	22	1	12	
77	23	1	23	
78	24	1	19	
79	25	1	20	
80	26	1	19	
81	27	1	19	
82	28	1	16	
83	29	1	13	

FIGURE 10.19
Boxplots for Exercise Data

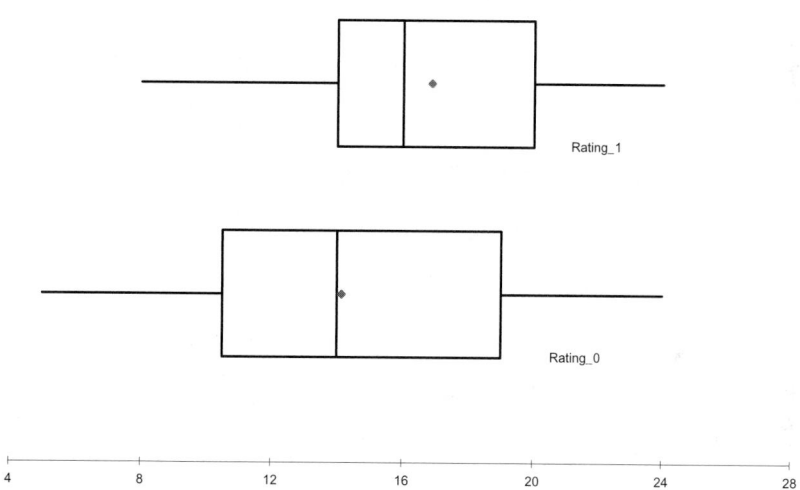

ciser populations. We use a one-tailed test, with the alternative of the "less than" variety, because the company expects higher ratings, on average, for the exercisers. The output for this test, along with a 95% confidence interval for $\mu_1 - \mu_2$, appears in Figure 10.20. We obtain it by using the StatPro/Statistical Inference/Two-Sample Analysis menu item and filling out the dialog boxes in the obvious way.[5]

If we can assume that the population standard deviations are at least approximately equal (and the values in cells G9 and H9 suggest that this assumption is plausible), then the output in the range G21:G27 is relevant. It shows that the observed sample mean difference, -2.725, is indeed negative. That is, the exercisers in the sample outperformed the nonexercisers by 2.725 rating points on average. The output also shows that (1) the standard error of the sample mean difference is 1.142, (2) the test statistic is -2.387, and

---

[5]StatPro's hypothesis-testing dialog box reminds you which difference is being tested. In this case, it is the mean rating for nonexercisers minus the mean rating for exercisers.

FIGURE 10.20

Analysis of Exercise Data

	E	F	G	H
3		**Two-sample analysis for Rating_0 minus Rating_1**		
4				
5		**Summary stats for two samples**		
6			Rating_0	Rating_1
7		Sample sizes	51	29
8		Sample means	14.137	16.862
9		Sample standard deviations	5.307	4.103
10				
11		**Confidence interval for difference between means**		
12		Confidence level	95.0%	
13		Sample mean difference	-2.725	
14		Pooled standard deviation	4.909	NA
15		Std error of difference	1.142	1.064
16		Degrees of freedom	78	71
17		Lower limit	-4.998	-4.847
18		Upper limit	-0.452	-0.603
19				
20		**Test of difference>=0 versus one-tailed alternative**		
21		Hypothesized mean difference	0.000	
22		Sample mean difference	-2.725	
23		Pooled standard deviation	4.909	NA
24		Std error of difference	1.142	1.064
25		Degrees of freedom	78	71
26		t-test statistic	-2.387	-2.560
27		p-value	0.010	0.006
28				
29		**Test of equality of variances**		
30		Ratio of sample variances	1.673	
31		p-value	0.073	

(3) the p-value for a one-tailed test is 0.010. In words, the data provide enough evidence to reject the null hypothesis at the 1% significance level. It is clear that exercisers perform better, in terms of mean ratings, than nonexercisers. A 95% confidence for this mean difference is all negative; it extends from -4.998 to -0.452.

This answers the first question we posed, but it doesn't answer the second. There is no way to be sure that the higher ratings for the exercisers are a direct result of exercise. It could be that employees who exercise are naturally more ambitious and hard-working people, and that this extra drive is responsible for *both* their exercising and their higher ratings. This study is an **observational study** because the company simply observes two randomly selected groups of employees and analyzes the results. It does not explicitly control for other factors, such as personality, that might be responsible for differences in ratings. Therefore, it can never be sure that there is a cause–effect relationship between exercise and performance ratings. All it can state is that exercisers appear, on average, to be more productive than nonexercisers—for whatever reason.

We are almost finished with this example, but not quite. What about the output in column H, and the test in rows 29–31? The test we just performed and the confidence interval we reported are based on the assumption of equal population standard deviations (or variances). As we discussed in Section 9.7.1, if this assumption is violated, then a slightly different form of analysis should be performed, and its results are reported in column H. As we see, the results are very similar to those in column G, although the p-value is slightly lower and the confidence interval is slightly narrower.

The test reported in rows 29–31 is a formal test of the hypothesis $H_0: \sigma_1^2/\sigma_2^2 = 1$ versus $H_a: \sigma_1^2/\sigma_2^2 \neq 1$, where the parameter being tested is the *ratio* of the two population variances. (The details behind this test are explained in the following subsection.) If we can reject this null hypothesis on the basis of a low p-value in cell G31, then we are fairly certain that the equal-variance assumption is *not* valid and that we should use the output in column H. Otherwise, we can use the output in column G. The p-value in cell G31,

0.073, suggests that the two population variances might not be equal, but the evidence is not overwhelming. Of course, the similarity of the outputs in columns G and H implies, especially from a practical point of view, that it doesn't really make much difference. In other examples it could be more critical. ●

Excel Tip *If the* p-value *for the test of equal variances is small, use the right column (here column H) for testing the difference between means. Otherwise, use the (traditional) left column (here column G).*

10.4.3 Hypothesis Test for Equal Population Variances

As we just saw, the two-sample procedure for a difference between population means depends on whether we can assume equal population variances.[6] Therefore, it is natural to test first for equal variances. We phrase this latter test in terms of the *ratio* of population variances, σ_1^2/σ_2^2. The null hypothesis is that this ratio is 1 (equal variances), whereas the alternative is that it is not 1 (unequal variances). The test statistic for this test is the ratio of sample variances:

$$F\text{-value} = s_1^2/s_2^2$$

The F distribution is a distribution of positive values and is always skewed to the right. It typically appears in tests of equal variances.

Assuming that the population variances are equal, this test statistic has an F distribution with $n_1 - 1$ and $n_2 - 1$ degrees of freedom.

The F distribution, named after the famous statistician R. A. Fisher, is another sampling distribution that arises frequently in statistical studies. (We will see it again when we study regression analysis in Chapters 11 and 12.) Because it always describes a ratio, there are two degrees of freedom parameters, one for the numerator and one for the denominator, and the numerator degrees of freedom is always quoted first.

Tables of the F distribution, for selected degrees of freedom, appear in many statistics books, but the necessary information can be obtained more easily with Excel's FDIST and FINV functions. The FDIST function takes the form

$$=\text{FDIST}(v, df1, df2)$$

This function returns the probability to the right of value v when the degrees of freedom are $df1$ and $df2$. Similarly, the FINV function takes the form

$$=\text{FINV}(p, df1, df2)$$

It returns the value with probability p to the right of it when the degrees of freedom are $df1$ and $df2$.

The F test for equal variances is performed as follows. We first create the test statistic by putting the *larger* of the two sample variances in the numerator. Hence, the test statistic is always greater than or equal to 1. Then we find the corresponding p-value with the FDIST function. For example, the formulas in cells G30 and G31 of Figure 10.20 are

$$=(G9/H9)\text{^}2$$

and

$$=\text{FDIST}(G30,G7-1,H7-1)$$

[6]The test in this section is traditionally stated in terms of variances, as we do here. It could also be stated in terms of standard deviations because equal variances imply equal standard deviations.

The StatPro Two-Sample procedure first checks that the larger sample standard deviation is in cell G9, so it puts the corresponding variance in the numerator of the test statistic. Then it finds the probability to the right of this test statistic with the FDIST function, where the degrees of freedom are $51 - 1$ and $29 - 1$. If the larger sample standard deviation had been in cell H9, then these formulas would have been

$$=(H9/G9)\wedge2$$

and

$$=FDIST(G30,H7\text{-}1,G7\text{-}1)$$

That is, the degrees of freedom would have been reversed.

For our purposes, the most important thing is the conclusion we draw from the test. If the p-value is small, we can conclude that the population variances are *not* equal. Otherwise, we can accept an equal-variance assumption. The p-value for the exercise data, 0.073, is in the "gray area." It provides some evidence of unequal variances, but the evidence is not overwhelming.

10.4.4 Hypothesis Tests for Differences Between Population Proportions

One of the most common uses of hypothesis testing is to test whether two population proportions are equal. Let p_1 and p_2 be the two population proportions, and let $\hat{p}_1$ and $\hat{p}_2$ be the corresponding sample proportions, based on sample sizes n_1 and n_2. The goal is to test whether the sample proportions differ enough to conclude that the *population* proportions are not equal. To base a test on the difference $\hat{p}_1 - \hat{p}_2$, we need its standard error. If the null hypothesis is true and $p_1 = p_2$, then it can be shown that the standard error of $\hat{p}_1 - \hat{p}_2$ is given by equation (10.5), where $\hat{p}_c$ is the pooled proportion from the two samples combined. For example, if $\hat{p}_1 = 20/85$ and $\hat{p}_2 = 34/115$, then $\hat{p}_c = (20 + 34)/(85 + 115) = 54/200$. (The reason we use this pooled estimate is that if the null hypothesis is true and the two population proportions are equal, then it makes sense to base an estimate of this common proportion on the *combined* sample of data.)

Standard Error for Difference Between Sample Proportions	$$\text{SE}(\hat{p}_1 - \hat{p}_2) = \sqrt{\hat{p}_c(1 - \hat{p}_c)(1/n_1 + 1/n_2)} \qquad \textbf{(10.5)}$$

Given this standard error, the rest is straightforward. Assuming that the sample sizes are reasonably large, the test statistic in equation (10.6) has (approximately) a standard normal distribution. Therefore, the corresponding p-value for the test can be found with Excel's NORMSDIST function, as illustrated in the next example.

Test Statistic for Difference Between Proportions	$$z\text{-value} = \frac{\hat{p}_1 - \hat{p}_2}{\text{SE}(\hat{p}_1 - \hat{p}_2)} \qquad \textbf{(10.6)}$$

Example 10.6 Employee Empowerment
at ArmCo Company

The ArmCo Company, a large manufacturer of automobile parts, has several plants in the United States. For years, ArmCo employees have complained that their suggestions for improvements in the manufacturing processes are ignored by upper management. In the spirit of employee empowerment, ArmCo management at the Midwest plant decided to initiate a number of policies to respond to employee suggestions. For example, a mailbox was placed in a central location, and employees were encouraged to drop suggestions into this box. No such initiatives were taken at the other ArmCo plants. As expected, there was a great deal of employee enthusiasm at the Midwest plant shortly after the new policies were implemented, but the question was whether life would revert to normal and the enthusiasm would dampen with time.

To check this, 100 randomly selected employees at the Midwest plant and 300 employees from other plants were asked to fill out a questionnaire 6 months after the implementation of the new policies at the Midwest plant. Employees were instructed to respond to each item on the questionnaire by checking either a "yes" box or a "no" box. Two specific items on the questionnaire were:

- Management at this plant is generally responsive to employee suggestions for improvements in the manufacturing processes.

- Management at this plant is more responsive to employee suggestions now than it used to be.

The results of the questionnaire for these two items appear in rows 5 and 6 of Figure 10.21. (See the file EMPOWER1.XLS.) Does it appear that the policies at the Midwest plant are appreciated? Should ArmCo implement these policies in its other plants?

Objective To use a test for the difference between proportions to see whether a program of accepting employee suggestions is appreciated by employees.

	A	B	C	D	E	F	G
1	**Employee empowerment results**						
2							
3	**Item 1: Management responds**				**Item 2: Things have improved**		
4		Midwest	Other			Midwest	Other
5	Yes	39	93		Yes	68	159
6	No	61	207		No	32	141
7	Totals	100	300		Totals	100	300
8							
9	Sample proportion yes	0.39	0.31		Sample proportion yes	0.68	0.53
10	Pooled proportion yes	0.33			Pooled proportion yes	0.5675	
11							
12	Difference between proportions	0.08			Difference between proportions	0.15	
13	Standard error of difference	0.054			Standard error of difference	0.057	
14	Test statistic	1.473			Test statistic	2.622	
15	p-value	0.070			p-value	0.004	
16							
17	Confidence interval for difference				Confidence interval for difference		
18	Confidence level	95%			Confidence level	95%	
19	Standard error of difference	0.056			Standard error of difference	0.055	
20	z-multiple	1.960			z-multiple	1.960	
21	Lower confidence limit	-0.029			Lower confidence limit	0.043	
22	Upper confidence limit	0.189			Upper confidence limit	0.257	

FIGURE 10.21 Results for Employee Empowerment Example

Solution

For either questionnaire item we let p_1 be the proportion of "yes" responses we would obtain at the Midwest plant if the questionnaire were given to all of its employees. We define p_2 similarly for the other plants. Management certainly hopes to find a larger proportion of "yes" responses (to either item) at the Midwest plant than at the other plants, so the appropriate test is one-tailed, with the hypotheses set up as H_0: $p_1 - p_2 \leq 0$ versus H_a: $p_1 - p_2 > 0$. (We could also write these as H_0: $p_1 \leq p_2$ versus H_a: $p_1 > p_2$, but this has no effect on the test.)

The data from this type of questionnaire are usually given as *counts* of "yes" and "no" responses, as in Figure 10.21, but these easily translate into sample proportions. (As we stated in Example 10.3, StatPro has no procedure for testing differences between proportions, but this example file EMPOWER1.XLS can be used as a "template" for any such test.) For the first questionnaire item (see columns B and C), the sample proportions of "yes" responses are $\hat{p}_1 = 39/100 = 0.39$ and $\hat{p}_2 = 93/300 = 0.31$, for a difference of $\hat{p}_1 - \hat{p}_2 = 0.08$. The standard error of this difference, under the assumption that $p_1 = p_2$, uses the pooled proportion $\hat{p}_c = (39 + 93)/(100 + 300) = 0.33$. This produces a standard error of 0.054, calculated in cell B13 with the formula

$$=\text{SQRT(B10*(1-B10)*(1/B7+1/C7))}$$

Then the test statistic is $0.08/0.054 = 1.473$, and the corresponding p-value for the test is the probability to the right of 1.473 in the standard normal distribution. Its value is 0.070, found in cell B15 with the formula

$$=\text{1-NORMSDIST(B14)}$$

A similar analysis for the second questionnaire item (see columns F and G) leads to a sample difference of $0.68 - 0.53 = 0.15$ and a p-value of 0.004.

These results should be fairly good news for management. There is moderate, but not overwhelming, support for the hypothesis that management at the Midwest plant is more responsive than at the other plants, at least as perceived by employees. There is convincing support for the hypothesis that things have improved more at the Midwest plant than at the other plants. Corresponding 95% confidence intervals for the differences between proportions appear in rows 21 and 22. Since they are almost completely positive, they reinforce the hypothesis-test findings. Moreover, they provide a range of plausible values for the differences between the population proportions.

The only real downside to these findings, from Midwest management's point of view, is the sample proportion $\hat{p}_1$ for the first item. Only 39% of the sampled employees at that plant believe that management generally responds to their suggestions, even though 68% believe things are better than they used to be. A reasonable conclusion by ArmCo management is that they are on the right track at the Midwest plant, and the policies initiated there ought to be initiated at other plants, but more must be done at *all* plants. ●

PROBLEMS

Level A

9. In the past, 60% of all undergraduate students enrolled at State University earned their degrees within 4 years of matriculation. A random sample of 36 students from the class that matriculated in the fall of 1998 was recently selected to test whether there has been a change in the proportion of students who graduate within 4 years. Administrators found that 15 of these 36 students graduated in the spring of 2002 (i.e., 4 academic years after matriculation).

a. Given the sample outcome, construct a 95% confidence interval for the relevant population proportion. Does this interval estimate suggest that there has been in a change in the proportion of students who graduate within 4 years? Why or why not?

b. Given the sample outcome, construct a 99% confidence interval for the relevant population proportion. Does this interval estimate suggest that there has been in a change in the proportion of students who graduate within 4 years? Why or why not?

10. Continuing the previous problem, suppose now that State University administrators want to test the claim made by faculty that the proportion of students who graduate within 4 years at State University has fallen *below* the historical value of 60% this year. Use the given sample proportion to test this claim. Report a p-value and interpret it in the context of this statistical hypothesis test.

11. The director of admissions of a distinguished (i.e., top-20) MBA program is interested in studying the proportion of entering students in similar graduate business programs who have achieved a composite score on the Graduate Management Admissions Test (GMAT) in excess of 630. In particular, the admissions director believes that the proportion of students entering top-rated programs with such composite GMAT scores is now 50%. To test this hypothesis, he has collected a random sample of MBA candidates entering his program in the fall of 1998. He believes that these students' GMAT scores are indicative of the scores earned by their peers in his program and in competitors' programs. The GMAT scores for these 25 individuals are given in the file P10_11.XLS. Test the admission director's claim at the 5% significance level and report your findings. Does your conclusion change when the significance level is increased to 10%?

12. A market research consultant hired by the Pepsi-Cola Co. is interested in determining the proportion of consumers who favor Pepsi-Cola over Coke Classic in a particular urban location. A random sample of 250 consumers from the market under investigation is provided in the file P09_17.XLS.

a. Construct a 99% confidence interval for the proportion of all consumers in this market who prefer Pepsi over Coke. Interpret this confidence interval for Pepsi's market researchers.

b. Does the confidence interval in part **a** support the claim made by one of Pepsi-Cola's marketing managers that more than half of the consumers in this urban location favor Pepsi over Coke? Explain your answer.

c. Comment on the sample size used by the market research consultant. Specifically, is the sample unnecessarily large? Why or why not?

13. The CEO of a medical supply company is committed to expanding the proportion of highly qualified

women in the organization's staff of salespersons. He claims that the proportion of women in similar sales positions across the country in 2002 is less than 50%. Hoping to find support for his claim, he directs his assistant to collect a random sample of salespersons employed by his company, which is thought to be representative of sales staffs of competing organizations in the industry. These data are listed in the file P10_13.XLS. Test this manager's claim using the given sample data and report a p-value. Do you find statistical support for his hypothesis that the proportion of women in similar sales positions across the country is less than 50%?

14. Management of a software development firm would like to establish a wellness program during the lunch hour to enhance the physical and mental health of its employees. Before introducing the wellness program, management must first be convinced that a sufficiently large majority of its employees are not already exercising at lunchtime. Specifically, it plans to initiate the program only if less than 40% of its personnel take time to exercise prior to eating lunch. To make this decision, management has surveyed a random sample of 100 employees regarding their midday exercise activities. The results of the survey are given in the file P10_14.XLS.

a. Using a 10% significance level, is there sufficient evidence for managers of this organization to initiate a corporate wellness program? Why or why not?

b. Using a 1% significance level, is there sufficient evidence for managers of this organization to initiate a corporate wellness program? Why or why not?

15. The managing partner of a major consulting firm is trying to assess the effectiveness of expensive computer skills training given to all new entry-level professionals. In an effort to make such an assessment, she administers a computer skills test immediately before and after the training program to each of 40 randomly chosen employees. The pretraining and post-training scores of these 40 individuals are recorded in the file P10_15.XLS.

a. Using a 10% level of significance, do the given sample data support the claim that the organization's training program is increasing the new employee's working knowledge of computing?

b. Using a 1% level of significance, do the given sample data support the claim that the organization's training program is increasing the new employee's working knowledge of computing?

16. A large buyer of household batteries wants to decide which of two equally priced brands to purchase. To do this, he takes a random sample of 100 batteries of each brand. The lifetimes, measured in hours, of the

randomly chosen batteries are recorded in the file P10_16.XLS.

a. Using the given sample data, generate a 95% confidence interval for the difference between the mean lifetimes of brand 1 and brand 2 batteries. Based on this confidence interval, which brand should the buyer purchase?

b. Using the given sample data, generate a 99% confidence interval for the difference in the mean lifetimes of brand 1 and brand 2 batteries. Based on this confidence interval, which brand should the buyer purchase?

c. How can your analyses in parts **a** and **b** be related to hypothesis testing?

17. The managers of a chemical manufacturing plant are interested in determining whether recent safety training workshops have reduced the weekly number of reported safety violations at the facility. The management team has randomly selected weekly safety violation reports for each of 25 weeks prior to the safety training and 25 weeks after the safety workshops. These data are provided in the file P10_17.XLS. Given this evidence, is it possible to conclude that the safety workshops have been effective in reducing the number of safety violations reported per week? Report a *p*-value and interpret your findings for the management team.

18. A real estate agent has collected a random sample of 75 houses that were recently sold in a suburban community. She is particularly interested in comparing the appraised value and recent selling price of the houses in this particular market. The values of these two variables for each of the 75 randomly chosen houses are provided in the file P09_24.XLS. Using these sample data, test whether there exists a statistically significant mean difference between the appraised values and selling prices of the houses sold in this suburban community. Report a *p*-value. For which levels of significance is it appropriate to conclude that *no* difference exists between these two values?

19. The owner of two submarine sandwich shops located in Gainesville, Florida, would like to know how the mean daily sales of the first shop (located in the downtown area) compares to that of the second shop (located on the southwest side of town). In particular, he would like to know whether the mean daily sales levels of these two restaurants are essentially equal. He records the sales (in dollars) made at each location for 30 randomly chosen days. These sales levels are given in the file P10_19.XLS. Construct a 99% confidence level for the mean difference between the daily sales of restaurant 1 and restaurant 2. Use this confidence interval to answer the following questions:

a. Is it possible to conclude that a statistically significant mean difference exists at the 1% level of significance? Explain why or why not.

b. Is it possible to conclude that a statistically significant mean difference exists at the 5% level of significance? Explain why or why not.

c. Is it possible to conclude that a statistically significant mean difference exists at the 10% level of significance? Explain why or why not.

20. Suppose that an investor wants to compare the risks associated with two different stocks. One way to measure the risk of a given stock is to measure the variation in the stock's daily price changes. The investor obtains a random sample of 25 daily price changes for stock 1 and 25 daily price changes for stock 2. These data are provided in the file P10_20.XLS. Show how this investor can compare the risks associated with the two stocks by testing the null hypothesis that the variances of the stocks are equal. Use $\alpha = 0.10$ and interpret the results of the statistical test.

21. A manufacturing company is interested in determining whether a significant difference exists between the variance of the number of units produced per day by one machine operator and the similar variance for another machine operator. The file P10_21.XLS contains the number of units produced by operator 1 and operator 2, respectively, on each of 25 days. Note that these two sets of days are not necessarily the same, so you can assume that the two samples are *independent* of one another.

a. Do these sample data indicate a statistically significant difference at $\alpha = 0.10$? Explain your answer.

b. If your conclusion in part **a** were incorrect, would you have committed a type I or type II error? Explain.

c. At which values of α could you *not* reject the null hypothesis?

22. A large buyer of household batteries wants to decide which of two equally priced brands to purchase. To do this, he takes a random sample of 100 batteries of each brand. The lifetimes, measured in hours, of the batteries are recorded in the file P10_16.XLS. Before testing for the difference between the mean lifetimes of these two batteries, he must first determine whether the underlying population variances are equal.

a. Perform a test for equal population variances. Report a *p*-value and interpret its meaning.

b. Based on your conclusion in part **a,** which test statistic should be used in performing a test for the existence of a difference between population means?

23. Do undergraduate business students who major in finance earn, on average, higher annual starting salaries than their peers who major in marketing? Before addressing this question through a statistical hypothesis test, we should determine whether the variances of annual starting salaries of the two types of majors are equal. The file P10_23.XLS contains the starting

salaries of 50 randomly selected finance majors and 50 randomly chosen marketing majors.

   a. Perform a test for equal population variances. Report a *p*-value and interpret its meaning.

   b. Based on your conclusion in part **a,** which test statistic should be used in performing a test for the existence of a difference between population means?

24. The CEO of a medical supply company is committed to expanding the proportion of highly qualified women in the organization's large staff of salespersons. Given the recent hiring practices of his human resources director, he claims that the company has increased the proportion of women in sales positions throughout the organization between 1997 and 2002. Hoping to find support for his claim, he directs his assistant to collect random samples of the salespersons employed by the company in 1997 and 2002. These data can be found in the file P10_13.XLS. Test the CEO's claim using the given sample data and report a *p*-value. Do you find statistical support for the efficacy of his committed strategy to hiring a greater proportion of female salespersons?

25. The director of admissions of a distinguished (i.e., top-20) MBA program is interested in studying the proportion of entering students in similar graduate business programs who have achieved a composite score on the Graduate Management Admissions Test (GMAT) in excess of 630. In particular, the admissions director believes that the proportion of students entering top-rated programs with such composite GMAT scores is higher in 2002 than it was in 1992. To test this hypothesis, he has collected random samples of MBA candidates entering his program in the fall of 2002 and in the fall of 1992. He believes that these students' GMAT scores are indicative of the scores earned by their peers in his program and in competitors' programs. The GMAT scores for the randomly selected students entering in each year are given in the file P10_11.XLS. Test the admission director's claim at the 5% significance level and report your findings. Does your conclusion change when the significance level is increased to 10%?

26. Managers of a software development firm have established a wellness program during the lunch hour to enhance the physical and mental health of their employees. Now, they would like to see whether the wellness program has increased the proportion of employees who exercise regularly during the lunch hour. To make this assessment, the managers surveyed a random sample of 100 employees about their noontime exercise habits *before* the wellness program was initiated. Later, *after* the program was initiated, another 100 employees were independently chosen and surveyed about their lunchtime exercise habits. The

results of these two surveys are given in the file P10_14.XLS.

   a. Construct a 99% confidence interval for the difference in the proportions of employees who exercise regularly during their lunch hour before and after the implementation of the corporate wellness program.

   b. Does the confidence interval found in part **a** support the belief that the wellness program has increased the proportion of employees who exercise regularly during the lunch hour? If so, at which levels of significance is this claim supported?

   c. Would your results in parts **a** and **b** differ if the *same* 100 employees surveyed before the program were also surveyed after the program? Explain.

Level B

27. An Environmental Protection Agency official asserts that more than 80% of the plants in the northeast region of the United States meet air pollution standards. An antipollution advocate does not believe the EPA's claim. She takes a random sample of 64 plants in the northeast region and finds that 56 meet the federal government's pollution standards.

   a. Does the sample information support the EPA's claim at the 5% level of significance?

   b. For which values of the sample proportion (based on a sample size of 64) would the sample data support the EPA's claim? Assume that $\alpha = 0.05$.

   c. Would the conclusion found in part **a** change if the sample proportion remained constant but the sample size increased to 124? Explain why or why not.

28. A television network decides to cancel one of its shows if it is convinced that less than 14% of the viewing public are watching this show.

   a. If a random sample of 1500 households with televisions is selected, what sample proportion values will lead to this show's cancellation? Assume a 5% significance level.

   b. What is the probability that this show will be canceled if 13.4% of all viewing households are watching it?

29. An economic researcher would like to know whether he can reject the null hypothesis, at the $\alpha = 0.10$ level, that no more than 20% of the households in Pennsylvania make more than $70,000 per year.

   a. If 200 Pennsylvania households are chosen at random, how many of them would have to be earning more than $70,000 per year for the researcher to reject the null hypothesis?

   b. Assuming that the true proportion of all Pennsylvania households with annual incomes of at least $70,000 is 0.217, find the probability of *not* rejecting a *false* null hypothesis when the sample size is 200.

30. Senior partners of an accounting firm are concerned about recent complaints by some female managers that they are paid less than their male counterparts. In response to these charges, the partners ask their human resources director to record the salaries of female and male managers with equivalent education, work experience, and job performance. A random sample of these pairs of managers is provided in the file P10_30.XLS.

   a. Do these data support the claim made by some female managers within this organization? Report and interpret a *p*-value.

   b. Assuming a 5% significance level, which values of the sample mean difference between the female and male salaries would support the claim of discrimination against female managers?

31. Do undergraduate business students who major in finance earn, on average, higher annual starting salaries than their peers who major in marketing? Address this question through a statistical hypothesis test. The file P10_23.XLS contains the starting salaries of 50 randomly selected finance majors and 50 randomly selected marketing majors.

   a. Is it appropriate to perform a paired-comparison *t*-test in this case? Explain why or why not.

   b. Perform an appropriate hypothesis test with a 1% significance level. Summarize your findings.

   c. How large would the difference between the mean starting salaries of finance and marketing majors have to be before you could conclude that finance majors earn more on average? Employ a 1% significance level in answering this question.

32. Consider a random sample of 100 households from a middle-class neighborhood that was the recent focus of an economic development study conducted by the local government. Specifically, for each of the 100 households, information was gathered on each of the following variables: family size, location of the household within the neighborhood, an indication of whether those surveyed owned or rented their home, gross annual income of the first household wage earner, gross annual income of the second household wage earner (if applicable), monthly home mortgage or rent payment, average monthly expenditure on utilities, and the total indebtedness (excluding the value of a home mortgage) of the household. The data are in the file P09_26.XLS.

   Test for the existence of a significant difference between the mean indebtedness levels of the households in the first (i.e., SW) and second (i.e., NW) sectors of this community. Perform similar hypothesis tests for the differences between the mean indebtedness levels of households from all other pairs of loca-

tions (i.e., first and third, first and fourth, second and third, second and fourth, and third and fourth). Summarize your findings.

33. Elected officials in a small Florida town are preparing the annual budget for their community. They would like to determine whether their constituents living across town are typically paying the same amount in real estate taxes each year. Given that there are over 3000 homeowners in this small community, officials have decided to sample a representative subset of taxpayers and thoroughly study their tax payments. A randomly selected set of 170 homeowners is given in the file P10_33.XLS. Specifically, the elected officials would like to test for the existence of a statistical difference between the mean real estate tax bill paid by residents of the *first* neighborhood of this town and each of the remaining five neighborhoods (i.e., neighborhoods 2–6).

   a. Before conducting any hypothesis tests on the difference between various pairs of mean real estate tax payments, perform a test for equal population variances for each pair of neighborhoods. For each pair, report a *p*-value and interpret its meaning.

   b. Based on your conclusions in part **a,** which test statistic should be used in performing a test for the existence of a difference between population means in each pair?

   c. Given your conclusions in part **b,** appropriately perform each of the tests for the existence of a difference between mean real estate tax payments in each pair of neighborhoods. For each pair, report a *p*-value and interpret its meaning.

34. Suppose that you sample two normal populations independently. The variances of these two populations are σ_1^2 and σ_2^2. You take random samples of sizes n_1 and n_2 and observe sample variances of s_1^2 and s_2^2.

   a. If $n_1 = n_2 = 21$, how large must the fraction s_1/s_2 be before you can reject the null hypothesis that σ_1^2 is no greater than σ_2^2 at the 5% significance level?

   b. Answer part **a** when $n_1 = n_2 = 41$.

   c. If s_1 is 25% greater than s_2, approximately how large must n_1 and n_2 be if you are able to reject the null hypothesis in part **a** at the 5% significance level? Assume that n_1 and n_2 are equal.

35. Two teams of workers assemble automobile engines at a manufacturing plant in Michigan. Quality control personnel inspect a random sample of the teams' assemblies and judge each assembly to be acceptable or unacceptable. A random sample of 127 assemblies from team 1 shows 12 unacceptable assemblies. A similar random sample of 98 assemblies from team 2 shows 5 unacceptable assemblies.

a. Construct a 90% confidence interval for the difference between the proportions of unacceptable assemblies generated by the two teams.

b. Based on a review of the confidence interval found in part **a,** is there sufficient evidence to conclude, at the 10% significance level, that the two teams differ with respect to their proportions of unacceptable assemblies?

c. For which values of the difference between these two sample proportions could you conclude that a statistically significant difference exists? Assume that $\alpha = 0.10$.

36. A market research consultant hired by the Pepsi-Cola Co. is interested in determining whether there is a difference between the proportions of female and male consumers who favor Pepsi-Cola over Coke Classic in a particular urban location. A random sample of 250 consumers from the market under investigation is provided in the file P09_17.XLS.

a. After separating the 250 randomly selected consumers by *gender,* perform the statistical test and report a *p*-value. At which levels of α will the market research consultant conclude that there is essentially no difference between the proportions of female and male consumers who prefer Pepsi to Coke in this urban area?

b. Marketing managers at the Pepsi-Cola Co. have asked their market research consultant to explore further the potential differences in the proportions of women and men who prefer drinking Pepsi to Coke Classic. Specifically, Pepsi managers would like to know whether the potential difference between the proportions of female and male consumers who favor Pepsi varies by the *age* of the consumers. Using the same random sample of consumers as in part **a,** assess whether this difference varies across the four given age categories: under 20, between 20 and 40, between 40 and 60, and over 60. Employ a 10% significance level in performing each of the *six* required hypothesis tests. Summarize your conclusions in detail. Finally, what recommendations would you make to the marketing managers in light of your statistical findings?

37. The employee benefits manager of a small private university would like to determine whether differences exist in the proportions of various groups of full-time employees who prefer adopting the second (i.e., plan B) of three available health care plans in the forthcoming annual enrollment period. A reliable frame of the university's employees and their tentative health care preferences are given in the file P08_25.XLS.

a. First, select a simple random sample of 25 employees from *each* of three employee classifica-

tions: administrative staff, support staff, and faculty.

b. Use the three simple random samples obtained in part **a** to perform tests for differences in the proportions of employees within respective classifications who favor plan B in the coming year. For instance, the first such test should examine the potential difference between the proportion of administrative employees who favor plan B and the proportion of the support staff who prefer plan B.

c. Report a *p*-value for each of your hypothesis tests and interpret your results. How might the benefits manager use the information you have derived from these statistical tests?

38. Consider a random sample of 100 households from a middle-class neighborhood that was the recent focus of an economic development study conducted by the local government. Specifically, for each of the 100 randomly selected households, information was gathered on each of the following variables: family size, location of the household within the neighborhood, an indication of whether those surveyed owned or rented their home, gross annual income of the first household wage earner, gross annual income of the second household wage earner (if applicable), monthly home mortgage or rent payment, average monthly expenditure on utilities, and the total indebtedness (excluding the value of a home mortgage) of the household. The data are provided in the file P09_26.XLS.

Researchers would like to use the available sample information to discern whether home ownership rates vary by household *location.* For example, is there a nonzero difference between the proportions of individuals who own their homes (as opposed to those who rent their homes) in households located in the first (i.e., SW) and second (i.e., NW) sectors of this community? Use the given sample to perform a test for the existence of a difference in home ownership rates in these two sectors as well as for those of other pairs of household locations. Assume that $\alpha = 0.05$. Interpret and summarize your results. (*Hint:* To be complete, you should construct and interpret a total of six hypothesis tests.)

39. For testing the difference between two proportions, we use $\sqrt{\hat{p}_c(1 - \hat{p}_c)(1/n_1 + 1/n_2)}$ as the approximate standard error of $\hat{p}_1 - \hat{p}_2$, where $\hat{p}_c$ is the pooled sample proportion. Explain why this is reasonable when the null-hypothesized value of $p_1 - p_2$ is zero. Why would this not be a good approximation when the null-hypothesized value of $p_1 - p_2$ is a nonzero number? What would you recommend using for the standard error of $\hat{p}_1 - \hat{p}_2$ in that case?

10.5 Tests for Normality

In this section we discuss several tests for normality. As we have already seen, many statistical procedures are based on the assumption that population data are normally distributed. The tests in this section allow us to test this assumption. The null hypothesis is that the population is normally distributed, whereas the alternative is that the population distribution is not normal. Therefore, the burden of proof is on showing that the population distribution is *not* normal. Unless there is sufficient evidence to this effect, we will accept the normal assumption.

The first test we discuss is called a *chi-square goodness-of-fit* test. It is quite intuitive. We form a histogram of the sample data and compare this to the *expected* histogram we would observe if the data were normally distributed with the same mean and standard deviation as the sample. If the two histograms are sufficiently similar, we accept the null hypothesis of normality. Otherwise, we reject it.

> The chi-square test for normality makes a comparison between the observed histogram and a histogram based on normality.

The test is based on a numerical measure of the difference between the two histograms. Let C be the number of categories in the histogram, and let O_i be the observed number of observations in category i. Also, let E_i be the expected number of observations in category i if the population were normal with the same mean and standard deviation as the sample. Then we use the goodness-of-fit measure in equation (10.7) as a test statistic. If the null hypothesis of normality is true, this test statistic has (approximately) a chi-square distribution with $C - 3$ degrees of freedom. Because *large* values of the test statistic indicate a poor fit—the O_i's do not match up well with the E_i's—the p-value for the test is the probability to the right of the test statistic in the chi-square distribution with $C - 3$ degrees of freedom.

Test Statistic for Chi-Square Test of Normality	
	$$\chi^2\text{-value} = \sum_{i=1}^{C} (O_i - E_i)^2/E_i \qquad \textbf{(10.7)}$$

(Here, χ is the Greek letter chi.)

Although it is possible to perform this test manually, it is certainly preferable to use StatPro, as we demonstrate in the following example.

Example 10.7 Distribution of Metal Strip Widths in Manufacturing

A company manufactures strips of metal that are supposed to have width 10 centimeters. For purposes of quality control, the manager plans to run some statistical tests on these strips. However, realizing that these statistical procedures assume normally distributed widths, he first tests this normality assumption on 90 randomly sampled strips. How should he proceed?

Objective To use the chi-square goodness-of-fit test to see whether the metal strip widths are normally distributed.

Solution

The sample data appear in Figure 10.22, where each width is measured to three decimal places. (See the file NORMTEST.XLS.) A number of summary measures also appear.

FIGURE 10.22
Data for Testing
Normality

	A	B	C	D	E	F	G
1	**Illustration of test for normality**						
2							
3	Part	Width			*Summary measures for selected variables*		
4	1	9.99				Width	
5	2	10.031			Count	90.000	
6	3	9.985			Mean	9.999	
7	4	9.983			Standard deviation	0.010	
8	5	10.004			Minimum	9.970	
9	6	10			Maximum	10.031	
10	7	9.992			First quartile	9.993	
11	8	9.996			Third quartile	10.006	
12	9	9.997			5th percentile	9.984	
13	10	9.993			95th percentile	10.014	
14	11	9.991					
89	86	9.977					
90	87	10.004					
91	88	10.007					
92	89	10.003					
93	90	9.996					

These summary measures help the manager to select "reasonable" categories for a histogram of the data. After observing them, the manager chooses 10 categories for the histogram. The extreme categories are "less than or equal to 9.980" and "greater than 10.020," and the middle eight categories each have length 0.005.[7]

To run the test, we select the StatPro/Tests for Normality/Chi-square Test menu item, which leads to the same dialog box as in the Histogram procedure. After specifying the histogram categories in the usual way (upper limit of 9.98 for first category, 10 categories in all, typical length 0.005), we obtain the message in Figure 10.23, the histograms in Figure 10.24 (page 518), and the numerical output in Figure 10.25. The histograms provide visual evidence of the goodness of fit. The solid bars represent the observed frequencies (the O_i's), and the hollow bars represent the expected frequencies for a normal distribution (the E_i's). The normal fit to the data appears to be quite good.

The message in Figure 10.23 (based on the output in Figure 10.25) confirms this statistically. In Figure 10.25, each value in column D is an E_i, calculated as the total number of observations multiplied by the normal probability of being in the corresponding category. Column E contains the individual $(O_i - E_i)^2/E_i$ terms, and cell H4 contains their sum, the chi-square test statistic. The corresponding p-value in cell H5, 0.814, is calculated with the formula

$$=\text{CHIDIST(H4,7)}$$

This large p-value provides no evidence whatsoever of nonnormality. It implies that if we repeated this procedure on many random samples, each taken from a population

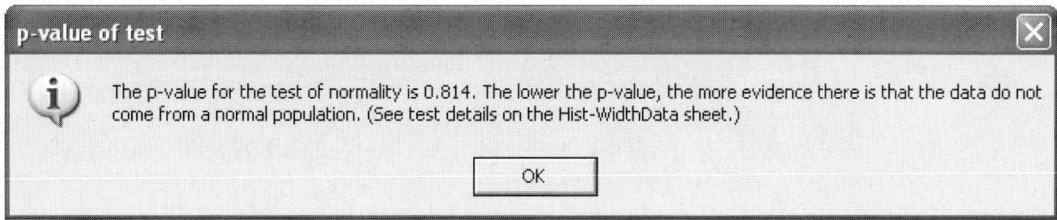

p-value of test ⊠

ⓘ The p-value for the test of normality is 0.814. The lower the p-value, the more evidence there is that the data do not come from a normal population. (See test details on the Hist-WidthData sheet.)

OK

FIGURE 10.23 StatPro Message for Chi-Square Test for Normality

[7]You might try defining the categories differently and rerunning the test. The category definitions *can* make a difference in the results.

FIGURE 10.24
Observed and
Normal
Histograms

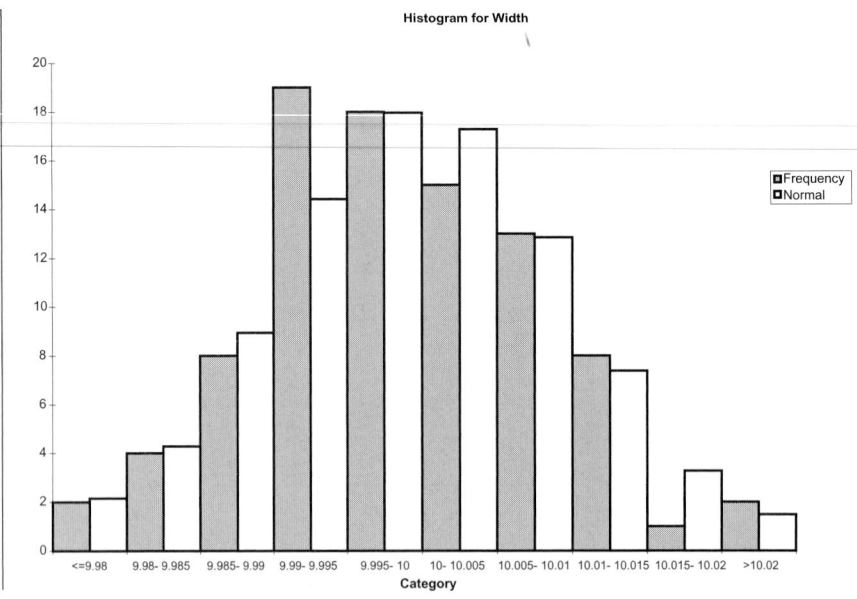

FIGURE 10.25 Chi-square Test of Normality

	A	B	C	D	E	F	G	H
1	Frequency table and normal test for Width							
2								
3	Upper limit	Category	Frequency	Normal	Distance measure		*Test of normal fit*	
4	9.98	<=9.98	2	2.150	0.010		Chi-square statistic	3.693
5	9.985	9.98- 9.985	4	4.277	0.018		p-value	0.814
6	9.99	9.985- 9.99	8	8.936	0.098			
7	9.995	9.99- 9.995	19	14.418	1.456			
8	10	9.995- 10	18	17.964	0.000			
9	10.005	10- 10.005	15	17.286	0.302			
10	10.01	10.005- 10.01	13	12.846	0.002			
11	10.015	10.01- 10.015	8	7.372	0.053			
12	10.02	10.015- 10.02	1	3.267	1.573			
13		>10.02	2	1.484	0.179			

known to be normal, we would obtain a fit at least this poor in about 81% of the samples. Stated differently, only about 19% of the fits would be *better* than the one we observed. Therefore, whatever statistical procedures the manager intends to use, he doesn't need to worry about the normality assumption. ●

We make three comments about this chi-square procedure. First, the test *does* depend on which (and how many) categories we use for the histogram. Reasonable choices are likely to lead to the same conclusion, but this is not guaranteed. Second, the test is not very effective unless the sample size is large, say, at least 80 or 100. Only then can we begin to see the true shape of the histogram and judge accurately whether it is normal. Finally, the test tends to be *too* sensitive if the sample size is really large. In this case any little "bump" on the observed histogram is likely to lead to a conclusion of nonnormality. This is one more example of practical versus statistical significance. With a large sample size we might be able reject the normality assumption with a high degree of certainty, but the practical difference between the observed and normal histograms might be unimportant.

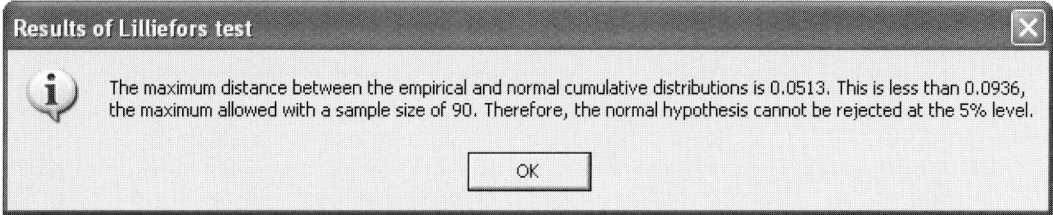

FIGURE 10.26 Lilliefors Test Results from StatPro

FIGURE 10.27
Normal and Empir-
ical Cumulative
Distribution Func-
tions

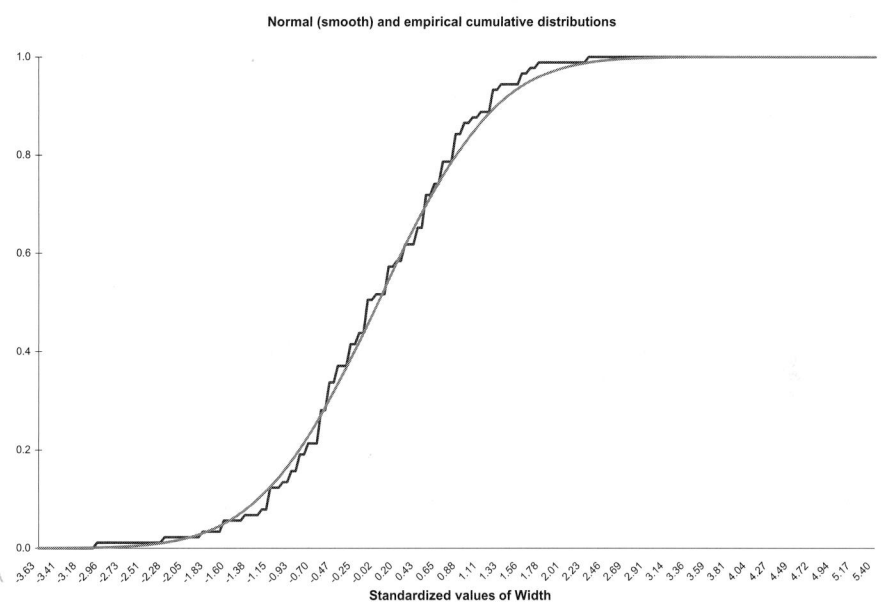

The chi-square test of normality is an intuitive one because it is based on histograms. However, it suffers from the first two points discussed in the previous paragraph. In particular, it is not as *powerful* as other available tests. This means that it is often unable to distinguish between normal and nonnormal distributions, and hence it often fails to reject the normal null hypothesis when it should be rejected. A more powerful test is called the *Lilliefors test*.[8] This test is based on the cumulative distribution function (cdf), which shows the probability of being less than or equal to any particular value. Specifically, the **Lilliefors test** compare two cdf's: the cdf from a normal distribution and the cdf corresponding to the given data. This latter cdf, called the **empirical cdf,** shows the fraction of observations less than or equal to any particular value. If the data come from a normal distribution, then the normal and empirical cdf's should be quite close. Therefore, the Lilliefors test compares the *maximum vertical distance* between the two functions and compares it to specially tabulated values. If this maximum vertical distance is sufficiently large, the null hypothesis of normality can be rejected.

To run the Lilliefors test in StatPro, we select the StatPro/Tests of Normality/Lilliefors Test menu item and the variable to be tested. StatPro then shows a message and a corresponding graph of the normal and empirical cdf's. These outputs for the Width variable in Example 10.7 appear in Figures 10.26 and 10.27. The message indicates that the

The Lilliefors test is based on a comparison of the cdf from the data and a normal cdf.

---

[8]This is actually a special case of the more general and widely known **Kolmogorov-Smirnoff (or K-S) test.**

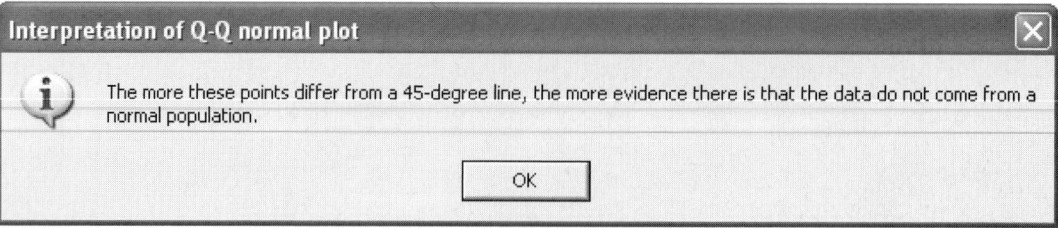

FIGURE 10.28 StatPro Message from Q-Q Plot Procedure

FIGURE 10.29
Q-Q Plot for
Width Data

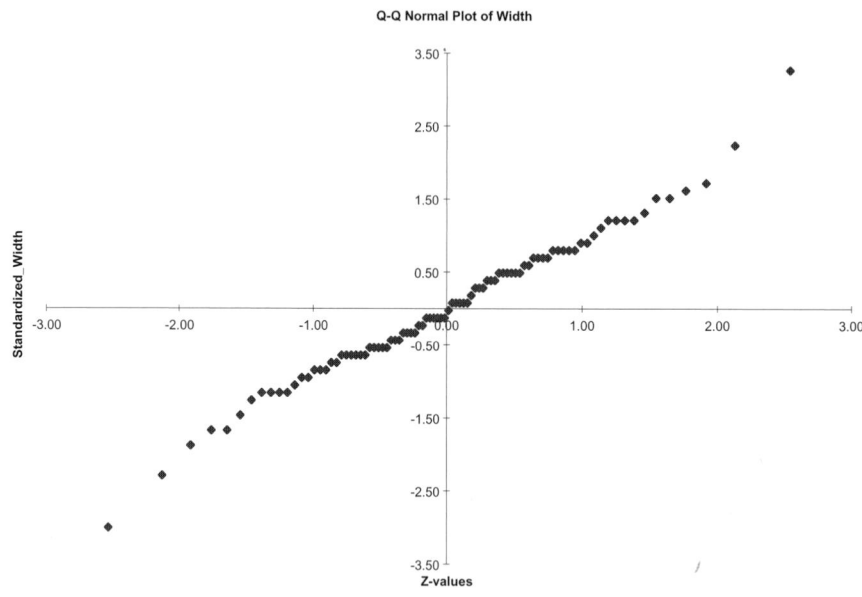

maximum vertical distance between the two curves is relatively small—not nearly large enough to reject the normal hypothesis. This conclusion agrees with the one based on the chi-square goodness-of-fit test, but the two tests do not agree on *all* data sets.

If data are normally distributed, the points on the corresponding Q-Q plot should be close to a 45° line.

We conclude this section with a popular, but informal, test of normality. This is based on a plot called a *quantile-quantile* (or *Q-Q*) *plot.* Although the technical details for forming this plot are somewhat complex, it is basically a scatterplot of the standardized values from the data set versus the values we would expect if the data were perfectly normally distributed (with the same mean and standard deviation as in the data set). If the data are, in fact, normally distributed, then the points in this plot will tend to cluster around a 45° line. Any large deviations from a 45° line signal some type of nonnormality. Again, however, this is not a formal test of normality. A Q-Q plot is usually used only to obtain a general idea of whether the data are normally distributed and, if they are not, what type of nonnormality exists. For example, if points on the right of the plot are well *above* a 45° line, this is an indication that the largest observations in the data set are larger than we would expect from a normal distribution. Therefore, these points might be high-end outliers and/or a signal of positive skewness.

To obtain a Q-Q plot in StatPro, we use the StatPro/Tests of Normality/Q-Q Plot menu item and select the variable to be tested. The output for the Width data in Example 10.7 appears in Figures 10.28 and 10.29. Although the points in this Q-Q plot do not all lie *ex-*

actly on a 45° line, they are about as close to doing so as we can expect from real data. Therefore, there is no reason to question the normal hypothesis for these data—the same conclusion we reached with the chi-square and Lilliefors tests.

PROBLEMS

Level A

40. A finance professor has just given a midterm examination in her corporate finance course. In particular, she is interested in determining whether the distribution of 100 exam scores is normally distributed. The data are in the file P02_05.XLS. Perform a chi-square goodness-of-fit test. Report and interpret the computed *p*-value. What can you conclude about normality?

41. The annual base salaries for 100 students graduating from a reputable MBA program this year are of interest to those in the admissions office who are responsible for marketing the program to prospective students. The data are in the file P10_41.XLS. Are these salaries normally distributed? Perform a chi-square goodness-of-fit test. Report and interpret the computed *p*-value.

42. The manager of a local fast-food restaurant is interested in improving the service provided to customers who use the restaurant's drive-up window. As a first step in this process, the manager asks his assistant to record the time (in minutes) it takes to serve 100 different customers at the final window in the facility's drive-up system. The given 100 customer service times are all observed during the busiest hour of the day for this fast-food operation. The data are in the file P10_42.XLS. Prior to performing some statistical tests on these data, the manager's assistant must know whether the given customer service times are normally distributed. Perform a chi-square goodness-of-fit test. Report and interpret the computed *p*-value.

43. A manufacturer is interested in determining whether it can claim that the boxes of detergent it sells contain, on average, more than 500 grams of detergent. From past experience the manufacturer assumes that the amount of detergent in the boxes is normally distributed. The firm takes a random sample of 100 boxes and records the amount of detergent (in grams) in each box. Based on the data in the file P10_02.XLS, is it still reasonable for the manufacturer to assume that the amount of detergent in these boxes is normally distributed? Perform a chi-square goodness-of-fit test. Report and interpret the computed *p*-value.

44. An aircraft manufacturer needs to buy aluminum sheets with an average thickness of 0.05 inch. The manufacturer collects a random sample of 100 sheets from a potential supplier. The thickness of each sheet in this sample is measured (in inches) and recorded in the file P10_07.XLS. Are these measurements normally distributed? Using a 5% significance level, perform a chi-square goodness-of-fit test. Summarize your results.

45. A U.S. Navy recruiting center knows from past experience that the mean height of its recruits is 68 inches. The recruiting center wants to test the claim that the average height of this year's recruits is greater than 68 inches. To do this, recruiting personnel take a random sample of 64 recruits from the past year and record their heights (in inches). These data are provided in the file P10_04.XLS. Before conducting an appropriate statistical test, the recruiters would like to check whether the given distribution of heights is normally distributed.
   a. On the basis of the available information, do the recruiters find support for the normality assumption at the 10% significance level? Explain.
   b. On the basis of the available information, do the recruiters find support for the normality assumption at the 1% significance level? Explain.

Level B

46. The chi-square test for normality discussed in this section is far from perfect. If the sample is too small, the test tends to accept the null hypothesis of normality for any population distribution even remotely bell shaped; that is, it is not **powerful** in detecting nonnormality. On the other hand, if the sample is very large, it will tend to reject the null hypothesis of normality for *any* data set.[9] Check this by using simulation. Simulate data from a normal distribution to illustrate that if the sample size is sufficiently large, there is a good chance that the null hypothesis will (wrongly) be rejected. Then simulate data from a nonnormal distribution (uniform or triangular, say) to illustrate that if the sample size is fairly small, there is a good chance that the null hypothesis will (wrongly) not be rejected. Summarize your findings in a short report.

[9]Actually, all of the tests for normality suffer from this latter problem.

10.6 Chi-Square Test for Independence

The test we discuss in this section, like one of the tests for normality from the previous section, uses the name "chi-square." However, this test, called the chi-square test for independence, has an entirely different objective. It is used in situations where a population is categorized in two different ways. For example, we might categorize people by their smoking habits and their drinking habits. The question then is whether these two attributes are *independent* in a probabilistic sense. They are **independent** if information on a person's drinking habits is of no use in predicting the person's smoking habits (or vice versa). In this particular example, however, we might suspect that these attributes are **dependent.** In particular, we might suspect that heavy drinkers are more likely (than nonheavy drinkers) to be heavy smokers, and we might suspect that nondrinkers are more likely (than drinkers) to be nonsmokers. The chi-square test for independence enables us to test this empirically.

Rejecting independence does not tell us the *form* of dependence. To see this, we must look more closely at the data.

The null hypothesis for this test is that the two attributes are independent. Therefore, statistically significant results are those that indicate some sort of dependence. As usual, this puts the burden of proof on the alternative hypothesis of dependence. In the smoking–drinking example, we would continue to believe that smoking and drinking habits are unrelated—that is, independent—unless there is sufficient evidence from the data that they are dependent. Furthermore, even if we are able to conclude that they are dependent, the test itself does not indicate the *form* of dependence. It could be that heavy drinkers tend to be nonsmokers, and nondrinkers tend to be heavy smokers. Although this is probably unlikely, it is definitely a form of dependence. The only way we can decide which form of dependence we have is to look closely at the data.

The data for this test consist of *counts* in various combinations of categories. We usually arrange these in a rectangular table called a **contingency table,** a **cross-tabs,** or, using Excel terminology, a pivot table.[10] For example, if there are three smoking categories and three drinking categories, the table would have three rows and three columns, for a total of nine cells. The data entry in a cell is the number of observations in that particular combination of categories. We illustrate this data setup and the resulting analysis in the following example.

Chi-Square Test for Independence

The **chi-square test for independence** is based on the counts in a contingency (or cross-tabs) table. It tests whether the counts for the row categories are probabilistically independent of the counts for the column categories.

Example 10.8 Relationship Between Demands for Desktops and Laptops at Big Office

Big Office, a chain of large office supply stores, sells an extensive line of desktop and laptop computers. Company executives want to know whether the demands for these two types of computers are related in any way. They might act as complementary products,

---

[10]Statisticians have long used the terms *contingency table* and *cross-tabs* for the tables we are discussing here. Pivot tables are more general—they can contain summary measures such as means and standard deviations, not just counts—but when they contain counts, they are equivalent to contingency tables and cross-tabs.

FIGURE 10.30
Counts of Daily
Demands for
Desktops and
Laptops

	A	B	C	D	E	F	G
1	Counts on 250 days of demands at Big Office						
2							
3			Desktops				
4			Low	MedLow	MedHigh	High	
5	Laptops	Low	4	17	17	5	43
6		MedLow	8	23	22	27	80
7		MedHigh	16	20	14	20	70
8		High	10	17	19	11	57
9			38	77	72	63	250

where high demand for desktops accompanies high demand for laptops (computers in general are hot), they might act as substitute products (demand for one takes away demand for the other), or their demands might be unrelated. Because of limitations in its information system, Big Office does not have the exact demands for these products. However, it does have daily information on categories of demand, listed in aggregate (that is, over all stores). These data appear in Figure 10.30. (See the file PCDEMAND.XLS.) Each day's demand for each type of computer is categorized as Low, Medium–Low, Medium–High, or High. The table is based on 250 days, so that the counts add to 250. The individual counts show, for example, that demand was high for both desktops *and* laptops on 11 of the 250 days. For convenience, we include row and column totals in the margins. Based on these data, can Big Office conclude that demands for these two products are independent?

Objective To use the chi-square test of independence to see whether demand for desktops is independent of demand for laptops.

Solution

The idea of the test is to compare the actual counts in the table with what we would *expect* them to be under independence. If the actual counts are sufficiently far from the expected counts, we can then reject the null hypothesis of independence. The "distance" measure used to check how far apart they are, shown in equation (10.8), is essentially the same chi-square statistic used in the chi-square test for normality. Here, O_{ij} is the actual count in cell i, j (row i, column j), E_{ij} is the expected count for that cell, and the sum is over all cells in the table. If this test statistic is sufficiently large, we reject the independence hypothesis. (We will provide more details of the test shortly.)

Test Statistic for Chi-Square Test for Independence	$$\text{Chi-square test statistic} = \sum_{ij} (O_{ij} - E_{ij})^2/E_{ij}$$	**(10.8)**

What do we expect under independence? The totals in row 9 indicate that demand for desktops was low on 38 of the 250 days. Therefore, if we had to estimate the probability of low demand for desktops, this estimate would be $38/250 = 0.152$. Now, if demands for the two products were independent, we should arrive at this *same* estimate from the data in any of rows 5–8. That is, a prediction about desktops should be the same regardless of the demand for laptops. The probability estimate of low desktop from row 5, for example, is $4/43 = 0.093$. Similarly, for rows 6, 7, and 8, it is $8/80 = 0.100$, $16/70 = 0.229$, and $10/57 = 0.175$. These calculations provide some evidence that desktops and laptops act as *substitute* products—the probability of low desktop demand is larger when laptop demand is medium–high or high than when it is low or medium–low.

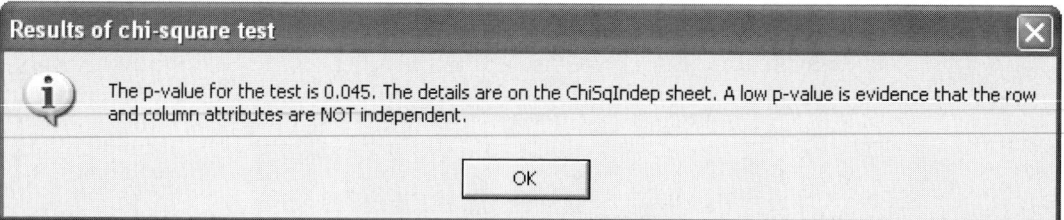

FIGURE 10.31 StatPro Message About the Test's Significance

This reasoning is the basis for calculating the E_{ij}'s. Specifically, it can be shown that the relevant formula for E_{ij} is given by equation (10.9), where R_i is the row total in row i, C_i is the total in column j, and N is the number of observations. For example, E_{11} for these data is 43(38)/250 = 6.536, which is slightly larger than O_{11} = 4.

Expected Counts Assuming Row and Column Independence		
	$$E_{ij} = R_i C_j / N$$	**(10.9)**

We can perform the calculations for the test easily with StatPro. We use the StatPro/Statistical Inference/Chi-square Independence Test menu item. There is only one dialog box, which asks for the range of the contingency table—*not* counting any labels or row or column totals that might surround the table. In this case, the relevant range in Figure 10.30 has been highlighted. StatPro then provides the message in Figure 10.31, and it appends a sheet named ChiSqIndep that contains the calculations. These appear in Figure 10.32.

We interpret the *p*-value of the test, 0.045, in the usual way. Specifically, we can reject the null hypothesis of independence at the 5% or 10% significance levels, but not at the 1% level. There is a good bit of evidence that the demands for the two products are not independent, but it is not overwhelming. (If you are interested in more details of the test, take a look at the formulas in rows 21–35 of Figure 10.32. In particular, the degrees of freedom parameter for the chi-square test is $(r - 1)(c - 1)$, where r is the number of row categories and c is the number of column categories.)

Tables of counts expressed as percentages of rows or of columns are useful for judging the form (and extent) of any possible dependence.

If we accept that there is some sort of dependence, we can use the output in Figure 10.32 to examine its form. The two tables in rows 8–19 are especially helpful. If the demands *were* independent, the rows of this first table should be identical, and the columns of the second table should be identical. This is because each row in the first table shows the distribution of desktop demand for each category of laptop demand, whereas each column in the second table shows the distribution of laptop demand for each category of desktop demand. A close study of these percentages again provides some evidence that the two products act as substitutes, but the evidence is not overwhelming. ●

It is important to realize that the table of counts necessary for the chi-square test of independence can be a pivot table. For example, the pivot table in Figure 10.33 shows counts of the Married and OwnHome attributes. (For Married, 1 means married, 0 means unmarried, and for OwnHome, 1 means a home owner, 0 means not a home owner. This pivot table is based on the data in the CATALOGS.XLS file we examined in Chapter 3.)

FIGURE 10.32
ChiSqIndep Sheet with Calculations

	A	B	C	D	E	F	G	H
1	Original counts, with row totals shown at right and column totals below in bold							
2		4	17	17	5	43		
3		8	23	22	27	80		
4		16	20	14	20	70		
5		10	17	19	11	57		
6		38	77	72	63	250		
7								
8	Shown as percentages of row totals							
9		9.3%	39.5%	39.5%	11.6%	100.0%		
10		10.0%	28.8%	27.5%	33.8%	100.0%		
11		22.9%	28.6%	20.0%	28.6%	100.0%		
12		17.5%	29.8%	33.3%	19.3%	100.0%		
13								
14	Shown as percentages of column totals							
15		10.5%	22.1%	23.6%	7.9%			
16		21.1%	29.9%	30.6%	42.9%			
17		42.1%	26.0%	19.4%	31.7%			
18		26.3%	22.1%	26.4%	17.5%			
19		100.0%	100.0%	100.0%	100.0%			
20								
21	Expected counts							
22		6.536	13.244	12.384	10.836			
23		12.160	24.640	23.040	20.160			
24		10.640	21.560	20.160	17.640			
25		8.664	17.556	16.416	14.364			
26								
27	Distances of observed from expected							
28		0.984	1.065	1.721	3.143			
29		1.423	0.109	0.047	2.321			
30		2.700	0.113	1.882	0.316			
31		0.206	0.018	0.407	0.788			
32								
33	Chi-square test statistic							
34		17.242						
35		0.045058						

FIGURE 10.33
Using a Pivot Table for a Chi-Square Test

	A	B	C	D
3	Count	OwnHome ▼		
4	Married ▼	0	1	Grand Total
5	0	307	191	498
6	1	177	325	502
7	Grand Total	484	516	1000

To see whether these two attributes are independent, we would perform the chi-square test on the table in the range B5:C6. You might want to check that the p-value for the test is 0.000, so that Married and OwnHome are *definitely* not independent.

PROBLEMS

Level A

47. The file P09_49.XLS contains data on 400 orders placed to the ElecMart company over the period of several months. For each order, the file lists the time of day, the type of credit card used, the region of the country where the customer resides, and others. Use a chi-square test for independence to see whether the following variables are independent. If the variables appear to be related, discuss the form of dependence you see.

a. Time and Region

b. Region and BuyCategory

c. Gender and CardType

48. The file P09_17.XLS categorizes 250 randomly selected consumers on the basis of their gender, their age, and their preference for Pepsi or Coke. Use a chi-square test for independence to see whether the drink preference is independent of gender; whether it is independent of age. If you find any dependence, discuss its nature.

49. The file P02_07.XLS contains data on 150 houses that were recently sold. Two variables in this data set are Price, the selling price of the house in $1000s, and Number_Bedrooms, the number of bedrooms in the house. We want to use a chi-square test for independence to see whether these two variables are independent. However, this test requires *categorical* variables, and Price is essentially continuous. Therefore, to run the test, first divide the prices into several categories: less than 120, 120 to 130, 130 to 140, and greater than 140. Then run the test and report your results.

Level B

50. The file P03_86.XLS contains salary data on almost 1800 NFL football players in the 2000 season. We want to use a chi-square test of independence to see whether Total (Salary plus Bonus) is related to other variables in the data set, such as the conference (NFC or AFC) or the player's position (quarterback, running back, and so on). Because the chi-square test requires *categorical* variables, you must first divide all salaries into several categories. Use these categories (in $1000s): less than 300, 300–600, 600–900, and greater than 900. Then test whether the following variables are independent. If they are not, discuss the type of dependence that appears to exist.

a. Salary and Conference
b. Salary and Off/Def
c. Salary and Position
d. Salary and Team

51. The file P03_83.XLS contains data on 1000 Marvak customers. The data set includes demographic variables for each customer as well as their salaries and the amounts they have spent at Marvak during the past year.

a. A lookup table in the file suggests a way to categorize the salaries. Use this categorization and chi-square tests of independence to see whether Salary is independent of (i) Age, (ii) Gender, (iii) Home, (iv) Married. Discuss any types of dependence you find.

b. Repeat part **a**, replacing Salary with AmtSpent. First you must categorize AmtSpent. Create four categories for AmtSpent based on the four quartiles. The first category is all values of AmtSpent below the first quartile of AmtSpent, the second category is between the first quartile and the median, and so on.

52. The file VIDEOS.XLS (the file that accompanies the case for Chapter 8) contains data on close to 10,000 videotape customers from several large cities in the United States. The variables include the customers' gender and their first choice among several types of movies. Perform chi-square tests of independence to see whether the following variables are related. If they are, discuss the form of dependence you are seeing.

a. State and FirstChoice
b. City and FirstChoice
c. Gender and FirstChoice

 # **One-Way ANOVA**

In Sections 9.7.1 and 10.4.2 we discussed the two-sample procedure for analyzing the difference between two population means. A natural extension is to *more* than two population means. The resulting procedure is commonly called *one-way analysis of variance,* or *one-way ANOVA.* There are two typical situations where one-way ANOVA is used. The first is when there are several distinct populations. For example, consider recent graduates with BS degrees in one of three disciplines: Business, Engineering, and Computer Science. We might sample randomly from each of these populations to discover whether there are any significant differences between them with respect to mean starting salary.

A second situation where one-way ANOVA is used is in randomized experiments. In this case a *single* population is treated in one of several ways. For example, a pharmaceutical company might select a group of people who suffer from allergies and randomly assign each person to a different type of allergy medicine currently being developed. Then the question is whether any of the treatments differ from one another with respect to the mean amount of symptom relief.

The data analysis in these two situations is identical; only the interpretation of the results differs. For the sake of clarity, we will phrase this discussion in terms of the first situation, where we randomly sample from each of several populations. Let I be the number of populations, and denote the means of these populations by μ_1 through μ_I. The null hypothesis is that the I means are all equal, whereas the alternative is that they are not all equal. Note that this alternative admits many possibilities. With $I = 4$, for example, we could have $\mu_1 = \mu_2 = \mu_3 = 5$ and $\mu_4 = 10$, or we could have $\mu_1 = \mu_2 = 5$ and $\mu_3 = \mu_4 = 10$, or we could have $\mu_1 = 5$, $\mu_2 = 7$, $\mu_3 = 9$, and $\mu_4 = 10$. The alternative hypothesis simply specifies that *the means are not all equal.*

Hypotheses for One-Way ANOVA	Null hypothesis: All means are equal Alternative hypothesis: At least one mean is different from others

The one-way ANOVA procedure is usually run in two stages. In the first stage we test the null hypothesis of equal means. If the resulting p-value is not sufficiently small, then there is not enough evidence to reject the equal-means hypothesis, and the analysis stops. However, if the p-value is sufficiently small, we can conclude with some assurance that the means are not all equal. Then the second stage is to discover which means are significantly different from which other means. This latter analysis is usually accomplished via confidence intervals.

If one-way ANOVA is basically a test of differences between means, why is it called analysis of *variance?* The answer to this question is the key to the procedure. Consider the boxplot in Figure 10.34. It corresponds to observations from four treatment levels with slightly different means and fairly large variances. (The large variances are indicated by the relatively wide boxes and long lines extending from them.) From these boxplots, would you conclude that the population means differ across the four treatment levels? Would your answer change if the data were instead as in Figure 10.35? We expect that it would.

The sample means in these two figures are virtually the same, but the variances *within* each treatment level in Figure 10.34 are quite large relative to the variance *between* the sample means. In contrast, there is very little variance within each treatment level in Figure 10.35. In the first case the large "within" variance makes it difficult to infer whether there are really any differences between population means, whereas the small "within" variance in the second case makes it easy to infer differences between population means.

FIGURE 10.34
Samples with Large Within Variation

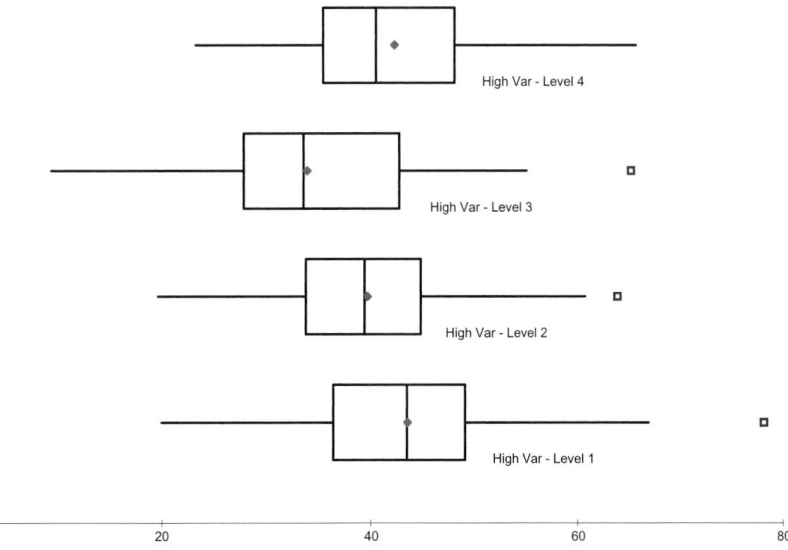

FIGURE 10.35
Samples with Small
Within Variation

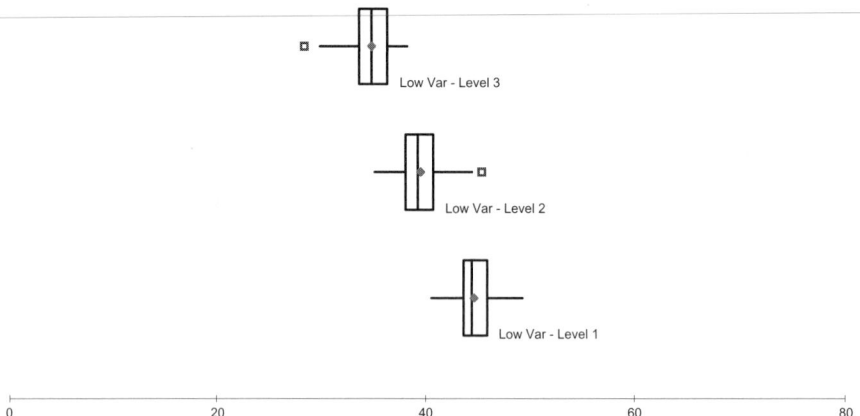

This is the essence of the ANOVA test. We compare variances *within* the individual samples to variance *between* the sample means. Only if the between variance is large relative to the within variance can we conclude with any assurance that there are differences between population means—and reject the equal-means hypothesis.

A test is robust *if its results are valid even when the assumptions behind it are not exactly true.*

The test itself is based on two assumptions: (1) the population variances are all equal to some common variance σ^2, and (2) the populations are normally distributed. These are analogous to the assumptions we made for the two-sample t test. Although these assumptions are never satisfied exactly in any application, we should keep them in mind and check for gross violations whenever possible. Fortunately, the test we present is fairly robust to violations of these assumptions, particularly when the sample sizes are large and roughly the same.

The between *variation measures how much the sample means differ from one another.*

To run the test, let $\overline{Y}_i$, s_i^2, and n_i be the sample mean, sample variance, and sample size from sample i. Also, let n and $\overline{\overline{Y}}$ be the combined number of observations and the sample mean of all n observations. (We call $\overline{\overline{Y}}$ the **grand mean.**) Then a measure of the between variation is *SSB* (sum of squares between):

$$SSB = \sum_{i=1}^{I} n_i \, (\overline{Y}_i - \overline{\overline{Y}})^2$$

The within *variation measures how much the observations within each sample differ from one another.*

Note that *SSB* is large if the individual sample means differ substantially from the grand mean $\overline{\overline{Y}}$, and this occurs only if they differ substantially from one another. A measure of the within variation is *SSW* (sum of squares within):

$$SSW = \sum_{i=1}^{I} (n_i - 1)s_i^2$$

This sum of squares is large if the individual sample variances are large. For example, *SSW* is much larger in Figure 10.34 than in Figure 10.35. However, *SSB* is the same in both figures.

Each of these sums of squares has an associated degrees of freedom, *dfB* and *dfW:*

$$dfB = I - 1$$

and

$$dfW = n - I$$

When we divide the sums of squares by their degrees of freedom, we obtain "mean squares," *MSB* and *MSW*:

$$MSB = \frac{SSB}{dfB}$$

and

$$MSW = \frac{SSW}{dfW}$$

Actually, it can be shown that *MSW* is a weighted average of the individual sample variances, where the sample variance s_i^2 receives weight $(n_i - 1)/(n - I)$. In this sense *MSW* is a pooled estimate of the common variance σ^2, just as in the two-sample procedure.

Finally, the ratio of these means squares, shown in equation (10.10), is the test statistic we use. Under the null hypothesis of equal population means, this test statistic has an *F* distribution with *dfB* and *dfW* degrees of freedom. If the null hypothesis is *not* true, then we would expect *MSB* to be large relative to *MSW*, as in Figure 10.35. Therefore, the *p*-value for the test is found by finding the probability to the *right* of the *F*-ratio in the *F* distribution with *dfB* and *dfW* degrees of freedom.

Test Statistic for One-Way ANOVA Test of Equal Means

$$F\text{-ratio} = \frac{MSB}{MSW} \tag{10.10}$$

The elements of this test are usually presented in an **ANOVA table,** as we will see shortly. The "bottom line" in this table is the *p*-value. If it is sufficiently small, we can conclude that the population means are not all equal. Otherwise, we cannot reject the equal-means hypothesis.

If we do reject the equal-means hypothesis, then it is customary to examine confidence intervals for the differences between all pairs of population means. This can lead to quite a few confidence intervals. For example, if there are $I = 5$ samples, then there are 10 pairs of differences (the number of ways 2 means can be chosen from 5 means). As usual, the confidence interval for any difference $\mu_i - \mu_j$ is of the form

$$\overline{Y}_i - \overline{Y}_j \pm \text{multiplier} \times \text{SE}(\overline{Y}_i - \overline{Y}_j)$$

The appropriate standard error is

$$\text{SE}(\overline{Y}_i - \overline{Y}_j) = s_p\sqrt{1/n_i + 1/n_j}$$

where s_p is the pooled standard deviation, calculated as $\sqrt{MSW}$.

There are several forms of these confidence intervals, some of which are implemented in StatPro, and the appropriate multiplier for the confidence intervals depends on which form is being used. We will not pursue the technical details here, except to say that the multiplier is sometimes chosen to be its "usual" value near 2 and is sometimes chosen to be considerably larger, say, around 3.5. The reason for the latter is that if we want to conclude with 95% confidence that *each* of these confidence intervals includes its corresponding mean difference, we must make the confidence intervals relatively wide.

For any of these confidence intervals that does *not* include the value 0, we infer that the corresponding means are not equal. But if a confidence interval does include 0, we cannot conclude that the corresponding means are unequal.

If the confidence interval for a particular difference does not include 0, we can conclude that these two means are different.

We have presented the formulas for one-way ANOVA because we believe they lend some insight into the procedure. However, StatPro's one-way ANOVA procedure takes care of all the tedious calculations, as illustrated in the following example.

Example 10.9 Employee Empowerment at ArmCo Company

We discussed the ArmCo Company in Example 10.6. It initiated an employee empowerment program at its Midwest plant, and the reaction from employees was basically positive. Let's assume now that ArmCo has initiated this policy in all five of its plants—in the South, Midwest, Northeast, Southwest, and West—and several months later it wants to see whether the policy is being perceived equally by employees across the plants. Random samples of employees at the five plants have been asked to rate the success of the empowerment policy on a scale of 1 to 10, 10 being the most favorable rating. The data appear in Figure 10.36.[11] (See the file EMPOWER2.XLS.) Is there any indication of mean differences across the plants? If so, which plants appear to differ from which others?

Objective To use one-way ANOVA to test whether the empowerment initiatives are appreciated equally across Armco's five plants.

Solution

One-way ANOVA does *not* require equal sample sizes.

First, note that the sample sizes are not equal. This could be because some employees opted not to cooperate or it could be due to other reasons. Fortunately, equal sample sizes are not necessary for the ANOVA test.

FIGURE 10.36
Data for Empowerment Example

	A	B	C	D	E
1	Empowerment results from several plants				
2					
3	South	Midwest	Northeast	Southwest	West
4	7	7	7	6	6
5	1	6	5	4	6
6	8	10	5	7	6
7	7	3	5	10	6
8	2	9	4	7	3
9	9	10	3	6	4
10	3	8	4	6	8
11	8	4	5	7	6
41	5	3	5	4	2
42	7	2	3	3	4
43	4	7	3	7	5
44		7	3	8	6
45		5	5	9	4
46		10	5	10	7
47		10		4	4
48		6		10	3
49		3		4	5
50		5		6	4
51		2			7
52		6			6
53		4			4
54		5			
55		2			
56		7			
57		8			
58		7			

[11]StatPro's One-Way ANOVA procedure accepts the data in stacked or unstacked form. The data in this example are unstacked.

FIGURE 10.37
StatPro Dialog
Box for ANOVA
Confidence
Intervals

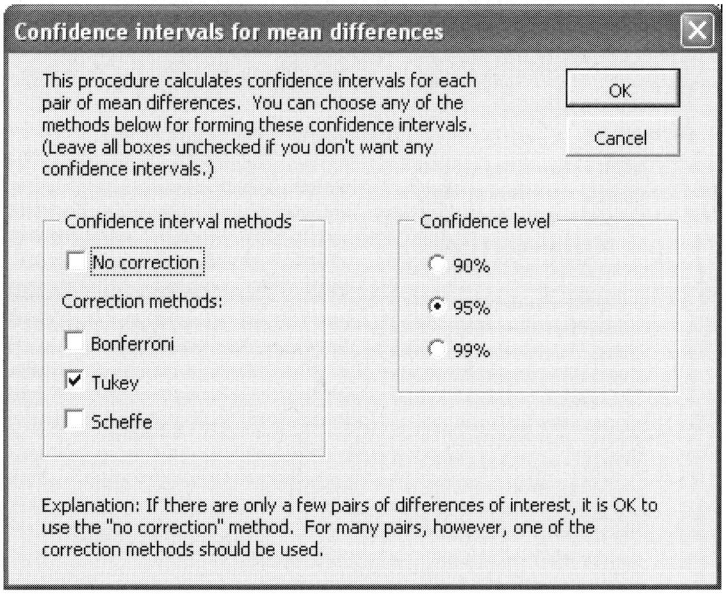

Confidence intervals for mean differences

This procedure calculates confidence intervals for each pair of mean differences. You can choose any of the methods below for forming these confidence intervals. (Leave all boxes unchecked if you don't want any confidence intervals.)

OK

Cancel

Confidence interval methods
☐ No correction

Correction methods:
☐ Bonferroni
☑ Tukey
☐ Scheffe

Confidence level
○ 90%
● 95%
○ 99%

Explanation: If there are only a few pairs of differences of interest, it is OK to use the "no correction" method. For many pairs, however, one of the correction methods should be used.

To run one-way ANOVA with StatPro on these data, select the StatPro/Statistical Inference/One-Way ANOVA menu item, check that these data are unstacked, select all five variables, and fill out the next dialog box as shown in Figure 10.37. This dialog box indicates that there are several "types" of confidence intervals available. Each of these uses a slightly different multiplier in the general confidence interval formula. Again, we will not pursue the differences between these confidence interval types, except to say that the default Tukey type is generally the most acceptable.

The resulting output in Figure 10.38 (page 532) consists of three basic parts: summary statistics, the ANOVA table, and confidence intervals. The summary statistics show that the Southwest has the largest mean rating, 6.745, and the Northeast has the smallest, 4.140, with the others in between. The sample standard deviations (or variances) vary somewhat across the plants, but not enough to invalidate the procedure. The side-by-side boxplots in Figure 10.39 illustrate these summary measures graphically. However, there is too much overlap between the boxplots to tell (graphically) whether the observed differences between plants are statistically significant.

The ANOVA table in rows 20–23 of Figure 10.38 shows the elements for the F test of equal means. All of it is based on the theory we developed above. The only part we didn't discuss is the Total variation in row 23. It is based on the total variation of all observations around the grand mean in cell I15, and is used mainly to check the calculations. (Note that SSB and SSW in cells I20 and I21 add up to the total sum of squares in cell I23. Similarly, the degrees of freedom add up in column J.) The F-ratio for the test is 10.480, in cell L21. Its corresponding p-value (to three decimal places) is 0.000. This leaves practically no doubt that the five population means are *not* all equal. Employees evidently do not perceive the empowerment policy equally across plants.

The main thing to remember from the ANOVA table is that a small p-value indicates that the population means are *not* all equal.

The 95% confidence intervals in rows 29–38 indicate which plants differ significantly from which others. For example, the mean for the Southwest plant is somewhere between 1.455 and 3.756 rating points above the mean for the Northeast plant. We see that the Southwest plant is rated significantly higher than the Northeast, West, and Midwest plants, and the South and Midwest plants are also rated significantly higher than the Northeast plant. Now it is up to ArmCo management to decide whether the magnitudes of these differences are *practically* significant, and, if so, what they can do to increase employee perceptions at the lower-rated plants.

FIGURE 10.38
Analysis of Empowerment Data

	G	H	I	J	K	L	M
3		**Results of one-way ANOVA**					
4							
5		**Summary stats for samples**					
6			South	Midwest	Northeast	Southwest	West
7		Sample sizes	40	55	43	47	50
8		Sample means	5.600	5.400	4.140	6.745	4.980
9		Sample standard deviations	2.073	2.469	1.820	1.687	1.635
10		Sample variances	4.297	6.096	3.313	2.846	2.673
11		Weights for pooled variance	0.170	0.235	0.183	0.200	0.213
12							
13		Number of samples	5				
14		Total sample size	235				
15		Grand mean	5.383				
16		Pooled variance	3.904				
17		Pooled standard deviation	1.976				
18							
19		**OneWay ANOVA table**					
20		Source	SS	df	MS	F	p-value
21		Between variation	163.653	4	40.913	10.480	0.0000
22		Within variation	897.879	230	3.904		
23		Total variation	1061.532	234			
24							
25		**Confidence intervals for mean differences**					
26		Confidence level	95.0%				
27		**Tukey method**					
28		Difference	Mean diff	Lower	Upper	Signif?	
29		South - Midwest	0.200	-0.933	1.333	No	
30		South - Northeast	1.460	0.263	2.658	Yes	
31		South - Southwest	-1.145	-2.318	0.028	No	
32		South - West	0.620	-0.537	1.777	No	
33		Midwest - Northeast	1.260	0.151	2.370	Yes	
34		Midwest - Southwest	-1.345	-2.428	-0.262	Yes	
35		Midwest - West	0.420	-0.645	1.485	No	
36		Northeast - Southwest	-2.605	-3.756	-1.455	Yes	
37		Northeast - West	-0.840	-1.974	0.294	No	
38		Southwest - West	1.765	0.657	2.872	Yes	

FIGURE 10.39
Boxplots for Empowerment Data

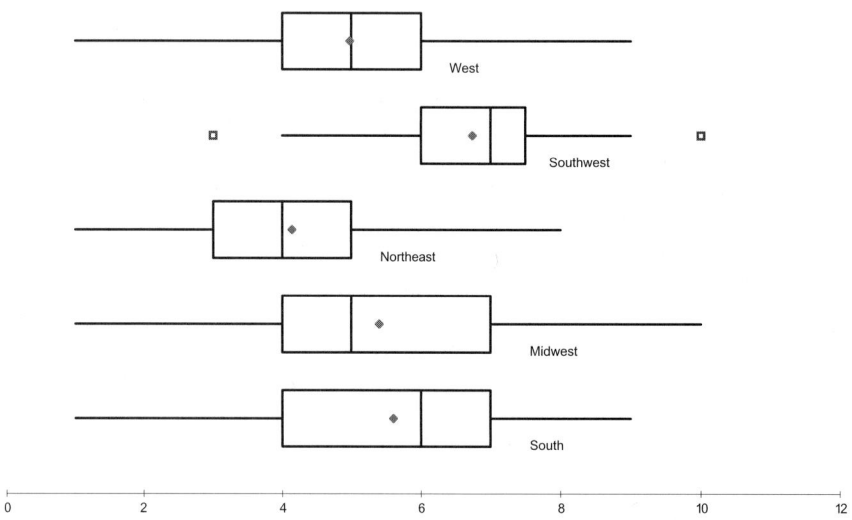

PROBLEMS

Level A

53. An automobile manufacturer employs sales representatives who make calls on dealers. The manufacturer wishes to compare the effectiveness of four different call-frequency plans for the sales representatives. Thirty-two representatives are chosen at random from the sales force and randomly assigned to the four call plans (eight per plan). The representatives follow their plans for 6 months, and their sales for the 6-month study period are recorded. These data are given in the file P10_53.XLS.

    a. Do the sample data support the hypothesis that at least one of the call plans helps produce a higher average level of sales than some other call plan? Perform an appropriate statistical test and report a *p*-value.

    b. If the sample data indicate the existence of mean sales differences across the call plans, which plans appear to produce different average sales levels? Construct 95% confidence levels for the differences between all pairs of means to help answer this question.

54. Consider a large chain of supermarkets that sell their own brand of potato chips in addition to many other name brands. Management would like to know whether the type of display used for the store brand has any effect on sales. Because there are four types of displays being considered, management decides to choose 24 similar stores to serve as experimental units. A random six of these are instructed to use display type 1, another random six are instructed to use display type 2, a third random six are instructed to use display type 3, and the final six stores are instructed to use display type 4. For a period of 1 month, each store keeps track of the *fraction* of total potato chips sales that are of the store brand. The data for the 24 stores are shown in the file P10_54.XLS. Note that one of the stores using display 3 is marked with an asterisk (*). This store did not follow instructions properly, so its observation is disregarded.

    a. Why do you think each store keeps track of the fraction of total potato chips sales that are of the store brand? Why do they not simply record the total amount of sales of the store brand potato chips?

    b. Do the data suggest different mean proportions of store brand sales at the 10% significance level? If so, construct 90% confidence intervals for the differences between all pairs of mean proportions to identify which of the display types are associated with higher fractions of sales.

55. National Airlines recently introduced a daily early-morning nonstop flight between Houston and Chicago. The vice president of marketing for National Airlines decided to perform a statistical test to see whether National's average passenger load on this new flight is different from that of each of its two major competitors (which we will call competitor 1 and competitor 2). Ten early-morning flights were selected at random from each of the three airlines and the percentage of *unfilled* seats on each flight was recorded. These data are stored in the file P10_55.XLS.

    a. Is there evidence that National's average passenger load on the new flight is different from that of its two competitors? Report a *p*-value and interpret the results of the statistical test.

    b. Select an appropriate significance level and construct confidence intervals for all pairs of differences between means. Which of these differences, if any, are statistically significant at the selected significance level?

56. Do graduates of undergraduate business programs with different majors tend to earn disparate average starting salaries? Consider the data given in the file P10_56.XLS.

    a. Is there any reason to doubt the equal-variance assumption made in the one-way ANOVA model in this particular case? Support your response to this question.

    b. Assuming that the variances of the four underlying populations are indeed equal, can you reject at the 10% significance level that the mean starting salary is the same for each of the given business majors? Explain why or why not.

    c. Generate 90% confidence intervals for all pairs of differences between means. Which of these differences, if any, are statistically significant at $\alpha = 0.10$?

10.8 Conclusion

The concepts and procedures we have discussed in this chapter occupy a cornerstone in both applied and theoretical statistics. Of particular importance is the interpretation of a *p*-value, especially since *p*-values are common outputs of all statistical software packages. A *p*-value summarizes the evidence in support of an alternative hypothesis, which is usually the hypothesis an analyst is trying to prove. Small *p*-values provide support for the alternative hypothesis, whereas large *p*-values provide little or no support for it.

Although hypothesis testing continues to be an important tool for analysts, it is important to note its limitations, particularly in business applications. First, given a choice between a confidence interval for some population parameter and a test of this parameter, we generally favor the confidence interval. For example, a confidence interval not only tells us whether a mean difference is 0, but it also gives us a plausible range for this difference. Second, many business *decision* problems cannot be handled adequately with hypothesis-testing procedures. Either they ignore important cost information or they treat the consequences of incorrect decisions (type I and type II errors) in an inappropriate way. Finally, the *statistical* significance at the core of hypothesis testing is sometimes quite different from the *practical* significance that is of most interest to business managers.

Summary of Key Terms

Term	Explanation	Excel	Page	Equation Number
Null hypothesis	Hypothesis that represents the current thinking or status quo.		481	
Alternative hypothesis	Typically, the hypothesis the analyst is trying to prove, also called the research hypothesis.		481	
One-tailed test	Test where values in only one direction will lead to rejection of null hypothesis.		483	
Two-tailed test	Test where values in both directions will lead to rejection of null hypothesis.		483	
Type I error	Error committed when null hypothesis is true but is rejected.		484	
Type II error	Error committed when null hypothesis is false but is not rejected.		484	
Rejection region	Sample results that lead to rejection of null hypothesis.		484	
Statistically significant results	Sample results that lead to rejection of null hypothesis.		484	
p-value	Probability of observing a sample result at least as extreme as the one actually observed.		485	

Summary of Key Terms (continued)

Term	Explanation	Excel	Page	Equation Number
Power	Probability of correctly rejecting null hypothesis when it is false.		487	
t test for a population mean	Test for a mean from a single population.	StatPro/ Statistical Inference/One-Sample Analysis	489	10.1
z test for a population proportion	Test for a proportion from a single population.	Must be done manually.	496	10.2
t test for difference between means from paired samples	Test for the difference between two population means when samples are paired in a natural way.	StatPro/ Statistical Inference/ Paired-Sample Analysis	498	10.3
t test for difference between means from independent samples	Test for the difference between two population means when samples are independent of one another.	StatPro/ Statistical Inference/Two-Sample Analysis	498	10.4
F test for equality of two variances	Test to check whether two population variances are equal, used to check an assumption of two-sample t test for difference between means.	StatPro/ Statistical Inference/Two-Sample Analysis	507	
F distribution	Skewed distribution useful for testing equality of variances.	= FDIST(*value, df1, df2*) = FINV(*prob, df1, df2*)	507	
z test for difference between proportions	Test for difference between similarly defined proportions from two populations.	Must be done manually	508	10.5, 10.6
Tests for normality	Tests to check whether a population is normally distributed; alternatives include chi-square test, Lilliefors test, and Q-Q plot.	StatPro/Tests for Normality	516–521	10.7
Chi-square test for independence	Test to check whether two attributes are probabilistically independent.	StatPro/ Statistical Inference/Chi-square Independence Test	522	10.8, 10.9
One-way ANOVA	Generalization of two-sample t test, used to test whether means from several populations are all equal, and if not, which are significantly different from which others.	StatPro/ Statistical Inference/One-Way ANOVA	526	10.10

PROBLEMS

Conceptual Exercises

C.1. Suppose that you wish to test a researcher's claim that the mean of a normally distributed population has increased from its commonly accepted value of 1.60. In order to carry out this test, you obtain a random sample of size 150 from this population. This sample yields a mean of 1.80 and a standard deviation of 1.30. What are the appropriate null and alternative hypotheses? Is this a one-tailed or two-tailed test?

C.2. Suppose that you wish to test a manager's claim that the proportion of defective items generated by a particular production process has decreased from its long-run historical value of 0.30. To carry out this test, you obtain a random sample of 300 items produced through this process. Computer-generated output indicates that the p-value for the sample is 0.01. Carefully interpret this p-value. At what levels of significance would you reject the null hypothesis?

C.3. A 99% confidence interval for the proportion (π) of all Lewisburg residents whose annual income exceeds $80,000 extends from 0.10 to 0.18. The confidence interval is based on a random sample of 150 Lewisburg residents. Using this information and a 1% level of significance, we wish to test the following hypotheses:

$$H_0: \quad \pi = 0.08$$
$$H_a: \quad \pi \neq 0.08$$

Based on the given information, what is the correct statistical decision in this case? Explain your recommendation.

C.4. Suppose that you are performing a one-tailed hypothesis test. "Assuming that everything else remains constant, a decrease in the test's level of significance (α) leads to a higher probability of rejecting the null hypothesis." Is this statement true or false? Explain your choice.

C.5. Can pleasant aromas help people work more efficiently? Researchers conducted an investigation to answer this question. Fifty students worked a paper-and-pencil maze ten times. On five attempts, the students wore a mask with floral scents. On the other five attempts, they wore a mask with no scent. The ten trials were done in random order and each used a different maze. The researchers found that the subjects took less time to complete the maze when wearing the scented mask. Is this an example of an *observational study?* Explain why or why not.

Level A

57. The file P10_57.XLS contains the number of days 44 mothers spent in the hospital giving birth (in the year 1995). Before health insurance rules were changed (the change was effective January 1, 1995), the average number of days spent in a hospital by a new mother was 2 days. For a 0.05 level of significance, do the data in the file indicate (the research hypothesis) that women are now spending less time in the hospital after giving birth than they were prior to 1995? Explain your answer in terms of the p-value for the test.

58. Eighteen readers took a speed-reading course. The file P10_58.XLS contains the number of words that they could read before and after the course. Test the alternative hypothesis at the 5% significance level that reading speeds have increased, on average, as a result of the course. Explain your answer in terms of the p-value. Do you need to assume that reading speeds (before and after) are normally distributed? Exactly what assumption do you need?

59. Statistics show that a child 0–4 years of age has a 0.0002 probability of getting cancer in any given year. Assume that during each of the last 7 years there have been 100 children ages 0–4 whose parents work in the business school. Four of these children have gotten cancer. Use this evidence to test whether the incidence of childhood cancer among children 0–4 whose parents work at the business school exceeds the national average. Write down your hypotheses and determine the appropriate p-value.

60. African-Americans in a St. Louis suburb sued the city claiming they were discriminated against in school-teacher hiring. Of the city's population, 5.7% were African-American; of 405 teachers in the school system, 15 were African-American. Set up appropriate hypotheses and determine whether African-Americans are underrepresented. Does your answer depend on whether you use a one-tailed or two-tailed test? In discrimination cases, the Supreme Court always uses a two-tailed test with $\alpha = .05$. (Source: U.S. Supreme Court Case Hazlewood versus City of St. Louis)

61. In the past, monthly sales for HOOPS, a small software firm, have averaged $20,000 with standard deviation $4,000. During the last year sales averaged $22,000 per month. Does this indicate that monthly sales have changed (in a statistically significant sense)? Use $\alpha = .05$. Assume monthly sales are normally distributed.

62. Twenty people have rated a new beer on a taste scale of 0 to 100. Their ratings are in the file P10_62.XLS. Marketing has determined that the beer will be a success if the average taste rating exceeds 76. If we use $\alpha = 0.05$, is there sufficient evidence to conclude that the beer will be a success? Discuss your result in terms of a p-value. Assume ratings are normally distributed.

63. We have asked 22 people to rate a competitive beer on a taste scale of 0 to 100. Another 22 people rated our beer on a taste scale of 0 to 100. The file P10_63.XLS contains the results. Do these data provide sufficient evidence to conclude, at the $\alpha = 0.01$ level, that people believe our beer tastes better than the competition? Assume ratings are normally distributed.

64. Callaway is thinking about entering the golf ball market. The company will make a profit if its market share is more than 20%. A market survey indicates that 140 of 624 golf ball purchasers will buy a Callaway golf ball.
   a. Is this enough evidence to persuade Callaway to enter the golf ball market?
   b. How would you make the decision if you were Callaway management?

65. Sales of a new product will be profitable if the average sales per store exceeds 100 per week. The product was test marketed for 1 week at 10 stores, with the results listed in the file P10_65.XLS. Assume that sales at each store follow a normal distribution.
   a. Is this enough evidence to persuade the company to market the new product?
   b. How would you make the decision if you were deciding whether to market the new product?

66. We are interested in determining whether the position of Coca-Cola in a store affects sales. Specifically, does Coke sell better when it is placed in the front or middle of an aisle? The file P10_66.XLS contains sales of Coke at 10 stores when Coke was placed in the front of an aisle and at another 10 stores when Coke was placed in the back of an aisle. What are reasonable null and alternative hypotheses to test? Use the data to test them. What can you conclude?

67. Target wants to know whether red or blue coats sell better. The sales of red and blue coats last winter were measured at several different stores. The file P10_67.XLS contains the results. What are reasonable null and alternative hypotheses to test? Use the data to test them. What can you conclude?

68. Marsh Supermarket wants to know whether putting a color flyer in the local paper has a greater effect on sales than putting a black and white flyer in the paper. For the last 9 times a color flyer was put in the paper, Marsh compared sales (in thousands of dollars) to the last week (at the same store) for which a black and white flyer was put in the paper. The file P10_68.XLS contains the data. Formulate reasonable hypotheses and test them. What do you conclude?

69. In the past, the Algood Company has produced an average of 12,000 good parts per day. Since the compensation system was changed 15 days ago, the average production has been 12,100 good parts per day. The sample standard deviation of the number of good parts produced during the last 15 days is 400. Is this sufficient evidence to conclude, at the $\alpha = 0.10$ level, that the new compensation plan has improved productivity? Justify your answer with a p-value.

Level B

70. You are trying to determine whether male and female Central Bank employees having equal qualifications receive different salaries. The file P10_70.XLS contains the salaries (in thousands of dollars) for 9 male and 9 female employees. Assume salaries are normally distributed.
   a. Assuming that each row of data represents paired observations, and using $\alpha = 0.05$, can you conclude that members of different genders are paid equally? Be sure to write down your hypotheses.
   b. How would you collect data to ensure that the observations are actually paired?

71. You are trying to determine whether male and female Indiana University grads having equal qualifications receive different starting salaries for their first job. The file P10_71.XLS contains the starting salaries for 10 male and 10 female IU grads.
   a. Assuming that each row of data represents paired observations, and using $\alpha = 0.05$, can you conclude that equally qualified people of different genders have the same starting salaries on average? Be sure to write down your null and alternative hypotheses.
   b. How would you collect data to ensure that the observations are actually paired?

72. A recent study concluded that children born to mothers who take Prozac tend to have more birth defects than children born to mothers who do not take Prozac.
   a. What do you think the null and alternative hypotheses were for this study?
   b. If you were a spokesperson for Eli Lilly (the company that produces Prozac), how would you rebut the conclusions of this study?

73. Suppose you are the state superintendent of Tennessee public schools. You want to know whether decreasing the class size in grades 1–3 will improve student performance. Explain how you would set up a test to determine whether decreased class size will improve student performance. What hypotheses would you use in this experiment. (This was actually done and smaller class size did help, particularly with minority students.)

74. Do chief executive officers of large corporations in different industries typically earn disparate annual salaries? Consider the data in the file P02_13.XLS, which came from a recent survey of chief executive officers of the nation's biggest businesses (*The Wall Street Journal,* April 9, 1998). In particular, we seek to discover whether significant differences exist between the mean levels of 1997 salaries earned by executives of *Technology* companies and those of each of the other seven company types. For instance, is the difference between the mean 1997 salaries of executives from *Technology* and *Basic Materials* companies significant?

a. Before conducting these hypothesis tests, perform a test for equal population variances for each pair of company types. For each pair, report a *p*-value and interpret it.

b. Based on your conclusions in part **a,** which test statistic should be used in performing a test for the existence of a difference between population means in each pair?

c. Given your conclusions in part **b,** perform a test for the existence of a difference in mean annual CEO salaries. For each pair of company types, report a *p*-value and interpret its meaning.

75. Is the overall cost of living higher or lower for urban areas in particular geographical regions of the United States? Consider the random sample of urban areas provided in the file P10_75.XLS (Source: *ACCRA Cost of Living Index*). In particular, determine whether the mean composite (cost of living) value in urban areas of the northeastern states is higher than the mean composite value in urban areas of each of the following: (a) southeastern states, (b) central states, (c) southwestern states, and (d) northwestern states. Note that you will need to assign each of the urban areas in the given sample to one of these five geographical regions before you can proceed further.

a. Before conducting any hypothesis tests on the difference between various pairs of mean composite values, perform a test for equal population variances in each pair of geographical regions. For each pair, report a *p*-value and interpret its meaning.

b. Based on your conclusions in part **a,** which test statistic should be used in performing a test for a difference between population means in each pair?

c. Given your conclusions in part **b,** perform a test for the difference between composite cost of living values in each pair of geographical regions. For each pair, report a *p*-value and interpret its meaning.

76. Consider a random sample of 100 households from a middle-class neighborhood that was the recent focus of an economic development study conducted by the local government. Specifically, for each of the 100 households in the sample, information was gathered on the gross annual income earned by the first wage earner of the household and on each of several other variables. The data are given in the file P09_26.XLS. Economic researchers would like to test for the existence of a significant difference between the mean an-

nual income levels of the first household wage earners in the first (i.e., SW) and second (i.e., NW) sectors of this community. In fact, they intend to perform similar hypothesis tests for the differences between the mean annual income levels of the first household wage earners from all other pairs of locations (i.e., first and third, first and fourth, second and third, second and fourth, and third and fourth).

a. Before conducting any hypothesis tests on the difference between various pairs of mean income levels, perform a test for equal population variances in each pair of locations. For each pair, report a *p*-value and interpret its meaning.

b. Based on your conclusions in part **a,** which test statistic should be used in performing a test for the existence of a difference between population means?

c. Given your conclusions in part **b,** perform a test for the existence of a difference in mean annual income levels in each pair of locations. For each pair, report a *p*-value and interpret its meaning.

77. A group of 25 husbands and wives were chosen randomly. Each person was asked to write the most he or she would be willing to pay for a new car (assuming they had decided to buy a new car). The results are shown in the file P10_77.XLS. Can you accept the alternative hypothesis that the husbands are willing to spend more, on average, than the wives at the 5% significance level? What is the associated *p*-value?

78. A company is concerned with the high cholesterol levels of many of its employees. To help combat the problem, it opens an exercise facility and encourages its employees to use this facility. After a year, it chooses a random 100 employees who claim they use the facility regularly, and another 200 who claim they don't use it at all. The cholesterol levels of these 300 employees are checked, with the results shown in the file P10_78.XLS.

a. Is this sample evidence "proof" that the exercise facility, when used, tends to lower the mean cholesterol level? Phrase this as a hypothesis-testing problem and do the appropriate analysis. Do you feel comfortable that your analysis answers the question definitively (one way or the other)? Why or why not?

b. Repeat part **a,** but replace "mean level" with "percentage with level over 215." (The company believes that any level over 215 is dangerous.)

79. Suppose that you are trying to compare two populations on some variable (GMAT scores of men versus women, for example). Specifically, you are testing the null hypothesis that the means of the two populations are equal versus a two-tailed hypothesis. Are the following statements correct? Why or why not?

a. A given difference (such as 5 points) between sample means from these populations will probably not be considered statistically significant if the sample sizes are small, but will probably be considered statistically significant if the sample sizes are large.

b. Virtually any difference between the population means will lead to statistically significant sample results if the sample sizes are sufficiently large.

80. Continuing the previous problem, analyze part **b** in Excel as follows. Start with hypothetical population mean GMAT scores for men and women, along with population standard deviations. Enter these at the top of a spreadsheet. You can make the two means as close as you like, but not identical. In column A simulate a sample of men's GMAT scores with your mean and standard deviation. Do the same for women in column B. The sample sizes do not have to be the same, but you can make them the same. Then run the test for the difference between two means. (The point of this problem is that if the population means are fairly close and you pick relatively small sample sizes, the sample mean differences probably won't be significant. If you find this, generate new samples of a larger sample size and redo the test. Now they might be significant. If not, try again with a still larger sample size. Eventually, you should get statistically significant differences.)

81. This problem concerns course scores (on a 0–100 scale) for a large undergraduate computer programming course. The class is composed of both underclassmen (freshmen and sophomores) and upperclassmen (juniors and seniors). Also, the students can be categorized according to their previous mathematical background from previous courses as "low" or "high" mathematical background. The data for these students are in the file P10_81.XLS. The variables are:
- Score: score on a 0–100 scale
- UpperCl: 1 for an upperclassman, 0 otherwise
- HighMath: 1 for a high mathematical background, 0 otherwise

For the following questions, assume that the students in this course represent a random sample from all college students who might take the course. This latter group is the "population."

a. Find a 90% confidence interval for the population mean score for the course. Do the same for the mean of all upperclassmen. Do the same for the mean of all upperclassmen with a high mathematical background.

b. The professor believes he has enough evidence to prove the research hypothesis that upperclassmen score at least 5 points better, on average, than lowerclassmen. Do you agree?

c. If we consider a "good" grade to be one that is at least 80, is there enough evidence to reject the null hypothesis that the fraction of good grades is the same for students with low math backgrounds as those with high math backgrounds?

82. A cereal company wants to see which of two promotional strategies, supplying coupons in a local newspaper or including coupons in the cereal package itself, is more effective. (In the latter case, there is a sign on the package indicating the presence of the coupon inside.) The company randomly chooses 80 Kroger's stores around the country—all of approximately the same size and overall sales volume—and promotes its cereal one way at 40 of these sites, and the other way at the other 40 sites. (All are at different geographical locations, so local newspaper ads for one of the sites should not affect sales at any other site.) Unfortunately, as in many business "experiments," there is a factor beyond the company's control, namely, whether its main competitor at any particular site happens to be running a promotion of its own. The file P10_82.XLS has 80 observations on three variables:
- Sales: number of boxes sold during the first week of the company's promotion
- PromType: 1 if coupons are in local paper, 0 if coupons are inside box
- CompProm: 1 if main competitor is running a promotion, 0 otherwise

a. Based on all 80 observations, find (1) the difference in sample mean sales between stores running the two different promotional types (and indicate which sample mean is larger), (2) the standard error of this difference, and (3) a 90% confidence interval for the population mean difference.

b. Test whether the population mean difference is 0 (the null hypothesis) versus a two-tailed alternative. State whether you should accept or reject the null hypothesis, and why.

c. Repeat part **b**, but now restrict the "population" to stores where the competitor is not running a promotion of its own.

d. Based on data from all 80 observations, can you accept the (alternative) hypothesis, at the 5% level, that the mean company sales drops by at least 30 boxes when the competitor runs its own promotion (as opposed to not running its own promotion)?

e. We often use the term "population" without really thinking what it means. If you talk about the population mean for the case, say, where coupons are put in boxes, explain in words exactly what this population mean refers to.

10.1 Regression Toward the Mean

In Chapters 11 and 12 we will study regression, a method for relating one variable to other explanatory variables. However, the term "regression" has sometimes been used in a slightly different way, meaning "regression toward the mean." The example often cited is of male heights. If a father is unusually tall, for example, his son will typically be taller than average but not as tall as the father. Similarly, if a father is unusually short, the son will typically be shorter than average but not as short as the father. We say that the son's height tends to regress toward the mean. This case will illustrate how regression toward the mean can occur.

Suppose a company administers an aptitude test to all of its job applicants. If an applicant scores below some value, he or she cannot be hired immediately but is allowed to retake a similar exam at a later time. In the interim the applicant can presumably study to prepare for the second exam. If we focus on the applicants who fail the exam the first time and then take it a second time, we would probably expect them to score better on the second exam. One plausible reason is that they are more familiar with the exam the second time. However, we will rule this out by assuming that the two exams are sufficiently different from one another. A second plausible reason is that the applicants have studied between exams, which has a beneficial effect. However, we will argue that even if studying has *no beneficial effect whatsoever,* these applicants will tend to do better the second time around. The reason is regression toward the mean. All of these applicants scored unusually low on the first exam, so they will tend to regress toward the mean on the second exam—that is, they will tend to score higher.

You can employ simulation to demonstrate this phenomenon, using the following model. Assume that the scores of *all* potential applicants are normally distributed with mean μ and standard deviation σ. Since we are assuming that any studying between exams has no beneficial effect, this distribution of scores is the *same* on the second exam as on the first. An applicant fails the first exam if his or her score is below some cutoff value L. Now, we would certainly expect scores on the two exams to be positively correlated, with some correlation ρ. That is, if everyone took both exams, then applicants who scored high on the first would tend to score high on the second, and those who scored low on the first would tend to score low on the second. (This isn't regression to the mean, but simply that some applicants are better than others.)

Given this model, you can proceed by simulating many pairs of scores, one pair for each applicant. The scores for each exam should be normally distributed with parameters μ and σ, but the trick is to make them correlated. You can use StatPro's BINORMAL_ function to do this. (Binormal is short for bivariate normal.) It takes a pair of means (both equal to μ), a pair of standard deviations (both equal to σ), and a correlation ρ as arguments, with the syntax =BINORMAL_(*means,stdevs, correlation*). To enter the formula, highlight two adjacent cells such as C6 and D6, type the formula, and press Ctrl-Shift-Enter. Then copy and paste to generate similar values for other applicants.

Once you have generated pairs of scores for many applicants, you should ignore all pairs except for those where the score on the first exam is less than L. (Sorting is suggested here, but "freeze" the random numbers first.) For these pairs, test whether the mean score on the second exam is *higher* than on the first, using a paired-samples test. If it is, you have demonstrated regression toward the mean. As you'll probably discover, however, the results will depend on the parameters you choose: μ, σ, ρ, and L. We encourage you to experiment with these. Assuming that you are able to demonstrate regression toward the mean, can you explain intuitively why it occurs?

10.2 Baseball Statistics

Baseball has long been the sport of statistics. Probably more statistics—both relevant and completely obscure—are kept on baseball games and players than for any other sport. During the early 1990s, the first author of this book was able to acquire an enormous set of baseball data.[12] It includes data on every at-bat for every player in every game for the 4-year period from 1987 to 1990. The majority of these data are on the CD-ROM that accompanies this book. (The bulk of the data are in eight large files with names such as 89AL.EXE—for the 1989 American League. See the BB_README.TXT file for detailed information about the files.) The files include data for approximately 500 player-years during this period. Each text file contains data for a particular player during an entire year, such as Barry Bonds during 1989, provided that the player had at least 500 at-bats during that year. Each record (row) of such a file lists the information pertaining to a single at-bat, such as whether the player got a hit, how many runners were on base, the earned-run-average (ERA) of the pitcher, and more.

The author analyzed these data to see whether batters tend to hit in "streaks," a phenomenon that has been debated for years among avid fans. [The results of this study are described in Albright (1993).] However, the data set is sufficiently rich to enable testing of any number of hypotheses. We challenge you to develop your own hypotheses and test them with these data.

[12]The data were collected by volunteers of a group called Project Scoresheet. These volunteers attended each game and kept detailed records of virtually everything that occurred. Such detail is certainly not available in newspaper box scores—it is not even available on the Web!

10.3 The Wichita Anti–Drunk Driving Advertising Campaign[13]

Each year drinking and driving behavior are estimated to be responsible for approximately 24,000 traffic fatalities in the United States. Data show that a preponderance of this problem is due to the behavior of young males. Indeed, a disproportionate number of traffic fatalities are young people between 15 and 24 years of age. Market research among young people has suggested that this perverse behavior of driving automobiles while under the influence of alcoholic beverages might be reduced by a mass media communications/advertising program based on an understanding of the "consumer psychology" of young male drinking and driving. There is some precedent for this belief. Reduction in cigarette smoking over the last 25 years is often attributed in part to mass antismoking advertising campaigns. There is also precedent for being less optimistic, because past experimental campaigns against drunk driving have shown little success.

Between March and August of 1986, an anti–drinking and driving advertising campaign was conducted in the city of Wichita, Kansas. In this federally sponsored experiment, several carefully constructed messages were aired on television and radio and also appeared in newspapers and on billboards. Unlike earlier and largely ineffective campaigns that depended on donated talent and media time, this test was sufficiently funded to create impressive anti–drinking and driving messages, and to place them so that the targeted audience would be reached. The messages were pretested before the program and the final version won an OMNI advertising award.

To evaluate the effectiveness of this anti–drinking and driving campaign, researchers collected before and after data (pre-program and post-program) of several types. In addition to data collection in Wichita, they also selected Omaha, Nebraska, as a "control" city. Omaha, another midwestern city on the Great Plains, was arguably similar to Wichita, but was not subjected to such an advertising campaign. The following tables contain some of the data gathered by researchers to evaluate the impact of the program.

Table 10.1 contains background demographics on the test and control cities. Table 10.2 contains data obtained from telephone surveys of 18-to-24-year-old

Table 10.1

Demographics for Wichita and Omaha

	Wichita	Omaha
Total population	411,313	483,053
Percentage 15–24 years	19.2	19.5
Race		
White	85	87
Black	8	9
Hispanic	4	2
Other	3	2
Percent high school graduates		
among those 18 years and older	75.4	79.9
Private car ownership	184,641	198,723

[13]This case was contributed by Peter Kolesar from Columbia University.

Table 10.2 **Telephone Survey of 18-to-24-Year-old Males**	Wichita		Omaha	
	Pre-program	**Post-program**	**Pre-program**	**Post-program**
Respondents	205	221	203	157
Drove after 4 drinks	71	61	77	69
Drove after 6 drinks	42	37	45	38

males in both cities. The surveys were done using a random telephone dialing technique. They had an 88% response rate during the pre-program survey and a 91% response rate during the post-program survey. Respondents were asked whether they had driven under the influence or 4 or more alcoholic drinks, or 6 or more alcoholic drinks, at least once in the previous month. The pre-program data were collected in September 1985, and the post-program data were collected in September 1986.

Table 10.3 contains counts of fatal or incapacitating accidents involving young people gathered from the Kansas and Nebraska Traffic Safety Departments during the spring and summer months of 1985 (pre-program) and 1986 (during the program). The spring and summer months were defined to be the period from March to August. These data were taken by the research team as "indicators" of driving under the influence of alcohol. Researchers at first proposed to also gather data on the blood alcohol content of drivers involved in fatal accidents. However, traffic safety experts pointed out that such data are often inconsistent and incomplete because police at the scene of a fatal accident have more pressing duties to per-

form than to gather such data. On the other hand, it is well established that alcohol is implicated in a major proportion of nighttime traffic fatalities, and for that reason, the data also focus on accidents at night among two classes of young people: the group of accidents involving 18-to-24-year-old males as a driver, and the group of accidents involving 15-to-24-year-old males and/or females as a driver.

The categories of accidents recorded were as follows:

- Total: total count of all fatal and incapacitating accidents in the indicated driver group
- Single vehicle: single vehicle fatal and incapacitating accidents in the indicated driver group
- Nighttime: nighttime (8 P.M. to 8 A.M.) fatal and incapacitating accidents in the indicated driver group

It was estimated that if a similar 6-month advertising campaign were run nationally, it would cost about $25 million. The Commissioner of the U.S. National Highway Safety Commission had funded a substantial part of the study and needed to decide what, if anything, to do next.

Table 10.3 **Average Monthly Number of Fatal and Incapacitating Accidents, March to August**			Wichita		Omaha	
	Driver Group	**Accident Type**	**1985**	**1986**	**1985**	**1986**
	18-to-24-Year-old Males	Total	68	55	41	40
		Single	13	13	13	14
		Night	36	35	25	26
	15-to-24-Year-old Males and Females	Total	117	97	59	57
		Single	22	17	16	20
		Night	56	52	34	38

10.4 Deciding Whether to Switch to a New Toothpaste Dispenser

John Jacobs works for the Fresh Toothpaste Company and has recently been assigned to investigate a new type of toothpaste dispenser. The traditional tube of toothpaste uses a screw-off cap. The new dispenser uses the same type of tube, but there is now a flip-top cap on a hinge. John believes this new cap is easier to use, although it is a bit messier than the screw-off cap—toothpaste tends to accumulate around the new cap. So far, the positive aspects appear to outweigh the negatives. In informal tests, consumers reacted favorably to the new cap. The next step was to introduce the new cap in a regional test market. The company has just conducted this test market for a 6-month period in 85 stores in the Cincinnati region. The results, in units sold per store, appear in Figure 10.40. (See the file TPASTE.XLS.)

John has done his homework on the financial side. Figure 10.41 shows a break-even analysis for the new dispenser relative to the current dispenser. The analysis is over the entire U.S. market, which consists of 9530 stores (of roughly similar size) that stock the product. Based on several assumptions that we'll soon discuss, John figures that to break even with the new

dispenser, the sales volume per store per 6-month period must be 3622 units. The question is whether the test market data support a decision to abandon the current dispenser and market the new dispenser nationally.

We first discuss the break-even analysis in Figure 10.41. The assumptions are listed in rows 4–8 and relevant inputs are listed in rows 11–17. In particular, the new dispenser involves an up-front investment of $1.5 million, and its unit cost is 2 cents higher than the unit cost for the current dispenser. However, the company doesn't plan to raise the selling price. Rows 22–26 calculate the net present value (NPV) for the next 4 years, assuming that the company does not switch to the new dispenser. Starting with *any* first-year sales volume in cell C30, rows 29–35 calculate the NPV for the next 4 years, assuming that the company does switch to the new dispenser. The goal of the break-even analysis is to find a value in cell C30 that makes the two NPVs (in cells B26 and B35) equal.

The trickiest part of the analysis concerns the depreciation calculations for the new dispenser. We find

FIGURE 10.40
Toothpaste Dispenser Data from Cincinnati Region

	A	B	C	D	E	F	G
1	Sales volumes in Cincinnati regional test market for 6 months						
2							
3	Store	Units sold					
4	1	4106					
5	2	2786					
6	3	3858					
7	4	3015					
8	5	3900					
9	6	3572					
10	7	4633					
11	8	4128					
12	9	3044					
13	10	2585					
85	82	1889					
86	83	6436					
87	84	4179					
88	85	3539					

FIGURE 10.41

Break-even Analysis
for Toothpaste
Example

	A	B	C	D	E	F
1	**Breakeven analysis for Stripe Toothpaste**					
2						
3	**Assumptions:**					
4	The planning horizon is 4 years					
5	Sales volume is expected to remain constant over the 4 years					
6	Unit selling prices and unit costs will remain constant over the 4 years					
7	Straight-line depreciation is used to depreciate the initial investment for the new dispenser					
8	Breakeven analysis is based on NPV for the four-year period					
9				Range names		
10	**Given data**			AfterTaxProfit: C33:F33		
11	Current volume (millions of units) using current dispenser		65.317	BeforeTaxContrib: C31:F31		
12	Initial investment ($ millions) for new dispenser		1.5	CashFlow: C34:F34		
13	Unit selling price (either dispenser)		$1.79	Deprec: C32:F32		
14	Unit cost (current dispenser)		$1.25	DiscRate: C17		
15	Unit cost (new dispenser)		$1.27	Invest: B34		
16	Tax rate		35%	TaxRate: C16		
17	Discount rate		16%			
18						
19	**Note:** From here on, all sales volumes are in millions of units, monetary values are in $ millions.					
20						
21	**Analysis of current dispenser**		Year 1	Year 2	Year 3	Year 4
22	Sales volume		65.317	65.317	65.317	65.317
23	Before-tax contribution		35.27	35.27	35.27	35.27
24	After-tax profit		22.926	22.926	22.926	22.926
25	Cash flow		22.926	22.926	22.926	22.926
26	NPV	$64.152				
27						
28	**Analysis of new dispenser**		Year 1	Year 2	Year 3	Year 4
29	Initial investment	$1.5				
30	Sales volume		69.027	69.027	69.027	69.027
31	Before-tax contribution		35.89	35.89	35.89	35.89
32	Depreciation		0.38	0.38	0.38	0.38
33	After-tax profit		23.087	23.087	23.087	23.087
34	Cash flow	($1.50)	23.462	23.462	23.462	23.462
35	NPV	$64.152				
36						
37	Number of stores nationally		9530			
38	Breakeven sales volume per store per 6 months		3622			

the before-tax contribution from sales in row 31 and subtract the depreciation each year (one-quarter of the investment) to figure the after-tax profit. For example, the formula in cell C33 is

$$=(BeforeTaxContrib-Deprec)*(1-TaxRate)$$

Then the depreciation is added back to obtain the cash flow, so that the formula in cell C34 is

$$=AfterTaxProfit+Deprec$$

Finally, we calculate the NPV for the new dispenser in cell B35 with the formula

$$=Invest+NPV(DiscRate,CashFlow)$$

Note that the initial investment, which is assumed to occur at the *beginning* of year 1, is not part of the NPV function, which includes only *end-of-year* cash flows.

We then use Excel's Goal Seek tool to force the NPVs in cells B26 and B35 to be equal. Again, we begin by entering *any* value for first-year sales volume with the new dispenser in cell C30. Then we use the Tools/Goal Seek menu item and fill out the dialog box as shown in Figure 10.42.

The file TPASTE.XLS does not yet contain the break-even calculations shown in Figure 10.41. Your first job is to enter the appropriate formulas, using any

FIGURE 10.42
Goal Seek Dialog Box

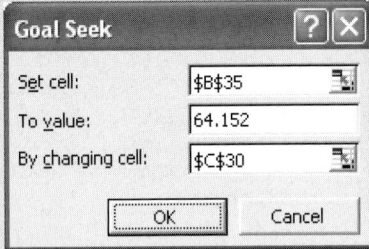

year 1 sales volume figure in cell C30. Next, you should use Excel's Goal Seek tool to find the break-even point. Finally, you should test the alternative hypothesis that the mean sales volume over all stores (for a 6-month period) will be large enough to warrant switching to the new dispenser. This hypothesis test should be based on the test market data from Cincinnati. Do you recommend that the company should switch to the new dispenser? Discuss whether this decision should be based on the results of a hypothesis test.

Regression Analysis: Estimating Relationships

© 2002 Harris Welles

Site Location of La Quinta Motor Inns

Regression analysis is an extremely flexible tool that can aid decision making in many areas. Kimes and Fitzsimmons (1990) describe how it has been used by La Quinta Motor Inns, a moderately priced hotel chain oriented toward serving the business traveler, to help make site location decisions. Location is one of the most important decisions for a lodging firm. All hotel chains search for ideal locations and often compete against each other for the same sites. A hotel chain that can select good sites more accurately and quickly than its competition has a distinct competitive advantage.

Kimes and Fitzsimmons, academics hired by La Quinta to model their site location decision process, used regression analysis. They collected data on 57 mature inns belonging to La Quinta during a 3-year business cycle. The data included profitability for each inn (defined as operating margin percentage—profit plus depreciation and interest expenses, divided by the total revenue), as well as a number of potential explanatory variables that could be used to predict profitability. These explanatory variables fell into five categories: competitive characteristics (such as number of hotel rooms in the vicinity and average room rates); demand generators (such as hospitals and office buildings within a 4-mile radius that might attract customers to the area); demographic characteristics (such as local population, unemployment rate, and median family income); market awareness (such as years inn has been open and state population per inn); and physical considerations (such as accessibility, distance to downtown, and sign visibility).

The analysts then determined which of these potential explanatory variables were most highly correlated (positively or negatively) with profitability and entered these variables into a regression equation for profitability. The estimated regression equation was

$$\text{Predicted Profitability} = 39.05 - 5.41\text{StatePop} + 5.81\text{Price}$$
$$-3.09\sqrt{\text{MedIncome}} + 1.75\text{ColStudents}$$

where StatePop is the state population (1000s) per inn, Price is the room rate for the inn, MedIncome is the median income ($1000s) of the area, ColStudents is the number of college students (1000s) within 4 miles, and all variables in this equation are standardized to have mean 0 and standard deviation 1. This equation predicts that profitability will increase when room rate and the number of college students *increase* and when state population and median income *decrease*. The R^2 value (to be discussed in this chapter) was a respectable 0.51, indicating a reasonable predictive ability. Using good statistical practice, the analysts validated this equation by feeding it explanatory variable data on a set of *different* inns, attempting to predict profitability for these new inns. The validation was a success—the regression equation predicted profitability fairly accurately for this new set of inns.

La Quinta management, however, was not as interested in predicting the exact profitability of inns as in predicting which would be profitable and which would be unprofitable. A cutoff value of 35% for operating margin was used to divide the profitable inns from the unprofitable inns. (Approximately 60% of the inns in the original sample were profitable by this definition.) The analysts were still able to use the regression equation they had developed. For any prospective site, they used the regression equation to predict profitability, and if the predicted value was sufficiently high, they predicted that this site would be profitable. They selected a decision rule—that is, how high was "sufficiently high"—from considerations of the two potential types of errors. One type of error, a false positive, was predicting that a site would be profitable when in fact it was headed for unprofitability. The opposite type of error, a false negative, was predicting that a site would be unprofitable (and rejecting the site) when in fact it would have been profitable. La Quinta management was more concerned about false positives, so it was willing to be conservative in its decision rule and miss a few potential opportunities for profitable sites.

Since the time of the study, La Quinta has implemented the regression model in spreadsheet form. For each potential site, it collects data on the relevant explanatory variables, uses the regression equation to predict the site's profitability, and applies the decision rule on whether to build. Of course, the model's recommendation is only that—a recommendation. Top management has the ultimate say on whether any site is used. As Sam Barshop, then chairman of the board and president of La Quinta Motor Inns stated, "We currently use the model to help us in our site-screening process and have found that it has raised the 'red flag' on several sites we had under consideration. We plan to continue using and updating the model in the future in our attempt to make La Quinta a leader in the business hotel market." ●

Introduction

Regression analysis is the study of relationships between variables. It is one of the most useful tools for a business analyst because it applies to so many situations. Some potential uses of regression analysis in business include the following:

- How do wages of employees depend on years of experience, years of education, and gender?

- How does the current price of a stock depend on its own past values, as well as the current and past values of a market index?

- How does a company's current sales level depend on its current and past advertising levels, the advertising levels of its competitors, the company's own past sales levels, and the general level of the market?

- How does the unit cost of producing an item depend on the total quantity of items that have been produced?

- How does the selling price of a house depend on such factors as the appraised value of the house, the square footage of the house, the number of bedrooms in the house, and perhaps others?

Each of these questions asks how a single variable, such as selling price or employee wages, depends on other relevant variables. If we can estimate this relationship, then we can not only better understand how the world operates, but we can also do a better job of predicting the variable in question. For example, we can not only understand how a company's sales are affected by its advertising, but we can also use the company's records of current and past advertising levels to predict future sales.

The branch of statistics that studies such relationships is called **regression analysis,** and is the subject of this chapter and the next. Regression analysis is one of the most pervasive of all statistical methods in the business world. This is because of its generality and applicability. Even when we restrict the analysis to a special case of regression analysis called **linear regression,** as we will do here, the number of potential applications is virtually unlimited.

There are several ways to categorize regression analysis. One categorization is based on the overall purpose of the analysis. As suggested previously, there are two potential objectives of regression analysis: to understand how the world operates and to make predictions. Either of these objectives might be paramount in any particular application. If the variable in question is employee wages and we are using variables such as years of experience, years of education, and gender to explain wage levels, then the purpose of the analysis is probably to understand how the world operates—that is, to explain how the variables combine in any given company to determine wages. More specifically, the purpose of the analysis might be to discover whether there is any gender discrimination in wages, after allowing for differences in work experience and education level.

On the other hand, the primary objective of the analysis might be prediction. A good example of this is when the variable in question is company sales, and variables such as advertising and past sales levels are used as explanatory variables. In this case it is certainly important for the company to know how the relevant variables impact its sales. But the company's primary objective is probably to predict *future* sales levels, given current and past values of the explanatory variables. The company might also use a regression model for a what-if analysis, where it predicts future sales for many conceivable patterns of advertising and then selects its advertising level on the basis of these predictions.

Fortunately, the same regression analysis enables us to solve both problems simultaneously. That is, it indicates how the world operates and it enables us to make predictions. So although the objectives of regression studies might differ, the same basic analysis always applies.

A second categorization of regression analysis is based on the type of data being analyzed. There are two basic types: cross-sectional data and time series data. **Cross-sectional data** are usually data gathered from approximately the same period of time from a cross section of a population. The housing and wage examples mentioned previously are typical cross-sectional studies. The first concerns a sample of houses, presumably sold during a short period of time, such as houses sold in Florida during the first couple of months of 1998. The second concerns a sample of employees observed at a particular point in time, such as a sample of automobile workers observed at the beginning of 1997.

Regression can be used to understand the way the world operates, or it can be used for prediction.

Regression can be used to analyze cross-sectional data or time series data.

In contrast, **time series data** involve one or more variables that are observed at several, usually equally spaced, points in time. The stock price example mentioned previously fits this description. We observe the price of a particular stock and possibly the price of a market index at the beginning of every week, say, and then try to explain the movement of the stock's price through time.

Regression analysis can be applied equally well to cross-sectional and time series data. In either case we might be attempting to understand how the world operates or make predictions. However, there are technical reasons for treating time series analysis somewhat differently. The primary reason is that time series variables are usually related to their own past values. This property of many time series variables is called **autocorrelation,** and it adds complications to the analysis that we will discuss briefly.

A third categorization of regression analysis involves the number of explanatory variables in the analysis. First, we need to introduce some terms. In every regression study there is a single variable that we are trying to explain or predict, called the **response** variable or the **dependent** variable. To help explain or predict the response variable, we use one or more **explanatory** variables. These explanatory variables are also called **predictor** variables or **independent** variables. If there is a single explanatory variable, the analysis is called **simple regression.** If there are several explanatory variables, it is called **multiple regression.**[1]

Response and Explanatory Variables

The **response** (or **dependent**) variable is the single variable being explained by the regression. The **explanatory** (or **independent**) variables are used to explain the response variable.

There are important differences between simple and multiple regression. The primary difference, as the name implies, is that simple regression is simpler. The calculations are simpler, the interpretation of output is somewhat simpler, and fewer complications can occur. We will begin with simple regression examples to introduce the ideas of regression. But we will soon see that simple regression is no more than a special case of multiple regression, and there is little need to single it out for separate discussion—especially when computer software is available to perform the calculations in either case.

Simple versus Multiple Regression

A **simple** regression includes a single explanatory variable, whereas a **multiple** regression can include any number of explanatory variables.

"Linear" regression allows us to estimate linear relationships as well as some nonlinear relationships. "Nonlinear" regression is a mathematical method we will not cover in this book.

A fourth and final categorization of regression analysis concerns linear versus nonlinear models. As mentioned previously, the only type of regression analysis we will study is *linear* regression. Generally, this means that the relationships between variables are *straight-line* relationships, whereas the term *nonlinear* implies curved relationships. By focusing on linear regression, it might appear that we are ignoring the many real-world relationships that are clearly nonlinear. Fortunately, linear regression can often be used to estimate nonlinear relationships. As we will see, the term *linear regression* is more general than it appears. Admittedly, many of the relationships we will study can be explained adequately by straight lines. But it is also true that many nonlinear relationships can be

---

[1]The traditional terms used in regression are *dependent* and *independent* variables. However, because these terms can cause confusion with probabilistic independence, a totally different concept, there has been an increasing use of the terms *response* and *explanatory* variables. We will use the latter terms in this book.

"linearized" by suitable mathematical transformations. Therefore, the only relationships we are ignoring in this book are those—and there are some—that cannot be transformed to linear. Such relationships can be studied, but only by advanced methods beyond the level of this book.

In this chapter we will focus on line-fitting and curve-fitting, that is, on estimating equations that describe relationships between variables. We will also discuss the interpretation of these equations, and we will provide a couple of numerical measures that indicate the goodness of fit of the equations we estimate. In the next chapter we will extend the analysis to statistical inference of regression output. As we will see, this chapter focuses only on a small part of the available regression output, but it is an extremely important part and one we should thoroughly understand before proceeding to the more complex issues in the next chapter.

11.2 Scatterplots: Graphing Relationships

A good way to begin any regression analysis is to draw one or more scatterplots. As we learned in Chapter 2, a scatterplot is a graphical plot of two variables, an X and a Y. Consider a data set with n observations, where for each observation there are at least two variables. We choose two of these variables and label them X and Y. Then for each of the observations, we plot the X and Y values as a point on a two-dimensional graph. The scatterplot is the resulting scatter of points. If there is any relationship between the two variables, it is usually apparent from the scatterplot.

The following example, which we will carry throughout this chapter, illustrates the usefulness of scatterplots. It is a typical example of cross-sectional data.

Example 11.1 Sales versus Promotions at Pharmex

Pharmex is a chain of drugstores that operate around the country. To see how effective its advertising and other promotional activities are, the company has collected data from 50 randomly selected metropolitan regions. In each region it has compared its own promotional expenditures and sales to those of the leading competitor in the region over the past year. There are two variables:

- Promote: Pharmex's promotional expenditures as a percentage of those of the leading competitor
- Sales: Pharmex's sales as a percentage of those of the leading competitor

Note that each of these variables is an "index," not a dollar amount. For example, if Promote equals 95 for some region, this tells us only that Pharmex's promotional expenditures in that region are 95% as large as those for the leading competitor in that region. The company expects that there is a positive relationship between these two variables, so that regions with relatively more expenditures have relatively more sales. However, it is not clear what the nature of this relationship is. The data are listed in the file PHARMEX.XLS. (See Figure 11.1 on page 552 for a partial listing of the data.) What type of relationship, if any, is apparent in a scatterplot?

Objective To use a scatterplot to examine the relationship between promotional expenses and sales at Pharmex.

FIGURE 11.1
Data for Drugstore Example

	A	B	C	D	E	F
1	**Data on drugstore promotional expenditures and sales**					
2						
3	Note: each value is a percentage of what the leading competitor did					
4						
5	Region	Promote	Sales			
6	1	77	85			
7	2	110	103			
8	3	110	102			
9	4	93	109			
10	5	90	85			
11	6	95	103			
12	7	100	110			
13	8	85	86			
14	9	96	92			
15	10	83	87			
53	48	100	98			
54	49	95	108			
55	50	96	87			

Solution

First, recall from Chapter 2 that there are two ways to create a scatterplot in Excel. We can use Excel's Chart Wizard to create an X–Y chart, or we can use StatPro's Scatterplot procedure. The advantage of the latter is that the X variable (the one on the horizontal axis) doesn't need to be to the left of the Y variable in the data set, so we generally favor its use in regression applications.

Which variable should be on the horizontal axis? In regression we always put the explanatory variable on the horizontal axis and the response variable on the vertical axis. In this example the store believes large promotional expenditures tend to "cause" larger values of sales, so we put Sales on the vertical axis and Promote on the horizontal axis. The resulting scatterplot appears in Figure 11.2.

This scatterplot indicates that there is indeed a positive relationship between Promote and Sales—the points tend to rise from bottom left to top right—but the relationship is

FIGURE 11.2
Scatterplot of Sales versus Promote

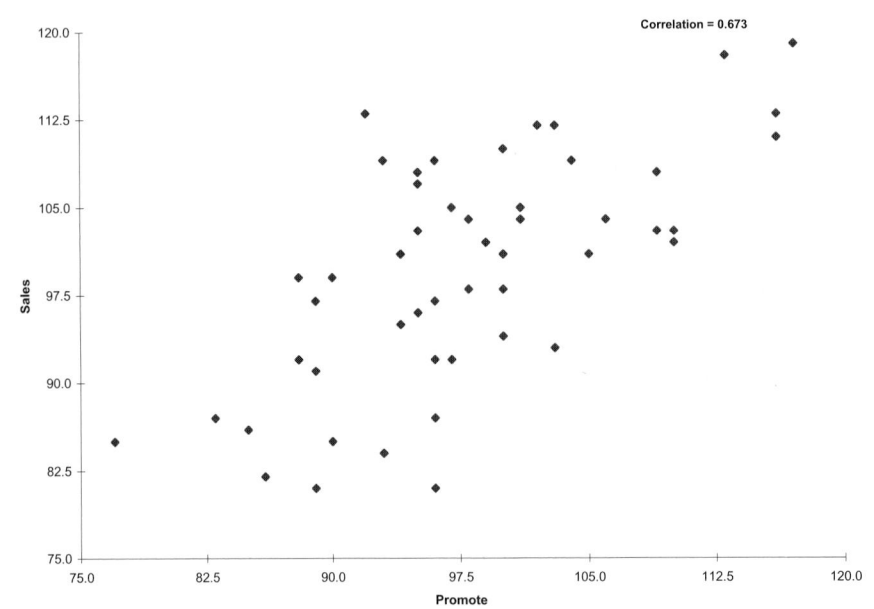

not perfect. If it were perfect, a given value of Promote would prescribe the value of Sales exactly. Clearly, this is not the case. For example, there are five regions with promotional values of 96 but different sales values. So the scatterplot indicates that while the variable Promote might be helpful for predicting the Sales value, it won't yield perfect predictions.

Note the correlation of 0.673 shown at the top of Figure 11.2. The StatPro add-in inserts this value automatically to indicate the strength of the linear relationship between the two variables. For now, just note that it is positive and its magnitude is moderately large. We will say more about correlations in the next section.

Finally, we say something about causation. There is a tendency for an analyst (such as a drugstore manager) to say that larger promotional expenses *cause* larger sales values. However, unless the data are obtained in a carefully controlled experiment—which is certainly not the case here—we can never make definitive statements about causation in regression analysis. The reason is that we can almost never rule out the possibility that some other variable is causing the variation in *both* of the observed variables. While this might be unlikely in this drugstore example, it is still a possibility. ●

The following example uses time series data to illustrate several other features of scatterplots. We will also follow this example throughout the chapter.

Example 11.2 Explaining Overhead Costs at Bendrix

The Bendrix Company manufactures various types of parts for automobiles. The manager of the factory wants to get a better understanding of overhead costs. These overhead costs include supervision, indirect labor, supplies, payroll taxes, overtime premiums, depreciation, and a number of miscellaneous items such as insurance, utilities, and janitorial and maintenance expenses. Some of these overhead costs are "fixed" in the sense that they do not vary appreciably with the volume of work being done, whereas others are "variable" and do vary directly with the volume of work. The fixed overhead costs tend to come from the supervision, depreciation, and miscellaneous categories, whereas the variable overhead costs tend to come from the indirect labor, supplies, payroll taxes, and overtime premiums categories. However, it is not easy to draw a clear line between the fixed and variable overhead components.

The Bendrix manager has tracked total overhead costs over the past 36 months. To help "explain" these, he has also collected data on two variables that are related to the amount of work done at the factory. These variables are:

- MachHrs: number of machine hours used during the month
- ProdRuns: the number of separate production runs during the month

The first of these is a direct measure of the amount of work being done. To understand the second, we note that Bendrix manufactures parts in fairly large batches. Each batch corresponds to a production run. Once a production run is completed, the factory must "set up" for the next production run. During this setup there is typically some downtime while the machinery is reconfigured for the part type scheduled for production in the next batch. Therefore, the manager believes both of these variables might be responsible (in different ways) for variations in overhead costs. Do scatterplots support this belief?

Objective To use scatterplots to examine the relationships among overhead, machine hours, and production runs at Bendrix.

FIGURE 11.3
Data for Bendrix
Overhead
Example

	A	B	C	D
1	Monthly data on manufacturing overhead costs			
2				
3	Month	MachHrs	ProdRuns	Overhead
4	1	1539	31	99798
5	2	1284	29	87804
6	3	1490	27	93681
7	4	1355	22	82262
8	5	1500	35	106968
9	6	1777	30	107925
10	7	1716	41	117287
11	8	1045	29	76868
12	9	1364	47	106001
13	10	1516	21	88738
37	34	1723	35	107828
38	35	1413	30	88032
39	36	1390	54	117943

Solution

The data appear in Figure 11.3. (See the BENDRIX1.XLS file.) Each observation (row) corresponds to a single month. We want to investigate any possible relationship between the Overhead variable and the MachHrs and ProdRuns variables, but because these are time series variables, we should also be on the lookout for any relationships between these variables and the Month variable. That is, we should investigate any time series behavior in these variables.

This data set illustrates, even with a modest number of variables, how the number of potentially useful scatterplots can grow quickly. (Remember that StatPro allows you to obtain many scatterplots in one pass.) At the very least, we should examine the scatterplot between each potential explanatory variable (MachHrs and ProdRuns) and the response variable (Overhead). These appear in Figures 11.4 and 11.5. We see that Overhead tends to increase as either MachHrs increases or ProdRuns increases. However, both relationships are far from perfect.

FIGURE 11.4
Scatterplot of
Overhead versus
Machine Hours

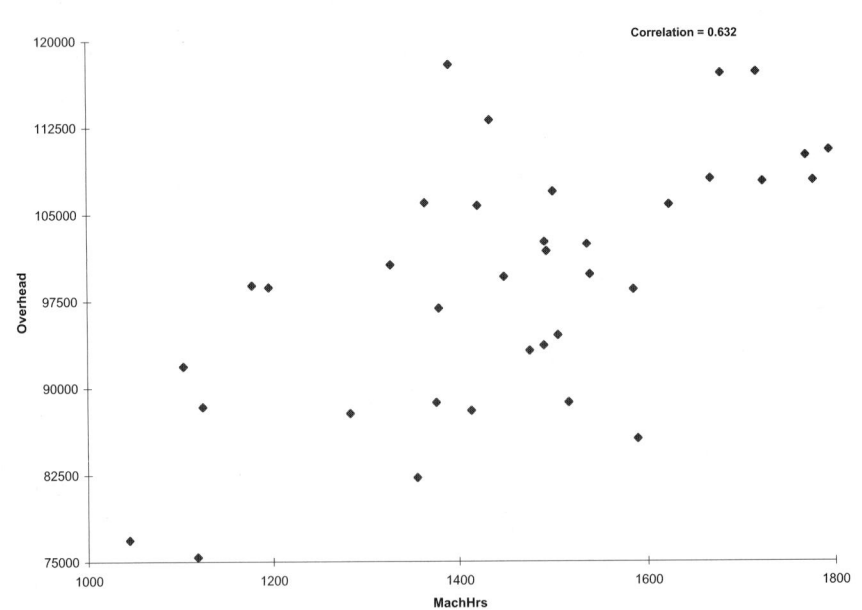

FIGURE 11.5

Scatterplot of
Overhead versus
Production Runs

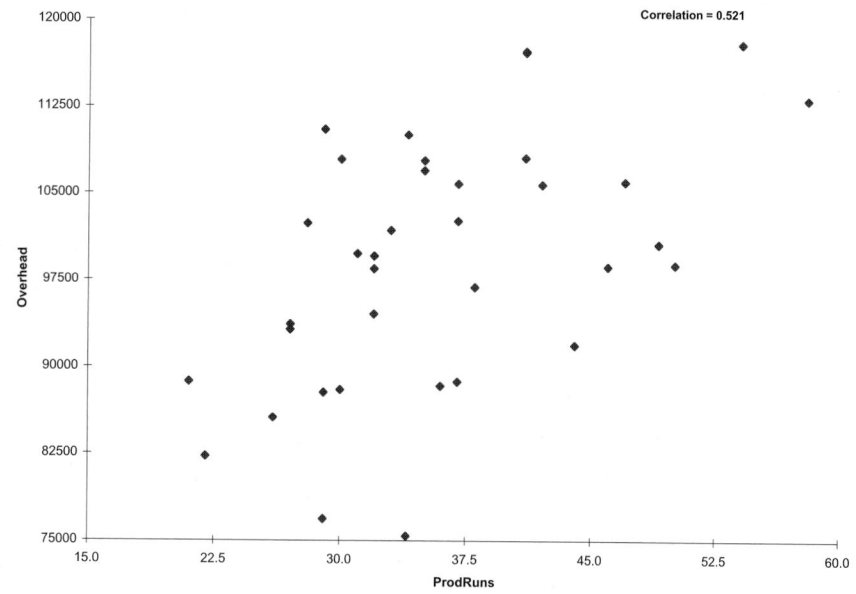

To check for possible time series patterns, we can also create a time series plot for
any of the variables. (Actually, this is equivalent to a scatterplot of the variable versus
Month, with the points joined by lines.) One of these, the time series plot for Overhead,
is in Figure 11.6. It shows a fairly random pattern through time, with no apparent upward
trend or other obvious time series pattern. You can check that time series plots of the
MachHrs and ProdRuns variables also indicate no obvious time series patterns.

Finally, when there are multiple explanatory variables, we can check for relationships
among them. The scatterplot of MachHrs versus ProdRuns appears in Figure 11.7. (Either variable could be chosen for the vertical axis.) This "cloud" of points indicates no
relationship worth pursuing.

FIGURE 11.6

Time Series Plot
of Overhead ver-
sus Month

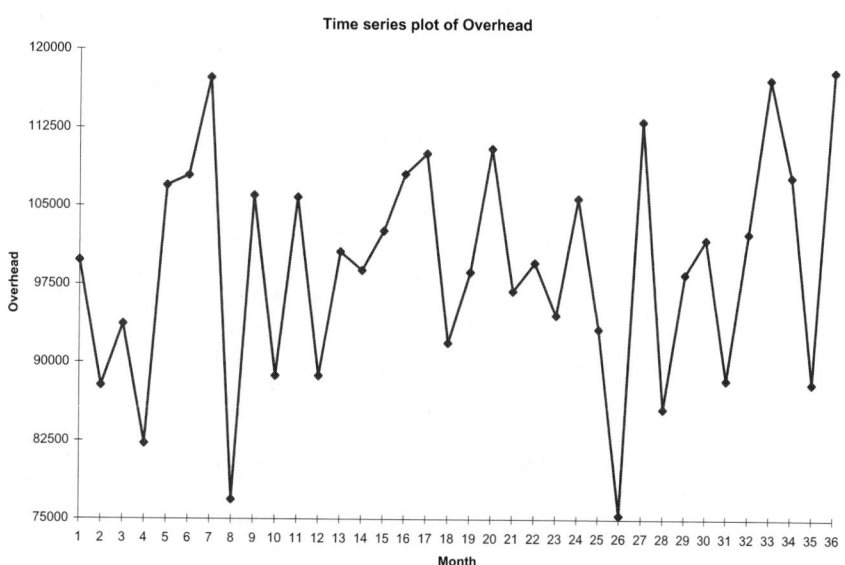

FIGURE 11.7
Scatterplot of
Machine Hours
versus Production
Runs

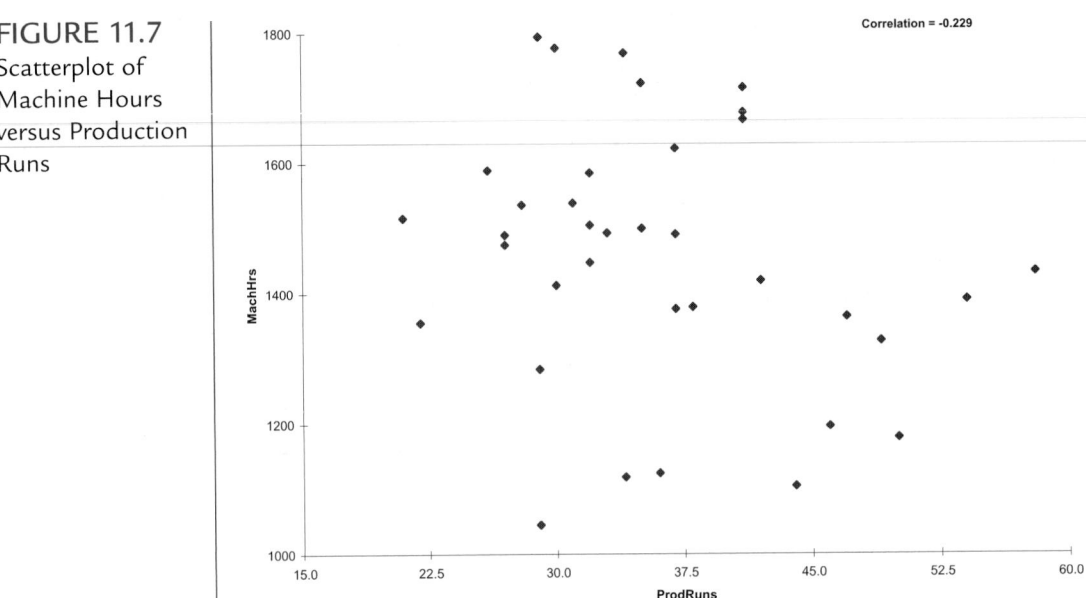

This is precisely the role of scatterplots: to give us a visual representation of relationships or the lack of relationships between variables.

In summary, the Bendrix manager should continue to explore the positive relationship between Overhead and each of the MachHrs and ProdRuns variables. However, none of the variables appears to have any time series behavior, and the two potential explanatory variables do not appear to be related to each other. ●

11.2.1 Linear Versus Nonlinear Relationships

Scatterplots are extremely useful for detecting behavior that might not be obvious otherwise. We illustrate some of these in the next few subsections. First, the typical relationship we hope to see is a straight-line, or *linear,* relationship. This doesn't mean that all points lie on a straight line—this is too much to expect in business data—but that the points tend to cluster around a straight line. The scatterplots in Figures 11.2, 11.4, and 11.5 all exhibit linear relationships, at least in the sense that no curvature is obvious.

The scatterplot in Figure 11.8, on the other hand, illustrates a relationship that is clearly nonlinear. The data in this scatterplot are 1990 data on over 100 countries. The variables listed are life expectancy (of newborns, based on current mortality conditions) and GNP per capita. The obvious curvature in the scatterplot can be explained as follows. For poor countries, a slight increase in GNP per capita has a large effect on life expectancy. However, this effect decreases for wealthier countries. A straight-line relationship is definitely not appropriate for these data. However, as we discussed previously *linear* regression—after an appropriate transformation of the data—still might be applicable.

11.2.2 Outliers

Scatterplots are especially useful for identifying **outliers,** observations that lie outside the typical pattern of points. The scatterplot in Figure 11.9 shows annual salaries versus years of experience for a sample of employees at a particular company. There is a clear linear relationship between these two variables—for all employees except one. Closer scrutiny of the data reveals that this one employee is the company president, whose salary is well above that of all the other employees!

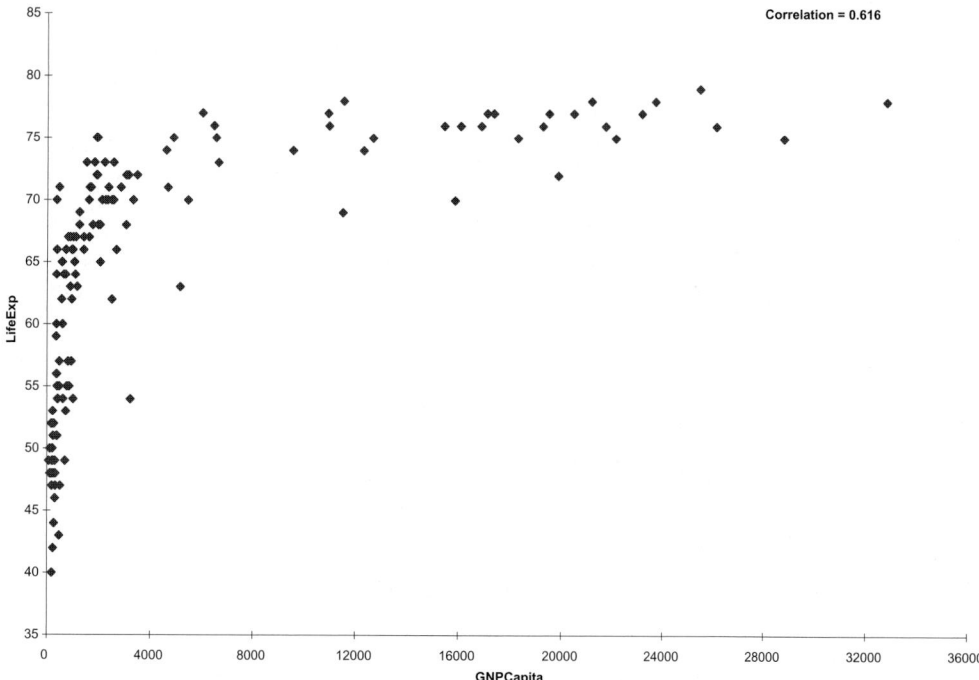

FIGURE 11.8 Scatterplot of Life Expectancy versus GNP per Capita

FIGURE 11.9
Scatterplot of
Salary versus Years
of Experience

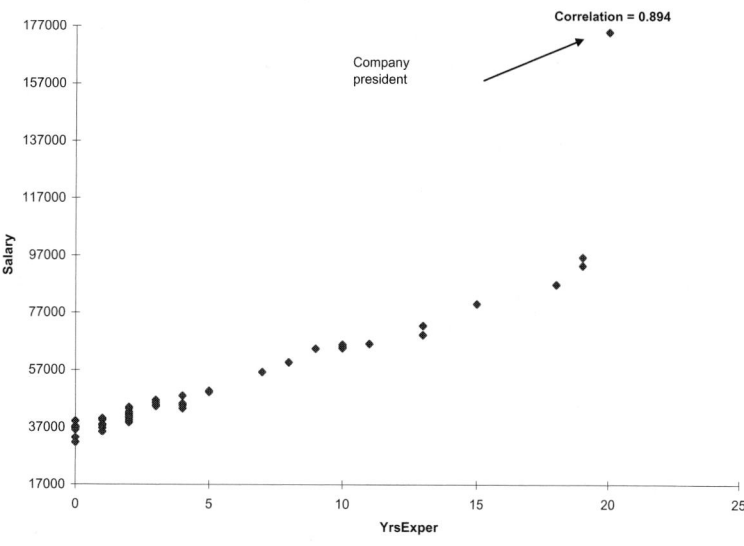

Outlier	An **outlier** is an observation that falls outside of the general pattern of the rest of the observations.

Although scatterplots are good for detecting outliers, they do not necessarily indicate what we ought to do about any outliers we find. This depends entirely on the particular situation. If we are attempting to investigate the salary structure for "typical" employees at a company, then we probably should not include the company president. First, the president's salary is not determined in the same way as the salaries for typical employees. Second, if we do include the president in the analysis, it can greatly distort the results for the mass of typical employees. In other situations, however, it might *not* be appropriate to eliminate outliers just to make the analysis come out more nicely.

It is difficult to generalize about the treatment of outliers, but the following points are worth noting.

- If an outlier is clearly not a member of the population of interest, then it is probably best to delete it from the analysis. This is the case for the company president in Figure 11.9.

- If it isn't clear whether outliers are members of the relevant population, we should run the regression analysis with them and without them. If the results are practically the same in both cases, then it is probably best to report the results with the outliers included. Otherwise, we should report both sets of results with a verbal explanation of the outliers.

11.2.3 Unequal Variance

Occasionally, there is a clear relationship between two variables, but the variance of the response variable depends on the value of the explanatory variable. We saw a good example of this in the catalog data in Example 3.11 of Chapter 3. Figure 11.10 reproduces one of the scatterplots from the data in that example. It shows AmountSpent versus Salary for the customers in the data set. There is a clear linear relationship, but the variability of AmountSpent increases as Salary increases. This is evident from the "fan" shape. As we will see in the next chapter, this unequal variance violates one of the assumptions of linear regression analysis, and there are special techniques to deal with it.

FIGURE 11.10

Unequal Variance of Response Variable in a Scatterplot

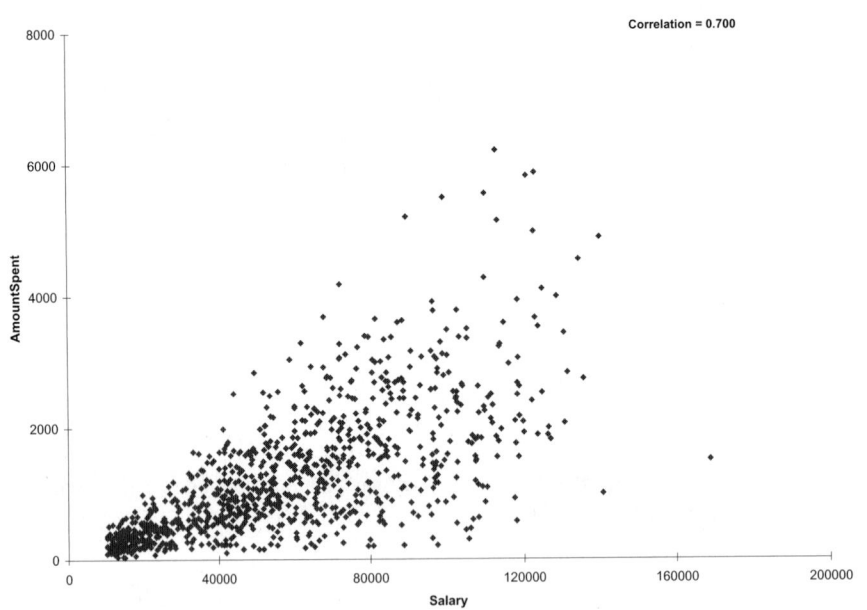

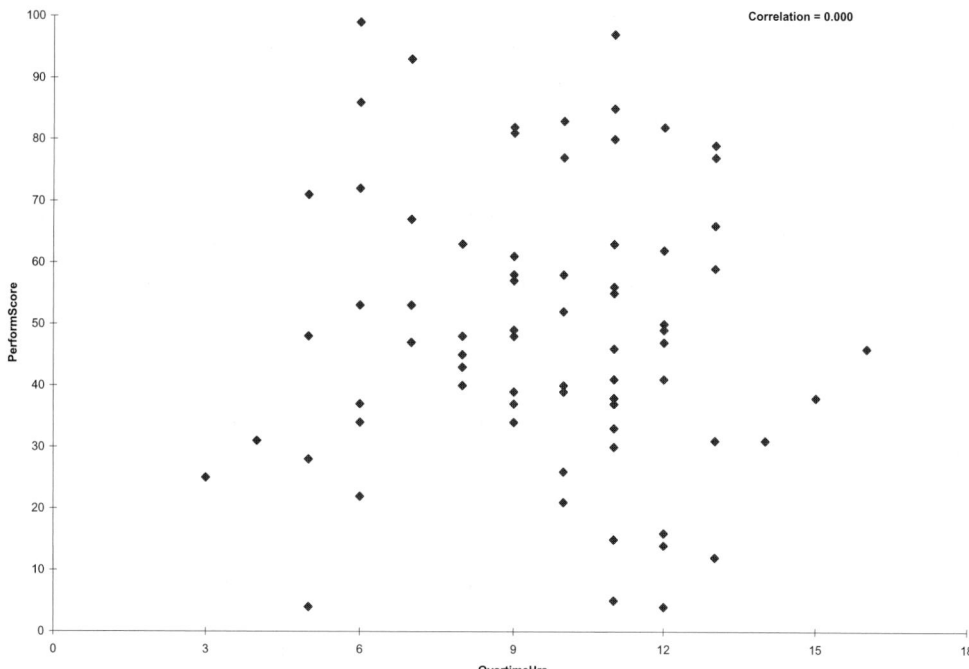

FIGURE 11.11 An Example of No Relationship

11.2.4 No Relationship

A scatterplot can provide one other useful piece of information—it can indicate that there is *no* relationship between a pair of variables, at least none worth pursuing. This is usually the case when the scatterplot appears as a shapeless swarm of points, as illustrated in Figure 11.11. Here the variables are an employee performance score and the number of overtime hours worked in the previous month for a sample of employees. There is virtually no hint of a relationship between these two variables in this plot, and, at least if these are the only two variables in the data set, the analysis could stop right here. Many people who use statistics evidently believe that a computer can perform magic on a set of numbers and find relationships that were completely hidden. Occasionally this is true, but when a scatterplot appears as in Figure 11.11, the variables are not related in any useful way, and that's all there is to it.

11.2.5 Other Scatterplot Features

Scatterplots are a useful way to begin nearly any regression analysis. In Excel they are fairly easy to construct, especially with an add-in such as StatPro. However, spreadsheet packages do have their limitations relative to special-purpose statistical software packages such as SPSS, Minitab, SAS, StatGraphics, and many others. The options for exploring data with scatterplots in some of these packages are very powerful and quite easy to use. We will briefly describe some of the options available. You can explore these depending on the software available to you.

- **Matrix of scatterplots.** We saw in Example 11.2 how the number of potentially relevant scatterplots can increase quickly as the number of variables increases. Rather than creating scatterplots one at a time, many software packages allow us to create a "matrix" of scatterplots all at once. For example, if there are five variables, there are ten possible *pairs* of variables. The matrix feature would create all ten of these with a single command, and we could see them all at once on the screen.

- **Coloring points based on a third variable.** Consider salary data such as those shown in Figure 11.9. Suppose we want to see whether the relationship between salary and years of experience is any different for men than for women. Some statistical software allows us to create a scatterplot of salary versus years of experience for all employees, but with the points for men colored (or shaped) differently than the points for women.

- **Brushing points based on a third variable.** The previous option is based on coloring points according to a variable with only two categories (for example, men and women). A similar idea is to color points based on a *continuous* third variable. For example, suppose we want to see how an employee's current performance score (given by a supervisor) affects the relationship between salary and years of experience. Some statistical software allows us to color the points in the scatterplot differently depending on a cutoff value for the performance score. For example, we might color all points blue for employees with performance scores less than 50 and color all other points red. In addition, it is usually easy to change the cutoff value and have the software recolor the points automatically.

- **Three-dimensional plots.** A final possibility is to show a "3-D" scatterplot of three variables simultaneously. Although these 3-D plots are often difficult to visualize on a two-dimensional computer screen, many software packages allow us to "rotate" the plot through various angles. A relationship that is not at all obvious from one angle might be apparent from another angle.

We are not claiming that a clever user could not accomplish these features with Excel—either with Excel's built-in capabilities or with its Visual Basic for Applications programming language. This is quite possible. The point, however, is that many special-purpose statistical software packages have this functionality built into them and are quite easy to use.

11.3 Correlations: Indicators of Linear Relationships

Scatterplots provide graphical indications of relationships, whether they be linear, nonlinear, or essentially nonexistent. **Correlations** are numerical summary measures that indicate the strength of relationships between pairs of variables.[2] A correlation between a pair of variables is a single number that summarizes the information in a scatterplot. A correlation can be very useful, but it has an important limitation: It can only measure the strength of a *linear* relationship. If there is a nonlinear relationship, as suggested by a scatterplot, the correlation can be completely misleading. With this important limitation in mind, let's look a bit more closely at correlations.

The usual notation for a correlation between two variables X and Y is r_{XY}. The subscripts can be omitted if the variables are clear from the context of the problem. The formula for r_{XY} is given by equation (11.1). Note that it is a sum of products in the numer-

---

[2]This section includes some material from Section 3.7, which is repeated here for convenience.

ator, divided by the product $s_X s_Y$ of the sample standard deviations of X and Y. This requires a considerable amount of computation, so that correlations are almost always computed by software packages.

Formula for Correlation	$$r_{XY} = \dfrac{\Sigma(X_i - \overline{X})(Y_i - \overline{Y})/(n - 1)}{s_X s_Y}$$	**(11.1)**

The numerator of equation (11.1) is also a measure of association between two variables X and Y, called the **covariance** between X and Y. Like a correlation, a covariance is a single number that measures the strength of the linear relationship between two variables. By looking at the sign of the covariance or correlation—plus or minus—we can tell whether the two variables are positively or negatively related. The drawback to a covariance, however, is that its magnitude depends on the units in which the variables are measured.

To illustrate, the covariance between Overhead and MachHrs in the Bendrix manufacturing data set is 1,333,138. (It can be found with Excel's COVAR function or with StatPro.) However, if we divide each overhead value by 1000, so that overhead costs are expressed in thousands of dollars, and we divide each value of MachHrs by 100, so that machine hours are expressed in hundreds of hours, the covariance decreases by a factor of 100,000 to 13.33138. This is in spite of the fact that the basic relationship between these variables has not changed and the revised scatterplot has exactly the same shape. For this reason it is often difficult to interpret the magnitude of a covariance, and we concentrate instead on correlations.

Unlike covariances, correlations have the attractive property that they are completely unaffected by the units in which the variables are measured. The rescaling described in the previous paragraph has absolutely no effect on the correlation between Overhead and MachHrs. In either case the correlation is 0.632. Moreover, all correlations are between -1 and $+1$, inclusive. The sign of a correlation, plus or minus, determines whether the linear relationship between two variables is positive or negative. In this respect, a correlation is just like a covariance. However, the strength of the linear relationship between the variables is measured by the absolute value, or magnitude, of the correlation. The closer this magnitude is to 1, the stronger the linear relationship is.

A correlation close to -1 or $+1$ indicates a strong linear relationship. A correlation close to 0 indicates virtually no linear relationship.

A correlation equal to zero or near zero indicates practically no linear relationship. A correlation with magnitude close to 1, on the other hand, indicates a strong linear relationship. At the extreme, a correlation equal to -1 or $+1$ occurs only when the linear relationship is perfect—that is, when all points in the scatterplot lie on a single straight line. Although such extremes practically never occur in business applications, "large" correlations, say, greater than 0.9 in magnitude, are not at all uncommon.

Looking back at the scatterplots for the Pharmex drugstore data in Figure 11.2, we see that the correlation between Sales and Promote is positive—as we would guess from the upward-sloping scatter of points—and that it is equal to 0.673. This is a moderately large correlation. It indicates what we see in the scatterplot, namely, that the points vary considerably around any particular straight line.

Similarly, the scatterplots for the Bendrix manufacturing data in Figures 11.4 and 11.5 indicate moderately large positive correlations, 0.632 and 0.521, between Overhead and MachHrs and between Overhead and ProdRuns. However, the correlation indicated in Figure 11.7 between MachHrs and ProdRuns, -0.229, is quite small and indicates almost no relationship between these two variables.

Correlations can be misleading when variables are related nonlinearly.

We must be a bit more careful when interpreting the correlations in Figures 11.8 and 11.9. The scatterplot between life expectancy and GNP per capita in Figure 11.8 is obviously nonlinear, and correlations are relevant descriptors only for *linear* relationships. If anything, the correlation of 0.616 in this example tends to underestimate the true strength of the relationship—the nonlinear one—between life expectancy and GNP per capita. In contrast, the correlation between salary and years of experience in Figure 11.9 is large, 0.894, but it is not nearly as large as it would be if the outlier were omitted. (It is then 0.992.) This example illustrates the considerable effect a single outlier can have on a correlation.

An obvious question is whether a given correlation is "large." This is a difficult question to answer directly. Clearly, a correlation such as 0.992 is quite large—the points tend to cluster very closely around a straight line. Similarly, a correlation of 0.034 is quite small—the points tend to be a shapeless swarm. But there is a continuum of in-between values, as exhibited in Figures 11.2, 11.4, and 11.5. We will give a more definite answer to this question when we examine the *square* of the correlation later in this chapter.

As for calculating correlations, there are two possibilities in Excel. To calculate a *single* correlation r_{XY} between variables X and Y, we can use Excel's CORREL function in the form

$$=\text{CORREL}(X\text{-range}, Y\text{-range})$$

Alternatively, we can use StatPro to obtain a whole table of correlations between a set of variables.

Finally, we reiterate the important limitation of correlations (and covariances), namely, that they apply only to *linear* relationships. If a correlation is close to zero, we cannot automatically conclude that there is no relationship between the two variables. We should look at a scatterplot first. The chances are that the points are a shapeless swarm and that no relationship exists. But it is also possible that the points cluster around some curve. In this case the correlation is misleading, and the nonlinear shape should be analyzed further.

Simple Linear Regression

Scatterplots and correlations are very useful for indicating linear relationships and the strengths of these relationships. But they do not actually *quantify* the relationships. For example, we know from the Pharmex drugstore data that sales are related to promotional expenditures. But from the knowledge presented so far, we do not know exactly what this relationship is. If the expenditure index for a given region is 95, what would we predict this region's sales index to be? Or if one region's expenditure index is 5 points higher than another region's, we would expect the former to have a larger sales index, but how much larger? To answer these questions, we need to quantify the relationship between the response variable Sales and the explanatory variable Promote.

Remember that "simple" linear regression does not mean "easy"; it means only that there is a *single* explanatory variable.

In this section we answer these types of questions for simple linear regression, where there is a *single* explanatory variable. We do so by fitting a straight line through the scatterplot of the response variable Y versus the explanatory variable X and then basing the answers to the questions on the fitted straight line. But which straight line? We address this issue next.

11.4.1 Least Squares Estimation

The scatterplot between Sales and Promote, repeated in Figure 11.12, hints at a linear relationship between these two variables. It would not be difficult to draw a straight line

FIGURE 11.12
Scatterplot with
Possible Linear Fit
Superimposed

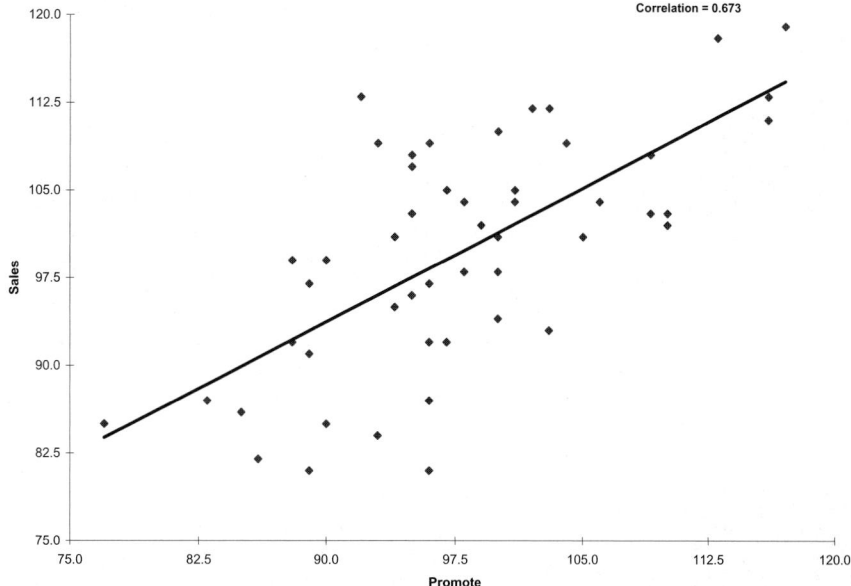

FIGURE 11.13
Fitted Values and
Residuals

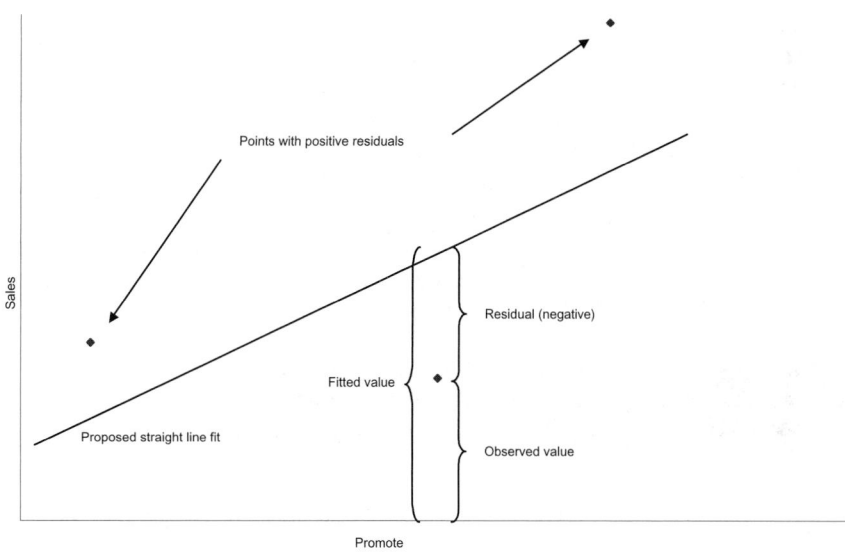

through these points to produce a reasonably good fit. In fact, a possible linear fit is indicated in the graph. But we will proceed more systematically than simply drawing lines freehand. Specifically, we will choose the line that makes the vertical distances from the points to the line as small as possible, as explained below.

Consider the magnified graph in Figure 11.13. Here we show several points in the scatterplot, along with a line drawn through them. Note that the vertical distance from the horizontal axis to any point, which is just the value of Sales for that point, can be decomposed into two parts: the vertical distance from the horizontal axis to the line, and the vertical distance from the line to the point. The first of these is called the **fitted value,** and the second is called the **residual.** The idea is very simple. By using a straight line to reflect the relationship between Sales and Promote, we expect a given Sales to be at the height of the line above any particular value of Promote. That is, we expect Sales to equal the fitted value. These terms are summarized in the box.

Fitted Values and Residuals	A **fitted value** is the predicted value of response variable. Graphically, it is the height of the line above a given explanatory value. The corresponding **residual** is the difference between the actual and fitted values of the response variable.

But the relationship is not perfect. Not all (perhaps not any) of the points lie exactly on the line. The differences are the residuals. They show how much the observed values differ from the fitted values. If a particular residual is positive, the corresponding point is above the line; if it is negative, the point is below the line. The only time a residual is zero is when the point lies directly on the line. The relationship between observed values, fitted values, and residuals is very general and is stated in equation (11.2).

Fundamental Equation for Regression	Observed Value = Fitted Value + Residual	(11.2)

We can now explain how to choose the "best-fitting" line through the points in the scatterplot. We choose the one with the *smallest sum of squared residuals*. The resulting line is called the **least squares line.** Why do we use the sum of *squared* residuals? Why not minimize some other measure of the residuals? First, we do not simply minimize the sum of the residuals because the positive residuals would cancel the negative residuals. In fact, the least squares line has the property that the sum of the residuals is always exactly zero. To adjust for this, we could minimize the sum of the *absolute values* of the residuals, and this is a perfectly reasonable procedure. However, for technical reasons it is not the procedure usually chosen. We settle on the sum of squared residuals because this method is deeply rooted in statistical tradition, and it works well.

Least Squares Line	The **least squares line** is the line that minimizes the sum of the squared residuals. It is the line quoted in regression outputs.

The minimization problem itself is a calculus problem that we will not discuss here. Virtually all statistical software packages perform this minimization automatically, so we need not be concerned with the technical details. However, we will provide the formulas for the least squares line.

Recall from basic algebra that the equation for any straight line can be written as

$$Y = a + bX$$

Here, a is the Y-intercept of the line, the value of Y when $X = 0$, and b is the slope of the line, the change in Y when X increases by 1 unit. Therefore, to specify the least squares line, all we need to specify is the slope and intercept. These are given by equations (11.3) and (11.4).

Equation for Slope in Simple Linear Regression	$$b = \frac{\Sigma(X_i - \overline{X})(Y_i - \overline{Y})}{\Sigma(X_i - \overline{X})^2} = r_{XY}\frac{s_Y}{s_X}$$	(11.3)

$$a = \overline{Y} - b\overline{X} \qquad \textbf{(11.4)}$$

We have presented these formulas primarily for conceptual purposes, not for hand calculations—the computer can take care of the calculations. From the right-hand formula for b, we see that it is closely related to the correlation between X and Y. Specifically, if we keep the standard deviations, s_X and s_Y, of X and Y constant, then the slope b of the least squares line varies directly with the correlation between the two variables. A relationship with a large correlation (negative or positive) has a steep slope, and a relationship with a small correlation has a shallow slope. At the extreme, a nonrelationship with a correlation of 0 has a slope of 0; that is, it results in a horizontal line. The effect of the formula for a is not quite as interesting. It simply forces the least squares line to go through the point of sample means, $(\overline{X}, \overline{Y})$.

It is easy to obtain the least squares line in Excel, either by using Excel's built-in Analysis ToolPak or StatPro's Simple Regression procedure. We illustrate the latter in the following continuations of Examples 11.1 and 11.2.

Example 11.1 Sales versus Promotions at Pharmex (continued)

Find the least squares line for the Pharmex drugstore data, using Sales as the response variable and Promote as the explanatory variable.

Objective To use StatPro's simple regression procedure to find the least squares line for sales as a function of promotional expenses at Pharmex.

Solution

We use the StatPro/Regression Analysis/Simple menu item. After specifying that Sales is the response (dependent) variable and that Promote is the explanatory (independent) variable, we see the dialog box in Figure 11.14. This gives us the option of creating several scatterplots involving the fitted values and residuals. We suggest checking the first two options, as shown.

The regression output includes three parts. The first two are a list of fitted values and residuals, placed in columns next to the data set, and any scatterplots selected from the dialog box in Figure 11.14. We will look at these shortly. The third part of the output is the most important. It is shown in Figure 11.15 on page 566.

FIGURE 11.14
Dialog Box for Scatterplot Options in Simple Regression Procedure

Diagnostic options for simple regression

Select from any of the following scatterplots:

☑ Fitted values versus actual Y values

☑ Residuals versus fitted values

☐ Residuals versus X values

OK

Cancel

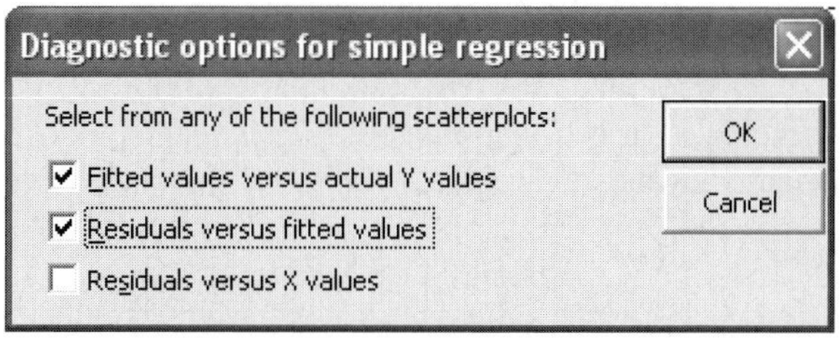

FIGURE 11.15

Regression Output for Drugstore Example

	A	B	C	D	E	F	G	H
1	**Results of simple regression for Sales**							
2								
3	**Summary measures**							
4		Multiple R	0.6730					
5		R-Square	0.4529					
6		StErr of Est	7.3947					
7								
8	**ANOVA table**							
9		Source	df	SS	MS	F	p-value	
10		Explained	1	2172.8804	2172.8804	39.7366	0.0000	
11		Unexplained	48	2624.7396	54.6821			
12								
13	**Regression coefficients**							
14			Coefficient	Std Err	t-value	p-value	Lower limit	Upper limit
15		Constant	25.1264	11.8826	2.1146	0.0397	1.2349	49.0179
16		Promote	0.7623	0.1209	6.3037	0.0000	0.5192	1.0054

We will eventually learn what all of the output in Figure 11.15 means, but for now, we will concentrate on only a small part of it. Specifically, we find the intercept and slope of the least squares line under the Coefficient label in cells C15 and C16. They imply that the equation for the least squares line is[3]

$$\text{Predicted Sales} = 25.1264 + 0.7623\text{Promote}$$

Excel Tip *The Simple Regression procedure in StatPro uses Excel formulas to calculate all of the regression output. It takes advantage of several built-in statistical functions available in Excel. Generate the StatPro output shown in Figure 11.15 and take a look at the formulas in cells C4, C5, C6, C15, and C16. You'll see how Excel's statistical functions CORREL, RSQ, STEYX, INTERCEPT, and SLOPE are used. Also, look at the Fitted Values column (which StatPro inserts to the right of the original data) to see how the TREND function is used. For now, note specifically that the slope and intercept of the least squares line can be calculated directly with the formulas*

$$=SLOPE(Y\text{-}range, X\text{-}range)$$

and

$$=INTERCEPT(Y\text{-}range, X\text{-}range)$$

*These formulas (with the appropriate X and Y ranges) can be entered anywhere in a spreadsheet to obtain the slope and intercept for a **simple** regression equation—no add-ins are necessary.*

We can interpret the regression equation for this example as follows. The slope, 0.7623, indicates that the sales index tends to increase by about 0.76 for each 1-unit increase in the promotional expenses index. Alternatively, if we compare two regions, where region 2 spends 1 unit higher than region 1, we predict the sales index for region 2 to be 0.76 larger than the sales index for region 1. The interpretation of the intercept is less important. It is literally the predicted sales index for a region that does no promotions. However, no region in the sample has anywhere near a zero promotional value. Therefore, in a situation like this, where the range of observed explanatory variable values does not include 0, it is best to think of the intercept term as an "anchor" for the least squares line that allows us to predict Y values for the range of *observed* X values.

In many applications it makes no sense to have the explanatory variable(s) equal to 0. Then the intercept term has no practical or economic meaning.

[3]We will always report the left side of the estimated regression equation as the *predicted* value of the response variable. It is not the *actual* value of the response variable because the observations do not all lie on the estimated regression line.

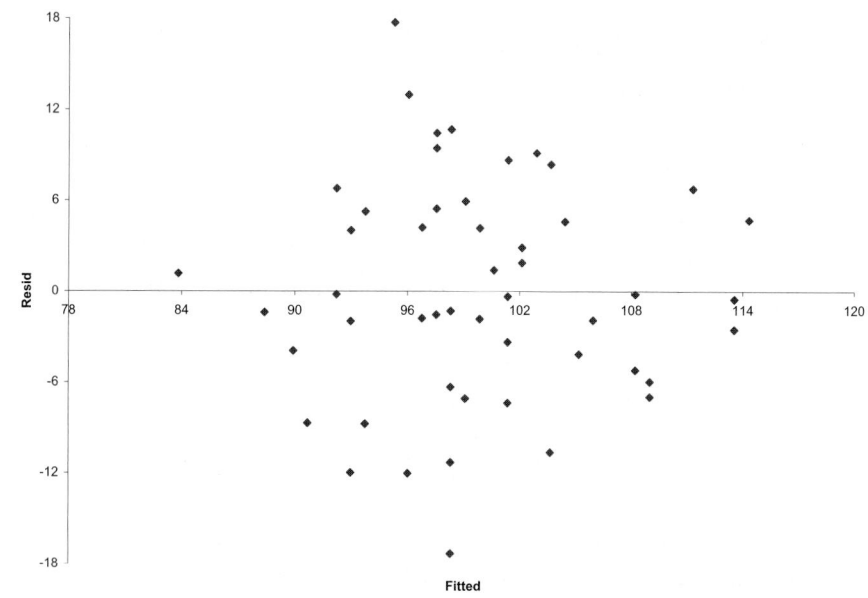

We typically like a shapeless "swarm" of points in a scatterplot of residuals versus fitted values.

A useful graph in almost any regression analysis is a scatterplot of residuals (on the vertical axis) versus fitted values. This scatterplot for the Pharmex data appears in Figure 11.16. We typically examine such a scatterplot for any striking patterns. A "good" fit not only has small residuals, but it has residuals scattered *randomly* around 0 with no apparent pattern. This appears to be the case for the Pharmex data. ●

Example 11.2 Explaining Overhead Costs at Bendrix (continued)

The Bendrix manufacturing data set has two potential explanatory variables, MachHrs and ProdRuns. Eventually, we will estimate a regression equation with *both* of these variables included. However, if we include only one at a time, what do they tell us about overhead costs?

Objective To use StatPro's simple regression procedure to regress overhead expenses at Bendrix against machine hours and then against production runs.

Solution

The regression output for Overhead with MachHrs as the single explanatory variable appears in Figure 11.17 (page 568). The output when ProdRuns is the only explanatory variable appears in Figure 11.18. The two least squares lines are therefore

$$\text{Predicted Overhead} = 48{,}621 + 34.7\text{MachHrs} \tag{11.5}$$

and

$$\text{Predicted Overhead} = 75{,}606 + 655.1\text{ProdRuns} \tag{11.6}$$

Clearly, these two equations are quite different, although each effectively breaks Overhead into a fixed component and a variable component. Equation (11.5) implies that

	A	B	C	D	E	F	G	H
1	**Results of simple regression for Overhead**							
2								
3	**Summary measures**							
4		Multiple R	0.6319					
5		R-Square	0.3993					
6		StErr of Est	8584.7394					
7								
8	**ANOVA table**							
9		Source	df	SS	MS	F	p-value	
10		Explained	1	1665463368.3416	1665463368.3416	22.5986	0.0000	
11		Unexplained	34	2505723491.6584	73697749.7547			
12								
13	**Regression coefficients**							
14			Coefficient	Std Err	t-value	p-value	Lower limit	Upper limit
15		Constant	48621.3546	10725.3327	4.5333	0.0001	26824.8705	70417.8388
16		MachHrs	34.7022	7.2999	4.7538	0.0000	19.8671	49.5374

FIGURE 11.17 Regression Output for Overhead versus MachHrs

	A	B	C	D	E	F	G	H
1	**Results of simple regression for Overhead**							
2								
3	**Summary measures**							
4		Multiple R	0.5205					
5		R-Square	0.2710					
6		StErr of Est	9457.2395					
7								
8	**ANOVA table**							
9		Source	df	SS	MS	F	p-value	
10		Explained	1	1130247999.2618	1130247999.2618	12.6370	0.0011	
11		Unexplained	34	3040938860.7382	89439378.2570			
12								
13	**Regression coefficients**							
14			Coefficient	Std Err	t-value	p-value	Lower limit	Upper limit
15		Constant	75605.5157	6808.6106	11.1044	0.0000	61768.7632	89442.2682
16		ProdRuns	655.0707	184.2747	3.5549	0.0011	280.5797	1029.5616

FIGURE 11.18 Regression Output for Overhead versus ProdRuns

the fixed component of overhead is about \$48,621. Bendrix can expect to incur this amount even if zero machine hours are used. The variable component is the 34.7MachHrs term. It implies that the expected overhead increases by about \$35 for each extra machine hour. Equation (11.6), on the other hand, breaks overhead down into a fixed component of \$75,606 and a variable component of about \$655 per each production run.

The difference between these two equations can be attributed to the fact that neither tells the whole story. If the manager's goal is to split overhead into a fixed component and a variable component, then the variable component should include *both* of the measures of work activity (and maybe even others) to give a more complete explanation of overhead. We will see how to do this when we reanalyze this example with multiple regression. ●

11.4.2 Standard Error of Estimate

We now reexamine fitted values and residuals to see how they lead to a useful summary measure for a regression equation. In a typical simple regression model, the expression

FIGURE 11.19
Fitted Values and Residuals for Pharmex Example

	A	B	C	D	E	F	G
1	Data on drugstore promotional expenditures and sales						
2							
3	Note: each value is a percentage of what the leading competitor did						
4							
5	Region	Promote	Sales			Fitted Values	Residuals
6	1	77	85			83.8232	1.1768
7	2	110	103			108.9790	-5.9790
8	3	110	102			108.9790	-6.9790
9	4	93	109			96.0200	12.9800
10	5	90	85			93.7331	-8.7331
11	6	95	103			97.5446	5.4554
12	7	100	110			101.3561	8.6439
13	8	85	86			89.9216	-3.9216
14	9	96	92			98.3069	-6.3069
15	10	83	87			88.3970	-1.3970
53	48	100	98			101.3561	-3.3561
54	49	95	108			97.5446	10.4554
55	50	96	87			98.3069	-11.3069

$a + bX$ is the fitted value of Y. Graphically, it is the height of the estimated line above the value X. We often denote it by $\hat{Y}$ (pronounced Y-hat):[4]

$$\hat{Y} = a + bX$$

Then a typical residual, denoted by e, is the difference between the observed value Y and the fitted value $\hat{Y}$ [a restatement of equation (11.2)]:

$$e = Y - \hat{Y}$$

We show some of the fitted values and associated residuals for the Pharmex drugstore example in Figure 11.19. (Recall that these columns are inserted automatically by StatPro's Simple Regression procedure.)

The magnitudes of the residuals provide a good indication of how useful the regression line is for predicting Y values from X values. However, because there are numerous residuals, it is useful to summarize them with a single numerical measure. This measure, called the **standard error of estimate** and denoted s_e, is essentially the standard deviation of the residuals. It is given by equation (11.7).

Formula for Standard Error of Estimate	$$s_e = \sqrt{\frac{\Sigma e_i^2}{n-2}}$$	**(11.7)**

Actually, because the average of the residuals from a least squares fit is always 0, this is identical to the standard deviation of the residuals except that we use the denominator $n - 2$ rather than the usual $n - 1$. As we'll see in more generality later on, the rule is to subtract the number of parameters being estimated from the sample size n to obtain the denominator. Here there are two parameters being estimated: the intercept a and the slope b.

About 2/3 of the fitted $\hat{Y}$ values are typically within 1 standard error of the actual Y values. About 95% are within 2 standard errors.

The usual empirical rules for standard deviations can be applied to the standard error of estimate. For example, we expect about 2/3 of the residuals to be within 1 standard error of their mean (which is 0). Stated another way, we expect about 2/3 of the observed Y values to be within 1 standard error of the corresponding fitted $\hat{Y}$ values. Similarly, we

---

[4]We could write Predicted Y instead of $\hat{Y}$, but the latter notation is more common in the statistics literature.

expect about 95% of the observed Y values to be within 2 standard errors of the corresponding fitted $\hat{Y}$ values.[5]

The standard error of estimate s_e is included in all regression outputs. Alternatively, it can be calculated directly with Excel's STEYX function in the form

$$=\text{STEYX}(Y\text{-range},X\text{-range})$$

In general, the standard error of estimate indicates the level of accuracy of predictions made from the regression equation. The smaller it is, the more accurate predictions tend to be.

The standard error for the Pharmex data appears in cell C6 of Figure 11.15. Its value, approximately 7.39, indicates the typical error we are likely to make when we use the fitted value (based on the regression line) to predict sales from promotional expenses. More specifically, if we use the regression equation to predict sales for many regions, based on the promotional expenses in each region, then about 2/3 of the predictions will be within 7.39 of the actual sales values, and about 95% of the predictions will be within 2 standard errors, or 14.78, of the actual sales values.

Is this level of accuracy good? One measure of comparison is the standard deviation of the sales variable, namely, 9.90. (This is obtained by the usual STDEV function applied to the observed sales values.) It can be interpreted as the standard deviation of the residuals around a *horizontal* line positioned at the mean value of Sales. This would be the relevant regression line if there were no explanatory variables—that is, if we ignored Promote. In other words, it is a measure of the prediction error we would make if we used the sample mean of Sales as the prediction for *every* region and ignored Promote. The fact that the standard error of estimate, 7.39, is not much less than 9.90 means that the Promote variable adds a relatively small amount to prediction accuracy. We can do nearly as well without it as with it. We would certainly prefer a standard error of estimate *well* below 9.90.

We can often use the standard error of estimate to judge which of several potential regression equations is the most useful. In the Bendrix manufacturing example we estimated two regression lines, one using MachHrs and one using ProdRuns. From Figures 11.17 and 11.18, their standard errors are approximately $8585 and $9457. These imply that MachHrs is a slightly better predictor of overhead. The predictions based on MachHrs will tend to be slightly more accurate than those based on ProdRuns. Of course, we might guess that predictions based on *both* predictors will yield even more accurate predictions, and this is definitely the case, as we will see when we discuss multiple regression.

11.4.3 R^2: The Coefficient of Determination

We now discuss another important measure of the goodness of fit of the least squares line: the *coefficient of determination,* or simply R^2. Along with the standard error of estimate s_e, it is the most frequently quoted measure in applied regression analysis. With a value always between 0 and 1, the **coefficient of determination** can be interpreted as the *fraction of variation of the response variable explained by the regression line.* (It is often expressed as a percentage, so that we talk about the *percentage* of variation explained by the regression line.)

Coefficient of Determination (R^2)	The **coefficient of determination (R^2)** is the percentage of variation of the response variable explained by the regression.

[5]This requires that the residuals be at least approximately normally distributed, a requirement we will discuss more fully in the next chapter.

To see more precisely what this means, we look into the derivation of R^2. In the previous section we suggested that one way to measure the regression equation's ability to predict is to compare the standard error of estimate, s_e, to the standard deviation of the response variable, s_Y. The idea is that s_e is (essentially) the standard deviation of the residuals, whereas s_Y is the standard deviation of the residuals that we would obtain from a horizontal regression line at height $\overline{Y}$, the response variable's mean. Therefore, if s_e is small compared to s_Y (that is, if s_e/s_Y is small), then the regression line has evidently done a good job in explaining the variation of the response variable.

The coefficient of determination is based on this idea. R^2 is defined by equation (11.8). (This value is obtained automatically with StatPro's regression procedure, or it can be calculated with Excel's RSQ function.) Equation (11.8) indicates that when the residuals are small, then R^2 will be close to 1, but when they are large, R^2 will be close to 0.

Formula for R^2

$$R^2 = 1 - \frac{\Sigma e_i^2}{\Sigma (Y_i - \overline{Y})^2} \tag{11.8}$$

R^2 measures the goodness of a linear fit. The better the linear fit is, the closer R^2 is to 1.

We see from cell C5 of Figure 11.15 that the R^2 measure for the Pharmex drugstore data is 0.453. In words, the single explanatory variable Promote is able to explain only 45.3% of the variation in the Sales variable. This is not particularly good—the same conclusion we made when we based goodness of fit on s_e. There is still 54.7% of the variation left unexplained. Of course, we would like R^2 to be as close to 1 as possible. Usually, the only way to increase it is to use better and/or more explanatory variables.

Analysts often compare equations on the basis of their R^2 values. We see from Figures 11.17 and 11.18 that the R^2 values using MachHrs and ProdRuns as single explanatory variables for the Bendrix overhead data are 39.9% and 27.1%. These provide one more piece of evidence that MachHrs is a slightly better predictor of Overhead than ProdRuns. Of course, they also suggest that the percentage of variation of Overhead explained could be increased by including *both* variables in a single equation. This is true, as we will see shortly.

In simple linear regression, R^2 is the square of the correlation between the response variable and the explanatory variable.

There is a good reason for the notation R^2. It turns out that R^2 is the square of the correlation between the observed Y values and the fitted $\hat{Y}$ values. This correlation appears in all regression outputs. For the Pharmex data it is 0.673, as seen in cell C4 of Figure 11.15. Aside from rounding error, the square of 0.673 is 0.453, the R^2 value right below it. In the case of simple linear regression, when there is only a single explanatory variable in the equation, the correlation between the Y variable and the fitted $\hat{Y}$ values is the same as the absolute value of the correlation between the Y variable and the explanatory X variable. For the Pharmex data we already saw that the correlation between Sales and Promote is indeed 0.673.

This interpretation of R^2 as the square of a correlation helps to decide the issue of when a correlation is "large." For example, if the correlation between two variables Y and X is ± 0.8, we know that the regression of Y on X will produce an R^2 of 0.64; that is, the regression with X as the only explanatory variable will explain 64% of the variation in Y. If the correlation drops to ± 0.7, this percentage drops to 49%; if the correlation increases to ± 0.9, the percentage increases to 81%. The point is that before a single variable X can explain a large percentage of the variation in some other variable Y, the two variables must be highly correlated—in *either* a positive or negative direction.

Level A

1. Explore the relationship between the selling prices (Y) and the appraised values (X) of the 150 homes in the file P02_07.XLS by estimating a simple linear regression model. Also, compute the standard error of estimate s_e and the coefficient of determination R^2 for the estimated least squares line. Interpret these measures and the least squares line for these data.
   a. Is there evidence of a *linear* relationship between the selling price and appraised value? If so, characterize the relationship (i.e., indicate whether the relationship is a positive or negative one, a strong or weak one, etc.).
   b. For which of the two remaining variables, the size of the home and the number of bedrooms in the home, is the relationship with the home's selling price *stronger*? Justify your choice with additional simple linear regression models.

2. What is the relationship between the number of short-term general hospitals (Y) and the number of general or family physicians (X) in U.S. metropolitan areas? Explore this question by estimating a simple linear regression model using the data in the file P02_17.XLS. Interpret your estimated regression model as well as the coefficient of determination R^2.

3. Motorco produces electric motors for use in home appliances. One of the company's production managers is interested in examining the relationship between the dollars spent per month in inspecting finished motor products (X) and the number of motors produced during that month that were returned by dissatisfied customers (Y). He has collected the data in the file P02_18.XLS to explore this relationship for the past 36 months. Generate a simple linear regression model using the given data and interpret it for this production manager. Also, compute and interpret s_e and R^2 for these data.

4. The owner of the Original Italian Pizza restaurant chain would like to understand which variable most strongly influences the sales of his specialty, deep-dish pizza. He has gathered data on the monthly sales of deep-dish pizzas at his restaurants and observations on other potentially relevant variables for each of his 15 outlets in central Pennsylvania. These data are provided in the file P11_04.XLS. Estimate a simple linear regression model between the quantity sold (Y) and each of the following candidates for the best explanatory variable: average price of deep-dish pizzas, monthly advertising expenditures, and disposable income per household in the areas surrounding the outlets. Which variable is *most* strongly associated with the number of pizzas sold? Be sure to explain your choice.

5. The human resources manager of DataCom, Inc. wants to examine the relationship between annual salaries (Y) and the number of years employees have worked at DataCom (X). These data have been collected for a sample of employees and are given in the file P11_05.XLS.
   a. Estimate the relationship between Y and X. Interpret the least squares line.
   b. How well does the estimated simple linear regression model fit the given data? Document your answer.

6. Consider the relationship between the size of the population (X) and the average household income level for residents of U.S. towns (Y). What do you expect the relationship between these two variables to be? Using the data in the file P02_24.XLS, produce and interpret a simple linear regression model involving these two variables. How well does the estimated model fit the given data?

7. Examine the relationship between the average utility bills for homes of a particular size (Y) and the average monthly temperature (X). The data in the file P11_07.XLS include the average monthly bill and temperature for each month of the past year.
   a. Use the given data to estimate a simple linear regression model. Interpret the least squares line.
   b. How well does the estimated regression model fit the given data? How might we do a better job of explaining the variation of the average utility bills for homes of a certain size?

8. The U.S. Bureau of Labor Statistics provides data on the year-to-year percentage changes in the wages and salaries of workers in private industries, including both "white-collar" and "blue-collar" occupations. Here we consider selected annual data in the file P02_56.XLS. Is there evidence of a strong relationship between the yearly changes in the wages and salaries of white-collar and blue-collar workers in the United States over the given time period? Answer this question by estimating and interpreting a simple linear regression model.

9. Management of a home appliance store in Charlotte would like to understand the growth pattern of the monthly sales of VCR units over the past 2 years. The managers have recorded the relevant data in an Excel spreadsheet, which can be found in the file P11_09.XLS. Have the sales of VCR units been growing linearly over the past 24 months? Using simple linear regression, explain why or why not.

10. Do the sales prices of houses in a given community vary systematically with their sizes (as measured in square feet)? Attempt to answer this question by esti-

mating a simple regression model where the sales price of the house is the response variable and the size of the house is the explanatory variable. Use the sample data given in the file P11_10.XLS. Interpret your estimated model and the associated coefficient of determination R^2.

11. The file P11_11.XLS contains annual observations of the American minimum wage. Has the minimum wage been growing at roughly a *constant* rate over this period? Use simple linear regression analysis to address this question. Explain the results you obtain.

12. Based on the data in the file P02_25.XLS from the U.S. Department of Agriculture, explore the relationship between the number of farms (X) and the average size of a farm (Y) in the United States. Specifically, generate a simple linear regression model and interpret it.

13. Estimate the relationship between monthly electrical power usage (Y) and home size (X) using the data in the file P11_13.XLS. Interpret your results. How well does a simple linear regression model explain the variation in monthly electrical power usage?

14. The *ACCRA Cost of Living Index* provides a useful and reasonably accurate measure of cost of living differences among a large number of urban areas. Items on which the index is based have been carefully chosen to reflect the different categories of consumer expenditures. The data are in the file P02_19.XLS. Use the given data to estimate simple linear regression models to explore the relationship between the composite index (i.e., response variable) and each of the various expenditure components (i.e., explanatory variable).
    a. Which expenditure component has the *strongest* linear relationship with the composite index?
    b. Which expenditure component has the *weakest* linear relationship with the composite index?

15. The management of Beta Technologies, Inc., is trying to determine the variable that best explains the variation of employee salaries using a sample of 52 full-time employees in the file P02_01.XLS. Estimate simple linear regression models to identify which of the following has the *strongest* linear relationship with annual salary: the employee's gender, age, number of years of relevant work experience prior to employment at Beta, number of years of employment at Beta, or number of years of post-secondary education. Provide support for your conclusion.

Multiple Regression

In general, there are two possible approaches to obtaining improved fits. The first is to examine a scatterplot of residuals for nonlinear patterns and then make appropriate modifications to the regression equation. We will discuss this approach later in this chapter. The second approach is much more straightforward: We simply add more explanatory variables to the regression equation. In the Bendrix manufacturing example we deliberately included only a single explanatory variable in the equation at a time so that we could keep the equations simple. But because scatterplots indicate that both explanatory variables are also related to Overhead, we ought to try including both in the regression equation. With any luck, the linear fit should improve.

When we include several explanatory variables in the regression equation, we move into the realm of *multiple* regression. Some of the concepts from simple regression carry over naturally to multiple regression, whereas some change considerably. The following list provides a starting point that we will expand on throughout this section.

Characteristics of Multiple Regression

- Graphically, we are no longer fitting a *line* to a set of points. If there are exactly two explanatory variables, then we are fitting a *plane* to the data in three-dimensional space. There is one dimension for the response variable and one for each of the two explanatory variables. Although we can imagine a flat plane passing through a swarm of points, it is difficult to graph this on a two-dimensional screen. If there are more than two explanatory variables, then we can only imagine the regression plane; drawing in four or more dimensions is impossible.

- The regression equation is still estimated by the least squares method—that is, by minimizing the sum of squared residuals. However, it is definitely not practical to implement this method by hand. A statistical software package such as StatPro is required.
- Simple regression is actually a special case of multiple regression—that is, an equation with a single explanatory variable can be considered as a "multiple" regression equation. Therefore, it is possible to use StatPro's Multiple Regression procedure in such situations.
- There is a "slope" term for each explanatory variable in the equation. The interpretation of these slope terms is somewhat more difficult than in simple regression, as we will discuss in the following subsection.
- The standard error of estimate and R^2 summary measures are almost exactly as in simple regression, as we will discuss in Section 11.5.2.
- Many *types* of explanatory variables can be included in the regression equation, as we will discuss in Section 11.6. To a large part, these are responsible for the wide applicability of multiple regression in the business world.

11.5.1 Interpretation of Regression Coefficients

A typical slope term measures the expected change in Y when the corresponding X increases by 1 unit.

If Y is the response variable and X_1 through X_k are the explanatory variables, then a typical multiple regression equation has the form shown in equation (11.9), where a is again the Y-intercept, and b_1 through b_k are the slopes. Collectively, we refer to a and the b's in equation (11.9) as the **regression coefficients.** The intercept a is the expected value of Y when all of the X's equal 0. Of course, this makes sense only if it is practical for all of the X's to equal 0. Each slope coefficient is the expected change in Y when this particular X increases by one unit and the other X's in the equation remain constant. For example, b_1 is the expected change in Y when X_1 increases by one unit and the other X's in the equation, X_2 through X_k, remain constant.

General Multiple Regression Equation

$$Y = a + b_1X_1 + b_2X_2 + \cdots + b_kX_k \qquad \text{(11.9)}$$

This extra proviso, "when the other X's in the equation remain constant," is very important for the interpretation of the regression coefficients. In particular, it means that the estimates of the b's depend on which other X's are included in the regression equation. We illustrate these ideas in the following continuation of the Bendrix manufacturing example.

Example (11.2) Explaining Overhead Costs at Bendrix (continued)

Estimate and interpret the equation for Overhead when both explanatory variables, MachHrs and ProdRuns, are included in the regression equation.

Objective To use StatPro's multiple regression procedure to estimate the equation for overhead costs at Bendrix as a function of machine hours and production runs.

FIGURE 11.20
Dialog Box of Options with Multiple Regression Procedure

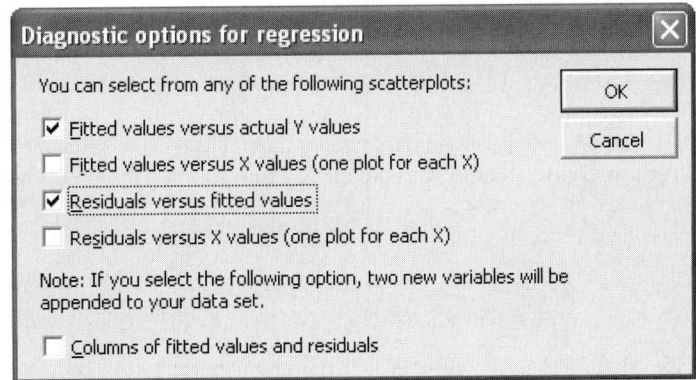

FIGURE 11.21 Multiple Regression Output for Bendrix Example

	A	B	C	D	E	F	G	H
1	*Results of multiple regression for Overhead*							
2								
3	*Summary measures*							
4		Multiple R	0.9308					
5		R-Square	0.8664					
6		Adj R-Square	0.8583					
7		StErr of Est	4108.9932					
8								
9	*ANOVA Table*							
10		Source	df	SS	MS	F	p-value	
11		Explained	2	3614020652.0000	1807010326.0000	107.0261	0.0000	
12		Unexplained	33	557166208.0000	16883824.4848			
13								
14	*Regression coefficients*							
15			Coefficient	Std Err	t-value	p-value	Lower limit	Upper limit
16		Constant	3996.6782	6603.6509	0.6052	0.5492	-9438.5612	17431.9176
17		MachHrs	43.5364	3.5895	12.1289	0.0000	36.2335	50.8393
18		ProdRuns	883.6179	82.2514	10.7429	0.0000	716.2760	1050.9598

Solution

Unlike the situation with simple regression, we do not even attempt to provide formulas for the regression coefficients in multiple regression. These require matrix algebra and are best left for a more technical course. Instead, we rely on computer software such as the StatPro add-in. To obtain the output, we use the StatPro/Regression Analysis/Multiple menu item, select Overhead as the response (dependent) variable, and select MachHrs and ProdRuns as the explanatory (independent) variables. The dialog box shown in Figure 11.20 then gives us options of which scatterplots to obtain and whether we want columns of fitted values and residuals placed next to the data set. We filled it in as shown in the figure, selecting scatterplots of the fitted values versus the observed values and residuals versus fitted values, and not requesting columns of fitted values and residuals.

The main regression output appears in Figure 11.21. The coefficients in the range C16:C18 indicates that the estimated regression equation is

$$\text{Predicted Overhead} = 3997 + 43.54\text{MachHrs} + 883.62\text{ProdRuns} \qquad \textbf{(11.10)}$$

The interpretation of equation (11.10) is that if the number of production runs is held constant, then the overhead cost is expected to increase by \$43.54 for each extra machine hour, and if the number of machine hours is held constant, the overhead cost is expected to increase by $883.62 for each extra production run. The Bendrix manager can interpret the intercept, $3997, as the fixed component of overhead. The slope terms involving MachHrs and ProdRuns are the variable components of overhead.

It is interesting to compare equation (11.10) with the separate equations for Overhead involving only a single variable each. From the previous section these are

$$\text{Predicted Overhead} = 48,621 + 34.7\text{MachHrs}$$

and

$$\text{Predicted Overhead} = 75,606 + 655.1\text{ProdRuns}$$

Note that the coefficient of MachHrs has increased from 34.7 to 43.5 and the coefficient of ProdRuns has increased from 655.1 to 883.6. Also, the intercept is now lower than either intercept in the single-variable equations. In general, it is difficult to guess the changes that will occur when we introduce more explanatory variables into the equation, but it is likely that changes will occur.

The reasoning is that when MachHrs is the only variable in the equation, we are obviously *not* holding ProdRuns constant—we are ignoring it—so in effect the coefficient 34.7 of MachHrs indicates the effect of MachHrs *and* the omitted ProdRuns on Overhead. But when we include both variables, then the coefficient 43.5 of MachHrs indicates the effect of MachHrs only, holding ProdRuns constant. Because the coefficients of MachHrs in the two equations have different *meanings,* it is not surprising that we obtain different numerical estimates of them. ●

> The estimated coefficient of any explanatory variable typically depends on which *other* explanatory variables are included in the equation.

11.5.2 Interpretation of Standard Error of Estimate and R-Square

The multiple regression output in Figure 11.21 is very similar to simple regression output.[6] In particular, cells C5 and C7 again show R^2 and the standard error of estimate s_e. Also, the square root of R^2 appears in cell C4. We interpret these quantities almost exactly as in simple regression. The standard error of estimate is essentially the standard deviation of residuals, but it is now given by equation (11.11), where n is the number of observations and k is the number of explanatory variables in the equation.

Formula for Standard Error of Estimate in Multiple Regression	$$s_e = \sqrt{\frac{\Sigma e_i^2}{n - (k + 1)}}$$	(11.11)

Fortunately, we interpret s_e exactly as before. It is a measure of the prediction error we are likely to make when we use the multiple regression equation to predict the response variable. In this example about 2/3 of the predictions should be within 1 standard error, or $4109, of the actual overhead cost. By comparing this with the standard errors from the single-variable equations for Overhead, $8585 and $9457, we see that the mul-

---

[6]One difference, however, is that the StatPro multiple regression output is not linked to the data by formulas. If the data change, you must rerun the Multiple Regression procedure to update the output.

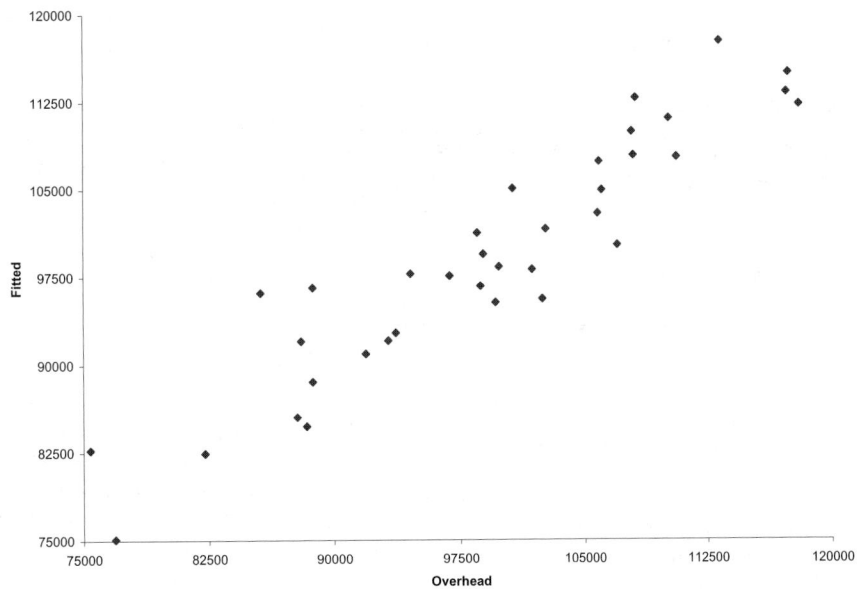

tiple regression equation is likely to provide predictions that are more than twice as accurate as the single-variable equations—quite an improvement!

The R^2 value is again the percentage of variation of the response variable explained by the combined set of explanatory variables. In fact, it even has the same formula as before [see equation (11.8)]. For the Bendrix data we see that MachHrs and ProdRuns combine to explain 86.6% of the variation in Overhead. This is a big improvement over the single-variable equations that were able to explain only 39.9% and 27.1% of the variation in Overhead. Remarkably, the combination of the two explanatory variables explains a larger percentage than the *sum* of their individual effects. This is not common, but as this example shows, it is possible.

The square root of R^2 shown in cell C4 of Figure 11.21 is again the correlation between the fitted values and the observed values of the response variable. For the Bendrix data the correlation between them is 0.931, quite high. A graphical indication of this high correlation can be seen in one of the scatterplots we requested, the plot of fitted versus observed values of Overhead. This scatterplot appears in Figure 11.22. If the regression equation gave *perfect* predictions, then all of the points in this plot would lie on a 45° line—each fitted value would *equal* the corresponding observed value. Although a perfect fit virtually never occurs, the closer the points are to a 45° line, the better the fit is, as indicated by R^2 or its square root.

R^2 is *always* the square of the correlation between the actual and fitted Y values—in both simple and multiple regression.

Although the R^2 value is one of the most frequently quoted values from a regression analysis, it does have one serious drawback: R^2 can only *increase* when extra explanatory variables are added to an equation. This can lead to "fishing expeditions," where we keep adding variables to an equation, some of which have no conceptual relationship to the response variable, just to inflate the R^2 value. To "penalize" the addition of extra variables that do not really belong, an **adjusted** R^2 value is typically listed in regression outputs. This adjusted value appears in cell C6 of Figure 11.21. Although it has no direct interpretation as "percentage of variation explained," it *can* decrease when extra explanatory variables that do not really belong are added to an equation. Therefore, it is a useful index that we can monitor. If we add variables and the adjusted R^2 *decreases,* then the extra variables are essentially not pulling their weight and should probably be omitted. We will have much more to say about the issue of which variables to include in the next chapter.

Adjusted R^2	The **adjusted R^2** is a measure that adjusts R^2 for the number of explanatory variables in the equation. It is used primarily to monitor whether extra explanatory variables really belong in the equation.

PROBLEMS

Level A

16. A trucking company wants to predict the yearly maintenance expense (Y) for a truck using the number of miles driven during the year (X_1) and the age of the truck (X_2, in years) at the beginning of the year. The company has gathered the data given in the file P11_16.XLS. Note that each observation corresponds to a particular truck.
  a. Formulate and estimate a multiple regression model using the given data. Interpret each of the estimated regression coefficients.
  b. Compute and interpret the standard error of estimate s_e and the coefficient of determination R^2 for these data.

17. DataPro is a small but rapidly growing firm that provides electronic data-processing services to commercial firms, hospitals, and other organizations. For each of the past 12 months, DataPro has tracked the number of contracts sold, the average contract price, advertising expenditures, and personal selling expenditures. These data are provided in P11_17.XLS. Assuming that the number of contracts sold is the response variable, estimate a multiple regression model with three explanatory variables. Interpret each of the estimated regression coefficients and the coefficient of determination R^2.

18. An antique collector believes that the price received for a particular item increases with its age and with the number of bidders. The file P11_18.XLS contains data on these three variables for 32 recently auctioned comparable items.
  a. Formulate and estimate a multiple regression model using the given data. Interpret each of the estimated regression coefficients. Is the antique collector correct in believing that the price received for the item increases with its age and with the number of bidders?
  b. Interpret the standard error of estimate s_e and the coefficient of determination R^2.

19. Stock market analysts are continually looking for reliable predictors of stock prices. Consider the problem of modeling the price per share of electric utility

stocks (Y). Two variables thought to influence this stock price are return on average equity (X_1) and annual dividend rate (X_2). The stock price, returns on equity, and dividend rates on a randomly selected day for 16 electric utility stocks are provided in the file P11_19.XLS.
  a. Formulate and estimate a multiple regression model using the given data. Interpret each of the estimated regression coefficients.
  b. Interpret the standard error of estimate s_e, the coefficient of determination R^2, and the adjusted R^2.

20. The manager of a commuter rail transportation system was recently asked by her governing board to determine which factors have a significant impact on the demand for rides in the large city served by the transportation network. The system manager has collected data on variables thought to be possibly related to the number of weekly riders on the city's rail system. The file P11_20.XLS contain these data.
  a. What are the expected signs of the coefficients of the explanatory variables in this multiple regression model? Provide reasoning for each of your stated expectations. (Answer this *before* using regression.)
  b. Formulate and estimate a multiple regression model using the given data. Interpret each of the estimated regression coefficients. Are the signs of the estimated coefficients consistent with your expectations as stated in part **a?**
  c. What proportion of the total variation in the number of weekly riders is *not* explained by this estimated multiple regression model?

21. Consider the enrollment data for *Business Week*'s top U.S. graduate business programs in the file P02_03.XLS. Use these data to estimate a multiple regression model to assess whether there is a systematic relationship between the total number of full-time students and the following explanatory variables: (i) the proportion of female students, (ii) the proportion of minority students, and (iii) the proportion of international students enrolled at these distinguished business schools.

a. Interpret the coefficients of your estimated regression model. Do any of these results surprise you? Explain.

b. How well does your estimated regression model fit the given data?

22. David Savageau and Geoffrey Loftus, the authors of *Places Rated Almanac* have ranked metropolitan areas in the United States with consideration of the following aspects of life in each area: cost of living, transportation, jobs, education, climate, crime, arts, health, and recreation. The data are in the file P02_55.XLS.

a. Use multiple regression analysis to explore the relationship between the metropolitan area's overall score and the set of potential explanatory variables.

b. Interpret each of the estimated coefficients in the regression model. Are the signs of the estimated coefficients consistent with your expectations? If not, can you explain any discrepancies between your findings and expectations?

c. Does the given set of explanatory variables do a good job of explaining changes in the overall score? Explain why or why not.

Level B

23. The owner of a restaurant in Bloomington, Indiana, has recorded sales data for the past 19 years. He has also recorded data on potentially relevant variables. The entire data set appears in the file P11_23.XLS.

a. Estimate a simple linear regression model involving annual sales (the response variable) and the size of the population residing within 10 miles of the restaurant (the explanatory variable). Interpret R^2.

b. Add another explanatory variable—annual advertising expenditures—to the regression model in part **a.** Estimate and interpret this expanded model. How does the R^2 value for this multiple regression model compare to that of the simple regression model estimated in part **a?** Explain any difference between the two R^2 values. Compute and interpret the *adjusted* R^2 value for the revised model.

c. Add one more explanatory variable to the multiple regression model estimated in part **b.** In particular,

estimate and interpret the coefficients of a multiple regression model that includes the *previous* year's advertising expenditure. How does the inclusion of this third explanatory variable affect the R^2 and adjusted R^2 values, in comparison to the corresponding values for the model of part **b?** Explain any changes in these values.

24. A regional express delivery service company recently conducted a study to investigate the relationship between the cost of shipping a package (Y), the package weight (X_1), and the distance shipped (X_2). Twenty packages were randomly selected from among the large number received for shipment, and a detailed analysis of the shipping cost was conducted for each package. These sample observations are given in the file P11_24.XLS.

a. Estimate a simple linear regression model involving shipping cost and package weight. Interpret the slope coefficient of the least squares line as well as the computed value of R^2.

b. Add another explanatory variable—distance shipped—to the regression model in part **a.** Estimate and interpret this expanded model. How does the R^2 value for this multiple regression model compare to that of the simple regression model estimated in part **a?** Explain any difference between the two R^2 values. Compute and interpret the *adjusted* R^2 value for the revised model.

25. Using the sample data given in the file P11_10.XLS, formulate a multiple regression model to predict the sales price of houses in a given community.

a. Add one explanatory variable at a time and estimate each partial regression equation. Report and explain changes in the standard error of estimate s_e, the coefficient of determination R^2, and the adjusted R^2 as each explanatory variable is added to the model.

b. Interpret each of the estimated regression coefficients in the full model.

c. What proportion of the total variation in the sales price is explained by the multiple regression model that includes all four explanatory variables?

 # 11.6 Modeling Possibilities

Once we move from simple to multiple regression, the floodgates open. All types of explanatory variables are potential candidates for inclusion in the regression equation. In this section we will examine several new types of explanatory variables. These include "dummy" variables, interaction variables, and nonlinear transformations. The techniques in this section provide us with many alternative approaches to modeling the relationship between a response variable and potential explanatory variables. In many applications these techniques produce much better fits than we could obtain without them.

As the title of this section suggests, these techniques are modeling *possibilities*. They provide a wide variety of explanatory variables to choose from. However, this does not mean that it is wise to include all or even many of these new types of explanatory variables in any particular regression equation. The chances are that only a few, if any, will significantly improve the linear fit. Knowing which explanatory variables to include requires a great deal of practical experience with regression, as well as a thorough understanding of the particular problem to be solved. The material in this section should *not* be an excuse for a mindless fishing expedition.

11.6.1 Dummy Variables

Some potential explanatory variables are categorical and cannot be measured on a quantitative scale. However, these categorical variables are often related to the response variable, so we need a way to include them in a regression equation. The trick is to use **dummy** variables, also called **indicator** or **0–1** variables. Dummy variables are variables that indicate the category a given observation is in. If a dummy variable for a given category equals 1, the observation is in that category; if it equals 0, the observation is not in that category.

Dummy Variable	A **dummy variable** is a variable with possible values 0 and 1. It equals 1 if the observation is in a particular category and 0 if it is not.

Categorical variables are used in two situations. The first and perhaps most common situation is when a categorical variable has only two categories. A good example of this is a "gender" variable that has the two categories "male" and "female." In this case we need only a *single* dummy variable, and we have the choice of assigning the 1's to either category. If we label the dummy variable Gender, then we can code Gender as 1 for males and 0 for females, or we can code Gender as 1 for females and 0 for males. We just need to be consistent and specify explicitly which coding scheme we are using.

The other situation is when there are more than two categories. A good example of this is when we have quarterly time series data and we want to treat the quarter of the year as a categorical variable with four categories, 1–4. Then we can create four dummy variables, Q1–Q4. For example, Q2 equals 1 for all second-quarter observations and equals 0 for all other observations. Although we can create four dummy variables, we will see that only three of them—*any* three—should be used in a regression equation.

The following example illustrates how we form, use, and interpret dummy variables in regression analysis.

Example 11.3 Possible Gender Discrimination in Salary at Fifth National Bank of Springfield

The Fifth National Bank of Springfield is facing a gender discrimination suit.[7] The charge is that its female employees receive substantially smaller salaries than its male employees. The bank's employee database is listed in the file BANK.XLS. For each of its 208 employees, the data set includes the following variables:

[7]This example and the accompanying data set are based on a real case. Only the bank's name has been changed.

FIGURE 11.23
Selected Data for
Bank Example

	A	B	C	D	E	F	G	H	I
1	Bank salary data								
2									
3	Employee	EducLev	JobGrade	YrHired	YrBorn	Gender	YrsPrior	PCJob	Salary
4	1	3	1	92	69	Male	1	No	32
5	2	1	1	81	57	Female	1	No	39.1
6	3	1	1	83	60	Female	0	No	33.2
7	4	2	1	87	55	Female	7	No	30.6
8	5	3	1	92	67	Male	0	No	29
9	6	3	1	92	71	Female	0	No	30.5
10	7	3	1	91	68	Female	0	No	30
11	8	3	1	87	62	Male	2	No	27
12	9	1	1	91	33	Female	0	No	34
13	10	3	1	86	64	Female	0	No	29.5
209	206	5	6	63	33	Male	0	No	88
210	207	5	6	60	36	Male	0	No	94
211	208	5	6	62	33	Female	0	No	30

- EducLev: education level, a categorical variable with categories 1 (finished high school), 2 (finished some college courses), 3 (obtained a bachelor's degree), 4 (took some graduate courses), 5 (obtained a graduate degree)

- JobGrade: a categorical variable indicating the current job level, the possible levels being 1–6 (6 is highest)

- YrHired: year employee was hired

- YrBorn: year employee was born

- Gender: a categorical variable with values "Female" and "Male"

- YrsPrior: number of years of work experience at another bank prior to working at Fifth National

- PCJob: a categorical yes/no variable depending on whether the employee's current job is computer-related

- Salary: current annual salary in thousands of dollars

Figure 11.23 lists a few of the observations. Do these data provide evidence that females are discriminated against in terms of salary?

Objective To use StatPro's Multiple Regression procedure to analyze whether the bank discriminates against females in terms of salary.

Solution

A naive approach to this problem compares the average female salary to the average male salary. This can be done with a pivot table, as in Chapter 2, or with a more formal hypothesis test, as in Chapter 10. Using these methods, we find that the average of all salaries is $39,922, the female average is $37,210, the male average is $45,505, and the difference between the male and female averages is statistically significant at any reasonable level of significance. The females are definitely earning less. But perhaps there is a reason for this. They might have lower education levels, they might have been hired more recently, they might be working at lower job grades, and so on. The question is whether the difference between female and male salaries is still evident after taking these other attributes into account. This is a perfect task for regression.

We first need to create dummy variables for the various categorical variables. We can do this manually with IF functions or we can use StatPro's Dummy Variable procedure.

To do it manually, we can create a dummy variable Female based on Gender in column J by entering the formula

$$=\text{IF}(F4=\text{"Female"},1,0)$$

in cell J4 and copying it down. Note that we are coding the females as 1's and the males as 0's. (The quotes are necessary when a nonnumerical value is used in an IF function.)

StatPro's Dummy Variable procedure is somewhat easier, especially when there are multiple categories. For example, to create five dummies, Ed_1 through Ed_5, for each of the education levels, we can use the StatPro/Data Utilities/Create Dummy Variables menu item, select the "Create several dummies from a categorical variable" option, select the EducLev variable to base the dummies on, and (after the dummies have been created) change their default names to Ed_1 through Ed_5. This procedure simply enters IF functions in the dummy cells, exactly as we would do manually. (StatPro automatically inserts the IF functions to create the dummy variables. It even checks for missing data.) We can follow the same procedure to create six dummies, Job_1 through Job_6, for the job grade categories.

We can add dummies to effectively collapse categories.

Sometimes we might want to collapse several categories. For example, we might want to collapse the five education categories into three categories: 1, (2,3), and (4,5). The new second category includes employees who have taken undergraduate courses or have completed a bachelor's degree, and the new third category includes employees who have taken graduate courses or have completed a graduate degree. It is easy to do this. We simply add the Ed_2 and Ed_3 columns to get the dummy for the new second category, and similarly add the Ed_4 and Ed_5 columns for the new third category.

Once the dummies have been created, we can run a regression analysis with Salary as the response variable, using any combination of numerical and dummy explanatory variables. However, there are two rules we must follow:

- We shouldn't use any of the *original* categorical variables, such as EducLev, that the dummies are based on.

- We should use *one less dummy* than the number of categories for any categorical variable.

Always include one less dummy than the number of categories. The omitted dummy corresponds to the *reference* category.

This second rule is a technical one. If we violate it, the statistical software will give us an error message. For example, if we want to use education level as an explanatory variable, we should enter only five of the six dummies Ed_1 through Ed_6. *Any* five of these can be used. The omitted dummy then corresponds to the **reference** category. As we will see, the interpretation of the dummy variable coefficients are all relative to this reference category. When there are only two categories, as with the gender variable, we typically name the variable with the category, such as Female, that corresponds to the 1's. If we create the dummy variables manually, we probably don't even bother to create a Male dummy. In this case "Male" automatically becomes the reference category.

To get used to dummy variables in regression, we will proceed in several stages in this example. We first estimate a regression equation with only one explanatory variable, Female. The output appears in Figure 11.24. The resulting equation is

$$\text{Predicted Salary} = 45.505 - 8.296\text{Female} \qquad \textbf{(11.12)}$$

To interpret this equation, recall that Female has only two possible values, 0 and 1. If we substitute Female=1 into equation (11.12), we obtain

$$\text{Predicted Salary} = 45.505 - 8.296(1) = 37.209$$

To interpret regression equations with dummy variables, it is useful to rewrite the equation for each category.

Since Female=1 corresponds to females, this equation simply indicates the average female salary. Similarly, if we substitute Female=0 into equation (11.12), we obtain

$$\text{Predicted Salary} = 45.505 - 8.296(0) = 45.505$$

FIGURE 11.24
Output for Bank
Example with a
Single Explanatory
Variable

FIGURE 11.24 Output for Bank Example with a Single Explanatory Variable

	A	B	C	D	E	F	G	H
1	**Results of multiple regression for Salary**							
2								
3	**Summary measures**							
4		Multiple R	0.3465					
5		R-Square	0.1201					
6		Adj R-Square	0.1158					
7		StErr of Est	10.5843					
8								
9	**ANOVA Table**							
10		Source	df	SS	MS	F	p-value	
11		Explained	1	3149.6346	3149.6346	28.1151	0.0000	
12		Unexplained	206	23077.4727	112.0266			
13								
14	**Regression coefficients**							
15			Coefficient	Std Err	t-value	p-value	Lower limit	Upper limit
16		Constant	45.5054	1.2835	35.4534	0.0000	42.9749	48.0360
17		Female	-8.2955	1.5645	-5.3024	0.0000	-11.3800	-5.2110

FIGURE 11.25 Regression Output with Two Numerical Explanatory Variables Included

	A	B	C	D	E	F	G	H
1	**Results of multiple regression for Salary**							
2								
3	**Summary measures**							
4		Multiple R	0.7016					
5		R-Square	0.4923					
6		Adj R-Square	0.4848					
7		StErr of Est	8.0794					
8								
9	**ANOVA Table**							
10		Source	df	SS	MS	F	p-value	
11		Explained	3	12910.6678	4303.5559	65.9279	0.0000	
12		Unexplained	204	13316.4395	65.2767			
13								
14	**Regression coefficients**							
15			Coefficient	Std Err	t-value	p-value	Lower limit	Upper limit
16		Constant	35.4917	1.3410	26.4661	0.0000	32.8476	38.1357
17		YrsPrior	0.1313	0.1809	0.7259	0.4687	-0.2254	0.4881
18		Female	-8.0802	1.1982	-6.7438	0.0000	-10.4426	-5.7178
19		YrsExper	0.9880	0.0809	12.2083	0.0000	0.8284	1.1476

Since Female=0 corresponds to males, this equation indicates the average male salary. Therefore, the interpretation of the −8.296 coefficient of the Female dummy variable is straightforward. It is the average female salary relative to the reference (male) category—females get paid $8296 less on average than males.

Obviously, equation (11.12) tells only part of the story. It ignores all information except for gender. We expand this equation by adding the experience variables YrsPrior and YrsExper, where YrsExper is years of experience with Fifth National and is calculated in a new column as 95 minus YrHired. (Remember that the data are from 1995.) The output with the Female dummy variable and these two experience variables appears in Figure 11.25. The corresponding regression equation is

Predicted Salary = 35.492 + 0.988YrsExper + 0.131YrsPrior − 8.080Female   **(11.13)**

It is again useful to write equation (11.13) in two forms: one for females (substituting Female=1) and one for males (substituting Female=0). After doing the arithmetic, they become

$$\text{Predicted Salary} = 27.412 + 0.988\text{YrsExper} + 0.131\text{YrsPrior}$$

and

$$\text{Predicted Salary} = 35.492 + 0.988\text{YrsExper} + 0.131\text{YrsPrior}$$

FIGURE 11.26
Regression Output with Education Dummies Included

	A	B	C	D	E	F	G	H
1	**Results of multiple regression for Salary**							
2								
3	**Summary measures**							
4		Multiple R	0.8030					
5		R-Square	0.6449					
6		Adj R-Square	0.6324					
7		StErr of Est	6.8244					
8								
9	**ANOVA Table**							
10		Source	df	SS	MS	F	p-value	
11		Explained	7	16912.6922	2416.0989	51.8787	0.0000	
12		Unexplained	200	9314.4150	46.5721			
13								
14	**Regression coefficients**							
15			Coefficient	Std Err	t-value	p-value	Lower limit	Upper limit
16		Constant	26.6134	1.7941	14.8335	0.0000	23.0755	30.1512
17		YrsPrior	0.3622	0.1581	2.2908	0.0230	0.0504	0.6740
18		Female	-4.5013	1.0858	-4.1458	0.0001	-6.6423	-2.3603
19		YrsExper	1.0329	0.0696	14.8404	0.0000	0.8957	1.1702
20		Ed_2	0.1602	1.6560	0.0968	0.9230	-3.1052	3.4257
21		Ed_3	4.7646	1.4734	3.2336	0.0014	1.8591	7.6700
22		Ed_4	7.3198	2.6942	2.7169	0.0072	2.0072	12.6325
23		Ed_5	11.7702	1.5102	7.7937	0.0000	8.7922	14.7482

Except for the intercept term, these equations are identical. We can now interpret the coefficient -8.080 of the Female dummy variable as the average salary disadvantage for females relative to males *after controlling for job experience.* Gender discrimination still appears to be a very plausible conclusion. However, note that the R^2 value is only 49.2%. Perhaps there is still more to the story.

We next add education level to the equation by including four of the five education level dummies. Although *any* four could be used, we use Ed_2 through Ed_5, so that the lowest level becomes the reference category. (We would expect this to lead to *positive* coefficients for these dummies, which are easier to interpret.) The resulting output appears in Figure 11.26. The estimated regression equation is now

$$\text{Predicted Salary} = 26.613 + 1.033\text{YrsExper} + 0.362\text{YrsPrior} - 4.501\text{Female} \\ + 0.160\text{Ed\_2} + 4.765\text{Ed\_3} + 7.320\text{Ed\_4} + 11.770\text{Ed\_5} \tag{11.14}$$

Now there are two categorical variables involved, gender and education level. However, we can still write a separate equation *for any combination* of categories by setting the dummies to the appropriate values. For example, the equation for females at the fifth education level is found by setting Female=1 and Ed_5=1, and setting the other job dummies equal to 0. After terms are combined, this equation is

$$\text{Predicted Salary} = 33.882 + 1.033\text{YrsExper} + 0.362\text{YrsPrior}$$

The intercept 33.882 is the intercept from equation (11.14), 26.613, plus the coefficients of Female and Ed_5.

We can interpret equation (11.14) as follows. For either gender and any education level, the expected increase in salary for one extra year of experience with Fifth National is $1033; the expected increase in salary for one extra year of prior experience with another bank is $362. The coefficients of the education dummies indicate the average increase in salary an employee can expect relative to the reference (lowest) education level. For example, an employee with education level 4 can expect to earn $7320 more than an employee with education level 1, all else being equal. Finally, the key coefficient, the $-$4501 for females, indicates the average salary disadvantage for females relative to males, given that they have the same experience levels *and* the same education levels. Note that the R^2 value is now 64.5%, quite a bit larger than when the education dummies were not included. We appear to be getting closer to the truth. In particular, we see that there appears to be gender discrimination in salaries, even after accounting for job experience and education level.

FIGURE 11.27
Pivot Table of Job
Grade Counts for
Bank Data

	A	B	C	D
1				
2				
3	Count	Gender ▼		
4	JobGrade ▼	Female	Male	Grand Total
5	1	34.29%	17.65%	28.85%
6	2	20.71%	19.12%	20.19%
7	3	25.71%	10.29%	20.67%
8	4	12.14%	16.18%	13.46%
9	5	6.43%	17.65%	10.10%
10	6	0.71%	19.12%	6.73%
11	Grand Total	100.00%	100.00%	100.00%

One further explanation for gender differences in salary might be job grade. Perhaps females tend to be in lower job grades, which would help explain why they get lower salaries on average. One way to check this is with a pivot table, as in Figure 11.27, where we put job grade in the row area, gender in the column area, and request counts, displayed as percentages of columns. Clearly, females tend to be concentrated at the lower job grades. For example, 28.85% of all employees are at the lowest job grade, but 34.29% of all females are at this grade and only 17.65% of males are at this grade. The opposite is true at the higher job grades. This certainly helps to explain why females get lower salaries on average.

We can go one step further to see the effect of job grade on salary by including the dummies for job grade in the equation, along with the other variables we have included so far. As with the education dummies, we use the lowest job grade as the reference category and include only the five dummies for the other categories. While we're at it, we include the other two potential explanatory variables to the equation: Age, coded as 95 minus YrBorn, and HasPCJob, a dummy based on the PCJob categorical variable. The regression output for this equation appears in Figure 11.28.

FIGURE 11.28
Regression Output with Other Variables Added

	A	B	C	D	E	F	G	H
1	Results of multiple regression for Salary							
2								
3	Summary measures							
4		Multiple R	0.8748					
5		R-Square	0.7652					
6		Adj R-Square	0.7482					
7		StErr of Est	5.6481					
8								
9	ANOVA Table							
10		Source	df	SS	MS	F	p-value	
11		Explained	14	20070.2508	1433.5893	44.9390	0.0000	
12		Unexplained	193	6156.8564	31.9008			
13								
14	Regression coefficients							
15			Coefficient	Std Err	t-value	p-value	Lower limit	Upper limit
16		Constant	29.6899	2.4900	11.9236	0.0000	24.7788	34.6011
17		YrsPrior	0.1677	0.1404	1.1943	0.2338	-0.1093	0.4447
18		Female	-2.5545	1.0120	-2.5242	0.0124	-4.5504	-0.5585
19		YrsExper	0.5156	0.0980	5.2621	0.0000	0.3223	0.7088
20		Ed_2	-0.4856	1.3987	-0.3472	0.7289	-3.2442	2.2731
21		Ed_3	0.5279	1.3575	0.3889	0.6978	-2.1496	3.2054
22		Ed_4	0.2852	2.4047	0.1186	0.9057	-4.4577	5.0281
23		Ed_5	2.6908	1.6209	1.6601	0.0985	-0.5061	5.8877
24		Job_2	1.5645	1.1858	1.3194	0.1886	-0.7742	3.9032
25		Job_3	5.2194	1.2624	4.1345	0.0001	2.7295	7.7092
26		Job_4	8.5948	1.4960	5.7451	0.0000	5.6442	11.5455
27		Job_5	13.6594	1.8743	7.2879	0.0000	9.9627	17.3561
28		Job_6	23.8324	2.7999	8.5119	0.0000	18.3101	29.3547
29		Age	-0.0090	0.0577	-0.1553	0.8767	-0.1228	0.1048
30		HasPCJob	4.9228	1.4738	3.3402	0.0010	2.0160	7.8297

As expected, the coefficients of the job grade dummies are all positive, and they increase as the job grade increases—it pays to be in the higher job grades. The effect of age appears to be minimal, and there appears to be a "bonus" of close to $5000 for having a PC-related job. The R^2 value has now increased to 76.5%, and the penalty for being a female has decreased to $2555—still large but not as large as before.

However, even if this penalty, the coefficient of Female in this last equation, is considered "small," is it convincing evidence against the argument for gender discrimination? We believe the answer is "no." We have used variations in job grades to reduce the penalty for being female. But the remaining question is then, Why are females predominantly in the low job grades? Perhaps this is the real source of gender discrimination. Perhaps management is not advancing the females as quickly as it should, which naturally results in lower salaries for females.

We conclude this example for now, but we will say more about it in the next two subsections. ●

11.6.2 Interaction Variables

Suppose that we regress a variable Y on a numerical variable X and a dummy variable D. If the estimated equation is of the form

$$\hat{Y} = a + b_1X + b_2D \qquad \text{(11.15)}$$

then, as in the previous section, we can break this equation down into two separate equations:

$$\hat{Y} = (a + b_2) + b_1X$$

and

$$\hat{Y} = a + b_1X$$

The first corresponds to $D = 1$, and the second corresponds to $D = 0$. The only difference between these two equations is the intercept term; the slope for each is b_1. Geometrically, they correspond to two *parallel* lines that are a distance b_2 apart. For example, if D corresponds to gender, then there is a female line and a parallel male line. The effect of X on Y is the same for females and males. When X increases by 1 unit, we predict Y to change by b_1 units for males or females.

In effect, when we include *only* a dummy variable in a regression equation, as in equation (11.15), we are allowing the intercepts of the two lines to differ (by an amount b_2), but we are *forcing* the lines to be parallel. Sometimes we want to allow them to have different slopes, in addition to possibly different intercepts. We can do this with an **interaction** variable. Algebraically, an interaction variable is the *product* of two variables. Its effect is to allow the effect of one of the variables on Y to depend on the value of the other variable.

Interaction Variable	An **interaction** variable is the product of two explanatory variables. We can include such a variable in a regression equation if we believe the effect of one explanatory variable on Y depends on the value of another explanatory variable.

Suppose we create the interaction variable XD (the product of X and D) and then estimate the equation

$$\hat{Y} = a + b_1X + b_2D + b_3XD$$

As usual, we rewrite this equation as two separate equations, depending on whether $D = 0$ or $D = 1$. If $D = 1$, we combine terms to write

$$\hat{Y} = (a + b_2) + (b_1 + b_3)X$$

If $D = 0$, the dummy and interaction variables drop out and we obtain

$$\hat{Y} = a + b_1X$$

The notation is not important. The important part is that the interaction term, b_3XD, allows the slope of the regression line to differ between the two categories.

The following continuation of the bank discrimination example illustrates one possible use of interaction variables.

Example 11.3 Possible Gender Discrimination in Salary at Fifth National Bank of Springfield (continued)

Earlier we estimated an equation for Salary using the numerical explanatory variables YrsExper and YrsPrior and the dummy variable Female. If we drop the YrsPrior variable from this equation (for simplicity) and rerun the regression, we obtain the equation

$$\text{Predicted Salary} = 35.824 + 0.981\text{YrsExper} - 8.012\text{Female} \qquad \textbf{(11.16)}$$

The R^2 value for this equation is 49.1%. If we decide to include an interaction variable between YrsExper and Female in this equation, what is its effect?

Objective To use multiple regression with an interaction variable to see whether the effect of years of experience on salary is different across the two genders.

Solution

We first need to form an interaction variable that is the product of YrsExper and Female. This can be done in two ways in Excel. We can do it manually by introducing a new variable that contains the product of the two variables involved, or we can use the StatPro/Data Utilities/Create Interaction Variable(s) menu item. For the latter, we select Female and YrsExper as the variables to be used to create an interaction variable, and we don't check either of the boxes in the next dialog box—we do *not* want either to be treated as a categorical variable.[8]

Once the interaction variable has been created, we include it in the regression equation in addition to the other variables in equation (11.16). The multiple regression output appears in Figure 11.29 (page 588). The estimated regression equation is

$$\text{Predicted Salary} = 30.430 + 1.528\text{YrsExper} + 4.098\text{Female}$$
$$- 1.248\text{YrsExper*Female}$$

(where YrsExper*Female is StatPro's default name for the interaction variable). As in the general discussion, it is useful to write this as two separate equations, one for females and one for males. The female equation (Female=1) is

$$\text{Predicted Salary} = (30.430 + 4.098) + (1.528 - 1.248)\text{YrsExper}$$
$$= 34.528 + 0.280\text{YrsExper}$$

[8]See the Help screen for this data utility. It explains the StatPro options for creating interaction variables.

	A	B	C	D	E	F	G	H
1	*Results of multiple regression for Salary*							
2								
3	*Summary measures*							
4		Multiple R	0.7991					
5		R-Square	0.6386					
6		Adj R-Square	0.6333					
7		StErr of Est	6.8163					
8								
9	*ANOVA Table*							
10		Source	df	SS	MS	F	p-value	
11		Explained	3	16748.8748	5582.9583	120.1620	0.0000	
12		Unexplained	204	9478.2324	46.4619			
13								
14	*Regression coefficients*							
15			Coefficient	Std Err	t-value	p-value	Lower limit	Upper limit
16		Constant	30.4300	1.2166	25.0129	0.0000	28.0314	32.8287
17		Female	4.0983	1.6658	2.4602	0.0147	0.8138	7.3827
18		YrsExper	1.5278	0.0905	16.8887	0.0000	1.3494	1.7061
19		Female*YrsExper	-1.2478	0.1367	-9.1296	0.0000	-1.5173	-0.9783

FIGURE 11.29 Regression Output with an Interaction Variable

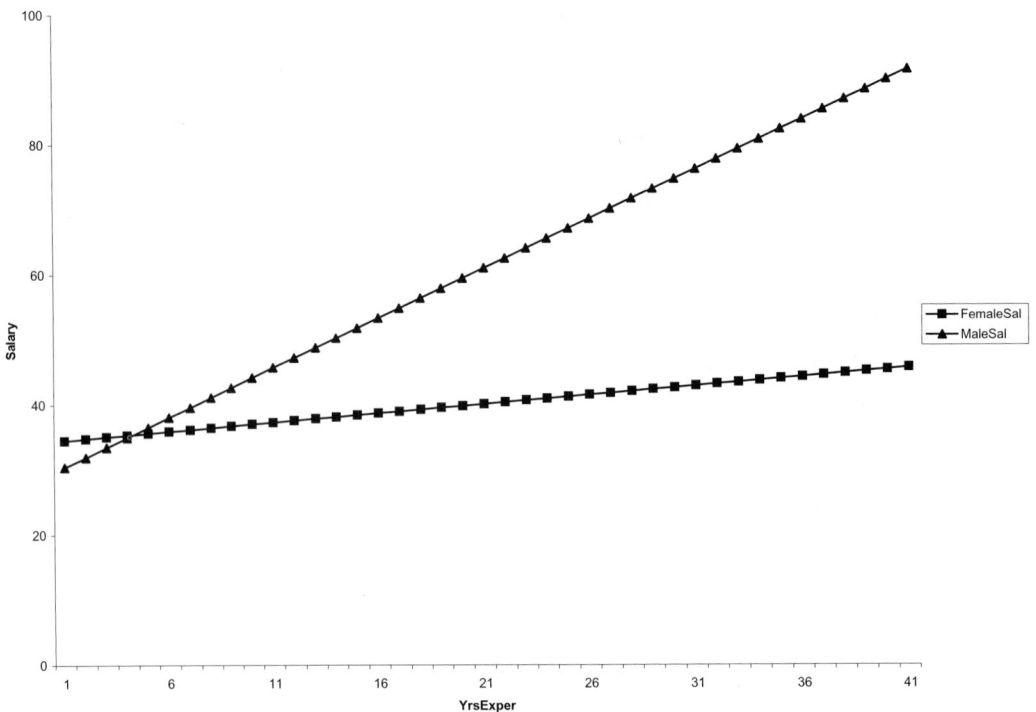

FIGURE 11.30 Nonparallel Female and Male Salary Lines

and the male equation (Female=0) is

$$\text{Predicted Salary} = 30.430 + 1.528\text{YrsExper}$$

Graphically, these equations appear as in Figure 11.30. The *Y*-intercept for the female line is slightly higher—females with no experience with Fifth National tend to start out slightly higher than males—but the slope of the female line is much lower. That is, males tend to move up the salary ladder much more quickly than females. Again, this provides another argument, although a somewhat different one, for gender discrimination against females. By the way, the R^2 value with the interaction variable has increased from 49.1% to 63.9%. The interaction variable has definitely added to the explanatory power of the equation. ●

This example illustrates just one possible use of interaction variables. The product of *any* two variables, a numerical and a dummy variable, two dummy variables, or even two numerical variables, can be used. The trick is to interpret the results correctly, and the easiest way to do this is the way we've been doing it—by writing several separate equations and seeing how they differ. To illustrate one further possibility (among many), suppose we include the variables YrsExper, Female, and HighJob in the equation for Salary, along with interactions between Female and YrsExper and between Female and HighJob. Here, HighJob is a new dummy variable that is 1 for job grades 4–6 and is 0 for job grades 1–3. (It can be calculated as the sum of the dummies Job_4 through Job_6.) The resulting equation is

$$\text{Predicted Salary} = 28.168 + 1.261\text{YrsExper} + 9.242\text{HighJob} + 6.601\text{Female}$$
$$-1.224\text{Female*YrsExper} + 1.564\text{Female*HighJob} \qquad \textbf{(11.17)}$$

and the R^2 value is now a hefty 76.6%.

The interpretation of equation (11.17) is quite a challenge because it is really composed of four separate equations, one for each combination of Female and HighJob. For females in the high job category, the equation becomes

$$\text{Predicted Salary} = (28.168 + 9.242 + 6.601 + 1.564) + (1.261 - 1.224)\text{YrsExper}$$
$$= 45.575 + 0.037\text{YrsExper}$$

and for females in the low job category it is

$$\text{Predicted Salary} = (28.168 + 6.601) + (1.261 - 1.224)\text{YrsExper}$$
$$= 34.769 + 0.037\text{YrsExper}$$

Similarly, for males in the high job category, the equation becomes

$$\text{Predicted Salary} = (28.168 + 9.242) + 1.261\text{YrsExper}$$
$$= 37.410 + 1.261\text{YrsExper}$$

and for males in the low job category it is

$$\text{Predicted Salary} = 28.168 + 1.261\text{YrsExper}$$

Putting this into words, we can interpret the various coefficients as follows:

Interpretation of Regression Coefficients

- The intercept 28.168 is the average *starting* salary (that is, with no experience at Fifth National) for males in the low job category.
- The coefficient 1.261 of YrsExper is the expected increase in salary per extra year of experience for males (in either job category).
- The coefficient 9.242 of HighJob is the expected salary "premium" for males starting in the high job category instead of the low job category.
- The coefficient 6.601 of Female is the expected starting salary premium for females relative to males, given that they start in the low job category.

- The coefficient -1.224 of Female*YrsExper is the penalty per extra year of experience for females relative to males—that is, male salaries increase this much more than female salaries each year.

- The coefficient 1.564 of Female*HighJob is the extra premium (in addition to the male premium) for females starting in the high job category instead of the low job category.

As we see, there are pros and cons to adding interaction variables. On the plus side, they allow for more complex and interesting models, and they can provide significantly better fits. On the minus side, they can become extremely difficult to interpret correctly. Therefore, we recommend that they be added only when there is good economic and statistical justification for doing so.

11.6.3 Nonlinear Transformations

The general linear regression equation has the form

$$\hat{Y} = a + b_1X_1 + b_2X_2 + \cdots + b_kX_k$$

We typically include nonlinear transformations in a regression equation because of economic considerations or curvature detected in scatterplots.

It is *linear* in the sense that the right-hand side of the equation is a constant plus a sum of products of constants and variables. However, there is no requirement that the response variable Y or the explanatory variables X_1 through X_k be *original* variables in the data set. Most often they are, but they are also allowed to be transformations of original variables. We already saw one example of this in the previous section with interaction variables. They are not original variables but are instead products of original (or even transformed) variables. We enter them in the same way as original variables; only the interpretation differs. In this section we will look at several nonlinear transformations of variables. These are often used because of curvature detected in scatterplots. They can also arise because of economic considerations. That is, economic theory often leads us to particular nonlinear transformations.

There are actually two cases we should distinguish. We can transform the response variable Y or we can transform any of the explanatory variables, the X's. We can also do both. In either case there are a few nonlinear transformations that are typically used. These include the natural logarithm, the square root, the reciprocal, and the square. The point of any of these is usually to "straighten out" the points in a scatterplot. If several different transformations straighten out the data equally well, then we prefer the one that is easiest to interpret.

We begin with a small example where only the X variable needs to be transformed.

Example 11.4 Demand and Cost for Electricity

The Public Service Electric Company produces different quantities of electricity each month, depending on the demand. The file POWER.XLS lists the number of units of electricity produced (Units) and the total cost of producing these (Cost) for a 36-month period. The data appear in Figure 11.31. How can regression be used to analyze the relationship between Cost and Units?

Objective To see whether the cost of supplying electricity is a nonlinear function of demand, and if it is, what form the nonlinearity takes.

FIGURE 11.31
Data for Electric
Power Example

	A	B	C	D
1	Data on cost versus production level			
2				
3	Month	Cost	Units	
4	1	45623	601	
5	2	46507	738	
6	3	43343	686	
7	4	46495	736	
8	5	47317	756	
9	6	41172	498	
10	7	43974	828	
11	8	44290	671	
12	9	29297	305	
13	10	47244	637	
37	34	46295	667	
38	35	45218	705	
39	36	45357	637	

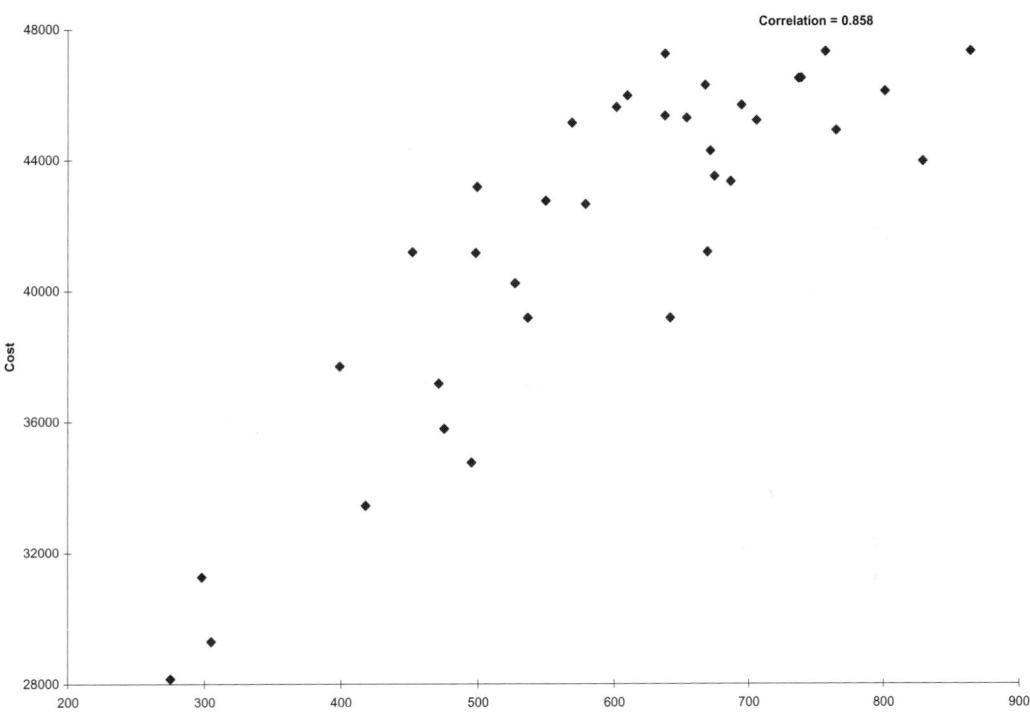

FIGURE 11.32 Scatterplot of Cost Versus Units for Electricity Example

Solution

A good place to start is with a scatterplot of Cost versus Units. This appears in Figure 11.32. It indicates a definite positive relationship and one that is nearly linear. However, there is also some evidence of curvature in the plot. The points increase slightly less rapidly as Units increases from left to right. In economic terms, there may be economies of scale, where the marginal cost of electricity decreases as more units of electricity are produced.

FIGURE 11.33
Residuals from a
Straight-Line Fit

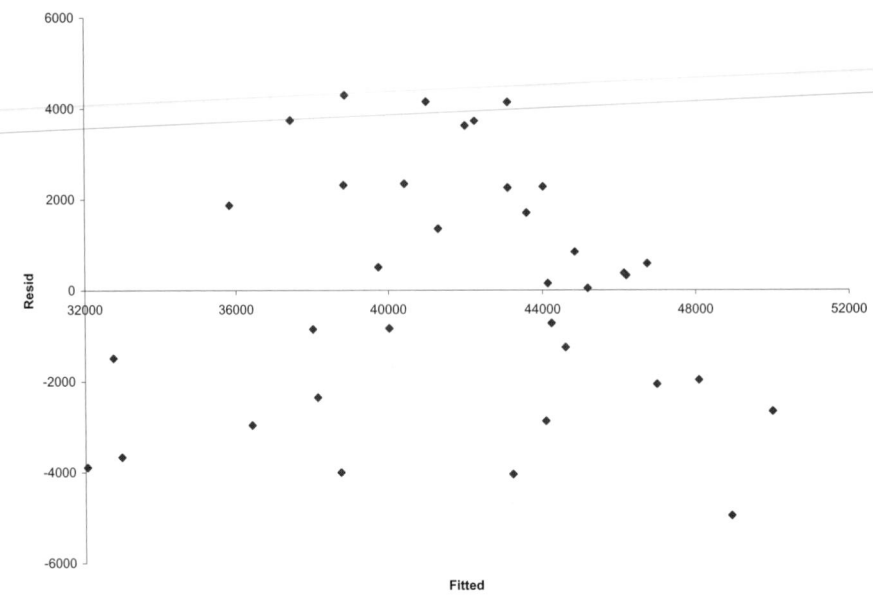

Nevertheless, we first use regression to estimate a *linear* relationship between Cost and Units. The resulting regression equation is

$$\text{Predicted Cost} = 23{,}651 + 30.53\text{Units}$$

The corresponding R^2 and s_e are 73.6% and \$2734. We also requested a scatterplot of the residuals versus the fitted values, always a good idea when nonlinearity is suspected. This plot is shown in Figure 11.33. The sign of nonlinearity in this plot is that the residuals to the far left and the far right are all negative, whereas the majority of the residuals in the middle are positive. Admittedly, the pattern is far from perfect—there are quite a few negative residuals in the middle—but the plot does hint at nonlinear behavior.

A scatterplot of residuals versus fitted values often indicates the need for a nonlinear transformation.

This negative–positive–negative behavior of residuals suggests a *parabola*—that is, a quadratic relationship with the *square* of Units included in the equation. We first create a new variable Sqr(Units) in the data set. This can be done manually (with the formula =C4^2 in cell D4, copied down) or with the StatPro/Data Utilities/Transform Variables menu item.[9] This latter method is easier to use and allows us to transform several variables simultaneously. Then we use multiple regression to estimate the equation for Cost with *both* explanatory variables, Units and Sqr(Units), included. The resulting equation, as shown in Figure 11.34, is

$$\text{Predicted Cost} = 5793 + 98.35\text{Units} - 0.0600\text{Sqr(Units)} \qquad \textbf{(11.18)}$$

Note that R^2 has increased to 82.2% and s_e has decreased to \$2281.

One way to see how this regression equation fits the scatterplot of Cost versus Units (in Figure 11.32) is to use Excel's trendline option. To do so, activate the scatterplot, use the Chart/Add Trendline menu item, click on the Type tab, and select the Polynomial type or order 2, that is, a quadratic. A graph of equation (11.18) is superimposed on the scatterplot, as shown in Figure 11.35. It shows a reasonably good fit, plus an obvious curvature.

[9]StatPro provides four nonlinear transformations: natural logarithm, square, square root, and reciprocal.

	A	B	C	D	E	F	G	H
1	*Results of multiple regression for Cost*							
2								
3	*Summary measures*							
4		Multiple R	0.9064					
5		R-Square	0.8216					
6		Adj R-Square	0.8108					
7		StErr of Est	2280.7998					
8								
9	*ANOVA Table*							
10		Source	df	SS	MS	F	p-value	
11		Explained	2	790511520.9722	395255760.4861	75.9808	0.0000	
12		Unexplained	33	171667568.0000	5202047.5152			
13								
14	*Regression coefficients*							
15			Coefficient	Std Err	t-value	p-value	Lower limit	Upper limit
16		Constant	5792.7983	4763.0586	1.2162	0.2325	-3897.7249	15483.3216
17		Units	98.3504	17.2369	5.7058	0.0000	63.2816	133.4192
18		Sqr(Units)	-0.0600	0.0151	-3.9806	0.0004	-0.0906	-0.0293

FIGURE 11.34 Regression Output with Squared Term Included

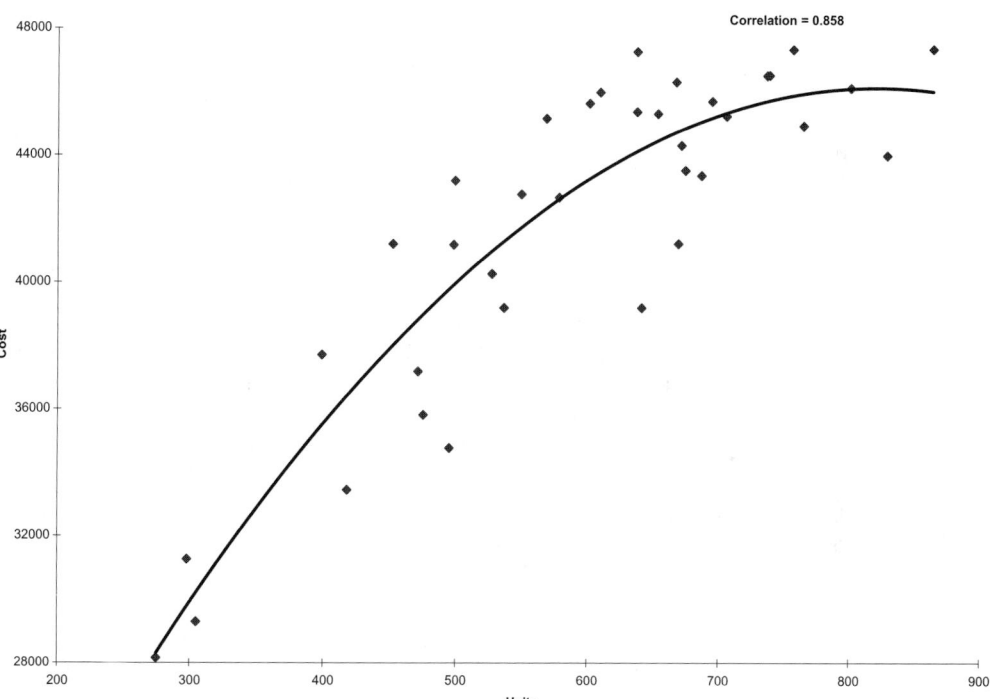

FIGURE 11.35 Quadratic Fit in Electricity Example

The main downside to a quadratic regression equation, as in equation (11.18), is that there is no easy interpretation of the coefficients of Units and Sqr(Units). For example, we can't conclude from the 98.35 coefficient of Units that Cost increases by 98.35 dollars when Units increases by 1. The reason is that when Units increases by 1, Sqr(Units) doesn't stay constant; it *also* increases. All we can say is that the terms in equation (11.18) combine to explain the nonlinear relationship between units produced and total cost.

A final note about this equation concerns the coefficient of Sqr(Units), −0.0600. First, the fact that it is negative makes the parabola bend "downward." This produces the

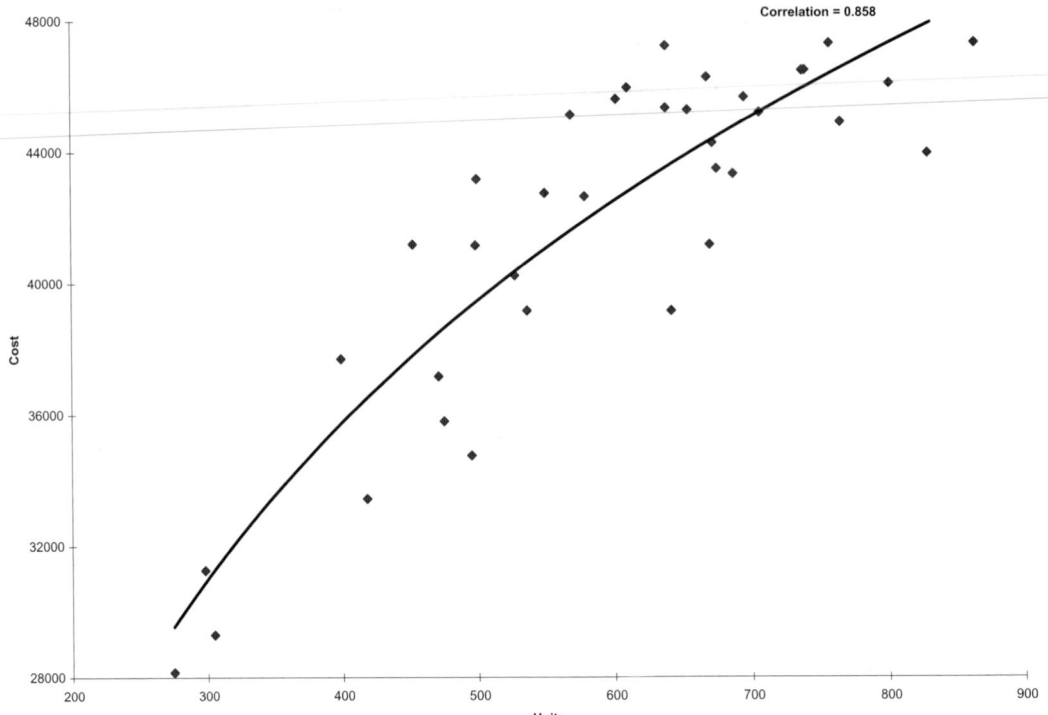

Correlation = 0.858

FIGURE 11.36 Logarithmic Fit to Electricity Data

decreasing marginal cost behavior, where every extra unit of electricity incurs a smaller cost. Actually, the curve described by equation (11.18) eventually goes downhill for large values of Units, but this part of the curve is irrelevant because the company evidently never produces such large quantities. Second, we shouldn't be fooled by the small magnitude of this coefficient. Remember that it is the coefficient of Units *squared,* which is a large quantity. Therefore, the effect of the product -0.0600Sqr(Units) is sizable.

There is at least one other possibility we might examine. Rather than a quadratic fit, we could try a logarithmic fit. In this case we create a new variable, Log(Units), the natural logarithm of Units, and then regress Cost against the *single* variable Log(Units). To create the new variable, we can either proceed manually with Excel's LN function or we can use StatPro/Data Utilities/Transform Variables menu item. Also, we can superimpose a logarithmic curve on the scatterplot of Cost versus Units by using Excel's trendline feature with the logarithmic option. This curve appears in Figure 11.36. To the naked eye, it appears to be similar, and about as good a fit, as the quadratic curve in Figure 11.35.

The resulting regression equation is

$$\text{Predicted Cost} = -63{,}993 + 16{,}654\text{Log(Units)} \qquad \textbf{(11.19)}$$

In general, if *b* is the coefficient of the log of *X*, then the expected change in *Y* when *X* increases by 1% is approximately 0.01 times *b*.

and the R^2 and s_e values are 79.8% and 2393. These latter values indicate that the logarithmic fit is not quite as good as the quadratic fit. However, the advantage of the logarithmic equation is that it is easier to interpret. In fact, one reason logarithmic transformations of variables are used as widely as they are in regression analysis is that they are fairly easy to interpret.

In the present case, where the log of an *explanatory* variable is used, we can interpret its coefficient as follows. Suppose that Units increases by 1%, for example, from 600 to

606. Then equation (11.19) implies that the expected Cost will increase by approximately $0.01(16,654) = 166.54$ dollars. In words, every 1% increase in Units is accompanied by an expected $166.54 increase in Cost. Note that for larger values of Units, a 1% increase represents a larger absolute increase (from 700 to 707 instead of from 600 to 606, say). But each such 1% increase entails the *same* increase in Cost. This is another way of describing the decreasing marginal cost property. ●

The electricity example has shown two possible nonlinear transformations of the *explanatory* variable (or variables) that we can use. All we need to do is create the transformed X's and run the regression. The interpretation of statistics such as R^2 and s_e is exactly the same as before; only the interpretation of the coefficients of the transformed X's changes. It is also possible to transform the response variable Y. Now, however, we must be careful when interpreting summary statistics such as R^2 and s_e, as we explain in the following examples.

A logarithmic transformation of Y is often useful when the distribution of Y values is skewed to the right.

Each of these examples transforms the response variable Y by taking its natural logarithm and then using the log of Y as the new response variable. This approach is taken in a wide variety of business applications. Essentially, it is often a good option when the distribution of Y is skewed to the right, with a few very large values and many small to medium values. The effect of the logarithm transformation is to spread the small values out and squeeze the large values together, making the distribution more symmetric. This is illustrated in Figures 11.37 and 11.38 (page 596) for a hypothetical distribution of household incomes. The histogram of incomes in Figure 11.37 is clearly skewed to the right. However, the histogram of the natural log of income in Figure 11.38 is much more nearly symmetric—and, for technical reasons, more suitable for use as a response variable in regression.

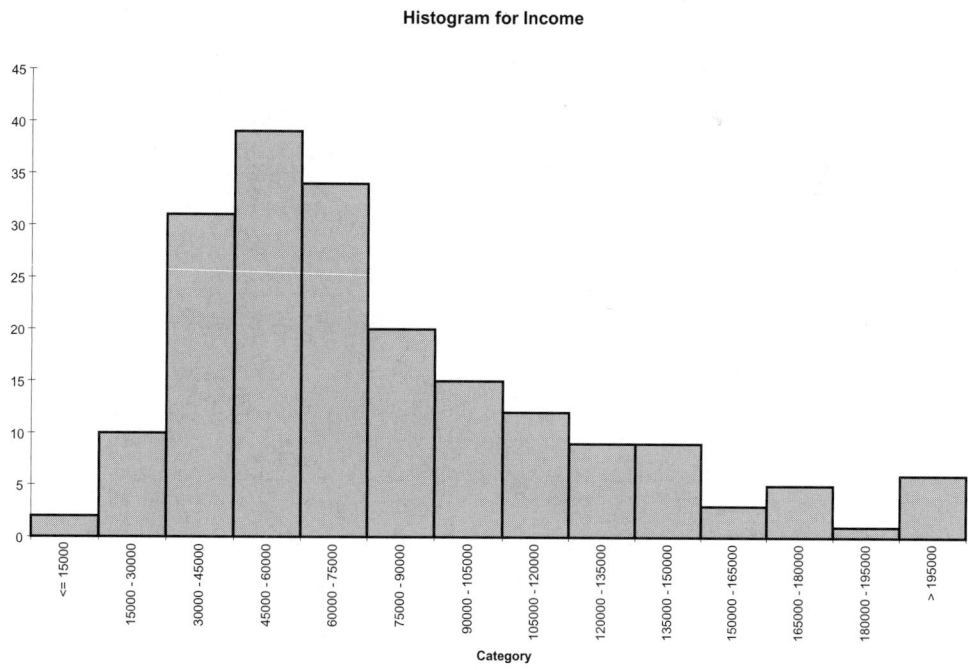

FIGURE 11.37 Skewed Distribution of Income

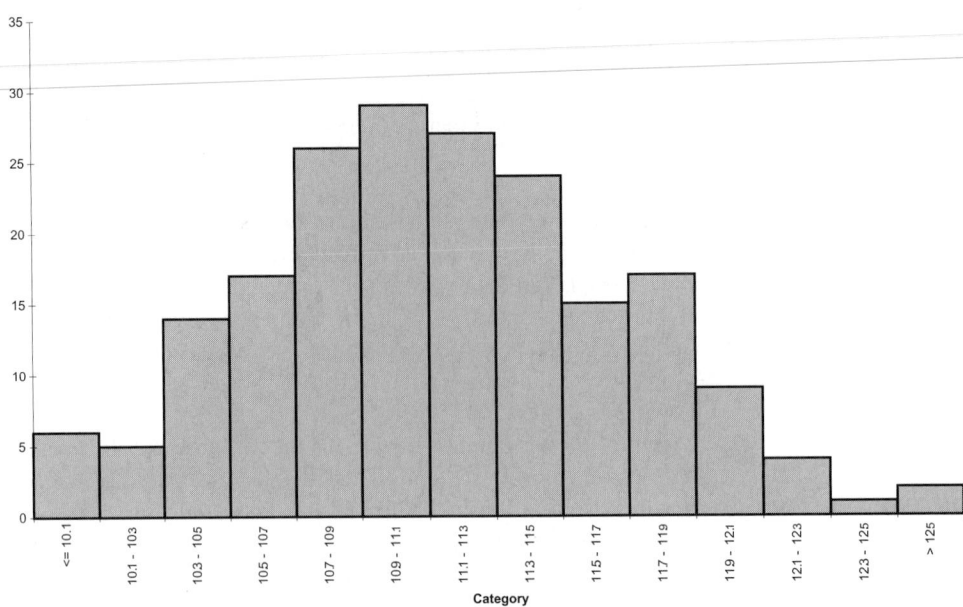

FIGURE 11.38 Symmetric Distribution of Log(Income)

Example 11.3 Possible Gender Discrimination in Salary at Fifth National Bank of Springfield (continued)

Returning to the bank discrimination example, a glance at the distribution of salaries of the 208 employees shows some skewness to the right—a few employees make substantially more than the majority of employees. Therefore, it might make sense to use the natural logarithm of Salary instead of Salary as the response variable. If we do this, how do we interpret the results?

Objective To reanalyze the bank salary data, now using the logarithm of salary as the response variable.

Solution

All of the analyses we did previously with this data set could be repeated except with Log(Salary) as the response variable. For the sake of discussion, we look only at the regression equation with Female and YrsExper as explanatory variables. After we create the Log(Salary) variable and run the regression, we obtain the output in Figure 11.39. The estimated regression equation is

$$\text{Predicted Log(Salary)} = 3.5829 + 0.0188\text{YrsExper} - 0.1616\text{Female} \quad \textbf{(11.20)}$$

The R^2 and s_e values are 42.4% and 0.1794. For comparison, when this same equation was estimated with Salary as the response variable, R^2 and s_e were 49.1% and 8.070.

We first interpret R^2 and s_e. Neither is directly comparable to the R^2 or s_e value with Salary as the response variable. Recall that R^2 in general is the percentage of the response variable explained by the regression equation. The problem here is that the two R^2 val-

	A	B	C	D	E	F	G	H
1	*Results of multiple regression for Log(Salary)*							
2								
3	*Summary measures*							
4		Multiple R	0.6514					
5		R-Square	0.4243					
6		Adj R-Square	0.4187					
7		StErr of Est	0.1794					
8								
9	*ANOVA Table*							
10		Source	df	SS	MS	F	p-value	
11		Explained	2	4.8613	2.4307	75.5556	0.0000	
12		Unexplained	205	6.5950	0.0322			
13								
14	*Regression coefficients*							
15			Coefficient	Std Err	t-value	p-value	Lower limit	Upper limit
16		Constant	3.5829	0.0280	128.0326	0.0000	3.5277	3.6381
17		Female	-0.1616	0.0265	-6.0936	0.0000	-0.2139	-0.1093
18		YrsExper	0.0188	0.0018	10.5556	0.0000	0.0153	0.0224

FIGURE 11.39 Regression Output with Log(Salary) as Response Variable

ues are percentages explained of *different* response variables, Log(Salary) and Salary. The fact that one is smaller than the other (42.4% versus 49.1%) does not necessarily mean that it corresponds to a "worse" fit. They simply aren't comparable.

The situation is even worse with s_e. Each s_e is a measure of a typical residual, but the residuals in the Log(Salary) equation are in log dollars, whereas the residuals in the Salary equation are in dollars. These units are completely different. For example, the log of $1000 is only 6.91. Therefore, it is no surprise that s_e for the Log(Salary) is *much* smaller than s_e for the Salary equation. If we want comparable standard error measures for the two equations, we should take antilogs of fitted values from the Log(Salary) equation to convert them back to dollars, subtract these from the original Salary values, and take the standard deviation of these "residuals." (The EXP function in Excel can be used to take antilogs.) You can check that the resulting standard deviation is 7.774.[10] This is somewhat smaller than s_e from the Salary equation, an indication of a slightly *better* fit.

> When logarithm of Y is used in the regression equation, the interpretations of s_e and R^2 are different because the units of the response variable are completely different.

Finally, we interpret equation (11.20) itself. Fortunately, this is fairly easy. When the response variable is Log(Y) and a term on the right-hand side of the equation is of the form bX, then whenever X increases by 1 unit, $\hat{Y}$ changes by a constant *percentage,* and this percentage is approximately equal to b (written as a percentage). For example, if $b = 0.035$, then when X increases by one unit, $\hat{Y}$ increases by approximately 3.5%. Applied to equation (11.20), this means that for each extra year of experience with Fifth National, an employee's salary can be expected to increase by about 1.88%. To interpret the Female coefficient, note that the only possible increase in Female is 1 unit (from 0 for male to 1 for female). When this occurs, the expected percentage *decrease* in salary is approximately 16.16%. In other words, equation (11.20) implies that females can expect to make about 16% less than men for comparable years of experience. ●

We are not necessarily claiming that the bank data are fit better with Log(Salary) as the response variable than with Salary—it appears to be a virtual toss-up. However, the lessons from this example are important in general. They are as follows.

[10]To make the two "standard deviations" comparable, we use the denominator $n - 3$ in each.

1. The R^2 values with Y and $\text{Log}(Y)$ as response variables are not directly comparable. They are percentages explained of *different* variables.

2. The s_e values with Y and $\text{Log}(Y)$ as response variables are usually of totally different magnitudes. To make the s_e from the log equation comparable, we need to go through the procedure described in the example, so that the residuals are in *original* units.

3. To interpret any term of the form bX in the log equation, we first express b as a percentage. For example, $b = 0.035$ becomes 3.5%. Then when X increases by 1 unit, the expected *percentage* change in Y is approximately this percentage b.

Remember these points, especially the third, when using the logarithm of Y as the response variable.

The log transformation of a response variable Y is used frequently. This is partly because it induces nice statistical properties (such as making the distribution of Y more symmetric). But an important advantage of this transformation is its ease of interpretation in terms of percentage changes.

Constant Elasticity Relationships A particular type of nonlinear relationship that has firm grounding in economic theory is called a *constant elasticity* relationship. It is also called a *multiplicative* relationship. It has the form shown in equation (11.21).

Formula for Multiplicative Relationship	$$Y = aX_1^{b_1}X_2^{b_2} \cdots X_k^{b_k}$$	**(11.21)**

One property of this type of relationship is that the effect of a change on any explanatory variable X_i on Y depends on the levels of the other X's in the equation. This is not true for the *additive* relationships

$$Y = a + b_1X_1 + b_2X_2 + \cdots + b_kX_k$$

that we have been discussing. For additive relationships, when any X_i increases by one unit, Y changes by b_i units, regardless of the levels of the other X's. Multiplicative relationships are defined in the box.

Multiplicative Relationship	In a **multiplicative** (or **constant elasticity**) **relationship,** the response variable is expressed as a *product* of explanatory variables raised to powers. When any explanatory variable changes by 1%, the response variable changes by a constant *percentage*.

The term *constant elasticity* comes from economics. Economists define the elasticity of Y with respect to X as the percentage change in Y that accompanies a 1% increase in X. Often this is in reference to a demand–price relationship. Then the *price elasticity* is the percentage decrease in demand when price increases by 1%. Usually, the elasticity depends on the current value of X. For example, the price elasticity when the price is \$35 might be different than when the price is \$50. However, when the relationship is of the form

$$Y = aX^b$$

then the elasticity is *constant,* the same for any value of X. Moreover, it is approximately equal to the exponent b. For example, if $Y = 2X^{-1.5}$, then the constant elasticity is approximately -1.5, so that when X increases by 1%, Y decreases by approximately 1.5%.

The constant elasticity property carries over to the multiple-X relationship in equation (11.21). Then each exponent is the approximate elasticity for its X. For example, if $Y = 2X_1^{-1.5}X_2^{0.7}$, then we can make the following statements:

- When X_1 increases by 1%, Y decreases by approximately 1.5%, regardless of the current values of X_1 and X_2.

- When X_2 increases by 1%, Y increases by approximately 0.7%, regardless of the current values of X_1 and X_2.

We can use linear regression to estimate the nonlinear relationship in equation (11.21) by taking natural logarithms of *all* variables. Here we exploit two properties of logarithms: (1) the log of a product is the sum of the logs, and (2) the log of X^b is b times the log of X. Therefore, taking logs of both sides of equation (11.21) gives

$$\text{Log}(Y) = \text{Log}(a) + b_1\text{Log}(X_1) + \cdots + b_k\text{Log}(X_k)$$

This equation is *linear* in the log variables $\text{Log}(Y)$ and $\text{Log}(X_1)$ through $\text{Log}(X_k)$, so it can be estimated in the usual way with multiple regression. We can then interpret the coefficients of the explanatory variables directly as elasticities. The following example illustrates the method.

Example 11.5 Factors Related to Sales of Domestic Automobiles

The file CARSALES.XLS contains annual data (1970–1999) on domestic auto sales in the United States. The data are listed in Figure 11.40 (page 600). The variables are defined as

- Sales: annual domestic auto sales (in number of units)
- PriceIndex: consumer price index of transportation
- Income: real disposable income
- Interest: prime rate of interest

Estimate and interpret a multiplicative (constant elasticity) relationship between Sales and PriceIndex, Income, and Interest.

Objective To use logarithms of variables in a multiple regression to estimate a multiplicative relationship for automobile sales as a function of price, income, and interest rate.

Solution

We first take natural logs of all four variables. (This can be done in one step with the StatPro/Data Utilities/Transform Variables menu item or we can use Excel's LN function.) We then use multiple regression, with Log(Quantity) as the response variable and Log(PriceIndex), Log(Income), and Log(Interest) as the explanatory variables. The resulting output is shown in Figure 11.41. The corresponding equation for Log(Quantity) is

Predicted Log(Sales) = 14.126 − 0.384Log(PriceIndex) + 0.388Log(Income) − 0.070Log(Interest)

If we like, we can convert this back to original variables, that is, back to multiplicative form, by taking antilogs. The result is

Predicted Sales = $1364048 \text{PriceIndex}^{-0.384} \text{Income}^{0.388} \text{Interest}^{-0.070}$

where the constant 1364048 is the antilog of 14.126 (and be calculated in Excel with the EXP function).

FIGURE 11.40
Data for Automobile Demand Example

	A	B	C	D	E
1	Car demand data				
2					
3	Year	Quantity	PriceIndex	Income	Interest
4	1970	7,115,270	37.5	2630	7.91%
5	1971	8,676,410	39.5	2745.3	5.72%
6	1972	9,321,310	39.9	2874.3	5.25%
7	1973	9,618,510	41.2	3072.3	8.03%
8	1974	7,448,340	45.8	3051.9	10.81%
9	1975	7,049,840	50.1	3108.5	7.86%
10	1976	8,606,860	55.1	3243.5	6.84%
11	1977	9,104,930	59	3360.7	6.83%
12	1978	9,304,250	61.7	3527.5	9.06%
13	1979	8,316,020	70.5	3628.6	12.67%
14	1980	6,578,360	83.1	3658	15.27%
15	1981	6,206,690	93.2	3741.1	18.87%
16	1982	5,756,610	97	3791.7	14.86%
17	1983	6,795,230	99.3	3906.9	10.79%
18	1984	7,951,790	103.7	4207.6	12.04%
19	1985	8,204,690	106.4	4347.8	9.93%
20	1986	8,222,480	102.3	4486.6	8.33%
21	1987	7,080,890	105.4	4582.5	8.21%
22	1988	7,526,334	108.7	4784.1	9.32%
23	1989	7,014,850	114.1	4906.5	10.87%
24	1990	6,842,733	120.5	5041.2	10.01%
25	1991	6,072,255	123.8	5033	8.46%
26	1992	6,216,488	126.5	5189.3	6.25%
27	1993	6,674,458	130.4	5261.3	6.00%
28	1994	7,181,975	134.3	5397.2	7.15%
29	1995	7,023,843	139.1	5539.1	8.83%
30	1996	7,139,884	143	5677.7	8.27%
31	1997	6,907,992	144.3	5854.5	8.44%
32	1998	6,756,804	141.6	6168.6	8.35%
33	1999	6,987,208	144.4	6320	8.00%

	A	B	C	D	E	F	G	H
1	Results of multiple regression for Log(Quantity)							
2								
3	Summary measures							
4		Multiple R	0.6813					
5		R-Square	0.4642					
6		Adj R-Square	0.4023					
7		StErr of Est	0.1053					
8								
9	ANOVA Table							
10		Source	df	SS	MS	F	p-value	
11		Explained	3	0.2496	0.0832	7.5073	0.0009	
12		Unexplained	26	0.2881	0.0111			
13								
14	Regression coefficients							
15			Coefficient	Std Err	t-value	p-value	Lower limit	Upper limit
16		Constant	14.1260	1.9838	7.1206	0.0000	10.0482	18.2037
17		Log(PriceIndex)	-0.3837	0.2091	-1.8351	0.0780	-0.8135	0.0461
18		Log(Income)	0.3881	0.3621	1.0720	0.2936	-0.3561	1.1324
19		Log(Interest)	-0.0698	0.0893	-0.7821	0.4412	-0.2534	0.1137

FIGURE 11.41 Regression Output for Multiplicative Relationship

In either form the equation implies that the elasticities are approximately equal to −0.384, 0.388, and −0.070. When PriceIndex increases by 1%, Sales tends to decrease by about 0.384%; when Income increases by 1%, Sales tends to increase by about 0.388%; and when Interest increases by 1%, Sales tends to decrease by about 0.070%.

Does this multiplicative equation provide a better fit to the automobile data than an additive relationship? Without doing considerably more work, it is difficult to answer this question with any certainty. As we discussed in the previous example, it is *not* sufficient to compare R^2 and s_e values for the two fits. Again, the reason is that one has Log(Sales) as the response variable, whereas the other has Sales, so the R^2 and s_e measures aren't comparable. We will simply state that the multiplicative relationship provides a reasonably good fit (for example, a scatterplot of its fitted values versus residuals shows no unusual patterns), and it makes sense economically.

Before leaving this example, we note that the results for this data set are not quite as clear as they might appear. (This is often the case with real data.) First, the correlation between Sales and Income, or between Log(Sales) and Log(Income), is negative, not positive. However, because of multicollinearity, a topic discussed in the next chapter, the regression coefficient of Log(Income) is positive. Second, most of the behavior appears to be driven by the early years. If you rerun the analysis from 1980 on, you will discover almost no relationship between Sales and the other variables. ●

One final example of a multiplicative relationship is the *learning curve* model. A **learning curve** relates the unit production time (or cost) to the cumulative volume of output since that production process first began. Empirical studies indicate that production times tend to decrease by a relatively constant *percentage* every time cumulative output doubles. To model this phenomenon, let Y be the time required to produce a unit of output, and let X be the *cumulative* amount of output that has been produced. If we assume that the relationship between Y and X is of the form

$$Y = aX^b$$

then it can be shown that whenever X doubles, Y decreases to a *constant* percentage of its previous value. This constant is often called the **learning rate.** For example, if the learning rate is 80%, then each doubling of cumulative production yields a 20% reduction in unit production time. It can be shown that the learning rate satisfies the equation

$$b = \ln(\text{learning rate})/\ln(2) \qquad\qquad \textbf{(11.22)}$$

(where "ln" refers to the natural logarithm). So once we estimate b, we can use equation (11.22) to estimate the learning rate.

The following example illustrates a typical application of the learning curve model.

Example (11.6) # The Learning Curve for Production of a New Product at Presario

The Presario Company produces a variety of small industrial products. It has just finished producing 22 batches of a new product (new to Presario) for a customer. The file LEARNING.XLS contains the times (in hours) to produce each batch. These data are listed in Figure 11.42 (page 602). Clearly, the times have tended to decrease as Presario has gained more experience in making the product. Does the multiplicative learning model apply to these data, and what does it imply about the learning rate?

Objective To use a multiplicative regression equation to estimate the learning rate for production time.

Solution

One way to check whether the multiplicative learning model is reasonable is to create the log variables Log(Time) and Log(Batch) in the usual way and then see whether a

FIGURE 11.42

Data for Learning
Curve Example

	A	B
1	**Learning curve effect**	
2		
3	Batch	Time
4	1	125.00
5	2	110.87
6	3	105.35
7	4	103.34
8	5	98.98
9	6	99.90
10	7	91.49
11	8	93.10
12	9	92.23
13	10	86.19
14	11	82.09
15	12	82.32
16	13	87.67
17	14	81.72
18	15	83.72
19	16	81.53
20	17	80.46
21	18	76.53
22	19	82.06
23	20	82.81
24	21	76.52
25	22	78.45

FIGURE 11.43

Scatterplot of Log
Variables with Lin-
ear Trend Super-
imposed

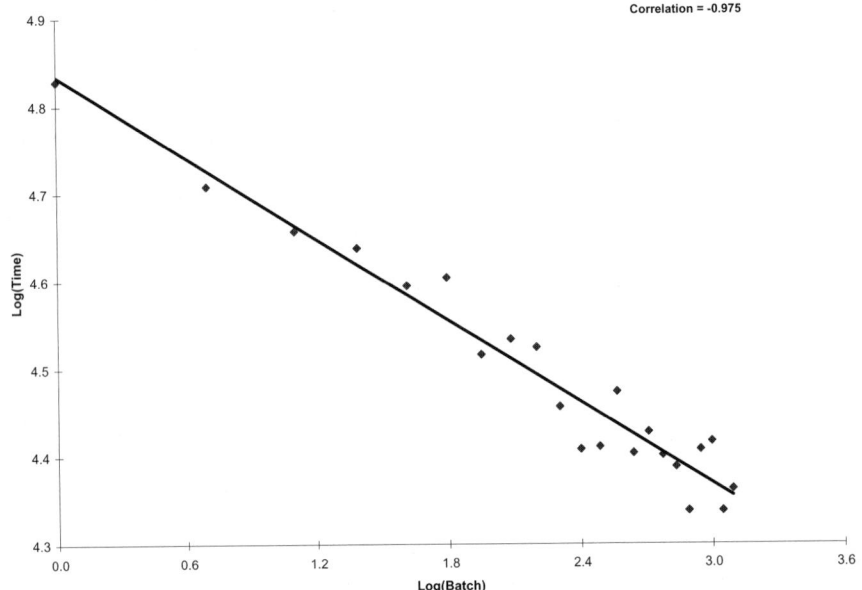

scatterplot of Log(Time) versus Log(Batch) is approximately *linear*. The multiplicative model implies that it should be. Such a scatterplot appears in Figure 11.43, along with a superimposed linear trend line. The fit appears to be quite good.

To estimate the relationship, we regress Log(Time) on Log(Batch). The resulting equation is

$$\text{Predicted Log(Time)} = 4.834 - 0.155\text{Log(Batch)} \qquad \textbf{(11.23)}$$

There are a couple of ways to interpret this equation. First, because it is based on a multiplicative relationship, we can interpret the coefficient -0.155 as an elasticity. That is, when Batch increases by 1%, Time tends to decrease by approximately 0.155%.

FIGURE 11.44 Using the Learning Curve Model for Predictions

Batch	Time	Log(Batch)	Log(Time)	Fitted Values	Residuals
1	125.00	0.00000	4.82831	4.834	-0.006
2	110.87	0.69315	4.70836	4.727	-0.018
3	105.35	1.09861	4.65729	4.664	-0.006
4	103.34	1.38629	4.63802	4.619	0.019
5	98.98	1.60944	4.59492	4.585	0.010
6	99.90	1.79176	4.60417	4.556	0.048
7	91.49	1.94591	4.51623	4.532	-0.016
8	93.10	2.07944	4.53367	4.512	0.022
9	92.23	2.19722	4.52429	4.493	0.031
10	86.19	2.30259	4.45655	4.477	-0.021
11	82.09	2.39790	4.40782	4.462	-0.055
12	82.32	2.48491	4.41061	4.449	-0.038
13	87.67	2.56495	4.47358	4.436	0.037
14	81.72	2.63906	4.40330	4.425	-0.022
15	83.72	2.70805	4.42748	4.414	0.013
16	81.53	2.77259	4.40097	4.404	-0.003
17	80.46	2.83321	4.38776	4.395	-0.007
18	76.53	2.89037	4.33768	4.386	-0.048
19	82.06	2.94444	4.40745	4.378	0.030
20	82.81	2.99573	4.41655	4.370	0.047
21	76.52	3.04452	4.33755	4.362	-0.025
22	78.45	3.09104	4.36246	4.355	0.008
23	77.32	3.13549		4.348
24	76.82	3.17805		4.341
25	76.33	3.21888		4.335
26	75.87	3.25810		4.329
27	75.43	3.29584		4.323
28	75.00	3.33220		4.318
29	74.60	3.36730		4.312
30	74.20	3.40120		4.307	Predictions
31	73.83	3.43399		4.302
32	73.47	3.46574		4.297
33	73.12	3.49651		4.292
34	72.78	3.52636		4.287
35	72.45	3.55535		4.283
36	72.14	3.58352		4.279
37	71.83	3.61092		4.274
	1115.18			Predicted total time for next 15 batches

Results of multiple regression for Log(Time)

Summary measures

Multiple R	0.9748
R-Square	0.9502
Adj R-Square	0.9477
StErr of Est	0.0299

ANOVA Table

Source	df	SS	MS	F	p-value
Explained	1	0.3410	0.3410	381.3880	0.0000
Unexplained	20	0.0179	0.0009		

Regression coefficients

	Coefficient	Std Err	t-value	p-value	Lower limit	Upper limit
Constant	4.8340	0.0186	259.7303	0.0000	4.7952	4.8728
Log(Batch)	-0.1550	0.0079	-19.5292	0.0000	-0.1715	-0.1384

Although this interpretation is correct, it is not as useful as the "doubling" interpretation we discussed previously. We know from equation (11.22) that the estimated learning rate satisfies

$$-0.155 = \ln(\text{learning rate})/\ln(2)$$

Solving for the learning rate (multiply through by $\ln(2)$ and then take antilogs), we find that it is 0.898, or approximately 90%. In words, whenever cumulative production doubles, the time to produce a batch decreases by about 10%.

Presario could use this regression equation to predict future production times. For example, suppose the customer places an order for 15 more batches of the same product. Note that Presario is already partway up the learning curve, that is, these batches are numbers 23–37, and the company already has experience producing the product. We can use equation (11.23) to predict the log of production time for each batch, then take their antilogs and sum them to obtain the total production time. The calculations are shown in rows 26–40 of Figure 11.44. We enter the batch numbers and calculate their logs in columns A and C. Then we substitute the values of Log(Batch) in column C into equation (11.23) to obtain the predicted values of Log(Time) in column F. Finally, we use Excel's EXP function to calculate the antilogs of these predictions in column B, and we calculate their sum in cell B41. The total predicted time to finish the order is about 1115 hours. ●

PROBLEMS

Level A

26. In a study of housing demand, a county assessor is interested in developing a regression model to estimate the selling price of residential properties within her jurisdiction. She randomly selects 15 houses and records the selling price in addition to the following values: the size of the house (in hundreds of square feet), the total number of rooms in the house, the age of the house, and an indication of whether the house has an attached garage. These data are stored in the file P11_26.XLS.

a. Estimate and thoroughly interpret a multiple regression model that includes the four potential explanatory variables.

b. Evaluate the estimated regression model's goodness of fit.

c. Use the estimated model to predict the sales price of a 3000-square-foot, 20-year-old home that has 7 rooms but no attached garage.

27. A manager of boiler drums wants to use regression analysis to predict the number of worker-hours needed to erect the drums in future projects. Consequently, data for 36 randomly selected boilers were collected. In addition to worker-hours (Y), the variables measured include boiler capacity, boiler design pressure, boiler type, and drum type. All of these measurements can be found in the file P11_27.XLS.

a. Formulate an appropriate multiple regression model to predict the number of worker-hours needed to erect boiler drums.

b. Estimate the formulated model using the given sample data, and interpret the estimated regression coefficients.

c. According to the estimated regression model, what is the difference between the mean number of worker-hours required for erecting industrial and utility field boilers?

d. According to the estimated regression model, what is the difference between the mean number of worker-hours required for erecting boilers with steam drums and those with mud drums?

e. Given the estimated regression model, predict the number of worker-hours needed to erect a utility-field, steam-drum boiler with a capacity of 550,000 pounds per hour and a design pressure of 1400 pounds per square inch.

f. Given the estimated regression model, predict the number of worker-hours needed to erect an industrial-field, mud-drum boiler with a capacity of 100,000 pounds per hour and a design pressure of 1000 pounds per square inch.

28. Suppose that a regional express delivery service company wants to estimate the cost of shipping a package (Y) as a function of cargo type, where cargo type includes the following possibilities: fragile, semifragile, and durable. Costs for 15 randomly chosen packages of approximately the same weight and same distance shipped, but of different cargo types, are provided in the file P11_28.XLS.

a. Formulate an appropriate multiple regression model to predict the cost of shipping a given package.

b. Estimate the formulated model using the given

sample data, and interpret the estimated regression coefficients.

c. According to the estimated regression model, which cargo type is the *most* costly to ship? Which cargo type is the *least* costly to ship?

d. How well does the estimated model fit the given sample data? How can the model's goodness of fit be improved?

e. Given the estimated regression model, predict the cost of shipping a package with semifragile cargo.

29. The file P11_11.XLS contains annual observations of the American minimum wage. Has the minimum wage been growing at roughly a *constant* rate over this period?

a. Generate a scatterplot diagram for these data. Comment on the observed behavior of the minimum wage over time.

b. Formulate and estimate an appropriate regression model to explain the variation of the American minimum age over the given time period. Interpret the estimated regression coefficients.

c. Analyze the estimated model's residuals. Is your estimated regression model adequate? If not, return to part **b** and revise your model. Continue to revise the model until your results are satisfactory.

30. Formulate a regression model that adequately estimates the relationship between monthly electrical power usage (Y) and home size (X) using the data in the file P11_13.XLS. Interpret your results. How well does your model explain the variation in monthly electrical power usage?

31. An insurance company wants to determine how its annual operating costs depend on the number of home insurance (X_1) and automobile insurance (X_2) policies that have been written. The file P11_31.XLS contains relevant information for 10 branches of the insurance company. The company believes that a multiplicative model might be appropriate because operating costs typically increase by a constant percentage as the number of either type of policy increases by a given percentage. Use the given data to estimate a multiplicative model for this insurance company. Interpret your results. Does a multiplicative model provide a good fit with these data?

32. Suppose that an operations manager is trying to determine the number of labor hours required to produce the ith unit of a certain product. Consider the data provided in the file P11_32.XLS. For example, the second unit produced required 517 labor hours, and the 600th unit required 34 labor hours.

a. Use the given data to estimate a relationship between the total number of units produced and the labor hours required to produce the last unit in the total set. Interpret your findings.

b. Use your estimated relationship to predict the number of labor hours that will be needed to produce the 800th unit.

Level B

33. The human resources manager of DataCom, Inc. wants to predict the annual salaries of given employees using the following explanatory variables: the number of years of prior relevant work experience, the number of years of employment at DataCom, the number of years of education beyond high school, the employee's gender, the employee's department, and the number of individuals supervised by the given employee. These data have been collected for a sample of employees and are given in the file P11_05.XLS.

    a. Formulate an appropriate multiple regression model to predict the annual salary of a given Data-Com employee.

    b. Estimate the formulated model using the given sample data, and interpret the estimated regression coefficients.

    c. According to the estimated regression model, is there a difference between the mean salaries earned by male and female employees at Data-Com? If so, how large is the difference?

    d. According to the estimated regression model, is there a difference between the mean salaries earned by employees in the sales department and those in the advertising department at DataCom? If so, how large is the difference?

    e. According to the estimated regression model, in which department are DataCom employees paid the *highest* mean salary? In which department are DataCom employees paid the *lowest* mean salary?

    f. Given the estimated regression model, predict the annual salary of a female employee who served in a similar department at another company for 10 years prior to coming to work at DataCom. This woman, a graduate of a 4-year collegiate business program, has been supervising 12 subordinates in the purchasing department since joining the organization 5 years ago.

34. Does the rate of violent crime acts vary across different regions of the United States?

    a. Using the data in the file P11_34.XLS, develop and estimate an appropriate regression model to explain the variation in acts of violent crime across the four established regions of the United States. Thoroughly interpret the estimated model. Rank the four regions from highest to lowest according to their mean violent crime rate.

    b. How would you modify the regression model in part **a** to account for possible differences in the violent crime rate across the various subdivisions of the given regions? Estimate your revised model

and interpret your findings. Rank the nine subdivisions from highest to lowest according to their mean violent crime rate.

35. Suppose that you are interested in predicting the price of a laptop computer based on its various features. The file P11_35.XLS contains observations on the sales price and a number of potentially relevant variables for a randomly chosen sample of laptop computers.

    a. Formulate a multiple regression model that includes all potential explanatory variables and estimate it with the given sample data.

    b. Interpret the estimated regression equation. Be sure to indicate the impact of each attribute on the computer's sales price. For example, what impact does the monitor type have on the average sales price of a laptop computer?

    c. How well does the estimated regression model fit the data given in the file?

    d. Use the estimated regression equation to predict the price of a laptop computer with the following features: a 60-megahertz processor, a battery that holds its charge for 240 minutes, 32-megabytes of RAM, a DX chip, a color monitor, a mouse pointing device, and a 24-hour, toll-free customer service hotline.

36. Continuing Problem 18, suppose that the antique collector believes that the *rate of increase* of the auction price with the age of the item will be driven upward by a large number of bidders. How would you revise the multiple regression model developed previously to model this feature of the problem?

    a. Estimate your revised model using the data in the file P11_18.XLS.

    b. Interpret each of the estimated coefficients in your revised model.

    c. Does this revised model fit the given data better than the original multiple regression model? Explain why or why not.

37. Continuing Problem 19, revise the multiple regression model developed previously to include an interaction term between the return on average equity (X_1) and annual dividend rate (X_2).

    a. Estimate your revised model using the data provided in the file P11_19.XLS.

    b. Interpret each of the estimated coefficients in your revised model. In particular, how do you interpret the coefficient for the interaction term in the revised model?

    c. Does this revised model fit the given data better than does the original multiple regression model? Explain why or why not.

38. Continuing Problem 24, suppose that one of the managers of this regional express delivery service company is trying to decide whether to add an interaction term involving the package weight (X_1) and the distance shipped (X_2) in the multiple regression model developed previously.

a. Why would the manager want to add such a term to the regression equation?
b. Estimate the revised model using the data given in the file P11_24.XLS.

c. Interpret each of the estimated coefficients in your revised model. In particular, how do you interpret the coefficient for the interaction term in the revised model?
d. Does this revised model fit the given data better than the original multiple regression model? Explain why or why not.

11.7 Validation of the Fit

The fit from a regression analysis is often overly optimistic. When we use the least squares procedure on a given set of data, we exploit all of the idiosyncrasies of the particular data to obtain the best possible fit. There is no guarantee that the fit will be as good when the estimated regression equation is applied to *new* data. In fact, it usually isn't. This is particularly important when our goal is to use the regression equation to predict new values of the response variable. The usual situation is that we use a given data set to estimate a regression equation. Then we gather new data on the *explanatory* variables and use these, along with the already-estimated regression equation, to predict the new (but unknown) values of the response variable.

One way to see whether this procedure will be successful is to split the original data set into two subsets: one subset for estimation and one subset for validation. We estimate the regression equation from the first subset. Then we substitute the values of explanatory variables from the second subset into this equation to obtain predicted values for the response variable. Finally, we compare these predicted values with the known values of the response variable in the second subset. If the agreement is good, there is reason to believe that the regression equation will predict well for new data.

This validation procedure is fairly simple to perform in Excel. We illustrate it for the Bendrix manufacturing data in Example 11.2. (See the file BENDRIX2.XLS.) There we used 36 monthly observations to regress Overhead on MachHrs and ProdRuns. For convenience, we repeat the regression output in Figure 11.45. In particular, it shows an R^2 value of 86.6% and an s_e value of $4109.

Now suppose that this data set is from one of Bendrix's two plants. The company would like to predict overhead costs for the other plant by using data on machine hours and production runs at the other plant. The first step is to see how well the regression

	F	G	H	I	J	K	L	M
3	**Results of multiple regression for Overhead**							
4								
5	**Summary measures**							
6		Multiple R	0.9308					
7		R-Square	0.8664					
8		Adj R-Square	0.8583					
9		StErr of Est	4108.9932					
10								
11	**ANOVA Table**							
12		Source	df	SS	MS	F	p-value	
13		Explained	2	3614020652.0000	1807010326.0000	107.0261	0.0000	
14		Unexplained	33	557166208.0000	16883824.4848			
15								
16	**Regression coefficients**							
17			Coefficient	Std Err	t-value	p-value	Lower limit	Upper limit
18		Constant	3996.6782	6603.6509	0.6052	0.5492	-9438.5612	17431.9176
19		MachHrs	43.5364	3.5895	12.1289	0.0000	36.2335	50.8393
20		ProdRuns	883.6179	82.2514	10.7429	0.0000	716.2760	1050.9598

FIGURE 11.45 Multiple Regression Output for Bendrix Example

FIGURE 11.46

Validation of Bendrix Regression Results

	A	B	C	D	E	F
1	Validation data					
2						
3	Coefficients from regression equation (based on original data)					
4		Constant	MachHrs	ProdRuns		
5		3996.6782	43.5364	883.6179		
6						
7	Comparison of summary measures					
8		Original	Validation			
9	R-square	0.8664	0.7733			
10	StErr of Est	4108.99	5251.53			
11						
12	Month	MachHrs	ProdRuns	Overhead	Fitted	Residual
13	1	1374	24	92414	85023	7391
14	2	1510	35	92433	100663	-8230
15	3	1213	21	81907	75362	6545
16	4	1629	27	93451	98775	-5324
17	5	1858	28	112203	109629	2574
18	6	1763	40	112673	116096	-3423
19	7	1449	44	104091	105960	-1869
20	8	1422	46	104354	106552	-2198
45	33	1534	38	104946	104359	587
46	34	1529	29	94325	96189	-1864
47	35	1389	47	98474	105999	-7525
48	36	1350	34	90857	92814	-1957

from Figure 11.45 fits data from the other plant. We perform this validation on the 36 months of data shown in Figure 11.46. The validation results also appear in this figure.

To obtain the results in this figure, we proceed as follows.

Procedure for Validating Regression Results

1. **Copy old results.** Copy the results from the original regression to the ranges B5:D5 and B9:B10.

2. **Calculate fitted values and residuals.** The fitted values are now the predicted values of overhead for the other plant, based on the original regression equation. We find these by substituting the new values of MachHrs and ProdRuns into the original equation. To do so, enter the formula

$$=\$B\$5+\text{SUMPRODUCT}(\$C\$5:\$D\$5,B13:C13)$$

in cell E13 and copy it down. Then calculate the residuals (prediction errors for the other plant) by entering the formula

$$=D13-E13$$

in cell F13 and copying it down.

3. **Calculate summary measures.** We see how well the original equation fits the new data by calculating R^2 and s_e values. Recall that R^2 in general is the square of the correlation between observed and fitted values. Therefore, enter the formula

$$=\text{CORREL}(E13:E48,D13:D48)^2$$

in cell C9. The s_e value is essentially the standard deviation of the residuals, but it uses the denominator $n - 3$ (when there are two explanatory variables) rather than $n - 1$. Therefore, enter the formula

$$=\text{SQRT}(35/33)*\text{STDEV}(F13:F48)$$

in cell C10.

The results in Figure 11.46 are typical. The validation results are usually not as good as the original results. The value of R^2 has decreased from 86.6% to 77.3%, and the value of s_e has increased from \$4109 to \$5252. Nevertheless, Bendrix might conclude that the original regression equation is adequate for making future predictions at either plant.

11.8 Conclusion

The material in this chapter has illustrated how to fit an equation to a set of points and how to interpret the resulting equation. We have also discussed two measures, R^2 and s_e, that indicate the goodness of fit of the regression equation. Although the general technique is called *linear* regression, we have seen how it can be used to estimate nonlinear relationships through suitable transformations of variables. We are not finished with our study of regression, however. In the next chapter we will make some statistical assumptions about the regression model and then discuss the types of inferences that can be made from regression output. In particular, we will discuss the accuracy of the estimated regression coefficients, the accuracy of predictions made from the regression equation, and the general topic of which explanatory variables "belong" in the regression equation.

Summary of Key Terms

Term	Symbol	Explanation	Excel	Page	Equation Number
Regression analysis		A general method for estimating the relationship between a response variable and one or more explanatory variables.		549	
Response (or dependent variable	Y	The variable being estimated or predicted in a regression analysis.		550	
Explanatory (or independent) variables	$X_1, X_2,$ and so on	The variables used to explain or predict the response variable.		550	
Simple regression		A regression model with a single explanatory variable.	StatPro/Regression Analysis/Simple	551, 562	
Multiple regression		A regression model with any number of explanatory variables.	StatPro/Regression Analysis/Multiple	551, 573	
Correlation	r_{XY}	A measure of strength of the linear relationship between two variables X and Y.	=CORREL(*range1*, *range2*), or StatPro/Summary Stats/Correlations, Covariances	561	11.1
Fitted value		The predicted value of response variable found by substituting explanatory values into the regression equation.		564	11.2
Residual		The difference between actual and fitted values of response variable.		564	11.2

Summary of Key Terms (continued)

Term	Symbol	Explanation	Excel	Page	Equation Number
Least squares line		The regression equation that minimizes the sum of squared residuals.	StatPro/Regression Analysis	564	11.3, 11.4
Standard error of estimate	s_e	Essentially, the standard deviation of the residuals; indicates the magnitude of the prediction errors.	StatPro/Regression Analysis	569, 576	11.7, 11.11
Coefficient of determination	R^2	The percentage of variation in the response variable explained by the regression model.	StatPro/Regression Analysis	570	11.8
Adjusted R^2		A measure similar to R^2, but adjusted for the number of explanatory variables in the equation.		578	
Regression coefficients	$b_1, b_2,$ and so on	The coefficients of the explanatory variables in a regression equation.	StatPro/Regression Analysis	574	11.9
Dummy variables		Variables coded as 0 or 1, used to capture categorical variables in a regression analysis.	StatPro/Data Utilities/Create Dummy Variable(s)	580	
Interaction variables		Products of explanatory variables, used when the effect of one on the response variable depends on the value of the other.	StatPro/Data Utilities/Create Interaction Variable(s)	586	
Nonlinear transformations		Variables created to capture nonlinear relationships in a regression model.	StatPro/Data Utilities/ Transform Variable(s)	590	
Quadratic model		A regression model with linear and squared explanatory variables.	StatPro/Regression Analysis	592	
Model with logarithmic transformations		A regression model using logarithms of Y and/or X's	StatPro/Regression Analysis	594–603	
Constant elasticity (or multiplicative relationship		A relationship where Y changes by a constant percentage when any X changes by 1%; requires logarithmic transformations.	StatPro/Regression Analysis	598	11.21
Learning curve		A particular multiplicative relationship used to indicate how cost or time in production decreases through time.	StatPro/Regression Analysis	601	11.22
Validation of fit		Checks how well a regression model based on one sample predicts a related sample.	StatPro/Regression Analysis	606	

PROBLEMS

Conceptual Exercises

C.1. Consider the relationship between yearly wine consumption (liters of alcohol from drinking wine, per person) and yearly deaths from heart disease (deaths per 100,000 people) in 19 developed countries. Suppose that you read a newspaper article in which the reporter states the following:

Researchers find that the correlation between yearly wine consumption and yearly deaths from heart disease is -0.84. Thus, it is reasonable to conclude that increased consumption of alcohol from wine causes fewer deaths from heart disease in industrialized societies.

Comment on the reporter's interpretation of the correlation measure in this case.

C.2. "It is generally appropriate to delete all outliers in the given data set when producing a scatterplot." Is this statement true or false? Explain your choice.

C.3. How does one interpret the relationship between two numerical variables when the estimated least squares regression line for them is essentially *horizontal* (i.e., flat)?

C.4. Suppose that you generate a scatterplot of residuals versus fitted values of the response variable for a given estimated regression model. Furthermore, you find the correlation between the residuals and fitted values to be 0.829. Does this provide a good indication that the estimated regression model is satisfactory? Explain why or why not.

C.5. Suppose that you have generated three alternative multiple regression models to explain the variation in a particular response variable. The regression output for each model can be summarized as follows:

	Model 1	Model 2	Model 3
No. indep. vars.	4	6	9
R^2	0.76	0.77	0.79
Adj. R^2	0.75	0.74	0.73

Which of these models would you select as "best"? Explain your choice.

Level A

39. Many companies manufacture products that are at least partially produced using chemicals (for example, paint, gasoline, and steel). In many cases, the quality of the finished product is a function of the temperature and pressure at which the chemical reactions take place. Suppose that a particular manufacturer wants to model the quality (Y) of a product as a function of the temperature (X_1) and the pressure (X_2) at which it is produced. The file P11_39.XLS contains data obtained from a carefully designed experiment involving these variables. Note that the assigned quality score can range from a maximum of 100 to a minimum of 0 for each manufactured product.

a. Formulate a multiple regression model that includes the two given explanatory variables. Estimate the model using the given sample data. Does the estimated model fit the data well?

b. Interpret each of the estimated coefficients in the multiple regression model.

c. Consider adding a term to model a possible interaction between the two explanatory models. Reformulate the model and estimate it again using the given data. Does the inclusion of the interaction term improve the model's goodness of fit?

d. Interpret each of the estimated coefficients in the revised model. In particular, how do you interpret the coefficient for the interaction term in the revised model?

40. Suppose that a power company located in southern Alabama wants to predict the peak power load (i.e., the maximum amount of power that must be generated each day to meet demand) as a function of the daily high temperature (X). A random sample of 25 summer days is chosen, and the peak power load and the high temperature are recorded on each day. The file P11_40.XLS contains these observations.

a. Generate a scatterplot diagram for these data. Comment on the observed relationship between the response variable and explanatory variable.

b. Formulate and estimate an appropriate regression model to predict the peak power load for this power company. Interpret the estimated regression coefficients.

c. Analyze the estimated model's residuals. Is your estimated regression model adequate? If not, return to part **b** and revise your model. Continue to revise the model until your results are satisfactory.

d. Use the final version of your model to predict the peak power load on a summer day with a high temperature of 100 degrees.

41. Management of a home appliance store in Charlotte would like to understand the growth pattern of the monthly sales of VCR units over the past 2 years. Managers have recorded the relevant data in an Excel spreadsheet, which can be found in the file P11_09.XLS. Have the sales of VCR units been growing *linearly* over the past 24 months?

a. Generate a scatterplot diagram for these data. Comment on the observed behavior of monthly VCR sales at this store over time.

b. Formulate and estimate an appropriate regression model to explain the variation of monthly VCR sales over the given time period. Interpret the estimated regression coefficients.

c. Analyze the estimated model's residuals. Is your estimated regression model adequate? If not, return to part **b** and revise your model. Continue to revise the model until your results are satisfactory.

42. Chipco, a small computer chip manufacturer, wants to be able to forecast monthly operating costs as a function of the number of units produced during a month. They have collected the 16 months of data in the file P11_42.XLS.

a. Determine an equation that can be used to predict monthly production costs from units produced. Are there any outliers?

b. How could the regression line obtained in part **a** be used to determine whether the company was efficient or inefficient during any particular month?

43. The file P11_43.XLS contains data on the following variables for several underdeveloped countries:
- Infant mortality rate
- Per capita GNP
- Percentage of people completing primary school
- Percentage of adults who can read (adult literacy)

Use these data to determine which of the given variables (by itself) best predicts infant mortality. Can you give an explanation for your answer?

44. The file P11_44.XLS contains data that relate the unit cost of producing a fuel pressure regulator to the cumulative number of fuel pressure regulators produced at the Ford plant in Bedford. For example, the 4000th unit cost $13.70 to produce.

a. Fit a learning curve to these data.

b. We would predict that doubling cumulative production reduces the cost of producing a regulator by what amount?

45. The "beta" of a stock is found by running a regression with the explanatory variable being the monthly return on a market index and the response variable being the monthly return on the stock. The beta of the stock is then the slope of this regression.

a. Explain why most stocks have a positive beta.

b. Explain why a stock with a beta with absolute value greater than 1 is more volatile than the market and a stock with a beta less than 1 (in absolute value) is less volatile than the market.

c. Use the data in the file P11_45.XLS to estimate the beta for Ford Motor Company.

d. What percentage of the variation in Ford's return is explained by market variation? What percentage is unexplained by market variation?

e. Verify (using Excel's COVAR and VARP functions) that the beta for Ford is given by

$$\frac{\text{Covariance between Ford and Market}}{\text{Variance of Market}}$$

Also, verify that the correlation between Ford return and Market return is the square root of R^2.

46. The file P11_46.XLS contains monthly returns on Anheiser-Busch (AB) and a market index. Use these data to answer the following questions:

a. What percentage of the variation in the return in AB is explained by variation in the market? What percentage is unexplained by variation in the market?

b. Predict the change in AB during a month in which the market goes up by 2%.

c. Use Excel's CORREL functions to determine the correlation between the return on AB and the market. Verify that this correlation between AB and the market is equal to the square root of R^2 from the regression output.

d. Estimate the beta (refer to Problem 45) for AB by using regression. Then verify that it can also be found (using Excel's COVAR and VARP functions) from

$$\frac{\text{Covariance between AB and Market}}{\text{Variance of Market}}$$

47. Investors are interested in knowing whether estimates of stock betas based on past history are good predictors of future betas (refer to Problem 45). How could you use a data set that gives monthly returns on several stocks over a 5-year period to see whether this is true?

48. The file P11_48.XLS contains monthly sales (in thousands) and price of a popular candy bar.

a. Describe the type of relationship between price and sales (linear, nonlinear, strong, weak).

b. What percentage of variation in monthly sales is explained by variation in price? What percentage is unexplained?

c. If the price of the candy bar is 55 cents, predict monthly candy bar sales.

d. Use the regression output to determine the correlation between price and candy bar sales.

e. Are there any outliers?

49. The file P11_49.XLS contains the amount of money spent advertising a product (in thousands of dollars) and the number of units sold (in millions) for 8 months.

a. Assume that the only factor influencing monthly sales is advertising. Fit the following three curves to these data: linear ($Y = a + bX$), exponential ($Y = ab^X$), and power ($Y = aX^b$). Which equation best fits the data?

b. Interpret the best-fitting equation.

c. Using the best-fitting equation, predict sales during a month in which $60,000 is spent on advertising.

50. Callaway Golf is trying to determine how the price of a set of clubs affects the demand for clubs. The file P11_50.XLS contains the price of a set of clubs (in dollars) and the monthly sales (in millions of sets sold).

a. Assume the only factor influencing monthly sales is price. Fit the following three curves to these data: linear ($Y = a + bX$), exponential ($Y = ab^X$), and power ($Y = aX^b$). Which equation best fits the data?

b. Interpret your best-fitting equation.

c. Using the best-fitting equation, predict sales during a month in which the price is $470.

51. The number of cars per 1000 people is known for virtually every country in the world. For many countries, however, per capita income is not known. Can you think of a way to estimate per capita income for countries where it is unknown?

52. The file P11_52.XLS contains the cost (in 1990 dollars) of making a 3-minute phone call from London to New York. Use regression to estimate how (or whether) the cost of a London to New York call has declined over time. Based on these data, predict the cost of a 3-minute phone call in the year 2000. (Source: *The Economist,* September 28, 1996)

53. The file P11_53.XLS contains the databit power per chip for computers. Use regression to estimate how databit power per chip has changed over time. (This result is called Moore's law.) Also predict the databit power per chip in the year 2000. Do you think Moore's law can continue indefinitely? (Source: *One World Ready or Not,* by William Greider, 1996)

54. The file P11_54.XLS contains the cost of building (in hundreds of millions of dollars) a plant to produce RAM chips for PCs. Use regression to estimate how (or whether) the cost of building a RAM plant has increased over time. Predict the cost of building a RAM plant in the year 2000. (Source: *One World Ready or Not,* by William Greider, 1996)

55. The file P11_55.XLS contains the unit cost of producing a unit of computer memory, as a function of the number of units of computer memory that have been produced to date. Use regression to analyze how (or whether) the price of a bit of memory has changed as more memory has been produced (Source: *Every Investor's Guide to High-Tech Stocks,* by Michael Murphy, 1998)

56. The Baker Company wants to develop a budget to predict how overhead costs vary with activity levels. Management is trying to decide whether direct labor hours (DLH) or units produced is the better measure of activity for the firm. Monthly data for the preceding 24 months appear in the file P11_56.XLS. Use regression analysis to determine which measure, DLH or Units (or both), should be used for the budget. How would the regression equation be used to obtain the budget for the firm's overhead costs?

57. The auditor of Kiely Manufacturing is concerned about the number and magnitude of year-end adjustments that are made annually when the financial statements of Kiely Manufacturing are prepared. Specifically, the auditor suspects that the management of Kiely Manufacturing is using discretionary write-offs to manipulate the reported net income. To check this, the auditor has collected data from 25 firms that are similar to Kiely Manufacturing in terms of manufacturing facilities and product lines. The cumulative reported third quarter income and the final net income reported are listed in the file P11_57.XLS for each of these 25 firms. If Kiely Manufacturing reported a cumulative third quarter income of $2,500,000 and a preliminary net income of $4,900,000, should the auditor conclude that the relationship between cumulative third quarter income and the annual income for Kiely Manufacturing differs from that of the 25 firms in this sample? Explain why or why not.

Level B

58. An economic development researcher wants to understand the relationship between the size of the monthly home mortgage or rent payment for households in a particular middle-class neighborhood and the following set of household variables: family size, approximate location of the household within the neighborhood, an indication of whether those surveyed own or rent their home, gross annual income of the first household wage earner, gross annual income of the second household wage earner (if applicable), average monthly expenditure on utilities, and the total indebtedness (excluding the value of a home mortgage) of the household. Observations on each of these variables for a large sample of households are recorded in the file P02_06.XLS.

a. Beginning with *family size,* iteratively add one explanatory variable and estimate the resulting regression equation to explain the variation in the monthly home mortgage or rent payment. If adding any explanatory variable causes the *adjusted R^2* measure to fall, do not include that variable in subsequent versions of the regression model. Otherwise, include the variable and consider adding the next variable in the set. Which variables are included in the final version of your regression model?

b. Interpret the final estimated regression equation you obtained through the process outlined in part **a.** Also, report and interpret the standard error of estimate s_e, the coefficient of determination R^2, and the adjusted R^2 for the final estimated model.

59. (Based on an actual court case in Philadelphia) In the 1994 congressional election, the Republican candidate outpolled the Democratic candidate by 400 votes (excluding absentee ballots). The Democratic candidate outpolled the Republican candidate by 500 absentee votes. The Republican candidate sued (and won), claiming that vote fraud must have played a role in the absentee ballot count. The Republican's lawyer ran a regression to predict (based on past elections) how the absentee ballot margin could be predicted from the votes tabulated on voting machines. Selected results are given in the file P11_59.XLS. Show how this regression could be used by the Republican to prove his claim of vote fraud. (*Hint:* Is the 1994 result an outlier?)

60. The file P11_60.XLS contains data on the price of new and used Taurus sedans. All prices for used cars are from 1995. For example, a new Taurus bought in 1985 cost $11,790 and the wholesale used price of that car in 1995 was $1700. A new Taurus bought in 1994 cost $18,680 and it could be sold used in 1995 for $12,600.

   a. You want to predict the resale value (as a percentage of the original price of the vehicle) as a function of the vehicle's age. Find an equation to do this. (You should try at least two different equations and choose the one with the best fit.)

   b. Suppose all police cars are Ford Tauruses. If you were the business manager for the New York Police Department, what use would you make of your findings from part **a?**

61. The data for this problem are fictitious, but they are not far off.) For each of the top 25 business schools, the file P11_61.XLS contains the average salary of a professor. Thus, for Indiana University (number 15 in the rankings), the average salary is $46,000. Use this information and regression to show that IU is doing a great job with its available resources.

62. Suppose the correlation between the average height of parents and the height of their firstborn male child is 0.5. You are also told that:
   - The average height of all parents is 66 inches.
   - The standard deviation of the average height of parents is 4 inches.

- The average height of all male children is 70 inches.
- The standard deviation of the height of all male children is 4 inches.

If a mother and father are 73 and 80 inches tall, respectively, how tall do you predict their son to be? Explain why this is called "regression toward the mean."

63. Do increased taxes increase or decrease economic growth? Table 11.1 gives tax revenues as a percentage of Gross Domestic Product (GDP) and the average annual percentage growth in GDP per capita for nine countries during the years 1970–1994. Do these data support or contradict the dictum of supply-side economics? (Source: *The Economist,* August 24, 1996)

64. For each of the four data sets in the file P11_64.XLS, calculate the least squares line. For which of these data sets would you feel comfortable in using the least squares line to predict Y? (Source: Frederic Anscombe, *The American Statistician*)

65. Suppose we run a regression on a data set of X's and Y's and obtain a least squares line of $Y = 12 - 3X$.
   a. If we double each value of X, what is the new least squares line?
   b. If we triple each value of Y, what is the new least squares line?
   c. If we add 6 to each value of X, what is the new least squares line?
   d. If we subtract 4 from each value of Y, what is the new least squares line?

66. The file P11_66.XLS contains monthly cost accounting data on overhead costs, machine hours, and direct material costs. This problem will help you explore the meaning of R^2 and the relationship between R^2 and correlations.
   a. Create a table of correlations between the individual variables.

Table **11.1**

Economic Data from Nine Countries	Country	Tax revenues as % of GDP	Average annual growth in GDP per capita
	Japan	26%	3.1%
	United States	27%	1.6%
	Italy	33%	2.5%
	Canada	34%	2.0%
	Switzerland	30%	1.0%
	Britain	36%	1.9%
	Germany	38%	2.2%
	France	42%	1.9%
	Sweden	49%	1.1%

b. If you ignored the two explanatory variables MachHrs and DirMatCost and predicted each OHCost as the *mean* of all OHCosts, then a typical "error" would be OHCost minus the mean of all OHCosts. Find the sum of squared errors using this form of prediction, where the sum is over all observations.

c. Now run three regressions: (1) OHCost versus MachHrs, (2) OHCost versus DirMatCost, and (3) OHCost versus both MachHrs and DirMatCost. (The first two are simple regressions, the third is a multiple regression.) For each, find the sum of squared residuals, and divide this by the sum of squared errors from part **b.** What is the relationship between this ratio and the associated R^2 for that equation? (Now do you see why R^2 is referred to as the percentage of variation explained?)

d. For the first two regressions in part **c,** what is the relationship between R^2 and the corresponding correlation between the response and explanatory variable? For the third regression it turns out that the R^2 can be expressed as a complicated function of all three correlations in part **a,** that is, not just the correlations between the response variable and each explanatory variable, but also the correlation between the explanatory variables. Note that this R^2 is not just the sum of the R^2 values from the first two regressions in part **c.** Why do you think this is true, intuitively? However, R^2 for the multiple regression is still the square of a correlation—namely, the correlation between the observed and predicted values of OHCost. Verify that this is the case for these data.

67. The file P11_67.XLS contains hypothetical starting salaries (in $1000's) for MBA students directly after graduation. The file also lists their years of experience prior to the MBA program and their class standing in the MBA program (on a 0–100 scale).

a. Estimate the regression equation with Salary as the response variable and Exper and Class as the explanatory variables. What does this equation imply? What does the standard error of estimate s_e tell you? What about R^2?

b. Repeat part **a,** but now include the interaction term Exper*Class (the product) in the equation as well as Exper and Class individually. Answer the same questions as in part **a.** What evidence is there that this extra variable (the interaction variable) is worth including? How do you interpret this regression equation?

68. In a study published in 1985 in *Business Horizons,* Platt and McCarthy employed multiple regression analysis to explain variations in compensations among the CEOs of large companies. Their primary objective was to discover whether levels of compensations are affected more by short-run considerations—"I'll earn more now if my company does well in the short run"—or long-run considerations—"My best method for obtaining high compensation is to stay with my company for a long time." The study used as its response variable the total compensation for each of the 100 highest paid CEOs in 1981. This variable was defined as the sum of salary, bonuses, and other benefits (measured in $1000's).

The following potential explanatory variables were considered. To capture short-run effects, the average of the company's previous 5 years' percentage changes in earnings per share (EPS) and the projected percentage change in next year's EPS were used. To capture the long-run effect, age and years as CEO, two admittedly correlated variables, were used. Dummy variables for the CEO's background (finance, marketing, and so on) were also considered. Finally, the researchers considered several nonlinear and interaction terms based on these variables. The best-fitting equation was the following:

$$\text{TotComp} = -3493 + 898.7(\text{Years as CEO})$$
$$+ 9.28(\text{Years as CEO})^2 - 17.19(\text{Years as CEO})(\text{Age})$$
$$+ 88.27\text{Age} + 867.4\text{Finance}$$

(The last variable represents a dummy variable, equal to 1 if the CEO had a finance background, 0 otherwise.) The corresponding R^2 was 19.4%.

a. Explain what this equation implies about CEO compensations.

b. The researchers drew the following conclusions. First, it appears that CEOs should indeed concentrate on long-run considerations—namely, those that keep them on their jobs the longest. Second, the absence of the short-run company-related variables from the equations helps to confirm the conjecture that CEOs who concentrate on earning the quick buck for their companies may not be acting in their best self-interest. Finally, the positive coefficient of the dummy variable may imply that financial people possess skills that are vitally important, and firms therefore outbid one another for the best financial talent. Based on the data given, do you agree with these conclusions?

c. Consider a CEO (other than those in the study) who has been in his position for 10 years and has a financial background. Predict his total yearly compensation (in $1000's) if he is 50 years old; if he is 55 years old. Explain why the difference between these two predictions is not 5(88.27), where 88.27 is the coefficient of the Age variable.

69. The Wilhoit Company has observed that there is a linear relationship between indirect labor expense (ILE) and direct labor hours (DLH). Data for direct labor hours and indirect labor expense for 18 months are given in the file P11_69.XLS. At the start of month 7, all cost categories in the Wilhoit Company increased by 10%, and they stayed at this level for months 7–12. Then at the start of month 13, another 10% across-the-board increase in all costs occurred, and the company operated at this price level for months 13–18.

a. Plot the data. Verify that the relationship between ILE and DLH is approximately linear within each 6-month period. Use regression (three times) to estimate the slope and intercept during months 1–6; during months 7–12; during months 13–18.

b. Use regression to fit a straight line to all 18 data points simultaneously. What values of the slope and intercept do you obtain?

c. Perform a price level adjustment to the data and re-estimate the slope and intercept using all 18 data points. Assuming no cost increases for month 19, what is your prediction for indirect labor expense if there are 35,000 direct labor-hours in month 19?

d. Interpret your results. What causes the difference in the linear relationship estimated in parts **b** and **c?**

70. The Bohring Company manufactures a sophisticated radar unit that is used in a fighter aircraft built by Seaways Aircraft. The first 50 units of the radar unit have been completed, and Bohring is preparing to submit a proposal to Seaways Aircraft to manufacture the next 50 units. Bohring wants to submit a competitive bid, but at the same time, it wants to ensure that all the costs of manufacturing the radar unit are fully covered. As part of this process, Bohring is attempting to develop a standard for the number of labor hours required to manufacture each radar unit. Developing a labor standard has been a continuing problem in the past. The file P11_70.XLS lists the number of labor hours required for each of the first 50 units of production. Bohring accountants want to see whether regression analysis, together with the concept of learning curves, can help solve the company's problem.

11.1 Quantity Discounts at the FirmChair Company

The FirmChair Company manufactures customized wood furniture and sells the furniture in large quantities to major furniture retailers. Jim Bolling has recently been assigned to analyze the company's pricing policy. He has been told that quantity discounts were usually given. For example, for one type of chair, the pricing changed at quantities of 200 and 400—that is, these were the quantity "breaks,"

where the marginal cost of the next chair changed. For this type of chair, the file FIRMCHAIR.XLS contains the quantity and total price to the customer for 81 orders. Use regression to help Jim discover the pricing structure that FirmChair evidently used. (*Note:* A linear regression of TotPrice versus Quantity will give you a "decent" fit, but you can do much better by introducing appropriate variables into the regression.)

11.2 Housing Price Structure in MidCity

Sales of single-family houses have been brisk in MidCity this year. This has especially been true in older, more established neighborhoods, where housing is relatively inexpensive compared to the new homes being built in the newer neighborhoods. Nevertheless, there are also many families who are willing to pay a higher price for the prestige of living in one of the newer neighborhoods. The file MIDCITY.XLS contains data on 128 recent sales in MidCity. For each sale, the file shows the neighborhood (1, 2, or 3) in which the house is located, the number of offers made on the house, the square footage, whether the house is made primarily of brick, the number of bathrooms, the number of bedrooms, and the selling price. Neighborhoods 1 and 2 are more traditional neighborhoods, whereas neighborhood 3 is a newer, more prestigious, neighborhood.

Use regression to estimate and interpret the pricing structure of houses in MidCity. Here are some considerations.

1. Is there a "premium" for a brick house, everything else being equal?

2. Is there a premium for a house in neighborhood 3, everything else being equal?

3. Is there an *extra* premium for a brick house in neighborhood 3, in addition to the usual premium for a brick house?

4. For purposes of estimation and prediction, could neighborhoods 1 and 2 be collapsed into a single "older" neighborhood?

Howie's Bakery is one of the most popular bakeries in town, and the favorite at Howie's is French bread. Each day of the week, Howie's bakes a number of loaves of French bread, more or less according to a daily schedule. To maintain its fine reputation, Howie's gives away to charity any loaves not sold on the day they are baked. Although this occurs frequently, it is also common for Howie's to run out of French bread on any given day—more demand than supply. In this case, no extra loaves are baked that day; the customers have to go elsewhere (or come back to Howie's the next day) for their French bread. Although French bread at Howie's is always popular, Howie's stimulates demand by running occasional 10% off sales.

Howie's has collected data for 20 consecutive weeks, 140 days in all. These data are listed in the file HOWIES.XLS. The variables are Day (Monday–Sunday), Supply (number of loaves baked that day), OnSale (whether French bread is on sale that day), and Demand (loaves actually sold that day). Howie's would like you to see whether regression can be used successfully to estimate Demand from the other data in the file. Howie reasons that if these other variables can be used to predict Demand, then he might be able to determine his daily supply (number of loaves to bake) in a more cost-effective way.

How successful is regression with these data? Is Howie correct that regression can help him determine his daily supply? Is any information "missing" that would be useful? How would you obtain it? How would you use it? Is this extra information *really* necessary?

11.4 Investing for Retirement

Financial advisors offer many types of advice to customers, but they generally agree that one of the best things people can do is invest as much as possible in tax-deferred retirement plans. Not only are the earnings from these investments exempt from income tax (until retirement), but the investment itself is tax-exempt. This means that if a person invests, say, $10,000 income of his $100,000 income a tax-deferred retirement plan, he pays income tax that year on only $90,000 of his income. This is probably the best method available to most people for avoiding tax payments. However, which group takes advantage of this attractive investment opportunity: everyone, people with low salaries, people with high salaries, or who?

The file RETIREPLAN.XLS lets you investigate this question. It contains data on 194 couples: number of dependent children, combined annual salary of husband and wife, current mortgage on home, average amount of other (nonmortgage) debt, and percentage of combined income invested in tax-deferred retirement plans (assumed to be limited to 15%, which is realistic). Using correlations, scatterplots, and regression analysis, what can you conclude about the tendency to invest in tax-deferred retirement plans in this group of people?

Regression Analysis: Statistical Inference

Predicting Movie Revenues

In the opener for Chapter 3, we discussed the article by Simonoff and Sparrow (2000) that examined movie revenues for 311 movies released in 1998 and late 1997. We saw that movie revenues were related to several variables, including genre, Motion Picture Association of America (MPAA) rating, country of origin, number of stars in the cast, whether the movie was a sequel, and whether the movie was released during a few "choice" times. In Chapter 3, we were limited to looking at summary measures and charts of the data. Now that we are studying regression, we can look further into the analysis performed by Simonoff and Sparrow. Specifically, they examined whether these variables, plus others, are effective in predicting movie revenues.

© 2002 Harris Welles

The authors actually report the results from three multiple regression models. All of these used the logarithm of the total U.S. gross revenue from the film as the response variable. (They used the *logarithm* because the distribution of gross revenues is very positively skewed.) The first model used only the "prerelease" variables listed in the previous paragraph. The values of these variables were all known prior to the movie's release. Therefore, the purpose of this model was to see how well revenues could be predicted *before* the movie was released.

The second model used the variables from model 1, along with two variables that could be observed after the first week of the movie's release: the first weekend gross, and the number of screens the movie opened on. (Actually, the logarithms of these latter two variables were used, again because of positive skewness. Also, the authors found it necessary to run two separate regressions at this stage—one for movies that opened on 10 or fewer screens, and another for movies that opened on more than 10 screens.) The idea here was that the success or failure of many movies depends to a large extent on how they do right after they are released. Therefore, it was expected that this information would add significantly to the predictive power of the regression model.

The third model built on the second by adding an additional explanatory variable: the number of Oscar nominations the movie received for key awards (Best Picture, Best Director, Best Actor, Best Actress, Best Supporting Actor, and Best Supporting Actress). This information is often not known until well after a movie's release, but it was hypothesized that Oscar nominations would lead to a significant increase in a movie's revenues, and that a regression model with this information could lead to very different predictions of revenue.

Simonoff and Sparrow found that the coefficients of the first regression model were in line with the boxplots we saw in Figure 3.1 of Chapter 3. For example, the variables that measured the number of "best" actors and actresses were both positive and significant, indicating that star power tends to lead to larger revenues. However, the predictive power of this model was poor. Given its standard error of prediction (and taking into account that the *logarithm* of revenue was the response variable), the authors stated that "the predictions of total grosses for an individual movie can be expected to be off by as much as a multiplicative factor of 100 high or low." It appears that there is no way to predict which movies will succeed and which will fail based on prerelease data only.

The second model added considerable predictive power. The regression equations indicated that gross revenue is positively related to first weekend gross and negatively related to the number of opening screens, both of these variables being significant. As for prediction, the factor of 100 mentioned in the previous paragraph decreased to a factor of 10 (for movies with 10 or fewer opening screens) or 2 (for movies with more than 10 opening screens). This is still not perfect—predictions of total revenue made after the movie's first weekend can still be pretty far off—but this additional information about initial success certainly helps.

The third model added only slightly to the predictive power, primarily because so few of the movies (10 out of 311) received Oscar nominations for key awards. However, the predictions for those that did receive nominations increased considerably. For example, the prediction for the multiple Oscar nominee *Saving Private Ryan*, based on the second model, was 194.622 (millions of dollars). Its prediction based on the third model increased to a whopping 358.237. (Interestingly, the prediction for this movie from the first model was only 14.791, and its actual gross revenue was 216.119. Perhaps the reason *Saving Private Ryan* did not make as much as the third model predicted was that the Oscar nominations were announced about 9 months after its release—too long to do much good.)

Simonoff and Sparrow then used their third model to predict gross revenues for 24 movies released in 1999—movies that were not in the data set used to estimate the regression model. They found that 21 out of 24 of the resulting 95% prediction intervals captured the actual gross revenues, which is about what we would expect. However, many of these prediction intervals were extremely wide, and several of the predictions were well above or below the actual revenues. The authors conclude by quoting Tim Noonan, a former movie executive: "Since predicting gross is extremely difficult, you have to serve up a [yearly] slate of movies and know that over time you'll have 3 or 4 to the left and 2 or 3 to the right. You must make sure you are doing things that mitigate your downside risk." ●

12.1 Introduction

In the previous chapter we learned how to fit a regression equation to a set of points by using the least squares method. The purpose of this regression equation is to provide a good fit to the points in the sample so that we can understand the relationship between a response variable and one or more explanatory variables. The entire emphasis of the discussion in the previous chapter was on finding a regression model that fits the observations in the sample. In this chapter we take a slightly different point of view: We assume that the observations in the sample are taken from some larger population. For example, the sample of 50 regions from the Pharmex drugstore example could represent a sample of all the regions where Pharmex does business. If that is the case, then we might be interested in the relationship between variables in the entire population, not just in the sample.

There are two basic problems we will discuss in this chapter. The first has to do with a *population regression model*. We want to infer its characteristics—that is, its intercept and slope term(s)—from the corresponding terms estimated by least squares. We also want to know which explanatory variables "belong" in the equation. We have seen that there are typically a large number of *potential* explanatory variables, and it is often not clear which of these do the best job of explaining variation in the response variable. In addition, we would like to infer whether there is any population regression equation worth pursuing. It might be that the potential explanatory variables provide very little explanation of the response variable, based on the sample data.

The second problem we will discuss in this chapter is prediction. We touched on the prediction problem in the previous chapter, primarily in the context of predicting the response variable for part of the sample held out for validation purposes. In reality, we had the values of the response variable for that part of the sample, so prediction was not really necessary. Now we will go beyond the sample and predict values of the response variable for *new* observations. There is no way to check the accuracy of these predictions, at least not right away, because the true values of the response variable are not yet known. However, we will provide prediction intervals to measure the accuracy of the predictions.

12.2 The Statistical Model

To perform statistical inference in a regression context, we must first make several assumptions about the population. Throughout the analysis these assumptions remain exactly that—they are only assumptions, not facts. These assumptions represent an idealization of reality, and as such, they are never likely to be entirely satisfied for the population in any real study. From a practical point of view, all we can ask is that they represent a close approximation to reality. If this is the case, then the analysis in this chapter is valid. But if the assumptions are grossly violated, we should be very suspicious of the statistical inferences that are based on these assumptions. Although we can never be entirely certain of the validity of the assumptions, there are ways to check for gross violations, and we will discuss some of these.

Regression Assumptions

1. There is a population regression line. It joins the *means* of the response variable for all values of the explanatory variables. For any fixed values of the explanatory variables, the mean of the errors is 0.

2. For any values of the explanatory variables, the standard deviation of the response variable is a constant, the same for all such values.

3. For any values of the explanatory variables, the response variable is normally distributed.

4. The errors are probabilistically independent.

Since these assumptions are so crucial to the regression analysis that follows, it is important to understand exactly what they mean. Assumption 1 is probably the most important. It implies that for some set of explanatory variables, there is an exact linear relationship in the population between the *means* of the response variable and the values of the explanatory variables.

Because of its importance, we discuss assumption 1 in more detail. Let Y be the response variable, and assume that there are k explanatory variables, X_1 through X_k. Let $\mu_{Y|X_1,\dots,X_k}$ be the mean of all Y's for any fixed values of the X's. Then assumption 1 implies that there is an exact linear relationship between the mean $\mu_{Y|X_1,\dots,X_k}$ and the X's. Specifically, it implies that there are coefficients α and β_1 through β_k such that the following equation holds for all values of the X's:

Population Regression Line Joining Means	$$\mu_{Y|X_1,\dots,X_k} = \alpha + \beta_1 X_1 + \cdots + \beta_k X_k$$	**(12.1)**

We commonly use Greek letters to denote *population* parameters and regular letters for their *sample* estimates.

In the terminology of the previous chapter, α is the intercept term, and β_1 through β_k are the slope terms. We use Greek letters for these coefficients to denote that they are *unobservable* population parameters. Assumption 1 implies the existence of a population regression equation and the corresponding α and β's. However, it tells us nothing about the values of these parameters. We still need to estimate them from sample data, and we will continue to use the least squares method to do so.

Equation (12.1) says that the *means* of the Y's lie on the population regression line. However, we know from a scatterplot that most *individual* Y's do not lie on this line. The vertical distance from any point to the line is called an **error term.** The error for any point, labeled ϵ, is the difference between Y and $\mu_{Y|X_1,\dots,X_k}$, that is,

$$Y = \mu_{Y|X_1,\dots,X_k} + \epsilon$$

By substituting the assumed linear form for $\mu_{Y|X_1,\dots,X_k}$, we obtain equation (12.2). This equation states that each value of Y is equal to a fitted part plus an error term. The fitted part is the linear expression $\alpha + \beta_1 X_1 + \cdots + \beta_k X_k$. The error term ϵ is sometimes positive, in which case the point is above the regression line, and sometimes negative, in which case the point is below the regression line. The last part of assumption 1 states that these errors average to 0 in the population, so that the positive errors cancel the negative errors.

Population Regression Line with Error Term	$$Y = \alpha + \beta_1 X_1 + \cdots + \beta_k X_k + \epsilon$$	**(12.2)**

Note that an error term ϵ is close to, but not the same as, a residual e. An error term is the vertical distance from a point to the (unobservable) population regression line. A residual is the vertical distance from a point to the estimated regression line. Residuals can be calculated from observed data; error terms cannot.

Assumption 2 concerns variation around the population regression line. Specifically, it states that the variation of the Y's about the regression line is the *same,* regardless of the values of the X's. A technical term for this property is **homoscedasticity.** We prefer a simpler term: **constant error variance.** In the Pharmex example (Example 11.1), constant error variance implies that the variation in Sales values is the same regardless of the

FIGURE 12.1

Illustration of Nonconstant Error Variance

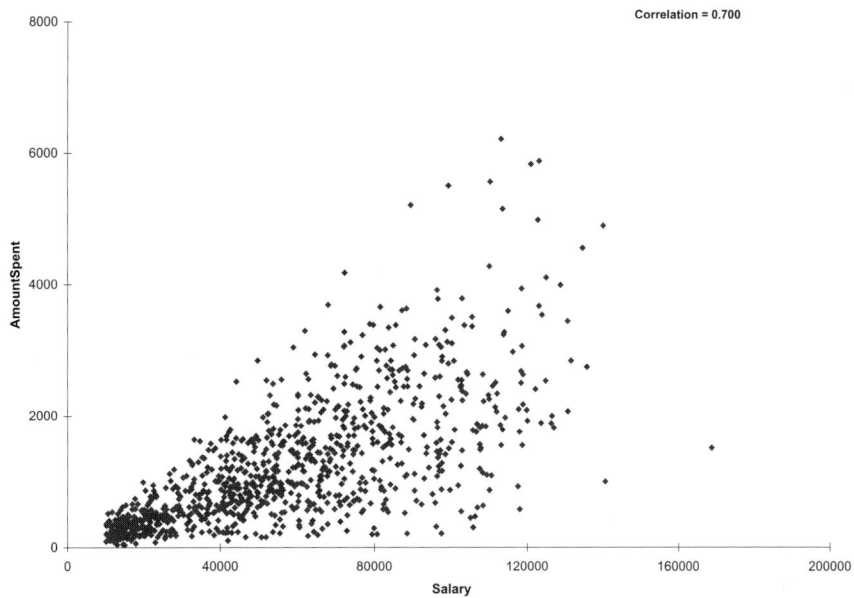

value of Promote. As another example, consider the Bendrix manufacturing example (Example 11.2). There we related overhead costs (Overhead) to the number of machine hours (MachHrs) and the number of production runs (ProdRuns). Constant error variance implies that overhead costs vary just as much for small values of MachHrs and ProdRuns as for large values—or any values in between.

There are many applications in which assumption 2 is questionable. The variation in Y often increases as X increases—a violation of assumption 2. We saw an example of this in Figure 11.10 (repeated here in Figure 12.1), which is based on the HyTex mail-order data in Example 3.11 from Chapter 3. This scatterplot shows AmountSpent versus Salary for a sample of HyTex's customers. Clearly, the variation in AmountSpent increases as Salary increases, which makes intuitive sense. Customers with small salaries have little disposable income, so they all tend to spend small amounts for mail-order items. Customers with large salaries have more disposable income. Some of them spend a lot of it on mail-order items and some spend only a little of it—hence a larger variation. Scatterplots with this "fan" shape are not at all uncommon in real studies, and they exhibit a clear violation of assumption 2.[1] We say that the data in this graph exhibit **heteroscedasticity,** or more simply, **nonconstant error variance.** These terms are summarized in the following box.

Homoscedasticity and Heteroscedasticity	**Homoscedasticity** means that the variability of Y values is the same for all X values. **Heteroscedasticity** means that the variability of Y values is larger for some X values than for others.

The easiest way to detect nonconstant error variance is through a visual inspection of a scatterplot. We draw a scatterplot of the response variable versus an explanatory variable X and see whether the points vary more for some values of X than for others. We can

[1]The fan shape in Figure 12.1 is probably the most common form of nonconstant error variance, but it is not the only possible form.

also examine the residuals with a residual plot, where residual values are on the vertical axis and some other variable (Y or one of the X's) is on the horizontal axis. If the residual plot exhibits a fan shape or other evidence of nonconstant error variance, this also indicates a violation of assumption 2.

Assumption 3 states that the errors are normally distributed. We can check this by forming a histogram or a Q–Q plot of the residuals. If assumption 3 holds, then the histogram should be approximately symmetric and bell shaped, and the points in the Q–Q plot should be close to a 45° line. But if there is an obvious skewness, too many residuals more than, say, 2 standard deviations from the mean, or some other nonnormal property, then this indicates a violation of assumption 3.

Finally, assumption 4 requires probabilistic independence of the errors. Intuitively, this assumption means that information on some of the errors provides no information on other errors. For example, if we are told that the overhead costs for months 1–4 are all above the regression line (positive residuals), we cannot infer anything about the residual for month 5 if assumption 4 holds.

Assumption 4 (independence of residuals) is usually suspect only for time series data.

For cross-sectional data there is generally little reason to doubt the validity of assumption 4 unless the observations are ordered in some particular way. For cross-sectional data we generally take assumption 4 for granted. However, for time series data, assumption 4 is often violated. This is because of a property called *autocorrelation*. For now, we simply mention that one output given automatically in many regression packages is the *Durbin-Watson statistic*. The Durbin–Watson statistic is one measure of autocorrelation and thus it measures the extent to which assumption 4 is violated. We can usually ignore it in cross-sectional studies, but it is important for time series data. We briefly discuss this Durbin-Watson statistic toward the end of this chapter and in Chapter 13.

One other assumption is important for numerical calculations. We must assume that no explanatory variable is an *exact* linear combination of any other explanatory variables. Another way of stating this is that there is no exact linear relationship between any set of explanatory variables. This would occur, for example, if one variable were an exact multiple of another, or if one variable were equal to the sum of several other variables. More generally, it occurs if one of the explanatory variables can be written as a weighted sum of several of the others.

If such a relationship holds, it means that there is *redundancy* in the data. One of the X's could be eliminated without any loss of information. Here is a simple example. Suppose that MachHrs1 is machine hours measured in hours, and MachHrs2 is machine hours measured in *hundreds* of hours. Then it is clear that these two variables contain exactly the same information, and either of them could (and should) be eliminated.

As another example, suppose that Ad1, Ad2, and Ad3 are the amounts spent on radio ads, television ads, and newspaper ads. Also, suppose that TotAd is the amount spent on radio, television, and newspaper ads combined. Then there is an exact linear relationship among these variables:

$$\text{TotAd} = \text{Ad1} + \text{Ad2} + \text{Ad3}$$

In this case there is no need to include TotAd in the analysis because it contains no information that is not already contained in the variables Ad1, Ad2, and Ad3.

Excel Tip *StatPro issues a warning and refuses to continue if it detects an exact linear relationship between explanatory variables in a regression model.*

Generally, it is fairly simple to spot an exact linear relationship such as these, and then to eliminate it by excluding the redundant variable from the analysis. However, if we do *not* spot the relationship and try to run the regression analysis with the redundant variable included, regression packages will typically respond with an error message. If the

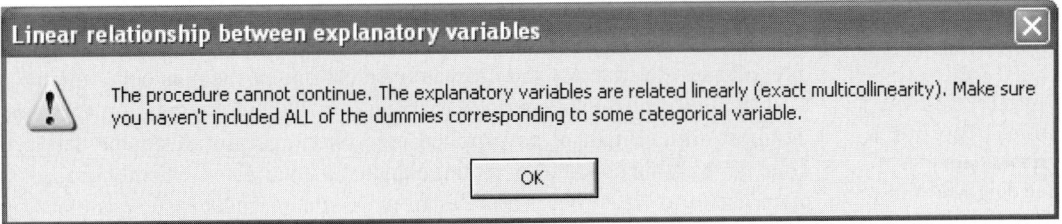

FIGURE 12.2 Error Message from StatPro Indicating Exact Multicollinearity

package interrupts the analysis with an error message containing the words "exact multi-collinearity" or "linear dependence," then we should look for a redundant explanatory variable. As an example, the message from StatPro in this case is shown in Figure 12.2. We got it by deliberately entering dummy variables from *each* category of a categorical variable—something we have warned *not* to do.

Although this problem can be a nuisance, it is usually caused by an oversight and can be fixed easily by eliminating a redundant variable. A more common and serious problem is **multicollinearity,** where explanatory variables are highly, but not exactly, correlated. A typical example is an employee's years of experience and age. Although these two variables are not equal for all employees, they are likely to be very highly correlated. If they are both included as explanatory variables in a regression analysis, the computer will not issue any error messages, but the estimates it produces might be unreliable. We will discuss multicollinearity in more detail later in this chapter.

Inferences About the Regression Coefficients

In this section we show how to make inferences about the population regression coefficients from sample data. We begin by making the assumptions discussed in the previous section. In particular, the first assumption states that there is a population regression line. Equation (12.2) for this line is repeated here:

$$Y = \alpha + \beta_1 X_1 + \cdots + \beta_k X_k + \epsilon$$

We refer to α and the β's collectively as the **regression coefficients.** Again, Greek letters are used to indicate that these quantities are unknown and unobservable. Actually, there is one other unknown constant in the model, the variance of the error terms. Regression assumption 2 states that these errors have a constant variance, the same for all values of the X's. We label this constant variance σ^2. Equivalently, the common standard deviation of the errors is σ.

This is how it looks in theory. There is a fixed set of explanatory variables, and given these variables, the problem is to estimate α, the β's, and σ. In practice, however, it is not usually this straightforward. In real regression applications the choice of relevant explanatory variables is almost never obvious. There are at least two guiding principles: relevance and data availability. We certainly want variables that are related to the response variable. The best situation is when there is an established economic or physical theory to guide us. For example, economic theory suggests that the demand for a product (response variable) is related to its price (possible explanatory variable). But there are not enough established theories to cover every situation. We often have to use the available data, plus some trial and error, to determine a *useful* set of explanatory variables. In this sense, it is usually pointless to search for one single "true" population regression equation. Instead, we typically estimate several competing models, each with a different set of explanatory variables, and ultimately select one of them as being the most useful.

Deciding which explanatory variables to include in a regression equation is probably the most difficult part of any applied regression analysis. Available data sets frequently offer an overabundance of potential explanatory variables. In addition, it is possible and often useful to create new variables from original variables, such as their logarithms. So where do we stop? Is it best to include every conceivable explanatory variable that might be related to the response variable? One overriding principle is **parsimony**—explaining the most with the least. For example, if we can explain a response variable just as well (or nearly as well) with two explanatory variables as with ten explanatory variables, then the principle of parsimony says to use only two. Models with fewer explanatory variables are generally easier to interpret, so we prefer them whenever possible.

> *Typically, the most difficult part of a regression analysis is deciding which explanatory variables to include in the regression equation.*

Parsimony	The principle of **parsimony** is to explain the most with the least. It favors a model with fewer explanatory variables, assuming that this model explains the response variable almost as well as a model with additional explanatory variables.

Before we can determine which equation has the "best" set of explanatory variables, however, we must be able to estimate the unknown parameters for a given equation. That is, for a given set of explanatory variables X_1 through X_k, we must be able to estimate α, the β's, and σ. We learned how to find point estimates of these parameters in the previous chapter. The estimates of α and the β's are the least squares estimates of the intercept and slope terms. For example, we used the 36 months of overhead data in the Bendrix example to estimate the equation

$$\text{Predicted Overhead} = 3997 + 43.54\text{MachHrs} + 883.62\text{ProdRuns}$$

This implies that the least squares estimates of α, β_1, and β_2 are 3997, 43.54, and 883.62. Furthermore, because the residuals are really estimates of the error terms, the standard error of estimate s_e is an estimate of σ. For the same overhead equation this estimate is $s_e = \$4109$.

However, we know from Chapters 8 and 9 that there is more to statistical estimation than finding point estimates of population parameters. Each potential sample from the population would typically lead to *different* point estimates. For example, if Bendrix estimated the equation for overhead from a different 36-month period, the results would almost certainly be different. Therefore, the question we now discuss is how these point estimates vary from sample to sample.

12.3.1 Sampling Distribution of the Regression Coefficients

The key idea is again sampling distributions. Recall that the sampling distribution of any estimate derived from sample data is the distribution of this estimate over all possible samples. This idea can be applied to the least squares estimate of a regression coefficient. For example, the sampling distribution of b_1, the least squares estimate of β_1, is the distribution of b_1's we would see if we observed many samples and ran a least squares regression on each of them.

Fortunately, mathematicians have used theoretical arguments to find the required sampling distributions. We state the main result as follows. Let β be any of the β's, and let

b be the least squares estimate of β. Then if the regression assumptions hold, the standardized value $(b - \beta)/s_b$ has a t distribution with $n - k - 1$ degrees of freedom, as given in the box. Here, k is the number of explanatory variables included in the equation, and s_b is the estimated standard deviation of the sampling distribution of b.

Sampling Distribution of a Regression Coefficient	If the regression assumptions are valid, the standardized value $$t = \frac{b - \beta}{s_b}$$ has a t distribution with $n - k - 1$ degrees of freedom.

This important result can be interpreted as follows. First, the estimate b is *unbiased* in the sense that its mean is β, the true but unknown value of the slope. If we calculated b's from repeated samples, some would underestimate β and others would overestimate β, but on average they would be on target.

Second, the estimated standard deviation of b is labeled s_b. It is usually called the **standard error of b.** This standard error is related to the standard error of estimate s_e, but it is not the same. Generally, the formula for s_b is quite complicated, and it is not shown here, but its value is printed in all standard regression outputs. It measures how much the b's would vary from sample to sample. A small value of s_b is preferred—it means that b is a more accurate estimate of the true coefficient β.

Finally, the shape of the distribution of b is symmetric and bell shaped. The relevant distribution is the t distribution with $n - k - 1$ degrees of freedom.

We have stated this result for a typical coefficient of one of the X's. These are usually the coefficients of most interest. However, exactly the same result holds for the intercept term α. Now we will see how to use this result.

Example 12.1 Explaining Overhead Costs at Bendrix

This example is a continuation of the Bendrix manufacturing example from the previous chapter. As before, the response variable is Overhead and the explanatory variables are MachHrs and ProdRuns. What inferences can we make about the regression coefficients?

Objective To use standard regression output to make inferences about the regression coefficients of machine hours and production runs in the equation for overhead costs.

Solution

When we use StatPro's Multiple Regression procedure, we obtain the output shown in Figure 12.3 on page 628. (See the file BENDRIX.XLS.) This output is practically identical to regression outputs from all other statistical software packages. We have already seen that the estimates of the regression coefficients appear under the label Coefficient in the range C16:C18. These values estimate the true, but unobservable, population coefficients. The next column, labeled Std Err, shows the s_b's. Specifically, 3.590 is the standard error of the coefficient of MachHrs, and 82.251 is the standard error of the coefficient of ProdRuns.

The b's represent point estimates of the β's, based on this particular sample. The s_b's indicate the accuracy of these point estimates. For example, the point estimate of β_1, the effect on Overhead of a 1-unit increase in MachHrs, is 43.536. We are about 95% confident that the true β_1 is within 2 standard errors of this point estimate, that is, from 36.357

	A	B	C	D	E	F	G	H
1	*Results of multiple regression for Overhead*							
2								
3	*Summary measures*							
4		Multiple R	0.9308					
5		R-Square	0.8664					
6		Adj R-Square	0.8583					
7		StErr of Est	4108.9932					
8								
9	*ANOVA Table*							
10		Source	df	SS	MS	F	p-value	
11		Explained	2	3614020652.0000	1807010326.0000	107.0261	0.0000	
12		Unexplained	33	557166208.0000	16883824.4848			
13								
14	*Regression coefficients*							
15			Coefficient	Std Err	t-value	p-value	Lower limit	Upper limit
16		Constant	3996.6782	6603.6509	0.6052	0.5492	-9438.5612	17431.9176
17		MachHrs	43.5364	3.5895	12.1289	0.0000	36.2335	50.8393
18		ProdRuns	883.6179	82.2514	10.7429	0.0000	716.2760	1050.9598

FIGURE 12.3 Regression Output for Bendrix Example

to 50.715. Similar statements can be made for the coefficient of ProdRuns and the intercept (Constant) term. ●

12.3.2 Confidence Intervals for the Regression Coefficients

As with any population parameters, we can use the sample data to obtain confidence intervals for the regression coefficients. For example, the preceding paragraph implies that an approximate 95% confidence interval for the coefficient of MachHrs extends from 36.357 to 50.715. More precisely, a confidence interval for any β is of the form

$$b \pm t\text{-multiple} \times s_b$$

where the t-multiple depends on the confidence level and the degrees of freedom (here $n - k - 1$). For example, the relevant t-multiple for the Bendrix data, assuming we want a 95% confidence interval, is the value that cuts off probability 0.025 of the t distribution with $36 - 2 - 1 = 33$ degrees of freedom. [It is 2.035 and can be found in Excel with the function TINV(0.05,33).][2] Using this multiple gives a 95% confidence interval from 36.234 to 50.839, as shown in Figure 12.3. StatPro always adds 95% confidence intervals for the regression coefficients automatically.

12.3.3 Hypothesis Tests for the Regression Coefficients

There is another important piece of information in regression outputs: the t-values for the individual regression coefficients. These are shown in the "t-value" column of the regression output in Figure 12.3. The formula for a t-value is simple. It is the ratio of the estimated coefficient to its standard error, as shown in equation (12.3). Therefore, it indicates how many standard errors the regression coefficient is from 0. For example, the t-value for MachHrs is about 12.13, so we know that the regression coefficient of

---

[2]If you look at the lower and upper limit cells in any StatPro multiple regression output, you can see how the TINV function is incorporated.

MachHrs, 43.536, is over 12 standard errors to the right of 0. Similarly, the coefficient of ProdRuns is more than 10 of its standard errors to the right of 0.

t-value for Test of Regression Coefficient	$t\text{-value} = b/s_b$	(12.3)

A *t*-value can be used in an important hypothesis test for the corresponding regression coefficient. To motivate this test, suppose that we want to decide whether a particular explanatory variable belongs in the regression equation. A sensible criterion for making this decision is to check whether the corresponding regression coefficient is 0. If a variable's coefficient is 0, there is no point in including this variable in the equation; the 0 coefficient will cancel its effect on the response variable.

Therefore, it is reasonable to test whether a variable's coefficient is 0. This is usually tested versus a *two-tailed* alternative. The null and alternative hypotheses are of the form $H_0:\beta = 0$ versus $H_a:\beta \neq 0$. If we can reject the null hypothesis and conclude that this coefficient is *not* 0, then we have an argument for including the variable in the regression equation. Conversely, if we cannot reject the null hypothesis, we might decide to eliminate this variable from the equation.

The *t*-value for a variable allows us to run this test easily. We simply compare the *t*-value in the regression output with a tabulated *t*-value and reject the null hypothesis only if the *t*-value from the computer output is greater in magnitude than the tabulated *t*-value. If the test is run at the 5% significance level, for example, then the appropriate tabulated *t*-value can be found in Excel with TINV($0.05,n - k - 1$), the same *t*-value used previously for confidence intervals.

Most computer packages, including StatPro, make this test even easier to run by reporting the corresponding *p*-value for the test. This eliminates the need for finding the tabulated *t*-value (or using the TINV function). The *p*-value is interpreted exactly as in Chapter 10. It is the probability (in both tails) of the relevant *t* distribution beyond the listed *t*-value. For example, referring again to Figure 12.3, the *t*-value for MachHrs is 12.13, and the associated *p*-value (rounded to four decimal places) is 0.0000. This means that there is virtually no probability beyond the observed *t*-value. In words, we are still not exactly sure of the true coefficient of MachHrs, but we are sure it is not 0. The same can be said for the coefficient of ProdRuns.

We will soon say even more about these *t*-values and how they can help to decide which variables to include or exclude in a regression equation. But we first make the following points about hypothesis tests for regression coefficients.

The test for whether a regression coefficient is 0 can be run by looking at the corresponding *p*-value: Reject the "equals 0" hypothesis if the *p*-value is small, say, less than 0.05. It can also be run by looking at the confidence interval for the coefficient: Reject the "equals 0" hypothesis if the confidence interval does *not* contain the value 0.

Hypothesis Tests and Regression Coefficients

1. A *t*-value is usually reported for the intercept (constant) term in the equation, as well as for the other coefficients. However, this information is often of little relevance. The reason is that there is often no practical reason for testing whether the intercept is 0. There are rare situations where an intercept equal to 0 has a meaningful interpretation, and in such situations the hypothesis test is relevant.

2. The test of $\beta = 0$ versus a two-tailed alternative at the 5% level, say, can also be run by calculating a 95% confidence interval for β and rejecting the null hypothesis if 0 is not within the confidence interval. That is, if a 95% confidence interval for β extends from a negative number to a positive number, we cannot reject the null hypothesis that $\beta = 0$.

3. The previous test, a two-tailed test of whether a particular β is 0, is only one of many hypothesis tests that can be run. For example, consider a sample of houses that have been sold recently. We would like to regress the selling prices of the houses on their appraised values, as obtained by a professional appraiser. Now, it is pretty clear, even before the data are observed, that selling prices will be *positively* related to appraised values. Therefore, there isn't much point in testing whether the coefficient of AppraisedValue is 0.

A more interesting test in this example is whether the coefficient of Appraised-Value is less than or greater than 1. If it is less than 1, say, then every extra dollar of appraised value contributes *less than* an extra dollar to the selling price. Therefore, we might run the one-tailed test of $H_0{:}\beta \geq 1$ versus $H_a{:}\beta < 1$. (We could also run a two-tailed test. It just depends on what we're trying to prove.) In this case we would base the test on the test statistic

$$t\text{-value} = \frac{b - 1}{s_b}$$

where b and s_b are the coefficient and standard error of AppraisedValue in the regression output. Its p-value could be calculated in Excel with the function TDIST(ABS(t-value), $n - 2$,1). This t-value and the corresponding p-value are *not* reported in computer outputs, but they are easy to obtain.

The point here is that most computer outputs provide the ingredients for a very natural test—whether a given regression coefficient is 0. Virtually no work is needed to perform this test because the t-value and p-value are given in the regression output. But other hypothesis tests on the coefficients are sometimes relevant, and they can be performed easily with Excel functions.

12.3.4 A Test for the Overall Fit: The ANOVA Table

The t-values for the regression coefficients allow us to see which of the potential explanatory variables are useful in explaining the response variable. But it is conceivable that *none* of these variables does a very good job. That is, it is conceivable that the entire group of explanatory variables explains only an insignificant portion of the variability of the response variable. Although this is the exception rather than the rule in most real applications, it can certainly happen. An indication of this is that we obtain a very small R^2 value. Because R^2 is the square of the correlation between the observed values of the response variable and the fitted values from the regression equation, another indication of a lack of fit is that this correlation (the "multiple R") is small. In this section we state a formal procedure for testing the overall fit, or explanatory power, of a regression equation.

Suppose that the response variable is Y and the explanatory variables are X_1 through X_k. Then the proposed population regression equation is

$$Y = \alpha + \beta_1 X_1 + \cdots + \beta_k X_k + \epsilon$$

To say that this equation has absolutely no explanatory power means that the same value of Y will be predicted regardless of the values of the X's. In this case it makes no difference which values of the X's we use because they all lead to the same predicted value of Y. But the only way this can occur is if all of the β's are 0. So the formal hypothesis we test in this section is $H_0{:}\beta_1 = \cdots = \beta_k = 0$ versus the alternative that at least one of the β's is not 0. In words, the null hypothesis is that this set of explanatory variables has no power to explain the variation in the response variable Y. If we can reject the null hypothesis, as we can in the majority of applications, this means that the explanatory variables *as a group* provide at least some explanatory power. This hypothesis is summarized in the box.

Hypotheses for ANOVA Test	The null hypothesis is that all coefficients of the explanatory variables are 0. The alternative is that at least one of these coefficients is not 0.

At first glance it might appear that we can test this null hypothesis by looking at the individual t-values. If they are all small (statistically insignificant), then we can accept the null hypothesis of no fit; otherwise, we can reject it. However, as we will see in the next section, it is possible, because of multicollinearity, to have small t-values even though the variables as a whole have significant explanatory power.

The alternative is to use an F test. This is sometimes referred to as the ANOVA (analysis of variance) test because the elements for calculating the required F-value are shown in an ANOVA table.[3] In general, an ANOVA table analyzes different sources of variation. In the case of regression, the variation in question is the variation of the response variable Y. The "total variation" of this variable is the sum of squared deviations about the mean and is labeled SST (sum of squares total)

$$SST = \sum (Y_i - \overline{Y})^2$$

The ANOVA table splits this total variation into two parts, the part *explained* by the regression equation, and the part left *unexplained*. The unexplained part is the sum of squared residuals, usually labeled SSE (sum of squared errors):

$$SSE = \sum e_i^2 = \sum (Y_i - \hat{Y}_i)^2$$

The explained part is then the difference between the total and unexplained variation. It is usually labeled SSR (sum of squares due to regression):

$$SSR = SST - SSE$$

The F test is a formal procedure for testing whether the explained variation is "large" compared to the unexplained variation. Specifically, each of these sources of variation has an associated degrees of freedom (df). For the explained variation, $df = k$, the number of explanatory variables. For the unexplained variation, $df = n - k - 1$, the sample size minus the total number of coefficients (including the intercept term). When we divide the explained or unexplained variation by its degrees of freedom, the result is called a mean square, or MS. The two mean squares we need are MSR and MSE, given by

$$MSR = \frac{SSR}{k}$$

and

$$MSE = \frac{SSE}{n - k - 1}$$

Note that MSE is the square of the standard error of estimate, that is,

$$MSE = s_e^2$$

---

[3]This ANOVA table is similar to the ANOVA table we discussed in the Chapter 10. However, we repeat the necessary material here for those who didn't cover the ANOVA section in Chapter 10.

Finally, the ratio of these mean squares is the required F-ratio for the test:

$$F\text{-ratio} = \frac{MSR}{MSE}$$

When the null hypothesis of no explanatory power is true, this F-ratio has an F distribution with k and $n - k - 1$ degrees of freedom. If the F-ratio is small, then the explained variation is small relative to the unexplained variation, and there is evidence that the regression equation provides little explanatory power. But if the F-ratio is large, then the explained variation is large relative to the unexplained variation, and we can conclude that the equation does have some explanatory power.

As usual, the F-ratio has an associated p-value that allows us to run the test easily. In this case the p-value is the probability to the *right* of the observed F-ratio in the appropriate F distribution. This p-value is reported in most regression outputs, along with the rest of the elements that lead up to it. If it is sufficiently small, less than 0.05, say, then we can conclude that the explanatory variables as a whole have at least some explanatory power.

Although this test is run routinely in most applications, there is often little doubt that the equation has some explanatory power; the only questions are how much, and which explanatory variables provide the best combination. In such cases the F-ratio from the ANOVA table is typically "off the charts" and the corresponding p-value is practically 0. On the other hand, F-ratios, particularly large ones, should not necessarily be used to choose between equations with different explanatory variables included.

For example, suppose that one equation with three explanatory variables has an F-ratio of 54 with an extremely small p-value—obviously very significant. Also, suppose that another equation that includes these three variables plus a few more has an F-ratio of 37 and also has a very small p-value. (When we say small, we mean *small*. These p-values are probably listed, to four decimal places, as 0.0000.) Is the first equation better because its F-ratio is higher? Not necessarily. The two F-ratios imply only that both of these equations have a good deal of explanatory power. It is better to look at their s_e values (or adjusted R^2 values) and their t-values to choose between them.

The ANOVA table is part of the StatPro output for any regression run. It appeared for the Bendrix example in Figure 12.3, which is repeated for convenience in Figure 12.4. The ANOVA table is in rows 10–12. We see the degrees of freedom in column C, the sums of squares in column D, the mean squares in column E, the F-ratio in cell F11, and

	A	B	C	D	E	F	G	H
1	Results of multiple regression for Overhead							
2								
3	Summary measures							
4		Multiple R	0.9308					
5		R-Square	0.8664					
6		Adj R-Square	0.8583					
7		StErr of Est	4108.9932					
8								
9	ANOVA Table							
10		Source	df	SS	MS	F	p-value	
11		Explained	2	3614020652.0000	1807010326.0000	107.0261	0.0000	
12		Unexplained	33	557166208.0000	16883824.4848			
13								
14	Regression coefficients							
15			Coefficient	Std Err	t-value	p-value	Lower limit	Upper limit
16		Constant	3996.6782	6603.6509	0.6052	0.5492	-9438.5612	17431.9176
17		MachHrs	43.5364	3.5895	12.1289	0.0000	36.2335	50.8393
18		ProdRuns	883.6179	82.2514	10.7429	0.0000	716.2760	1050.9598

FIGURE 12.4 Regression Output for Bendrix Example

its associated *p*-value in cell G11. As predicted, this *F*-ratio is "off the charts," and the *p*-value is practically 0.

This information wouldn't be much comfort for the Bendrix manager who is trying to understand the causes of variation in overhead costs. This manager already *knows* that machine hours and production runs are related positively to overhead costs—everyone in the company knows that! What he really wants is a set of explanatory variables that yields a high R^2 and a low s_e. The low *p*-value in the ANOVA tables does not guarantee these. All it guarantees is that MachHrs and ProdRuns are of "some help" in explaining variations in Overhead.

As this example indicates, the ANOVA table can be used as a screening device. If the explanatory variables do not explain a significant percentage of the variation in the response variable, then we can either discontinue the analysis or search for an entirely new set of explanatory variables. But even if the *F*-ratio in the ANOVA table is extremely significant, there is no guarantee that the regression equation provides a good enough fit for practical uses. This depends on other measures such as s_e and R^2.

PROBLEMS

Level A

1. Explore the relationship between the selling prices (*Y*) and the appraised values (*X*) of the 150 homes in the file P02_07.XLS by estimating a simple linear regression model. Construct a 95% confidence interval for the model's slope (i.e., β_1) parameter. What does this confidence interval tell you about the relationship between *Y* and *X* for these data?

2. The owner of the Original Italian Pizza restaurant chain would like to predict the sales of his specialty, deep-dish pizza. He has gathered data on the monthly sales of deep-dish pizzas at his restaurants and observations on other potentially relevant variables for each of his 15 outlets in central Pennsylvania. These data are provided in the file P11_04.XLS.
   a. Estimate a multiple regression model between the quantity sold (*Y*) and the following explanatory variables: average price of deep-dish pizzas, monthly advertising expenditures, and disposable income per household in the areas surrounding the outlets.
   b. Is there evidence of any violations of the key assumptions of regression analysis in this case?
   c. Which of the variables in this model have regression coefficients that are statistically different from 0 at the 5% significance level?
   d. Given your findings in part **c**, which variables, if any, would you choose to remove from the model estimated in part **a**? Explain your decision.

3. The *ACCRA Cost of Living Index* provides a useful and reasonably accurate measure of cost of living differences among a large number of urban areas. Items on which the index is based have been carefully chosen to reflect the different categories of consumer expenditures. The data are in the file P02_19.XLS.

   a. Use multiple regression to explore the relationship between the composite index (response variable) and the various expenditure components (explanatory variables).
   b. Is there evidence of any violations of the key assumptions of regression analysis?
   c. Which of the variables in this model have regression coefficients that are statistically different from 0 at the 5% significance level?
   d. Given your findings in part **c**, which variables, if any, would you choose to remove from the model estimated in part **a**? Explain your decision.

4. A trucking company wants to predict the yearly maintenance expense (*Y*) for a truck using the number of miles driven during the year (X_1) and the age of the truck (X_2, in years) at the beginning of the year. The company has gathered the information given in the file P11_16.XLS. Note that each observation corresponds to a particular truck.
   a. Formulate and estimate a multiple regression model using the given data.
   b. Does autocorrelation, multicollinearity, or heteroscedasticity appear to be a problem?
   c. Construct 95% confidence intervals for the regression coefficients of X_1 and X_2. Based on these interval estimates, which variables, if any, would you choose to remove from the model estimated in part **a**? Explain your decision.

5. Based on the data in the file P02_25.XLS from the U.S. Department of Agriculture, explore the relationship between the number of farms (*X*) and the average size of a farm (*Y*) in the United States.
   a. Use the given data to estimate a simple linear regression model.

b. Test whether there is sufficient evidence to conclude that the slope parameter (i.e., β_1) is *less than* 0. Use a 5% significance level.

c. Based on your finding in part **b**, is it possible to conclude that a linear relationship exists between the number of farms and the average farm size between 1950 and 1997? Explain.

6. An antique collector believes that the price received for a particular item increases with its age and the number of bidders. The file P11_18.XLS contains data on these three variables for 32 recently auctioned comparable items.

a. Estimate an appropriate multiple regression model using the given data.

b. Interpret the ANOVA table for this model. In particular, does this set of explanatory variables provide at least some power in explaining the variation in price? Report a *p*-value for this hypothesis test.

7. Consider the enrollment data for *Business Week*'s top U.S. graduate business programs in the file P02_03.XLS. Use these data to estimate a multiple regression model to assess whether there is a systematic relationship between the total number of full-time students and the following explanatory variables: (i) the proportion of female students, (ii) the proportion of minority students, and (iii) the proportion of international students enrolled at these distinguished business schools. Next, interpret the ANOVA table for this model. In particular, does this set of explanatory variables provide at least some power in explaining the variation in total full-time enrollment at the top graduate business programs? Report a *p*-value for this hypothesis test.

8. The U.S. Bureau of Labor Statistics provides data on the year-to-year percentage changes in the wages and salaries of workers in private industries, including both "white-collar" and "blue-collar" occupations. Here we consider these data in the file P02_56.XLS. Is there evidence of a linear relationship between the yearly changes in the wages and salaries of "white-collar" (*Y*) and "blue-collar" (*X*) workers in the United States over the given time period? Begin to answer this question by estimating a simple linear regression model.

a. Construct a 95% confidence interval for the model's slope (i.e., β_1) parameter. Interpret this interval estimate to answer the question posed above.

b. Interpret the ANOVA table for this model. In particular, does the explanatory variable included in this simple regression model provide at least some power in explaining the variation in the response variable? Report a *p*-value for this hypothesis test.

c. What is the relationship between the computed *t*-ratio for the estimated coefficient of the explanatory variable and the *F*-ratio found in the ANOVA section of the output? Do these two test statistic values provide the same indication regarding a possible relationship between yearly changes in the wages and salaries of white-collar and blue-collar workers? Explain why or why not.

9. Suppose that a regional express delivery service company wants to estimate the cost of shipping a package (*Y*) as a function of cargo type, where cargo type includes the following possibilities: fragile, semifragile, and durable. Costs for 15 randomly chosen packages of approximately the same weight and same distance shipped, but of different cargo types, are provided in the file P11_28.XLS.

a. Estimate an appropriate multiple regression model to predict the cost of shipping a given package.

b. Interpret the ANOVA table for this model. In particular, do the explanatory variables included in your model formulated in part **a** provide at least some power in explaining the variation in the cost of shipping a package? Report a *p*-value for this hypothesis test.

10. A simple linear regression with 11 observations yielded the ANOVA table in Table 12.1

a. Complete this ANOVA table.

b. Using $\alpha = 0.05$, test the hypothesis of no linear regression.

Level B

11. Consider the relationship between the size of the population (*X*) and the average household income level (*Y*) for residents of U.S. towns.

Table **12.1**

ANOVA Table

	Degrees of Freedom	Sum of Squares
Regression		1000
Error		
Total		2500

Table **12.2**

ANOVA Table

	Degrees of Freedom	Sum of Squares
SSR		20
SSE	4	
SST		100

a. Using the data in the file P02_24.XLS, estimate a regression model involving these two variables.
b. Does autocorrelation, multicollinearity, or heteroscedasticity appear to be a problem?
c. Test whether there is sufficient evidence to conclude that the slope parameter (i.e., β_1) is *greater than* 0.0035. Use a 5% significance level.
d. Based on your finding in part **c,** is it possible to conclude that a linear relationship exists between the size of the population and the average house-hold income level for residents of U.S. towns? Explain.

12. Suppose you find the ANOVA table shown in Table 12.2 for a simple linear regression.
a. Find the correlation between X and Y. Assume the slope of the least squares line is negative.
b. Find the p-value for the test of the hypothesis of no linear regression.

12.4 Multicollinearity

Recall that the coefficient of any variable in a regression equation indicates the effect of this variable on the response variable, provided that the other variables in the equation remain constant. Another way of stating this is that the coefficient represents the effect of this variable on the response variable *in addition to* the effects of the other variables in the equation. For example, if MachHrs and ProdRuns are included in the equation for Overhead, then the coefficient of MachHrs indicates the *extra* amount MachHrs explains about variation in Overhead, in addition to the amount already explained by ProdRuns. Similarly, the coefficient of ProdRuns indicates the extra amount ProdRuns explains about variation in Overhead, in addition to the amount already explained by MachHrs. Therefore, the relationship between an explanatory variable X and the response variable Y is not always accurately reflected in the coefficient of X; it depends on which *other X*'s are included or not included in the equation.

This is especially true when there is a linear relationship between two or more *explanatory* variables, in which case we have *multicollinearity*. By definition, **multicollinearity** is the presence of a fairly strong linear relationship between two or more explanatory variables, and it can make estimation difficult.

Multicollinearity

Multicollinearity occurs when there is a nearly linear relationship among a set of explanatory variables.

Consider the following example. It is a rather trivial example, but it is useful for illustrating the potential effects of multicollinearity.

Example 12.2 Height as a Function of Foot Length

We want to explain a person's height by means of foot length. The response variable is Height, and the explanatory variables are Right and Left, the length of the right foot and the left foot, respectively. What can occur when we regress Height on *both* Right and Left?

Objective To illustrate the problem of multicollinearity when both foot length variables are used in a regression for height.

Solution

Admittedly, there is no need to include both Right and Left in an equation for Height—either one of them would do—but we include them both to make a point. Now, it is likely that there is a large correlation between height and foot size, so we would expect this regression equation to do a good job. For example, the R^2 value will probably be large. But what about the coefficients of Right and Left? Here we have a problem. The coefficient of Right indicates the right foot's effect on Height in addition to the effect of the left foot. This additional effect is probably minimal. That is, after the effect of Left on Height has already been taken into account, the extra information provided by Right is probably minimal. But it goes the other way also. The extra effect of Left, in addition to that provided by Right, is probably minimal.

To show what can happen numerically, we generated a hypothetical data set of heights and left and right foot lengths. (See the file HEIGHT.XLS.) We did this so that, except for random error, height is approximately 32 plus 3.2 times foot length (all expressed in inches). As shown in Figure 12.5, the correlation between Height and either Right or Left in our data set is quite large, and the correlation between Right and Left is very close to 1.

The regression output when both Right and Left are entered in the equation for Height appears in Figure 12.6. This tells a somewhat confusing story. The multiple R and the corresponding R^2 are about what we would expect, given the correlations between Height and either Right or Left in Figure 12.5. In particular, the multiple R is close to the correlation between Height and either Right or Left. Also, the s_e value is quite good. It implies

FIGURE 12.5
Correlations in Example of Height versus Foot Length

	E	F	G	H	I
3	**Table of correlations**				
4			Height	Right	Left
5		Height	1.000		
6		Right	0.903	1.000	
7		Left	0.900	0.999	1.000

FIGURE 12.6
Regression Output for Height versus Foot Length Example

	A	B	C	D	E	F	G	H
1	**Results of multiple regression for Height**							
2								
3	**Summary measures**							
4		Multiple R	0.9042					
5		R-Square	0.8176					
6		Adj R-Square	0.8140					
7		StErr of Est	2.0041					
8								
9	**ANOVA Table**							
10		Source	df	SS	MS	F	p-value	
11		Explained	2	1836.3845	918.1923	228.6003	0.0000	
12		Unexplained	102	409.6916	4.0166			
13								
14	**Regression coefficients**							
15			Coefficient	Std Err	t-value	p-value	Lower limit	Upper limit
16		Constant	31.7603	1.9595	16.2087	0.0000	27.8737	35.6469
17		Right	6.8229	3.4285	1.9901	0.0493	0.0226	13.6233
18		Left	-3.6448	3.4411	-1.0592	0.2920	-10.4701	3.1806

that predictions of height from this regression equation will typically be off by only about 2 inches.

However, the coefficients of Right and Left are not at all what we might expect, given that we generated heights as approximately 32 plus 3.2 times foot length. In fact, the coefficient of Left is the wrong sign—it is *negative*! Besides this "wrong" sign, the tip-off that there is a problem is that the *t*-value of Left is quite small and the corresponding *p*-value is quite large. Judging by this, we might conclude that Height and Left are either not related or are related negatively. But we know from Figure 12.5 that both of these conclusions are false. In contrast, the coefficient of Right has the "correct" sign, and its *t*-value and associated *p*-value do imply statistical significance, at least at the 5% level. However, this happened mostly by chance. Slight changes in the data could change the results completely—the coefficient of Right could become negative and insignificant, or both coefficients could become insignificant.

The problem is that although both Right and Left are clearly related to Height, it is impossible for the least squares method to distinguish their *separate* effects. Note that the regression equation does estimate the combined effect fairly well—the sum of the coefficients of Right and Left is $6.823 + (-3.645) = 3.178$. This is close to the coefficient 3.2 we used to generate the data. Also, the estimated intercept 31.760 is close to the intercept 32 we used to generate the data. Therefore, the estimated equation will work well for predicting heights. It just does not have reliable estimates of the individual coefficients of Right and Left.

To see what happens when either Right or Left is excluded from the regression equation, we show the results of *simple* regression. When Right is the only variable in the equation, it becomes

$$\text{Predicted Height} = 31.546 + 3.195\text{Right}$$

The R^2 and s_e values are 81.6% and 2.005, and the *t*-value and *p*-value for the coefficient of Right are now 21.34 and 0.0000—very significant. Similarly, when Left is the only variable in the equation, it becomes

$$\text{Predicted Height} = 31.526 + 3.197\text{Left}$$

The R^2 and s_e values are 81.1% and 2.033, and the *t*-value and *p*-value for the coefficient of Left are 20.99 and 0.0000—again very significant. Clearly, both of these equations tell almost identical stories, and they are much easier to interpret than the equation with both Right and Left included. The message, therefore, is that when two variables are very highly correlated, only one of them should be included in the regression equation. ●

This example illustrates an extreme form of multicollinearity, where two explanatory variables are very highly correlated. In general, there are various degrees of multicollinearity. In each of them, there is a linear relationship between two or more explanatory variables, and this relationship makes it difficult to estimate the individual effect of the X's on the response variable. The symptoms of multicollinearity can be "wrong" signs of the coefficients, smaller-than-expected *t*-values, and larger-than-expected (insignificant) *p*-values. In other words, variables that are really related to the response variable can look like they aren't related, based on their *p*-values. The reason is that their effects on Y are already explained by other X's in the equation.

Sometimes multicollinearity is easy to spot and treat. For example, it would be silly to include both Right and Left foot length in the equation for Height. They are obviously very highly correlated and only one is needed in the equation for Height. We should exclude one of them—either one—and reestimate the equation. However, multicollinearity is not usually this easy to treat or even diagnose.

Multicollinearity often causes regression coefficients to have the "wrong" sign, t-values to be too small, and p-values to be too large.

Multicollinearity typically causes unreliable estimates of regression coefficients, but it does not generally cause poor predictions.

Moderate to extreme multicollinearity poses a problem in many regression applications. Unfortunately, there are usually no easy solutions.

Suppose, for example, that we want to use regression to explain variations in salary. Three potentially useful explanatory variables are age, years of experience in the company, and years of experience in the industry. It is very likely that each of these is positively related to salary, and it is also very likely that they are very closely related to each other. However, it isn't clear which, if any, we should exclude from the regression equation. If we include all three, we are likely to find that at least one of them is insignificant (high p-value), in which case we might consider excluding it from the equation. If we do so, the s_e and R^2 values will probably not change very much—the equation will provide equally good predicted values—but the coefficients of the variables that remain in the equation could change considerably.

PROBLEMS

Level A

13. Using the data given in P11_10.XLS, estimate a multiple regression equation to predict the sales price of houses in a given community. Employ all available explanatory variables. Is there evidence of multicollinearity in this model? Explain why or why not.

14. Consider the enrollment data for *Business Week*'s top U.S. graduate business programs in the file P02_03.XLS. Use these data to estimate a multiple regression model to assess whether there is a systematic relationship between the total number of full-time students and the following explanatory variables: (i) the proportion of female students, (ii) the proportion of minority students, and (iii) the proportion of international students enrolled at these distinguished business schools.

   a. Determine whether each of the regression coefficients for the explanatory variables in this model is statistically different from 0 at the 5% significance level. Summarize your findings.

   b. Is there evidence of multicollinearity in this model? Explain why or why not.

15. The manager of a commuter rail transportation system was recently asked by her governing board to determine the factors that have a significant impact on the demand for rides in the large city served by the transportation network. The system manager has collected data on variables that might be related to the number of weekly riders on the city's rail system. The file P11_20.XLS contains these data.

   a. Estimate a multiple regression model using all of the available explanatory variables. Perform a test of significance for each of the model's regression coefficients. Are the signs of the estimated coefficients consistent with your expectations?

   b. Is there evidence of multicollinearity in this model? Explain why or why not. If multicollinearity appears to be present, explain what you would do to eliminate this problem.

Level B

16. The human resources manager of DataCom, Inc., wants to examine the relationship between annual salaries (Y) and the number of years employees have worked at DataCom (X). These data have been collected for a sample of employees and are given in the file P11_05.XLS.

   a. Estimate the relationship between Y and X using simple linear regression analysis. Is there evidence to support the hypothesis that the coefficient for the number of years employed is statistically different from 0 at the $\alpha = 0.05$ level?

   b. Next, formulate a multiple regression model to explain annual salaries of DataCom employees with X and X^2 as explanatory variables. Estimate this model using the given data. Perform relevant hypothesis tests to determine the significance of the regression coefficients of these two variables. Let $\alpha = 0.05$. Summarize your findings.

   c. How do you explain your findings in part **b** in light of the results found in part **a?**

17. The owner of a restaurant in Bloomington, Indiana, has recorded sales data for the past 19 years. He has also recorded data on potentially relevant variables. The data appear in the file P11_23.XLS.

   a. Estimate a multiple regression equation that includes annual sales as the response variable and the following explanatory variables: year, size of the population residing within 10 miles of the restaurant, annual advertising expenditures, and advertising expenditures in the *previous* year.

   b. Which of the explanatory variables have significant effects on sales at the 10% significance level? Do any of these results surprise you? Explain why or why not.

   c. Exclude all insignificant explanatory variables from the full model and estimate the reduced model. Comment on the significance of each remaining explanatory variable. Again, use a 10% significance level.

d. Based on your analysis of this problem, does mul-

ticollinearity appear to be present in the original or reduced versions of the model? Provide the reasoning behind your response.

 # Include/Exclude Decisions

In this section we make further use of the t-values of regression coefficients. In particular, we will see how they can be used to make include/exclude decisions for explanatory variables in a regression equation. From Section 12.3 we know that a t-value can be used to test whether a population regression coefficient is 0. But does this mean that we should automatically include a variable if its t-value is significant and automatically exclude it if its t-value is not significant? The decision is not always this simple.

The bottom line is that we are always trying to get the best fit possible, and because of the principle of parsimony, we want to use the fewest number of variables. This presents a trade-off, where there are often no easy answers. On the one hand, more variables certainly increase R^2 and they usually reduce the standard error of estimate s_e. On the other hand, fewer variables are better for parsimony. Therefore, we present several guidelines. These guidelines are not hard and fast rules, and they are sometimes contradictory. In real applications there are often several equations that are equally good for all practical purposes, and it is rather pointless to search for a single "true" equation.

Guidelines for Including/Excluding Variables in a Regression Equation

1. Look at a variable's t-value and its associated p-value. If the p-value is above some accepted significance level, such as 0.05, then this variable is a candidate for exclusion.

2. Check whether a variable's t-value is less than 1 or greater than 1 in magnitude. If it is less than 1, then s_e will decrease (and adjusted R^2 will increase) if this variable is excluded from the equation. If it is greater than 1, the opposite will occur. These are mathematical facts. Because of them, some statisticians advocate excluding variables with t-values less than 1 and including variables with t-values greater than 1.

3. Look at t-values and p-values, rather than correlations, when making include/exclude decisions. An explanatory variable can have a fairly high correlation with the response variable, but because of *other* variables included in the equation, it might not be needed. This would be reflected in a low t-value and a high p-value, and this variable could possibly be excluded for reasons of parsimony. This often occurs in the presence of multicollinearity.

4. When there is a group of variables that are in some sense logically related, it is sometimes a good idea to include all of them or exclude all of them. In this case, their individual t-values should not be used. Instead, the "partial F test" discussed in Section 12.7 should be used.

5. Use economic and/or physical theory to decide whether to include or exclude variables, and put less reliance on t-values and/or p-values. The idea is that some variables might really *belong* in an equation because of their theoretical relationship with the response variable, and their low t-values, possibly the result of an unlucky sample, should not disqualify them from being in the equation. Similarly, a variable that has no economic or physical relationship with the response variable might have gotten a significant t-value just by chance. This does not necessarily mean that it should be included in the equation. We should not use a computer package blindly to hunt for "good" explanatory variables. We should have some idea, before running the package, which variables belong and which do not.

Again, these guidelines can give contradictory signals. Specifically, guideline 2 bases the include/exclude decision on whether the magnitude of the t-value is greater or less than 1. However, analysts who base the decision on statistical significance at the usual 5% level, as in guideline 1, typically exclude a variable from the equation unless its t-value is at least 2 (approximately). This latter approach is more stringent—fewer variables will be retained—but it is probably the more popular approach. However, either approach is likely to result in "similar" equations for all practical purposes.

We illustrate how these guidelines can be used in the following example. It uses a slightly modified version of the data set on HyTex's mail-order customers from Chapter 3.

Example 12.3 Explaining Spending Amounts at HyTex

The file CATALOGS.XLS contains data on 1000 customers who purchased mail-order products from the HyTex Company in the current year. Recall from Example 3.11 in Chapter 3 that HyTex is a direct marketer of stereo equipment, personal computers, and other electronic products. HyTex advertises entirely by mailing catalogs to its customers, and all of its orders are taken over the telephone. The company spends a great deal of money on its catalog mailings, and it wants to be sure that this is paying off in sales. For each customer there are data on the following variables:

- Age: age of the customer at the end of the current year
- Gender: coded as 1 for males, 0 for females
- Own_Home: coded as 1 if customer owns a home, 0 otherwise
- Married: coded as 1 if customer is currently married, 0 otherwise
- Close: coded as 1 if customer lives reasonably close to a shopping area that sells similar merchandise, 0 otherwise
- Salary: combined annual salary of customer and spouse (if any)
- Children: number of children living with customer
- PrevCust: coded as 1 if customer purchased from HyTex during the previous year, 0 otherwise
- PrevSpent: total amount of purchases made from HyTex during the previous year
- Catalogs: number of catalogs sent to the customer this year
- AmountSpent: total amount of purchases made from HyTex this year

Estimate and interpret a regression equation for AmountSpent based on all of these variables.

Objective To see which potential explanatory variables are useful for explaining current year spending amounts at HyTex with multiple regression.

Solution

First, if you compare this data set to the data set in Chapter 3, you'll see that we made the following modifications to simplify the regression analysis.

- Age is now a continuous variable, not a categorical variable with three categories.
- Before, we had a History variable with four categories, depending on how much, if any, the customer purchased from HyTex in the previous year. Now we use the dummy variable PrevCust to indicate whether the customer purchased anything from

FIGURE 12.7

Regression Output with All Explanatory Variables Included

	A	B	C	D	E	F	G	H
1	**Results of multiple regression for AmountSpent**							
2								
3	**Summary measures**							
4		Multiple R	0.8893					
5		R-Square	0.7908					
6		Adj R-Square	0.7820					
7		StErr of Est	423.8584					
8								
9	**ANOVA Table**							
10		Source	df	SS	MS	F	p-value	
11		Explained	10	162299316.1832	16229931.6183	90.3390	0.0000	
12		Unexplained	239	42937764.0000	179655.9163			
13								
14	**Regression coefficients**							
15			Coefficient	Std Err	t-value	p-value	Lower limit	Upper limit
16		Constant	257.3477	132.9876	1.9351	0.0542	-4.6298	519.3251
17		Age	0.1884	1.7626	0.1069	0.9150	-3.2839	3.6607
18		Gender	-124.0805	55.7627	-2.2252	0.0270	-233.9295	-14.2315
19		Own Home	62.2752	60.7581	1.0250	0.3064	-57.4144	181.9648
20		Married	49.8426	70.1742	0.7103	0.4782	-88.3963	188.0815
21		Close	-282.7266	71.7762	-3.9390	0.0001	-424.1213	-141.3319
22		Salary	0.0143	0.0017	8.4930	0.0000	0.0110	0.0177
23		Children	-155.2858	31.5902	-4.9156	0.0000	-217.5166	-93.0550
24		PrevCust	-729.7213	92.3670	-7.9002	0.0000	-911.6786	-547.7639
25		PrevSpent	0.4725	0.0782	6.0447	0.0000	0.3185	0.6264
26		Catalogs	42.5806	4.3503	9.7880	0.0000	34.0108	51.1504

HyTex in the previous year. We also use the continuous variable PrevSpent for the amount purchased the previous year. Of course, if PrevCust equals 0, so does PrevSpent.

With this much data, 1000 observations, we can certainly afford to set aside part of the data set for validation, as discussed in Section 11.7. Although any split can be used, we decided to base the regression on the first 250 observations and use the other 750 for validation. Remember to select only the range through row 253 when using StatPro. Otherwise, all 1000 observations will be used.

We begin by entering all of the potential explanatory variables. Our goal is then to exclude variables that aren't necessary, based on their t-values and p-values. The multiple regression output with all explanatory variables appears in Figure 12.7. It indicates a fairly good fit. The R^2 value is 79.1% and s_e is about $424. When we consider that the actual amounts spent in the current year vary from a low of under $50 to a high of over $5500, with a median of about $950, a typical prediction error of around $424 is decent but not great.

From the p-value column, we see that there are three variables, Age, Own_Home, and Married, that have p-values well above 0.05. These are the obvious candidates for exclusion from the equation. We could rerun the equation with all three of these variables excluded, but it is better to proceed one step at a time. It is possible that when one of these variables is excluded, another one of them will become significant (the Right–Left foot phenomenon).

Actually, this did not happen. We first excluded the variable with the largest p-value, Age, and reran the regression. At this point, Own_Home and Married still had large p-values, and all other variables had small p-values. Next, we excluded Married, the variable with the largest remaining p-value, and reran the regression. Now, only Own_Home had a large p-value, so we ran one more regression with this variable excluded. The resulting output appears in Figure 12.8 on page 642. As we see, the R^2 and s_e values of 79.0% and $423 are practically as good as they were with all variables included, and all of the t-values are now large (well above 2 in absolute value) and the p-values are all small (well below 0.05).

FIGURE 12.8
Regression Output with Insignificant Variables Excluded

	A	B	C	D	E	F	G	H
1	**Results of multiple regression for AmountSpent**							
2								
3	**Summary measures**							
4		Multiple R	0.8885					
5		R-Square	0.7895					
6		Adj R-Square	0.7834					
7		StErr of Est	422.5169					
8								
9	**ANOVA Table**							
10		Source	df	SS	MS	F	p-value	
11		Explained	7	162035108.1832	23147872.5976	129.6650	0.0000	
12		Unexplained	242	43201972.0000	178520.5455			
13								
14	**Regression coefficients**							
15			Coefficient	Std Err	t-value	p-value	Lower limit	Upper limit
16		Constant	269.8643	108.5596	2.4859	0.0136	56.0218	483.7067
17		Gender	-130.3226	55.2112	-2.3604	0.0190	-239.0784	-21.5668
18		Close	-287.5537	70.8671	-4.0576	0.0001	-427.1488	-147.9586
19		Salary	0.0154	0.0014	11.1924	0.0000	0.0127	0.0181
20		Children	-158.4511	31.3378	-5.0562	0.0000	-220.1809	-96.7214
21		PrevCust	-724.0651	91.5870	-7.9058	0.0000	-904.4746	-543.6556
22		PrevSpent	0.4699	0.0777	6.0452	0.0000	0.3168	0.6230
23		Catalogs	42.6638	4.3204	9.8751	0.0000	34.1535	51.1741

We can interpret this final regression equation as follows:

Interpretation of Regression Equation

- The coefficient of Gender implies that an average male customer spent about $130 less than an average female customer, all other variables being equal. Similarly, an average customer living close to stores with this type of merchandise spent about $288 less than an average customer living far from such stores.

- The coefficient of Salary implies that, on average, about 1.5 cents of every extra salary dollar was spent on HyTex merchandise.

- The coefficient of Children implies that about $158 *less* was spent for every extra child living at home.

- The PrevCust and PrevSpent terms are somewhat more difficult to interpret. First, both of these terms are 0 for customers who didn't purchase from HyTex in the previous year. For those who did, the terms become $-724 + 0.47$PrevSpent. The coefficient 0.47 implies that each extra dollar spent the previous year can be expected to contribute an extra 47 cents in the current year. The -724 literally means that if we compare a customer who didn't purchase from HyTex last year to another customer who purchased only a tiny amount, the latter would be expected to spend about $724 less than the former this year. However, none of the latter customers were in the data set. A look at the data shows that of all customers who purchased from HyTex last year, almost all spent at least $100 and most spent considerably more. In fact, the median amount spent by these customers last year was about $900 (the median of all positive values for the PrevSpent variable). If we substitute this median value into the expression $-724 + 0.47$PrevSpent, we obtain -301. Therefore, this "median" spender from last year can be expected to spend about $301 less this year than the previous year nonspender.

- The coefficient of Catalogs implies that each extra catalog can be expected to generate about $43 in extra spending.

We conclude this example with a couple of cautionary notes. First, when we validate this final regression equation on the other 750 customers, using the procedure from Section 11.7, we find R^2 and s_e values of 71.8% and $522. Actually, these aren't bad. They show only a little deterioration from the values based on the original 250 customers. Second, we haven't tried all possibilities yet. We haven't tried nonlinear or interaction variables, nor have we looked at different coding schemes (such as treating Catalogs as a categorical variable and using dummy variables to represent it); we haven't checked for nonconstant error variance (remember that Figure 12.1 is based on this data set) or looked at the potential effects of outliers. ●

PROBLEMS

Level A

18. David Savageau and Geoffrey Loftus, the authors of *Places Rated Almanac* have ranked metropolitan areas in the United States with consideration of the following aspects of life in each area: cost of living, transportation, jobs, education, climate, crime, arts, health, and recreation. The data are in the file P02_55.XLS. Use multiple regression analysis to explore the relationship between the metropolitan area's overall score and the set of aforementioned numerical factors. Which explanatory variables should be included in a final version of this regression model? Justify your choices.

19. A manager of boiler drums wants to use regression analysis to predict the number of worker-hours needed to erect the drums in future projects. Consequently, data for 36 randomly selected boilers were collected. In addition to worker-hours (Y), the variables measured include boiler capacity, boiler design pressure, boiler type, and drum type. All of these measurements are listed in the file P11_27.XLS. Estimate an appropriate multiple regression model to predict the number of worker-hours needed to erect given boiler drums using all available explanatory variables. Which explanatory variables should be included in a final version of this regression model? Justify your choices.

20. An economic development researcher wants to understand the relationship between the size of the monthly home mortgage or rent payment for households in a particular middle-class neighborhood and the following set of household variables: family size, approximate location of the household within the neighborhood, an indication of whether those surveyed own or rent their home, gross annual income of the first household wage earner, gross annual income of the second household wage earner (if applicable), average monthly expenditure on utilities, and the total indebtedness (excluding the value of a home mortgage) of the household. Observations on each of these variables for a large sample of households are recorded in the file P02_06.XLS.
   a. In an effort to explain the variation in the size of the monthly home mortgage or rent payment, estimate a multiple regression model that includes all of the aforementioned household explanatory variables.
   b. Using your regression output, determine which of the explanatory variables should be *excluded* from the regression equation. Explain why you decide to remove each such variable.

21. Managers at Beta Technologies, Inc., have collected current annual salary figures and potentially related data for a random sample of 52 of the company's full-time employees. The data are in the file P02_01.XLS. These data include each selected employee's gender, age, number of years of relevant work experience prior to employment at Beta, the number of years of employment at Beta, and the number of years of post-secondary education.
   a. Estimate a multiple regression model to explain the variation in employee salaries at Beta Technologies using all of the potential explanatory variables.
   b. Using your regression output, determine which of the explanatory variables should be *excluded* from the regression equation. Provide reasoning for your decision to remove each such variable.

22. Stock market analysts are continually looking for reliable predictors of stock prices. Consider the problem of modeling the price per share of electric utility stocks (Y). Two variables thought to influence such a stock price are return on average equity (X_1) and annual dividend rate (X_2). The stock price, returns on equity, and dividend rates on a randomly selected day for 16 electric utility stocks are provided in the file P11_19.XLS.
   a. Estimate a multiple regression model using the given data. Include linear terms as well as an interaction term involving the return on average equity (X_1) and annual dividend rate (X_2).
   b. Which of the three explanatory variables (X_1, X_2, and X_1X_2) should be included in a final version of this regression model? Explain. Does your conclusion make sense in light of your knowledge of corporate finance?

12.6 Stepwise Regression[4]

Multiple regression represents an improvement over simple regression because it allows any number of explanatory variables to be included in the analysis. Sometimes, however, the large number of potential explanatory variables makes it difficult to know which variables to include. Many statistical packages provide some assistance by including automatic equation-building options. These options estimate a series of regression equations by successively adding (or deleting) variables according to prescribed rules. Generically, the methods are referred to as **stepwise regression.**

Before discussing how stepwise procedures work, consider a naive approach to the problem. We have already looked at correlation tables for indications of linear relationships. Why not simply include all explanatory variables that have large correlations with the response variable? There are two reasons for not doing this. First, although a variable is highly correlated with the response variable, it might also be highly correlated with other explanatory variables. Therefore, this variable might not be needed in the equation once the other explanatory variables have been included.

Second, even if a variable's correlation with the response variable is small, its contribution when it is included with a number of other explanatory variables can be greater than anticipated. Essentially, this variable can have something unique to say about the response variable that none of the other variables provides, and this fact might not be apparent from the correlation table.

For these reasons it is sometimes useful to let the computer discover the best combination of variables by means of a stepwise procedure. There are a number of procedures for building equations in a stepwise manner, but they all share a basic idea. Suppose that we have an existing regression equation and we want to add another variable to this equation from a set of variables not yet included. At this point, the variables already in the equation have explained a certain percentage of the variation of the response variable. The residuals represent the part still unexplained. Therefore, in choosing the next variable to enter the equation, we pick the one that is most highly correlated with the current residuals. If none of the remaining variables is highly correlated with the residuals, we might decide to quit. This is the essence of stepwise regression. However, besides adding variables to the equation, a stepwise procedure might delete a variable. This is sometimes reasonable because a variable entered early in the procedure might no longer be needed, given the presence of other variables that have entered since.

Stepwise regression (and its variations) can be helpful in discovering a useful regression model, but it should not be used mindlessly.

Many statistical packages have three types of equation-building procedures: *forward, backward,* and *stepwise.* A **forward** procedure begins with no explanatory variables in the equation and successively adds one at a time until no remaining variables make a significant contribution. A **backward** procedure begins with all potential explanatory variables in the equation and deletes them one at a time until further deletion would do more harm than good. Finally, a true **stepwise** procedure is much like a forward procedure, except that it also considers possible deletions along the way. All of these procedures have the same basic objective—namely, to find an equation with a small s_e and a large R^2 (or adjusted R^2). There is no guarantee that they will all produce exactly the same final equation, but in most cases their final results are very similar. The important thing to realize is that the equations estimated along the way, including the final equation, are estimated exactly as before—by least squares. Therefore, none of these procedures produces any new results. They merely take the burden off the user of having to decide ahead of time which variables to include in the equation.

---

[4]This section can be omitted without any loss of continuity.

The StatPro add-in implements each of the forward, backward, and stepwise procedures. To use them, we select the response variable and a set of *potential* explanatory variables. Then we specify the criterion for adding and/or deleting variables from the equation. This can be done in two ways, with an *F*-value or a *p*-value. We suggest using *p*-values because they are easier to understand, but either method is easy to use. In the *p*-value method, we select a *p*-value such as 0.05. If the regression coefficient for a potential entering variable would have a *p*-value less than 0.05 (if it were entered), then it is a candidate for entering (if the forward or stepwise procedure is used). The procedure selects the variable with the *smallest p*-value as the next entering variable. Similarly, if any currently included variable has a *p*-value greater than some value such as 0.05, then (with the stepwise and backward procedures) it is a candidate for leaving the equation. The methods stop when there are no candidates (according to their *p*-values) for entering or leaving the current equation.

The following continuation of the HyTex mail-order example illustrates these stepwise procedures.

Example 12.3 Explaining Spending Amounts at HyTex (continued)

The analysis of the HyTex mail-order data (for the first 250 customers in the data set) resulted in a regression equation that included all potential explanatory variables except for Age, Own_Home, and Married. We excluded these because their *t*-values were large and their *p*-values were small (less than 0.05). Do forward, backward, and stepwise procedures produce the same regression equation for the amount spent in the current year?

Objective To use StatPro's Stepwise Regression procedure to analyze the HyTex data.

Solution

Each of these options is found under the StatPro/Regression Analysis menu item. In each, we specify AmountSpent as the response variable and select all of the other variables (besides Customer) as *potential* explanatory variables. We then see a dialog box as in Figure 12.9 where we choose *p*-values or *F*-values as the appropriate criterion. (We chose

FIGURE 12.9
Dialog Box for Choosing Criterion in Stepwise Procedure

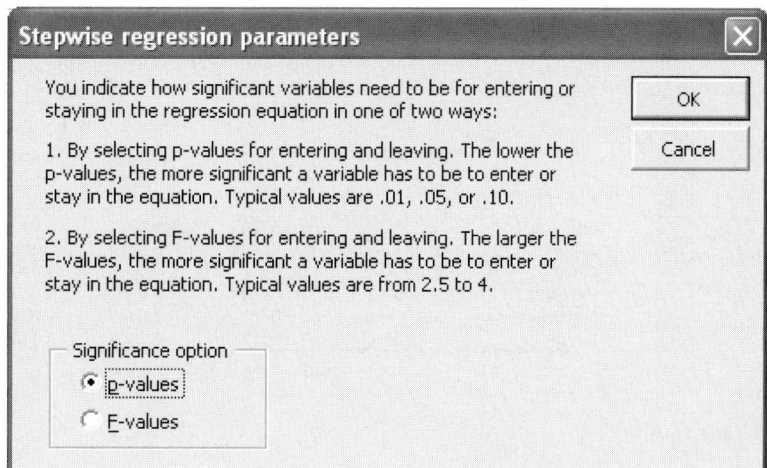

FIGURE 12.10

Dialog Box for Choosing *p*-values for Entering and Leaving

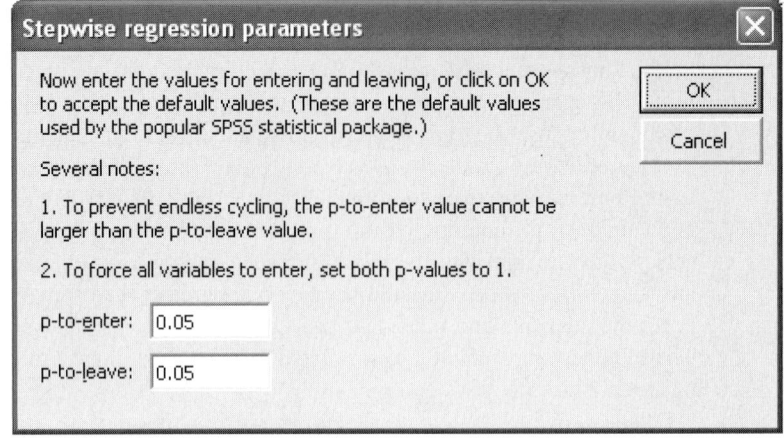

Stepwise regression parameters

Now enter the values for entering and leaving, or click on OK to accept the default values. (These are the default values used by the popular SPSS statistical package.)

Several notes:

1. To prevent endless cycling, the p-to-enter value cannot be larger than the p-to-leave value.

2. To force all variables to enter, set both p-values to 1.

p-to-enter: 0.05

p-to-leave: 0.05

OK

Cancel

FIGURE 12.11

Regression Output from Stepwise Procedure

	A	B	C	D	E	F	G	H
1	**Results of stepwise regression for AmountSpent**							
2								
3	**Step 1 - Entering variable: Salary**							
4								
5	**Summary measures**							
6		Multiple R	0.6624					
7		R-Square	0.4387					
8		Adj R-Square	0.4365					
9		StErr of Est	681.5285					
10								
11	**ANOVA Table**							
12		Source	df	SS	MS	F	p-value	
13		Explained	1	90045768.1832	90045768.1832	193.8631	0.0000	
14		Unexplained	248	115191312.0000	464481.0968			
15								
16	**Regression coefficients**							
17			Coefficient	Std Err	t-value	p-value	Lower limit	Upper limit
18		Constant	57.1089	91.6991	0.6228	0.5340	-123.4992	237.7170
19		Salary	0.0203	0.0015	13.9235	0.0000	0.0174	0.0232
20								
126	**Step 7 - Entering variable: Gender**							
127								
128	**Summary measures**			Change	% Change			
129		Multiple R	0.8885	0.0027	0.3%			
130		R-Square	0.7895	0.0048	0.6%			
131		Adj R-Square	0.7834	0.0041	0.5%			
132		StErr of Est	422.5169	-3.9560	-0.9%			
133								
134	**ANOVA Table**							
135		Source	df	SS	MS	F	p-value	
136		Explained	7	162035108.1832	23147872.5976	129.6650	0.0000	
137		Unexplained	242	43201972.0000	178520.5455			
138								
139	**Regression coefficients**							
140			Coefficient	Std Err	t-value	p-value	Lower limit	Upper limit
141		Constant	269.8643	108.5596	2.4859	0.0136	56.0218	483.7067
142		Salary	0.0154	0.0014	11.1924	0.0000	0.0127	0.0181
143		Catalogs	42.6638	4.3204	9.8751	0.0000	34.1535	51.1741
144		Children	-158.4511	31.3378	-5.0562	0.0000	-220.1809	-96.7214
145		Close	-287.5537	70.8671	-4.0576	0.0001	-427.1488	-147.9586
146		PrevCust	-724.0651	91.5870	-7.9058	0.0000	-904.4746	-543.6556
147		PrevSpent	0.4699	0.0777	6.0452	0.0000	0.3168	0.6230
148		Gender	-130.3226	55.2112	-2.3604	0.0190	-239.0784	-21.5668

p-values.) Next we see a dialog box as in Figure 12.10 where we choose the particular *p*-value or *F*-value for entering or leaving. (This particular dialog box is for the stepwise procedure, so we must choose a *p*-value for entering *and* for leaving. The latter cannot be less than the former, but they can be equal, as we show here. Note that the *default* *p*-value for leaving is 0.10; we selected 0.05 instead.)

It turns out that each procedure produces a *final* equation that is exactly the same as we obtained previously, with all variables except Age, Own_Home, and Married included. This often happens, but not always. The stepwise and forward procedures add the variables in the order Salary, Catalogs, Children, Close, PrevCust, PrevSpent, and Gender. The backward procedure, which starts with *all* variables in the equation, eliminates variables in the order Age, Married, and Own_Home.

A sample of the stepwise output appears in Figure 12.11. For each step of the procedure, we see which variable enters or leaves the equation. We also see the usual regression output for all the variables in the equation at that step, along with percentage changes in the key summary measures from the previous step to the current step. Again, however, the final equation's output is *exactly* the same as if we used multiple regression with these particular variables. ●

Stepwise regression or any of its variations can be very useful for narrowing down the set of all possible explanatory variables to a set that is useful for explaining a response variable. However, these procedures should not be used as a substitute for thoughtful analysis. With the availability of such procedures in statistical software packages, there is sometimes a tendency to turn the analysis over to the computer and accept its output. A good analyst does not just collect as much data as possible, throw it into a computer package, and blindly report the results. There should always be some rationale, whether it be based on economic theory, business experience, or common sense, for the variables that we use to explain a given response variable. A thoughtless use of stepwise regression can sometimes capitalize on chance to obtain an equation with a reasonably large R^2 but no useful or practical interpretation.

PROBLEMS

Level A

23. Suppose that you are interested in predicting the price of a laptop computer based on its various features. The file P11_35.XLS contains observations on the sales price and a number of potentially relevant variables for a randomly chosen sample of laptop computers. Employ stepwise regression to decide which explanatory variables to include in a regression equation. Use the *p*-value method with a cutoff value of 0.05 for entering and leaving. Summarize your findings.

24. Does the rate of violent crime acts vary across different regions of the United States? Using the data in P11_34.XLS and a stepwise regression procedure, develop an appropriate regression model to explain the variation in acts of violent crime across the United States. Use the *p*-value method with a cutoff value of 0.05 for entering and leaving. Summarize your results.

25. In a study of housing demand, a county assessor is interested in developing a regression model to estimate the selling price of residential properties within her ju-risdiction. She randomly selects 15 houses and records the selling price in addition to the following values: the size of the house (in hundreds of square feet), the total number of rooms in the house, the age of the house, and an indication of whether the house has an attached garage. These data are stored in the file P11_26.XLS.

a. Use stepwise regression to decide which explanatory variables should be included in the assessor's statistical model. Use the *p*-value method with a cutoff value of 0.05 for entering and leaving. Summarize your findings.

b. How do your results in part **a** change when the critical *p*-value for entering and leaving is increased to 0.10? Explain any differences between the regression equation obtained here and the one found in part **a**.

26. Continuing Problem 2, employ stepwise regression to evaluate your conclusions regarding the specification of a regression model to predict the sales of deep-dish pizza by the Original Italian Pizza restaurant chain.

Sample observations on all potentially relevant variables are provided in P11_04.XLS. Use the *p*-value method with a cutoff value of 0.05 for entering and leaving. Compare your conclusions in Problem 2 with those derived from a stepwise regression procedure in completing this problem.

27. Continuing Problem 3, employ stepwise regression to evaluate your conclusions regarding the specification of a regression model to explain the variation in values of the ACCRA Cost of Living Index. Data on potentially relevant expenditure components (i.e., explanatory variables) are provided in the file P02_19.XLS. Use the *p*-value method with a cutoff value of 0.05 for entering and leaving. Compare your conclusions in Problem 3 with those derived from a stepwise regression procedure in completing this problem.

Level B

28. What factors are truly useful in predicting a chief executive officer's annual base salary? Explore this question by employing a stepwise regression procedure on potentially relevant variables for which survey data have been collected and recorded in the file P02_13.XLS. Assess only those variables that make economic sense in predicting CEO base salaries. Also, consider incorporating a set of categorical variables to account for any potential variation in the base salaries that is explained by the CEO's company type. Use the *p*-value method with a cutoff value of 0.10 for entering and leaving. Summarize your findings.

12.7 The Partial *F* Test[5]

There are many situations where a set of explanatory variables form a logical group. It is then common to include all of the variables in the equation or exclude all of them. An example of this is when one of the explanatory variables is categorical with more than two categories. In this case we model it by including dummy variables—one less than the number of categories. If we decide that the categorical variable is worth including, we might want to keep all of the dummies. Otherwise, we might exclude all of them. We will look at an example of this type subsequently.

For now, consider the following general situation. We have already estimated an equation that includes the variables X_1 through X_j, and we are proposing to estimate a larger equation that includes X_{j+1} through X_k in addition to the variables X_1 through X_j. That is, the larger equation includes all of the variables from the smaller equation, but it also includes $k - j$ extra variables. These extra variables are the ones that form a group. We assume that it makes logical sense to include all of them or none of them.

The complete equation always contains all of the explanatory variables in the reduced equation, plus some more.

In this section we describe a test to determine whether the extra variables provide enough *extra* explanatory power as a group to warrant their inclusion in the equation. The test is called the partial *F* test. The original equation is called the **reduced** equation, and the larger equation is called the **complete** equation. In simple terms, the partial *F* test tests whether the complete equation is significantly better than the reduced equation.[6]

The test itself is intuitive. We use the output from the ANOVA tables of the reduced and complete equations to form an *F*-ratio. This ratio measures how much the sum of squared residuals, *SSE, decreases* by including the extra variables in the equation. It *must* decrease by some amount because the sum of squared residuals cannot increase when extra variables are added to an equation. But if it is does not decrease sufficiently, then the extra variables might not explain enough to warrant their inclusion in the equation, and we should probably exclude them. The *F*-ratio measures this. If it is sufficiently large, then we can conclude that the extra variables are worth including; otherwise, we can safely exclude them.

To state the test formally, we first state the relevant hypotheses. Let β_{j+1} through β_k be the coefficients of the extra variables in the complete equation. Then the null hypoth-

---

[5]This section is somewhat more advanced and can be omitted without any loss of continuity.
[6]StatPro does not run the partial *F* text, but it provides all of the ingredients.

esis is that these extra variables have no effect on the response variable; that is, $H_0: \beta_{j+1} = \cdots = \beta_k = 0$. The alternative is that at least one of the extra variables has an effect on the response variable, so that at least one of these β's is not 0. The hypotheses are summarized in the box.

Hypotheses for the Partial F Test	The null hypothesis is that the coefficients of all the extra explanatory variables in the complete equation are 0. The alternative is that at least one of these coefficients is not 0.

To run the test, we estimate both the reduced and complete equations and look at the associated ANOVA tables. Let SSE_R and SSE_C be the sums of squared errors from the reduced and complete equations, respectively. Also, let MSE_C be the mean square error for the complete equation. All of these quantities appear in the ANOVA tables. Next, we form the F-ratio in equation (12.4).

Test Statistic for Partial F Test	$$F\text{-ratio} = \frac{(SSE_R - SSE_C)/(k - j)}{MSE_C}$$	**(12.4)**

Note that the numerator includes the reduction in sum of squared errors discussed previously. If the null hypothesis is true, then this F-ratio has an F distribution with $k - j$ and $n - k - 1$ degrees of freedom. If it is sufficiently large, we reject H_0. As usual, the best way to run the test is to find the p-value corresponding to this F-ratio. This is the probability beyond the calculated F-ratio in the F distribution with $k - j$ and $n - k - 1$ degrees of freedom. In words, we reject the hypothesis that the extra variables have no explanatory power if this p-value is sufficiently small, less than 0.05, say.

This F-ratio and corresponding p-value are *not* part of the StatPro regression output. However, they are fairly easy to obtain. We run two regressions, one for the reduced equation and one for the complete equation, and use the appropriate values from their ANOVA tables to calculate the F-ratio in equation (12.4). Then we use Excel's FDIST function in the form FDIST(F-ratio,$k - j$,$n - k - 1$) to calculate the corresponding p-value. The procedure is illustrated in the following example. It uses the bank discrimination data from Example 11.3 of the previous chapter.

Example 12.4 Possible Gender Discrimination in Salary at Fifth National Bank of Springfield

Recall from Example 11.3 that Fifth National Bank has 208 employees. The data for these employees are stored in the file BANK.XLS. In the previous chapter we ran several regressions for Salary to see whether there is convincing evidence of salary discrimination against females. We will continue this analysis here. First, we'll regress Salary versus the Female dummy, YrsExper, and the interaction between Female and YrsExper, labeled Female*YrsExper. This will be the reduced equation. Then we'll see whether the JobGrade dummies Job_2 to Job_6 add anything significant to the reduced equation. If so, we will then see whether the interactions between the Female dummy and the JobGrade dummies, labeled Job2*Female to Job6*Female, add anything significant to what we already have. If so, we'll finally see whether the education dummies Ed_2 to Ed_5 add anything significant to what we already have.

FIGURE 12.12

Reduced Equation for Bank Example

	A	B	C	D	E	F	G	H
1	**Results of multiple regression for Salary**							
2								
3	**Summary measures**							
4		Multiple R	0.7991					
5		R-Square	0.6386					
6		Adj R-Square	0.6333					
7		StErr of Est	6.8163					
8								
9	**ANOVA Table**							
10		Source	df	SS	MS	F	p-value	
11		Explained	3	16748.8748	5582.9583	120.1620	0.0000	
12		Unexplained	204	9478.2324	46.4619			
13								
14	**Regression coefficients**							
15			Coefficient	Std Err	t-value	p-value	Lower limit	Upper limit
16		Constant	30.4300	1.2166	25.0129	0.0000	28.0314	32.8287
17		Female	4.0983	1.6658	2.4602	0.0147	0.8138	7.3827
18		YrsExper	1.5278	0.0905	16.8887	0.0000	1.3494	1.7061
19		Female*YrsExper	-1.2478	0.1367	-9.1296	0.0000	-1.5173	-0.9783

Objective To use several partial F tests to see whether various groups of explanatory variables should be included in a regression equation for salary, given that other variables are already in the equation.

Solution

First, note that we created all of the dummies and interaction variables with StatPro's Data Utilities procedures. These could be entered directly with Excel functions (see the formulas in the AnalysisData sheet of the BANK.XLS file for details), but StatPro makes the process much quicker and easier. Also, note that we have used three sets of dummies, for gender, job grade, and education level. When we use these in a regression equation, the dummy for one category of each should always be excluded; it is the reference category. The reference categories we have used are "male," job grade 1, and education level 1.

The output for the "smallest" equation, using Female, YrsExper, and Female*YrsExper as explanatory variables, appears in Figure 12.12. (We put this output in a sheet called Reduced.) We're off to a good start. These three variables already explain 63.9% of the variation in Salary.

The output for the next equation, which adds the explanatory variables Job_2 to Job_6, appears in Figure 12.13. (We put this output in a sheet called Complete.) This equation appears to be much better. For example, R^2 has increased to 81.1%. We check whether it is *significantly* better with the partial F test in rows 26–30. (This part of the output is not given by StatPro; we have to enter it manually.) The degrees of freedom in cell C27 is 5, the number of *extra* variables. The degrees of freedom in cell C28 is the same as the value in cell C12, the degrees of freedom for *SSE*. Then we calculate the F-ratio in cell C29 with the formula

$$=((Reduced!D12-Complete!D12)/Complete!C27)/Complete!E12$$

where Reduced!D12 refers to *SSE* for the reduced equation from the Reduced sheet. Finally, we calculate the corresponding p-value in cell C30 with the formula

$$=FDIST(C29,C27,C28)$$

It is practically 0, so there is no doubt that the job grade dummies add significantly to the explanatory power of the equation.

FIGURE 12.13

Equation with Job
Dummies Added

	A	B	C	D	E	F	G	H
1		*Results of multiple regression for Salary*						
2								
3		*Summary measures*						
4		Multiple R	0.9005					
5		R-Square	0.8109					
6		Adj R-Square	0.8033					
7		StErr of Est	4.9916					
8								
9		*ANOVA Table*						
10		Source	df	SS	MS	F	p-value	
11		Explained	8	21268.7391	2658.5924	106.7004	0.0000	
12		Unexplained	199	4958.3682	24.9164			
13								
14		*Regression coefficients*						
15			Coefficient	Std Err	t-value	p-value	Lower limit	Upper limit
16		Constant	26.1042	1.1054	23.6143	0.0000	23.9243	28.2841
17		Female	6.0633	1.2663	4.7881	0.0000	3.5662	8.5605
18		YrsExper	1.0709	0.1020	10.4975	0.0000	0.8697	1.2720
19		Female*YrsExper	-1.0211	0.1187	-8.6001	0.0000	-1.2552	-0.7869
20		Job_2	2.5965	1.0101	2.5705	0.0109	0.6046	4.5884
21		Job_3	6.2214	0.9982	6.2328	0.0000	4.2530	8.1898
22		Job_4	11.0720	1.1726	9.4423	0.0000	8.7597	13.3842
23		Job_5	14.9466	1.3402	11.1521	0.0000	12.3037	17.5895
24		Job_6	17.0974	2.3907	7.1517	0.0000	12.3831	21.8117
25								
26		*Partial F test for including job dummies*						
27		df numerator	5					
28		df denominator	199					
29		F-ratio	36.2802					
30		p-value	0.0000					

Do the interactions between the Female dummy and the job dummies add anything more? We again use the partial F test, but now the previous *complete* equation becomes the new *reduced* equation, and the equation that includes the new interaction terms becomes the new complete equation. The output for this new complete equation appears in Figure 12.14 (page 652). (We put this output in a sheet called MoreComplete.) We perform the partial F test in rows 31–35 exactly as before. For example, the formula for the F-ratio in cell C34 is

$$=((\text{Complete!D12}-\text{MoreComplete!D12})/\text{MoreComplete!C32})/\text{MoreComplete!E12}$$

Note how the SSE_R term in equation (12.4) now comes from the Complete sheet since this sheet contains the current *reduced* equation. As we see, the terms "reduced" and "complete" are relative. What is complete in one stage becomes reduced in the next stage. In any case, the p-value in cell C35 is again extremely small, so there is no doubt that the interaction terms add significantly to what we already had (even though R^2 has increased from 81.1% to only 84.0%).

Finally, we add the education dummies. The resulting output is shown in Figure 12.15 (page 653). (We put this output in a sheet called StillMoreComplete.) Again, we see how the terms reduced and complete are relative. This output now corresponds to the complete equation, and the previous output corresponds to the reduced equation. The formula in cell C38 for the F-ratio is now

$$=((\text{MoreComplete!D12}-$$
$$\text{StillMoreComplete!D12})/\text{StillMoreComplete!C36})/\text{StillMoreComplete!E12}$$

Its SSE_R value comes from the MoreComplete sheet. Note that the increase in R^2 is from 84.0% to only 84.7%. Also, the p-value is not extremely small. According to the partial F test, it is not quite enough to qualify for statistical significance at the 5% level. Based

FIGURE 12.14

Regression Output with Interaction Terms Added

	A	B	C	D	E	F	G	H
1	**Results of multiple regression for Salary**							
2								
3	**Summary measures**							
4		Multiple R	0.9163					
5		R-Square	0.8396					
6		Adj R-Square	0.8289					
7		StErr of Est	4.6564					
8								
9	**ANOVA Table**							
10		Source	df	SS	MS	F	p-value	
11		Explained	13	22020.7615	1693.9047	78.1242	0.0000	
12		Unexplained	194	4206.3457	21.6822			
13								
14	**Regression coefficients**							
15			Coefficient	Std Err	t-value	p-value	Lower limit	Upper limit
16		Constant	26.5155	1.4324	18.5112	0.0000	23.6904	29.3406
17		Female	4.7245	1.7354	2.7225	0.0071	1.3019	8.1470
18		YrsExper	0.9608	0.1042	9.2214	0.0000	0.7553	1.1663
19		Female*YrsExper	-0.8060	0.1303	-6.1845	0.0000	-1.0630	-0.5490
20		Job_2	3.3410	1.8642	1.7922	0.0747	-0.3357	7.0177
21		Job_3	7.8720	2.2149	3.5540	0.0005	3.5035	12.2404
22		Job_4	10.6919	1.9567	5.4643	0.0000	6.8328	14.5510
23		Job_5	13.1464	1.9931	6.5958	0.0000	9.2154	17.0774
24		Job_6	20.9794	2.7676	7.5803	0.0000	15.5210	26.4379
25		Job_2*Female	-0.9434	2.1640	-0.4359	0.6634	-5.2114	3.3246
26		Job_3*Female	-1.9350	2.4414	-0.7926	0.4290	-6.7502	2.8801
27		Job_4*Female	0.4338	2.3750	0.1827	0.8553	-4.2503	5.1179
28		Job_5*Female	4.8734	2.6232	1.8578	0.0647	-0.3002	10.0470
29		Job_6*Female	-27.3274	5.7700	-4.7361	0.0000	-38.7074	-15.9474
30								
31	**Partial F test for including job dummies**							
32		df numerator	5					
33		df denominator	194					
34		F-ratio	6.9368					
35		p-value	0.0000					

on this evidence, there is not much to gain from including the education dummies in the equation, so we would probably elect to exclude them.

Note that the results could be very different if we entered groups in a different order. For example, you might try entering the education dummies, and then interactions between these dummies and Female, *before* entering the job grade dummies. The results will be quite different. Again, remember that because of potential multicollinearity, what is significant can depend on what *other* variables are already in the equation.

Before leaving this example, we make several comments. First, the partial test is *the* formal test of significance for an extra set of variables. Many users look only at the R^2 and/or s_e values to check whether extra variables are doing a "good job." For example, they might cite that R^2 went from 81.1% to 84.0% or that s_e went from 4.992 to 4.656 as evidence that extra variables provide a "significantly" better fit. Although these are important indicators, they are not the basis for a formal hypothesis test.

Second, if the partial F test shows that a block of variables is significant, it does not imply that each variable in this block is significant. Some of these variables can have low t-values. Consider Figure 12.14, for example. We are able to conclude that the Female/Job interactions as a whole are significant. But three of these interactions, Job_2*Female to Job_4*Female are clearly not significant, and Job_5*Female is borderline. In fact, Job_6*Female is the only one that is clearly significant. Some analysts favor excluding the *individual* variables that aren't significant, whereas others favor keeping the whole block or excluding the whole block. We lean toward the latter but recognize that either approach is valid—and the results are nearly the same either way.

Third, producing all of these outputs and doing the partial F tests is a lot of work. Therefore, we included a "Block" option in StatPro to make life easier. To run the analy-

FIGURE 12.15
Regression Output with Education Dummies Added

	A	B	C	D	E	F	G	H
1	*Results of multiple regression for Salary*							
2								
3	*Summary measures*							
4		Multiple R	0.9205					
5		R-Square	0.8473					
6		Adj R-Square	0.8336					
7		StErr of Est	4.5914					
8								
9	*ANOVA Table*							
10		Source	df	SS	MS	F	p-value	
11		Explained	17	22221.6888	1307.1582	62.0060	0.0000	
12		Unexplained	190	4005.4185	21.0811			
13								
14	*Regression coefficients*							
15			Coefficient	Std Err	t-value	p-value	Lower limit	Upper limit
16		Constant	26.0205	1.6784	15.5027	0.0000	22.7097	29.3313
17		Female	4.3738	1.7247	2.5360	0.0120	0.9718	7.7758
18		YrsExper	1.0024	0.1045	9.5878	0.0000	0.7961	1.2086
19		Female*YrsExper	-0.7608	0.1299	-5.8584	0.0000	-1.0170	-0.5047
20		Ed_2	-0.6648	1.1204	-0.5933	0.5537	-2.8748	1.5453
21		Ed_3	0.6124	1.0823	0.5658	0.5722	-1.5225	2.7473
22		Ed_4	0.0491	1.9616	0.0250	0.9800	-3.8201	3.9184
23		Ed_5	2.8082	1.3035	2.1543	0.0325	0.2370	5.3793
24		Job_2	2.6973	1.8757	1.4380	0.1521	-1.0025	6.3971
25		Job_3	6.8626	2.2492	3.0511	0.0026	2.4259	11.2993
26		Job_4	8.7459	2.0547	4.2565	0.0000	4.6930	12.7989
27		Job_5	10.5796	2.1800	4.8530	0.0000	6.2794	14.8797
28		Job_6	18.2024	2.9402	6.1908	0.0000	12.4027	24.0021
29		Job_2*Female	-0.7138	2.1484	-0.3323	0.7401	-4.9515	3.5239
30		Job_3*Female	-1.7529	2.4305	-0.7212	0.4717	-6.5472	3.0415
31		Job_4*Female	1.0232	2.3839	0.4292	0.6683	-3.6791	5.7255
32		Job_5*Female	5.2410	2.6231	1.9980	0.0471	0.0669	10.4151
33		Job_6*Female	-29.3752	5.7539	-5.1053	0.0000	-40.7249	-18.0255
34								
35	*Partial F test for including job dummies*							
36		df numerator	4					
37		df denominator	190					
38		F-ratio	2.3828					
39		p-value	0.0530					

sis in this example in one step, we use the StatPro/Regression Analysis/Block menu item. After selecting Salary as the response variable, we see the dialog box in Figure 12.16 (page 654) and fill it out as shown. We want four blocks of explanatory variables, and we want a given block to enter only if it passes the partial F test at the 5% level. In later dialog boxes, we then specify the explanatory variables in each block. Block 1 has Female, YrsExper, and Female*YrsExper, block 2 has the job grade dummies, and so on.

Once we have specified all of this, the regression calculations are done in stages. At each stage, the partial F test checks whether a block is significant. If so, the variables in this block enter and we progress to the next stage. If not, the process ends; neither this block nor any later blocks are entered.

The output from this procedure appears in Figures 12.17 and 12.18. (It wouldn't all fit in a single figure.) This output is basically a repetition of Figures 12.12, 12.13, and 12.14. However, in rows 26–29 and 51–54 note the percentage changes in the key summary statistics from one block to the next. Also, note that the output with the final block, the education dummies, does not appear at all. This block did not pass the partial F test at the 5% level, so it does not appear in the output.

Finally, we have concentrated on the partial F test and statistical significance in this example. We don't want you to lose sight, however, of the bigger picture. Once we have decided on a "final" regression equation, say, the one in Figure 12.14, we need to analyze its implications for the problem at hand. In this case the bank is interested in possible salary discrimination against females, so we should interpret this final equation in these

FIGURE 12.16

Dialog Box for
StatPro's Block
Regression Option

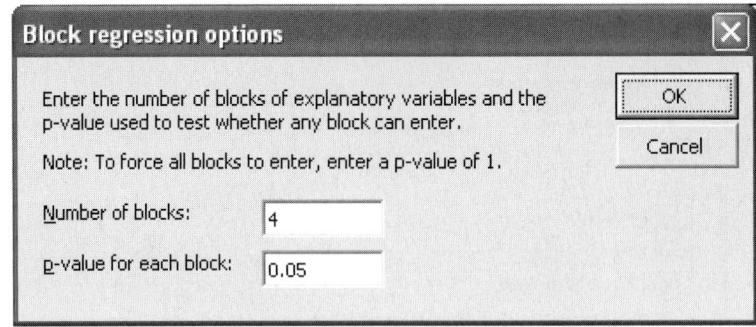

Block regression options

Enter the number of blocks of explanatory variables and the
p-value used to test whether any block can enter.

Note: To force all blocks to enter, enter a p-value of 1.

OK

Cancel

Number of blocks: 4

p-value for each block: 0.05

FIGURE 12.17

First Part of
Block Regression
Output

	A	B	C	D	E	F	G	H
1	*Results of block regression for Salary*							
2								
3	*Block 1 enters (corresponding variables shown in bold)*							
4								
5	*Summary measures*							
6		Multiple R	0.7991					
7		R-Square	0.6386					
8		Adj R-Square	0.6333					
9		StErr of Est	6.8163					
10								
11	*ANOVA Table*							
12		Source	df	SS	MS	F	p-value	
13		Explained	3	16748.8748	5582.9583	120.1620	0.0000	
14		Unexplained	204	9478.2324	46.4619			
15								
16	*Regression coefficients*							
17			Coefficient	Std Err	t-value	p-value	Lower limit	Upper limit
18		Constant	30.4300	1.2166	25.0129	0.0000	28.0314	32.8287
19		**Female**	4.0983	1.6658	2.4602	0.0147	0.8138	7.3827
20		**YrsExper**	1.5278	0.0905	16.8887	0.0000	1.3494	1.7061
21		**Female*YrsExper**	-1.2478	0.1367	-9.1296	0.0000	-1.5173	-0.9783
22								
23	*Block 2 enters (corresponding variables shown in bold)*							
24								
25	*Summary measures*			Change	% Change			
26		Multiple R	0.9005	0.1014	12.7%			
27		R-Square	0.8109	0.1723	27.0%			
28		Adj R-Square	0.8033	0.1700	26.9%			
29		StErr of Est	4.9916	-1.8247	-26.8%			
30								
31	*ANOVA Table*							
32		Source	df	SS	MS	F	p-value	
33		Explained	8	21268.7391	2658.5924	106.7004	0.0000	
34		Unexplained	199	4958.3682	24.9164			
35								
36	*Regression coefficients*							
37			Coefficient	Std Err	t-value	p-value	Lower limit	Upper limit
38		Constant	26.1042	1.1054	23.6143	0.0000	23.9243	28.2841
39		Female	6.0633	1.2663	4.7881	0.0000	3.5662	8.5605
40		YrsExper	1.0709	0.1020	10.4975	0.0000	0.8697	1.2720
41		Female*YrsExper	-1.0211	0.1187	-8.6001	0.0000	-1.2552	-0.7869
42		**Job_2**	2.5965	1.0101	2.5705	0.0109	0.6046	4.5884
43		**Job_3**	6.2214	0.9982	6.2328	0.0000	4.2530	8.1898
44		**Job_4**	11.0720	1.1726	9.4423	0.0000	8.7597	13.3842
45		**Job_5**	14.9466	1.3402	11.1521	0.0000	12.3037	17.5895
46		**Job_6**	17.0974	2.3907	7.1517	0.0000	12.3831	21.8117

FIGURE 12.18
Second Part of Block Regression Output

	A	B	C	D	E	F	G	H
48	**Block 3 enters (corresponding variables shown in bold)**							
49								
50	**Summary measures**			Change	% Change			
51		Multiple R	0.9163	0.0158	1.8%			
52		R-Square	0.8396	0.0287	3.5%			
53		Adj R-Square	0.8289	0.0255	3.2%			
54		StErr of Est	4.6564	-0.3352	-6.7%			
55								
56	**ANOVA Table**							
57		Source	df	SS	MS	F	p-value	
58		Explained	13	22020.7615	1693.9047	78.1242	0.0000	
59		Unexplained	194	4206.3457	21.6822			
60								
61	**Regression coefficients**							
62			Coefficient	Std Err	t-value	p-value	Lower limit	Upper limit
63		Constant	26.5155	1.4324	18.5112	0.0000	23.6904	29.3406
64		Female	4.7245	1.7354	2.7225	0.0071	1.3019	8.1470
65		YrsExper	0.9608	0.1042	9.2214	0.0000	0.7553	1.1663
66		Female*YrsExper	-0.8060	0.1303	-6.1845	0.0000	-1.0630	-0.5490
67		Job_2	3.3410	1.8642	1.7922	0.0747	-0.3357	7.0177
68		Job_3	7.8720	2.2149	3.5540	0.0005	3.5035	12.2404
69		Job_4	10.6919	1.9567	5.4643	0.0000	6.8328	14.5510
70		Job_5	13.1464	1.9931	6.5958	0.0000	9.2154	17.0774
71		Job_6	20.9794	2.7676	7.5803	0.0000	15.5210	26.4379
72		**Job_2*Female**	-0.9434	2.1640	-0.4359	0.6634	-5.2114	3.3246
73		**Job_3*Female**	-1.9350	2.4414	-0.7926	0.4290	-6.7502	2.8801
74		**Job_4*Female**	0.4338	2.3750	0.1827	0.8553	-4.2503	5.1179
75		**Job_5*Female**	4.8734	2.6232	1.8578	0.0647	-0.3002	10.0470
76		**Job_6*Female**	-27.3274	5.7700	-4.7361	0.0000	-38.7074	-15.9474

terms. We will not go through this exercise again here—we did similar interpretations in the previous chapter. Our point is simply that you shouldn't get so caught in the details of statistical significance that you lose sight of the original purpose of the analysis! ●

PROBLEMS

Level A

29. A regional express delivery service company recently conducted a study to investigate the relationship between the cost of shipping a package (Y), the package weight (X_1), and the distance shipped (X_2). Twenty packages were randomly selected from among the large number received for shipment and a detailed analysis of the shipping cost was conducted for each package. These sample observations are given in the file P11_24.XLS.

   a. Estimate a multiple regression model involving the two given explanatory variables. Using the ANOVA table, perform and interpret the result of an F test. Use a 5% significance level in making the statistical decision in this case.

   b. Is it worthwhile to add the terms X_1^2 and X_2^2 to the regression equation of part a? Base your decision here on a partial F test. Once again, employ a 5% significance level in performing this test.

   c. Is it worthwhile to add the term X_1X_2 to the most appropriate reduced equation as determined in part

b? Again, perform a partial F test with a 5% significance level.

   d. Based on the previous findings, what regression equation should this company use in predicting the cost of shipping a package? Defend your recommendation.

30. Suppose you are interested in predicting the price of a laptop computer based on its features. The file P11_35.XLS contains observations on the sales price and a number of potentially relevant variables for a randomly chosen sample of laptop computers.

   a. Estimate a multiple regression model that predicts the price of a laptop computer using the following quantitative variables: the speed of the computer's CPU, the length of time the computer's battery maintains its charge, and the size of the computer's RAM. Assess this set of explanatory variables with an F test, and report a p-value.

   b. Do explanatory variables that model the computer's chip type and monitor type contribute significantly to the prediction of the laptop's sales price? Let the equation estimated in part **a** serve as the reduced equation in a partial F test. Employ a 5% significance level in conducting the appropriate hypothesis test in this case.

c. Do explanatory variables that model the computer's pointing device and the availability of a help line for buyers contribute significantly to the prediction of the laptop's sales price? Let the most appropriate equation found from the analysis in part **b** serve as the reduced equation in a partial F test. Again, employ a 5% significance level in conducting the appropriate hypothesis test in this case.

31. Many companies manufacture products that are at least partially produced using chemicals (for example, paint, gasoline, and steel). In many cases, the quality of the finished product is a function of the temperature and pressure at which the chemical reactions take place. Suppose that a particular manufacturer wants to model the quality (Y) of a product as a function of the temperature (X_1) and the pressure (X_2) at which it is produced. The file P11_39.XLS contains data obtained from a designed experiment involving these variables. Note that the assigned quality score can range from a minimum of 0 to a maximum of 100 for each manufactured product.

a. Estimate a multiple regression model that includes the two given explanatory variables. Assess this set of explanatory variables with an F test, and report a p-value.

b. Conduct a partial F test to decide whether it is worthwhile to add second-order terms (i.e., X_1^2, X_2^2, and X_1X_2) to the multiple regression equation estimated in part **a.** Employ a 5% significance level in conducting this hypothesis test.

c. Which regression equation is the most appropriate one for modeling the quality of the given product? Bear in mind that a good statistical model is usually parsimonious.

Level B

32. Continuing Problem 6, we'll refer to the original multiple regression model (i.e., the one that includes the age of the auctioned item and the number of bidders as explanatory variables) as the *reduced* equation. Suppose now that the antique collector believes that

the *rate of increase* of the auction price with the age of the item will be driven upward by a large number of bidders.

a. Revise the multiple regression model developed previously to model this additional feature of the problem. Estimate this larger regression equation, which we call the *complete* equation, using the sample data in the file P11_18.XLS.

b. Perform a partial F test to check whether the complete equation is significantly better than the reduced equation. Use a 5% level of significance.

33. An economic development researcher wants to understand the relationship between the size of the monthly home mortgage or rent payment for households in a particular middle-class neighborhood and the following set of household variables: family size, approximate location of the household within the neighborhood, an indication of whether those surveyed owned or rented their home, gross annual income of the first household wage earner, gross annual income of the second household wage earner (if applicable), average monthly expenditure on utilities, and the total indebtedness (excluding the value of a home mortgage) of the household. Observations on these variables for a large sample of households are recorded in the file P02_06.XLS.

a. To explain the variation in the size of the monthly home mortgage or rent payment, formulate a multiple regression model that includes all of the *quantitative* household variables in the aforementioned set. Estimate this model using the given sample data. Perform an F test of the model's overall significance, and report a p-value.

b. Determine whether the *qualitative* (i.e., categorical) variable that models the location of the household within the neighborhood adds significantly to explaining the variation in the size of the monthly home mortgage or rent payment. Use a 5% significance level in conducting this hypothesis test.

c. Determine whether it is worthwhile to add a variable that models whether the home is owned or rented to the most appropriate regression equation from part **b.** Again, use a 5% significance level in conducting this hypothesis test.

12.8 Outliers

In all of the regression examples we have analyzed to this point, we have ignored the possibility of outliers. Unfortunately, in many real applications we cannot afford to ignore outliers. They are often present, and they can often have a substantial effect on the results. In this section we will briefly discuss outliers in the context of regression—how to detect them and what to do about them.

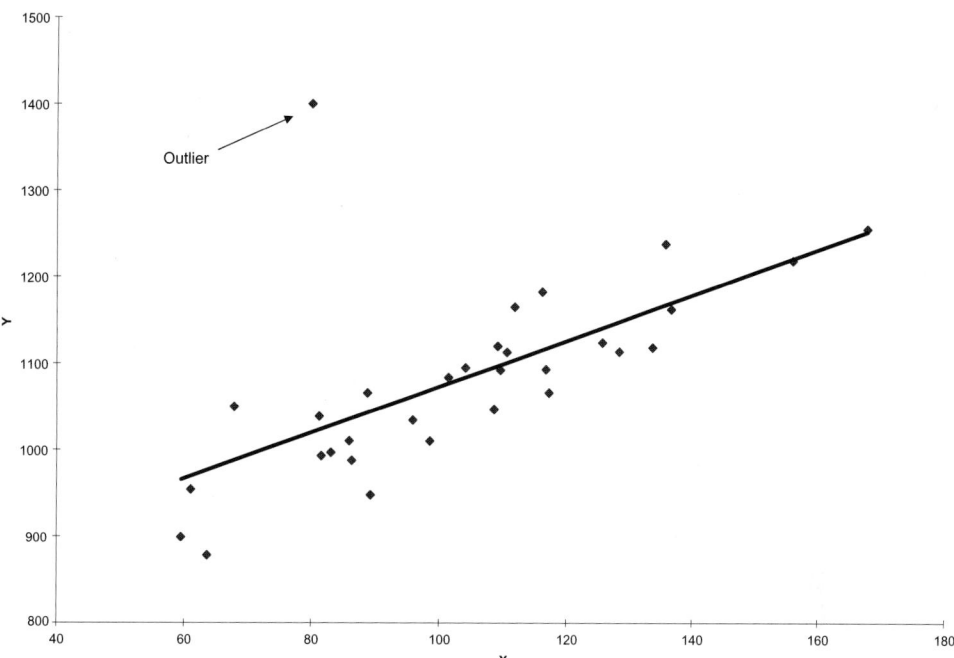

FIGURE 12.19 Outlier with a Large Residual

We tend to think of an outlier as an observation that has an extreme value for at least one variable. For example, if salaries in a data set are mostly in the $40,000–$80,000 range, but one salary is $350,000, then this observation is a clear outlier with respect to salary. However, in a regression context outliers are not always this obvious. In fact, an observation can be considered an outlier for several reasons, and some types of outliers can be difficult to detect. An observation can be an outlier for one or more of the following reasons.

Characteristics of an Outlier

Outliers can come in several forms, as indicated in this list.

1. It has an extreme value for one or more variables.

2. Its value of the response variable is much larger or smaller than predicted by the regression line, and its residual is abnormally large in magnitude. An example appears in Figure 12.19. The line in this scatterplot fits most of the points, but it misses badly on the one obvious outlier. This outlier has a large positive residual, but its Y value is not abnormally large. Its Y value is only large relative to points with the same X value that it has.

3. Its residual is not only large in magnitude, but this point "tilts" the regression line toward it. An example appears in Figure 12.20 on page 658. The two lines shown are the regression lines with the outlier and without it. If we keep the outlier, it makes a big difference on the slope and intercept of the regression line. This type of outlier is called an **influential** point, for the obvious reason.

4. Its values of individual explanatory variables are not extreme, but they fall outside the general pattern of the other observations. An example appears in Figure 12.21. Here, we assume that the two variables shown, YrsExper (years of experience) and Rating (an employee's performance rating) are both explanatory variables for some other response variable (Salary) that isn't shown in the plot. The obvious outlier does not have

FIGURE 12.20
Outlier That Tilts the Regression Line

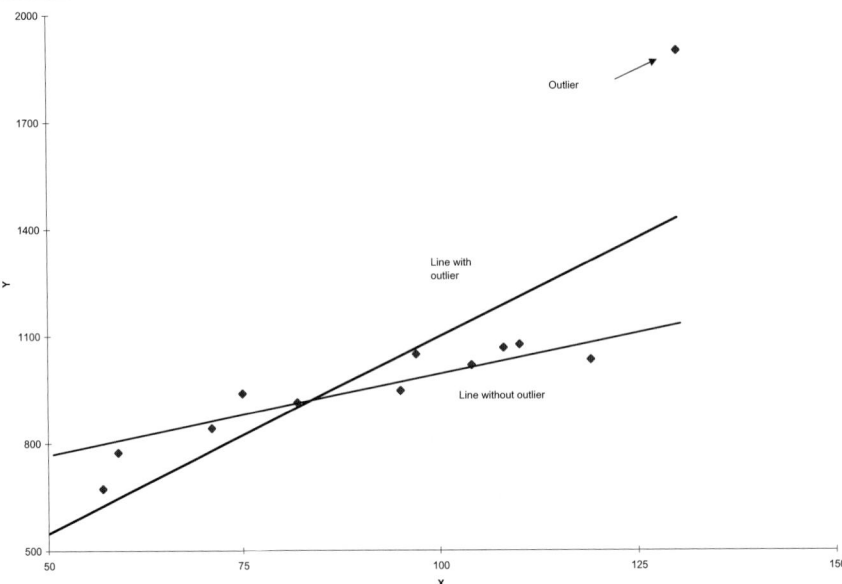

FIGURE 12.21
Outlier Outside Pattern of Explanatory Variables

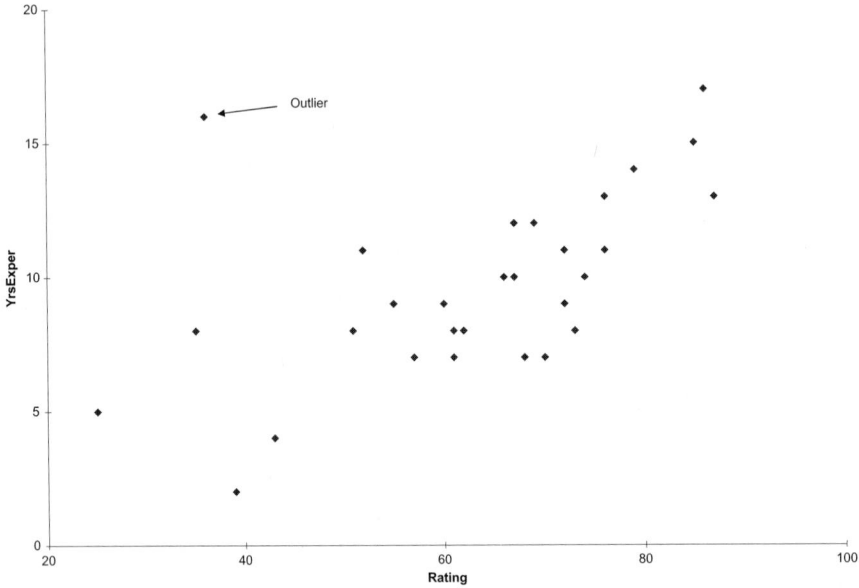

an abnormal value of either YrsExper or Rating, but it falls well outside the pattern of most employees.

Once we have identified outliers, there is still the thorny problem of what to do with them. In most cases the regression output will look "nicer" if we delete outliers, but this is not necessarily appropriate. If we can argue that the outlier isn't really a member of the relevant population, then it is appropriate and probably best to delete it. But if no such argument can be made, then it is not really appropriate to delete the outlier just to make the analysis come out better. Perhaps the best advice in this case is the advice we gave in the previous chapter. Run the analysis with the outliers and without them. If the key outputs do not change much, then it does not really matter whether the outliers are included or not. If the key outputs do change substantially, then report the results both with and without the outliers, along with a verbal explanation.

We illustrate this procedure in the following continuation of the bank discrimination example.

Example (12.4) **Possible Gender Discrimination in Salary at Fifth National Bank of Springfield (continued)**

Of the 208 employees at Fifth National Bank, are there any obvious outliers? In what sense are they outliers? Does it matter to the regression results, particularly those concerning gender discrimination, whether the outliers are removed?

Objective To locate possible outliers in the bank salary data, and to see to what extent they affect the regression model.

Solution

There are several places we could look for outliers. An obvious place is the Salary variable. The boxplot in Figure 12.22 shows that there are several employees making substantially more in salary than most of the employees. (See the file BANK.XLS.) We could consider these outliers and remove them, arguing perhaps that these are senior managers who shouldn't be included in the discrimination analysis. We leave it to you to check whether the regression results are any different with these high-salary employees than without them.

Another place to look is at a scatterplot of the residuals versus the fitted values. This type of plot (offered as an option by StatPro) shows points with abnormally large residuals. For example, we ran the regression with Female, YrsExper, Female*YrsExper, and the five job grade dummies, and we obtained the output and scatterplot in Figures 12.23 and 12.24 (page 660). This scatterplot has several points that could be considered outliers, but we focus on the point identified in the figure. The residual for this point is approximately -21. Given that s_e for this regression is approximately 5, this residual is over

FIGURE 12.22
Boxplot of Salaries for Bank Data

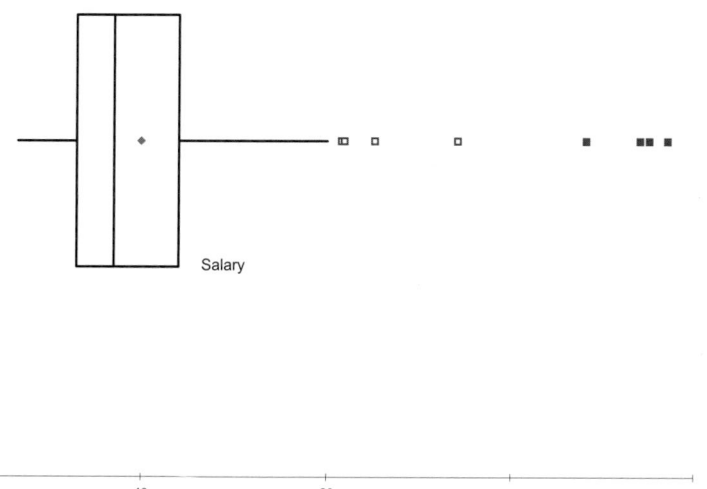

FIGURE 12.23
Regression Output with Outlier Included

	A	B	C	D	E	F	G	H
1	**Results of multiple regression for Salary**							
2								
3	**Summary measures**							
4		Multiple R	0.9005					
5		R-Square	0.8109					
6		Adj R-Square	0.8033					
7		StErr of Est	4.9916					
8								
9	**ANOVA Table**							
10		Source	df	SS	MS	F	p-value	
11		Explained	8	21268.7391	2658.5924	106.7004	0.0000	
12		Unexplained	199	4958.3682	24.9164			
13								
14	**Regression coefficients**							
15			Coefficient	Std Err	t-value	p-value	Lower limit	Upper limit
16		Constant	26.1042	1.1054	23.6143	0.0000	23.9243	28.2841
17		Female	6.0633	1.2663	4.7881	0.0000	3.5662	8.5605
18		YrsExper	1.0709	0.1020	10.4975	0.0000	0.8697	1.2720
19		Female*YrsExper	-1.0211	0.1187	-8.6001	0.0000	-1.2552	-0.7869
20		Job_2	2.5965	1.0101	2.5705	0.0109	0.6046	4.5884
21		Job_3	6.2214	0.9982	6.2328	0.0000	4.2530	8.1898
22		Job_4	11.0720	1.1726	9.4423	0.0000	8.7597	13.3842
23		Job_5	14.9466	1.3402	11.1521	0.0000	12.3037	17.5895
24		Job_6	17.0974	2.3907	7.1517	0.0000	12.3831	21.8117

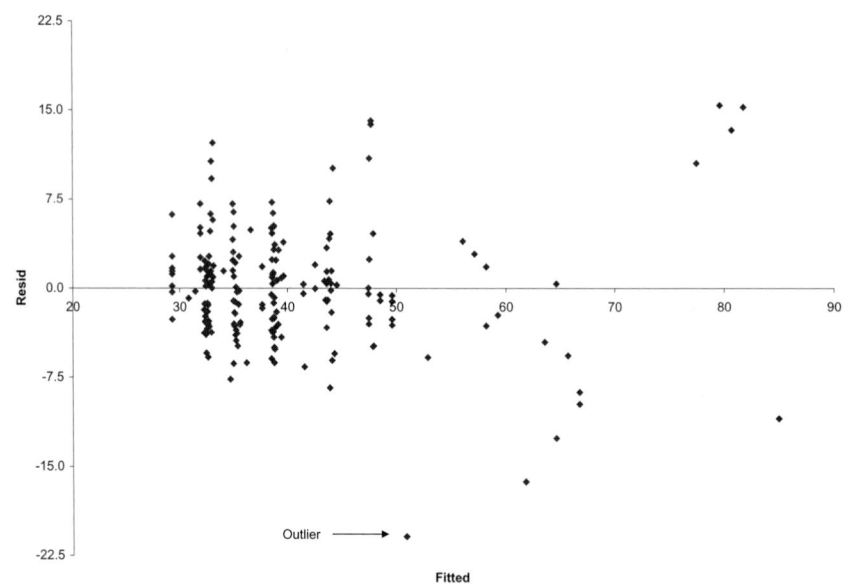

four standard errors below 0—quite a lot. When we examine this point more closely, we see that it corresponds to employee 208, who is a 62-year-old female employee in the highest job grade. She has 33 years of experience with Fifth National, she has a graduate degree, and she earns only $30,000. She is clearly an unusual employee, and there are probably special circumstances that can explain her small salary.

In any case, if we delete this employee and rerun the regression with the same variables, we obtain the output in Figure 12.25.[7] Now, recalling that gender discrimination is the key issue in this example, we compare the coefficients of Female and Female*

[7]As it turns out, this employee is the last observation in the data set. An easy way to run the regression (with StatPro) without this employee is to insert a blank row above her row and then position the cursor anywhere above this blank row.

FIGURE 12.25

Regression Output with Outlier Excluded

	A	B	C	D	E	F	G	H
1	**Results of multiple regression for Salary**							
2								
3	**Summary measures**							
4		Multiple R	0.9130					
5		R-Square	0.8336					
6		Adj R-Square	0.8269					
7		StErr of Est	4.6857					
8								
9	**ANOVA Table**							
10		Source	df	SS	MS	F	p-value	
11		Explained	8	21780.9962	2722.6245	124.0064	0.0000	
12		Unexplained	198	4347.1909	21.9555			
13								
14	**Regression coefficients**							
15			Coefficient	Std Err	t-value	p-value	Lower limit	Upper limit
16		Constant	26.7103	1.0440	25.5840	0.0000	24.6515	28.7691
17		Female	4.3531	1.2321	3.5331	0.0005	1.9234	6.7828
18		YrsExper	0.8977	0.1012	8.8675	0.0000	0.6980	1.0973
19		Female*YrsExper	-0.7206	0.1252	-5.7578	0.0000	-0.9674	-0.4738
20		Job_2	2.7179	0.9485	2.8655	0.0046	0.8474	4.5883
21		Job_3	6.2572	0.9370	6.6777	0.0000	4.4093	8.1050
22		Job_4	10.9838	1.1008	9.9777	0.0000	8.8129	13.1547
23		Job_5	15.4645	1.2619	12.2547	0.0000	12.9759	17.9530
24		Job_6	22.3234	2.4530	9.1004	0.0000	17.4861	27.1608

YrsExper in the two outputs. The coefficient of Female has dropped from 6.063 to 4.353. In words, the Y-intercept for the female regression line used to be about $6000 higher than for the male line; now it's only about $4350 higher. More importantly, the coefficient of Female*YrsExper has changed from -1.021 to -0.721. This coefficient indicates how much less steep the female line for Salary versus YrsExper is than the male line. So a change from -1.021 to -0.721 indicates *less* discrimination against females now than before. In other words, this unusual female employee accounts for a good bit of the discrimination argument—although a strong argument still exists even without her. ●

PROBLEMS

Level A

34. The file P12_34.XLS contains the sales, Y, in thousands of dollars per week, for randomly selected fast-food outlets in each of four cities. Furthermore, this data set includes the traffic flow, in thousands of cars, through each of the selected fast-food outlets.

a. Use the given data to estimate a model for predicting sales as a function of traffic flow. This regression model should account for city-to-city variations that might be due to size or other market conditions. Assume that the level of mean sales will differ from city to city, but that the change in mean sales per unit increase in traffic flow will remain the same for all cities (i.e., traffic flow and city factors do not interact).

b. Perform an F test of the overall significance of the multiple regression model estimated in part **a,** and report a p-value.

c. How do you explain the result of your statistical hypothesis test in part **b**? What, if anything, would you do to obtain more satisfactory results?

35. A manufacturing firm wants to determine whether a relationship exists between the number of work-hours an employee misses per year (Y) and the employee's annual wages (X). The data provided in the file P12_35.XLS are based on a random sample of 15 employees from this organization.

a. Estimate a simple linear regression model using the sample data. How well does the estimated model fit the sample data?

b. Perform an F test for the existence of a linear relationship between Y and X. Use a 5% level of significance.

c. How do you explain the results you have found in parts **a** and **b?**

d. Suppose you learn that the 10th worker in the sample has been fired for missing an excessive number of work-hours during the past year. In light of this information, how would you proceed to estimate the relationship between the number of work-hours an employee misses per year and the employee's annual wages, using the available information? If you decide to revise your estimate of this regression equation, repeat parts **a** and **b**.

Level B

36. Statistician Frank J. Anscombe created a data set to illustrate the importance of doing more than just examining the standard regression output. These data are provided in the file P12_36.XLS.

    a. Regress Y_1 on X. How well does the estimated model fit the data? Is there evidence of a linear relationship between Y_1 and X at the 5% significance level?

    b. Regress Y_2 on X. How well does the estimated model fit the data? Is there evidence of a linear relationship between Y_2 and X at the 5% significance level?

    c. Regress Y_3 on X. How well does the estimated model fit the data? Is there evidence of a linear relationship between Y_3 and X at the 5% significance level?

    d. Regress Y_4 on X_4. How well does the estimated model fit the data? Is there evidence of a linear relationship between Y_4 and X_4 at the 5% significance level?

    e. Compare these four simple linear regression models (i) in terms of goodness of fit and (ii) in terms of overall statistical significance.

    f. How do you explain these findings, considering that each of the regression equations is based on a *different* set of variables?

    g. What role, if any, do outliers have on each of these estimated regression models?

12.9   Violations of Regression Assumptions

Much of the theoretical research in the area of regression has dealt with violations of the regression assumptions in Section 12.2. There are three issues: how to detect violations of the assumptions, what goes wrong if we ignore violations, and what to do about them if they are detected. Detection is usually relatively easy. We can look at scatterplots, histograms, and time series plots for visual signs of violations, and there are a number of numerical measures (many not covered here) that have been developed for diagnostic purposes. The second issue, what goes wrong if we ignore violations, depends on the type of violation and its severity. The third issue is the most difficult. There are some relatively easy fixes and some that are well beyond the level of this book. In this section we will briefly discuss some of the most common violations and a few possible remedies for them.

12.9.1 Nonconstant Error Variance

The second regression assumption states that the variance of the errors should be *constant* for all values of the explanatory variables. This is a lot to ask, and it is almost always violated to some extent. Fortunately, mild violations do not have much effect on the validity of the regression output, so we can usually ignore them.

    However, one particular form of nonconstant error variance occurs fairly often and should be dealt with. This is the "fan shape" we saw in the scatterplot of AmountSpent versus Salary in Figure 12.1. As salaries increase, the variability of amounts spent also increases. Although this fan shape appears in the scatterplot of the response variable AmountSpent versus the explanatory variable Salary, it also appears in the scatterplot of residuals versus fitted values when we regress AmountSpent versus Salary. If we ignore this nonconstant error variance, then the standard error of the regression coefficient of Salary is inaccurate, so that a confidence interval for this coefficient or a hypothesis test concerning it can be misleading.

    There are at least two ways to deal with this fan-shape phenomenon. The first is to use a different estimation method than least squares. It is called *weighted least squares,*

A fan shape can cause an incorrect value for the standard error of estimate, so that confidence intervals and hypothesis tests for the regression coefficients are not valid.

FIGURE 12.26
Scatterplot without
Fan Shape

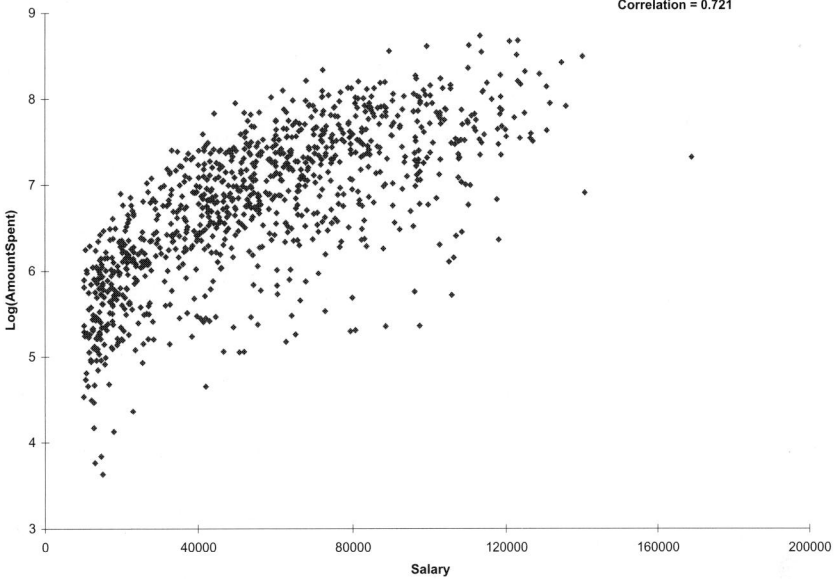

and it is an option available in many statistical software packages. However, it is fairly advanced and it is not available with Excel (or StatPro), so we won't discuss it here.

The second method is simpler. When we see a fan shape, where the variability increases from left to right in a scatterplot, we can try a logarithmic transformation of the response variable. The reason this often works is that the logarithmic transformation squeezes the large values closer together and pulls the small values farther apart. The scatterplot of the log of AmountSpent versus Salary is in Figure 12.26. Clearly, the fan shape evident in Figure 12.1 is gone.

This logarithmic transformation is not a magical cure for all instances of nonconstant error variance. For example, it appears to have introduced some curvature into the plot in Figure 12.26. However, as we discussed in the previous chapter, whenever the distribution of the response variable is heavily skewed to the right, as it often is, the logarithmic transformation is worth exploring.

A logarithmic transformation of Y can sometimes cure the fan-shape problem.

12.9.2 Nonnormality of Residuals

The third regression assumption states that the error terms are normally distributed. We can check this assumption fairly easily by forming a histogram of the residuals. We can even perform a formal test of normality of the residuals by using the procedures discussed in Section 10.5 of Chapter 10. However, unless the distribution of the residuals is severely nonnormal, the inferences we make from the regression output are still approximately valid. In addition, a form of nonnormality often encountered is skewness to the right, and this can often be remedied by the same logarithmic transformation of the response variable that remedies nonconstant error variance.

12.9.3 Autocorrelated Residuals

The fourth regression assumption states that the error terms are probabilistically independent. This assumption is usually valid for cross-sectional data, but it is often violated for time series data. The problem with time series data is that the residuals are often correlated with nearby residuals, a property called **autocorrelation.** The most frequent type of autocorrelation is positive autocorrelation. For example, if residuals separated by 1 month are autocorrelated—called **lag 1 autocorrelation**—in a positive direction, then an overprediction in January, say, will likely lead to an overprediction in February, and an underprediction in January will likely lead to an underprediction in February. If this autocorrelation is large, then serious prediction errors can occur if it isn't dealt with appropriately.

A numerical measure has been developed to check for lag 1 autocorrelation. It is called the **Durbin–Watson statistic** (after the two statisticians who developed it), and it is quoted automatically in the regression output of many statistical software packages. The Durbin–Watson (DW) statistic is scaled to be between 0 and 4. Values close to 2 indicate very little lag 1 autocorrelation, values below 2 indicate positive autocorrelation, and values above 2 indicate negative autocorrelation.

Since *positive* autocorrelation is the usual culprit, the question becomes how much below 2 the DW statistic must be before we should react. There is a formal hypothesis test for answering this question, and a set of tables appears in many statistics texts. Without going into the details, we will simply state that when the number of time series observations, n, is about 30 and the number of explanatory variables is fairly small, say, 1 to 5, then any DW statistic less than 1.2 should get our attention. If n increases to around 100, then we shouldn't be concerned unless the DW statistic is below 1.5.

If e_i is the ith residual, then the formula for the DW statistic is

A Durbin–Watson statistic below 2 signals that nearby residuals are *positively* correlated with one another.

$$DW = \frac{\Sigma_{i=2}^{n}(e_i - e_{i-1})^2}{\Sigma_{i=1}^{n}e_i^2}$$

This is obviously not very attractive for hand calculation, so we have included the DW function in the StatPro add-in. To use it, run any regression and check the option to supply columns of fitted values and residuals. (This option is automatic with the StatPro's Simple Regression procedure.) Then enter the formula

=DW(ResidRange)

in any cell, substituting the actual range of residuals for "ResidRange."

The following continuation of Example 12.1 with the Bendrix manufacturing data— the only time series data set we've analyzed with regression—checks for possible lag 1 autocorrelation.

Example 12.1 Explaining Overhead Costs at Bendrix (continued)

Is there any evidence of lag 1 autocorrelation in the Bendrix data when Overhead is regressed on MachHrs and ProdRuns?

Objective To use the Durbin–Watson statistic to check whether there is any serious autocorrelation in the residuals from the Bendrix regression model for overhead costs.

Solution

We run the usual multiple regression and check that we want columns of fitted values and residuals appended to the data set. The results (with some rows hidden) are in Figure 12.27. (See the file BENDRIX.XLS.) The residuals are listed in column G. Each repre-

	A	B	C	D	E	F	G	H	I	J
1	**Monthly data on manufacturing overhead costs**									
2										
3	Month	MachHrs	ProdRuns	Overhead		Fitted Values	Residuals		Durbin-Watson	
4	1	1539	31	99798		98391.351	1406.649		1.3131	
5	2	1284	29	87804		85522.333	2281.667			
6	3	1490	27	93681		92723.595	957.405			
7	4	1355	22	82262		82428.092	-166.092			
8	5	1500	35	106968		100227.903	6740.097			
9	6	1777	30	107925		107869.395	55.605			
10	7	1716	41	117287		114933.472	2353.528			
11	8	1045	29	76868		75117.134	1750.866			
12	9	1364	47	106001		104910.368	1090.632			
13	10	1516	21	88738		88553.834	184.166			
37	34	1723	35	107828		109936.520	-2108.520			
38	35	1413	30	88032		92022.147	-3990.147			
39	36	1390	54	117943		112227.640	5715.360			

FIGURE 12.27 Regression Output with Residuals and DW Statistic

FIGURE 12.28
Time Series Graph
of Residuals

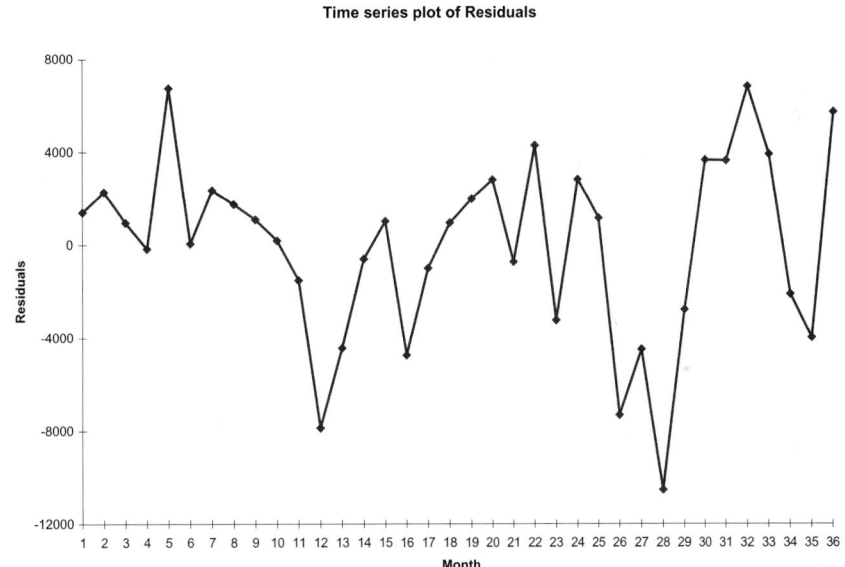

sents how much the regression overpredicts (if negative) or underpredicts (if positive) the overhead cost for that month. We can check for lag 1 autocorrelation in two ways, with the DW statistic and by examining the time series graph of the residuals in Figure 12.28. We calculate the DW statistic in cell I4 of Figure 12.27 with the formula

$$=DW(G4:G39)$$

(Remember that DW is *not* a built-in Excel function. It is available only if StatPro is loaded.) Based on our guidelines for DW values, 1.3131 suggests positive autocorrelation—it is less than 2—but not enough to cause concern.[8] This general conclusion is supported by the time series graph. Serious autocorrelation of lag 1 would tend to show long runs of residuals alternating above and below the horizontal axis—positives would tend to follow positives, and negatives would tend to follow negatives. There is some indication of this behavior in the graph but not an excessive amount. ●

---

[8]A more formal test, using Durbin–Watson tables, supports this conclusion.

What should we do if the DW statistic signals significant autocorrelation? Unfortunately, the answer to this question would take us much deeper into time series analysis than we can go in this book. Suffice it to say that time series analysis in the context of regression can become very complex, and there are no easy fixes for the autocorrelation that often occurs.

PROBLEMS

Level A

37. Motorco produces electric motors for use in home appliances. One of the company's production managers is interested in examining the relationship between the dollars spent per month in inspecting finished motor products (X) and the number of motors produced during that month that were returned by dissatisfied customers (Y). He has collected the data in the file P02_18.XLS to explore this relationship for the past 36 months.
    a. Generate a simple linear regression model using the given data and interpret it for this production manager.
    b. Conduct an appropriate hypothesis test for the existence of a linear relationship between Y and X in this case, and report a p-value.
    c. Examine the residuals of the estimated regression equation. Do you see evidence of any violations of the assumptions regarding the errors of the regression model?
    d. Conduct a Durbin–Watson test on the model's residuals. Interpret the result of this test for the production manager.
    e. In light of your result in part d, do you recommend modifying the original regression model? If so, how would you revise it?

38. Examine the relationship between the average utility bills for homes of a particular size (Y) and the average monthly temperature (X). The data in the file P11_07.XLS include the average monthly bill and temperature for each month of the past year.

    a. Use the given data to estimate a simple linear regression model. How well does the estimated regression model fit the given data?
    b. Conduct an appropriate hypothesis test for the existence of a linear relationship between Y and X, and report a p-value.
    c. Examine the residuals of the estimated regression equation. Do you see evidence of any violations of the assumptions regarding the errors of the regression model?
    d. Conduct a Durbin–Watson test on the model's residuals. Interpret the result of this test.
    e. In light of your result in part d, do you recommend modifying the original regression model? If so, how would you revise it?

39. The manager of a commuter rail transportation system was recently asked by her governing board to predict the demand for rides in the large city served by the transportation network. The system manager has collected data on variables thought to be related to the number of weekly riders on the city's rail system. The file P11_20.XLS contains these data.
    a. Estimate a multiple regression model using all of the available explanatory variables.
    b. Conduct and interpret the result of an F test on the given model. Employ a 5% level of significance in conducting this statistical hypothesis test.
    c. Is there evidence of autocorrelated residuals in this model? Explain why or why not.

12.10 Prediction

Once we have estimated a regression equation from a set of data, we might want to use this equation to predict the value of the response variable for *new* observations. As an example, suppose that a retail chain is considering opening a new store in one of several proposed locations. It naturally wants to choose the location that will result in the largest revenues. The problem is that the revenues for the new locations are not yet known. They can be observed only after stores are opened in these locations, and the chain cannot afford to open more than one store at the current time. An alternative is to use regression analysis. Using data from *existing* stores, the chain can run a regression of the response variable revenue on several explanatory variables such as population density, level of wealth in the vicinity, number of competitors nearby, ease of access given the existing roads, and so on.

Assuming that the regression equation has a reasonably large R^2 and, even more important, a reasonably small s_e, the chain can then use this equation to predict revenues for the proposed locations. Specifically, it will gather values of the explanatory variables for each of the proposed locations, substitute these into the regression equation, and look at the predicted revenue for each proposed location. All else being equal, the chain will probably choose the location with the highest predicted revenue.

As another example, suppose that we are trying to explain the starting salaries for undergraduate college students. We want to predict the *mean* salary of all graduates with certain characteristics, such as all male marketing majors from state-supported universities. To do this, we first gather salary data from a sample of graduates from various universities. Included in this data set are relevant explanatory variables for each graduate in the sample, such as the type of university, the student's major, GPA, years of work experience, and so on. We then use these data to estimate a regression equation for starting salary and substitute the relevant values of the explanatory variables into the regression equation to obtain the required prediction.

These two examples illustrate two types of prediction problems in regression. The first problem, illustrated by the retail chain example, is probably the more common of the two. Here we are trying to predict the value of the response variable for one or more *individual* members of the population. In this specific example we are trying to predict the future revenue for several potential locations of the new store. In the second problem, illustrated by the salary example, we are trying to predict the *mean* of the response variable for all members of the population with certain values of the explanatory variables. In the first problem we are predicting an individual value; in the second problem we are predicting a mean.

The second problem is inherently easier than the first in the sense that the resulting prediction is bound to be more accurate. The reason should be intuitive. Recall that the mean of the response variable for any fixed values of the explanatory variables lies on the population regression line. Therefore, if we can accurately estimate this line—that is, if we can accurately estimate the regression coefficients—we can accurately predict the required mean. In contrast, most individual points do *not* lie on the population regression line. Therefore, even if our estimate of the population regression line is perfectly accurate, we still cannot predict exactly where an individual point will fall.

Stated another way, when we predict a mean, there is a single source of error: the possibly inaccurate estimates of the regression coefficients. But when we predict an individual value, there are two sources of error: the inaccurate estimates of the regression coefficients and the inherent variation of individual points around the regression line. Actually, this second source of error often dominates the first.

We illustrate these comments in Figure 12.29 on page 668. For the sake of illustration, the response variable is salary and the single explanatory variable is years of experience with the company. Let's suppose that we want to predict either the salary for a particular employee with 10 years of experience or the mean salary of all employees with 10 years of experience. The two lines in this graph represent the population regression line (which in reality is unobservable) and the estimated regression line. For each prediction problem the point prediction—the best guess—is the value above 10 on the estimated regression line. The error in predicting the mean occurs because the two lines in the graph are not the same, that is, the estimated line is not quite correct. The error in predicting the individual value (the point shown in the graph) occurs because the two lines are not the same and also because this point does not lie on the population regression line.

Regression can be used to predict Y for a single observation, or it can be used to predict the mean Y for many observations, all with the same X values.

FIGURE 12.29

Prediction Errors
for an Individual
Value and a Mean

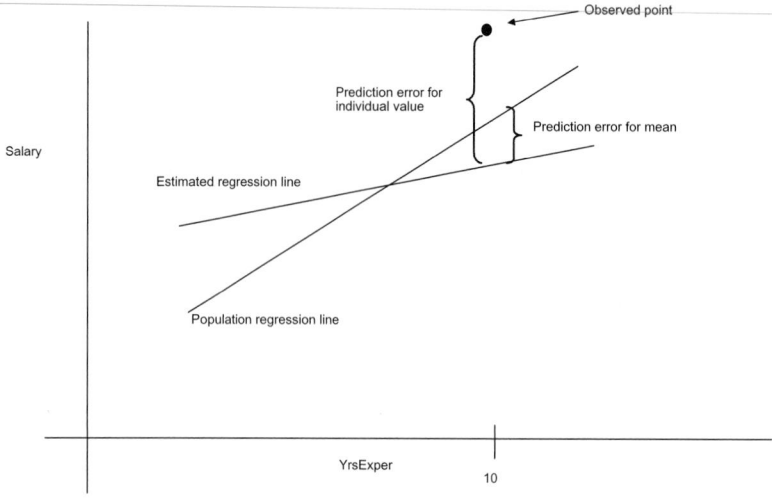

One general aspect of prediction becomes apparent by looking at this graph. If we let X's denote the explanatory variables, predictions for values of the X's close to their means are likely to be more accurate than predictions for X's far from their means. In the graph, the mean years of experience is about 7. (This is approximately where the two lines cross.) Because the slopes of the two lines are different, they get farther apart as YrsExper gets farther away from 7 (on either side). As a result, predictions tend to become less accurate.

This phenomenon shows up as higher standard errors of prediction as the X's get farther away from their means. However, for extreme values of the X's, there is another problem. Suppose, for example, that all values of YrsExper in the data set are between 1 and 15, and we attempt to predict the salary for an employee with 25 years of experience. This is called **extrapolation;** we are attempting to predict beyond the limits of the sample.

The problem here is that there is no guarantee, and sometimes no reason to believe, that the relationship within the range of the sample is valid outside of this range. It is perfectly possible that the effect of years of experience on salary is considerably different in the 25-year range than in the range of the sample. If it is, then extrapolation is bound to yield inaccurate predictions. In general, we should try to avoid extrapolation whenever possible. If we really want to predict the salaries of employees with 25-plus years of experience, then we should include some employees of this type in the original sample.

We now discuss how to make predictions and how to estimate their accuracy, both for individual values and for means. To keep it simple, we first assume that there is a single explanatory variable X. We choose a fixed "trial" value of X, labeled X_0, and predict the value of a single Y or the mean of all Y's when $X = X_0$. For both prediction problems the **point prediction,** or best guess, is found by substituting into the right-hand side of the estimated regression equation. Graphically, this is the height of the estimated regression line above X_0.

It is more difficult to predict for extreme X's than for X's close to the mean. Trying to predict for X's beyond the range of the data set (extrapolation) is quite risky.

Point Prediction (of single Y or mean of Y's)	To calculate the **point prediction,** substitute the given values of the X's into the estimated regression equation.

The standard error of prediction for a single Y is approximately equal to the standard error of estimate.

To measure the accuracy of these point predictions, we calculate a standard error for each prediction. These standard errors can be interpreted in the usual way. For example, we are about 68% certain that the actual values will be within 1 standard error of the point predictions, and we are about 95% certain that the actual values will be within 2 standard errors of the point predictions. For the individual prediction problem, the standard error is labeled s_{ind} and is given by equation (12.5). As indicated by the approximate equality on the right, when the sample size n is large and X_0 is fairly close to $\bar{X}$, the last two terms inside the square root are relatively small, and this standard error of prediction can be approximated by s_e, the standard error of estimate.

Standard Error of Prediction for a Single Y	$$s_{\text{ind}} = s_e \sqrt{1 + \frac{1}{n} + \frac{(X_0 - \bar{X})^2}{\sum_{i=1}^{n}(X_i - \bar{X})^2}} \simeq s_e$$	(12.5)

For the prediction of the mean, the standard error is labeled s_{mean} and is given by equation (12.6). Here, if X_0 is fairly close to $\bar{X}$, then the last term inside the square root is relatively small, and this standard error of prediction is approximately $s_e/\sqrt{n}$.

Standard Error of Prediction for the Mean Y	$$s_{\text{mean}} = s_e \sqrt{\frac{1}{n} + \frac{(X_0 - \bar{X})^2}{\sum_{i=1}^{n}(X_i - \bar{X})^2}} \simeq s_e/\sqrt{n}$$	(12.6)

The standard error of prediction for a mean of Y's is approximately equal to the standard error of estimate divided by the square root of the sample size.

These standard errors can be used to calculate a 95% prediction interval for an individual value and a 95% confidence interval for a mean value. Exactly as in Chapter 9, we go out a t-multiple of the relevant standard error on either side of the point prediction. The t-multiple is the value that cuts off 0.025 probability in the right-hand tail of a t distribution with $n - 2$ degrees of freedom.

The term *prediction* interval (rather than confidence interval) is used for an individual value because an individual value of Y is not a population *parameter*. However, the interpretation is basically the same. If we calculate a 95% prediction interval for many members of the population, we expect their actual Y values to fall within the corresponding prediction intervals about 95% of the time.

To see how all of this can be implemented in Excel, we revisit the Pharmex drugstore data set, where a sales index is regressed on a promotional expenditures index for a sample of 50 regions.

Example 12.5 Sales versus Promotions at Pharmex

Besides the 50 regions in the data set, Pharmex does business in five other regions, which have promotional expenses indexes of 114, 98, 84, 122, and 101. Find the predicted Sales and a 95% prediction interval for each of these regions. Also, find the *mean* Sales for all regions with each of these values of Promote, along with 95% confidence intervals for these means.

Objective To predict sales at several Pharmex locations from the promotional expenses at these locations.

	A	B	C	D	E	F	G	H	I	J	K	L
1	Pharmex drugstore data											
2												
3	Region	Promote	Sales			Results of multiple regression for Sales						
4	1	77	85									
5	2	110	103			Summary measures						
6	3	110	102			Multiple R	0.6730					
7	4	93	109			R-Square	0.4529					
8	5	90	85			Adj R-Square	0.4415					
9	6	95	103			StErr of Est	7.3947					
10	7	100	110									
11	8	85	86			ANOVA Table						
12	9	96	92			Source	df	SS	MS	F	p-value	
13	10	83	87			Explained	1	2172.8805	2172.8805	39.7366	0.0000	
14	11	88	99			Unexplained	48	2624.7395	54.6821			
15	12	94	101									
16	13	104	109			Regression coefficients						
17	14	89	81				Coefficient	Std Err	t-value	p-value	Lower limit	Upper limit
18	15	95	107			Constant	25.1264	11.8826	2.1146	0.0397	1.2349	49.0179
19	16	94	95			Promote	0.7623	0.1209	6.3037	0.0000	0.5192	1.0054
20	17	96	109									
21	18	92	113			Predictions for individual customers			Approximate std err:		7.3947	
22	19	93	84			Region	Promote	Pred Sales	Std Err	Lower limit	Upper limit	
23	20	98	98			1	114	112.03	7.7185	96.51	127.55	
24	21	103	112			2	98	99.83	7.4683	84.82	114.85	
25	22	95	96			3	84	89.16	7.6546	73.77	104.55	
26	23	103	93			4	122	118.13	8.0177	102.01	134.25	
27	24	89	97			5	101	102.12	7.4778	87.08	117.15	
28	25	97	92									
29	26	97	105			Predictions of means			Approximate std err:		1.0458	
30	27	99	102			Region	Promote	Pred Sales	Std Err	Lower limit	Upper limit	
31	28	101	105			1	114	112.03	2.2122	107.58	116.48	
32	29	113	118			2	98	99.83	1.0459	97.73	101.93	
33	30	86	82			3	84	89.16	1.9776	85.18	93.14	
34	31	100	101			4	122	118.13	3.0986	111.90	124.36	
35	32	96	97			5	101	102.12	1.1118	99.88	104.35	
36	33	105	101									
52	49	95	108									
53	50	96	87									

FIGURE 12.30 Prediction in Simple Regression

Solution

This example can be solved partly with StatPro and partly with Excel's built-in functions. The end results appear in Figure 12.30. We first use StatPro's multiple regression procedure to obtain the usual regression output for sales as a function of promotional expenses. (We could use the simple regression option, but it is just as easy to use multiple regression, even when there is a single explanatory variable.) We then calculate the point predictions and prediction intervals in rows 23–27 (for individual values) and rows 31–35 (for means).

To calculate any point prediction in column H, we substitute the given promotional expense into the regression equation. For example, the formula in cell H23, which is then copied down, is

$$=\$G\$18+\$G\$19*G23$$

The more difficult part is calculating the standard errors of prediction from equations (12.5) and (12.6). Fortunately, Excel has a built-in function, DEVSQ, for calculating the sum of squared deviations of the X's around their mean—the sum in the denominator of each equation. In addition, the standard error of estimate has already been calculated in cell G9 by StatPro. Therefore, we can implement equations (12.5) and (12.6) in cells I23 and I31 with the formulas

$$=\$G\$9*SQRT(1+1/50+(G23-AVERAGE(\$B\$4:\$B\$53))^2/DEVSQ(\$B\$4:\$B\$53))$$

and

$$=\$G\$9*SQRT(1/50+(G31-AVERAGE(\$B\$4:\$B\$53))^2/DEVSQ(\$B\$4:\$B\$53))$$

(The trick is to get the parentheses correct!) Each of these formulas is then copied down.

Finally, the lower and upper limits of the prediction intervals are found in the usual way, by subtracting and adding a t-multiple of the appropriate standard error to the point prediction. This t-multiple is found from the Excel function TINV(0.05,48), which evaluates to approximately 2.

We have gone through these rather tedious calculations to make several points. First, the approximate standard errors in equations (12.5) and (12.6), s_e and $s_e/\sqrt{n}$, are more or less accurate, depending on how near the given X value is to the mean of the X's in the sample. For practical purposes, the approximations usually suffice. This is fortunate because the exact standard errors are difficult to calculate and are not always given in statistical software packages. This is particularly the case in *multiple* regression, which we will discuss shortly. Second, a simple rule of thumb for calculating individual 95% prediction intervals is to go out an amount $2s_e$ on either side of the predicted values. Again, this is not exactly correct, but as the calculations in this example indicate, it works quite well.

Finally, we see from the wide prediction intervals how much uncertainty remains. For example, the prediction interval for a region with Promote equal to 114 extends from 96.51 to 127.55. This implies that the point prediction for this region, 112.03, contains a considerable amount of uncertainty. The reason is the relatively large standard error of estimate, s_e. If we could halve the value of s_e, the length of the prediction interval would be only half as large. Contrary to what you might expect, this is not a sample size problem. That is, a larger sample size would almost surely *not* produce a smaller value of s_e. The whole problem is that Promote is not highly correlated with Sales. The only way to decrease s_e and get more accurate predictions is to find other explanatory variables that are more closely related to Sales. ●

The situation is only slightly different for multiple regression. Again, to find a point prediction in either prediction problem, we substitute the trial values of the X's into the right-hand side of the estimated regression equation. The formulas for the standard errors of prediction, however, are considerably more complex and are not given here. Some (but not all) statistical packages provide these standard errors. Fortunately, we can again approximate the standard error of an individual value by s_e and the standard error of a mean by $s_e/\sqrt{n}$.

PROBLEMS

Level A

40. The human resources manager of DataCom, Inc., wants to predict the annual salaries of given employees using the following explanatory variables: (1) the number of years of prior relevant work experience, (2) the number of years of employment at DataCom, (3) the number of years of education beyond high school, (4) the employee's gender, (5) the employee's department, and (6) the number of individuals supervised by the given employee. These data have been collected for a sample of employees and are given in the file P11_05.XLS.

a. Estimate an appropriate multiple regression model to predict the annual salary of a given DataCom employee.

b. Conduct and interpret the result of an F test on the given model. Employ a 5% level of significance in conducting this statistical hypothesis test.

c. Given the estimated regression model, predict the annual salary of a male employee who served in a similar department at another company for 5 years prior to coming to work at DataCom. This man, a graduate of a 4-year collegiate business program, has been supervising 6 subordinates in the sales department since joining the organization 7 years ago.

d. Find a 95% prediction interval for the salary earned by a DataCom employee as characterized in part **c.**

e. Find a 95% confidence interval for the mean salary earned by all DataCom employees sharing the characteristics provided in part **c.**

f. How do you explain the difference between the widths of the intervals in parts **d** and **e?**

41. Suppose you are interested in predicting the price of a laptop computer based on its various features. The file P11_35.XLS contains observations on the sales price and on a number of potentially relevant variables for a randomly chosen sample of laptop computers.

a. Estimate a multiple regression model that includes all available explanatory variables.

b. Conduct and interpret the result of an F test on the given model. Employ a 5% level of significance in conducting this statistical hypothesis test.

c. Use the estimated regression equation to predict the price of a laptop computer with the following features: a 50-megahertz processor, a battery that holds its charge for 180 minutes, 20 megabytes of RAM, a DX chip, a color monitor, a trackball pointing device, and a 24-hour, toll-free customer service hotline.

d. Find a 99% prediction interval for the price of a laptop computer as characterized in part **c.**

e. Find a 99% confidence interval for the average price of all laptop computers sharing the characteristics provided in part **c.**

f. How do you explain the difference between the widths of the intervals in parts **d** and **e?**

42. Suppose that a power company located in southern Alabama wants to predict the peak power load (i.e., Y, the maximum amount of power that must be generated each day to meet demand) as a function of the daily high temperature (X). A random sample of 25 summer days is chosen, and the peak power load and the high temperature are recorded on each day. The file P11_40.XLS contain these observations.

a. Use the given data to estimate a simple linear regression model. How well does the estimated regression model fit the given data?

b. Conduct an appropriate hypothesis test for the existence of a linear relationship between Y and X, and report a p-value.

c. Examine the residuals of the estimated regression equation. Do you see evidence of any violations of the assumptions regarding the errors of the regression model?

d. Conduct a Durbin–Watson test on the model's residuals. Interpret the result of this test.

e. Given your result in part **d,** do you recommend modifying the original regression model in this case? If so, how would you revise it?

f. Use the final version of your regression model to predict the peak power load on a summer day with a high temperature of 90 degrees.

g. Find a 95% prediction interval for the peak power load on a summer day with a high temperature of 90 degrees.

h. Find a 99% confidence interval for the average peak power load on all summer days with a high temperature of 90 degrees.

12.11 Conclusion

In these two chapters on regression, we have seen how useful regression analysis can be for a variety of business applications and how statistical software such as the StatPro add-in in Excel enables us to obtain relevant output—both graphical and numerical—with very little effort. However, we have also seen that there are many concepts that need to be understood well before regression analysis can be used appropriately. Given the user-friendly software currently available, it is all too easy to generate enormous amounts of regression output and then misinterpret or misuse much of it.

At the very least, you should (1) be able to interpret the standard regression output, including statistics on the regression coefficients, summary measures such as R^2 and s_e, and the ANOVA table, (2) know what to look for in the many scatterplots available, (3) know how to use dummy variables, interaction terms, and nonlinear transformations to improve a fit, and (4) be able to spot clear violations of the regression assumptions. However, we haven't covered everything. Indeed, many entire books are devoted exclusively to regression analysis. Therefore, you should recognize when you *don't* know enough to handle a regression problem such as nonconstant error variance or autocorrelation appropriately. In this case you should consult a statistical expert.

Summary of Key Terms

Term	Symbol	Explanation	Excel	Page	Equation Number
Statistical model for regression		A theoretical model, including several assumptions, that must be satisfied, at least approximately, for inferences from regression output to be valid.		621	12.1
Homoscedasticity (and heteroscedasticity)		Equal (and unequal) variance of the response variable for different values of the explanatory variables.		623	
Autocorrelation of residuals		Lack of independence in the series of residuals, especially relevant for time series data.		624	
Parsimony		The concept of explaining the most with the least.		626	
Standard error of regression coefficient	s_b	Measures how much the estimates of a regression coefficient vary from sample to sample.	StatPro/ Regression Analysis	627	
t-value for regression coefficient	t	The ratio of the estimate of a regression coefficient to its standard error, used to test whether the coefficient is 0.	StatPro/ Regression Analysis	629	12.3
Confidence interval for regression coefficient		An interval likely to contain the population regression coefficient.	StatPro/ Regression Analysis	628	
Hypothesis test for regression coefficient		Typically, a two-tailed test, where the null hypothesis is that the regression coefficient is 0.	StatPro/ Regression Analysis	629	
ANOVA table for regression		Used to test whether the explanatory variables, as a whole, have any significant explanatory power.	StatPro/ Regression Analysis	630	
Multicollinearity		Occurs when there is a fairly strong linear relationship between explanatory variables.		635	
Include/exclude decisions		Guidelines for deciding whether to include or exclude potential explanatory variables.		639	
Stepwise regression		A class of "automatic" equation-building methods, where variables are added (or deleted) in order of their importance.	StatPro/ Regression Analysis/ Stepwise (or Forward or Backward)	644	

Summary of Key Terms (continued)

Term	Symbol	Explanation	Excel	Page	Equation Number
Partial F test		Tests whether a set of extra explanatory variables adds any explanatory power to an existing regression equation.	Must be done manually.	648	12.4
Outliers		Observations that lie outside the general pattern of points and can have a substantial effect on the regression model.		656	
Influential point		A point that can "tilt" the regression line.		657	
Durbin–Watson statistic		A measure of the autocorrelation between residuals, especially useful for time series data.	=DW(*range*), a StatPro function	664	
Standard errors of prediction	s_{ind}, s_{mean}	Measures of the accuracy of prediction when predicting Y for an individual observation, or predicting the mean of all Y's for fixed values of the explanatory variables.	StatPro/ Regression Analysis (provides approximate values only)	669	12.5, 12.6

PROBLEMS

Conceptual Exercises

C.1. Suppose a regression output produces the following 99% confidence interval for one of the regression coefficients:

$$[-32.47, -16.88]$$

Given this information, should an analyst reject the null hypothesis that this population regression coefficient is equal to 0? Explain your answer.

C.2. Explain why it is not possible to estimate a linear regression model that contains *all* dummy variables associated with a particular categorical explanatory variable.

C.3. Suppose that you are serving as the mentor for a summer intern in your organization. As a first assignment, you direct this undergraduate student to generate a multiple regression model that does a good job of explaining the variation in the monthly sales of one of the products your company manufactures. Shortly thereafter, the intern submits a report that recommends a particular estimated equation. The student tells you that she found the "best" model by enumerating all possible explanatory variables, gathering a random sample for each possible variable, and using a statistical software package that automatically finds a model containing only those variables

with statistically significant regression coefficients. What feedback would you give to the intern based on the overall approach she has taken in completing this statistical assignment?

C.4. Distinguish between the test of significance of an individual regression coefficient and the ANOVA test. When, if ever, are these two statistical tests essentially equivalent?

C.5. Which of these intervals based on the same estimated regression equation with fixed values of the explanatory variables would be *wider*: (i) a 95% prediction interval for an individual value of Y or (ii) a 95% confidence interval for the mean value of Y? Explain your answer. How would you interpret the wider of these two intervals in words?

Level A

43. For 12 straight weeks you have observed the sales (in number of cases) of canned tomatoes at Mr. D's. Each week you kept track of the following:
- Was a promotional notice placed in all shopping carts for canned tomatoes?
- Was a coupon given for canned tomatoes?
- Was a price reduction (none, 1, or 2 cents off) given?

The file P12_43.XLS contains these data.

a. Use multiple regression to determine how these factors influence sales.

b. Discuss whether your final equation has any problems with autocorrelation, heteroscedasticity, or multicollinearity.

c. Predict sales of canned tomatoes during a week in which Mr. D's uses a shopping cart notice, a coupon, and a 1-cent price reduction.

44. The file P12_44.XLS contains data on pork sales. Price is in dollars per hundred pounds, quantity sold is in billions of pounds, per capita income is in dollars, U.S. population is in millions, and GNP is in billions of dollars.

a. Use the data to develop a regression equation that could be used to predict the quantity of pork sold during future periods. Does heteroscedasticity, autocorrelation, or multicollinearity appear to be a problem?

b. Suppose that during each of the next two quarters, price is 45, U.S. population is 240, GNP is 2620, and per capita income is 10,000. (These are in the units described previously.) Predict the quantity of pork sold during each of the next two quarters.

45. The file P12_45.XLS contains monthly sales (in thousands of dollars) for a photography studio and the price charged per portrait during each month. Suppose we try to predict the current month's sales from last month's sales and the current month's price.

a. If the price of a portrait during month 21 is $10, predict month 21 sales.

b. Does autocorrelation, multicollinearity, or heteroscedasticity appear to be a problem?

46. The file P12_46.XLS contains data on a motel chain's revenue and advertising.

a. Use the data and multiple regression to make predictions for the motel chain's revenues during the next four quarters. Assume that advertising during each of the next four quarters is $50,000.

b. Does autocorrelation appear to be a problem?

47. The file P12_47.XLS contains the quarterly revenues (in millions of dollars) of Washington Gas and Light for the years 1992–1998. We want to use these data to build a multiple regression model that can be used to forecast future revenues.

a. Which variables should be included in the regression? Explain your rationale for including or excluding variables.

b. Interpret the coefficients of your final equation.

c. Make a forecast for revenues during the first quarter of 1999. Also, estimate the probability that 1999 Quarter 1 revenues will be at least $150 million dollars. (*Hint:* Use the standard error of prediction and the fact that the errors are approximately normally distributed.)

48. The file P11_43.XLS contains the following data for several underdeveloped countries:

- Infant mortality rate
- Adult literacy rate
- Percentage of students finishing primary school
- Per capita GNP

a. Use these data to develop an equation that can be used to predict the infant mortality rate. Justify your equation.

b. Are there any outliers? If so, what happens if you omit them? *Should* they be omitted?

c. Interpret the coefficients in your equation.

d. Does heteroscedasticity or multicollinearity appear to be a problem?

e. Why is autocorrelation not important in this problem?

f. Within what amount should 95% of our predictions for the infant mortality rate be accurate?

g. For a country with a $2000 GNP, 90% adult literacy, and 80% finishing primary school, the regression implies that there is a 1% chance of infant mortality exceeding what value? (*Hint:* Use the standard error of prediction and the fact that the errors are approximately normally distributed.)

49. The file P12_49.XLS contains data on 128 recent home sales in MidCity. For each sale, the file shows the neighborhood (1, 2, or 3) in which the house is located, the number of offers made on the house, the square footage, whether the house is made primarily of brick, the number of bathrooms, the number of bedrooms, and the selling price. Neighborhoods 1 and 2 are more traditional neighborhoods, whereas neighborhood 3 is a newer, more prestigious, neighborhood. For each part below, use StatPro to estimate the relevant regression equation for selling price.

a. Base this first equation on all variables (other than Home in column A), treating information on brick and neighborhood as categorical and treating all other variables as regular quantitative variables. From this equation, explain what the "premium" is for a house being made of brick, all else being equal. What is the premium for a house being in neighborhood 3, all else being equal? Also, explain the effect of the other variables (besides brick and neighborhood) on price. Are they all significant? (For now, don't eliminate any variables, even if they are insignificant.)

b. Is there an *extra* premium for a brick house in neighborhood 3, in addition to the usual premium for being brick? Answer by allowing an interaction effect between brick and neighborhood. (This equation should have two extra variables in addition to those in the equation part **a.**)

c. Starting with the equation in part **b,** remove any variables with *p*-values greater than 0.1. Explain exactly what this latter equation says about the effect of neighborhood on price. Does it provide much different information than the equation in part **b?**

50. Recall the movie star data from Chapter 2 (in the file ACTORS.XLS).
    a. Determine an equation to predict salary on the basis of gender, domestic gross, and foreign gross. Make sure all variables in your equation are significant at the 0.15 level.
    b. Interpret the coefficients in your equation.
    c. Does your equation exhibit any autocorrelation, heteroscedasticity, or multicollinearity?
    d. Identify and interpret any outliers.

51. You are trying to determine how the marketing mix influences the sale of Cornpone cereal. The file P12_51.XLS contains the following information for 17 consecutive weeks. (*Note:* Weekly sales are in millions of boxes.)
    • Was price cut during the week?
    • Was there a prize in the package?
    • Was there a coupon in the package?
    a. Use these data to determine an equation that can be used to predict weekly Cornpone sales. (Ignore any possible effect of trend.) Make sure all variables in your equation are significant at the 0.15 level.
    b. Interpret the coefficients in your equation.
    c. Are there any outliers?
    d. Is either multicollinearity or autocorrelation a problem?
    e. During a week in which there is a price cut and both a prize and a coupon are in the package, what is the probability that sales will be less than 50 million boxes? You may assume that heteroscedasticity and autocorrelation are not problems. (*Hint:* Use the standard error of prediction and the fact that the errors are approximately normally distributed.)

52. The belief that larger majorities for a president in a presidential election help the president's party increase its representation in the House and Senate is called the "coat-tail" effect. The file P12_52.XLS gives the percentage by which each president since 1948 won the election and the number of seats in the House and Senate gained (or lost) during each election. Are these data consistent with the idea of presidential coat-tails? (Source: *Wall Street Journal,* September 10, 1996)

53. The file P12_53.XLS lists the U.S. unemployment rate, the percentage growth in the U.S. economy (in real terms), and the percentage growth in prices for years 1960–2001. Determine how changes in unemployment, economic growth, and price changes are related. (Source: *Wall Street Journal Almanac*)

54. The file P12_54 contains the golf handicap and an index of their company's stock performance over the last 3 years for 50 CEOs. A higher index indicates a better stock performance, whereas a lower handicap indicates better golfing ability. For example, Jerry Choate, the CEO of Allstate, has a 10.1 golf handicap, and his company's stock performance index is 83. (The maximum possible stock performance index is 100.) The May 31, 1998, *New York Times* reported that these data indicate that better golfers make better CEOs. What do you think?

55. When potential workers apply for a job that requires extensive manual assembly of small intricate parts, they are initially given three different tests to measure their manual dexterity. The ones who are hired are then periodically given a performance rating on a 0–100 scale that combines their speed and accuracy in performing the required assembly operations. The file P12_55.XLS lists the test scores and performance ratings for a randomly selected group of employees. It also lists their seniority (months with the company) at the time of the performance rating.
    a. Look at a matrix of correlations. Can you say with certainty (based only on these correlations) that the R^2 value for the regression will be at least 35%? Why or why not?
    b. Is there any evidence (from the correlation matrix) that multicollinearity will be a problem? Why or why not?
    c. Run the regression of JobPerf versus all four independent variables. List the equation, the value of R^2, and the value of s_e. Do all of the coefficients have the signs you would expect? Briefly explain.
    d. Referring to the equation in part **c,** if a worker (outside of the 80 in the sample) has 15 months of seniority and test scores of 57, 71, and 63, give a prediction and an approximate 95% prediction interval for this worker's JobPerf score.
    e. One of the *t*-values for the coefficients in part **c** is less than 1. Explain briefly why this occurred. Does it mean that this variable is not related to JobPerf?
    f. Arguably, the three test measures provide overlapping (or redundant) information. For the sake of parsimony (explaining "the most with the least"), it might be sensible to regress JobPerf versus only two explanatory variables, Sen and AvgTest, where AvgTest is the average of the three test scores— that is, AvgTest = (Test1 + Test2 + Test3)/3. Run this regression and report the same measures as in part **c:** the equation itself, R^2, and s_e. Would you argue that this equation is "just as good as" the equation in part **c?** Explain briefly.

56. Nicklaus Electronics manufactures electronic components used in the computer and space industries. The annual rate of return on the market portfolio and the annual rate of return on Nicklaus Electronics stock for

the last 36 months are shown in the file P12_56.XLS. The company wants to calculate the "systematic risk" of its common stock. (It is systematic in the sense that it represents the part of the risk that Nicklaus shares with the market as a whole.) The rate of return Y_t in period t on a security is hypothesized to be related to the rate of return m_t on a market portfolio by the equation

$$Y_t = a + bm_t + e_t$$

Here, a is the risk-free rate of return, b is the security's systematic risk, and e_t is an error term. Using the data available, estimate the systematic risk of the common stock of Nicklaus Electronics. Would you say that Nicklaus stock is a "risky" investment? Why or why not?

57. The auditor of Kaefer Manufacturing uses regression analysis during the analytical review stage of the firm's annual audit. The regression analysis attempts to uncover relationships that exist between various account balances. Any such relationship is subsequently used as a preliminary test of the reasonableness of the reported account balances. The auditor wants to determine whether a relationship exists between the balance of accounts receivable at the end of the month and that month's sales. The file P12_57.XLS contains data on these two accounts for the last 36 months. It also shows the sales levels 2 months before month 1.
   a. Is there any statistical evidence to suggest a relationship between the monthly sales level and accounts receivable?
   b. Referring to part a, would the relationship be described any better by including this month's sales and the previous month's sales (called lagged sales) in the equation for accounts receivable? What about adding the sales from more than a month ago to the equation? For this problem, why might it make accounting sense to include lagged sales variables in the equation? How do you interpret their coefficients?
   c. During month 37, which is a fiscal year-end month, the sales were $1,800,000. The reported accounts receivable balance was $3,000,000. Does this reported amount seem consistent with past experience? Explain.

58. A company gives prospective managers four separate tests for judging their potential. For a sample of 30 managers, the test scores and the subsequent job effectiveness ratings (JobEff) given 1 year later are listed in the file P12_58.XLS.
   a. Look at scattergrams and the table of correlations for these five variables. Does it appear that a multiple regression equation for JobEff, with the test scores as explanatory variables, will be successful? Can you foresee any problems in obtaining accu-

rate estimates of the individual regression coefficients?
   b. Estimate the regression equation that includes all four test scores, and find 95% confidence intervals for the coefficients of the explanatory variables. How can you explain the negative coefficient of Test3, given that the correlation between JobEff and Test3 is positive?
   c. Can you reject the null hypothesis that these test scores, as a whole, have no predictive ability for job effectiveness at the 1% level? Why or why not?
   d. If a new prospective manager has test scores of 83, 74, 65, and 77, what do you predict his job effectiveness rating will be in 1 year? What is the approximate standard error of this prediction?

Level B

59. Confederate Express is attempting to determine how its monthly shipping costs depend on the number of units shipped during a month. The file P12_59.XLS contains the number of units shipped and total shipping costs for the last 15 months.
   a. Use regression to determine a relationship between units shipped and monthly shipping costs.
   b. Plot the errors for the predictions in order of time sequence. Is there any unusual pattern?
   c. We have been told that there was a trucking strike during months 11–15, and we believe that this might have influenced shipping costs. How could the answer to part a be modified to account for the effects of the strike? After accounting for the effects of the strike, does the unusual pattern in part b disappear?

60. You are trying to determine the effects of three packaging displays (A, B, and C) on sales of toothpaste. The file P12_60.XLS contains the number of cases of toothpaste sold for 9 consecutive weeks. The type of store (GR = grocery, DI = discount, and DE = department store) and the store location (U = urban, S = suburban, and R = rural) are also listed.
   a. Run a multiple regression to determine how the type of store, display, and store location influence sales. Which potential explanatory variables should be included in the equation? Explain your rationale for including or excluding variables.
   b. What type of store, store location, and display appears to maximize sales?
   c. For the type of store in your part b answer, estimate the probability that 80 or more cases of toothpaste will be sold during a week. (*Hint:* Use the standard error of prediction and the fact that the errors are approximately normally distributed.)

d. Does multicollinearity or autocorrelation seem to be a problem?

61. You want to determine the variables that influence bus usage in major American cities. For 24 cities, the following data are listed in the file P12_61.XLS:
- Bus travel (annual, in thousands of hours)
- Income (average per capita income)
- Population (in thousands)
- Land area (in square miles)

a. Use these data to fit the equation

$$\text{BusTravel} = \alpha\text{Income}^{\beta_1}\text{Population}^{\beta_2}\text{LandArea}^{\beta_3}$$

b. Are all variables significant at the 0.05 level?

c. Interpret the values of β_1, β_2, and β_3.

62. The file P12_62.XLS contains data on 80 managers at a large (fictitious) corporation. The variables are Salary (current annual salary), YrsExper (years of experience in the industry), YrsHere (years of experience with this company), and MglLevel (current level in the company, coded 1 to 4). You want to regress Salary on the potential explanatory variables. What is the "best" ways to do so? Specifically, how should you handle Mg1Level? Also, should you include both YrsExper and YrsHere or only one of these, and if only one, which one? Present your results, and explain them and your reasoning behind them.

63. Mattel has assigned you to analyze the factors influencing Barbie sales. The number of Barbie dolls sold (in millions) during the last 23 years is given in the file P12_63.XLS. Year 23 is last year, year 22 is the year before that, and so on. The following factors are thought to influence Barbie sales:
- Was there a recession?
- Were Barbies on sale at Christmas?
- Was there an upward trend over time?

a. Determine an equation that can be used to predict annual Barbie sales. Make sure that all variables in your equation are significant at the 0.15 level.

b. Interpret the coefficients in your equation.

c. Are there any outliers?

d. Is heteroscedasticity or autocorrelation a problem?

e. During the current year (year 24), a recession is predicted and Barbies will be put on sale at Christmas. There is a 1% chance that sales of Barbies will exceed what value? You may assume here that heteroscedasticity and autocorrelation are not a problem. (*Hint:* Use the standard error of prediction and the fact that the errors are approximately normally distributed.)

64. The capital asset pricing model (CAPM) is a cornerstone of finance. To apply the CAPM, we assume that each stock has a risk measure (called the beta of the stock) associated with it. Then the CAPM asserts that
- The expected return on $1 invested in a stock is a linear function of the stock's beta.
- $1 invested in a stock with a 0 beta will earn an annual return equal to the risk-free interest rate (r_f) on 90-day treasury bills.
- $1 invested in a stock with a beta of 1 will yield an annual return equal to the annual return (r_m) on the market portfolio.

a. Formulate a population regression model incorporating these features of the CAPM. The explanatory variable is the stock's beta and the response variable is the annual return on $1 invested in the stock.

b. Given the data in Table 12.3, test the adequacy of the model developed in part **a.** Assume $r_f = 0.09$ and $r_m = 0.18$.

65. How does inflation in a country affect changes in exchange rates? The file P12_65.XLS contains the following information for 11 countries.
- Ratio of percentage increase in prices in local country to percentage increase in U.S. prices from 1973 to 1995.
- Ratio of 1995 units of local currency per dollar to 1973 units of local currency per dollar.

Use these data to explain how inflation affects exchange rates. Do you have an explanation for these results? (Source: *The Economist,* January 20, 1996)

66. The file P12_66.XLS shows the "yield curve" (at monthly intervals). For example, in January 1985 the annual rate on a 3-month T-bill was 7.76% and the annual rate on a 30-year government bond was 11.45%. Use regression to determine which interest rates tend to move together most closely. (Source: International Investment and Exchange Database Devel-

Table **12.3**

Stock Returns and Betas	Company	Beta	Annual Return
	AT&T	0.56	0.14
	IBM	1.07	0.19
	GM	0.76	0.16
	Polaroid	2.17	0.28
	Chrysler	1.04	0.18

oped by Craig Holden, Indiana University School of Business)

67. The Keynesian school of macroeconomics believes that increased government spending leads to increased growth. The file P12_67.XLS contains the following annual data:
    - Government spending as percentage of GDP (gross domestic product)
    - Percentage annual growth in annual GDP

    Are these data consistent with the Keynesian school of economics? (Source: *Wall Street Journal*)

68. The June 1997 issue of *Management Accounting* gave the following rule for predicting your current salary if you are a managerial accountant. Take $31,865. Next, add $20,811 if you are top management, add $3604 if you are senior management, or subtract $11,419 if you are entry management. Then add $1105 for every year you have been a managerial accountant. Add $7600 if you have a master's degree or subtract $12,467 if you have no college degree. Add $11,257 if you have a professional certification. Finally, add $8667 if you are male.
    a. How do you think the journal derived this "method" of estimating an accountant's current salary? Be specific.
    b. How could a managerial accountant use this information to determine whether he or she is significantly underpaid?

69. Suppose you are trying to use regression to predict the current salary of a major league baseball player. What variables might you use?

70. The file P12_70.XLS contains sample data on annual sales for Prozac, a drug produced by Ely Lilly. For each year, the file lists the price per day of therapy (DOT) charged for Prozac and total Prozac sales (in millions of DOT) for the year. Assuming that price is the only factor influencing Prozac sales, determine the number of DOT of Prozac that Lilly should produce for the year to ensure that there is only a 1% chance that Lilly runs out of Prozac. Assume the current price of Prozac is $1.75. (*Hint:* Use the standard error of prediction and the fact that the errors are approximately normally distributed.)

71. A business school committee was charged with studying admissions criteria to the school. Until that time, only juniors were admitted. Part of the committee's task was to see whether freshman courses would be equally good predictors of success as freshman and sophomore courses combined. Here, we'll take "success" to mean doing well in A-core (a combination of the junior level finance, marketing, and production courses, F301, M301, and P301). The file P12_71.XLS contains data on 250 students who had just completed A-core. For each student, the file lists their grades in the following courses:

- M118 (freshman)—finite math
- M119 (freshman)—calculus
- K201 (freshman)—computers
- W131 (freshman)—writing
- E201, E202 (sophomore)—micro- and macroeconomics
- L201 (sophomore)—business law
- A201, A202 (sophomore)—accounting
- E270 (sophomore)—statistics
- A-core (junior)—finance, marketing, and production

    Except for A-core, each value is a grade point for a specific course (such as 3.7 for an A−). For A-core, each value is the average grade point for the three courses comprising A-core.
    a. The A-core grade point will be the eventual response variable in a regression analysis. Look at the correlations between all variables. Is multicollinearity likely to be a problem? Why or why not?
    b. Run a multiple regression using all of the potential explanatory variables. Now, eliminate the variables as follows. (This is a reasonable variation of the procedures discussed in the chapter.) Look at 95% confidence intervals for their coefficients (as usual, not counting the intercept term). Any variable whose confidence interval contains the value 0 is a candidate for exclusion. For all such candidates, eliminate the variable with the t-value lowest in magnitude. Then rerun the regression, and use the same procedure to possibly exclude another variable. Keep doing this until 95% confidence intervals of the coefficients of all remaining variables do *not* include 0. Report this final equation, its R^2 value, and its standard error of estimate s_e.
    c. Give a quick summary of the properties of the equation in part **b.** Specifically, (i) do the variables have the "correct" signs, (ii) which courses tend to be the best predictors, (iii) are the predictions from this equation likely to be much good, and (iv) are there any obvious violations of the regression assumptions?
    d. Redo part **b,** but now use as your potential explanatory variables only courses taken in the freshman year. As in part **b,** report the final equation, its R^2, and its standard error of estimate s_e.
    e. Briefly, do you think there is enough predictive power in the freshman courses, relative to the freshman and sophomore courses combined, to change to a sophomore admit policy? (Answer only on the basis of the regression results; don't get into other merits of the argument.)

72. The file P12_72.XLS has data on several countries. The variables are listed here.
    - Country: name of country
    - GNPCapita: GNP per capita

- PopGrowth: average annual percentage change in population, 1980–1990
- Calorie: daily per capita calorie content of food used for domestic consumption
- LifeExp: average life expectancy of newborn given current mortality conditions
- Fertility: births per woman given current fertility rates

   With data such as these, cause and effect are difficult to determine. For example, does low LifeExp cause GNPCapita to be low, or vice versa? Therefore, the purpose of this problem is to experiment with the following sets of response and explanatory variables. In each case, look at scatterplots (and use economic reasoning) to find and estimate the best form of the equation, using only linear and logarithmic variables. Then interpret precisely what each equation is saying.
   a. Response: LifeExp; Explanatories: Calorie, Fertility
   b. Response: LifeExp; Explanatories: GNPCapita, PopGrowth
   c. Response: GNPCapita; Explanatories: PopGrowth, Calorie, Fertility

73. Suppose that an economist has been able to gather data on the relationship between demand and price for a particular product. After analyzing scatterplots and using economic theory, the economist decides to estimate an equation of the form $Q = aP^b$, where Q is quantity demanded and P is price. An appropriate regression analysis is then performed, and the estimated parameters turn out to be $a = 1000$ and $b = -1.3$. Now consider two scenarios: (1) the price increases from $10 to $12.50; (2) the price increases from $20 to $25.
   a. Do you expect the percentage decrease in demand to be the same in scenario (1) as in scenario (2)? Why or why not?
   b. What is the expected percentage decrease in demand in scenario (1); in scenario (2)? Be as exact as possible. (*Hint:* Remember from economics that an elasticity shows directly what happens for a "small" percentage change in price. These changes aren't that small, so you'll have to do some calculating.)

74. A human resources analyst believes that in a particular industry, the wage rate ($/hr) is related to seniority by an equation of the form $W = ae^{bS}$, where W equals wage rate and S equals seniority (in years). However, the analyst suspects that both parameters, a and b, might depend on whether the workers belong to a union. Therefore, the analyst gathers data on a number of workers, both union and nonunion, and estimates the following equation with regression:

$$\ln(W) = 2.14 + 0.027S + 0.12U + 0.006SU$$

Here $\ln(W)$ is the natural log of W, U is 1 for union workers and 0 for nonunion workers, and SU is the product of S and U.
   a. According to this model, what is the predicted wage rate for a nonunion worker with 0 years of seniority? What is it for a union worker with 0 years of seniority?
   b. Explain exactly what this equation implies about the predicted effect of seniority on wage rate for a nonunion worker; for a union worker.

75. A company has recorded its overhead costs, machine hours, and labor hours for the past 60 months. The data are in the file P12_75.XLS. The company decides to use regression to explain its overhead hours linearly as a function of machine hours and labor hours. However, recognizing good statistical practice, it decides to estimate a regression equation for the first 36 months, then validate this regression with the data from the last 24 months. That is, it will substitute the values of machine and labor hours from the last 24 months into the regression equation that is based on the first 36 months and see how well it does.
   a. Run the regression for the first 36 months. Explain briefly why the coefficient of labor hours is not significant.
   b. For this part, use the regression equation from part **a** with both variables still in the equation (even though one was insignificant). Fill in the fitted and residual columns for months 37–60. Then do relevant calculations to see whether the R^2 (or multiple R) and the standard error of estimate s_e are as good for these 24 months as they are for the first 36 months. Explain your results briefly. (*Hint:* Remember the meaning of the multiple R and the standard error of estimate.)

76. Pernavik Dairy produces and sells a wide range of dairy products. Because most of the dairy's costs and prices are set by a government regulatory board, most of the competition between the dairy and its competitors takes place through advertising. The controller of Pernavik has developed the sales and advertising levels for the last 52 weeks. These appear in the file P12_76.XLS. Note that the advertising levels for the 3 weeks prior to week 1 are also listed. The controller wonders whether Pernavik is spending too much money on advertising. He argues that the company's contribution-margin ratio is about 10%. That is, 10% of each sales dollar goes toward covering fixed costs. This means that each advertising dollar has to generate at least $10 of sales or the advertising is not cost-effective. Use regression to determine whether advertising dollars are generating this type of sales response. (*Hint:* It is very possible that the sales value in any week is affected not only by advertising this week, but also by advertising levels in the past 1, 2,

or 3 weeks. These are called "lagged" values of advertising. Try regression models with lagged values of advertising included, and see whether you get better results.)

77. The Pierce Company manufactures drill bits. The production of the drill bits occurs in lots of 1000 units. Due to the intense competition in the industry and the correspondingly low prices, Pierce has undertaken a study of the manufacturing costs of each of the products it manufactures. One part of this study concerns the overhead costs associated with producing the drill bits. Senior production personnel have determined that the number of lots produced, the direct labor hours used, and the number of production runs per month might help to explain the behavior of overhead costs. The file P12_77.XLS contains the data on these variables for the past 36 months.
    a. See how well you can predict overhead costs on the basis of these variables with a linear regression equation. Why might you be disappointed with the results?
    b. A production supervisor believes that labor hours and the number of production run setups affect overhead because Pierce uses a lot of supplies when it is working on the machines and because the machine setup time for each run is charged to overhead. As he says, "When the rate of production increases, we use overtime until we can train the additional people that we require for the machines. When the rate of production falls, we incur idle time until the surplus workers are transferred to other parts of the plant. So it would seem to me that there will be an additional overhead cost whenever the level of production changes. I would also say that because of the nature of this rescheduling process, the bigger the change in production, the greater the effect of the change in production on the increase in overhead." How might you use this information to find a better regression equation than in part **a?** (*Hint:* Develop a new explanatory variable, and use the fact that the number of lots produced in the month preceding month 1 was 5964.)

78. Danielson Electronics manufactures color television sets for sale in a highly competitive marketplace. Recently Ron Thomas, the marketing manager of Danielson Electronics, has been complaining that the company is losing market share because of a poor-quality image, and he has asked that the company's major product, the 25-inch console model, be redesigned to incorporate a higher quality level. The company general manager, Steve Hatting, is considering the request to improve the product quality but is not convinced that consumers will be willing to pay the additional expense for improved quality.

As the company controller, you are in charge of determining the cost effectiveness of improving the quality of the television sets. With the help of the marketing staff, you have obtained a summary of the average retail price of the company's television set and the prices of 29 competitive sets. In addition, you have obtained from *The Shoppers' Guide,* a magazine that evaluates and reports on various consumer products, a quality rating of the television sets produced by Danielson Electronics and its competitors. The file P12_78.XLS summarizes these data. According to *The Shoppers' Guide,* the quality rating, which varies from 0 to 10 (10 being the highest level of quality), considers such factors as the quality of the picture, the frequency of repair, and the cost of repairs. Discussions with the product design group suggest that the cost of manufacturing this type of television set is $125 + Q^2$, where Q is the quality rating.
    a. Regress AvgPrice versus QualityRating. Does the regression equation imply that customers are willing to pay a premium for quality? Explain.
    b. Given the results from part **a,** is there a preferred level of quality for this product? Assume that the quality level will affect only the price charged and not the level of sales of the product.
    c. How might you answer part **b** if the level of sales is also affected by the quality level (or alternatively, if the level of sales is affected by price)?

79. The file P12_79.XLS contains data on gasoline consumption and several economic variables. The variables are gasoline consumption for passenger cars (GasUsed), service station price excluding taxes (SSPrice), retail price of gasoline including state and federal taxes (RPrice), Consumer Price Index for all items (CPI), Consumer Price Index for public transportation (CPIT), number of registered passenger cars (Cars), average miles traveled per gallon (MPG), and real per capita disposable income (DispInc). (Sources: *Basic Petroleum Data Book,* published by the American Petroleum Institute, 2001, and *Economic Report of the President,* 2002)
    a. Regress GasUsed linearly versus CPIT, Cars, MPG, DispInc, and DefRPrice, where DefRPrice is the deflated retail price of gasoline (RPrice divided by CPI). What signs would you expect the coefficients to have? Do they have these signs? Which of the coefficients are statistically significant at the 0.05 level?
    b. Suppose the government makes the claim that for every 1 cent of tax on gasoline, there will be a $1 billion increase in tax revenue. Use the estimated equation in part **a** to support or refute the government's claim.

80. On October 30, 1995, the citizens of Quebec went to the polls to decide the future of their province. They

were asked to vote "Yes" or "No" to whether Quebec, a predominantly French-speaking province, should secede from Canada and become a sovereign country. The "No" side was declared the winner, but only by a thin margin. Immediately following the vote, however, allegations began to surface that the result was closer than it should have been. (Source: Cawley and Sommers (1996)). In particular, the ruling separatist Parti Quebecois, whose job was to decide which ballots were rejected, was accused by the "No" voters of systematic electoral fraud by voiding thousands of "No" votes in the predominantly allophone and anglophone electoral divisions of Montreal. (An allophone refers to someone whose first language is neither English nor French. An anglophone refers to someone whose first language is English.)

Cawley and Sommers examined whether electoral fraud had been committed by running a regression, using data from the 125 electoral divisions in the October 1995 referendum. The response variable was REJECT, the percentage of rejected ballots in the electoral division. The explanatory variables were as follows:

- ALLOPHONE: percentage of allophones in the electoral division
- ANGLOPHONE: percentage of anglophones in the electoral division
- REJECT94: percentage of rejected votes from that electoral division during a similar referendum in 1994
- LAVAL: dummy variable equal to 1 for electoral divisions in the Laval region, 0 otherwise
- LAV_ALL: interaction (i.e., product) of LAVAL and ALLOPHONE

The estimated regression equation (with t-values in parentheses) is

$$\text{Prediced REJECT} = 1.112 + \underset{(5.68)}{0.020} \text{ ALLOPHONE} \underset{(4.34)}{}$$

$$+ \underset{(0.12)}{0.001} \text{ ANGLOPHONE} + \underset{(2.64)}{0.223} \text{ REJECT94}$$

$$- \underset{(-8.61)}{3.773} \text{ LAVAL} + \underset{(15.62)}{0.387} \text{ LAV\_ALL}$$

The R^2 value was 0.759. Based on this analysis, Cawley and Sommers state that, "The evidence presented here suggests that there were voting irregularities in the October 1995 Quebec referendum, especially in Laval." Discuss how they came to this conclusion.

81. Suppose we are trying to explain variations in salaries for technicians in a particular field of work. The file P12_81.XLS contains annual salaries for 200 technicians. It also shows how many years of experience each technician has, as well as his or her education level. There are four education levels, as explained in the comment in cell D3. Three suggestions are put forth for the relationship between Salary and these two explanatory variables:

- We should regress Salary linearly versus the two given variables, YrsExper and EducLev.
- All that really matters in terms of education is whether the person got a college degree or not. Therefore, we should regress Salary linearly versus YrsExper and a dummy variable indicating whether he or she got a college degree.
- Each level of education might result in different jumps in salary. Therefore, we should regress Salary linearly versus YrsExper and dummy variables for the different education levels.

a. Run the indicated regressions for each of these three suggestions. Then (i) explain what each equation is saying and how the three are different (focus here on the coefficients), (ii) which you prefer, and (iii) whether (or how) the regression results in your preferred equation contradict the average salary results shown in the PivTab sheet of the file.

b. Consider the four workers shown on the Predict sheet of the file. (These are four new workers, not among the original 200.) Using your preferred equation, calculate a predicted salary and an approximate 95% prediction interval for each of these four workers.

c. It turns out (you don't have to check this) that the interaction between years of experience and education level is *not* significant for this data set. In general, however, argue why we might expect an interaction between them for salary data of technical workers. What form of interaction would you suspect? (There is not necessarily one right answer, but argue convincingly one way or the other, that is, for a positive or a negative interaction.)

12.1 The Artsy Corporation[9]

The Artsy Corporation has been sued in the U.S. Federal Court on charges of sex discrimination in employment under Title VII of the Civil Rights Act of 1964.[10] The litigation at contention here is a class-action lawsuit brought on behalf of all females who were employed by the company, or who had applied for work with the company, between 1979 and 1987. Artsy operates in several states, runs four quite distinct businesses, and has many different types of employees. The allegations of the plaintiffs deal with issues of hiring, pay, promotions, and other "conditions of employment."

In such large class-action employment discrimination lawsuits, it has become common for statistical evidence to play a central role in the determination of guilt or damages. In an interesting twist on typical legal procedures, a precedent has developed in these cases that plaintiffs may make a prima facie case purely in terms of circumstantial statistical evidence. If that statistical evidence is reasonably strong, the burden of proof shifts to the defendants to rebut the plaintffs' statistics with other data, other analyses of the same data, or nonstatistical testimony. In practice, statistical arguments often dominate the proceedings of such Equal Employment Opportunity (EEO) cases. Indeed, in this case the statistical data used as evidence filled numerous computer tapes, and the supporting statistical analysis comprised thousands of pages of printouts and reports. We will work here with a typical subset that pertains to one contested issue at one of the company's locations.

The data in the file ARTSY.XLS relate to the pay of 256 employees on the hourly payroll at one of the company's production facilities. The data include an identification number (ID) that would permit us to identify the person by name or social security number; the person's gender (Gender), where 0 denotes female and 1 denotes male; the person's job grade in 1986 (Grade); the length of time (in years) the person had been in that job grade as of December 31, 1986 (TInGrade); and the person's weekly pay rate as of December 31, 1986 (Rate). These data permit a statistical examination of one of the issues in the case—fair pay for female employees. We deal with one of three pay classes of employees—those on the bi-weekly payroll, and at one of the company's locations at Pocahantus, Maine.

The plaintiffs' attorneys have proposed settling the pay issues in the case for this group of female employees for a "back pay" lump payment to female employees of 25% of their pay during the period 1979–1987. It is our task to examine the data statistically for evidence in favor of, or against, the charges. We are to advise the lawyers for the company on how to proceed. Consider the following issues as they have been laid out to us by the attorneys representing the firm:

1. Overall, how different is pay by gender? Are the differences in pay statistically significant? Does a statistical significance test have meaning in a case like this? If so, how should it be performed? Lay out as succinctly as possible the arguments that you anticipate the plaintiffs will make with this data set.

2. The company wishes to argue that a legitimate explanation of the pay rate differences may be the difference in job grades. (In this analysis, we will tacitly assume that each person's job grade is, in fact, appropriate for him or her, even though the plaintiffs' attorneys have charged that females have been unfairly kept in the lower grades. Other statistical data, not available here, are used in that analysis.) The lawyers ask, "Is there a relatively easy way to understand, analyze, and display the pay differences by job grade? Is it easy enough that it could be presented to an average jury without confusing them?" Again, use the data to anticipate the possible arguments of the plaintiffs. To what extent does job grade appear to explain the pay rate differences between the genders? Propose and carry out appropriate hypothesis tests or

---

[9]This case was contributed by Peter Kolesar from Columbia University.

[10]Artsy is an actual corporation, and the data given in this case are real, but the name has been changed to protect the firm's true identity.

confidence intervals to check whether the difference in pay between genders is statistically significant within each of the grades.

3. In the actual case, the previous analysis suggested to the attorneys that differences in pay rates are due, at least in part, to differences in job grades. They had heard that in another EEO case, the dependence of pay rate on job grade had been investigated with regression analysis. Perform a simple linear regression of pay rate on job grade for them. Interpret the results fully. Is the regression significant? How much of the variability in pay does job grade account for? Carry out a full check of the quality of your regression. What light does this shed on the pay fairness issue? Does it help or hurt the company? Is it fair to the female employees?

4. It is argued that seniority within a job grade should be taken into account because the company's written pay policy explicitly calls for the consideration of this factor. How different are times in grade by gender? Are they enough to matter?

5. The Artsy legal team wants an analysis of the simultaneous influence of grade and time in grade on pay. Perform a multiple regression of pay rate versus grade and time in grade. Is the regression significant? How much of the variability in pay rates is explained by this model? Will this analysis help your clients? Could the plaintiffs effectively attack it? Consider residuals in your analysis of these issues.

6. Organize your analyses and conclusions in a brief report summarizing your findings for your client, the Artsy Corporation. Be complete but succinct. Be sure to advise them on the settlement issue. Be as forceful as you can be in arguing "the Artsy Case" without misusing the data or statistical theory. Apprise your client of the risks they face by developing the forceful and legitimate counterargument the female plaintiffs could make.

12.2 Heating Oil at Dupree Fuels Company[11]

Dupree Fuels Company is facing a difficult problem. Dupree sells heating oil to residential customers. Given the amount of competition in the industry, both from other home heating oil suppliers and from electric and natural gas utilities, the price of the oil supplied and the level of service are critical in determining a company's success. Unlike electric and natural gas customers, oil customers are exposed to the risk of running out of fuel. Home heating oil suppliers therefore have to guarantee that the customer's oil tank will not be allowed to run dry. In fact, Dupree's service pledge is, "50 free gallons on us if we let you run dry." Beyond the cost of the oil, however, Dupree is concerned about the perceived reliability of his service if a customer is allowed to run out of oil.

To estimate customer oil use, the home heating oil industry uses the concept of "degree days." A degree day is equal to the difference between the average daily temperature and 68 degrees Fahrenheit. So if the average temperature on a given day is 50, the degree days for that day will be 18. (If the degree day calculation results in a negative number, the degree days number is recorded as 0.) By keeping track of the number of degree days since the customer's last oil fill, by knowing the size of the customer's oil tank, and by estimating the customer's oil consumption as a function of the number of degree days, the oil supplier can estimate when the customer is getting low on fuel and then resupply the customer.

Dupree has used this scheme in the past but is disappointed with the results and the computational burdens it places on the company. First, the system requires that a consumption-per-degree-day figure be estimated for each customer to reflect that customer's consumption habits, size of home, quality of home insulation, and family size. Because Dupree has over 1500 customers, the computational burden of keeping track of all of these customers is enormous. Second, the system is crude and unreliable. The consumption per degree day for each customer is computed by dividing the oil consumption during the preceding year by the degree days during the preceding year. Customers have tended to use less fuel than estimated during the colder months and more fuel than estimated during the warmer months. This means that Dupree is making more deliveries than necessary during the colder months and customers are running out of oil during the warmer months.

Dupree wants to develop a consumption estimation model that is practical and more reliable. The following data are available in the file DUPREE.XLS:

- The number of degree days since the last oil fill and the consumption amounts for 67 customers.

- The number of people residing in the homes of each of the 67 customers. Dupree thinks that this might be important in predicting the oil consumption of customers using oil-fired hot water heaters because it provides an estimate of the hot-water requirements of each customer. Each of the customers in this sample uses an oil-fired hot water heater.

- An assessment, provided by Dupree sales staff, of the home type of each of these 67 customers. The home type classification, which is a number between 1 and 5, is a composite index of the home size, age, exposure to wind, level of insulation, and furnace type. A low index implies a lower oil consumption per degree day, and a high index implies a higher consumption of oil per degree day. Dupree thinks that the use of such an index will allow them to estimate a consumption model based on a sample data set and then to apply the same model to predict the oil demand of each of his customers.

Use regression to see whether a statistically reliable oil consumption model can be estimated from the data.

[11]Case Studies 12.2–12.4 are based on problems from *Advanced Management Accounting,* 2nd edition, by Robert S. Kaplan and Anthony A. Atkinson, Prentice Hall, 1989. We thank them for allowing us to adopt their problems.

12.3 Developing a Flexible Budget at the Gunderson Plant

The Gunderson Plant manufactures the industrial product line of FGT Industries. Plant management wants to be able to get a good, yet quick, estimate of the manufacturing overhead costs that can be expected each month. The easiest and simplest method to accomplish this task is to develop a flexible budget formula for the manufacturing overhead costs. The plant's accounting staff has suggested that simple linear regression be used to determine the behavior pattern of the overhead costs. The regression data can provide the basis for the flexible budget formula. Sufficient evidence is available to conclude that manufacturing overhead costs vary with direct labor hours. The actual direct labor hours and the corresponding manufacturing overhead costs for each month of the last 3 years have been used in the linear regression analysis.

The 3-year period contained various occurrences not uncommon to many businesses. During the first year, production was severely curtailed during 2 months due to wildcat strikes. In the second year, production was reduced in 1 month because of material shortages, and increased significantly (scheduled overtime) during 2 months to meet the units required for a one-time sales order. At the end of the second year, employee benefits were raised significantly as the result of a labor agreement. Production during the third year was not affected by any special circumstances. Various members of Gunderson's accounting staff raised some issues regarding the historical data collected for the regression analysis. These issues were as follows.

- Some members of the accounting staff believed that the use of data from all 36 months would provide a more accurate portrayal of the cost behavior. While they recognized that any of the monthly data could include efficiencies and inefficiencies,

they believed these efficiencies and inefficiencies would tend to balance out over a longer period of time.

- Other members of the accounting staff suggested that only those months that were considered normal should be used so that the regression would not be distorted.

- Still other members felt that only the most recent 12 months should be used because they were the most current.

- Some members questioned whether historical data should be used at all to form the basis for a flexible budget formula.

The accounting department ran two regression analyses of the data—one using the data from all 36 months and the other using only the data from the last 12 months. The information derived from the two linear regressions is shown below (t-values shown in parentheses). The 36-month regression is

$$OH_t = 123,810 + 1.60 \text{ DLH}_t, \quad R^2 = 0.32$$
$$(1.64)$$

The 12-month regression is

$$OH_t = 109,020 + 3.00 \text{ DLH}_t, \quad R^2 = 0.48$$
$$(3.01)$$

Questions

1. Which of the two results (12 months versus 36 months) would you use as a basis for the flexible budget formula?

2. How would the four specific issues raised by the members of Gunderson's accounting staff influence your willingness to use the results of the statistical analyses as the basis for the flexible budget formula? Explain your answer.

Wagner Printers performs all types of printing including custom work, such as advertising displays, and standard work, such as business cards. Market prices exist for standard work, and Wagner Printers must match or better these prices to get the business. The key issue is whether the existing market price covers the cost associated with doing the work. On the other hand, most of the custom work must be priced individually. Because all custom work is done on a job-order basis, Wagner routinely keeps track of all the direct labor and direct materials costs associated with each job. However, the overhead for each job must be estimated. The overhead is applied to each job using a predetermined (normalized) rate based on estimated overhead and labor hours. Once the cost of the prospective job is determined, the sales manager develops a bid that reflects both the existing market conditions and the estimated price of completing the job.

In the past, the normalized rate for overhead has been computed by using the historical average of overhead per direct labor hour. Wagner has become increasingly concerned about this practice for two reasons. First, it hasn't produced accurate forecasts of overhead in the past. Second, technology has changed the printing process, so that the labor content of jobs has been decreasing, and the normalized rate of overhead per direct labor hour has steadily been increasing. The file WAGNER.XLS shows the overhead data that Wagner has collected for its shop for the past 52 weeks. The average weekly overhead for the last 52 weeks is $54,208, and the average weekly number of labor hours worked is 716. Therefore, the normalized rate for overhead that will be used in the upcoming week is about $76 (= 54,208/716) per direct labor hour.

Questions

1. Determine whether you can develop a more accurate estimate of overhead costs.

2. Wagner is now preparing a bid for an important order that may involve a considerable amount of repeat business. The estimated requirements for this project are 15 labor hours, 8 machine hours, $150 direct labor cost, and $750 direct material cost. Using the existing approach to cost estimation, Wagner has estimated the cost for this job as $2040 (= 150 + 750 + (76 × 15)). Given the existing data, what cost would you estimate for this job?

13

Time Series Analysis and Forecasting

Forecasting Labor Requirements at Taco Bell

How much quantitative analysis occurs at fast-food restaurants? At Taco Bell, a lot! An article by Huerter and Swart (1998) explains the approach to labor management that has occurred at Taco Bell restaurants over the past decade. Labor is a large component of costs at Taco Bell. Approximately 30% of every sales dollar goes to labor. However, the unique characteristics of fast-food restaurants make it difficult to plan labor utilization efficiently. In particular, the Taco Bell product—food— cannot be inventoried; it must be made fresh at the time the customer orders it. Because of shifting demand throughout any given day, where the lunch period accounts for approximately 52% of a day's sales and as much as 25% of a day's sales can occur during the busiest hour, labor requirements vary greatly throughout the day. If too many workers are on hand during slack times, they are paid for doing practically nothing. Worse than that, however, are the lost sales (and unhappy customers) that occur if too few workers are on hand during peak times. Prior to 1988, Taco Bell made very little effort to manage the labor problem in an efficient, centralized manner. The company simply allocated about 30% of each store's sales to the store managers and let them allocate it as best they could—not always with good results.

In 1988 Taco Bell initiated its "value meal" deals, where certain meals were priced as low as 59 cents. This increased demand to the point where management could no longer ignore the labor allocation problem. Therefore, in-store computers were installed, data from all stores were collected, and a team of analysts was assigned the task of developing a cost-efficient labor allocation system. This system, which has now been fully integrated into all Taco Bell stores since 1993, is composed of three subsystems: (1) a forecasting subsystem that, for each store, forecasts the arrival rate of customers by 15-minute interval by day of week; (2) a simulation subsystem that, for each store, simulates the congestion and number

© Getty Images

of lost customers that will occur for any customer arrival rate, given a specific number (and deployment) of workers; and (3) an optimization subsystem that, for each store, indicates the minimum cost allocation of workers, subject to various constraints, such as a minimum service level and a minimum shift length for workers. Although all three of these subsystems are important, the forecasting subsystem is where it all starts. Each store must have a reasonably accurate forecast of future customer arrival rates, broken down by small time intervals (such as 11:15 A.M. to 11:30 A.M. on Friday), before labor requirements can be predicted and labor allocations can be made in an intelligent manner. Like many real-world forecasting systems, Taco Bell's has two important characteristics: (1) it requires extensive data, which have been made available by the in-store computer systems, and (2) the eventual forecasting method used is mathematically a fairly simple one, namely, 6-week moving averages, which we will study in this chapter.

Simple or not, the forecasts, as well as the other system components, have enabled Taco Bell to cut costs and increase profits considerably. In its first 4 years, 1993–1996, the labor management system is estimated to have saved Taco Bell approximately $40.34 million in labor costs. Because the number of Taco Bell stores is constantly increasing, the annual company-wide savings from the system will certainly grow in the future. In addition, the focus on quantitative analysis has produced other side benefits for Taco Bell. Its service is now better and more consistent across stores, with many fewer customers leaving because of slow service. Also, the quantitative models developed have enabled Taco Bell to evaluate the effectiveness of various potential productivity enhancements, including self-service drink islands, customer-activated touch screens for ordering, and smaller kitchen areas. So the next time you order food from Taco Bell, you can be assured that there is definitely a method to the madness! ●

13.1 Introduction

Many decision-making applications depend on a forecast of some quantity. Here are several examples.

Examples of Forecasting Applications

- When a service organization, such as a fast-food restaurant, plans its staffing over some time period, it must forecast the customer demand as a function of time. This might be done at a very detailed level, such as the demand in successive half-hour periods, or at a more aggregate level, such as the demand in successive weeks.

- When a company plans its ordering or production schedule for a product it sells to the public, it must forecast the customer demand for this product so that it can stock appropriate quantities—neither too much nor too little.

- When an organization plans to invest in stocks, bonds, or other financial instruments, it typically attempts to forecast movements in stock prices and interest rates.

- When government representatives plan policy, they attempt to forecast movements in macroeconomic variables such as inflation, interest rates, and unemployment.

Unfortunately, forecasting is a very difficult task, both in the short run and in the long run. Typically, we base forecasts on observations made in the past. We investigate past behavior, search for patterns or relationships, and then we make forecasts. There are two problems with this approach. The first is that it is not always easy to uncover historical patterns or relationships. In particular, it is often difficult to separate the noise, or random behavior, from the underlying patterns. Some forecasts can even overdo it, by attributing importance to patterns that are in fact random variations and are unlikely to repeat themselves.

The second problem is that there are no guarantees that past patterns will continue in the future. The OPEC countries could raise their oil prices again, a company's competitor could introduce a new product into the market, the bottom could fall out of the stock market, and so on. Each of these shocks to the system being studied could drastically alter the future in a highly unpredictable way. This partly explains why forecasts are almost always wrong. Unless they have inside information to the contrary, forecasters must assume that history will repeat itself. But we all know that history does *not* always repeat itself. Therefore, there are many famous forecasts that turned out to be way off the mark, even though the forecasters made reasonable assumptions and used standard forecasting techniques. Nevertheless, forecasts are required constantly, so fear of failure is no excuse for not giving it our best effort.

13.2 Forecasting Methods: An Overview

There are many forecasting methods available, and all practitioners have their favorites. To say the least, there is little agreement among practitioners or theoreticians as to the best forecasting method. The methods can generally be divided into three groups: (1) *judgmental* methods, (2) *extrapolation* (or *time series*) methods, and (3) *econometric* (or *causal*) methods. The first of these is basically nonquantitative and will not be discussed here; the last two are quantitative. In this section we will describe extrapolation and econometric methods in some generality. In the rest of the chapter, we will go into more detail, particularly about the extrapolation methods.

13.2.1 Extrapolation Methods

Extrapolation methods are quantitative methods that use past data of a time series variable—and nothing else, except possibly time itself—to forecast future values of the variable. The idea is that we can use past movements of a variable, such as company sales or U.S. exports to Japan, to forecast its future values. Many extrapolation methods are available, including trend-based regression, exponential smoothing, moving averages, and autoregression models. Some of these methods are relatively simple, both conceptually and in terms of the calculations required, whereas others are quite complex. Also, as the names imply, some of these methods use the same regression methods we have discussed in the previous two chapters, whereas others do not.

All of these extrapolation methods search for *patterns* in the historical series and then extrapolate these patterns into the future. Some try to track long-term upward or downward trends and then project these. Some try to track the seasonal patterns (sales up in November and December, down in other months, for example) and then project these. Basically, the more complex the method, the more closely it tries to track historical patterns. Researchers have long believed that it is an asset of a method to be able to track the ups and downs—the zigzags on a graph—of a time series. This has led to voluminous research and increasingly complex methods. But is complexity always better?

Surprisingly, empirical evidence shows that it is sometimes worse. This is documented in the quarter-century review article by Armstrong (1986) and the article by Schnarrs and Bavuso (1986). They document a number of empirical studies on literally thousands of time series forecasts where complex methods fared no better, and often worse, than simple methods. In fact, the Schnarrs and Bavuso article presents evidence that a naive forecast from a "random walk" model often outperforms all of the more sophisticated extrapolation methods. With this naive model we forecast that next period's value will be the same as this period's value. So if today's closing stock price is 51.375, we forecast that tomorrow's closing stock price will be 51.375. We will discuss random walks in more detail in Section 13.5.

The evidence in favor of simpler models is not accepted by everyone, particularly not those who have spent years investigating complex models, and complex models continue to be studied and used. However, there is a very plausible reason why simple models might provide better forecasts. The whole idea behind extrapolation methods is to extrapolate historical patterns into the future. But it is often difficult to determine which patterns are real and which represent noise, that is, random ups and downs that are not likely to repeat themselves. Also, if something important changes (a competitor introduces a new product or interest rates increase, for example), it is always possible that the historical patterns will change. A potential problem with complex methods is that they often track the historical series *too* closely. That is, they often track patterns that are really noise. Simpler methods, on the other hand, track only the most basic underlying patterns and therefore can be more flexible and accurate in forecasting the future.

13.2.2 Econometric Models

Econometric models, also called **causal** models, use regression to forecast a time series variable by using other explanatory time series variables. For example, a company might use a causal model to regress future sales on its advertising level, the population income level, the interest rate, and possibly others. In one sense, regression analysis involving time series variables is similar to the regression analysis discussed in the previous two chapters. For example, it is still possible to use the same least squares approach and the same multiple regression software in many time series regression models. In fact, several examples and problems in the previous two chapters used time series data.

However, causal regression models for time series data present new mathematical challenges that go well beyond the level of this book. To get a glimpse of the potential difficulties, suppose a company wants to use a causal regression model to forecast its monthly sales for some product, based on two other time series variables: its monthly advertising levels for the product and its main competitor's monthly advertising levels for a competing product. We could simply estimate a regression equation of the form

$$Y_t = \alpha + \beta_1 X_{1t} + \beta_2 X_{2t} \tag{13.1}$$

Here, Y_t is the company's sales in month t, and X_{1t} and X_{2t} are the company's and the competitor's advertising levels in month t. We might learn something useful from this regression model, but we should be aware of the following problems.

One problem is that we must decide on the appropriate "lags" for the regression equation. Do sales this month depend only on advertising levels *this* month, as specified in equation (13.1), or also on advertising levels in the previous month, the previous two months, and so on? A second problem is whether to include lags of the *sales* variable in the regression equation as explanatory variables. Presumably, sales in one month might depend on the level of sales in previous months (as well as on advertising levels). A third problem is that the two advertising variables can be *autocorrelated* and *cross-correlated*. Autocorrelation means, for example, that the company's advertising level in one month can depend on its advertising levels in previous months. Cross-correlation means, for example, that the company's advertising level in one month can be related to the competitor's advertising levels in previous months, or vice versa, that the competitor's advertising in one month can be related to the company's advertising levels in previous months.

These are difficult issues, and the way in which they are addressed can make a big difference in the usefulness of the resulting regression model. We will examine several regression-based models in this chapter, but we will avoid situations such as the one just described, where one time series variable Y is regressed one or more time series X's. [Pankratz (1991) is a good reference for these latter types of models.]

13.2.3 Combining Forecasts

There is one other general forecasting method that is worth mentioning. In fact, it has attracted a lot of attention in recent years, and many researchers believe that it has great potential for increasing forecast accuracy. The method is simple—combine two or more forecasts to obtain the final forecast. The reasoning behind this method is also simple—the forecast errors from different forecasting methods might cancel one another. The forecasts that are combined can be of the same general type—extrapolation forecasts, for example—or they can be of different types, such as judgmental and extrapolation. The *number* of forecasts to combine and the *weights* to use in combining them have been the subject of several research studies.

Although the findings are not entirely consistent, it appears that the marginal benefit from each individual forecast after the first two or three is minor. Also, there is not much evidence to suggest that the simplest weighting scheme—weight each forecast equally, that is, average them—is any less accurate than more complex weighting schemes.

13.2.4 Components of Time Series Data

In Chapter 2 we discussed time series plots, a useful graphical means of depicting time series data. We now use these time series plots to help explain and identify four important components of a time series. These components are called the *trend* component, the *seasonal* component, the *cyclic* component, and the *random* (or noise) component.

Let's start by looking at a very simple time series. This is a time series where every observation is the same. Such a series is shown in Figure 13.1. The graph in this figure shows time (t) on the horizontal axis and the observation values (Y) on the vertical axis. It is assumed that Y is measured at regularly spaced intervals, usually days, weeks, months, quarters, or years, with Y_t being the value of the observation at time period t. As indicated in Figure 13.1, the individual observation points are usually joined by straight

FIGURE 13.1
The Base Series

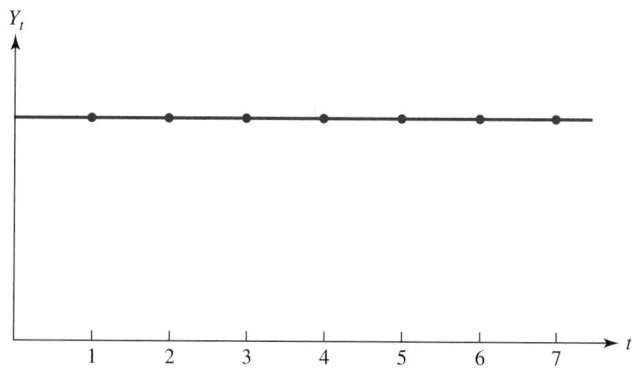

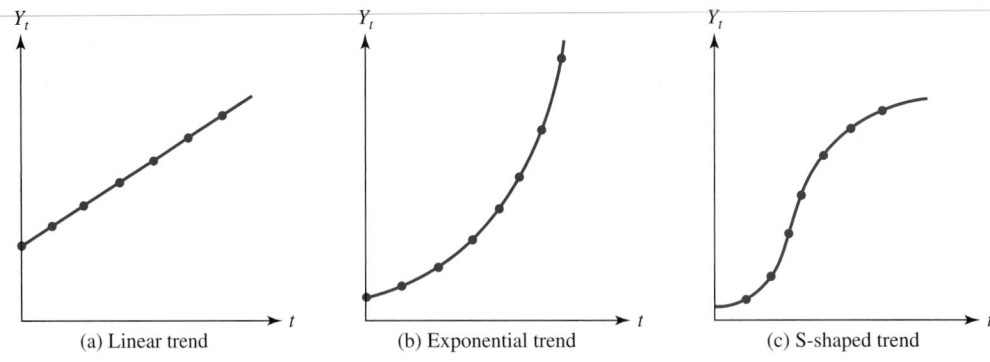

FIGURE 13.2 Series with Trends

lines to make any patterns in the time series more apparent. Since all observations in this time series are equal, the resulting time series plot is a horizontal line. We refer to this time series as the *base* series. We will now illustrate more interesting time series built from this base series.

If the observations increase or decrease regularly through time, we say that the time series has a **trend** (sometimes called a secular trend). The graphs in Figure 13.2 illustrate several possible trends. The *linear* trend in Figure 13.2a occurs if a company's sales, for example, increase by the same amount from period to period. This constant per period change is then the slope of the linear trend line. The curve in Figure 13.2b is an *exponential* trend curve. It occurs in a business such as the personal computer business, where sales are increasing at a tremendous rate. For this type of curve, the *percentage* increase in Y_t from period to period remains constant. The curve in Figure 13.2c is an *S-shaped* trend curve. For example, this type of trend curve is appropriate for a new product that takes a while to catch on, then exhibits a rapid increase in sales as the public becomes aware of it, and finally tapers off to a fairly constant level. The curves in Figure 13.2 all represent *upward* trends. Of course, we could just as well have *downward* trends of the same types.

Many time series have a **seasonal** component. For example, a company's sales of swimming pool equipment increase every spring, then stay relatively high during the summer, and then drop off until next spring, at which time the yearly pattern repeats itself. An important aspect of the seasonal component is that it tends to be predictable from one year to the next. That is, the *same* seasonal pattern tends to repeat itself every year.

In Figure 13.3 we show two possible seasonal patterns. In Figure 13.3a there is nothing but the seasonal component. That is, if there were no seasonal variation, we would have the base series from Figure 13.1. In Figure 13.3b we show a seasonal pattern superimposed on a linear trend line.

The third component of a time series is the **cyclic** component. By studying past movements of many business and economic variables, it becomes apparent that there are business cycles that affect many variables in similar ways. For example, during a recession housing starts generally go down, unemployment goes up, stock prices go down, and so on. But when the recession is over, all these variables tend to move in the opposite direction.

We know that the cyclic component exists for many time series because we are able to see it as periodic swings in the levels of the time series graphs. However, the cyclic component is harder to predict than the seasonal component. The reason is that seasonal variation is much more regular. For example, swimming pool supplies sales *always* start to increase during the spring. Cyclic variation, on the other hand, is more irregular for the

FIGURE 13.3
Series with
Seasonality

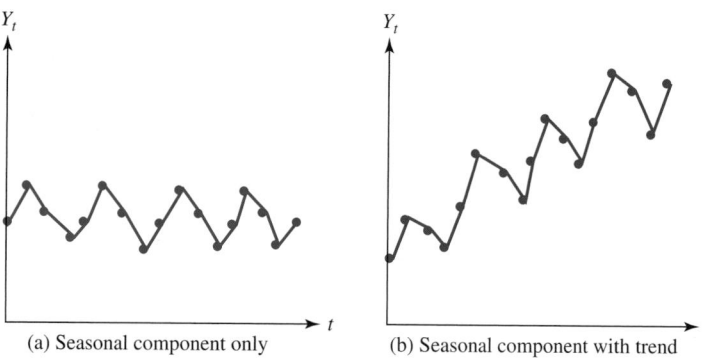

(a) Seasonal component only   (b) Seasonal component with trend

FIGURE 13.4
Series with Cyclic
Component

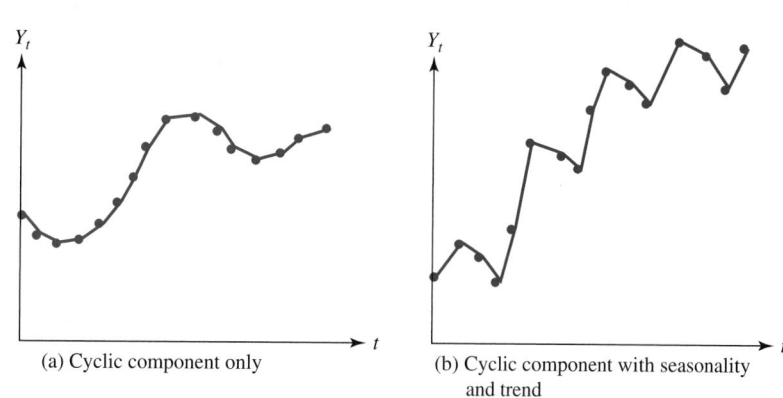

(a) Cyclic component only   (b) Cyclic component with seasonality
and trend

simple reason that the "business cycle" does not always have the same length. A further distinction is that the length of a seasonal cycle is generally one year; the length of a business cycle is generally much longer than one year.

The graphs in Figure 13.4 illustrate the cyclic component of a time series. In Figure 13.4a cyclic variation is superimposed on the base series from Figure 13.1. In Figure 13.4b this same cyclic variation is superimposed on the series from Figure 13.3b. The resulting graph has trend, seasonal variation, and cyclic variation.

The final component in a time series is called the **random** component, or simply **noise.** This unpredictable component gives most time series graphs their irregular, jagged-edge appearances. Usually, a time series can be determined only to a certain extent by its trend, seasonal, and cyclic components. Then other factors determine the rest. These other factors may be inherent randomness, unpredictable "shocks" to the system, the unpredictable behavior of human beings who interact with the system, and probably others. Whatever these factors are, however, there is no doubt that they exist and that they combine to create a certain amount of unpredictability in almost all time series.

Figure 13.5 and 13.6 (page 696) show the effect that noise can have on a time series graph. The graph on the left of each figure shows the random component only, superimposed on the base series. Then on the right of each figure, the random component is superimposed on the trend-with-seasonal-component graph from Figure 13.3b. The difference between Figures 13.5 and 13.6 is the relative magnitude of the noise. When it is small, as in Figure 13.5, the other components emerge fairly clearly; they are not disguised by the noise. But if the noise is large in magnitude, as in Figure 13.6, the noise makes it very difficult to distinguish the other components.

FIGURE 13.5
Series with Noise

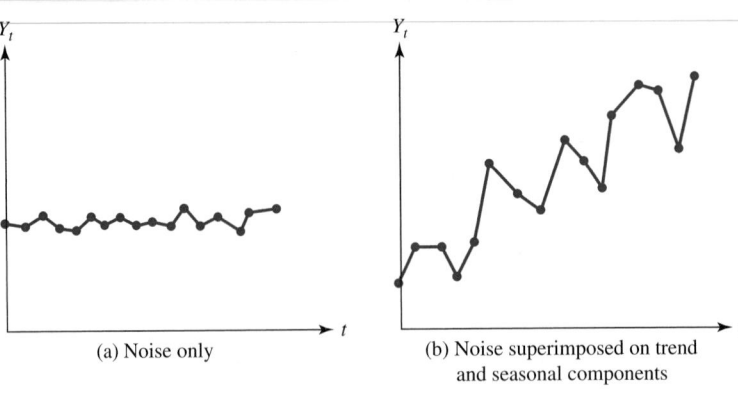

(a) Noise only

(b) Noise superimposed on trend
and seasonal components

FIGURE 13.6
Series with More
Noise

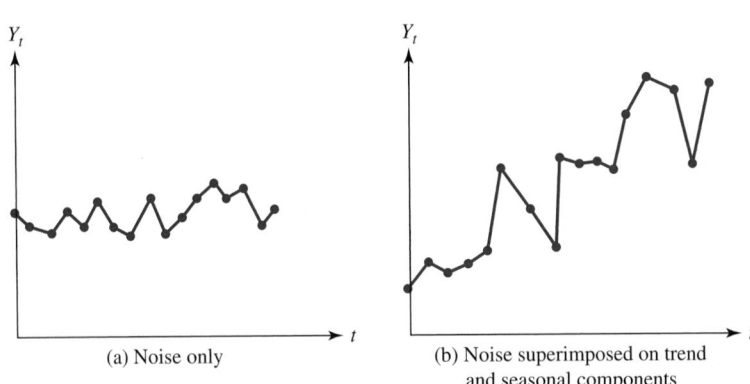

(a) Noise only

(b) Noise superimposed on trend
and seasonal components

13.2.5 General Notation and Formulas

We now introduce a bit of notation and discuss some aspects common to most forecasting methods. In general, we let Y denote the variable we want to forecast. Then Y_t denotes the observed value of Y at time t. Typically, the first observation (the most distant one) corresponds to period $t = 1$, and the last observation (the most recent one) corresponds to period $t = T$, so that T denotes the number of historical observations of Y. The periods themselves might be weeks, months, quarters, years, or any other convenient unit of time.

Suppose we have just observed Y_{t-k} and want to make a "k-period-ahead" forecast; that is, we want to use the information through time $t - k$ to forecast Y_t. Then we denote the resulting forecast by $F_{t-k,t}$. The first subscript denotes the period in which the forecast is made, and the second subscript denotes the period being forecasted. As an example, if the data are monthly and September 2000 corresponds to $t = 67$, then a forecast of Y_{69}, the value in November 2000, would be labeled $F_{67,69}$. The **forecast error** is the difference between the actual value and the forecast. It is denoted by E with appropriate subscripts. Specifically, the forecast error associated with $F_{t-k,t}$ is

$$E_{t-k,t} = Y_t - F_{t-k,t}$$

This double-subscript notation is necessary to specify when the forecast is being made and which period is being forecasted. However, the former is often clear from context. Therefore, to simplify the notation, we will usually drop the first subscript and write F_t and E_t to denote the forecast of Y_t and the error in this forecast.

There are actually two steps in any forecasting procedure. The first step is to build a model that fits the historical data well. The second step is to use this model to forecast the future. Most of the work goes into the first step. For any trial model we see how well it "tracks" the known values of the time series. Specifically, we calculate the one-period-ahead forecasts F_t (or more precisely, $F_{t-1,t}$) from the model and compare these to the known values, Y_t, for each t in the historical time period. We attempt to find a model that produces small forecast errors, E_t. We expect that if the model forecasts the *historical* data well, it will also forecast *future* data well.

Forecasting software packages typically report several summary measures of the forecast errors. The most important of these are MAE (mean absolute error), RMSE (root mean square error), and MAPE (mean absolute percentage error). These are defined in equations (13.2), (13.3), and (13.4). Fortunately, models that make any one of these measures small tend to make the others small, so that we can choose whichever measure we want to minimize. In the following formulas, N denotes the number of terms in each sum. This value is typically slightly less than T, the number of historical observations, because it is not usually possible to provide a forecast for each historical period.

Mean Absolute Error

$$\text{MAE} = \left(\sum_{t=1}^{N} |E_t| \right) / N \qquad \text{(13.2)}$$

Root Mean Square Error

$$\text{RMSE} = \left(\sum_{t=1}^{N} E_t^2 \right) / N \qquad \text{(13.3)}$$

Mean Absolute Percentage Error

$$\text{MAPE} = 100\% \times \left(\sum_{t=1}^{N} |E_t/Y_t| \right) / N \qquad \text{(13.4)}$$

RMSE is similar to a standard deviation in that the errors are squared; because of the square root, its units are the same as those of the forecasted variable. The MAE is similar to the RMSE, except that absolute values of errors are used instead of squared errors. The MAPE is probably the most easily understood measure because it does not depend on the units of the forecasted variable; it is always stated as a percentage. For example, the statement that the forecasts are off on average by 2% has a clear meaning, even if you do not know the units of the variable being forecasted.

Forecasting software packages are sometimes able to choose the best model from a given class (such as the best exponential smoothing model) by minimizing MAE, RMSE, or MAPE. However, because the details of the package are not always given, you might not be absolutely sure which of these measures is being minimized. Fortunately, this is probably not too important. For example, a model with a small RMSE typically has a small MAPE and MAE. In any case, small values of these measures only guarantee that the model forecasts the *historical* observations well. There is still no guarantee that the model will forecast *future* values accurately.

We will now examine a number of useful forecasting models. You should be aware that more than one of these models can be appropriate for any particular time series data. For example, a random walk model and an autoregression model might be equally effective for forecasting stock price data. (Remember also that we can combine forecasts from more than one model to obtain a possibly better forecast.) We will try to give some insights into choosing the best type of model for various types of time series data. But ultimately the choice depends on the experience of the user.

13.3 Testing for Randomness

All forecasting models we build have the general form shown in equation (13.5). The fitted value in this equation is the part we calculate from past data and any other available information (such as the identity of the current season), and it is used as a forecast for Y. The residual is the forecast error, the difference between the observed value of Y and its forecast:

$$Y_t = \text{Fitted Value} + \text{Residual} \tag{13.5}$$

In a time series context the terms *residual* and *forecast error* can be used interchangeably.

For time series data, there is a residual for each historical period, that is, for each value of t. We want this time series of residuals to be random "noise," as discussed in Section 13.2.4. The reason is that if this series of residuals is not noise, then it can be modeled further. For example, if the residuals trend upwardly, then we can refine our model to include this trend component in the *fitted* value. The point is that we want the fitted value to include all components of the original series that can possibly be forecasted, and we want the leftover residuals to be noise.

The question we now examine is how to determine whether a time series of residuals is random noise (which we usually abbreviate to "random"). The simplest method, but not always a reliable one, is to examine time series plots of residuals visually. This often enables us to detect nonrandom patterns. For example, the time series plots in Figures 13.7–13.11 illustrate some common nonrandom patterns. In Figure 13.7, there is an upward trend. In Figure 13.8, the variance increases through time (larger zigzags to the right). Figure 13.9 exhibits seasonality, where observations in certain months are consistently larger than those in other months. There is a "meandering" pattern in Figure 13.10, where large observations tend to be followed by other large observations, and small observations tend to be followed by other small observations. Finally, the opposite behavior of Figure 13.10 is illustrated in Figure 13.11. Here, there are too *many* zigzags—large

FIGURE 13.7
A Series with Trend

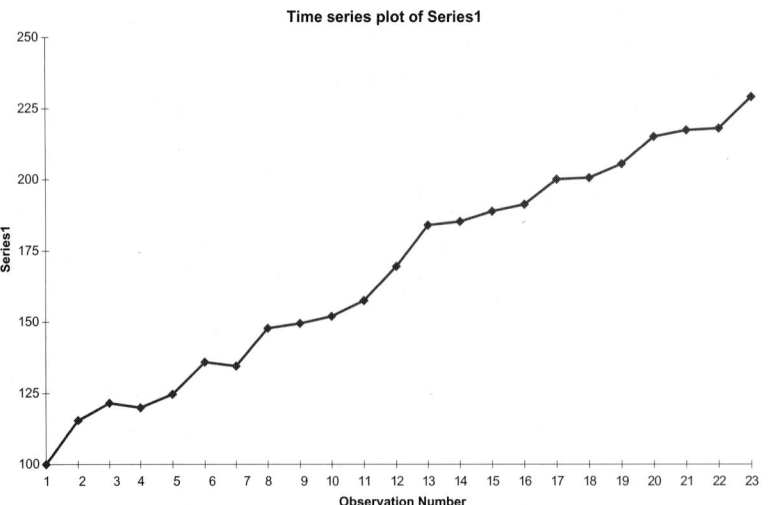

FIGURE 13.8
A Series with
Increasing Variance
Through Time

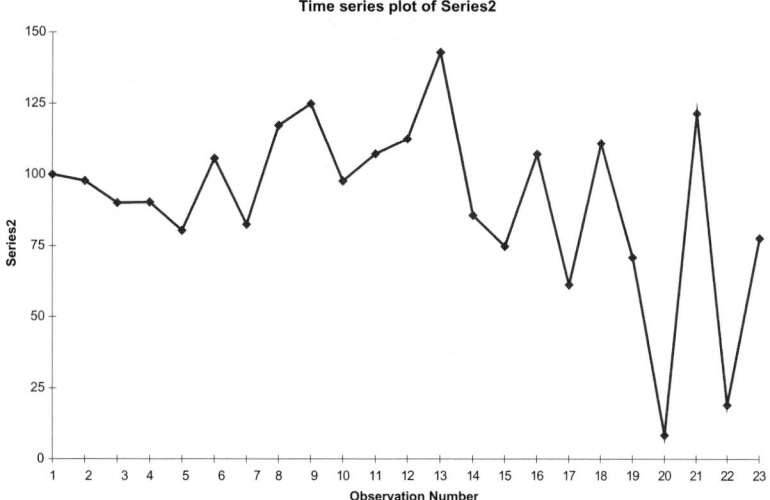

FIGURE 13.9
A Series with
Seasonality

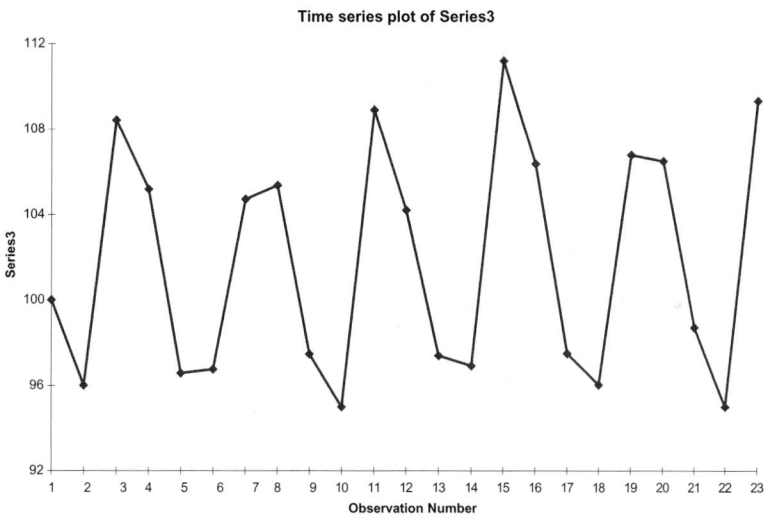

FIGURE 13.10
A Series That
Meanders

FIGURE 13.11

A Series That Oscillates Frequently

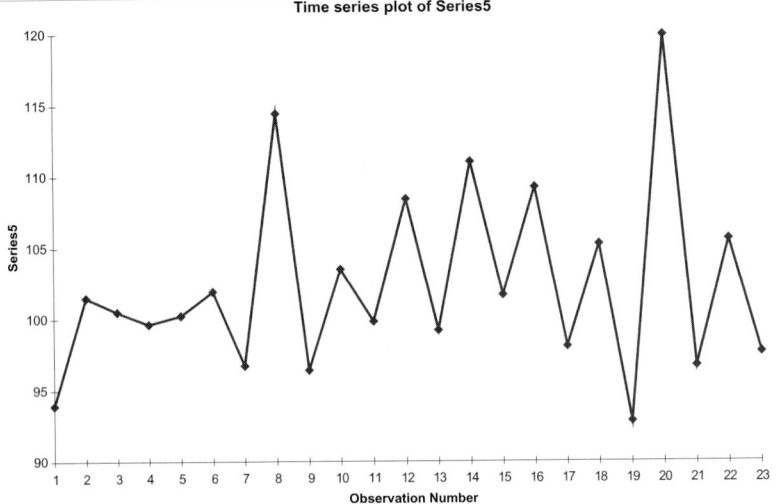

Time series plot of Series5

observations tend to follow small observations and vice versa. None of the time series in these figures can be considered random.

13.3.1 The Runs Test

This runs test can be used on *any* time series, not just a series of residuals.

It is not always easy to detect randomness or the lack of it from the visual inspection of a graph. Therefore, we mention two more formal methods of testing for randomness. The first is called the *runs test*. We first choose a base value, which could be the average value of the series, the median value, or even some other value. Then we define a **run** as a consecutive series of observations that remain on one side of this base level. For example, if the base level is 0 and we observe the series 1, 5, 3, −3, −2, −4, −1, 3, 2, then there are three runs: 1, 5, 3; −3, −2, −4, −1; and 3, 2. The idea behind the runs test is that a random series should have a number of runs that is neither too large nor too small. If the series has too few runs, then it could be trending (as in Figure 13.7) or it could be meandering (as in Figure 13.10). If the series has too many runs, then it is zigzagging too often (as in Figure 13.11).

Runs Test	The **runs test** is a formal test of the null hypothesis of randomness. If there are too many or too few runs in the series, then we conclude that the series is not random.

We will not provide the mathematical details of the runs test, but we will illustrate how it is implemented in StatPro in the following example.

Example 13.1 Forecasting Monthly Stereo Sales

Monthly sales for a chain of stereo retailers are listed in the file STEREO.XLS. They cover the period from the beginning of 1998 to the end of 2001, during which there was no upward or downward trend in sales and no clear seasonal peaks or valleys. This behavior is apparent in the time series plot of sales in Figure 13.12. Therefore, a simple forecast model of sales is to use the *average* of the series, 182.67, as a forecast of sales for each month. Do the resulting residuals represent random noise?

FIGURE 13.12
Time Series Plot
of Stereo Sales

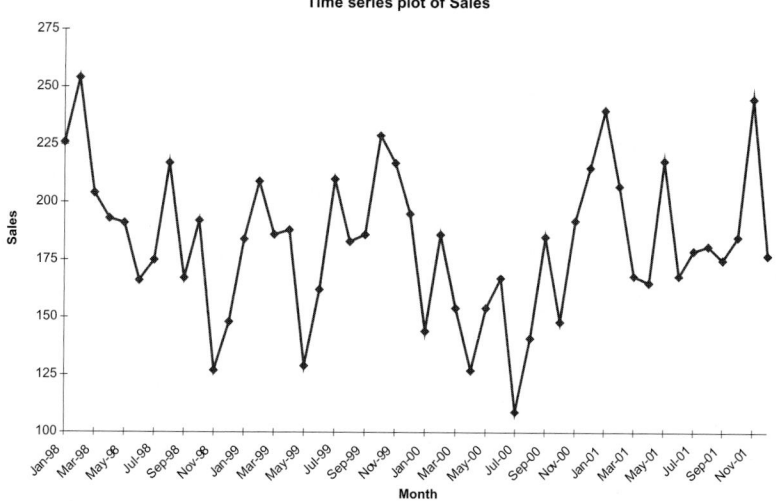

Objective To use StatPro's Runs Test procedure to check whether the residuals from this simple forecasting model represent random noise.

Solution

To obtain the residuals for this forecasting model, we subtract the average, 182.67, from each observation. Therefore, the plot of the residuals, shown in Figure 13.13, has exactly the same shape as the plot of sales. The only difference is that it is shifted down by 182.67 and has mean 0. We now use the runs test to check whether there are too many or too few runs around the base value of 0 in this residual plot. To do so, we use the StatPro/Statistical Inference/Runs Test for Randomness menu item, choose Residual as the variable to analyze, and choose "mean of the series" as the cutoff value. (This corresponds to the horizontal line at 0 in Figure 13.13.) This produces the output in columns F–K of Figure 13.14.

StatPro inserts columns F and G to count the runs. You can check the formulas in these columns, but basically, a 1 in column G indicates the beginning of a new run.

FIGURE 13.13
Time Series Plot
of Residuals

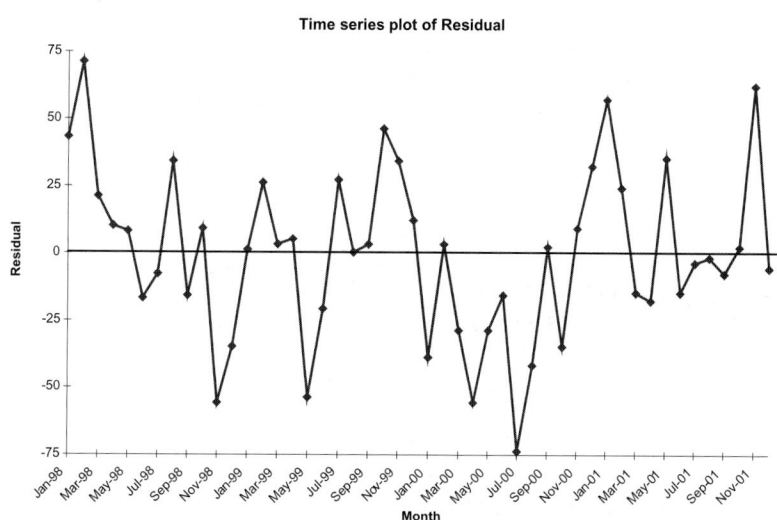

	A	B	C	D	E	F	G	H	I	J	K
1	Monthly stereo sales (in $1,000s)										
2											
3	**Month**	**Sales**	**Forecast**	**Residual**		**Residual_High**	**Residual_NewRun**			*Runs Test Results for Residual*	
4	Jan-98	226	182.67	43.33		1	1				
5	Feb-98	254	182.67	71.33		1	0			Number of obs	48
6	Mar-98	204	182.67	21.33		1	0			Number above cutoff	26
7	Apr-98	193	182.67	10.33		1	0			Number below cutoff	22
8	May-98	191	182.67	8.33		1	0			Number of runs	20
9	Jun-98	166	182.67	-16.67		0	1				
10	Jul-98	175	182.67	-7.67		0	0			E(R)	24.833
11	Aug-98	217	182.67	34.33		1	1			Stdev(R)	3.403
12	Sep-98	167	182.67	-15.67		0	1			Z-value	-1.420
13	Oct-98	192	182.67	9.33		1	1			p-value (2-tailed)	0.155
14	Nov-98	127	182.67	-55.67		0	1				
15	Dec-98	148	182.67	-34.67		0	0				
16	Jan-99	184	182.67	1.33		1	1				
17	Feb-99	209	182.67	26.33		1	0				
18	Mar-99	186	182.67	3.33		1	0				

FIGURE 13.14 Runs Test for Randomness

Columns J and K then report the results of the runs test. Without going into all of the details, the important points are the following:

- The number of observed runs is 20, in cell K8.

- The number of runs *expected* under an assumption of randomness is 24.833, in cell K10. Therefore, the series of residuals has too *few* runs. Positive values tend to follow positive values, and negative values tend to follow negative values.

- The z-value in cell K12, (1.42, indicates how many standard errors the observed number of runs is below the expected number of runs. The corresponding p-value indicates how extreme this z-value is. It should be interpreted just like other p-values for hypothesis tests. If it is small, say, less than 0.05, then we can reject the null hypothesis of randomness and conclude that the series of residuals is *not* random noise. However, the p-value for this example is only 0.155. Therefore, there is not convincing evidence of nonrandomness in the residuals, and we can conclude that the residuals represent noise. ●

A small p-value provides evidence of nonrandomness.

13.3.2 Autocorrelation

Like the runs test, autocorrelations can be calculated for *any* time series, not just a series of residuals.

In this section we discuss one further way to check for randomness of a time series of residuals—we examine its *autocorrelations*. The "auto" means that successive observations are correlated with one other. For example, in the most common form of autocorrelation, *positive* autocorrelation, large observations tend to follow large observations, and small observations tend to follow small observations. In this case the runs test is likely to pick it up, because there will be fewer runs than expected. Another way to check for the same nonrandomness property is to calculate the autocorrelations of the time series.

Autocorrelation	An **autocorrelation** is a type of correlation used to measure whether values of a time series are related to their own past values.

To understand autocorrelations it is first necessary to understand what it means to *lag* a time series. This concept is easy to understand in spreadsheets. We'll again use the monthly stereo sales data for 1998–2001 in the STEREO.XLS file. To lag by 1 month, we simply "push down" the series by one row. See column E of Figure 13.15. Note that

	A	B	C	D	E	F	G	H	I	J	K	L
1	Monthly stereo sales (in $1,000s)											
2												
3	Month	Sales	Forecast	Residual	Residual_Lag1	Residual_Lag2	Residual_Lag3			Autocorrelations for Residual		
4	Jan-98	226	182.67	43.33 *	*	*				Lag	Autocorr	StErr
5	Feb-98	254	182.67	71.33	43.33 *	*				1	0.3492	0.1443
6	Mar-98	204	182.67	21.33	71.33	43.33 *				2	0.0772	0.1443
7	Apr-98	193	182.67	10.33	21.33	71.33	43.33			3	0.0814	0.1443
8	May-98	191	182.67	8.33	10.33	21.33	71.33			4	-0.0095	0.1443
9	Jun-98	166	182.67	-16.67	8.33	10.33	21.33			5	-0.1353	0.1443
10	Jul-98	175	182.67	-7.67	-16.67	8.33	10.33			6	0.0206	0.1443
11	Aug-98	217	182.67	34.33	-7.67	-16.67	8.33					
12	Sep-98	167	182.67	-15.67	34.33	-7.67	-16.67					
13	Oct-98	192	182.67	9.33	-15.67	34.33	-7.67					
14	Nov-98	127	182.67	-55.67	9.33	-15.67	34.33					
15	Dec-98	148	182.67	-34.67	-55.67	9.33	-15.67					
16	Jan-99	184	182.67	1.33	-34.67	-55.67	9.33					
17	Feb-99	209	182.67	26.33	1.33	-34.67	-55.67					
18	Mar-99	186	182.67	3.33	26.33	1.33	-34.67					
19	Apr-99	188	182.67	5.33	3.33	26.33	1.33					
20	May-99	129	182.67	-53.67	5.33	3.33	26.33					
21	Jun-99	162	182.67	-20.67	-53.67	5.33	3.33					
22	Jul-99	210	182.67	27.33	-20.67	-53.67	5.33					
23	Aug-99	183	182.67	0.33	27.33	-20.67	-53.67					

FIGURE 13.15 Lags and Autocorrelations for Stereo Sales

there is a blank cell at the top of the lagged series (in cell E4). We can continue to push the series down one row at a time to obtain other lags. For example, the lag 3 version of the series appears in column G. Now there are three missing observations at the top. Note that in December 1998, say, the first, second, and third lags correspond to the observations in November 1998, October 1998, and September 1998, respectively. That is, lags are simply previous observations, removed by a certain number of periods from the present time. These lagged columns can be obtained by copying and pasting the original series or by using the StatPro/Data Utilities/Create Lagged Variables menu item.

Then the autocorrelation of lag k, for any integer k, is essentially the correlation between the original series and the lag k version of the series. For example, in Figure 13.15 the lag 1 autocorrelation is the correlation between the observations in columns D and E. Similarly, the lag 2 autocorrelation is the correlation between the observations in columns D and F.[1]

We have shown the lagged versions of Sales in Figure 13.15, and we have explained autocorrelations in terms of these lagged variables, to help motivate the concept of autocorrelation. However, we can use StatPro's Autocorrelation procedure directly, *without* forming the lagged variables, to calculate autocorrelations. This is illustrated in the following continuation of Example 13.1.

Example 13.1 Forecasting Monthly Stereo Sales (continued)

The runs test on the stereo sales data suggests that the pattern of sales is not completely random. Large values tend to follow large values, and small values tend to follow small values. Do autocorrelations support this conclusion?

Objective To examine the autocorrelations of the residuals from the forecasting model for evidence of nonrandomness.

---

[1] We'll ignore the exact details of the calculations here. Just be aware that the formula for autocorrelations that is usually used differs slightly from the correlation formula in Chapter 3. However, the difference is very slight and of little practical importance.

FIGURE 13.16
Correlogram for
Residuals

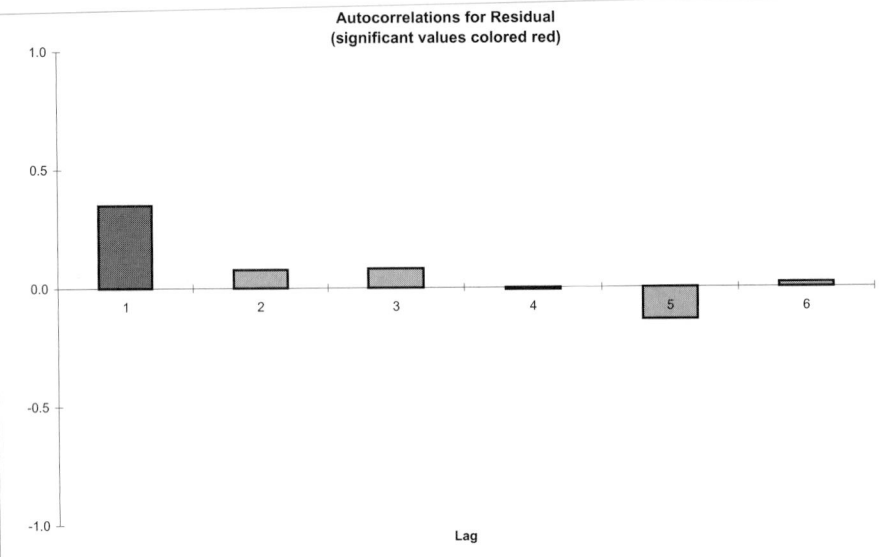

Autocorrelations for Residual
(significant values colored red)

Solution

We use StatPro's Autocorrelation procedure, found under the StatPro/Summary Stats/ Autocorrelations menu item. It requires us to specify the time series variable (Residual), the number of lags we want (we chose 6), and whether we want a chart of the autocorrelations. This chart is called a **correlogram.** The resulting autocorrelations and correlogram appear in Figures 13.15 and 13.16, respectively. A typical autocorrelation of lag k indicates the relationship between observations k periods apart. For example, the autocorrelation of lag 3, 0.0814, indicates that there is very little relationship between residuals separated by 3 months.

How large is a "large" autocorrelation? Under the assumption of randomness, it can be shown that the standard error of any autocorrelation is approximately $1/\sqrt{T}$, in this case $1/\sqrt{48} = 0.1443$. (Recall that T denotes the number of observations in the series.) If the series is truly random, then only an occasional autocorrelation should be larger than 2 standard errors in magnitude. Therefore, any autocorrelation that *is* larger than 2 standard errors in magnitude is worth our attention. These significantly nonzero autocorrelations are boldfaced in the numerical output and shown in red in the chart. The only "large" autocorrelation for the residuals is the first, or lag 1, autocorrelation of 0.3492. The fact that it is *positive* indicates once again that there is some tendency for large residuals to follow large residuals and for small to follow small. The autocorrelations for other lags are less than two standard errors in magnitude and can be ignored. ●

Typically, we can ask for autocorrelations up to as many lags as we like. However, there are several practical considerations to keep in mind. First, it is common practice to ask for no more lags than 25% of the number of observations. For example, if there are 48 observations, we should ask for no more than 12 autocorrelations (lags 1–12).

Second, the first few lags are typically the most important. Intuitively, if there is any relationship between successive observations, it is likely to be between nearby observations. The June 1999 observation is more likely to be related to the May 1999 observation than to the October 1998 observation. Sometimes there is a fairly large spike in the correlogram at some large lag, such as lag 9. However, this can often be ignored as a random "blip" unless there is some obvious reason for its occurrence. A similarly large autocorrelation at lag 1 or 2 should usually be taken more seriously. The one exception to this is a *seasonal* lag. For example, for monthly data an autocorrelation at lag 12 corresponds to a relationship between observations a year apart, such as May 1999 and May 1998. If this autocorrelation is significantly large, it probably should not be ignored.

We will not examine autocorrelations much further in this book. However, we mention that many advanced forecasting techniques are based largely on the examination of the autocorrelation structure of time series. This autocorrelation structure tells us how a series is related to its own past values through time, which can be very valuable information for forecasting *future* values.

PROBLEMS

Level A

1. The file P13_01.XLS contains the monthly number of airline tickets sold by the CareFree Travel Agency. Is this time series *random?* Perform a runs test and compute a few autocorrelations to support your answer.

2. The file P13_02.XLS contains the weekly sales at a local bookstore for each of the past 25 weeks. Is this time series *random?* Perform a runs test and compute a few autocorrelations to support your answer.

3. The number of employees on the payroll at a food processing plant is recorded at the start of each month. These data are provided in the file P13_03.XLS. Perform a runs test and compute a few autocorrelations to determine whether this time series is random.

4. The quarterly numbers of applications for home mortgage loans at a branch office of Northern Central Bank are recorded in the file P13_04.XLS. Perform a runs test and compute a few autocorrelations to determine whether this time series is random.

5. The number of reported accidents at a manufacturing plant located in Flint, Michigan, was recorded at the start of each month. These data are provided in the file P13_05.XLS. Is this time series *random?* Perform a runs test and compute a few autocorrelations to support your answer.

6. The file P13_06.XLS contains the weekly sales at the local outlet of WestCoast Video Rentals for each of the past 36 weeks. Perform a runs test and compute a few autocorrelations to determine whether this time series is random.

Level B

7. Determine whether the RAND() function in Excel actually generates a random stream of numbers. Generate at least 100 random numbers to perform this test. Summarize your findings.

8. Use a runs test and calculate autorrelations to decide whether the random series explained in each part **a–c** are random. For each part, generate at least 100 random numbers in the series.

   a. A series of independent normally distributed values, each with mean 70 and standard deviation 5.

   b. A series where the first value is normally distributed with mean 70 and standard deviation 5, and each succeeding value is normally distributed with mean equal to the *previous* value and standard deviation 5. (For example, if the fourth value is 67.32, then the fifth value will be normally distributed with mean 67.32.)

   c. A series where the first value, Y_1, is normally distributed with mean 70 and standard deviation 5, and each succeeding value, Y_t, is normally distributed with mean $(1 + a_t)Y_{t-1}$ and standard deviation $5(1 + a_t)$, where the a_t's are independent, normally distributed values with mean 0 and standard deviation 0.2. (For example, if $Y_{t-1} = 67.32$ and $a_t = -0.2$, then Y_t will be normally distributed with mean $0.8(67.32) = 53.856$ and standard deviation $0.8(5) = 4$.)

13.4 Regression-Based Trend Models

Many time series follow a long-term trend except for random variation. This trend can be upward or downward. A straightforward way to model this trend is to estimate a regression equation for Y_t, using time t as the *single* explanatory variable. In this section we will discuss the two most frequently used trend models, *linear* trend and *exponential* trend.

13.4.1 Linear Trend

A linear trend means that the time series variable changes by a constant *amount* each time period. The relevant equation is given by equation (13.6), where, as in previous regression equations, a is the intercept, b is the slope, and ϵ_t is an error term.

Linear Trend Model		
	$$Y_t = a + bt + \epsilon_t$$	(13.6)

The interpretation of b is that it represents the expected change in the series from one period to the next. If b is positive, the trend is upward; if b is negative, the trend is downward. The intercept term a is less important. It literally represents the expected value of the series at time $t = 0$. If time t is coded so that the first observation corresponds to $t = 1$, then a is where we expect the series to have been one period before we started observing. However, it is possible that time is coded in another way. For example, we might have annual data that start in 1985. Then the first value of t might be entered as 1985, which means that the intercept a corresponds to a period 1985 years earlier! Clearly, we would not take its value literally in this case.

As always, the graph of the time series is a good place to start. It indicates whether a linear trend model is likely to provide a good fit. Generally, the graph should rise or fall at approximately a constant rate through time, without too much random variation. But even if there is a lot of random variation—a lot of zigzags—fitting a linear trend to the data might still be a good starting point. Then the *residuals* from this trendline, which should have no remaining trend, could possibly be modeled by some other method in this chapter.

Example 13.2 Quarterly Sales of Johnson & Johnson

The file JOHNSON&JOHNSON.XLS contains quarterly sales data for Johnson & Johnson from second quarter 1991 through first quarter 2001 (in millions of dollars). The time series plot of these data appears in Figure 13.17. Sales increase from $3031 million in the initial quarter to $7791 million in the final quarter. How well does a linear trend fit these data? Are the residuals from this fit random?

Objective To fit a linear trend line to Johnson & Johnson's quarterly sales and examine its residuals for randomness.

Solution

The plot in Figure 13.17 indicates a clear upward trend with little or no curvature. Therefore, a linear trend is certainly plausible. To estimate it with regression, we first need a *numeric* time variable—labels such as Q2-91 will not do. We construct this time variable in column C of the data set, using the consecutive values 1–40. (See Figure 13.18.) We then run a simple regression of Sales versus Time, with the results shown. The estimated linear trend line is

$$\text{Forecasted Sales} = 2585 + 122.77\text{Time}$$

FIGURE 13.17
Time Series Plot of
Johnson & Johnson
Sales

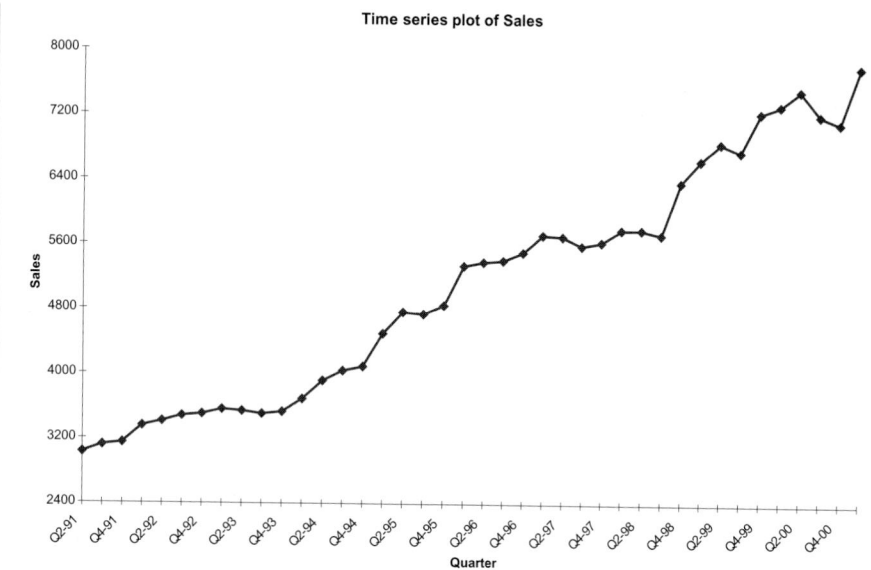

Time series plot of Sales

FIGURE 13.18 Regression Output for Linear Trend

	A	B	C	D	E	F	G	H	I	J	K	L	M	N	O
1	Quarterly sales at Johnson & Johnson (in $ millions)														
2															
3	Quarter	Sales	Time		Fitted Values	Residuals			Results of simple regression for Sales						
4	Q2-91	3031	1		2707.7524	323.2476									
5	Q3-91	3119	2		2830.5241	288.4759			Summary measures						
6	Q4-91	3148	3		2953.2958	194.7042			Multiple R	0.9851					
7	Q1-92	3357	4		3076.0674	280.9326			R-Square	0.9705					
8	Q2-92	3413	5		3198.8391	214.1609			StErr of Est	253.4685					
9	Q3-92	3480	6		3321.6108	158.3892									
10	Q4-92	3503	7		3444.3825	58.6175			ANOVA table						
11	Q1-93	3560	8		3567.1541	-7.1541			Source	df	SS	MS	F	p-value	
12	Q2-93	3541	9		3689.9258	-148.9258			Explained	1	80338465.8779	80338465.8779	1250.4766	0.0000	
13	Q3-93	3506	10		3812.6975	-306.6975			Unexplained	38	2441358.5221	64246.2769			
14	Q4-93	3531	11		3935.4691	-404.4691									
15	Q1-94	3690	12		4058.2408	-368.2408			Regression coefficients						
16	Q2-94	3916	13		4181.0125	-265.0125					Coefficient	Std Err	t-value	p-value	Lower limit	Upper limit
17	Q3-94	4038	14		4303.7841	-265.7841			Constant		2584.9808	81.6807	31.6474	0.0000	2419.6269	2750.3346
18	Q4-94	4090	15		4426.5558	-336.5558			Time		122.7717	3.4718	35.3621	0.0000	115.7433	129.8001
19	Q1-95	4496	16		4549.3275	-53.3275									
20	Q2-95	4762	17		4672.0992	89.9008									

This equation implies that we expect sales to increase by $122.77 million per quarter. (The 2585 value in this equation is what we would predict sales to be at time 0—quarter 1 of 1991.) To use this equation to forecast future sales, we start with the final sales figure, 7791 in Q1 of 2001, and add 122.77 times the number of quarters in the future being forecasted. For example, the forecast for Q1 of 2002 is

$$\text{Forecasted Sales Q1-02} = 7791 + 122.77(4) = 8282.08$$

Excel supplies an easier way to obtain this trend line. Once the plot in Figure 13.17 is constructed, we can use the Chart/Add Trendline menu item. This gives us several types of trend lines to choose from, and we select the linear option for this example. We can also click on the Options tab in the Add Trendline dialog box and select the options to show the regression equation and its R^2 value on the chart, as we have done in Figure 13.19. This superimposed trend line indicates a reasonably good fit.

However, the fit is not perfect, as the plot of the residuals in Figure 13.20 indicates. These residuals tend to "meander," staying positive for a while, then negative, then positive, and so on. You can check that the runs test for these residuals produces a z-value of −3.204, with a corresponding p-value of 0.001, and that its first two autocorrelations

FIGURE 13.19
Time Series Plot
with Linear Trend
Superimposed

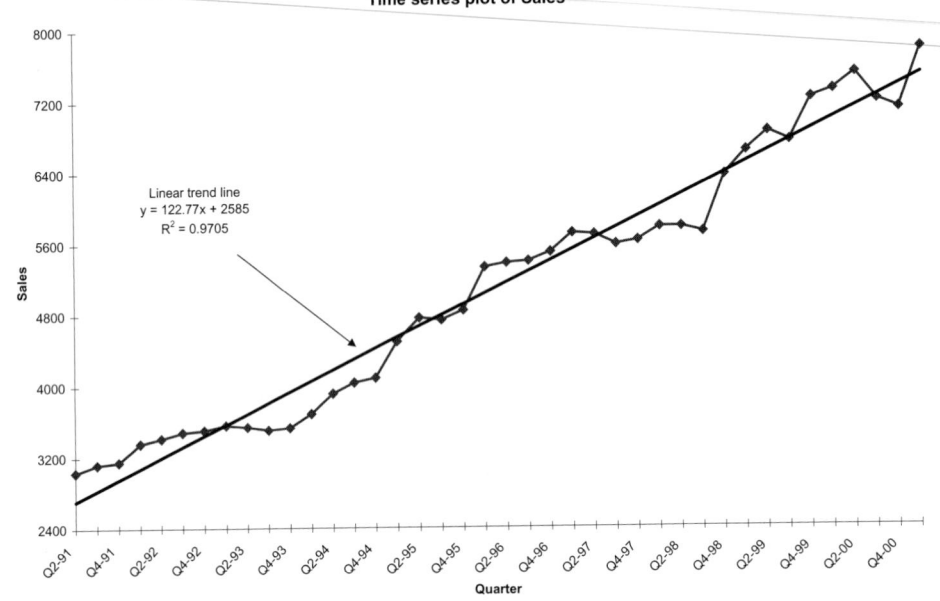

Time series plot of Sales

Linear trend line
y = 122.77x + 2585
R² = 0.9705

FIGURE 13.20
Time Series Plot
of Residuals

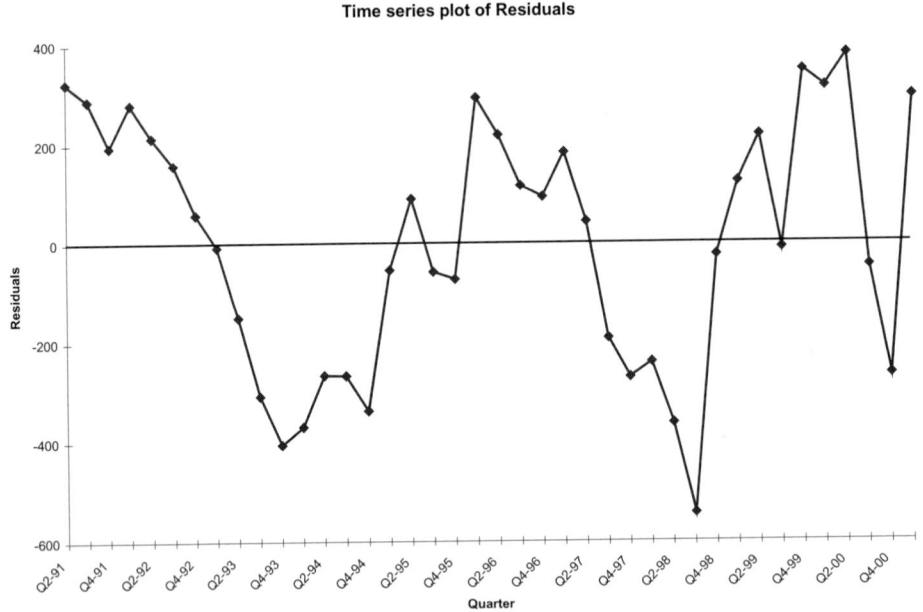

Time series plot of Residuals

are significantly positive. In short, these residuals are *not* random noise, and they could be modeled further. However, we do not present the tools to do so in this book. ●

13.4.2 Exponential Trend

In contrast to a linear trend, an exponential trend is appropriate when the time series changes by a constant *percentage* (as opposed to a constant dollar amount) each period. Then the appropriate regression equation is given in equation (13.7), where *c* and *b* are constants, and u_t represents a *multiplicative* error term.

| Exponential Trend Model | $$Y_t = ce^{bt}u_t$$ | **(13.7)** |

An exponential trend for Y is equivalent to a linear trend for the logarithm of Y.

Equation (13.7) is useful for understanding how an exponential trend works, as we will discuss, but it is not useful for estimation. For that, we require a *linear* equation. Fortunately, we can achieve linearity by taking natural logarithms of both sides of equation (13.7). (The key, as usual, is that the logarithm of a product is the sum of the logarithms.) The result appears in equation (13.8), where $a = \ln(c)$ and $\epsilon_t = \ln(u_t)$. This equation represents a *linear* trend, but the response variable is now the logarithm of the original Y_t. This implies the following important fact: If a time series exhibits an exponential trend, then a plot of its logarithm should be approximately linear.

| Equivalent Linear Trend for Logarithm of Y | $$\ln(Y_t) = a + bt + \epsilon_t$$ | **(13.8)** |

Because the computer does the calculations, our main responsibility is to interpret the final result. This is not too difficult. It can be shown that the coefficient b (expressed as a percentage) is approximately the percentage change per period. For example, if $b = 0.05$, then the series is increasing by approximately 5% per period.[2] On the other hand, if $b = -0.05$, then the series is decreasing by approximately 5% per period.

An exponential trend can be estimated with StatPro's regression procedure (simple or multiple, whichever you prefer), but only after the log transformation has been made on Y_t. We illustrate this in the following example.

Example 13.3 Quarterly Sales at Intel

The file INTEL.XLS contains quarterly sales data (in millions of dollars) for the chip manufacturing firm Intel from the first quarter of 1986 through the first quarter of 2001. Are Intel's sales growing exponentially through this entire period?

Objective To estimate Intel's exponential growth and to see whether it has been maintained during the entire period from 1986 until early 2001.

Solution

We will first estimate and interpret an exponential trend for the years 1986–1996. Then we will see how well the projection of this trend into the future fits the data after 1996. The time series plot through 1996 appears in Figure 13.21 on page 710. We have used Excel's Chart/Add Trendline menu item, with the Exponential option, to superimpose an exponential trend line on this plot. The fit is evidently quite good. Equivalently, Figure 13.22 illustrates the time series of log sales for this same period, with a *linear* trend line superimposed. Its fit is equally good.

We can also use StatPro's regression procedure to estimate this exponential trend, as shown in Figure 13.23. We first add a time variable in column C and make a logarithmic transformation of Sales in column D. Then we regress Log(Sales) on Time (remembering

[2]A more accurate estimate of this percentage change is $e^b - 1$. For example, when $b = 0.05$, this is $e^b - 1 = 5.13\%$.

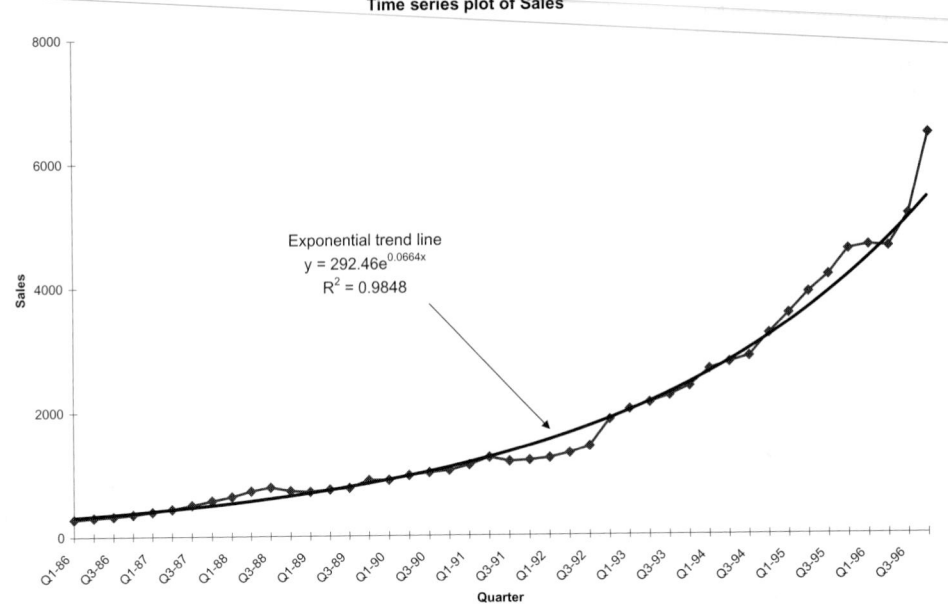

Time series plot of Sales

Exponential trend line
$y = 292.46e^{0.0664x}$
$R^2 = 0.9848$

FIGURE 13.22
Time Series Plot
of Log Sales with
Linear Trend
Superimposed

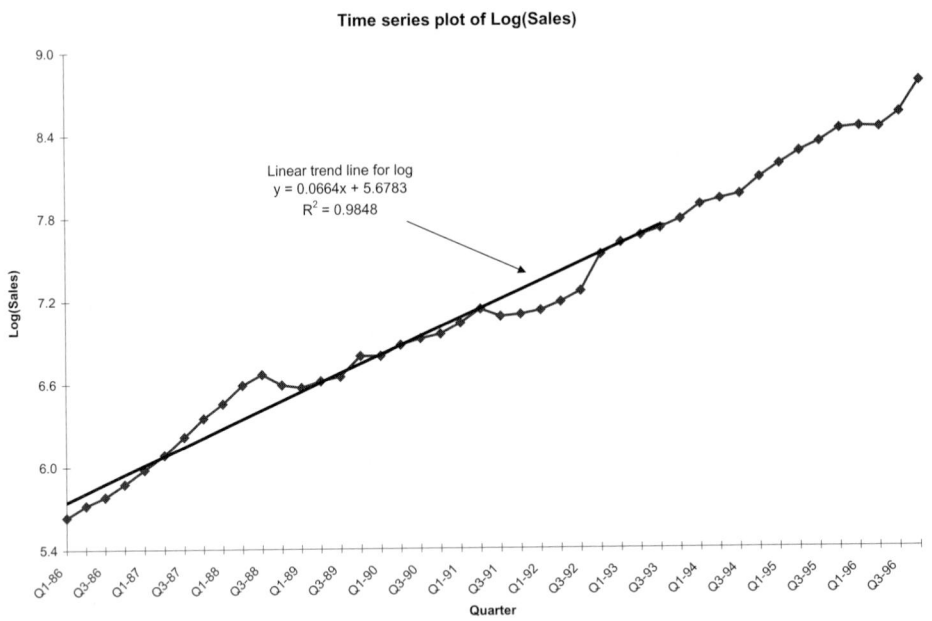

Time series plot of Log(Sales)

Linear trend line for log
$y = 0.0664x + 5.6783$
$R^2 = 0.9848$

to use the data through 1996 only) to obtain the regression output. Note that its two coefficients in cells H18 and H19 are the same as those shown for the linear trend in Figure 13.22. If we take the antilog of the constant in cell H21 (with the formula = EXP(H18)), we obtain the constant *multiple* shown in Figure 13.21. It corresponds to the constant c in equation (13.7).

What does it all mean? The estimated equation (13.7) is

$$\text{Forecasted Sales} = 292.46e^{0.0664t}$$

FIGURE 13.23 Regression Output for Estimating Exponential Trend

	A	B	C	D	E	F	G	H	I	J	K	L	M
1	Quarterly data on Intel sales ($ millions)												
2													
3	Quarter	Sales	Time	Log(Sales)			Results of multiple regression for Log(Sales) (through 1996 only)						
4	Q1-86	280.05	1	5.63498									
5	Q2-86	305.18	2	5.72090			Summary measures						
6	Q3-86	324.14	3	5.78117			Multiple R	0.9924					
7	Q4-86	355.64	4	5.87392			R-Square	0.9848					
8	Q1-87	394.53	5	5.97770			Adj R-Square	0.9845					
9	Q2-87	438.96	6	6.08440			StErr of Est	0.1071					
10	Q3-87	501.13	7	6.21686									
11	Q4-87	572.49	8	6.34999			ANOVA Table						
12	Q1-88	635.81	9	6.45490			Source	df	SS	MS	F	p-value	
13	Q2-88	726.68	10	6.58849			Explained	1	31.2785	31.2785	2725.0426	0.0000	
14	Q3-88	784.94	11	6.66561			Unexplained	42	0.4821	0.0115			
15	Q4-88	727.34	12	6.58939									
16	Q1-89	713.08	13	6.56960			Regression coefficients						
17	Q2-89	747.34	14	6.61652				Coefficient	Std Err	t-value	p-value	Lower limit	Upper limit
18	Q3-89	771.44	15	6.64826			Constant	5.6783	0.0329	172.7969	0.0000	5.6120	5.7447
19	Q4-89	894.97	16	6.79680			Time	0.0664	0.0013	52.2019	0.0000	0.0638	0.0690
20	Q1-90	894.46	17	6.79622									
21	Q2-90	968.30	18	6.87554			Antilog of constant	292.46					
22	Q3-90	1012.44	19	6.92012									
23	Q4-90	1046.07	20	6.95280									

	A	B	C	D	E	F	G	H	I
1	Quarterly data on Intel sales ($ millions)								
2									
3	Quarter	Sales	Time	Log(Sales)	Forecast		Results of multiple regression for Log(S		
4	Q1-86	280.05	1	5.63498	312.54				
5	Q2-86	305.18	2	5.72090	334.00		Summary measures		
6	Q3-86	324.14	3	5.78117	356.93		Multiple R		0.9924
7	Q4-86	355.64	4	5.87392	381.43		R-Square		0.9848
8	Q1-87	394.53	5	5.97770	407.62		Adj R-Square		0.9845
9	Q2-87	438.96	6	6.08440	435.60		StErr of Est		0.1071
10	Q3-87	501.13	7	6.21686	465.50				
11	Q4-87	572.49	8	6.34999	497.46		ANOVA Table		
12	Q1-88	635.81	9	6.45490	531.61		Source		df
13	Q2-88	726.68	10	6.58849	568.11		Explained		1
14	Q3-88	784.94	11	6.66561	607.11		Unexplained		42
15	Q4-88	727.34	12	6.58939	648.79				
16	Q1-89	713.08	13	6.56960	693.33		Regression coefficients		
17	Q2-89	747.34	14	6.61652	740.92				Coefficient
18	Q3-89	771.44	15	6.64826	791.79		Constant		5.6783
19	Q4-89	894.97	16	6.79680	846.15		Time		0.0664
20	Q1-90	894.46	17	6.79622	904.23				
21	Q2-90	968.30	18	6.87554	966.31		Antilog of constant		292.46
22	Q3-90	1012.44	19	6.92012	1032.65				
23	Q4-90	1046.07	20	6.95280	1103.54				

FIGURE 13.24 Creating Forecasts of Sales

The most important constant in this equation is the regression coefficient of Time, $b = 0.0664$. Expressed as a percentage, this coefficient implies that Intel's sales were increasing by approximately 6.64% per quarter throughout this 11-year period. (The constant multiple, $c = 292.46$, is our forecast of sales at time 0—in quarter 4 of 1985.) To use this equation for forecasting into the future, we start with the final observation, 6440 in quarter 4 of 1996, and multiply by 1.0664 for as many quarters as we are forecasting ahead. For example, the forecast of the last quarter of 1998 is

$$\text{Forecasted Sales in Q4-98} = 6440(1.0664)^8 = 10{,}771$$

Has this exponential growth continued beyond 1996 at Intel? As you might guess, it has *not*, due to slumping sales in the computer industry and increased competition from other chip manufacturers. We checked this by creating the Forecast column in Figure 13.24 (enter the formula =$I$21*EXP($I$19*C4) in cell E4 and copy the whole way

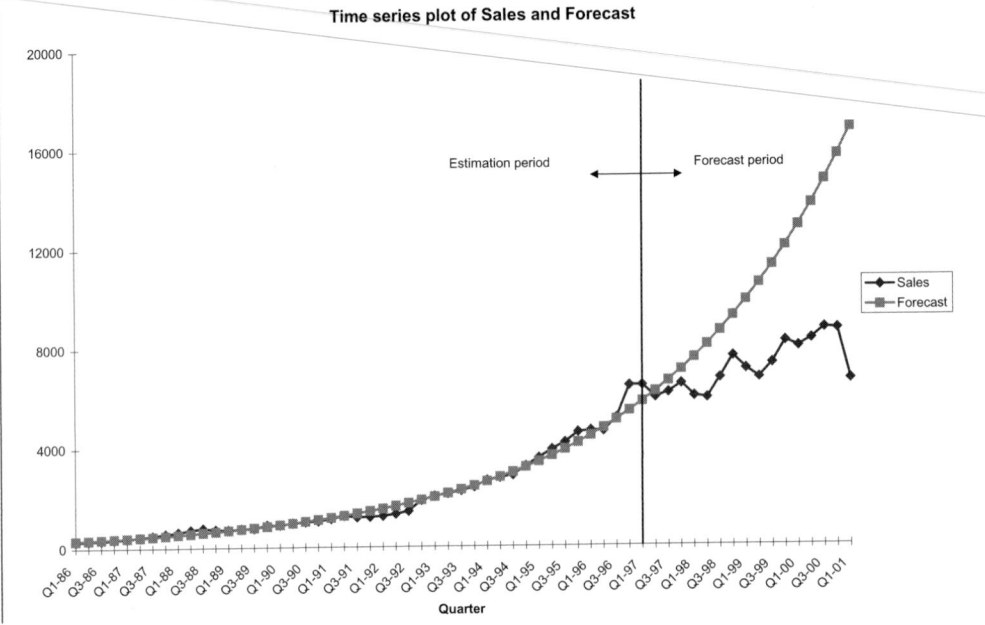

FIGURE 13.25
Time Series Plot of Forecasts Superimposed on Sales

FIGURE 13.26 Measures of Forecast Errors

	A	B	C	D	E	F	G	H	I	J	K	L	M
1	Quarterly data on Intel sales ($ millions)												
2													
3	Quarter	Sales	Time	Log(Sales)	Forecast	SqError	AbsError	AbsPctError			Results of multiple regression for Log(Sales) (through		
4	Q1-86	280.05	1	5.63498	312.54	1055.49	32.49	11.60%					
5	Q2-86	305.18	2	5.72090	334.00	830.63	28.82	9.44%			Summary measures		
6	Q3-86	324.14	3	5.78117	356.93	1075.24	32.79	10.12%			Multiple R	0.9924	
7	Q4-86	355.64	4	5.87392	381.43	665.08	25.79	7.25%			R-Square	0.9848	
8	Q1-87	394.53	5	5.97770	407.62	171.18	13.08	3.32%			Adj R-Square	0.9845	
9	Q2-87	438.96	6	6.08440	435.60	11.26	3.36	0.76%			StErr of Est	0.1071	
10	Q3-87	501.13	7	6.21686	465.50	1269.08	35.62	7.11%					
11	Q4-87	572.49	8	6.34999	497.46	5629.04	75.03	13.11%			ANOVA Table		
12	Q1-88	635.81	9	6.45490	531.61	10856.59	104.19	16.39%			Source	df	SS
13	Q2-88	726.68	10	6.58849	568.11	25146.18	158.58	21.82%			Explained	1	31.2785
14	Q3-88	784.94	11	6.66561	607.11	31623.70	177.83	22.66%			Unexplained	42	0.4821
15	Q4-88	727.34	12	6.58939	648.79	6170.60	78.55	10.80%					
16	Q1-89	713.08	13	6.56960	693.33	390.32	19.76	2.77%			Regression coefficients		
17	Q2-89	747.34	14	6.61652	740.92	41.13	6.41	0.86%				Coefficient	Std Err
18	Q3-89	771.44	15	6.64826	791.79	414.15	20.35	2.64%			Constant	5.6783	0.0329
19	Q4-89	894.97	16	6.79680	846.15	2384.33	48.83	5.46%			Time	0.0664	0.0013
20	Q1-90	894.46	17	6.79622	904.23	95.59	9.78	1.09%					
21	Q2-90	968.30	18	6.87554	966.31	3.96	1.99	0.21%			Antilog of constant	292.46	
22	Q3-90	1012.44	19	6.92012	1032.65	408.31	20.21	2.00%					
23	Q4-90	1046.07	20	6.95280	1103.54	3302.19	57.46	5.49%			Measures of forecast error (through 1996 only)		
24	Q1-91	1132.78	21	7.03243	1179.30	2164.14	46.52	4.11%			RMSE	208.43	
25	Q2-91	1252.69	22	7.13305	1260.26	57.33	7.57	0.60%			MAE	118.68	
26	Q3-91	1187.70	23	7.07978	1346.78	25304.01	159.07	13.39%			MAPE	7.49%	
27	Q4-91	1205.45	24	7.09461	1439.23	54654.64	233.78	19.39%					

down). This implements our estimate of equation (13.7). We then used StatPro's time series plot procedure to plot the two series Sales and Forecast, shown in Figure 13.25. It is clear that sales in the forecast period remained rather constant—nowhere near the 6.64% growth they exhibited in the estimation period. As Intel clearly realizes, nothing that good lasts forever.

Before leaving this example, we comment briefly on the standard error of estimate shown in cell I9 of Figure 13.24. This value, 0.1071, is in *log* units, not original dollar units. Therefore, it is a totally misleading indicator of the forecast errors we might make from the exponential trend equation. To obtain more meaningful measures, we first obtain the forecasts of sales, as explained previously. Then we can easily obtain any of the three forecast error measures discussed previously in equations (13.2), (13.3), and (13.4). The results appear in Figure 13.26. The squared errors, absolute errors, and absolute percentage errors are first calculated with the formulas =(B4-E4)^2, =ABS(B4-E4), and =G4/B4 in cells F4, G4, and H4, which are then copied down. The error measures (for the data through 1996 only) then appear in cells L24, L25, and L26. The corresponding formulas for RMSE, MAE, and MAPE are straightforward. RMSE is the square root of the average of the squared errors in column F, and MAE and MAPE are the averages of the values in columns G and H, respectively. The latter is particularly simple to interpret. Forecasts for the 11-year estimate period were off, on average, by about 7.5%. (Of course, as you can check, forecasts for the quarters *after* 1996 were off by much more!) ●

Whenever we observe a time series that is increasing at an increasing rate (or decreasing at a decreasing rate), an exponential trend model is worth trying. The key to the analysis is to regress the *logarithm* of the time series variable versus time. The coefficient of time, written as a percentage, is then the percentage increase (if positive) or decrease (if negative) per period.

PROBLEMS

Level A

9. The file P13_01.XLS contains the monthly number of airline tickets sold by the CareFree Travel Agency.
   a. Does a linear trend appear to fit these data well? If so, estimate and interpret the linear-trend model for this time series. Also, interpret the R^2 and s_e values.
   b. Provide an indication of the typical forecast error generated by the estimated model in part **a**.
   c. Is there evidence of some seasonal pattern in these sales data? If so, characterize the seasonal pattern.

10. The file P13_10.XLS contains the daily closing prices of Wal-Mart stock for a 1-year period. Does a linear or exponential trend fit these data well? If so, estimate and interpret the best trend model for this time series. Also, interpret the R^2 and s_e values.

11. The file P13_11.XLS contains annual data on the amount of life insurance in force in the United States. Fit an exponential growth curve to these data. Write a short report to summarize your findings.

12. The file P13_12.XLS contains 5 years of monthly data on sales (number of units sold) for a particular company. The company suspects that except for random noise, its sales are growing by a constant *percentage* each month and that they will continue to do so for at least the near future.
   a. Explain briefly whether the plot of the series visually supports the company's suspicion.
   b. Fit the appropriate regression model to the data. Report the resulting equation and state explicitly what it says about the percentage growth per month.
   c. What are the RMSE and MAPE for the forecast model in part **b**? In words, what do they measure?

Considering their magnitudes, does the model seem to be doing a good job?
   d. In words, how does the model make forecasts for future months? Specifically, given the forecast value for the last month in the data set, what simple arithmetic could you use to obtain forecasts for the next few months?

13. The file P13_13.XLS contains quarterly data on GDP. (The data are expressed in billions of current dollars, they are seasonally adjusted, and they represent annualized rates.)
   a. Look at a time series plot of GDP. Does it suggest a linear relationship; an exponential relationship?
   b. Use regression to estimate a linear relationship between GDP and Time. Interpret the associated "constant" term and the "slope" term. Would you say that the fit is good?

Level B

14. The file P13_14.XLS contains monthly time series data on corporate bond yields. These are averages of daily figures, and each is expressed as an annual rate. The variables are:
   • YieldAAA: average yield on AAA bonds
   • YieldBAA: average yield on BAA bonds

   If you examine either Yield variable, you will notice that the autocorrelations of the series are not only large for many lags, but that the lag 1 autocorrelation of the *differences* is significant. This is very common. It means that the series is not a random walk and that it is probably possible to provide a better forecast than the "naive" forecast from the random walk model. Here is the idea. The large lag 1 autocorrelation of the differences means that the differences are related to the first lag of the differences. This relationship can be estimated by creating the difference variable and a lag of it, then regressing the former on the latter, and finally using this information to forecast the original Yield variable.

a. Verify that the autocorrelations are as described, and form the difference variable and the first lag of it. Call these DYield and L1DYield (where D is for difference, L1 is for first lag).

b. Run a regression with DYield as the response variable and L1DYield as the single explanatory variable. In terms of the original variable Yield, this equation can be written as

$$\text{Yield}_t - \text{Yield}_{t-1} = a + b(\text{Yield}_{t-1} - \text{Yield}_{t-2})$$

Solving for Yield_t is equivalent to the following equation that can be used for forecasting:

$$\text{Yield}_t = a + (1 + b)\text{Yield}_{t-1} - b\text{Yield}_{t-2}$$

Try it—that is, try forecasting the next month from the known last 2 months' values. How might you forecast values 2 or 3 months from the last observed month? (*Hint:* If you do not have an *observed* value to use in the right side of the equation, use a forecasted value.)

c. The autocorrelation structure led us to the equation in part **b.** That is, the autocorrelations of the original series took a long time to die down, so we looked at the autocorrelations of the differences,

and the large spike at lag 1 led to regressing L1DYield on DYield. In turn, this led ultimately to an equation for Yield_t in terms of its first two lags. Now see what you would have obtained if you had tried regressing Yield_t on its first two lags in the first place—that is, if you had used regression to estimate the equation

$$\text{Yield}_t = a + b_1\text{Yield}_{t-1} + b_2\text{Yield}_{t-2}$$

When you use multiple regression to estimate this equation, do you get the same equation as in part **b?**

15. The unit sales of a new drug for the first 25 months after its introduction to the marketplace are recorded in the file P13_15.XLS.

a. Estimate a linear trend equation using the given data. How well does the linear trend fit these data? Are the residuals from this linear trend model *random?*

b. If the residuals from this linear trend model are *not* random, propose another regression-based trend model that more adequately explains the long-term trend in this time series. Estimate the alternative model(s) using the given data. Check the residuals from the model(s) for randomness. Summarize your findings.

c. Given the best estimated model of the trend in this time series, interpret R^2 and s_e.

13.5 The Random Walk Model

Random series are sometimes building blocks for other time series models. The model we now discuss, the **random walk** model, is an example of this. In a random walk model the series itself is not random. However, its *differences*—that is, the changes from one period to the next—are random. This type of behavior is typical of stock price data (as well as various other time series data). For example, the graph in Figure 13.27 shows monthly closing prices for Caterpillar stock from January 1995 through April 2001. (See the file CATERPILLAR.XLS.) This series is not random, as can be seen from its gradual upward trend at the beginning and the general meandering behavior throughout. (Although the runs test and autocorrelations are not shown for the series itself, they confirm that the series is not random. There are significantly *fewer* runs than expected, and the autocorrelations are significantly *positive* for many lags.)

If we were standing in April 2001 and were asked to forecast Caterpillar prices for the next few months, it is intuitive that we would not use the average of the historical values as our forecast. This forecast would probably be too low because the series has an upward trend. Instead, we might base our forecast on the most recent observation. This is exactly what the random walk model does.

Equation (13.9) for the random walk model is given in the box, where μ is a constant and ϵ_t is a random series (noise) with mean 0 and some standard deviation σ that remains *constant* through time.

FIGURE 13.27

Time Series Plot of Caterpillar Stock Prices

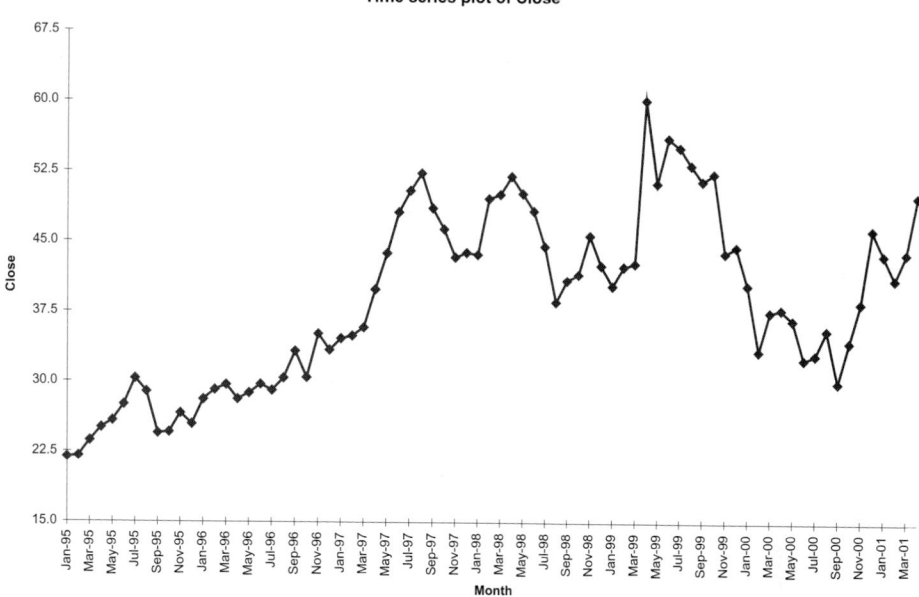

Time series plot of Close

Random Walk Model

$$Y_t = Y_{t-1} + \mu + \epsilon_t \qquad \text{(13.9)}$$

If we let $DY_t = Y_t - Y_{t-1}$, the change in the series from time t to time $t-1$ (where D stands for difference), then we can write the random walk model as in equation (13.10). This implies that the differences form a random series with mean μ and standard deviation σ. An estimate of μ is the average of the differences, labeled $\overline{Y}_D$, and an estimate of σ is the sample standard deviation of the differences, labeled s_D.

Difference Form of Random Walk Model

$$DY_t = \mu + \epsilon_t \qquad \text{(13.10)}$$

In words, a series that behaves according to this random walk model has random differences, and the series tends to trend upward (if $\mu > 0$) or downward (if $\mu < 0$) by an amount μ each period. If we are standing in period t and want to make a forecast F_{t+1} of Y_{t+1}, then a reasonable forecast is given by equation (13.11). That is, we add the estimated trend to the current observation to forecast the next observation.

One-Step-Ahead Forecast for Random Walk Model

$$F_{t+1} = Y_t + \overline{Y}_D \qquad \text{(13.11)}$$

We illustrate this method in the following example.

Example (13.4) Random Walk Model
of Stock Prices at Caterpillar

The monthly closing prices of Caterpillar stock from January 1995 through April 2001, shown in Figure 13.27, indicate some upward trend. Do they follow a random walk model with an upward drift? If so, how should future values of these stock prices be forecasted?

Objective To check whether monthly closing prices at Caterpillar follow a random walk model with an upward drift, and to see how future prices can be forecasted.

Solution

We have already seen that the closing price series itself is not random, due to the upward trend. To check for the adequacy of a random walk model, we need the *differenced* series. Each value in the differenced series is that month's closing price minus the previous month's closing price. This series can be calculated easily with an Excel formula, or it can be generated automatically with the StatPro/Data Utilities/Create Difference Variables menu item. (When asked for the *number* of difference variables, accept the default value of 1.) This differenced series appears in column C of Figure 13.28. This figure also shows the mean and standard deviation of the differences, 0.373 and 3.851, which will be used in forecasting. Finally, Figure 13.28 shows several autocorrelations of the differences, all of which are insignificant. A runs test for the differences, not shown here, has a large *p*-value, 0.668, which supports the conclusion that the differences are random.

The plot of the differences appears in Figure 13.29. A visual inspection of the plot also supports the conclusion of random differences, although these differences do not vary around a mean of 0. Rather, they vary around a mean of 0.373, indicated by the horizontal line. This positive value measures the upward drift—the closing prices increase, on average, by 0.373 per month. Finally, the variability in this figure is fairly constant (except for the two wide swings in 1999). Specifically, the zigzags do not tend to get appreciably wider through time. Therefore, we can conclude that the random walk model with an upward drift fits these data quite well.

FIGURE 13.28
Differences of
Closing Prices

	A	B	C	D	E	F	G	H
1	Monthly closing prices of Caterpillar stock							
2								
3	Month	Close	Close_Diff1			Summary measures for selected variables		
4	Jan-95	21.9177	*				Close_Diff1	
5	Feb-95	22.0243	0.1066			Count	75.000	
6	Mar-95	23.6775	1.6532			Mean	0.373	
7	Apr-95	25.0685	1.391			Standard deviation	3.851	
8	May-95	25.8184	0.7499					
9	Jun-95	27.5325	1.7141			Autocorrelations for Close_Diff1		
10	Jul-95	30.3007	2.7682				Lag	Autocorr	StErr
11	Aug-95	28.9014	-1.3993				1	-0.1421	0.1155
12	Sep-95	24.4882	-4.4132				2	0.1189	0.1155
13	Oct-95	24.6015	0.1133				3	-0.0848	0.1155
14	Nov-95	26.6065	2.005				4	-0.0230	0.1155
15	Dec-95	25.4686	-1.1379				5	0.0075	0.1155
16	Jan-96	28.0882	2.6196				6	-0.0528	0.1155
17	Feb-96	29.1245	1.0363				7	-0.0031	0.1155
18	Mar-96	29.6699	0.5454				8	-0.0034	0.1155
19	Apr-96	28.1245	-1.5454				9	0.0257	0.1155
20	May-96	28.7823	0.6578				10	-0.1709	0.1155
21	Jun-96	29.7143	0.932				11	-0.0177	0.1155
22	Jul-96	29.0688	-0.6455				12	-0.0295	0.1155
23	Aug-96	30.3927	1.3239						

FIGURE 13.29
Time Series Plot
of Differences

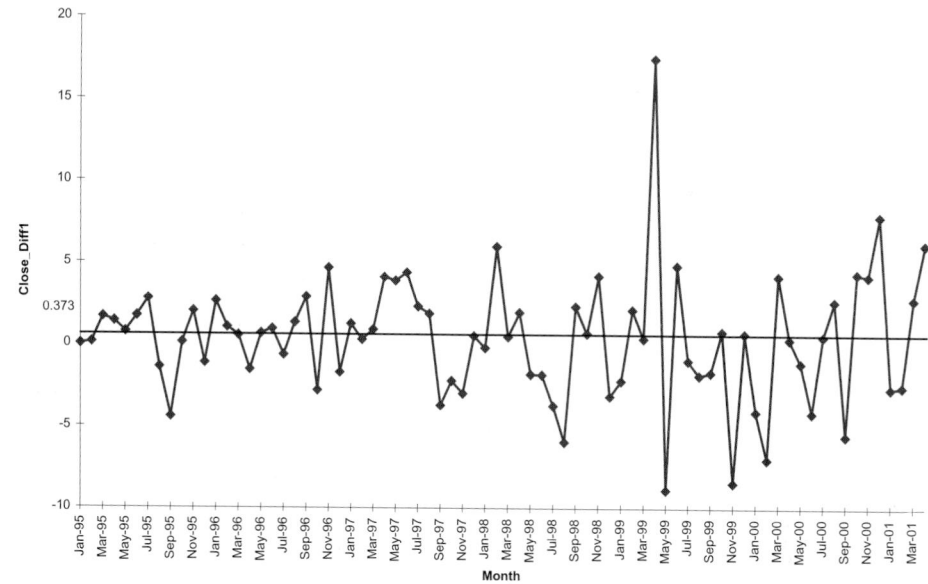

Time series plot of Close_Diff1

To forecast future closing prices, we add the number of months ahead being forecasted times the mean difference to the final closing price (49.872 in April 2001). For example, a forecast of the closing price for September 2001 is as follows:

$$\text{Forecasted Closing Price for 9/01} = 49.872 + 0.373(5) = 51.737$$

As a rough measure of the accuracy of this forecast, we can use the standard deviation of the differences, 3.851. Specifically, it can be shown that the standard error for forecasting k periods ahead is the standard deviation of the differences multiplied by the square root of k. In this case, the standard error is 8.611. As usual, we can be 95% confident that the actual closing price in September will be no more than 2 standard errors from the forecast. Unfortunately, this results in a wide interval—from 34.155 to 68.599. This simply reflects the fact that it is very difficult to make accurate forecasts for a series with this much variability. ●

PROBLEMS

Level A

16. The file P13_16.XLS contains the daily closing prices of American Express stock for a 1-year period.
    a. Use the random walk model to forecast the closing price of this stock on the next trading day.
    b. We can be about 95% certain that the forecast made in part **a** is off by no more than how many dollars?

17. The closing value of the AMEX Airline Index for each trading day during a 1-year period, is given in the file P13_17.XLS.
    a. Use the random walk model to forecast the closing price of this stock on the next trading day.

b. We can be about 68% certain that the forecast made in part **a** is off by no more than how many dollars?

18. The file P13_18.XLS contains the daily closing prices of ChevronTexaco stock for a 1-year period.
    a. Use the random walk model to forecast the closing price of this stock on the next trading day.
    b. We can be about 99.7% certain that the forecast made in part **a** is off by no more than how many dollars?

19. The closing value of the Dow Jones Industrial Average for each trading day for a 1-year period, is provided in the file P13_19.XLS.

a. Use the random walk model to forecast the closing price of this stock on the next trading day.

b. Would it be wise to use the random walk model to forecast the closing price of this stock for a trading day approximately *one month* after the next trading day? Explain why or why not.

20. Continuing the previous problem, consider the differences between consecutive closing values of the Dow Jones Industrial Average for the given set of trading days. Do these differences form a random series? Demonstrate why or why not.

21. The closing price of a share of J.P. Morgan's stock for each trading day during a 1-year period, is recorded in the file P13_21.XLS.

a. Use the random walk model to forecast the closing price of this stock on the next trading day.

b. We can be about 68% certain that the forecast made in part **a** is off by no more than how many dollars?

22. The purpose of this problem is to get you used to the concept of autocorrelation in a time series. You could do it with any time series, but here you should use the series of Wal-Mart daily stock prices in the file P13_10.XLS.

a. First, do it the "easy" way. Use the Autocorrelations procedure in StatPro to get a list of autocorrelations and a corresponding correlogram of the closing prices. You can choose the number of lags.

b. Now do it the "hard" way. Create columns of lagged versions of the Close variable—3 or 4 lags will suffice. Next, look at scatterplots of Close versus its first few lags. If the autocorrelations are large, you should see fairly tight scatters—that's what autocorrelation is all about. Also, generate a correlation matrix to see the correlations between Close and its first few lags. These should be approximately the same as the autocorrelations from

part **a.** (Autocorrelations are calculated slightly differently than regular correlations, which accounts for any slight discrepancies you might notice, but these discrepancies should be minor.)

c. Create the first differences of Close in a new column. (You can do this manually with formulas, or you can use StatPro's Differences procedure under Utilities.) Now repeat parts **a** and **b** with the differences instead of the original closing prices— that is, examine the autocorrelations of the differences. They should be small, and the scattergrams of the differences versus lags of the differences should be "swarms." This illustrates what happens when the differences of a time series variable have "insignificant" autocorrelations.

d. Write a short report of your findings.

Level B

23. Consider a random walk model with the following equation: $Y_t = Y_{t-1} + 500 + \epsilon_t$, where ϵ_t is a normally distributed random series with mean 0 and standard deviation 10.

a. Use Excel to generate a time series that behaves according to this random walk model.

b. Use the time series you constructed in part **a** to forecast the next observation.

24. The file P13_24.XLS contains the daily closing prices of Procter & Gamble stock for a 1-year period. Use only the data from 2001 to estimate the trend component of the random walk model. Next, use the estimated random walk model to forecast the behavior of the time series for the 2002 dates in the series. Comment on the accuracy of the generated forecasts over this period. How could you improve the forecasts as you progress through these 2002 trading days?

13.6 Autoregression Models[3]

A regression-based extrapolation method is to regress the current value of the time series on past (lagged) values. This is called **autoregression,** where the "auto" means that the explanatory variables in the equation are lagged values of the response variable, so that we are regressing the response variable on lagged versions of itself. This procedure is fairly straightforward on a spreadsheet. We first create lags of the response variable and then use a regression procedure to regress the original column on the lagged columns. Some trial and error is generally required to see how many lags are useful in the regression equation. The following example illustrates the procedure.

[3]This section can be omitted without any loss of continuity.

Example (13.5) Forecasting Hammer Sales

A retailer has recorded its weekly sales of hammers (units purchased) for the past 42 weeks. (See the file HAMMERS.XLS.) A graph of this time series appears in Figure 13.30. It reveals a "meandering" behavior. The values begin high and stay high awhile, then get lower and stay lower awhile, then get higher again. (This behavior could be caused by any number of things, including the weather, increases and decreases in building projects, and possibly others.) How useful is autoregression for modeling these data and how would it be used for forecasting?

Objective To use autoregression, with the appropriate number of lagged terms, to forecast hammer sales.

Solution

A good place to start is with the autocorrelations of the series. These indicate whether the Sales variable is linearly related to any of its lags. The first six autocorrelations are shown in Figure 13.31 (page 720). The first three of them are significantly positive, and then they decrease. Based on this information, we create three lags of Sales and run a regression of Sales versus these three lags. The output from this regression appears in Figure 13.32. We see that R^2 is fairly high, about 57%, and that s_e is about 15.7. However, the p-values for lags 2 and 3 are both quite large. It appears that once the first lag is included in the regression equation, the other two are not really needed.

It is generally best to begin with plenty of lags and then delete the higher numbered lags that aren't necessary.

Therefore, we reran the regression with only the first lag included. (Actually, we first omitted only the third lag. But the resulting output showed that the second lag was still insignificant.) The regression output with only the first lag included appears in Figure 13.33 (page 721). In addition, a graph of the response and fitted variables, that is, the original Sales variable and its forecasts, appears in Figure 13.34. (This latter graph was formed from the Week, Sales, and Fitted columns.) The estimated regression equation is

$$\text{Forecasted Sales}_t = 13.763 + 0.793\text{Sales}_{t-1}$$

FIGURE 13.30
Time Series Plot of Sales of Hammers

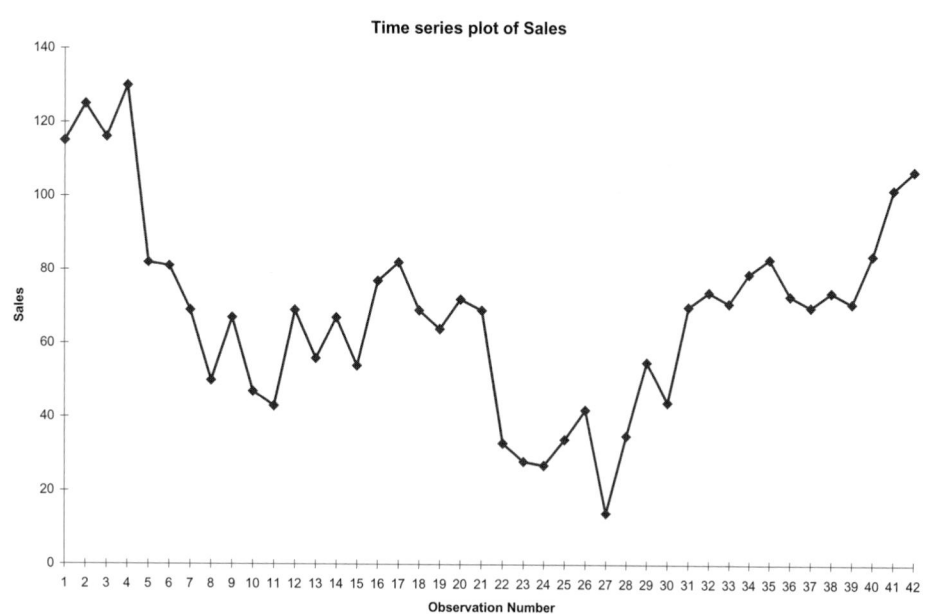

FIGURE 13.31
Correlogram for
Hammer Sales
Data

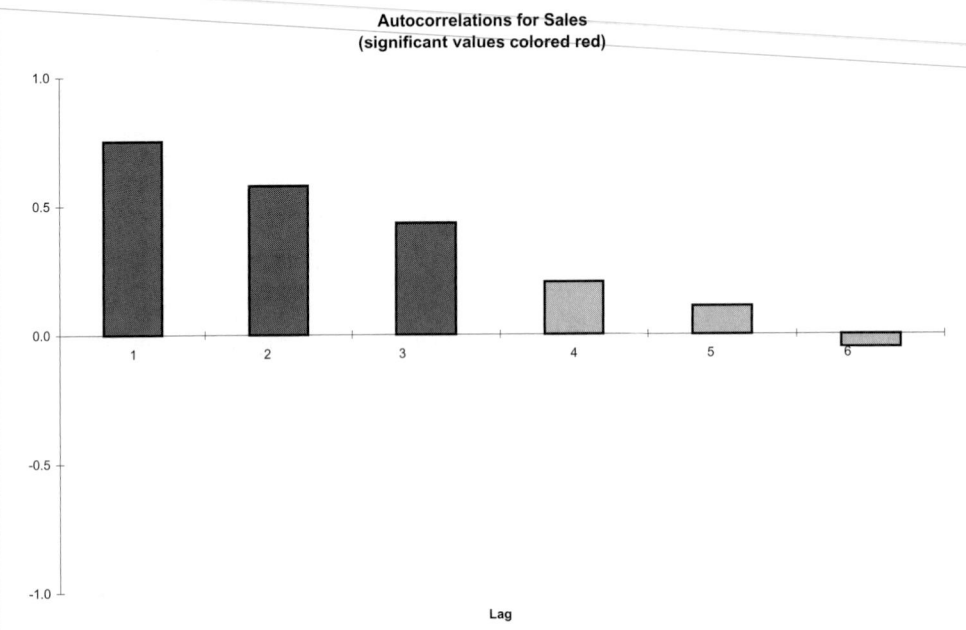

FIGURE 13.32
Autoregression
Output with
Three Lagged
Variables

	A	B	C	D	E	F	G	H
1	**Results of multiple regression for Sales**							
2								
3	**Summary measures**							
4		Multiple R	0.7573					
5		R-Square	0.5736					
6		Adj R-Square	0.5370					
7		StErr of Est	15.7202					
8								
9	**ANOVA Table**							
10		Source	df	SS	MS	F	p-value	
11		Explained	3	11634.2001	3878.0667	15.6927	0.0000	
12		Unexplained	35	8649.3896	247.1254			
13								
14	**Regression coefficients**							
15			Coefficient	Std Err	t-value	p-value	Lower limit	Upper limit
16		Constant	15.4986	7.8820	1.9663	0.0572	-0.5027	31.5000
17		Sales_Lag1	0.6398	0.1712	3.7364	0.0007	0.2922	0.9874
18		Sales_Lag2	0.1523	0.1987	0.7665	0.4485	-0.2510	0.5556
19		Sales_Lag3	-0.0354	0.1641	-0.2159	0.8303	-0.3686	0.2977

The associated R^2 and s_e values are approximately 65% and 15.4. The R^2 value is a measure of the reasonably good fit we see in Figure 13.34, whereas s_e is a measure of the likely forecast error for short-term forecasts. It implies that a short-term forecast could easily be off by as much as 2 standard errors, or about 31 hammers.

To use the regression equation for forecasting *future* sales values, we substitute known or forecasted sales values in the right-hand side of the equation. Specifically, the forecast for week 43, the first week after the data period, is

$$\text{Forecasted Sales}_{43} = 13.763 + 0.793\text{Sales}_{42} = 13.763 + 0.793(107) \approx 98.6$$

FIGURE 13.33
Autoregression
Output with a
Single Lagged
Variable

	A	B	C	D	E	F	G	H
1	**Results of multiple regression for Sales**							
2								
3	**Summary measures**							
4		Multiple R	0.8036					
5		R-Square	0.6458					
6		Adj R-Square	0.6367					
7		StErr of Est	15.4476					
8								
9	**ANOVA Table**							
10		Source	df	SS	MS	F	p-value	
11		Explained	1	16969.9761	16969.9761	71.1146	0.0000	
12		Unexplained	39	9306.5117	238.6285			
13								
14	**Regression coefficients**							
15			Coefficient	Std Err	t-value	p-value	Lower limit	Upper limit
16		Constant	13.7634	6.7906	2.0268	0.0496	0.0281	27.4987
17		Sales_Lag1	0.7932	0.0941	8.4329	0.0000	0.6029	0.9834

FIGURE 13.34
Forecasts from
Autoregression

The two curves in
this figure look
pretty close to one
another. But if you
compare the *vertical*
distances between
pairs of points,
you'll see that they
are not so close
after all.

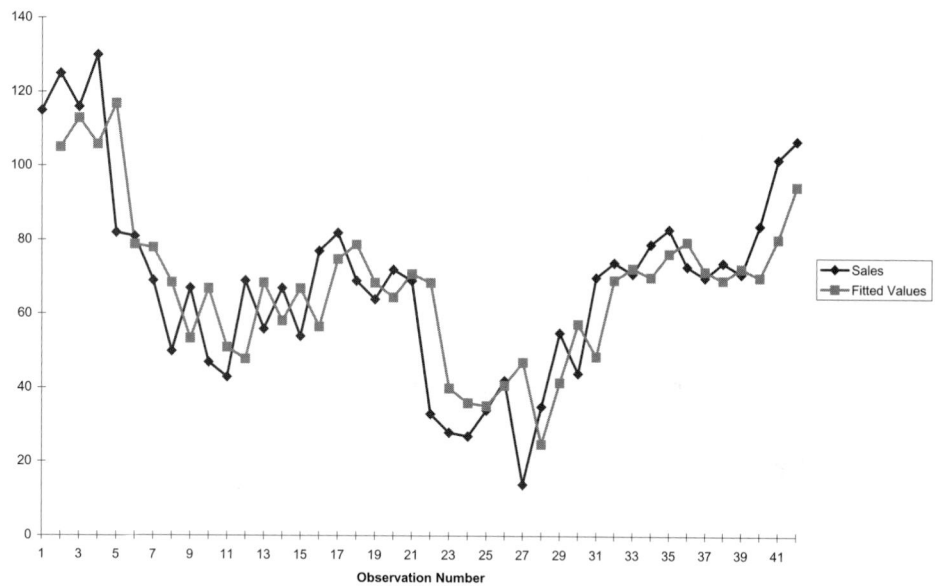

Here we use the *known* value of sales in week 42. However, the forecast for week 44 requires the *forecasted* value of sales in week 43:

$$\text{Forecasted Sales}_{44} = 13.763 + 0.793\text{Forecasted Sales}_{43}$$
$$= 13.763 + 0.793(98.6) \approx 92.0$$

Perhaps these two forecasts of future sales values are on the mark, and perhaps they are not. The only way we'll know for certain is by observing future sales values. However, it is interesting that in spite of the *upward* movement in the series in the last 3 weeks, the forecasts for weeks 43 and 44 are for *downward* movements. This is a combination of two properties of the regression equation. First, the coefficient of Sales_{t-1}, 0.793, is positive. Therefore, the equation forecasts that large sales will be followed by large sales, that is, positive autocorrelation. Second, however, this coefficient is less than 1, and this provides a dampening effect. The equation forecasts that a large will follow a large, but not *that* large. ●

Sometimes an autoregression model can be virtually equivalent to another forecasting model. As an example, suppose we find that the following equation adequately models a time series variable Y:

$$Y_t = 75.65 + 0.976Y_{t-1}$$

The coefficient of the lagged term, 0.976, is nearly equal to 1. If this coefficient were 1, we could subtract the lagged term from both sides of the equation and write that the *difference* series is a constant—that is, we would have a random walk model. As you can see, a random walk model is a special case of an autoregression model. However, autoregression models are much more general. Unfortunately, a more thorough study of them would take us well beyond the level of this book.

PROBLEMS

Level A

25. Consider the Consumer Price Index (CPI), which provides the annual percentage change in consumer prices. The data are in the file P02_26.XLS.
    a. Compute the first six autocorrelations of this time series.
    b. Use the results of part **a** to specify one or more "promising" autoregression models. Estimate each model with the available data. Which model provides the best fit to the given data?
    c. Use the best autoregression model from part **b** to produce a forecast of the CPI in the next year. Also, provide a measure of the likely forecast error.

26. The Consumer Confidence Index (CCI) attempts to measure people's feelings about general business conditions, employment opportunities, and their own income prospects. The file P02_28.XLS contains the annual average values of the CCI.
    a. Compute the first six autocorrelations of this time series.
    b. Use the results of part **a** to specify one or more "promising" autoregression models. Estimate each model with the available data. Which model provides the best fit to the given data?
    c. Use the best autoregression model from part **b** to produce a forecast of the CCI in the next year. Also, provide a measure of the likely forecast error.

27. Consider the proportion of Americans under the age of 18 living below the poverty level. The data are in the file P02_29.XLS.
    a. Compute the first six autocorrelations of this time series.
    b. Use the results of part **a** to specify one or more "promising" autoregression models. Estimate each model with the available data. Which model provides the best fit to the given data?

    c. Use the best autoregression model from part **b** to produce a forecast of the proportion of American children living below the poverty level in the next year. Also, provide a measure of the likely forecast error.

28. Examine the trend in the annual average values of the discount rate. The data are in the file P02_30.XLS.
    a. Specify one or more "promising" autoregression models based on autocorrelations of this time series. Estimate each model with the available data. Which model provides the best fit to given data?
    b. Use the best autoregression model from part **a** to produce forecasts of the discount rate in the next 2 years.

29. The file P02_34.XLS contains the annual cigar consumption per capita in the United States for selected years.
    a. Specify one or more "promising" autoregression models based on autocorrelations of this time series. Estimate each model with the available data. Which model provides the best fit to the given data?
    b. Use the best autoregression model from part **a** to produce forecasts of the cigar consumption per capita in the next 3 years.

30. Consider the average annual interest rates on 30-year fixed mortgages in the United States. The data are recorded in the file P02_35.XLS.
    a. Specify one or more "promising" autoregression models based on autocorrelations of this time series. Estimate each model with the available data. Which model provides the best fit to the given data?
    b. Use the best autoregression model from part **a** to produce forecasts of the average annual interest rates on 30-year fixed mortgages in the next 3 years.

31. The file P13_31.XLS lists the monthly unemployment rates for several years. A common way to forecast time series is by using regression with lagged variables.

  a. Predict future monthly unemployment rates using some combination of the unemployment rates for the last 4 months. For example, you might use last month's unemployment rate and the unemployment rate from 3 months ago as explanatory variables. Make sure all variables that you finally decide to keep in your equation are significant at the 0.15 level.

  b. Do the residuals in your equation exhibit any autocorrelation?

  c. Predict the next month's unemployment rate.

  d. There is a 5% chance that the next month's unemployment rate will be less than what value?

  e. What is the probability the next month's unemployment rate will be less than 6%?

Level B

32. The unit sales of a new drug for the first 25 months after its introduction to the marketplace are recorded in the file P13_15.XLS. Specify one or more "promising" autoregression models based on autocorrelations of this time series. Estimate each model with the available data. Which model provides the best fit to the given data? Use the best autoregression model you found to forecast the sales of this new drug in the 26th month.

33. The file P13_02.XLS contains the weekly sales at a local bookstore for each of the past 25 weeks.

  a. Specify one or more "promising" autoregression models based on autocorrelations of this time series. Estimate each model with the available data. Which model provides the best fit to the given data?

  b. What general result emerges from your analysis in part **a**? In other words, what is the most appropriate autoregression model for any given *random* time series?

  c. Use the best autoregression model from part **a** to produce forecasts of the weekly sales at this bookstore for the next 3 weeks.

34. The file P13_24.XLS contains the daily closing prices of Procter & Gamble stock for a 1-year period.

  a. Use only the data from 2001 to estimate an appropriate autoregression model.

  b. Next, use the estimated autoregression model to forecast the behavior of the time series for the 2002 dates of the series. Comment on the accuracy of the generated forecasts over this period.

  c. How well does the autoregression model perform in comparison to the random walk model with respect to the accuracy of these forecasts? Explain any significant differences between the forecasting abilities of the two models.

 # Moving Averages

Perhaps the simplest and one of the most frequently used extrapolation methods is the method of **moving averages.** To implement the moving averages method, we first choose a **span,** the number of terms in each moving average. Let's say the data are monthly and we choose a span of 6 months. Then the forecast of next month's value is the average of the most recent 6 months' values. For example, we average January–June to forecast July, we average February–July to forecast August, and so on. This procedure is the reason for the term *moving* averages.

Moving Average and Span	A **moving average** is the average of the observations in the past few periods, where the number of terms in the average is the **span.**

A moving averages model with a span of 1 is a random walk model with a mean trend of 0.

The role of the span is important. If the span is large—say, 12 months—then many observations go into each average, and extreme values have relatively little effect on the forecasts. The resulting series of forecasts will be much smoother than the original series. (For this reason, the moving average method is called a **smoothing** method.) In contrast, if the span is small—say, 3 months—then extreme observations have a larger effect on the forecasts, and the forecast series will be much less smooth. In the extreme, if the span is 1, there is no smoothing effect at all. The method simply forecasts next month's value to be the same as the current month's value. This is often called the **naive** forecasting model. It is a special case of the random walk model we discussed previously, with the mean difference equal to 0.

What span should we use? This requires some judgment. If we believe the ups and downs in the series are random noise, then we don't want future forecasts to react too quickly to these ups and downs, and we should use a relatively large span. But if we want to track every little zigzag—under the belief that each up or down is predictable—then we should use a smaller span. We shouldn't be fooled, however, by a plot of the (smoothed) forecast series superimposed on the original series. This graph will almost always look better when a small span is used, because the forecast series will appear to track the original series better. Does this mean it will always provide better future forecasts? Not necessarily. There is little point in tracking random ups and downs closely if they represent unpredictable noise.

The following example illustrates the use of moving averages.

Example 13.6 Houses Sold in the Midwest

The file HOUSESALES.XLS contains monthly data on the number of houses sold in the Midwest (in thousands) from January 1994 through May 2001. (These data are seasonally adjusted.)[4] A time series plot of the data appears in Figure 13.35. Does a moving averages model fit this data set well? What span should be used?

Objective To see whether a moving averages model with an appropriate span fits the housing sales data, and to see how StatPro implements this method.

Solution

Although the moving averages method is quite easy to implement in Excel—we just form an average of the appropriate span and copy it down—it can be tedious. Therefore,

FIGURE 13.35
Time Series Plot of Monthly House Sales

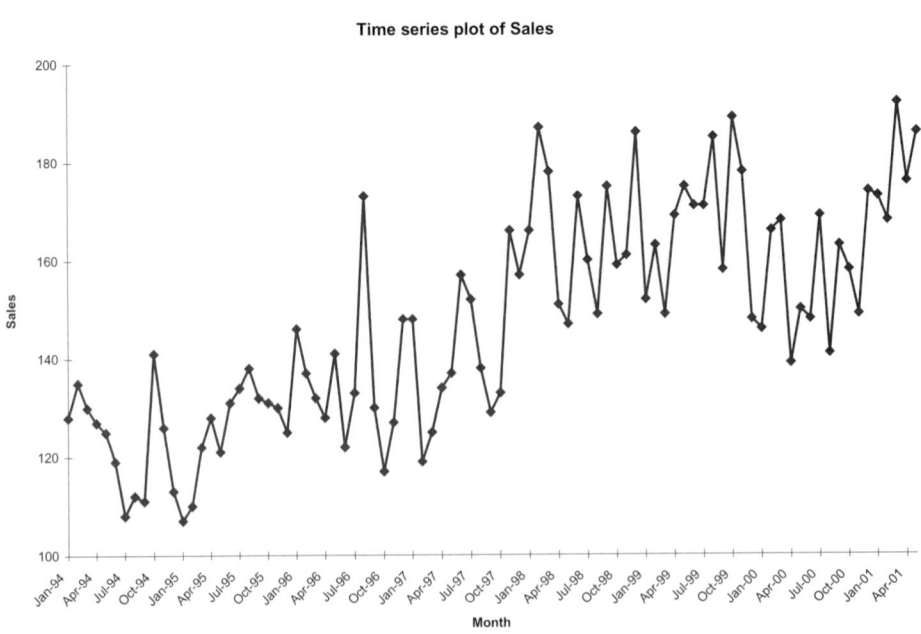

[4]We will discuss seasonal adjustment in Section 13.9. Government data are often reported in seasonally adjusted form, with the seasonality removed, to make any trends more apparent.

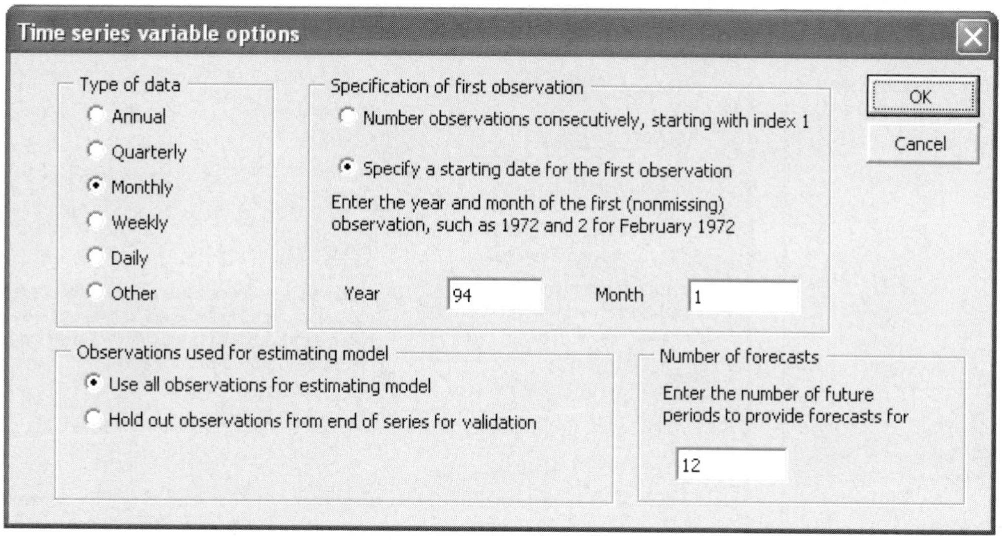

FIGURE 13.36 Timing Dialog Box for StatPro Forecasting Procedure

we call on the forecasting procedure of StatPro. Actually, this procedure is fairly general in that it allows us to forecast with several methods, either with or without taking seasonality into account. Since this is our first exposure to this procedure, we'll go through it in some detail in this example. In later examples, we'll mention some of its other capabilities.

To use the StatPro Forecasting procedure, the cursor needs to be in a data set with time series data. The data set can contain a "date" variable (for labeling charts), but this is not required. We use the StatPro/Forecasting menu item and eventually choose Sales as the variable to analyze. We then see several dialog boxes, the first of which appears in Figure 13.36. Here we specify the timing. Are the data annual, quarterly, and so on, and (if relevant) what year and period do they begin? We can also elect to "hold out" a subset of the data for validation purposes, and we can specify how many periods to forecast into the future. (If we hold out several periods at the end of the data set for validation, then any model that is built is estimated only for the nonholdout observations, and summary measures are reported for the nonholdout and holdout subsets separately.) Figure 13.36 shows that the Dow data are monthly, they begin in the first month of 1994, we do not hold out any data for validation, and we want to forecast 12 months into the future.

In the next dialog box, shown in Figure 13.37, we specify which forecasting method to use and any parameters of that method. Here we are using the moving averages method with a span of 3. (A span of 3 is StatPro's default value, but you can request any reasonable span.) Also, we indicate in this dialog box that the data are *not* seasonal. We next see a dialog box that allows us to request various time series plots, and finally we get the usual choice of whether to report the output on the current worksheet or a new worksheet.

The output consists of several parts, as shown in Figures 13.38–13.41 (pages 726–728). Figures 13.38 and 13.40 are for a span of 3, whereas the other two figures are for a span of 12. (We also obtained output for a span of 6, with results similar to those for a span of 12.)[5] First, the forecasts and forecast errors are shown for the historical period of the data. Actually, with moving averages we lose some forecasts at the beginning of the pe-

---

[5]You'll notice that StatPro has its own minor Y2K problem. When the dates extend from the 90s into the next century, the labels are incorrect, as in Jan-100 rather than Jan-00. For now, you can change these labels manually.

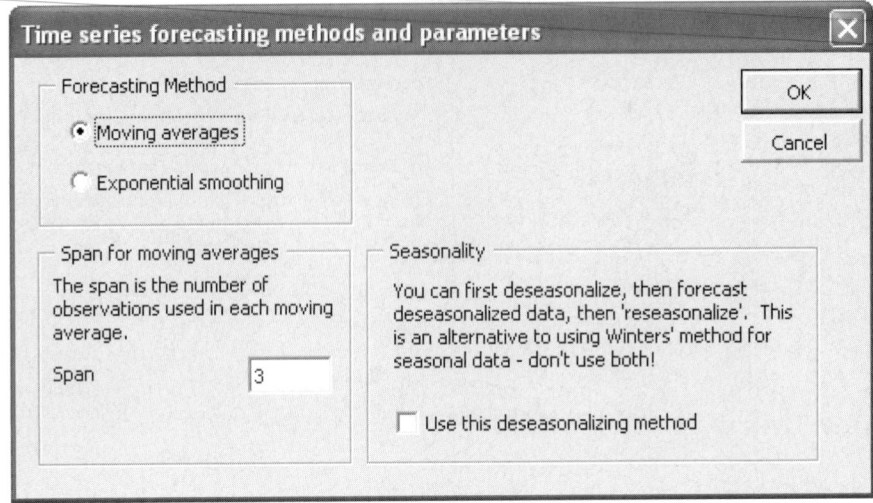

FIGURE 13.37
Method Dialog
Box for StatPro
Forecasting
Procedure

FIGURE 13.38
Moving Averages
Output with
Span 3

	A	B	C	D	E	F	G
1	**Forecasting results for Sales**			Date	Observation	Forecast	Error
2				Jan-94	128.000		
3	**Moving averages**			Feb-94	135.000		
4				Mar-94	130.000		
5	Span	3		Apr-94	127.000	131.000	-4.000
6				May-94	125.000	130.667	-5.667
7	**Estimation period**			Jun-94	119.000	127.333	-8.333
8				Jul-94	108.000	123.667	-15.667
9	MAE	12.2636		Aug-94	112.000	117.333	-5.333
10	RMSE	14.9558		Sep-94	111.000	113.000	-2.000
11	MAPE	8.37%		Oct-94	141.000	110.333	30.667
12				Nov-94	126.000	121.333	4.667
13				Dec-94	113.000	126.000	-13.000
89				Apr-01	176.000	177.667	-1.667
90				May-01	186.000	178.667	7.333
91				Jun-01		184.667	
92				Jul-01		182.222	
93				Aug-01		184.296	
94				Sep-01		183.728	
95				Oct-01		183.416	
96				Nov-01		183.813	
97				Dec-01		183.652	
98				Jan-02		183.627	
99				Feb-02		183.698	
100				Mar-02		183.659	
101				Apr-02		183.661	
102				May-02		183.673	

riod. For example, we lose three when the span is 3 because we don't have enough *previous* data to calculate these early averages. If we ask for *future* forecasts, they are shown in red at the bottom of the data series. Of course, there are no accompanying forecast errors because we don't yet have observations for these future periods. To the left of all this, we see the summary measures MAE, RMSE, and MAPE of the forecast errors. Finally, if we ask for any time series plots, these appear on separate sheets.

The essence of the forecasting method is very simple and is captured in column F of Figure 13.38 (for a span of 3). It uses the formula

$$=AVERAGE(\$E2:\$E4)$$

in cell F5, which is then copied down. The forecast errors are then just the differences between columns E and F. For the future periods, the forecast formulas use observations when they are available. If they are not available, previous forecasts are used. For exam-

FIGURE 13.39
Moving Averages
Output with
Span 12

	A	B	C	D	E	F	G
1	*Forecasting results for Sales*			Date	Observation	Forecast	Error
2				Jan-94	128.000		
3	**Moving averages**			Feb-94	135.000		
4				Mar-94	130.000		
5	Span	12		Apr-94	127.000		
6				May-94	125.000		
7	**Estimation period**			Jun-94	119.000		
8				Jul-94	108.000		
9	MAE	12.4253		Aug-94	112.000		
10	RMSE	15.6188		Sep-94	111.000		
11	MAPE	8.07%		Oct-94	141.000		
12				Nov-94	126.000		
13				Dec-94	113.000		
89				Apr-101	176.000	160.333	15.667
90				May-101	186.000	163.417	22.583
91				Jun-101		166.417	
92				Jul-101		167.951	
93				Aug-101		167.864	
94				Sep-101		170.103	
95				Oct-101		170.695	
96				Nov-101		171.752	
97				Dec-101		173.648	
98				Jan-102		173.619	
99				Feb-102		173.671	
100				Mar-102		174.143	
101				Apr-102		172.655	
102				May-102		172.377	

FIGURE 13.40
Moving Averages
Forecasts with
Span 3

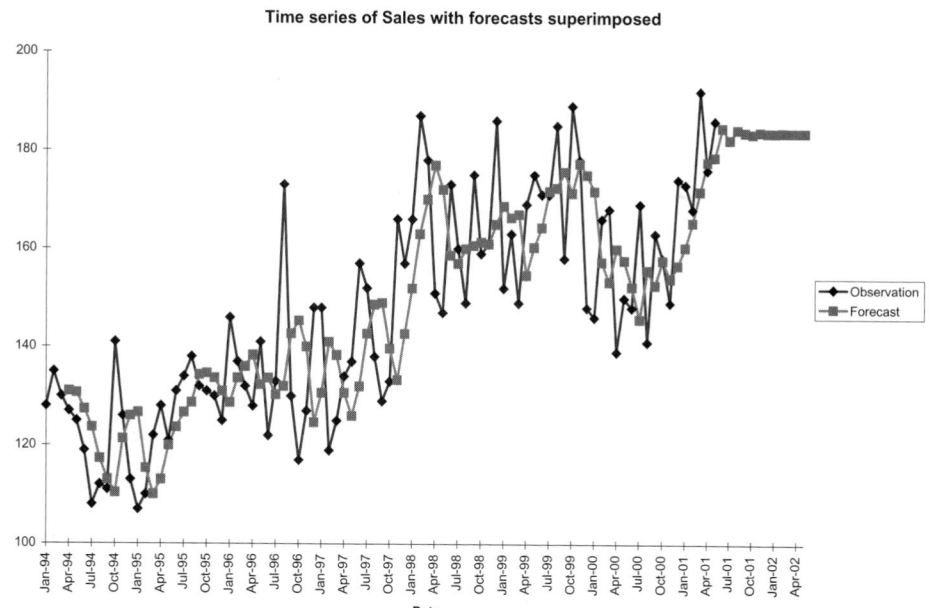

ple, the value in cell F92, the forecast for July 2001, is the average of the *observed* values in April and May and the *forecasted* value in June. Finally, the summary formulas in the range B9:B11 implement equations 13.2, 13.3, and 13.4.[6]

The plots in Figures 13.40 and 13.41 show the behavior of the forecasts. The forecasted series with span 3 follows the ups and downs of the actual series fairly closely, whereas the forecasted series with span 12 is much smoother and doesn't react nearly as

[6]If you want to learn some interesting features of Excel, take a close look at the formulas in these three cells. The formulas with curly brackets around them are *array* formulas. To enter them, type the formula without the curly brackets and then press Ctrl-Shift-Enter.

FIGURE 13.41

Moving Averages
Forecasts with
Span 12

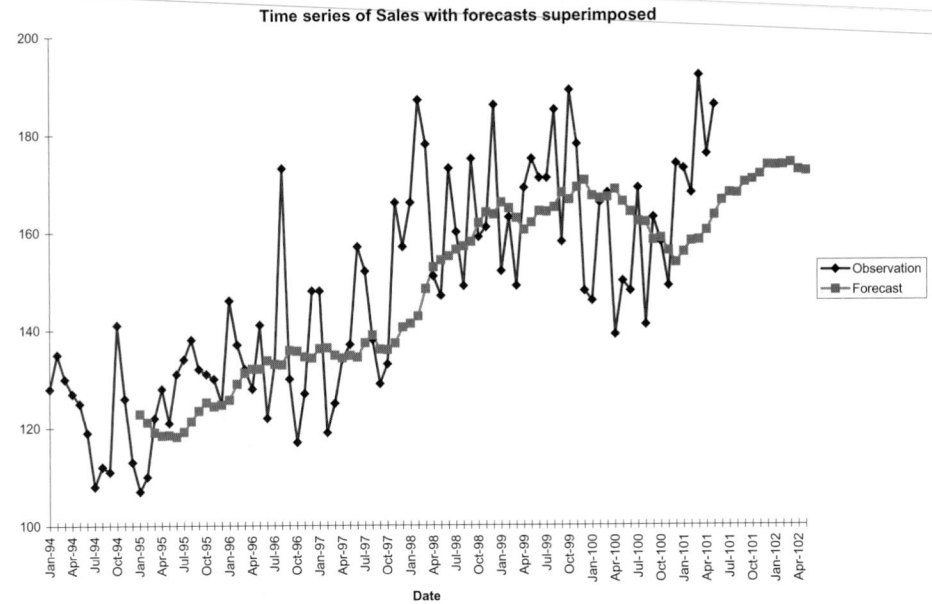

Time series of Sales with forecasts superimposed

much to these ups and downs. Which of these is better? The error summary measures indicate that it is a virtual toss-up. The MAPE with span 12 is slightly lower, indicating that these forecasts are off, on average, by about 8.1%, but the RMSE with span 3 is slightly lower. Note that the *future* forecasts are considerably lower with span 12 than with span 3.

At this point, how to proceed is up to the judgment of the forecaster, who presumably has some knowledge of the housing sales market in the Midwest. If she believes that the ups and downs in the original series are largely unpredictable noise, then she will probably trust the smooth forecasts from a span of 12. Otherwise, she might use a span of 3 (or some other intermediate span, such as 6). ●

The moving average method we have presented is the simplest of a group of moving average methods used by professional forecasters. We *smoothed* exactly once; that is, we took moving averages of several observations at a time and used these as forecasts. More complex methods smooth more than once, basically to get rid of random noise. They take moving averages, then moving averages of these moving averages, and so on for several stages. This can become quite complex, but the objective is quite simple—to smooth the data so that we can see underlying patterns.

PROBLEMS

Level A

35. The file P13_16.XLS contains the daily closing prices of American Express stock for a 1-year period.
   a. Using a span of 3 days, forecast the price of this stock for the next trading day with the moving average method. How well does this method with span 3 forecast the known observations in this data set?
   b. Repeat part **a** with a span of 10.
   c. Which of these two spans appears to be more appropriate? Explain your choice.

36. The closing value of the AMEX Airline Index for

each trading day during a 1-year period is given in the file P13_17.XLS.

a. How well does the moving average method track this series when the span is 4 days; when the span is 12 days?

b. Using the more appropriate span, forecast the closing value of this index on the next trading day with the moving average method.

37. The closing value of the Dow Jones Industrial Average for each trading day during a 1-year period is provided in the file P13_19.XLS.

a. Using a span of 2 days, forecast the price of this stock on the next trading day with the moving average method. How well does the moving average method with span 2 forecast the known observations in this data set?

b. Repeat part **a** with a span of 5 days; with a span of 15 days.

c. Which of these three spans appears to be most appropriate? Explain your choice.

38. The file P13_10.XLS contains the daily closing prices of Wal-Mart stock during a 1-year period. Use the moving average method with a carefully chosen span to forecast this time series for the next 3 trading days. Defend your choice of the span used.

39. The Consumer Confidence Index (CCI) attempts to measure people's feelings about general business conditions, employment opportunities, and their own income prospects. The file P02_28.XLS contains the annual average values of the CCI. Use the moving average method with a carefully chosen span to forecast this time series in the next 2 years. Defend your choice of the span used here.

Level B

40. Consider the file P02_37.XLS, which contains total monthly U.S. retail sales data. While retaining the final 6 months of observations for validation purposes, use the method of moving averages with a carefully chosen span to forecast U.S. retail sales in the next year. Comment on the performance of your model. What makes this time series more challenging to forecast?

41. Consider a random walk model with the following equation: $Y_t = Y_{t-1} + \epsilon_t$, where ϵ_t is a random series with mean 0 and standard deviation 1. Specify a moving average model that is equivalent to this random walk model. In particular, what is the appropriate size of the span in the equivalent moving average model? Describe the smoothing effect of this span choice.

 # Exponential Smoothing

There are two possible criticisms of the moving averages method. First, it puts equal weight on each value in a typical moving average when making a forecast. Many people would argue that if next month's forecast is to be based on the previous 12 months' observations, then more weight ought to be placed on the more recent observations. The second criticism is that the moving averages method requires a lot of data storage. This is particularly true for companies that routinely make forecasts of hundreds or even thousands of items. If 12-month moving averages are used for 1000 items, then 12,000 values are needed for next month's forecasts. This may or may not be a concern considering today's relatively inexpensive computer storage capabilities.

Exponential smoothing is a method that addresses both of these criticisms. It bases its forecasts on a weighted average of past observations, with more weight put on the more recent observations, and it requires very little data storage. In addition, it is not difficult for most business people to understand, at least conceptually. Therefore, this method finds widespread use in the business world, particularly when frequent and automatic forecasts of many items are required.

There are many versions of exponential smoothing. The simplest is called, surprisingly enough, *simple* exponential smoothing. It is relevant when there is no pronounced trend or seasonality in the series. If there is a trend but no seasonality, then *Holt's* method is applicable. If, in addition, there is seasonality, then *Winters'* method can be used. This does not exhaust the list of exponential smoothing models—researchers have invented many other variations—but these three models will suffice for us.

Exponential Smoothing Models	**Simple** exponential smoothing is appropriate for a series without a pronounced trend or seasonality. **Holt's** method is appropriate for a series with trend but no seasonality. **Winters'** method is appropriate for a series with seasonality (and possibly trend).

In this section we will examine simple exponential smoothing and Holt's model for trend. Then in the next section we will examine Winters' model when we focus on seasonal models in general.

13.8.1 Simple Exponential Smoothing

We now examine simple exponential smoothing in some detail. We first introduce two new terms. Every exponential model has at least one **smoothing constant,** which is always between 0 and 1. Simple exponential smoothing has a single smoothing constant denoted by α. (Its role will be discussed shortly.) The second new term is L_t, called the **level** of the series at time t. This value is not observable but can only be estimated. Essentially, it is where we think the series would be at time t if there were no random noise. Then the simple exponential smoothing method is defined by the following two equations, where F_{t+k} is the forecast of Y_{t+k} made at time t:

Simple Exponential Smoothing Formulas	$$L_t = \alpha Y_t + (1 - \alpha)L_{t-1} \qquad \text{(13.12)}$$ $$F_{t+k} = L_t \qquad \text{(13.13)}$$

Even though you usually won't have to substitute into these equations manually, you should understand what they say. Equation (13.12) shows how to update the estimate of the level. It is a weighted average of the current observation, Y_t, and the previous level, L_{t-1}, with respective weights α and $1 - \alpha$. Equation (13.13) shows how forecasts are made. It says that the k-period-ahead forecast, F_{t+k}, made of Y_{t+k} in period t is the most recently estimated level, L_t. This is the *same* for any value of $k \geq 1$. The idea is that in simple exponential smoothing, we believe that the series is not really going anywhere. So as soon as we estimate where the series ought to be in period t (if it weren't for random noise), we forecast that this is where it will also be in any future period.

The smoothing constant α is analogous to the span in moving averages. There are two ways to see this. The first way is to rewrite equation (13.12), using the fact that the forecast error, E_t, made in forecasting Y_t at time $t - 1$ is $Y_t - F_t = Y_t - L_{t-1}$. A bit of algebra then gives equation (13.14).

Equivalent Formula for Simple Exponential Smoothing	$$L_t = L_{t-1} + \alpha E_t \qquad \text{(13.14)}$$

This equation says that the next estimate of the level is adjusted from the previous estimate by adding a multiple of the most recent forecast error. This makes sense. If our previous forecast was too high, then E_t is negative, and we adjust the estimate of the level downward. The opposite is true if our previous forecast was too low. However, equation (13.14) says that we do not adjust by the entire magnitude of E_t, but only by a fraction of it. If α is small, say, $\alpha = 0.1$, then the adjustment is minor; if α is close to 1, the adjustment is large. So if we want to react quickly to movements in the series, we choose a large α; otherwise, we choose a small α.

Another way to see the effect of α is to substitute recursively into the equation for L_t. If you are willing to go through some algebra, you can verify that L_t satisfies equation (13.15), where the sum extends back to the first observation at time $t = 1$.

$$L_t = \alpha Y_t + \alpha(1 - \alpha)Y_{t-1} + \alpha(1 - \alpha)^2 Y_{t-2} + \alpha(1 - \alpha)^3 Y_{t-3} + \cdots \quad \textbf{(13.15)}$$

Small smoothing constants provide forecasts that respond slowly to changes in the data. Large smoothing constants do the opposite.

Equation (13.15) shows how the exponentially smoothed forecast is a weighted average of previous observations. Furthermore, because $1 - \alpha$ is less than 1, the weights on the Y's decrease from time t backward. Therefore, if α is close to 0, then $1 - \alpha$ is close to 1 and the weights decrease very slowly. In other words, observations from the distant past continue to have a large influence on the next forecast. This means that the graph of the forecasts will be relatively smooth, just as with a large span in the moving averages method. But when α is close to 1, the weights decrease rapidly, and only very recent observations have much influence on the next forecast. In this case forecasts react quickly to sudden changes in the series.

What value of α should we use? There is no universally accepted answer to this question. Some practitioners recommend always using a value around 0.1 or 0.2. Others recommend experimenting with different values of α until a measure such as RMSE or MAPE is minimized. Some packages even have an optimization feature to find this optimal value of α. (This is the case with StatPro.) But just as we discussed in the moving averages section, the value of α that tracks the historical series most closely does not necessarily guarantee the most accurate *future* forecasts.

Example 13.6 Houses Sold in the Midwest (continued)

Previously, we used the moving averages method to forecast monthly housing sales in the Midwest. (See the HOUSESALES.XLS file.) How well does simple exponential smoothing work with this data set? What smoothing constant should we use?

Objective To see how well a simple exponential smoothing model, with an appropriate smoothing constant, fits the housing sales data, and to see how StatPro implements this method.

Solution

We will use StatPro to implement the simple exponential smoothing model, specifically equations (13.12) and (13.13). We do this again with the StatPro/Forecasting menu item. We first specify that the data are monthly, beginning in January 1994, we do not hold out any of the data for validation, and we ask for 12 months of future forecasts. We then fill out the next dialog box as shown in Figure 13.42. That is, we select the exponential

FIGURE 13.42

Method Dialog
Box for StatPro
Forecasting
Procedure

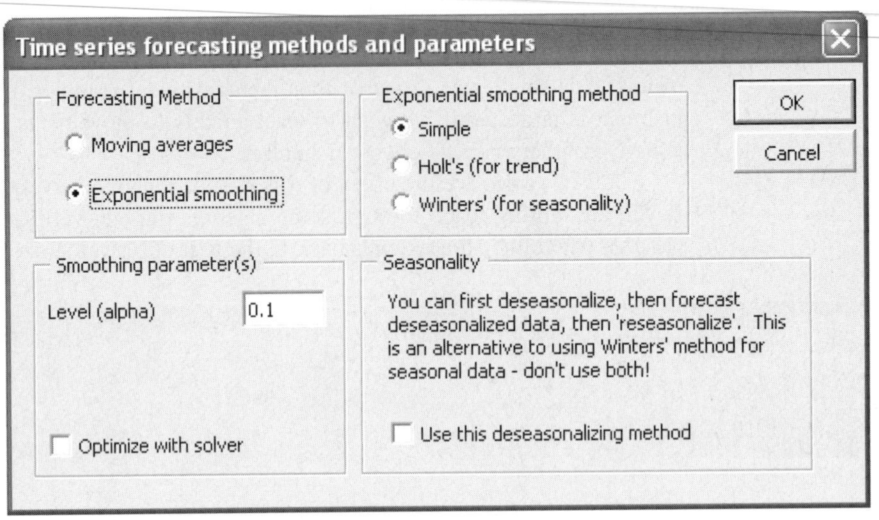

smoothing option in the upper left, select the Simple option, choose a smoothing constant (0.1 was chosen here, but any other value could be chosen) and elect not to optimize, and specify that the data are not seasonal. On the next dialog sheet we ask for time series charts of the series with the forecasts superimposed and the series of forecast errors.

The results appear in Figures 13.43 and 13.44. The heart of the method takes place in columns F, G, and H of Figure 13.43. Column F calculates the smoothed levels (L_t) from equation (13.12), column G calculates the forecasts (F_t) from equation (13.13), and column H calculates the forecast errors (E_t) as the observed values minus the forecasts. For example, the formulas in row 6 are

$$=\text{Alpha*E6+(1-Alpha)*F5}$$
$$=\text{F5}$$

and

$$=\text{E6-G6}$$

One exception to this scheme is in row 2. Every exponential smoothing method requires *initial* values, in this case the initial smoothed level in cell F2. There is no way to calculate this value, L_1, from equation (13.12) because the *previous* value, L_0, is unknown. Different implementations of exponential smoothing initialize in different ways. We have simply set L_1 equal to Y_1 (in cell E2). The effect of initializing in different ways is usually minimal because any effect of early data is usually washed out as we forecast into the future. In the present example, data from 1994 have little effect on forecasts of 2001 and beyond.

You might object that this series *does* have an upward trend. This is true, and in the next subsection we will use Holt's method on this series to see whether it captures the trend better than simple exponential smoothing.

Note that the 12 future forecasts (rows 91 down) are all equal to the last calculated smoothed level, the one for May 2001 in cell F90. The fact that these remain constant is a consequence of the assumption behind *simple* exponential smoothing, namely, that the series is not really going anywhere. Therefore, the last smoothed level is the best indication of future values of the series we have.

Figure 13.44 shows the forecast series superimposed on the original series. We see the obvious smoothing effect of a relatively small α level. The forecasts don't track the series very well, but if the various zigzags in the original series are really random noise, then perhaps we don't want the forecasts to track these random ups and downs too closely. Perhaps we instead prefer a forecast series that emphasizes the basic underlying pattern.

	A	B	C	D	E	F	G	H
1	*Forecasting results for Sales*			Date	Observation	SmLevel	Forecast	Error
2				Jan-94	128.000	128.000		
3	**Simple exponential smoothing**			Feb-94	135.000	128.700	128.000	7.000
4				Mar-94	130.000	128.830	128.700	1.300
5	Smoothing constant(s)			Apr-94	127.000	128.647	128.830	-1.830
6	Level	0.100		May-94	125.000	128.282	128.647	-3.647
7				Jun-94	119.000	127.354	128.282	-9.282
8	**Estimation period**			Jul-94	108.000	125.419	127.354	-19.354
9				Aug-94	112.000	124.077	125.419	-13.419
10	MAE	11.9332		Sep-94	111.000	122.769	124.077	-13.077
11	RMSE	15.0820		Oct-94	141.000	124.592	122.769	18.231
12	MAPE	7.91%		Nov-94	126.000	124.733	124.592	1.408
13				Dec-94	113.000	123.560	124.733	-11.733
89				Apr-01	176.000	165.708	164.565	11.435
90				May-01	186.000	167.737	165.708	20.292
91				Jun-01			167.737	
92				Jul-01			167.737	
93				Aug-01			167.737	
94				Sep-01			167.737	
95				Oct-01			167.737	
96				Nov-01			167.737	
97				Dec-01			167.737	
98				Jan-02			167.737	
99				Feb-02			167.737	
100				Mar-02			167.737	
101				Apr-02			167.737	
102				May-02			167.737	

FIGURE 13.43 Simple Exponential Smoothing Output

FIGURE 13.44
Plot of Forecasts from Simple Exponential Smoothing

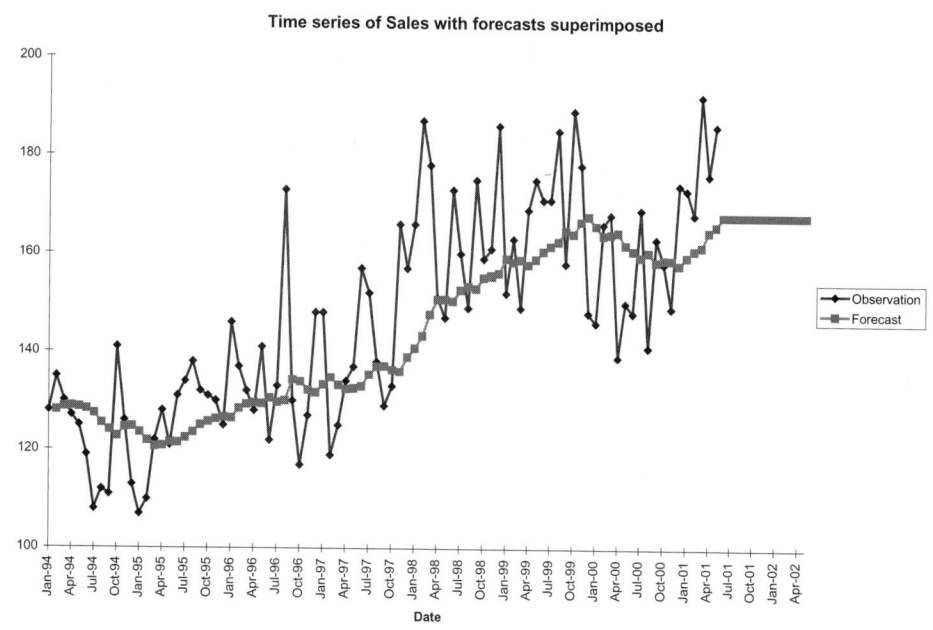

We see several summary measures of the forecast errors from Figure 13.43. The RMSE and MAE indicate that the forecasts from this model are typically off by a magnitude of about 12 to 15 thousand, and the MAPE indicates that they are off by about 7.9%. (These are similar to the errors we obtained with moving averages.) These imply fairly sizable errors. One way to reduce the errors is to use a different smoothing method. We will try this in the next subsection with Holt's method. Another way to reduce the errors is to use a different smoothing constant. There are two methods you can use. First,

FIGURE 13.45

Plot of Forecasts with an Optimal Smoothing Constant

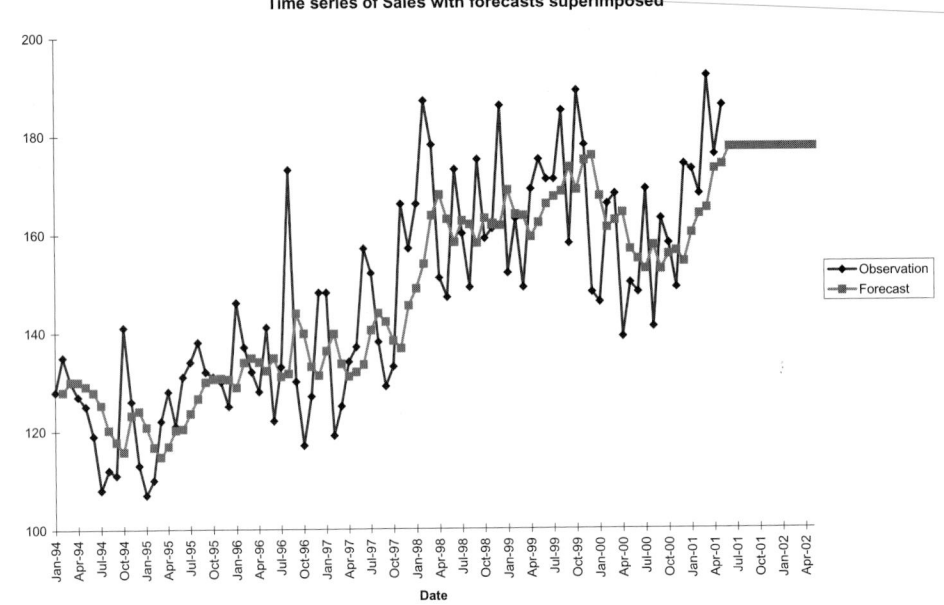

Time series of Sales with forecasts superimposed

you can simply enter different values in cell B6 of Figure 13.43. All formulas, including those for MAE, RMSE, and MAPE, will update automatically.

Second, you can check the Optimize with Solver box in Figure 13.42. This automatically runs the Excel Solver to find the smoothing constant that minimizes RMSE. We tried this for the housing data and obtained the forecasts in Figure 13.45 (from a smoothing constant of 0.295). The corresponding MAE, RMSE, and MAPE are 11.4, 14.1, and 7.7%, respectively—slightly better than before. This larger smoothing constant produces a less smooth forecast curve and slightly better error measures. However, there is no guarantee that *future* forecasts made with this optimal smoothing constant will be any better than with a smoothing constant of 0.1. ●

13.8.2 Holt's Model for Trend

The simple exponential smoothing model generally works well if there is no obvious trend in the series. But if there is a trend, then this method consistently lags behind it. For example, if the series is constantly increasing, simple exponential smoothing forecasts will be consistently low. Holt's method rectifies this by dealing with trend explicitly. In addition to the level of the series, L_t, Holt's method includes a trend term, T_t, and a corresponding smoothing constant β. The interpretation of L_t is exactly as before. The interpretation of T_t is that it represents an estimate of the change in the series from one period to the next. The equations for Holt's model are as follows.

Formula's for Holt's Exponential Smoothing Method	$L_t = \alpha Y_t + (1 - \alpha)(L_{t-1} + T_{t-1})$	**(13.16)**
	$T_t = \beta(L_t - L_{t-1}) + (1 - \beta)T_{t-1}$	**(13.17)**
	$F_{t+k} = L_t + kT_t$	**(13.18)**

These equations are not as bad as they look. (And don't forget that the computer typically does all of the calculations for you.) Equation (13.16) says that the updated level is a weighted average of the current observation and the previous level plus the estimated change. Equation (13.17) says that the updated trend term is a weighted average of the difference between two consecutive levels and the previous trend term. Finally, equation (13.18) says that the k-period-ahead forecast made in period t is the estimated level plus k times the estimated change per period.

Everything we said about α for simple exponential smoothing applies to both α and β in Holt's model. The new smoothing constant β controls how quickly the method reacts to perceived changes in the trend. If β is small, the method reacts slowly. If it is large, the method reacts more quickly. Of course, there are now two smoothing constants to select. Some practitioners suggest using a small value of α (0.1 to 0.2, say) and setting β equal to α. Others suggest using an optimization option (available in StatPro) to select the "best" smoothing constants. We illustrate the possibilities in the following continuation of the housing sales example.

Example 13.6 Houses Sold in the Midwest (continued)

We again examine the monthly data on housing sales in the Midwest. (See the file HOUSESALES.XLS.) In the previous subsection, we saw that simple exponential smoothing, even with an optimal smoothing constant, does only a "fair" job of forecasting housing sales. Given that there is an upward trend in housing sales over this period, we might expect Holt's method to perform better. Does it? What smoothing constants are appropriate?

Objective To see whether Holt's method, with appropriate smoothing constants, captures the trend in the housing sales data better than simple exponential smoothing (or moving averages).

Solution

We implement Holt's method in StatPro almost exactly like we did in simple exponential smoothing. The only difference is that we can now choose *two* smoothing constants, as shown in Figure 13.46 (page 736). They can have different values, although we have chosen them to be their default values of 0.1.

The StatPro outputs in Figure 13.47 and 13.48 are also very similar to the simple exponential smoothing outputs. The only difference is that there is now a trend column, column G, in the numerical output. You can check that the formulas in columns F, G, and H implement equations (13.16), (13.17), and (13.18). As before, there is an initialization problem in row 2. These require values of L_1 and T_1 to get the method started. Different implementations of Holt's method obtain these initial values in slightly different ways, but the effect is fairly minimal in most cases. (You can check cells F2 and G2 to see how StatPro does it.)

Somewhat surprisingly, the error measures for this implementation of Holt's method are no better than for simple exponential smoothing, even though this series exhibits a gradual upward trend. Perhaps this is because 0.1 and 0.1 are not the *optimal* smoothing constants. Therefore, we ran Holt's method a second time, checking the Optimize with Solver box. This resulted in somewhat better results and the forecasts shown in Figure 13.49. The optimal smoothing constants were $\alpha = 0.252$ and $\beta = 0.000$, and the MAE, RMSE, and MAPE values were 11.2, 13.9, and 7.7%—almost identical to simple exponential smoothing with an optimal smoothing constant.

We make two final comments about this example. First, the optimal smoothing constant for trend, $\beta = 0.000$, does *not* mean that the method ignores trend. We can see from

FIGURE 13.46
StatPro Dialog Box
for Holt's Method

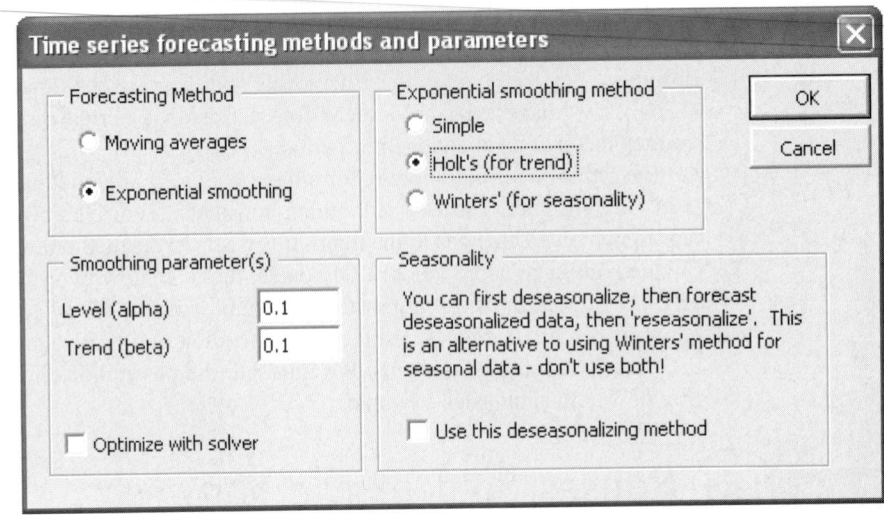

FIGURE 13.47 Output from Holt's Method

	A	B	C	D	E	F	G	H	I
1	*Forecasting results for Sales*			Date	Observation	SmLevel	SmTrend	Forecast	Error
2				Jan-94	128.000	128.000	0.652		
3	**Holt's exponential smoothing**			Feb-94	135.000	129.287	0.715	128.652	6.348
4				Mar-94	130.000	130.002	0.715	130.002	-0.002
5	Smoothing constant(s)			Apr-94	127.000	130.345	0.678	130.717	-3.717
6	Level	0.100		May-94	125.000	130.421	0.618	131.023	-6.023
7	Trend	0.100		Jun-94	119.000	129.835	0.497	131.038	-12.038
8				Jul-94	108.000	128.099	0.274	130.332	-22.332
9	**Estimation period**			Aug-94	112.000	126.736	0.110	128.373	-16.373
10				Sep-94	111.000	125.261	-0.048	126.846	-15.846
11	MAE	12.0491		Oct-94	141.000	126.792	0.110	125.213	15.787
12	RMSE	15.0762		Nov-94	126.000	126.811	0.101	126.902	-0.902
13	MAPE	8.23%		Dec-94	113.000	125.521	-0.038	126.912	-13.912
14				Jan-95	107.000	123.634	-0.223	125.483	-18.483
89				Apr-01	176.000	165.268	0.028	164.076	11.924
90				May-01	186.000	167.366	0.235	165.296	20.704
91				Jun-01				167.601	
92				Jul-01				167.836	
93				Aug-01				168.071	
94				Sep-01				168.306	
95				Oct-01				168.541	
96				Nov-01				168.776	
97				Dec-01				169.011	
98				Jan-02				169.246	
99				Feb-02				169.481	
100				Mar-02				169.716	
101				Apr-02				169.951	
102				May-02				170.186	

Figure 13.49 that it does indeed include trend because of the upward-trending *future* forecasts at the right of the graph. The zero smoothing constant simply means that our initial estimate of trend (in cell G2 of Figure 13.47) is never updated. Second, you should not conclude from this example that Holt's method is never superior to simple exponential smoothing. Holt's method is often able to react quickly to a sudden upswing or downswing in the data, whereas simple exponential smoothing typically has a delayed reaction to such a change. It just happened in this example that the trend was gradual, so that both methods were able to react to it in equivalent ways.

FIGURE 13.48
Forecasts from
Holt's Method
with Nonoptimal
Smoothing
Constants

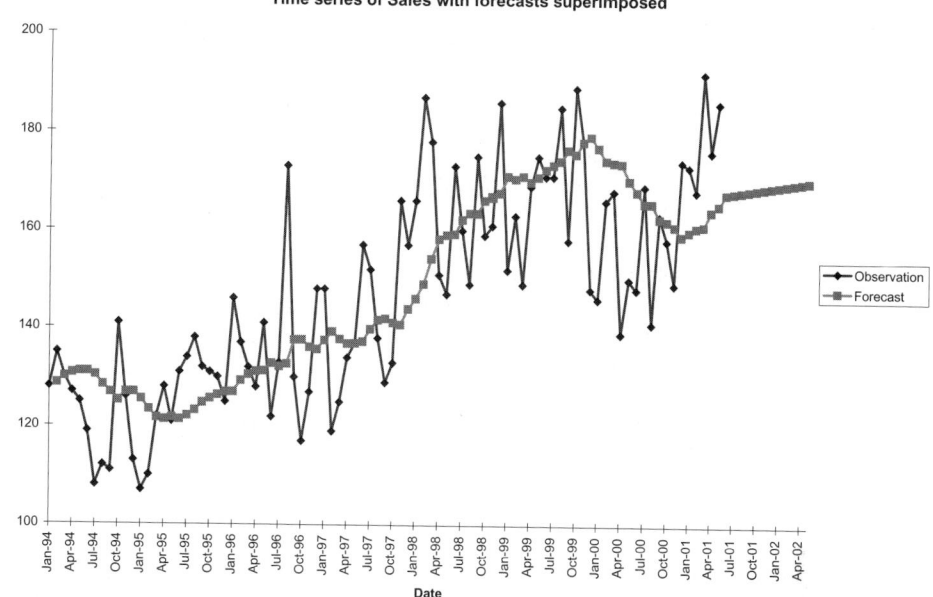

FIGURE 13.49
Forecasts from
Holt's Method
with Optimal
Smoothing
Constants

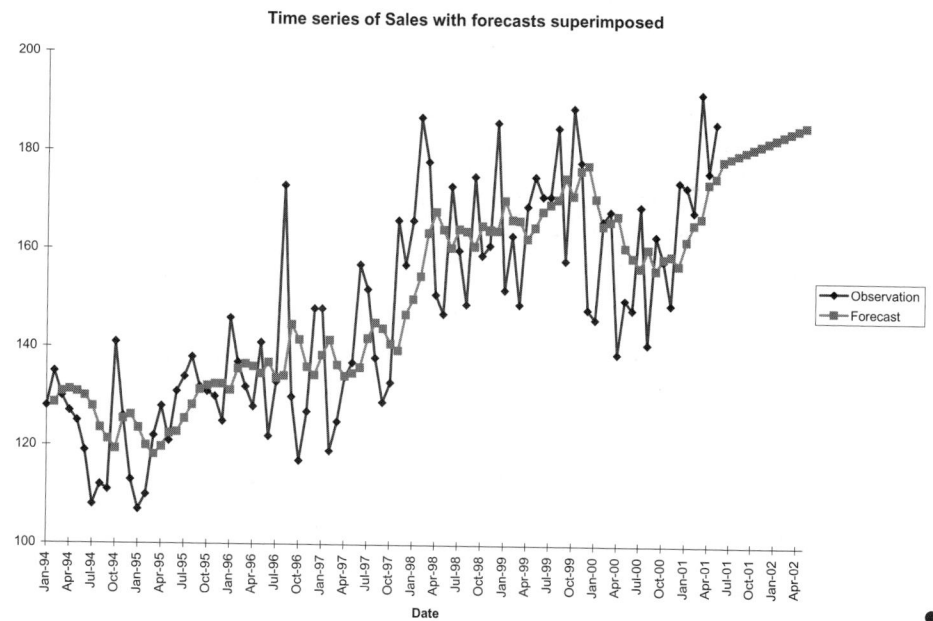

P R O B L E M S

Level A

42. Consider the airline ticket data in the file
P13_01.XLS.
  a. Create a time series chart of the data. Based on
what you see, which of the exponential smoothing
models do you think should be used for forecasting? Why?
  b. Use simple exponential smoothing to forecast
these data, using no holdout period and requesting
12 months of future forecasts. Use the default
smoothing constant of 0.1.

c. Repeat part **b,** optimizing the smoothing constant. Does it make much of an improvement?

d. Write a short report to summarize your results.

43. Consider the applications for home mortgages data in the file P13_04.XLS.

   a. Create a time series chart of the data. Based on what you see, which of the exponential smoothing models do you think should be used for forecasting? Why?

   b. Use simple exponential smoothing to forecast these data, using no holdout period and requesting 4 quarters of future forecasts. Use the default smoothing constant of 0.1.

   c. Repeat part **b,** optimizing the smoothing constant. Does it make much of an improvement?

   d. Write a short report to summarize your results.

44. Consider the American Express closing price data in the file P13_16.XLS. Focus only on the closing prices.

   a. Create a time series chart of the data. Based on what you see, which of the exponential smoothing models do you think should be used for forecasting? Why? (*Note:* The data are currently sorted from most recent to most distant in the past. Sort them in the opposite order first.)

   b. Use Holt's exponential smoothing to forecast these data, using no holdout period and requesting 20 days of future forecasts. Use the default smoothing constants of 0.1.

   c. Repeat part **b,** optimizing the smoothing constant. Does it make much of an improvement?

   d. Repeat parts **a** and **b,** this time using a holdout period of 50 days.

   e. Write a short report to summarize your results.

45. Consider the poverty level data in the file P02_29.XLS. Focus only on the Percent variable.

   a. Create a time series chart of the data. Based on what you see, which of the exponential smoothing models do you think should be used for forecasting? Why?

   b. Use simple exponential smoothing to forecast these data, using no holdout period and requesting 3 years of future forecasts. Use the default smoothing constants of 0.1.

   c. Repeat part **b,** optimizing the smoothing constant. Make sure you request a chart of the series with the forecasts superimposed. Does the optimal smoothing constant make much of an improvement?

   d. Write a short report to summarize your results. Considering the chart in part **c,** would you say the forecasts are "good"?

Problems 46–48 ask you to apply the exponential smoothing formulas. These do not require StatPro. In

fact, they do not even require Excel. You can do them with a hand calculator (or with Excel).

46. TOD Chevy is using Holt's method to forecast weekly car sales. Currently, the level is estimated to be 50 cars per week, and the trend is estimated to be 6 cars per week. During the current week, 30 cars are sold. After observing the current week's sales, forecast the number of cars 3 weeks from now. Use $\alpha = \beta = 0.3$.

47. You have been assigned to forecast the number of aircraft engines ordered each month by Commins Engine Company. At the end of February, the forecast is that 100 engines will be ordered during April. Then during March, 120 engines are actually ordered.

   a. Using $\alpha = 0.3$, determine a forecast (at the end of March) for the number of orders placed during April; during May. Use simple exponential smoothing.

   b. Suppose MAE = 16 at the end of March. At the end of March, Commins can be 68% sure that April orders will be between what two values, assuming normally distributed forecast errors? (*Hint:* The standard deviation of forecast errors is approximately 1.25 times MAE.)

48. Simple exponential smoothing with $\alpha = 0.3$ is being used to forecast sales of radios at Lowland Appliance. Forecasts are made on a monthly basis. After August radio sales are observed, the forecast for September is 100 radios.

   a. During September, 120 radios are sold. After observing September sales, what do we forecast for October radio sales? For November radio sales?

   b. It turns out that June sales were recorded as 10 radios. Actually, however, 100 radios were sold in June. After correcting for this error, develop a forecast for October radio sales.

Level B

49. Holt's method assumes an additive trend. For example, a trend of 5 means that the level will increase by 5 units per period. Suppose there is actually a **multiplicative trend.** For example, if the current estimate of the level is 50 and the current estimate of the trend is 1.2, we would predict demand to increase by 20% per period. So we would forecast the next period's demand to be 50(1.2) and forecast the demand 2 periods in the future to be $50(1.2)^2$. If we want to use a multiplicative trend in Holt's method, we should use the following equations:

$$L_t = \alpha Y_t + (1 - \alpha)(I)$$
$$T_t = \beta(II) + (1 - \beta)T_{t-1}$$

a. Determine (*I*) and (*II*).

b. Suppose we are working with monthly data and month 12 is December, month 13 is January, and so on. Also suppose that $L_{12} = 100$ and $T_{12} = 1.2$. Suppose $Y_{13} = 200$. At the end of month 13, what is the prediction for Y_{15}? Assume $\alpha = \beta = 0.5$ and a multiplicative trend.

50. A version of simple exponential smoothing can be used to predict the outcome of sporting events. To illustrate, consider pro football. We first assume that all games are played on a neutral field. Before each day of play, we assume that each team has a rating. For example, if the rating for the Bears is $+10$ and the rating for the Bengals is $+6$, we predict the Bears to beat the Bengals by $10 - 6 = 4$ points. Suppose that the Bears play the Bengals and win by 20 points. For this game, we "underpredicted" the Bears' performance by $20 - 4 = 16$ points. Assuming that the best α for pro football is $\alpha = 0.10$, we would increase the Bears' rating by $16(0.1) = 1.6$ and decrease the Bengals' rating by 1.6 points. In a rematch, the Bears would then be favored by $(10 + 1.6) - (6 - 1.6) = 7.2$ points.

a. How does this approach relate to the equation $L_t = L_{t-1} + \alpha e_t$?

b. Suppose that the home field advantage in pro football is 3 points; that is, home teams tend to outscore visiting teams by an average of 3 points a game. How could the home field advantage be incorporated into this system?

c. How might we determine the *best* α for pro football?

d. How might we determine ratings for each team at the beginning of the season?

e. Suppose we apply this method to predict pro football (16-game schedule), college football (11-game schedule), college basketball (30-game schedule), and pro basketball (82-game schedule). Which sport do you think would have the smallest optimal α; the largest optimal α? Why?

f. Why might this approach yield poor forecasts for major league baseball?

 Seasonal Models

So far we have said practically nothing about seasonality. Seasonality is the consistent month-to-month (or quarter-to-quarter) differences that occur each year. For example, there is seasonality in beer sales—high in the summer months, lower in other months. Toy sales are also seasonal, with a huge peak in the months preceding Christmas. In fact, if you start thinking about time series variables that you are familiar with, the majority of them probably have some degree of seasonality.

How do we know whether there is seasonality in a time series? The easiest way is to check whether a plot of the time series has a *regular* pattern of ups and/or downs in particular months or quarters. Although random noise can sometimes obscure such a pattern, the seasonal pattern is usually fairly obvious.

Some time series software packages have special types of graphs for spotting seasonality, but we won't discuss these here.

There are basically three methods for dealing with seasonality. First, we can use Winters' exponential smoothing model, as we will discuss here. It is similar to simple exponential smoothing and Holt's method, except that it includes another component (and smoothing constant) to capture seasonality. Second, we can *deseasonalize* the data, then use any of our forecasting methods to model the deseasonalized data, and finally "reseasonalize" these forecasts. Finally, we can use multiple regression with dummy variables for the seasons.

As we saw with the housing sales data, government agencies often perform part of the second method for us—that is, they deseasonalize the data.

Seasonal models are usually classified as *additive* or *multiplicative*. Suppose that we have monthly data, and that the average of the 12 monthly values for a typical year is 150. An **additive** model finds seasonal indexes, one for each month, that we *add* to the monthly average, 150, to get a particular month's value. For example, if the index for March is 22, then we expect a typical March value to be $150 + 22 = 172$. If the seasonal index for September is -12, then we expect a typical September value to be $150 - 12 = 138$. A **multiplicative** model also finds seasonal indexes, but we *multiply* the monthly average by these indexes to get a particular month's value. Now if the index for March is 1.3, we expect a typical March value to be $150(1.3) = 195$. If the index for September is 0.9, then we expect a typical September value to be $150(0.9) = 135$. These models are summarized here.

Additive Seasonal Model	In an **additive** seasonal model, we add an appropriate seasonal index to a "base" forecast. These indexes, one for each season, typically average to 0.

Multiplicative Seasonal Model	In a **multiplicative** seasonal model, we multiply a "base" forecast by an appropriate seasonal index. These indexes, one for each season, typically average to 1.

Either an additive or a multiplicative model can be used to forecast seasonal data. However, because multiplicative models are somewhat easier to interpret (and have worked well in applications), we will focus on them. Note that the seasonal index in a multiplicative model can be interpreted as a percentage. Using the figures in the previous paragraph as an example, March tends to be 30% above the monthly average, whereas September tends to be 10% below it. Also, the seasonal indexes in a multiplicative model should average to 1. Computer packages typically ensure that this happens.

13.9.1 Winters' Exponential Smoothing Model

We now turn to Winters' exponential smoothing model. It is very similar to Holt's model—it again has level and trend terms and corresponding smoothing constants α and β—but it also has seasonal indexes and a corresponding smoothing constant γ (gamma). This new smoothing constant γ controls how quickly the method reacts to perceived changes in the pattern of seasonality. If γ is small, the method reacts slowly. If it is large, the method reacts more quickly. As with Holt's model, there are equations for updating the level and trend terms, and there is one extra equation for updating the seasonal indexes. For completeness, we list these equations in the accompanying box, but they are clearly too complex for hand calculation and are best left to the computer. In equation (13.21), S_t refers to the multiplicative seasonal index for period t. In equations (13.19), (13.21), and (13.22), M refers to the number of seasons ($M = 4$ for quarterly data, $M = 12$ for monthly data).

Formulas for Winters' Exponential Smoothing Model	$$L_t = \alpha \frac{Y_t}{S_{t-M}} + (1 - \alpha)(L_{t-1} + T_{t-1})$$	(13.19)
	$$T_t = \beta(L_t - L_{t-1}) + (1 - \beta)T_{t-1}$$	(13.20)
	$$S_t = \gamma \frac{Y_t}{L_t} + (1 - \gamma)S_{t-M}$$	(13.21)
	$$F_{t+k} = (L_t + kT_t)S_{t+k-M}$$	(13.22)

To see how the forecasting in equation (13.22) works, suppose we have observed data through June and want a forecast for the coming September, that is, a 3-month-ahead forecast. (In this case t refers to June and $t + k = t + 3$ refers to September.) Then we first add 3 times the current trend term to the current level. This gives a forecast for September that would be appropriate if there were no seasonality. Next, we multiply this forecast by the most recent estimate of September's seasonal index (the one from the previous September) to get the forecast for September. Of course, the computer does all of the arithmetic, but this is basically what it is doing. We illustrate the method in the following example.

Example 13.7 Quarterly Sales at Coca-Cola

The data in the COCACOLA.XLS file represent quarterly sales (in millions of dollars) for Coca-Cola from quarter 1 of 1986 through quarter 1 of 2001. As we might expect, there has been an upward trend in sales during this period, and there is also a fairly regular seasonal pattern, as shown in Figure 13.50. Sales in the warmer quarters, 2 and 3, are consistently higher than in the colder quarters, 1 and 4. How well can Winters' method track this upward trend and seasonal pattern?

Objective To see how well Winters' method, with appropriate smoothing constants, can forecast Coca-Cola's seasonal sales.

Solution

To use Winters' method with StatPro, we proceed exactly as with any of the other exponential smoothing methods. However, for a change (and because we have so many years of data), we use StatPro's option of holding out some of the data for validation. Specifically, we fill out the first large dialog box as shown in Figure 13.51 (page 742), basing the model on the data through quarter 1, 1999, holding out 8 quarters of data (quarter 2, 1999, through quarter 1, 2001), and forecast 4 quarters into the future. Then we fill in the second large dialog box as shown in Figure 13.52. Note that the "Use this deseasonalizing method" option at the bottom right has been disabled. It wouldn't make much sense to deseasonalize *and* use Winters' method; we use one or the other. Also, we have elected to optimize the smoothing constants, but this is optional.

Parts of the output are shown in Figure 13.53. The following points are worth noting. (1) The optimal smoothing constants (those that minimize RMSE) are $\alpha = 1.0$, $\beta = 0.0$, and $\gamma = 0.139$. Intuitively, these mean that we react right away to changes in level, we never react to changes in trend, and we react fairly slowly to changes in the seasonal pattern. (2) If we ignore seasonality, the series is trending upward at a rate of 57.265 per quarter (see column G). This is our initial estimate of trend and, because $\beta = 0$, it never

FIGURE 13.50
Time Series Plot of Coca-Cola Sales

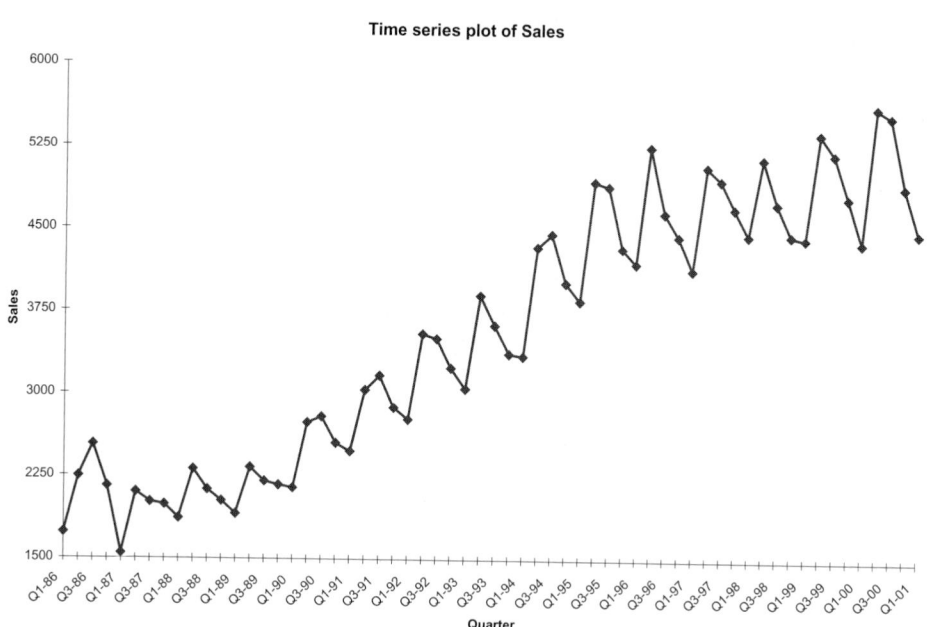

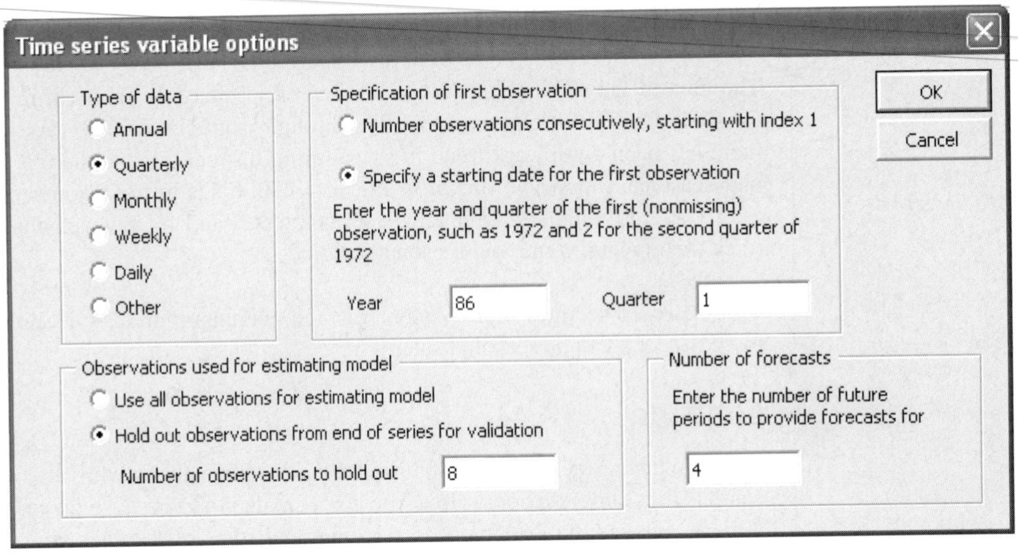

FIGURE 13.51 Using StatPro's Holdout Option

FIGURE 13.52
Method Dialog
Box from StatPro
Forecasting Pro-
cedure

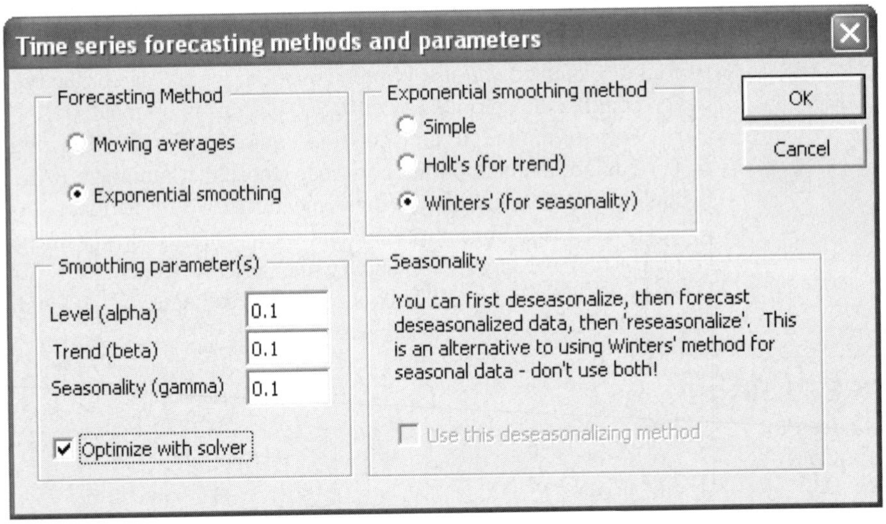

changes. (3) The seasonal pattern stays constant throughout this 10-year period. The seasonal indexes, shown in column H, are 0.887, 1.097, 1.055, and 0.961. For example, quarter 1 is about 11% below the yearly average, and quarter 2 is almost 10% above the yearly
average. (4) The forecast series tracks the actual series quite well during the nonholdout
period. For example, MAPE is 3.95%, meaning that on average our forecasts are only
about 4% in error. Even for the holdout period, MAPE is only slightly larger, at 4.17%.

The plot of the forecasts superimposed on the original series, shown in Figure 13.54,
indicates that Winters' method clearly picks up the seasonal pattern and the upward trend
and projects both of these into the future. In later examples, we will investigate whether
other seasonal forecasting methods can do this well.

One final comment is that we are not obligated to find the *optimal* smoothing constants. Some analysts might suggest using more "typical" values such as $\alpha = \beta = 0.2$
and $\gamma = 0.5$. (We often choose γ larger than α and β because each season's seasonal

You can check that
if we hold out 3
years of data, the
MAPE for the hold-
out period increases
quite a lot. It is
common for the fit
to be considerably
better in the estima-
tion period than in
the holdout period.

	A	B	C	D	E	F	G	H	I	J
1	*Forecasting results for Sales*			Date	Observation	SmLevel	SmTrend	SmSeason	Forecast	Error
2				Q1-86	1734.827	1955.059	57.265	0.887		
3	**Winters' exponential smoothing**			Q2-86	2244.961	2046.287	57.265	1.097	2207.701	37.260
4				Q3-86	2533.805	2402.289	57.265	1.055	2218.713	315.092
5	Smoothing constant(s)			Q4-86	2154.963	2242.857	57.265	0.961	2363.168	-208.205
6	Level	1.000		Q1-87	1547.819	1744.311	57.265	0.887	2041.020	-493.201
7	Trend	0.000		Q2-87	2104.412	1918.176	57.265	1.097	1976.491	127.921
8	Seasonality	0.139		Q3-87	2014.363	1909.809	57.265	1.055	2083.589	-69.226
9				Q4-87	1991.747	2072.984	57.265	0.961	1889.987	101.760
10	**Estimation period**			Q1-88	1869.050	2106.321	57.265	0.887	1890.283	-21.233
11				Q2-88	2313.632	2108.881	57.265	1.097	2373.649	-60.017
12	MAE	120.2047		Q3-88	2128.320	2017.851	57.265	1.055	2284.734	-156.414
13	RMSE	162.6609		Q4-88	2026.829	2109.497	57.265	0.961	1993.795	33.034
14	MAPE	3.95%		Q1-89	1910.604	2153.151	57.265	0.887	1922.683	-12.079
15				Q2-89	2331.165	2124.862	57.265	1.097	2425.025	-93.860
16	**Holdout period**			Q3-89	2206.550	2092.020	57.265	1.055	2301.590	-95.040
17				Q4-89	2173.968	2262.638	57.265	0.961	2065.058	108.910
18	MAE	203.3919		Q1-90	2148.278	2420.997	57.265	0.887	2058.572	89.706
19	RMSE	218.2511		Q2-90	2739.308	2496.885	57.265	1.097	2718.877	20.431
20	MAPE	4.17%		Q3-90	2792.754	2647.798	57.265	1.055	2693.980	98.774
21				Q4-90	2556.010	2660.262	57.265	0.961	2599.055	-43.045
54				Q1-99	4428.000	4990.124	57.265	0.887	4167.979	260.021
55				Q2-99	5379.000				5537.440	-158.440
56				Q3-99	5195.000				5384.114	-189.114
57				Q4-99	4803.000				4959.632	-156.632
58				Q1-00	4391.000				4631.258	-240.258
59				Q2-00	5621.000				5788.741	-167.741
60				Q3-00	5543.000				5625.716	-82.716
61				Q4-00	4903.000				5179.717	-276.717
62				Q1-01	4479.000				4834.517	-355.517
63				Q2-01					6040.043	
64				Q3-01					5867.318	
65				Q4-01					5399.801	
66				Q1-02					5037.775	

FIGURE 13.53 Output from Winters' Method

FIGURE 13.54

Plot of Forecasts from Winters' Method

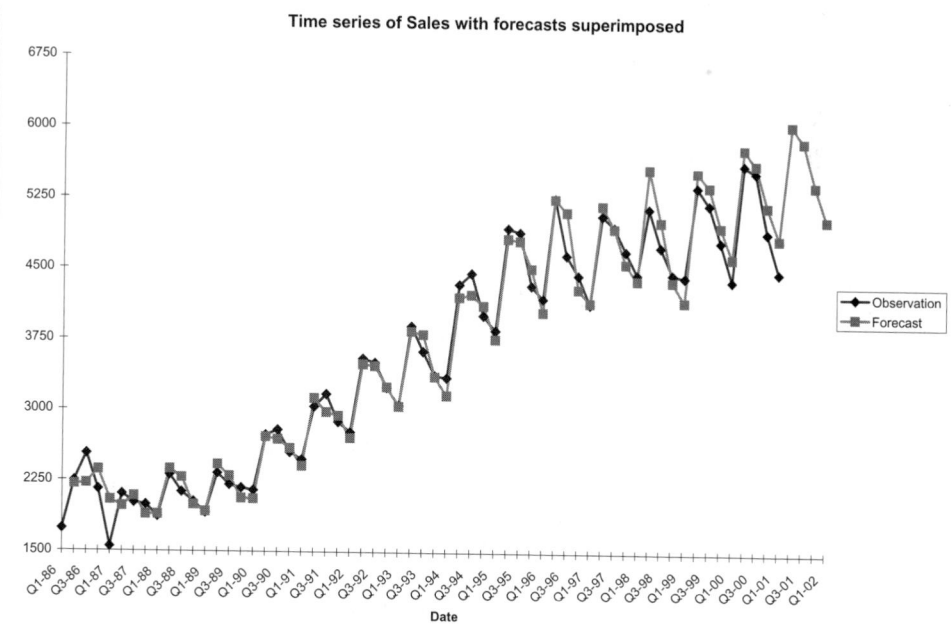

index gets updated only once per year.) To see how these smoothing constants would affect the results, we can simply substitute their values in the range B6:B8 of Figure 13.53. As we would expect, MAE, RMSE, and MAPE all get worse (they increase to 163, 217 and 5.43%, respectively, for the estimation period), but a plot of the forecasts superimposed on the original sales data still indicates a very good fit. ●

The three exponential smoothing methods we have examined are not the only ones available. For example, there are linear and quadratic models available in some software packages. These are somewhat similar to Holt's model except that they use only a single smoothing constant. There are also adaptive exponential smoothing models, where the smoothing constants themselves are allowed to change through time. Although these more complex models have been studied thoroughly in the academic literature and are used by some practitioners, they typically offer only marginal gains in forecast accuracy over the models examined here.

13.9.2 Deseasonalizing: The Ratio-to-Moving-Averages Method

You have all probably seen references to time series data that have been *deseasonalized*. In this section we will discuss why this is done and how it is done. We will also see how it can be used to forecast seasonal time series. First, data are often published in deseasonalized form so that readers can spot trends more easily. For example, if we see a time series of sales that has not been deseasonalized, and it shows a large increase from November to December, we might not be sure whether this represents a real increase in sales or a seasonal phenomenon (Christmas sales). However, if this increase is really just a seasonal effect, then the deseasonalized version of the series will show no such increase in sales.

Government economists and statisticians have a variety of sophisticated methods for deseasonalizing time series data, but they are all variations of the **ratio-to-moving-averages** method described here. This method is applicable when we believe that seasonality is multiplicative, as described in the previous section. Our job is to find the seasonal indexes, which can then be used to deseasonalize the data. For example, if we estimate the index for June to be 1.3, this means that June's values are typically about 30% larger than the average for all months. Therefore, to deseasonalize a June value, we *divide* it by 1.3 (to make it smaller). Similarly, if February's index is 0.85, then February's values are 15% below the average for all months. So to deseasonalize a February value, we again divide it by 0.85 (to make it larger).

Deseasonalizing	To **deseasonalize** an observation (assuming a multiplicative model of seasonality), *divide* it by the appropriate seasonal index.

To find the seasonal index for June 91 (or any other month) in the first place, we essentially divide June's observation by the average of the 12 observations surrounding June. (This is the reason for the term "ratio" in the name of the method.) There is one minor problem with this approach. June 91 isn't exactly in the middle of any 12-month sequence. If we use the 12 months from January 91 to December 91, June 91 is in the *first* half of the sequence; if we use the 12 months from December 90 to November 91, June 91 is in the *last* half of the sequence. Therefore, it is best to compromise by averaging the January-to-December and December-to-November averages. This is called a **centered** average. Then the seasonal index for June is June's observation divided by this centered average. The following equation shows more specifically how it works.

$$\text{Jun91 index} = \frac{\text{Jun91}}{\left(\frac{\text{Dec90}+\cdots+\text{Nov91}}{12} + \frac{\text{Jan91}+\cdots+\text{Dec91}}{12}\right)/2}$$

The only remaining question is how to combine all of the indexes for any specific month such as June. After all, if we have data for several years, the above procedure produces several June indexes, one for each year. The usual way to combine them is simply to average them. This single average index for June is then used to deseasonalize *all* of the June observations.

Once the seasonal indexes are obtained, we divide each observation by its seasonal index to deseasonalize the data. The deseasonalized data can then be forecasted by any of the methods we have described (other than Winters' method, which wouldn't make much sense). For example, we could use Holt's method or the moving average method to forecast the deseasonalized data. Finally, we "reseasonalize" the forecasts by *multiplying* them by the seasonal indexes.

As this description suggests, the method is not meant for hand calculations! However, it is straightforward to implement in StatPro, as we illustrate in the following example.

Example 13.7 Quarterly Sales at Coca-Cola (continued)

We return to the Coca-Cola sales data. (See the file COCACOLA.XLS.) Is it possible to obtain the same forecast accuracy with the ratio-to-moving-averages method as we obtained with Winters' method?

Objective To use the ratio-to-moving-averages method to deseasonalize the Coca-Cola data and then forecast the deseasonalized data.

Solution

The answer to this question depends on which forecasting method we use to forecast the *deseasonalized* data. The ratio-to-moving-averages method only provides a means for deseasonalizing the data and providing seasonal indexes. Beyond this, any method can be used to forecast the deseasonalized data, and some methods obviously work better than others. For this example, we compared two possibilities: the moving averages method with a span of 4 quarters, and Holt's exponential smoothing method optimized. However, we show the results only for the latter. Because the deseasonalized series still has a clear upward trend, we would expect Holt's method to do well, and we would expect the moving averages forecasts to lag behind the trend. This is exactly what occurred. For example, the values of MAPE for the two methods are 6.08% (moving averages) and 3.85% (Holt's). (To make a fair comparison with the Winters' method output for these data, we again held out 8 quarters. The MAPE values reported are for the non-holdout period.)

To implement this latter method in StatPro, we proceed exactly as before, but this time we check the "Use this deseasonalizing method" box in the dialog box in Figure 13.55 (page 746). Note that when the Holt's option is checked, the deseasonalizing option is enabled. When we check this option, we get a larger selection of optional charts. We can see charts of the deseasonalized data and/or the original "reseasonalized" data.

Selected outputs are shown in Figures 13.56–13.59. Figures 13.56 and 13.57 show the numerical output. In particular, Figure 13.57 shows the seasonal indexes from the ratio-to-moving averages method in column G. These are virtually identical to the seasonal indexes we found using Winters' method, although the methods are mathematically different. Column H contains the deseasonalized sales (column F divided by column G),

FIGURE 13.55
Checking the
Deseasonalizing
Option

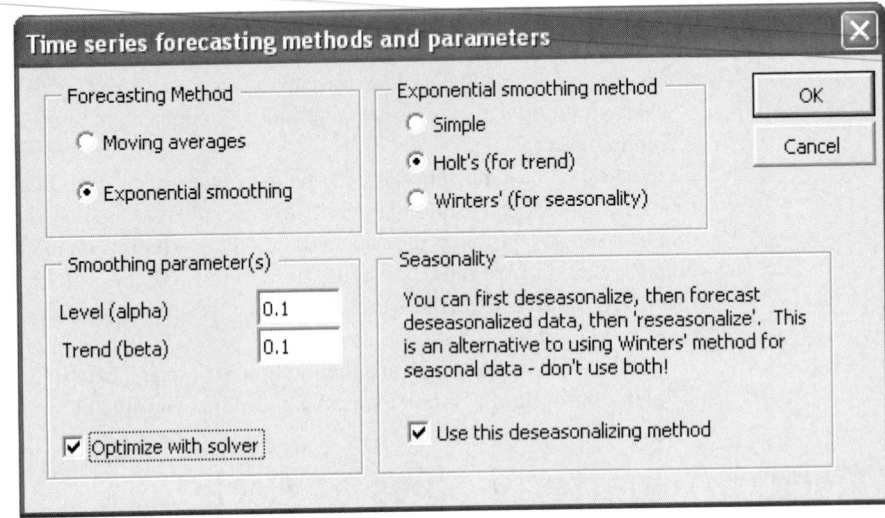

Time series forecasting methods and parameters

Forecasting Method
- ○ Moving averages
- ● Exponential smoothing

Exponential smoothing method
- ○ Simple
- ● Holt's (for trend)
- ○ Winters' (for seasonality)

[OK] [Cancel]

Smoothing parameter(s)

Level (alpha) 0.1
Trend (beta) 0.1

☑ Optimize with solver

Seasonality

You can first deseasonalize, then forecast deseasonalized data, then 'reseasonalize'. This is an alternative to using Winters' method for seasonal data - don't use both!

☑ Use this deseasonalizing method

FIGURE 13.56
Summary Measures for Forecast Errors

	A	B	C
1	*Forecasting results for Sales*		
2			
3	**Holt's exponential smoothing**		
4			
5	Smoothing constant(s)		
6	Level	0.925	
7	Trend	0.000	
8			
9	**Estimation period**		
10		Deseas	Actual
11	MAE	117.9743	117.9189
12	RMSE	163.6374	162.4921
13	MAPE	3.85%	3.85%
14			
15	**Holdout period**		
16		Deseas	Actual
17	MAE	187.2492	180.9577
18	RMSE	210.5482	414.2752
19	MAPE	3.73%	3.73%

columns I–L implement Holt's method on the deseasonalized data, and columns M and N are the "reseasonalized" forecasts and errors.

The deseasonalized data, with forecasts superimposed, appear in Figure 13.58. Here we see only the smooth upward trend with no seasonality, which Holt's method is able to track very well. Then Figure 13.59 shows the results of reseasonalizing. Again, the forecasts track the actual sales data very well. In fact, we see that the summary measures of forecast errors (in Figure 13.56, range C11:C13) are quite comparable to those from Winters' method. The reason is that both arrive at virtually the same seasonal pattern.

	Date	Observation	SeasIndex	DeseasObs	SmLevel	SmTrend	DeseasFCast	DeseasError	Forecast	Error
1	Date	Observation	SeasIndex	DeseasObs	SmLevel	SmTrend	DeseasFCast	DeseasError	Forecast	Error
2	Q1-86	1734.827	0.887	1955.059	1955.059	57.265				
3	Q2-86	2244.961	1.097	2046.287	2043.727	57.265	2012.324	33.963	2207.701	37.260
4	Q3-86	2533.805	1.055	2402.289	2379.577	57.265	2100.993	301.296	2216.014	317.791
5	Q4-86	2154.963	0.961	2242.857	2257.480	57.265	2436.843	-193.985	2341.346	-186.383
6	Q1-87	1547.819	0.887	1744.311	1787.310	57.265	2314.745	-570.434	2053.995	-506.176
7	Q2-87	2104.412	1.097	1918.177	1912.629	57.265	1844.575	73.601	2023.665	80.747
8	Q3-87	2014.363	1.055	1909.808	1914.338	57.265	1969.894	-60.086	2077.738	-63.375
9	Q4-87	1991.747	0.961	2072.984	2065.342	57.265	1971.603	101.381	1894.339	97.408
10	Q1-88	1869.050	0.887	2106.321	2107.549	57.265	2122.608	-16.286	1883.502	-14.452
11	Q2-88	2313.632	1.097	2108.881	2113.097	57.265	2164.814	-55.933	2374.996	-61.364
12	Q3-88	2128.320	1.055	2017.850	2029.347	57.265	2170.363	-152.512	2289.182	-160.862
13	Q4-88	2026.829	0.961	2109.497	2107.772	57.265	2086.612	22.885	2004.841	21.988
14	Q1-89	1910.604	0.887	2153.151	2154.047	57.265	2165.037	-11.887	1921.152	-10.548
15	Q2-89	2331.165	1.097	2124.863	2131.379	57.265	2211.312	-86.449	2426.008	-94.843
16	Q3-89	2206.550	1.055	2092.020	2099.303	57.265	2188.644	-96.625	2308.464	-101.914
17	Q4-89	2173.968	0.961	2262.638	2254.642	57.265	2156.569	106.069	2072.056	101.912
18	Q1-90	2148.278	0.887	2420.997	2412.774	57.265	2311.907	109.089	2051.477	96.801
19	Q2-90	2739.308	1.097	2496.886	2494.862	57.265	2470.039	26.847	2709.855	29.453
20	Q3-90	2792.754	1.055	2647.797	2640.586	57.265	2552.128	95.670	2691.847	100.907
54	Q1-99	4428.000	0.887	4990.124	4967.690	57.265	4692.509	297.615	4163.910	264.090
55	Q2-99	5379.000	1.097	4902.972			5024.955	-121.984	5512.827	-133.827
56	Q3-99	5195.000	1.055	4925.356			5082.221	-156.865	5360.453	-165.453
57	Q4-99	4803.000	0.961	4998.900			5139.486	-140.586	4938.077	-135.077
58	Q1-00	4391.000	0.887	4948.427			5196.751	-248.324	4611.351	-220.351
59	Q2-00	5621.000	1.097	5123.555			5254.017	-130.462	5764.128	-143.128
60	Q3-00	5543.000	1.055	5255.293			5311.282	-55.989	5602.054	-59.054
61	Q4-00	4903.000	0.961	5102.978			5368.547	-265.569	5158.162	-255.162
62	Q1-01	4479.000	0.887	5047.598			5425.813	-378.215	4814.610	-335.610
63	Q2-01		1.097				5483.078		6015.429	
64	Q3-01		1.055				5540.344		5843.656	
65	Q4-01		0.961				5597.609		5378.247	
66	Q1-02		0.887				5654.874		5017.868	

FIGURE 13.57 Ratio-to-Moving-Averages Output

FIGURE 13.58
Forecast Plot of
Deseasonalized
Series

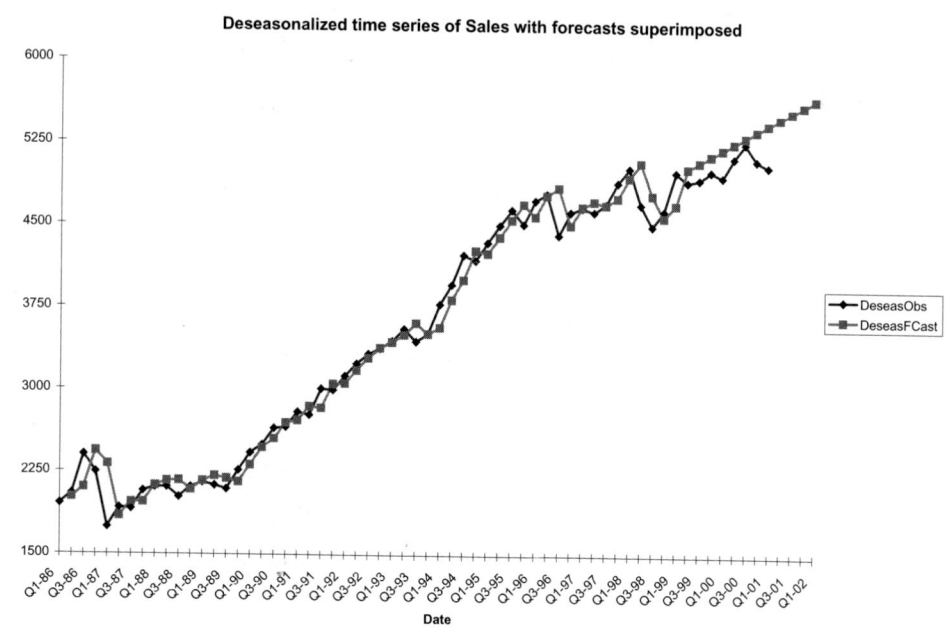

Deseasonalized time series of Sales with forecasts superimposed

FIGURE 13.59

Forecast Plot of
Reseasonalized
(Original) Series

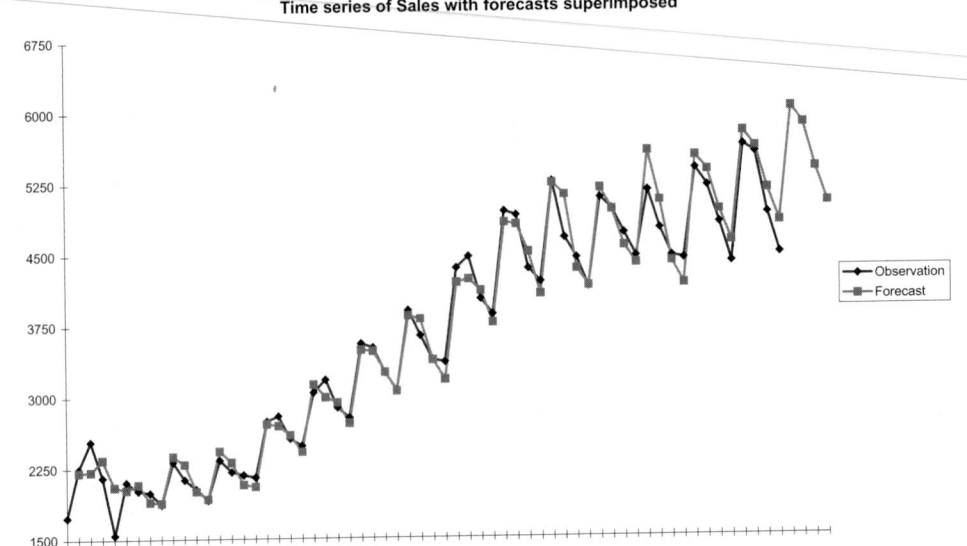

Time series of Sales with forecasts superimposed

13.9.3 Estimating Seasonality with Regression

We now examine a regression approach to forecasting seasonal data that uses dummy variables for the seasons. Depending on how we write the regression equation, we can create either an additive or a multiplicative seasonal model.

As an example, suppose that the data are quarterly data with a possible linear trend. Then we can introduce dummy variables Q_1, Q_2, and Q_3 for the first three quarters (using quarter 4 as the reference quarter) and estimate the additive equation

$$\hat{Y}_t = a + bt + b_1Q_1 + b_2Q_2 + b_3Q_3$$

Then the coefficients of the dummy variables, b_1, b_2 and b_3, indicate how much each quarter differs from the reference quarter, quarter 4.

For example, if the estimated equation is

$$\hat{Y}_t = 130 + 25t + 15Q_1 + 5Q_2 - 20Q_3$$

then the average increase from one quarter to the next is 25 (the coefficient of t). This is the trend effect. However, quarter 1 averages 15 units higher than quarter 4, quarter 2 averages 5 units higher than quarter 4, and quarter 3 averages 20 units lower than quarter 4. These coefficients indicate the effect of seasonality.

As discussed in Chapter 11, it is also possible to estimate a *multiplicative* model using dummy variables for seasonality (and possibly time for trend). Then we would estimate the equation

$$\hat{Y}_t = ae^{bt}e^{b_1Q_1}e^{b_2Q_2}e^{b_3Q_3}$$

or, after taking logs,

$$\ln \hat{Y}_t = \ln a + bt + b_1Q_1 + b_2Q_2 + b_3Q_3$$

One advantage of this approach is that it provides a model with *multiplicative* seasonal factors. It is also fairly easy to interpret the regression output, as illustrated in the following continuation of the Coca-Cola example.

Example 13.7 Quarterly Sales at Coca-Cola (continued)

Returning to the Coca-Cola sales data (see the file COCACOLA.XLS), does a regression approach provide forecasts that are as accurate as those provided by the other seasonal methods in this chapter?

Objective To use a multiplicative regression equation, with dummy variables for seasons and a time variable for trend, to model Coca-Cola sales.

Solution

We illustrate the multiplicative approach, although an additive approach is also possible. Figure 13.60 illustrates the data setup. Besides the Sales and Time variables, we need dummy variables for three of the four quarters (these were created manually), and a Log(Sales) variable. The other output in this figure will be discussed below. We then use multiple regression, with Log(Sales) as the response variable, and Time, Q1, Q2, and Q3 as the explanatory variables.

The regression output appears in Figure 13.61. (Again, to make a fair comparison with previous methods, we base the regression only on the data through quarter 1 of 1999. That

FIGURE 13.60
Data Setup for Multiplicative Model with Dummies

	A	B	C	D	E	F	G
1	Coca Cola quarterly sales (millions of dollars)						
2							
3	Quarter	Sales	Time	Q1	Q2	Q3	Log(Sales)
4	Q1-86	1734.83	1	1	0	0	7.45866
5	Q2-86	2244.96	2	0	1	0	7.71644
6	Q3-86	2533.80	3	0	0	1	7.83748
7	Q4-86	2154.96	4	0	0	0	7.67553
8	Q1-87	1547.82	5	1	0	0	7.34460
9	Q2-87	2104.41	6	0	1	0	7.65179
10	Q3-87	2014.36	7	0	0	1	7.60806
11	Q4-87	1991.75	8	0	0	0	7.59677
12	Q1-88	1869.05	9	1	0	0	7.53319
13	Q2-88	2313.63	10	0	1	0	7.74657
14	Q3-88	2128.32	11	0	0	1	7.66309
15	Q4-88	2026.83	12	0	0	0	7.61423

FIGURE 13.61
Regression Output for Multiplicative Model

	A	B	C	D	E	F	G	H
1	Results of multiple regression for Log(Sales)							
2								
3	Summary measures							
4		Multiple R	0.9696					
5		R-Square	0.9401					
6		Adj R-Square	0.9351					
7		StErr of Est	0.0890					
8								
9	ANOVA Table							
10		Source	df	SS	MS	F	p-value	
11		Explained	4	5.9594	1.4899	188.2487	0.0000	
12		Unexplained	48	0.3799	0.0079			
13								
14	Regression coefficients							
15			Coefficient	Std Err	t-value	p-value	Lower limit	Upper limit
16		Constant	7.4421	0.0333	223.3590	0.0000	7.3751	7.5091
17		Time	0.0214	0.0008	26.7027	0.0000	0.0197	0.0230
18		Q1	-0.0776	0.0343	-2.2645	0.0281	-0.1465	-0.0087
19		Q2	0.1391	0.0349	3.9827	0.0002	0.0689	0.2093
20		Q3	0.0954	0.0349	2.7337	0.0087	0.0252	0.1656

	A	B	C	D	E	F	G	H	I	J	K	L	M	N
1	Coca Cola quarterly sales (millions of dollars)													
2													Estimation period	
3	Quarter	Sales	Time	Q1	Q2	Q3	Log(Sales)	Forecast	AbsError	SqError	AbsPctError			
4	Q1-86	1734.83	1	1	0	0	7.45866	1612.94	121.885	14856	7.03%		MAE	223.348
5	Q2-86	2244.96	2	0	1	0	7.71644	2046.53	198.436	39377	8.84%		RMSE	294.8479
6	Q3-86	2533.80	3	0	0	1	7.83748	2001.30	532.508	283565	21.02%		MAPE	6.78%
7	Q4-86	2154.96	4	0	0	0	7.67553	1858.43	296.529	87930	13.76%			
8	Q1-87	1547.82	5	1	0	0	7.34460	1756.77	208.947	43659	13.50%		Holdout period	
9	Q2-87	2104.41	6	0	1	0	7.65179	2229.01	124.599	15525	5.92%			
10	Q3-87	2014.36	7	0	0	1	7.60806	2179.75	165.387	27353	8.21%		MAE	1035.368
11	Q4-87	1991.75	8	0	0	0	7.59677	2024.15	32.400	1050	1.63%		RMSE	1050.642
12	Q1-88	1869.05	9	1	0	0	7.53319	1913.41	44.364	1968	2.37%		MAPE	20.76%
13	Q2-88	2313.63	10	0	1	0	7.74657	2427.77	114.136	13027	4.93%			
14	Q3-88	2128.32	11	0	0	1	7.66309	2374.11	245.794	60415	11.55%			
56	Q1-99	4428.00	53	1	0	0	8.39570	4896.15	468.146	219160	10.57%			
57	Q2-99	5379.00	54	0	1	0	8.59026	6212.30	833.302	694392	15.49%			
58	Q3-99	5195.00	55	0	0	1	8.55545	6075.01	880.011	774419	16.94%			
59	Q4-99	4803.00	56	0	0	0	8.47700	5641.34	838.343	702819	17.45%			
60	Q1-00	4391.00	57	1	0	0	8.38731	5332.73	941.727	886850	21.45%			
61	Q2-00	5621.00	58	0	1	0	8.63426	6766.24	1145.243	1311581	20.37%			
62	Q3-00	5543.00	59	0	0	1	8.62029	6616.71	1073.710	1152852	19.37%			
63	Q4-00	4903.00	60	0	0	0	8.49760	6144.37	1241.373	1541006	25.32%			
64	Q1-01	4479.00	61	1	0	0	8.40716	5808.24	1329.238	1766873	29.68%			
65	Q2-01		62	0	1	0		7369.58						
66	Q3-01		63	0	0	1		7206.71						
67	Q4-01		64	0	0	0		6692.26						
68	Q1-02		65	1	0	0		6326.15						

FIGURE 13.62 Forecast Errors and Summary Measures

is, we again hold out the last 8 quarters.) Of particular interest are the coefficients of the explanatory variables. Recall that for a log response variable, these coefficients can be interpreted as *percentage* changes in the original sales variable. Specifically, the coefficient of Time means that deseasonalized sales increase by about 2.1% per quarter. Also, the coefficients of Q1, Q2, and Q3 mean that sales in quarters 1, 2, and 3 are, respectively, about 7.8% below, 13.9% above, and 9.6% above sales in the reference quarter, quarter 4. This pattern is quite comparable to the pattern of seasonal indexes we saw in previous models for these data.

To compare the forecast accuracy of this method with earlier models, we must go through several steps manually. (See Figure 13.62 for reference.) We first calculate the forecasts in column H by entering the formula

=EXP(MultRegr!$C$16+MMULT(Data!C4:F4,MultRegr!$C$17:$C$20))

in cell H4 and copying down. (This formula assumes the regression output is in a sheet named MultRegr. It uses Excel's MMULT function to sum the products of explanatory values and regression coefficients. You can replace this by "writing out" the sum of products if you like. The formula then takes EXP of the resulting sum to convert the log sales value back to the original sales units.) Next, we calculate the absolute errors, squared errors, and absolute percentage errors in columns I, J, and K, and we summarize them in the usual way, both for the estimation period and the holdout period, in column N.

Note that these summary measures are considerably larger for this regression model than for the previous seasonality models, especially in the holdout period. We can get some idea why the holdout period does so poorly by looking at the plot of observations versus forecasts in Figure 13.63. The multiplicative regression model with Time included really implies *exponential* growth (as in Section 13.4.2), with seasonality superimposed. However, Coca-Cola's growth tapered off in the last couple of years and did not keep up with the exponential growth curve. In short, the dummy variables do a good job of tracking seasonality, but the underlying exponential trend curve outpaces actual sales. We conclude that this regression model is *not* as good for forecasting Coca-Cola sales as Winters' method or Holt's method on the deseasonalized data.

FIGURE 13.63
Plot of Forecasts
for Multiplicative
Model

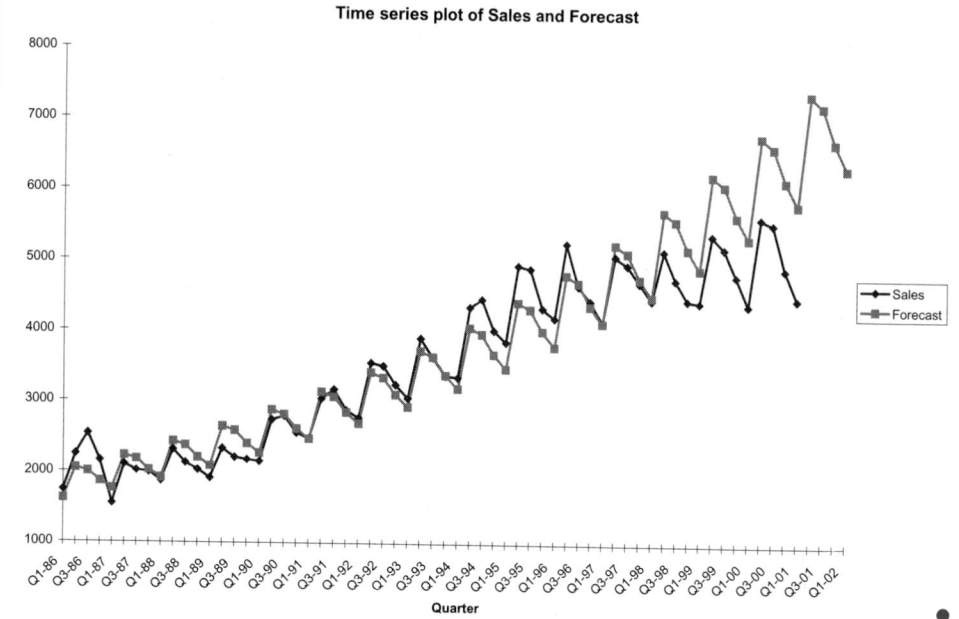

Time series plot of Sales and Forecast

This method of detecting seasonality by using dummy variables in a regression equation is always an option. The other variables included in the regression equation could be time t, lagged versions of Y_t, and/or current or lagged versions of other independent variables. These variables would capture any time series behavior other than seasonality. Just remember that there is always one less dummy variable than the number of seasons. If the data are quarterly, then three dummies are needed; if the data are monthly, then eleven dummies are needed. If the coefficients of any of these dummies turn out to be statistically insignificant, they can be omitted from the equation. Then the omitted terms are in effect combined with the reference season. For example, if the Q_1 term were omitted, then quarters 1 and 4 would essentially be combined and treated as the reference season, and the other two seasons would be compared to them through their dummy variable coefficients.

PROBLEMS

Level A

51. The University Credit Union is open Monday through Saturday. Winters' method is being used (with $\alpha = \beta = \gamma = 0.5$) to predict the number of customers entering the bank each day. After incorporating the arrivals of 16 October 2002, $L_t = 200$, $T_t = 1$, and the "seasonalities" are as follows: Monday, 0.90; Tuesday, 0.70; Wednesday, 0.80; Thursday, 1.1; Friday, 1.2; Saturday, 1.3. For example, the number of customers entering the bank on a typical Monday is 90% of the number of customers entering the bank on an average day. On Tuesday, 17 October 2002, 182 customers enter the bank. At the close of

business on 17 October 2002, forecast the number of customers who will enter the bank on 25 October 2002.

52. Last National Bank is using Winters' method (with $\alpha = 0.2$, $\beta = 0.1$, and $\gamma = 0.5$) to forecast the number of customers served each day. The bank is open Monday through Friday. At present, the following "seasonalities" have been estimated: Monday, 0.80; Tuesday, 0.90; Wednesday, 0.95; Thursday, 1.10; Friday, 1.25. A seasonality of 0.80 for Monday means that on a Monday, the number of customers served by the bank tends to be 80% of the average daily value. Currently, the level is estimated to be 20 customers, and the trend is estimated to equal 1 customer. After observing that 30 customers are served by the bank on Monday, forecast the number of customers who will be served by the bank on Wednesday.

53. Suppose that Winters' method is used to forecast quarterly U.S. retail sales (in billions of dollars). At the end of the first quarter of 2002, $L_t = 300$, $T_t = 30$, and the seasonal indexes are as follows: quarter 1, 0.90; quarter 2, 0.95; quarter 3, 0.95; quarter 4, 1.20. During the second quarter of 2002, retail sales are $360 billion. Assume $\alpha = 0.2$, $\beta = 0.4$, and $\gamma = 0.5$.
    a. At the end of the second quarter of 2002, develop a forecast for retail sales during the fourth quarter of 2002.
    b. At the end of the second quarter of 2002, develop a forecast for the second quarter of 2003.

54. The file P02_38.XLS contains monthly retail sales of U.S. liquor stores.
    a. Is seasonality present in these data? If so, characterize the seasonality pattern and then deseasonalize this time series using the ratio-to-moving-average method.
    b. If you decided to deseasonalize this time series in part a, forecast the deseasonalized data for each month of the next year using the moving average method with an appropriate span.
    c. Does Holt's exponential smoothing method, with optimal smoothing constants, outperform the moving average method employed in part b? Demonstrate why or why not.

55. Continuing Problem 54, how do your responses to the questions change when you employ Winters' method to handle seasonality in this time series? Explain. Which forecasting method do you prefer, Winters' method or a method used in the previous problem? Defend your choice.

56. The file P02_39.XLS contains monthly time series data for total U.S. retail sales of building materials (which includes retail sales of building materials, hardware and garden supply stores, and mobile home dealers).
    a. Is seasonality present in these data? If so, characterize the seasonality pattern and then deseasonalize this time series using the ratio-to-moving-average method.
    b. If you decided to deseasonalize this time series in part a, forecast the deseasonalized data for each month of the next year using the moving average method with an appropriate span.
    c. Does Holt's exponential smoothing method, with optimal smoothing constants, outperform the moving average method employed in part b? Demonstrate why or why not.

57. The file P02_40.XLS consists of the monthly retail sales levels of U.S. gasoline service stations.

a. Is there a seasonal pattern in these data? If so, how do you explain this seasonal pattern? Also, if necessary, deseasonalize these data using the ratio-to-moving-average method.
    b. Forecast this time series for the first 4 months of the next year using the most appropriate method for these data. Defend your choice of forecasting method.

58. The number of employees on the payroll at a food processing plant is recorded at the start of each month. These data are provided in the file P13_03.XLS.
    a. Is there a seasonal pattern in these data? If so, how do you explain this seasonal pattern? Also, if necessary, deseasonalize these data using the ratio-to-moving-average method.
    b. Forecast this time series for the first 4 months of the next year using the most appropriate method. Defend your choice of forecasting method.

59. Consider the file P02_37.XLS, which contains total monthly U.S. retail sales data. Compare the effectiveness of Winters' method with that of the ratio-to-moving-average method in deseasonalizing this time series. Using the deseasonalized time series generated by each of these two methods, forecast U.S. retail sales with the most appropriate method. Defend your choice of forecasting method.

60. Suppose that a time series consisting of 6 years (1998–2003) of quarterly data exhibits definite seasonality. In fact, assume that the seasonal indexes turn out to be 0.75, 1.45, 1.25, and 0.55.
    a. If the last four observations of the series (the four quarters of 2003) are 2502, 4872, 4269, and 1924, calculate the deseasonalized values for the four quarters of 2003.
    b. Suppose that a plot of the deseasonalized series shows an upward linear trend, except for some random noise. Therefore, a linear regression of this series versus time is estimated, and it produces the equation

$$\text{Predicted deseasonalized value} = 2250 + 51\text{Quarter}$$

Here the time variable "Quarter" is coded so that Quarter $= 1$ corresponds to first quarter 1998, Quarter $= 24$ corresponds to fourth quarter 2003, and the others fall in between. Forecast the actual (not deseasonalized) values for the four quarters of 2004.

61. The file P13_61.XLS contains monthly data on federal receipts of taxes. There are two variables: IndTax (taxes from individuals) and CorpTax (corporate taxes). For this problem, work only with IndTax.
    a. What evidence is there that seasonality is important in this series? Find seasonal indexes (by any method you like) and state briefly what they mean.

b. Forecast the next 12 months by using a linear trend on the seasonally adjusted data. State briefly the steps you use to get this type of forecast, give the final RMSE, MAPE, and forecast for the next month. Then show numerically how you could replicate this forecast, i.e., show on paper how the package uses its estimated model to get the next month forecast. (Substitute in specific numbers, and do the arithmetic.)

62. Quarterly sales for a department store over a 6-year period are given in the file P13_62.XLS.

    a. Use multiple regression to develop a model that can be used to predict future quarterly sales. (*Hint:* Use dummy variables and an explanatory variable for the quarter number, 1–24.)

    b. Letting Y_t be the sales during quarter number t, discuss how to fit the following model to these data.

$$Y_t = \alpha \beta_1^t \beta_2^{X_1} \beta_3^{X_2} \beta_4^{X_3}$$

    Here $X_1 = 1$ if t is a first quarter, 0 otherwise; $X_2 = 1$ if t is a second quarter, 0 otherwise; and $X_3 = 1$ if t is a third quarter, 0 otherwise.

    c. Interpret the answer to part **b.**

    d. Which model appears to yield better predictions for sales, the one in part **a** or the one in part **b?**

63. Confederate Express Service is attempting to determine how its shipping costs for a month depend on the number of units shipped during a month. The number of units shipped and total shipping cost for the last 15 months are given in the file P13_63.XLS.

    a. Determine a relationship between units shipped and monthly shipping cost.

    b. Plot the errors for the predictions in order of time sequence. Is there any unusual pattern?

    c. We have been told that there was a trucking strike during months 11–15, and we believe that this might have influenced shipping costs. How could the answer to part **a** be modified to account for the effect of the strike? After accounting for this effect, does the unusual pattern in part **b** disappear?

Level B

64. In our discussion of Winters' method, a monthly seasonality of 0.80 for January, say, means that during January, air conditioner (AC) sales are expected to be 80% of the sales during an average month. An alternative approach to modeling seasonality, called an **additive model,** is to let the seasonality factor for each month represent how far above average AC sales will be during the current month. For instance, if $S_{Jan} = -50$, then AC sales during January are expected to be 50 less than AC sales during an average month. (This is 50 ACs, not 50%.) If $S_{July} = 90$, then AC sales during July are expected to be 90 more than AC sales during an average month. Let

S_t = Seasonality for month t after observing month t demand

L_t = Estimate of level after observing month t demand

T_t = Estimate of trend after observing month t demand

Then the Winters' method equations given in the text should be modified as follows:

$$L_t = \alpha(I) + (1 - \alpha)(L_{t-1} + T_{t-1})$$
$$T_t = \beta(L_t - L_{t-1}) + (1 - \beta)T_{t-1}$$
$$S_t = \gamma(II) + (1 - \gamma)S_{t-12}$$

    a. Determine (I) and (II).

    b. Suppose that month 13 is January, $L_{12} = 30$, $T_{12} = -3$, $S_1 = -50$, and $S_2 = -20$. Let $\alpha = \gamma = \beta = 0.5$. Suppose 12 ACs are sold during month 13. At the end of month 13, what is the prediction for AC sales during month 14 using an additive model?

65. Winters' method assumes a multiplicative seasonality but an additive trend. For example, a trend of 5 means that the level will increase by 5 units per period. Suppose that there is actually a *multiplicative* trend. Then (ignoring seasonality) if the current estimate of the level is 50 and the current estimate of the trend is 1.2, we would predict demand to increase by 20% per period. So we would forecast the next period's demand to be 50(1.2) and forecast the demand 2 periods in the future to be $50(1.2)^2$. If we want to use a multiplicative trend in Winters' method, we should use the following equations (assuming a period is a month):

$$L_t = \alpha \left(\frac{Y_t}{S_{t-12}} \right) + (1 - \alpha)(I)$$
$$T_t = \beta(II) + (1 - \beta)\,T_{t-1}$$
$$S_t = \gamma \left(\frac{Y_t}{L_t} \right) + (1 - \gamma)S_{t-12}$$

    a. Determine (I) and (II).

    b. Suppose that we are working with monthly data and month 12 is December, month 13 is January, and so on. Also suppose that $L_{12} = 100$, $T_{12} = 1.2$, $S_1 = 0.90$, $S_2 = 0.70$, and $S_3 = 0.95$. Also, suppose $Y_{13} = 200$. At the end of month 13, what is the prediction for Y_{15} using $\alpha = \beta = \gamma = 0.5$ and a multiplicative trend?

66. Consider the file P02_37.XLS, which contains total monthly U.S. retail sales data. Does a regression approach for estimating seasonality provide forecasts that are as accurate as those provided by (i) Winters' method and (ii) the ratio-to-moving-average method? Compare the summary measures of forecast errors associated with each method for deseasonalizing the given time series. Summarize your findings after performing these comparisons.

67. The file P02_39.XLS contains monthly time series data for total U.S. retail sales of building materials (which includes retail sales of building materials, hardware and garden supply stores, and mobile home dealers). Does a regression approach for estimating seasonality provide forecasts that are as accurate as those provided by (i) Winters' method and (ii) the ratio-to-moving-average method? Compare the summary measures of forecast errors associated with each method for deseasonalizing the given time series. Summarize your findings after performing these comparisons.

 13.10 Conclusion

We have covered a lot of ground in this chapter. Because forecasting is such an important activity in business, it has received a tremendous amount of attention by both academicians and practitioners. All of the methods discussed in this chapter—and more—are actually used, often on a day-to-day basis. There is really no point in arguing which of these methods is best. All of them have their strengths and weaknesses. The most important point is that when they are applied properly, they have all been found to be useful in real business applications.

Summary of Key Terms

Term	Explanation	Excel	Page	Equation Number
Extrapolation methods	Forecasting methods where only past values of a variable (and possibly time itself) are used to forecast future values.		691	
Causal (or econometric) methods	Forecasting methods based on regression, where other time series variables are used as explanatory variables.		692	
Trend	A systematic increase or decrease of a time series variable through time.		694	
Seasonality	A regular pattern of ups and downs based on the season of the year, typically months or quarters.		694	
Cyclic variation	An irregular pattern of ups and downs caused by business cycles.		694	
Noise (or random variation)	The unpredictable ups and downs of a time series variable.		695	
Mean absolute error (MAE)	The average of the absolute forecast errors.	StatPro/ Forecasting	697	13.2
Root mean square error (RMSE)	The square root of the average of the squared forecast errors.	StatPro/ Forecasting	697	13.3
Mean absolute percentage error (MAPE)	The average of the absolute percentage forecast errors.	StatPro/ Forecasting	697	13.4
Runs test	A test of whether the forecast errors are random noise.	StatPro/Statistical Inference/Runs Test for Randomness	700	
Autocorrelations of residuals	Correlations of forecast errors with themselves, used to check whether they are random noise.	StatPro/ Summary Stats/ Autocorrelations	702	

Summary of Key Terms (continued)

Term	Explanation	Excel	Page	Equation Number
Correlogram	A bar chart of autocorrelations at different lags.	StatPro/ Summary Stats/ Autocorrelations	704	
Linear trend model	A regression model where a time series variable changes by a constant amount each time period.	StatPro/ Regression Analysis	706	13.6
Exponential trend model	A regression model where a time series variable changes by a constant percentage each time period.	StatPro/ Regression Analysis	709	13.7
Random walk model	A model indicating that the differences between adjacent observations of a time series variable are constant except for random noise.		715	13.9–13.11
Autoregression model	A regression model where the only explanatory variables are lagged values of the response variable.	StatPro/ Regression Analysis	718	
Moving averages model	A forecasting model where the average of several past observations is used to forecast the next observation.	StatPro/ Forecasting	723	
Span	The number of observations in each average of the moving averages model.	StatPro/ Forecasting	723	
Exponential smoothing models	A class of forecasting models where forecasts are based on weighted averages of previous observations, giving more weight to more recent observations.	StatPro/ Forecasting	729	
Smoothing constants	Constants between 0 and 1 that prescribe the weight attached to previous observations and hence the smoothness of the series of forecasts.	StatPro/ Forecasting	730	
Simple exponential smoothing	An exponential smoothing model useful for time series with no prominent trend or seasonality.	StatPro/ Forecasting	730	13.12, 13.13
Holt's method	An exponential smoothing model useful for time series with trend but no seasonality.	StatPro/ Forecasting	734	13.16–13.18
Winters' method	An exponential smoothing model useful for time series with seasonality (and possibly trend).	StatPro/ Forecasting	740	13.19–13.22
Deseasonalizing	A method for removing the seasonal component from time series data.	StatPro/ Forecasting	744	
Ratio-to-moving-averages method	A method for deseasonalizing a time series, so that some other method can then be used to forecast the deseasonalized series.	StatPro/ Forecasting	744	
Dummy variables for seasonality	A regression-based method for forecasting seasonality, where dummy variables are used for the seasons.	StatPro/ Regression analysis	748	

Conceptual Exercises

C.1. "A truly random series will likely have a very small number of runs." Is this statement true or false? Explain your choice.

C.2. Distinguish between a *correlation* and an *autocorrelation*. How are these measures similar? How are these measures different?

C.3. What is the relationship between the random walk model and the autoregression model?

C.4. Under what conditions would you prefer a simple exponential smoothing model to the moving averages method for forecasting a time series?

C.5. Is it more appropriate to use an *additive* or a *multiplicative* model to forecast seasonal data? Summarize the difference(s) between these two types of seasonal models.

Level A

68. The file P13_68.XLS contains quarterly revenues of Toys 'R Us. Discuss the seasonal and trend components of the growth of Toys 'R Us revenues. Also, use any reasonable forecasting method to forecast quarterly revenues for the next year. Explain your choice of forecasting method.

69. The file P13_69.XLS contains quarterly revenues and earnings per share (EPS) for the following companies: Mattel, McDonald's, Eli Lilly, General Motors, Microsoft, AT&T, Nike, GE, Coca-Cola, and Ford.
   a. For each company, use a regression model with trend and seasonal components to forecast revenues and EPS.
   b. For each company, use Winters' method to forecast revenues and EPS.
   c. For each company, which method appears to be more accurate?

70. The file P13_70.XLS contains the sales in (millions of dollars) for Sun Microsystems.
   a. Use these data to predict the company's sales for the next 2 years. You need consider only a linear and exponential trend, but you should justify the equation you choose.
   b. In words, how do your predictions of sales increase from year to year?
   c. Are there any outliers?

71. The file P13_71.XLS contains the sales in (millions of dollars) for Procter & Gamble.
   a. Use these data to predict Procter & Gamble sales for the next 2 years. You need consider only a linear and exponential trend, but you should justify the equation you choose.

b. Use your answer from part **a** to explain how your predictions of Procter & Gamble sales increase from year to year.
   c. Are there any outliers?
   d. We can be approximately 95% sure that Procter & Gamble sales in the year following next year will be between what two values?

72. The file P13_72.XLS lists the sales of Nike. Forecast sales in the next 2 years with a linear or exponential trend. Are there any outliers in your predictions for the observed period?

73. The file P12_44.XLS contains data on pork sales. Price is in dollars per hundred pounds sold, quantity sold is in billions of pounds, per capita income is in dollars, U.S. population is in millions, and GNP is in billions of dollars.
   a. Use these data to develop a regression equation that could be used to predict the quantity of pork sold during future periods. Is autocorrelation, heteroscedasticity, or multicollinearity a problem?
   b. Suppose that during each of the next two quarters, price is $45, U.S. population is 240, GNP is 2620, and per capita income is $10,000. (All of these are expressed in the units described above.) Predict the quantity of pork sold during each of the next 2 quarters.
   c. We expect our prediction of pork sales to be accurate within what value 68% of the time?
   d. Use Winters' method to develop a forecast of pork sales during the next 2 quarters.

74. The file P13_74.XLS contains data on a motel chain's revenue and advertising.
   a. Use these data and multiple regression to make predictions of the motel chain's revenues during the next 4 quarters. Assume that advertising during each of the next 4 quarters is $50,000. (*Hint:* Try using advertising, lagged by 1 quarter, as an explanatory variable.)
   b. Use simple exponential smoothing to make predictions for the motel chain's revenues during the next 4 quarters.
   c. Use Holt's method to make forecasts for the motel chain's revenues during the next 4 quarters.
   d. Use Winters' method to determine predictions for the motel chain's revenues during the next 4 quarters.
   e. Which of these forecasting methods would you expect to be the most reliable for these data?

75. The file P13_75.XLS contains data on monthly U.S. housing sales (in thousands of houses).
   a. Using Winters' method, find values of α, β, and γ that yield an RMSE as small as possible.

b. Although we have not discussed autocorrelation for smoothing methods, good forecasts derived from smoothing methods should exhibit no autocorrelation. Do the forecast errors for this problem exhibit autocorrelation?

c. At the end of the observed period, what is the forecast of housing sales during the next months?

76. Let Y_t be the sales during month t (in thousands of dollars) for a photography studio, and let P_t be the price charged for portraits during month t. The data are in the file P12_45.XLS. Use regression to fit the following model to these data:

$$Y_t = \alpha + \beta_1 Y_{t-1} + \beta_2 P_t + \epsilon_t$$

This equation indicates that last month's sales and the current month's price are explanatory variables. The last term, ϵ_t, is an error term.

a. If the price of a portrait during month 21 is $10, what would we predict for sales in month 21?

b. Does there appear to be a problem with autocorrelation, heteroscedasticity, or multicollinearity?

77. The file P13_77.XLS gives quarterly auto sales, GNP, interest rates and unemployment rates.

a. With all but the most recent 2 years of data, use regression to forecast auto sales. Carefully interpret the coefficients in your final equation.

b. Use all but the most recent 2 years of data to develop an exponential smoothing model to forecast future auto sales.

c. To *validate* your model, determine which model does the best job of forecasting for the most recent 2 years of data. It is usually recommended to hold back some of your data to validate any forecast model. This helps avoid "overfitting."

Level B

78. The file P13_78.XLS gives monthly exchange rates (dollars per unit of local currency) for 25 countries. Technical analysts believe that by charting past changes in exchange rates, it is possible to predict future changes of exchange rates. After analyzing the autocorrelations for these data, do you believe that technical analysis has potential?

79. The file P13_79.XLS contains 5 years of monthly data for a particular company. The first variable is Time (1–60). The second variable, Sales1, has data on sales of a product. Note that Sales1 increases linearly throughout the period, with only a minor amount of "noise." (The third variable, Sales2, will be discussed and used in the next problem.) For this problem use the Sales1 variable to see how the following forecasting methods are able to track a linear trend.

a. Forecast this series with the moving average method with various spans such as 3, 6, and 12. What can you conclude?

b. Forecast this series with simple exponential smoothing with various smoothing constants such as 0.1, 0.3, 0.5, and 0.7. What can you conclude?

c. Now repeat part **b** with Holt's exponential smoothing method, again for various smoothing constants. Can you do significantly better than in parts **a** and **b?**

d. What can you conclude from your findings in parts **a, b,** and **c** about forecasting this type of series?

80. The Sales2 variable in the file from the previous problem was created from the Sales1 variable by multiplying by monthly seasonal factors. Basically, the summer months are high and the winter months are low. This might represent the sales of a product that has a linear trend and seasonality.

a. Repeat parts **a, b,** and **c** from the previous problem to see how well these forecasting methods can deal with trend *and* seasonality.

b. Now use Winters' method, with various values of the three smoothing constants, to forecast the series. Can you do much better? Which smoothing constants work well?

c. Use the ratio-to-moving-average method, where you first do the seasonal decomposition and then forecast (by any appropriate method) the deseasonalized series. Does this do as well as, or better than, Winters' method?

d. What can you conclude from your findings in parts **a, b,** and **c** about forecasting this type of series?

81. The file P13_81.XLS contains monthly time series data on federal expenditures in various categories. All values are in billions of current dollars. The variables are:

- Defense: expenditures on national defense
- Science: expenditures on science, space, and technology
- Energy: expenditures on energy
- Environ: expenditures on natural resources and environment
- Trans: expenditures on transportation

Analyze the Science variable by (i) simple exponential smoothing, (ii) Holt's method, (iii) simple exponential smoothing on the trend-adjusted data (the residuals from regressing linearly versus time), and (iv) moving averages on the adjusted or unadjusted data. Experiment with the smoothing constants [or span in (iv)], or use the optimize feature. Do any of these methods produce significantly better fits than the others as measured by RMSE or MAPE?

82. The data in the file P13_82.XLS represent annual changes in the average surface air temperature of the earth. (The source doesn't say exactly how this was measured.) A look at the time series shows a gradual upward trend, starting with negative values and ending with (mostly) positive values. This might be used to support the theory of global warming.

a. Is this series a random walk? Explain.

b. Regardless of your answer in part **a,** use a random walk model to forecast the next value of the series. What is your forecast, and what is an approximate 95% forecast interval?

c. Forecast the series in three ways: (i) simple exponential smoothing ($\alpha = 0.35$), (ii) Holt's method ($\alpha = 0.5$, $\beta = 0.1$), and (iii) simple exponential smoothing ($\alpha = 0.3$) on trend-adjusted data, that is, the residuals from regressing linearly versus time. (These smoothing constants are close to "optimal.") For each of these, list the MAPE, the RMSE, and the forecast for next year. Also, comment on any "problems" with forecast errors from any of these three approaches. Finally, compare the "qualitative" features of the three forecasts (for example, how do their short-run or longer-run forecasts differ?). Is any one of the methods clearly superior to the others?

d. Does your analysis predict convincingly that global warming would occur during the observed years? Explain.

13.1 Arrivals at the Credit Union

The Indiana University Credit Union Eastland Plaza Branch was having trouble getting the correct staffing levels to match customer arrival patterns. On some days, the number of tellers was too high relative to the customer traffic, so that tellers were often idle. On other days, the opposite occurred. Long customer waiting lines formed because the relatively few tellers could not keep up with the number of customers. The credit union manager, James Chilton, knew that there was a problem, but he had little of the quantitative training he believed would be necessary to find a better staffing solution. James figured that the problem could be broken down into three parts. First, he needed a reliable forecast of each day's number of customer arrivals. Second, he needed to translate these forecasts into staffing levels that would make an adequate trade-off between teller idle-

ness and customer waiting. Third, he needed to translate these staffing levels into individual teller work assignments—who should come to work when.

The last two parts of the problem require analysis tools (queueing and scheduling) that we have not covered. However, you can help James with the first part—forecasting. The file CREDITUNION.XLS lists the number of customers entering this credit union branch each day of the past year. It also lists other information: the day of the week, whether the day was a staff or faculty payday, and whether the day was the day before or after a holiday. Use this data set to develop one or more forecasting models that James could use to help solve his problem. Based on your model(s), make any recommendations about staffing that appear reasonable.

13.2 Forecasting Weekly Sales at Amanta

Amanta Appliances sells two styles of refrigerators at over 50 locations in the Midwest. The first style is a relatively expensive model, whereas the second is a standard, less expensive model. Although weekly demand for these two products is fairly stable from week to week, there is enough variation to concern management at Amanta. There have been relatively unsophisticated attempts to forecast weekly demand, but they haven't been very successful. Sometimes demand (and the corresponding sales) are lower than forecasted, so that inventory costs are high. Other times the forecasts are too low. When this happens and on-hand inventory is not sufficient to meet customer demand, Amanta requires expedited shipments to keep customers happy—and this nearly wipes out Amanta's profit margin on the expedited units.[7] Profits at Amanta would almost certainly increase if demand could be forecasted more accurately.

Data on weekly sales of both products appear in the file AMANTA.XLS. A time series chart of the two sales variables indicates what Amanta management expected—namely, there is no evidence of any upward or downward trends or of any seasonality. In fact, it might appear that each series is an unpredictable sequence of random ups and downs. But is this really true? Is it possible to forecast either series, with some degree of accuracy, with an extrapolation method (where only past values of *that* series are used to forecast current and future values)? What method appears to be best? How accurate is it? Also, is it possible, when trying to forecast sales of one product, to somehow incorporate current or past sales of the *other* product in the forecast model? After all, these products might be "substitute" products, where high sales of one go with low sales of the other, or they might be complementary products, where sales of the two products tend to move in the *same* direction.

[7]Because Amanta uses expediting when necessary, its sales each week are equal to its customer demands. Therefore, we use the terms *demand* and *sales* interchangeably.

Analysis of Variance and Experimental Design

Using Controlled Experiments in Business

Can statistical principles and careful experimentation lead to improved products and lower costs? They certainly can, argues Rita Koselka in the article "The New Mantra: MVT" in *Forbes* magazine [Roselka (1996)]. Most products are a result of several controllable inputs. The question is, Which combination of input settings results in the best quality and lowest cost? This is where experimentation enters the picture. For years, companies experimented (if they experimented at all) by changing the level of one input at a time. This one-at-a-time method of experimentation is not only costly and time-consuming, but it often fails to identify the best *combination* of input settings. As we will discuss in Section 14.5, the input factors often interact, so that the best setting of one factor might depend on the settings of other factors, and one-at-a-time testing will probably not discover this fact. A better alternative is to test multiple factors *simultaneously*. This is called **multiple variable testing,** or **MVT,** and it is quickly becoming regarded as one of the most important statistical techniques for product improvement. If you have never heard of MVT, you probably will. It is a natural outgrowth of the quality control movement that has been so pervasive in the past two decades. As the article states, "It [MVT] doesn't just tell you how to raise the quality of your output. It tells you how to do that cost-effectively."

The potential improvements with MVT can occur in traditional manufacturing and service industries. The following are several examples discussed in the article.

- Five years ago a subsidiary of Raychem Corp. called Elo TouchSystems was losing $3 million annually making touch-sensitive computer screens for products like automatic teller machines. The problem was a bubbling between the screen and the coating, and it resulted in a disastrous 25% reject rate. Raychem had spent 18 months on quality

© James Nazz/CORBIS

improvement efforts, but they hadn't worked. Then the company hired statistical consultants who experimented with MVT. Their solution, which would never have surfaced with one-at-a-time testing, was to change three things at the same time: the type of polyester, the coversheet shaping process, and the adhesive. Within months, the reject rate had decreased to less than 1%, many fewer quality inspectors were required, and the company was breaking even. Today, it is making $15 million on $50 million in sales.

- For years, Boise Cascade had been experimenting, with limited success, with small variations in its pulping process at a Louisiana paper mill. After using MVT with eight variables, it came to the counterintuitive conclusion that the mill could maintain its paper quality while switching to a cheaper grade of wood. The result was paper of at least as high a quality level and a savings of $3 million per year.

- Saint Luke's Hospital in Kansas City was concerned about the misuse of warfarin, an anti-bloodclotting drug that can be fatal if used improperly. In 1992 the hospital worked with statistical consultants to experiment with ways to keep patients from misusing the drug. They tested seven variables to better educate patients and provide emergency access to nurses. They found that having a standardized instruction sheet and having the pharmacist discuss the drug with patients yielded a 68% improvement in patient understanding of how to use the drug appropriately.

- A shoe company selling sneakers in over 100 stores was considering a proposal to increase sales with a costly, high-tech display. Before doing so, it hired statistical consultants, who persuaded the company to experiment with a whole range of possible changes: in sales techniques, advertising, separation of shoes by color, and various discounts, as well as the displays. The findings were surprising. Although the new display had the potential to increase sales, its impact would not be nearly as great as a simple combination of using the old display case and arranging the shoes by color. Using this suggestion, the company did not spend money on new displays, and it was still able to increase sales by 33%.

- In 1994 Southwestern Bell was having problems. At the same time it was trying to reduce errors on service orders, it was getting complaints about response times to its service calls from the Texas state government in Austin, one of its biggest customers. Statistical consultants prescribed simultaneous 5-month tests of 15 changes in the way the company did business. These included (1) whether the company should continue to have weekly meetings for its customer service representatives, (2) whether it should adopt performance incentives, (3) whether the service reps or a separate typing pool should type in orders, (4) whether the service reps or lower-paid clerks should answer the phones, (5) whether the length of employee training should be changed, and so on. The results of the experiment were that the weekly meetings should be retained, the clerks should do the typing but not the phone-answering, and employee training should be shortened.

MVT is not a new scientific method. Statisticians have studied and applied experimental designs for years, particularly in the physical sciences. However, it is relatively new to business. For the most part, the well-known strategic consultants and the big accounting firms have been strangers to MVT. Fortunately, this is changing. For example, experimental design is now being taught formally at Motorola University in Illinois. As Roselka concludes, "The design of experiments involves some cleverness. It may cost a lot of money to shut down a production line and rearrange it; it may take precious months for a billing department or a mail-order operation to see whether a novel way of doing things will pay off. MVT is the science of gleaning the most amount of information from the least amount of costly testing." ●

14.1 Introduction

One of the most frequent applications of statistics is the comparison of several populations on some characteristic. We discussed the simplest version of this comparison problem in Sections 9.7.1 and 10.4.2 when we discussed the two-sample procedure for analyzing the difference between two population means. A natural extension is to *more* than two population means, which is the topic of this chapter. The resulting procedure is commonly called **analysis of variance,** or ANOVA. Actually, we discussed a special case of this method, one-way ANOVA, briefly in Chapter 10. We will now discuss it in more depth. For completeness, we will repeat some of the discussion from Chapter 10 here.[1]

There are two typical situations where ANOVA is used. The first is when there are several distinct populations. For example, consider recent graduates with BS degrees in one of three disciplines: Business, Engineering, and Computer Science. We might sample randomly from each of these populations to discover whether there are any significant differences among them with respect to mean starting salary. A second situation where ANOVA is used is in randomized experiments. In this case a *single* population is treated in one of several ways. For example, a pharmaceutical company might select a group of people who suffer from allergies and randomly assign each person to a different type of allergy medicine currently being developed. Then the question is whether any of the treatments differ from one another with respect to the mean amount of symptom relief.

These two examples illustrate two basic situations where ANOVA is used. The comparison of recent graduates is called an **observational study.** In this case we analyze the data that are already available to us, that is, the starting salaries of recent graduates from the three disciplines. Unfortunately we don't first get to choose which students should major in which disciplines. It might be nice to do so because it would help us to rule out other possible causes besides discipline, such as unequal academic abilities, that might affect starting salaries. But we don't get to make these choices. The students themselves choose their disciplines, and all we can do is analyze the resulting data on starting salaries.

> **Observational Study**
>
> In an **observational study,** we analyze data already available to us. The disadvantage is that it is difficult or impossible to rule out factors over which we have no control for the effects we observe.

In contrast, the allergy example illustrates a **designed experiment.** The researchers in this example are interested in whether different allergy medicines cause different amounts of symptom relief. Therefore, they will select the subjects for the experiment so that the subjects receiving one allergy medicine are as much alike, in every characteristic that might matter—age, medical history, gender mix, and so on—as the subjects receiving any other allergy medicine. In this way, if there are any differences across groups with respect to symptom relief, the researchers will be able to attribute the differences to the types of medicine, not some extraneous factor.

> **Designed Experiment**
>
> In a **designed experiment,** we control for various factors such as age, gender, or socioeconomic status so that we can learn more precisely what is responsible for the effects we observe.

---

[1] We included the ANOVA section in Chapter 10 for those who want only a quick introduction to the topic. For those who want a more in-depth treatment, we advise skipping the section in Chapter 10 and studying this chapter instead.

It should be clear from this discussion that designed experiments are generally preferable to observational studies. In a carefully designed experiment, where we can "control for" extraneous factors such as age or gender that are not of direct interest, we can be fairly sure that any differences across groups with respect to some measurement variable are due to the variables that we purposely manipulate. This ability to infer causal relationships is never possible with observational studies. For example, if recent Business graduates are found to make more, on average, than Computer Science graduates, we can never be sure whether this is a result of being a *Business* graduate rather than a *Computer Science* graduate or whether, say, it is due to the fact that the Business graduates in our study have more work experience than the Computer Science graduates. We didn't control for work experience, so we cannot rule out the possibility that it might have had an effect.

ANOVA has been used in many disciplines. In fact, it began in agricultural studies, where researchers wanted to learn, for example, which types of wheat produce the greatest yield per acre. Because the results from such an experiment can take many months to obtain, the agricultural researchers have had to design their experiments very carefully, so that they could obtain the most *information* from the resulting data. This idea of obtaining the most useful information from a limited amount of data continues to be crucial in ANOVA studies and has spawned a whole area of scientific research called **experimental design.** The essential goal of experimental design is to decide which observations to make, given a limited budget (in time and/or money), to maximize the chances of seeing differences across groups that actually exist. For example, the allergy researchers want to design their experiment so that if there really are differences across medicine types, the analysis will have a good chance of detecting them.

Experimental Design

Experimental design is the science (and art) of setting up an experiment so that we obtain the most information for the time and money involved.

We will concentrate on the most common and basic experimental designs in this chapter, leaving more complex designs to specialized textbooks. However, because our audience is mostly *business* students, it is important to note that the use of designed experiments in business situations is probably less prevalent than in, say, medicine or agriculture. Business managers do not always have the luxury of being able to design a controlled experiment for obtaining data. Instead, they often have to rely on whatever data are available, that is, observational data. Nevertheless, as the introductory vignette to this chapter attests, there are many potentially profitable uses of experimental design in the business world, and many intelligent companies are beginning to use experimentation for competitive advantage.

It is no coincidence that some of this terminology is the same as that used in regression analysis. We will see why later in this chapter, when we investigate the relationship between ANOVA and regression.

Before proceeding, there is some general terminology we should introduce. In all of our examples, there is a variable of primary interest that we wish to measure. It is called the **response variable** (or sometimes the **dependent** or **criterion** variable) and is the variable we measure to detect differences among groups. The groups themselves are determined by one or more **factors** (sometimes called **independent** or **explanatory variables**), each varied at several **levels** (or **treatment levels**). The number of factors determines the type of ANOVA. If there is a single factor, we refer to **one-way ANOVA;** if there are two factors, we refer to **two-way ANOVA;** if there are three factors, we refer to **three-way ANOVA;** and so on. The only types we will discuss in this book are the two most common types, one-way and two-way ANOVA. It is best to think of a factor as a categorical variable, with the possible categories being its levels. Finally, the "entities" measured at each treatment level (or combination of levels) are called **experimental units.** Some examples will help to clarify this terminology.

One-Way ANOVA	In **one-way ANOVA,** a single response variable is measured at various levels of a single factor. Each experimental unit is assigned to one of these levels.

Two-Way ANOVA	In **two-way ANOVA,** a single response variable is measured at various combinations of the levels of two factors. Each experimental unit is assigned to one of these combinations of levels.

Consider the observational study on graduates of Business, Engineering, and Computer Science. The response variable is starting salary, the experimental units are the individual graduates, and the *single* factor is the student's major discipline. This factor has three levels: Business, Engineering, and Computer Science, and each student is "assigned" to one of these levels. If we also wanted to see how gender affects starting salary, we could introduce a second factor, gender, at the two levels "male" and "female." Then each student would be "assigned" to one of the combinations of levels, such as a female in Business.

For the study on allergy medicines, the response variable is the amount of relief from allergy symptoms, the experimental units are the individual patients, and the single factor is medicine type. Its levels are the various types of medicines used in the experiment. Each patient in the experiment receives exactly one of these types of medicine. We could also introduce a second factor here. For example, if we purposely wanted to see whether age has an effect on the response variable (or on which medicine type is most effective), we could introduce a second factor, age, with levels such as 5 to 20, 20 to 35, 35 to 50, and 50 or older. Then each patient would be at some combination of the two factors, such as a person 20 to 35 years old receiving the second type of medicine.

Although the experimental units in both of these two examples are people, this is not always the case. Suppose a company wants to see whether five different shelf layouts for its product lead to different levels of sales. The company could choose a sample of 50 supermarkets that sell its product, try the first layout in 10 of them, the second layout in another 10, and so on. Then the response variable is sales level, the single factor is shelf layout, varied at five levels, and the experimental units are the 50 supermarkets. Note that in this example, each of the experimental units, that is, each supermarket, is chosen (probably in some random way) to "receive" one of the five treatments and each treatment is applied to a separate subset of 10 supermarkets. When there are an equal number of experimental units assigned to each treatment level (or combination of levels, for a two-factor or multi-factor design), we call this a **balanced design.** Balanced designs are somewhat easier to analyze, and we prefer them whenever possible. In fact, the only two-factor design we will discuss in this book is a balanced design.

Balanced Design	In a **balanced design,** an equal number of experimental units is assigned to each combination of treatment levels.

One-Way ANOVA

We begin our discussion with the simplest design to analyze, the one-factor design. As discussed in the introduction, there are basically two situations. First, the data could be observational data, in which case the levels of the single factor might best be considered as "subpopulations" of an overall population—graduates of Business, Engineering, and Computer Science, for example. Second, the data could be generated from a designed experiment, where a single population of experimental units, allergy patients, say, is treated by different types of allergy medicine. Fortunately, the data analysis is basically the same in either case. We normally ask two questions. First, are there any significant differences in the mean of the response variable across the different groups? If the answer to this question is "yes," then we typically ask the second question: Which of the groups differs significantly from which others, again with respect to the mean of the response variable?

14.2.1 The Equal-Means Test

We set up the first question as a hypothesis test. Let J be the number of levels of the single factor, and let μ_j be the mean of the response variable for level j. (As usual, this Greek letter is used as a "population" mean, the mean of the response variable if *all* experimental units received treatment level j.) The null hypothesis is that there are no differences in population means across treatment levels:

$$H_0: \quad \mu_1 = \mu_2 = \cdots = \mu_J$$

The alternative is then the opposite, namely, that at least one pair of μ's are not equal. If we can reject this null hypothesis at some typical level of significance (usually the 5% or 10% level), then we hunt further to see which means are different from which others. To do this, we typically calculate confidence intervals for differences between pairs of means and see which of these confidence intervals do *not* include zero. For example, if the confidence interval for the difference $\mu_2 - \mu_4$ extends from 5.35 to 9.31, we would conclude that μ_2 and μ_4 are *not* equal (and that μ_2 is in fact larger than μ_4).

This is the general plan. Now we'll see how to put it into action. First, we ask an obvious question: If ANOVA is basically a test of differences between means, why is it called analysis of *variance* and not analysis of *means?* The answer to this question is the key to the procedure. Consider the boxplot in Figure 14.1. It corresponds to observations from four treatment levels with slightly different means and fairly large variances. (The large variances are indicated by the relatively wide boxes and long lines extending from them.) From these boxplots, would you conclude that the population means differ across the four treatment levels? Would your answer change if the data were instead as in Figure 14.2?[2] We expect that it would.

The sample means in these two figures are virtually the same, but the variation *within* each treatment level in Figure 14.1 is quite large relative to the variation *between* the sample means. In contrast, there is very little variation within each treatment level in Figure 14.2. In the first case, the large "within" variation makes it difficult to infer whether there are really any differences across population means, whereas the small "within" variation in the second case makes it clearer that differences across population means probably exist.

This is the essence of the ANOVA procedure. We compare variation *within* the individual treatment levels to variation *between* the sample means. Only if the between vari-

---

[2]Note that we keep the horizontal scale the same in both charts for a fair comparison.

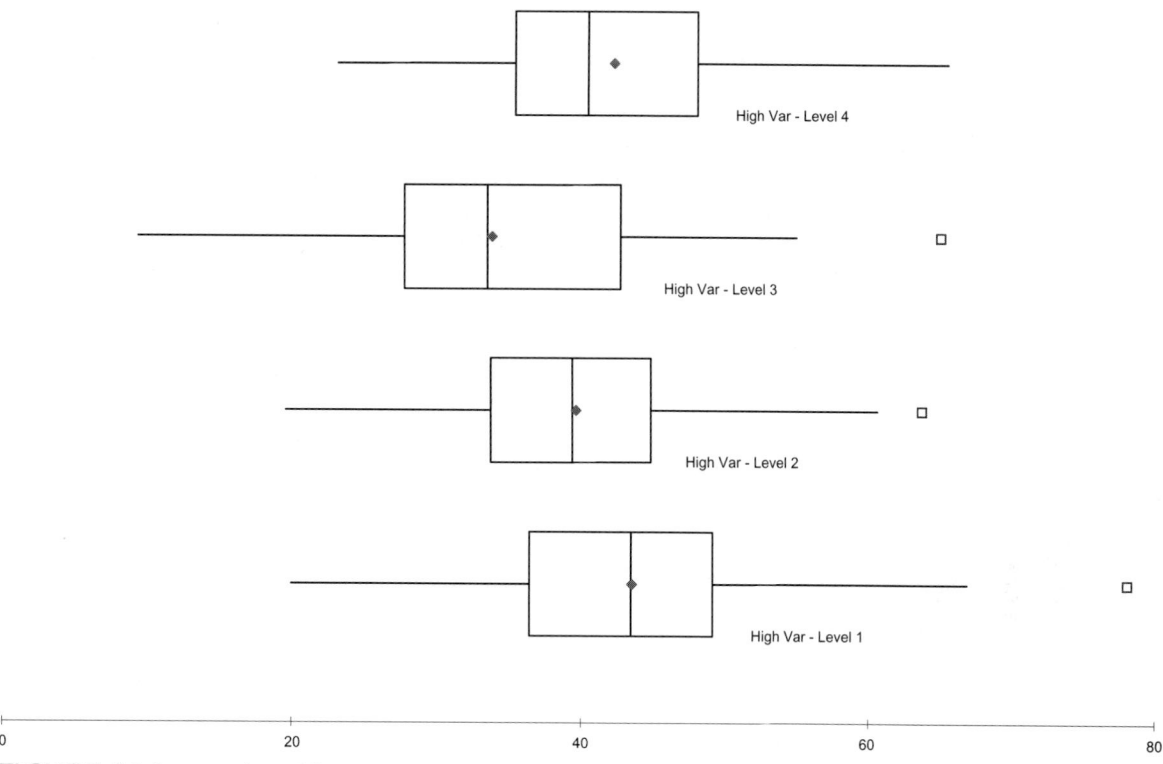

FIGURE 14.1 Samples with Large Within Variation

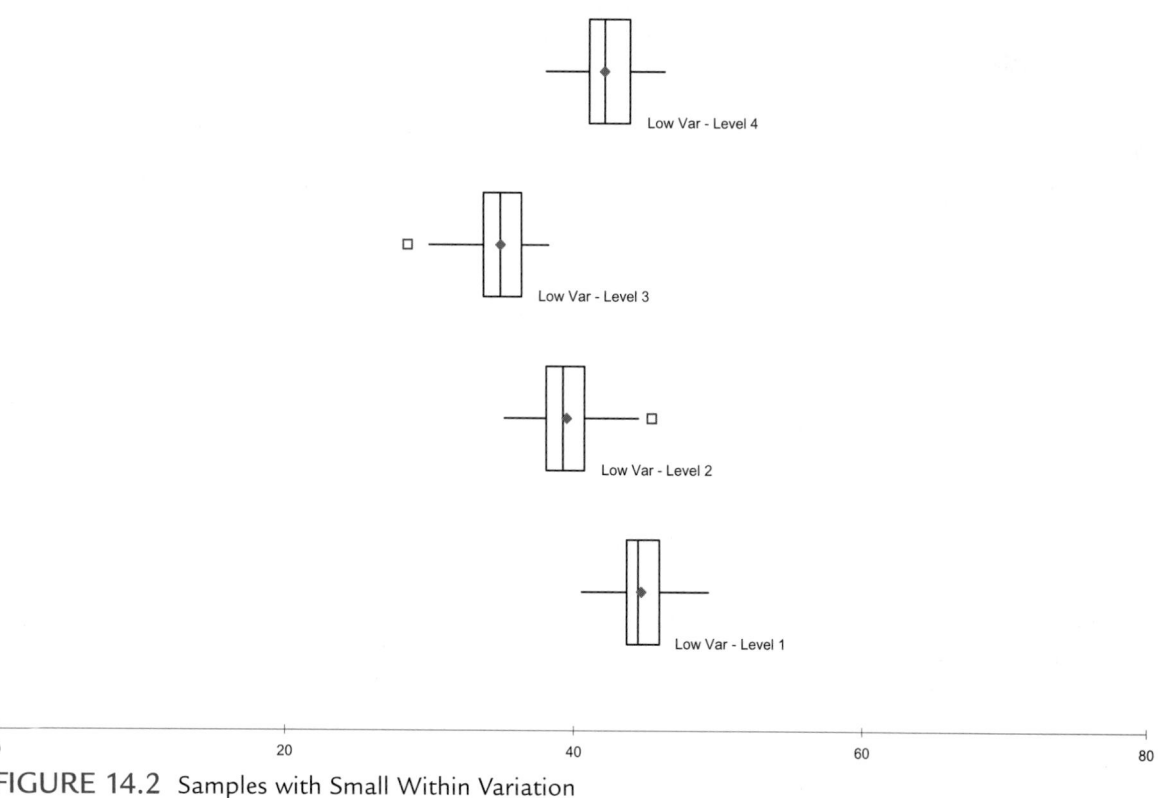

FIGURE 14.2 Samples with Small Within Variation

ation is large relative to the within variation can we conclude with any assurance that there are differences across population means—and reject the equal-means hypothesis.

The test itself is based on two assumptions: (1) The population variances are all equal to some common variance σ^2, and (2) the populations are normally distributed. These are analogous to the assumptions we made for the two-sample procedures in Chapters 9 and 10. Although these assumptions are never satisfied exactly in any real application, we should keep them in mind and check for gross violations whenever possible. Fortunately, the test we present is fairly robust to violations of these assumptions, particularly when the sample sizes are large and roughly the same.

> Remember that a *robust* test is one in which the conclusions are approximately valid even when the assumptions behind it are violated to some extent.

To run the test, let $\bar{Y}_j$, s_j^2, and n_j be the sample mean, sample variance, and sample size from treatment level j. Also, let n and $\bar{Y}$ be the combined number of observations and the sample mean of all n observations. (We call $\bar{Y}$ the **grand mean.**) Then a measure of the between variance is *MSB* (mean square between), given in equation (14.1). Note that *MSB* is large if the individual sample means differ substantially from the grand mean $\bar{Y}$, and this occurs only if they differ substantially from one another.

Measure of Between Variation

$$MSB = \frac{\sum_{j=1}^{J} n_j(\bar{Y}_j - \bar{\bar{Y}})^2}{J - 1}$$ (14.1)

A measure of the within variance is *MSW* (mean square within), given in equation (14.2). This value is really just a weighted average of the individual sample variances, where the sample variance s_j^2 receives weight $(n_j - 1)/(n - J)$. In fact, *MSW* is the average of the sample variances if the sample sizes, the n_j's, are equal. In this sense, *MSW* is a pooled estimate of the (assumed) common variance σ^2, just as in the two-sample procedures from Chapters 9 and 10. Therefore, *MSW* is large if the individual sample variances are large. For example, *MSW* is much larger in Figure 14.1 than in Figure 14.2. However, *MSB* is the same in both figures.

Measure of Within Variation

$$MSW = \frac{\sum_{j=1}^{J} (n_j - 1)s_j^2}{n - J}$$ (14.2)

The numerators of equations (14.1) and (14.2) are called **sums of squares** (often labeled *SSB* and *SSW*), and the denominators are called **degrees of freedom** (often labeled *dfB* and *dfW*). As we will see, they are always reported in ANOVA output. Finally, the ratio of the mean squares is the test statistic we use, the *F*-ratio in equation (14.3). Under the null hypothesis of equal population means, this test statistic has an *F* distribution with *dfB* and *dfW* degrees of freedom. If the null hypothesis is *not* true, then we would expect *MSB* to be large relative to *MSW*, as in Figure 14.2. Therefore, the *p*-value for the test is found by finding the probability to the *right* of the *F*-ratio in the *F* distribution with *dfB* and *dfW* degrees of freedom.

F-ratio for ANOVA Test

$$F\text{-ratio} = \frac{MSB}{MSW}$$ (14.3)

The elements of this test are usually presented in an **ANOVA table,** as we will see shortly. The "bottom line" in this table is the p-value for the F-ratio. If the p-value is sufficiently small, we can conclude that the population means are not all equal. Otherwise, we cannot reject the equal-means hypothesis.

14.2.2 Confidence Intervals for Differences Between Means

If we cannot reject the equal-means hypothesis, then there is little incentive to examine differences between individual pairs of means. However, if we *can* reject the equal-means hypothesis, then it is customary to form confidence intervals for the differences between pairs of population means. This can lead to quite a few confidence intervals. For example, if there are $J = 5$ treatment levels, then there are 10 pairs of differences (the number of ways 2 means can be chosen from a total of 5 means). The confidence interval for any difference $\mu_i - \mu_j$ is of the form shown in expression (14.4).

Confidence Interval for Difference Between Means	$\bar{Y}_i - \bar{Y}_j \pm \text{multiplier} \times \sqrt{MSW(1/n_i + 1/n_j)}$	(14.4)

As we will discuss in Section 14.4, there are several possibilities for the appropriate multiplier in this expression. Regardless of which multiplier we use, however, we are always looking for confidence intervals that do *not* include 0. If the confidence interval for $\mu_i - \mu_j$ is all positive, for example, then we can conclude with high confidence that these two means are not equal and that μ_i is indeed *larger* than μ_j. However, if the confidence interval for $\mu_i - \mu_j$ includes 0, that is, if it extends from a negative number to a positive number, we cannot conclude that these two means are different.

We have presented the formulas for one-way ANOVA because they lend some insight into the procedure. However, StatPro's one-way ANOVA procedure takes care of all the tedious calculations, as we illustrate in the following example.

Example 14.1 The Effect of Shelf Height on Cereal Sales at Midway

Does it matter which shelf a popular brand is placed on? It certainly might, because we tend to purchase items that are easiest to see. To test this, suppose that Midway is a large chain of supermarket stores with many stores in many locations. Midway selects 125 of these stores for an experiment. Specifically, it selects these particular 125 stores to be as alike as possible, so that store size, amount of customer traffic, types of customers, and other characteristics are as similar across stores as possible. Each store stocks cereal in a similar location in the store on five-shelf displays. In the experiment, 25 randomly selected stores place a particular popular brand of cereal—we'll call it Brand X—on the lowest shelf for a month. Another randomly selected 25 stores place Brand X on the next-to-lowest shelf, another 25 place it on the middle shelf, another 25 place it on the next-to-highest shelf, and the final 25 place it on the highest shelf. Then the number of boxes of Brand X sold is recorded at each of the stores for the last two weeks of the experiment.

FIGURE 14.3
Data for Cereal
Experiment

	A	B	C	D	E
1	Testing the effect of shelf height on cereal sales				
2					
3	Lowest	Next-to-lowest	Middle	Next-to-highest	Highest
4	340	347	444	456	358
5	376	428	281	471	427
6	378	219	378	484	325
7	371	431	425	448	428
8	395	377	485	330	522
9	332	238	353	405	455
10	307	368	332	375	315
11	333	364	453	546	466
12	239	529	466	489	341
13	301	399	377	502	204
14	298	505	471	373	317
15	358	412	178	486	342
16	373	430	301	513	326
17	387	328	504	346	371
18	351	431	388	319	331
19	235	541	423	475	387
20	307	459	426	242	416
21	278	318	418	424	422
22	455	302	442	425	479
23	346	394	327	274	351
24	355	225	354	358	330
25	202	374	381	411	449
26	389	345	284	564	461
27	417	329	349	395	375
28	250	374	346	546	399

(The first two weeks allow customers to get used to the shelving arrangement.) The resulting data are in the file CEREAL.XLS, as shown in Figure 14.3. Does shelf height appear to make a difference in sales?

Objective To use one-way ANOVA to see whether shelf height makes any difference in mean sales of Brand X, and if so, to discover which shelf heights outperform the others.

Solution

First, the sample sizes are equal—this is a balanced design. This is not absolutely necessary in an experiment of this type, but since Midway is able to specify which stores use which shelving heights, it makes sense to use a balanced design. Second, this is a designed experiment, not an observational study. Midway deliberately chose the 125 stores in the experiment to be alike in as many ways as possible. This helps to ensure that any differences in sales across the five groups can be attributed to differences in shelf heights and not to other extraneous factors. Of course, it is virtually impossible to control for *all* other factors in an experiment such as this—the 125 stores are certainly not identical in *all* of their characteristics—but Midway has tried its best to keep them similar. Also, it has *randomly* assigned the stores to treatment levels (shelf heights), rather than arbitrarily assigning them. By using a random assignment, Midway avoids any possible bias it might have unconsciously introduced with a nonrandom assignment.

To analyze the data, we use the StatPro/Statistical Inference/One-Way ANOVA menu item. In the second and third dialog boxes, we check the Unstacked option because there is a separate sales column for each of the shelf heights, and then we select all five variables for analysis. (Note that we would use the Stacked option if there were two columns of length 125 each, one with the shelf height and the other with sales.) In the next dialog

FIGURE 14.4
StatPro Dialog Box
for Confidence
Intervals

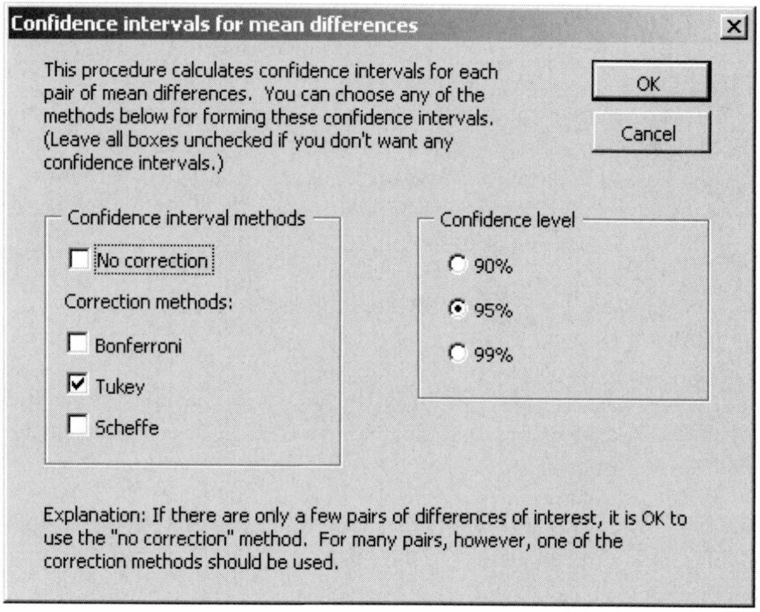

Confidence intervals for mean differences

This procedure calculates confidence intervals for each pair of mean differences. You can choose any of the methods below for forming these confidence intervals. (Leave all boxes unchecked if you don't want any confidence intervals.)

| OK |
| Cancel |

Confidence interval methods

☐ No correction

Correction methods:

☐ Bonferroni

☑ Tukey

☐ Scheffe

Confidence level

○ 90%

◉ 95%

○ 99%

Explanation: If there are only a few pairs of differences of interest, it is OK to use the "no correction" method. For many pairs, however, one of the correction methods should be used.

	A	B	C	D	E	F	G
1	*Results of one-way ANOVA*						
2							
3	*Summary stats for samples*						
4			Lowest	Next-to-lowest	Middle	Next-to-highest	Highest
5		Sample sizes	25	25	25	25	25
6		Sample means	334.920	378.680	383.440	426.280	383.880
7		Sample standard deviations	61.043	84.081	75.625	85.054	69.619
8		Sample variances	3726.243	7069.560	5719.173	7234.210	4846.777
9		Weights for pooled variance	0.200	0.200	0.200	0.200	0.200
10							
11		Number of samples	5				
12		Total sample size	125				
13		Grand mean	381.440				
14		Pooled variance	5719.193				
15		Pooled standard deviation	75.625				
16							
17	*OneWay ANOVA table*						
18		Source	SS	df	MS	F	p-value
19		Between variation	104807.680	4	26201.920	4.581	0.0018
20		Within variation	686303.120	120	5719.193		
21		Total variation	791110.800	124			
22							
23	*Confidence intervals for mean differences*						
24		Confidence level	95.0%				
25	*Tukey method*						
26		Difference	Mean diff	Lower	Upper	Signif?	
27		Lowest - Next-to-lowest	-43.760	-103.050	15.530	No	
28		Lowest - Middle	-48.520	-107.810	10.770	No	
29		Lowest - Next-to-highest	-91.360	-150.650	-32.070	Yes	
30		Lowest - Highest	-48.960	-108.250	10.330	No	
31		Next-to-lowest - Middle	-4.760	-64.050	54.530	No	
32		Next-to-lowest - Next-to-highest	-47.600	-106.890	11.690	No	
33		Next-to-lowest - Highest	-5.200	-64.490	54.090	No	
34		Middle - Next-to-highest	-42.840	-102.130	16.450	No	
35		Middle - Highest	-0.440	-59.730	58.850	No	
36		Next-to-highest - Highest	42.400	-16.890	101.690	No	

FIGURE 14.5 One-Way ANOVA Output

box, shown in Figure 14.4, we accept the default settings, with a check next to the Tukey option for confidence intervals. (We will discuss these confidence interval options in Section 14.4.)

The one-way ANOVA output is shown in Figure 14.5. The summary statistics at the top indicate that the next-to-highest shelf height has the largest average sales, 426.3, almost 100 boxes larger than the lowest shelf height, which has the smallest average sales. This information is confirmed by the side-by-side boxplots in Figure 14.6. (Although these boxplots are not created as part of the ANOVA output, they are always a useful addition.) The sample standard deviations vary from about 61 to 85 over the five treatment levels. Although these tend to indicate unequal variances, the equal-variance assumption is almost never satisfied exactly in any study, and this much discrepancy in the standard deviations is nothing to worry about—it certainly does not invalidate the analysis.

The effect of unequal variances is mitigated by having equal, or nearly equal, sample sizes for the treatment levels. This is another reason for using a balanced design.

It appears from the summary statistics and the boxplots that mean sales differ for different shelf heights, but are the differences significant? The test of equal means answers this question. It appears in rows 18–21 of the output. The values in this ANOVA table are based on equations (14.1)–(14.3). (The only part we didn't discuss is the Total variation in row 21. It is based on the total variation of all observations around the grand mean in cell C13 and is used mainly as a check of the calculations. Note that *SSB* and *SSW* in cells C19 and C20 add up to the total sum of squares in cell C21. Similarly, the degrees of freedom add up in column D.) The *F*-ratio in cell F19 is 4.581, the ratio of the mean squares in cells E19 and E20. Its corresponding *p*-value is 0.0018, nearly zero. This leaves practically no doubt that the five population means are *not* all equal. Shelf height evidently does make a significant difference in sales.

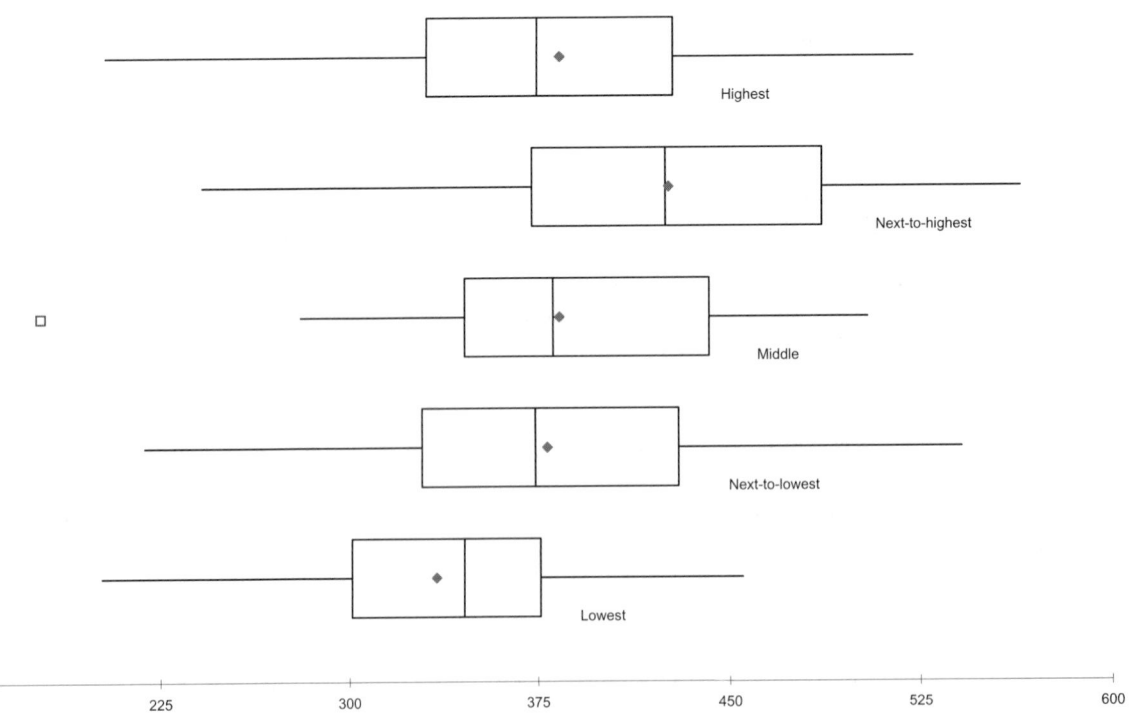

FIGURE 14.6 Side-by-Side Boxplots of Sales

The 95% confidence intervals in rows 27–36 indicate which shelf heights differ significantly from which others. There is a "Yes" next to any difference whose confidence interval does *not* include 0. In this example, there is only one such difference, the one between the next-to-highest height and the lowest height. Not surprisingly, these are the treatment levels with the largest and smallest average sales. None of the other differences are significant. For example, even though the difference between the next-to-highest and next-to-lowest heights is 47.6, the corresponding confidence interval extends from a negative number to a positive number. Therefore, we cannot declare this difference to be statistically significant.

The main conclusion from this example is that shelf height definitely appears to make a difference in mean sales, at least for the population of stores similar to the ones in the study. Customers tend to purchase fewer boxes of cereal when they are placed on the bottom shelf, and they tend to purchase more when they are placed on the next-to-highest shelf—presumably right around eye level. ●

14.2.3 Using a Logarithmic Transformation

Recall that the inferences based on the ANOVA procedure rely on two assumptions: equal variances across treatment levels and normally distributed data. Although these assumptions are never met *exactly* in any real study, we should check whether they are at least approximately valid. Often a look at side-by-side boxplots, as in Figure 14.6, can tell us whether there are serious violations of these assumptions. For example, the boxplots in this figure are reasonably symmetric and indicate reasonably similar variances, so that the ANOVA results should be valid. If the assumptions are seriously violated, however, we should not blindly report the ANOVA results. In some cases, a transformation of the data will help, as illustrated in the next example.

Example (14.2) Payments for Orders at Rebco

Rebco is a manufacturing company that supplies parts to many other manufacturing companies, its customers. Rebco is concerned about the time it takes these customers to pay for their orders. The file REBCO.XLS contains data (a subset of which is shown in Figure 14.7 on page 774) on the most recent payment from 91 of its customers. The customers are categorized as small, medium, and large. For each customer we see the number of days it took the customer to pay and the amount of the payment. Are there any differences in the mean time to pay across the three customer sizes? What about differences across the mean payment amounts?

Objective To see how a logarithm transformation can be used to ensure the validity of the ANOVA assumptions, and to see how the resulting output should be interpreted.

Solution

Unlike Example 14.1, this is a one-factor observational study, where the single factor is customer size at three levels: small, medium, and large. The experimental units are the bills for the orders, and there are two response variables, days until payment and payment amount, that we will examine. Focusing first on the days until payment, we see from the side-by-side boxplots in Figure 14.8 that whatever differences there are appear to be slight. Perhaps the large customers pay, on average, a bit more promptly, but it is difficult to see from the plots whether the apparent differences are significant. Therefore, we turn

FIGURE 14.7
Data for Rebco
Example

	A	B	C	D
1	**Payments to Rebco**			
2				
3	Customer	CustSize	Days	Amount
4	1	Large	12	1352
5	2	Small	21	274
6	3	Small	20	267
7	4	Small	21	229
8	5	Large	14	1870
9	6	Small	29	246
10	7	Large	14	840
11	8	Medium	19	743
12	9	Large	18	1245
13	10	Small	18	195
14	11	Medium	19	608
15	12	Large	14	994
16	13	Small	17	215
17	14	Small	19	131
18	15	Medium	17	583
19	16	Medium	13	501
20	17	Medium	17	391
21	18	Medium	25	741
92	89	Medium	22	372
93	90	Large	23	1045
94	91	Medium	15	671

to the numerical results. The summary results and the ANOVA table in Figure 14.9 show that the differences between the sample means are not even close to being statistically significant. The *p*-value for the test is only 0.318. Rebco cannot reject the null hypothesis that customers of all sizes take, on average, the same number of days to pay.

The analysis of the *amounts* these customers pay is quite different. This is immediately evident from the side-by-side boxplots in Figure 14.10. Actually, two things are clear. First, there is little doubt that small customers tend to have lower bills than medium-size customers, who in turn tend to have lower bills than large customers. Second, however, we see that the equal-variance assumption is grossly violated. There is very little

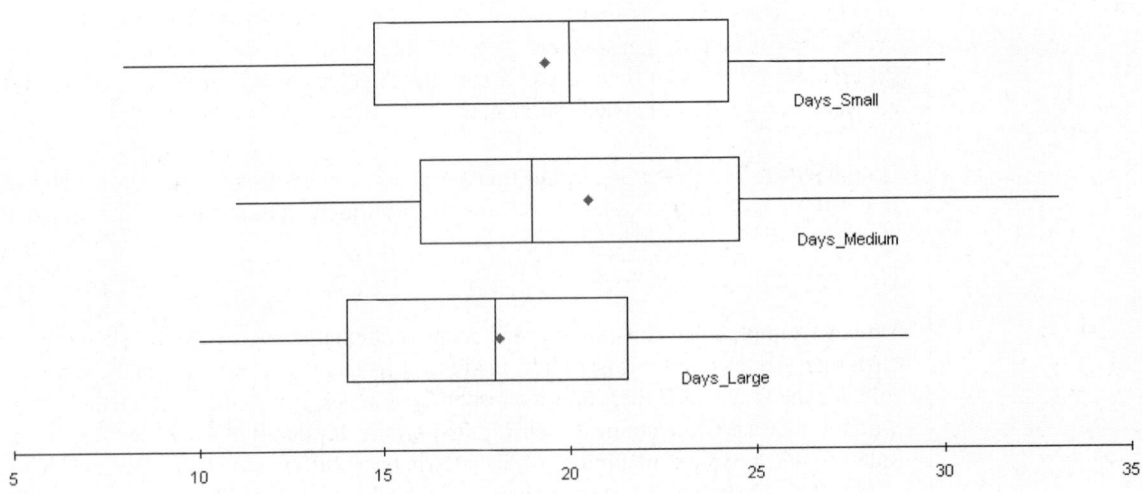

FIGURE 14.8 Boxplots for Days Until Payment

FIGURE 14.9
ANOVA Results for
Days Until Payment

	A	B	C	D	E	F	G
1	Results of one-way ANOVA						
2							
3	Summary stats for samples						
4			Days_Large	Days_Medium	Days_Small		
5		Sample sizes	20	39	32		
6		Sample means	18.100	20.487	19.375		
7		Sample standard deviations	4.887	5.707	6.318		
8		Sample variances	23.884	32.572	39.919		
9		Weights for pooled variance	0.216	0.432	0.352		
10							
11		Number of samples	3				
12		Total sample size	91				
13		Grand mean	19.571				
14		Pooled variance	33.285				
15		Pooled standard deviation	5.769				
16							
17	One-Way ANOVA table						
18		Source	SS	df	MS	F	p-value
19		Between variation	77.242	2	38.621	1.160	0.3181
20		Within variation	2929.044	88	33.285		
21		Total variation	3006.286	90			

variation in payment amounts from small customers and a large amount of variation from large customers. We should remedy this situation before running any formal ANOVA.

One common method for equalizing variances is to take logarithms of the response variable and then use the transformed variable as the new response variable. This log transformation tends to spread apart small values and compress together large values—exactly what we want in this example. After taking the logarithms of the payment amounts, we obtain the boxplots in Figure 14.11 on page 776. The log transformation retains the ordering, so that logs of small amounts are still less than logs of large amounts, but the variances are now much closer to being equal. The resulting ANOVA on the log variable appears in Figure 14.12. The p-value in the ANOVA table is again the key for checking whether we can reject the equal-means hypothesis. The fact that it is virtually 0 assures us that the means of the log variables are *not* equal.

This graph in Figure 14.10 indicates definite violations of the ANOVA equal-variance assumption.

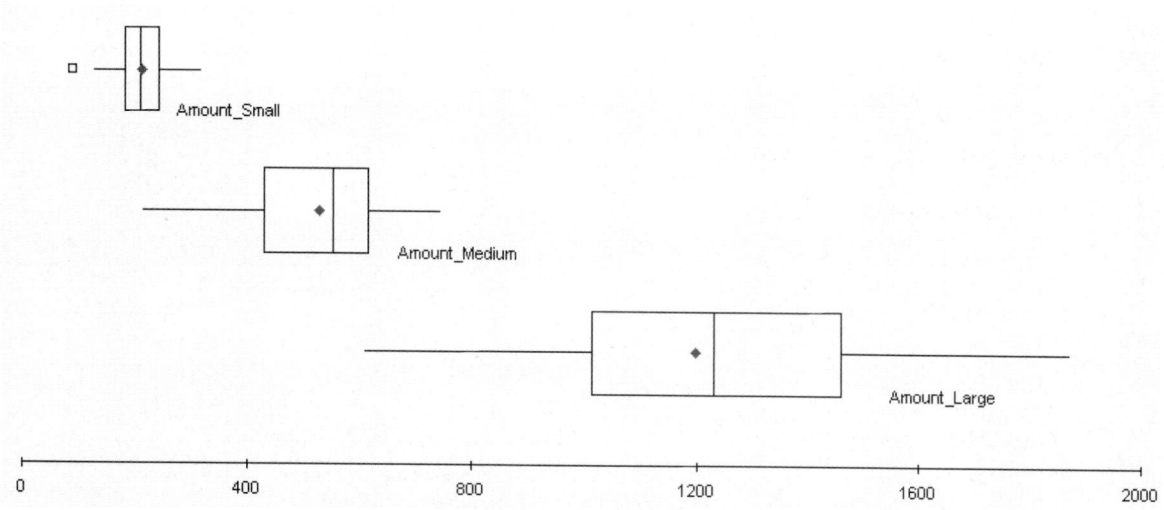

FIGURE 14.10 Boxplots for Payment Amounts

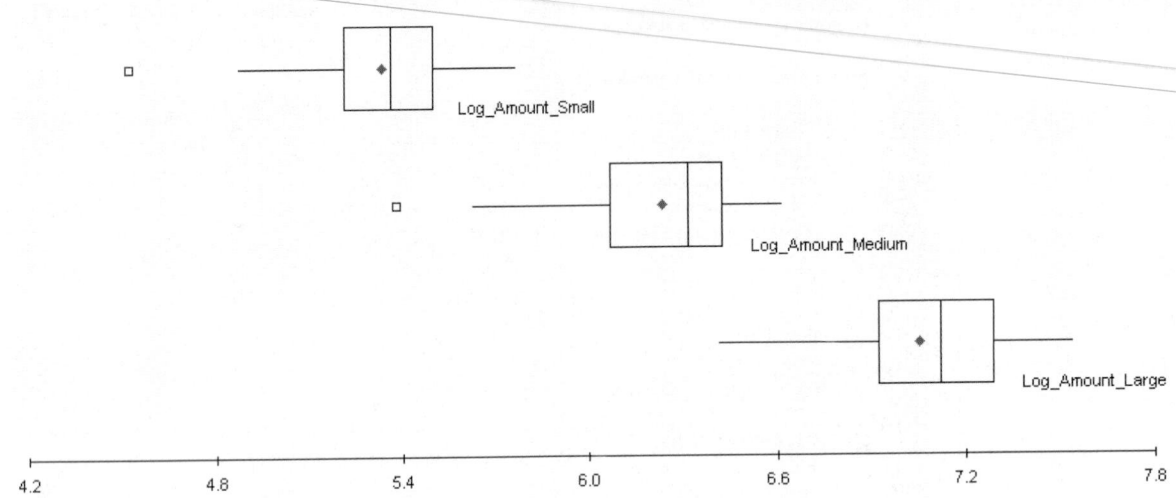

FIGURE 14.11 Boxplots of Log-Transformed Amounts

	A	B	C	D	E	F	G
1	*Results of one-way ANOVA*						
2							
3	*Summary stats for samples*						
4			Log_Amount_Large	Log_Amount_Medium	Log_Amount_Small		
5		Sample sizes	20	39	32		
6		Sample means	7.047	6.235	5.332		
7		Sample standard deviations	0.303	0.286	0.252		
8		Sample variances	0.092	0.082	0.064		
9		Weights for pooled variance	0.216	0.432	0.352		
10							
11		Number of samples	3				
12		Total sample size	91				
13		Grand mean	6.096				
14		Pooled variance	0.078				
15		Pooled standard deviation	0.279				
16							
17	*One-Way ANOVA table*						
18		Source	SS	df	MS	F	p-value
19		Between variation	37.519	2	18.760	241.540	0.0000
20		Within variation	6.835	88	0.078		
21		Total variation	44.354	90			
22							
23	*Confidence intervals for mean differences*						
24		Confidence level	95.0%				
25	*Usual (no correction) method*						
26		Difference	Mean diff	Lower	Upper	Signif?	
27		Log_Amount_Large - Log_Amount_Medium	0.812	0.660	0.965	Yes	
28		Log_Amount_Large - Log_Amount_Small	1.715	1.557	1.873	Yes	
29		Log_Amount_Medium - Log_Amount_Small	0.903	0.771	1.035	Yes	
30							
31	**Antilogs**						
32		Amount_Large - Amount_Medium	2.253	1.935	2.624		
33		Amount_Large - Amount_Small	5.557	4.745	6.507		
34		Amount_Medium - Amount_Small	2.466	2.161	2.815		
35							
36		For example, our best guess is that the median amount for large customers is 2.253 times as large as the median					
37		amount for medium customers, and we are 95% confident that this factor is between 1.935 and 2.624.					
38							

FIGURE 14.12 ANOVA Results for Log-Transformed Amounts in Example 14.2

What does this say about the original variables? The bottom part of the output in Figure 14.12 answers this question, although we have to be very careful when interpreting the results. First, when we ran StatPro's one-way ANOVA procedure on the log of the Amount variable, we requested the confidence intervals in rows 27–29.[3] However, each of these is a confidence interval for the difference between means of the *log-transformed* variables. Because they are in log units, these numbers have little practical meaning. The trick is to take their antilogarithms (with the EXP function), as shown in rows 32–34, and then interpret the antilogs correctly. It can be shown that the correct interpretation is the one given at the bottom of the spreadsheet—namely, each antilog is a *ratio of medians* for the respective treatment levels. (If the populations are reasonably symmetric, the antilogs can also be interpreted as approximate ratios of *means*.) For example, our best guess is that the median amount paid by large customers is 2.253 times as large as the median amount paid by medium-sized customers, and we are 95% confident that this ratio is between 1.935 and 2.624. Since the populations are reasonably symmetric (see the boxplots in Figure 14.10), this same statement applies, at least approximately, to the means.

The bottom line for Rebco is that its large customers have bills that are typically over twice as large as those for medium-sized customers, which in turn are typically over twice as large as those for small customers. Even though all customers currently tend to take about the same number of days to pay, there is a greater incentive to get the large customers to pay early—more money is at stake. ●

The "ratio of medians" interpretation discussed in this example is the correct interpretation in any comparison problem where a log transformation is used (probably to equalize variances) on the response variable. It applies not only to ANOVA studies such as Rebco's but to the two-sample t procedures discussed in Chapters 9 and 10.

PROBLEMS

Level A

1. An automobile manufacturer employs sales representatives who make calls on dealers. The manufacturer wishes to compare the effectiveness of four different call-frequency plans for the sales representatives. Thirty-two representatives are chosen at random from the sales force and randomly assigned to the four call plans (eight per plan). The representatives follow their plans for 6 months, and their sales for the 6-month study period are recorded. These data are given in P14_01.XLS.
   a. Do the sample data support the hypothesis that at least one of the call plans helps produce a higher average level of sales? Perform an appropriate statistical test and report a p-value.
   b. If the sample data indicate the existence of mean sales differences across the call plans, which plans appear to produce different average sales levels? Construct 95% confidence levels for the differences between all pairs of population means in this case.

2. Consider a large chain of supermarkets that sell their own brand of potato chips in addition to many other name brands. Management would like to know whether the type of display used for the store brand has any effect on sales. Since there are four types of displays being considered, management decides to choose 24 similar stores to serve as experimental units. A random six of these are instructed to use display type 1, another random six are instructed to use display type 2, a third random six are instructed to use display type 3, and the final six stores are instructed to use display type 4. For a period of one month, each store keeps track of the *fraction* of total potato chips sales that are of the store brand. The data for the 24 stores are shown in the file P14_02.XLS. Note that one of the stores using display 3 is marked with a "*". This store did not follow instructions properly, so its observation is disregarded.
   a. Why do you think each store keeps track of the fraction of total potato chips sales that are of the store brand? Why do they not simply record the total amount of sales of the store brand potato chips?

---

[3] Again, we will discuss the type of confidence interval method shown here in Section 14.4.

b. Do the data suggest different mean proportions of store brand sales at the 10% significance level? If so, construct 90% confidence intervals for the differences between all pairs of mean proportions to identify which of the display types are associated with higher fractions of sales.

3. National Airlines recently introduced a daily (i.e., early morning) nonstop flight between Houston and Chicago. The vice president of marketing for National Airlines decided to perform a statistical test to see whether National's average passenger load on this new flight was different from that of each of its two major competitors (which we will call competitor 1 and competitor 2). Ten early-morning flights were selected at random from each of the three airlines and the percent of *unfilled* seats on each flight was recorded. These data are stored in the file P14_03.XLS.
   a. Is there evidence that National's average passenger load on the new flight is different from that of its two competitors? Report a *p*-value and interpret the results of the statistical test.
   b. Select an appropriate significance level and construct confidence intervals for all pairs of differences between means. Which of these differences, if any, are statistically significant at the selected significance level?

4. A hotel manager would like to know whether customers who pay by different methods have different-sized bills. She divides all customers into four categories: those who pay by check or cash, those who pay with a VISA or MasterCard, those who pay with an American Express card, and those who use some other type of credit card. She then collects data on daily bills, which are given in the file P14_04.XLS. Note that these bills contain the room charge, plus any other charges to the customer's account.
   a. Test whether the different categories of customers have different-sized bills at the 10% significance level.
   b. Compute 90% confidence intervals for all pairs of differences between means. Which of these differences, if any, are significantly nonzero at the 10% significance level?

5. A company sells identical soap in four different packages at the same price. The sales of each package type for 12 months are provided in the file P14_05.XLS. Is there any indication of differences in the mean sales of this brand of soap across the various package types? Perform an appropriate statistical test and report a *p*-value.

Level B

6. Do graduates of undergraduate business programs with different majors tend to earn disparate average starting salaries? Consider the data given in the file P14_06.XLS.
   a. Is there any reason to doubt the equal-variance assumption made in the one-way ANOVA model in this particular case? Support your response to this question.
   b. Assuming that the variances of the four underlying populations are indeed equal, can you reject at the 10% significance level that the mean starting salary is the same for each of the given business majors? Explain why or why not.
   c. Generate 90% confidence intervals for all pairs of differences between means. Which of these differences, if any, are statistically significant at the 10% significance level?

7. A company that employs a large number of salespeople is interested in determining which of the following subsets of the sales staff sells, on average, the most: (1) those whose compensation consists of a fixed salary, (2) those whose compensation is based strictly on commission, and (3) those whose compensation is based on a smaller fixed portion *and* a reduced commission rate. Sales data (in dollars) from the previous quarter are collected for randomly selected salespeople who are compensated according to one of the three aforementioned schemes. The data are given in the file P14_07.XLS.
   a. Is there any reason to doubt the equal-variance assumption made in the one-way ANOVA model in this particular case? Support your response to this question.
   b. Can you reject at the 5% significance level that the mean sales are the same for each of the three groups of salespeople? Explain why or why not.
   c. Generate 95% confidence intervals for all pairs of differences between means. Which of these differences, if any, are statistically significant at the 5% significance level?

14.3 Using Regression to Perform ANOVA

The method we discussed in the previous section for performing ANOVA—calculating sums of squares by rather complex formulas and showing the results in an ANOVA table—is the traditional way of implementing ANOVA. Indeed, it is the method implemented in most statistical software packages, and it can be extended to many experimental designs besides one-way ANOVA. However, it is worth knowing that much of the same ANOVA results can be obtained by multiple regression analysis, as we will briefly discuss in this section. The advantage of using regression is that many people understand regression better than the sometimes obscure formulas used in traditional ANOVA. The disadvantage is that some of the traditional ANOVA output, such as the confidence intervals for mean differences, can be obtained with regression only with some difficulty—they are not standard parts of the regression output. Therefore, regression is not a perfect substitute for traditional ANOVA, but it can supplement the analysis.

To perform ANOVA with regression, we simply run a regression with the same response variable as in ANOVA and use dummy variables for the treatment levels as the *only* explanatory variables. For example, if there is a single factor with 5 treatment levels, we create 4 dummy variables, one for each of the treatment levels except a designated "reference" level, and we run the regression with these 4 dummies as the only explanatory variables. In the resulting regression output, the ANOVA table will be *exactly* the same as the ANOVA table we obtain from traditional ANOVA, and the coefficients of the dummy variables will be estimates of the mean differences between the corresponding treatment levels and the reference level.

For example, if we have 5 treatment levels and we designate level 5 as the reference level, then the regression coefficients will estimate the mean differences $\mu_1 - \mu_5$, $\mu_2 - \mu_5$, $\mu_3 - \mu_5$, and $\mu_4 - \mu_5$. Therefore, the reported confidence intervals for these coefficients are really confidence intervals for these mean differences. However, we do not automatically obtain confidence intervals for other mean differences such as $\mu_2 - \mu_3$. Also, the confidence intervals we obtain are not of the "Tukey" type we obtained with ANOVA. (They are of the "no correction" type we will discuss in the next section.) On the plus side, however, the regression output provides an R^2 value, the percentage of the variation of the response variable explained by the various treatment levels of the single factor. This R^2 value is *not* part of the traditional ANOVA output.

To see how this works, we revisit Midway's cereal experiment from Example 14.1.

Example 14.1 The Effect of Shelf Height on Cereal Sales at Midway (continued)

Recall that the Midway supermarket chain ran a study on 125 stores to see whether shelf height, set at five different levels, has any effect on sales of a popular brand of cereal. (See the file CEREAL.XLS.) Does Midway get the same results as before if it analyzes the data with regression?

Objective To see how Midway can analyze its data with regression, using only dummy variables for the treatment levels.

Solution

Before we can run a regression, we must first reorganize the data. Recall that the original data in the CEREAL.XLS file are in unstacked form—one sales column for each shelf height. For regression, the data must be in stacked form. This is easy to accomplish with StatPro. We use the StatPro/Data Utilities/Stack Variables menu item. When asked for "codes" for the different levels, we specified the numbers 1 through 5, although any codes would do. This produces two long columns, one for the codes and one for the sales values. We then use StatPro's dummy variable procedure to create dummies for the different shelf heights, based on the code column. The results for the first few stores appear in Figure 14.13 on page 780.

	G	H	I	J	K	L	M	N
1	**Data for regression (obtained with StatPro's Stack Variables and Create Dummy Variables procedures)**							
2								
3	Shelf height	Sales	Shelf height_1	Shelf height_2	Shelf height_3	Shelf height_4	Shelf height_5	
4	1	340	1	0	0	0	0	
5	1	376	1	0	0	0	0	
6	1	378	1	0	0	0	0	
7	1	371	1	0	0	0	0	
8	1	395	1	0	0	0	0	
9	1	332	1	0	0	0	0	
10	1	307	1	0	0	0	0	
11	1	333	1	0	0	0	0	
12	1	239	1	0	0	0	0	
13	1	301	1	0	0	0	0	
14	1	298	1	0	0	0	0	
15	1	358	1	0	0	0	0	
16	1	373	1	0	0	0	0	
17	1	387	1	0	0	0	0	
18	1	351	1	0	0	0	0	
19	1	235	1	0	0	0	0	
20	1	307	1	0	0	0	0	
21	1	278	1	0	0	0	0	
22	1	455	1	0	0	0	0	
23	1	346	1	0	0	0	0	
24	1	355	1	0	0	0	0	
25	1	202	1	0	0	0	0	
26	1	389	1	0	0	0	0	
27	1	417	1	0	0	0	0	
28	1	250	1	0	0	0	0	
29	2	347	0	1	0	0	0	
30	2	428	0	1	0	0	0	
31	2	219	0	1	0	0	0	

FIGURE 14.13 Stacked Variables and Dummy Variables

We now run a multiple regression with the Sales variable as the response variable and the dummies Shelf height_2 through Shelf height_5 as the explanatory variables. We used Shelf height_1 (the lowest height) as the reference level, although any level could have been used. The regression output is shown in Figure 14.14.

The first thing to notice is that the ANOVA table from the regression output is *identical* to the ANOVA table from traditional ANOVA. (See Figure 14.5.) This will always be the case. We can infer, because of the extremely low *p*-value in this table, that the population regression coefficients are not all 0. However, because these regression coefficients are really mean differences between the various levels and the reference level, we can infer that these mean differences are not all 0. Specifically, at least one of the upper heights differs from the lowest height. The estimates of the mean differences, given in the range C17:C20, are simply the observed average differences in sales between upper heights and the lowest height. Also, the constant in cell C16 is simply the observed average sales for the lowest height.

If you compare the confidence intervals in the range G17:H20 of the regression output to the corresponding confidence intervals for the ANOVA output in Figure 14.5, you will see that they are somewhat different. For example, the confidence interval for $\mu_2 - \mu_1$ from Figure 14.14 extends from 1.41 to 86.11, whereas the similar confidence interval in Figure 14.5 extends from -15.53 to 103.05. (We had to reverse the signs to get the confidence interval for $\mu_2 - \mu_1$, not $\mu_1 - \mu_2$.) In particular, the confidence interval from

	A	B	C	D	E	F	G	H
1	**Results of multiple regression for Sales**							
2								
3	**Summary measures**							
4		Multiple R	0.3640					
5		R-Square	0.1325					
6		Adj R-Square	0.1036					
7		StErr of Est	75.6253					
8								
9	**ANOVA Table**							
10		Source	df	SS	MS	F	p-value	
11		Explained	4	104807.6750	26201.9188	4.5814	0.0018	
12		Unexplained	120	686303.1250	5719.1927			
13								
14	**Regression coefficients**							
15			Coefficient	Std Err	t-value	p-value	Lower limit	Upper limit
16		Constant	334.9200	15.1251	22.1434	0.0000	304.9735	364.8666
17		Shelf height_2	43.7600	21.3901	2.0458	0.0430	1.4092	86.1108
18		Shelf height_3	48.5200	21.3901	2.2683	0.0251	6.1692	90.8708
19		Shelf height_4	91.3600	21.3901	4.2711	0.0000	49.0092	133.7108
20		Shelf height_5	48.9600	21.3901	2.2889	0.0238	6.6092	91.3108

FIGURE 14.14 Regression Output for Cereal Example

regression, although centered around the same mean difference, is much narrower. In fact, it is entirely positive, leading us to conclude that this mean difference is significant. The ANOVA output led us to the opposite conclusion. The reason for this apparent discrepancy is the subject of the next section. It is basically because the Tukey intervals quoted in the ANOVA output are more "conservative" and typically lead to *fewer* significant differences.

One final comment about the regression output regards its R^2 value. We see that differences in the shelf height account for 13.2% of the variation in sales. This means that although shelf height has some effect on sales, there is a lot of "random" variation in sales across stores that cannot be accounted for by shelf height. ●

PROBLEMS

Level A

8. For the National Airlines data in Problem 3 (see the file P14_03.XLS), perform a regression analysis using dummy variables for the airlines. Comment on the meaning of the regression output. How does it compare with the ANOVA output from Problem 3? Does it give you any extra insights?

9. For the soap data in Problem 5 (see the file P14_05.XLS), perform a regression analysis using dummy variables for the different packages. Comment on the meaning of the regression output. How does it compare with the ANOVA output from problem 8? Does it give you any extra insights?

Level B

10. In Problem 6, the salary data for graduates of the undergraduate business programs (see the file P14_06.XLS) represent an unbalanced design—there are more students from some majors than others. Run a regression on starting salaries, using dummies for majors as the explanatory variables. Do you get the same results as with the ANOVA output from Problem 6? Specifically, consider the constant term and the regression coefficients in the output. Is the constant equal to the average starting salary for the reference major? (You can designate any of the majors as the reference major.) Are the regression coefficients equal to average differences between the various majors and the reference major?

14.4 The Multiple Comparison Problem

In many statistical analyses, including ANOVA studies, we want to make statements about *multiple* unknown parameters. For example, in the cereal study (Example 14.1), we wanted to create confidence intervals for differences between each pair of means—10 confidence intervals in all.[4] Any time we make such a statement, there is a chance that we will be wrong; that is, there is a chance that the true value will not be inside the confidence interval. If we create a 95% confidence interval, say, then the error probability is 0.05. In fact, as we explained in Chapter 9, the endpoints of the confidence interval are chosen so that the error probability will be 1 minus the confidence level we choose. What do we mean by "error probability," however, when we make several statements based on the same data? This is the issue we will address in this section.

We can use simulation to get an idea of the problem. In the file MULTIPLECOMPARISON.XLS, we simulate a data set very much like those encountered in one-way ANOVA. (See the range A10:H70 of Figure 14.15. Note that a lot of rows have been hidden so that the figure fits on the page.) The data in this range correspond to data from a one-factor design with 8 treatment levels and 60 observations per level—480 observations all together. However, we entered the same formula, =NORMINV(Rand(),0,1), in each cell. This means that each treatment level is generating normal data with mean 0 and standard deviation 1. Therefore, we *know* that the equal-means hypothesis of ANOVA is true—all population means are equal to 0. Nevertheless, we calculate 95% confidence intervals for each of the 28 possible differences between means in rows 84–111 (many are hidden because of space considerations), using the two-sample procedure described in Chapter 9. If a confidence interval does *not* include 0, we indicate it as a significant difference by putting "Yes" in column D. Then if *at least one* of these 28 confidence intervals is significant, we record "Yes" in cell D113. (At least one of the hidden rows evidently has "Yes" for these particular random numbers.) Finally, we replicate this procedure 500 times in columns J and K, each time recording the value in cell D113, and we report the percentage of replications with "Yes" in cell K5. As we see, the reported percentage (in cell K5) is over 50%.

What does this prove? Recall that all of the population means equal 0. This is how we simulated the random numbers in the first place. Therefore, if any one of the 28 confidence intervals in rows 84–111 turns out to be significant and we report it as such, we are making an error. That is, we are reporting that these two population means are not equal, when in fact they both equal 0. Because *none* of the 28 confidence intervals in rows 84–111 should be significant, we will have a perfect record only if we report "No" in cell D113. Of course, a perfect record cannot always be obtained, but by using a 95% confidence level, we might expect a perfect record in 95% of the replications. Unfortunately, the simulation shows that we are not even close to this. We get a perfect record less than half of the time! In statistical terms, if we run each confidence interval at the 95% level, the *overall* confidence level (of having *all* 28 statements correct) is much less than 95%. (Worse yet, we are never really sure how much less.) This is usually called the **multiple comparison problem.** It says that if we make a lot of statements, each at a given confidence level such as 95%, then the chance of making at least one wrong statement is much greater than 5%.

The question is how to get the overall confidence level equal to the desired value, such as 95%. Or in the simulation in Figure 14.15, how can we get the error rate in cell K5 to

---

[4]Note that if a confidence interval for a difference such as $\mu_1 - \mu_2$ is reported, the confidence interval for its opposite, $\mu_2 - \mu_1$, is *not* reported. To get the latter, however, just take the negative of the former.

Simulation of multiple comparison problem

In this simulation, samples from several normal populations, all with the *same* mean and standard deviation, will be simulated, and the usual 95% confidence interval will be constructed for *each* mean difference. Then a data table will be used to check what percentage of replications have at least one confidence interval that does not include 0. It will probably be larger than 5%.

Percent of replications with any significant
(from data table below)

50.2%

Simulation of data and confidence intervals

	Sample1	Sample2	Sample3	Sample4	Sample5	Sample6	Sample7	Sample8
	0.687256261	-0.599591	1.221754	-1.1708835	0.286768	-0.300112	0.78126	1.616168
	0.338887958	2.200759	-1.56836	0.89505875	-1.547232	-0.566299	1.10708	-0.221773
	-0.009360974	1.46119	0.651398	0.28793806	-0.401756	-1.713015	1.41968	1.14449
	-1.810403774	-0.582618	-0.412329	-0.1017406	1.836361	3.179011	0.368991	0.520039
	-1.352527761	0.951588	0.990105	0.75851176	0.319209	1.716076	-0.592736	-2.949819
	-0.066420398	-0.740249	-0.185373	0.2527122	-0.107927	-0.914201	-0.061922	0.942741
	0.435021548	1.074109	0.144885	-0.2472223	0.431783	2.485976	-0.658473	-1.296826
	-0.623854248	0.756877	0.565556	-0.580327	0.272715	-0.043411	1.473159	-0.833288
	-1.044306828	0.805396	1.286739	-0.2292677	-1.202779	0.46545	-0.636473	-0.250632
	-0.242787337	1.278522	-2.024681	-0.6794767	-2.015431	0.230222	0.274786	1.403696
	0.616757916	1.185492	1.125798	1.43218131	0.204714	0.907046	1.457997	0.056477
	-0.416378043	-0.831308	-0.682631	-0.3546938	-0.787918	1.334797	1.811213	1.14876

Means
-0.060752711	0.205536	-0.096191	-0.1095192	-0.009159	0.206538	0.272333	0.127728

Stdevs
1.071654353	0.992258	1.071104	0.99391675	0.931993	1.109639	1.06644	0.988844

Pooled stdev	1.029755
StErr of diff	0.188007
t-multiple	2.001716

95% confidence intervals for differences between means

Difference	Lower limit	Upper limit	Significant?
1 minus 2	-0.642624	0.110048	No
1 minus 3	-0.340897	0.411775	No
1 minus 4	-0.32757	0.425103	No
5 minus 7	-0.657828	0.094844	No
5 minus 8	-0.513224	0.239448	No
6 minus 7	-0.442131	0.310541	No
6 minus 8	-0.297526	0.455146	No
7 minus 8	-0.231731	0.520941	No
	Any significant?		Yes

Data table for whether any conf ints are significant

Replication	Any Signif?
	Yes
1	Yes
2	Yes
3	No
4	No
5	No
6	No
55	No
56	Yes
57	Yes
58	Yes
59	Yes
60	No
61	No
62	Yes
63	Yes
64	No
65	Yes
66	No
67	No
68	Yes
69	Yes
70	No
71	No
72	No
73	Yes
74	Yes
75	Yes
96	Yes
97	No
98	No
99	No
100	Yes
101	No
102	No
103	Yes
104	No
497	Yes
498	Yes
499	Yes
500	Yes

FIGURE 14.15 Simulation to Illustrate the Multiple Comparison Problem

be approximately 5%? The answer is that we need to *correct* the individual confidence intervals, so that we do *not* calculate them exactly as described in Chapter 9. Several corrections have been proposed by statisticians, and StatPro includes three of the most popular correction methods in its one-way ANOVA procedure: the Bonferroni, Tukey, and Scheffé methods. (They can be chosen from the dialog box shown in Figure 14.16 on page 784.) Although the details of these methods are beyond the scope of this book, they are all methods for coping with the multiple comparison problem. They differ only in the multiplier they use in the typical confidence interval formula for a difference between means:

$$\bar{Y}_i - \bar{Y}_j \pm \text{multiplier} \times \sqrt{MSW(1/n_i + 1/n_j)}$$

Recall that the multiplier used for the usual "no-correction" method from Chapter 9 is a t-value that, for a 95% confidence level, is approximately equal to 2 (at least for reasonably large sample sizes). The correction methods all use multiples that are *larger* than this. The idea is that by using a larger multiplier, we get a wider confidence interval. This decreases the chance that the confidence interval will fail to include the true mean dif-

FIGURE 14.16
Dialog Box for Multiple Confidence Intervals in ANOVA

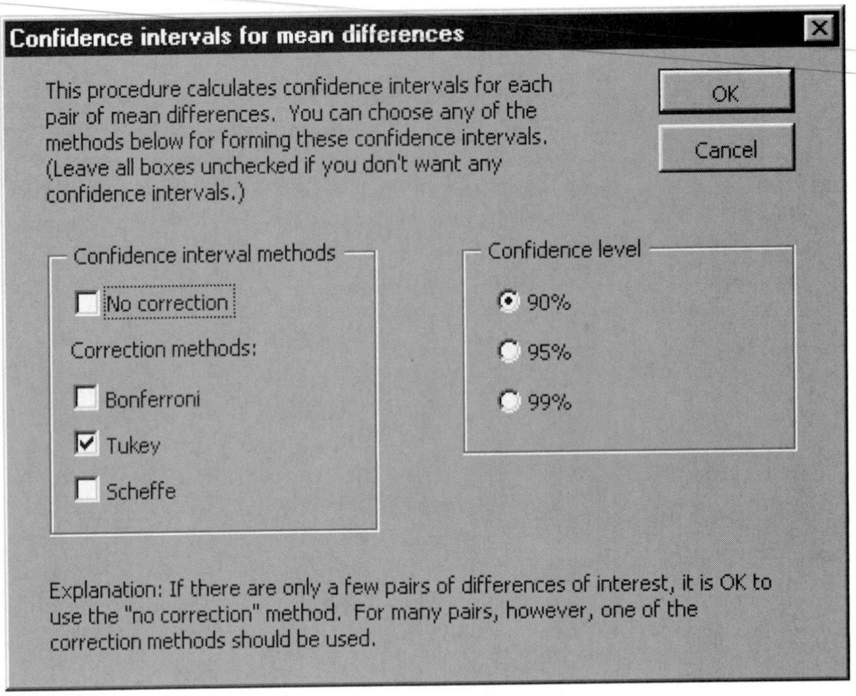

Confidence intervals for mean differences

This procedure calculates confidence intervals for each pair of mean differences. You can choose any of the methods below for forming these confidence intervals. (Leave all boxes unchecked if you don't want any confidence intervals.)

Confidence interval methods

☐ No correction

Correction methods:

☐ Bonferroni

☑ Tukey

☐ Scheffe

Confidence level

⦿ 90%

○ 95%

○ 99%

OK

Cancel

Explanation: If there are only a few pairs of differences of interest, it is OK to use the "no correction" method. For many pairs, however, one of the correction methods should be used.

ference, which in turn decreases the chance that at least one of several such confidence intervals will fail to include its true mean difference. The larger the multiplier is, the more conservative the confidence intervals will be (where "conservative" means wider intervals). Scheffé's and Bonferroni's methods tend to be the most conservative, whereas Tukey's method strikes a balance between being too conservative and not conservative enough. It is the method favored by many researchers when the focus is on many confidence intervals for mean differences, as in Example 14.1.

To follow our simulation one step further, the multiplier used in the individual confidence intervals in rows 84–111 of Figure 14.15 is approximately equal to 2, as shown in cell B80. Using appropriate formulas (not presented here), it can be shown that the multipliers for the Tukey, Bonferroni, and Scheffé methods are 3.04, 3.27, and 3.77, respectively. Furthermore, if the Tukey multiplier is used in the simulation, the percentage in cell K5 becomes approximately 5%, exactly what we want.

To see how these correction methods might affect results, we report all four types of confidence intervals for the cereal data of Example 14.1 in Figure 14.17. (We reported only the Tukey intervals earlier.) You should note the following. First, the confidence intervals get wider as we move from no correction (from Chapter 9) to Tukey to Bonferroni to Scheffé. Second, all three correction methods report exactly the same significant differences. Specifically, they all report that the only significant difference is between the next-to-highest and lowest shelf heights. The three correction methods do not agree exactly in all data sets, but they usually produce similar results. In contrast, the no-correction method finds 7 of the 10 differences to be significant, a very different result. This is typical. Because this method does not correct for the number of confidence intervals being reported, it tends to find *too many* significant differences.

At this point, it is natural to ask why there are so many methods. The reason has to do with the purpose of the study. A researcher who initiates a study might have a particular interest in a few specific differences. For example, the analyst in Example 14.1 might be particularly interested in the differences between the lowest height and each of the

FIGURE 14.17
Confidence Intervals from Different Methods

	A	B	C	D	E	F
23		**Confidence intervals for mean differences**				
24		Confidence level	95.0%			
25		**Usual (no correction) method**				
26		Difference	Mean diff	Lower	Upper	Signif?
27		Lowest - Next-to-lowest	-43.760	-86.111	-1.409	Yes
28		Lowest - Middle	-48.520	-90.871	-6.169	Yes
29		Lowest - Next-to-highest	-91.360	-133.711	-49.009	Yes
30		Lowest - Highest	-48.960	-91.311	-6.609	Yes
31		Next-to-lowest - Middle	-4.760	-47.111	37.591	No
32		Next-to-lowest - Next-to-highest	-47.600	-89.951	-5.249	Yes
33		Next-to-lowest - Highest	-5.200	-47.551	37.151	No
34		Middle - Next-to-highest	-42.840	-85.191	-0.489	Yes
35		Middle - Highest	-0.440	-42.791	41.911	No
36		Next-to-highest - Highest	42.400	0.049	84.751	Yes
37		**Bonferroni method**				
38		Difference	Mean diff	Lower	Upper	Signif?
39		Lowest - Next-to-lowest	-43.760	-104.932	17.412	No
40		Lowest - Middle	-48.520	-109.692	12.652	No
41		Lowest - Next-to-highest	-91.360	-152.532	-30.188	Yes
42		Lowest - Highest	-48.960	-110.132	12.212	No
43		Next-to-lowest - Middle	-4.760	-65.932	56.412	No
44		Next-to-lowest - Next-to-highest	-47.600	-108.772	13.572	No
45		Next-to-lowest - Highest	-5.200	-66.372	55.972	No
46		Middle - Next-to-highest	-42.840	-104.012	18.332	No
47		Middle - Highest	-0.440	-61.612	60.732	No
48		Next-to-highest - Highest	42.400	-18.772	103.572	No
49		**Tukey method**				
50		Difference	Mean diff	Lower	Upper	Signif?
51		Lowest - Next-to-lowest	-43.760	-103.050	15.530	No
52		Lowest - Middle	-48.520	-107.810	10.770	No
53		Lowest - Next-to-highest	-91.360	-150.650	-32.070	Yes
54		Lowest - Highest	-48.960	-108.250	10.330	No
55		Next-to-lowest - Middle	-4.760	-64.050	54.530	No
56		Next-to-lowest - Next-to-highest	-47.600	-106.890	11.690	No
57		Next-to-lowest - Highest	-5.200	-64.490	54.090	No
58		Middle - Next-to-highest	-42.840	-102.130	16.450	No
59		Middle - Highest	-0.440	-59.730	58.850	No
60		Next-to-highest - Highest	42.400	-16.890	101.690	No
61		**Scheffe method**				
62		Difference	Mean diff	Lower	Upper	Signif?
63		Lowest - Next-to-lowest	-43.760	-110.684	23.164	No
64		Lowest - Middle	-48.520	-115.444	18.404	No
65		Lowest - Next-to-highest	-91.360	-158.284	-24.436	Yes
66		Lowest - Highest	-48.960	-115.884	17.964	No
67		Next-to-lowest - Middle	-4.760	-71.684	62.164	No
68		Next-to-lowest - Next-to-highest	-47.600	-114.524	19.324	No
69		Next-to-lowest - Highest	-5.200	-72.124	61.724	No
70		Middle - Next-to-highest	-42.840	-109.764	24.084	No
71		Middle - Highest	-0.440	-67.364	66.484	No
72		Next-to-highest - Highest	42.400	-24.524	109.324	No

other four heights. The whole study is intended to study these specific differences. In this case, the differences of interest are called **planned comparisons.** On the other hand, the analyst might initiate the study just to see what differences there are. This analyst will examine all pairwise differences to see which are significant. Here we talk about **unplanned comparisons** because the analyst does not specify which differences to focus on *before* collecting the data.

In the case of planned comparisons, if there are only a few differences of interest, it is usually acceptable to report confidence intervals for these differences using the no-correction method. If there are more than a few planned comparisons (even trained statisticians do not always agree on the interpretation of "a few"), then it is better to report Bonferroni intervals. In the case of unplanned comparisons, the Tukey method is usually the preferred method. It keeps the overall confidence level close to the desired level (such as 95%) without making the intervals overly wide. More important, it keeps the entire study from becoming a "fishing expedition," where a few differences become significant just by the luck of the draw (as occurred in the simulation in Figure 14.15).

The Scheffé method can be used for planned or unplanned comparisons. It tends to produce the widest intervals because it is intended not only for differences between means, such as $\mu_2 - \mu_4$, but also for more general **contrasts,** where a contrast is the difference between weighted averages of means. For example, if the analyst in Example 14.1 is interested in how the lowest height compares to the *average* of the other four heights, then the difference $\mu_1 - (\mu_2 + \mu_3 + \mu_4 + \mu_5)/4$ would be of interest. Although the analysis of general contrasts such as this is deferred until the next section, we note that Scheffé's method was developed specifically to deal with them. If we are interested only in simple differences like $\mu_2 - \mu_4$, then Tukey's method should be used instead.[5]

PROBLEMS

Level A

11. Consider again the one-way ANOVA hypothesis test described in Problem 1. Address the multiple comparison problem by applying the Bonferroni, Tukey, and Scheffé methods to obtain an *overall* confidence level of approximately 95%. Summarize your results. Recall that the relevant data are given in the file P14_01.XLS.

12. Consider again the one-way ANOVA hypothesis test described in Problem 2. Address the multiple comparison problem by applying the Bonferroni, Tukey, and Scheffé methods to obtain an *overall* confidence level of approximately 90%. How do these results compare to the uncorrected 90% confidence intervals? Recall that the relevant data are given in the file P14_02.XLS.

13. Consider again the one-way ANOVA hypothesis test described in Problem 3. Address the multiple comparison problem by applying the Bonferroni, Tukey, and Scheffé methods to obtain an *overall* confidence level of approximately 99%. Compare the widths of the confidence intervals generated with each of these methods with those of uncorrected 99% confidence intervals. Explain your findings. Recall that the relevant data are given in the file P14_03.XLS.

14. Consider again the one-way ANOVA hypothesis test described in Problem 4. Address the multiple comparison problem by applying the Bonferroni, Tukey, and Scheffé methods to obtain an *overall* confidence level of approximately 90%. Summarize your results. Re-

call that the relevant data are given in the file P14_04.XLS.

15. Consider again the one-way ANOVA hypothesis test described in Problem 5. Address the multiple comparison problem by applying the Bonferroni, Tukey, and Scheffé methods to obtain an *overall* confidence level of approximately 99%. How do these results compare to the uncorrected 99% confidence intervals? Recall that the relevant data are given in the file P14_05.XLS.

Level B

16. Consider again the one-way ANOVA hypothesis test described in Problem 6. Suppose that we are interested in comparing the mean starting salary of accounting students with that of each of the other three majors (i.e., marketing, finance, and management). Recall that the relevant data are provided in the file P14_06.XLS.
   a. Which method for generating confidence intervals for mean differences is most appropriate in this situation? Explain your choice.
   b. Apply the method identified in part **a** to estimate the mean differences of interest. Briefly interpret your findings.

17. Consider again the one-way ANOVA hypothesis test described in Problem 7. Suppose that we are interested in comparing the mean sales achieved by salespeople with varying compensation schemes. In other words, we are interested in making all possible com-

---

[5]See Chapter 17 of the textbook by Neter et al. (1996) for a much more complete discussion of multiple comparisons in ANOVA.

parisons at this point. Recall that the relevant data are provided in the file P14_07.XLS.

a. Which method for generating confidence intervals for mean differences is most appropriate in this situation? Explain your choice.

b. Apply the method identified in part **a** to estimate the mean differences of interest. Briefly interpret your findings.

14.5 Two-Way ANOVA

The examples discussed so far in this chapter have been single-factor designs. There is a single factor, such as shelf height in Example 14.1 or customer size in Example 14.2, that we observe at several levels. The question then is whether the mean of a response variable is equal across all levels. In this section we allow two factors, each at several levels. As we will see, some of the ideas from one-way ANOVA carry over to two-way ANOVA. However, there are differences in the data setup, the analysis itself, and, perhaps most important, the types of questions we ask. Because an abstract discussion of two-way ANOVA can be difficult to follow, we immediately introduce the following example.

Example **Driving Distances for Golf Ball Brands**

If you are a golfer, or even if you have ever seen golf ball commercials on television, you know that a number of golf ball manufacturers claim to have the "longest ball," that is, the ball that goes the farthest on drives. This example illustrates how these claims might be tested. We assume that there are five major brands, labeled A through E. A consumer testing service runs an experiment where 60 balls of each brand are driven under three temperature conditions. The first 20 are driven in cool weather (about 40 degrees), the next 20 are driven in mild weather (about 65 degrees), and the last 20 are driven in warm weather (about 90 degrees). The goal is to see whether some brands differ significantly, on average, from other brands and what effect temperature has on mean differences between brands. For example, it is possible that brand A is the longest ball in warm weather but some other brand is longest in cool temperatures.

Objective To use two-way ANOVA to analyze the effects of golf ball brands and temperature on driving distances.

Solution

This example represents a controlled experiment. The consumer testing service decides exactly how to run the experiment, namely, by assigning 20 randomly chosen balls of each brand to each of three temperature levels. In our general terminology, the experimental units are the individual golf balls and the response variable is the length (in yards) of each drive. There are two factors: brand and temperature. The brand factor has five treatment levels, A through E, and temperature has three levels, cool, mild, and warm. The design is balanced because the same number of balls, 20, is used at each of the $5 \times 3 = 15$ treatment level combinations. In fact, balanced designs are the only two-way designs we will discuss in this book. (The analysis of unbalanced designs is more complex and is best left to a more advanced book.) There is one further piece of terminology. We call this a **full factorial** two-way design because we test golf balls at *each* of the 15 possible treatment level combinations. If, for example, we decided not to test any brand A balls at a temperature of 65 degrees, then the resulting experiment would be called an design. We will discuss incomplete designs briefly in the next section—and why they are sometimes used—but we prefer full factorial designs whenever possible.

Full Factorial Design	In a **full factorial design,** we assign experimental units to *each* treatment level combination.

Incomplete Design	In an **incomplete design,** we assign experimental units to some of the treatment level combinations but not to all of them.

How should the consumer testing service actually carry out the experiment? One possibility is to have 15 golfers, each of approximately the same skill level, hit 20 balls each. Golfer 1 could hit 20 brand A balls in cool weather, golfer 2 could hit 20 brand B balls in cool weather, and so on. You can probably see the downside of this design. Brand A might come out the longest ball just because the golfers assigned to brand A have good days. Therefore, if the consumer testing company decides to use human golfers, it should spread them evenly among brands and weather conditions. For example, it could employ 10 golfers to hit two balls of each brand at each of the weather conditions. Even here, however, the use of different golfers introduces an unwanted source of variation: the different abilities of the golfers (or how well they happen to be driving that day). Is the solution, then, to use a *single* golfer for all 300 balls? This has its own downside—namely, that the golfer might get tired in the process of hitting this many balls. Even if he hits the brands in random order, the fatigue factor could play a role in the results.

These are the types of things designers of experiments must consider. They must attempt to eliminate as many unwanted sources of variation as possible, so that any differences across the factor levels of interest can be attributed to these factors and not to extraneous factors. In this example, we suspect that the best option for the consumer testing service is to employ a "mechanical" golf ball driving machine to hit all 300 balls. This should reduce the inevitable random variation that would occur by using human golfers. Still, there will be some random variation. Even a mechanical device, hitting the same brand under the same weather conditions, will not hit every drive exactly the same length.

Data for StatPro's two-way ANOVA procedure cannot be in unstacked form. Also, a balanced design must be used.

Once the details of the experiment have been decided and the golf balls have been hit, we will have 300 observations (yardages) at various conditions. The usual way to enter the data in Excel—and the *only* way StatPro's two-way ANOVA procedure will accept it—is in the stacked form shown in Figure 14.18. (See the file GOLFBALL.XLS.) There must be two "code" variables that represent the levels of the two factors (Brand and Temp) and a measurement variable that represents the response variable (Yards). Although many rows are hidden in this figure, there are actually 300 rows of data, 20 for each of the 15 combinations of Brand and Temp. Again, this is a *balanced* design, which is what StatPro expects for its two-way ANOVA procedure. (StatPro will issue an error message if it finds an unbalanced design, that is, unequal numbers of observations at the various treatment level combinations.)

Now that we have the data, what can we learn from them? In fact, which questions should we ask? Here it helps to look at a table of sample means, such as in Figure 14.19 (page 790). (This table is part of the output from StatPro's two-way ANOVA procedure. Alternatively, it can be obtained easily with the Excel pivot table tool, as we did here.[6]) Prompted by this table, here are some questions we might ask.

[6]Note that the default label Excel uses in cells L10 and P4 is Grand Total—and you cannot change them. However, they are really "grand averages."

FIGURE 14.18
Data for Golf Ball
Example

	A	B	C
1	Golf ball driving distance study		
2			
3	Brand	Temp	Yards
4	A	Cool	220.6
5	A	Cool	204.0
6	A	Cool	233.6
7	A	Cool	229.1
8	A	Cool	214.6
9	A	Cool	208.8
10	A	Cool	204.6
11	A	Cool	219.2
12	A	Cool	208.0
13	A	Cool	214.3
14	A	Cool	219.6
15	A	Cool	240.9
16	A	Cool	212.1
17	A	Cool	215.3
18	A	Cool	240.0
19	A	Cool	221.9
20	A	Cool	229.7
21	A	Cool	216.7
22	A	Cool	215.3
23	A	Cool	208.1
24	A	Mild	248.6
25	A	Mild	217.0
26	A	Mild	230.8
27	A	Mild	237.5
28	A	Mild	237.1
29	A	Mild	224.9
30	A	Mild	238.1

1. Looking at column P, do any brands average significantly more yards than any others (where these averages are averages over all temperatures)?

2. Looking at row 10, do average yardages differ significantly across temperatures (where these averages are across all brands)?

3. Looking at the *columns* in the range M5:O9, do differences among averages of brands depend on temperature? For example, does one brand dominate in cool weather and another in warm weather?

4. Looking at the *rows* in the range M5:O9, do differences among averages of temperatures depend on brand? For example, are some brands very sensitive to changes in temperature, while others are not?

It is useful to characterize the type of information these questions are seeking. Question 1 is asking about the **main effect** of the brand factor. If we ignore the temperature (by averaging over the various levels of it), do some brands tend to go farther than some others? This is obviously a key question for the study. Question 2 is also asking about a main effect, the main effect of the temperature factor. If we ignore the brand (by averaging over all brands), do balls tend to go farther in some temperatures than others? The answer to this question is obvious to golfers. They all know that balls compress better, and hence go farther, in warm temperatures than cool temperatures. Therefore, this is not a

FIGURE 14.19

Table of Sample
Means in Golf Ball
Example

	L	M	N	O	P
1	Table of sample mean yardages				
2					
3	Average of Yards	Temp			
4	Brand	Cool	Mild	Warm	Grand Total
5	A	218.8	236.5	258.4	237.9
6	B	224.1	245.1	258.3	242.5
7	C	228.0	242.7	263.0	244.6
8	D	215.0	237.6	256.1	236.2
9	E	224.8	255.7	270.9	250.5
10	Grand Total	222.1	243.5	261.4	242.3

key question for the study, although we would certainly expect the study to confirm what experience tells us.

Main Effects

Main effects indicate whether there are different means for treatment levels of one factor when averaged over the levels of the other factor.

Questions 3 and 4 are asking about **interactions** between the two factors. These interactions are often the most interesting results of a two-way study. Essentially, interactions (if there are any) provide information that could not be guessed by knowing the main effects alone. In this example, interactions are patterns of the averages in the range M5:O9 that could not be guessed by looking only at the "main effect" averages in column P and row 10. Specifically, the order of brands in column P, from largest to smallest average yardages, is E, C, B, A, D. If there were no interactions at all, this ordering would hold at each temperature. For these data, it is close. At the cool temperature, the ordering is C, E, B, A, D; for mild, it is E, B, C, D, A; for warm, it is E, C, A, B, D. Actually, having no interactions implies even more than the preservation of these rankings. It implies that the difference between any two brands' averages is the same at any of the three temperature levels. For example, the differences between brands E and D at the three temperatures are $224.8 - 215.0 = 9.8$, $255.7 - 237.6 = 18.1$, and $270.9 - 256.1 = 14.8$. If there were no interactions at all, these three differences would be equal.

Interactions

Interactions indicate patterns of differences in means that could not be guessed from the main effects alone. They exist when the effect of one factor on the response variable depends on the level of the other factor.

Remember that the interactions become stronger as the lines in either of these graphs become more nonparallel.

The concept of interaction is much easier to understand by looking at graphs. The graphs in Figures 14.20 and 14.21, which are both outputs from StatPro's two-way ANOVA procedure, represent two ways of looking at the pattern of averages for different combinations of brand and temperature—that is, the averages in the range M5:O9 of Figure 14.19. The first of these shows a line for each brand, where each point on the line corresponds to a different temperature. The second shows the same information with the roles of brand and temperature reversed. Neither graph is "better" than the other; they simply show the same data from different perspectives. The key to either is whether the lines are *parallel*. If they are, then there are no interactions—the effect of one factor on average yardage is the *same* regardless of the level of the other factor. The more nonparallel they are, however, the stronger the interactions are. The lines in either of these graphs

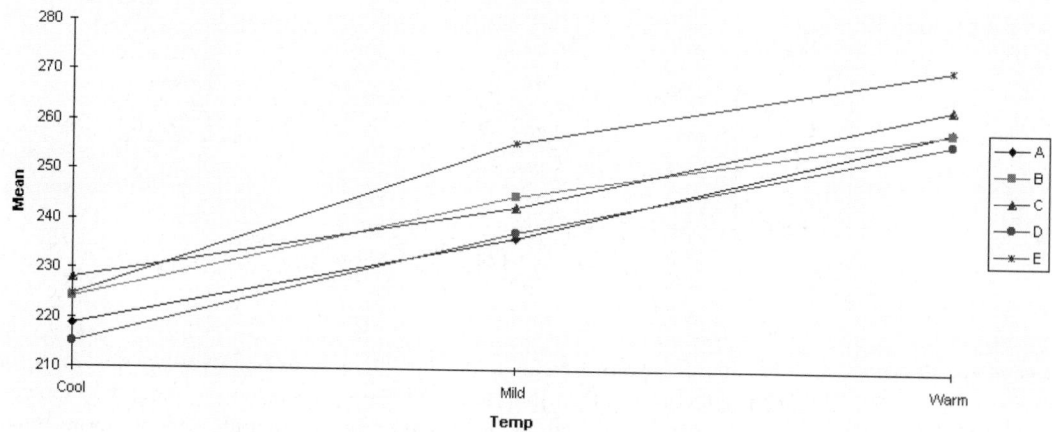

Chart of Cell Means to Check for Interactions

FIGURE 14.20 One View of Interactions in Golf Ball Example

are not exactly parallel, but they are nearly so. This implies that there is very little interaction between brand and temperature in these data.

In general, interactions can be of several types. We show two contrasting types in Figures 14.22 and 14.23 on page 792. (For simplification, these focus on two brands only. They are based on *different* data from those used in the GOLFBALL.XLS file.) In Figure 14.22, brand A dominates at all temperatures. However, there is an interaction because the difference between brands increases as temperature increases. In this situation the interaction effect is interesting, but the main effect of brand—brand A is better when averaged over all temperatures—is also interesting. The situation is quite different in Figure 14.23, where there is a "crossover." Brand A is somewhat better at cool temperatures, but brand B is better at mild and warm temperatures. In this case the interaction is the most interesting finding, and the main effect of brand is much less interesting. In simple terms, if you are a golfer, you'd buy brand A in cool temperatures and brand B otherwise, and you wouldn't care very much which brand is better when averaged over *all* temperatures.

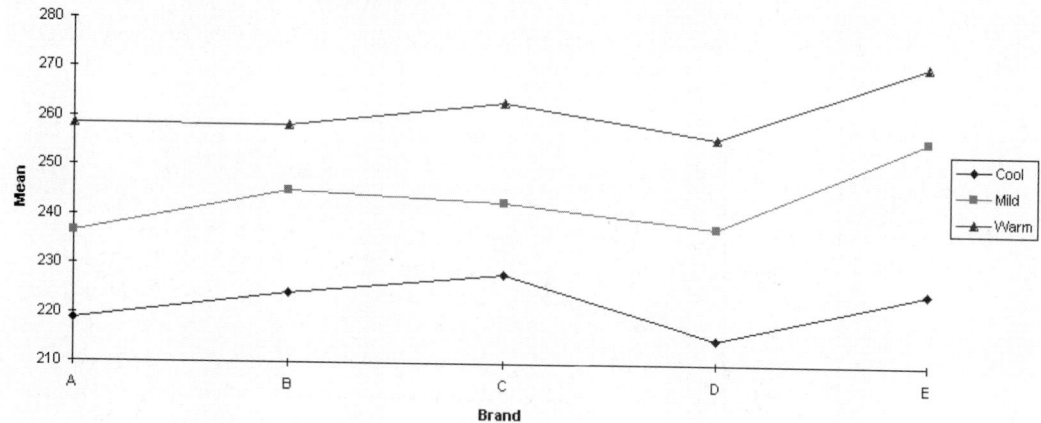

Chart of Cell Means to Check for Interactions

FIGURE 14.21 Another View of Interactions in Golf Ball Example

FIGURE 14.22
One Possible Pattern of Interactions

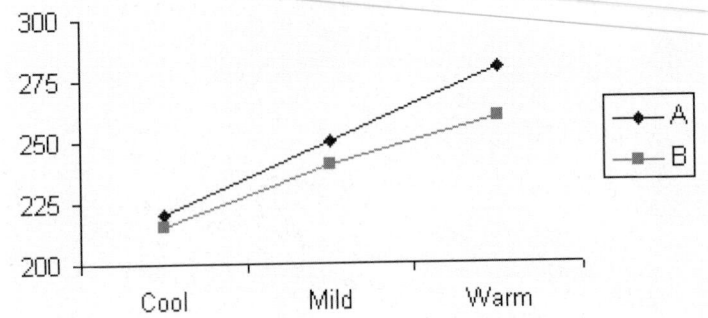

For these reasons, we check *first* for interactions in a two-way design. If there are significant interactions, then the main effects might not be as interesting. However, if there are no significant interactions, then main effects generally become more important.

Summing up what we have seen so far, main effects are differences in averages across the levels on one factor, where these averages are averages over *all* levels of the other factor. In a table of sample means, such as in Figure 14.19, we can check for main effects by looking at the averages in the "Grand Total" column and row. In contrast, the interactions are patterns of averages in the main body of the table and are best shown graphically, as in Figures 14.20 and 14.21. They indicate whether the effect of one factor depends on the level of the other factor.

The next question is whether the main effects and interactions we see in a table of sample means are statistically significant. As in one-way ANOVA, this is answered by an ANOVA table. However, instead of having just two sources of variation, within and between, as in one-way ANOVA, there are now four sources of variation: one for the main effect of each factor, one for interactions, and one for variation within treatment level combinations. For the golf ball data, two-way ANOVA separates the total variation across all 300 observations into four sources. First, there is variation due to different brands producing different average yardages. Second, there is variation due to different average yardages at different temperatures. Third, there is variation due to the interactions we saw in the interaction graphs. Finally, there is the same type of "within" variation as in one-way ANOVA. This is the variation that occurs because yardages for the 20 balls of the same brand hit at the same temperature are not all identical. (This within variation is usually called the "error" variation in statistical software packages.)

Two-way ANOVA collects this information about the different sources of variation, using fairly complex formulas, in an ANOVA table as shown in Figure 14.24. (This is the output from StatPro, using the StatPro/Statistical Inference/Two-way ANOVA menu item,

FIGURE 14.23
Another Possible Pattern of Interactions

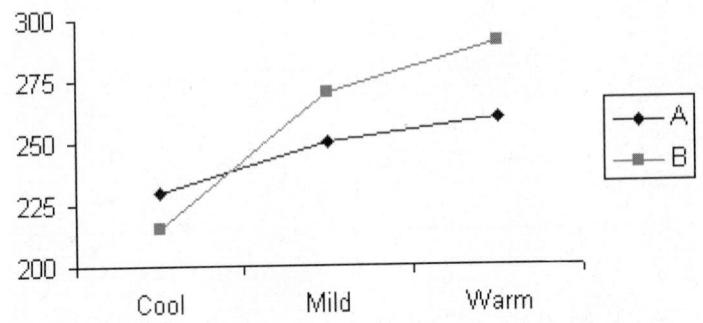

FIGURE 14.24
StatPro Two-Way
ANOVA Output for
Golf Ball Data

	E	F	G	H	I	J	K
3	TwoWay ANOVA results for Yards						
4							
5	Sample sizes for treatments (Brand down, Temp across)						
6			Cool	Mild	Warm		
7		A	20	20	20	60	
8		B	20	20	20	60	
9		C	20	20	20	60	
10		D	20	20	20	60	
11		E	20	20	20	60	
12			100	100	100		
13							
14	Sample means for treatments (Brand down, Temp across)						
15			Cool	Mild	Warm		
16		A	218.820	236.450	258.435	237.902	
17		B	224.145	245.130	258.270	242.515	
18		C	227.995	242.715	263.040	244.583	
19		D	214.995	237.620	256.110	236.242	
20		E	224.785	255.745	270.940	250.490	
21			222.148	243.532	261.359		
22							
23	Sample standard deviations for treatments (Brand down, Temp across)						
24			Cool	Mild	Warm		
25		A	10.904	8.826	11.009	19.222	
26		B	11.696	9.801	8.931	17.365	
27		C	10.849	14.251	7.084	18.149	
28		D	13.644	10.176	12.127	20.691	
29		E	10.665	10.960	9.052	21.836	
30			12.285	12.778	10.977		
31							
32	TwoWay ANOVA Table						
33		Source	SS	df	MS	F	p-value
34		Brand	7702.436	4	1925.609	16.466	0.000
35		Temp	77085.997	2	38542.998	329.584	0.000
36		Interaction	1999.966	8	249.996	2.138	0.032
37		Error	33329.126	285	116.944		
38		Total	120117.526	299			

selecting Brand and Temp as the "code" variables and Yards as the "measurement" variable. The output includes tables of sample sizes, sample means, and sample standard deviations, as well as the ANOVA table.) The four sources of variation appear in rows 34–37. Rows 34 and 35 are for the main effects of brand and temperature, row 36 is for interactions, and row 37 is for the within (or error) variation. Each source has a sum of squares and a degrees of freedom. Also, each has a mean square, the ratio of the sum of squares to the degrees of freedom. Finally, the first three sources have an F-ratio and an associated p-value, where each F-ratio is the ratio of the mean square in that row to the mean square error in cell I37.

We test whether main effects or interactions are statistically significant in the usual way—by examining p-values. Specifically, we claim statistical significance if the corresponding p-value is sufficiently small, less than 0.05, say. Looking first at the interactions, the p-value is about 0.03, which says that the lines in the interaction graphs are significantly nonparallel, at least at the 5% significance level. We might dispute whether this nonparallelism is *practically* significant, but there is statistical evidence that at least some interaction between brand and temperature exists. The two p-values for the main effects in cells K34 and K35 are practically 0, meaning that there *are* differences across brands and across temperatures. Of course, the main effect of temperature was a foregone conclusion—we already knew that balls do not go as far in cold temperatures—but the main effect of brand is more interesting. According to the evidence, some brands definitely go farther, on average, than some others. ●

14.5.1 Confidence Intervals for Contrasts

If we find that main effects and/or interactions are significant, then we will probably want to check which factor levels, or factor level combinations, produce significantly larger means than which others. Recall that StatPro's one-way ANOVA procedure provides the option of giving confidence intervals for differences between each pair of means. We do not provide this same option in two-way ANOVA because there would typically be too much output to digest, and much of it would probably not be very useful. Given the purposes of any particular study, there are usually a few comparisons we would like to make, and this can be done fairly easily after the StatPro two-way ANOVA procedure has been run. We will illustrate the methods in this section.

First, we recall that a **contrast** is any difference between weighted averages of means. An example of a simple contrast is the difference between two means, such as $\mu_3 - \mu_1$. We would study this contrast if we were interested in whether μ_3 is different from μ_1. An example of a more complex contrast is $(\mu_1 + \mu_2)/2 - (\mu_3 + \mu_4 + \mu_5)/3$. We would study this contrast if we were interested in whether the average of μ_1 and μ_2 is different from the average of μ_3, μ_4, and μ_5. Note that the coefficients of these contrasts sum to 0. For example, $(\frac{1}{2} + \frac{1}{2}) - (\frac{1}{3} + \frac{1}{3} + \frac{1}{3}) = 0$. All contrasts have this property. Obviously, many contrasts could be constructed. The ones that we construct for any particular study depend entirely on what we are interested in. We might be interested in several simple contrasts or one or two more complex contrasts.

Contrast	A **contrast** is any linear combination of means (sum of coefficients multiplied by means) such that the sum of the coefficients is 0. It is typically used to compare one weighted average of means to another.

StatPro's two-way ANOVA procedure finds *MSW* for this formula. However, you must calculate the other ingredients manually with Excel formulas.

Once we have used StatPro to run a two-way ANOVA, we can then form confidence intervals for any contrasts of interest. The general form of the confidence interval is given by expression (14.5). Here, the point estimate of the contrast is formed by substituting sample means for the μ's in the contrast, *MSW* is the mean square error from the ANOVA table, n_j is the sample size corresponding to any particular mean in the contrast, c_j is the coefficient of the corresponding μ_j in the contrast, and the summation is over all terms in the contrast.

Confidence Interval for Contrast

$$\text{Point estimate of contrast} \pm \text{multiplier} \times \sqrt{MSW \sum_j c_j^2/n_j} \tag{14.5}$$

As an example, if the contrast is a simple difference between means such as $\mu_1 - \mu_4$, then the point estimate is $\bar{Y}_1 - \bar{Y}_4$, and the c_j's are $c_1 = 1$ and $c_4 = -1$, so that $c_1^2 = 1$ and $c_4^2 = 1$. Therefore, the confidence interval becomes

$$\bar{Y}_1 - \bar{Y}_4 \pm \text{multiplier} \times \sqrt{MSW(1/n_1 + 1/n_4)}$$

which is the same as the formula given in Sections 14.2 and 14.4 for one-way ANOVA.

As in one-way ANOVA, the multiplier in the confidence interval can be chosen in several ways to handle the multiple comparison problem appropriately. We indicate typical possibilities in the following continuation of the golf ball example.

Example 14.3 Driving Distances for Golf Ball Brands (continued)

One golf ball retail shop would like to test the claims that (1) brand C beats the average of the other four brands in cool weather and (2) brand E beats the average of the other four brands when it is not cool. Are these two claims supported by the data in the GOLF-BALL.XLS file?

Objective To form and test contrasts for the golf ball data, and to interpret the results.

Solution

Let $\mu_{C,Cool}$ be the mean yardage for brand C balls hit in cool weather, and define similar means for the other brands and temperatures. Then the first claim concerns the contrast

$$\mu_{C,Cool} - \frac{\mu_{A,Cool} + \mu_{B,Cool} + \mu_{D,Cool} + \mu_{E,Cool}}{4}$$

and the second concerns the contrast

$$\frac{\mu_{E,Mild} + \mu_{E,Warm}}{2} - \frac{\frac{\mu_{A,Mild} + \mu_{A,Warm}}{2} + \frac{\mu_{B,Mild} + \mu_{B,Warm}}{2} + \frac{\mu_{C,Mild} + \mu_{C,Warm}}{2} + \frac{\mu_{D,Mild} + \mu_{D,Warm}}{2}}{4}$$

$$= \frac{\mu_{E,Mild} + \mu_{E,Warm}}{2} - \frac{\mu_{A,Mild} + \mu_{A,Warm} + \mu_{B,Mild} + \mu_{B,Warm} + \mu_{C,Mild} + \mu_{C,Warm} + \mu_{D,Mild} + \mu_{D,Warm}}{8}$$

(Note in this second contrast how we average over the mild and warm temperatures. This is because the second claim just specifies "not cool.") A good way to handle the calculations in Excel is illustrated in Figure 14.25 on page 796. (This output is right below the output illustrated in Figure 14.24, so you should refer to that output as well.) For either contrast, we first record the coefficients of the means in the contrast. These appear in the ranges G46:I50 and G59:I63. (Note that the sum of the values in each of these ranges is 0. This is required for contrasts, as we discussed previously.) Then the point estimate of a contrast is the SUMPRODUCT of the sample means and these coefficients. For example, we calculate the point estimate of the first contrast in cell G52 with the formula

=SUMPRODUCT(G46:I50,G16:I20)

The multiplier for these confidence intervals is always a thorny issue, but most statisticians agree that if only a small number of confidence intervals are being formed, as we are doing here, then we can use the usual t-value, where the degrees of freedom is the one corresponding to the error (within) variation. Therefore, the multiplier for each contrast is approximately 2, found with the formula

=TINV(1-G41,H37)

FIGURE 14.25

Confidence Intervals for Contrasts in Golf Ball Example

	E	F	G	H	I	J
40	**Confidence intervals for contrasts**					
41		Confidence level	95%			
42						
43		Comparing brand C against average of others in cool weather				
44		Matrix of coefficients				
45			Cool	Mild	Warm	
46		A	-0.25	0	0	
47		B	-0.25	0	0	
48		C	1	0	0	
49		D	-0.25	0	0	
50		E	-0.25	0	0	
51		Multiplier	1.968			
52		Point estimate	7.31			
53		Lower limit	1.99			
54		Upper limit	12.63			
55						
56		Comparing brand E against average of others in non-cool weather				
57		Matrix of coefficients				
58			Cool	Mild	Warm	
59		A	0	-0.125	-0.125	
60		B	0	-0.125	-0.125	
61		C	0	-0.125	-0.125	
62		D	0	-0.125	-0.125	
63		E	0	0.5	0.5	
64		Multiplier	1.968			
65		Point estimate	13.62			
66		Lower limit	9.86			
67		Upper limit	17.38			

Then since each sample size is 20 (the value in cell G7), we can find the lower and upper limits of the confidence intervals from expression (14.5). For example, the confidence interval for the first contrast is found with the formulas

$$=\text{G52-G51*SQRT(I37*SUMSQ(G46:I50)/G7)}$$

and

$$=\text{G52+G51*SQRT(I37*SUMSQ(G46:I50)/G7)}$$

in cells G53 and G54.

As we see, both claims are supported. The confidence intervals for the two contrasts extend from 1.99 to 12.63 and from 9.86 to 17.38—all positive yardages. It looks like brand C beats the average of the competition by at least 1.99 yards in cool weather, and brand E beats the average of the competition by at least 9.86 yards in weather that was not cool. ●

> Remember that Excel's SUMSQ function sums the squares of the values in a given range.

If we want to examine a lot of contrasts, then we should use one of the other confidence interval methods discussed in Section 14.4, the two preferred methods being the Bonferroni and Scheffé methods. The only difference is in the multiplier used. For the Bonferroni method, suppose we want to form k confidence intervals. Then rather than using the *t*-value that has probability α in the tails, we use the *t*-value that has probability α/k in the tails. For example, if we want to form $k = 2$ confidence intervals at the 95% confidence level, as above, then $\alpha = 0.05$ and $\alpha/k = 0.025$, so we put probability 0.025 in the tails rather than 0.05. The effect is that each of the two confidence intervals is constructed separately at the 97.5% level. The multiplier in the golf ball example would use the formula

$$=\text{TINV}((1\text{-}G41)/2,H37)$$

which evaluates to 2.253. This larger multiplier would result in slightly wider confidence intervals.

The Scheffé method is the most "conservative" method in the sense that it generally produces the widest confidence intervals. However, the relevant multiplier for this method is rather complex and will not be discussed here.

14.5.2 Assumptions of Two-Way ANOVA

The assumptions for the two-way ANOVA procedure are basically the same as for one-way ANOVA. If we focus on any particular combination of factor levels, such as brand A golf balls hit in cool weather, then we assume that (1) the distribution of values (yardages) for this combination is normal, and (2) the variance of values at this combination is the same as at any other combination. It is always wise to check for at least gross violations of these assumptions, especially the equal-variance assumption. The StatPro output provides an informal check by providing a table of standard deviations for the factor level combinations. For the golf ball example, this table is shown in Figure 14.26. Obviously, these standard deviations are not all exactly equal, but we would never expect *exact* equality in any real study. Because these standard deviations are of similar magnitude, there is no reason to worry about the equal-variance assumption for these data. Besides, the equal-variance assumption is less important when the design is balanced, as this one is.

As we demonstrated in Example 14.2, however, the log transformation is often useful when variances are far from equal across factor level combinations. At least, this transformation is often worth trying. If it works—that is, if it tends to equalize the variances and maybe even make the data more normal—then two-way ANOVA can be carried out on the log-transformed data exactly as we demonstrated in Example 14.2.

FIGURE 14.26
Checking the Equal-Variance Assumption

			Cool	Mild	Warm	
23	*Sample standard deviations for treatments (Brand down, Temp across)*					
24			Cool	Mild	Warm	
25		A	10.904	8.826	11.009	19.222
26		B	11.696	9.801	8.931	17.365
27		C	10.849	14.251	7.084	18.149
28		D	13.644	10.176	12.127	20.691
29		E	10.665	10.960	9.052	21.836
30			12.285	12.778	10.977	

PROBLEMS

Level A

18. Suppose a company that sells residential carpet cleaning equipment wants to judge the sales effectiveness of two factors: factor *A*, the type of sales presentation used by its salespeople, and factor *B*, the type of previous experience or training its salespeople have had in selling this type of equipment. Specifically, there are two types of presentations the company wishes to test. These two levels of factor *A* are the "hard-sell" approach, level 1, and the more relaxed "soft-sell" approach, level 2. The company also differentiates among four levels of past experience/training. These levels of factor *B* are labeled 1 through 4 and are defined as follows: (1) no past experience as a salesperson and no formal training in how to be a salesperson, (2) no past experience and some formal training, (3) some past experience and no formal training, and (4) some past experience and some formal training.

To see how presentation and experience/training affect sales, the company runs an experiment with 80 of its recently hired salespeople, 20 of whom fall into each of the four experience/training levels described above. Within each group of 20 salespeople at a given experience/training level, 10 are told to use a hard-sell approach and the other 10 are told to use a soft-sell approach. Of course, the hard-sellers and soft-sellers are instructed very carefully in the types of presentation they are supposed to use. During a 4-month period, the number of sales for each of the 80 salespeople is recorded. The data are provided in the file P14_18.XLS. The company wishes to infer from these data whether the different presentations and experience/training backgrounds cause significant differences in sales.

  a. Assess the main effect of the presentation approach factor upon sales.
  b. Assess the main effect of the previous experience and training factor upon sales.
  c. Do you find evidence of significant interactions between the two factors in this case? Explain.

19. A study is performed on a sample of residential homes to discover whether the size of the monthly heating bill depends on the type of heat or the type of home. In particular, three types of heat are examined: electric, natural gas, and oil. Also, all homes are classified into two types: those on a single level and those with at least two stories. In a single community, ten houses of each type, using each type of heat, are located and their heating bills for February of the past year are observed. These data are recorded in the file P14_19.XLS. Assume that the homes in this study are approximately equivalent in terms of overall square footage and level of insulation.

  a. Do you find evidence of a significant main effect for the heat type factor in this case? Explain.
  b. Do you find evidence of a significant main effect for the home type factor in this case? Explain.
  c. Do you find evidence of significant interactions between the two factors in this case? Explain.

20. A Chrysler dealer would like to know whether the amount of money spent on a new automobile depends on (1) the age of the buyer, and (2) whether the buyer is accompanied by his or her spouse. Data on 60 recent new vehicle purchases, including purchase prices, are provided in the file P14_20.XLS. Test for any significant main effects and interactions at the 10% level, and briefly summarize your findings.

Level B

21. Consider again the two-way ANOVA hypothesis test described in Problem 18. Construct a 95% confidence interval for each possible contrast between means in this case. Interpret your results. Recall that the relevant data are given in the file P14_18.XLS.

22. Consider again the two-way ANOVA hypothesis test described in Problem 19. Recall that the relevant data are given in the file P14_19.XLS.

  a. A natural gas supplier claims that homes that use gas heat generate an average February heating bill that is *less than* the average February heating bill of all other homes that use electricity or heating oil. Is this claim supported by the given data? Explain.
  b. A heating oil supplier claims that homes that use heating oil generate an average February heating bill that is *less than* the average February heating bill of all other homes that use electricity or natural gas. Is this claim supported by the given data? Explain.

23. The file P14_23.XLS provides data for a two-way ANOVA in which each of the two factors has two levels. Note that there are 25 observations for each of the four treatment combinations in this case.

  a. Are the assumptions of two-way ANOVA met in this case? If not, do what you can to correct any problem(s).
  b. Test for any significant main effects and interactions at the 5% level. Briefly summarize your results.
  c. Construct a 95% confidence interval for each possible contrast between means in this case. Interpret your results.

14.6 More About Experimental Design

The purpose of this chapter is to introduce key ideas and analysis techniques for the most common (and simple) single-factor and multi-factor models. In each of these models, we analyze how a response variable varies when one or more factors are varied at several levels. Although the same analysis can be used in observational studies or in designed experiments, we focus now on designed experiments. This is particularly important because many businesses are just now beginning to see the potential of designed experiments for reducing cost, increasing profit, and producing higher-quality items. (We saw examples of this in the introductory vignette to this chapter.)

We can break up the topic of experimental design into two parts: (1) the actual design of the experiment, and (2) the analysis of the resulting data. In this section we will expand on these, mostly on the design but to some extent on the analysis, just to provide a sense of what is possible. Specifically, we will discuss some key issues in experimental design and some of the more popular designs. However, the discussion in this section is by no means complete. Many books have been written about experimental design and the required statistical analysis [see Berger and Maurer (2002), Schmidt and Launsby (1994), and DeVor et al. (1992), for example, for very readable accounts of the topic], so we can barely scratch the surface.

Experimental design, as opposed to the statistical methods for analyzing the resulting data, has to do with the selection of factors, the choice of the treatment levels, the way experimental units are assigned to the treatment level combinations, and the conditions under which the experiment is run. These decisions must be made *before* the experiment is performed, and they should be made very carefully. Experiments are typically costly and time-consuming, so we want to ensure that the experiment is designed (and performed) in a way that will give us the most useful information possible. Unfortunately, proper experimental design is by no means intuitive. We know we want the most for our money, but it is usually not clear how to achieve it. Therefore, a whole science of experimental design has developed through the years. We will summarize some of its most important results here.

14.6.1 Randomization

The purpose of most experiments is to see which of several factors have an effect on a response variable. The factors in question are chosen as those that are controllable and are most likely (among all possible factors) to have some effect. Often, however, there are "nuisance" factors that we cannot control, at least not directly. If nothing is done about these nuisance factors, they can possibly mask the effect of the "important" factors, so that we do not achieve the results we hoped for. One important method for dealing with such nuisance factors is **randomization,** where we attempt to spread out randomly the levels of the nuisance factors to the various levels of the experimental factors. We illustrate this extremely important idea in the following example.

Randomization	**Randomization** is the process of randomly assigning experimental units so that nuisance factors are spread uniformly across treatment levels.

Example 14.4 Testing for Sharpness in Three Brands of Inkjet Printers

A computer magazine company regularly tests products from different manufacturers for differences in various aspects of quality. For its next issue, it would like to test sharpness of printed image across three popular brands of inkjet printers. It purchases one printer of each brand, prints several pages on each printer, and measures the sharpness of image on a 0–100 scale for each page. The data and analysis appear in Figure 14.27. (See the file PRINTERS.XLS.) They indicate that printer A is best on average and C is worst. Why might these results be misleading?

Objective To use randomization of paper types to see whether differences in sharpness are really due to different brands of printers.

Solution

This is a single-factor design, where the single factor, brand of printer, is varied at three levels. Suppose, however, that there is another factor, type of paper, that is not the primary focus of the study but might affect the sharpness of image. For the sake of discussion, suppose further that all type 1 paper is used in printer A, all type 2 paper is used in printer B, and all type 3 paper is used in printer C. Then it is very possible that the apparent effect of printer is really an effect of paper type. Specifically, it is possible that type 1 paper tends to produce the sharpest image, *regardless* of the printer used. We can't know this for sure, but it is certainly possible given our (flawed) experimental design. The solution is to randomize over paper type. For each sheet of paper to be printed by any

	A	B	C	D	E	F	G	H	I	J
1	Testing printers for sharpness of image									
2										
3	Printer brand	Sharpness			Results of one-way ANOVA					
4	A	92								
5	A	85			Summary stats for samples					
6	A	88				Sharpness_A	Sharpness_B	Sharpness_C		
7	A	92			Sample sizes	10	10	10		
8	A	85			Sample means	88.800	83.700	81.800		
9	A	87			Sample standard deviations	3.011	2.669	4.022		
10	A	89			Sample variances	9.067	7.122	16.178		
11	A	92			Weights for pooled variance	0.333	0.333	0.333		
12	A	86								
13	A	92			Number of samples	3				
14	B	84			Total sample size	30				
15	B	88			Grand mean	84.767				
16	B	81			Pooled variance	10.789				
17	B	80			Pooled standard deviation	3.285				
18	B	83								
19	B	83			One-Way ANOVA table					
20	B	87			Source	SS	df	MS	F	p-value
21	B	81			Between variation	262.067	2	131.033	12.145	0.0002
22	B	86			Within variation	291.300	27	10.789		
23	B	84			Total variation	553.367	29			
24	C	80								
25	C	84								
26	C	77								
27	C	83								
28	C	83								
29	C	77								
30	C	85								
31	C	76								
32	C	86								
33	C	87								

FIGURE 14.27 Results from an Experiment Before Randomizing

FIGURE 14.28

Experimental
Design Using
Randomization

	A	B	C	D	E	F	G	H
1	Testing printers for sharpness of image							
2								
3	Printer brand	Random number	Paper type	Sharpness		Distribution of paper type		
4	A	0.266195164	1			Type	Pct	
5	A	0.664517239	2			1	50%	
6	A	0.435129345	1			2	35%	
7	A	0.61661941	2			3	15%	
8	A	0.030550167	1					
9	A	0.244512381	1					
10	A	0.567424817	2					
11	A	0.890457789	3					
12	A	0.397271654	1					
13	A	0.816237506	2					
14	B	0.479874446	1					
15	B	0.058263785	1					
16	B	0.419962876	1					
17	B	0.666729799	2					
18	B	0.164683134	1					
19	B	0.564385205	2					
20	B	0.038422271	1					
21	B	0.556448142	2					
22	B	0.088533026	1					
23	B	0.694174721	2					
24	C	0.701259602	2					
25	C	0.28187615	1					
26	C	0.516994921	2					
27	C	0.618448763	2					
28	C	0.996632788	3					
29	C	0.141936342	1					
30	C	0.168143457	1					
31	C	0.54821817	2					
32	C	0.261978154	1					
33	C	0.098773533	1					

printer, we randomly select a paper type. This will tend to even out the paper types across the printers. Then if the average sharpness of image from printer A is still higher than the averages from the other two brands, we will have more confidence that this is due to differences in printers, not types of paper. Note that it is *not* necessary to use equal numbers of sheets of each paper type in the experiment. For example, if paper type 1 is the most used paper type by actual users, then we might use more of it in the experiment. The important point is that no printer is fed a much higher proportion of any paper type than any other printer.

We illustrate how this might be implemented with random numbers in Figure 14.28. Based on actual usage, suppose that approximately 50% of the paper used in the experiment is of type 1, 35% is of type 2, and 15% is of type 3. This information is entered in columns F and G. Then to randomize paper types across printers, we enter random numbers in column B with the RAND() function, enter the formula

$$=IF(B4<\$G\$5,1,IF(B4<\$G\$5+\$G\$6,2,3))$$

in cell C4, and copy down column C. Of course, Figure 14.28 shows only the experimental design. Now it is up to the company to run the experiment with the printers and paper types shown (one piece of paper per row), measure the sharpness levels, and perform the same statistical analysis as described in Section 14.2. That is, after we randomize and collect the data, the analysis is the usual one-way ANOVA. This time, however, because we have randomized over paper types, we can be more confident that any observed differences across printers are indeed due to the printers themselves and not differences in paper. ●

14.6.2 Blocking

Randomization is one method for eliminating the effects of one or more nuisance factors. Another method is called **blocking.** Like randomization, blocking is extremely important and is used in many applications. Actually, we have already seen perhaps the simplest form of blocking in Chapters 9 and 10 when we studied the paired-sample procedure. In a study of differences between pretest and posttest performance scores, for example, each person is defined as a "block." The idea is that pretest and posttest tend to be correlated—some people do well on both, whereas some do poorly on both—so by using a paired-sample procedure, we "block out" the differences among people and are able to focus on the differences between the two tests.

There are many forms of blocking designs, but we will describe only the simplest: the **randomized block** design with a single experimental factor and a single blocking variable. Suppose there are T treatment levels of the single factor and B blocks. Then we use $T \times B$ experimental units and assign T of these to each block. If it is possible (or makes sense), we also randomize the T experimental units in any block to the T treatment levels. We illustrate the typical setup in the following example.

Randomized Block Design	In a **randomized block design,** we divide the experimental units into several "similar" blocks. Then each experimental unit within a given block is randomly assigned a different treatment level.

Example 14.5 · The Effect of Soap Dispenser on Sales of Liquid Soap

The SoftSoap Company is introducing a new product into the market: liquid soap for washing hands. Four types of soap dispensers are being considered. SoftSoap has no idea which of these four dispensers will be perceived as the most attractive or easy to use, so it runs an experiment. It chooses eight supermarkets that have traditionally carried Soft-Soap products, and it asks each supermarket to stock all four versions of its new product for a 2-week test period. It records the number of items purchased of each type at each store during this period. (See the file SOAP.XLS.) How might we describe (and analyze) this experiment?

Objective To use a blocking design with store as the blocking variable to see whether type of dispenser makes a difference in sales of liquid soap.

Solution

As this example illustrates, the blocking variable—and even the decision whether to block—depends entirely on the specific problem.

At first glance, this might look exactly like a one-way design as described in Section 14.2. There is a single factor, dispenser type, varied at four levels, and there are eight observations at each level. For example, we obtain a count of sales for dispenser type 1 at each of eight stores. However, it is very possible that the response variable, number of sales, is correlated with store. That is, some stores might sell a lot of *each* dispenser type, whereas others might not sell many of any dispenser type. (For example, stores in areas where there are a lot of manual labor jobs might sell a lot more hand soap than stores in a university area.) Therefore, we treat each store as a block, so that the experimental design appears as in Figure 14.29. Each treatment level (dispenser type) is assigned exactly

SoftSoap dispenser experiment

Store	Dispenser	Sales
1	1	68
1	2	82
1	3	94
1	4	72
2	1	72
2	2	96
2	3	104
2	4	78
3	1	70
3	2	73
3	3	76
3	4	59
4	1	49
4	2	56
4	3	60
4	4	61
5	1	66
5	2	84
5	3	94
5	4	75
6	1	48
6	2	54
6	3	56
6	4	43
7	1	57
7	2	75
7	3	81
7	4	70
8	1	65
8	2	77
8	3	80
8	4	81

Two-Way ANOVA results for Sales

Sample sizes for treatments (Store down, Dispenser across)

	1	2	3	4	
1	1	1	1	1	4
2	1	1	1	1	4
3	1	1	1	1	4
4	1	1	1	1	4
5	1	1	1	1	4
6	1	1	1	1	4
7	1	1	1	1	4
8	1	1	1	1	4
	8	8	8	8	

Sample means for treatments (Store down, Dispenser across)

	1	2	3	4	
1	68.000	82.000	94.000	72.000	79.000
2	72.000	96.000	104.000	78.000	87.500
3	70.000	73.000	76.000	59.000	69.500
4	49.000	56.000	60.000	61.000	56.500
5	66.000	84.000	94.000	75.000	79.750
6	48.000	54.000	56.000	43.000	50.250
7	57.000	75.000	81.000	70.000	70.750
8	65.000	77.000	80.000	81.000	75.750
	61.875	74.625	80.625	67.375	

Sample standard deviations for treatments (Store down, Dispenser across)

	1	2	3	4	
1	0.000	0.000	0.000	0.000	11.605
2	0.000	0.000	0.000	0.000	15.000
3	0.000	0.000	0.000	0.000	7.416
4	0.000	0.000	0.000	0.000	5.447
5	0.000	0.000	0.000	0.000	12.010
6	0.000	0.000	0.000	0.000	5.909
7	0.000	0.000	0.000	0.000	10.210
8	0.000	0.000	0.000	0.000	7.365
	9.372	14.040	16.724	12.478	

Two-Way ANOVA Table

Source	SS	df	MS	F	p-value
Store	4313.500	7	616.214	17.751	0.000
Dispenser	1617.000	3	539.000	15.527	0.000
Error	729.000	21	34.714		
Total	6659.500	31			

FIGURE 14.29 Randomized Block Design for Soap Example

once to each block (store). As a practical matter, if each dispenser type is stocked on a different shelf in the store, we could also use randomization, where we instruct each store to randomize the order of dispenser types from top shelf to bottom shelf.

We can analyze these data essentially the same way we analyze a two-factor design, that is, with two-way ANOVA. There are two differences, one technical and one of interpretation. The technical difference is that because there is only one observation in each combination of treatment level and block, it is impossible to estimate an interaction effect and a "within" error variance simultaneously. Therefore, we *assume* there are no important interaction effects between treatment levels and blocks, and we attribute all variation other than that from main effects to error variation. StatPro takes care of this automatically, as shown in Figure 14.29. To obtain this output, we use StatPro's two-way ANOVA procedure, with Store and Dispenser as the two "code" variables. StatPro recognizes that there is only one observation per store/dispenser combination, so it sets the ANOVA table up with no Interaction row. Nevertheless, it still provides interaction charts, one of which appears in Figure 14.30 (page 804), to check for the no-interaction assumption. If there were no interactions, the lines in this chart (one for each store) would be parallel. Although these lines are not *exactly* parallel, it appears that the effect of

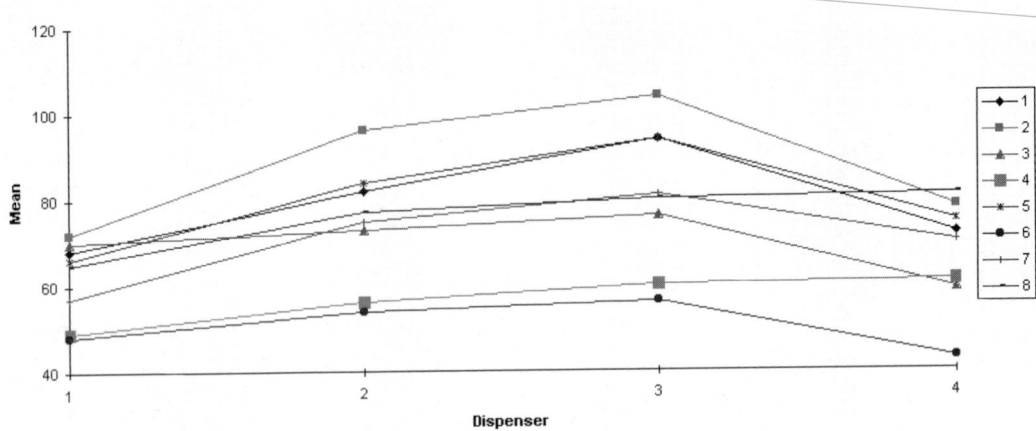

FIGURE 14.30 Interaction Chart for Soap Example

dispenser type is approximately the same at each store. Therefore, the no-interaction assumption appears to be confirmed.

There are two *F*-values and corresponding *p*-values in the ANOVA table in Figure 14.29. One is for the main effect of dispenser type, whereas the other is for the main effect of store. We are clearly more interested in the former because this is the focus of the experiment. Its *p*-value is essentially 0, meaning that there *are* significant differences across dispenser types. In fact, judging by the sample means, the ranking of dispenser types in decreasing order is 3, 2, 4, 1, and there is a considerable gap between each of these. If SoftSoap had to market only one dispenser type, it would almost certainly select type 3. The *p*-value for the main effect of store is also essentially 0, which means that the stores differ significantly with respect to average sales. This is not as interesting a finding—in fact, we used a block design precisely because we suspected such an effect—but it does confirm that a block design was a good idea.

FIGURE 14.31
Results for Soap
Example Using
One-Way ANOVA

	E	F	G	H	I	J	K
3	*Results of one-way ANOVA*						
4							
5	*Summary stats for samples*						
6			Sales_1	Sales_2	Sales_3	Sales_4	
7		Sample sizes	8	8	8	8	
8		Sample means	61.875	74.625	80.625	67.375	
9		Sample standard deviations	9.372	14.040	16.724	12.478	
10		Sample variances	87.839	197.125	279.696	155.696	
11		Weights for pooled variance	0.250	0.250	0.250	0.250	
12							
13		Number of samples	4				
14		Total sample size	32				
15		Grand mean	71.125				
16		Pooled variance	180.089				
17		Pooled standard deviation	13.420				
18							
19	*One Way ANOVA table*						
20		Source	SS	df	MS	F	p-value
21		Between variation	1617.000	3	539.000	2.993	0.0477
22		Within variation	5042.500	28	180.089		
23		Total variation	6659.500	31			

We can also confirm that blocking was useful by running a one-way ANOVA on the data, using Dispenser as the *single* factor and ignoring Store. The results appear in Figure 14.31. The differences across dispenser type are still significant at the 5% level (the p-value is still less than 0.05), but they are not as significant as when a blocking variable is used. By comparing the ANOVA tables in Figures 14.29 and 14.31, we see that the error (within) sum of squares in the latter, 5042.5, is split into two parts in the former: the block sum of squares, 4313.5, and the error sum of squares, 729. By having a lower error sum of squares, we obtain a more powerful test for dispenser differences. The point is that when differences across stores are ignored, they tend to mask the differences across dispenser types.

Blocking is one of the most powerful methods in experimental design. It allows us to "control" for a variable, such as store, that is not of primary interest but could introduce an unwanted source of variation. Experimental designers should always be on the lookout for possible blocking variables. They generally result in more powerful tests. ●

14.6.3 Incomplete Designs

Recall that the two-factor designs we discussed in Section 14.5 are called *full factorial* designs. In a full factorial design we obtain one or more observations for *each* combination of treatment levels. For example, if there are two factors with 5 and 7 treatment levels, respectively, then we replicate the experiment at each of the $5 \times 7 = 35$ treatment level combinations. If there are three factors with 3, 5, and 7 treatment levels, respectively, then we replicate at each of the $3 \times 5 \times 7 = 105$ combinations. By running an experiment in this way, we can estimate all main effects and interactions. A full factorial design is the preferred way to run an experiment from a statistical point of view, but it can be very expensive, even infeasible, if there are more than a few factors.

In industrial settings, there are often a *large* number of input factors that can be varied to produce a product. (Think, for example, of the number of factors that might be varied in an attempt to produce a car door that doesn't rattle.) Each of these factors might have a main effect on some response variable of interest, and there might also be important interactions between input factors. The question is how to design an experiment so that we get as much useful information as possible and stay within budget (either time or money). To get an idea of the problem, suppose there are 12 input factors. Even if we use only two treatment levels ("low" and "high") for each factor, there are $2^{12} = 4096$ treatment level combinations in a full factorial design—probably many more than could be tested.

Because this is very common in real applications, statisticians have devised **incomplete, or fractional factorial,** designs that test only a fraction of the possible treatment level combinations. Obviously, we lose something by not gaining information on *all* of the possible combinations. Specifically, different effects are **confounded,** which means that we cannot estimate them independently. As an example, the main effect of factor D might be confounded with the three-way interaction effect of factors A, B, and C. In this case it is impossible to tell, because of our design, whether a particular set of observed differences is due to factor D or to the interaction of factors A, B, and C. We would probably conclude that the differences are due to factor D, just because three-way interaction effects are typically not very important, but we cannot be absolutely sure.

This is a fairly difficult topic, and we will not be able to cover it in much detail. However, just to give you a taste of what is involved, we illustrate a "half-fractional" design with four factors, each at two levels, in Figure 14.32 on page 806. (See the file FRACTIONAL.XLS.) If this were a full factorial design, there would be $2^4 = 16$ combinations of treatment levels. The "half-fractional" design means that we use only half,

FIGURE 14.32
A Half-Fractional design with Four Factors

	A	B	C	D
1	Half-factorial design with 4 factors			
2				
3	A	B	C	D
4	1	1	1	1
5	1	1	-1	-1
6	1	-1	1	-1
7	1	-1	-1	1
8	-1	1	1	-1
9	-1	1	-1	1
10	-1	-1	1	1
11	-1	-1	-1	-1

or eight, of these. When using only two levels for each factor, it is customary to label the lower level with a -1 and the higher level with a $+1$. Therefore, each row in the figure represents one of eight combinations of the factor levels. For example, in the first row we would use the higher level of each factor. (Then when implementing the experiment, we would assign several experimental units to each combination, so that we would have several observations per row.)

To see how the confounding works, it is useful to create new columns by multiplying the appropriate original A–D columns. For example, the AC column is the product, row by row, of the A and C columns. As in usual algebra, the result is $+1$ if there are an even number of -1's, and -1 if there are an odd number of -1's. The results appear in Figure 14.33. Note that we have created a column for each possible two-way and three-way interaction. Now compare these columns. You'll notice that they come in pairs. For example, the A column has exactly the same pattern as the BCD column, the AB column has the same pattern as the CD column, and so on. When two columns are identical, we say that one is the **alias** of the other. The practical impact is that if two effects are aliases of one another, it is impossible to estimate their *separate* effects. Therefore, we try to design the experiment so that only one of these is likely to be important and the other is likely to be insignificant. In this particular design, each main effect (single letter) is aliased with a three-way interaction—A with BCD, B with ACD, and so on. If three-way interactions are unlikely to be important, then we can attribute any significant findings to main effects, not three-way interactions. But note that the two-way interactions are confounded with each other—AB with CD, AC with BD, and AD with BC. It will probably be difficult to unravel these.

As we have indicated, there is a whole science devoted to creating incomplete designs such as the one in Figure 14.32, and to analyzing the resulting data. [We again refer to Schmidt and Launsby (1994) and DeVor et al. (1992) for introductory accounts of the topic.] The usual approach, especially when there are a large number of potentially important input factors, is to run a highly fractional experiment (a small fraction of all possible treatment level combinations) to "screen" for the relatively few factors that have im-

	A	B	C	D	E	F	G	H	I	J	K	L	M	N
1	Half-factorial design with 4 factors													
2							Two-way interactions					Three-way interactions		
3	A	B	C	D	AB	AC	AD	BC	BD	CD	ABC	ABD	ACD	BCD
4	1	1	1	1	1	1	1	1	1	1	1	1	1	1
5	1	1	-1	-1	1	-1	-1	-1	-1	1	-1	-1	1	1
6	1	-1	1	-1	-1	1	-1	-1	1	-1	-1	1	-1	1
7	1	-1	-1	1	-1	-1	1	1	-1	-1	1	-1	-1	1
8	-1	1	1	-1	-1	-1	1	1	-1	-1	-1	1	1	-1
9	-1	1	-1	1	-1	1	-1	-1	1	-1	1	-1	1	-1
10	-1	-1	1	1	1	-1	-1	-1	-1	1	1	1	-1	-1
11	-1	-1	-1	-1	1	1	1	1	1	1	-1	-1	-1	-1

FIGURE 14.33 Counfounding Effects in an Incomplete Design

portant effects. Having found these, we can then run a more detailed experiment, perhaps even a full factorial experiment, to investigate the few important factors more fully. As the introductory vignette to this chapter explains, the results are often very impressive. These experiments can lead to lower costs, higher sales, higher reliability, and higher customer satisfaction—in short, to better products.

PROBLEMS

Level A

24. Suppose that a producer of single-room air conditioners wishes to test four prototype air conditioning units. The response variable is the number of days an air conditioner will function properly before its motor needs major repair. In this case the producer is interested in only one factor, the type of air conditioner, at four different levels. However, the manufacturer suspects that the type of use might affect the time until major repair. Specifically, these air conditioning units are used in three environments: (1) in residential homes located in northern climates, where they are used only on an occasional basis during the summer months; (2) in residential homes located in moderate climates, where they are used frequently during the summer months and seldom during other seasons of the year; and (3) in residential homes in southern climates, where they are used frequently throughout the year except during the cooler winter months. The producer suspects that these different environments may tend to obscure real differences among the four types of air conditioners.

To conduct this experiment, the producer has allocated 20 air conditioners of each type. Provided that the air conditioner producer is interested primarily in how the type of unit affects the time until major re-

pair, how can the company control for the type of environment? Assume that approximately 10% of all single-room air conditioners produced by this company are used in homes located in northern climates, 25% are used in homes located in moderate climates, and 65% are used in homes located in southern climates. Explain, in detail, how the producer should set up this experiment.

25. Consider again the one-way ANOVA hypothesis test described in Problem 6. How could blocking be employed to control for a factor that is not of primary interest yet could introduce an unwanted source of variation in this case?

26. Consider again the one-way ANOVA hypothesis test described in Problem 7. How could blocking be employed to control for a factor that is not of primary interest yet could introduce an unwanted source of variation in this case?

Level B

27. Following the example presented in Section 14.6.3, illustrate a half-fractional design with *five* factors, each at two levels. Specifically, generate figures similar to Figures 14.32 and 14.33 to support your verbal explanation. Identify the aliases.

Conclusion

This chapter has focused on the design of experiments and the statistical analysis of the resulting data, called analysis of variance. This methodology has long played an important role in agriculture and many natural sciences, particularly the medical sciences. The business world is just beginning to realize the importance of designed experiments for designing and producing better products, and this trend will undoubtedly continue as more people receive training in the techniques of experimental design and analysis of variance. It is important to keep sight of the overall goal: to see whether variations in one or more factors have significant effects on a response variable of interest. The role of experimental design is to set up experiments in a way—using randomization, blocking, fractional factorial designs, or whatever—to get as much information from the resulting data as possible. Then the techniques of ANOVA allow us to say whether any main effects or interactions are significant. If there are significant effects, we can then form confidence intervals to measure the magnitudes of specific differences between means or other contrasts. The goal of good experimental design is to identify important factor effects when they exist.

Summary of Key Terms

Term	Explanation	Excel	Pages	Equation Number
Analysis of variance (ANOVA)	A collection of methods for testing for differences in means across subpopulations (or across a single population treated in different ways).		763	
Observational study	A study that uses readily available information.		763	
Designed experiment	A study in which data are obtained under controlled experimental conditions.		763	
Experimental design	The plan that determines how many observations to obtain at which combinations of experimental conditions.		764	
Response variable	The variable that is measured in an ANOVA study.		764	
Factors	The categorical variables that serve as the explanatory variables in an ANOVA study.		764	
Treatment levels	The possible values of a factor.		764	
Experimental units	The people, machines, or whatever, that are measured in an ANOVA study.		765	
Balanced design	An experimental design where the same number of experimental units is assigned to each treatment level combination.		765	
One-way ANOVA	An ANOVA study with a single factor.	StatPro/ Statistical Inference/ One-Way ANOVA	766	
ANOVA table	A table that includes the ingredients (sums of squares, degrees of freedom, mean squares, F-ratio, and p-value) for tests of equal means.	StatPro/ Statistical Inference/ One-Way (or Two-Way) ANOVA	769	14.1, 14.2, 14.3
Confidence intervals in ANOVA	Confidence intervals for differences between pairs of means (or contrasts).	StatPro/ Statistical Inference/ One-Way ANOVA	769	14.4, 14.5
Multiple comparison problem	The problem that when many statements are made, each with a stated level of confidence, the probability that at least one will be wrong is much larger than anticipated.		782	
Contrast	A weighted combination of means where the weights sum to 0; used to contrast one combination of means with another.		786, 794	

Summary of Key Terms (continued)

Term	Explanation	Excel	Pages	Equation Number
Bonferroni, Tukey, Scheffé methods	Methods that expand confidence interval lengths to correct for the multiple comparison problem.	StatPro/ Statistical Inference/ One-Way ANOVA	783	
Two-way ANOVA	An ANOVA study with two factors.	StatPro/ Statistical Inference/ Two-Way ANOVA	787	
Full factorial design	An experimental design in which observations are made at each combination of factor levels.		787	
Incomplete (or fractional) design	An experimental design in which observations are made only at a selected subset of the combinations of factor levels.		788	
Main effects	Indications of differences across levels of one factor (when averaged over the levels of the other factor).	Statistical Inference/ Two-Way ANOVA	789	
Interactions	Indications of differences in means that could not be anticipated from main effects alone.	Statistical Inference/ Two-Way ANOVA	790	
Randomization	The random assignment of experimental units to various levels of factors.		799	
Blocking	A technique of assigning experimental units to similar blocks of experimental units to decrease error variation.	Statistical Inference/ Two-Way ANOVA	802	
Confounding	The (unavoidable) confusion of some effects with others in an incomplete experimental design.		805	

PROBLEMS

Level A

28. Although four similar-sized small-car models exhibit similar miles per gallon (mpg) sticker ratings, there is some skepticism as to whether their mean mpg values are really equal. To test this equal-means hypothesis, several cars of each model are driven for 10,000 miles under nearly identical driving conditions. The observed mpg values are listed in the file P14_28.XLS. Use one-way ANOVA to help decide whether the different models have equal mean mpg values, and write a short report to summarize your findings.

29. A professional golf association wants to compare the mean distances traveled by four brands of golf balls when struck by the same driver. Specifically, a robotic golfer uses a driver to hit a random sample of 80 balls (i.e., 20 balls of each brand). Note that the 80 balls are hit in random order. The distance is recorded for each hit, and the results are provided in the file P14_29.XLS.

a. Is there any indication of differences in the mean distances traveled by the four types of balls? Perform an appropriate statistical test and report a p-value.

b. Select an appropriate significance level and construct confidence intervals for all pairs of differences between means. Which of these differences, if any, are statistically significant at the selected significance level?

30. Boxes of Cheerios cereal are filled by five identical machines at a local General Mills plant. Independent samples are randomly drawn from a large number of Cheerios boxes filled by each machine, and the number of ounces of cereal in each selected box is recorded in the file P14_30.XLS. Use one-way ANOVA to help decide whether the five machines are yielding essentially equivalent average fills (in ounces). Briefly summarize your findings.

31. Assume that we gather independent random samples from large batches of each of three different brands of lightbulbs. We then record the lifetime of each selected bulb in the file P14_31.XLS.
a. Test whether the different brands of lightbulbs have equal average lifetimes at the 10% significance level.
b. Compute 90% confidence intervals for all pairs of differences between means. Which of these differences, if any, are significantly nonzero at the 10% significance level?

32. Consider again the one-way ANOVA hypothesis test described in Problem 28. Address the multiple comparison problem by applying the Bonferroni, Tukey, and Scheffé methods to obtain an *overall* confidence level of approximately 95%. How do these results compare to the uncorrected 95% confidence intervals? Recall that the relevant data are given in the file P14_28.XLS.

33. Consider again the one-way ANOVA hypothesis test described in Problem 29. Address the multiple comparison problem by applying the Bonferroni, Tukey, and Scheffé methods to obtain an *overall* confidence level of approximately 99%. Compare the widths of the confidence intervals generated with each of these methods with those of uncorrected 99% confidence intervals. Explain your findings. Recall that the relevant data are given in the file P14_29.XLS.

34. Consider again the one-way ANOVA hypothesis test described in Problem 30. Address the multiple comparison problem by applying the Bonferroni, Tukey, and Scheffé methods to obtain an *overall* confidence level of approximately 95%. Summarize your results. Recall that the relevant data are given in the file P14_30.XLS.

35. Consider again the one-way ANOVA hypothesis test described in Problem 31. Address the multiple comparison problem by applying the Bonferroni, Tukey, and Scheffé methods to obtain an *overall* confidence level of approximately 90%. Compare the widths of the confidence intervals generated with each of these methods with those of uncorrected 90% confidence intervals. Explain your findings. Recall that the relevant data are given in the file P14_31.XLS.

36. A commuter airline wants to determine the combination of advertising medium (four levels) and advertising agency (two levels) that would produce the largest increase in ticket sales per advertising dollar spent. Each of the two advertising agencies has prepared advertisements in formats required for distribution by each of the media (including television, radio, newspaper, and Web site). Forty small towns of roughly the same size have been selected for this experiment. Furthermore, groups of five of these small towns have been assigned to receive an advertisement prepared and distributed by each of the eight agency–medium combinations. The dollar increases in ticket sales per advertising dollar spent, based on a 1-month period, are given in the file P14_36.XLS. Test for any significant main effects and interactions at the 5% level, and briefly summarize your results.

37. The file P14_37.XLS gives the miles per gallon for each of three different octanes (Octane A, Octane B, and Octane C) of gasoline and three types of vehicles (light, medium, and heavy). Subsets of ten vehicles of each type have been randomly assigned to each octane level.
a. Do you find evidence of a significant main effect for the octane factor in this case? Explain.
b. Do you find evidence of a significant main effect for the vehicle type factor in this case? Explain.
c. Do you find evidence of significant interactions between the two factors in this case? Explain.

38. In an effort to increase unit sales of particular products in the short run, many supermarkets reduce the price of these products and increase their display space. Consider three levels of each factor: for the price factor, (1) normal price, (2) moderately reduced price, and (3) heavily reduced price; and for the display factor, (1) normal display space, (2) moderately increased display space, and (3) heavily increased display space. Suppose that each of these nine treatment combinations was applied five times to a specific product at a particular supermarket. Each treatment application lasted 7 days, and the response variable was unit sales for the week. The data for this experiment can be found in the file P14_38.XLS. Test for any significant main effects and interactions at the 1% level, and briefly summarize your results.

39. Consider again the one-way ANOVA hypothesis test described in Problem 29. Suppose now that the professional golf association wants to compare the mean distances traveled by four brands of golf balls using *human* golfers instead of a robotic golfer. Each human golfer who participates in the experiment will employ the same type of driver to hit a subset of the 80 balls.

a. Explain how a randomized experimental design could be used to perform this one-way ANOVA.

b. Explain how a randomized block design could be used to perform this one-way ANOVA.

Level B

40. A production manager believes that the time required to assemble a particular product depends on the type of training that workers on the line receive. Four different training programs have been administered to workers of roughly equal experience at the local plant during the past year. To test her hypothesis, the production manager gathers assembly time data for randomly selected subsets of workers who have participated in one of the four training programs. These times are recorded in the file P14_40.XLS. Use one-way ANOVA to help decide whether the different training programs yield equivalent average assembly times, and write a short report to summarize your findings.

41. Consider again the one-way ANOVA hypothesis test described in Problem 40. Address the multiple comparison problem by applying the Bonferroni, Tukey, and Scheffé methods to obtain an *overall* confidence level of approximately 95%. Briefly summarize your results. Recall that the relevant data are given in the file P14_40.XLS.

42. Consider again the two-way ANOVA hypothesis test described in Problem 36. Recall that the relevant data are given in the file P14_36.XLS.

a. Advertising agency 1 claims that the mean increase in ticket sales per advertising dollar owing to its Web site-based advertisements is *greater than* the average increase in sales arising from other types of ads prepared by its firm. Is this claim supported by the given data? Explain.

b. Advertising agency 2 claims that the mean increase in ticket sales per advertising dollar owing to its Web site-based advertisements is *greater than* the average increase in sales arising from other ads prepared by its firm or the competition. Is this claim supported by the given data? Explain.

43. Consider again the two-way ANOVA hypothesis test described in Problem 37. Recall that the relevant data are given in the file P14_37.XLS.

a. Is the average miles per gallon for all vehicles using Octane A different from the average miles per gallon for all vehicles that use other types of octane? Support your conclusion with one or more appropriate confidence intervals.

b. Is the average miles per gallon for all medium-type vehicles different from the average miles per gallon for all other types of vehicles? Support your conclusion with one or more appropriate confidence intervals.

14.1 Krentz Appraisal Services

Nancy Krentz, the owner and manager of a property appraisal service based in York, Pennsylvania, is concerned that her four appraisers (Allen, Felan, Maloy, and Nelson) are producing appraisals of comparable properties that are generally not equivalent. She wants to conduct an investigation to determine whether her concerns are valid. Nancy directs her administrative assistant, Katie Shaffer, to identify 40 similar properties in the York area for use in the study. Given the sample of comparable properties, Nancy then arbitrarily divides the 40 properties into four subsets of ten. Next, she randomly assigns each subset to one of the four appraisers for assessment. The appraisals of the given 40 properties are recorded in the file KRENTZ.XLS. Given Nancy's limited background in statistical analysis, she has asked for your expert assistance in evaluating the data that her assistant has compiled. She recalls that at one point in her business studies she learned a systematic method, called analysis of variance, for comparing the averages of related groups of quantitative data. However, she cannot recall the assumptions that must be met to apply this methodology, nor the procedures for implementing the appropriate method and correctly interpreting the results. Nancy has prepared the following list of questions that she would like for you to help her answer:

1. What requirements must be met to apply analysis of variance? Is it appropriate to use analysis of variance in this case?

2. Assuming that it is appropriate to apply a form of analysis of variance here, how can she use the right method to analyze the data that have been collected and recorded in the file KRENTZ.XLS?

3. Does the statistical analysis confirm her suspicion that there are individual differences among the four appraisers? If so, which of the four appraisers are typically generating evaluations that are larger or smaller than those of the others?

4. Has the statistical test been formulated in the best manner? In particular, was it appropriate for Nancy to divide the 40 selected properties into four subsets of ten and then assign each subset to one of the appraisers? If not, how could the design of the study be modified to elicit the most useful information in evaluating the appraisal staff at Krentz? Be as specific as possible.

5. In light of the results of this data analysis, what steps, if any, should Nancy take to improve the situation in her organization?

15

Data Mining Techniques: Discriminant Analysis, Logistic Regression, and OLAP

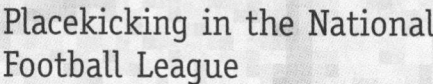

© Alan Schein Photography/CORBIS

Placekicking in the National Football League

One of the most important players on every National Football League (NFL) team is the placekicker. This sometimes overlooked player is responsible for the points after touchdowns (PATs) and field goals (FGs), he scores in nearly every game, and he often decides the outcome of the game. As background, the objective in all placekicks is to kick the ball over a crossbar and through the two uprights of the goalposts to score points. A PAT is attempted, almost always from 20 yards away, after each touchdown. If successful, it is worth one point. A field goal can be attempted from any distance, and if successful, it is worth three points. In the 1995–1996 NFL season, for example, attempted FGs varied in length from 18 to 66 yards. All placekicks are "all or nothing." There is no reward for a placekick that just barely misses, and a placekick that is just barely inside the uprights is as good as one right through the middle.

Bilder and Loughin (1998) studied placekicking results for the 1995–1996 NFL season. Their objective was to discover the factors that help determine the results of placekicks. Therefore, they gathered data (from the NFL web site, http://www.nfl.com, and other sources) on the circumstances surrounding each of over 1700 placekicks during the season. They assumed that these circumstances—and *not* differences in abilities across placekickers—determine the probability of a successful placekick. Because a placekick is all or nothing, the authors chose logistic regression, one of the topics in this chapter, as the method for modeling the probability of a successful kick. Specifically, they regressed the "logit" of the success probability against a linear expression of the explanatory variables. This logit is the logarithm of the odds ratio—the probability of success divided by the probability of failure—and can be interpreted in a fairly straightforward way, as we will discuss shortly.

The authors considered a variety of explanatory variables. Some of these turned out to be significant and some, surprisingly, did not. The nonsignificant variables, those that did not have a significant effect on the probability of a successful place-kick, included: Altitude (the elevation of the city where the game is played), Dome (a dummy variable indicating whether the game is inside a dome), Home (a binary variable indicating whether the placekicker is on his home field), Precipitation (a dummy variable indicating whether there is any precipitation at game time), Surface (a dummy variable indicating whether the game is on artificial turf), Temperature (the temperature at game time), and Week (a variable that ranges from 1 to 17 and indicates the week of the season).

The significant variables included: Change (a dummy variable indicating whether a successful placekick would create a lead change), Distance (the distance of the placekick in yards), PAT (a dummy variable indicating whether the placekick was a PAT versus a FG), Wind (a dummy variable indicating whether the wind speed was greater than 15 miles per hour at game time), and an interaction between Distance and Wind. The estimated regression equation using these variables was

$$\text{Estimated logit} = 4.498 + 1.259\text{PAT} - 0.331\text{Change} - 0.081\text{Distance}$$
$$+ 2.878\text{Wind} - 0.091\text{Distance} * \text{Wind}$$

The interaction term makes this equation a bit tricky to interpret, so we rewrite it for two possible cases: when Wind = 0 and when Wind = 1. When Wind = 0 (little or no wind), the equation becomes

$$\text{Estimated logit} = 4.498 + 1.259\text{PAT} - 0.331\text{Change} - 0.081\text{Distance}$$

When Wind = 1 (fairly heavy wind), the equation becomes

$$\text{Estimated logit} = (4.498 + 2.878) + 1.259\text{PAT} - 0.331\text{Change}$$
$$- (0.081 + 0.091)\text{Distance}$$

From these equations, we see that the odds of a successful placekick are: (1) greater for a PAT than for a FG, all other factors being constant; (2) less when a successful kick would create a lead change than when it would not; and (3) less as the distance of the kick increases. Also, the presence of wind magnifies the negative effect of distance on the odds of success.

These interpretations come directly from the *signs* of the coefficients of the explanatory variables. As in any good logistic regression analysis, however, Bilder and Loughin then explore the meaning of the regression equation in more depth. Specifically, if $\hat{\beta}$ is any estimated regression coefficient, then $\exp(\hat{\beta})$ can be interpreted as the change in the odds of success if the corresponding explanatory variable increases by one unit. For example, $\exp(-0.331) = 0.72$. This means that the odds of success when Change = 1 are 72% as great as when Change = 0. Similarly, $\exp(1.259) = 3.52$, so that the odds of success of a PAT are 3.52 times as large as for a FG. (This latter result is not as intuitive as it first appears. It compares PATs and FGs under the *same* conditions, including the same distance. So 20-yard FGs are evidently missed more often than PATs.) Finally, $\exp(-10 \times (0.081 + 0.091)) = 0.18$. This implies that the odds of success are only 18% as great for each 10-yard increase in distance under windy conditions. For example, if the odds of success were 3 (that is, 3 to 1) at 30 yards, they would be only 0.54 (that is, about 1 to 2) at 40 yards.

The regression equation can also be used to estimate the probability of success for any specific conditions. To do so, remember that the estimated logit is really $\log(\hat{p}/(1 - \hat{p}))$, so the regression equation can be "solved" for $\hat{p}$. The authors do this calculation for a famous situation in the 1996 playoff game between the Kansas City Chiefs and the Indianapolis Colts. Lin Elliott, Kansas City's placekicker, missed a FG at the end of the game, and Kansas City's season ended with a three-point loss. The distance of this FG was 42 yards, there was little wind, it was a FG as opposed to a PAT, and it *would* have caused a lead change. (The authors consider coming from behind to tie a "lead change.") Substituting these values into the regression equation, we obtain $\hat{p} = 0.69$, or about 2 out of 3. So if you were one of the outraged Kansas City fans calling for Elliott to be traded, just remember that *any* placekicker would have about one chance out of three of missing under these conditions! ●

15.1   Introduction

There are many situations in business where observations can be separated into two or more well-defined groups. These observations might be people, cities, universities, countries, or others. The purpose of a statistical analysis might then be to classify each observation into the appropriate group, given data on related variables. Here are some typical examples.

- A company that markets a product divides customers into two groups: those who have tried the product and those who have not. The company would like to understand the differences between these two groups with respect to variables such as age, income, hours of television watched, and others.

- A direct marketing company mails brochures to thousands of potential customers. It also collects data on these customers, such as their age, income, how often they have purchased from this company, the time of their most recent purchase from the company, and others. The company wants to learn the best predictors for which customers respond to mailings and which do not. By doing so, it can better target the customers most likely to respond.

- Some companies go bankrupt and others do not. By looking at a number of financial variables on a number of companies, is it possible to predict which are about to go bankrupt?

- The sales representatives for a textbook publishing company encourage university professors to adopt one of its books. Some universities choose not to adopt the book, some adopt it and then stop using it, and others adopt it and continue to use it. What are the differences among these three groups of universities? Does it have to do with enrollment, average SAT scores, or what?

- Most people who borrow from a lending company repay their loans, but a significant minority default on their loans. Lending companies would certainly like to be able to predict whether a potential customer will default *before* granting that customer a loan. These companies need to know what data to collect on potential customers and how to use the data to predict the probability of default.

- Soft drink companies would like to know what variables separate customers who prefer low calorie drinks from those who prefer the "full calorie" drinks. By learning how these two groups differ, the companies can target them more successfully in their advertisements.

All of these situations are similar in that there are two or more well-defined groups, and an analyst wants to find variables that have the ability to explain or predict which group each observation belongs to. In some cases the *explanatory* objective is the primary one, as in the bankruptcy example. There, financial analysts might simply want to know which financial indicators lead some companies to bankruptcy and others to solvency. It would be interesting to predict whether a particular company will go bankrupt, but this might not be the analysts' primary objective. In other cases the *predictive* objective is primary. This is true in the loan-granting example. The lending company certainly wants to understand why some people default and others do not, but the company's primary objective is to predict who will default so that likely defaulters will not be granted loans in the first place.

In this chapter we will discuss two statistical methods for solving these problems. The first method, discriminant analysis, attempts to find one or more linear functions of potential explanatory variables that *discriminate between,* or separate, the groups. (We will see what this means graphically in the next section.) Using these functions, the method then classifies each observation into one of the groups. If the explanatory variables do a good job of discriminating between the groups, then most of the classifications should be correct; otherwise, there will be many misclassifications. The second method, logistic regression, appears to be quite different from discriminant analysis, but its objectives and even its results are usually quite similar. As the name suggests, logistic regression is a variation of the regression analysis we discussed in Chapters 11 and 12. However, it differs fundamentally from the regression models we studied earlier because it uses a *categorical* response variable, and its output shows how the *probability* of being in either of the groups depends on the explanatory variables.

Discriminant analysis and logistic regression can both be considered *data mining* techniques. More generally, a data mining technique is any method that attempts to extract information from a large data set. These techniques are becoming increasingly important in today's business world, where companies are able to amass huge *data warehouses* with all sorts of raw data. The problem is one of converting the raw data into useful information—information that gives a company a better knowledge of its customers and hence a competitive advantage in its business. Besides discriminant analysis and logistic regression, there are many data mining techniques, some of which require specialized (and expensive) software to implement.

We will look at one of the simpler data mining techniques toward the end of this chapter. It is called OLAP, or *online analytical processing*. OLAP is essentially an extension of the pivot tables we discussed in Chapters 2–4. It allows us to break data down by various categories and by hierarchies within categories. For example, OLAP might enable a manager of a supermarket chain to break sales down by product category—not just by individual products, but by product families, then by product departments, then by product categories, and so on. In other words, it allows the manager to "drill down" through a hierarchy of product levels to understand the company's sales better. Although OLAP technology is typically performed with specialized software, a somewhat scaled-down version of it can be implemented in Excel, as we will demonstrate here.

Before proceeding, we note that discriminant analysis and logistic regression are typically found only in more advanced multivariate statistics books. These methods are not easy, and a thorough coverage of their intricacies is well beyond the level of this book. However, because of the importance of these methods for data mining in today's business world, we believe some introduction to them is warranted. Fortunately, StatPro performs the calculations for us, so that we can focus on interpreting the results. In contrast, there is nothing mathematically difficult about OLAP. It is simply a generalization of pivot tables, which, as we know, summarize data with sums, averages, counts, and so on. The power of OLAP is that it can perform this summarization in a matter of seconds, even when the underlying data source is huge. Therefore, there is no mystery in what OLAP does; it is just a matter of learning the software to implement the technique.

15.2 Discriminant Analysis

Discriminant analysis is sometimes described as a method for *discriminating* between, or separating, two or more groups, and it is sometimes described as a method for *classifying* observations to one of several groups. Actually, it accomplishes both objectives. Given a set of observations, each of which is in one of several well-defined groups (such as customers who have either tried a product or have not) and a set of potential explanatory variables (such as the customers' ages and incomes), the method first tries to see whether it is possible to discriminate between these groups with respect to the explanatory variables. For example, we might find that customers who have tried a product tend to be high-income and older customers, whereas customers who have not tried it tend to be lower-income and younger. As we will see, this discrimination problem is not as simple as it might appear, due to overlap among the groups and correlation among the explanatory variables. However, the result is typically a linear function of the explanatory variables (actually, several linear functions if there are more than two groups) that best discriminates between the groups.

Once we have this linear function, we use it to classify observations. Specifically, in the two-group problem, we calculate the value of the function for each observation and compare it to a "cutoff" value. If the function value is greater than the cutoff, then we classify the observation as group 1; otherwise, we classify it as group 2. Because we know which group each observed value belongs to, we can then see how well our classification procedure has done. If it classifies most observations correctly, then we have some confidence that the procedure will classify *new* observations correctly. If it makes a lot of incorrect classifications, however, we have no such confidence.

Although discriminant analysis requires many detailed calculations and produces voluminous numerical output, the idea behind the method is fairly simple and can be understood best through a graphical approach. It revolves around the concept of *statistical distance,* which we discuss next.

15.2.1 Statistical Distance and Separation

Suppose you take an exam and score 750 out of 1000 possible points. The mean on the exam is 700. Then one measure of your "distance" from the class average is 50, the difference between 750 and 700. However, this distance measure might not really reflect how far above the average your score is because it doesn't take variation into account. If everyone other than you scored between 675 and 725, then your 750 is amazingly high. On the other hand, if the scores range from 500 to 1000, then your 750 isn't all that far from the mean.

A better measure of distance from the mean, called **statistical distance,** is the standardized score $(750 - 700)/s$, where s is the standard deviation of all scores. If $s = 20$, say, then your distance from the mean is 2.5 (2.5 standard deviations above the mean), whereas if $s = 100$, your distance from the mean is only 0.5 (0.5 standard deviation above the mean). We used this concept of statistical distance in several previous chapters. In fact, we used it every time we standardized a value by subtracting the mean and dividing the difference by the standard deviation. It is an extremely important concept in statistics.

Statistical Distance (Univariate Case)	In the univariate (single variable) case, the **statistical distance** of an observation from the mean is the number of standard deviations it differs from the mean.

Now suppose you take two such exams. Your scores are 750 and 680, the class means are 700 and 720, and the class standard deviations are 50 and 80. On the first exam, your statistical distance from the mean is $(750 - 700)/50 = 1.0$, and on the second exam it is $(680 - 740)/80 = -0.75$. The question, however, is how we might combine this information to see how far you are from the class mean on both exams combined. That is, we need a *multivariate* version of statistical distance. The following example illustrates how we can obtain this, at least in the case of two variables. Fortunately, the idea extends to more than two variables in a natural way.

Example 15.1 Measuring Statistical Distance Based on Expenditures for Culture and Sports

The file STATDIST1.XLS contains data on 40 randomly selected families. Each row of the data set indicates how much the family spent on cultural events and sporting events during the past year. A partial listing of the data appears in Figure 15.1. Also, data for three new families are listed in the range F19:G21. How far are these three new families from the "average" of the 40 original families?

Objective To develop a measure of statistical distance for families from a mean, given their expenditures on cultural and sporting events.

Solution

The key to understanding the analysis is the scatterplot in Figure 15.2. The diamond-shaped points represent the original 40 families, the triangle-shaped point in the middle represents the sample means for these 40 families (the values in the range F10:G10 of Figure 15.1), and the three square-shaped points represent the new families. Clearly, there is a negative correlation between spending on culture and spending on sports, and we emphasize this by drawing tilted ellipses around the sample mean point in the middle. (We

	A	B	C	D	E	F	G	H	I	J
1	**Illustration of statistical distance**									
2										
3	Family	Culture	Sports		*Multivariate summary statistics*					
4	1	400	320						**Range names:**	
5	2	200	610		*Sample size*				Data - B4:C43	
6	3	260	440			40			SampSize - F6	
7	4	280	550						Means - F10:F11	
8	5	250	500		*Sample means*				CovarMat - G14:H15	
9	6	150	740			Culture	Sports			
10	7	260	600			298.25	495.75			
11	8	340	650							
12	9	330	640		*Matrix of sample variances and covariances*					
13	10	340	520				Culture	Sports		
14	11	250	400			Culture	5563.53	-5871.73		
15	12	500	380			Sports	-5871.73	15189.17		
16	13	240	480							
17	14	320	300		**New observations**					
18	15	300	530			Family	Culture	Sports	Distance	
19	16	310	560			41	370	700	3.11	
20	17	380	300			42	480	250	2.50	
21	18	300	510			43	200	470	1.90	
22	19	350	580							
42	39	270	610							
43	40	360	370							

FIGURE 15.1 Data to Illustrate Statistical Distance

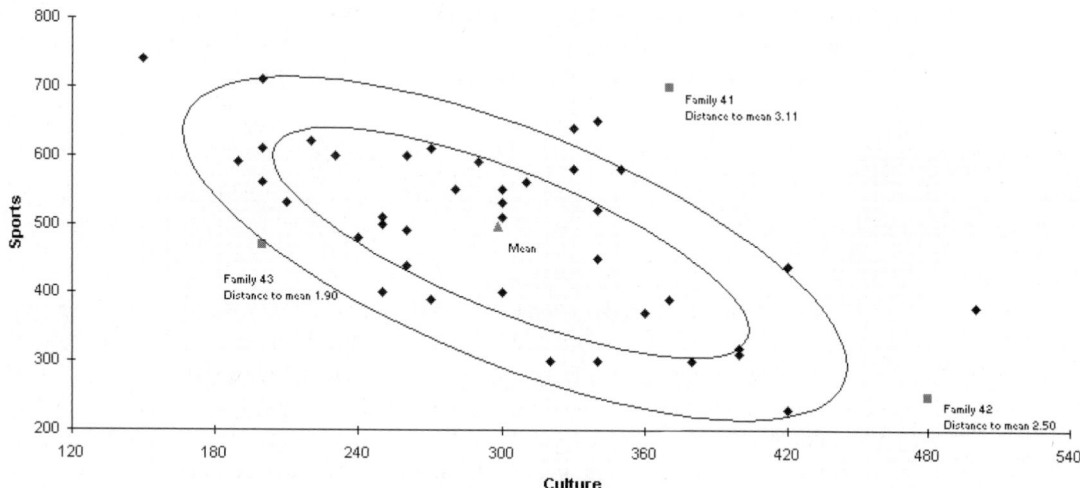

FIGURE 15.2 Scatterplot to Illustrate Statistical Distance

This chart and similar charts in this chapter were created with Excel's drawing tools. There is no StatPro option for creating them.

Because of the elliptical shape of the scatter, straight-line distance is not the same (or as useful) as statistical distance.

have drawn only two ellipses, but you can imagine a whole continuum of them emanating from the mean point.) Each ellipse represents an equally likely set of points. Equivalently, each ellipse represents a particular statistical distance from the mean point. Larger ellipses correspond to points that are farther from the mean point; in this sense, they correspond to more "unusual" points.

The statistical distances of the three new families from the mean point are shown in the chart. They might surprise you. Families 41 and 43 appear to be about the same "straight-line" distance from the mean point, but family 43's *statistical* distance is much smaller. This is because family 43 is a more "typical" family. That is, its pattern of spending is similar to the other 40 families, whereas family 41's pattern of high spending for *both* culture and sports is unusual. Similarly, family 42 appears to be farther away from the mean point than family 41, but family 42's statistical distance is considerably smaller. (This might not be obvious from the graph, but compare cells I19 and I20.) Again, this is because family 42's pattern of spending is more in line with the typical pattern than family 41's.

Scatterplots are very useful for two variables, but they are more difficult to see for three variables, and they can only be imagined for more than three variables. Fortunately, there is a numerical approach that works for any number of variables. Recall that statistical distance for a single variable is the standardized value $(X - \bar{X})/s$. Alternatively, we could write the square of the distance as $(X - \bar{X})^2/s^2$, or in the somewhat strange, but equivalent, form shown in equation (15.1):

$$D^2 = (X - \bar{X})(s^2)^{-1}(X - \bar{X}) \qquad (15.1)$$

Here, D represents distance, and the -1 exponent means "reciprocal." The multivariate version of statistical distance is a direct translation of equation (15.1) that uses vectors and matrices. It is given by equation (15.2). Here, X and $\bar{X}$ are (row) vectors of observations and sample means, T means "transpose" (make the row vector a column vector), S is the matrix of sample variances and covariances (called the **sample covariance matrix**), and the -1 exponent denotes matrix inverse.

Multivariate Formula for Statistical Distance

$$D^2 = (X - \bar{X})S^{-1}(X - \bar{X})^T \qquad (15.2)$$

The *S* matrix itself could be calculated as a series of individual variances and covariances, but it can be calculated all at once by using the matrix formula in equation (15.3), where *n* is the sample size and *Data* is the matrix of original data.

Formula for Covariance Matrix	$$S = (Data^T Data - n\bar{X}^T\bar{X})/(n - 1) \qquad (15.3)$$

As usual, range names are not required, but we have used them here to make the Excel formulas more readable.

Unless you have had a course in matrix algebra, equations (15.2) and (15.3) probably make little sense, but they are key formulas in multivariate analysis. Fortunately, it is fairly easy to implement them in Excel, using Excel's built-in transpose function, TRANSPOSE, its matrix multiplication function, MMULT, and its matrix inverse function, MINVERSE. Although StatPro will take care of the details in discriminant analysis, we illustrate the formulas here, and you might want to try them out. We first calculate the sample covariance matrix in the range G14:H15. To do so, highlight the range G14:H15, type the formula

= (MMULT(TRANSPOSE(Data),Data)-SampSize*MMULT(TRANSPOSE(Means),Means))/(SampSize-1)

and press Ctrl-Shift-Enter. This implements equation (15.3). To obtain the statistical distance for family 41 from the means, type the formula

= SQRT(MMULT((G19:H19-Means),MMULT(MINVERSE(CovarMat),TRANSPOSE(G19:H19-Means))))

in cell I19 and press Ctrl-Shift-Enter. Then copy this formula to the range I20:I21 for the other two families. This implements equation (15.2) (where the SQRT gives us distance, not squared distance).[1]

As you can see in the range H19:H21, family 43 is indeed closest to the means calculated from the original 40 families, and family 41 is farthest away. This confirms what we saw in the scatterplot. The actual distances can be interpreted as the multivariate analogs of "standard deviations from the mean." For example, family 43 is, in multivariate terms, 1.90 standard deviations from the mean. ●

So what does this all have to do with discriminating between groups? The following example indicates the role of statistical distance more clearly when multiple groups are involved.

Example 15.2 Classifying Families as University or Non-university Families on the Basis of Their Expenditures on Culture and Sports

The file STATDIST2.XLS contains data similar to the data in Example 15.1. Now, however, there are two groups: university families and non-university families. (See Figure 15.3 for a partial listing of the data.) There are also data on three new families in the range

---

[1]Excel's matrix functions, MMULT (matrix multiplication), MINVERSE (matrix inverse), MDETERM (matrix determinant), and TRANSPOSE (make a row into a column or vice versa), are quite powerful. In general, to use a matrix function, some special steps are required. If the result of the function will be a matrix or vector, then you must first select the destination range of the correct size before entering the function. After you type the function, you must press Ctrl-Shift-Enter (all three keys at once). Pressing just Enter will not work.

	A	B	C	D	E	F	G	H	I	J	K
1	**Illustration of closest statistical distance**										
2											
3	Family	Culture	Sports	Group			*Multivariate summary statistics*			**Range names:**	
4	1	540	340	Univ						Data1 - B4:C38	
5	2	340	390	Univ			*Sample sizes*			Data2 - B39:C83	
6	3	550	310	Univ			Univ	35		SampSize1 - H6	
7	4	620	150	Univ			NonUniv	45		SampSize2 - H7	
8	5	330	430	Univ						Means1 - H11:I11	
9	6	600	190	Univ			*Sample means*			Means2 - H12:I12	
10	7	530	110	Univ				Culture	Sports	CovarMat1 - H16:I17	
11	8	600	260	Univ			Univ	402.00	431.71	CovarMat2 - H20:I21	
12	9	460	330	Univ			NonUniv	297.33	594.44	CovarMatP - H24:I25	
13	10	500	300	Univ							
14	11	510	260	Univ			*Matrices of sample variances and covariances*				
15	12	410	330	Univ			**Univ**	Culture	Sports		
16	13	580	170	Univ			Culture	16257.65	-20127.06		
17	14	390	360	Univ			Sports	-20127.06	29773.45		
18	15	490	310	Univ							
19	16	470	380	Univ			**NonUniv**	Culture	Sports		
20	17	410	490	Univ			Culture	5965.45	-2735.61		
21	18	540	250	Univ			Sports	-2735.61	4720.71		
22	19	470	360	Univ							
23	20	460	310	Univ			**Pooled**	Culture	Sports		
24	21	350	660	Univ			Culture	10451.79	-10316.50		
25	22	340	540	Univ			Sports	-10316.50	15641.13		
26	23	290	660	Univ							
27	24	440	540	Univ			**New observations**				
28	25	270	580	Univ			Family	Culture	Sports	Distance to 1	Distance to 2
29	26	250	590	Univ			81	320	450	1.17	1.67
30	27	270	530	Univ			82	420	600	2.52	2.09
31	28	230	620	Univ			83	550	500	3.24	3.24
32	29	330	520	Univ							
38	35	230	660	Univ							
39	36	250	560	NonUniv							
40	37	140	710	NonUniv							
41	38	350	580	NonUniv							
81	78	280	590	NonUniv							
82	79	370	510	NonUniv							
83	80	340	620	NonUniv							

FIGURE 15.3 Data for Two Groups of Families

G29:H31, but we are not told whether these are university or non-university families. How might we classify these new families?

Objective To understand how statistical distance can be used to classify families into one of two groups, based on their expenditures on culture and sports.

Solution

A scatterplot is again very useful for illustrating the ideas. The scatterplot in Figure 15.4 (page 822) shows four sets of points, as described in the legend. (We have enclosed the majority of university and non-university points in ellipses, simply to indicate the general swarm of points for each group.) It is clear that the university families tend to spend more on culture and less on sports, and vice versa for the non-university families, although there is considerable overlap between the two groups. Also, the scatter for each group has the same type of elliptical shape as in the previous example. Therefore, one idea for classification is to find the statistical difference from each new family's point to each of the group means and then classify the family to the group with the smallest distance. We indicate this in the scatterplot. Although it is not necessarily obvious visually, family 81 is

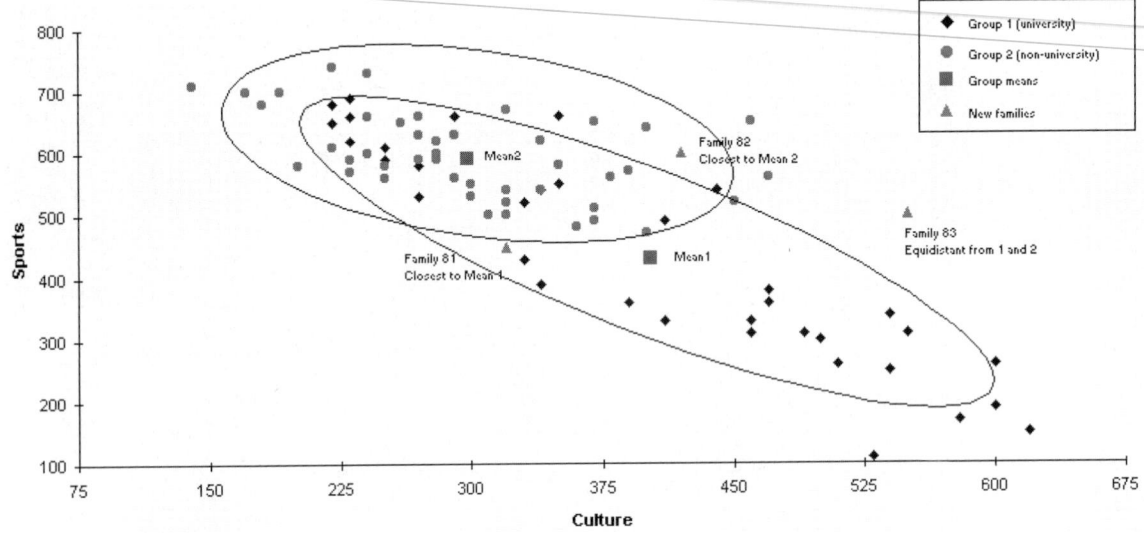

FIGURE 15.4 Illustrating Closest Statistical Distance with a Scatterplot

closest to group mean 1, family 82 is closest to group mean 2, and family 83 is equidistant from the two group means. We would classify family 81 as a university family, we would classify family 82 as a non-university family, and we could go either way with family 83.

The calculations for these statistical distances appear on the right of Figure 15.3. They are essentially the same as in the previous example, except that we now repeat the calculations for each group. Note that there is a sample covariance matrix for each group. The usual assumption in discriminant analysis is that the *population* covariance matrices are the same for all groups, so we "pool" the sample covariance matrices to obtain an estimate of this common population covariance matrix, and we use this pooled estimate in all further calculations.[2] The pooled matrix appears in the range H24:I25. To calculate it, we highlight the range H24:I25, type the formula

$$= ((\text{SampSize1-1}) * \text{CovarMat1} + (\text{SampSize2-1}) * \text{CovarMAT2})/(\text{SampSize1} + \text{SampSize2-2})$$

and press Ctrl-Shift-Enter. Then the distances in the range J29:J31 are calculated from equation (15.2), using the pooled sample covariance matrix for S. As we see, family 81 is indeed closer to group mean 1, family 82 is closer to group mean 2, and family 83 is equally distant from the two group means. ●

This example illustrates the basic ideas behind discriminant analysis. It is essentially a classification method that uses statistical distance as the basis for classification. Actually, there are several equivalent ways of implementing discriminant analysis. One way is this one—calculate the statistical distance from an observation to each group mean and classify it according to the smallest statistical distance. We will illustrate a slightly different approach for the two-group problem in the next subsection. Although this new approach has certain advantages, it is basically equivalent to the statistical distance approach.

[2]This assumption is analogous to the equal-variance assumption we made for the two-sample procedures in Chapters 9 and 10. However, it is a stronger assumption. There are methods for testing whether the assumption is satisfied, at least approximately, but we will not present these methods here.

15.2.2 Analysis of Two Groups

This paragraph is the key to a conceptual understanding of discriminant analysis.

In this section we will study the two-group discriminant problem in some depth. Recall that one objective of discriminant analysis is to see whether a set of explanatory variables can be used to discriminate between the two groups. If this is possible, we can turn to the second objective, classification. The original method for discrimination, developed by the famous statistician Sir Ronald Fisher, was to search for a linear *discriminant function* of the explanatory variables that "maximally separates" the two groups. The idea is simple. Given a scatterplot of the two groups, is there a line such that most of the points in one group lie on one side of the line, and most of the points in the other group lie on the other side of the line? If so, then it is possible to discriminate between the groups. In addition, we can use the line for classification—any new point is classified according to which side of the line it is on. We illustrate the method in the following example.

Example 15.3 Classifying People as *Wall Street Journal* Subscribers or Nonsubscribers

The file WSJ1.XLS contains observations on 84 people, each of whom either subscribes or does not subscribe to *The Wall Street Journal* (WSJ). These are the two groups. The variables that we believe might be useful for discriminating between these two groups are Income (the person's annual income) and InvestAmt (the total amount the person has invested in stocks and bonds). A partial listing of the data appears in Figure 15.5. How well do these two variables discriminate between the subscribers and nonsubscribers? How well can we classify these 84 people on the basis of these two variables?

Objective To use discriminant analysis to see whether we can discriminate between WSJ subscribers and nonsubscribers on the basis of income and level of investment.

FIGURE 15.5
Data for WSJ Example

	A	B	C	D	E	F
1	Discriminant analysis: two groups and two explanatory variables					
2						
3	Person	Income	InvestAmt	WSJSubscriber		
4	1	66400	26900	No		
5	2	68000	7100	No		
6	3	54900	21500	No		
7	4	50600	19300	No		
8	5	54100	16700	No		
56	53	60900	25800	No		
57	54	88900	28600	No		
58	55	68200	12300	No		
59	56	88400	34500	No		
60	57	66600	32200	No		
61	58	77800	48500	Yes		
62	59	86600	66600	Yes		
63	60	72900	39400	Yes		
64	61	90900	63800	Yes		
65	62	64300	50100	Yes		
66	63	53900	36400	Yes		
84	81	74100	36700	Yes		
85	82	78500	46000	Yes		
86	83	75200	51100	Yes		
87	84	100700	58800	Yes		

Solution

As usual, it helps to start with a scatterplot, as shown in Figure 15.6. The legend shows that most of the subscriber points are above and to the right of the nonsubscriber points. Subscribers tend to have larger incomes and larger amounts invested. Fisher's method is to find a line that separates the two groups of points as much as possible. As in our discussion of regression, we will not discuss the technical details for finding this line, but it is clear that the line shown in the figure does a good job. (You might see whether you can find a better line.) Most of the subscriber points are above this line, and most of the nonsubscriber points are below it. Because of the overlap between the two sets of points, no line can separate the two groups *perfectly,* but the one shown is fairly successful at separating the two groups. Furthermore, if we observed a new person, we could plot his or her point and classify it according to which side of the line it is on.

The numerical calculations (on which the line in the scatterplot is based) appear in Figure 15.7. This output is available from StatPro by selecting the StatPro/Classification Analysis/Discriminant Analysis menu item and working through the dialog boxes. (The procedure is straightforward. You must indicate a "code" variable that specifies which group each person is in, and you must indicate any explanatory variables you want to use. There are also some optional settings you can select. We will discuss them shortly). As in our discussion of statistical distance, the procedure starts by calculating a row of means for each group, a covariance matrix for each group, and a pooled covariance matrix. The next step is to calculate the coefficients of the linear "separating" function in the range I29:J29. (These coefficients are calculated from matrix formulas that we will not present here.)

Given these coefficients, the method then calculates a *discriminant score* for each person in column H. Specifically, the formula in H4 is

$$= \text{SUMPRODUCT}(\$L\$29{:}\$M\$29,B4{:}C4)$$

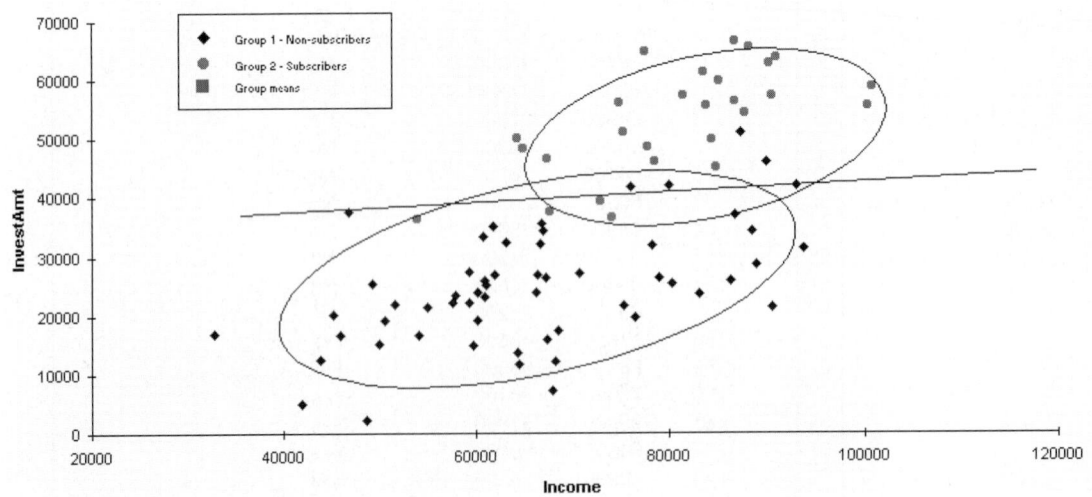

FIGURE 15.6 The Line of Maximal Separation

	F	G	H	I	J	K	L	M	N	O	P
3	StatDist1	StatDist2	Discriminant	Classification			**Discriminant Analysis**				
4	0.229005023	2.769086119	-5.115462701	No							
5	2.363424601	5.232611294	1.974062536	No			**Sample sizes**				
6	0.84174592	3.258632671	-3.968084673	No			No	57			
7	1.1526726	3.496346953	-3.47526607	No			Yes	27			
8	0.982441143	3.762475179	-2.327645579	No							
9	0.936638963	2.553357587	-6.101988807	No			**Sample means**				
10	0.154394118	3.130470253	-4.035156788	No					Income	InvestAmt	
11	1.702584371	4.216609761	-1.482657848	No			No	66042.10526	24952.63158	
12	2.194452945	4.96524554	0.995859757	No			Yes	80485.18519	53000	
13	0.468008148	2.887145418	-4.864871499	No							
14	1.586104663	4.531384548	0.085698944	No			**Matrices of sample variances and covariances**				
15	0.811604171	2.658922372	-5.717576593	No			**No**	Income	InvestAmt		
16	1.498825874	3.779795893	-2.902971023	No			Income	209308552.6	79303458.65	
17	1.051153082	4.037750481	-1.323906946	No			InvestAmt	79303458.65	99485394.74	
18	1.783079745	1.215415818	-9.7742289	Yes							
19	2.285171145	1.522254218	-10.37553465	Yes			**Yes**	Income	InvestAmt		
20	2.584398307	4.112206434	-3.807596444	No			Income	118875156.7	64698846.15	
21	0.656529285	3.063131908	-4.447286826	No			InvestAmt	64698846.15	80470769.23	
22	1.151117028	1.927512098	-7.728043781	No							
23	2.694566838	0.783608978	-12.24648373	Yes			**Pooled**	Income	InvestAmt		
24	1.068687563	3.268971623	-4.151118834	No			Income	180634549	74672727.86	
25	1.555385495	2.973282909	-5.712566408	No			InvestAmt	74672727.86	93456367.14	
26	1.255593415	3.47388108	-3.677492549	No							
27	2.499871629	4.57232058	-1.594781354	No			**Coefficients of discriminant function**				
28	0.635006685	3.18985884	-4.037177052	No				Income	InvestAmt	
29	1.594025006	2.385751663	-7.347712378	No				6.586E-05	-0.000352735	
30	1.140249151	3.228452886	-4.361790063	No							
31	1.085977277	3.962671787	-1.661449493	No			**Prior probabilities**				
32	1.419896921	3.994688488	-1.952445593	No			No	0.5		
33	1.773809943	1.347916166	-9.58792188	Yes			Yes	0.5		
34	1.537078161	1.894001335	-8.310844125	No							
35	2.344238423	5.304145289	2.396091714	No			**Misclassification costs (actual along side, predicted along top)**				
36	0.597009637	3.537897725	-2.843010114	No				No	Yes	
37	1.129307847	3.705601273	-2.695087728	No			No	0	1	
38	0.348948079	2.844629057	-4.938085164	No			Yes	1	0	
39	0.536874514	2.71930283	-5.3699732	No							
40	2.15596176	3.571478008	-4.869517991	No			**Cutoff value for classification**				
41	1.638349227	3.485691684	-4.190230856	No			-8.923160017				
42	2.134951307	2.238615689	-8.696468458	No							
43	0.472353892	3.011263985	-4.500863723	No			**Classification matrix (actual along side, predicted along top)**				
44	0.149718273	2.845333196	-4.8864073	No				No	Yes	Pct Correct
45	1.775737117	3.754375365	-3.452113982	No			No	52	5	91.2%
46	1.122616242	2.122151741	-7.301529624	No			Yes	2	25	92.6%
47	0.376297422	3.118732273	-4.130714397	No							
48	1.637949222	3.839573794	-2.893435384	No			**Summary of overall classification results**				
49	1.253200361	3.899890451	-2.103842826	No			Pct correct	91.7%		
50	1.219426186	4.122411997	-1.169519792	No			Base	67.9%		
51	1.337628046	4.294208942	-0.597669195	No			Improvement	74.1%		
52	1.307471303	1.799993927	-8.157911553	No							
86	2.907536738	0.394886835	-13.07207715	Yes							
87	3.569000736	1.538400361	-14.10870531	Yes							

FIGURE 15.7 Discriminant Analysis for WSJ Example

and this is copied down column H.[3] This formula simply substitutes each person's values into the linear discriminant function defined by the coefficients. For example, the score for the first person is

$$0.00006586(66400) - 0.0003527(26900) = -5.115$$

Finally, the method calculates a cutoff score in cell L41 (more about this value below) and compares each discriminant score to the cutoff score for classification in column I. The formula in cell I4 is

$$= \text{IF}(\text{H4} > \$\text{L}\$41, \text{"No"}, \text{"Yes"})$$

[3]StatPro doesn't enter this formula in column H; it enters only the numerical result of the calculation. The reason is that in general, the explanatory variables might not be in contiguous columns.

which is copied down column I. For example, the first person is classified (correctly) as a nonsubscriber because this person's discriminant score, -5.115, is greater than the cut-off score.

The output from row 43 down checks the accuracy of the 84 classifications. As we see, 52 of the 57 nonsubscribers (91.2%) are correctly classified, and 25 out of the 27 subscribers (92.6%) are correctly classified. The Summary section indicates that 91.7% of the 84 people are classified correctly. Is this very good? One rather naive answer is that it is good because it is considerably greater than 50%. However, note that there are many more nonsubscribers than subscribers (57 versus 27). Therefore, we could use a very simple classification scheme—classify *everyone* as nonsubscribers—and be correct in 57 out of 84 cases (or 67.9%). This "base" percentage is shown in cell M50. The real question, then, is whether discriminant analysis does much better than this base percentage. The *best* it could do is 100% correct, so cell M51 reports the percentage of the gap from 67.9% to 100% that the discriminant analysis fills. In words, 91.7% is 74.1% of the way from 67.9% to 100%. Obviously, there is room for improvement—7 of the people are classified incorrectly—but the discriminant analysis represents a big improvement over the simple classification scheme.

The StatPro output confirms that classifying according to discriminant scores is equivalent to classifying according to statistical distance (when there are equal prior probabilities and equal misclassification costs).

Recall that statistical distance can also be used as a criterion for classification. Therefore, StatPro calculates the statistical distance of each observation to each of the group mean vectors and reports these in columns F and G. For example, the value in cell F4 is the statistical distance of the first person's explanatory scores to the mean vector for all nonsubscribers, whereas the value in G4 is this person's statistical distance to the mean vector for all subscribers. You can check that each person's classification, based on the discriminant score, is the *same* as if we had used statistical distance, classifying each according to the group with the smallest statistical distance.

Finally, we look at the cutoff value used for classification. It is usually selected so that classification based on the linear discriminant function is equivalent to classification based on smallest statistical distance. However, it can be modified in two ways. First, based on historical data, we might know that one group is much more likely than another. Then we can enter *prior probabilities* of the two groups, as shown in the range M32:M33. (Actually, StatPro gives us this option in one of its dialog boxes, but new prior probabilities can also be entered manually.) These are the probabilities that a "typical" member of the population will fall in the various groups, based on any prior knowledge we might have. The default values are those shown, 0.5 each, but we can "bias" the classification by entering unequal prior probabilities. For example, if we believe from prior knowledge that only 20% of all people subscribe to WSJ, then the bottom part of the output changes as shown in Figure 15.8. (Note that the discriminant coefficients do not change; only the cutoff value changes.) Now we are more likely than before to classify someone as a non-subscriber. Therefore, we make fewer misclassifications of one kind, but we make more of the other. Also, the new classification is *not* equivalent to smallest statistical distance.

The second modification concerns the *costs* of misclassification, shown in cells N37 and M38. Depending on the context and purpose of the analysis, incorrectly misclassifying a group 1 observation might be much more costly than incorrectly misclassifying a group 2 observation (or vice versa). This is probably not an issue in this example, but it certainly could be in the classification, say, of patients as having or not having cancer. Specifically, classifying a patient with cancer as a healthy person is probably much more costly than classifying a healthy person as having cancer. By using appropriate misclassification costs in cells N38 and M37, the cutoff value is chosen so that the expected cost of misclassification is minimized.[4] The default costs are 1, but StatPro allows you to en-

---

[4]The formula for the cutoff value is complicated and is not given here. However, it is automatically implemented by StatPro.

FIGURE 15.8

Classification with
Unequal Prior
Probabilities

	K	L	M	N	O	P
31	*Prior probabilities*					
32		No	0.8			
33		Yes	0.2			
34						
35	*Misclassification costs (actual along side, predicted along top)*					
36			No	Yes		
37		No	0	1		
38		Yes	1	0		
39						
40	*Cutoff value for classification*					
41		-10.30945438				
42						
43	*Classification matrix (actual along side, predicted along top)*					
44			No	Yes	Pct Correct	
45		No	55	2	96.5%	
46		Yes	5	22	81.5%	
47						
48	*Summary of overall classification results*					
49		Pct correct	91.7%			
50		Base	67.9%			
51		Improvement	74.1%			

ter other costs. Actually, one of these can always be chosen as 1 because only the *ratio* of the costs is important. For example, if one of the misclassification costs is 4 times as large as another, it doesn't matter whether these costs are 4 and 1, 100 and 25, or 4000 and 1000; only the ratio, 4 to 1, matters.

Figure 15.9 illustrates the case where the prior probabilities are equal but the cost of misclassifying a subscriber as a nonsubscriber is 4 times as large as the opposite misclassification cost. Now there are *no* misclassifications of the more costly type, but there are more of the less costly type. Again, this modified classification is *not* equivalent to smallest statistical distance.

These illustrations of unequal prior probabilities and/or unequal misclassification costs show that we can bias the classification one way or the other, but we cannot get something for nothing. Any time we manipulate the cutoff value to reduce the number of misclassifications of one type, we are likely to increase the number of misclassifications of the other type.

FIGURE 15.9

Classification with
Unequal Misclas-
sification Costs

	K	L	M	N	O	P
31	*Prior probabilities*					
32		No	0.5			
33		Yes	0.5			
34						
35	*Misclassification costs (actual along side, predicted along top)*					
36			No	Yes		
37		No	0	1		
38		Yes	4	0		
39						
40	*Cutoff value for classification*					
41		-7.536865656				
42						
43	*Classification matrix (actual along side, predicted along top)*					
44			No	Yes	Pct Correct	
45		No	47	10	82.5%	
46		Yes	0	27	100.0%	
47						
48	*Summary of overall classification results*					
49		Pct correct	88.1%			
50		Base	67.9%			
51		Improvement	63.0%			

Discriminant analysis can use any number of explanatory variables for discrimination. These can be continuous "measurement" variables or they can be dummy 0–1 variables. The next example illustrates how a typical analysis might proceed.

Example 15.4 Understanding the Characteristics of Great Italian Company's Lasagna Customers

The Great Italian Company makes upscale frozen Italian entrees that people can put in the microwave for instant dinners. The company has collected data on 856 people, 495 of whom have tried the company's lasagna dinner and 361 of whom have never tried it. (See the file LASAGNA1.XLS.) The data include the person's age, weight, income, gender (0 for females, 1 for males), and whether he or she lives alone (0 if no, 1 if yes). A partial listing of the data appears in Figure 15.10. Great Italian wants to know whether it can correctly classify triers and nontriers on the basis of these (or some subset of these) explanatory variables.

Objective To use discriminant analysis to classify people as triers or nontriers of the company's lasagna, based on demographic data.

Solution

If we use StatPro's discriminant analysis procedure with all five explanatory variables (and equal prior probabilities and equal misclassification costs), we obtain the output in Figure 15.11. Note that because of multiple explanatory variables, we cannot rely on scatterplots; scatterplots are useful only for examining two variables at a time. However, the

FIGURE 15.10
Data for Lasagna
Example

	A	B	C	D	E	F	G
1	Discriminant analysis: two groups and more than two explanatory variables						
2							
3	Person	Age	Weight	Income	Gender	LiveAlone	HaveTried
4	1	27	167	40600	1	0	No
5	2	40	219	55200	1	0	No
6	3	31	190	48500	1	1	No
7	4	44	186	72100	0	0	No
8	5	45	152	60800	0	0	No
360	357	57	225	68100	1	0	No
361	358	41	210	38400	1	1	No
362	359	56	244	105000	0	0	No
363	360	47	168	57800	1	0	No
364	361	40	183	50700	0	1	No
365	362	40	167	78900	0	0	Yes
366	363	47	184	67000	0	1	Yes
367	364	39	174	37400	1	0	Yes
368	365	40	158	55500	1	0	Yes
851	848	31	223	86600	0	0	Yes
852	849	29	187	71100	0	0	Yes
853	850	40	178	80700	1	0	Yes
854	851	37	163	79300	0	0	Yes
855	852	34	158	64400	1	0	Yes
856	853	30	215	64600	0	0	Yes
857	854	36	171	53700	1	1	Yes
858	855	30	206	53400	1	1	Yes
859	856	50	231	67500	0	0	Yes

FIGURE 15.11
Discriminant
Analysis with All
Five Explanatory
Variables

	N	O	P	Q	R	S	T
3	*Discriminant Analysis*						
4							
5	*Sample sizes*						
6		No	361				
7		Yes	495				
8							
9	*Sample means*						
10			Age	Weight	Income	Gender	LiveAlone
11		No	42.52354571	190.7174515	58404.70914	0.495844875	0.102493
12		Yes	36.05454545	194.0787879	71892.92929	0.563636364	0.218182
13							
14	*Matrices of sample variances and covariances*						
15		**No**	Age	Weight	Income	Gender	LiveAlone
16		Age	94.73347184	21.02611573	27095.0277	-0.071429671	0.304524
17		Weight	21.02611573	572.1088335	14271.88981	-0.748399508	0.417929
18		Income	27095.0277	14271.88981	229017061.1	-326.7859341	-350.7618
19		Gender	-0.071429671	-0.748399508	-326.7859341	0.250677131	0.004594
20		LiveAlone	0.304524469	0.417928593	-350.7617729	0.004593721	0.092244
21							
22		**Yes**	Age	Weight	Income	Gender	LiveAlone
23		Age	73.14479205	7.6576371	23395.1233	0.084578579	0.309937
24		Weight	7.6576371	638.1982333	-3245.190774	0.097202797	-0.247994
25		Income	23395.1233	-3245.190774	367962561.2	-23.94184763	-207.5635
26		Gender	0.084578579	0.097202797	-23.94184763	0.246448289	-0.003791
27		LiveAlone	0.309937431	-0.247994111	-207.5634891	-0.003790946	0.170924
28							
29		**Pooled**	Age	Weight	Income	Gender	LiveAlone
30		Age	82.24540648	13.29306135	24954.80197	0.018813977	0.307656
31		Weight	13.29306135	610.3385332	4139.058653	-0.259257191	0.032723
32		Income	24954.80197	4139.058653	309390687.6	-151.6044602	-267.9281
33		Gender	0.018813977	-0.259257191	-151.6044602	0.248230939	-0.000256
34		LiveAlone	0.30765562	0.03272272	-267.9281052	-0.000256426	0.137757
35							
36	*Coefficients of discriminant function*						
37		Age	Weight	Income	Gender	LiveAlone	
38		0.100290861	-0.007408144	-5.2753E-05	-0.321858987	-1.165228701	
39							
40	*Prior probabilities*						
41		No	0.5				
42		Yes	0.5				
43							
44	*Misclassification costs (actual along side, predicted along top)*						
45			No	Yes			
46		No	0	1			
47		Yes	1	0			
48							
49	*Cutoff value for classification*						
50		-1.279108461					
51							
52	*Classification matrix (actual along side, predicted along top)*						
53			No	Yes	Pct Correct		
54		No	272	89	75.3%		
55		Yes	142	353	71.3%		
56							
57	*Summary of overall classification results*						
58		Pct correct	73.0%				
59		Base	57.8%				
60		Improvement	36.0%				

These sample means do not necessarily tell the whole story because they ignore correlations between the variables, but they point us in the right direction.

sample means in the numerical output provide some indication of how the two groups differ. The people who have tried the lasagna tend to be younger and slightly heavier, they have more income, they are primarily male, and they tend to live alone. The rest of the output is exactly as in the previous example. As we see, the five explanatory variables do a reasonably good job of classifying—73% of the people are classified correctly, a considerable improvement over the naive base case of 57.8% (found by classifying everyone as triers).

As with multiple regression, it is possible to "fine-tune" the discriminant analysis by eliminating explanatory variables that do not really help with the classification. However, we will not pursue this direction here because it would not really help the company discriminate better between triers and nontriers. If the company really needs a better correct classification percentage, then—exactly as in regression—its best alternative is to use more (or different) explanatory variables. (Can you think of any potentially good ones?) ●

15.2.3 Analysis of More Than Two Groups

For multiple groups, we use only the statistical distance criterion for classification.

Discriminant analysis can also be performed when there are more than two groups, but the analysis is somewhat more complex. As with two groups, we can base the classification on statistical distance or on linear functions of the explanatory variables. Although these two methods are essentially equivalent, we believe the latter is more difficult to understand. The problem is that with more than two groups, there is not a *single* linear function and cutoff value that can be used for classification. We need *multiple* linear functions because the variables that are most useful for discriminating between groups 1 and 2, say, might not be the most useful for discriminating between groups 1 and 3. Therefore, we present only the statistical distance method. This method is simple to state for any number of groups. For any observation, we calculate its statistical distance to each of the group mean vectors, and we classify it in the group with the *smallest* statistical distance. StatPro makes this procedure straightforward, as illustrated in the following example.

Example 15.5 Understanding the Market for One of JBH's Textbooks

The JBH Publishing Company specializes in university textbooks for Business Schools. One of its current best-selling books is a computer tools book that is aimed at large introductory computer courses. It covers spreadsheets and database programs with a hands-on approach. Although this book has gained a fairly large share of the lucrative computer tools market, JBH would like to know why some universities have never adopted it and others who adopted it have since switched to a competing book. How might the company analyze this situation?

Objective To use multiple-group discriminant analysis to classify potential customers for a particular computer tools textbook.

Solution

The file TEXTBOOK.XLS contains data on 119 universities that JBH considers its potential customers. Each university falls into one of three groups: (1) Never (the school has never adopted the book), (2) NoMore (the school adopted the book but has since switched to a competing book), and (3) Still (the school adopted the book and is still using it). The potential explanatory variables are

- Enrollment—total enrollment at the university
- AvgSAT—average SAT score of all entering freshmen to the university
- PctTenure—percentage of courses in the Business School taught by tenure-track faculty
- PCLabs—number of PCs in public university computer labs available to Business School classes
- PctOwnPC—percentage of Business School students who own a PC
- Tuition—average yearly tuition charged by the university

FIGURE 15.12
Group Means for
Textbook Example

		Enrollment	AvgSAT	PctTenure	PCLabs	PctOwnPC	Tuition
3	**Discriminant Analysis**						
4							
5	**Sample sizes**						
6	Never	40					
7	NoMore	37					
8	Still	42					
9							
10	**Sample means**						
11		Enrollment	AvgSAT	PctTenure	PCLabs	PctOwnPC	Tuition
12	Never	14799.1	1134.3	80.6%	148.2	54.9%	14997.5
13	NoMore	14888.5	921.7	59.9%	101.6	44.0%	9878.4
14	Still	19575.6	950.1	59.5%	153.3	52.0%	9681.0

To get some understanding of its customers, JBH might begin by looking at the group means listed in Figure 15.12. These indicate some variables that might be good for discriminating between groups. For example, the Never group has larger mean average SAT scores and tuition than the other two groups. Perhaps the more exclusive schools consider JBH's book too elementary. We can see from the scatterplot in Figure 15.13 (which again uses ellipses to help us see the general swarms of points) that AvgSAT and Tuition indeed do a good job of separating the Never group from the combined NoMore and Still groups. However, the group means suggest that these two variables do *not* discriminate well between the NoMore and Still groups. To discriminate between these, we might try (again based on the means) Enrollment, PCLabs, and/or PctOwnPC. For example, the scatterplot in Figure 15.14 (page 832) indicates that PCLabs and PctOwnPC separate these two

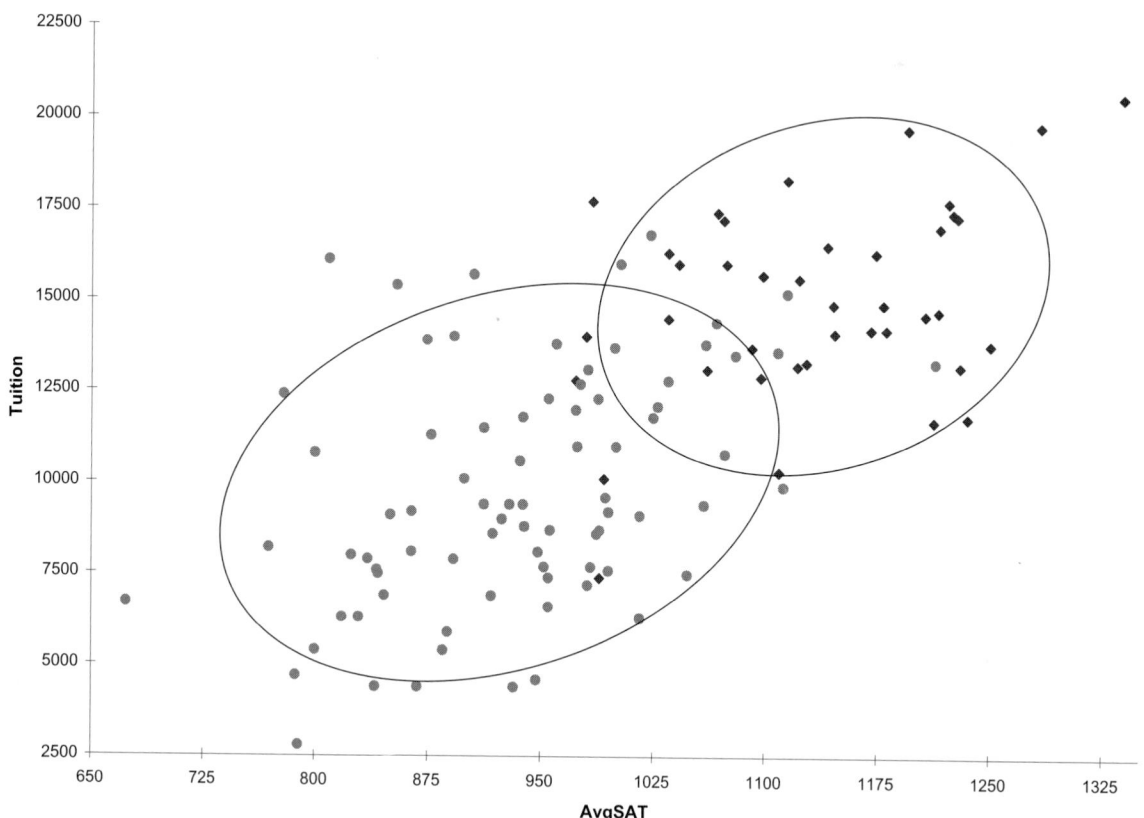

FIGURE 15.13 Discriminating Between Never and Combined NoMore and Still Groups

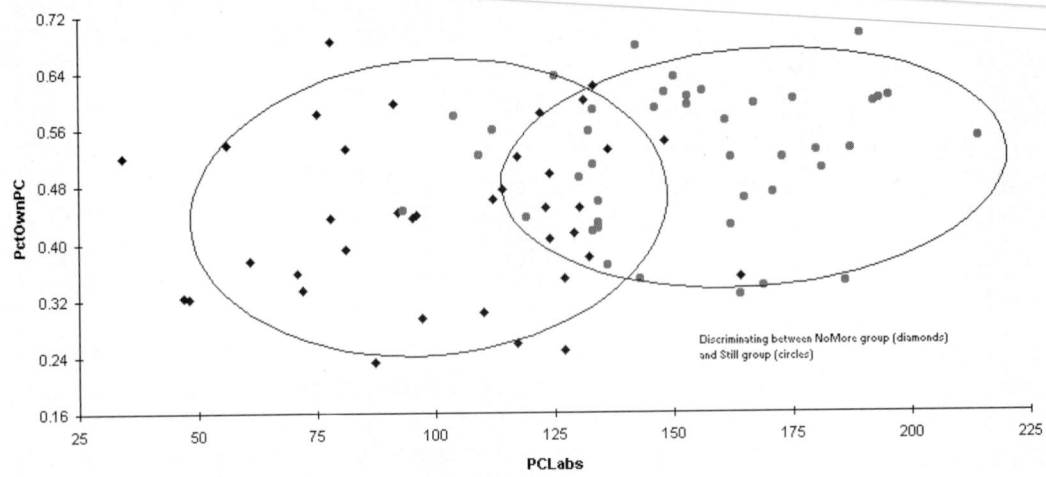

FIGURE 15.14 Discriminating Between NoMore and Still Groups

groups fairly well. Specifically, the Still group tends to have more available lab PCs and more students who own PCs than the NoMore group.

This analysis indicates that accurate classification is probably possible, but it is still unclear how we should classify the schools. For example, a school with relatively high values of PCLabs and PctOwnPC should not be classified as NoMore, but it is not clear whether it should be classified as Never or Still. This is the reason for using statistical distance. It takes *all* of the variables into account, and it uses information on their variances and covariances, not just means. Fortunately, StatPro takes care of the details. For each university, it calculates three statistical distances, one to each of the three group means, and reports these next to the data, as shown for a few schools in Figure 15.15. Then it classifies each school depending on which distance is the smallest. For the data shown, only university 4 is classified incorrectly. It evidently breaks the pattern of "typical" schools in the Never group. For example, its AvgSAT is abnormally low. However, the output in Figure 15.16 shows that the vast majority of schools are classified correctly.

JBH can now use this analysis to target its customers more effectively. It appears that this book is considered too low-level by the more exclusive schools. Maybe JBH should hire an author to write a somewhat higher-level computer tools textbook. Also, this book is being dropped by schools whose students do not have access to as many PCs. This is a more difficult problem for JBH. It cannot simply give these schools more PCs, but it might consider publishing another textbook that is less hands-on.

	A	B	C	D	E	F	G	H	I	J	K	L	M
1	Discriminant analysis: more than two groups, more than two explanatory variables												
2													
3	University	Enrollment	AvgSAT	PctTenure	PCLabs	PctOwnPC	Tuition	Status		StatDist1	StatDist2	StatDist3	Classification
4	1	17455	1068	79.3%	154	46.5%	17400	Never		2.26616	4.453118	4.407021	Never
5	2	14445	1173	84.6%	162	50.6%	16300	Never		1.269066	4.875467	5.143399	Never
6	3	14773	1122	80.9%	158	68.0%	15600	Never		1.857446	4.285611	3.775915	Never
7	4	16138	992	63.4%	133	54.2%	10100	Never		3.229906	1.933789	1.771154	Still
8	5	16717	1141	80.7%	123	64.4%	16500	Never		2.355581	4.204571	4.019527	Never
9	6	18002	1109	84.5%	127	50.9%	10300	Never		2.629935	5.226973	4.797633	Never
10	7	14030	1180	85.2%	146	56.5%	14200	Never		1.140585	4.714975	4.938711	Never
11	8	17453	1146	79.0%	144	57.6%	14100	Never		1.612812	4.152255	3.579208	Never
12	9	13931	1127	78.8%	126	60.0%	13300	Never		1.311491	3.430159	3.917526	Never
13	10	13046	984	71.8%	143	41.7%	17700	Never		3.271203	3.450484	4.91474	Never
14	11	17171	1121	76.8%	148	49.1%	13200	Never		1.329401	3.860794	3.631125	Never
15	12	14944	1230	80.7%	161	70.0%	13200	Never		2.391374	4.860317	4.190698	Never

FIGURE 15.15 Classification on the Basis of Smallest Statistical Distance

FIGURE 15.16
Classification
Results for
Textbook Example

		Never	NoMore	Still	Pct Correct
49	**Classification matrix (actual along side, predicted along top)**				
50		Never	NoMore	Still	Pct Correct
51	Never	39	0	1	97.5%
52	NoMore	0	34	3	91.9%
53	Still	0	3	39	92.9%
54					
55	**Summary of overall classification results**				
56	Pct correct	94.1%			
57	Base	35.3%			
58	Improvement	90.9%			

PROBLEMS

Level A

1. The file P15_01.XLS contains data on 100 consumers who drink beer. A portion of these individuals prefer to drink light beer, whereas another subset prefers to drink regular beer. In particular, a major beer producer believes that the following variables might be useful in discriminating between these two groups: the individual's gender, marital status, annual income level, and age.

   a. Assess how well the beer producer can classify these 100 beer drinkers on the basis of the four given variables.

   b. What other variables might be useful in discriminating between these two groups of beer drinkers?

2. Producers of the late-night television news program "Nightline" are trying to increase the show's viewership. In an effort to understand who is currently watching their program, the producers collect data on four potentially useful variables for discriminating between viewers and nonviewers of "Nightline": the individual's gender, marital status, annual income level, and age. Such information is collected for 75 randomly selected late-night television viewers and recorded in the file P15_02.XLS.

   a. Assess how well the producers can classify these 75 individuals on the basis of the four given variables.

   b. What other variables might be useful in discriminating between these two groups of late-night television viewers?

3. The manager of a rapidly growing health club is interested in determining who is most likely to purchase a membership at the club. She proceeds to survey 100 randomly selected individuals who live in the surrounding community. Specifically, she collects data on each of the following: the individual's gender, an indication of whether the individual exercises regularly, the individual's annual income level, and the individual's age. These data are provided in the file P15_03.XLS.

   a. How well can the health club manager discriminate between those who are members and those who are not members on the basis of the given variables?

   b. How does your answer to the question in part **a** change when the prior probability of not being a member changes from 0.50 to 0.60? Explain.

   c. How does your answer to the question in part **a** change when the cost of misclassifying a member as a nonmember becomes *twice* as large as that of the other possible misclassification? Explain.

4. The owner of a new Chinese restaurant in Centerville, Ohio, is trying to determine who is most likely to have tried his new establishment. He directs his manager to survey randomly selected members of the town and then to collect the following information for each individual in the sample: an indication of whether the respondent has received a coupon for a 20% discount on a dinner at the restaurant, an indication of whether the respondent has seen an advertisement for the opening of the restaurant in one of the local newspapers, the respondent's annual income, and the number of days in a typical week that the respondent eats dinner outside the home. The data collected for a sample of 100 individuals are given in the file P15_04.XLS.

   a. How well can the restaurant owner discriminate between those who have tried his new restaurant and those who have not on the basis of the given variables?

   b. How does your answer to the question in part **a** change when the prior probability of not having tried the restaurant changes from 0.50 to 0.60? Explain.

   c. How does your answer to the question in part **a** change when the cost of misclassifying a person who has not tried the restaurant as someone who has becomes *twice* as large as that of the other possible misclassification? Explain.

5. Admissions directors of graduate business programs constantly review the criteria by which they make admission decisions. Suppose that the director of a particular top-20 MBA program is trying to understand how she and her staff discriminate between those who are admitted to her institution's program and those who are rejected. In an effort to do so, she collects data on each of the following variables for 100 randomly selected individuals who applied in the past academic year: an indication of whether the applicant graduated in the top 10% of his or her undergraduate class, an indication of whether the admissions office interviewed the applicant in person or over the telephone, the applicant's overall GMAT score (where the highest possible score is 800), and the applicant's undergraduate grade-point average (standardized on a four-point scale). These data are provided in the file P15_05.XLS.

a. How well can this admissions director discriminate between those who are admitted to her program and those who are not on the basis of the given variables?

b. How does your answer to the question in part **a** change when the prior probability of not admitting an applicant changes from 0.50 to 0.75? Explain.

c. How does your answer to the question in part **a** change when the cost of misclassifying an applicant who has been admitted as someone who has been rejected becomes *three* times as large as that of the other possible misclassification? Explain.

6. Who tends to own a personal computer in his or her home? Consider the data provided in the file P15_06.XLS. Specifically, we have collected observations on each of the following variables for 100 randomly chosen consumers: an indication of the consumer's educational attainment (i.e., a high school education or less, or more than a high school education), an indication of whether the consumer uses a computer in his or her work, the consumer's annual income, and the consumer's age.

a. How well can we discriminate between those who own personal computers and those who do not on the basis of the four given variables?

b. How does your answer to the question in part **a** change when the prior probability of not owning a PC changes from 0.50 to 0.30? Explain.

c. How does your answer to the question in part **a** change when the cost of misclassifying a consumer who does not own a PC as someone who does becomes *three* times as large as that of the other possible misclassification? Explain.

Level B

7. The owner of a restaurant in Gainesville, Florida, has recorded sales data for the past 19 years. He has also recorded data on potentially relevant variables. The entire data set appears in the file P15_07.XLS. Use discriminant analysis to assess how well this restaurant owner can use the given explanatory variables to classify yearly sales into one of *three* levels: high, medium, and low.

8. A regional express delivery service company recently conducted a study to investigate how to distinguish among various cost levels for shipping packages. The company believes that it can classify the cost of shipping a given package into one of *three* levels (high, medium, and low) by knowing the weight of the package and the distance to the package's destination. Using the data in the file P15_08.XLS, perform an appropriate discriminant analysis and briefly summarize your findings. Should this company be encouraged by your findings?

9. Consider again the classification problem described in Problem 1. Recall that the relevant data set can be found in the file P15_01.XLS.

a. How should the beer producer classify an additional beer drinker who is 44 years old, male, married, and earns $40,000 per year?

b. How should the beer producer classify an additional beer drinker who is 44 years old, female, single, and earns $40,000 per year?

10. Consider again the classification problem described in Problem 2. Recall that the relevant data set can be found in the file P15_02.XLS.

a. How should the producers of "Nightline" classify an additional late-night TV viewer who is 46 years old, male, single, and earns $49,000 per year?

b. How should the producers of "Nightline" classify an additional late-night TV viewer who is 46 years old, female, married, and earns $49,000 per year?

 15.3 # Logistic Regression

Logistic regression is an alternative to discriminant analysis for trying to decide which groups individuals are in, given values of a set of explanatory variables. However, instead of using statistical distance or a linear function for classifying customers, it estimates the *probability* that an individual is in a particular group. As its name implies, logistic regression is somewhat similar to the regression analysis we discussed in Chapters 11 and 12, but its approach is quite different. As we explain next, it uses a *nonlinear* function of the explanatory variables for prediction.

15.3.1 The Model

Logistic regression is essentially regression with a dummy (0–1) response variable. For the two-group problem, the only version of logistic regression we will discuss, the dummy variable indicates whether an observation is in group 0 or group 1.[5] (Because it is natural to code the response variable as 0 and 1, we will refer to the groups as group 0 and group 1, rather than group 1 and group 2.) One approach to the classification problem, an approach that is sometimes actually used, is to run the usual multiple regression on the data, using the dummy variable as the response variable. However, this approach has two serious drawbacks. First, it violates the regression assumption that the error terms should be normally distributed. It is easy to show that normality is impossible if the response variable has only two possible values, 0 and 1. The second problem is that the predicted values of the response variable can be between 0 and 1, less than 0, or greater than 1. If we want a predicted value to estimate a *probability*, then values less than 0 or greater than 1 make no sense.

We therefore take a slightly different approach. Suppose that X_1 through X_k are the potential explanatory variables. If we form a linear function of them, $\beta_0 + \beta_1 X_1 + \cdots + \beta_k X_k$, there is no guarantee that this linear function will be between 0 and 1, and hence that it will qualify as a probability. But the nonlinear function

$$\frac{1}{1 + e^{-(\beta_0 + \beta_1 X_1 + \cdots + \beta_k X_k)}}$$

is *always* between 0 and 1. In fact, the function $f(x) = 1/(1 + e^{-x})$ is an "s-shaped" curve, as shown in Figure 15.17. For large negative values of x, the function approaches 0, and for large positive value of x, it approaches 1.

FIGURE 15.17
Graph of the
Function Used in
Logistic Regression

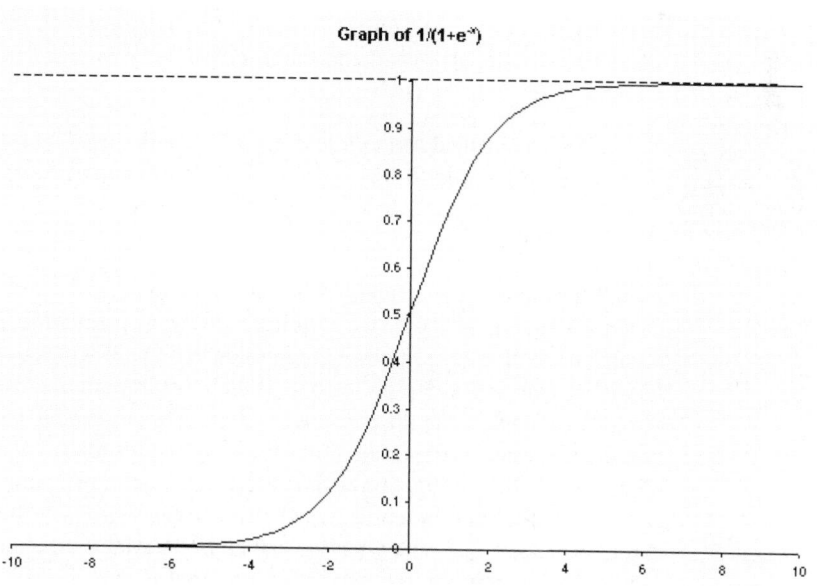

Graph of $1/(1+e^{-x})$

The logistic regression model uses this function to estimate the probability that any observation will be in group 1. Specifically, if p is the probability of being in group 1, we estimate the model in equation (15.4).

---

[5]Logistic regression can be extended to more than two groups, but the calculations and interpretation become considerably complex. Therefore, we will not discuss this extension here.

Model for Probability p of Being in Group 0	$$p = \frac{1}{1 + e^{-(\beta_0 + \beta_1 X_1 + \cdots + \beta_k X_k)}}$$	**(15.4)**

By manipulating equation (15.4) algebraically, we obtain the equivalent form in equation (15.5). This equation says that the natural logarithm of $p/(1 - p)$ is a *linear* function of the explanatory variables. The ratio $p/(1 - p)$ is called the **odds ratio.**

Model for Logarithm of Odds Ratio (Logit)	$$\ln\left(\frac{p}{1 - p}\right) = \beta_0 + \beta_1 X_1 + \cdots + \beta_k X_k$$	**(15.5)**

The odds ratio is frequently used in everyday language. Suppose, for example, that the probability p of a company going bankrupt is 0.25. Then the odds that the company will go bankrupt are $p/(1 - p) = 0.25/0.75 = 1/3$. In words, the odds are 1 to 3 that the company will go bankrupt. Odds ratios are probably most common in sports. If we read that the odds of Duke winning the NCAA basketball championship are 2 to 1, this means that the probability of Duke winning the championship is 2/3. Or if we read that the odds *against* Indiana winning the championship are 99 to 1, then the probability that Indiana will win is only 1/100.

The logarithm of the odds ratio, the quantity on the left side of equation (15.5), is called the **logit.** Therefore, the logistic regression model states that the logit is a linear function of the explanatory variables. Although this is probably a bit mysterious and there is no easy way to justify it intuitively, logistic regression *does* produce useful results in many applications. Therefore, our goals here are to explain how the regression coefficients, the β's, are estimated and, more importantly, how to interpret the results.

Logit and Odds Ratio	The **logit** is defined as the logarithm of the **odds ratio,** the ratio of the probability of being in group 1 to the probability of being in group 0.

We turn first to the estimation of the regression coefficients. Because of the nonlinear nature of equation (15.4), we require a different estimation procedure from the least squares method used in multiple regression. The new procedure is called *maximum likelihood estimation* and represents one of the most important ideas in statistical theory. The details are complex, but the concept is simple. For any model, we create a likelihood function that is basically the probability of seeing the results we actually saw. Then we maximize this likelihood function with respect to the unknown parameters, in this case the regression coefficients. In words, we choose the regression coefficients so that the resulting model is most in line with what we actually observed. Fortunately, Excel's Solver add-in can be used to perform the maximization. That is, we use Solver (behind the scenes) in StatPro's implementation of logistic regression.

15.3.2 Interpretation of the Results

Once we obtain the estimates of the β's, which we label as b's, we need to interpret them. To do so, we take the antilogarithm of both sides of equation (15.5) to obtain

$$\frac{p}{1-p} = e^{b_0 + b_1 X_1 + \cdots + b_k X_k}$$

Now suppose that X_1, say, increases by an amount Δ and the other X's do not change. What happens to the odds ratio $p/(1 - p)$? Using the laws of exponents, we see that the odds ratio changes by a *factor* of $e^{b_1 \Delta}$. In particular, if $\Delta = 1$, this factor is e^{b_1}.

As an example, suppose that group 1 represents people who have had a heart attack and group 0 represents people who have not. We run a logistic regression with several explanatory variables, one of which is Cholesterol, the person's cholesterol level, and its estimated coefficient turns out to be 0.083. Then for every unit increase in cholesterol level, the odds of having a heart attack increase by a factor of $e^{0.083} = 1.087$. Because 1 unit is so small, we might prefer to think in terms of 10-unit increases. Then the appropriate factor is $e^{0.083(10)} = 2.293$. In words, if a person's cholesterol level increases by 10 points, we estimate that his or her odds of having a heart attack increase by a factor of 2.293.

But what does it really mean that the odds of having a heart attack increase by a factor of 2.293? This does *not* mean that the *probability* of having a heart attack increases by this factor. To see this, suppose that the probability of the person having a heart attack, before the cholesterol increase, was $p = 0.25$. Then the person's odds of having a heart attack were $p/(1 - p) = 0.25/0.75 = 1/3$ (or, as we would say, 1 to 3). After a 10-unit increase in cholesterol, the odds ratio increases to $2.293(1/3) = 0.764$. Now, if $\tilde{p}$ is the new probability of a heart attack, we know that $\tilde{p}/(1 - \tilde{p}) = 0.764$. Solving for $\tilde{p}$, we get $\tilde{p} = 0.764/(1 + 0.764) = 0.433$.[6] If this person experiences another 10-unit increase in cholesterol level, the odds ratio again increases by a factor of 2.293, to $2.293(0.764) = 1.752$, and the new probability of a heart attack is $1.752/(1 + 1.752) = 0.637$.

We now see how logistic regression can be implemented in Excel.

Example 15.6 Classifying People as *Wall Street Journal* Subscribers or Nonsubscribers

This example continues the analysis of the WSJ data in Example 15.3. Recall that each person either subscribes (group 1) or does not subscribe (group 0) to the WSJ. The explanatory variables are Income and InvestAmt, the annual income and total amount invested in stocks and bonds for each person. The data are listed in the file WSJ2.XLS. (The only difference between the files WSJ1.XLS and WSJ2.XLS is that the two groups are now coded as 0 or 1, rather than No or Yes.) Does logistic regression produce different results from the discriminant analysis results we saw earlier?

Objective To use logistic regression to see whether we can classify people as WSJ subscribers or nonsubscribers on the basis of income and level of investment.

Solution

If the original data are coded as "Yes" and "No," then we need to create a dummy variable based on this categorical variable.

First, it is important that the response variable be coded as a 0–1 dummy variable before running StatPro. This is not necessary with discriminant analysis, but it is with logistic regression because the latter performs *numerical* calculations on the response variable. To run the analysis, select the StatPro/Classification Analysis/Logistic Regression menu item. After a couple of initial dialog boxes, you will see the dialog box in Figure 15.18 (page 838). Fill it in as shown. (Toward the end of this section, we will explain the purpose of the "count variable" option.) Then fill out the remaining dialog boxes, selecting

[6]In general, if r represents the odds ratio, so that $r = p/(1 - p)$, then algebra can be used to obtain $p = r/(1 + r)$.

FIGURE 15.18
Dialog Box for
Logistic Regression
Options

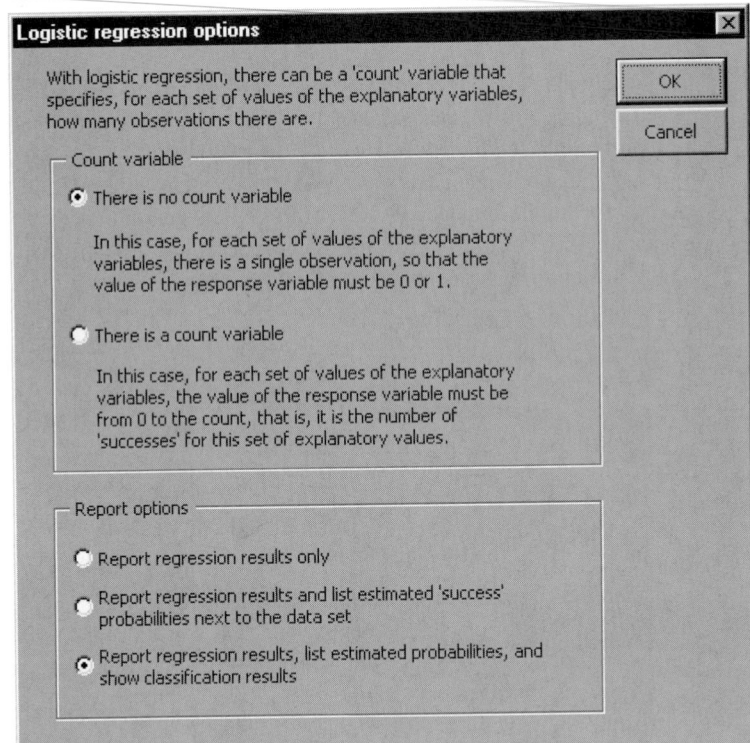

Logistic regression options

With logistic regression, there can be a 'count' variable that specifies, for each set of values of the explanatory variables, how many observations there are.

OK

Cancel

Count variable

⦿ There is no count variable

In this case, for each set of values of the explanatory variables, there is a single observation, so that the value of the response variable must be 0 or 1.

○ There is a count variable

In this case, for each set of values of the explanatory variables, the value of the response variable must be from 0 to the count, that is, it is the number of 'successes' for this set of explanatory values.

Report options

○ Report regression results only

○ Report regression results and list estimated 'success' probabilities next to the data set

⦿ Report regression results, list estimated probabilities, and show classification results

WSJSubscriber as the response variable, and Income and InvestAmt as the explanatory variables. Note that once you fill in all of the dialog boxes, StatPro launches a numerically intensive search for the maximum likelihood estimates of the regression coefficients. Depending on the speed of your computer, this can take a while.

The output for the WSJ data appears in Figure 15.19. Note the resemblance in rows 11–15 between logistic regression output and the usual multiple regression output. The only differences are that the values in column M are now called Wald statistics rather than t-values, but they are calculated exactly the same and have exactly the same interpretation.[7] Also, we append the values in column Q, the antilogs of the regression coefficients of Income and InvestAmt.

Regarding interpretation, we make several points. The negative coefficient of Income implies that as income increases, the probability of being a WSJ subscriber decreases slightly. In contrast, the positive coefficient of InvestAmt implies that as the amount invested increases, the probability of being a WSJ subscriber increases. The p-values corresponding to the regression coefficients can be interpreted in the usual way. Specifically, the relatively large p-value for Income (even though it *is* less than 0.10) means that Income does not have as great an effect on the response variable as InvestAmt.

The values in column Q can be interpreted as in our general discussion of logistic regression. For example, the value in cell Q15, 1.000352, is the factor by which the odds of being a WSJ subscriber change when InvestAmt increases by \$1 and Income stays constant. (Note that the odds are always interpreted in terms of the group coded as 1, in this case, the subscribers.) Because \$1 is such a small change, we might also calculate the appropriate factor for, say, a \$1000 increase in InvestAmt. This is $e^{0.000352(1000)} = 1.422$. (It can be calculated in Excel with the formula =EXP(1000*J15).) Therefore, if a person's

Keep in mind that p in the logistic regression model is the probability of being in group 1— here the probability of being a subscriber.

[7]They are named for the famous statistician A. Wald, who conducted some of the seminal research in logistic regression.

	I	J	K	L	M	N	O	P	Q
3	**Results of logistic regression for WSJSubscriber**								
4									
5	**Summary measures**								
6		Null deviance	105.4942						
7		Model deviance	26.8361						
8		Improvement	78.6581						
9		p-value	0.0000						
10									
11	**Regression coefficients**								
12			Coefficient	Std Err	Wald	p-value	Lower limit	Upper limit	Exp(Coeff)
13		Constant	-7.9303	3.1000	-2.5582	0.0105	-14.0064	-1.8543	
14		Income	-0.000092	0.0001	-1.7346	0.0828	-0.0002	0.0000	0.999908
15		InvestAmt	0.000352	0.0001	3.6623	0.0003	0.0002	0.0005	1.000352
16									
17	**Cutoff value for classification**								
18		0.5							
19									
20	**Classification matrix (actual along side, predicted along top**								
21			0	1	Pct Correct				
22		0	54	3	94.7%				
23		1	4	23	85.2%				
24									
25	**Summary of overall classification results**								
26		Pct correct	91.7%						
27		Base	67.9%						
28		Improvement	74.1%						

FIGURE 15.19 Logistic Regression Output for WSJ Example

amount invested increases by $1000 (and income stays constant), the odds of being a WSJ subscriber increase by a factor of 1.422.

By checking the bottom option in Figure 15.18, we also obtain the estimated probabilities of subscribing for all people in the data set (as well as classification results). A few of these probabilities are shown in Figure 15.20. We would expect small probabili-

FIGURE 15.20
Estimated Probabilities and Classifications for WSJ Example

	A	B	C	D	E	F	G
1	**Logistic regression: two groups and two explanatory variables**						
2							
3	Person	Income	InvestAmt	WSJSubscriber		Est Probs	Classification
4	1	66400	26900	0		0.0102	0
5	2	68000	7100	0		0.0000	0
6	3	54900	21500	0		0.0044	0
7	4	50600	19300	0		0.0030	0
8	5	54100	16700	0		0.0009	0
9	6	78200	31900	0		0.0199	0
10	7	66200	23800	0		0.0035	0
11	8	43900	12400	0		0.0005	0
12	9	41900	5000	0		0.0000	0
13	10	61100	25200	0		0.0092	0
14	11	64500	11800	0		0.0001	0
15	12	59400	27300	0		0.0221	0
16	13	45900	16800	0		0.0019	0
17	14	59700	14900	0		0.0003	0
18	15	76000	41900	0		0.4557	0
19	16	89900	46200	0		0.5142	1
20	17	32700	16900	0		0.0067	0
83	80	85100	59800	1		0.9949	1
84	81	74100	36700	1		0.1379	0
85	82	78500	46000	1		0.7379	1
86	83	75200	51100	1		0.9582	1
87	84	100700	58800	1		0.9706	1

ties for nonsubscribers and large probabilities for subscribers. We can then use these estimated probabilities to classify each person. The usual cutoff value for classification is 0.5—classify as a subscriber if the estimated probability is at least 0.5 and as a nonsubscriber otherwise—although trial and error can be attempted (in cell I18 of Figure 15.19) to achieve more accurate classifications. With the cutoff of 0.5, we see in Figure 15.19 that 91.7% of the people are classified correctly. This is the same percentage we obtained with discriminant analysis in Example 15.3—each method misclassifies 7 people. However, a closer comparison shows that discriminant analysis misclassifies 5 nonsubscribers (persons 15, 16, 20, 30, and 51) and 2 subscribers (persons 77 and 81), whereas logistic regression misclassifies 3 nonsubscribers (persons 16, 20, and 51) and 4 subscribers (persons 60, 63, 77, and 81). The agreement between the two methods is close, but it is not perfect. ●

As with usual multiple regression, the explanatory variables in logistic regression can be quantitative variables or categorical variables coded as dummy variables. The following example illustrates how to incorporate dummy variables and interpret the output.

Example 15.7 Understanding the Characteristics of Great Italian Company's Lasagna Customers

This example continues the Great Italian Company's analysis of its lasagna customers from Example 15.4. Recall that each customer has either tried the company's frozen lasagna dinner or has not. The explanatory variables include the customer's age, weight, income, gender, and whether he or she lives alone. The data are in the file LASAGNA2.XLS. (Again, the only difference between this file and the earlier file is that the response variable HaveTried is now coded as 1 or 0, rather than Yes or No.) Do all of these variables help to predict whether a customer has tried the lasagna dinner?

Objective To use logistic regression to estimate the probability of people trying the company's lasagna, based on demographic data.

Solution

We run StatPro's logistic regression procedure exactly as in the previous example, except that we now select all five of the potential explanatory variables. The output appears in Figure 15.21. From the signs of the coefficients, we see that the probability of having tried the lasagna increases as each explanatory variable increases, except for Age—older people are *less* likely to have tried the dinner. Also, all of the *p*-values of the regression coefficients are small (one is slightly above 0.05, but we'll count it as significant), so that all of these variables contribute significantly to the prediction.

To understand a dummy variable's coefficient, we use the same basic interpretation as before. For example, the antilog of the coefficient of LiveAlone is 3.714 (cell T18). This means that as the value of this variable increases by 1 unit—that is, from 0 to 1—the odds of the person having tried the dinner increase by a factor of 3.714. In other words, the odds for a person living alone are 3.714 times greater than for a similar person (same age, weight, income, and gender) not living alone.

If we request the estimated probabilities of having tried the lasagna (not shown here) and then classify each person based on the cutoff probability of 0.5, the output in Figure 15.21 shows that we are correct on about 75% of the classifications. The same percent-

FIGURE 15.21
Logistic Regression
Output for Lasagna
Example

	L	M	N	O	P	Q	R	S	T
3	**Results of logistic regression for Have Tried**								
4									
5	**Summary measures**								
6	Null deviance	1165.6048							
7	Model deviance	885.5894							
8	Improvement	280.0154							
9	p-value	0.0000							
10									
11	**Regression coefficients**								
12		Coefficient	Std Err	Wald	p-value	Lower limit	Upper limit	Exp(Coeff)	
13	Constant	-1.2911	0.7614	-1.6956	0.0900	-2.7835	0.2013		
14	Age	-0.1027	0.0099	-10.3791	0.0000	-0.1221	-0.0833	0.9024	
15	Weight	0.0085	0.0034	2.4984	0.0125	0.0018	0.0151	1.0085	
16	Income	0.0001	0.0000	10.6307	0.0000	0.0000	0.0001	1.0001	
17	Gender	0.3123	0.1647	1.8966	0.0579	-0.0104	0.6351	1.3666	
18	LiveAlone	1.3121	0.2434	5.3909	0.0000	0.8351	1.7892	3.7140	
19									
20	**Cutoff value for classification**								
21	0.5								
22									
23	**Classification matrix (actual along side, predicted along top)**								
24			0	1	Pct Correct				
25	0		239	122	66.2%				
26	1		94	401	81.0%				
27									
28	**Summary of overall classification results**								
29	Pct correct	74.8%							
30	Base	57.8%							
31	Improvement	40.2%							

age when using discriminant analysis was 73%, so the two methods give very similar results. It is interesting that logistic regression misclassifies a much larger percentage of the non-triers than the triers, whereas discriminant analysis misclassifies about the same percentage of each group. However, there is nothing sacred about the cutoff probability 0.5 used for classification. By changing it to 0.56, we get almost identical results to the discriminant analysis. ●

15.3.3 Goodness-of-Fit Measures

In multiple regression we used R^2 as a measure of the goodness of fit of a regression equation. There is no exact analog of R^2 for logistic regression, but several goodness-of-fit measures have been developed by researchers. We report some of these at the top of the logistic regression output from StatPro. (See rows 5–9 of Figures 15.19 and 15.21.[8]) These measures all concern "deviances," which are similar to sum of squared residuals for regular regression. The first deviance reported is for a "null" model that includes *no* explanatory variables. This model is typically very poor, so it usually has a large deviance. The second deviance reported is the deviance for the proposed model. If the model is doing a good job, this deviance should be considerably less than the null deviance. Therefore, we report the difference between these deviances and a *p*-value for the difference. This *p*-value can be interpreted similarly to the *p*-value in a regression's ANOVA table—if it is small, then the model has at least *some* explanatory power. The *p*-values for both the WSJ and lasagna examples are essentially 0, leaving no doubt that the explanatory variables we are using have some predictive power.

The model deviance can also be used for the same purpose as the partial *F* test in multiple regression. (See Section 12.7 of Chapter 12.) Specifically, suppose we estimate a

[8]We also include cell comments for these measures to help you remember what they mean.

	I	J	K	L	M	N	O	P	Q
3	*Results of logistic regression for HaveTried (all variables included)*								
4									
5	*Summary measures*								
6		Null deviance	1165.6048						
7		Model deviance	885.5894						
8		Improvement	280.0154						
9		p-value	0.0000						
10									
11	*Regression coefficients*								
12			Coefficient	Std Err	Wald	p-value	Lower limit	Upper limit	Exp(Coeff)
13		Constant	-1.2911	0.7614	-1.6956	0.0900	-2.7835	0.2013	
14		Age	-0.1027	0.0099	-10.3791	0.0000	-0.1221	-0.0833	0.9024
15		Weight	0.0085	0.0034	2.4984	0.0125	0.0018	0.0151	1.0085
16		Income	0.0001	0.0000	10.6306	0.0000	0.0000	0.0001	1.0001
17		Gender	0.3123	0.1647	1.8966	0.0579	-0.0104	0.6351	1.3666
18		LiveAlone	1.3121	0.2434	5.3909	0.0000	0.8351	1.7892	3.7140
19									
20	*Results of logistic regression for HaveTried (two variables omitted)*								
21									
22	*Summary measures*								
23		Null deviance	1165.6048						
24		Model deviance	895.3028						
25		Improvement	270.3020						
26		p-value	0.0000						
27									
28	*Regression coefficients*								
29			Coefficient	Std Err	Wald	p-value	Lower limit	Upper limit	Exp(Coeff)
30		Constant	0.4366	0.4197	1.0403	0.2982	-0.3860	1.2591	
31		Age	-0.1004	0.0097	-10.3150	0.0000	-0.1195	-0.0813	0.9044
32		Income	0.0001	0.0000	10.6177	0.0000	0.0000	0.0001	1.0001
33		LiveAlone	1.3087	0.2422	5.4042	0.0000	0.8340	1.7833	3.7012
34									
35	**Comparison of models**								
36		Drop in deviance	9.7134						
37		p-value	0.008						

FIGURE 15.22 Using Model Deviance to Compare Two Models

"complete" model and obtain a model deviance of D_C. Then suppose we omit $j \geq 1$ of the explanatory variables to form a "reduced" model and obtain a model deviance of D_R. The question is whether the reduced model is just as good as the complete model. To answer this question formally, we formulate a null hypothesis and an alternative. The null hypothesis states that the coefficients of the extra j variables are all 0, so that these variables are of no use in the logistic regression equation. The alternative is that at least one of these coefficients is *not* 0. Then it can be shown that under the null hypothesis, the difference $D_R - D_C$ has (approximately) a chi-square distribution with j degrees of freedom. Therefore, if this difference is large, we can reject the null hypothesis—that is, we can conclude that the extra j variables *are* useful for prediction and should not be omitted from the equation. However, if this difference is small, we can probably omit these variables from the analysis.

We illustrate how this model comparison might proceed in Figure 15.22, again using the lasagna data. The top output is the same as we saw previously, with all five explanatory variables included. We then reran the logistic regression with the two least significant variables, Weight and Gender, omitted. (These variables are both significant in the complete model, so we would normally not omit them. We do so here only for the sake of illustration.) The output for the reduced model begins in row 20. Then we manually calculate the difference between model deviances in cell K36 with the formula

=K24-K7, and we find its *p*-value in cell K37 with the formula **=CHIDIST(K36,2).** (The second argument is 2 because we omitted 2 variables to obtain the reduced model.) Because this *p*-value is so small, we conclude that it is *not* a good idea to omit Weight and Gender from the equation; they are evidently useful predictors.

15.3.4 The Count Variable Option

Recall that the StatPro dialog box for logistic regression (see Figure 15.18) includes a "count variable" option. We provide a brief explanation of this option in this section. Usually, the cases (or rows) in a logistic regression setup correspond to individual "trials." For example, a person with certain characteristics has either tried or has not tried a lasagna dinner. Sometimes, however, each case corresponds to a *set* of trials. Then the response variable is no longer a 0–1 dummy variable. Instead, it indicates the number of trials in the set that result in one of the two possible outcomes. The following example illustrates one possibility.

Example (15.8) Classifying Categories of People as Good or Bad Credit Risks

In an investigation of credit default, a consumer organization gathers data on several categories of people. Each of the people in these categories is classified as a "good" credit risk or a "bad" credit risk, where those in the latter group have defaulted on loan payments at least once. The people are categorized according to age (younger or older) and income (low, middle, or high). These categorizations essentially define the explanatory variables. For each joint category, such as older middle-income people, the organization records the number of people in the category and the number who are considered bad credit risks. How should the analysis proceed?

Objective To use the "count variable" option in StatPro to estimate the probability of being a bad credit risk for various categories of people.

Solution

The file CREDIT.XLS contains the data. (See Figure 15.23.) There is one row for each joint category. Each row specifies the characteristics of the category, including the number of people in it and the number who are bad credit risks. You can imagine expanding this data set so that there is a separate row for each *person* (1800 rows total!), but the setup shown here contains all of the required information and is much more compact. The Number variable in column D, which records the number of people in each joint category, is the "count" variable referred to in StatPro. For example, it indicates that 300 younger

FIGURE 15.23
Data Set with a Count Variable

	A	B	C	D	E	F	G	H
1	Logistic regression with a count variable							
2								
3	Joint category	Age	Income	Number	NumberBad	Young	LowInc	MidInc
4	1	Young	Low	500	25	1	1	0
5	2	Young	Middle	300	9	1	0	1
6	3	Young	High	200	2	1	0	0
7	4	Old	Low	200	8	0	1	0
8	5	Old	Middle	400	4	0	0	1
9	6	Old	High	200	1	0	0	0

Est Probs		Results of logistic regression for NumberBad							
0.0535									
0.0245		Summary measures							
0.0095		Null deviance	23.4869						
0.0313		Model deviance	1.5019						
0.0141		Improvement	21.9850						
0.0055		p-value	0.0001						
		Regression coefficients							
			Coefficient	Std Err	Wald	p-value	Lower limit	Upper limit	Exp(Coeff)
		Constant	-5.2037	0.6183	-8.4161	0.0000	-6.4156	-3.9918	
		Young	0.5602	0.3388	1.6536	0.0982	-0.1038	1.2242	1.7510
		LowInc	1.7703	0.6094	2.9050	0.0037	0.5759	2.9647	5.8726
		MidInc	0.9588	0.6443	1.4880	0.1368	-0.3041	2.2216	2.6084

FIGURE 15.24 Logistic Regression XLS Output for Credit Example

middle-income people were observed. The next variable indicates that 9 of these are bad credit risks.

Because the only explanatory variables available are categorical, we first need to convert them to dummies. For age, we create a Young dummy, and for income, we create LowInc and MidInc dummies. These appear in columns F–H of Figure 15.23. (Just as in regular regression, there should always be one less dummy than the number of categories.) We then run StatPro's logistic regression procedure, specifying that there *is* a count variable, the count variable is Number, the response variable is NumberBad, and the explanatory variables are Young, LowInc, and MidInc. The results appear in Figure 15.24, and we interpret them exactly as in previous examples. Perhaps the most useful part of the output is column J. There we see the estimated probability of being a bad credit risk for each joint category of people. For example, the group most likely to default on loans are the younger low-income people. The estimated probability that a person in this category is a bad credit risk is 0.0535. In contrast, the estimated probability of an older high-income person defaulting is only 0.0055.[9] ●

The key to knowing whether you have a count variable in logistic regression is checking whether each row corresponds to a single observation or is a summary of several observations. The former situation is probably more prevalent in business applications, but the latter occurs occasionally.

15.3.5 A Computational Possibility

If you run a logistic regression on data where the two groups are perfectly separated by one or more explanatory variables, you will get a surprising error message. This is not a bug in StatPro; it is the nature of the maximum likelihood procedure used to estimate the logistic regression coefficients. We illustrate how this can happen in the following example.

Example 15.9 Estimating the Probability of Getting a High Grade in a Course

The file GRADES.XLS contains data on students in a college course. For each student, it lists the student's GPA coming into the course, the student's self-reported number of

---

[9]In case you were wondering, we could include interactions between Young and LowInc and between Young and MidInc in the regression analysis. But then we would get a perfect fit.

FIGURE 15.25
Data Set with
Perfect Separation
Between Groups

	A	B	C	D
1	**Example where logistic regression doesn't work**			
2				
3	Student	GPA	Hours	GradeA
4	1	2.62	7	0
5	2	2.42	4	0
6	3	2.62	3	0
7	4	2.95	7	0
8	5	3.19	3	0
9	6	3.16	4	0
10	7	2.71	3	0
11	8	2.75	3	0
12	9	2.78	6	0
13	10	2.69	5	0
14	11	2.94	3	0
15	12	2.44	4	0
16	13	3.05	6	0
17	14	2.85	2	0
18	15	2.67	5	0
19	16	2.70	2	0
20	17	2.90	6	0
21	18	2.67	6	0
22	19	2.83	4	0
23	20	3.07	3	0
24	21	2.98	2	0
25	22	2.86	7	0
26	23	2.65	4	0
27	24	2.75	3	0
28	25	2.76	5	0
29	26	2.74	3	0
30	27	2.79	5	0
31	28	3.44	10	1
32	29	3.68	11	1
33	30	3.64	10	1
34	31	3.77	11	1
35	32	3.83	11	1
36	33	3.47	12	1
37	34	3.63	9	1
38	35	3.65	10	1
39	36	3.75	11	1
40	37	3.34	10	1
41	38	3.82	12	1
42	39	3.57	10	1
43	40	3.72	10	1

hours spent on the course per week, and whether the student got at least an A- in the course (the "A group"). As the listing in Figure 15.25 shows, both the GPA and the number of hours are higher for *every* student in the A group than for the students in the other group. That is, there is no overlap at all between these two groups with respect to the two explanatory variables. In this situation, what does logistic regression tell us?

Objective To illustrate an "error message" that can occur in logistic regression when two groups are separated perfectly by a set of explanatory variables.

Solution

When we run StatPro's logistic regression procedure with GradeA as the response variable and GPA and Hours as the explanatory variables, we get no numerical output. All we get is the message in Figure 15.26 on page 846. And it would not do any good to try this on another statistical software package. We would get a similar error message. The

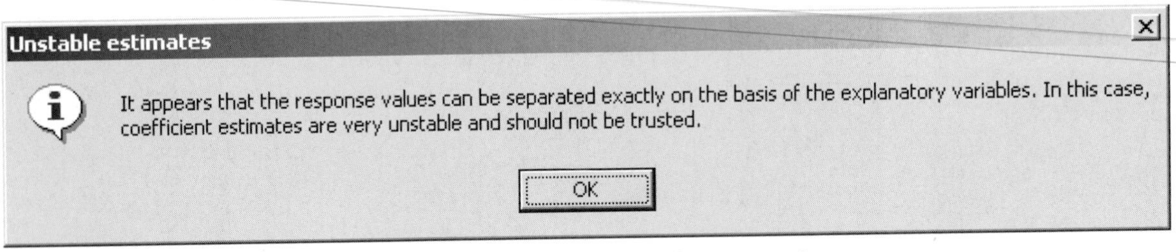

FIGURE 15.26 StatPro's Error Message When There Is Perfect Separation

problem is a technical one, where the method is unable to find a maximum in the maximum likelihood procedure. Admittedly, this will not happen often, because two groups are usually not separated perfectly by a set of explanatory variables, but you should be aware that it can happen. Furthermore, as the message indicates, if the groups are *nearly,* but not perfectly, separated, the estimates of the regression coefficients tend to become unstable. ●

15.3.6 Comparison with Discriminant Analysis

Before concluding this section, we say a few words about which classification method, discriminant analysis or logistic regression, you should choose in any application. There are no hard and fast guidelines, and it might come down to a matter of taste (or which software is available). The strengths of discriminant analysis are that it is based on statistical distance, an intuitive concept that can be shown graphically, it can incorporate unequal prior probabilities and/or misclassification costs, and the linear discriminant function(s) can sometimes provide insights into which explanatory variables are most important for discrimination. The logistic regression method, with its reliance on the logit function, is probably somewhat less intuitive. However, its strengths are that it provides estimates of the *probabilities* of being in the groups, and its output is more in line with the familiar multiple regression output.

Fortunately, as we have seen in the WSJ and lasagna examples, the two methods usually provide very similar results. Given the availability of software for each, such as in StatPro, it is a good idea to try both methods and compare their results.

PROBLEMS

Level A

11. The file P15_01.XLS contains data on 100 consumers who drink beer. A portion of these individuals prefer to drink light beer, whereas another subset prefers to drink regular beer. In particular, a major beer producer believes that the following variables might be useful in classifying a given beer consumer: the individual's gender, marital status, annual income level, and age. Use logistic regression to determine which of these variables are useful in classifying beer drinkers. How well does your estimated model perform in classifying the given consumers?

12. Producers of the late-night television news program "Nightline" are trying to increase the show's viewer-

ship. In an effort to understand who is currently watching their program, the producers collect data on four potentially useful variables for discriminating between viewers and nonviewers of "Nightline": the individual's gender, marital status, annual income level, and age. Such information is collected for 75 randomly selected late-night television viewers and recorded in the file P15_02.XLS. Use logistic regression to determine which of these variables are useful in classifying the late-night TV viewers. How well does your estimated model perform in classifying the given television viewers?

13. The manager of a rapidly growing health club is interested in determining who is most likely to purchase a membership at the club. She proceeds to survey 100

randomly selected individuals who live in the surrounding community. Specifically, she collects data on each of the following: the individual's gender, an indication of whether the individual exercises regularly, the individual's annual income level, and the individual's age. These data are provided in the file P15_03.XLS. Use logistic regression to assess how well the health club manager can discriminate between those who are members and those who are not members on the basis of the given variables. Refer to your computer-generated output in answering each of the following questions:
a. Interpret each of the estimated regression coefficients.
b. Which of the given explanatory variables appear to play a significant role in classifying the given individuals?
c. Evaluate the estimated model's goodness of fit.

14. The owner of a new Chinese restaurant in Centerville, Ohio, is trying to determine who is most likely to have tried his new establishment. He directs his manager to survey randomly selected members of the town and then to collect the following information for each individual in the sample: an indication of whether the respondent has received a coupon for a 20% discount on a dinner at the restaurant, an indication of whether the respondent has seen an advertisement for the opening of the restaurant in one of the local newspapers, the respondent's annual income, and the number of days in a typical week that the respondent eats dinner outside the home. The data collected for a sample of 100 individuals are given in the file P15_04.XLS. Use logistic regression to assess how well the restaurant owner can discriminate between those who have tried his new restaurant and those who have not on the basis of the given variables. Refer to your computer-generated output in answering each of the following questions.
a. Interpret each of the estimated regression coefficients.
b. Which of the given explanatory variables appear to play a significant role in classifying the given individuals?
c. Evaluate the estimated model's goodness of fit.

15. Who tends to own a personal computer in his or her home? Consider the data provided in the file P15_06.XLS. Specifically, we have collected observations on each of the following variables for 100 randomly chosen consumers: an indication of the consumer's educational attainment (i.e., a high school education or less, or more than a high school education), an indication of whether the consumer uses a computer in his or her work, the consumer's annual income, and the consumer's age. Use logistic regression to assess how well we can discriminate between those who own personal computers and those who do not on the basis of the four given variables. Refer to

your computer-generated output in answering each of the following questions:
a. Interpret each of the estimated regression coefficients.
b. Which of the given explanatory variables appear to play a significant role in classifying the given consumers?
c. Evaluate the estimated model's goodness of fit.

16. Admissions directors of graduate business programs constantly review the criteria by which they make admission decisions. Suppose that the director of a particular top-20 MBA program is trying to understand how she and her staff discriminate between those who are admitted to her institution's program and those who are rejected. In an effort to do so, she collects data on each of the following variables for 100 randomly selected individuals who applied in the past academic year: an indication of whether the applicant graduated in the top 10% of his or her undergraduate class, an indication of whether the admissions office interviewed the applicant in person or over the telephone, the applicant's overall GMAT score (where the highest possible score is 800), and the applicant's undergraduate grade-point average (standardized on a four-point scale). These data are provided in the file P15_05.XLS. How useful is logistic regression in discriminating between those who are admitted to this MBA program and those who are not on the basis of the given variables? Explain.

17. In a study to measure the effectiveness of various price reductions on a given product, a manufacturer sent a coupon offering one of several possible price reductions (2%, 4%, 6%, 8%, or 10%) to each of 500 households. In other words, each of the households was randomly assigned to one of the five price reduction levels. The manufacturer then observed which of the 500 coupons were redeemed within 4 months. A count of the households assigned to each possible price reduction level and the associated number who choose to purchase the product are provided in the file P15_17.XLS. Employ logistic regression to determine whether the price reduction is useful in classifying the households in this case. Interpret your computer-generated results.

18. The directors of a professional association are interested in assessing the effect of an increase in the organization's annual dues upon the number of membership renewals in the coming year. To determine whether a proposed increase in dues would have a significant impact on the number of renewals, the directors prepare a survey that asks each respondent to indicate whether she or he will likely renew her or his membership if the dues are increased by one of several possible amounts ($15, $30, $45, $60, or $75). Each of 175 selected members of the association is randomly assigned to one of the possible increases in

annual dues. A count of the members assigned to each possible increase level and the associated number who choose to renew their membership are provided in the file P15_18.XLS. Employ logistic regression to determine whether the increase in dues is useful in classifying the members in this case. Interpret your computer-generated results.

Level B

19. Consider again the classification problem described in Problem 13. Estimate a logistic regression model that includes all potential explanatory variables using the data given in the file P15_03.XLS. Next, remove all variables with p-values in excess of 0.10. Estimate a new logistic regression model with the remaining explanatory variables. Finally, perform an appropriate statistical hypothesis test to determine whether the omitted variables are truly useful for prediction in this case. Interpret the statistical result.

20. Consider again the classification problem described in Problem 14. Estimate a logistic regression model that includes all potential explanatory variables using the data given in the file P15_04.XLS. Next, remove all variables with p-values in excess of 0.10. Estimate a new logistic regression model with the remaining explanatory variables. Finally, perform an appropriate statistical hypothesis test to determine whether the omitted variables are truly useful for prediction in this case. Interpret the statistical result.

21. Consider again the classification problem described in Problem 15. Estimate a logistic regression model that includes all potential explanatory variables using the data given in the file P15_06.XLS. Next, remove all variables with p-values in excess of 0.05. Estimate a new logistic regression model with the remaining explanatory variables. Finally, perform an appropriate statistical hypothesis test to determine whether the omitted variables are truly useful for prediction in this case. Interpret the statistical result.

15.4 Online Analytical Processing (OLAP)

In Chapters 2–4, we saw how pivot tables allow us to break data down by categories. In this section we will take a brief look at an exciting new data mining technique called *online analytical processing,* or OLAP, that takes the ideas from pivot tables one step further. OLAP not only allows us to break down data by categories—which it calls dimensions—but it allows hierarchies within dimensions so that we can "drill down" to finer and finer levels of the dimensions. OLAP is one of the most conceptually simple but powerful data mining methods available.

To understand the concept of OLAP, it is first necessary to understand the type of database structure that OLAP requires. The required structure is called a *star* structure, as we will illustrate in a typical example using supermarket sales data. For this example, we assume that a chain of supermarkets throughout North America keeps track of every sale of every product sold in each of its stores. For each sale, it records information about (1) the customer who purchased the product, (2) the product that was purchased, (3) the store where the product was purchased, and (4) the date of the purchase. Each of these four items is called a *dimension*. There is a table in the database for each dimension. We label these tables Customers, Products, Stores, and Dates. For example, the Products table includes a row for each product the company sells. The fields in this table list any relevant information about the product. Similarly, the Dates table includes a row for every date the supermarket is open.

The supermarket database records data about each sale in a *facts* table. This table has a record for each sale of each product at each store. A given record includes two numerical measures for each sale: the revenue from the sale and the number of units sold. It also includes the index of the customer, the product, the store, and the date. These indexes link to the dimension tables. For example, if the customer index is 351, we can look up this index in the Customers table to see who made this purchase and where this customer lives. In database terminology, these indexes are called *foreign keys* to the various dimension tables. Therefore, the Facts table includes measures about a particular sale (revenue and units sold) and foreign keys to the dimension tables. In real applications, the dimension tables are relatively small (a few hundred or a few thousand records), whereas the Facts table is typically huge (potentially millions of records).

<table>
<tr>
<td>Star Structure, Dimension Tables, and Facts Table</td>
<td>A database with a star structure has links radiating from a facts table to various dimension tables. The dimension tables contain information about various categories, such as products, customers, or stores. The facts table contains information about numeric measures, such as revenue or quantity sold, for each transaction, as well as foreign keys to the dimension tables.</td>
</tr>
</table>

A relationship diagram for this type of database appears in Figure 15.27. (It is taken from the FOODMART.MDB Access file that we will use throughout this section.) We see why the term *star* structure is used. The dimension tables radiate like stars from the central Facts table.

In addition to the dimensions themselves, we see that there a number of natural *hierarchies* in the dimensions. Perhaps the most obvious is the product hierarchy. Based on the way this company categorizes its products, the top level of the hierarchy is the product family (drink, food, nonconsumable). Within each of these, there are product departments, then product categories, then product subcategories, then brand names, and finally individual product names. For example, one product's name is Fabulous Berry Juice. Its brand name is Fabulous, its subcategory is Juice, its category is Pure Juice Beverages, its department is Beverages, and its family is Drink.

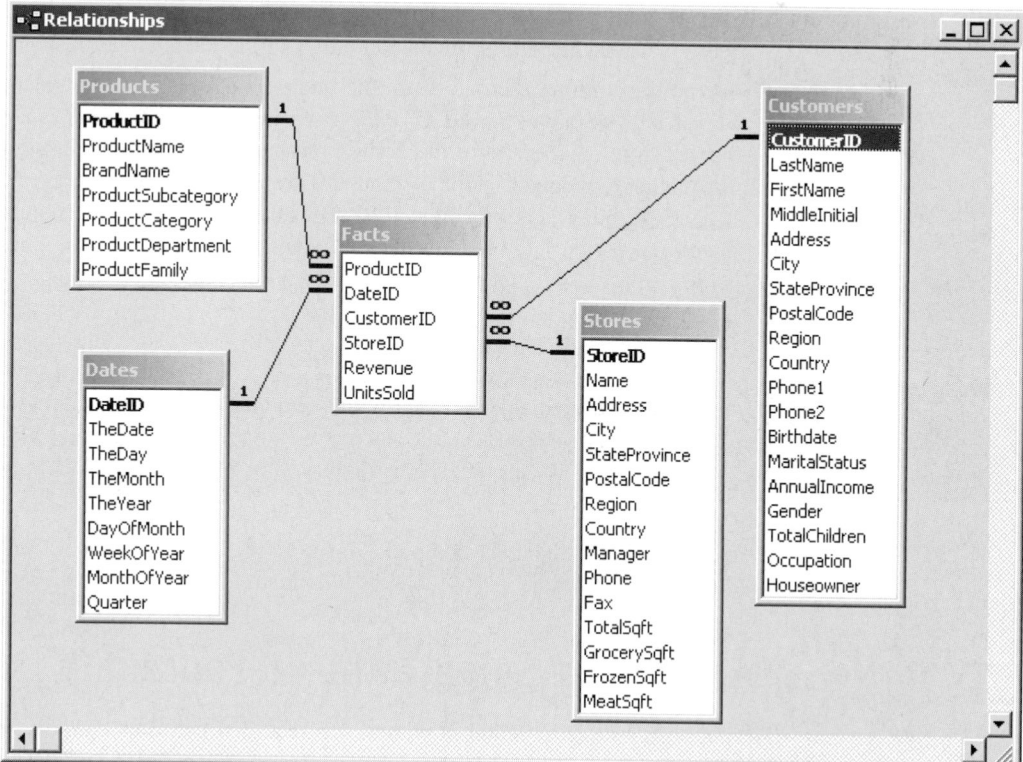

FIGURE 15.27 Star Structure of Foodmart Database

Hierarchies are also possible for the Customers and Stores dimensions. For each, we could use a geographical hierarchy, starting with Country, then Region, then StateProvince, then City, and then PostalCode. For the Dates dimension in any such database, there is always a natural hierarchy: Year, Quarter, Month, Week, Day. We also point out that some fields in a dimension table might not be part of any hierarchy. For example, the Gender field in the Customers table is not part of a hierarchy, but we can still use it in the analysis to break sales down by gender.

The importance of hierarchies in OLAP is that we can drill down to any level of the hierarchy of interest to a manager. For example, a manager might be particularly interested in sales of products in the Beverages department, broken down by Region for stores and by quarter for 1997. This becomes extremely easy to implement, as we will soon see.

OLAP is all about creating pivot tables, and drilling down through dimensions and their hierarchies, to obtain useful summary information. There are a number of (expensive) software products available to perform this type of analysis. Microsoft entered the OLAP market in a big way with its OLAP Services software, bundled with its enterprise SQL Server 7 database package. One version later, with its newest SQL Server 2000 package, the company renamed its OLAP software as Analysis Services (because the term OLAP was evidently considered too esoteric). For large data mining projects, Microsoft expects customers to use Analysis Services in SQL Server 2000. It is powerful, easy to use, and relatively inexpensive. Fortunately, a subset of this technology is available within Excel, as we will demonstrate here. All we need is Excel 2000 (or later) and a database with a star structure. (We should mention that Microsoft is not the only developer of OLAP software. There are a number of others, but their packages tend to be quite expensive.)

To implement OLAP within Excel, we use the following general steps. (These build on the knowledge from Section 4.4 of Chapter 4, which you might want to review at this time.)

1. Starting with a database with the proper structure, use Microsoft Query to create a query that links all of the dimension tables to the Facts table.

2. Instead of returning the data from the query to Excel (which might not even *fit* in a worksheet), we create an *OLAP cube* from the query data. This is a special type of file that stores aggregates of the desired measures. For example, it might store, among many other aggregates, sums of revenues from sales of all beverages to males in the state of Washington during June 1997. By storing only aggregates, cube files are considerably smaller than the data files on which they are based.

3. Use Excel to create a pivot table in the usual way, but base it on the OLAP cube from step 2.

OLAP Cube	An **OLAP cube** is a specially formatted file that contains preprocessed aggregate measures such as sums or counts for various combinations of the categories in the dimension tables. It can be used as the basis for Excel pivot tables.

If this procedure sounds complex, don't worry. It is not much more difficult than building a "plain vanilla" pivot table, as we illustrate in the following example.

Example 15.10 Analyzing Sales Data at Foodmart

The supermarket chain called Foodmart has recorded the sales data illustrated in Figure 15.27 for the years 1997 and 1998 for its stores in North America. These data are stored in the Access database file FOODMART.MDB. (This is a rather large database, although still not large by real standards. It has data on 1560 products, 10,281 customers, 24 stores, 730 dates, and 251,395 individual sales. It is a subset of the database that is packaged with SQL Server 2000.) Create an OLAP cube file from this database, and use it to analyze the data with Excel pivot tables.

Objective To illustrate how to create an OLAP cube with Microsoft Query and then use Excel pivot tables to obtain useful information from the cube.

Solution

Step 1: Create a query The first step is to use Microsoft Query in the usual way. We first create a new data source called Foodmart (exactly as in Section 4.4) and then form a query such as the one shown in Figure 15.28. At this stage, we double-click on each field from each table that we might want to include in pivot tables. The entire list doesn't fit in Figure 15.28. We included the following fields, although you could include others (and they can be entered in any order).

Excel Tip *With this large a database, you can speed things up by clicking on the (!) button on the Microsoft Query toolbar (so that it isn't depressed). The effect is that it won't look up the data values as you double-click on the various fields. After you have double-clicked on all fields you want in the query, you can click on the ! button to obtain the data.*

- Customers table: Country, Region, StateProvince, City, PostalCode, Gender, House-Owner, TotalChildren
- Stores table: Country, Region, StateProvince, City

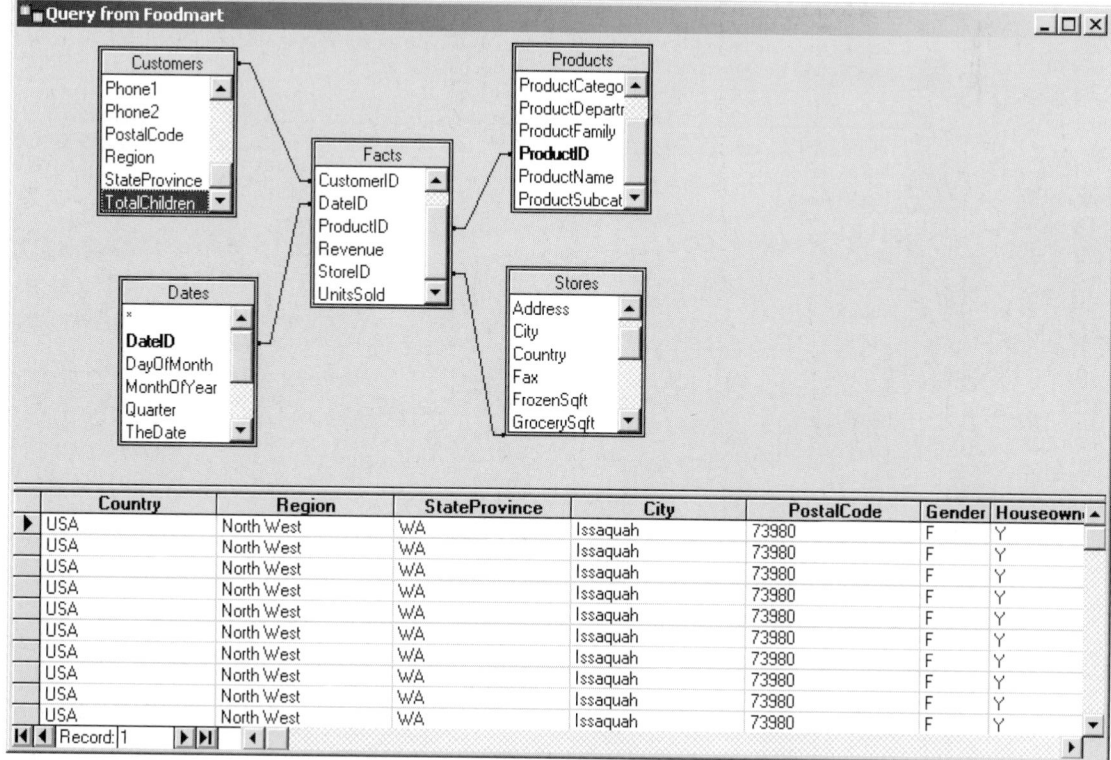

FIGURE 15.28 Query to Capture Required Data for Cube

- Products table: ProductFamily, ProductDepartment, ProductCategory, ProductSubcategory, BrandName, ProductName
- Dates table: TheDate
- Facts table: Revenue, UnitsSold

Step 2: Create a cube file The second step is to create the cube file. Fortunately, a three-step wizard walks you through the procedure. In Microsoft Query, select the File/Create OLAP Cube menu item. You will see an introductory screen. Click on Next to see the dialog box in Figure 15.29. This is where you select any measures you might want to summarize in a pivot table. The wizard automatically checks all numeric fields, as shown in the figure. However, we want to summarize only Revenue and UnitsSold, so remove the checks from all but these two fields.

In addition, we can summarize these measures by using four possible aggregate functions: Sum, Count, Max, and Min. Sum is the default aggregate measure, and we will use it here. However, if you want to summarize by using any of the other three, you would do so by selecting from a dropdown list in the Summarize By column. (If you want to summarize a given measure by more than one aggregate measure, you would need to double-click on this field multiple times when you create the query in step 1.)

Click on Next to go to the second step of the wizard. This is where you specify dimensions and hierarchies. You begin with the dialog box in Figure 15.30, with all fields from your query (other than the measures already selected in the previous screen) listed on the left. You can now drag these to the right to create dimensions and hierarchies. Start by dragging ProductFamily to the right. You will see ProductFamily twice. Right-click on the top version and rename it Product. This indicates that you now have a dimension

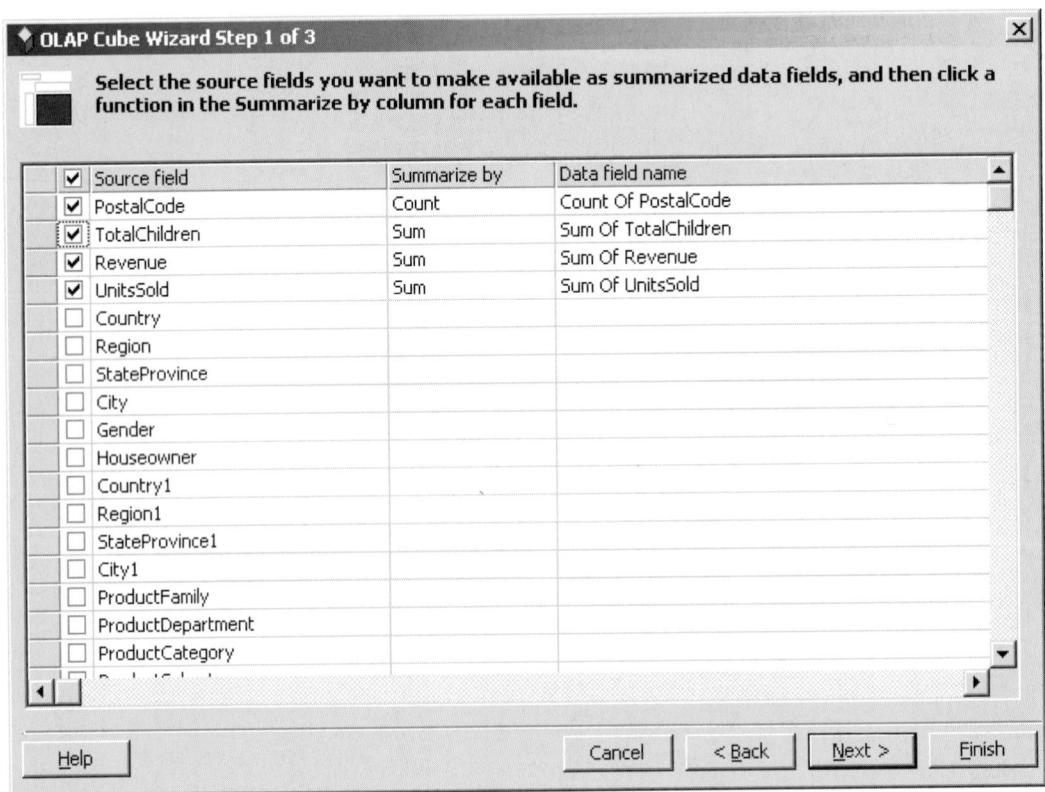

FIGURE 15.29 Cube Wizard Step 1

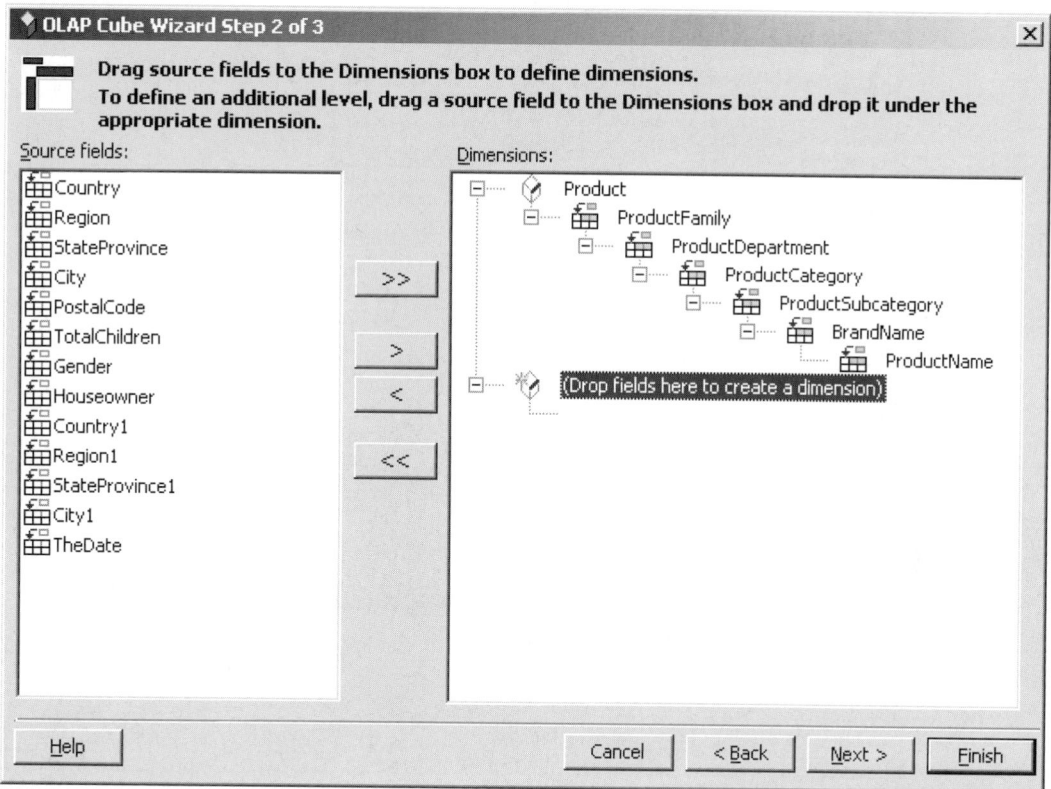

Drag source fields to the Dimensions box to define dimensions.
To define an additional level, drag a source field to the Dimensions box and drop it under the appropriate dimension.

Source fields:

- Country
- Region
- StateProvince
- City
- PostalCode
- TotalChildren
- Gender
- Houseowner
- Country1
- Region1
- StateProvince1
- City1
- TheDate

Dimensions:

- Product
  - ProductFamily
    - ProductDepartment
      - ProductCategory
        - ProductSubcategory
          - BrandName
            - ProductName
- (Drop fields here to create a dimension)

>>   >   <   <<

Help   Cancel   < Back   Next >   Finish

FIGURE 15.30 Product Dimension and Its Hierarchy

called Product, and the top level of the Product hierarchy is ProductFamily. Next, drag ProductDepartment on top of ProductFamily. The effect is that ProductDepartment will be one level down from ProductFamily. Finish the Products dimension by dragging ProductCategory on top of ProductDepartment, then ProductSubcategory on top of ProductCategory, then BrandName on top of ProductSubcategory, and finally ProductName on top of BrandName. This spells out the entire Product hierarchy. At this point, the dialog box should appear as in Figure 15.30.

Next, create a new dimension called Gender by dragging the Gender field to the right. Similarly, create another new dimension called Houseowner. Each of these should be a separate dimension, even though they both come from the Customers table, because there is no hierarchy implied by the Gender, Houseowner combination.

Next, create a store dimension. Start by dragging the Country1 field to the right. (Note that there are two Country fields, one from the Customers table and one from the Stores table. If you selected the Customer fields first when you built the query, like we did, then Country will refer to the Customers table and Country1 will refer to the Stores table.) As before, you will see two versions of Country1. Rename the top one Store and the bottom one Country. Then drag Region1 on top of Country, then StateProvince1 on top of Region1, and then City1 on top of StateProvince1 (and eventually remove the 1's from the names).

Finally, create a Date dimension simply by dragging TheDate field to the right. The OLAP wizard is intelligent about dates. It automatically creates the implied hierarchy for you.

The bottom of your dialog box should appear as in Figure 15.31 on page 854. (If you messed up, try dragging fields to the positions where they should be—the wizard is pretty

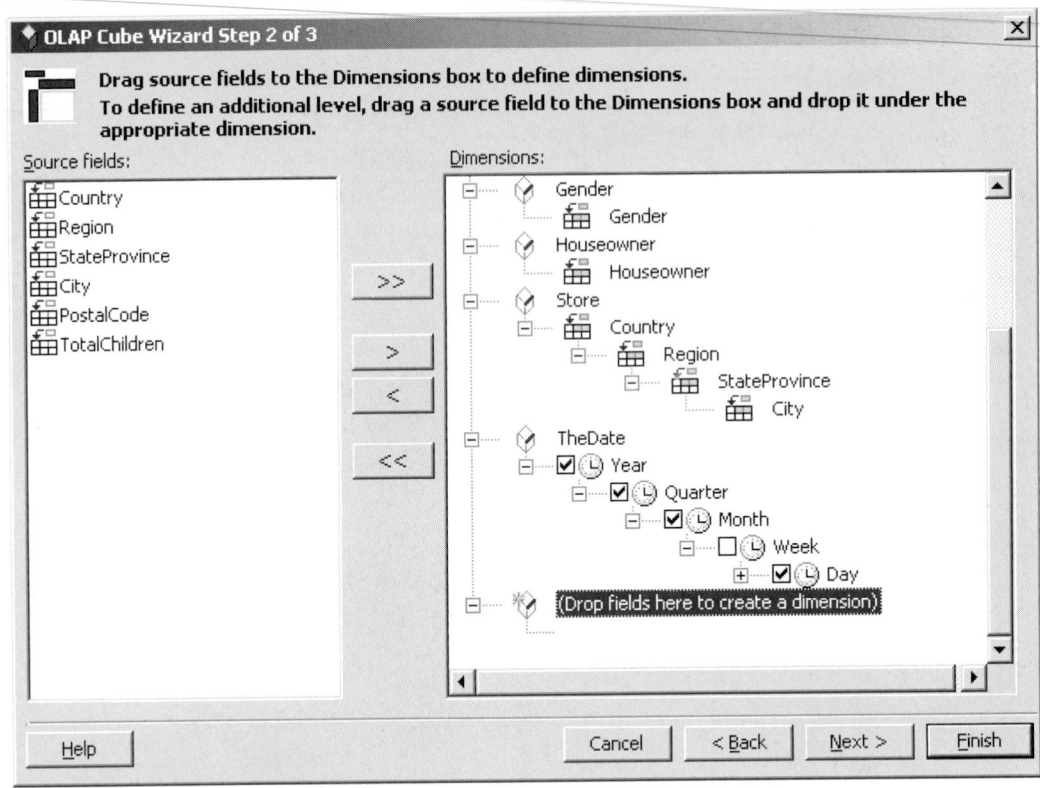

OLAP Cube Wizard Step 2 of 3

Drag source fields to the Dimensions box to define dimensions.
To define an additional level, drag a source field to the Dimensions box and drop it under the appropriate dimension.

Source fields:
- Country
- Region
- StateProvince
- City
- PostalCode
- TotalChildren

`>>` `>` `<` `<<`

Dimensions:
- Gender
  - Gender
- Houseowner
  - Houseowner
- Store
  - Country
    - Region
      - StateProvince
        - City
- TheDate
  - ☑ Year
    - ☑ Quarter
      - ☑ Month
        - ☐ Week
          - ☑ Day
- (Drop fields here to create a dimension)

Help   Cancel   < Back   Next >   Finish

FIGURE 15.31 Other Dimensions and Their Hierarchies

forgiving. If this doesn't work, you can always start over.) Note that some of the fields on the left have not been used. We could use them to create new customer dimensions, but we have chosen not to, simply to illustrate the various options.

This finishes the second step of the wizard, so click on Next. This brings up the dialog box in Figure 15.32, where you can indicate how and where you want to save the cube file. You can accept the default options, although you might store the cube file (FOODMART.CUB) in a folder of your choice. Then click on Finish. You will be prompted to store the *definition* of the cube in an .OQY file (FOODMART.OQY). Again, you might want to change the default folder where this is stored.

After you make this choice, the computer goes to work, calculating and storing all of the aggregate information in the cube. For a large database, this can take a minute or two. (The precise way it calculates and stores all of this information need not concern us, but you should be aware that it is performing a *lot* of arithmetic!) When it is finished, you are asked whether you want to return the data to Excel. If you respond "Yes," you are taken directly to the third step of the pivot table wizard. Although this is the option you are likely to choose in many situations, please respond "No" for now. We will illustrate the pivot table creation separately.

We now have the OLAP cube (.CUB) file and its associated query (.OQY) file. The final step of the overall procedure is to use these to create a pivot table. If necessary, open a blank Excel workbook and select the Data/PivotTable and PivotChart Report menu item. In step 1 of the wizard, select the External data source option. In step 2, click on Get Data, then click on the OLAP cube tab, and Browse for your Foodmart query (.OQY)

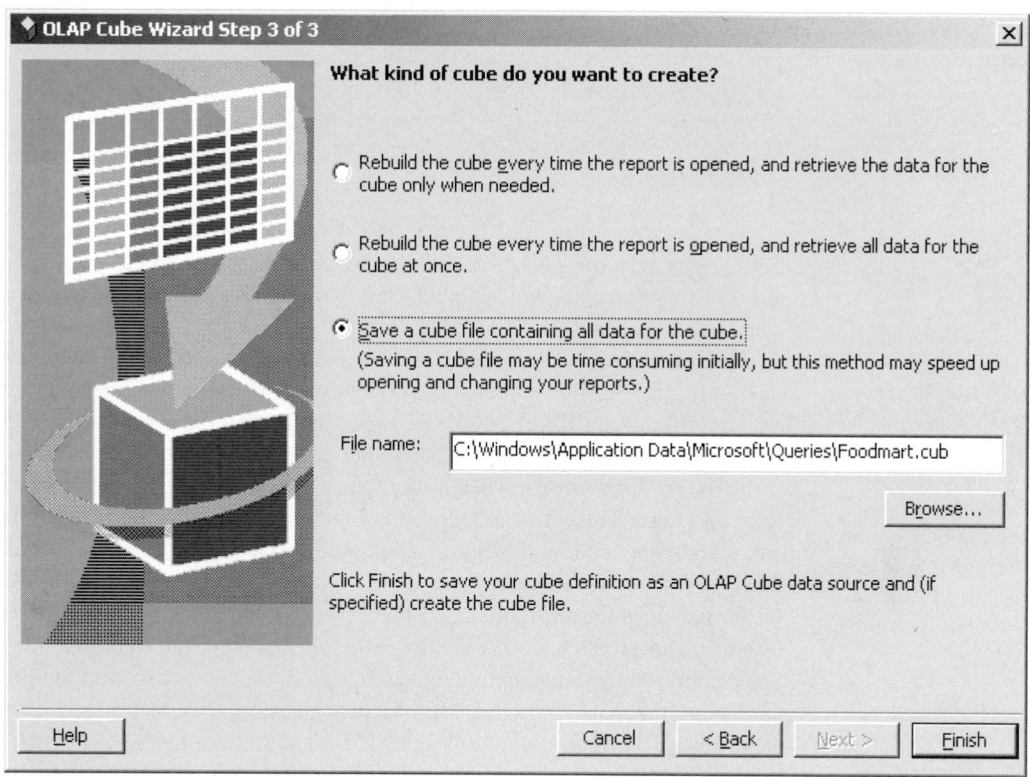

FIGURE 15.32 Final Step of Cube Wizard

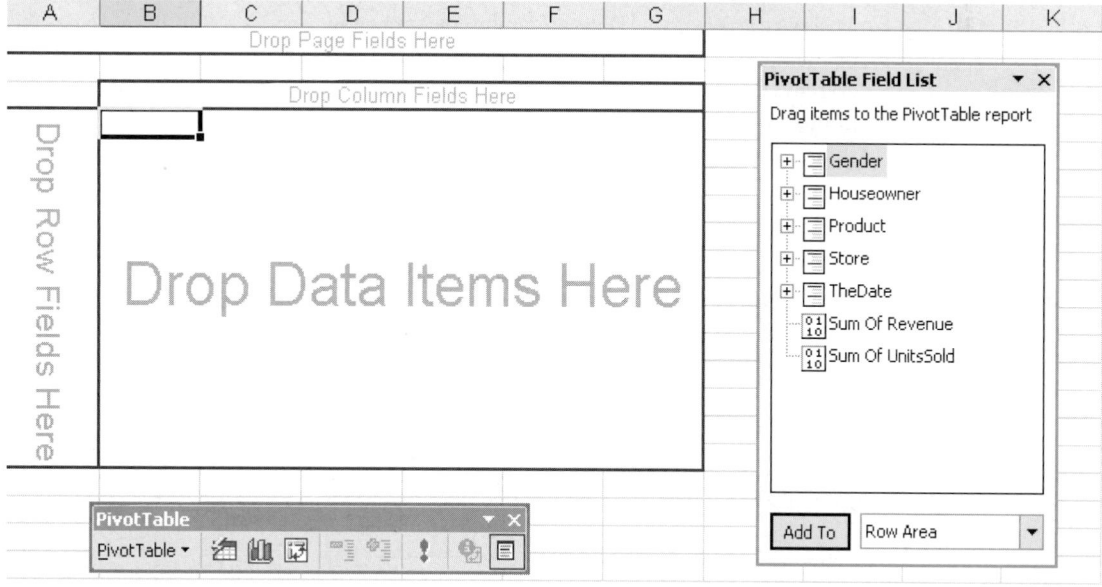

FIGURE 15.33 Pivot Table Toolbar

FIGURE 15.34
Basic Pivot Table

	A	B	C	D	E
1	Store	All			
2					
3	Sum Of Revenue	ProductFamily			
4	Year	Drink	Food	Non-Consumable	Grand Total
5	1997	$48,836	$409,036	$107,366	$565,238
6	1998	$93,742	$778,136	$207,270	$1,079,147
7	Grand Total	$142,578	$1,187,171	$314,636	$1,644,386

If you click some-
where on the sheet
and the field list dis-
appears, just click on
the pivot table to
make it reappear. If
the toolbar should
disappear, you can
recover it with the
View/Toolbars menu
item.

file.[10] After selecting this file, click on OK and then on Next to go to step 3 of the wizard. For this example, we will skip the Layout dialog box, so click directly on Finish to complete the wizard.

You should see a blank pivot table, a field list, and the pivot table toolbar, as shown in Figure 15.33. (The screenshot in Figure 15.33 is from Excel XP. The interface is slightly different in Excel 2000, where the fields are on the toolbar and there is no separate field list window.) Note that there are two types of icons in the field list. The fields at the top are dimensions, whereas the two at the bottom are aggregates of measures (revenue and units sold). You can drag the dimensions only to the Row, Column, and Page areas, and you can drag the aggregate measures only to the Data area. (This makes sense, doesn't it?)

Now comes the fun (and easy) part. We will summarize total revenue by date, store location, and product. To do so, drag the Sum of Revenue button to the Data area, the Store button to the Page area, the Date button to the Row area, and the Product button to the Column area. (This is just one of many possible ways to proceed. Don't be afraid to experiment.) The pivot table in Figure 15.34 shows total revenue for each item at the top hierarchy levels of the Date and Product dimensions for all stores. (We also used the Field Settings option to change the number format to currency.)

This is just the start. To "drill down," try the following. First, select the 1997 and 1998 cells and click on the Show Detail button of the pivot table toolbar. (Remember that if your toolbar ever disappears, you can get it back through the View/Toolbars menu item.) The next level of the Date hierarchy, Quarter, now appears. Similarly, select the Drinks cell and click on the Show Detail button to drill down to the next level of the Product hierarchy, ProductDepartment, for the Drinks family. Finally, click on the dropdown arrow in the Page area and select USA to see revenue totals only for the USA. The pivot table should now appear as in Figure 15.35.

	A	B	C	D	E	F	G	H	I
1	Store	USA							
2									
3	Sum Of Revenue		ProductFamily	ProductDepartment					
4			Drink			Drink Total	Food	Non-Consumable	Grand Total
5	Year	Quarter	Alcoholic Beverages	Beverages	Dairy				
6	1997	1	$3,082	$6,771	$1,733	$11,586	$101,261	$26,781	$139,628
7		2	$3,506	$6,772	$1,637	$11,915	$95,436	$25,316	$132,666
8		3	$3,450	$6,882	$1,662	$11,994	$101,808	$26,470	$140,272
9		4	$3,990	$7,324	$2,027	$13,342	$110,531	$28,799	$152,672
10	1997 Total		$14,029	$27,749	$7,059	$48,836	$409,036	$107,366	$565,238
11	1998	1	$3,772	$7,226	$1,759	$12,757	$107,621	$28,025	$148,403
12		2	$3,618	$7,120	$1,969	$12,708	$105,197	$27,589	$145,494
13		3	$3,507	$7,431	$2,060	$12,997	$104,277	$27,884	$145,158
14		4	$2,779	$5,378	$1,563	$9,720	$79,971	$22,062	$111,753
15	1998 Total		$13,676	$27,155	$7,350	$48,182	$397,065	$105,561	$550,808
16	Grand Total		$27,705	$54,904	$14,409	$97,018	$806,101	$212,927	$1,116,047

FIGURE 15.35 Expanded Pivot Table

[10]When you create the .CUB and .OQY files in the first place, Excel stores a link in the .OQY file to the folder where the .CUB file is stored. If you move the .CUB file to a different folder, the link will no longer work and you will have to browse to find the .CUB file.

We'll let you explore it from here. You are limited only by your imagination. If you want to drill further down a hierarchy, just select one or more cells, such as 1 and 2 under quarters of 1998, and click on the Show Detail button. To drill back up a hierarchy, just select a cell such as Drink in Figure 13.35 and click on the Hide Detail button of the toolbar. If you would rather see the sum of units sold instead of the sum of revenue, simply drag the Sum of Revenue button off the pivot table and drag the Sum of UnitsSold button to the vacated cell.

The only drawback to this procedure is that you can include only those aggregate measures and dimensions in your pivot table that you included in your cube. If you suddenly notice that you need a customer location dimension, or you'd rather report the *maximum* revenue instead of the *total* revenue in each data cell, you must go back and build a new cube. In real applications, this means that analysts must think carefully about the pivot table reports they need *before* going through all the work of building their cubes. ●

We have illustrated many useful data analysis tools in this book, but one could argue that none of these tools is more powerful, flexible, and easy to use than the OLAP cube/pivot table combination we have demonstrated here. These tools are a manager's dream!

PROBLEMS

Level A

22. Use the same .CUB and .OQY files we developed in this section for the Foodmart data to summarize *both* numerical measures (revenue and units sold) in the same pivot table. Break these down by houseowner and gender. Make sure you create a corresponding pivot chart. Summarize (in words) the information you see in this pivot table/pivot chart combination.

23. Use the same .CUB and .OQY files we developed in this section for the Foodmart data to summarize revenue with a pivot table and a corresponding pivot chart. Do it so that the resulting pivot table shows only monthly results for the first half of 1997 broken down by the various brands in the Soda subcategory of the Carbonated Beverages category of the Beverages family of the Drink category for the Product dimension. Then refine this further so that it shows only the monthly results for the first half of 1997 for the individual products in the Washington brand. Summarize (in words) the information you see in these pivot table/pivot chart combinations.

24. Using the same Foodmart database, create a *new* OLAP cube that includes the same numeric measures and the same dimensions as before but also includes a Customer dimension with the hierarchy Country, Region, StateProvince, City, PostalCode. Then create a pivot table/pivot chart combination that shows total revenue, broken down by quarter of year for 1997 only and by customer city for all cities in the United States only. Summarize (in words) the information you see in these pivot table/pivot chart combinations.

Level B

25. Using the same Foodmart database, create a *new* OLAP cube that includes the same dimensions as before but now summarizes only one of the two numeric measures (you can choose *either* revenue or units sold) by Count (not Sum). Then use it to create a pivot table/pivot chart combination that shows counts broken down by months during 1997 only and by product families. Summarize (in words) the information you see in these pivot table/pivot chart combinations. In particular, indicate what the numbers in the pivot table are counting.

26. Using the same Foodmart database, create a *new* OLAP cube that includes the same dimensions as before but now summarizes revenue in two ways: by Sum and by Max. (When you create the original query in MS Query, double-click on the Revenue field twice to create two versions of it, one of which you'll use for Sum and the other for Max.) Then use it to create a pivot table/pivot chart combination that shows sums and maximums of revenue broken down by months during 1997 only and by product families. Summarize (in words) the information you see in these pivot table/pivot chart combinations.

⑮.⑤ Conclusion

To put the material in this chapter in some perspective, recall that in Chapters 9 and 10, we studied the two-sample procedure for distinguishing between two groups. There we saw how to test whether the sample means of a specified variable are different across the two groups. Then in Chapter 12 we used ANOVA to expand this group comparison problem to more than two groups, still using a single measurement variable for the comparison of means. The material in this chapter is essentially a multivariate extension of our earlier analysis. Now we measure one *or more* variables for each of two or more groups and ask whether these variables can be used for classifying observations into the groups. Obviously, the classification will be successful only if there are significant differences between the means of the groups—otherwise, there will be too much overlap between groups for accurate classification. But assuming there *are* significant differences between group means, we need one of the methods in this chapter, either discriminant analysis or logistic regression, to perform the classification.

Discriminant analysis and logistic regression are probably the most complex methods discussed in this book. They involve multivariate analysis and nonlinear optimization, and these are inherently difficult concepts. However, we have tried to explain the key ideas as intuitively as possible, and we have provided the software in StatPro to perform the complex calculations. Above all, we have tried to provide examples that suggest how *useful* these methods can be in the real business world.

Summary of Key Terms

Term	Explanation	Excel	Page	Equation Number
Data mining	A collection of methods used to extract meaningful information from (typically large) databases.		816	
Data warehouse	A (typically large) database specially designed for data mining.		816	
Discriminant analysis	A method for discriminating between two or more groups, or for classifying members into these groups, based on a number of explanatory variables.	StatPro/ Classification Analysis/ Discriminant Analysis	817	
Statistical distance	A measure of the distance from one observation to the mean of observations that takes variances and correlations into account.	Use MMULT, MINVERSE, and TRANSPOSE functions	817	15.1, 15.2
Sample covariance matrix	Matrix of variances (on diagonal) and covariances (off diagonal) used in calculating statistical distance.	Use MMULT and TRANSPOSE functions	819	15.3
Discriminant function	Linear function of explanatory variables used to discriminate between two groups in discriminant analysis.	StatPro/ Classification Analysis/ Discriminant Analysis	823	
Discriminant score	Value for each sample member found by substituting the member's explanatory variable values into the discriminant function; used to classify members into one of two groups.	StatPro/ Classification Analysis/ Discriminant Analysis	824	
Logistic regression	A type of regression where the response variable is binary (0 or 1).	StatPro/ Classification Analysis/ Logistic Regression	834	
Model for probability of being in one group	Nonlinear expression involving a linear function of the explanatory variables		836	15.4
Model for logit	Relates logit to a linear function of the explanatory variables		836	15.5
Odds ratio	Ratio of the probability of being in one group to the probability of being in the other group.		836	
Logit	Logarithm of odds ratio.		836	

Term	Explanation	Excel	Page	Equation Number
Maximum likelihood estimation	Method used (in this chapter for logistic regression) to estimate the model parameters that are most likely to have generated the observed data.	Implemented by Solver in StatPro's logistic regression procedure	836	
Deviance	Goodness-of-fit measure used in logistic regression to indicate how badly the model fits the data.	StatPro/ Classification Analysis/ Logistic Regression	841	
Count variable option	Option in logistic regression when observations are summarized rather being expressed in 0–1 form.	StatPro/ Classification Analysis/ Logistic Regression	843	
Online analytical processing (OLAP)	Data mining method for breaking data down by various dimensions with pivot tables.		848	
Dimension tables	Database tables that list members of categories or category hierarchies to be used in OLAP.		849	
Facts tables	Database tables that include foreign keys to the dimension tables and various numerical measures to be summarized.		849	
Star structure	Database structure that links dimension tables to a central facts table.		849	
OLAP cube	Special type of file that contains pre-aggregated measures of interest; used to enable quick generation of pivot tables in OLAP.		850	

PROBLEMS

Level A

27. Senior managers of a particular firm have been accused of discriminating on the basis of gender in their recent hiring practices. The file P15_27.XLS contains observations on each of the following variables for 80 randomly selected recent applicants: gender, an indication of whether the applicant works better individually or in a group, the number of years of relevant full-time work experience, and the number of years of higher education completed.
  a. Do these data support the charge of gender-based discrimination in this firm's recent hiring decisions? Explain.

  b. How well do all of the given variables perform in discriminating between the applicants? Briefly summarize your findings.

28. A computer programming expert would like to know the roles played by experience (in months of full-time experience) and education (in years of higher education) in a programmer's ability to complete a complex task in a specified amount of time. In particular, the expert is interested in discriminating between those programmers who are able to complete the given task within the specified time and those who are not using these two variables. To conduct this study, the expert gathers observations on each of these variables for 75 randomly selected programmers who attempted to complete the given task. These data are provided in the file P15_28.XLS.

a. How well can the programming expert distinguish between those who are able to complete the given task within the specified time and those who are not on the basis of the given variables?

b. How does your answer to the question in part **a** change when the prior probability of not completing the task successfully changes from 0.50 to 0.80? Explain.

c. How does your answer to the question in part **a** change when the cost of misclassifying a programmer who cannot complete the task as someone who can becomes *two and one-half* times as large as that of the other possible misclassification? Explain.

29. In a study to measure the effectiveness of various price reductions on a given product, a manufacturer sent a coupon offering one of several possible price reductions (2%, 4%, 6%, 8%, or 10%) to each of 500 households. In other words, each of the households was randomly assigned to one of the five price reduction levels. The manufacturer then observed which of the 500 coupons were redeemed within 4 months. These data are recorded in the file P15_29.XLS. Based on a discriminant analysis of these data, can we conclude that the price reduction level adequately distinguishes between those households that purchase the product (i.e., redeem the coupon) and those that do not? If not, how could we proceed to do a better job of discriminating between these two groups of households? Explain.

30. The directors of a professional association are interested in assessing the effect of an increase in the organization's annual dues upon the number of membership renewals in the coming year. To determine whether a proposed increase in dues would have a significant impact on the number of renewals, the directors prepare a survey that asks each respondent to indicate whether she or he will likely renew her or his membership if the dues are increased by one of several possible amounts ($15, $30, $45, $60, or $75). Each of 175 selected members of the association is randomly assigned to one of the possible increases in annual dues. These assignments, along with the respondents' intentions regarding the upcoming membership renewal, are provided in the file P15_30.XLS. Based on a discriminant analysis of these data, can we conclude that the amount of the increase in annual dues adequately distinguishes between those members who renew their membership and those who do not? If not, how could we proceed to do a better job of discriminating between these two groups of members? Explain.

31. The DaimlerChrysler Corporation wants to know who typically buys Jeep vehicles. Market researchers at

Chrysler survey 1000 randomly selected automobile owners and gather observations on each of the following: an indication of whether the individual currently owns a Jeep, an indication of whether the individual has recently test-driven a Jeep, the individual's annual income, the current mileage of the primary vehicle driven by the individual, and an indication of whether the individual plans to buy a Jeep in the next 6 months. The survey results are given in the file P15_31.XLS. How well can Chrysler discriminate between those individuals who purchase a Jeep and those who do not on the basis of the given variables? Briefly explain your answer to this question.

32. The credit managers of a large department store chain routinely must decide to whom they should extend credit for purchases. The file P15_32.XLS contains observations on each of several potentially useful variables for discriminating between those customers whose credit applications are approved and those whose credit applications are rejected. In particular, the file contains data for 100 customers who have recently applied for credit within the store. How well can the management of this department store chain discriminate between those credit applicants who are approved and those who are not approved? Briefly explain your answer.

33. A decision sciences professor, who teaches a first-year quantitative methods course for MBA students, is interested in finding a set of variables that will help her classify students by the final grade they earn in the course. In particular, this professor would like to distinguish between those MBA students who earn a final course grade of "A−" or higher from those who earn a final grade of "B+" or lower. The file P15_33.XLS contains observations on each of three potentially useful variables for each of 80 randomly selected MBA students who completed this professor's quantitative methods course during the previous semester.

a. Assess how well the professor can classify these 80 students on the basis of the given variables.

b. What other variables might be useful in discriminating between these two groups of students?

34. The *Places Rated Almanac* ranked metropolitan areas in the United States with consideration of the following aspects of life in each area: cost of living, transportation, jobs, education, climate, crime, arts, health, and recreation. These data are provided in the file P15_34.XLS. How well do these variables perform in discriminating between those metropolitan areas with overall ratings in the top 25% of the distribution and those metropolitan areas with overall ratings below the top quartile? Explain your findings.

35. Senior managers of a particular firm have been accused of discriminating on the basis of gender in their recent hiring practices. The file P15_27.XLS contains

observations on each of the following variables for 80 randomly selected recent applicants: gender, an indication of whether the applicant works better individually or in a group, the number of years of relevant full-time work experience, and the number of years of higher education completed. How useful is logistic regression in classifying these applicants on the basis of the given explanatory variables? Explain.

36. A computer programming expert would like to know the roles played by experience (in months of full-time experience) and education (in years of higher education) in a programmer's ability to complete a complex task in a specified amount of time. In particular, the expert is interested in discriminating between those programmers who are able to complete the given task within the specified time and those who are not using these two variables. To conduct this study, the expert gathers observations on each of these variables for 75 randomly selected programmers who attempted to complete the given task. These data are provided in the file P15_28.XLS. Use logistic regression to assess how well the expert can distinguish between those who are able to complete the given task within the specified time and those who are not on the basis of the given variables. Refer to your computer-generated output in answering each of the following questions.
    a. Interpret each of the estimated regression coefficients.
    b. Interpret the estimated probability and classification for the *first* programmer in the data set.

37. The DaimlerChrysler Corporation wants to know who typically buys Jeep vehicles. Market researchers at Chrysler survey 1000 randomly selected automobile owners and gather observations on each of the following: an indication of whether the individual currently owns a Jeep, an indication of whether the individual has recently test-driven a Jeep, the individual's annual income, the current mileage of the primary vehicle driven by the individual, and an indication of whether the individual plans to buy a Jeep in the next 6 months. The survey results are given in the file P15_31.XLS. Use logistic regression to assess how well Chrysler can discriminate between those individuals who purchase a Jeep and those who do not on the basis of the given variables. Refer to your computer-generated output in answering each of the following questions.
    a. Interpret each of the estimated regression coefficients.
    b. Interpret the estimated probability and classification for the *third* automobile owner in the data set.

38. A decision sciences professor, who teaches a first-year quantitative methods course for MBA students, is interested in finding a set of variables that will help her classify students by the final grade they earn in the course. In particular, this professor would like to dis-

tinguish between those MBA students who earn a final course grade of "A−" or higher from those who earn a final grade of "B+" or lower. The file P15_33.XLS contains observations on each of three potentially useful variables for each of 80 randomly selected MBA students who completed this professor's quantitative methods course during the previous semester. Use logistic regression to assess how well the professor can classify these 80 students on the basis of the given variables. Refer to your computer-generated output in answering each of the following questions.
    a. Interpret each of the estimated regression coefficients.
    b. Interpret the estimated probability and classification for the *second* student in the data set.

39. The *Places Rated Almanac* ranked metropolitan areas in the United States with consideration of the following aspects of life in each area: cost of living, transportation, jobs, education, climate, crime, arts, health, and recreation. These data are provided in the file P15_34.XLS. Use logistic regression to assess how well these variables perform in discriminating between those metropolitan areas with overall ratings in the top 25% of the distribution and those metropolitan areas with overall ratings below the top quartile. Refer to your computer-generated output in answering each of the following questions.
    a. Interpret each of the estimated regression coefficients.
    b. Interpret the estimated probability and classification for Abilene (TX), the *first* metropolitan area in the data set.

40. A real estate agent is interested in classifying the sales prices of homes in a particular area on the basis of four potentially relevant variables: home size (in square feet), lot size (in acres), total number of rooms in the home, and number of bathrooms in the home. Observations on each of these variables have been collected for 150 randomly selected homes in the area. These data have been recorded in the file P15_40.XLS. Assume that the real estate agent wants to discriminate between those homes with sales prices in the top half of the relevant distribution and those homes with sales prices in the bottom half of the distribution. Use logistic regression to assess how well the real estate agent can classify these 150 homes on the basis of the given variables. Refer to your computer-generated output in answering each of the following questions.
    a. Interpret each of the estimated regression coefficients.
    b. Interpret the estimated probability and classification for the *fifth* home in the data set.

41. Consider again the classification problem described in Problem 33. Using the same set of explanatory variables, the professor would now like to classify each of

her MBA students into one of *three* groups: those who earn a final course grade of "A−" or greater, those who earn a final course grade between "B+" and "B−", and those who earn a final course grade of "C+" or lower. The data are given in the file P15_41.XLS. Assess how well the professor can now classify the given students on the basis of these variables.

42. A real estate agent is interested in classifying the sales prices of homes in a particular area on the basis of four potentially relevant variables: home size (in square feet), lot size (in acres), total number of rooms in the home, and number of bathrooms in the home. Observations on each of these variables have been collected for 150 randomly selected homes in the area. These data have been recorded in the file P15_40.XLS. Assume that the real estate agent wants to discriminate between those homes with sales prices in the top half of the relevant distribution and those homes with sales prices in the bottom half of the distribution.
    a. Assess how well the real estate agent can classify these 150 homes on the basis of the given variables.
    b. How should the real estate agent classify an additional home with 1800 square feet, seven rooms, two bathrooms, and a one-acre lot?

43. Consider again the classification problem described in Problem 12. Estimate a logistic regression model that includes all potential explanatory variables using the data given in the file P15_02.XLS. Next, remove all variables with p-values in excess of 0.05. Estimate a new logistic regression model with the remaining explanatory variables. Finally, perform an appropriate statistical hypothesis test to determine whether the omitted variables are truly useful for prediction in this case. Interpret the statistical result.

44. Consider again the classification problem described in Problem 38. Compare the classification results of logistic regression with those of discriminant analysis in this case. Which method do you find preferable here? Justify your choice.

45. Consider again the classification problem described in Problem 40. Compare the classification results of logistic regression with those of discriminant analysis in this case. Which method do you find preferable here? Justify your choice.

46. The database in the Access file P15_46.mdb has a star structure. (To see this structure, open the file in Access and click on the Relationships button on the main toolbar.) The database contains a small set of data such as those that might be obtained by an automobile manufacturer. There are five dimensions: Time, Dealer (a record for each dealer), Customer (a record for each type of customer, where the customers have been categorized according to various demographic characteristics), Product (a record for each type of car manufactured), and FinanceMethod (a record for each method customers use to finance car purchases). There is also a SalesFacts table, which contains a foreign key to each dimension and various numerical measures for each car sale. (To see the definitions of these measures, open the SalesFacts table in Design view in Access.) Formulate at least three interesting questions that the manufacturer or the dealers might ask and be able to answer, based on a database set up in this way. Then create an OLAP cube and answer your questions with appropriate pivot table/pivot chart combinations.

47. The database in the Access file P15_47.mdb has a star structure. (To see this structure, open the file in Access and click on the Relationships button on the main toolbar.) The database contains a small set of data such as those that might be obtained by a hotel chain. There are three dimensions: Day (a record for each day), Hotel (a record for each hotel in the chain), and RoomType (a record for each type of hotel room). There is also an OccupancyFacts table, which contains a foreign key to each dimension and various numerical measures for each type of room at each hotel on each day. (To see the definitions of these measures, open the OccupancyFacts table in Design view in Access.) Formulate at least three interesting questions that a hotel manager might ask and be able to answer, based on a database set up in this way. Then create an OLAP cube and answer your questions with appropriate pivot table/pivot chart combinations.

Understanding Cereal Brand Preferences

Stephen Michael, the marketing manager for a well-known brand of raisin bran cereal (which we henceforth call Brand A), would like to explain why some customers in the midwest region of the United States prefer to buy a locally produced brand of raisin bran cereal (which we shall refer to as Brand B) instead of the major brand manufactured by Stephen's employer. Stephen is concerned inasmuch as Brand B's market share in the midwest region has been steadily increasing, while Brand A's share has been consistently declining, over the past year. He would like to implement some changes in his marketing campaign for Brand A in hopes of preventing further decline in the brand's share of the market. Stephen has directed his staff to collect data on 500 households who have routinely purchased either Brand A or Brand B at grocery stores in representative communities across the region. The database for all households included in the sample is provided in the file CEREAL.XLS. For each of the 500 households, the data set includes the following variables:

- **DisplayA:** a dummy variable that indicates whether Brand A was on display in the grocery store at the time of the given sale (value = 1) or not (value = 0).

- **DisplayB:** a dummy variable that indicates whether Brand B was on display in the grocery store at the time of the given sale (value = 1) or not (value = 0).

- **PriceA:** the price (in dollars) of Brand A at the time of the given sale.

- **PriceB:** the price (in dollars) of Brand B at the time of the given sale.

- **DiscountA:** the published discount (in dollars) in the selling price of Brand A (relative to the regular selling price of Brand A) at the time of the given sale.

- **DiscountB:** the published discount (in dollars) in the selling price of Brand B (relative to the regular selling price of Brand B) at the time of the given sale.

- **LoyaltyA:** the value of an index that measures the extent to which this household has exhibited loyalty to Brand A in the past (this measure is based on raisin bran cereal purchases during the past 18 months, and higher values of this index are associated with greater levels of customer loyalty to Brand A) at the time of the given sale.

- **LoyaltyB:** the value of an index that measures the extent to which this household has exhibited loyalty to Brand B in the past (this measure is based on raisin bran cereal purchases during the past 18 months, and higher values of this index are associated with greater levels of customer loyalty to Brand B) at the time of the given sale.

- **Brand Preference:** the brand of raisin bran cereal selected by the household at the time of the given sale.

Use the classification methods presented in this chapter, along with other appropriate data analysis techniques, to help Stephen and his colleagues understand why some customers have been selecting Brand B instead of their own brand. What step(s) should Stephen take to make Brand A more competitive in the future? Interpret your findings in a carefully written and well-supported memorandum. Provide technical support for your verbal conclusion in the form of one or more attachments to your memo.

16

Statistical Process Control

© Jim Zuckerman/CORBIS

Statistical Process Control at Nortel

Northern Telecom (Nortel) is a leading global provider of digital network hardware and software solutions for communications, information, entertainment, education, and commerce. As of 1996, it had annual world-wide revenues of $12.85 billion. Due to a 1995 U.S. Federal Communications Commission action, the auctioning of a section of frequency bandwidth reserved for advanced digital cellular communications, Nortel saw the opportunity to move into a new technology, personal communications services (PCS). The company decided to build a new manufacturing operation, Wireless Networks Raleigh (WNR) in Raleigh, North Carolina. The new plant was dedicated to manufacturing advanced cell-site base-station and antenna equipment, with many of the designs for the plant's products coming from Nortel operations in France and the United Kingdom. From the outset, senior managers recognized that state-of-the-art manufacturing practices were necessary for WNR. Among other things, this included computer-based data collection from the manufacturing processes, statistical process control on literally hundreds of product characteristics, and the ability to act on the information in a timely manner, both in North Carolina and in European sites.

A team of analysts created a decision support system to achieve these objectives. Their work is described in the article by Brinkley et al. (1998). A key component of the system is its statistical process control capability. The communications products developed at WNR contain circuit boards and other assembly items that must meet stringent quality specifications. To ensure that this happens, the system continually monitors hundreds of product characteristics and produces real-time control charts. Typical illustrations of these appear in Figures 16.1 and 16.2 on page 866. (Much of this chapter deals with the meaning and interpretation of such charts, so we urge you to take another look at them after you have studied this chapter.) The basic objectives of these charts are to show everyone involved—workers

FIGURE 16.1
X̄ and R Charts for
Nortel System

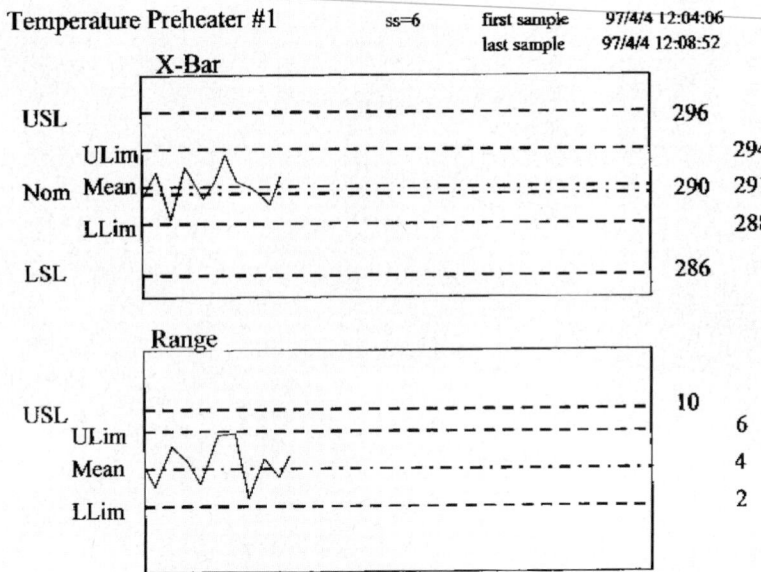

on the shop floor, design engineers, and senior managers—how the manufacturing process is currently operating and to suggest corrective actions in case the process is not behaving as desired. Such charts are in use in most manufacturing companies. However, an interesting aspect of the system developed by Nortel is that these charts (and the spreadsheet data on which they are based) are available instantly through the company's intranet—over the World Wide Web. For example, a design engineer in Paris can immediately see current data from Raleigh simply by logging onto the intranet. This real-time capability cuts the time for data analysis and decision making tremendously, compared to the traditional methods of report generation and distribution.

The analysts developed and implemented this decision support system within a 12-month period at a cost of $500,000. The annual costs of supporting the system

FIGURE 16.2
Process Capability
Chart for Nortel
System

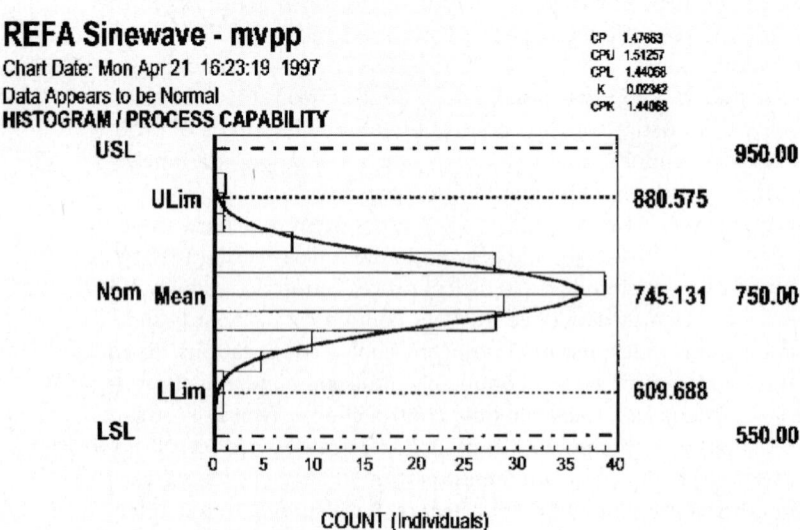

are approximately $800,000. Given that revenues in 1996 at WNR were already $320 million, this low system cost is very impressive. The company estimates that the direct benefits of the system, primarily from reduced work-in-process and reduced rework and scrap production, are over $1 million annually. However, there are other more qualitative benefits from the system. It has had a definite positive impact on customer satisfaction due to improved quality and reliability, and employee satisfaction has improved as a result of increased knowledge and empowerment. •

16.1 Introduction

One of the areas where statistics has had the largest impact in the business world is the area of quality. For many years quality was not emphasized, especially by U.S. companies. Gradually, spurred on by foreign competition, high levels of quality became a competitive advantage for the best companies, and today most companies are simply not competitive unless they have excellent quality. One of the best examples of this change is in the automobile industry. All we need to do is compare U.S. cars produced in the 1970s with those produced now to see the tremendous improvements in quality that have been achieved in the past decade. U.S. automobile manufacturers can now compete successfully with their Japanese and European counterparts, and much of this is due to the improvements in quality in U.S. manufacturing plants. Of course, this improvement in quality became *necessary* in the U.S. automobile industry. Without quality improvements, U.S. market shares would have continued to fall into the hands of foreign competitors.

The quality movement in the United States (and abroad) has taken on almost a religious fervor in recent years. It has spawned a number of acronyms, including TQM (total quality management), QC (quality control), SPC (statistical process control), QFD (quality function deployment), and others. Some of the best-known consultants and researchers in the area, including Joseph Juran, Genichi Taguchi, Philip Crosby, and W. Edwards Deming, have become business heroes and are commonly referred to as "quality gurus." Since its inception in 1987, there has been fierce competition for the Malcolm Baldrige National Quality Award, an award given to U.S. manufacturing and service companies on the basis of their superior quality initiatives and performance.

In the larger context, the quality movement comprises much more than just statistical or quantitative methods. It is also about leadership, worker empowerment, producer/supplier relationships, interest in the customer, and other broad business issues. However, a large part of the success of the quality movement is due to the increased use of quantitative methods, both in manufacturing and in service industries. Our focus in this chapter is on a set of quantitative tools generally referred to as **statistical process control** (or **SPC**).

There are various themes to SPC, but perhaps the two most important themes can be summarized as:

- Get it right the first time.
- Reduce variation.

In the past, quality was often synonymous with *inspection*. Completed parts or assemblies were routinely inspected for problems, and those that failed inspection were scrapped or sent back for rework. This emphasis on inspection is good in that it tries to keep faulty products from reaching the customer, but it is expensive. Time and resources used to fix mistakes could be used more profitably if there were no mistakes to fix. The quality gurus argued that it is much better to catch mistakes early in the production process, where they are less costly to fix, than to wait for final inspection. Besides, by focusing on the sources of problems early in the process, the *causes* of the problems become more apparent, and future problems can be prevented. Therefore, getting it right the first time is the focus of many quality program initiatives.

The second theme is to reduce variability. As Deming and others have preached, variability is the main culprit that hurts quality, and everything that is possible must be done to eliminate variability. However, to do so, we must be able to measure it and give workers a way to reduce it. This is exactly the objective of control charts, the statistical tool we will study throughout much of this chapter. Control charts enable an operator on the shop floor to see what a production process is currently doing. Then by following established guidelines, this operator and/or management can often root out the causes of variability and produce a better product—*before* it passes on to final inspection.

The quality movement and the statistical tools for quality improvement are usually discussed in the context of manufacturing. Indeed, much of the impetus for SPC came from manufacturing industries. However, the same ideas that have resulted in dramatic improvements in manufacturing apply to service industries. The banking industry, the hotel industry, and others also need to reduce variability and get it right the first time. Fortunately, these service industries can also benefit from control charts and other SPC tools. A bit more creativity is sometimes required to apply the tools appropriately, but it can be done.

16.2 Deming's 14 Points

W. Edwards Deming is probably more responsible than any other single individual for the recent emphasis on quality.[1] A statistician by trade, Deming took his theories and statistical tools to Japan shortly after World War II. At that time, of course, Japan was just starting to rebound from the devastation of the war. Deming taught the newly emerging Japanese industries the principles of quality management, for which they are now well known. He became a legend in Japan. In fact, the quality award Japan gives that is comparable to our Baldrige Award is called the Deming Prize.

However, Deming (and his teachings) remained largely unknown in the United States until the early 1980s. At that time, U.S. manufacturing industries were floundering with poor-quality products, and Deming—along with a few other quality gurus—began teaching them the statistical principles they would need to compete successfully. Today Deming's philosophy greatly influences decision making at many major U.S. companies, including Procter and Gamble, Xerox, Ford, and General Motors. In fact, the operation of the Saturn division of GM is based largely on Deming's philosophy.

Deming is perhaps best remembered for his famous 14 points, a list of precepts he taught in all of his seminars. We present and discuss these points here to give an overview of quality management. [They are adapted from Deming (1986).] As will be obvious, some of these 14 points are more quantitatively oriented than others. However, we reiterate that Deming was first and foremost a statistician who strongly believed in taking constant measurements and applying appropriate statistical techniques to reduce variation. He was no idle armchair philosopher!

1. **CONSTANCY OF PURPOSE. Create constancy of purpose toward improvement of product and service, allocating resources to provide for long-range needs rather than only short-term profitability, with a plan to become competitive, stay in business, and provide jobs.**

[1]Deming died at the age of 93 in 1993.

It is tempting for U.S. companies, whose attention has traditionally been on short-term financial measures, to adopt quality initiatives (with a lot of fanfare!) and then abandon them whenever they conflict with short-term objectives. This defeats the purpose of the quality initiatives. Besides, it leads employees to believe that this or that quality initiative is simply the "gimmick of the month." The only way to adopt Deming's teachings successfully is for management to "walk the talk" and show a constancy of purpose that the workers can appreciate.

2. **THE NEW PHILOSOPHY. Adopt the new philosophy. We are in a new economic age, created in Japan. We can no longer live with commonly accepted levels of delays, mistakes, defective materials, and defective workmanship. Transformation of Western management style is necessary to halt the continued decline of industry.**

The type of change Deming envisioned is not simply a few quick changes that can be adopted overnight. It is an entirely new philosophy, and most companies will have to completely rethink the way they do things. However, unless companies strive continually to move in this direction, they are doomed to lose out to their more enlightened competitors.

3. **CEASE DEPENDENCE ON MASS INSPECTION. Eliminate the need for mass inspection as a way to achieve quality by building quality into the product in the first place. Require statistical evidence of built-in quality in both manufacturing and purchasing functions.**

Companies should concentrate on *preventing* defects rather than detecting them. As we stated earlier, the traditional emphasis in American companies was on final inspection. Unfortunately, final inspection of a defective product does not always provide information on how to improve product quality. It is better to use control charts and other statistical tools to monitor product quality at each step of a process. Then quality problems can be corrected as they occur. Besides, 100% inspection is costly and delays deliveries, and it is not always accurate. To illustrate that 100% inspection is not always accurate, simply ask several people to count the number of "f"s on this page. (Most will miss the "f"s in "of.") So even if the inspectors know what they're looking for, they won't always find it.

4. **END LOWEST-TENDER CONTRACTS. End the practice of awarding business solely on the basis of price tag.**

It is sad but true that the lowest-price vendor may also have the lowest quality. Saving one cent on each "widget" placed in a car may later cost a company millions of dollars in repair costs and lawsuits due to accidents caused by the faulty part. Deming believed a company should use as few suppliers as possible. Ideally he believed a company should use a single supplier (called **single sourcing**). He argued that single sourcing leads to a trusting relationship that will increase the quality of goods produced. In accordance with Deming's teachings, the automobile companies have in recent years greatly reduced the number of suppliers they use. In this way they and their suppliers no longer play an adversarial role but instead forge long-term relationships and work together to improve quality. This cooperation has become increasingly possible with Web technology and enterprise resource planning (ERP) software, such as SAP, that enables companies to share information with their suppliers in real time.

5. **IMPROVE EVERY PROCESS. Improve constantly and forever the system of production and service, to improve quality and productivity, and thus constantly decrease costs.**

Deming believed that unenlightened managers often respond only to crises—that is, they wait until things get so bad that they are impossible to ignore. It is much better to make improvements continually and thereby try to keep crises from occurring in the first place. The statistical tools we will discuss are intended to do this. By using them continually, companies are able to detect a problem almost immediately when it occurs, and they are generally able to learn more about their processes so that lasting improvements can be made.

6. INSTITUTE TRAINING. Institute modern methods of training for everybody's job, including management, to make better use of every employee.

Employees need to understand the entire process, not just their own jobs. Most Japanese companies, including Toyota, have adopted this philosophy. Deming believed that training is an investment for the long term in the company's most important assets, its employees. For example, in its early days, Motorola U (Motorola's employee training center) estimated a 3300% return on training at plants where managers supported the quality policy. Unfortunately, some companies fail to see the long-term value of worker training and scrap training programs as soon as financial resources become tight. Deming argued strongly that this is exactly the wrong strategy to take.

7. INSTITUTE LEADERSHIP OF PEOPLE. Adopt and institute leadership aimed at helping people to do a better job.

Supervisors should not focus on the negative. They should promote teamwork, not divisiveness, and they should stress quality, not quantity. One of Deming's most controversial views is his rejection of performance appraisals. He believed they undermine teamwork. His reasons are that (1) most variation is a part of the *system,* over which the workers have little or no control, so it makes little sense to reward or penalize workers for something they cannot control; (2) individual and departmental targets and objectives destroy cooperation between departments; and (3) reliance on pay as a motivator destroys pride in work and individual creativity. Deming saw the ideal leader as a coach and teacher, not a watchman. He believed that the primary objective of leadership should be to motivate people to work to their maximum level of performance. When they complain about "worker attitudes" as the cause of poor quality, they usually miss the real cause: the system itself. It is actually worse than this: When workers are placed under conditions that force them to do a poor job, they are likely to stop caring, in which case they will probably do an even worse job.

8. DRIVE OUT FEAR. Encourage effective two-way communication and other means to drive out fear throughout the organization so that everybody can work effectively and more productively for the company.

When all relationships between different levels of employees (workers versus supervisors, middle management versus upper management, and so on) are adversarial and based on fear, it is difficult to make improvements. Workers attempt to "hide" from their supervisors rather than cooperate with them, and each level of management is fed what the level below thinks it wants to hear. Fortunately, this style of management, once so prevalent in the United States, is now beginning to give way to one based on trust and cooperation.

9. BREAK DOWN BARRIERS. Break down barriers between departments and staff areas.

It is usually counterproductive to the company as a whole if the individual departments have little knowledge of one another's roles, and it is even worse if individual departments work only for their own good, not the good of the company. Fortunately, we are now seeing more cross-functional teams, often formed to tackle specific problems. Obviously, this brings different areas of expertise to the solution of problems, and it encourages cooperation.

10. ELIMINATE EXHORTATIONS. Eliminate the use of slogans, posters, and exhortations for the workforce, demanding zero defects and new levels of productivity, without providing methods.

Deming's point here is that it does little good to exhort workers to "do it right the first time" if the *system* prevents them from doing so. It actually discourages workers, since they are being exhorted to do something beyond their power to achieve. Management would be wiser to improve the system so that workers can do the jobs they are being asked to do.

11. ELIMINATE ARBITRARY NUMERICAL TARGETS. Eliminate work standards that prescribe quotas for the workforce and numerical goals for people in management.

Establishing arbitrary standards for workers can be counterproductive. On the one side, if the standard is set too low, workers will do just enough to meet the standard, when in fact they *could* accomplish more. On the other side, if the standard is set too high, workers will either become discouraged by the impossibility of meeting the standard or they will cut corners—and produce lower quality—in their attempt to turn out the numbers. A better approach is for management to provide helpful leadership and coaching so that workers can turn out better quality *and* achieve higher productivity. After all, which worker is more productive: a worker who produces 50 parts per hour, 10 of which require rework, or a worker who produces 45 parts per hour, all of which meet specifications?

12. PERMIT PRIDE OF WORKMANSHIP. Remove the barriers that rob hourly workers, and people in management, of their right to pride of workmanship.

This is similar to the previous point. Managers must provide the workers with the motivation and tools to perform their jobs as well as possible. Most workers don't want just to serve time at their jobs; they want to have pride in what they accomplish. Deming argued that when workers are placed in an environment where they can do their jobs properly, then productivity and quality will both improve, and the workers will be happier at what they do.

13. ENCOURAGE EDUCATION. Institute a vigorous program of education, and encourage self-improvement for everyone.

Whereas point 6 is concerned primarily with job training, this point is broader. It refers to self-improvement in any dimension. Deming's idea is that a better educated workforce is a more valuable one. It is better able to evolve with today's constantly changing technology, and it is more aware of broader business issues.

14. TOP MANAGEMENT COMMITMENT AND ACTION. Clearly define top management's permanent commitment to ever-improving quality and productivity, and their obligation to implement all of these principles.

This is essentially a summation of the other points. It says that if a company plans to initiate the quality improvements that are necessary to compete successfully in today's business world—as spelled out by the previous 13 points—then top management must lead the way. Otherwise, it won't happen.

Deming's 14 points are both a philosophy for becoming a quality leader and a prescription for how to do so. For the past several decades, Deming and his disciples have "spread the word" to numerous U.S. (and foreign) companies, and in many cases top management has taken the advice. We customers who now purchase high-quality cars, refrigerators, computers, and a host of other consumer items should thank Deming for his insights. They have resulted in lower prices, better and more useful features, and considerably longer times between repairs.

16.3 Basic Ideas Behind Control Charts

We now discuss control charts, one of the most important statistical tools available for reducing variability and improving quality. These types of charts were originally developed by Dr. Walter A. Shewhart of the Bell Telephone Laboratories in the 1920s, and they are still in widespread use today. They are generally easy to use, even for people not specifically trained in statistics, and they provide a wealth of information about a process.

To understand the reasoning behind control charts, we need to discuss two types of variability in a process. No process, whether it be in a manufacturing or a service company, ever produces outputs with *exactly* the same characteristics from item to item. There is always some variability. The question is whether the variability is an inherent part of the process or can be attributed to **assignable causes.** If the current variability in the output of a process is due entirely to the inherent nature of the process, then we say that its variability is due to **common causes** and that the process is **in statistical control,** or simply, **in control.** On the other hand, if some of the current variability of the process is due to specific **assignable causes,** such as a bad batch of raw materials, an improperly adjusted machine, a new operator unfamiliar with the process, or others, then we say that the process is **out of control.**

Types of Variation	**Common cause** variation is the inherent variation in an in-control process. **Assignable cause** variation is the extra variation observed when a process goes out of control—which could be for any number of reasons.

One of the main purposes of control charts is to monitor a process so that we can see when a process goes from an in-control condition to an out-of-control condition. When such a transition is discovered, then a person knowledgeable about the process (the operator of a machine, for example) can search for an assignable cause that led to the out-of-control condition and fix it, thereby bringing the process back into control.

It is important to realize that a process in control is not necessarily a good process. It could be making a lot of items "out of specs." There are two reasons we want to distinguish between in-control and out-of-control processes. First, an in-control process is at least *predictable,* regardless of whether it is any good. By measuring an in-control process, we can estimate the ability of the process to produce quality items. An out-of-control process, on the other hand, might not only be producing a lot of faulty items, but it is also *unpredictable*. It is difficult to change a process in some sensible way if we don't know exactly how good or bad it is.

The second reason is that the assignable causes that produce out-of-control behavior can often be corrected by the workers on the shop floor; that is, they generally do not require management intervention to correct. Once corrected, the process is brought back into control, and the charting can continue.

Unfortunately, there is little workers can do to improve an in-control process that has unacceptable variability. Control charts allow them to *measure* the amount of variability, but there is generally no way they can *reduce* the amount of variability without guidance from management. Essentially, workers are stuck with the current process, and (compare with Deming's point 10) no amount of encouragement or exhortation can enable them to produce a higher-quality product unless management makes a fundamental improvement in the process itself. Control charts can measure how well an in-control process is doing, but they can't improve a faulty process. That is the job of management.

Before looking at control charts in detail, we list the primary reasons they have become so popular.

1. **They improve productivity and lower costs.** Here we define productivity as the number of *good* items produced per hour. Control charts typically allow mistakes to be found (or prevented) early in the process—before they result in poor finished products. Therefore, instead of having workers spend a lot of time producing items that are eventually scrapped or require rework, control charts enable them to get it right the first time. Obviously, when workers spend their time producing good items, they are more productive and the costs of scrap and rework are minimized.

2. **They prevent unnecessary process adjustments.** Even an in-control process contains a certain amount of inherent common-cause variability. Without the guidance of control charts, a human operator is likely to respond to every observed up and down in the process. If the diameter of one part is too high, a downward adjustment of a machine might be made; if a subsequent diameter is too low, an upward adjustment might be made. However, if the process is actually in control and the observed ups and downs are really just "normal" amounts of variability, then such adjustments can actually make the process *worse*. Deming called this "tampering" and advised never to tamper with an in-control process. Control charts allow the operator to see when a process is really in need of an adjustment—because it has gone out-of-control. As long as the process is in control, adjustments shouldn't be necessary.

3. **They provide diagnostic information about the process.** Control charts are similar to medical diagnostic tests. They not only signal when something is wrong, but they provide clues as to the cause of the problem. An experienced operator will monitor control charts for telltale signs of various problems. When these are spotted, then a search for assignable causes can be made, and in many cases problems can be fixed before they become major.

4. **They provide information about process capability.** Process capability is defined as the ability of a process to produce outputs that meet specifications. Obviously, managers want a high level of process capability. However, to achieve this, they need to know how well their *current* process is doing. Control charts help provide this information, at least when the process is in control. For example, a control chart might indicate that under current in-control conditions, approximately 6 out of every 1000 parts fail to meet specs. This is the process capability. Armed with this information, management can decide whether fundamental changes in the process are warranted.

Tampering (making many unnecessary adjustments) with an in-control process will typically make it *worse*.

16.4 Control Charts for Variables

There are two basic types of control charts: control charts for variables and control charts for attributes. Charts for variables are relevant when there is a measurable quantity, such as a diameter, a weight, or a thickness, that can be monitored. In this case the purpose of the chart is to see how this quantity varies through time. On the other hand, in many situations an item is judged to conform to specifications or not—it is either a good item or a bad one—in which case a control chart for *attributes* is appropriate. This type of chart tracks the proportion of conforming (or nonconforming) parts through time. An attributes chart is also appropriate for tracking the number of *defects* (such as paint blemishes, scratches, and so on) through time. We will discuss control charts for attributes in the next section.

In this section we will illustrate two of the most common types of variables control charts: the $\bar{X}$ chart and the R chart. Consider any product that has some measurable characteristic such as a diameter or a thickness. To produce $\bar{X}$ and R charts for this product, we typically proceed as follows. Every so often, say, every half hour, we randomly sample a small number of items and measure the characteristic. This "small number" is usually from 3 to 6, and the resulting sample of measurements is called a **subsample.** For the $\bar{X}$ chart we calculate the average of the measurements in the subsample, and for the R chart we calculate the range (maximum minus minimum) of the measurements in the subsample. Then we plot the sequence of averages ($\bar{X}$'s) and the sequence of ranges (R's) in time series plots.

$\bar{X}$ and R charts	An $\bar{X}$ **chart** charts the averages of small subsamples through time. Its purpose is to see how the mean of the process is changing through time. An R **chart** measures the ranges (maximum minus minimum) of small subsamples through time. Its purpose is to see how the variability of the process is changing through time.

The resulting time series plots are more informative when we add centerlines and control limits. A centerline indicates the average value that the $\bar{X}$'s (or R's) vary around. Control limits place upper and lower bounds on where the $\bar{X}$'s (or R's) should be for a process in control. The following example provides an illustration of these charts. We will follow it up with more details on how the charts are formed.

Example 16.1 Measuring the Volumes of "12-ounce" Soda Cans

The file SODACANS.XLS contains data on the number of ounces of soda in cans labeled "12-ounce" cans. Every half hour, five cans of soda from a production process were measured for fill volume. This was done for 70 consecutive half-hour periods. Create and interpret the $\bar{X}$ and R charts.

Objective To use $\bar{X}$ and R charts to check whether the process of filling soda cans is performing as it should.

Solution

Although $\bar{X}$ and R charts are quite easy to create by hand—this is the way they are usually created on the shop floor—the process is tedious and better suited for computer implementation. We have done so in StatPro and will explain the steps here. First, the SODACANS.XLS file is set up in the appropriate way for StatPro. There are five adjacent columns for the five observations taken each half hour. (These columns need not be adjacent, but it's natural for them to be.)

To use StatPro, place the cursor anywhere inside the data set and select the StatPro/Quality Control/XBar, R Charts menu item. When prompted for variables, select the five adjacent observation variables, Obs1–Obs5. The next dialog box, shown in Figure 16.3, provides several options for building the charts. Fill it out as shown. We will plot data for only the first 30 half hours and base the control limits on these. Then provide a name (we used Wt1) for the sheets that StatPro will create.

FIGURE 16.3
Dialog Box for $\bar{X}$, R
Chart Options

X-Bar, R chart options

Observations
You can base the chart on all 30 of the subsamples, or on just the subsamples specified below.

○ Plot all of the observations
⦿ Plot only those within the limits below

Starting index: 1   Ending index: 30

Control limits
You can base the control limits for the charts from all or part of the current data, or you can use control limits from previous data.

○ Control limits based on all of the current data
⦿ Control limits based on only the subsamples specified below

Starting index: 1   Ending index: 30

○ Control limits based on previous data

Previous subsample size   5
Previous X-DoubleBar:
Previous R-Bar:

OK
Cancel

Extra lines on X-Bar chart
Besides the control limits (which are at about 3 sigmas from the center line), you can add lines at 1 and 2 sigmas from the mean on the X-Bar chart.

☐ Include extra lines on X-Bar chart

FIGURE 16.4
$\bar{X}$ Chart for Soda
Can Fill Volumes

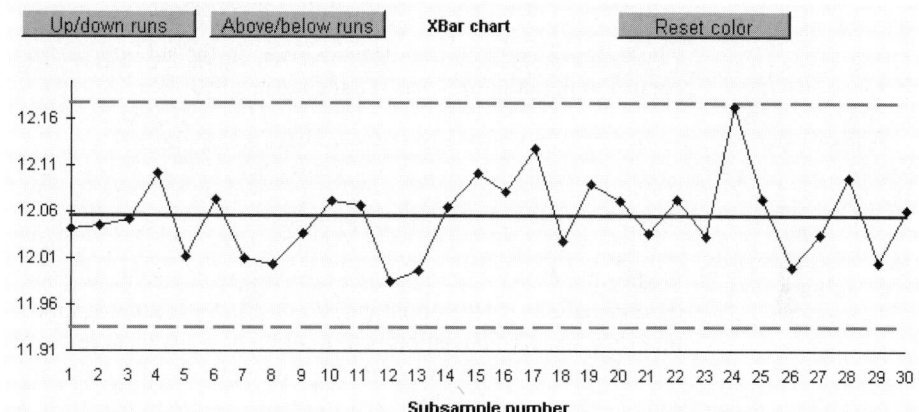

FIGURE 16.5
R Chart for Soda
Can Fill Volumes

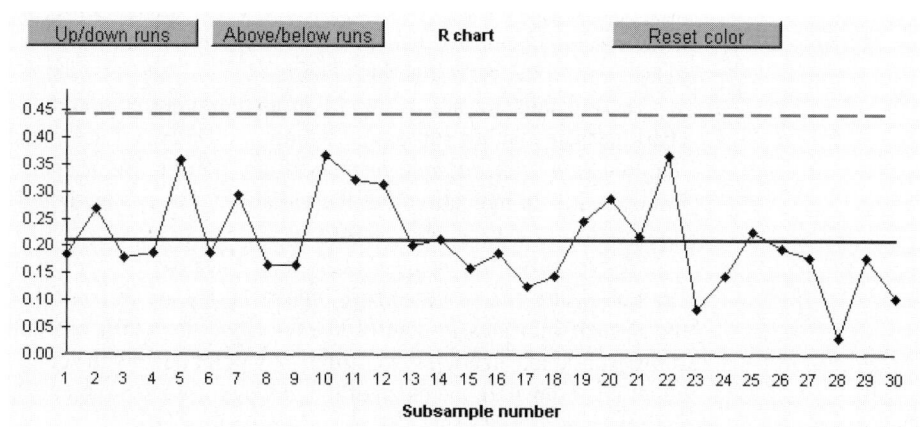

StatPro creates three new sheets. One (named XBar,R Chart Data) contains the data that the control charts are based on. The other two (named XBar Chart and R Chart) are chart sheets that contain the $\bar{X}$ and R charts. These appear in Figures 16.4 and 16.5. (We will soon discuss the buttons at the tops of these chart sheets.) On each chart we see that the points vary around a centerline and stay within upper and lower control limits (although one point on the $\overline{X}$ chart is very close to the upper limit). The behavior we see in these charts is typical *in-control* behavior. No points are outside of the control limits, and there is no obvious "nonrandom" behavior, such as an upward trend through time. Therefore, this process appears to be in control. If there are specifications on the soda cans—for example, the fill volume of a can should be between 11.88 and 12.20 ounces—then we could use the data (and the fact that the process is in control, that is, predictable) to estimate the percentage of *all* cans within specs. ●

We now discuss in some detail how these charts are formed. The $\overline{X}$ chart is a plot of the subsample averages, that is, the individual $\overline{X}$'s. The centerline for this plot is the average of all $\overline{X}$'s, denoted $\overline{\overline{X}}$. The lower and upper control limits, denoted *LCL* and *UCL*, are approximately three standard deviations (of $\overline{X}$) on either side of the centerline, where the standard deviation of $\overline{X}$ is $\sigma/\sqrt{n}$ and n is the subsample size. However, it has been traditional to measure variability by *ranges* rather than standard deviations. (This is easier for nontechnical operators.) Fortunately, there is a simple relationship between them. If $\hat{\sigma}$ is an estimate of the unknown standard deviation σ and $\overline{R}$ is the average of all R's (one from each subsample), then

$$\hat{\sigma} = \overline{R}/d_2$$

where d_2 is a constant that depends only on the subsample size n.[2] The lower and upper control limits in the $\overline{X}$ chart are then given by

$$LCL = \overline{\overline{X}} - 3\hat{\sigma}/\sqrt{n}$$

and

$$UCL = \overline{\overline{X}} + 3\hat{\sigma}/\sqrt{n}$$

> The main thing to remember—for all control charts we will discuss—is that the lower and upper control limits are approximately three standard deviations below and above the centerline.

For the R chart, we use the average $\overline{R}$ as the centerline and again go out three standard deviations (of R) on either side to form the control limits. It turns out that the appropriate standard deviation is a multiple of $\overline{R}$, so that we can write the lower and upper control limits as $D_3\overline{R}$ and $D_4\overline{R}$. The constants D_3 and D_4 again depend only on the subsample size n and are tabulated in quality control books. We simply note that the "natural" lower control limit (three standard deviations below the centerline) can sometimes be negative. Since a negative range value doesn't make sense, we instead set $LCL = 0$ in these cases. This means that the upper and lower control limits are not always the same distance from the centerline in R charts.

These details are not as important as the interpretation and use of the charts. The R chart measures within-subsample variation over time. Each R measures the variability in the process *at a given point in time*. If any point in the R chart goes beyond the control limits, this is an indication that the variability has changed, and we can begin searching for a reason—an assignable cause.

We typically look at the R chart first. Because the control limits for the $\overline{X}$ chart depend on $\overline{R}$, they make little sense unless the R's are in control. Assuming, however, that the R chart indicates in-control behavior, we then shift our attention to the $\overline{X}$ chart. It shows subsample averages over time. Any point beyond the control limits suggests a shift, either up or down, in the mean of the process. If we see such a point, we can begin searching for an assignable cause.

---

[2]This and other constants here are tabulated in books on quality control. They have been incorporated automatically into StatPro.

The description to this point might suggest that control charting is a static, one-time procedure. However, it is actually a dynamic, ongoing procedure. Typically, a company periodically recalculates (or calculates for the first time, if this is a new process) the control limits for $\overline{X}$ and R by using a fresh set of subsamples. (At least 20 or 25 subsamples are recommended.) If any points are beyond the control limits, a search for assignable causes begins. If the search is successful, the problems are fixed. In any case, the points beyond the control limits are usually eliminated, and the remaining subsamples are used to reestimate new control limits. The purpose in this stage is to bring the process into control, if necessary.

Once the process is in control, we "continue" the control charts by plotting new points but using the just-established centerlines and control limits. We continue to do this until some type of out-of-control behavior is observed, at which time we search for assignable causes. As time proceeds, we learn more and more about the process. This learning not only enables us to keep the process from slipping out of control for extended periods of time, but it also enables us to make lasting improvements to the process. The following continuation of Example 16.1 illustrates how events might unfold.

StatPro gives us the flexibility to plot only certain subsamples and/or to base the control limits on certain subsamples or even on historical values. This enables us to mimic the way control charts are used on the shop floor. We illustrate some of these possibilities in the following continuation of the soda can example.

Example 16.1 Measuring the Volumes of "12-ounce" Soda Cans (continued)

We now assume that the first 30 subsamples of soda can fill volumes, the ones used previously in Example 16.1, were used to determine centerlines and control limits for the $\overline{X}$ and R charts. When we plot the other 40 subsamples, using the *same* centerlines and control limits, what do we learn about the process?

Objective To continue the $\overline{X}$ and R charts to learn whether the soda can process stays in control beyond the subsamples on which the original charts were based.

Solution

As the dialog box in Figure 16.3 indicates, StatPro allows us to choose the subsamples to plot, as well as the subsamples to base the control limits on. Here we will plot all of the subsamples but base the control limits (and centerlines) only on subsamples 1–30. The dialog box should be filled out as shown in Figure 16.6 (page 878). The resulting $\overline{X}$ and R charts appear in Figures 16.7 and 16.8. (StatPro replaces the previous sheets with these new ones. If you want to keep the previous charts, rename their sheets first.)

We first look at the R chart. It shows that the process stayed in control for at least 10 more half-hour periods beyond subsample 30. However, beginning shortly after subsample 40, the process variability appears to have increased (many points above the centerline), and finally two points, subsamples 49 and 53, jumped above the upper control limit. As if this weren't enough evidence of an upward shift in variability, we can also search for "runs" of at least eight points above or below the centerline by clicking on the "Above/below runs" button on the chart. (It is very unlikely to see a run this long in an in-control process.) If we click on this button, points 47–54 turn red, indicating a long run above the centerline.

FIGURE 16.6
Dialog Box for
Continuation of
Control Charts

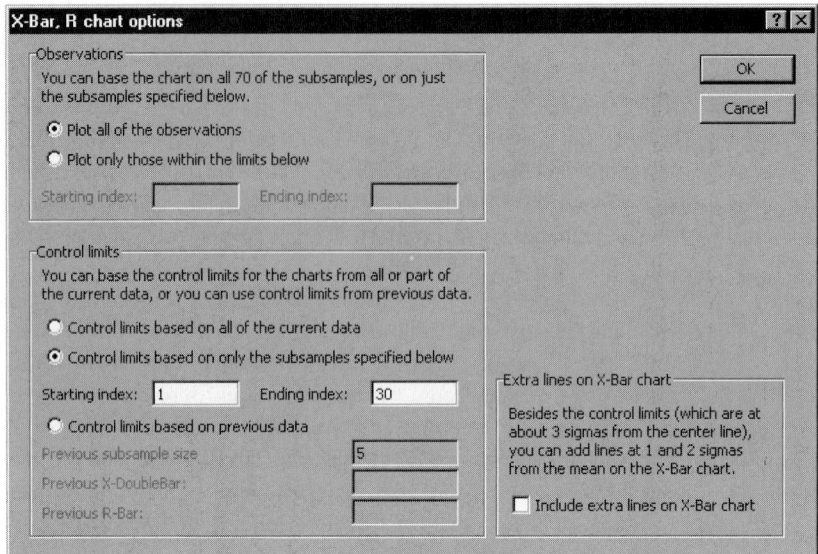

FIGURE 16.7
$\bar{X}$ Chart for
Continuation of
Subsamples

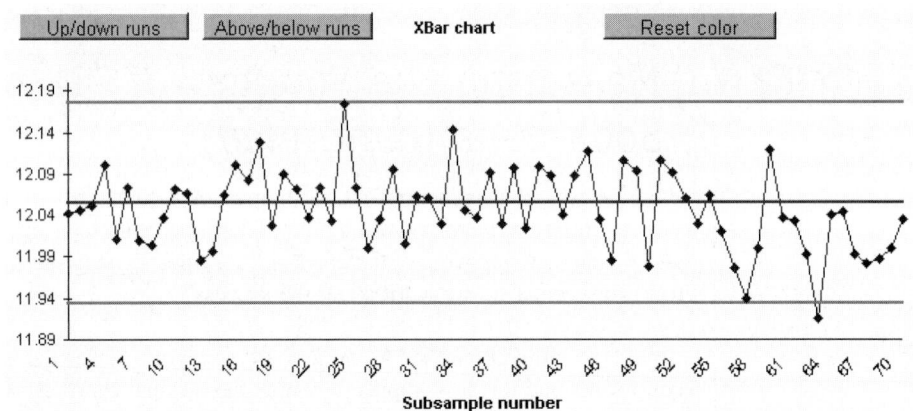

FIGURE 16.8
R Chart for
Continuation of
Subsamples

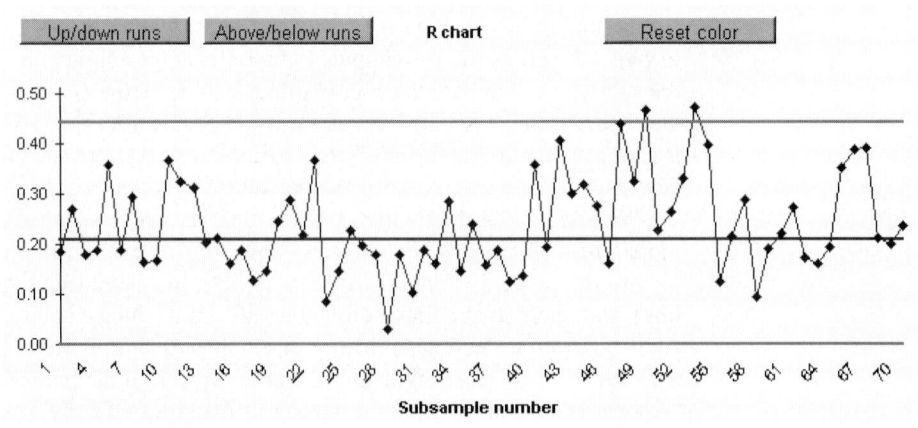

Presumably, the operator of the process discovered the problem that was causing abnormally high variation and fixed it at around the time of subsample 55. After that point, the R chart goes back into control (where "in control" is relative to the first 30 subsamples). However, at about this same time, the $\bar{X}$ chart suggests a downward shift in the process mean. Many points are below the centerline, and one finally crosses the lower control limit on subsample 63. Many machines have a mechanism for adjusting the mean to some target level, such as 12.05 ounces. In the present case it appears that this machine simply needs to be readjusted to bring its mean back up to the previous level. After this is done, both of the control charts should indicate an in-control process—at least until some other assignable causes force it out of control again. ●

This example illustrates how control charts allow an operator to monitor a process continuously and react quickly when problems are indicated. Without this continuous monitoring, out-of-control conditions could persist indefinitely, causing poor quality and higher costs.

16.4.1 Control Charts and Hypothesis Testing

It is enlightening to think of control charts in the context of hypothesis testing. We let the null and alternative hypotheses correspond to in-control and out-of-control conditions, respectively. As we monitor the process with control charts, there are two types of errors we can make. The first, a type I error, is when we react to an out-of-control indication when in fact the process is still in control. We call this a false alarm. For example, there is some chance that a process operating in control will produce a point beyond the control limits. In this case we might begin a search for assignable causes when there are none. We might also make an unnecessary adjustment to the process to bring it back into control (unnecessary because it is still *in* control).

We want to make the probability of a type I error fairly small. If it is too large, we react to too many false alarms and, in Deming's terminology, we tamper with the process. This could not only be costly, but it could actually cause an *increase* in the variability of the process. Therefore, we set the control limits fairly far apart—typically three standard deviations from the centerline—so that the chance of observing a point beyond them is very small.

The *ARL* indicates the mean length of time between false alarms in an in-control process.

To pursue this a bit further, assume that the $\overline{X}$'s are normally distributed. (Since each $\overline{X}$ is an average of several observations, the central limit theorem suggests that this normality assumption is reasonable.) Then we know that the probability of any $\overline{X}$ being more than three standard deviations from the mean is 0.0027. From this, we can calculate the mean number of subsamples, called the **average run length,** or *ARL*, until an in-control process produces a point beyond the control limits. It is simply[3]

$$ARL = 1/0.0027 \simeq 370$$

In other words, false alarms will be few and far between if the process remains in control.

Of course, the flip side is a type II error. This means that the process has gone out of control but the control charts do not indicate it. As usual, it is difficult to calculate the probability of a type II error because there are many types of out-of-control conditions that *could* occur. However, let's concentrate on one possible type of out-of-control condition, where the process variation remains constant but the mean shifts from μ to $\mu +$ $k\sigma$, where k is some fixed constant. For example, if $k = 1$, then the process mean has

[3]An analogy is how long, on average, we would have to wait to roll double sixes with two dice. Since there are 36 possible outcomes for the two dice, the probability of double sixes on a single toss is 1/36. Therefore, the expected number of tosses until double sixes occurs is $1/(1/36) = 36$.

shifted upward by one standard deviation. We would like to spot this shift immediately, but we won't spot it until an $\overline{X}$ falls above the upper control limit. How long, on average, will this take?

Assuming that the $\overline{X}$ chart has centerline μ, the upper control limit is $\mu + 3\sigma/\sqrt{n}$, and the mean of the process has shifted up to $\mu + \sigma$, we first calculate the probability that an $\overline{X}$ is above the upper control limit. Since $\overline{X}$ now has mean $\mu + \sigma$ and standard deviation $\sigma/\sqrt{n}$, the calculation is a typical normal probability calculation, where we subtract the mean and then divide by the standard deviation:

$$P(\overline{X} > \mu + 3\sigma/\sqrt{n}) = P\left(Z > \frac{(\mu + 3\sigma/\sqrt{n}) - (\mu + \sigma)}{\sigma/\sqrt{n}}\right) = P(Z > 3 - \sqrt{n})$$

Here, Z is normal with mean 0 and standard deviation 1. In the soda can example, $n = 5$, so this probability is

$$P(Z > 3 - \sqrt{5}) = P(Z > 0.764) = 0.222$$

Therefore, there is less than 1 chance in 4 that any particular $\overline{X}$ will be beyond the upper control limit. Another way of looking at it is to calculate the *ARL*, the expected number of subsamples until the out-of-control behavior is spotted:

$$ARL = 1/0.222 \approx 4.5$$

For example, if subsamples are taken every half hour, it will take, on average, over 2 hours to realize that the process has gone out of control.

We would like to keep both type I and type II errors to a minimum. That is, we would like to minimize the number of false alarms, but at the same time we would like to spot out-of-control conditions quickly. One strategy is to sample more frequently. Instead of sampling every half hour, we could sample every 15 minutes. Another strategy is to increase the subsample size n from, say, 5 to 10. Both of these strategies are intended to decrease the *ARL* when the process goes out of control.

For example, if we use $n = 10$ instead of $n = 5$ in the above calculations, we obtain

$$P(Z > 3 - \sqrt{10}) = P(Z > -0.162) = 0.564$$

and

$$ARL = 1/0.564 \approx 1.77$$

Now, assuming that we are still sampling every half hour, the average time to spot the out-of-control condition is less than an hour. Alternatively, if we keep $n = 5$ but sample every 15 minutes, then the previous *ARL* of 4.5 now translates to only slightly more than 1 hour.

16.4.2 Other Out-of-Control Indications

To this point, the only formal indication of an out-of-control process is a point beyond the control limits. There are a number of other possible indications of "nonrandom" behavior that we might want to react to. The usual ones that have been suggested include:

1. At least 8 upward (or downward) consecutive changes

2. At least 8 consecutive points above (or below) the centerline

3. At least 2 of 3 consecutive points beyond two standard deviations from the centerline (where both are on the *same* side of the centerline); usually applied only to $\overline{X}$ charts

4. At least 4 of 5 consecutive points beyond one standard deviation from the centerline (where all 4 are on the *same* side of the centerline); usually applied only to $\overline{X}$ charts

We can sample more frequently or we can take larger subsamples to detect out-of-control behavior more quickly.

FIGURE 16.9
Zone in an $\overline{X}$ Chart

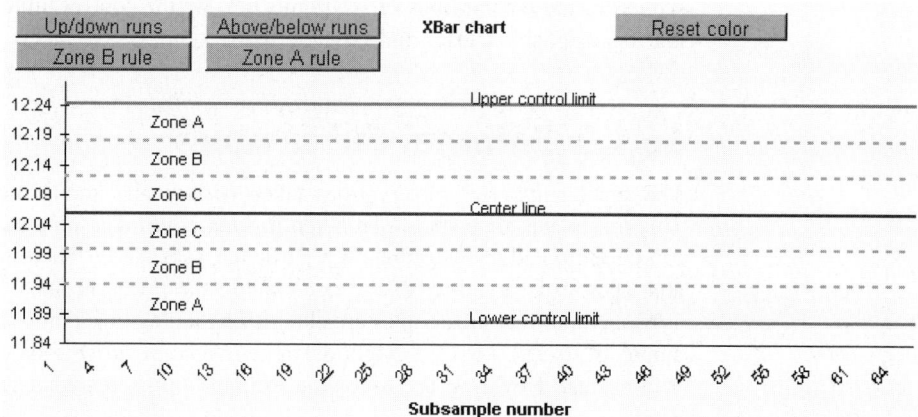

For these last two conditions it is common to divide the region between the center-line and either control limit into three "zones" of width one standard deviation each, as indicated in Figure 16.9. Then condition 3 is called the **Zone A rule,** and condition 4 is called the **Zone B rule.** In either case the idea is that although points within zone A and zone B are within the control limits, it is unlikely that an in-control process would have this many nearby points in zone A or B.

StatPro places buttons on the charts to check for these four conditions. When any button is clicked, any offending points are colored red. To check conditions 3 and 4 on the $\overline{X}$ chart, the "Include extra lines on X-bar chart" box must be checked in the dialog box in Figure 16.6. If this is done for the soda can data in the continuation of Example 16.1, you can check that none of conditions 1, 2, or 3 hold for the $\overline{X}$ chart, but condition 4 holds for two different sets of five consecutive points, as indicated in Figure 16.10. The fourth point in each five-point set in zone B is colored red. Since the five-point sets can overlap, there is actually just one red point on the chart, the point corresponding to subsample 69. (Subsamples 65–69 and 66–70 both satisfy condition 4, and subsample 69 is the fourth point in zone B of each of these five-point sequences.)

We do not want to overemphasize these (or any other) possible indications of out-of-control behavior. The more such conditions we check for, the more likely we will find false alarms. In a real situation an experienced operator is likely to give different emphasis to different out-of-control indications. For example, if he sees any of conditions 1–4, but no points beyond the control limits, he might start sampling more frequently—every 15 minutes instead of every half hour, say. If he then continues to see more in-

FIGURE 16.10
Illustration of
Zone B Rule

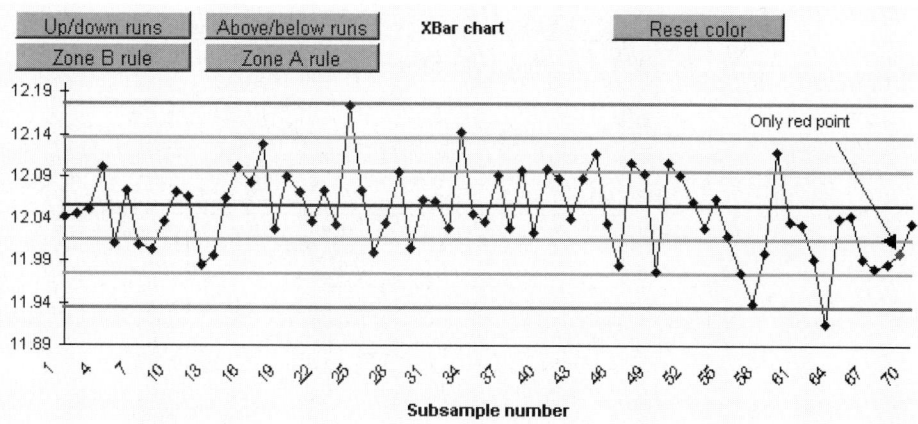

stances of these conditions or see points beyond the control limits, he might start searching for assignable causes and possible fixes.

16.4.3 Rational Subsamples

The small number of observations taken periodically should be **rational subsamples.** This means that they should be taken in such a way that only common-cause variability can be attributed to the points in a particular subsample. There shouldn't be any assignable causes of variability that affect some of the points in the subsample and not others. Typically, rational subsamples are obtained by taking observations nearby in time. For example, every half hour we might examine five *consecutive* soda cans coming off the production line. However, the following example illustrates what can happen if we are not careful.

Example 16.2 Checking the Quality of Gaskets Made on Different Machines

In a manufacturing process for gaskets, two parallel production machines produce identical types of gaskets. A crucial dimension of the gaskets is their thickness, measured in millimeters. The file GASKETS.XLS contains data from this process. Every 15 minutes, four gaskets were sampled, two from each machine. What can we learn from the $\bar{X}$ and R charts?

Objective To see how nonrational samples can produce misleading information of $\bar{X}$ and R charts.

Solution

The observations that comprise a particular subsample are labeled (in the file) M1Obs1, M1Obs2, M2Obs1, and M2Obs2. The first two are from machine 1; the last two are from machine 2. The $\bar{X}$ and R charts for these data, where the centerline and control limits are based on all 50 subsamples, appear in Figures 16.11 and 16.12. The R chart looks perfectly well within control, and the $\bar{X}$ chart looks even better. In fact, it looks suspiciously *too* good, with almost no points outside the one standard deviation band, let alone the

FIGURE 16.11
$\bar{X}$ Chart for Gasket Data from Both Machines

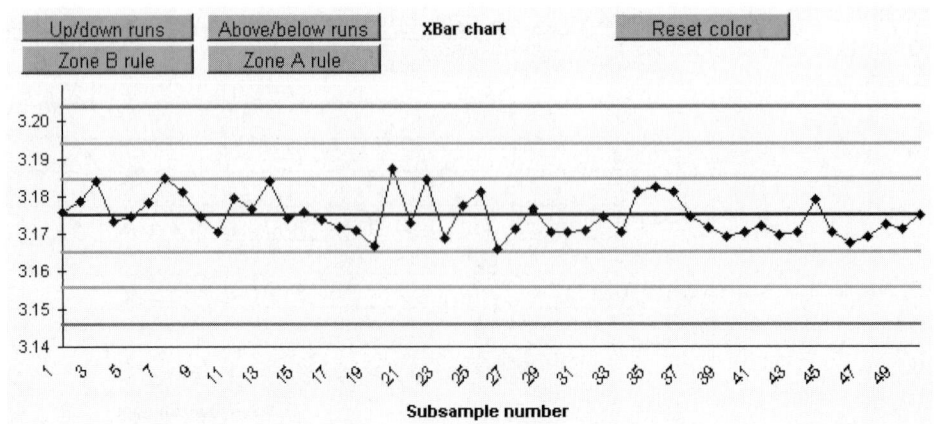

FIGURE 16.12

R Chart for Gasket
Data from Both
Machines

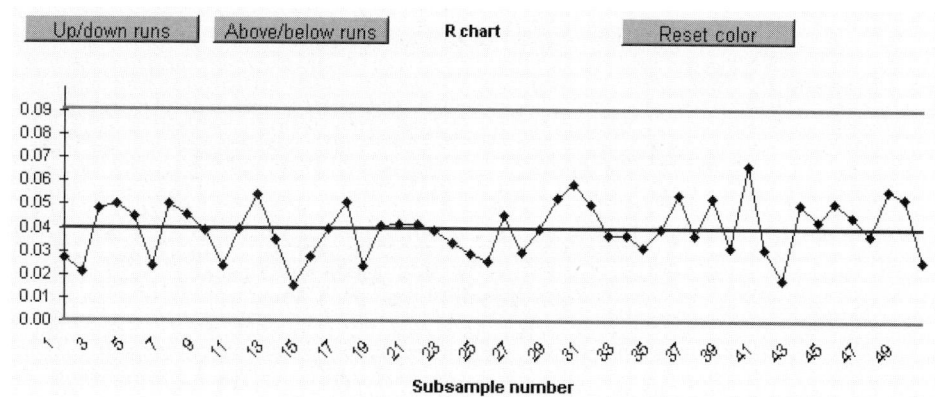

three standard deviation band. The process appears to be in control, but is the small amount of variation in the $\overline{X}$ chart (relative to the control limits) telling us something?

A simple look at the data shows that the observations from machine 1 are consistently below those from machine 2. The variability in the data from each machine is roughly the same, but they are varying around *different means*. Think of what this does to the control charts. First, each *R* is probably a large value from machine 2 minus a small value from machine 1. So the *R*'s are fairly large. This causes the control limits on the $\overline{X}$ chart to be fairly far apart. However, each $\overline{X}$ is an average of two typical machine 1 observations and two typical machine 2 observations. Such averages are not only fairly stable through time, but the highs tend to cancel out the lows. The result is the unusually low variability we see in Figure 16.11.

For the sake of illustration, we assume *four* observations were taken from each machine each half hour. (These are labeled M1Obs1–M1Obs4 and M2Obs1–M2Obs4 in the file.) Only the first two observations from each machine were used in the above control charts. A *rational* subsample philosophy would suggest separate control charts for each machine. It turns out (you can check this) that the control charts for machine 1, based on the subsamples of size 4, indicate perfect in-control behavior. The charts for machine 2, again based on subsamples of size 4, appear in Figures 16.13 and 16.14 (page 884). As we see from the *R* chart, the variability in machine 2 suddenly increased shortly after subsample 25. This causes one out-of-control point in the $\overline{X}$ chart and nearly another. Machine 2 should be checked for assignable causes!

FIGURE 16.13

$\overline{X}$ Chart for Gaskets
from Machine 2

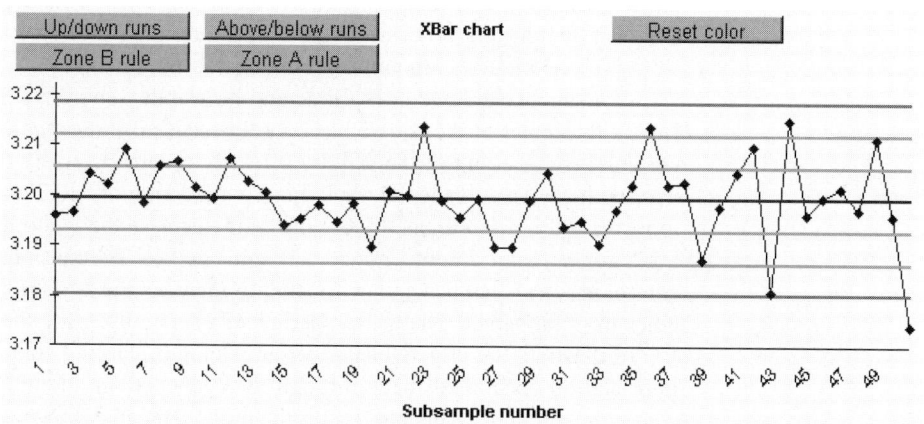

FIGURE 16.14
R Chart for Gaskets from Machine 2

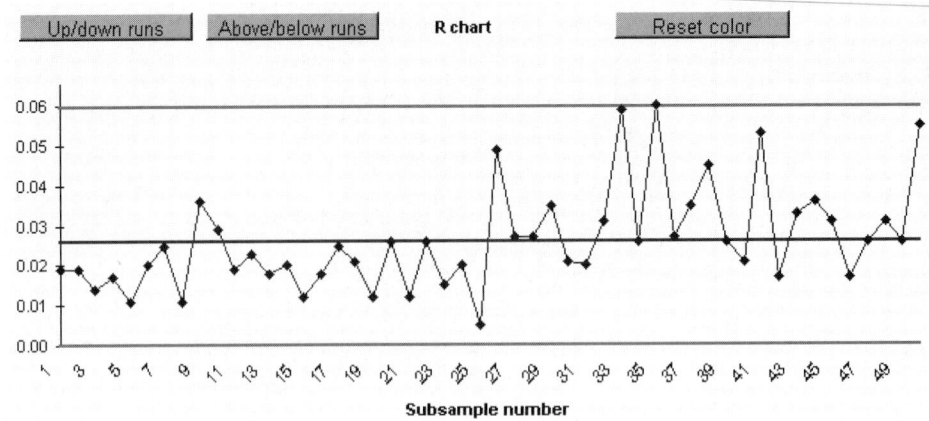

The problem here is that when we combine observations from the two machines into subsamples, the out-of-control behavior is masked by the mixing of highs and lows. We are unable to learn about each machine separately. Therefore, the lesson from this example is that observations within any particular subsample should come from a *single* process, not the mixture of two or more processes. ●

16.4.4 Deming's Funnel Experiment and Tampering

In the quest for reduced variability, it is tempting to make frequent small adjustments to a system. However, if the system is already in control, these adjustments can actually make a system *worse*. Deming called this "tampering" and often demonstrated it in his seminars with the following funnel experiment.

To illustrate the idea, suppose that we are in the business of drilling a tiny hole in the exact center of a square piece of wood. In the past, the holes we have drilled have averaged being in the center of the wood with both the x- and y-coordinates having a standard deviation of 0.1 inch. Also, the drilling process has been in control. Specifically, the deviations from the center of the square (measured in each of the x- and y-coordinates) follow a normal distribution with mean 0 and standard deviation 0.1 inch. This means, for example, that 68% of the holes have their x-coordinate within 0.1 inch of the center, 95% of the holes have their x-coordinate within 0.2 inch of the center, and 99.7% of the holes have their x-coordinate with 0.3 inch of the center. This describes the *inherent* variability in the drilling process. Without changing the process by which holes are drilled, we must live with this amount of common cause variation.

Now suppose that we drill a hole and its x- and y-coordinates are $x = 0.1$ and $y = 0$ [where the center of the square has coordinates (0, 0)]. A natural reaction is to reduce (if possible) the x-setting of the drill by 0.1 inch to correct for the fact that the x-coordinate was too high. Then if the next hole has coordinates $x = -0.2$ and $y = 0.1$, we might try to increase the x-coordinate by 0.2 inch and decrease the y-coordinate by 0.1 inch. Deming's funnel experiment shows that this method of continually readjusting an in-control process—tampering—will actually *increase* the variability of the distance of the holes from the target. That is, tampering will generally make the process worse!

To illustrate the effects of tampering, Deming placed a funnel above a target on the floor and dropped small balls through the funnel in an attempt to hit the target. As he demonstrated, many balls did *not* hit the target. His goal, however, was to make the balls fall as close to the target as possible. Deming proposed four rules for adjusting the position of the funnel.

Rules for Funnel Experiment

1. Never move the funnel.

2. After each ball is dropped, move the funnel—*relative to its previous position*—to compensate for any error. To illustrate, suppose the funnel begins directly over the target, at coordinates (0, 0). If the ball lands at (0.5, 0.1) on the first drop, we compensate by repositioning the funnel at $(-0.5, -0.1)$. If the second drop has coordinates $(1, -2)$, we now reposition the funnel at $(-0.5 - 1, -0.1 - (-2)) = (-1.5, 1.9)$.

3. Move the funnel—*relative to its original position at* (0, 0)—to compensate for any error. If the ball lands at (0.5, 0.1) on the first drop, we compensate by repositioning the funnel at $(-0.5, -1)$. If the second drop has coordinates $(1, -2)$, we now reposition the funnel at $(0 - 1, 0 - (-2)) = (-1, 2)$.

4. Always reposition the funnel directly over the last drop. Thus if the first ball lands at (0.5, 1), we reposition the funnel to (0.5, 1). If the second drop has coordinates (1, 2), we reposition the funnel to (1, 2). This rule might be followed, for example, by an automobile manufacturer's painting department. With each new batch of paint, they attempt to match the color of the previous batch, regardless of whether the previous color was "correct."

To see how these rules work, we run a simulation in Excel. We assume that the *x*-coordinate on each drop is normally distributed with a mean equal to the *x*-coordinate of the funnel position and a standard deviation of 1. A similar statement holds for the *y*-coordinate. Also, we assume that the *x*- and *y*-coordinates are selected independently of one another. These assumptions describe the inherent variability in the process of dropping the balls.

We now develop a spreadsheet to simulate the four rules. For each rule we simulate 50 consecutive drops of the ball and then use a data table to replicate the distance from the 50th drop to the target 100 times. That is, we replicate the experiment of 50 drops 100 times, and we record the distance of the 50th drop from the target on each replication. A good rule should have a small average distance, and the standard deviation of the distances (across the replications) should also be small.

It helps to introduce some notation. Let $P_{x,t}$ and $P_{y,t}$ be the *x*- and *y*-coordinates of the position of the funnel just before drop *t*, where $P_{x,1}$ and $P_{y,1}$, the coordinates of the initial position, are both set to 0 for all of the rules. Also, let X_t and Y_t be the coordinates where drop *t* *actually* falls. Our assumptions imply that X_t and Y_t are normally distributed with means $P_{x,t}$ and $P_{y,t}$ and standard deviations 1. The four rules determine the coordinates of the *next* funnel position, $P_{x,t+1}$ and $P_{y,t+1}$, as follows:

$$P_{x,t+1} = P_{x,t}, \quad P_{y,t+1} = P_{y,t} \qquad \textbf{(Rule 1)}$$

$$P_{x,t+1} = P_{x,t} - X_t, \quad P_{y,t+1} = P_{y,t} - Y_t \qquad \textbf{(Rule 2)}$$

$$P_{x,t+1} = 0 - X_t = -X_t, \quad P_{y,t+1} = 0 - Y_t = -Y_t \qquad \textbf{(Rule 3)}$$

$$P_{x,t+1} = X_t, \quad P_{y,t+1} = Y_t \qquad \textbf{(Rule 4)}$$

These equations allow us to simulate 50 consecutive drops for any of the four rules very easily in Excel. We illustrate this in Figure 16.15 (page 886) for rule 2. (See the file FUNNEL.XLS.) After entering zeros in cells B7 and C7, we enter the formula

$$=\text{NORMINV}(\text{RAND}(),\text{B7},1)$$

in cell D7 and copy it to the range D7:E56 to generate normal random numbers with the appropriate means and standard deviation 1. Then we enter the formula

$$=\text{B7-D7}$$

FIGURE 16.15
Simulation of Rule
2 for Funnel Experi-
ment

	A	B	C	D	E
1	Deming's funnel experiment: Rule 2				
2					
3	Move funnel relative to its last position to compensate for error.				
4		Funnel positioned at:		Drop lands at:	
~					
6	Drop	Xpos	Ypos	Xdrop	Ydrop
7	1	0	0	-0.13	0.57
8	2	0.13	-0.57	1.98	-3.43
9	3	-1.84	2.86	-1.18	1.62
10	4	-0.67	1.24	-1.24	0.12
11	5	0.58	1.12	1.16	1.41
12	6	-0.59	-0.28	1.44	0.07
13	7	-2.03	-0.35	-1.18	-0.80
14	8	-0.85	0.45	-1.30	1.59
54	48	2.68	-0.11	3.14	0.83
55	49	-0.46	-0.95	-1.74	1.20
56	50	1.28	-2.15	1.13	-1.81

in cell B8 and copy it to the range B8:C56. This implements the rule 2 positioning equations. The formulas for the other three rules are similar and follow directly from the positioning equations.

After implementing each of the four rules for 50 drops, we use a data table, as shown in Figure 16.16, to replicate 100 times the distance from the 50th drop to the target for each rule. (Each distance is the square root of the sum of squares of the coordinates of the 50th drop.) The average, standard deviation, and maximum of the 100 distances appear in rows 5–7. Since we want these distances to be *small,* we see that rule 1 is performing best, with rule 2 following fairly close behind, and rules 3 and 4 performing terribly.

This behavior is reinforced by the histograms of the 100 replicated distances for each rule in Figures 16.17–16.20. (They are all shown on the same scale to facilitate comparisons.) As we see, most of the distances for rule 1 are within 2 units of the target and most of the distances for rule 2 are within 3 units of the target, but most of the distances for rules 3 and 4 are more than 9 units from the target. As Deming predicted, tampering with an in-control system never helps—and it can have very negative consequences.

We conclude this discussion of the funnel experiment by noting that the system obtained by using rule 1, the leave-it-alone rule, is not necessarily a *good* system. It may indeed require improvement. The point, though, is that continual tampering with this sys-

FIGURE 16.16
Distances from
Drop 50 to Target
for Four Rules

	A	B	C	D	E	F
1	Data table for replicating distance from center of 50th drop					
2						
3	Summary measures for replications below					
4		Rule1	Rule2	Rule3	Rule4	
5	Average	1.28	1.87	9.11	9.58	
6	Stdev	0.65	0.88	4.98	4.99	
7	Maximum	3.60	4.36	24.96	25.91	
8						
9	Replication	Rule1	Rule2	Rule3	Rule4	
10	0	1.61	1.51	12.61	17.18	
11	1	1.08	3.11	9.73	16.56	
12	2	0.92	0.59	13.34	11.70	
13	3	0.38	1.84	6.91	1.88	
14	4	0.94	1.42	12.93	6.12	
15	5	0.92	3.63	3.01	13.16	
106	96	1.59	1.11	8.69	11.64	
107	97	0.23	0.98	16.46	1.33	
108	98	0.76	0.45	21.46	6.80	
109	99	2.49	2.90	13.93	2.42	
110	100	0.72	4.24	14.25	9.60	

FIGURE 16.17
Histogram of Distances for Rule 1

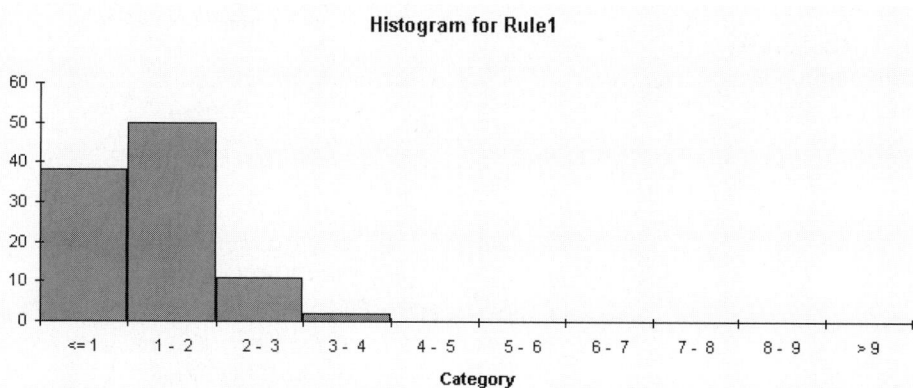

FIGURE 16.18
Histogram of Distances for Rule 2

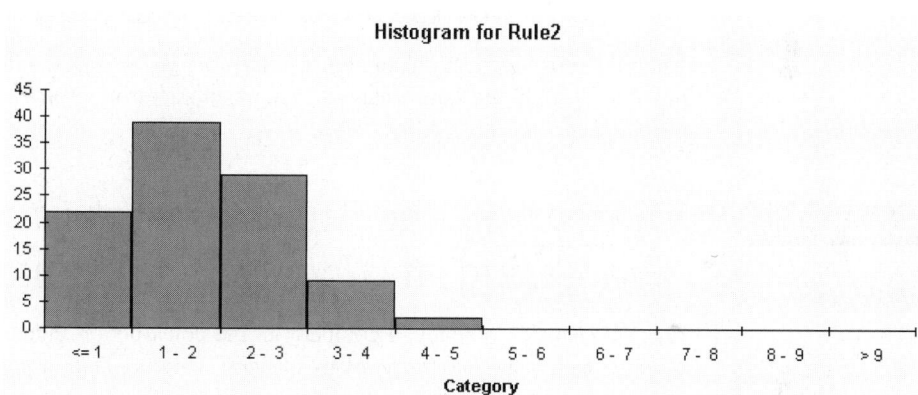

FIGURE 16.19
Histogram of Distances for Rule 3

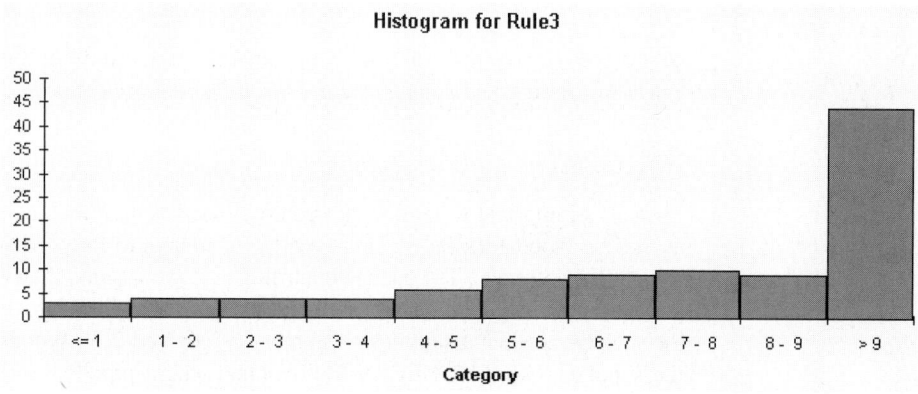

tem will not produce the required improvement; it will only tend to make things worse. The only way to make a lasting improvement to the system is for management to change it fundamentally. Workers on the shop floor do not typically have the knowledge or authority to make such a fundamental change. It must come from management.

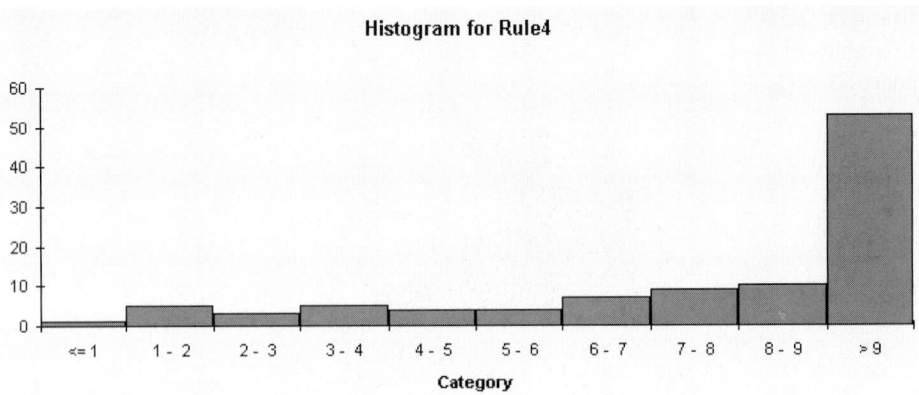

FIGURE 16.20
Histogram of Distances for Rule 4

16.4.5 A Nonmanufacturing Example

Although most applications of control charts are in the manufacturing area, it is certainly possible to apply the same analysis to nonmanufacturing problems. The following example illustrates one possibility.

Example 16.3 Measuring the Timeliness of Check Processing at Woodstock Company

The Woodstock Company, a company in the construction industry, had recently experienced considerable expansion of its business volume. Due in part to this expansion, the finance department of the company was having difficulty processing checks to suppliers in a timely manner. Many of its suppliers were being paid beyond the normal 30-day period. This was not only making the suppliers unhappy, but Woodstock was also failing to obtain the discounts many suppliers offered for prompt payments. How could control charts help Woodstock solve its problem?

Objective To see how control charts can help Woodstock find the reasons for untimely check processing and suggest ways of decreasing check processing time.

Solution

First, it is important to realize that control charts cannot magically solve a problem such as the one Woodstock faced. However, they can help to show what is happening and point to possible solutions. To produce control charts, we assume that Woodstock measured the processing times for five checks completed each day. Each processing time is defined as the time from when a supplier's shipment is received until Woodstock sends the check to the supplier. The file CHECKS.XLS contains these processing times for 60 consecutive business days. Observations for the first 30 days were used to form control limits. The R chart (not shown here) for these 30 days is well within control, but the $\overline{X}$ chart, shown in Figure 16.21, indicates out-of-control points on days 7 and 10.

Upon closer examination, Woodstock learned that on day 7 the people in finance, trying to improve a process with high variability and large processing times, implemented a

FIGURE 16.21
X̄ Chart for First 30 Days

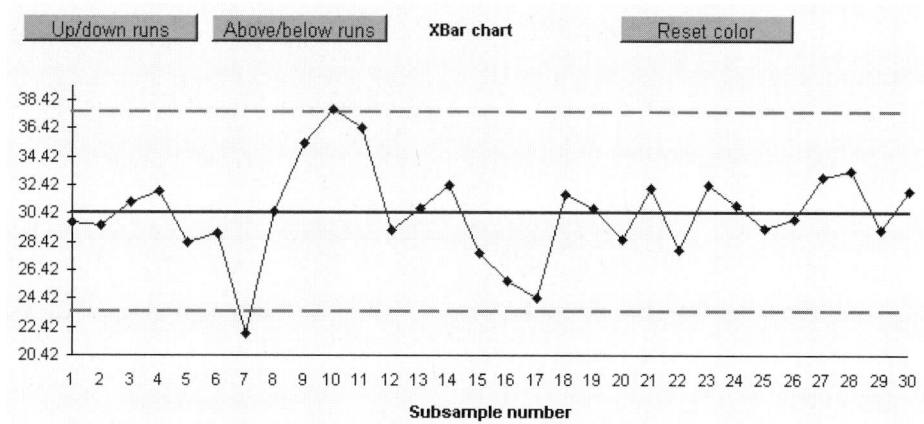

change in the check preparation process. However, this change backfired—it actually made things worse—and was eliminated after 5 days. This change is a clear example of an *assignable cause*. The points we observe in Figure 16.21 are actually the result of two separate processes, those without the change and those (points 7–11) with the change. To understand the original process, Woodstock needed to eliminate points 7–11 and form new charts. This was done, and the plots of days 1–6 and 12–30 (not shown here) showed statistical control.[4]

The process was now in statistical control, but this was no place to stop! Woodstock was alarmed at the high average processing times (about 30 days) and the high variability (average R's of nearly 12 days). Management took a closer look at the check preparation process and discovered several unnecessary steps—duplicate paperwork and excessive "hand-offs" from one person to another. They took steps to streamline the process, and they continued to plot, using the control limits and centerlines from days 1–6 and 12–30. The $\bar{X}$ and R charts through day 60 (again, with days 7–11 eliminated) appear in Figures 16.22 and 16.23 (page 890).

FIGURE 16.22
X̄ Chart for Days 1–60 (with Days 7–11 Eliminated)

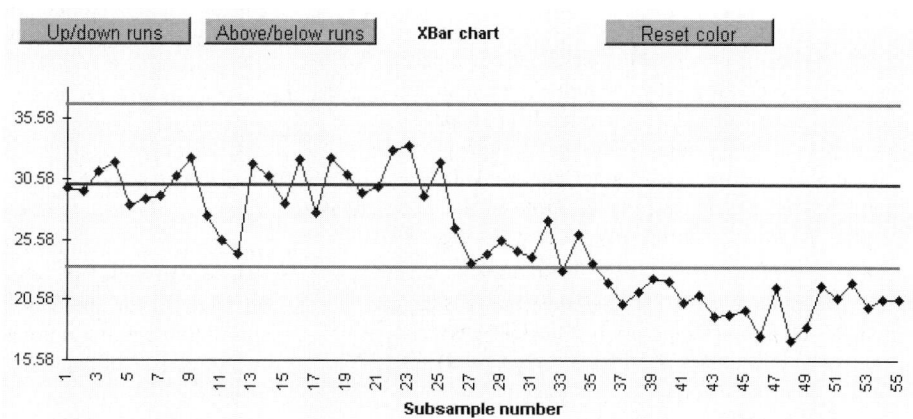

---

[4]To do this in Excel, we copied the original data sheet to a new data sheet, deleted the rows corresponding to days 7–11, and formed control charts from the first 25 rows of this new data set.

FIGURE 16.23
R Chart for Days 1–60 (with Days 7–11 Eliminated)

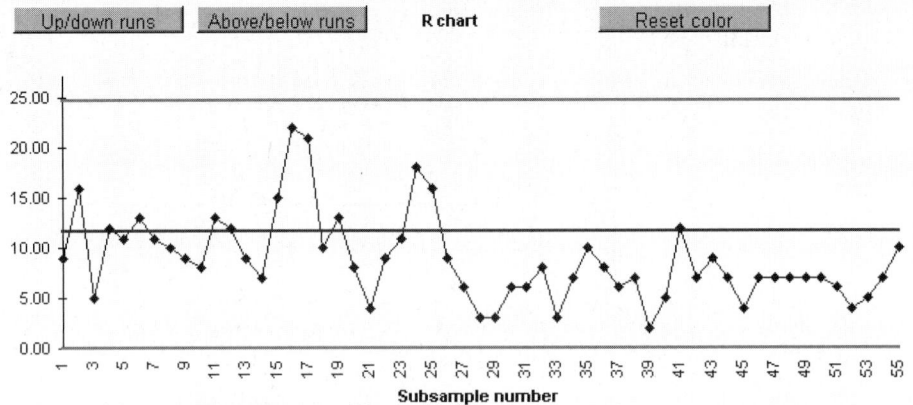

These control charts both indicate out-of-control behavior, but of the kind Woodstock is happy to see. The *R* chart indicates a lower level of variability, and the $\overline{X}$ chart indicates a decreased average time to process checks. The *R*'s are now averaging about 6.5 days, and the average check processing times are about 20 days. These improvements are a direct result of Woodstock's management interventions, but these interventions were prompted by observing control charts and trying to understand what was causing them.

Even after day 60, Woodstock should not rest on its laurels. First, it should recalculate control limits and centerlines, based on new data, say, from days 51–80. It could use these to check whether the improved process is in control with respect to the new limits. At least as importantly, it should continue to search for potential improvements in the process. If the average check preparation time could be reduced from 30 days to about 20 days, and the variability could be reduced as well, who's to say that further improvements are not possible? ●

PROBLEMS

Level A

1. The file P16_01.XLS contains data on the amount of soda (in ounces) placed in aluminum cans by a particular filling process. Ideally, the filling process should place 12 ounces of soda in each can. Every hour, 4 cans of soda were randomly selected from the production process and measured for amount of fill. This was repeated for 25 consecutive hours. Generate and interpret $\overline{X}$ and *R* charts for the given data. Does this filling process appear to be in control?

2. The data in the file P16_02.XLS consist of 25 subsamples of 4 observations each on the diameters (measured in centimeters) of ball bearings produced by a manufacturing process. The target diameter of these ball bearings is 4 centimeters.
   a. Generate and interpret $\overline{X}$ and *R* charts for the given data. Based on these charts, does this manufacturing process appear to be in control?
   b. Are there any other indications that the given process may be out of control? If so, explain.

3. The data in the file P16_03.XLS consist of 25 subsamples of 6 observations each on the fill weights of cans of paint. The ideal fill weight of these paint cans is 20 pounds.
   a. Generate and interpret $\overline{X}$ and *R* charts for the given data. Does this filling process appear to be in control?
   b. Are there any other indications that the given process may be out of control? If so, explain.

4. Continuing the previous problem, operators have made an adjustment that they hope will improve the functioning of this process. The file P16_04.XLS contains 25 subsamples of 6 observations each on the fill weights of cans of paint, taken after the process was modified.
   a. Generate and interpret $\overline{X}$ and *R* charts for the given data. Which control limits do you believe are most appropriate? Explain. Does this filling process appear to be in control now?
   b. Are there any other indications that the given process may be out of control now? If so, explain.

c. What advice would you give to the operators of this filling process based on your analysis of the latest sample information?

5. Producers of a particular brand of ready-to-eat breakfast cereal place, in theory, 15 ounces of cereal in each box of this product. In an effort to assess this stage of the manufacturing process, an operations manager gathers 25 subsamples of 4 observations each on the amount (measured in ounces) of cereal in selected boxes. These data are given in the file P16_05.XLS.

a. Generate and interpret $\overline{X}$ and R charts for the given data. Does this process appear to be in control?

b. Are there any other indications that the given process may be out of control? If so, explain.

c. If your analysis reveals that a problem does exist with the process, what advice would you give to the manager of this operation?

6. The data in the file P16_06.XLS consist of 25 subsamples of 4 observations each on the lengths of particular bolts manufactured for use in large aircraft. The target length of these bolts is 37 centimeters.

a. Generate and interpret $\overline{X}$ and R charts for the given data. Does this production process appear to be in control?

b. Are there any other indications that the given process may be out of control? If so, explain.

7. Continuing the previous problem, managers have made an adjustment that they hope will improve the functioning of this process. The file P16_07.XLS contains 25 subsamples of 4 observations each on selected bolt lengths, taken after the production process was modified.

a. Generate and interpret $\overline{X}$ and R charts for the given data. Which control limits do you believe are most appropriate? Explain. Does this production process appear to be in control now?

b. Are there any other indications that the given process may be out of control now? If so, explain.

c. What advice would you give to the operators of this production process based on your analysis of the latest sample information?

8. The operations manager of an airline check-in counter is interested in evaluating the service provided to the company's customers. In particular, she would like to make sure that customers are not waiting excessively long prior to being served at the check-in facility. Ideally, she would like to see that customers wait, on average, less than 7 minutes prior to being served at the check-in counter. The file P16_08.XLS contains the waiting times (in minutes) of 5 randomly selected passengers, observed during the same hour on each of 25 different days.

a. Construct and interpret $\overline{X}$ and R charts for the given data. Based on these charts, does this service process appear to be in control?

b. Are there any other indications that the given service process may be out of control? If so, explain.

c. What specific advice would you give to the operations manager for improving customer service, at least in the short run?

9. Continuing the previous problem, the operations manager of this airline check-in counter has recently made some changes to the operation of the counter that she hopes will reduce customer waiting times. Again, she observes the waiting times of 5 randomly selected passengers during the same hour on each of another 25 days. These observations are given in the file P16_09.XLS. Characterize the impact of the operations manager's refinements on the performance of this service operation. Does this facility, with respect to customer waiting times, appear to be operating well now? Explain why or why not.

10. The file P16_10.XLS contains the breaking strengths (measured in pounds) of randomly selected pieces of a certain welded material. In particular, these data consist of 25 subsamples of 5 observed breaking strengths each. The target breaking strength of this welded material is 300 pounds.

a. Generate and interpret $\overline{X}$ and R charts for the given data. Does this production process appear to be in control?

b. Are there any other indications that the given process may be out of control? If so, explain.

c. If the process is out of control, which subsamples could be eliminated to achieve in-control behavior in the R chart? Why might it be legitimate to eliminate these subsamples?

11. The file P16_11.XLS contains data on the amount of liquid detergent (in ounces) placed in plastic containers by a particular filling process. Ideally, the filling process should place 100 ounces of detergent in each container. Every hour, 6 detergent containers were randomly selected from the production process and measured for amount of fill. This was repeated for 25 consecutive hours.

a. Generate and interpret $\overline{X}$ and R charts for the given data. Does this filling process appear to be in control?

b. Are there any other indications that the given process may be out of control? If so, explain.

c. Given your analysis of the sample information, what steps, if any, should be taken to adjust this filling process?

12. Management of a local bank is interested in assessing the process used in opening new checking accounts for bank customers. In particular, management would like to examine the time required to process a customer's request to open a new checking account. Currently, managers believe that it should typically take about 7 minutes to process such a request. The file P16_12.XLS contains the time required to process new checking account requests for each of 6 customers selected randomly on a given day. A different subsample of 6 customer requests was collected on each of 25 days. Generate and interpret $\overline{X}$ and R charts for the given data. Does this customer service process appear to be in control? Explain why or why not.

13. The manager of a supermarket would like to evaluate the effectiveness of a large freezer unit currently used to store excess supplies of various frozen food items in the supermarket's inventory. Specifically, the manager wants to determine whether the current freezer is maintaining the valuable store inventory at a roughly constant temperature of 7 degrees. To make this evaluation, he asks his assistant to take 6 temperature readings within the freezer at various points in the day for a total of 30 days. These measurements are given in file P16_13.XLS.
   a. Generate and interpret $\overline{X}$ and R charts for the given data. Do these data indicate the presence of one or more problems with the operation of this freezer? Explain.
   b. How might you explain the trend in the R chart for subsamples 21 through 30?
   c. What advice would you give to the supermarket manager regarding this freezer?

Level B

14. Consider a situation where a given process goes out of control but the relevant control charts do not indicate so. Assume that the process variation remains constant but the mean shifts from μ to $\mu + k\sigma$, where k is a fixed constant.
   a. Provided that $k = 1.5$ and $n = 5$, what is the mean number of subsamples required for $\overline{X}$ to rise above the upper control limit?
   b. Provided that $k = 1.5$ and $n = 9$, what is the mean number of subsamples required for $\overline{X}$ to rise above the upper control limit?
   c. Provided that $k = 2.0$ and $n = 5$, what is the mean number of subsamples required for $\overline{X}$ to rise above the upper control limit?

15. Consider a situation where a given process goes out of control but the relevant control charts do not indicate so. Assume that the process variation remains constant but the mean shifts from μ to $\mu - k\sigma$, where k is a fixed constant.
   a. Provided that $k = 1.0$ and $n = 4$, what is the mean number of subsamples required for $\overline{X}$ to fall below the lower control limit?
   b. Provided that $k = 1.0$ and $n = 8$, what is the mean number of subsamples required for $\overline{X}$ to fall below the lower control limit?
   c. Provided that $k = 0.5$ and $n = 4$, what is the mean number of subsamples required for $\overline{X}$ to fall below the lower control limit?

16.5 Control Charts for Attributes

Often there are no explicit measurements available. We are simply able to check whether each item produced conforms to specifications or not. For example, a computer chip either works as it should or it doesn't. An item that fails to conform to specifications is called a **nonconforming** (or **defective**) item. When items can be classified only as conforming or nonconforming, then we typically chart the proportions that are conforming during consecutive periods of time. The resulting chart is called a *p* **chart.** It is one of several types of charts called **attributes** charts, where the term *attribute* indicates an "on/off" type of measurement: The item either has the attribute or it does not.

StatPro's online help indicates how your data should be set up to use *c* charts and *u* charts.

There are other types of attributes charts called *c* **charts** and *u* **charts.** These are used to chart the number (or rate) of defects in successive items, where a defect is any flaw in an item, such as a paint blemish on a car door, a defective weld in a pipeline, a broken rivet on an airplane wing, and so on. Clearly, different types of defects vary in their seriousness, and any combination of them could cause an item (a car door, for example) to be classified as nonconforming. We will not discuss *c* charts and *u* charts in this book, but they are very similar to the other control charts we discuss, and they can be formed easily with StatPro's Quality Control procedures.

16.5.1 The p Chart

We now discuss p charts in some detail. During consecutive periods of time, we sample a number of items from a process and label each of these as conforming or nonconforming. Specifically, suppose we sample n_i items during period i, and k_i of these fail to conform to specifications. The number n_i either could be *all* of the items produced during period i, or it could represent a sample of all items produced. Also, these sample sizes could be constant for all periods, or they could differ. In any case, we let $\hat{p}_i$ be the proportion of nonconforming items in sample i:

$$\hat{p}_i = k_i/n_i$$

A p chart is then a time series plot of the $\hat{p}_i$'s.

p Chart	A *p* chart plots the proportions of items that are nonconforming (defective) through time. It is a type of attributes chart.

The idea behind p charts is exactly the same as with $\overline{X}$ and R charts. We place a centerline and control limits on the chart in such a way that the $\hat{p}_i$'s for an in-control process vary randomly around the centerline and almost never cross the control limits. The centerline is placed at the overall proportion of nonconforming items. This value, denoted by $\overline{p}$, is given by

$$\text{Centerline} = \overline{p} = \frac{\Sigma_i\, k_i}{\Sigma_i\, n_i} = \frac{\text{number of nonconforming items}}{\text{number of items produced}}$$

If the n_i's are constant across samples, then $\overline{p}$ is the average of the $\hat{p}_i$'s.

To specify the control limits, we first examine the special case where the n_i's are constant and equal to a common value n. If we can assume that each item is nonconforming with some constant probability p, then we know from Chapter 6 that the number of non-conforming items in sample i is binomially distributed with mean np and standard deviation $\sqrt{np(1-p)}$. Equivalently, the sample proportion $\hat{p}$ has mean p and standard deviation $\sqrt{p(1-p)/n}$. Because p is typically unknown, we use the value $\overline{p}$ as an estimate of p to form the following control limits:

Again, the main thing to remember is that the lower and upper control limits are three standard deviations below and above the centerline, where the centerline is the proportion of all items that are nonconforming.

$$LCL = \overline{p} - 3\sqrt{\overline{p}(1-\overline{p})/n}$$

and

$$UCL = \overline{p} + 3\sqrt{\overline{p}(1-\overline{p})/n}$$

That is, we go out three standard deviations (of $\hat{p}$) on each side of the centerline value $\overline{p}$.

If the sample sizes are *not* equal, then each sample has its own control limits of the form

$$\overline{p} \pm 3\sqrt{\overline{p}(1-\overline{p})/n_i}$$

Now the denominator n_i varies from sample to sample instead of being constant. The effect is that the control limits vary through time and are not straight lines on the control chart. This is an annoying feature, so in practice the *average* of the n_i's is often used in place of the n_i's as a "common sample size" unless the n_i's differ greatly from one another. StatPro gives the user this option of using a common sample size even if the individual sample sizes are not equal.

The following example illustrates how a p chart can be constructed and interpreted. Although some of the details are different, the basic interpretation and use of p charts are exactly the same as with $\overline{X}$ and R charts. Specifically, they are monitored through time to provide a better understanding of a process and suggestions for possible improvement.

Example 16.4 Measuring the Proportion of Defective Chips at SoundTech

SoundTech is a company that manufactures electronic chips for sound systems in personal computers. Each chip is classified as conforming or nonconforming. The nonconforming chips cannot be used and are discarded. Each hour 75 chips are tested for conformance. These 75 chips represent a random sample of all chips that are produced in a given hour. The file CHIPS1.XLS lists the number of nonconforming chips (out of 75) for 25 consecutive hours. Is the process currently in control? Is it behaving well?

Objective To use p charts to see whether the chip manufacturing process at SoundTech is in control and is producing a "small" number of nonconforming chips.

Solution

The mechanics of constructing a p chart with StatPro are very similar to those for $\overline{X}$ and R charts. The main difference is that the data can be set up in several ways. First, the data can either list the *numbers* of nonconforming items or the *fractions* of nonconforming items. (CHIPS1.XLS lists the former.) Second, a variable that lists the sample sizes, the n_i's, can either be present or absent. (There is such a variable in CHIPS1.XLS.) If the sample size variable is absent, then it is assumed that the sample sizes are constant, and you must enter this common value in a dialog box.

To create the p chart, we use StatPro's Quality Control/P Chart menu item. After the usual opening message, the dialog box in Figure 16.24 appears. For this example it should be filled out as shown. (Alternatively, the bottom option could be checked, in which case the value 75 should be entered manually in the box.) Next, the procedure prompts for the variable containing the numbers nonconforming and the sample size variable. (The variables Nonconforming and SampSize should be selected for this example.) Next, the dia-

FIGURE 16.24
Dialog Box for Specifying Variables for p Chart

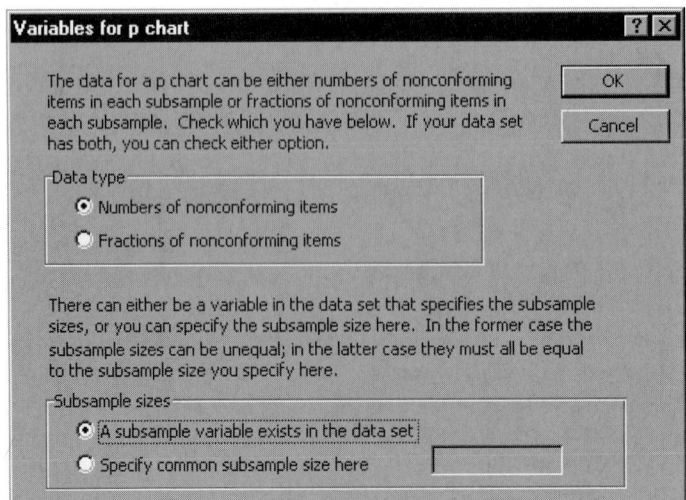

FIGURE 16.25
Dialog Box for
Other *p* Chart
Options

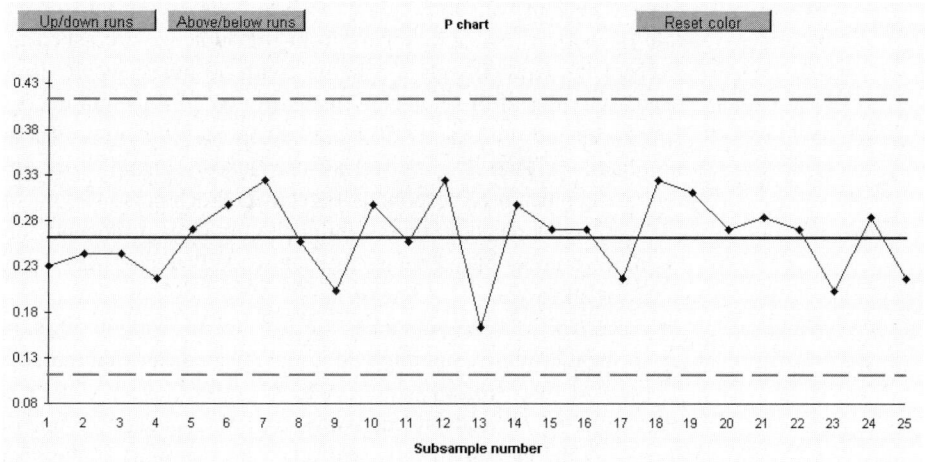

p chart options ? X

Observations
You can base the chart on all 25 of the subsamples, or on just the subsamples specified below.
⦿ Plot all of the observations
○ Plot only those within the limits below
Starting index: [   ]   Ending index: [   ]

Control limits
You can base the control limits for the chart from all or part of the current data, or you can base control limits on previous data.
⦿ Control limits based on all of the current data
○ Control limits based on only the subsamples specified below
Starting index: [   ]   Ending index: [   ]
○ Control limits based on previous data
Previous subsample size [   ]
Previous p-bar (average defectives per sample): [   ]

Extra lines on p chart
Besides the control limits (which are at about 3 sigmas from the center line), you can add lines at 1 and 2 sigmas from the mean on the p chart.
☐ Include extra lines on p chart
Note that this option isn't available unless the subsample sizes are equal.

OK   Cancel

log box shown in Figure 16.25 appears. It is almost exactly the same as for $\overline{X}$ and R charts and should be filled in as shown. Two sheets are then created, one with the *p* chart and one with the data used to build it.

The *p* chart appears in Figure 16.26. We see that the points, each of which indicates a proportion nonconforming, vary randomly around a centerline of $\overline{p} = 0.255$. The control limits are at 0.104 and 0.407, and no points are beyond the control limits. Therefore, the current process appears to be in control. But is it any good? We would argue that an average percent nonconforming of about 25% is *not* very good. As usual, an in-control process is *predictable* but not necessarily acceptable. SoundTech management should begin searching for improvements to its process. For example, they might select different suppliers of raw material, purchase new machinery, or institute better worker training. Then by charting future values, the company can see whether any improvements it employs have the desired effect.

For the sake of comparison, we illustrate a variation of this example where SoundTech samples a *different* number of chips each hour. For example, it might actually sample *all* chips produced, but production quantities might vary considerably from hour to hour. The

Remember, "in control" does not necessarily mean *good;* it just means *predictable.*

FIGURE 16.26
p Chart for Non-
conforming Chips

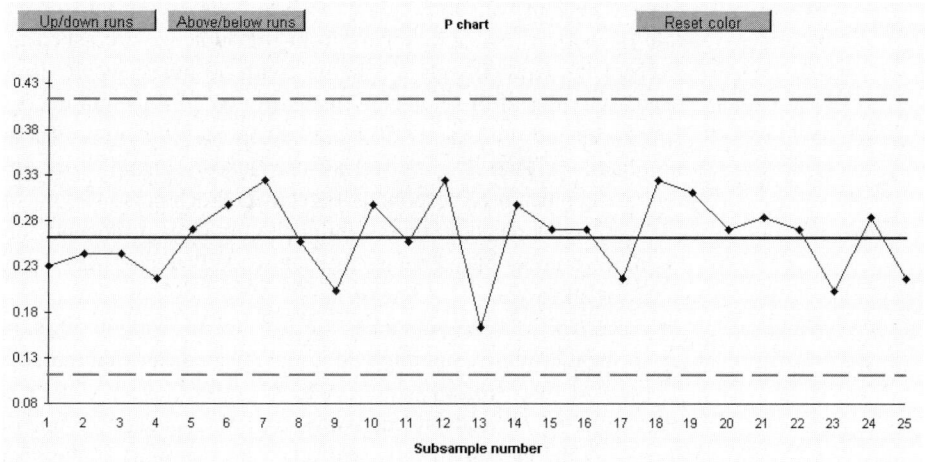

| Up/down runs | Above/below runs | P chart | Reset color |

Subsample number

FIGURE 16.27
A *p* Chart with Un-
equal Sample Sizes

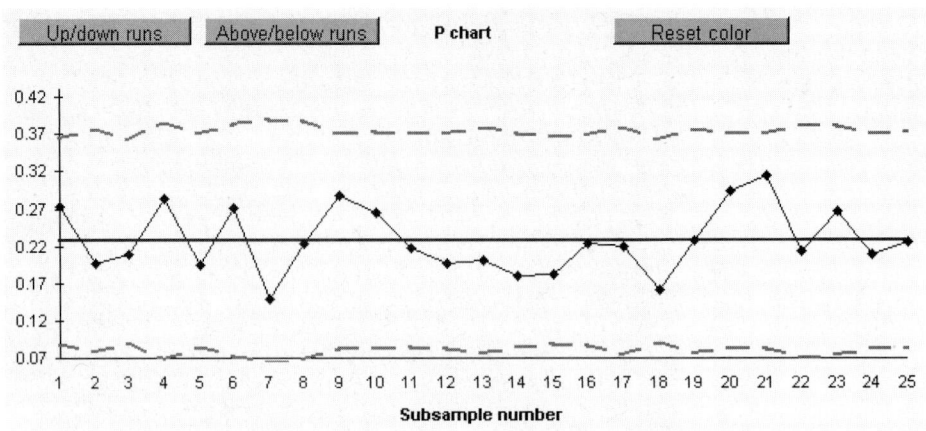

file CHIPS2.XLS contains the data. The only difference is that the SampSize variable in this file is not constant. We have two options. We can fill out the dialog box in Figure 16.24 exactly as before, or we can check the bottom box and enter an "average" sample size. If we select the former option, the resulting *p* chart appears in Figure 16.27. As we see, the nonconstant sample sizes result in uneven control limits. Although we are still looking for points beyond the control limits, the bumpiness of these limits is somewhat distracting. Therefore, SoundTech might decide to base the chart on the average sample size (about 75). Fortunately, the practical difference between these two approaches is usually minor. ●

16.5.2 The Red Bead Experiment

Recall that several of Deming's 14 points concern the role of management in helping workers to do a better job. Deming believed that it is *not* management's role simply to exhort workers to do a better job. Management needs to change a system that prevents workers from performing up to standards. To illustrate this concept, Deming often used the following "red bead experiment." It illustrates clearly that in a system subject only to common-cause variation, some workers are bound to be the "best" on some days and "worst" on others, for no particular reasons (such as slacking off or working harder). It also illustrates how all workers can fail to live up to standards, through no fault of their own, if the system is not designed correctly.

The experiment is very simple. There is a large container of beads, 20% of which are red and 80% of which are white. Several people from the audience are asked to play the role of workers, whereas others from the audience are asked to help out as inspectors. Each of the workers gets a "paddle" with 50 holes, where each hole can hold a single bead. The rules of the game are that each worker must put his or her paddle into the container and pull out exactly 50 beads. Each such draw corresponds to one day's production quantity. That is, each worker "produces" exactly 50 beads per day. They are also told that red beads correspond to defectives. Each person's job is to produce no more than two defectives per day; their continued employment depends on it. The inspectors then count the number of red beads for each worker for each day's production and tally the results for all to see.

Let's say the workers' names are Jim, Tricia, Tom, and Lisa. Several things about the experiment are fairly obvious. First, the mechanics of the process make it impossible for workers to "fish" for all white beads. Therefore, every worker gets a random sample of 50 red and white beads on each draw from the container. Some days Jim will—totally by luck—get the most red beads, and other days he will get the fewest red beads. The same applies for the other workers. Certainly, there is no reason to reprimand Jim in the first

FIGURE 16.28
Simulation Results
for Red Bead Exper-
iment

	Day	Jim	Tricia	Tom	Lisa	Total	Best	Worst	Winner	Loser
Red bead experiment										
Simulation of 30 days										
	Day	Jim	Tricia	Tom	Lisa	Total	Best	Worst	Winner	Loser
	1	10	8	11	10	39	8	11	Tricia	Tom
	2	9	14	13	13	49	9	14	Jim	Tricia
	3	13	11	6	8	38	6	13	Tom	Jim
	4	6	20	11	17	54	6	20	Jim	Tricia
	5	15	9	10	5	39	5	15	Lisa	Jim
	6	13	6	11	9	39	6	13	Tricia	Jim
	7	6	13	14	6	39	6	14	Jim	Tom
	8	11	7	12	12	42	7	12	Tricia	Tom
	9	9	10	9	13	41	9	13	Jim	Lisa
	10	9	8	10	8	35	8	10	Tricia	Tom
	11	10	4	5	11	30	4	11	Tricia	Lisa
	12	6	9	11	12	38	6	12	Jim	Lisa
	13	9	8	14	13	44	8	14	Tricia	Tom
	14	8	14	13	8	43	8	14	Jim	Tricia
	15	10	11	9	13	43	9	13	Tom	Lisa
	16	5	16	14	11	46	5	16	Jim	Tricia
	17	11	9	12	11	43	9	12	Tricia	Tom
	18	12	6	15	2	35	2	15	Lisa	Tom
	19	17	13	8	9	47	8	17	Tom	Jim
	20	10	7	12	9	38	7	12	Tricia	Tom
	21	6	10	11	10	37	6	11	Jim	Tom
	22	8	8	13	10	39	8	13	Jim	Tom
	23	10	8	9	8	35	8	10	Tricia	Jim
	24	12	9	13	15	49	9	15	Tricia	Lisa
	25	11	11	12	16	50	11	16	Jim	Lisa
	26	10	9	11	9	39	9	11	Tricia	Tom
	27	11	9	7	11	38	7	11	Tom	Jim
	28	7	10	11	11	39	7	11	Jim	Tom
	29	5	13	10	12	40	5	13	Jim	Tricia
	30	14	10	12	8	44	8	14	Lisa	Jim
Tally of winners and losers										
		Wins	Losses							
Jim		12	7							
Tricia		11	5							
Tom		4	12							
Lisa		3	6							
Total		30	30							

case or reward him in the second. But Deming says that this is exactly what occurs in many job settings.

Second, the experiment is stacked against the workers. It is impossible for them, on most days, to draw two or fewer red beads. On average, each draw will result in 20%, or 10, red beads. For them to do their job as instructed, the *system* must change. For example, management could remove a lot of the red beads from the container. Even though all of this was obvious to the "workers" in Deming's experiments, it is interesting that many of them nevertheless tried their best to perform as instructed, and many were genuinely frustrated when they continued to draw too many red beads.

We can illustrate the red bead experiment with an Excel simulation, together with a p chart. The REDBEAD.XLS file contains the results. We first simulate the number of red beads for each worker on 30 successive days. (See Figure 16.28.) For each worker and each day, the number of red beads is a binomial random value based on 50 trials and probability 0.2 of "success" (a red bead) on each trial. To simulate such a value, we can use the CRITBINOM function in Excel.[5] Specifically, we enter the formula

$$=\text{CRITBINOM}(50,0.2,\text{RAND}())$$

in cell B6 and copy it to the range B6:E35.

---

[5]Alternatively, we could use StatPro's built-in function BINOMIAL_ in the form =BINOMIAL_(50,0.2).

We then sum the number of red beads per day in column F, and in columns G–J we record the "best" and "worst" workers each day. For example, the formulas in cells G6 and I6 are

$$=MIN(B6:E6)$$

and

$$=IF(G6=\$B6,"Jim",IF(G6=\$C6,"Tricia",IF(G6=\$D6,"Tom","Lisa")))$$

(Note that if there is a tie for best or worst, only one of the workers in the tie is listed.) Finally, we tally the winners and losers in the range B39:C42. The formula in B39 is

$$=COUNTIF(I\$6:I\$35,\$A39)$$

which is then copied to the range B39:C42.

Using the daily production quantities of red beads in column F, we can also create a *p* chart, as shown in Figure 16.29. It shows a process well in control. In particular, it shows how the daily proportion of red beads varies randomly around 0.2. Of course, this doesn't mean the workers are producing what management *wants* them to produce (no more than two red beads per day), but it certainly is not the workers' fault, and there is nothing they can do about it until the system changes.

If you look at this REDBEAD file, you'll see that all of the random numbers are "live," so that they change any time the spreadsheet recalculates. Therefore, you'll see

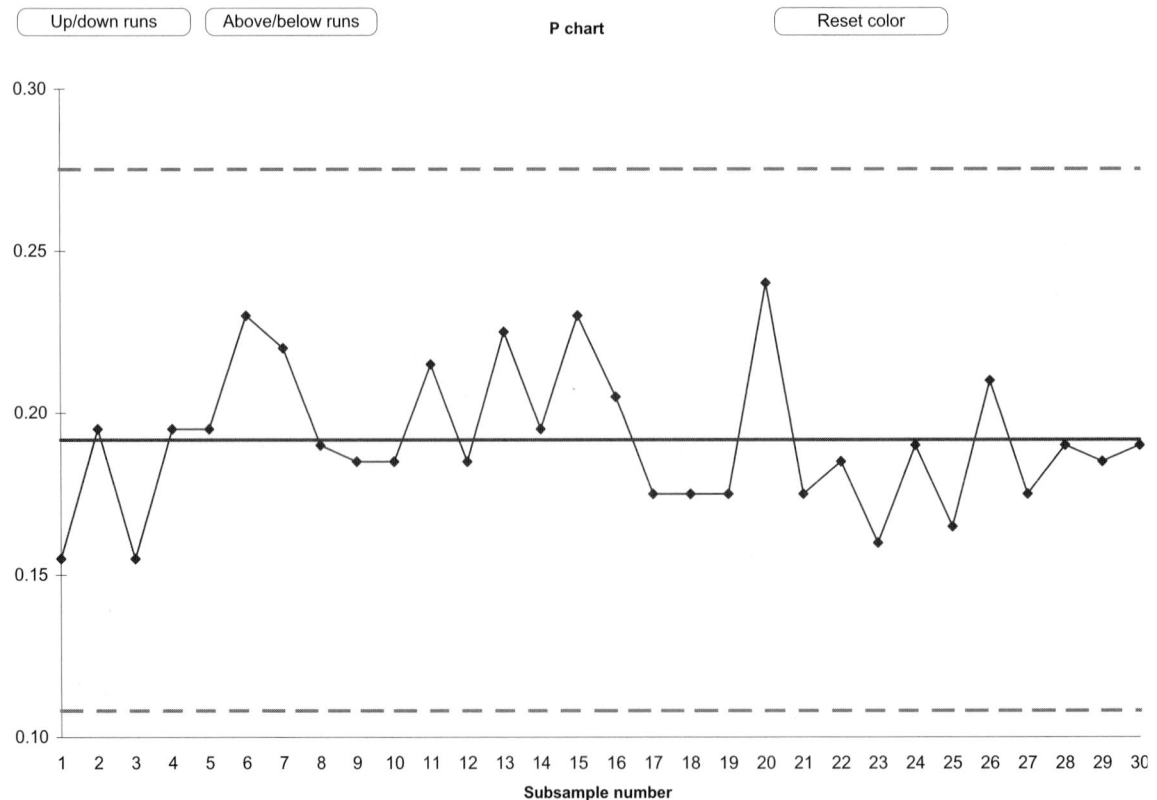

FIGURE 16.29 *p* Chart for the Red Bead Experiment

values different from those in Figures 16.28 and 16.29. However, they should all tell approximately the same story, as summarized here:

- Variation is inherent in any process.
- The system determines workers' performance; until it changes, workers are typically unable to improve their performance.
- Only management can change the system.
- Given an in-control process, some workers will *appear* to be best or worst on different days, but at least part of this is a matter of luck, not skill or working harder. When this is the case, rewards or reprimands are likely to do more harm than good.

PROBLEMS

Level A

16. Suppose that a manufacturer of electronic computer chips classifies each chip as either defective or nondefective. Each hour 100 electronic chips are randomly selected from a very large batch of chips and tested for possible defectiveness. The file P16_16.XLS lists the number of defective chips (out of the 100 sampled) found during each of 25 hours of production. Based on the given sample data, is this production process currently in control? Provide support for your conclusions.

17. Construct and interpret a *p* chart for the data provided in the file P16_17.XLS. These data consist of the number of defective units found in each of 25 samples with a common size of 200 units. Is this production process currently in control? Is it behaving "well"?

18. Continuing the previous problem, suppose that the given production process has been modified in an effort to reduce the variation in the proportion of defective items. An additional 25 samples that share a common size of 200 are gathered, and the number of defective items in each sample is recorded in the file P16_18.XLS. Have the refinements to this system achieved the desired result? Explain why or why not.

19. Construct and interpret a *p* chart for the data provided in the file P16_19.XLS. These data consist of the number of defective units found in each of 25 samples with a common size of 100 units. Is this production process currently in control? Is it behaving "well"?

20. Suppose that a manufacturer of a particular automotive part classifies each unit produced as either defective or nondefective. Each hour 200 parts are randomly selected from a very large batch of manufactured items and tested for possible defectiveness. The file P16_20.XLS lists the number of defective parts (out of the 200 sampled) found during each of 30 consecutive hours of production.
  a. Based on the given sample data, is this production process currently in control? Provide support for your conclusions.

b. What advice, if any, would you give to the managers of this production process?

21. A new Internet-based bookstore ships thousands of books to its customers each week. The director of this firm's shipping department is concerned about a seemingly large number of complaints from customers regarding errors in shipments. The managers of the organization know that the future profitability of this virtual bookstore depends essentially on its ability to fill customer orders quickly and accurately. In an effort to investigate this problem further, the shipping director collects data on the number of shipments not conforming to corresponding customer orders for 25 days in a row. The file P16_21.XLS contains the number of reported nonconforming shipments out of the 75 shipments randomly selected on each of the 25 days. Do these data indicate a high average proportion of shipment errors? Is there an excessive amount of variability in the error rate of the company's shipments? If so, what can managers do in both the short run and the long run to improve the situation?

22. Continuing the previous problem, the firm's shipping director has implemented a new quality control program within her department. She is interested in determining whether this program is reducing the mean error rate of the company's shipments. She also hopes that the quality control program will serve to reduce the variability in the proportion of book shipments not conforming to customer orders. To assess the efficacy of the new program, she obtains another set of sample data. The file P16_22.XLS contains the number of reported nonconforming shipments out of the 75 shipments randomly selected on each of 25 consecutive days *after* the implementation of the quality control program. Based on these data, does the program appear to be meeting the shipping director's goals? Explain why or why not.

23. Construct and interpret a *p* chart for the data provided in the file P16_23.XLS. These data consist of the proportion of defective heating control units found in each of 30 samples with a common size of 150 units.

Is this production process currently in control? Is it behaving "well"?

24. Construct and interpret a *p* chart for the data provided in the file P16_24.XLS. These data consist of the proportion of defective lightbulbs found in each of 30 samples with a common size of 100 units. Is this production process currently in control? Is it behaving "well"?

25. Continuing the previous problem, the managers of this manufacturing process have implemented more stringent quality control procedures to reduce the variability of the lightbulbs' defective rate. The file P16_25.XLS contains the proportion of defective lightbulbs found in each of 30 samples (again, with a common size of 100 units) taken after the implementation of the new quality control procedures. Use these sample observations to assess the impact of the managers' corrective actions? Support your conclusions with a *p* chart.

26. Management of a new credit card company has recorded the number of nonconforming customer bills found in random samples of 150 bills obtained during the first 30 weeks of the firm's operation. These observations are provided in the file P16_26.XLS. Given this sample information, how would you evaluate the performance of this company with respect to the accuracy of produced customer bills? Support your assessment with a relevant control chart.

27. A mail-order clothing retailer is interested in improving the accuracy of its customer service agents who enter customer orders into the firm's computerized record system. To monitor the accuracy of the agents' data entry activities, a manager records the number of data entry errors detected in 150 randomly selected customer orders placed over the course of 1 month (i.e., 30 days). These data are stored in the file P16_27.XLS. Construct a *p* chart for the given sample data.

a. Based on the given sample data, is this data entry process currently in control? Provide support for your conclusion here.

b. What advice, if any, would you give to the managers of this process?

28. Construct and interpret a *p* chart for the data provided in the file P16_28.XLS. These data consist of the proportion of defective radar detectors found in each of 30 samples with a common size of 200 units. Is this production process currently in control? Is it behaving "well"? Comment on any discernible trend(s) in the *p* chart based on the given sample data.

29. Explain the difference in the actions a manager must take on a process when a point on a *p* chart exceeds the upper control limit versus when a point falls below the lower control limit.

Level B

30. Provided that $\bar{p} = 0.01$, determine a sample size large enough to avoid the construction of a *p* chart with a *negative* lower control limit. How does this required sample size change when $\bar{p}$ increases to 0.10?

31. To establish the subsample size *n* for a *p* chart, a probability of 0.95 is specified for finding at least one nonconforming item in any subsample of *n* items. If the process has an average nonconformance rate of 5%, what subsample size *n* should be used? If the process has an average nonconformance rate of 1%, what subsample size *n* should be used? (*Hint:* A binomial random variable *X* can be approximated by a Poisson random variable with parameter $\lambda = np$, where *n* is the subsample size and *p* is the underlying proportion of nonconforming items.)

32. For a fixed subsample size *n*, what values of $\bar{p}$ lead to a positive lower control limit on a *p* chart?

33. Use Excel to simulate Deming's red bead experiment, described in Section 16.5.2 of this text, with *six* workers instead of four. How do the results change when the number of workers is increased to six? What new conclusions, if any, emerge from your revision of this simulation?

Process Capability

Recall that one of the main goals of control charts is to bring a process into control so that it is predictable. In this section we assume that a process is in control, and we predict how *capable* it is of producing outputs that meet specifications. These specifications are typically set outside of a process from considerations of what an "acceptable" product is. For example, a team of engineers might determine that a machined rod can function properly only if its diameter is between 20.80 and 20.95 millimeters. Or a manager might decide that its check processing department is operating within acceptable limits only if check processing times are no more than 30 days. These examples illustrate what we *want* outputs to be. The question then is whether the current process is *capable* of meeting these specifications. When we analyze whether a process is able to meet set specifications, we call it a **process capability analysis.**

<table>
<tr><td>Process Capability</td><td>**Process capability** measures the ability of an in-control process to produce items that meet specifications, which are typically defined by lower and upper specification limits.</td></tr>
</table>

In a process capability analysis we are typically given lower and upper specification limits, denoted *LSL* and *USL,* and we want to calculate the proportion of outputs from a given process that fall within these limits.[6] Based on data generated from the process, we perform a probability calculation to see how capable the current process is of producing outputs within the specification limits. The following example illustrates a typical calculation.

Example 16.5 Checking Whether Diameters of Manufactured Rods Meet Specifications

A manufacturing process produces rods for a mechanical device. Engineers have determined that the diameters of the rods must be between 20.80 and 20.95 millimeters; rods with diameters outside these limits are unusable. As part of the standard control charting the company does, the data in the file RODS.XLS have been collected. Here diameters of six randomly selected rods were measured every half hour for several production shifts. How capable is this process of meeting engineering specifications?

Objective To use control charts to check whether the manufacturing process is in control, and if it is, to use standard statistical procedures to estimate the proportion of rods that meet specs.

Solution

The Data sheet in the RODS.XLS file is set up exactly as we have seen in previous control chart examples. That is, there is a separate column for each observation in the subsamples of size 6. Therefore, we can—and should—examine $\overline{X}$ and R charts as a first step to see whether the current process is in control. If it *isn't* in control, then it lacks the predictability necessary to judge whether it is capable of meeting specifications. Fortunately, control charts show that the current process *is* in control. We show the $\overline{X}$ chart in Figure 16.30 (page 902). Note that its centerline is 20.897 mm, and its lower and upper control limits are 20.867 mm and 20.928 mm. These values indicate how the process *is* operating. They might or might not bear any relationship to how we would *like* it to be operating.

Now that we know the process is in control, hence predictable, we estimate the proportion of rods that fall within the specification limits *LSL* = 20.80 and *USL* = 20.95. One obvious way to do this is to count the number of *observed* rods with diameters within the limits. There are 45(6) = 270 rods in the 45 subsamples, and a simple tally shows that none of the 270 diameters are less than the *LSL*, whereas four are greater than the *USL*. Therefore, the proportion within the limits is 266/270 = 0.985.

[6]In some cases, such as in the check processing example, only one of the limits is relevant. In that example, *USL* = 30, but there is no lower limit of interest.

FIGURE 16.30
$\underline{X}$ Control Chart for
Rod Diameters

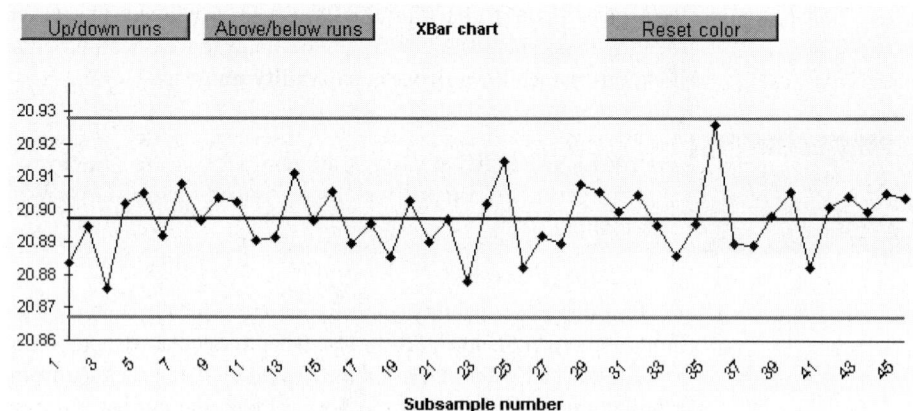

The specification lim-
its, *LSL* and *USL,* are
not shown on the
control chart. This is
the standard—and
preferred—practice.

However, this calculation uses only observed rods. What about other rods the process has been producing and will produce? To answer this question, we use a probability model. We assume the distribution of rod diameters is some standard distribution, typically the normal distribution, and then we do a probability calculation based on the estimated parameters of this distribution. We will illustrate this procedure in a couple of steps.

First, is it reasonable to assume that rod diameters are normally distributed? We check this by creating a histogram of rod lengths. (To use StatPro to do this, we first need to obtain one long variable of 270 diameters. This can be done either manually by copying and pasting or with StatPro's Stack procedure.) The resulting histogram in Figure 16.31 indicates a reasonably bell-shaped distribution of diameters, so that a normal probability model is reasonable. The histogram also indicates that none of the diameters are near the *LSL*, and that only a few are above the *USL*.

Next, we use the normal probability model to calculate the probabilities of falling outside the specification limits. We assume a typical rod has a diameter that is normally distributed with mean and standard deviation equal to the *observed* mean and standard deviation from the sample. These are $\overline{X} = 10.897$ and $s = 0.025$. We then use the NORMDIST function to calculate the probability below the *LSL* and above the *USL*. (See Figure 16.32.) The formulas in cells N11 and N13 are

$$=\text{NORMDIST(LSL,SampMean,SampStdev,1)}$$

and

$$=\text{1-NORMDIST(USL,SampMean,SampStdev,1)}$$

FIGURE 16.31
Histogram of Rod
Diameters

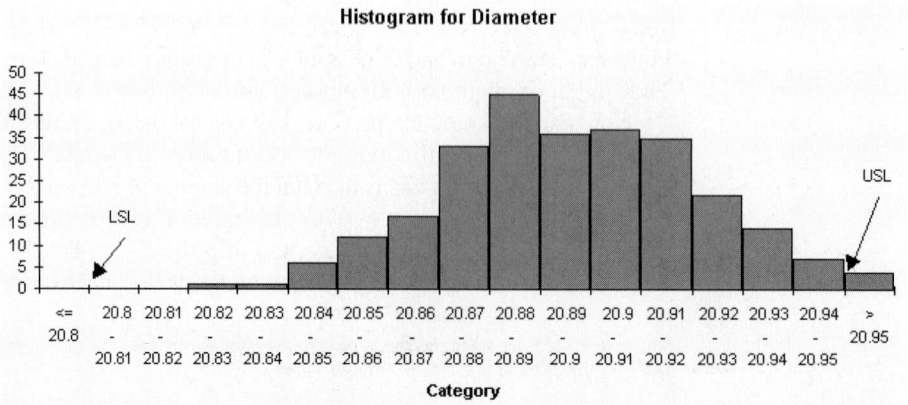

FIGURE 16.32
Capability Analysis
for Rod Diameters

	L	M	N	O	P	Q
3	*Summary measures for selected variables*					
4			Diameter			
5		Mean	20.897			
6		Standard deviation	0.025			
7						
8	**Normal probability calculations**					
9		LSL	20.80			
10		USL	20.95			
11		P(below LSL)	5.966E-05			
12		Per million	60			
13		P(above USL)	0.0181			
14		Per million	18093			
15						
16	**Capability indices**					
17		Cp	0.990			
18		Cpk	0.698			

Named ranges:
SampMean: N5
SampStdev: N6
LSL: N9
USL: N10

We see that there is almost no probability of being below the *LSL*, but the probability of being above the *USL* is just below 0.02. Therefore, slightly more than 98% of rods should meet specifications if the process continues to operate as it is currently operating.

Although the probabilities of not meeting specifications are quite small, it is common to project the results to a *large* number of items. Specifically, the capability of a process is often quoted in parts per million (ppm). We have done this in cells N12 and N14 by multiplying each of the probabilities by 1,000,000. Surprisingly, the extremely small probability in cell N11 still implies that 60 ppm will fall below the *LSL*, and the probability in cell N13 implies that over 18,000 ppm will fall above the *USL*. Perhaps this process isn't as capable as we initially thought! Two other capability indices, denoted C_p and C_{pk}, are also listed in Figure 16.32. We will discuss them in the next subsection. ●

Before continuing, we make one important point about control charts and specification limits. The specification limits, *LSL* and *USL*, should *not* be shown on a control chart. If lines are drawn at these limits, we get the impression that the purpose of the charts is to get all of the points within these limits. Remember, however, that the real purpose of control charts is to show us what the process *is* doing, not what we *want* it to do. As Deming preached, anyone who draws specification limits on a control chart doesn't really understand the purpose of the chart.

16.6.1 Process Capability Indexes

If the outputs from an in-control process are approximately normally distributed with mean μ and standard deviation σ, then we know that almost all of the items produced will be within three standard deviations of the mean, that is, within the interval $\mu \pm 3\sigma$. Note that this interval has length 6σ. On the other hand, we *want* the items to be within the interval from *LSL* to *USL*, an interval of length $USL - LSL$. One way to judge the capability of a process is to compare the lengths of these two intervals. Specifically, the capability index denoted by C_p is defined by equation (16.1).

C_p Capability Index (Used When Process Mean Is on Target)

$$C_p = \frac{USL - LSL}{6\sigma}$$

(16.1)

FIGURE 16.33
Distribution for a
Process with $C_p = 1$

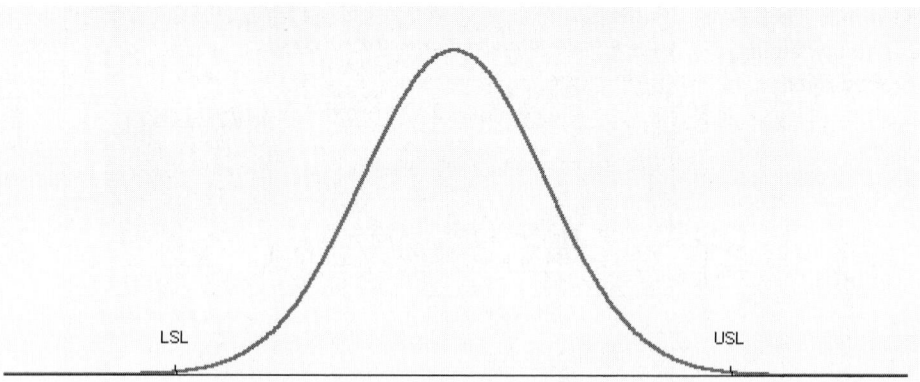

To understand C_p, assume that the ideal output value, called the "target," is halfway between the *LSL* and the *USL*. Also, assume that the current mean μ of the process is equal to the target, and that the distance from the target to either specification limit is 3σ. Then $C_p = 1$, and we have the picture in Figure 16.33. In words, the current process is making just about what needs to be made, and the probability of falling outside the specification limits is only 0.0027 (the probability that a normally distributed random value is more than three standard deviations from the mean). Alternatively, the ppm beyond the specification limits is 2,700 [=0.0027(1,000,000)].

Now consider changes from this "baseline" situation. There are essentially three ways we could make the process more or less capable: (1) we could change the specification limits, *LSL* and *USL*, (2) we could change the mean μ so that it is not equal to the target, and (3) we could increase or decrease the variability in the process, as measured by σ. For the time being, we'll assume that *LSL* and *USL* are fixed because of engineering requirements and that μ continues to equal the target, so that only σ varies.

Figures 16.34 and 16.35 indicate two possible changes in σ. Figure 16.34 represents a *more* variable process, where the distance from the target to either specification limit is only 2σ. The specification limits haven't changed, but σ has increased. In this case $C_p = 4\sigma/6\sigma = 0.667$, and the probability of falling outside the specification limits is 0.045392 (45,392 ppm). On the other hand, Figure 16.35 represents a *less* variable process, where the distance from the target to either specification limit is now 4σ. Again, the specification limits haven't changed, but σ has decreased. In this case $C_p = 8\sigma/6\sigma = 1.333$, and the probability of falling outside the specification limits is 0.000064 (64 ppm). Clearly,

FIGURE 16.34
Distribution for a
Process with $C_p = 0.667$

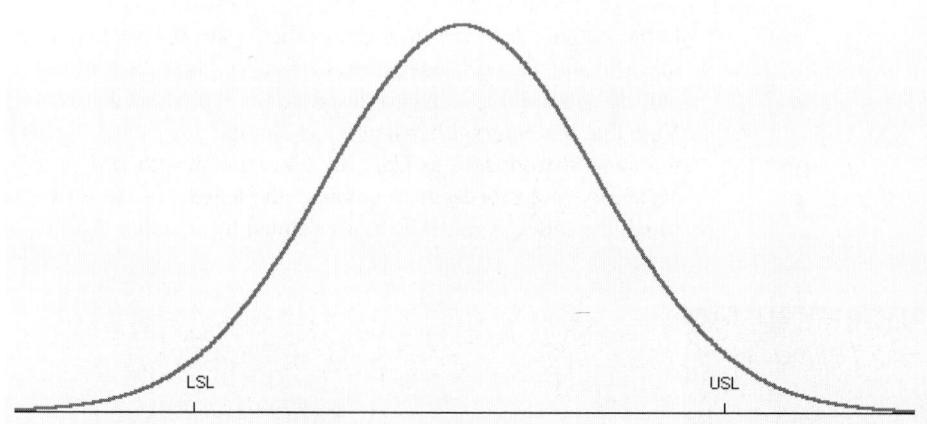

FIGURE 16.35
Distribution for a
Process with C_p =
1.333

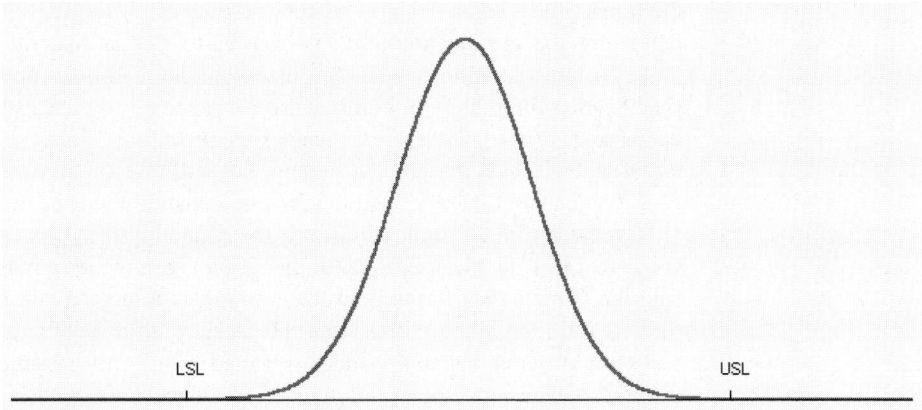

the relationship of σ to the distance between the specification limits is crucial in determining the capability of the process, and we want C_p to be as large as possible. Most world-class manufacturers attempt to achieve a C_p of at least 1.333, and many even try to improve on this.

When the target is midway between the specification limits, the process mean is equal to the target, and the process is normally distributed, it can be shown that the probability of falling outside the specification limits is

$$P(\text{beyond specification limits}) = 2P(Z < -3C_p) \qquad \textbf{(16.2)}$$

where Z is normal with mean 0 and standard deviation 1. We use equation (16.2) in Figure 16.36 to show the effect of C_p. (See the file CAPINDEX.XLS.) We treat C_p in cell B10 (range-named Cp) as an input. The formula in cell B12 is

$$=2*NORMSDIST(-3*Cp)$$

We then create the data table in columns D–F. Clearly, the ppm outside the specification limits decreases dramatically as C_p increases.

An equivalent way of thinking about C_p is by considering its reciprocal. For example, suppose that C_p = 1.333, so that its reciprocal is 3/4. Specifically, let's assume that LSL = 92, USL = 108, μ = 100 (the target), and σ = 2. (Check that C_p = 1.333 for these parameters.) Then we know that the interval $\mu \pm 3\sigma$, which contains almost all of

FIGURE 16.36

Effect of C_p on ppm
Beyond Specification Limits

	A	B	C	D	E	F
1	**Cp index**					
2						
3	**Assumptions:**					
4	Spec limits are fixed					
5	Target is midway between spec limits					
6	Process mean equals target					
7	Process distribution is normal					
8						
9	**Typical calculation**			Data table		
10	Cp	1			Cp P(beyond)	ppm
11					0.0027	2700
12	P(beyond spec limits)	0.0027		0.333	0.3178	317795
13	ppm beyond spec limits	2700		0.667	0.0454	45392
14				1	0.0027	2700
15				1.333	0.0001	64
16				1.667	0.0000	1
17				2	0.0000	0

the items produced, is from 94 to 106, an interval of length 12. In contrast, the interval defined by the specification limits is from 92 to 108, an interval of length 16. Therefore, the $\pm 3\sigma$ interval within which the process outputs fall takes up only $12/16 = 3/4$ of the specification-limit interval. In this sense the process is quite capable. Of course, if we can decrease σ, then C_p will increase and its reciprocal will decrease. Therefore, the $\pm 3\sigma$ interval within which the process outputs fall will take up an even smaller fraction of the specification-limit interval, and the process capability will be even greater.

Essentially, the C_p index indicates the capability of a process if it is centered properly—that is, if its mean is equal to the target. Then the only reason for parts falling outside the specification limits is excess variation—a large σ—in the process. But what if there is variation in the process *and* the process mean is off target? In this case we need a slightly different capability index, denoted by C_{pk}, to measure how close the process mean is to the nearest specification limit. We concentrate on the *nearest* specification limit because this is where most of the problems are likely to occur. The "baseline" case is now where the distance from the process mean to the nearest specification limit is 3σ, for in this case the process will be producing just about what we want it to produce. We define C_{pk} so that it is equal to 1 in this baseline case, as shown in equation (16.3).

C_{pk} Capability Index (Used When Process Mean Is Not on Target)	$$C_{pk} = \min\left\{\frac{USL - \mu}{3\sigma}, \frac{\mu - LSL}{3\sigma}\right\}$$	(16.3)

An illustration of C_{pk} similar to that for C_p in Figure 16.36 appears in Figure 16.37. (This is also in the CAPINDEX.XLS file.) Here we assume that the process is nearer to *USL* than to *LSL*. Then it can be shown that the probability of being beyond *USL* is $P(Z > 3C_{pk})$. Therefore, for any trial value of C_{pk} in cell B10 (range-named Cpk), we enter the formulas

$$=1\text{-NORMSDIST}(3*\text{Cpk})$$

and

$$1000000*\text{B12}$$

FIGURE 16.37

Effect of C_{pk} on ppm Beyond the Nearest Specification Limit

	A	B	C	D	E	F
1	**Cpk index**					
2						
3	**Assumptions:**					
4	Spec limits are fixed					
5	Target is midway between spec limits					
6	Process mean doesn't equal the target (here we'll assume it's closer to USL than to LSL)					
7	Process distribution is normal					
8						
9	**Typical calculation**			**Data table**		
10	Cpk	1		Cpk	P(above USL)	ppm above USL
11					0.00135	1350
12	P(above USL)	0.00135		0.333	0.15890	158897
13	ppm above USL	1350		0.667	0.02270	22696
14				1	0.00135	1350
15				1.333	0.00003	32
16				1.667	0.00000	0
17				2	0.00000	0
18						
19						
20						
21				Because we're assuming the mean is closer		
22				to USL than to LSL, P(below LSL) and ppm		
23				below LSL are even smaller than these values.		

in cells B12 and B13. Using these, we form a data table in columns D–F to show how the process capability varies as C_{pk} varies. Of course, this shows only half of the story, the probability of being beyond specifications on the *high* side. But the probability of being beyond specifications on the *low* side is even smaller, since we assumed the process mean is closer to *USL* than to *LSL*.

To illustrate C_{pk} with data, we look again at Figure 16.32 from the example on rod diameters. Since *LSL* = 20.80 and *USL* = 20.95, the target is their midpoint, 20.875. However, the process mean, estimated by $\overline{X}$ = 20.897, appears to be closer to the *USL*. Therefore, it is likely that most rods beyond the specification limits will be *above* the *USL*. We estimate σ for this process with the sample standard deviation, s = 0.025. Then the (estimated) C_{pk} is

$$C_{pk} = \frac{USL - \overline{X}}{3s} = \frac{20.95 - 20.897}{3(0.025)} = 0.698$$

This is considerably below the baseline case where C_{pk} = 1, so that, as we see in Figure 16.32, the current process has a fairly large ppm beyond specifications—almost all on the high side.

If the C_{pk} is unacceptably small—and again, most world-class manufacturers try to achieve a value of at least 1.333—then there are two possibilities.[7] First, we could try to "center" the process by adjusting the process mean to the target. In this case C_p and C_{pk} coincide. If we could do this for the rod example, we could achieve a C_p value of 0.990, as shown in cell N17 in Figure 16.32. This would be much more acceptable than the current process. Alternatively, we could try to reduce the process variation, with or without a shift in the mean. By reducing σ, we automatically reduce C_{pk} (and C_p), regardless of whether the mean is on target.

Both C_p and C_{pk} are simply *indexes* of process capability. The larger they are, the more capable the process is. An equivalent descriptive measure is the "number of sigmas" of a process. A *k*-**sigma process** is one for which the distance from the process mean to the nearest specification limit is $k\sigma$. For example, a 3-sigma process is one where C_{pk} = 1, since this is exactly how C_{pk} is defined. (In case the process mean is on target, C_p also equals 1 for a 3-sigma process.) As we will discuss in more detail below, Motorola has become famous for its 6-sigma processes, where C_{pk} = 2. This is remarkable quality! It implies that almost *no* items out of specifications per million items are produced. This is because an item is out of specifications in a 6-sigma process only if it is beyond six standard deviations from the mean, an extremely unlikely event. Motorola and other world-class companies have achieved this by reducing variation to a bare minimum—and by continually searching for ways to reduce it even further.

k-sigma Process	A *k*-**sigma process** is one in which the distance from the process mean to the nearest specification limit is $k\sigma$. Here, σ is the standard deviation of the process.

We can summarize the ideas in this section as follows.

1. The C_p index is appropriate for processes in which the mean is equal to the target value (midway between the specification limits). Processes with C_p = 1 produce about 2700 out-of-specification items per million, but this number decreases dramatically as C_p increases.

---

[7]This assumes that changing the specification limits is *not* an option.

2. The C_{pk} index is appropriate for all processes, but it is especially useful when the mean is off target. (In case the mean is on target, C_p and C_{pk} are equivalent.) Processes with $C_{pk} = 1$ produce about 1350 out-of-specification items per million on the side nearest the target (and less on the other side), and again this number decreases dramatically as C_{pk} increases.

3. Both C_p and C_{pk} are only indices of process capability. However, they imply the probability of an item being beyond specifications (and the ppm beyond specifications), as illustrated in Figures 16.36 and 16.37.

4. A 3-sigma process has $C_{pk} = 1$, whereas a 6-sigma process has $C_{pk} = 2$. In general, the distance from the process mean to the nearest specification limit in a k-sigma process is $k\sigma$. Quality improves dramatically as k increases.

16.6.2 More on Motorola and 6-Sigma

We defined a k-sigma process as one where the distance from the target to the nearest specification limit is $k\sigma$. Until the 1990s most companies were very content to achieve a 3-sigma process, that is, $C_p = 1$. Assuming that each part's measurement is normally distributed, they reasoned that 99.73% of all parts would be within specifications. Motorola questioned this wisdom on two counts:

- Products are made of many parts. The probability that a product is acceptable is the probability that *all* parts making up the product are acceptable.

- When using control charts to monitor quality, shifts of 1.5 standard deviations or less in the process mean are difficult to detect. Therefore, when we are producing a product, there is a reasonable chance that the process mean will shift by as much as 1.5σ up or down without being detected (at least in the short run).

Given that the process mean might be as far as 1.5σ from the target and that a product is made up of *many* parts, a 3-sigma process might not be as good as we originally stated. Just how good is it?

Suppose a product is composed of m parts. We will calculate the probability that all m parts are within specifications when the process mean is 1.5σ above the target and the distance from the target to either specification limit is $k\sigma$.[8] That is, we are considering a k-sigma process with a process mean off center by an amount 1.5σ. Let X be the measurement for a typical part, and let p be the probability that X is within the specification limits, that is, $p = P(LSL < X < USL)$. If p_m is the probability that all m parts are within the specification limits, then assuming that all parts are identical and probabilistically independent, the multiplication rule for probability implies that $p_m = p^m$.

To calculate $p = P(LSL < X < USL)$, we need to standardize each term inside the probability by subtracting the process mean μ and dividing the difference by σ. Let T be the target. Then we have $LSL = T - k\sigma$ and $USL = T + k\sigma$ (since the process is a k-sigma process) and $\mu = T + 1.5\sigma$ (since the mean has shifted upward by an amount 1.5σ). Therefore, the standardized specification limits are

$$\frac{LSL - \mu}{\sigma} = \frac{(T - k\sigma) - (T + 1.5\sigma)}{\sigma} = -k - 1.5$$

and

$$\frac{USL - \mu}{\sigma} = \frac{(T + k\sigma) - (T + 1.5\sigma)}{\sigma} = k - 1.5$$

[8]The case where the process mean is 1.5σ *below* the target is completely analogous.

This implies equation (16.4), where Z is normal with mean 0 and standard deviation 1.

$$p = P(-k - 1.5 < Z < k - 1.5) = P(Z < k - 1.5) - P(Z < -k - 1.5) \quad (16.4)$$

We can easily implement this in Excel, as shown in Figure 16.38. (See the file MULTPART.XLS.) Equation (16.4) is implemented in cell B10 with the formula

$$=NORMSDIST(A10-1.5)-NORMSDIST(-A10-1.5)$$

and this probability is raised to the 10th, 100th, and 1000th powers in cells C10 to E10. All of this is then copied down for other values of k. As we see, a 3-sigma process (row 11) is not all that great. Almost 7% of its individual parts are out of specifications, about half of its 10-part products are out of specifications, and almost *all* of its 100-part and 1000-part products are out of specifications. In contrast, a 6-sigma process (row 14) is extremely capable, with only 0.34% of its 1000-part products out of specifications. No wonder Motorola's 5-year goal (as of 1992) was to achieve "6-sigma capability in everything we do."

By the way, if 1 minus the probability in cell B14 is multiplied by 1,000,000, the result is 3.4. (The *exact* value in cell B14 is 0.9999966; its format does not show it, however.) This value has become very well known in the quality world. It says that if a process is a 6-sigma process with the mean off target by an amount 1.5σ, then the process will produce only 3.4 ppm out of specifications. Again, this is remarkable quality!

The previous analysis shows how we can calculate the capability of a process if we *know* it is a k-sigma process for any specific k. We conclude this section by asking a slightly different question. If a company has produced many parts and has observed a certain fraction to be out of specifications, what is the estimated value of k? For example, suppose that after monitoring thousands of gaskets produced on its machines, a company has observed that 0.545% of them are out of specifications. Is this company's process a 3-sigma process, a 4-sigma process, or what?

To answer this question, we again assume a "worst-case" scenario where the mean is above the target by an amount 1.5σ. Then, from equation (16.4), we know that the probability of being within specifications is

$$p = P(Z < k - 1.5) - P(Z < -k - 1.5)$$

if the process is a k-sigma process. However, we now know p from observed data, and we want to estimate k. This can be done with Excel's Goal Seek tool, as shown in Figure 16.39. (See the file KSIG.XLS.) First, we enter the observed fractions in and out of

FIGURE 16.38
Probability of Multipart Products Meeting Specifications

	A	B	C	D	E
1	Process capability for multiple-part products				
2					
3	Assumptions:				
4	Each product has m identical, probabilistically independent parts				
5	The process is a k-sigma process				
6	The process mean is 1.5 stdevs above the target				
7					
8	Calculations				
9	k	p	p_{10}	p_{100}	p_{1000}
10	2	0.69123	0.02490	0.00000	0.00000
11	3	0.93319	0.50084	0.00099	0.00000
12	4	0.99379	0.93961	0.53638	0.00197
13	5	0.99977	0.99768	0.97700	0.79239
14	6	1.00000	0.99997	0.99966	0.99660

FIGURE 16.39
Finding *k* for a
k-sigma Process

	A	B	C
1	**Finding k for a k-sigma process**		
2			
3	Assumption:		
4	Mean is 1.5 sigmas above target		
5			
6	Observed fractions		
7	Out of specs	0.00545	
8	Within specs	0.99455	
9			
10	Trial value of k	4.077	
11			
12	P(within specs)	0.99501	

FIGURE 16.40
Goal Seek Dialog
Box Settings

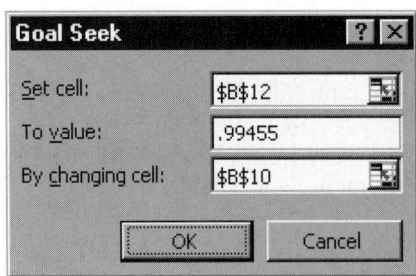

specifications in cells B7 and B8. Next, we enter *any* trial value of *k* in cell B10 and use it to calculate the probability of being within specifications in cell B12 with the formula

$$=\text{NORMSDIST}(B10-1.5)-\text{NORMSDIST}(-B10-1.5)$$

Finally, we use the Tools/Goal Seek menu item and fill out the dialog box as in Figure 16.40. It immediately shows that this process is slightly better than a 4-sigma process.[9]

PROBLEMS

Level A

34. The file P16_01.XLS contains data on the amount of soda (in ounces) placed in aluminum cans by a particular filling process. The filling process should place between *LSL* = 11.95 ounces and *USL* = 12.05 ounces of soda in each can. Every hour, 4 cans of soda were randomly selected from the production process and measured for amount of fill. This was repeated for 25 consecutive hours.
  a. Based on the given sample data, calculate the probabilities that this process will yield soda cans (i) falling below the lower specification limit and (ii) exceeding the upper specification limit.
  b. Represent your results found in part **a** in parts per million (ppm).

  c. Estimate C_p and C_{pk} in this case. If a difference exists between these two capability indexes, explain it.

35. The data in the file P16_07.XLS consist of 25 subsamples of 4 observations each on the lengths of particular bolts manufactured for use in large aircraft. The target length of these bolts is 37 centimeters. Furthermore, the bolt manufacturer has established *LSL* = 36.95 centimeters and *USL* = 37.05 centimeters.
  a. Estimate C_p and the probability of meeting specifications. Evaluate C_p in this case. Is this a highly capable production process?
  b. Estimate C_{pk} in this case. Does your estimate of C_{pk} differ from that of C_p computed in part **a?** If so, how do you explain this difference?

[9]Note that Goal Seek's solution in cell B12 doesn't match the probability in cell B8 exactly, but it is close enough for all practical purposes.

36. Management of a local bank is interested in assessing the process used in opening new checking accounts for bank customers. In particular, management would like to examine the time required to process a customer's request to open a new checking account. Currently, managers believe that it should take about 7 minutes to process such a request. Furthermore, they believe that the time required to process this type of request should be between $LSL = 5.5$ minutes and $USL = 8.5$ minutes. The file P16_12.XLS contains the time required to process new checking account requests for each of 6 customers selected randomly on a given day. A different subsample of 6 customer requests was collected on each of 25 days.

    a. Estimate C_p and the probability of meeting specifications. Evaluate C_p in this case. Is this a highly capable production process?

    b. Estimate C_{pk} in this case. Does your estimate of C_{pk} differ from that of C_p computed in part **a?** If so, how do you explain this difference?

37. A computer printout shows that a certain process has a C_p of 1.50 and a C_{pk} of 0.80. Assuming that this process is in control, what do these two index values indicate about the capability of the process?

Level B

38. For a given process, can C_{pk} ever exceed C_p? Provide a mathematical and/or verbal argument to support your answer.

39. Suppose that a product is composed of 25 identical and probabilistically independent parts. Assume that this product is manufactured through the use of a k-sigma process with a process mean that is 1.25 standard deviations above the target mean. Assuming that $k = 3$, what proportion of these multipart products are *not* within specification limits? Answer this question again for the case where $k = 6$. Explain the difference between these two probabilities.

40. Suppose that a product is composed of m identical and probabilistically independent parts. Assume that this product is manufactured through the use of a 6-sigma process with a process mean that is 1.5 standard deviations below the target mean. Assuming that $m = 5$, what proportion of these multipart products are *not* within specification limits? Answer this question again for the cases where $m = 50$ and $m = 500$. Explain the differences among your three computed probabilities.

41. Suppose that after monitoring a large number of electronic computer chips, a manufacturer observes that 1% of them are out of specifications. Assume that the process mean exceeds the target mean by 1.25 standard deviations. Given this information, it is possible to conclude that this manufacturer is employing a k-sigma process. What is k approximately?

42. Suppose that after monitoring thousands of ball bearings, a manufacturer observes that 0.27% of them are out of spec. Assume that the process mean is 1.5 standard deviations below the target mean. Given this information, it is possible to conclude that this manufacturer is employing a k-sigma process. What is k approximately?

 # Conclusion

Some critics have claimed that the quality movement, with all of its acronyms, is a fad that will eventually lose favor in the business world. We do not believe this is true. In the past decade or two, many companies in the United States and abroad have embraced the teachings of Deming and others to gain a competitive advantage with superior quality. By now, quality has improved to such a level in many industries that superior quality no longer ensures a competitive advantage; it is a prerequisite for staying in business!

In this chapter we have discussed two quantitative tools from the quality movement: control charts and process capability analysis. It is clear from Deming's 14 points that there is much more to achieving quality than crunching numbers—for example, good management and worker training are crucial—but the use of proven statistical techniques is key to any program of continual improvement. The fanfare surrounding the quality movement may indeed die down in the future, but the careful monitoring of processes, together with the use of the statistical tools we have discussed, will still be required elements for companies that want to remain competitive.

Summary of Key Terms

Term	Symbol	Explanation	Excel	Pages	Equation Number
Statistical process control (SPC)		A set of statistical tools, including control charts, for monitoring quality.		867	
Deming's 14 points		A list of rules espoused by Deming for creating quality improvements in manufacturing and service operations.		868	
Assignable cause variability		The inherent variability in an in-control process.		872	
In-control process		Process that is predictable and is subject only to common cause variability.		872	
Special cause variability		The extra variability observed when a process goes out of control.		872	
Out-of-control process		Process that is unpredictable because of special cause variability.		872	
Control chart		Graphical device for monitoring a process to see whether and when it goes out of control.	StatPro/ Quality Control	873	
Subsample		A small set of observations (usually 3 to 6) from a process taken at approximately the same moment in time.		874	
$\bar{X}$ chart		Plot of subsample averages through time, used to monitor the process mean.	StatPro/ Quality Control/X-bar, R Charts	874	
R chart		Plot of subsample ranges (maximum minus minimum) through time, used to monitor process variability.	StatPro/ Quality Control/ X-bar, R Charts	874	
Lower and upper control limits	LCL, UCL	Lines on a control chart, based on the observed data, typically three standard deviations below and above the centerline.	StatPro/ Quality Control	876	
Average run length	ARL	Mean time until an out-of-control point occurs on a control chart; reciprocal of the probability of an out-of-control point.		879	
Other out-of-control conditions		Excessive runs in one direction or on one side of the centerline, or sequences satisfying "zone" rules that would be unexpected by a process in control.	StatPro/ Quality Control	880	

Summary of Key Terms (continued)

Term	Symbol	Explanation	Excel	Pages	Equation Number
Rational subsample		A subsample such that its observations are subject to the same common cause variability.		882	
Funnel experiment		Deming's experiment used to illustrate the harmful effects of tampering with an in-control process.		884	
p chart		Plot of proportions of nonconforming (defective) items through time.	StatPro/ Quality Control/ P Chart	892	
Red bead experiment		Deming's experiment used to illustrate how workers' performance is limited by the system.		896	
Process capability		Measures the ability of an in-control process to produce items that meet specifications.		900	
Lower and upper specification limits	LSL, USL	Limits that prescribe where measurements must fall for items to function as they should.		901	
Process capability index	C_p	Index of process capability, used when the process mean is on target; the larger the better.	Must be calculated manually	903	16.1
Process capability index	C_{pk}	Index of process capability, used when the process mean is not on target; the larger the better.	Must be calculated manually	906	16.3
k-sigma process		Process in which the distance to the nearest specification limit is $k\sigma$; the larger k is, the better.		907	
Motorola 6-sigma		The goal of Motorola and other companies to produce no more than 3.4 nonconforming parts per million.		908	

PROBLEMS

Level A

43. To monitor product quality, Wintel inspects 30 chips each week and determines the fraction of defective chips. The resulting data are in the file P16_43.XLS. Construct a P chart based on 12 weeks of data. Is week 5 out of control? Why or why not? Is week 6 out of control? Why or why not? What do these charts tell you about Wintel's production process?

44. Eleven samples of size 3 were taken in an effort to monitor the voltage held by a regulator. The data are

in the file P16_44.XLS. A regulator meets specifications if it can hold between 40 and 60 volts.
 a. Construct $\bar{X}$ and R charts for the data. Is the process in control?
 b. Estimate C_p and the probability of meeting specifications.

45. You are the manager of a hospital emergency room. You are interested in analyzing the time patients wait to see a physician. For 25 samples of 5 patients each, the file P16_45.XLS contains the time each patient had to wait before seeing a physician. Construct and

interpret $\overline{X}$ and R charts for this situation. Also answer the following questions:
  a. We are 68% sure that a patient will wait between what two values (in minutes) to see a physician?
  b. We are 95% sure that a patient will wait between what two values (in minutes) to see a physician?
  c. We are 99.7% sure that a patient will wait between what two values (in minutes) to see a physician?

46. A mail-order company (Seas Beginning) processes 100 invoices per day. For each of several days, they have kept track of the number of invoices that contain errors. The data are in the file P16_46.XLS. Use these data to construct a p chart and then interpret the chart.

47. The file P16_47.XLS contains the measured diameters (in inches) reported by the production foreman of 500 rods produced by Rodco. A rod is considered acceptable if it is at least 1.0 inch in diameter. In the past the diameter of the rods produced by Rodco has followed a symmetric distribution.
  a. Construct a histogram of these measurements.
  b. Comment on any unusual aspects of the histogram.
  c. Can you guess what might have caused the unusual aspect(s) of the histogram? (*Hint:* One of Deming's 14 points is to "Drive Out Fear.")

48. John makes 20 computers per day for Pathway computer. Production data appear in the file P16_48.XLS. Construct a control chart based on the number of defective computers produced during each of the last 30 days. Explain as fully as possible what you learn from this control chart.

49. For the data in the file P16_49.XLS, suppose $USL = 1.06$ inches and $LSL = 0.94$ inch.
  a. Is the process in control? If it is out of control, describe any observed out-of-control pattern.
  b. If possible, estimate C_p, C_{pk}, and the probability of meeting specifications.

50. For the data in the file P16_50.XLS, suppose that $LSL = 190$ and $USL = 210$.
  a. Is the process in control? If it is out of control, describe any observed out-of-control pattern.
  b. If possible, estimate C_p, C_{pk}, and the probability of meeting specifications.

51. For the data in the file P16_51.XLS, suppose that $LSL = 195$ and $USL = 205$.
  a. Is the process in control? If it is out of control, describe any observed out-of-control pattern.
  b. If possible, estimate C_p, C_{pk}, and the probability of meeting specifications.

52. Consider a k-sigma process.
  a. If $k = 4$, determine the fraction of all parts that meet specifications and calculate C_p. Now suppose that a car consists of 1000 parts, each of which is governed by a 4-sigma process. What fraction of all cars will be perfect (meaning all 1000 parts meet specs)?

  b. Repeat part **a** for a 6-sigma process. Now can you see why companies like Motorola are not satisfied with anything less than a 6-sigma process?

53. A part is considered within specifications if its tensile strength is between 180 and 200. For 20 straight hours the tensile strength of 4 randomly chosen parts was measured. The data are in the file P16_53.XLS.
  a. Is the process in control?
  b. What is C_p?
  c. What is the probability that the specifications will be met on a typical part?

54. Twelve samples of size 4 were taken in an effort to monitor the voltage held by a regulator. The data are in the file P16_54.XLS. A regulator meets specifications if it can hold between 18 and 56 volts. Estimate C_p and the probability of meeting specifications.

55. Suppose that the employees of D&D's each service 100 accounts per week. The file P16_55.XLS contains the number of accounts that each employee "messed up" during the week. Do these data indicate that Jake should receive a raise and Billy should be fired? Discuss.

Level B

56. Continuing the previous problem, the file P16_56.XLS contains 2 more weeks of data for D&D's.
  a. By plotting each employee's weekly fraction of "mess ups," does it appear that Amanda is a problem? Discuss.
  b. Now combine the 2 weeks of data for each employee and answer the question in part **a**.

57. A company has a 3-sigma process. What must the company do to change it to a 6-sigma process? Be as specific as possible.

58. For Ford to designate a supplier as Q-1 (its highest designation), Ford requires that the supplier have a C_p equal to 1.33. Currently your firm has $C_p = 1$. What must you do to increase your C_p to 1.33? Be as specific as possible.

59. How does continuous improvement manifest itself on a p chart? What about on $\overline{X}$ and R charts?

60. A screw manufacturer produces screws that are supposed to be 0.125 inch in diameter. A screw is deemed satisfactory if its diameter is between 0.124 inch and 0.126 inch. The company quality manager therefore uses an $\overline{X}$ chart with a centerline of 0.125 inch, a UCL of 0.126 inch, and an LCL of 0.124 inch. Why is this incorrect?

61. A sudden change in a particular production process has *lowered* the process mean by 1 standard deviation. It has been determined that the weight of the product being measured is approximately normally distributed. Furthermore, it is known that the recent change had virtually no effect on the variability of this process.

What proportion of points is expected to fall outside the control limits on the $\bar{X}$ chart if the subsample size is 4? Compute this proportion again for the case where the subsample size is 9. Provide an explanation for the difference between these two proportions.

62. The SteelCo company manufactures steel rods. The specification limits on the lengths of these rods are from 95.6 inches to 95.7 inches. The process that produces these rods currently yields lengths that are normally distributed with mean 95.66 inches and standard deviation 0.025 inch.
    a. What is the probability that a single rod will be within specification limits?
    b. What is the probability that at least 90 of 100 rods will be within specification limits?
    c. SteelCo's best customer currently buys 200 of these rods each day and pays the company $20 apiece. However, it gets a $40 refund for each rod that doesn't meet specifications. What is SteelCo's current expected profit per day? How small would its standard deviation need to be before it would net an expected $3900 per day?

63. Continuing the previous problem, suppose that SteelCo can pay money to reduce the standard deviation of the process. It costs e^{1000d} dollars to reduce the standard deviation from 0.025 to 0.025 − d. (This reflects the fact that small reductions are fairly cheap, but large reductions are quite expensive.) If the company wants to make sure that at least 99% of all rods meet specifications, how much will it have to spend? (Remember that you evaluate e^x in Excel with the EXP function.)

64. We reconsider SteelCo from the previous two problems from a different point of view. Now we assume SteelCo doesn't know its process mean and standard deviation, so it uses sampling. The file P16_64.XLS lists 150 randomly sampled rod lengths.
    a. Calculate a 95% confidence interval for the population mean length of all rods produced.
    b. Continuing part a, find a 95% confidence interval for the population proportion that meet the specifications listed at the top of the spreadsheet (the same as in the previous problem).
    c. Using the sample standard deviation found in part a as a best guess for the population standard deviation, how large a sample size is required to achieve a 95% confidence interval for the mean of the form "point estimate plus or minus 0.002"?

65. Simulation is useful to see how long it might take before an out-of-control condition is recognized by an $\bar{X}$ or R control chart. Proceed as follows:
    a. Generate 30 subsamples of size $n = 5$ each, where each observation comes from a normal

distribution with a given mean μ and standard deviation σ. (You can choose μ and σ.) Then "freeze" these values (with the Copy and Paste-Special/Values commands), and form $\bar{X}$ and R control charts based on all 30 subsamples.
    b. Below the subsamples in part a, generate 30 more subsamples of size $n = 5$ each, where each observation comes from a normal distribution with mean $\mu + k_1\sigma$ and standard deviation $k_2\sigma$. That is, the mean has shifted by an amount $k_1\sigma$, and the standard deviation has been multiplied by a factor k_2. The values k_1 and k_2 should be entered as input parameters that you can change easily. Initially, set $k_1 = 1.5$ and $k_2 = 1$, although you can try other values in a follow-up sensitivity analysis. Do *not* freeze the observations in these 30 new subsamples. Create new $\bar{X}$ and R control charts that plot all 60 subsamples but have the *same* control limits from part a. By pressing the F9 key and/or changing the values in the k_1 and k_2 input cells, you change the behavior of the control charts.
    c. Write up your results. In particular, indicate how long it takes for the control charts to realize that the process is out of control with respect to the *original* control limits, and how this depends on k_1 and k_2.

66. Are all "capable" processes the same? Consider the data in the file P16_66.XLS. The data come from two processes that produce the same type of part. These parts should be within the specification limits 10.45 inches to 10.55 inches, with a target of 10.50 inches. Are both processes capable of staying within the specification limits? If you were a manufacturer and had to select one of these processes as a supplier of parts, which would you choose? Why?

67. Continuing the previous problem, one of the leading quality gurus, Genichi Taguchi of Japan, suggested the idea of a **quadratic loss function** when judging quality. Rather than saying that a part is "good" when its measurement falls within specification limits and is "bad" otherwise, the quadratic loss function estimates the part's quality as $(x - T)^2$, where x is the part's measurement and T is the target measurement.
    a. Using this loss function, estimate the average quality of parts from process 1, given the data in the file P16_66.XLS and a target of 10.50 inches. Do the same for process 2. Which process appears to be better?
    b. Explain intuitively why a quadratic loss function might be preferable to a simple 0–1 function (where a part either meets specifications or it doesn't) when assessing quality.

68. A manufacturer supplies a certain type of assembly to customers. The manufacturer recognizes the advantages of control charts and uses them consistently. In the past month, its $\bar{X}$ and R charts for the assembly in-

dicated a process well within control. For these charts, 100 subsamples of size $n = 5$ each were used. In addition, the manufacturer compared the 100 $\bar{X}$'s to the specification limits set by one of its customers. Only 2 out of these 100 averages were outside the specification limits. This was good news because the customer was willing to accept orders with no more than 5% out of specifications. However, the manufacturer was shocked when the customer rejected an order for 1000 assemblies. The customer claimed that it inspected 50 of the 1000 assemblies, and 10% (i.e., 5) of them were out of specifications. What is going on here? Is it likely that the customer could see so many bad assemblies, given what the manufacturer observed in its process? Perform appropriate calculations, and write up your results in a short report. Make whatever assumptions you deem relevant.

69. A type of assembly is produced by gluing 5 identical wafers together in a sandwich-like arrangement. The critical dimension of this assembly is its width, the sum of the widths of the 5 wafers. The specifications for the assembly width are from 4.95 inches to 5.05 inches, with a target of 5.00 inches. The manufacturer wants at least 99.5% of the assemblies to meet specifications. Based on a lot of evidence, the individual wafer widths are normally distributed with mean 1 inch and some standard deviation σ.

a. One engineer, Bob Smith, argues that the company should try to achieve a value of σ (through appropriate changes in the process) such that 99.5% of all *individual* wafer widths are between 0.99 inch and 1.01 inches. What value of σ is necessary to achieve this?

b. A second engineer, Ed Jones, argues that Smith is solving the wrong problem. Instead, he says, the company should try to achieve a value of σ such that the *sum* of 5 normally distributed random values has only a 0.5% chance of falling outside the interval from 4.95 inches to 5.05 inches. What value of σ is necessary to achieve this?

c. Which of the engineers is solving the "correct" problem? If the company follows the advice of the wrong engineer, will it err on the high side (too many assemblies out of specifications) or the low side? Is there any disadvantage to erring on the low side?

16.1 The Lamination Process at Intergalactica[10]

Charlie Hobbs, production manager of the 120-employee Rock Isle Plant of the Plastron Division of Intergalactica Chemicals Ltd., stared at the tables of reject data his people had gathered over the last 3 months. Hobbs wanted to use these data to help justify the purchase of a new lamination press. Past proposals for a new press had been rejected by Intergalactica Corporate Finance. But now that Intergalactica had embarked on a new Total Quality Management program, perhaps some fancy quality statistics would dazzle the people at headquarters.

The Plastron Division with $32 million in sales and $8 million in net revenues is the smallest and most profitable of Intergalactica Chemicals Ltd.'s 14 business units. Plastron manufactures a variety of laminated resin-impregnated paper products. Its lead product is a laminate board, trademarked under the name "Plastfoam," that has a wide variety of applications in commercial art and photography. The product is a very high-quality hard-surfaced, rigid but lightweight, multilayered polystyrene foam core board that is used for direct photographic and lithographic printing as well as for mounting of displays, photographs, and the like. Plastfoam facing is made by first saturating a special heavy absorbent paper in a proprietary blend of plastic resins, and then drying the paper to form a rigid yet somewhat rough-surfaced material. Next, these rough-faced resin-impregnated sheets are pressed under heat (calendered) to form a very smooth and rather hard material similar to the well-known "Formica" surface that is frequently used as countertops. Then, two of these smooth faces are glued to the top and bottom of a polystyrene foam core to form a "sandwich." The final step in the production process is edge trimming and packaging. The resulting Plastfoam product is considered the "Cadillac" of the industry and commands a premium price that makes it Plastron's most profitable product line by far. As noted earlier, the Plastron Division also manufactures a va-

riety of resin-impregnated paper-base materials that are used as facings in some economy brands of furniture. These products share some of the same production facilities and personnel with the Plastfoam product. Production scheduling and product changeovers are serious issues in the plant.

Key purchased input materials include the specialty absorbent paper that gets impregnated, several types of resins that Plastron blends to create its unique recipe for the saturation process, the large styrofoam blocks that are cut to size in-house, glues, and dyes. A rough schematic of the main production processes is shown in Figures 16.41 and 16.42. The physical production facilities employed include two resin mixing tanks, the paper saturator—similar in appearance

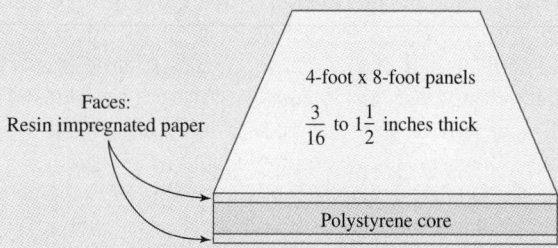

FIGURE 16.41 Plastron Product

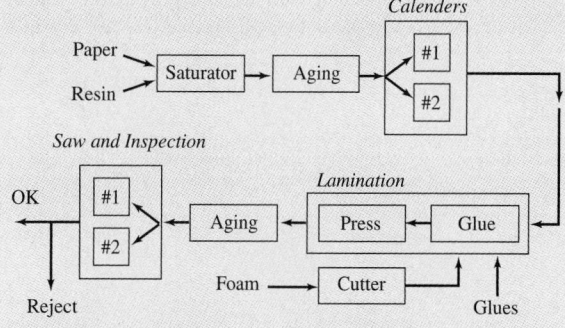

FIGURE 16.42 Plastron Production Process

---

[10]These cases were contributed by Professor Peter Kolesar at Columbia University.

to a paper machine and 150 feet long—two converted multiopening retrofitted plywood presses used for calendering, two hot wire styrofoam cutters, a hot glue applicator and hot glue press, two edge trimmers, and inspection/packing stands.

The production line uses technology that was adapted by Plastron from other purposes, and it has been in place essentially unchanged for about 14 years. The typical flow time of product through the system is 19 days. Of this time, about 5 days are designed for "curing" the product between saturation and calendering and 1 day for setting the glue after the glue press. Over the last 6 months, actual flow times from saturation to packing have varied from a minimum of 11 days to a maximum of 31 days.

Intergalactica Chemicals Ltd., the parent of the Plastron Division, is a global enterprise with 14 very diverse chemical process business units. Overall, Intergalactica has $14 billion in sales and 73,000 employees on three continents. The corporation recently embarked on a corporation-wide quality improvement program called the Intergalactica Quality Process. Twelve weeks ago, just before he began the collection of detailed end-of-the-line reject data, Charlie Hobbs had completed an intensive statistically focused 1-week quality improvement training session at the new "Intergalactica Corporate Quality College."

At Quality College, Hobbs was exposed to the basic principles and tools of statistical process control. The course aim was to enable the participants to help move the company toward its 5-year corporate goals of "total process control and process capability." It emphasized the concept that measurement and data are the keys to quality and productivity improvement.

On returning to the Rock Isle Plant, Hobbs realized that the available historical data on product quality at Plastron were quite limited—they kept only overall monthly reject rates by product. Therefore, as his first step, Hobbs instituted a systematic recording of detailed counts of panels rejected at the end of the production process—by cause of rejection.

Just before packing, each (4 ft by 8 ft) finished Plastfoam panel is inspected. There, under special lighting, the panels are rotated by two workers and each side is examined visually for defects. Thus, all output is subjected to 100% visual inspection before shipment. Although the Plastfoam product line is quite profitable and, as mentioned previously, is thought of as the "Cadillac" of the industry, historical reject rates have been very high—in some months as bad as 20%.

So, Charlie Hobbs had 3 months of defect data available (see the file PLASTRON.XLS), and he had to decide what to do next. His data confirmed that delaminations had been the most frequent or second most frequent cause of rejects in each of the last 3 months. (Delamination means that the Plastfoam sandwich had partially separated.) Hobbs had expected that delaminations would be near the top of the charts. Indeed, that was why he wanted a new lamination press. Now, he wanted to do more with the numbers. One of the articles Hobbs had read at Intergalactica Quality College quoted W. Edwards Deming, the dean of American quality experts, as saying that "85% of the problems are with the system!" Hobbs read that as "85% of the delaminations are due to the lamination press."

What kind of case could Hobbs make with these data for a new press?

Paper Production for Fornax at the Pluto Mill

Ed Michaels, the recently appointed director of quality assurance at Great Western Papelco's Pluto Mill, did not much like what he had just heard at the mill's morning meeting. This meeting, a long-standing Pluto Mill tradition, was attended by the mill manager, his direct reports, and most of their direct reports—typically 17 persons. It was held each day at the start of the first shift. The nominal purpose of the morning meeting was general communication and planning, but, in fact, most of the discussion usually focused on things that had gone wrong during the previous day, firefighting efforts under way, and the like. The wrap-up of the morning meeting was a report on the previous day's production and shipment figures. An enormous chart plotting daily tons shipped dominated the conference room wall, and before the meeting started, the production manager was obliged to have yesterday's figures posted. The mill manager was obliged to telephone these figures personally to Salt Lake City headquarters as soon as the meeting ended—usually to the White Paper Group vice president himself.

This morning there had been a special topic—the upcoming run of Fornax reproduction paper, the first to happen at the Pluto Mill in more than 6 weeks. The schedule called for 2400 tons of paper with production to start on Monday morning, if the paper makers could get the paper machine tuned up in time. The group vice president of White Papers, George Philliston, was leaning heavily on the Pluto mill manager, Rich Johnson, not to repeat the disastrous customer rejection that had occurred on the last run of Fornax paper. At the morning meeting, Ed and the Pluto Mill quality improvement lead team had been charged to help "do it right this time," and he was worried.

This was a crucial time for Great Western and the Pluto Mill. The Great Western Papelco Company, a large, fully integrated forest products company, was just 8 months into a massive, company-wide total quality management effort that had been named "Quality Is Everything." To kick off this effort, all mill management had been through a 2½-day quality awareness training, new quality posters were on the walls of the company conference rooms, and Ed Michaels had been the first person appointed to hold the new position of mill director of quality assurance. In this job, Ed reported directly to Rich Johnson, the mill manager. Ed had, of course, gone through the 2½ day course and, in addition, had just returned from a 1-week training course on statistical methods for quality improvement, run by nearby Erehwon University.

Many new and very technical ideas had been covered in those 4½ days at Erehwon. Ed came back to Pluto with a binder of notes that was fully 3 inches thick, a new calculator, and a diskette with statistical process control (SPC) software. Pareto charts, control charts, C_p ratios, Ishikawa diagrams, and standard deviations filled his mind. In 3 months Ed would be going back for a second week of training that promised to include regression analysis, experimental design, and hypothesis testing. (It had been 17 years since his last math course in college.) Some of Ed's colleagues in the mill, and indeed Rich Johnson, the mill manager, expected this new methodology of SPC to be the silver bullet that would magically solve Pluto's serious quality problems. Ed wryly noted that Rich, a crafty old-timer, had scheduled himself to be out of town during the original quality awareness training. "Ed, I'm counting on you to lead this quality effort," were his parting words.

Indeed, there were many serious quality problems. A task force consisting of outside consultants had, at the outset of the Quality Is Everything program, estimated that quality nonconformance costs at the Pluto Mill were an incredible 19% of revenues. Fully 70% of all plant overtime had been traced to specific quality problems, many shipments of even commodity-grade product had nonconformances on one or more specifications, Monday morning absenteeism was 11%, and the list goes on.

Members of the senior management team of Great Western Papelco were already talking enthusiastically about the positive impact on corporate profits if even half of these quality-related costs could be reduced through SPC. But Ed Michaels, no newcomer himself to Great Western Papelco or to the Pluto Mill, had seen a series of quality programs introduced over the years with great fanfare, but little follow-up or long-term effect. There had been a Zero Defects program in the late 1960s, then a Quality Circle initiative in 1978. The latest program had been an Overhead Value Analysis in 1982. Though each program had brought initial benefits, none had delivered on its long-term promises and, in fact, had left most employees cynical or worse. "At least half the mill probably thinks Quality Is Everything will be another fiasco," mused Ed, "and this time I'm the point man." Ed knew that senior management had discussed the shortcomings of the past programs and was stating forcefully that Quality Is Everything would be different, but he wondered how real their commitment would be as the paper market continued to soften, and both prices and sales volumes dropped.

The Pluto Mill, one of seven paper mills in the White Paper Group of Great Western Papelco, manufactures a variety of white papers for high-speed printing and reprographic applications. The Pluto Mill, indeed Great Western itself, has a very fragile relationship with the Fornax Corporation, a major producer of reprographic equipment. Pluto produces $8\frac{1}{2} \times 11$ inch paper for the Fornax Corporation, which is sold under the Fornax Company label, and is used in the Fornax Fourth-Generation Super-Duper Hi-Speed Laser-Phasor Publishing System. Recently, Pluto made a large run of "Fornax" paper that, according to the tests done at the paper mill, met Fornax quality specifications. Nevertheless, the shipment was rejected by the customer when it was inspected upon receipt at the Fornax warehouse. The rejection was based on physical testing done according to a military standard statistical sampling plan. The tests were done at Fornax's own lab by the Fornax Quality Assurance Department.

This situation caused a crisis at the Pluto Mill, as the return of the entire Fornax shipment cost Great Western Papelco well over $100,000. (What do you do with over 800,000 reams of paper in Fornax wrapping and labels?) Of course, Fornax itself lost $25,000 and more than a little customer goodwill when the shipment was rejected. This loss was particularly acute as Fornax's just-in-time inventory system depends on timely and high-quality shipments. Unable to meet its own customers' needs, Fornax had to special-order paper from one of its other suppliers. To make matters worse, this recent rejection was just one in a long series of similar incidents between Great Western Papelco and Fornax.

Great Western Papelco has ten paper machines in its system that are nominally capable of producing Fornax paper, but only one, machine C at Pluto, has been "qualified" to supply Fornax, and though qualified, it frequently has trouble meeting the specifications. The operators at the Pluto Mill often take more than a day to change over to Fornax production from the company's own "Great Western High-Speed Reproduction Paper," whose quality specifications, essentially those of the industry at large, are looser than those of the Fornax brand. A paper machine operates continuously so that during these changeovers, a great deal of off-quality paper is made as the operators literally fight the paper machine to bring all 24 key quality characteristics within the customer's specifications. It is true that, although some changeovers have gone remarkably smoothly, on one occasion last year it actually took over 4 days to begin the Fornax run! In contrast, the quality of several other Fornax paper suppliers is apparently high enough that many of their paper machines are qualified, and their quality history is so high that their product is only "skip-lot" inspected. Under such skip-lot plans, Fornax does much less testing, essentially showing confidence in the suppliers' abilities to consistently meet the Fornax needs and specifications.

At the Pluto Mill, "Fornax" paper is produced continuously on paper machine C, at a rate of about 20 tons per hour. The linear speed of the machine is about 20 miles per hour. As the paper comes off the dry end of the machine, it is rolled onto huge reels, each of which is about 20 tons in weight, 20 miles long, and 20 feet wide.[11] Thus, it takes about 1 hour

---

[11]The size of this machine alone was seen by the paper makers as a potential problem. No other paper machine close to this size was qualified to run Fornax paper.

to produce such a reel of paper. After production, reels are rewound and cut, across the paper machine direction, into 10-mile-long "sets" (still 20 feet wide) and then rewound again and slit four times, again along the paper machine direction, into 5-foot-wide rolls from which the paper is finally cut into $8\frac{1}{2} \times 11$ inch sheets. This last operation is done on a machine called the "WilSheeter," which handles 6 rolls at a time. At the back end of the sheeter, the paper is packaged into 500-sheet reams, labeled, and then packed for Fornax in 20-ream cartons. The run in question consisted of 132 reels, or 2640 tons of paper.

Many of the more than 20 quality characteristics measured on each reel of paper are of concern to Fornax and have caused the Pluto Mill problems in the past. The key issues lately, however, appear to have been moisture content, smoothness, and curl. Moisture content was the chief complaint on the last Fornax run. It affects both the printability of the paper and the speed and ease with which it goes through the complex Fornax machines. These state-of-the-art reproduction devices take ream paper in at one end and produce completely bound reports at the other. Their high speed, high quality, and versatility are important competitive edges in Fornax Company's own struggles to regain share in the market it once dominated. The high quality and low cost of the duplication machines offered by several competitors had nearly knocked Fornax out of the market.

Part of the Fornax corporate strategy to regain market share is an intensive total quality management effort incorporating a thorough vendor qualification program. Fornax engineers have often visited the Pluto Mill, sometimes during Fornax runs, and it was they who had first introduced some key statistical and quality improvement ideas to Pluto personnel. Although Fornax has provided help and encouragement to Pluto for some time, it appears that their patience is running very thin. Fornax procurement personnel rather bluntly reminded Great Western Papelco's vice president, George Philliston, of their intention to reduce the number of suppliers by half over a 3-year period. By the way, as Great Western well knows, Fornax is able to market reproduction paper wrapped under its own Fornax brand name at a premium price that is well above that of Great Western's own products. Sales of this product make a handsome contribution to Fornax's profitability. Ironically, due to Pluto's difficulty in changing over to Fornax product and its frequent necessity to cull out "off-spec" paper made during the run, it was suspected by some in the mill that they actually lost money on many Fornax runs. On the other hand, it was a help to Great Western sales people to tell other customers for commodity grades that machine C was Fornax qualified. If only they really knew how tenuous that qualification was!

Fornax insists on 90% compliance to its $4.0 \pm 0.5\%$ moisture specifications (3.5% to 4.5% moisture). The Pluto Mill paper makers claim that achieving compliance to moisture specifications on their own is not too difficult. The problem, they claim, is that moisture is frequently adjusted to bring other characteristics into compliance. Smoothness has been another long-standing concern, and the paper makers say that adjustments in moisture, caliper, and basis weight are continually being made to get smoothness within specifications. ("Caliper" is a measure of paper thickness in thousandths of an inch, whereas "basis weight," recorded in pounds per 500 sheets, is essentially a measure of density.) These quality characteristics are dependent on many factors that are under the nominal control of the paper makers, including the machine speed, a variety of temperatures, pressures and nip clearances, and the distribution and quality of the pulp as it is laid down on the moving wire web at the head of the paper machine. Although there are many parameters under their influence, selecting an optimal control strategy has proven to be far from easy.

The effects of many parameters often go in opposite directions. For example, increasing machine speed decreases basis weight while simultaneously increasing moisture. Increasing steam flow after the size press increases both caliper and smoothness, but it decreases moisture. In addition, the physical characteristics of the incoming pulp are crucial and can vary considerably from batch to batch. Paper makers around Great Western frequently say, "Making paper is one-third art, one-third science, and one-third luck with a pinch of black magic. The day nature produces cylindrical, knot-free identical trees is the day paper making becomes pure science!"

As part of the organizational architecture of Great Western's Quality Is Everything company-wide total quality program, each paper mill has formed a quality improvement lead team. In his role as director of quality assurance, Ed Michaels is the Pluto Mill's lead team leader/facilitator. However, he and his team have, as yet, had very little hands-on experience using the tools of statistical problem solving. No specific quality improvement tasks had yet been selected for attack when mill manager Rich Johnson charged Ed and his team with taking on the Fornax problem. Ed is uncomfortable with starting on such a high-impact, high-visibility problem that has defied the efforts of so many others at the mill for so long, but he has little choice. The Pluto Mill culture is such that when Rich Johnson says "jump" the only question is "how high?"

The Pluto quality improvement lead team will meet in the morning to consider the Fornax moisture problem in light of the warnings from the Great Western sales department: Fornax management is so unhappy with recent quality that Great Western is in danger of losing the valuable Fornax account if another lot is rejected. Indeed, it is widely rumored that Fornax continues to give business to Pluto largely to keep them as a back-up against the possibility of supply disruption due to the volatile labor relations in the paper industry. With a large Fornax run scheduled for next week, group vice president Philliston wants to know what actions the Mill is planning to take so that this run meets Fornax specifications.

1. Taking advantage of the statistics training that he just went through, Ed Michaels wants to see what he can learn from the Pluto RRDB (Really Reel Data Base) about past and potential quality problems. Since the last run was rejected because of low moisture, Ed is focusing attention on the data in the file PLUTO.XLS, which gives the moisture measurements from the RRDB for the 132 reels from the rejected run. The moisture measurements contained in this data set are the result of physical tests conducted on a single, 12-inch square sample cut off the end of each reel—at its center "across the reel." It takes approximately half an hour to get the sample to the lab, conduct the moisture test, and report the results back to the machine operator. (Moisture is measured by weighing the sample sheet, baking it for 8 minutes at 100°C and then weighing it again. The weight loss is presumed to be due to evaporation of water.)

Your assignment is to perform a statistical analysis on this data set that might be useful to Ed and his team in understanding and resolving their problem. The following issues are among those that Ed thinks might be relevant to an understanding of Great Western's problem:

a. Great Western's management has traditionally dealt with quality issues in terms of averages such as moisture averaged over a reel, averaged over a run, or averaged over a month. How well does the average reel moisture over the run conform to the Fornax specifications of 3.5% to 4.5%? How well do the individual reel moistures conform to the Fornax specifications? What should the specification apply to anyway—averages, individuals, samples, or the whole run? What would senior management at Great Western think? At Fornax?

b. Pluto Mill management has reminded the Great Western sales department that traditional industry standards on this grade of paper are 4.0 ± 1.0%. How well do the individual reel moistures conform to these weaker standards? The Pluto Mill manager has complained, "Isn't the problem the inconsistency between Fornax and the rest of the industry?"

c. The next Fornax run is scheduled to be 120 reels. How many of those reels can Ed expect to meet Fornax specifications? How far off could he be in this prediction? Could he give the Great Western sales department a range on the number of "off-spec" reels that might be produced? If the mill were to run an extra number of reels and cull out the off-spec paper, how many reels would they have to run to be certain of getting 120 good ones?

d. Dr. D. Vader of the Fornax quality assurance department devised the sampling inspection plan that was used by Fornax when it rejected the recent run. Vader's starting point was the frequently used MIL-STD 105D plan (which he modified because Pluto ships in FIFO order and Fornax and Pluto are collaborating on a "just in time" inventory program). Thus, the testing at Fornax is done up front and just in time to decide whether the rest of the shipment should be unloaded from the rail cars. In

Vader's plan, one carton of paper is selected at random from each of the first 14 reels of paper received from Pluto. Then, 5 reams are selected at random from each carton, and 5 sheets are selected at random from each ream. On the basis of these data (only $14 \times 5 \times 5 = 350$ sheets of paper out of the more than 429 million sheets in the run), Fornax rejected the entire shipment! Mill management is asking Ed how such data could have led to the lot rejection. "Didn't our own tests, made before shipment, show that we met specs?" The Fornax Company has not given Great Western any specifics on their data. All Ed knows is that, according to Vader, 58 of the 350 sampled sheets were below and 1 sheet was above the moisture specs. Aren't these data, Ed mused, inconsistent with the numbers in the RRDB table? What insight can Ed get from his own RRDB records? Is such a small sample enough to assure that Fornax is getting 90% compliance on such a large production run? Are the results—as political pollsters put it—within an allowable margin of error? Might Fornax have made an error in testing? What could explain the differences?

2. As mentioned previously, the next run will be starting on Monday morning—if the paper makers can manage the changeover in a timely fashion. Besides working harder and putting forth "best efforts," what can Pluto Mill do to assure that there will not be a repetition of last month's disastrous product rejection? Ed Michaels would like to put his statistical training to work to do some root-cause problem solving, but how?

3. Great Western sales and marketing are chagrined at the potential loss of Fornax as a customer. Indeed, they expect that quality standards will be even stricter in the future, and if Great Western is ever to hold on in this product line, or penetrate new markets, real quality improvement will be necessary. It is clear to senior management that the company's quality status has slipped and that Great Western is far from being the quality leader it once was in the industry. Perhaps the Fornax problem is a blessing in disguise. Pluto Mill, to-

gether with corporate engineering, has developed a proposal to spend $23 million to upgrade the wet end of the paper machine in an attempt to cure a variety of performance problems, most particularly "curl." Curl also affects how easily paper runs through a duplicator without jamming. It has been a frequent problem with Fornax. Bigabeta, Inc., a leading supplier of paper machine equipment, assures Great Western that the proposed upgrade to machine C will, among other things, ease the Fornax moisture and smoothness problems. Unfortunately, the recent history of such capital projects at Great Western has been disappointing, to say the least. Over the last 10 years, fully 60% of funded projects have failed to live up to their projected impacts. The Great Western board of directors is uneasy about the efficacy of their capital funding process. Something appears to be broken in how these projects are approved or implemented, but what is it?

a. What light do the data in the file PLUTO.XLS cast on the worthiness of this capital proposal for the moisture problem? Of course, in reality a variety of financial and marketing analyses would be necessary, but in the spirit of the case, focus on the process performance itself.

b. In addition to the types of financial information that usually accompany such capital requests, what other data would you request in evaluating this proposal?

c. Ed himself is no longer as confident as he once was that technology itself will be a cure to the problems on machine C. He wonders, "Is this paper machine potentially capable of meeting the Fornax specifications? Is the problem in the paper machine, or in how the paper machine is operated? Could the problem be (as claimed by the machine operators) in the pulping process upstream? And how does all this relate to the smoothness, abrasion, and curl problems they have had on Fornax paper over the past several months?"

4. From the Fornax view, what do you think about the Fornax specifications and testing? Do you have any suggestions for improvement?

Statistical Reporting

A.1 Introduction

By now, you have learned a wide variety of statistical tools, ranging from simple charts and descriptive measures to more complex tools such as regression and time series analysis. We suspect that all of you will be required to use some of these tools in your later coursework and in your eventual jobs. This means that you will not only need to understand the statistical tools and apply them correctly, but you will also have to write reports of your analyses for someone else—an instructor, a boss, or a client—to read. Unfortunately, the best statistical analysis is worth little if the report is written poorly. A good report must be accurate from a statistical point of view, but maybe even more important, it must be written in clear, concise English.[1]

As instructors, we know from experience that statistical report writing is the downfall of many students in statistics courses. Many students appear to believe that they will be evaluated entirely on whether the numbers are right and that the quality of the write-up is at best secondary. This is simply not true. It is not true in an academic environment, and it is certainly not true in a business environment. Managers and executives in business are very busy people who have little time or patience to wade through poorly written reports. In fact, if a report starts out badly, the remainder will probably not be read at all. Only when it is written clearly, concisely, and accurately will it have a chance of making any impact. Stated simply, a statistical analysis is often worthless if not reported well.

The goals of this brief appendix are to list several suggestions for writing good reports and to provide examples of good reports based on analyses we have done in previous chapters. You have undoubtedly had many classes in writing throughout your school years, and we cannot hope to make you a good writer if you have not already developed good basic writing skills. However, we can do three things to make you a competent statistical report writer. First, we can motivate you to spend time on your report writing by stressing how important it is in the business world. Indeed, we believe that poor writing often occurs because writers do not believe the quality of their writing makes any difference to anyone. However, we promise you that it does make a difference in the business world—your job might depend on it! Second, we can list several suggestions for improving your statistical report writing. Once you believe that good writing is really important, these tips might be all you need to help you improve you report writing

---

[1]This appendix discusses report *writing*. However, we acknowledge that oral presentation of statistical analysis is also very important. Fortunately, virtually all of our suggestions for good report writing carry over to making effective presentations. Also, we focus here on *statistical* reporting. The same comments are relevant for other quantitative reports, such as those dealing with optimization models or simulation.

significantly. Finally, we can provide examples of good reports based on previous examples from this book. Some people learn best by example, so these "templates" should come in very handy.

There is no single best way to write a statistical report. Just as there are many different methods for writing a successful novel or a successful biography, there are many different methods for writing a successful statistical report. The examples we provide look good to us, but you might want to change them according to your own tastes—or maybe even improve on them. Nevertheless, there are some bad habits that practically all readers will object to, and there are some good habits that will make your writing more effective. We list several suggestions here and expand on them in the next section.

Planning
- Clarify the objective
- Develop a clear plan
- Give yourself enough time

Developing a Report
- Write a quick first draft
- Edit and proofread
- Give your report a professional look

Guidelines for Effective Reporting
Be clear
- Provide sufficient background information
- Tailor statistical explanations to your audience
- Place charts and tables in the body of the report

Be concise
- Let charts do the talking
- Be selective in the computer outputs you include

Be precise
- List assumptions and potential limitations
- Limit the decimal places
- Report the results fairly
- Get advice from an expert

 # Suggestions for Good Statistical Reporting

To some extent, the habits that make someone a good statistical report writer are the same habits that make someone a good writer in general. Good writing is good writing! However, there are some specific aspects of good statistical reporting that do not apply to other forms of writing. In this section we will list several suggestions for becoming a good writer in general and for becoming a good statistical report writer in particular.

A.2.1 Planning

Clarify the objective When you write a statistical report, you are probably writing it *for* someone—an instructor, a boss, or maybe even a client. Make sure you know exactly what this other person wants, so that you do not write the wrong report (or perform the wrong statistical analysis). If there is any doubt in your mind about the objective of the report, clarify it with the other person before proceeding. Do not just assume that coming close to the target objective will be good enough.

Develop a clear plan Before you dive into writing the report, make a plan for how you are going to organize it. This can be a mental plan, especially if the report is short and straightforward, or it can be a more formal written outline. Think about the "best" length for the report. It should be long enough to cover the important points, but it should not be verbose. Think about the overall organization of the report and how you can best divide it into sections (if separate sections are appropriate). Think about the computer outputs you need to include (and those you can exclude) to make your case as strong as possible. Think about the audience for whom you are writing and what level of detail they will demand or will be able to comprehend. If you have a clear plan before you begin writing, the writing itself will flow much more smoothly and easily than if you make up a plan as you go. Most effective statistical reports essentially follow the outline below. We recommend that you try it.

- Executive summary
- Problem description
- Data description
- Statistical methodology
- Results and conclusions

Give yourself enough time Need we say it? If you want to follow the suggestions we list here, you need to give yourself time to do the job properly. If the report is due first thing Monday morning and you begin writing it on Sunday evening, your chances of producing anything of high quality are slim. Get started early, and don't worry if your first effort is not perfect. If you produce *something* a week ahead of time, you'll have plenty of time to polish it in time for the deadline.

A.2.2 Developing a Report

Write a quick first draft We have all seen writers in movies who agonize over the first sentence of a novel, and we suspect many of you suffer the same problem when writing a report. You want to get it exactly right the first time through, so you agonize over every word. We suggest writing the first draft as quickly as possible—just get *something* down in writing—and then worry about fixing it up with careful editing later on. The worst thing many of us face as writers is a blank piece of paper (or a blank computer document). Once there is something written, even if it is only in preliminary form, the hard part is over and the perfecting can begin.

Edit and proofread The secret of good writing is rewriting! We believe this suggestion (when coupled with the previous suggestion) can have the most immediate impact on the quality of your writing—and it is relatively easy to do. With today's software, there is no excuse for not editing and checking thoroughly, yet we are constantly amazed at how many people fail to do so. Spell checkers and grammar checkers are available in all of the popular word processors, and although they will not catch all errors, they should definitely be used. Then the real editing task can begin. A report that contains no spelling or grammatical errors is not necessarily well written. We believe a good practice, given enough time and planning, is to write a report and then reread it with a critical eye in a day or two. Better yet, get a knowledgeable friend to read it. Often the wording you thought was fine the first time around will sound awkward or confusing on a second reading. If this is the case, rewrite it! And don't just change a word or two. If a sentence sounds really awkward or a paragraph does not get your point across, don't be afraid to delete the whole thing and explore better ways of structuring it. Finally, proofread the final copy at least once, preferably more than once. Just remember that this report has *your* name on it, and any careless spelling or grammar mistakes will reflect badly on you. Admittedly, this editing and proofreading process can be time-consuming, but it can also be very rewarding when you realize how much better the final report reads.

Give your report a professional look We are not necessarily fans of all of the glitz that today's software enables (fancy colored fonts, 3-D charts, and so on), and we suspect that many writers spend too much time on glitz as opposed to substance. Nevertheless, it is important to give your reports a professional look. If nothing else, an attractive report makes a good first impression, and a first impression matters. It indicates to the reader that you have spent some time on the report and that there *might* be something inside worth reading. Of course, the fanciest report in the world cannot overcome a lack of substance, but at least it will gain you some initial respect. A sloppy report, even if it presents a great statistical analysis, might never be read at all! In any case, leave the glitz until last. Spend sufficient time to ensure that your report reads well and makes the points you want to make. Then you can have some fun "dressing it up."

A.2.3 Guidelines for Effective Reporting

We divide the following guidelines for effective reporting into three categories: be clear, be concise, and be precise.

Be clear How many times have you read a passage from a book, only to find that you need to read it again—maybe several times—because you keep losing your train of thought? It could be that you were daydreaming about something else, but it could also be that the writing itself is not clear. If a report is written clearly, the chances are that you will pick up its meaning on the first reading. Therefore, strive for clarity in your own writing. Avoid long, involved sentence structure. Don't beat around the bush, but come right out and say what you mean to say. Make sure each paragraph has a single theme that hangs together. Don't use jargon (unless you define it explicitly) that your intended readers are unlikely to understand. And, of course, read and reread what you have written—that is, edit it—to ensure that your writing is as clear as you thought.

Provide sufficient background information After working on a statistical analysis for weeks or even months, you might lose sight of the fact that others are not as familiar with the project as you are. Make sure you include enough background information to bring the reader up to speed on the context of your report. As instructors, we have read through the fine details of many student reports without knowing exactly what the overall report is all about. Don't put your readers in this position.

Tailor statistical explanations to your audience Once you begin writing the "analysis" section of a statistical report, you will probably start wondering how much explanation you need to include. For example, if you are describing the results of a regression analysis, you certainly want to mention the R^2 value, the standard error of estimate, and the regression coefficients, but do you need to explain the *meaning* of these statistical concepts? This depends entirely on your intended audience. If this report is for a statistics class, your instructor is certainly familiar with the statistical concepts, and you do not need to define them in your report. But if your report is for a nontechnical boss who knows very little about statistics beyond means and medians, a bit of explanation is certainly useful. Even in this case, however, keep in mind that your task is not to write a statistics textbook; it is to analyze a particular problem for your boss. So keep the statistical explanations brief, and get on with the analysis.

Place charts and tables in the body of the report This is a personal preference and can be disputed, but we favor placing charts and tables in the body of the report, right next to where they are referenced, rather than at the back of the report in an appendix. This way, when readers see a reference to Figure 3 or Table 2 in the body of the report, they do not have to flip through pages to find Figure 3 or Table 2. Given the options in today's word processors, this can be done in a visually attractive manner with very little extra work.

A.2.4 Be Concise

Statistical report writing is not the place for the flowery language used in novels. Your readers want to get straight to the point, and they typically have no patience for verbose reports. Make sure each paragraph, each sentence, and even each word has a purpose, and delete everything that is extraneous. This is the time where you can put critical editing to good use. Just remember that many professionals have a one-page rule—they refuse to read anything that does not fit on a single page. You might be surprised at how much you can say on a single page once you realize that this is the limit of your allotted space.

Let charts do the talking After writing this book, we are the first to admit that it can sometimes be very difficult to explain a statistical result in a clear, concise, and precise manner. It is sometimes easy to get mired in a tangle of words, even when the statistical concepts are fairly simple. This is where charts can help immensely. A well-constructed chart can be a great substitute for a long drawn-out sentence or paragraph. For example, we have seen many confusing discussions of interaction effects in regression or two-way ANOVA studies, although an accompanying chart of interactions makes the results clear and simple to understand. Do not omit the accompanying verbal explanations completely, but keep them short and refer instead to the charts.

Be selective in the computer outputs you include With today's statistical software, it is easy to produce masses of numerical outputs and accompanying charts. There is unfortunately a tendency to include everything the computer spews out—often in an appendix to the report. Worse yet, there are often no references to some of these outputs in the body of the report; the outputs are just there, supposedly self-explanatory to the intended reader. This is a bad practice. Be selective in the outputs you include in your report, and don't be afraid to alter them (with a text processor or a graphics package, say) to help clarify your points. Also, if you believe a table or chart is really important enough to include in the report, be sure to refer to it in some way in your write-up. For example, you might say, "We can see from the chart in Figure 3 that men over 50 years old are much more likely to try our product than women under 50 years old." This observation is probably clear from the chart in Figure 3—this is probably why you *included* Figure 3—but you should definitely bring attention to it in your write-up.

A.2.5 Be Precise

Statistics is a science, as well as an art. The way a statistical concept or result is explained can affect its meaning in a critical way. Therefore, try to use very precise language in your statistical reports. If you are unsure of the most precise wording, look at the wording used in this book (or another statistics book) for guidance. For example, if you are reporting a confidence interval, don't report that, "The probability is 95% that the sample mean is between 97.3 and 105.4." This might sound good enough, but it is not really correct. A more precise statement is, "We are 95% confident that the true but unobserved population mean is between 97.3 and 105.4." Of course, you must understand a statistical result (and sometimes the theory behind it) before you can report it precisely, but we suspect that imprecise statements are often due to laziness, not lack of understanding. Make the effort to phrase your statistical statements as precisely as possible.

List assumptions and potential limitations Many of the statistical procedures we have discussed rely on certain assumptions to hold. For example, in standard regression analysis there are assumptions about equal error variance, lack of residual autocorrelation, and normality of the residuals. If your analysis relies on certain assumptions for validity, mention these in your report, especially when there is evidence that they are violated. In fact, if they appear to be violated, warn the reader about the possible limitations of your results. For example, a confidence interval reported at the 95% level might really, due to the violation of an equal variance assumption, be valid at the 80% or 85% level. Don't just ignore assumptions—with the implication that they do not matter.

Limit the decimal places We are continually surprised at the number of students who quote statistical results (directly from computer outputs, of course) to 5–10 decimal places, even when the original data are given with much less precision. For example, when forecasting sales a year from now, given historical sales data like $3440, $4120, and so on, some people will quote a forecast like $5213.2345. Who are they kidding! Statistical methods are exact only up to a certain limit. If you quote a forecast like $5213.2345, just because this is what appears in your computer output, you are not gaining precision; you are showing your lack of understanding of the limits of the statistical methodology. If you instead quote a forecast of "about $5200," you will probably gain more respect from critical readers.

Report the results fairly We have all heard statements such as, "It is easy to lie with statistics." It is true that the *same* data can often be analyzed and reported by two different analysts to support diametrically opposite points of view. Certain results can be omitted, the axes of certain charts can be distorted, important assumptions can be ignored, and so on. This is partly a statistical issue and partly an ethical issue. There is not necessarily anything wrong with two competent analysts using different statistical methods to arrive at different conclusions. For example, in a case where gender discrimination in salary has been charged, honest statisticians might very well disagree as to the legitimacy of the charges, depending on how they analyze the data. The world is not always black and white, and statistical analysts often find themselves in the gray areas. However, you are ethically obligated to report your results as fairly as possible. You should not *deliberately* try to lie with statistics.

Get advice from an expert Even if you have read and understood every word in this book, you are still not an expert in statistics. You know a lot of useful techniques, but there are many specific details and nuances of statistical analysis that we have not had time to cover. A good example is violation of assumptions. At several places we have discussed how to *detect* violations of assumptions, but we have usually not discussed possible remedies because they typically require advanced methods. If you become stuck on how to write a specific part of your report because you lack the statistical knowledge, don't be afraid to consult someone with more statistical expertise. For example, try e-mailing former instructors. They might be flattered that you remember them and value their knowledge—and they can probably provide the information you need.

A.3 Examples of Statistical Reports

Because many of you probably learn better from *examples* of report writing than from lists of suggestions, we now present several example reports. Each of these is based on a statistical analysis we performed in a previous chapter. (We embellish these with details that were not given earlier to bring them more alive.) Again, our reports represent just one possible style of writing, and other styles might be equally good or even better. But we have attempted to follow the suggestions listed in the previous section. In particular, we have strived for clarity, conciseness, and precision—and the final reports you see here are the result of much editing!

Example A.1

This example is adapted from Example 3.9. I am working for the Spring Mills Company, and my boss, Sharon Sanders, has asked me to report on the accounts receivable problem our company is currently experiencing. My task is to describe data on our customers, analyze the magnitude of interest lost because of late payments from our customers, and suggest a solution for remedying the problem. Ms. Sanders knows basic statistics, but she might need a refresher on the meaning of boxplots. ●

SPRING MILLS COMPANY
ZANESVILLE, OHIO

To: Sharon Sanders
From: Chris Albright
Subject: Report on accounts receivable
Date: July 6, 2001

Executive summary
Our company produces and distributes a wide variety of manufactured goods. Due to this variety, we have a large number of customers. We have classified our customers as small, medium, or large depending on the amount of business they do with us. Recently, we have had problems with accounts receivable. We are not getting paid as promptly as we would like, and we sense that it is costing our company a good deal of money in potential interest. You assigned me to investigate the magnitude of the problem and to suggest a strategy for fixing it. This report discusses my findings.

Data set

I collected data on 280 customer accounts. The breakdown by size is: 150 small customers, 100 medium customers, and 30 large customers. For each account, my data set includes the number of days since the customer was originally billed (Days) and the amount the customer currently owes (Amount). If necessary, we can identify any of these accounts by name, although specific names will not appear in this report. However, my data and analysis are in the file Receive.xls. I have attached this file to my report in case you want to see further details.

Software

My analysis was performed entirely in Excel XP, using the well-known StatPro add-in where necessary.

Analysis

Given the objectives of the analysis, my analysis is broken down by customer size. Exhibit 1 shows summary statistics for the Days and Amount for each customer size. (Small, medium, and large are coded throughout as 1, 2, and 3. For example, Days1 refers to the Days variable for small customers). We see, not surprisingly, that larger customers tend to owe larger amounts. The median amounts for small, medium, and large customers are $250, $470, and $1395, and the mean amounts follow a similar pattern. In contrast, medium and large companies tend to delay payments about equally long (median days delayed about 19–20), whereas small companies tend to delay only about half this long. The standard deviations in this exhibit indicate some variation across companies of any size, although this variation is considerably smaller for the amounts owed by small companies.

Graphical comparisons of these different size customers appear in Exhibits 2 and 3. Each of these shows side-by-side boxplots (the first of Amount, the second of Days) for easy visual comparison. (For any boxplot, recall that the box contains the middle 50% of the observations, the line and the dot inside the box represent the median and mean, and individual points outside the box represent extreme observations.) These boxplots graphically confirm the patterns we observed in Exhibit 1.

Exhibits 1–3 describe the variables Days and Amount individually, but they do not indicate whether there is a relationship between them. Do our customers who owe large amounts tend to delay longer? To investigate this, we created scatterplots of Amount versus Days for each customer size. The scatterplot for small customers (not shown) indicates no relationship whatsoever; the correla-

EXHIBIT A.1

Summary Measures for Different Size Customers

	A	B	C	D	E	F	G	H
1	Summary measures for selected variables							
2			Days1	Amount1	Days2	Amount2	Days3	Amount3
3		Count	150.000	150.000	100.000	100.000	30.000	30.000
4		Sum	1470.000	38180.000	2055.000	48190.000	577.000	43630.000
5		Mean	9.800	254.533	20.550	481.900	19.233	1454.333
6		Median	10.000	250.000	20.000	470.000	19.000	1395.000
7		Standard deviation	3.128	49.285	6.622	99.155	6.191	293.888
8		Minimum	2.000	140.000	8.000	280.000	3.000	930.000
9		Maximum	17.000	410.000	39.000	750.000	32.000	2220.000

EXHIBIT A.2
Boxplots of Days by Different Size Customers

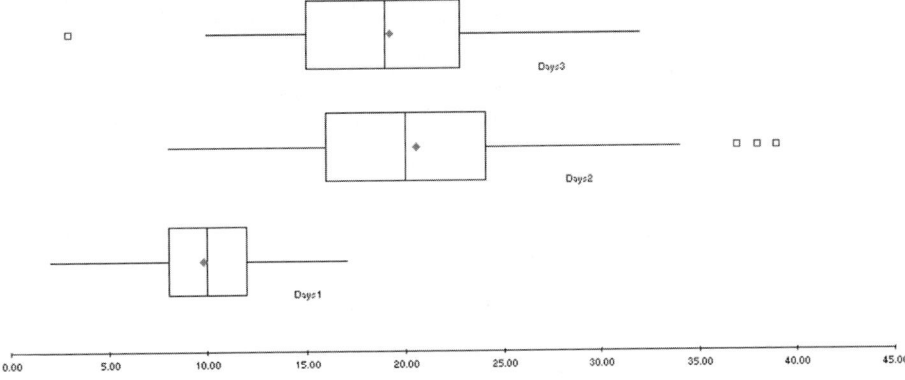

EXHIBIT A.3
Boxplots of Amount by Different Size Customers

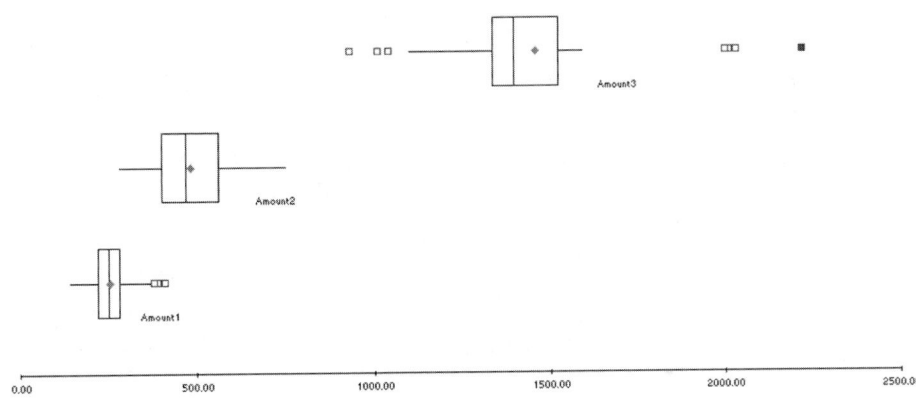

tion between Days and Amount is a negligible −0.044. However, the scatterplots for medium and large customers both indicate a fairly strong positive relationship. We show the scatterplot for medium-size customers in Exhibit 4. (The one for large customers is similar, only with many fewer points.) The correlation is a hefty 0.612, and the upward sloping (and reasonably linear) pattern is clear: the larger the delay, the larger the amount owed—or vice versa.

EXHIBIT A.4
Scatterplot of Amount versus Days for Medium Customers

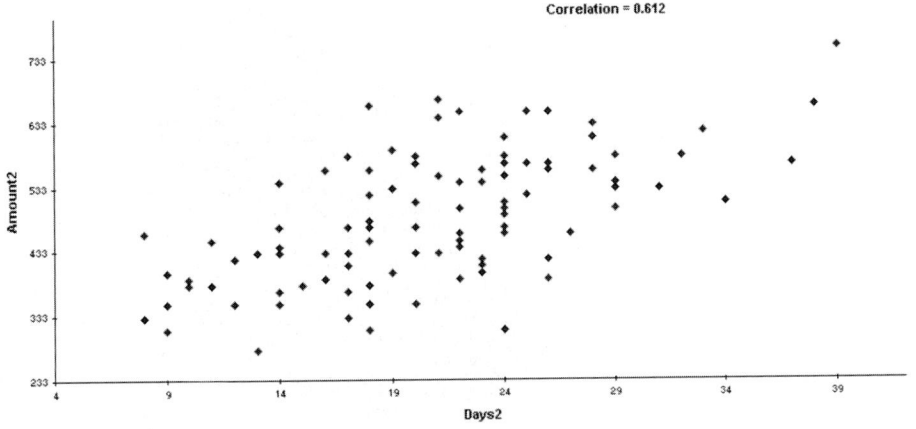

EXHIBIT A.5
Table of Lost
Interest

	A	B	C	D	E	F	G	H	I
1	Interest lost								
2									
3	Summary measures for selected variables								
4		Lost1	Lost2	Lost3					
5	Sum	$122.68	$338.65	$287.25					
6									
7	Annual interest rate		12%						
8									
9	Days1	Amount1	Lost1	Days2	Amount2	Lost2	Days3	Amount3	Lost3
10	7	$180	$0.41	17	$470	$2.63	19	$1,330	$8.31
11	8	$210	$0.55	22	$540	$3.91	20	$1,400	$9.21
12	10	$210	$0.69	28	$560	$5.16	14	$1,550	$7.13
13	8	$150	$0.39	24	$470	$3.71	15	$1,460	$7.20
14	9	$300	$0.89	26	$650	$5.56	23	$2,030	$15.35
15	5	$240	$0.39	29	$530	$5.05	19	$1,520	$9.49
16	4	$330	$0.43	21	$550	$3.80	15	$1,330	$6.56
17	10	$290	$0.95	33	$620	$6.73	17	$1,520	$8.50

The analysis to this point describes our customer population, but it does not directly answer our main concerns: How much potential interest are we losing and what can we do about it? The analysis in Exhibit 5 and accompanying pie chart in Exhibit 6 address the first of these questions. To create Exhibit 5, I assumed that we can earn an annual rate of 12% on excess cash. Then for each customer, I calculated the interest lost by not having a payment made for a certain number of days. (These calculations are in rows 10 down; data are shown for only a few of the customers.) Then I summed these lost interest amounts to obtain the totals in row 5 and created a pie chart from the totals.

The message from the pie chart is fairly clear. We do not need to worry about our many small customers; the interest we are losing because of them is relatively small. However, we might want to put some pressure on the medium

EXHIBIT A.6
Pie Chart of Lost
Interest

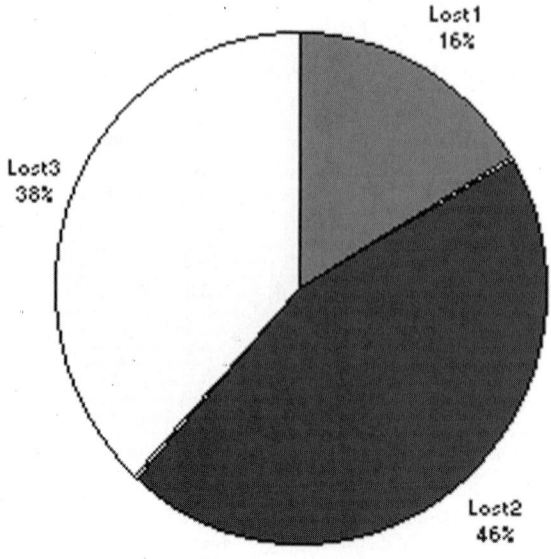

Interest Lost by Customer Type

and large customers. I would suggest targeting the large customers first, especially those with large amounts due. There are fewer of them, so that we can concentrate our efforts more easily. Also, remember that amounts due and days delayed are positively correlated for the large customers. Therefore, the accounts with large amounts due are where we are losing the most potential interest.

Attachment: Receive.xls

Example A.2

This example is adapted from Example 9.9. I am a student in a required MBA statistics course. For the statistical inference part of the course, each student has been assigned to gather data in a real setting that can be used to find a suitably narrow confidence interval for a population parameter. Although the instructor, Rob Jacobs, certainly knows statistics well, he has asked us to include explanations of relevant statistical concepts in our reports, just to confirm that *we* know what we are talking about. Professor Jacobs has made it clear that he does not want a lot of "padding." He wants reports that are short and to the point. ●

Report on Confidence Intervals for Professor Rob Jacobs
Managerial Statistics, S540 – Spring semester, 2001
Submitted by Wayne Winston

Executive summary
This report summarizes my findings on potential differences between husbands and wives in their ratings of automobile presentations. I chose this topic because my uncle manages a Honda dealership in town, and he enabled me to gain access to the data for this report. The report contains the following: (1) an explanation of the overall study, (2) a rationale for the sample size I chose, (3) the data, (4) the statistical methodology, and (5) a summary of my results.

The study
We tend to associate automobiles with males—horsepower, dual cams, and V-6 engines are macho terms. I decided to investigate whether husbands, when shopping for new cars with their wives, tend to react more favorably to salespersons' presentations than their wives. (My bias that this is true is bolstered by the fact that all salespeople I have seen, including all of those in this study, are men.) To test this, I asked a sample of couples at the Honda dealership to rate the sales presentation they had just heard on a 1 to 10 scale, 10 being the most favorable. The husbands and wives were asked to give independent ratings. I then used these data to calculate a confidence interval for the mean *difference* between the husbands' and wives' ratings. If my initial bias was correct, this confidence interval should be predominantly positive.

The sample size
Before I could conduct the study, I had to choose a sample size: the number of couples to sample. The eventual sample was based on two considerations: the time I could devote to the study and the length of the confidence interval I

desired. For the latter consideration, I used StatPro's sample size determination procedure to get an estimate of the required sample size. This procedure requests a confidence level (I chose the usual 95% level), a desired confidence interval half-length, and a standard deviation of the differences. I suspected that most of the differences (husband rating minus wife rating) would be from -1 to $+3$, so I (somewhat arbitrarily) chose a desired half-length of 0.25 and guessed a standard deviation of 0.75. StatPro reported that this would require a sample size of 35 couples. I decided that this was reasonable, given the amount of time I could afford, so I used this sample size and proceeded to gather data from 35 husbands and wives. Of course, I realized that if the *actual* standard deviation of differences turned out to be larger than my guess, then my confidence interval would not be as narrow as I specified.

The data

The data I collected includes a husband and a wife rating for each of the 35 couples in the sample. Exhibit 1 presents data for the first few couples, together with several summary statistics for the entire data set. As the sample means and medians indicate, husbands do tend to rate presentations somewhat higher than their wives, but this comparison of means and medians is only preliminary. The statistical *inference* is discussed next.

Statistical methodology

My goal is to compare two means: the mean rating of husbands and the mean rating of wives. There are two basic statistical methods for comparing two means: the two-sample method and the paired-sample method. I chose the latter. The two-sample method assumes that the observations from the two samples are *independent*. Although I asked each husband–wife pair to evaluate the presentation independently, I suspected that husbands and wives, by the very fact that they live together and tend to think alike, would tend to give positively correlated ratings. The data confirmed this. The correlation between the husband and wife ratings was a fairly large and positive 0.44. When data come in natural pairs and are positively correlated, then the paired-sample method for comparing means is preferred. The reason is that it takes advantage of the positive correlation to provide a narrower confidence interval than the two-sample method.

EXHIBIT A.1
Data and Summary Measures

	A	B	C	D	E	F	G	H
3	Pair	Husband	Wife			Summary measures for selected variables		
4	1	6	3				Husband	Wife
5	2	7	8			Mean	6.914	5.286
6	3	8	5			Median	7.000	5.000
7	4	6	4			Standard deviation	1.222	1.792
8	5	8	5					
9	6	7	6					
10	7	8	5					
11	8	6	7					
12	9	7	8					
13	10	7	5					

	F	G	H	I
3	*Paired-sample analysis for Husband minus Wife*			
4				
5	*Summary measures for Husband-Wife*			
6		Sample size	35	
7		Sample mean	1.629	
8		Sample standard deviation	1.664	
9				
10	*Confidence interval for mean*			
11		Confidence level	95.0%	
12		Sample mean	1.629	
13		Std error of mean	0.281	
14		Degrees of freedom	34	
15		Lower limit	1.057	
16		Upper limit	2.200	

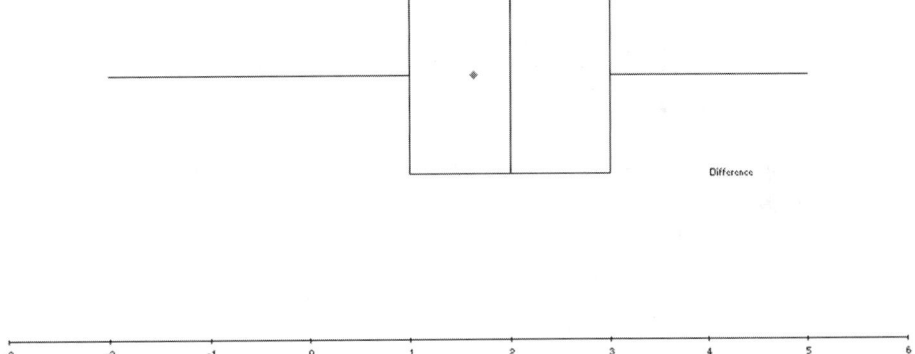

Results To obtain the desired confidence interval, I used StatPro's paired-sample procedure. This creates a new column of differences and then analyzes these differences. Exhibit 2 contains the StatPro output. The summary measures at the top of this output provide one more indication that husbands react, on average, more favorably to presentations than their wives. The mean difference is about 1.6 rating points. A graphical illustration of this difference appears in Exhibit 3, which includes a boxplot of the "husband minus wife" differences. We see that the vast majority of the differences are positive.

The bottom part of Exhibit 2 contains the statistical inference, including the 95% confidence interval for the mean difference. This interval extends from approximately 1 to 2.2. To understand how it is formed, the method first calculates the standard error of the sample mean difference in cell H13. This is the standard deviation of the differences divided by the square root of the sample size. Then it goes out approximately two standard errors on either side of the sample mean difference to form the limits of the confidence interval.

Because the confidence interval includes only positive values (and the lower limit is not even close to 0), there is little doubt that husbands, on average, react more positively to sales presentations than their wives. Note, though, that the confidence interval is not nearly as narrow as I specified in the sample size section. This is because the standard deviation of differences turned out to be considerably larger than I guessed (1.66 versus 0.75). If I really wanted a narrower confidence interval, then I would need a considerably larger sample. Given that I have essentially proved my conjecture that the mean difference is positive, however, a larger sample does not appear to be necessary.

Example A.3

This example is adapted from Example 11.2. I am a statistical consultant, and I have been hired by the Bendrix Company, a manufacturing company, to analyze its overhead data. The company has supplied me with historical monthly data from the past three years on overhead expenses, machine hours, and the number of production runs. My task is to develop a method for forecasting overhead expenses in future months, given estimates of the machine hours and number of production runs that are expected in these months. My contact, Dave Clements, is in the company's finance department. He obtained an MBA degree about 10 years ago, and he vaguely remembers some of the statistics he learned at that time. However, he does not profess to be an expert. The more I can write my report in laymen's terms, the more he will appreciate it. ●

CHRISTOPHER ZAPPE STATISTICAL CONSULTING SERVICES
BLOOMINGTON, INDIANA

To: Dave Clements, financial manager
Subject: Forecasting overhead
Date: July 20, 2001

Dave, here is the report you requested. (See also the attached Excel XP file, Overhead.xls, that contains the details of my analysis. By the way, it was done with the help of the StatPro add-in for Excel. If you plan to do any further statistical analysis, I would strongly recommend purchasing this add-in.) As I will explain in this report, regression analysis is the best-suited statistical methodology for your situation. It fits an equation to historical data, it uses this equation to forecast future values of overhead, and it provides a measure of accuracy of these forecasts. I believe you will be able to "sell" this analysis to your colleagues. The theory behind the regression analysis is admittedly complex, but the outputs I will provide are quite intuitive, even to nonstatisticians.

Objectives and data
To ensure that we are on the same page, I will briefly summarize my task. You supplied me with Bendrix monthly data, from July 1999 through June 2001, on three variables: Overhead (total overhead expenses during the month), MachHrs (number of machine hours used during the month), and ProdRuns (number of separate production runs during the month). You suspect that Overhead is directly related to MachHrs and ProdRuns, and you want me to quantify this relationship so that you can forecast *future* overhead expenses on

EXHIBIT A.1
Scatterplot of Overhead versus MachHrs

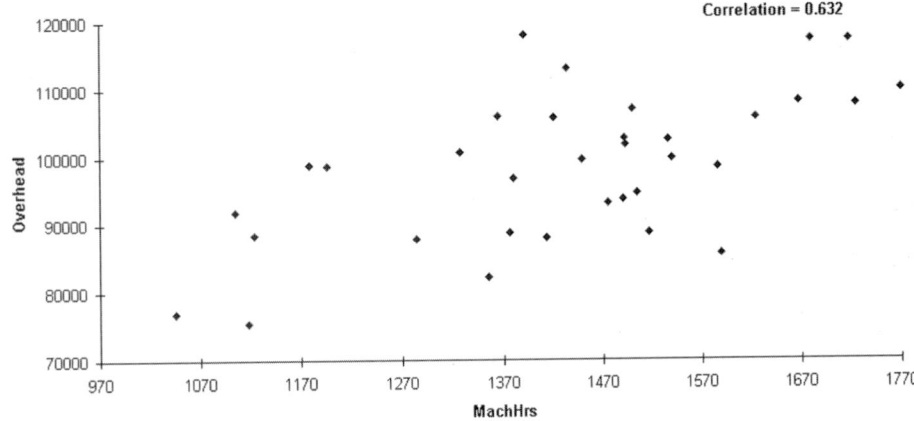

the basis of (estimated) future values of MachHrs and ProdRuns. Although you did not state this explicitly in your requirements, I assume that you would also like a measure of the accuracy of the forecasts.

Statistical methodology
Fortunately, there is a natural methodology for solving your problem: regression analysis. Regression analysis was developed specifically to quantify the relationship between a single *response* variable and one or more *explanatory* variables (assuming that there is a relationship to quantify). In your case, the response variable is Overhead, the explanatory variables are MachHrs and ProdRuns, and from a manufacturing perspective, there is every reason to believe that Overhead is related to MachHrs and ProdRuns. The outcome of the regression analysis will be a regression equation that can be used to forecast future values of Overhead and provide a measure of the accuracy of these forecasts. There are a lot of calculations involved in regression analysis, but statistical software such as StatPro takes care of these calculations easily, allowing you to focus on the interpretation of the results.

Preliminary analysis of the data
Before diving into the regression analysis itself, it is always a good idea to check graphically for relationships between the variables. The best type of chart for your problem is a scatterplot, which shows the relationship between any pair of variables. The scatterplots in Exhibits 1 and 2 illustrate how Overhead varies with MachHrs and with ProdRuns. In both charts the points follow a reasonably linear pattern from bottom left to upper right. That is, Overhead tends to increase linearly with MachHrs and with ProdRuns, which is probably what you suspected. The correlations in the upper right corners of these plots indicate the strength of the linear relationships. The magnitudes of these correlations, 0.632 and 0.521, are fairly large. (The maximum possible correlation is 1.0.) They provide hope that regression analysis will yield reasonably accurate forecasts of overhead expenses.

Before moving to the regression analysis, there are two other charts you should consider. First, you ought to check whether there is a relationship between the two explanatory variables, MachHrs and ProdRuns. If the correlation between these variable is high (negative or positive), then you have a phenomenon called *multicollinearity*. This is not necessarily bad, but it complicates the

EXHIBIT A.2
Scatterplot of Over-
head versus
ProdRuns

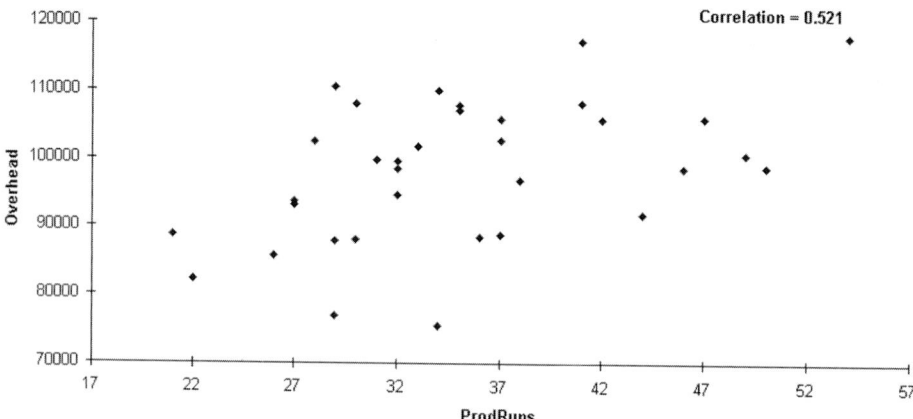

EXHIBIT A.3
Scatterplot of
MachHrs versus
ProdRuns

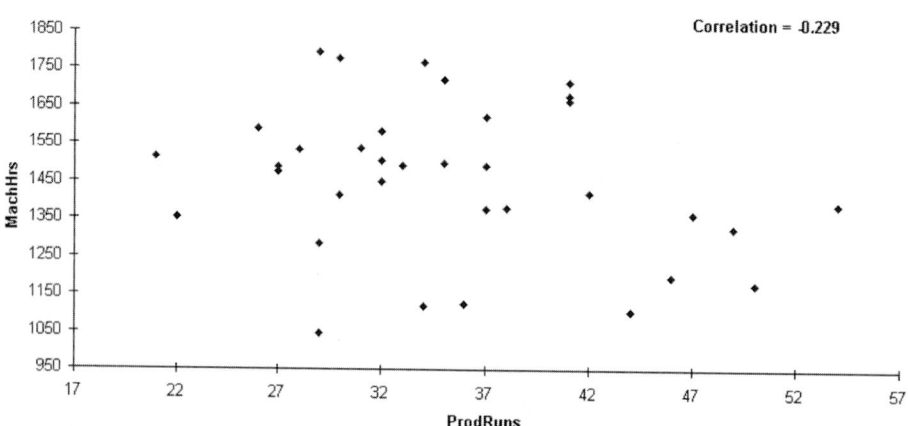

interpretation of the regression equation. Fortunately, as Exhibit 3 indicates, there is virtually no relationship between MachHrs and ProdRuns, so multicollinearity is not a problem for you.

You should also check the time series nature of your overhead data. For example, if your overhead expenses are trending upward over time, or if there is a seasonal pattern to your expenses, then MachHrs and ProdRuns, by themselves, would probably not be adequate to forecast future values of Overhead. However, as illustrated in Exhibit 4, a time series plot of Overhead indicates no obvious trends or seasonal patterns.

Regression analysis
The plots in Exhibits 1–4 provide some confidence that regression analysis for Overhead, using MachHrs and ProdRuns as the explanatory variables, will yield useful results. Therefore, I used StatPro's multiple regression procedure to estimate the regression equation. As you may know, the regression output from practically any software package, including StatPro, can be a bit intimidating. For this reason, I will report only the most relevant outputs. (You can see the rest in the Overhead.xls file if you like.) The estimated regression equation is

$$\text{Forecasted Overhead} = 3997 + 43.54\text{MachHrs} + 883.62\text{ProdRuns}$$

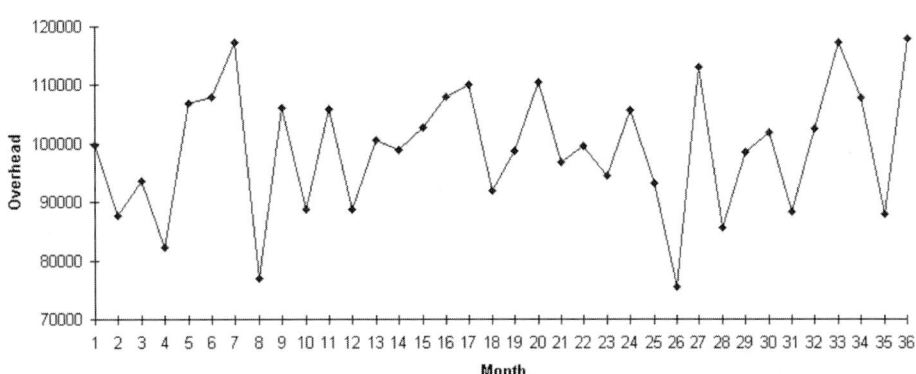

Time series chart of Overhead

Two important summary measures in any regression analysis are *R*-square and the standard error of estimate. Their values for this analysis are 93.1% and $4109.

Now let's turn to interpretation. The two most important values in the regression equation are the coefficients of MachHrs and ProdRuns. For each extra machine hour your company uses, the regression equation predicts that an extra $43.54 in overhead will be incurred. Similarly, each extra production run is predicted to add $883.62 in overhead. Of course, these values should be considered approximate only, but they provide a sense of how much extra machine hours and extra production runs add to overhead. (Don't spend too much time trying to interpret the constant term, 3997. Its primary use is to get the forecasts to the correct "level.")

The *R*-square value indicates that 93.1% of the variation in overhead expenses you observed during the past 36 months can be "explained" by the values of MachHrs and ProdRuns your company used. Alternatively, 6.9% of the variation in overhead has still not been explained. To explain this remaining variation, you would probably need data on one or more *other* relevant variables. However, 93.1% is quite good. In statistical terms, you have a good fit.

For forecasting purposes, the standard error of estimate is even more important than *R*-square. It indicates the approximate magnitude of forecast errors you can expect when you base your forecasts on the regression equation. This standard error can be interpreted much like a standard deviation. Specifically, there is about a 68% chance that a forecast will be off by no more than one standard error, and there is about a 95% chance that a forecast will be off by no more than two standard errors.

Forecasting

Your forecasting job is now quite straightforward. Suppose, for example, that you expect 1525 machine hours and 45 production runs next month. (These values are in line with your historical data.) Then you simply plug these values into the regression equation to obtain the forecasted overhead:

Forecasted overhead = 3997 + 43.54(1525) + 883.62(45) = $101,158

Given the standard error of estimate of $4109, you can be about 68% confident that this forecast will be off by no more than $4109 on either side, and you can be about 95% confident that it will be off by no more than $8218 on either side. Of course, I'm sure you know better than to take any of these values too literally, but I believe this level of forecasting accuracy should be useful to your company.

One last recommendation I have is to update the analysis as time moves on. As you observe future values of the variables, incorporate them into the data set (and remove old values if you believe they are obsolete), and rerun the regression analysis. You can do this easily with the same Overhead.xls file I have attached.

If you have any questions, feel free to call me at any time. You have my number.

 # Conclusion

Many people believe that statistical analysis is heavy-duty number crunching and little else. As many of our former students have told us, however, this is definitely not true. They continually testify to the importance of written reports (and oral presentations) in their jobs. In fact, we believe that many of you will be judged more by the quality of your *writing* (and speaking) than by the quality of your quantitative analysis. Therefore, keep the suggestions and examples in this chapter handy—you might need them more than you realize. Just remember that well-designed studies and careful statistical analysis are often useless unless they are communicated clearly and effectively to the audience that needs them.

References

Aarvik, O., and P. Randolph. "The Application of Linear Programming to the Determination of Transmission Line Fees in an Electrical Power Network." *Interfaces* 6 (1975): 17–31.

Albright, S. C. "A Statistical Analysis of Hitting Streaks in Baseball." *Journal of the American Statistical Association* 88, no. 424 (1993): 1175–1196.

Altman, E. *Handbook of Corporate Finance.* New York: Wiley, 1986.

Appleton, D., J. French, and M. Vanderpump. "Ignoring a Covariate: An Example of Simpson's Paradox." *The American Statistician* 50 (1996): 340–341.

Armstrong, S. "Forecasting by Extrapolation: Conclusions from 25 Years of Research." *Interfaces* 14, no. 6 (1984): 52–66.

Armstrong, S. *Long-Range Forecasting.* New York: Wiley, 1985.

Armstrong, S. "Research on Forecasting: A Quarter-Century Review, 1960–1984." *Interfaces* 16, no. 1 (1986): 89–103.

Arntzen, B., G. Brown, T. Harrison, and L. Trafton. "Global Supply Chain Management at Digital Equipment Corporation." *Interfaces* 25, no. 1 (1995): 69–93.

Babich, P. "Customer Satisfaction: How Good Is Good Enough?" *Quality Progress* 25 (1992): 65–68.

Balson, W., J. Welsh, and D. Wilson. "Using Decision Analysis and Risk Analysis to Manage Utility Environmental Risk." *Interfaces* 22, no. 6 (1992): 126–139.

Barnett, A. "Genes, Race, IQ, and *The Bell Curve.*" *ORMS Today* 22, no. 1 (1995): 18–24.

Bean, J., C. Noon, S. Ryan, and G. Salton. "Selecting Tenants in a Shopping Mall." *Interfaces* 18, no. 2 (1988): 1–10.

Bean, J., C. Noon, and G. Salton. "Asset Divestiture at Homart Development Company." *Interfaces* 17, no. 1 (1987): 48–65.

Benninga, S. *Numerical Methods in Finance.* Cambridge, MA: MIT Press, 1989.

Berger, P., and R. Maurer. *Experimental Design.* Belmont, CA: Duxbury, 2002.

Bilder, C. R., and T. M. Loughin. " 'It's Good!' An Analysis of the Probability of Success for Placekicks." *Chance* 11, no. 2 (1998): 20–30.

Black, F., and M. Scholes. "The Pricing of Options and Corporate Liabilities." *Journal of Political Economy* 81 (1973): 637–654.

Blyth, C. "On Simpson's Paradox and the Sure-Thing Principle." *Journal of the American Statistical Association* 67 (1972): 364–366.

Borison, A. "Oglethorpe Power Corporation Decides about Investing in a Major Transmission System." *Interfaces* 25, no. 2 (1995): 25–36.

Boykin, R. "Optimizing Chemical Production at Monsanto." *Interfaces* 15, no. 1 (1985): 88–95.

Brigandi, A., D. Dargon, M. Sheehan, and T. Spencer. "AT&T's Call Processing Simulator (CAPS) Operational Design for Inbound Call Centers." *Interfaces* 24, no. 1 (1994): 6–28.

Brinkley, P., D. Stepto, J. Haag, K. Liou, K. Wang, and W. Carr. "Nortel Redefines Factory Information Technology: An OR-Driven Approach." *Interfaces* 28, no. 1 (1998): 37–52.

Brown, G., et al. "Real-Time Wide Area Dispatch of Mobil Tank Trucks." *Interfaces* 17, no. 1 (1987): 107–120.

Cawley, J., and P. Sommers, "Voting Irregularities in the 1995 Referendum on Quebec Sovereignty." *Chance,* 9, no. 4 (Fall 1996): 29–30.

Cebry, M., A. DeSilva, and F. DiLisio. "Management Science in Automating Postal Operations: Facility and Equipment Planning in the United States Postal Service." *Interfaces* 22, no. 1 (1992): 110–130.

Charnes, A., and L. Cooper. "Generalization of the Warehousing Model." *Operational Research Quarterly* 6 (1955): 131–172.

Citro, C. "Window on Washington." *Chance* 12, no. 1 (1999): 38–41.

Cox, J., S. Ross, and M. Rubenstein. "Option Pricing: A Simplified Approach." *Journal of Financial Economics* 7 (1979): 229–263.

Deming, E., *Out of the Crisis.* Cambridge, MA: MIT Center for Advanced Engineering Study, 1986.

DeVor, R., T. Chang, and J. Sutherland. *Statistial Quality Design and Control.* New York: Macmillan, 1992.

DeWitt, C., L. Lasdon, A. Waren, D. Brenner, and S. Melhem. "OMEGA: An Improved Gasoline Blending System for Texaco." *Interfaces* 19, no. 1 (1989): 85–101.

Eaton, D., et al. "Determining Emergency Medical Service Vehicle Deployment in Austin, Texas." *Interfaces* 15, no. 1 (1985): 96–108.

Efroymson, M., and T. Ray. "A Brand-Bound Algorithm for Plant Location." *Operations Research* 14 (1966): 361–368.

Engemann, K., and H. Miller. "Operations Risk Management at a Major Bank." *Interfaces* 22, no. 6 (1992): 140–149.

Eppen, G., K. Martin, and L. Schrage. "A Scenario Approach to Capacity Planning." *Operations Research* 37, no. 4 (1989): 517–527.

Fabian, T. "A Linear Programming Model of Integrated Iron and Steel Production." *Management Science* 4 (1958): 415–449.

Feinstein, C. "Deciding Whether to Test Student Athletes for Drug Use." *Interfaces* 20, no. 3 (1990): 80–87.

Fitzsimmons, J., and L. Allen. "A Warehouse Location Model Helps Texas Comptroller Select Out-of-State Audit Offices." *Interfaces* 13, no. 5 (1983): 40–46.

GeneHunter. Ward Systems Group, Frederick, Maryland, 1995.

Glover, F., G. Jones, D. Karney, D. Klingman, and J. Mote. "An Integrated Production, Distribution, and Inventory System." *Interfaces* 9, no. 5 (1979): 21–35.

Glover, F., et al. "The Passenger-Mix Problem in the Scheduled Airlines." *Interfaces* 12 (1982): 873–880.

Graddy, K. "Do Fast-Food Chains Price Discriminate on the Race and Income Characteristics of an Area?" *Journal of Business & Economic Statistics* 15, no. 4 (1997): 391–401.

Grossman, S., and O. Hart. "An Analysis of the Principal Agent Problem." *Econometrica* 51 (1983): 7–45.

Hauser, J., and S. Gaskin. "Application of the Defender Consumer Model." *Marketing Science* 3, no. 4 (1984): 327–351.

Herrnstein, R., and C. Murray. *The Bell Curve.* New York: The Free Press, 1994.

Hertz, D. "Risk Analysis in Capital Investment." *Harvard Business Review* 42 (Jan.–Feb. 1964): 96–108.

Hess, S. "Swinging on the Branch of a Tree: Project Selection Applications." *Interfaces* 23, no. 6 (1993): 5–12.

Holmer, M. "The Asset-Liability Management Strategy System at Fannie Mae." *Interfaces* 24, no. 3 (1994): 3–21.

Hoppensteadt, F., and C. Peskin. *Mathematics in Medicine and the Life Sciences.* New York: Springer-Verlag, 1992.

Howard, R. "Heathens, Heretics, and Cults: The Religious Spectrum of Decision Aiding." *Interfaces* 22, no. 6 (1992): 15–27.

Huerter, J., and W. Swart. "An Integrated Labor-Management System for Taco Bell," *Interfaces* 28, no. 1 (1998): 75–91.

Kauffman, J., B. Matsik, and K. Spencer. *Beginning SQL Programming.* Birmingham, UK: Wrox Press Ltd, 2001.

Kelly, J. "A New Interpretation of Information Rate." *Bell System Technical Journal* 35 (1956): 917–926.

Kimes, S., and J. Fitzsimmons. "Selecting Profitable Hotel Sites at La Quinta Motor Inns. *Interfaces* 20, no. 2 (1990): 12–20.

Kirkwood, C. "An Overview of Methods for Applied Decision Analysis" *Interfaces* 22, no. 6 (1992): 28–39.

Klingman, D., N. Phillips, D. Steiger, and W. Young. "The Successful Deployment of Management Science throughout Citgo Petroleum Corporation." *Interfaces* 17, no. 1 (1987): 4–25.

Kovar, M. "Four Million Adolescents Smoke: Or Do They?" *Chance* 13, no. 2 (2000): 10–14.

Krajewski, L., L. Ritzman, and P. McKenzie. "Shift Scheduling in Banking Operations: A Case Application." *Interfaces* 10, no. 2 (1980): 1–8.

Krumm, F., and C. Rolle. "Management and Application of Decision and Risk Analysis in Du Pont." *Interfaces* 22, no. 6 (1992): 84–93.

Lanzenauer, C., E. Harbauer, B. Johnston, and D. Shuttleworth. "RRSP Flood: LP to the Rescue." *Interfaces* 17, no. 4 (1987): 27–40.

Levy, P., and S. Lemeshow. *Sampling of Populations: Methods and Applications,* 3rd ed. New York: Wiley, 1999.

Littlechild, S. "Marginal Pricing with Joint Costs." *Economic Journal* 80 (1970): 323–334.

Love, R., and J. Hoey. "Management Science Improves Fast Food Operations." *Interfaces* 20, no. 2 (1990): 21–29.

Magoulas, K., and D. Marinos-Kouris. "Gasoline Blending LP." *Oil and Gas Journal* (July 1988): 44–48.

Marcus, A. "The Magellan Fund and Market Efficiency." *Journal of Portfolio Management* (Fall 1990): 85–88.

Martin, C., D. Dent, and J. Eckhart. "Integrated Production, Distribution, and Inventory Planning at Libbey-Owens-Ford." *Interfaces* 23, no. 3 (1993): 68–78.

McDaniel, S., and L. Kinney. "Ambush Marketing Revisited: An Experimental Study of Perceived Sponsorship Effects on Brand Awareness, Attitude Toward the Brand, and Purchase Intention." *Journal of Promotion Management* 3 (1996): 141–167.

Mellichamp, J., D. Miller, and O. Kwon. "The Southern Company Uses a Probability Model for Cost Justification of Oil Sample Analysis." *Interfaces* 23, no. 3 (1993): 118–124.

Miser, H., "Avoiding the Corrupting Lie of a Poorly Stated Problem." *Interfaces* 23, no. 6 (1993): 114–119.

Morrison, D., and R. Wheat. "Pulling the Goalie Revisited." *Interfaces* 16, no. 6 (1984): 28–34.

Mulvey, J. "Reducing the U.S. Treasury's Taxpayer Data Base by Optimization." *Interfaces* 10 (1980): 101-111.

Neter, J., M. Kutner, C. Nachtsheim, and W. Wasserman. *Applied Linear Statistical Models,* 4th ed. Chicago: Irwin, 1996.

Norton, R. "A New Tool to Help Managers." *Fortune* (May 30, 1994): 135–140.

Oliff, M., and E. Burch. "Multiproduct Production Scheduling at Owens-Corning Fiberglass." *Interfaces* 15, no. 5 (1985): 25–34.

Pankratz, A., *Forecasting with Dynamic Regression Models.* New York: Wiley, 1991.

Pass, S. "Digging for Value in a Mountain of Data." *ORMS Today* 24, no. 5 (1997): 24–28.

Peterson, R., and E. Silver. *Decision Systems for Inventory Management and Production Planning.* 2nd ed. New York: Wiley, 1985.

Press, S. J. "Sample-Audit Tax Assessment for Businesses: What's Fair?" *Journal of Business & Economic Statistics* 13, no. 3 (1995): 357–359.

Ramsey, F., and D. Schafer. *The Statistical Sleuth: A Course in Methods of Data Analysis.* Belmont, CA: Duxbury Press, 1997.

Robichek, A., D. Teichroew, and M. Jones. "Optimal Short-Term Financing Decisions." *Management Science* 12 (1965): 1–36.

Robinson, P., L. Gao, and S. Muggenborg. "Designing an Integrated Distribution System at DowBrands, Inc." *Interfaces* 23, no. 3 (1993): 107–117.

Rohn, E. "A New LP Approach to Bond Portfolio Management." *Journal of Financial and Quantitative Analysis* 22 (1987): 439–467.

Roselka, Rita. "The New Mantra: MVT." *Forbes* (March 11, 1996): 114–118.

Rothstein, M. "Hospital Manpower Shift Scheduling by Mathematical Programming." *Health Services Research* (1973).

Salkin, H., and C. Lin. "Aggregation of Subsidiary Firms for Minimal Unemployment Compensation Payments via Integer Programming." *Management Science* 25 (1979): 405–408.

Schindler, S., and T. Semmel. "Station Staffing at Pan American World Airways." *Interfaces* 23, no. 3 (1993): 91–106.

Schmidt, S., and R. Launsby. *Understanding Industrial Designed Experiments,* 4th ed. Colorado Springs: Air Academy Press, 1994.

Schnarrs, S., and J. Bavuso. "Extrapolation Models on Very Short-Term Forecasts." *Journal of Business Research* 14 (1986): 27–36.

Simonoff, J., and I. Sparrow. "Predicting Movie Grosses: Winners and Losers, Blockbusters and Sleepers." *Chance* 13, no. 3 (2000): 15–24.

Smith, S. "Planning Transistor Production by Linear Programming." *Operations Research* 13 (1965): 132–139.

Stanley, T., and W. Danko. *The Millionaire Next Door.* Atlanta, GA: Longstreet Press, 1996.

Strong, R. "LP Solves Problem: Eases Duration Matching Process." *Pension and Investment Age* 17, no. 26 (1989): 21.

Swart, W., and L. Donno. "Simulation Modeling Improves Operations, Planning and Productivity of Fast-Food Restaurants." *Interfaces* 11, no. 6 (1981): 35–47.

Ulvila, J. "Postal Automation (ZIP+4) Technology: A Decision Analysis." *Interfaces* 17, no. 2 (1987): 1–12.

Volkema, R., "Managing the Process of Formulating the Problem." *Interfaces* 25, no. 3 (1995): 81–87.

Walkenbach, J. *Microsoft Excel 2000 Bible.* Foster City, CA: IDG Books Worldwide, Inc., 1999.

Walker, W. "Using the Set Covering Problem to Assign Fire Companies to Firehouses." *Operations Research* 22 (1974): 275–277.

Westbrooke, I. "Simpson's Paradox: An Example in a New Zealand Survey of Jury Composition." *Chance* 11, no. 2 (1998): 40–42.

Westerberg, C., B. Bjorklund, and E. Hultman. "An Application of Mixed Integer Programming in a Swedish Steel Mill." *Interfaces* 7, no. 2 (1977): 39–43.

Winston, W. L. *Operations Research: Applications and Algorithms.* 3rd ed. Belmont, CA: Duxbury Press, 1994.

Zahavi, J. "Franklin Mint's Famous AMOS." *ORMS Today* 22, no. 5 (1995): 18–23.

Zangwill, W. "The Limits of Japanese Production Theory." *Interfaces* 22, no. 5 (1992): 14–25.

Index